Handbook of Biological Wastewater Treatment

Handbook of Biological Wastewater Treatment

Design and Optimisation of Activated Sludge Systems

Second Edition

A.C. van Haandel and
J.G.M. van der Lubbe
www.wastewaterhandbook.com

IWA
Publishing
London • New York

Published by **IWA Publishing**
Alliance House
12 Caxton Street
London SW1H 0QS, UK
Telephone: +44 (0)20 7654 5500
Fax: +44 (0)20 7654 5555
Email: publications@iwap.co.uk
Web: www.iwapublishing.com

First Published in Hardback 2012, Reprinted 2012
Published in Paperback 2015

British Library Cataloguing in Publication Data
A CIP catalogue record for this book is available from the British Library

ISBN: 9781780407753 (Paperback)
ISBN: 9781780400808 (eBook)

Contents

Chapter 3

Chapter 4

Chapter 5

Chapter 6

Chapter 7

Chapter 13

Chapter 14

Preface

In this book the authors seek to present the state-of-the-art theory concerning the various aspects of the activated sludge system and to develop procedures for optimized cost based design- and operation. The book has been written for students at MSc or PhD level, as well as for engineers in consulting firms and environmental protection agencies.

Since its conception almost a century ago, the activated sludge system evolved as the most popular configuration for wastewater treatment. Originally this was due to its high efficiency at removing suspended solids and organic material, which at that time was considered as the most important treatment objective.

The earliest design principles for activated sludge systems date back to the second half of the 20th century, almost fifty years after the first systems were constructed and many further developments have occurred since. As nitrogen is one of the key components in eutrophication of surface water, in the 1970s nitrogen removal became a requirement and this resulted in the incorporation of nitrification- and denitrification processes in the activated sludge system. An important subsequent development was the introduction of chemical- and biological phosphorus removal in the 1980s and 1990s.

Over the last decades the predominance of the activated sludge system has been consolidated, as cost-efficient and reliable biological removal of suspended solids, organic material and the macro-nutrients nitrogen and phosphorus has consistently been demonstrated. This versatility is also shown in the continuous development of new configurations and treatment concepts, such as anaerobic pre-treatment, membrane bioreactors, granular aerobic sludge and innovative systems for nitrogen removal. It is therefore scarcely surprising that many books have been dedicated to the subject of wastewater treatment and more specifically to one or more aspects of the activated sludge system. So why should you consider buying this particular book? The two main reasons why this book is an invaluable resource for everybody working in the field of wastewater treatment are the following:

- The scope of this book is extremely broad and deep, as not only the design of the activated sludge system, but also that of auxiliary units such as primary and final settlers, pre-treatment units, sludge thickeners and digesters is extensively discussed;

– The book offers a truly integrated design method, which can be easily implemented in spreadsheets and thus may be adapted to the particular needs of the user.

In this text, the theory related to the different processes taking place in activated sludge systems is presented. It is demonstrated that the sludge age is the main design parameter for both aerobic and anaerobic systems. A steady-state model is developed that will prove extremely useful for the design and optimisation of activated sludge systems. This model describes the removal of organic material in the activated sludge system and its consequences for the principal parameters determining process performance: effluent quality, excess sludge production and oxygen consumption.

The design guidelines for biological and chemical nutrient removal are integrated with those of other main treatment units, such as final settlers, primary settlers and anaerobic pre-treatment units, sludge thickeners and -digesters. Finally, the text will also deal with operational issues: for example sludge settling and -bulking, oxygen transfer, maintenance of an adequate pH, sludge digestion and methane production.

Visit us at our website www.wastewaterhandbook.com for more information, the latest updates and free Excel design tools, or contact us at info@wastewaterhandbook.com.

Notes on the second edition

This significantly revised and updated second edition expands upon our earlier work. Valuable feedback was received from the wastewater treatment courses, based on this handbook, given in the period 2007 to 2011. This welcome feedback has been incorporated in the book in order to improve the didactic qualities. Where needed the book structure was adapted to make it more intuitively understandable by the reader, while many additional examples have been introduced to clarify the text. Finally, obsolete text has been removed and a number of obvious errors corrected. The main additions/changes with regards to the book contents are:

Chapter 3 – Organic material removal

First of all, a new section has been written that explains the model notation used in this book in much more detail. Additional examples facilitate the readers understanding about the way the steady state model for COD removal is constructed and how it can be used. The difference between true and apparent yield is explained, while also the section on the F/M ratio, and especially the reasons not to use it, has been expanded.

Chapter 4 – Aeration

The section on aeration, previously part of Chapter 3, has been updated and moved to a separate Chapter.

Chapter 5 – Nitrogen removal

The effect of the oxygen recycle to the anoxic zones on the denitrification capacity is now explicitly included in the model. Furthermore, the concept of available nitrate, i.e. the flux of nitrate to the pre-D and post-D zones is explained in more detail. The design procedure for nitrification has been elaborated and several extensive examples for optimized design of nitrogen removal have been added.

Chapter 6 – Innovative systems for nitrogen removal

As the developments on the subject of innovative nitrogen removal are so rapid, this section has been significantly rewritten and expanded and now merits it own chapter.

Chapter 7 – Phosphorus removal

Several examples on the design of chemical phosphorus removal systems have been added.

Chapter 8 – Sludge settling

To explain the theory better, several examples have been added. The section on sludge thickening was expanded with an alternative empirical design approach and has been moved to Chapter 12 – Sludge Treatment and Disposal.

Chapter 9 – Sludge bulking and scum formation

The section on sludge separation problems has been rewritten and expanded to include the latest theories and experimental findings on the development and prevention of both sludge bulking- and scum formation.

Chapter 10 – Membrane bioreactors

The chapter on new system configurations is now devoted to MBR only, as the section on aerobic granulated sludge has been updated based on the return of experience from full-scale installations and is moved to Appendix A9. Several new examples detail the design of both cross-flow and submerged membrane configurations.

Chapter 11 – Moving bed biofilm reactors

A new chapter about a technology that has become popular due to its compactness and its potential for upgrading of existing activated sludge systems.

Chapter 12 – Sludge treatment and disposal

The chapter is expanded with a section on sludge thickening: both the solids flux design method and an empirical design approach are presented.

Chapter 13 – Anaerobic pre-treatment

This part has been completely rewritten based on the experiences obtained from an extensive review of large full-scale UASB based sewage treatment plants. The main design and operational issues in UASB treatment are discussed, while new sections have been introduced on the subject of the loss of methane with the effluent, the impact on greenhouse gas emissions and the problems related to the presence of sulphate in the raw sewage.

The anaerobic design model has been expanded to include the presence of sulphate in the influent and that of suspended solids in the effluent. A new section has been introduced that deals with the methodology of UASB reactor design. The section on combined anaerobic-aerobic treatment has been adapted to reflect the latest findings on the extent of nitrogen removal possible after anaerobic pre-treatment. Some interesting new treatment configurations are presented, combining anaerobic pre-treatment with innovative nitrogen removal. Finally a thoroughly updated section on industrial anaerobic reactors has been included, based on the authors experiences within Biothane Systems International.

Chapter 14 – Integrated cost-based design and operation

The section on cost calculation now contains several examples of the calculation of investment-, operational and annualized costs. Furthermore the chapter is expanded with two extensive integrated design examples: (I) combined nitrogen and phosphorus removal in which bio-P removal is compared with pre- and simultaneous precipitation and (II) MBR in which the system configurations for submerged and cross-flow membranes are evaluated.

List of model parameters

Complementary to the section on model notation, a comprehensive list of all parameters used throughout the book has been compiled and added for easy reference.

New appendices

- Appendix A5 - determination of denitrification kinetics
- Appendix A7 - empiric methods for final settler sizing
- Appendix A8 - denitrification in the final settler
- Appendix A9 - aerobic granulated sludge

About the authors

Adrianus van Haandel (1948) holds an MSc degree from the Technical University of Eindhoven – The Netherlands and a PhD from the University of Cape Town – South Africa. He has worked at the University of Campina Grande in Brazil since 1971, where he coordinates research on biological wastewater treatment. He has extensive experience as an independent consultant and is involved with a number of international expert committees. Together with other authors he has written several books about different aspects of wastewater treatment including "Anaerobic sewage treatment in regions with a hot climate" and "Advanced biological treatment processes for industrial wastewaters: principles and applications". Adrianus can be contacted at prosab@uol.com.br.

Jeroen van der Lubbe (1971) is a senior process & product development engineer at Biothane Systems International, part of Veolia Water – Solutions and Technologies. Apart from process design and consultancy, he has been responsible for the development of the Upthane™ – Veolia's municipal UASB solution while currently he is product development manager of the anaerobic MBR – Memthane™ and involved in the first European implementation. He graduated in 1995 at the Environmental Department of the Wageningen University – The Netherlands and since then has been involved extensively in the design, engineering and operation of both industrial and municipal wastewater treatment plants. Before joining Biothane, he worked at Fontes & Haandel Engenharia Ambiental, Raytheon Engineers & Constructors, DHV Water and Tebodin Consultants and Engineers. Jeroen can be contacted at info@wastewaterhandbook.com.

Acknowledgements

This book reflects the experience of the authors with different aspect of biological wastewater treatment. Insofar as the theory of biological processes is concerned, it has very much been influenced by the ideas developed by the research group lead by Professor Gerrit Marais at the University of Cape Town – South Africa. Another important input was the ongoing cooperative research program at several Brazilian universities, PROSAB, financed by the federal government through its agency FINEP. The experimental results generated by this group and the discussions, especially with Professors Pedro Alem and Marcos von Sperling, constituted important contributions.

In the Netherlands, the following persons are acknowledged for their input: Merle de Kreuk at the Technical University Delft and Tom Peeters from DHV BV – for their input to and review of the section on aerobic granular sludge, Wouter van der Star at the Technical University Delft and Tim Hülsen of Paques BV – for their review of the section on innovative nitrogen removal, Darren Lawrence at Koch Membrane Systems and Hans Ramaekers at Triqua BV – for their contribution to the section on MBR technology, Hallvard Ødegaard, professor emeritus at the Department of Hydraulic and Environmental Engineering of the Norwegian University of Science and Technology in Trondheim, for his extensive input to the chapter on MBBR, Sybren Gerbens at the Friesland Water Authority – for his input on construction and treatment costs while he also provided several photos used in this book, André van Bentem at DHV BV and Joost de Haan at the Delfland Water Board who supplied many interesting photos and finally Barry Heffernan for licensing photos and proofreading.

Finally a special word of thanks to the author's wives, Paula Frassinetti and Lotje van de Poll, for their unfailing support during the long incubation period in which this book....and the second edition was written. Not to mention the time it took to develop the course material...

Symbols, parameters and abbreviations

In this book a naming convention is used in which (I) the number of characters required to identify a unique parameter is minimized and (II) the description of the parameter can be deducted in a logical way from its individual constituents. Thus in general a parameter is constructed from a combination of one or more main identifiers (either in capital- or in normal font) followed by one or more subscripts (capital- or normal font).

The main identifiers indicate the class of the parameter, such as daily applied load or production (M), substrate (S), solids (X) or constants (K), while the subscripts specify the type involved, such as ($_v$)= volatile, ($_t$)=total, et cetera. Thus for example MS_{ti} is defined as the total ($_t$) daily applied mass (M) of organic material (S) in the influent ($_i$). In most cases a specific letter can therefore have more than one meaning. However, it should be easy to deduct what it refers to from the context where it is used. As such the amount of characters required to uniquely identify a specific parameter is reduced to the minimum.

In the remainder of this section the list of abbreviations and the list of symbols and parameters are presented. The latter contains in alphabetical order all of the parameters used in the second edition of the Handbook, including a short description and the unit of measure. Subsequently, after a number of key parameters have been introduced in the main text, the model notation used in this book will be explained in much more detail in Section 3.2.

LIST OF ABBREVIATIONS

AF = anaerobic filter
AIC = annualized investment costs
Anammox = anaerobic (anoxic) ammonium oxidation
APT = activated primary tank
AT = aeration tank
ATU = allyl-thio-urea
ATV = abwasser technik verband
AF = anaerobic filter
BABE = bio-augmentation batch enhanced

BAS	= biofilm activated sludge system
BDP	= Bardenpho
BOD	= biological oxygen demand
CANON	= completely autotrophic nitrogen removal over nitrite
CAS	= conventional activated sludge system
CF	= cross-flow
CHP	= combined heat and power
CIP	= cleaning in place
COD	= chemical oxygen demand
CSTR	= completely stirred tank reactor (completely mixed reactor)
DEMON	= de-ammonification
DSVI	= diluted sludge volume index
DWF	= dry weather flow
EGSB	= expanded granular sludge bed
EPA	= environmental protection agency
FSS	= fixed suspended solids
GLS	= gas-liquid-solids
GSBR	= granulated sludge bed reactor
HUSB	= hydrolysis upflow sludge blanket
IC	= internal circulation
IFAS	= integrated fixed film activated sludge system
ISS	= inert suspended solids
LPCF	= low pressure cross-flow
MBR	= membrane bioreactor
MBBR	= moving bed biofilm reactor
MF	= micro-filtration
OGF	= oil, grease and fat
OLAND	= oxygen limited autrotrophic nitrification – denitrification
OUR	= oxygen uptake rate
PAO	= phosphate accumulating organisms
PE	= people equivalent
PF	= plug flow
PHB	= poly-hydroxy-butyrates
RWF	= rainy weather flow rate
SBR	= sequencing batch reactor
SHARON	= single reactor for high activity ammonium removal over nitrite
$SSVI_{3.5}$	= stirred sludge volume index (determined at $3.5\ g \cdot l^{-1}$)
STORA	= stichting toegepast onderzoek naar de reiniging van afvalwater
STOWA	= stichting toegepast onderzoek waterbeheer
SVI	= sludge volume index
TAC	= total annualised costs
TIC	= total investment costs
TKN	= total Kjeldahl nitrogen
TMP	= trans-membrane pressure
TOC	= total operational costs

TOC	= total organic carbon
TS	= total solids
TSS	= total suspended solids
UASB	= upflow anaerobic sludge blanket
UCT	= university of Cape Town
UF	= ultra-filtration
VFA	= volatile fatty acids
VS	= volatile solids
VSS	= volatile suspended solids
WRC	= water research council
ZSV	= zone settling velocity

LIST OF SYMBOLS AND PARAMETERS

Par.	Short description	UoM
a	= projected width of a gas collection plate	m
a	= mixed liquor recirculation factor (from nitrification zone to pre-D zone)	(–)
A_a	= total area occupied by apertures in a UASB reactor	m^2
A_d	= surface area of final settler	m^2
A_{dmin}	= minimum final settler surface area	m^2
$a_{i,n}$	= annualisation factor	(–)
AIC	= annualized investment costs	$US\$ \cdot yr^{-1}$
Alk	= alkalinity	$mg\ CaCO_3 \cdot l^{-1}$
Alk_∞	= final alkalinity after complete decay of active sludge in aerobic digester	$mg\ CaCO_3 \cdot l^{-1}$
Alk_d	= alkalinity consumed in the aerobic digester	$mg\ CaCO_3 \cdot l^{-1}$
Alk_i	= initial alkalinity concentration (aerobic digestion)	$mg\ CaCO_3 \cdot l^{-1}$
Alk_e	= final alkalinity concentration (aerobic digestion)	$mg\ CaCO_3 \cdot l^{-1}$
A_m	= membrane surface area	m^2
A_{mod}	= membrane surface area in a module	m^2
A_o	= overflow area in UASB reactor	m^2
a_{th}	= specific thickener surface area	$m^2 \cdot d \cdot kg^{-1}$ COD
A_{th}	= thickener surface area	m^2
A_u	= surface area of UASB reactor	m^2
A_{umin}	= minimum UASB surface area	m^2
b	= projected height of a gas collection plate	m
b_{an}	= anaerobic decay rate	d^{-1}
b_h	= decay rate for heterotrophic bacteria (non bio-P)	d^{-1}
b_{hT}	= decay rate for heterotrophic bacteria (non bio-P) at temperature T	d^{-1}

B_n	= mass balance recovery factor for nitrogenous material	(–)
b_n	= decay rate for nitrifiers	d^{-1}
B_o	= mass balance recovery factor for COD	(–)
BOD_{vss}	= BOD value of a unit of organic sludge (aerobic digestion)	$mg\ BOD \cdot mg^{-1}\ VSS$
B_p	= mass balance recovery factor for phosphorus	(–)
b_p	= decay rate of bio-P organisms	d^{-1}
b_v	= apparent decay constant of heterotrophic bacteria (non bio-P)	d^{-1}
C_{ae}	= unit construction costs of aeration system	$US\$ \cdot kW^{-1}$
C_d	= unit volume construction costs of final settler	$US\$ \cdot m^{-3}$
C_{d1}	= unit volume construction costs of the primary settler	$US\$ \cdot m^{-3}$
C_{da}	= unit volume construction costs of aerobic digester	$US\$ \cdot m^{-3}$
C_{di}	= unit volume construction costs of anaerobic digester	$US\$ \cdot m^{-3}$
C_{dl}	= costs of discharge to sewer (levies)	$US\$ \cdot PE^{-1}$
C_{el}	= price of electricity	$US\$ \cdot kWh^{-1}$
C_{gen}	= unit construction cost of power generation	$US\$ \cdot kW^{-1}$
C_h	= costs of heating (e.g. with gas or oil)	$US\$ \cdot m^{-3}$ or kg^{-1} fuel
$[CH_4]_{eq}$	= equilibrium methane concentration	$mg\ CH_4 \cdot l^{-1}$
c_p	= proportionality constant between stirred and diluted sludge volume index	(–)
C_r	= unit volume construction costs of the aeration tank	$US\$ \cdot m^{-3}$
C_r	= specific active biomass production per unit mass daily applied biodegradable COD	$mg\ VSS \cdot d \cdot mg^{-1}\ COD$
C_{rh}	= specific active biomass production of heterotrophic organisms per unit mass daily applied biodegradable COD	$mg\ VSS \cdot d \cdot mg^{-1}\ COD$
C_{rn}	= specific active nitrifiers production of per unit mass of daily applied nitrifiable nitrogen	$mg\ VSS \cdot d \cdot mg^{-1}\ N$
C_{rp}	= specific active biomass production of bio-P organisms per unit mass daily applied biodegradable COD	$mg\ VSS \cdot d \cdot mg^{-1}\ COD$
C_{sd}	= costs of sludge disposal	$US\$ \cdot ton^{-1}\ TSS$
C_{th}	= unit volume construction costs of a sludge thickener	$US\$ \cdot m^{-3}$
C_u	= unit volume construction costs of a UASB reactor	$US\$ \cdot m^{-3}$
D_c	= denitrification capacity	$mg\ N \cdot l^{-1}$ influent
D_{c1}	= denitrification capacity in pre-D zone	$mg\ N \cdot l^{-1}$ influent
D_{c1p}	= denitrification capacity from utilization of slowly biodegradable COD	$mg\ N \cdot l^{-1}$ influent
D_{c1s}	= denitrification capacity from utilization of easily biodegradable COD	$mg\ N \cdot l^{-1}$ influent
D_{c3}	= denitrification capacity in post-D zone	$mg\ N \cdot l^{-1}$ influent
D_{cd}	= denitrification capacity in the final settler	$mg\ N \cdot l^{-1}$ influent
D_d	= diameter of final settler	m

DO_{av}	= average oxygen concentration during OUR test	$mg\ O_2 \cdot l^{-1}$
DO_l	= oxygen concentration in the liquid phase	$mg\ O_2 \cdot l^{-1}$
DO_m	= oxygen concentration measured by oxygen sensor	$mg\ O_2 \cdot l^{-1}$
DO_{mt}	= oxygen concentration in the membrane tank	$mg\ O_2 \cdot l^{-1}$
DO_s	= saturation concentration of dissolved oxygen in the mixed liquor at pressure "p"	$mg\ O_2 \cdot l^{-1}$
DO_{s20}	= saturation concentration of dissolved oxygen at 20°C	$mg\ O_2 \cdot l^{-1}$
DO_{sa}	= saturation concentration of dissolved oxygen under actual conditions	$mg\ O_2 \cdot l^{-1}$
DO_{sp}	= saturation concentration of dissolved oxygen at standard pressure	$mg\ O_2 \cdot l^{-1}$
DO_{ss}	= saturation concentration of dissolved oxygen at 20°C and 1 atm (9.1 $mg \cdot l^{-1}$)	$mg\ O_2 \cdot l^{-1}$
DO_{sT}	= saturation concentration of dissolved oxygen at temperature T	$mg\ O_2 \cdot l^{-1}$
f	= fraction of the influent flow discharged to the first reactor in step feed systems	(–)
f	= endogenous residue	$mg\ VSS \cdot mg^{-1}\ VSS$
F	= fouling factor	(–)
F	= solids flux	$kg\ TSS \cdot m^{-2} \cdot d^{-1}$
F/P	= feed to permeate ratio	(–)
$f_{a(N-1)}$	= active sludge fraction in the sludge entering the Nth digester	$mg\ VSS \cdot mg^{-1}\ VSS$
f_{ac}	= fraction of construction costs required for construction of additional (non-specified) units	$mg\ VSS \cdot mg^{-1}\ VSS$
f_{ae}	= active sludge concentration in aerobic digester	$mg\ VSS \cdot mg^{-1}\ VSS$
f_{aer}	= aerobic sludge mass fraction	$kg\ TSS \cdot kg^{-1}\ TSS$
f_{ai}	= initial active sludge concentration (aerobic digestion)	$mg\ VSS \cdot mg^{-1}\ VSS$
f_{aN}	= active sludge fraction in the sludge leaving the Nth aerobic digester	$mg\ VSS \cdot mg^{-1}\ VSS$
f_{an}	= anaerobic sludge mass fraction	$kg\ TSS \cdot kg^{-1}\ TSS$
f_{at}	= active fraction of sludge	$mg\ VSS \cdot mg^{-1}\ TSS$
f_{av}	= active fraction of organic sludge	$mg\ VSS \cdot mg^{-1}\ VSS$
f_{av1}	= active fraction of organic sludge from primary settling	$mg\ VSS \cdot mg^{-1}\ VSS$
f_{av2}	= active fraction of organic sludge from activated sludge system	$mg\ VSS \cdot mg^{-1}\ VSS$
f_{ave}	= active fraction of organic stabilised sludge	$mg\ VSS \cdot mg^{-1}\ VSS$
f_{avu}	= active fraction of organic UASB sludge	$mg\ VSS \cdot mg^{-1}\ VSS$
f_{bh}	= fraction of S_{bi} consumed by normal heterotrophic biomass	$mg\ COD \cdot mg^{-1}\ COD$
f_{bp}	= fraction of S_{bi} sequestered by bio-P organisms	$mg\ COD \cdot mg^{-1}\ COD$
f_{bp}	= slowly biodegradable (particulate) COD fraction in the raw wastewater	$mg\ COD \cdot mg^{-1}\ COD$

f'_{bp}	= slowly biodegradable (particulate) COD fraction in the pre-settled wastewater	$mg\ COD \cdot mg^{-1}\ COD$
f_{bpu}	= biodegradable particulate fraction of organic COD in anaerobic effluent	$mg\ COD \cdot mg^{-1}\ COD$
f_{bs}	= easily biodegradable (soluble) COD fraction in the raw wastewater	$mg\ COD \cdot mg^{-1}\ COD$
f'_{bs}	= easily biodegradable (soluble) COD fraction in the pre-settled wastewater	$mg\ COD \cdot mg^{-1}\ COD$
f_{bsh}	= fraction of S_{bsi} consumed by normal heterotrophic bacteria	$mg\ COD \cdot mg^{-1}\ COD$
f_{bsp}	= fraction of S_{bsi} sequestered by bio-P organisms	$mg\ COD \cdot mg^{-1}\ COD$
f_{bsu}	= biodegradable soluble fraction of organic COD in anaerobic effluent	$mg\ COD \cdot mg^{-1}\ COD$
f_{cv}	= proportionality constant between bacterial mass and mass of COD	$mg\ COD \cdot mg^{-1}\ VSS$
f_d	= activity factor for a bivalent ion	(–)
f_{dn}	= denitrification constant = $(1 - f_{cv} \cdot Y)/2.86$	(–)
f_{ep}	= endogenous residue of bio-P organisms	$mg\ VSS \cdot mg^{-1}\ VSS$
f_{h2s}	= inorganic H_2S-COD in UASB effluent expressed as fraction of influent COD	$mg\ COD \cdot mg^{-1}\ COD$
f_{h2su}	= inorganic H_2S-COD fraction in anaerobic effluent	$mg\ COD \cdot mg^{-1}\ COD$
f_i	= additional investment costs (non-construction related)	(–)
F_l	= limiting solids flux	$kg\ TSS \cdot m^{-2} \cdot d^{-1}$
F_m	= membrane flux	$l \cdot m^{-2} \cdot h^{-1}$
f_m	= maximum anoxic sludge fraction allowed for selected sludge age (when $N_{ae} = N_{ad}$)	$m^3 \cdot m^{-3}$
f_m	= activity coefficient for a monovalent ion in the mixed liquor	(–)
f_{max}	= maximum allowed anoxic mass fraction	$kg\ TSS \cdot kg^{-1}\ TSS$
f_{mi}	= mineral fraction influent	$mg\ ISS \cdot mg^{-1}\ COD$
f_{min}	= minimum required anoxic sludge mass fraction	$kg\ TSS \cdot kg^{-1}\ TSS$
f_n	= nitrogen fraction in organic biomass	$mg\ N \cdot mg^{-1}\ VSS$
f'_{np}	= inert particulate COD fraction after primary settling	$mg\ COD \cdot mg^{-1}\ COD$
f_{np}	= inert particulate influent COD fraction	$mg\ COD \cdot mg^{-1}\ COD$
f_{npu}	= inert particulate fraction of COD in anaerobic effluent	$mg\ COD \cdot mg^{-1}\ COD$
f'_{ns}	= inert soluble COD fraction after primary settling	$mg\ COD \cdot mg^{-1}\ COD$
f_{ns}	= non biodegradable, soluble influent COD fraction	$mg\ COD \cdot mg^{-1}\ COD$
f_{nsu}	= non biodegradable, soluble COD fraction in anaerobic effluent	$mg\ COD \cdot mg^{-1}\ COD$
f_p	= phosphorus fraction in organic biomass	$mg\ P \cdot mg^{-1}\ VSS$
f_{pd}	= fraction of bio-P organisms capable of denitrification	(–)
f_{pp}	= maximum poly-P fraction of bio-P organisms	$mg\ P \cdot mg^{-1}\ VSS$
f_{pr}	= phosphorus release constant	$mg\ P \cdot mg^{-1}\ COD$

f_{pu}	= putrescible fraction of anaerobic sludge	$mg\ VSS \cdot mg^{-1}\ VSS$
f_r	= average frequency of exposure at the chlorine injection point	d^{-1}
F_s	= applied solids load (drying beds)	$kg\ TSS \cdot m^{-2}$
f'_{sb}	= fraction of biodegradable COD that is easily biodegradable remaining after primary settling	$mg\ COD \cdot mg^{-1}\ BCOD$
f_{sb}	= fraction of biodegradable COD that is easily biodegradable	$mg\ COD \cdot mg^{-1}\ BCOD$
F_{sol}	= solids loading rate	$kg\ TSS \cdot m^2 \cdot d^{-1}$
F_t	= total solids flux in final settler	$kg\ TSS \cdot m^2 \cdot d^{-1}$
F_u	= solids flux due to sludge abstraction	$kg\ TSS \cdot m^2 \cdot d^{-1}$
f_v	= organic sludge fraction = ratio between volatile and total sludge concentration	$mg\ VSS \cdot mg^{-1}\ TSS$
F_v	= solids flux due to sludge settling	$kg\ TSS \cdot m^2 \cdot d^{-1}$
f_{ve}	= organic sludge fraction in stabilised sludge	$mg\ VSS \cdot mg^{-1}\ TSS$
f_{vp}	= organic sludge fraction of bio-P organisms	$mg\ VSS \cdot mg^{-1}\ TSS$
f_{vu}	= organic sludge fraction anaerobic sludge	$mg\ VSS \cdot mg^{-1}\ TSS$
f_x	= total anoxic sludge mass fraction	$kg\ TSS \cdot kg^{-1}\ TSS$
f_{x1}	= pre-D anoxic sludge mass fraction	$kg\ TSS \cdot kg^{-1}\ TSS$
f_{x3}	= post-D anoxic sludge mass fraction	$kg\ TSS \cdot kg^{-1}\ TSS$
f_{xd}	= sludge mass fraction located in final settler	$kg\ TSS \cdot kg^{-1}\ TSS$
f_{xvd}	= fraction of final settler volume filled with sludge	$m^3 \cdot m^{-3}$
g	= gravitational acceleration constant	$m \cdot s^{-2}$
h	= liquid height above base of V-notch or above perforation	m
H_1	= thickener inlet zone / thickening zone (ATV)	m
H_2	= thickener clarification zone / sludge storage zone (ATV)	m
H_3	= thickener compression zone / separation zone (ATV)	m
H_4	= thickener sludge removal zone / clear water zone (ATV)	m
H_d	= height of final settler	m
H_{dav}	= average depth of final settler	m
H_{db}	= height of the sludge buffer zone	m
H_{df}	= deflector height	m
H_{dif}	= level of air diffusers above reactor bottom	m
H_{dig}	= height of digestion zone in UASB reactor	m
H_{fb}	= height of freeboard of UASB reactor	m
H_{gb}	= liquid height of gas box	m
H_{gls}	= liquid GLS height	m
H_{liq}	= liquid height UASB reactor	m
H_{th}	= height of sludge thickener	m
H_u	= total height of UASB reactor	m
i	= interest rate	%

Symbol	Description	Unit
I	= investment costs	US$
I_{dsv}	= diluted sludge volume index	$ml \cdot g^{-1}$ TSS
I_{ssv}	= stirred sludge volume index	$ml \cdot g^{-1}$ TSS
k	= Vesilind constant	$l \cdot g^{-1}$ TSS
K_1	= rate constant for denitrification on easily biodegradable organic material	$mg\ N \cdot g^{-1}\ X_a\text{-VSS} \cdot d^{-1}$
k_1	= equilibrium constant for CO_2 dissociation	$mol \cdot l^{-1}$
k_1^*	= "real" equilibrium constant for CO_2 dissociation, corrected for ionic activity	$mol \cdot l^{-1}$
K_2	= rate constant for denitrification on slowly biodegradable organic material	$mg\ N \cdot g^{-1}\ X_a\text{-VSS} \cdot d^{-1}$
k_2	= equilibrium constant for bicarbonate dissociation	$mol \cdot l^{-1}$
k_2^*	= "real" equilibrium constant of the bicarbonate dissociation, corrected for ionic activity	$mol \cdot l^{-1}$
K_3	= rate constant for denitrification due to endogenous respiration	$mg\ N \cdot g^{-1}\ X_a\text{-VSS} \cdot d^{-1}$
K_a	= adsorption rate constant	$litre \cdot mg^{-1}\ X_a \cdot d^{-1}$
k_{abs}	= adsorption constant	h^{-1}
K_{ap}	= adsorption saturation constant	$mg\ COD \cdot mg^{-1}\ X_a$
K_c	= fermentation constant	$l \cdot mg^{-1}\ X_a\text{-VSS} \cdot d^{-1}$
K_h	= Henry constant	atm or $mg \cdot l^{-1} \cdot atm^{-1}$
k_{la}	= oxygen transfer coefficient	h^{-1}
k_{laa}	= oxygen transfer coefficient under actual conditions	h^{-1}
k_{las}	= oxygen transfer constant at 20°C	h^{-1}
k_{laT}	= oxygen transfer constant at T°C	h^{-1}
K_m	= specific utilisation rate constant	$mg\ COD \cdot mg^{-1}\ X_a \cdot d^{-1}$
K_{mp}	= specific utilisation rate of slowly bio-degradable (adsorbed) organic material	$mg\ COD \cdot mg^{-1}\ X_a \cdot d^{-1}$
K_{ms}	= specific utilisation rate of easily biodegradable organic material	$mg\ COD \cdot mg^{-1}\ X_a \cdot d^{-1}$
K_n	= saturation constant for nitrifiers	$mg\ N \cdot l^{-1}$
K_o	= half saturation constant for aerobic processes	$mg\ O_2 \cdot l^{-1}$
k_r	= relaxation constant	h^{-1}
K_s	= saturation constant (Monod)	$mg\ COD \cdot l^{-1}$
K_{sp}	= saturation constant (Monod) for growth on slowly biodegradable, adsorbed substrate	$mg\ COD \cdot mg^{-1}\ X_a$
K_{ss}	= saturation constant (Monod) for growth on easily biodegradable substrate	$mg\ COD \cdot l^{-1}$
k_w	= equilibrium constant for the dissociation of water	$mol^2 \cdot l^{-2}$
k_w^*	= "real" equilibrium constant for the dissociation of water, corrected for ionic activity	$mol^2 \cdot l^{-2}$

L_e	= height of water layer remaining at end of drying period	mm
L_i	= height of initial water layer applied to sludge bed	mm
L_u	= length of UASB reactor	m
m	= maintenance costs	% of TIC per year
m_{civ}	= maintenance costs for civil part of plant	% of TIC per year
MC_{rd}	= construction costs of aeration tank and final settler	US$
MC_{thdi}	= total construction costs of thickener and anaerobic digester	US$
MD_{c1}	= total pre-D denitrification capacity	$kg\ N \cdot d^{-1}$
MD_{c3}	= total post-D denitrification capacity	$kg\ N \cdot d^{-1}$
ME_{chem}	= total chemical excess sludge production	$kg\ TSS \cdot d^{-1}$
mE_d	= specific digested sludge mass	$kg\ VSS \cdot kg^{-1}\ COD$
ME_d	= digested sludge mass	$kg\ VSS \cdot d^{-1}$
ME_{meoh}	= chemical excess sludge production (metal oxides)	$kg\ TSS \cdot d^{-1}$
ME_{mep}	= chemical excess sludge production (metal phosphates)	$kg\ TSS \cdot d^{-1}$
mE_t	= specific excess sludge production (equal to apparent yield Y_{ap})	$mg\ TSS \cdot mg^{-1}\ COD$
ME_t	= excess sludge production	$kg\ TSS \cdot d^{-1}$
mE_{t1}	= specific primary excess sludge production	$mg\ TSS \cdot mg^{-1}\ COD$
ME_{t1}	= primary excess sludge production	$kg\ TSS \cdot d^{-1}$
mE_{t2}	= specific secondary excess sludge production	$mg\ TSS \cdot mg^{-1}\ COD$
ME_{t2}	= secondary excess sludge production	$kg\ TSS \cdot d^{-1}$
mE_{te}	= specific stabilised excess sludge production	$mg\ TSS \cdot mg^{-1}\ COD$
ME_{te}	= stabilised excess sludge production	$kg\ TSS \cdot d^{-1}$
mE_{tu}	= specific anaerobic excess sludge production	$mg\ TSS \cdot mg^{-1}\ COD$
ME_{tu}	= anaerobic excess sludge production	$kg\ TSS \cdot d^{-1}$
ME_{tx}	= total (secondary) excess sludge production corrected for loss of suspended solids in the effluent	$kg\ TSS \cdot d^{-1}$
mE_v	= specific organic sludge production (apparent yield Y_{ap})	$mg\ VSS \cdot mg^{-1}\ COD$
ME_v	= volatile or organic excess sludge production	$kg\ VSS \cdot d^{-1}$
ME_{v1}	= organic primary excess sludge production	$kg\ VSS \cdot d^{-1}$
ME_{v2}	= organic secondary excess sludge production	$kg\ VSS \cdot d^{-1}$
mE_{ve}	= specific stabilised organic excess sludge production	$mg\ VSS \cdot mg^{-1}\ COD$
ME_{ve}	= stabilised organic excess sludge production	$kg\ VSS \cdot d^{-1}$
ME_{vu}	= organic anaerobic excess sludge production	$kg\ VSS \cdot d^{-1}$
mE_{vxa}	= specific active excess sludge production	$mg\ VSS \cdot mg^{-1}\ COD$
mE_{xvna}	= specific inactive excess sludge production	$mg\ VSS \cdot mg^{-1}\ COD$
mM_d	= specific methane production	$kg\ CH_4 \cdot kg^{-1}\ COD$
MM_d	= methane production	$kg\ CH_4 \cdot d^{-1}$
M_{ME}	= consumption of metal salts	$kg \cdot d^{-1}$

$m_{me\&i}$	= maintenance costs for mechanical, electrical and instrumentation part of plant	% of TIC
MN_{av1}	= mass of nitrate available in (i.e. returned to) the pre-D zone	$kg\ N \cdot d^{-1}$
MN_d	= mass of denitrified nitrogen	$kg\ N \cdot d^{-1}$
MN_{d1}	= mass of nitrate denitrified in the pre-D reactor	$kg\ N \cdot d^{-1}$
MN_{d3}	= mass of nitrate denitrified in the post-D reactor	$kg\ N \cdot d^{-1}$
MN_{dd}	= mass of nitrate denitrified in the final settler	$kg\ N \cdot d^{-1}$
MN_{dp}	= denitrification due to consumption of slowly biodegradable COD	$kg\ N \cdot d^{-1}$
MN_{ds}	= denitrification due to consumption of easily biodegradable COD	$kg\ N \cdot d^{-1}$
mN_l	= specific nitrogen discharge with the excess sludge	$mg\ N \cdot mg^{-1}\ COD$
MN_l	= nitrogen removal with produced excess sludge	$kg\ N \cdot d^{-1}$
mN_{ld}	= specific nitrogen release in digester	$mg\ N \cdot mg^{-1}\ COD$
MN_{ld}	= mass of nitrogen released in digester	$kg\ N \cdot d^{-1}$
mN_{le}	= specific nitrogen removal due to discharge with the stabilised excess sludge	$mg\ N \cdot mg^{-1}\ COD$
MN_{le}	= mass of nitrogen removed with stabilised excess sludge	$kg\ N \cdot d^{-1}$
MN_{lx}	= mass of nitrogen removed with the excess sludge corrected for the loss of organic nitrogen with the effluent	$kg\ N \cdot d^{-1}$
MN_{te}	= nitrogen load in effluent	$kg\ N \cdot d^{-1}$
MN_{ti}	= nitrogen load in influent	$kg\ N \cdot d^{-1}$
MO_c	= oxygen demand for COD oxidation (= MS_o)	$kg\ O_2 \cdot d^{-1}$
MO_{eq}	= equivalent oxygen demand (recovered oxygen from denitrification)	$kg\ O_2 \cdot d^{-1}$
MO_n	= oxygen demand for nitrification	$kg\ O_2 \cdot d^{-1}$
MO_t	= total oxygen demand	$kg\ O_2 \cdot d^{-1}$
MP_{chem}	= mass of phosphorus removed by chemical precipitation	$kg\ P \cdot d^{-1}$
mP_{el}	= specific power production	$kWh \cdot kg^{-1}\ COD$
MP_l	= phosphorus removal with excess sludge production	$kg\ P \cdot d^{-1}$
mP_l	= specific phosphorus discharge with the excess sludge	$mg\ P \cdot mg^{-1}\ COD$
MP_{l1}	= mass of phosphorus removed with the primary excess sludge	$kg\ P \cdot d^{-1}$
mP_{le}	= specific phosphorus removal due to discharge with the stabilised excess sludge	$mg\ P \cdot mg^{-1}\ COD$
MP_{le}	= mass of phosphorus removed with stabilised excess sludge	$kg\ P \cdot d^{-1}$
MP_{lex2}	= mass of phosphorus removed with the secondary excess sludge, corrected for the loss of organic nitrogen with the effluent	$kg\ P \cdot d^{-1}$
MP_{lx}	= mass of phosphorus removed with the excess sludge, corrected for loss of organic phosphorus in the effluent	$kg\ P \cdot d^{-1}$

MP_{te}	= phosphorus load in the effluent	$kg\ P \cdot d^{-1}$
mq_1	= specific primary excess sludge flow rate	$m^3 \cdot kg^{-1}\ COD$
mq_2	= specific secondary excess sludge flow rate	$m^3 \cdot kg^{-1}\ COD$
mq_{th}	= specific thickened sludge production	$m^3 \cdot kg^{-1}\ COD$
mS_{bu}	= fraction of total COD present as biodegradable COD in UASB effluent	$mg\ COD \cdot mg^{-1}\ COD$
mS_d	= fraction of influent COD that is digested	$mg\ COD \cdot mg^{-1}\ COD$
MS_d	= mass of COD digested in the system	$kg\ COD \cdot d^{-1}$
MS_{da}	= COD mass digested in UASB and emitted to atmosphere	$kg\ COD \cdot d^{-1}$
mS_{du}	= fraction of influent COD digested in UASB	$mg\ COD \cdot mg^{-1}\ COD$
MS_{du}	= COD mass digested in UASB	$kg\ COD \cdot d^{-1}$
mS_e	= fraction of influent COD leaving the system with the effluent (soluble COD only)	$mg\ COD \cdot mg^{-1}\ COD$
mS_{eu}	= fraction of influent COD ending up as non-settleable COD in the UASB effluent	$mg\ COD \cdot mg^{-1}\ COD$
MS_{eu}	= non settleable COD load in UASB effluent	$kg\ COD \cdot d^{-1}$
mS_{mb}	= fraction of influent COD metabolized	$mg\ COD \cdot mg^{-1}\ COD$
MS_{mb}	= metabolized sludge mass	$kg\ COD \cdot d^{-1}$
mS_o	= fraction of influent COD that is oxidized	$mg\ COD \cdot mg^{-1}\ COD$
MS_o	= mass of COD oxidized in the system (= MO_c)	$kg\ COD \cdot d^{-1}$
mS_{od}	= fraction of influent COD oxidized in aerobic digester	$mg\ COD \cdot mg^{-1}\ COD$
MS_{seq}	= mass of COD sequestered by bio-P organisms	$kg\ COD \cdot d^{-1}$
mS_{te}	= fraction of influent COD leaving the system with the effluent (includes particulate COD)	$mg\ COD \cdot mg^{-1}\ COD$
MS_{te}	= COD load in the effluent	$kg\ COD \cdot d^{-1}$
MS_{ti}	= applied COD load	$kg\ COD \cdot d^{-1}$
MS_{xv}	= mass of COD discharged from the system in the excess sludge	$kg\ COD \cdot d^{-1}$
mS_{xv}	= fraction of influent COD discharged from the system in the excess sludge	$mg\ COD \cdot mg^{-1}\ COD$
mS_{xv1}	= fraction of influent COD leaving the system in the primary excess sludge	$kg\ COD \cdot d^{-1}$
mS_{xv2}	= fraction of influent COD discharged from the system in the secondary excess sludge	$kg\ COD \cdot d^{-1}$
mS_{xve}	= fraction of influent COD leaving the system with stabilised excess sludge	$mg\ COD \cdot mg^{-1}\ COD$
MS_{xve}	= mass of COD discharged from the system in the stabilised excess sludge	$kg\ COD \cdot d^{-1}$
mS_{xvu}	= influent COD fraction converted into anaerobic excess sludge	$mg\ COD \cdot mg^{-1}\ COD$

MS_{xvu}	= COD mass discharged as anaerobic excess sludge from the UASB	$kg\ COD \cdot d^{-1}$
mw_{meoh}	= molecular weight metal hydroxide	$g \cdot mol^{-1}$
mw_{mp}	= molecular weight metal phosphate	$g \cdot mol^{-1}$
mw_{ms}	= molecular weight metal salt	$g \cdot mol^{-1}$
mX_a	= active sludge mass per unit mass daily applied COD	$mg\ VSS \cdot d \cdot mg^{-1}\ COD$
MX_a	= total active sludge mass in system	kg VSS
MX_{ah}	= total active heterotrophic sludge mass in system	kg VSS
MX_{an}	= total active nitrifier sludge mass in system	kg VSS
MX_{ap}	= total mass of active bio-P organisms in system	kg VSS
mX_{au}	= active anaerobic sludge mass per unit mass daily applied COD	$mg\ VSS \cdot d \cdot mg^{-1}\ COD$
MX_{au}	= total active anaerobic sludge mass in system	kg VSS
mX_{bpu}	= non-degraded biodegradable sludge mass per unit mass daily applied COD	$mg\ VSS \cdot d \cdot mg^{-1}\ COD$
MX_{bpu}	= total mass of non-degraded biodegradable sludge mass in system	kg VSS
MX_{chem}	= total mass of chemical sludge in system	kg TSS
mX_e	= endogenous sludge mass per unit mass daily applied COD	$mg\ VSS \cdot mg^{-1}\ COD \cdot d^{-1}$
MX_e	= total mass of endogenous sludge in system	kg VSS
MX_{en}	= total mass of endogenous nitrifier sludge in system	kg VSS
MX_{ep}	= total mass of endogenous bio-P sludge in system	kg VSS
mX_{eu}	= endogenous anaerobic sludge mass per unit mass daily applied COD	$mg\ VSS \cdot d \cdot mg^{-1}\ COD$
MX_{eu}	= total mass of endogenous anaerobic sludge	kg VSS
mX_i	= inert sludge mass per unit mass daily applied COD	$mg\ VSS \cdot d \cdot mg^{-1}\ COD$
MX_i	= total mass of inert sludge in system	kg VSS
mX_{iu}	= non-biodegradable particulate anaerobic sludge mass per unit mass daily applied COD	$mg\ VSS \cdot d \cdot mg^{-1}\ COD$
MX_{iu}	= total mass of non-biodegradable particulate anaerobic sludge in system	kg VSS
mX_{mu}	= inorganic anaerobic sludge mass per unit mass daily applied COD	$mg\ ISS \cdot d \cdot mg^{-1}\ COD$
MX_{mu}	= total mass of inorganic anaerobic sludge in system	kg VSS
MX_n	= total nitrifier mass in system	kg VSS
mX_t	= total sludge mass per unit mass daily applied COD	$mg\ TSS \cdot d \cdot mg^{-1}\ COD$
MX_t	= total sludge mass in system	kg TSS
MX_{tba}	= available sludge mass storage capacity in final settler	kg TSS
MX_{tbr}	= required sludge mass storage capacity in final settler	kg TSS
MX_{td}	= total sludge mass in final settler	kg TSS
mX_{tu}	= anaerobic sludge mass per unit mass daily applied COD	$mg\ TSS \cdot mg^{-1}\ COD \cdot d^{-1}$

MX_{tu}	= total mas of anaerobic sludge in system	kg TSS
mX_v	= volatile sludge mass per unit mass daily applied COD	$mg\ VSS \cdot mg^{-1}\ COD \cdot d^{-1}$
MX_v	= total volatile sludge mass in system	kg VSS
MX_{vh}	= total organic heterotrophic biomass in system	kg VSS
mX_{vu}	= anaerobic organic sludge per unit mass daily applied COD	$mg\ VSS \cdot mg^{-1}\ COD \cdot d^{-1}$
MX_{vu}	= total anaerobic organic sludge mass in system	kg VSS
n	= economical lifetime	years
n	= number of gas boxes	(–)
n	= insurance costs	% of TIC per year
N	= number of UASB reactors	(–)
N	= number of aerobic digesters	(–)
N_{ad}	= desired/required effluent ammonium concentration	$mg\ N \cdot l^{-1}$
N_{ae}	= ammonium effluent concentration	$mg\ N \cdot l^{-1}$
N_{av1}	= nitrate available in pre-D zone	$mg\ N \cdot l^{-1}$ influent
N_{av3}	= nitrate available in post-D zone	$mg\ N \cdot l^{-1}$ influent
N_c	= nitrification capacity (= nitrified ammonium concentration)	$mg\ N \cdot l^{-1}$ influent
N_c/S_{bi}	= ratio between nitrification capacity and biodegradable influent COD	mg N/mg COD
$(N_c/S_{bi})_l$	= limiting ratio between nitrification capacity and biodegradable influent COD for the Bardenpho process	$mg\ N \cdot mg^{-1}\ COD$
$(N_c/S_{bi})_o$	= maximum ratio between nitrification capacity and biodegradable influent COD allowing full nitrogen removal	$mg\ N \cdot mg^{-1}\ COD$
N_d	= denitrified nitrogen concentration	$mg\ N \cdot l^{-1}$ influent
N_{dd}	= concentration of nitrate that will be denitrified in the return sludge stream per passage through the final settler	$mg\ N \cdot l^{-1}$
N_{ddmax}	= maximum allowable production of nitrogen gas in the return sludge flow during its passage through the final settler to the abstraction point	$mg\ N \cdot l^{-1}$
N_{dp}	= denitrification due to consumption of slowly biodegradable COD	$mg\ N \cdot l^{-1}$ influent
N_{ds}	= denitrification due to consumption of easily biodegradable COD	$mg\ N \cdot l^{-1}$ influent
N_{ke}	= effluent Kjeldahl nitrogen concentration	$mg\ N \cdot l^{-1}$
N_{ki}	= influent Kjeldahl nitrogen concentration	$mg\ N \cdot l^{-1}$
N_l	= nitrogen concentration removed with the excess sludge	$mg\ N \cdot l^{-1}$ influent
N_{ld}	= nitrogen concentration released in digester	$mg\ N \cdot l^{-1}$ influent
N_{le}	= nitrogen concentration removed with the stabilised excess sludge	$mg\ N \cdot l^{-1}$ influent
N_{lh}	= nitrogen concentration removed with the heterotrophic excess sludge	$mg\ N \cdot l^{-1}$ influent

N_{ln}	= nitrogen concentration removed with the nitrifier excess sludge	$mg\ N \cdot l^{-1}$ influent
N_{lx}	= nitrogen concentration discharged with excess sludge (corrected for loss of organic nitrogen in the effluent)	$mg\ N \cdot l^{-1}$ influent
N_{N2eq}	= equilibrium dissolved nitrogen gas concentration at the maximum liquid depth of the final settler, assuming an atmosphere of 100% nitrogen	$mg\ N \cdot l^{-1}$
N_{N2in}	= dissolved nitrogen gas concentration in the incoming mixed liquor flow	$mg\ N \cdot l^{-1}$
$N_{n\infty}$	= nitrate concentration when decay of active sludge is complete (aerobic digestion)	$mg\ N \cdot l^{-1}$
N_{nd}	= nitrate production in the aerobic digester	$mg\ N \cdot l^{-1}$
N_{ne}	= nitrate/nitrate effluent concentration	$mg\ N \cdot l^{-1}$
N_{ni}	= initial nitrate concentration (aerobic digestion)	$mg\ N \cdot l^{-1}$
N_{ni}	= influent nitrate/nitrite concentration	$mg\ N \cdot l^{-1}$
N_{oe}	= organic nitrogen in effluent	$mg\ N \cdot l^{-1}$
N_{oi}	= influent organic nitrogen concentration	$mg\ N \cdot l^{-1}$
N_{ope}	= particulate organic nitrogen in effluent	$mg\ N \cdot l^{-1}$
N_{ose}	= soluble organic nitrogen in effluent	$mg\ N \cdot l^{-1}$
N_p	= nitrification potential (= maximum ammonium concentration that can be nitrified)	$mg\ N \cdot l^{-1}$ influent
N_{te}	= effluent total nitrogen concentration	$mg\ N \cdot l^{-1}$
$N_{te,max}$	= maximum nitrogen effluent concentration (all released nitrogen recycled to aeration tank)	$mg\ N \cdot l^{-1}$
$N_{te,min}$	= minimum nitrogen effluent concentration (no recycle of released nitrogen to aeration tank)	$mg\ N \cdot l^{-1}$
N_{ti}	= influent Kjeldahl nitrogen concentration	$mg\ N \cdot l^{-1}$
$(N_{ti}/S_{ti})_l$	= limiting ratio between influent TKN and total influent COD for the applicability of the Bardenpho process	$mg\ N \cdot mg^{-1}$ COD
$(N_{ti}/S_{ti})_o$	= maximum ratio between influent TKN and total influent COD allowing full nitrogen removal	$mg\ N \cdot mg^{-1}$ COD
o	= operational costs	% of TIC per year
O_c	= oxygen uptake rate (respiration) for COD oxidation	$mg\ O_2 \cdot l^{-1} \cdot d^{-1}$
O_{en}	= endogenous respiration rate	$mg\ O_2 \cdot l^{-1} \cdot d^{-1}$
O_{eq}	= oxygen recovery rate (equivalent oxygen uptake rate) due to denitrification	$mg\ O_2 \cdot l^{-1} \cdot d^{-1}$
O_{ex}	= exogenous respiration rate	$mg\ O_2 \cdot l^{-1} \cdot d^{-1}$
$O_{ex,sbp}$	= exogenous respiration rate due to consumption of slowly biodegradable (adsorbed) substrate	$mg\ O_2 \cdot l^{-1} \cdot d^{-1}$
$O_{ex,sbs}$	= exogenous respiration rate due to consumption of easily biodegradable substrate	$mg\ O_2 \cdot l^{-1} \cdot d^{-1}$

O_n	= oxygen uptake rate for nitrification	$mg\ O_2 \cdot l^{-1} \cdot d^{-1}$
O_t	= total oxygen uptake rate	$mg\ O_2 \cdot l^{-1} \cdot d^{-1}$
$OT_{4,5}$	= oxygen transfer efficiency at 4.5 m submergence	%
OT_a	= actual oxygen transfer efficiency	$kg\ O_2 \cdot kWh^{-1}$ or %
O_{td}	= total oxygen uptake rate (aerobic digester)	$mg\ O_2 \cdot l^{-1} \cdot d^{-1}$
OT_s	= standard oxygen transfer efficiency	$kg\ O_2 \cdot kWh^{-1}$ or %
OUR	= oxygen uptake rate	$mg\ O_2 \cdot l^{-1} \cdot h^{-1}$
OUR_a	= apparent OUR	$mg\ O_2 \cdot l^{-1} \cdot h^{-1}$
OUR_{abs}	= rate of change of oxygen concentration in reactor due to hydraulic effects	$mg\ O_2 \cdot l^{-1} \cdot h^{-1}$
OUR_{en}	= endogenous respiration rate	$mg\ O_2 \cdot l^{-1} \cdot h^{-1}$
OUR_h	= rate of change of oxygen concentration in reactor due to adsorption of atmospheric oxygen	$mg\ O_2 \cdot l^{-1} \cdot h^{-1}$
OUR_m	= maximum oxygen uptake rate due to nitrification	$mg\ O_2 \cdot l^{-1} \cdot h^{-1}$
p	= personnel costs	% of TIC per year
p	= atmospheric pressure	bar
P	= static point	(–)
P_{aer}	= required aeration power	kW
P_{aerm}	= installed aeration power	kW
p_{ch4}	= partial methane pressure	atm
P_{chem}	= concentration of phosphorus to be chemically removed	$mg\ P \cdot l^{-1}$ influent
p_{dis}	= discharge pressure	bar or m liquid
P_{diss}	= dissipated power	$W \cdot m^{-3}$
P_{el}	= power production	kW
P_{el}	= electrical power consumption (pumps)	kW
PE_{res}	= residual pollution load in wastewater after treatment	$US\$ \cdot PE^{-1}$
P_h	= required heating power	m^3 gas or kg fuel $\cdot d^{-1}$
P_l	= influent phosphorus concentration removed with the excess sludge	$mg\ P \cdot l^{-1}$ influent
P_{ld}	= influent phosphorus concentration in digested sludge (i · e · released to liquid phase)	$mg\ P \cdot l^{-1}$ influent
P_{le}	= influent phosphorus concentration removed with the stabilised excess sludge	$mg\ P \cdot l^{-1}$ influent
P_{lx}	= phosphorus concentration discharged with excess sludge (corrected for loss of organic phosphorus with the effluent)	$mg\ P \cdot l^{-1}$ influent
P_{min}	= minimum required energy required to keep sludge in suspension	$W \cdot m^{-3}$
p_{o2}	= partial oxygen pressure	atm
P_{ope}	= particulate organic phosphorus in effluent	$mg\ P \cdot l^{-1}$
P_{ose}	= soluble organic phosphorus in effluent	$mg\ P \cdot l^{-1}$
P_{pe}	= phosphate concentration in effluent	$mg\ P \cdot l^{-1}$

p_s	= standard pressure	bar
P_{td}	= desired/required total phosphorus concentration in the effluent	$mg\ P \cdot l^{-1}$
P_{te}	= effluent total phosphorus concentration	$mg\ P \cdot l^{-1}$
$P_{te,max}$	= maximum phosphorus effluent concentration (all released phosphorus recycled to aeration tank)	$mg\ P \cdot l^{-1}$
$P_{te,min}$	= minimum phosphorus effluent concentration (no recycle of released phosphorus to aeration tank)	$mg\ P \cdot l^{-1}$
P_{ti}	= influent phosphorus concentration	$mg\ P \cdot l^{-1}$
p_w	= water vapor pressure	bar
Q	= flow rate	$m^3 \cdot h^{-1}$ or $m^3 \cdot s^{-1}$
q	= excess sludge flow	$m^3 \cdot d^{-1}$
q_1	= primary excess sludge flow	$m^3 \cdot d^{-1}$
q_2	= secondary excess sludge flow	$m^3 \cdot d^{-1}$
Q_{air}	= air flow	$kg \cdot h^{-1}$ or $Nm^3 \cdot h^{-1}$
Q_{ch4}	= methane gas flow rate	$Nm^3 \cdot h^{-1}$
Q_{bg}	= biogas flow rate	$Nm^3 \cdot h^{-1}$
Q_f	= module feed flow (cross-flow membranes)	$m^3 \cdot h^{-1}$
Q_i	= influent flow rate	$m^3 \cdot d^{-1}$ or $m^3 \cdot h^{-1}$
Q_p	= permeate flow rate	$m^3 \cdot h^{-1}$
Q_{pf}	= influent peak flow rate	$m^3 \cdot h^{-1}$
Q_{rec}	= recirculation flow (cross-flow MBR)	$m^3 \cdot h^{-1}$
q_{th}	= thickened excess sludge flow	$m^3 \cdot d^{-1}$
q_w	= excess sludge flow	$m^3 \cdot d^{-1}$
r	= recirculation factor from pre-D zone to anaerobic zone	(–)
R	= gas constant	$kJ \cdot mol^{-1} \cdot K^{-1}$
r_a	= adsorption rate of slowly biodegradable material	$mg\ COD \cdot l^{-1} \cdot d^{-1}$
r_d	= decay rate	$mg\ VSS \cdot l^{-1} \cdot d^{-1}$
r_d	= denitrification rate	$mg\ N \cdot l^{-1} \cdot d^{-1}$
R_d	= retention time in aerobic digester	days
R_{di}	= retention time in anaerobic digester	days
R_{dmin}	= theoretical minimum total aerobic digestion retention time for $N \rightarrow \infty$	days
r_{dp}	= denitrification rate on slowly biodegradable COD	$mg\ N \cdot l^{-1} \cdot d^{-1}$
r_{ds}	= denitrification rate on easily biodegradable COD	$mg\ N \cdot l^{-1} \cdot d^{-1}$
R_{dtot}	= minimum total aerobic digestion retention time	days
r_g	= growth rate	$mg\ VSS \cdot l^{-1} \cdot d^{-1}$
R_h	= hydraulic retention time	days
R_{h1}	= hydraulic retention time in pre-D reactor	days
r_{hi}	= hydrolysis rate of stored slowly biodegradable material	$mg\ COD \cdot l^{-1} \cdot d^{-1}$

R_{hth}	= thickening time final settler (ATV)	days
R_{hu}	= hydraulic retention time UASB	hr
R_{min}	= minimum retention time for complete utilisation of the S_{bs} present in the influent in the pre-D reactor	days
r_n	= nitrification rate	$mg\ N \cdot l^{-1} \cdot d^{-1}$
R_N	= retention time in N^{th} aerobic digester	days
R_{rel}	= relative evaporation rate of water in the exposed sludge batch	(–)
R_s	= sludge age	days
R_{sa}	= true sludge age (including sludge mass present in final settler)	days
r_{sbp}	= net production of slowly biodegradable material	$mg\ COD \cdot l^{-1} \cdot d^{-1}$
r_{sbs}	= net production of easily biodegradable material	$mg\ COD \cdot l^{-1} \cdot d^{-1}$
r_{sbs}	= feeding rate of easily biodegradable material to the pre-D reactor	$mg\ COD \cdot l^{-1} \cdot d^{-1}$
R_{sm}	= minimum sludge age required to achieve desired effluent ammonium concentration	days
R_{sn}	= minimum sludge age required for nitrification	days
r_{spa}	= net production of adsorbed biodegradable material	$mg\ COD \cdot l^{-1} \cdot d^{-1}$
R_{su}	= anaerobic sludge age	days
r_u	= utilisation rate of organic material	$mg\ COD \cdot l^{-1} \cdot d^{-1}$
r_{us}	= utilisation rate of easily biodegradable influent organic material	$mg\ COD \cdot l^{-1} \cdot d^{-1}$
r_v	= decay rate of volatile solids	$mg\ VSS \cdot l^{-1} \cdot d^{-1}$
R_w	= water evaporation rate	$mm \cdot day^{-1}$
r_{xa}	= net production of active sludge	$mg\ VSS \cdot l^{-1} \cdot d^{-1}$
r_{xe}	= production rate of endogenous residue	$mg\ VSS \cdot l^{-1} \cdot d^{-1}$
s	= sludge recycle factor	(–)
S_{bh}	= biodegradable COD consumed by normal heterotrophic biomass	$mg\ COD \cdot l^{-1}$ influent
S_{bi}	= biodegradable influent COD concentration	$mg\ COD \cdot l^{-1}$ influent
S_{bp}	= biodegradable COD sequestered by bio-organisms	$mg\ COD \cdot l^{-1}$ influent
S_{bp}	= slowly biodegradable COD concentration (reactor)	$mg\ COD \cdot l^{-1}$ influent
S_{bs}	= easily biodegradable COD concentration (reactor)	$mg\ COD \cdot l^{-1}$ influent
S_{bsh}	= easily biodegradable COD consumed by normal heterotrophic biomass	$mg\ COD \cdot l^{-1}$ influent
S'_{bsi}	= influent concentration of easily biodegradable material after correction for denitrification in the anaerobic zone	$mg\ COD \cdot l^{-1}$ influent
S_{bsi}	= easily biodegradable influent COD concentration	$mg\ COD \cdot l^{-1}$ influent
S_{bsN}	= residual concentration of the easily biodegradable material in the effluent of the Nth reactor of a series	$mg\ COD \cdot l^{-1}$
S_{bsp}	= easily biodegradable COD sequestered by bio-P organisms	$mg\ COD \cdot l^{-1}$ influent

s_c	= critical sludge recirculation factor	(–)
s_d	= safety factor used to allow for locally increased dissolved nitrogen gas concentrations	(–)
S_{eu}	= soluble (non settleable) COD concentration in UASB effluent	$mg\ COD \cdot l^{-1}$ influent
s_{fd}	= safety factor used in design final settler	(–)
s_{fth}	= safety factor used in design sludge thickener	(–)
S_{hab}	= COD discharge per capita	$g\ COD \cdot inhab^{-1}$
s_{min}	= minimum value of sludge recirculation flow (MBR)	(–)
S_{ni}	= non biodegradable influent COD concentration	$mg\ COD \cdot l^{-1}$ influent
S_{nsi}	= non biodegradable soluble influent COD concentration	$mg\ COD \cdot l^{-1}$ influent
S_{pa}	= concentration of absorbed slowly biodegradable material (reactor)	$mg\ COD \cdot l^{-1}$ influent
s_{pf}	= return sludge ratio during peak flow (ATV)	(–)
S_{pi}	= particulate influent COD concentration	$mg\ COD \cdot l^{-1}$ influent
S_{te}	= total effluent COD concentration	$mg\ COD \cdot l^{-1}$
S_{ti}	= total influent COD concentration	$mg\ COD \cdot l^{-1}$ influent
S_{tu}	= total UASB effluent COD concentration	$mg\ COD \cdot l^{-1}$
S_{VFA}	= VFA concentration	$mg\ COD \cdot l^{-1}$ influent
t	= aerobic digestion time	days
T	= sewage temperature	°C
t_1	= time required for preparation of the sludge bed and application onto the bed of the sludge to be dried	days
t_2	= time required for percolation	days
t_3	= time required for evaporation	days
t_4	= time required for removal of the dried sludge and cleaning of the bed for the next batch	days
t_c	= total drying cycle time	days
TCC	= total construction costs	US$
t_{comp}	= compression time (thickener)	days
T_{dig}	= temperature in the anaerobic digester	°C
TIC	= total investment costs	US$
T_{in}	= blower inlet temperature	°C
T_{max}	= maximum reactor temperature	°C
T_{min}	= minimum reactor temperature (often equal to design temperature)	°C
TOC	= total operational costs	US$
t_p	= duration of primary phase (denitrification)	d
T_s	= hydraulic loading rate	$m \cdot h^{-1}$
T_{sm}	= maximum allowable hydraulic loading rate	$m \cdot h^{-1}$

T_{spf}	= hydraulic loading rate during peak flow (ATV)	$m \cdot h^{-1}$
T_{vx}	= sludge volume loading rate	$l \cdot m^{-2} \cdot h^{-1}$
T_{vxm}	= maximum sludge volume loading rate	$l \cdot m^{-2} \cdot h^{-1}$
u	= downward liquid velocity in settler	$m \cdot h^{-1}$
U	= humidity	%
U_e	= final humidity	%
U_i	= initial humidity	%
v_0	= Vesilind constant	$m \cdot d^{-1}$ or $m \cdot h^{-1}$
v_a	= liquid velocity in UASB apertures	$m \cdot h^{-1}$
V_{aer}	= volume aerobic zone	m^3
V_{an}	= volume of anaerobic zone (bio-P removal)	m^3
V_c	= volume of settler cone	m^3
v_d	= hydraulic retention time in final settler	days
V_d	= volume of final settler	m^3
V_{d1}	= volume of primary settler	m^3
V_{da}	= aerobic digestion volume	m^3
v_{da}	= specific aerobic digestion volume	$m^3 \cdot d \cdot kg^{-1}$ COD
V_{db}	= available volume for sludge buffering in the final settler	m^3
V_{di}	= anaerobic digester volume	m^3
V_{hab}	= reactor volume required per capita	$m^3 \cdot inhab^{-1}$
v_l	= liquid upflow velocity in UASB reactor	$m \cdot h^{-1}$
v_o	= liquid overflow velocity in UASB reactor	$m \cdot h^{-1}$
V_r	= volume of aeration tank	m^3
v_r	= biological reactor volume	$m^3 \cdot d \cdot kg^{-1}$ COD
V_t	= total volume	m^3
v_{th}	= specific thickener volume	$m^3 \cdot d \cdot kg^{-1}$ COD
V_{th}	= thickener volume	m^3
V_{tp}	= molar gas volume at actual temperature and pressure	$liter \cdot mol^{-1}$
V_u	= UASB volume	m^3
v_x	= sludge volume	$ml \cdot l^{-1}$
V_{x1}	= volume pre-D zone	m^3
V_{x3}	= volume post-D zone	m^3
W_a	= width of single aperture in UASB reactor	m
W_{gb}	= outer width of gas box	m
W_u	= width of UASB reactor	m
X_a	= active sludge concentration in reactor	kg $VSS \cdot m^{-3}$
$X_{a(N-1)}$	= active sludge concentration in (N-1)th digester and its effluent (aerobic digestion)	kg $VSS \cdot m^{-3}$
X_{ad}	= digested active sludge concentration (aerobic digestion)	kg $VSS \cdot m^{-3}$

X_{ae}	= active sludge concentration in digester and its effluent (aerobic digestion)	$kg\ VSS \cdot m^{-3}$
$X_{ah,an}$	= active heterotrophic sludge concentration in anaerobic zone	$kg\ VSS \cdot m^{-3}$
X_{ai}	= initial or incoming active sludge concentration (aerobic digestion)	$kg\ VSS \cdot m^{-3}$
X_{aN}	= active sludge concentration in N-th digester and its effluent (aerobic digestion)	$kg\ VSS \cdot m^{-3}$
X_{an}	= active nitrifier concentration	$kg\ VSS \cdot m^{-3}$
X_{av}	= average concentration at which sludge will accumulate in the final settler	$kg\ TSS \cdot m^{-3}$
X_{bpu}	= non degraded biodegradable solids concentration	$kg\ VSS \cdot m^{-3}$
X_c	= critical sludge concentration	$kg\ TSS \cdot m^{-3}$
x_{ch4}	= mol fraction of dissolved methane gas in water	$mol \cdot mol^{-1}$
X_{d1}	= primary sludge concentration	$kg\ TSS \cdot m^{-3}$
X_e	= concentration of endogenous residue in reactor	$kg\ VSS \cdot m^{-3}$
X_{ee}	= endogenous sludge concentration formed in aerobic digester	$kg\ VSS \cdot m^{-3}$
X_{en}	= concentration of endogenous residue from nitrifiers	$kg\ VSS \cdot m^{-3}$
X_f	= average sludge concentration on settler bottom (ATV)	$kg\ TSS \cdot m^{-3}$
X_i	= inert organic sludge concentration in reactor	$kg\ VSS \cdot m^{-3}$
X_l	= limiting sludge concentration	$kg\ TSS \cdot m^{-3}$
X_m	= minimum sludge concentration	$kg\ TSS \cdot m^{-3}$
X_{mi}	= concentration of inorganic solids in influent	$mg\ ISS \cdot l^{-1}$
X_{mu}	= inorganic sludge concentration in reactor	$kg\ ISS \cdot m^{-3}$
X_{nae}	= inactive sludge concentration in digester (aerobic digestion)	$kg\ VSS \cdot m^{-3}$
X_{nai}	= initial or incoming inactive sludge concentration	$kg\ VSS \cdot m^{-3}$
X_r	= return sludge concentration	$kg\ TSS \cdot m^{-3}$
X_{rm}	= maximum return sludge concentration	$kg\ TSS \cdot m^{-3}$
X_{rmax}	= maximum allowed sludge concentration in membrane tank	$kg\ TSS \cdot m^{-3}$
X_t	= total sludge concentration in reactor	$kg\ TSS \cdot m^{-3}$
X_{t1}	= sludge concentration in first reactor (step feed systems)	$kg\ TSS \cdot m^{-3}$
X_{t2}	= sludge concentration in second reactor (step feed systems)	$kg\ TSS \cdot m^{-3}$
X_{te}	= total stabilised sludge concentration	$kg\ TSS \cdot m^{-3}$
X_{te}	= effluent total solids concentration	$mg\ TSS \cdot l^{-1}$
X_{th}	= thickened excess sludge concentration	$kg\ TSS \cdot m^{-3}$
X_{thl}	= limiting thickening sludge concentration	$kg\ TSS \cdot m^{-3}$
X_{tpf}	= sludge concentration in the reactor during peak flow	$kg\ TSS \cdot m^{-3}$
X_{tu}	= average UASB sludge concentration in reactor	$kg\ TSS \cdot m^{-3}$
X_{tud}	= average UASB sludge concentration in digestion zone	$kg\ TSS \cdot m^{-3}$

Symbol	Description	Unit
X_v	= volatile sludge concentration in reactor	$kg\ VSS \cdot m^{-3}$
$X_{v\infty}$	= final volatile sludge concentration when decay of active sludge is complete (aerobic digestion)	$kg\ VSS \cdot m^{-3}$
X_{vd}	= digested organic sludge concentration	$kg\ VSS \cdot m^{-3}$
X_{ve}	= stabilised organic sludge concentration	$kg\ VSS \cdot m^{-3}$
X_{vi}	= initial volatile sludge concentration (aerobic digestion)	$kg\ VSS \cdot m^{-3}$
X_{vu}	= organic anaerobic sludge concentration	$kg\ VSS \cdot m^{-3}$
X_w	= waste sludge concentration	$kg\ TSS \cdot m^{-3}$
Y_{an}	= anaerobic yield	$mg\ VSS \cdot mg^{-1}\ COD$
Y_{ao}	= yield of ammonia oxidisers	$mg\ VSS \cdot mg^{-1}\ N$
Y_{ap}	= apparent yield	$mg\ VSS\ or\ TSS \cdot mg^{-1}\ COD$
Y or Y_h	= heterotrophic yield	$mg\ VSS \cdot mg^{-1}\ COD$
Y_n	= nitrifier yield	$mg\ VSS \cdot mg^{-1}\ N$
Y_{no}	= yield of nitrite oxidisers	$mg\ VSS \cdot mg^{-1}\ N$
α	= inclination	$mm \cdot m^{-1}$
α	= ratio of the oxygen transfer rate in mixed liquor and in pure water	(–)
α	= plate inclination or angle of base of V-notch	°
β	= ratio of the saturation concentration of DO in mixed liquor and in pure water	(–)
ΔAlk_{am}	= alkalinity change from ammonification	$mg\ CaCO_3 \cdot l^{-1}$ influent
ΔAlk_d	= alkalinity change from denitrification	$mg\ CaCO_3 \cdot l^{-1}$ influent
ΔAlk_n	= alkalinity change from nitrification	$mg\ CaCO_3 \cdot l^{-1}$ influent
ΔAlk_t	= total alkalinity change	$mg\ CaCO_3 \cdot l^{-1}$ influent
ΔD_{c1}	= reduction in pre-D denitrification capacity due to recycle of oxygen to pre-D zone	$mg\ N \cdot l^{-1}$
ΔD_{c3}	= reduction in post-D denitrification capacity due to influx of oxygen in post-D zone	$mg\ N \cdot l^{-1}$
ΔL	= height of water layer removed during drying period	mm
ΔMX_t	= mass of sludge transferred from the reactor to the final settler during peak flow	kg TSS
ΔN_a	= variation of ammonium concentration	$mg\ N \cdot l^{-1}$ influent
ΔN_{am}	= ammonified nitrogen concentration in the activated sludge process	$mg\ N \cdot l^{-1}$ influent
ΔN_n	= variation of nitrate concentration	$mg\ N \cdot l^{-1}$ influent
Δp	= differential pressure	bar
Δp_{mod}	= differential pressure over a membrane module	bar
Δp_{TM}	= trans membrane pressure	bar
ΔX_t	= change in reactor sludge concentration during peak flow	$g\ TSS \cdot l^{-1}$
η_{aer}	= efficiency of blower	%

η_{ch4}	= methane fraction in biogas	%
η_{COD}	= COD removal efficiency	%
η_d	= efficiency factor to account for short circuiting between inlet- and outlet of final settler (ATV)	(–)
η_{dn}	= maximum solids removal efficiency of inactive sludge fraction	%
η_{dp}	= maximum solids removal efficiency of active sludge fraction	%
η_{el}	= electrical efficiency of pump, biogas engine and gas motor	%
η_m	= ratio between net and gross membrane flux	(–)
η_{sb}	= sludge drying bed productivity	$kg\ TSS \cdot m^{-2} \cdot d^{-1}$
η_{xv}	= fraction of solids converted in digester	%
η_1	= COD removal efficiency of primary settler	%
η_{dn}	= degree of solids conversion inert and endogenous sludge	(–)
η_{dp}	= degree of solids conversion active sludge	(–)
η_{x1}	= solids removal efficiency of primary settler	%
μ'_m	= (apparent) maximum specific nitrifier growth rate in systems with non aerated zones	d^{-1}
μ	= specific growth rate of nitrifiers	d^{-1}
μ_m	= maximum specific nitrifier growth rate	d^{-1}
$\nu_{T/T,ref}$	= sweet water viscosity at process temperature or T = 15°C	cP
ρ	= density	$kg \cdot m^{-3}$
Φ	= membrane permeability	$litre \cdot m^{-2} \cdot h^{-1} \cdot bar^{-1}$
$\Phi_{T/T,ref}$	= membrane permeability at process temperature/at T = 15°C	$litre \cdot m^{-2} \cdot h^{-1} \cdot bar^{-1}$
θ	= temperature dependency coefficient (Arrhenius)	(–)
ω	= contraction coefficient	(–)

Chapter 1
Scope of the text

1.0 INTRODUCTION

Suspended solids are the most visible of all impurities in wastewater and may be either organic or inorganic in nature. It is therefore not surprising that the first wastewater treatment systems, introduced by the end of the 19th century, were designed as units for the separation of solids and liquid by means of gravity settling: a process known as the primary treatment of wastewater. When the first efficient and reliable treatment units entered into operation, it soon became clear that these could treat wastewaters only partially for a simple reason: a large fraction of the organic material in wastewater is not settleable and therefore is not removed by primary treatment.

With the objective of improving the treatment efficiency of wastewater treatment plants, secondary treatment was introduced in the first decades of the 20th century. Secondary treatment is characterised by the use of biological methods to remove the organic material present in the wastewater. In search of an efficient wastewater treatment system, the activated sludge process was developed in 1914 by Lockett and Ardern at the University of Manchester. They noted that aeration of municipal sewage resulted in an increased removal rate of organic material, while at the same time the formation of macroscopic flocs was observed, which could be separated from the liquid phase by settling, forming a biological sludge. The important contribution made by Lockett and Ardern was the observation that the addition of this sludge to a new batch of wastewater tremendously accelerated the removal rate of the organic material. The capacity of the sludge to increase the removal rate of organic material led to the common denomination "activated sludge".

In its original version, the activated sludge process was operated as a batch process: wastewater was introduced into a biological reactor containing settled sludge, the reactor contents were then aerated, resulting in the removal of organic material from the liquid phase. Subsequently, the aeration was interrupted and the sludge was then separated from the treated influent by settling. After discharging the treated water as effluent, a new batch of wastewater was introduced into the reactor and a new cycle was initiated. Although this "ancient" activated sludge process has been replaced gradually by other configurations, nevertheless it has survived in the form of the Sequential Batch Reactor (SBR), which has regained popularity over the last decades, especially for application to smaller wastewater streams. Furthermore, a new variant of the SBR has been developed recently, in which a granular sludge is cultivated that settles very well, resulting in a significant reduction of required reactor volume.

The basic principle of the activated sludge process has not changed since the first application: organic material is still placed in contact with activated sludge in an aerobic environment. However, in the decades that followed the introduction of the activated sludge process, many researchers made important contributions, which improved the performance of the activated sludge process both in terms of organic material removal efficiency and of treatment capacity. In addition operational stability was increased as well.

1.1 ADVANCES IN SECONDARY WASTEWATER TREATMENT

The first important advance in the development of the activated sludge process was the transformation of the original sequential batch process into a continuous process, through the addition of a settling tank after the biological reactor. Figure 1.1 shows the basic configuration of a continuous activated sludge process designed for both primary and secondary wastewater treatment.

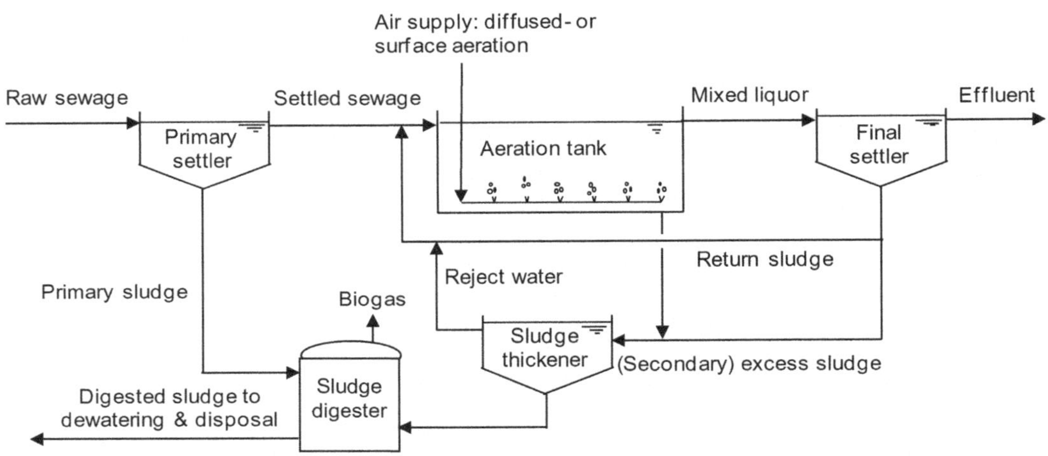

Figure 1.1 Representation of the basic configuration of the activated sludge system (configuration for primary and secondary wastewater treatment)

The biological reactor or aeration tank is filled with a mixture of activated sludge and influent, known as "mixed liquor". The aeration equipment (either surface aerators or compressors connected to submerged air diffusers) transfers the oxygen necessary for the oxidation of organic material into the reactor, while simultaneously introducing enough turbulence to keep the sludge flocs in suspension.

The continuous introduction of new influent results in a continuous discharge of mixed liquor to the final settler (or secondary clarifier), where phase separation of solids and liquid takes place. The liquid leaves the system as treated effluent, whereas the sludge is recirculated to the aeration tank and for that reason is called "return sludge". A primary settler (or primary clarifier) may be introduced to remove part of the suspended solids present in the influent. This reduces the organic load to the biological reactor.

The settled suspended solids ("primary sludge") are often sent to an anaerobic digester, together with the activated sludge that is discharged from the biological reactor: the excess sludge. In the anaerobic digester, the volatile suspended solids in the excess sludge are partly degraded, in the absence of oxygen, into methane and carbon dioxide.

Without the discharge of excess sludge, there would be a continuous growth of sludge in the reactor and consequently, an increase of the sludge concentration in the process. In practice the activated sludge concentration must not be allowed exceed a certain maximum value in order to guarantee proper functioning of the final settler (secondary clarifier). For concentrations beyond the maximum, sludge will escape together with the effluent. A constant sludge mass is maintained when the rate of sludge production is equal to the rate of sludge wastage, where this wastage may be unintentional (in the effluent) or intentional (as excess sludge). In practice, excess sludge is discharged either directly from the aeration tank or from the return sludge line, although the first option is advantageous, as will be explained later in this book. The production of excess sludge adds an extra dimension to the activated sludge process: apart from the wastewater treatment process, there is also a need to find a suitable method for the treatment and final disposal of the produced excess sludge. In practice the sludge is generally submitted to a biological stabilisation process with the objective to reduce the fraction of biodegradable material (biomass and organic material) and as such to avoid putrification. After removing a large part of the water fraction of the sludge, a solid end product (sludge cake) is obtained that may be used in agriculture, disposed into a landfill or sent to an incinerator.

The importance of a controlled rate of sludge wastage was only recognized in the 1950s, when the first models to quantitatively describe the activated sludge process were developed. In these models the concept of sludge age was defined as the ratio between the sludge mass present in the process and the rate of sludge wastage. Physically the sludge age is equal to the mean retention time of the sludge in the process. In this text the sludge age will be identified as the most important operational- and design variable of the activated sludge process.

In the 1950s, additional to the organic material removal, nitrification was introduced in the activated sludge process. Nitrification is a two-step biological oxidation of ammonium, using oxygen as an oxidant: the first step is the oxidation of ammonium to nitrite, while the second step is the oxidation of nitrite to nitrate. Nitrification was initially applied only to reduce the effluent oxygen demand. In the case of municipal wastewater, the oxygen demand for nitrification makes up about half of the demand for organic material removal. It was noted that nitrification in the activated sludge process was perfectly feasible if the applied sludge age was long enough. This requirement was due to the relatively slow growth rate of the nitrifying bacteria.

1.2 TERTIARY WASTEWATER TREATMENT

Once it was possible to produce an effluent with a very low oxygen demand, it became clear that this alone was not always sufficient for significant improvement of the quality of the receiving water body. It was established that the presence of mineral compounds in the effluent, especially the so-called macro-nutrients nitrogen and phosphorus, could cause a serious disruption of the ecological equilibrium in the receiving water. This phenomenon, called eutrophication, was due to an excessive growth of the aquatic life that was able to develop because of the availability of the nutrients. To protect the water quality in the receiving water bodies, it became necessary to develop tertiary treatment systems in which, in addition to the removal of suspended solids (primary treatment) and organic material (secondary treatment), also the macro-nutrients nitrogen and phosphorus were eliminated.

Biological nitrogen removal is obtained when the processes of nitrification and denitrification are applied sequentially. Denitrification is the reduction of nitrate (or nitrite) to nitrogen gas, using organic material as a reductor. Denitrification only develops in an anoxic environment, which is characterized by the presence of nitrate or nitrite and the absence of dissolved oxygen. In the first units constructed for biological nitrogen removal, the nitrified effluent from an activated sludge process was discharged in a second reactor, operated without aeration. Organic material, usually in the form of methanol, was added to the second reactor. Thus, the treatment system was composed of two reactors with different sludges, the first one being for organic material removal and nitrification and the second one for denitrification. However, soon it was established that the organic material present in the wastewater could very well be used for nitrate reduction, with the double advantage that neither external organic material was needed nor a separate unit with denitrifying sludge. These "single sludge" processes have unaerated zones for denitrification and aerated zones where nitrification takes place together with organic material removal. Figure 1.2 shows an areal view of a large modern wastewater treatment plant for tertiary treatment.

Figure 1.2 Aerial view of the large Harnaschpolder STP under construction (1.3 million P.E), located near Delft in the Netherlands. Courtesy of Delfland Waterboard – picture taken by Aeroview – Rotterdam

Recently several new processes have been developed that optimise the nitrogen removal process further. However, application is currently restricted to wastewaters with a high ammonium content, such as reject water produced during the dewatering of digested sludge. Combined nitritation – denitritation (e.g. the SHARON process) is an example of such a new process. The reactor is operated under conditions where the second nitrification step, oxidation of nitrite to nitrate, is not allowed to occur. The advantages are: (I) a reduction in oxygen demand and (II) a reduction in the consumption of organic material. The latter is an advantage as many wastewaters contain insufficient organic material for complete nitrate removal.

A second innovation is the process of anaerobic ammonium oxidation, where a recently discovered bacterial species (Anammox) is used to remove ammonium, using nitrite as the oxidant instead of oxygen. Strictly speaking this process is named inappropriately, as the term anaerobic indicates that both dissolved oxygen and nitrate/nitrite are absent. In contrast to the conventional removal of nitrite or nitrate by denitrification, no organic material is required. A logical next step, which has recently been applied at full-scale, is the combination of nitritation and anaerobic ammonium oxidation, either in a single reactor or as a two reactor system. In the latter configuration the nitritation reactor is used to produce a suitable feed for the Anammox reactor: i.e. an effluent containing ammonium and nitrite in approximately the same ratio.

The second macro-nutrient, phosphorus, can be removed with biological- and chemical methods. Chemical precipitation with metal salts or lime results in the formation of an insoluble metal-phosphate complex, which is removed together with the excess sludge. Disadvantages are the large increase in excess sludge production and the costs of the chemicals. Biological phosphorus removal (bio-P removal) depends on the artificial increase of the phosphorus content of the activated sludge. Again, the phosphorus removal mechanism is disposal with the excess sludge. Bio-P removal is enhanced when an anaerobic zone is introduced in the biological reactor. The mixed liquor is exposed first to the anaerobic environment and subsequently to either an anoxic- or an aerobic phase. Phosphate is removed from the liquid phase and stored as poly-phosphates inside the bacterial cell, increasing the phosphorus content of the sludge. The need for an anaerobic zone implies that in general nitrogen removal is a prerequisite for biological phosphorus removal, as the removal of nitrate is required.

1.3 TEMPERATURE INFLUENCE ON ACTIVATED SLUDGE DESIGN

Currently numerous full-scale activated sludge systems for tertiary treatment are in operation and the majority of these discharge an effluent substantially free of organic material and nutrients. Most of these wastewater treatment plants have been constructed in regions with a temperate climate, notably in Europe and North America. South Africa is the only nation with a large number of tertiary wastewater treatment plants located in regions with a hot climate. Other countries in the tropics and subtropics have usually built activated sludge processes for secondary treatment only.

In many cases the performance of activated sludge processes in regions with a warm climate has been less than satisfactory, especially when these are designed for secondary treatment only. This can be attributed partially to the lack of financial means for proper operation, but in many cases the problem is mainly due to the fact that inadequate design criteria are used. Often these criteria are adaptations from those developed in regions with a colder climate, where the vast majority of the activated sludge processes have been constructed. However, the difference in temperature has such an important influence on the activated sludge behaviour, that some of the design criteria developed in regions with a temperate climate have only a limited applicability in the tropics and subtropics. A clear example is the process of nitrification.

In regions with a cold climate, nitrification will develop only when the activated sludge system is specifically designed for it, through application of a long sludge age. In contrast, in the tropics the growth rate of the nitrifiers is so fast that nitrification is practically unavoidable, even when the applied sludge age is very short. Thus in the tropics, nitrification will develop at least partially. If the aeration capacity of the process is insufficient for organic material removal and nitrification together, there will be competition for the available oxygen by the different bacteria, with the result that both processes develop only partially. The resulting effluent quality will be poor, containing both organic material and ammonium.

Frequently, the low dissolved oxygen concentration in the reactor will lead to the development of a sludge that exhibits extremely poor settling behaviour (filamentous sludge), resulting in the discharge of suspended solids together with the effluent. In that case, even primary treatment quality cannot always be guaranteed.

If the activated sludge process is designed for nitrification but not for denitrification, the latter process is likely to occur spontaneously in the final settler, in the absence of dissolved oxygen. Microscopic nitrogen gas bubbles will appear, predominantly inside the sludge flocs, causing them to rise to the liquid surface where they will form a layer of floating sludge on the surface of the final settler, which will eventually be discharged with the effluent. This loss of sludge may lead to serious disruption of the treatment process: not only will the effluent quality be poor due to the presence of suspended solids, but also the remaining sludge mass may be too small to metabolise the applied organic load. Thus, the absence of provisions for tertiary treatment in regions with a warm climate will tend to cause a decrease in the efficiency of both primary- and secondary treatment processes. It is concluded that in the tropics and subtropics, tertiary treatment in activated sludge processes is not really optional: if biological nitrogen removal is not applied, the performance of the process will be far below the usual level obtained in regions with a temperate climate.

The inclusion of biological nitrogen removal in the treatment process has important repercussions on the design of activated sludge processes. Often it will be necessary to operate the process at a relatively long sludge age, which is achieved by reducing sludge wastage. As a consequence, the sludge mass in the system will increase and hence the reactor volume will be larger. On the other hand, the unit for excess sludge treatment will then be relatively small.

Sludge stabilisation is another aspect of the activated sludge process that is profoundly affected by temperature. The objective of sludge stabilisation is to reduce the fraction of biodegradable material in the sludge and thus to improve its hygienic quality and rheological properties. This stabilisation process is carried out in a separate biological reactor, the sludge digester. If the digester is aerated, the active sludge mass will decrease due to natural decay. If the digester is not aerated, an anaerobic sludge will develop, that uses the wasted sludge as a substrate.

Anaerobic sludge digestion has the advantage that oxygen is not required, but on the other hand, it develops very slowly at temperatures below 15 to 18°C. For this reason, anaerobic digesters operating in a cold climate usually are heated, which reduces the attractiveness of this process. Under these circumstances aerobic sludge digestion, which is feasible at very low temperatures, may be an interesting option, especially for small wastewater treatment plants. However, in regions with a warm climate anaerobic digestion can be carried out at high rate without the need for artificial heating. Thus, in the tropics it is always advantageous to apply anaerobic digestion, unless the process cannot be applied, for instance due to the presence of toxic material in the wasted sludge, as may be the case for plants treating industrial wastewater. In regions with a hot climate, the applicability of the anaerobic digestion process is not limited to the stabilisation of the excess sludge or treatment of high-strength wastewaters. In many cases municipal sewage can be submitted to anaerobic digestion, followed by complementary treatment in an activated sludge process. Under favourable conditions, the combined anaerobic-aerobic process offers great advantages compared to the conventional activated sludge process: a high quality effluent can be obtained at substantially lower investment and operational costs, due to large reductions in both required reactor volume and oxygen demand. However, if nitrogen removal is required, this configuration may be less attractive as the availability of organic material for denitrification will be reduced. This text is mainly a reflection of experimental work in countries with a warm climate, and for that reason, much attention is paid to the particular problems and opportunities that a high average wastewater temperature offers.

1.4 OBJECTIVE OF THE TEXT

The main objective of this text is to offer the reader the tools required for the design and optimisation of activated sludge processes, for both municipal- and industrial wastewater. Nowadays, this will in general include tertiary treatment and anaerobic sludge digestion.

A simplified quantitative steady state model for COD removal is presented that will prove very useful in the design and optimisation of activated sludge systems. The model describes the removal of organic material in the activated sludge system and its consequences for the principal parameters of the process: effluent quality, excess sludge production and oxygen consumption. It has been extended with modules for both nitrogen- and phosphorus removal. An unique feature is the integrated design of biological reactor and final settler, allowing optimisation in terms of lowest total cost design.

The validity of the steady state model has been thoroughly tested during experimental research at bench-, pilot- and full-scale processes, treating different wastewaters under very diverse operational conditions. Most of the concepts presented in this book have been developed at the University of Cape Town (UCT) in South Africa and form the backbone of the Activated Sludge Models I to III as developed by the specialist group of the International Water Association (IWA; 1987, 1994 and 2000). However, the format and contents of the IWA models do not make them particularly suitable for application as a design tool. One should consider that the main objective of these models is to simulate activated sludge system behaviour under varying (dynamic) conditions. For this purpose, a large number of variables and parameters are included. These are indispensable when studying system reactions to disturbances or to process control measures, but can be considered as unnecessary ballast from a design viewpoint. In fact, the IWA models are of such a complexity that an analytical optimised design solution is not possible.

An example is the dissolved oxygen (DO) concentration, which is included in the IWA models as one of many state variables. Each state variable has its own mass balance. Furthermore, the concentration of dissolved oxygen is included in nearly all reaction rate equations, in the form of a switching function. This Monod type switching function is either of the form DO/(K+DO) or (K+DO)/DO and thus "switches" a particular process on or -off, depending on the dissolved oxygen concentration. This is a crucial feature when simulating the behaviour of activated sludge systems. However, it is not required for system design, where sufficient availability of oxygen in the aerobic reactors and the absence of oxygen in anoxic- and anaerobic reactors are presupposed. Proper aeration design and -control, including installation of sufficient aeration capacity and a suitable process control system, will ensure that oxygen will be present at the right time, location and quantity.

Another issue is that most models only take into consideration the processes that develop in the biological reactor, such as metabolisation of organic material and nitrogen removal. The design of auxiliary units such as final settlers, thickeners and digesters is either excluded or not integrated with that of the biological units.

In this book an integrated cost-based design approach is presented that includes all the main treatment units of the activated sludge system: biological reactors, final settler, sludge thickener, sludge digester and optionally pre-treatment units such as the primary settler and the UASB reactor. In various detailed examples, the use of this design approach will be demonstrated in a step-by-step determination of the optimal activated sludge system configuration. Finally, this text will also deal with operational problems of activated sludge systems: e.g. sludge settling and bulking problems, oxygen transfer, maintenance of an adequate pH, sludge digestion and methane production.

Chapter 2

Organic material and bacterial metabolism

2.0 INTRODUCTION

The organic compounds present in wastewater are of particular interest in sanitary engineering. A great variety of micro-organisms – which may be present in the wastewater or in the receiving water body – will interact with organic compounds, using these either as an energy source or as a material source for synthesis of new cellular material. The utilisation of organic material by micro-organisms is called metabolism. The biochemical reactions that produce energy result in the dissimilation of the organic compounds and the production of stable end products, a process called catabolism. Finally, the synthesis of new cellular matter is called anabolism. In order to be able to describe the metabolic processes that occur in the activated sludge process, it is necessary to:

- Determine a quantitative parameter that adequately describes the concentration of organic compounds present in wastewater;
- Establish the different catabolic – and anabolic processes that may occur.

Both these aspects will be discussed in this chapter.

2.1 MEASUREMENT OF ORGANIC MATERIAL

In view of the enormous number of different compounds present in most wastewaters, it is totally impractical, if not impossible, to determine these individually. For this reason the concept of organic material is introduced, which is indicative for the combined concentration of all the organic compounds present in a wastewater. To quantify the mass or concentration of organic material, it is possible to use the properties that practically all organic compounds have in common: (I) they can be oxidised and (II) they contain organic carbon. In sanitary engineering the property that organic material can be oxidised has found the widest application. There are two standard tests based on this property: the biological oxygen demand (BOD) and the chemical oxygen demand (COD) tests. Both have standardised procedures that are described in several specialised texts (for example Standard Methods, 2002). The experimental details will not be discussed here.

2.1.1 The COD test

In both the COD and BOD tests, the organic material concentration is calculated from the oxidant consumption necessary for the oxidation of the organic material. The main differences are the oxidant that is used and the operational conditions during the tests. In the case of COD, a sample of wastewater containing organic material is placed in contact with a very strong inorganic oxidant, a mixture of dichromate and sulphuric acid with silver sulphate as a catalyst. The temperature is increased to the point of ebullition of the mixture, resulting in an increase of the oxidation rate. After two hours (the standard duration of the test) oxidation of the organic compounds is virtually complete. The resulting COD value can be determined by means of titration or with the aid of a spectrophotometer by reading the concentration of formed chromium (Cr^{3+}) concentration. The theoretical COD value of a specific compound can be calculated from stoichiometric considerations. If this theoretical value corresponds to the experimental value, it is concluded that the oxidation of the organic material is complete.

The theoretical COD of a compound with a structural formula $C_xH_yO_z$ can be determined from the two redox equations that describe the overall reaction.

(a) Oxidation reaction

$$C_xH_yO_z + (2x - z)H_2O \longrightarrow x\,CO_2 + (4x + y - 2z)H^+ + (4x + y - 2z)e^- \quad \text{or}$$
$$1/(4x + y - 2z)C_xH_yO_z + (2x - z)/(4x + y - 2z)H_2O \longrightarrow x/(4x + y - 2z)CO_2 + H^+ + e^- \tag{2.1a}$$

(b) Reduction reaction

$$e^- + H^+ + \tfrac{1}{4}\,O_2 \rightarrow \tfrac{1}{2}\,H_2O \tag{2.1b}$$

After combining Eqs. (2.1a and 2.1b) and rearranging one finds:

$$C_xH_yO_z + \tfrac{1}{4}\cdot(4x + y - 2z)O_2 \rightarrow x\,CO_2 + \tfrac{y}{2}\,H_2O \tag{2.1}$$

From Eq. (2.1) it can be noted that the theoretical COD (i.e. the theoretical oxygen demand) of 1 mole of a compound $C_xH_yO_z$ amounts to ¼ · $(4x + y - 2z)$ moles of O_2. Knowing that the molar mass of $C_xH_yO_z$ can be expressed as $(12x + y + 16z)$ g · mol^{-1} and the molar mass for oxygen is 32 grams, it is concluded that the COD of $(12x + y + 16z)$ grams of the compound $C_xH_yO_z$ is equal to ¼ · $(4x + y - 2z) \cdot 32 = 8 \cdot (4x + y - 2z)$ gram O_2. Hence the theoretical COD per unit mass of $C_xH_yO_z$ is given by:

$$COD_t = 8 \cdot (4x + y - 2z)/(12x + y + 16z)\text{g COD} \cdot g^{-1}\,C_xH_yO_z \tag{2.2}$$

When the procedure for the COD test is strictly followed, for almost all compounds the experimental result will not differ more than a few percent from the theoretical value. This leads to the conclusion that (I) during the COD test the organic material is completely oxidized and (II) the precision and reproducibility of the test are good.

Equation (2.2) can be used to calculate the theoretical COD per unit mass for different structural formulas $C_xH_yO_z$. Table 2.1 shows the COD values for some selected compounds. It can be observed that the COD_t value varies considerably, with a minimum value of 0.18 g COD · g^{-1} $C_xH_yO_z$ in the case of oxalic acid and a maximum of 4.0 g COD · g^{-1} for methane. These figures indicate very clearly, that the mass of an organic compound is not a *priori* indicative for its COD. Hence, the expression “mass of organic material” in the case of COD does not really reflect the mass of the organic compounds, but rather the mass of oxygen required for their complete oxidation.

Table 2.1 Theoretical values of COD and TOC per unit mass for selected compounds (I = COD content; II = TOC content and III = COD/TOC ratio)

Compound	X	Y	Z	I mg COD · mg^{-1} $C_xH_yO_z$	II mg TOC · mg^{-1} $C_xH_yO_z$	III mg COD · mg^{-1} TOC
Oxalic acid	2	2	4	0.18	0.27	0.67
Formic acid	1	2	2	0.35	0.26	1.33
Citric acid	6	8	7	0.75	0.38	2.00
Glucose	6	12	6	1.07	0.40	2.67
Lactic acid	3	6	3	1.07	0.40	2.67
Acetic acid	2	4	2	1.07	0.40	2.67
Glycerine	3	8	3	1.22	0.39	3.11
Phenol	6	6	1	2.38	0.77	3.11
Ethyl. glycol	2	6	2	1.29	0.39	3.33
Benzene	6	6	0	3.08	0.92	3.33
Acetone	3	6	1	2.21	0.62	3.56
Palmitic acid	16	32	2	2.88	0.75	3.83
Cyclohexane	6	12	0	3.43	0.86	4.00
Ethylene	2	4	0	3.43	0.86	4.00
Ethanol	2	6	1	2.09	0.52	4.00
Methanol	1	4	1	1.50	0.38	4.00
Ethane	2	6	0	3.73	0.80	4.67
Methane	1	4	0	4.00	0.75	5.33

It can also be concluded that if oxygen is consumed for the oxidation of organic material in a biological treatment plant, then by definition the mass of consumed oxygen will be equal to the mass of oxidised COD. The oxidation of organic material results it its transformation into stable, inorganic compounds like carbon dioxide and water. Hence the mass of oxidised organic material (expressed as COD) can be measured directly from the consumption of oxygen required for this oxidation. This is the basis of respirometrics, the study of biological processes through measurement of the rate of oxygen consumption.

EXAMPLE 2.1

What is the theoretical COD value of a solution of 1 g · l^{-1} of glucose ($C_6H_{12}O_6$)?

Solution

From Eq. (2.2) and knowing that x = 6; y = 12 and z = 6, one has:

$$COD_t = 8 \cdot (4 \cdot 6 + 12 - 2 \cdot 6)/(12 \cdot 6 + 12 + 16 \cdot 6) = 192/180 = 1.067\ \text{mg COD} \cdot \text{mg}^{-1} C_6H_{12}O_6.$$

Hence the solution with 1 g · l^{-1} of glucose has a theoretical COD value of 1067 mg · l^{-1}.

EXAMPLE 2.2

In the traditional COD test (open reflux), a mixture of 10 ml of wastewater sample, 5 ml of 0.25 N potassium dichromate and 15 ml of sulphuric acid is utilised. What is the highest value of the COD concentration that still can be determined?

Solution

In the initial mixture the available quantity of dichromate $= 5 \cdot 0.25 = 1.25$ meq. If the oxidant is entirely used during the COD test, this would mean that 1.25 meq of organic material is consumed. This is equivalent to $1.25 \cdot 8 = 10$ mg O_2, as the equivalent weight of oxygen is $32/4 = 8$ gram $\cdot$ eq^{-1}, see Eq. (2.1b). As the 10 mg of organic material (expressed as COD) were present in a 10 ml wastewater sample, its concentration was 10 mg per 10 ml or 1000 mg $\cdot$ l^{-1}.

It is concluded it is impossible to measure a COD concentration higher than 1000 mg $\cdot$ l^{-1}, because there would be no residual dichromate left. In practice it will be attempted to dilute the sample so that the expected COD concentration is about equal to 500 mg $\cdot$ l^{-1}. Note that the traditional open reflux COD test is more and more being replaced by the use of rapid (but expensive) test-kits.

2.1.2 The BOD test

In the BOD test oxygen is used for the biological oxidation of organic material and therefore this process requires the presence of micro-organisms. If the wastewater does not contain sufficient micro-organisms, they must be added (seeded) at the beginning of the test, together with mineral nutrients and a buffer to maintain a neutral pH.

While in the COD test the oxidation of organic material is essentially complete in less than two hours, in the BOD test the oxidation rate is very slow and complete oxidation will take several weeks. As it is impractical to wait such a long time for the result of the test, a standard test duration of 5 days has been selected, even though it is well known that this is insufficient for complete oxidation. According to folklore, a five-day period was selected because in the UK, where the BOD test was originally developed, it will take the longest river about 5 days to reach the sea... Because temperature has been found to affect the oxidation rate, a standard temperature of 20°C is used. Hence, unless differently stated, BOD means the $BOD_{5,20}$ i.e. the BOD after 5 days of incubation at 20°C.

Some organic compounds (especially small molecules) can be metabolised immediately by micro-organisms. On the other hand, most wastewaters also contain suspended solids, colloidal material and macro molecules. These compounds need to be hydrolysed into smaller molecules prior to metabolisation. Other organic compounds have a very low rate of metabolism, thus requiring little oxygen during the five day test period for BOD.

The organic material metabolised during the test is determined by the oxygen consumption and is called the biodegradable material. Organic compounds that cause no measurable oxygen consumption are called non-biodegradable and are therefore not detected with this test. In the case of biodegradable material, the oxidation will not be complete after 5 days of incubation. Therefore it is not possible to calculate a *priori* a theoretical BOD value for a solution of a known composition, as was done above for the COD test. Before the start of the test it is not known which proportion of the organic material metabolised by the micro-organisms will be oxidised (hence contributing to the BOD) and which part will be incorporated in the cell mass.

As a result of the decay of micro-organisms, part of the influent COD will in the end remain as an inert endogenous residue (refer also to Section 2.3.1) and will *not* exhibit an oxygen demand. Therefore the value of BOD_∞ will always be lower than the biodegradable COD value (BCOD). It will be demonstrated in Example 2.5 that BOD_∞ is about 86% of BCOD.

An often-used empirical equation for the consumption of oxygen and hence for the BOD in a solution of biodegradable material is:

$$BOD_{t,20} = BOD_{\infty,20} \cdot [1 - e^{(-k_{BOD} \cdot t)}] \qquad (2.3)$$

where:

$BOD_{\infty,20}$ = ultimate BOD i.e. the BOD after a long incubation time (>3 weeks) at 20°C, when oxidation of the biodegradable material is assumed to be complete
$BOD_{t,20}$ = BOD after an incubation time of "t" days at 20°C
k_{BOD} = degradation constant for organic material (d^{-1} at 20°C)
t = duration of test in days.

The value of the organic material degradation constant k_{BOD} is variable and depends on the type of wastewater used. Roeleveld et al. (2002) found that in the Netherlands, for municipal wastewater the k_{BOD} value varied between 0.15 to 0.8 d^{-1}. However, often a typical value of 0.23 d^{-1} at 20°C is assumed. With the aid of Eq. (2.3) the ratio of the BOD after an incubation time of 5 days and the BOD after a long (infinite) incubation period is given by:

$$BOD_{5,20}/BOD_{\infty,20} = 1 - e^{(-5 \cdot k_{BOD})} = 0.68 \qquad (2.4)$$

Eq. (2.4) indicates that, for a k_{BOD} value of 0.23 d^{-1}, 68% of the biodegradable material is oxidised during the incubation period of 5 days. It is important to note that Eqs. (2.3 and 2.4) are empirical relationships, developed for sewage and that they do not apply necessarily to other wastewaters. Although labour intensive, it is possible to calculate the k_{BOD} value from a series of BOD determinations, as is demonstrated in Example 2.4

EXAMPLE 2.3

For practical reasons (non working weekends), a 7 days incubation period is used in Sweden instead of the traditional 5 days. What is the additional BOD that may be expected during the extra two days? Assume $k_{BOD} = 0.23 \cdot d^{-1}$.

Solution

From Eq. (2.3) one has $BOD_7/BOD_5 = (1 - e^{(-7 \cdot k_{BOD})})/(1 - e^{(-5 \cdot k_{BOD})}) = 0.80/0.68 = 1.18$. Hence after 7 days of incubation the BOD will be 18% higher than after 5 days.

It is clear that the k_{BOD} value has a significant effect on the values of BOD_5 and BOD_∞ that will be reported. It is evident that the use of a fixed ratio to relate BOD_5 to BOD_∞ can easily lead to large errors, when the wastewaters are different in composition. To illustrate this fact, consider a wastewater with a BOD_∞ value of

EXAMPLE 2.4

For a certain wastewater the $BOD_{t,20}$ value is determined as a function of the incubation time for a period of 20 days. The results are listed in Table 2.2. Determine the value of k_{BOD}.

Table 2.2 $BOD_{t,20}$ values as a function of the incubation time

Incubation time (days)	$BOD_{t,20}$ ($mg\ O_2 \cdot l^{-1}$)	Incubation time (days)	$BOD_{t,20}$ ($mg\ O_2 \cdot l^{-1}$)
1	95	7	318
2	165	8	350
3	206	9	354
4	242	10	365
5	260	15	400
6	293	20	405

Solution

In Figure 2.1 the data points are plotted. Using Eq. (2.3), theoretical curves of $BOD_{t,20}$ as a function of incubation time are generated for different values of k_{BOD}. A good fit is obtained for $k_{BOD} = 0.23\ d^{-1}$. At higher incubation times, the BOD_t tends towards a value of 400 mg $O_2 \cdot l^{-1}$, which is taken as the value of $BOD_{\infty,20}$. The BCOD and total COD values are indicated as well.

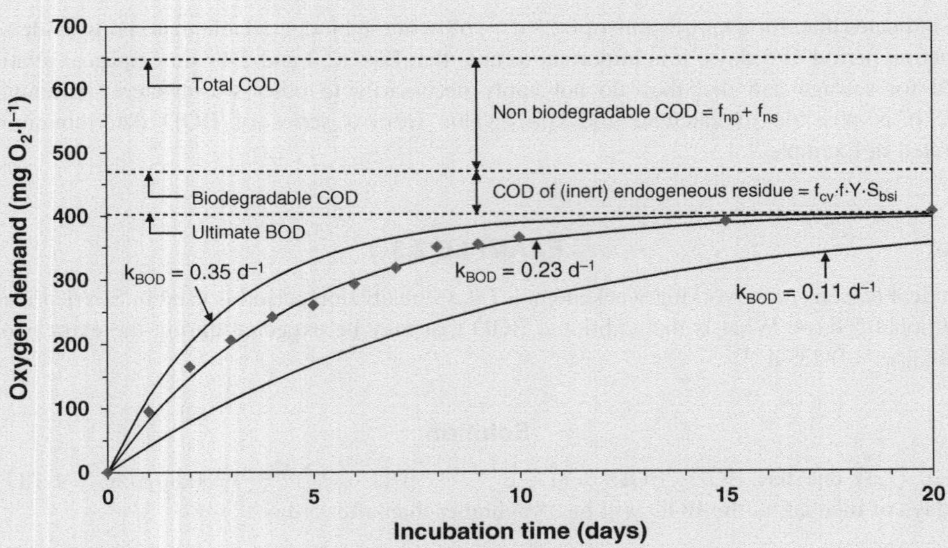

Figure 2.1 $BOD_{t,20}$ as a function of the incubation time for different k_{BOD} values. f_{ns} / f_{np} are inert soluble resp. inert particulate influent COD fraction, Y = biomas yield, f = endogenous residue remaining upon decay and f_{cv} = ratio between COD and VSS of cell mass

400 mg $O_2 \cdot l^{-1}$, as in Example 2.4. Now evaluate what happens if the actual k_{BOD} value is different from the typical k_{BOD} value of 0.23 d^{-1}. For instance, if the true k_{BOD} value is 0.11 d^{-1}, the measured BOD_5 value would have been only 168 mg $O_2 \cdot l^{-1}$, as can be observed in Figure 2.1. Using the BOD_5/BOD_∞ ratio of 0.68 as determined previously with Eq. (2.4), this yields an expected BOD_∞ value of 249 mg $O_2 \cdot l^{-1}$, a difference of 81mg $O_2 \cdot l^{-1}$ or more than 30%. On the other hand, when the true k_{BOD} value is 0.35 d^{-1}, a BOD_∞ of 486 mg $O_2 \cdot l^{-1}$ is calculated.

It can be concluded that the reproducibility of the BOD test is much lower than that of the COD test. The data in Table 2.3 (Heukelian, 1958) are an another example. The BOD determination of several solutions of single, biodegradable compounds with known concentrations was carried out. The observed standard deviations ranged from 13 to 62% of the average values: this is much larger than those determined for the COD test. Table 2.3 also shows clearly that after 5 days the biological oxidation of organic material is still incomplete. In the last column, the ratio between the experimental BOD_5 value and the theoretical COD concentration is calculated. The experimental oxygen demand was only 36% (ethyl acetate) to 75% (glucose) of the demand for complete oxidation.

Table 2.3 Experimental BOD_5 values for selected compounds, the standard deviation and the ratio between experimental BOD_5 and theoretical COD

Compound	No. of tests	BOD_5	Standard deviation		BOD_5/COD_t
Acetic acid	9	0.62	0.18	29%	0.58
Sodium acetate	7	0.33	0.18	54%	0.42
Methyl alcohol	11	0.86	0.11	13%	0.57
Ethyl alcohol	12	1.25	0.23	18%	0.60
Glycerine	6	0.75	0.14	19%	0.62
Formaldehyde	5	0.57	0.30	53%	0.44
Acetone	9	0.89	0.55	62%	0.40
Glucose	10	0.80	0.45	56%	0.75
Ethyl acetate	6	0.66	0.29	44%	0.36
Phenol	5	0.76	0.25	14%	0.74

2.1.3 The TOC test

In the total organic carbon (TOC) test the production of carbon dioxide is measured upon complete oxidation of organic material through combustion at high temperature. The carbon dioxide mass is indicative of the mass of organic carbon initially present in the sample. The equipment for the TOC test is rather sophisticated but it has the advantage of taking only a few minutes, so that it permits virtually on line control. In the case of a known compound ($C_xH_yO_z$) the theoretical TOC value is easily calculated from stoichiometry: as indicated by Eq. (2.1), upon oxidation of 1 mol of $C_xH_yO_z$, i.e. $(12x + y + 16z)$ grams of $C_xH_yO_z$, x moles of CO_2 are formed, so that the TOC is given by 12 x grams. Hence the theoretical TOC per unit mass is calculated as:

$$TOC_t = 12x/(12x + y + 16z) \qquad (2.5)$$

Eq. (2.5) has been used to calculate the TOC for the compounds in Table 2.1. It can be noted that the value of the TOC per unit mass varies significantly for different compounds. Table 2.1 also shows the theoretical COD/TOC ratio. This ratio can be calculated from Eqs. (2.2 and 2.5):

$$COD_t/TOC_t = 8 \cdot (4x + y - 2z)/12x \tag{2.6}$$

Table 2.1 shows that the COD/TOC ratio is not constant. This leads to the conclusion that if one parameter is a good indicator for the organic material concentration, the other one is not. In the following section it will be established that the COD is the correct parameter to evaluate the organic material concentration. The TOC can only be used when the composition of the organic material of the wastewater will be essentially constant. In those cases an experimental COD/TOC ratio can be determined and the COD concentration may be estimated from the measured TOC value.

2.2 COMPARISON OF MEASUREMENT PARAMETERS

In this section, we will attempt to relate the parameters used to quantify the concentration of organic material (COD and TOC) to the chemical energy contained in the material. To do so, it will be necessary to introduce a basic thermodynamic concept, free energy, defined as the amount of useful energy released during a chemical reaction, for example during oxidation of organic material. Values of the free energy release of many compounds can be found in thermodynamic tables. Table 2.4 shows that the values of the released free energy, expressed as $kJ \cdot mol^{-1}$, vary enormously for different chemical compounds. However, when the released free energy per unit mass of theoretical COD is calculated, the value is more or less constant for the different compounds. The only substantially different values are those for the first two from the list: oxalic acid ($21.6\,kJ \cdot g^{-1}$ COD and formic acid ($18.0\,kJ \cdot g^{-1}$ COD). If these two compounds are excluded, an average value of 13.7 kJ is calculated for all other compounds and none of the individual values deviates more than 10% from this average value.

The large difference observed for oxalic and formic acid can be explained by taking into account the oxidation state of these compounds, which is reflected by the number of electrons transferred per carbon atom during the oxidation process. In Eq. (2.1a) the number of electron equivalents transferred during the oxidation of 1 mol of $C_xH_yO_z$ (N_{eq}) is given by:

$$N_{eq} = 4x + y - 2z \tag{2.7}$$

As 1 mol of $C_xH_yO_z$ contains x moles of carbon, it can be calculated that the number of electron equivalents per mol C or the number of electrons per carbon atom N_{el} is equal to:

$$N_{el} = (4x + y - 2z)/x \tag{2.8}$$

The values of N_{eq} and N_{el} are presented in Table 2.4 as well. Figure 2.2 is a graphical representation of the released free energy of the organic compounds in Table 2.4 as a function of the number of electrons released per C-atom.

There is a tendency for the free energy release to decrease as the number of transferred electrons increases, but for numbers above $N_{el} = 3$ electrons per C-atom, this tendency is not very significant and an approximately constant value of $13.7\,kJ \cdot g^{-1}$ COD is maintained. The exceptions to the rule are oxalic and formic acid, with N_{el} values of 1 and 2 respectively. The overwhelming majority of the compounds in wastewaters have N_{el} values in the range of 4 electrons per C-atom (carbohydrates, proteins) to 6 electrons per C-atom (lipids). Hence it can be justified to assume that for organic material in wastewaters the free energy content will be 13.7 $kJ \cdot g^{-1}$ COD $\pm 10\%$.

The compounds in Table 2.4 and Figure 2.2 represent the entire spectrum from the most reduced organic material (methane, N_{el} = 8 electrons per C atom) to the most oxidised organic material (oxalic acid, $N_{el} = 1$ electron per C atom). Purposely, the compounds were chosen for their very different nature: the series has saturated aliphatics (alkanes, alcohols, aldehydes and fatty acids) and unsaturated aliphatics (alkenes), aromatic compounds and a carbohydrate. The objective of such a diverse selection is to show that the released free energy per unit mass of oxidised COD is practically constant. Based on this data it can be concluded that the COD is a good parameter to estimate the chemical energy present in organic material.

Table 2.4 Free energy release/electron transfer upon oxidation of selected compounds

No.	Compound	I ($kJ \cdot mol^{-1}$)	II ($kJ \cdot g^{-1}$ COD_t)	III ($kJ \cdot g^{-1}$ TOC_t)	IV N_{eq}	V N_{el}
1	Oxalic acid	344.4	21.55	14.45	2	1
2	Formic acid	285.6	18.02	23.98	2	2
3	Citric acid	722.4	15.04	30.07	6	3
4	Glucose	2881.2	14.99	40.03	24	4
5	Lactic acid	1369.2	14.24	38.01	12	4
6	Acetic acid	869.4	13.57	36.20	8	4
7	Glycerine	1625.4	14.49	39.44	14	4⅔
8	Phenol	3036.6	13.52	42.04	28	4⅔
9	Ethyl. glycol	1180.2	14.74	49.10	10	5
10	Benzene	3196.2	13.31	44.31	30	5
11	Acetone	1722	13.44	51.16	16	5⅓
12	Palmitic acid	9819.6	13.36	51.16	92	5¾
13	Cyclohexane	3784.2	13.10	52.42	36	6
14	Ethylene	1318.8	13.73	54.94	12	6
15	Ethanol	1310.4	13.65	54.60	12	6
16	Methanol	693	14.45	57.79	6	6
17	Ethane	1444.8	12.89	60.19	14	7
18	Methane	802.2	12.52	65.44	8	8

Headings: I = free energy content; II = energy content per g COD; III = energy content per g TOC, IV = number of electron equivalents per mol compound and V = number of electrons transferred per C-atom.

2.3 METABOLISM

The term metabolism refers to the utilisation of a substrate such as organic material by micro-organisms. Invariably part of the metabolised organic material is transformed into chemically stable end products, which is an energy releasing process. The chemical transformation of the organic material is called catabolism or dissimilation. A second process, occurring simultaneously with catabolism is anabolism, the process of synthesis of new cellular mass. Depending on the type of micro-organisms involved, the source material for synthesis may be organic material (heterotrophic micro-organisms) or carbon dioxide (autotrophic micro-organisms).

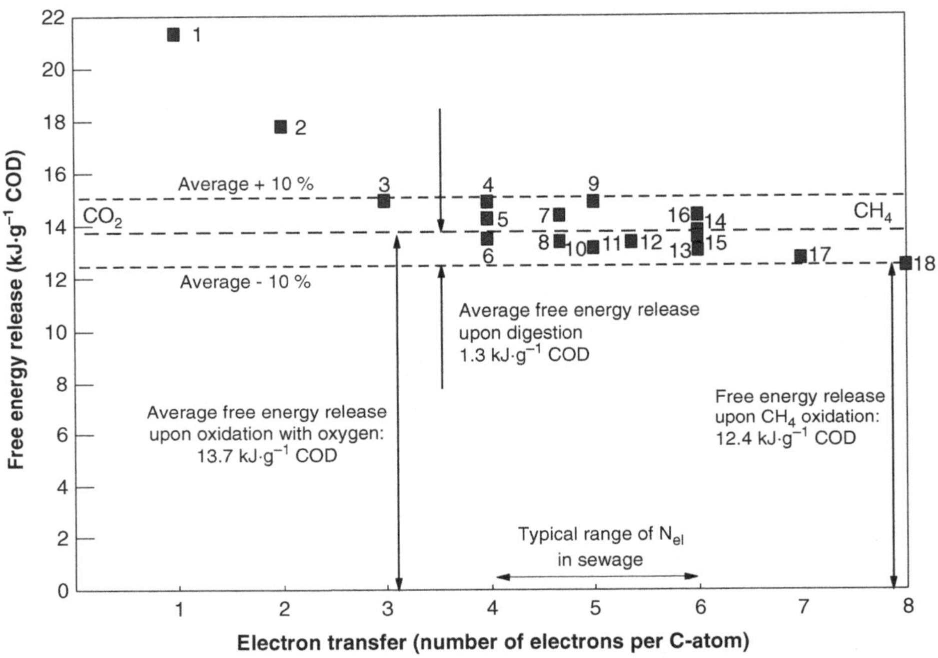

Figure 2.2 Value of the released free energy as a function of the number of transferred electrons per C-atom. The numbers in Figure 2.2 correspond to the compounds listed in Table 2.4

2.3.1 Oxidative metabolism

First the metabolism of heterotrophs in an aerobic environment will be considered. In this case the catabolic process will be the oxidation of organic material by oxygen, also called aerobic respiration. The anabolic process will be the synthesis of cellular material from organic material. It is concluded that the organic material represents both an energy source and a material source for the micro-organisms. Figure 2.3 schematically displays the metabolism of organic material by heterotrophic bacteria in an aerobic environment.

The processes of catabolism and anabolism are interdependent: without anabolism it is not possible to maintain a mass of living micro-organisms and consequently metabolism itself would be impossible. On the other hand, anabolism is an energy demanding process and micro-organisms obtain this energy from catabolic activity. Hence anabolism is necessary for catabolism and vice-versa. The processes of catabolism and anabolism result in measurable phenomena. Oxidation of organic material leads to consumption of dissolved oxygen and this consumption can be measured by respirometric techniques (Spanjers et al., 1996). Furthermore the generated microbiological mass can be detected by the increase in (volatile) suspended solids content (gram VSS).

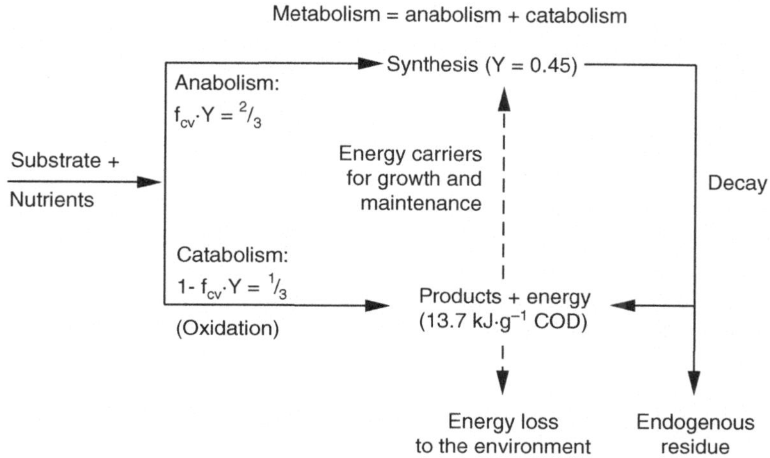

Figure 2.3 Metabolism of organic material in an aerobic environment

This parameter can be determined as the difference between the weight of a dried sample at 103°C (total suspended solids, TSS) and the weight of the same sample after combustion of the organic material at a temperature of 550°C (fixed of inorganic suspended solids, FSS). The mass difference is equal to the mass of volatile suspended organic material. Experimental details of the determination of volatile suspended solids can be found in Standard Methods (2002).

Experimental results indicate that the production of bacterial mass is in the range of 0.35 g to 0.52 g VSS per gram of metabolised COD. An average value of 0.45 g VSS $\cdot$ g^{-1} COD has been reported many times and will be adopted in this text. The ratio between the synthesised microbial mass and the metabolised COD mass is called the yield coefficient Y. To determine which fraction of the metabolised COD is used for anabolism, the COD value of a unit mass of micro-organisms (f_{cv}) must be determined. Several researchers suggested empirical structural formulae and calculated on that basis experimental values of f_{cv} as shown in Table 2.5. Marais and Ekama (1976) found an average value of 1.48 mg COD $\cdot$ mg^{-1} VSS in a very extensive research project. Another investigation in Brazil led to a f_{cv} value of 1.50 mg COD $\cdot$ mg^{-1} VSS for micro-organisms generated from treatment of raw sewage (Dias, 1981).

Accepting the sludge mass parameters $Y = 0.45$ mg VSS $\cdot$ mg^{-1} COD and $f_{cv} = 1.5$ mg COD $\cdot$ mg^{-1} VSS, the fraction of organic material that is anabolised in the aerobic environment can now be calculated. Upon metabolism of 1 gram of COD, the obtained micro-organism mass is Y gram with a COD mass of $f_{cv} \cdot Y$ gram. Hence, the remainder $(1 - f_{cv} \cdot Y)$ g COD will be catabolised. By definition, the required oxygen mass for this catabolism is equal to $(1 - f_{cv} \cdot Y)$. Numerically one has $f_{cv} \cdot Y = 1.48 \cdot 0.45 = 0.67$ and $(1 - f_{cv} \cdot Y) = 0.33$. It is concluded that in an aerobic environment a fraction of 67% or ⅔rd of the metabolised organic material is anabolised, whereas a fraction of 33% or ⅓rd is oxidised.

Table 2.5 Calculated values of the f_{cv} ratio for different empirical formulas of microbial composition (McCarty, 1964)

Formula	Molar weight	COD per gram	Reference
$C_5H_7O_2N$	113	1.42	Hoover and Porges (1952)
$C_5H_9O_3N$	131	1.22	Speece and McCarty (1964)
$C_7H_{10}O_3N$	156	1.48	Sawyer (1956)
$C_5H_8O_2N$	114	1.47	Symons and McKinney (1958)

Figure 2.3 shows an aspect of metabolism that has not yet been discussed. The cellular mass itself contains biodegradable organic material and can be oxidised, at least partially. The oxygen consumption due to oxidation of the cellular material is called endogenous respiration, to distinguish it from the consumption for oxidation of extra-cellular material denominated exogenous respiration. In Figure 2.3 it is indicated that only a part of the cellular mass is oxidised. The remaining non biodegradable solids, called the endogenous residue, are a fraction $f = 0.2$ of the decayed micro-organism mass. In Chapters 3 and 12 the stoichiometric and kinetic aspects of the degradation of cellular mass and the consequential oxygen consumption and endogenous residue generation are discussed in detail.

EXAMPLE 2.5

Calculate the ultimate BOD value of a solution that contains a theoretical COD concentration of $1\ g \cdot l^{-1}$, composed of biodegradable organic material.

Solution

During the metabolism of $1\ g\ COD \cdot l^{-1}$ of biodegradable organic material, there is a synthesis of microbial mass of $Y = 0.45$ g VSS. After endogenous respiration, an endogenous residue of $f \cdot Y = 0.2 \cdot 0.45 = 0.09$ g VSS will remain. This residue will have a COD mass of $f_{cv} \cdot f \cdot Y = 1.5 \cdot 0.2 \cdot 0.45 = 0.135$ g COD, so that $1 - 0.135 = 0.865$ g COD was oxidised. For this oxidation an oxygen mass of 0.865 g O_2 was required.

It is concluded that in a solution with $1\ g \cdot l^{-1}$ of biodegradable organic material (present as COD), the oxygen consumption after completing endogenous respiration is $0.865\ g \cdot l^{-1}$ or equivalently, the BOD is $0.865\ g \cdot l^{-1}$ or $865\ mg \cdot l^{-1}$. Hence, even in the case that all the biological oxygen demand of biodegradable organic material is satisfied, the theoretical oxygen consumption will only be 86.5% of the chemical oxygen demand.

2.3.2 Anoxic respiration

Until now it was assumed that the organic material is metabolised in an aerobic environment, i.e. in the presence of oxygen. However, oxygen may not be available to the micro-organisms and in that case other compounds may serve as an alternative oxidant. In wastewater treatment plants nitrate (NO_3^-), nitrite (NO_2^-) and sulphate (SO_4^{2-}) are possible substitutes for oxygen. Most bacteria in activated sludge can use nitrate or nitrite if no oxygen is available (facultative bacteria). In contrast, sulphate reducers are micro-organisms that cannot survive in an aerobic environment. The half reactions of the oxidants can be written as:

Oxygen:	$e^- + H^+ + \frac{1}{4}\ O_2$	$\rightarrow \frac{1}{2}\ H_2O$	(2.9a)
Nitrate:	$e^- + \frac{6}{5}\ H^+ + \frac{1}{5}\ NO_3^-$	$\rightarrow \frac{1}{10}\ N_2 + \frac{3}{5}\ H_2O$	(2.9b)
Nitrite:	$e^- + \frac{4}{3}\ H^+ + \frac{1}{3}\ NO_2^-$	$\rightarrow \frac{1}{6}\ N_2 + \frac{2}{3}\ H_2O$	(2.9c)
Sulphate:	$e^- + H^+ + \frac{1}{8}\ SO_4^{2-}$	$\rightarrow \frac{1}{8}\ S^{2-} + \frac{1}{2}\ H_2O$	(2.9d)

The equations show the equivalence between ¼ mol O_2 (= 8 g O_2), ⅕ mol NO_3^- (= 14/5 = 2.8 g N) and ⅛ mol SO_4^{2-} (= 32/8 = 4 g S). Hence stoichiometrically 1 g NO_3-N is equivalent to 8/2.8 = 2.86 g O_2 and 1 g SO_4-S is equivalent to 8/4 = 2 g O_2.

It is interesting to consider that, contrary to common perception, it is not the oxygen atom in the alternative oxidant that represents the oxidative potential, but instead the nitrogen- or sulphur atom. The oxidation number of the oxygen atom does not change upon reduction of the alternative oxidant (the value remains -2), whereas those of sulphur and nitrogen are reduced (e.g. from +5 to 0 in the case of nitrate).

EXAMPLE 2.6

If sulphite is used by bacteria, reducing it to sulphide, how many grams of COD can be oxidised per gram of SO_3^{2-}-S?

Solution

The half equation for the reduction of sulphite is:

$$e^- + H^+ + \frac{1}{6}\ SO_3{}^{2-} \rightarrow \frac{1}{6}\ S^{2-} + \frac{1}{2}\ H_2O \qquad (2.9e)$$

Thus ⅙ mol of sulphite (32/6 = 5.33 g S) is equivalent to ¼ mol of O_2, so that it can oxidise 8 g of COD. Hence, the oxidation of 1 g of COD requires 5.33/8 = 0.67 g SO_3-S. Stated differently, it takes 8/5.33 = 1.50 g COD to reduce 1 g of SO_3-S.

In the activated sludge process the reduction of nitrate to molecular nitrogen is called denitrification. This is a process of great importance in wastewater treatment, as it is required for the biological removal of nitrogen from wastewater. Nitrite is an intermediate in the nitrification process ($NH_4^+ \rightarrow NO_2^- \rightarrow NO_3^-$), but as the oxidation of nitrite to nitrate proceeds (in general) faster than that of ammonium to nitrite, its concentration is very low under normal circumstances.

The reduction of sulphate generates hydrogen sulphide gas with its characteristic bad odour. This normally does not take place in the activated sludge process, but the process may develop under anaerobic conditions, for example in an excess sludge digester or in pre-treatment units such as primary clarifiers and sand traps.

2.3.3 Anaerobic digestion

In the preceding sections some aspects of aerobic (or oxic) metabolism have been discussed. However, there are also micro-organisms that can metabolise organic material even in the absence of an oxidant, a process that is called fermentation. It results in a rearrangement of the electrons in molecules of the metabolised compound in such a fashion that at least two new molecules are formed. Sometimes only one particular type of molecule is formed, but in general different types of molecules are produced, one being more oxidised and the other one being more reduced than the original molecule.

Fermentations are of very great importance in the food industry (e.g. for the production of cheese, yoghurt and beer). In sanitary engineering, the fermentation of particular interest is anaerobic digestion. This fermentation is characterised by the fact that the end products are methane and carbon dioxide. The particularity is that methane cannot be further reduced and carbon dioxide cannot be further oxidised, so that anaerobic digestion is the most complete of all fermentation processes. For a compound $C_xH_yO_z$ the anaerobic digestion process (excluding biomass growth) can be written as:

$$C_xH_yO_z + \tfrac{1}{4}\cdot(4\cdot x - y - 2\cdot z)H_2O \rightarrow \tfrac{1}{8}\cdot(4\cdot x - y + 2\cdot z)CO_2 + \tfrac{1}{8}\cdot(4\cdot x + y - 2\cdot z)CH_4 \qquad (2.10)$$

EXAMPLE 2.7

A beer brewery considers anaerobic digestion of waste ethanol. What will be the theoretical composition of the biogas generated by the fermentation of ethanol?

Solution

For ethanol one has $x = 2$, $y = 6$ and $z = 1$, thus per mol digested $C_2H_6O_1$ an amount of $⅛\cdot(4\cdot 2 - 6 + 2\cdot 1) = ½$ mol CO_2 and $⅛\cdot(4\cdot 2 + 6 - 2\cdot 1) = 1½$ mol CH_4 are formed. Hence, the theoretical biogas composition will be equal to 75% methane and 25% carbon dioxide. In practice, the gas will be richer in methane, because of the higher solubility of carbon dioxide in water, so less will escape to the gas phase.

Another equation to describe anaerobic digestion of organic matter is the Buswell equation, which can be used if the digested organic matter contains nitrogen:

$$\begin{aligned} C_xH_yO_zN_w &+ \tfrac{1}{4}\cdot(4\cdot x - y - 2\cdot z + 3\cdot w)H_2O \\ &\rightarrow \tfrac{1}{8}\cdot(4\cdot x - y + 2\cdot z + 3\cdot w)CO_2 + \tfrac{1}{8}\cdot(4\cdot x + y - 2\cdot z - 3\cdot w)CH_4 + w\,NH_3 \end{aligned} \qquad (2.11)$$

In the anaerobic digestion process there is no oxidation of organic material, as no oxidant is available. Thus the electron transfer capacity does not change and will end up in the formed methane, which has a chemical oxygen demand of 4 g $COD\cdot g^{-1}$ CH_4. Therefore, it can be concluded that in order to produce 1 gram of methane, the mass of organic material to be digested also equals 4 gram COD. Therefore the mass of generated methane will be a quarter of the digested COD mass.

An aspect of great importance concerns the energy released in anaerobic digestion. Different from the oxidative catabolism that results in the destruction of organic material, fermentation only converts the organic material and a large proportion of the chemical energy is transferred to the formed methane. It was shown in Section 2.3 that the free energy release upon oxidation of "normal" organic material is approximately 13.7 $kJ \cdot g^{-1}$ COD (see Figure 2.2). Furthermore, in the same figure, it can be seen that the free energy release for the oxidation of methane is 12.4 $kJ \cdot g^{-1}$ COD. Thus, it can be concluded that the anaerobic digestion of organic material results in an average free energy release of only 13.7 – 12.4 = 1.3 $kJ \cdot g^{-1}$ COD. Hence, the free energy release of the anaerobic digestion process is much smaller (at about 10%) of the energy release from the oxidation of organic material.

Figure 2.4 Overview of the rectangular final settlers of the 270.000 $m^3 \cdot d^{-1}$ wastewater treatment plant Houtrust – The Hague in the Netherlands. Courtesy of Delfland Water Board

The consequence of this small energy release is that more organic material must be converted into methane for the bacteria to obtain the same amount of energy required for anabolism. This leads to the conclusion that the proportion of catabolised material to anabolised material will be much larger in the case of anaerobic digestion than in the case of aerobic metabolism. In practice about 95% of the digested organic material is transformed into methane and only 5% is synthesized ($Y_{an} = 0.05$ mg $VSS \cdot mg^{-1}$ COD). In contrast, in aerobic metabolism only 33% of the organic material is catabolised and 67% is synthesized. In the activated sludge process anaerobic digestion can be applied to reduce the mass of excess sludge produced and/or as a pre-treatment process to reduce the organic load to the activated sludge process. These applications of anaerobic digestion are discussed in Chapters 12 and 13.

Chapter 3
Organic material removal

3.0 INTRODUCTION

In this chapter, a quantitative steady state model is developed that describes the removal of organic material in the activated sludge system and its consequences for the principal parameters of the process: effluent quality, excess sludge production and oxygen consumption. The validity of the model has been thoroughly tested during experimental work at bench-, pilot- and full-scale processes, treating different wastewaters under very diverse operational conditions. In all cases the correlation between the experimental values of the process variables and the values predicted by the model was excellent. The model is applicable to all aerobic suspended growth treatment systems, which include the different versions of the conventional activated sludge system, sequential batch reactors, carrousels and aerated lagoons.

In Chapter 13 it will be shown that the model remains valid when it is used for sewage that has received anaerobic pre-treatment, while in Chapters 5 and 7 it is demonstrated that the model can be extended to include nitrogen- and phosphorus removal. Most of the concepts presented in this chapter have been developed at the University of Cape Town (UCT) in South Africa and form the backbone of the activated sludge models as developed later by the specialist group of the International Water Association (IWA, 1986, 1994 and 2000).

The general, dynamic model presented in this chapter is capable of predicting the variation in space and in time of all measurable parameters related to organic material removal in activated sludge systems with reactors in series and operated under variable flow and load conditions. To use the general model, it will be necessary to use a computer program. However, such a dynamic model, written in the form of differential equations, is very suitable for simulating system behaviour but less so for optimised design. Advanced and often costly simulation software is available, but in general requires calibration of a large number of kinetic and stoichiometric parameters. On the other hand, the main parameters of interest in biological wastewater treatment, i.e. effluent quality, sludge production, average oxygen demand and required treatment volume, can be calculated very well with the steady state model. In summary, the main advantages of this steady state model are:

- It allows for easy design optimisation of activated sludge systems;
- It can be extended with nitrogen- and phosphorus removal (Chapters 5 and 7);

– Design of auxiliary systems such as final settlers (Chapter 8), sludge thickeners and digesters (Chapter 12), primary clarifiers and UASB reactors (Chapter 13) can all be included. This allows for truly optimised design.

3.1 ORGANIC MATERIAL AND ACTIVATED SLUDGE COMPOSITION

3.1.1 Organic material fractions in wastewater

In this text COD will be used as the parameter for organic material measurement. Some advantages of its use over its alternatives BOD_5, BOD_{20} and TOC have been mentioned already in the previous chapter. In Section 3.1.3 another important advantage will be presented: the possibility to verify if the organic material mass balance closes. The concentration and composition of the organic material depends on the origin of the wastewater. For the purpose of modelling the activated sludge system, it is necessary to divide the influent organic material into four different fractions.

In Chapter 2 a first distinction was made between biodegradable and non-biodegradable material, the former being susceptible to metabolism by the bacterial mass, whereas the latter was not affected by the biochemical actions of the micro-organisms. For a more refined description of the activated sludge system, both the biodegradable and the non-biodegradable fractions are divided into a dissolved part and a particulate part. The subdivision takes into consideration the physical size of the organic material.

In most wastewaters, the particles of the organic material show a large variation in size: part is present as a true solution, another part may be present as a colloidal suspension and the remainder as a suspension with macroscopic particles. With respect to the metabolism of organic material by micro-organisms, a distinction is made between dissolved and particulate material (colloids and macroscopic particles), which is a simplification of a more complex reality. However, it will be shown that this simple approach leads to a surprisingly precise description of activated sludge behaviour, even under extreme operational conditions.

The activated sludge flocs act as a strong coagulant, resulting in the removal of particulate organic material by physical processes: the sludge flocs can capture the particles by screening, enmeshment or adsorption, making them part of the solid (sludge) phase. These physical processes remove both biodegradable- and non-biodegradable particles. This leads to the conclusion that the behaviour of organic material of the dissolved non-biodegradable fraction will be different from that of the particulate non-biodegradable fraction: the former will not be affected by the presence of sludge and will leave the process without modifications, whereas the latter will accumulate in the solid phase, until it is discharged as part of the excess sludge.

For the biodegradable organic material there is also a difference between the organic material of dissolved- and particulate origin. Due to the small size of the molecules in the dissolved fraction, these penetrate through the cell membranes and thus will be metabolised directly. In contrast, the particulate organic material can only be metabolised after several preparatory processes that may include flocculation, adsorption on the cell wall and hydrolysis of the adsorbed material with the consequential production of soluble organic material. It is concluded that the physical removal of the organic material occurs at a high rate, but in the case of particulate material, metabolism will not be immediate. In this chapter it will be shown that it may take several hours before organic influent material is actually metabolised in the activated sludge system. Thus a distinction can be made between easily (dissolved) and slowly (particulate) biodegradable materials.

As the division of the influent organic material in the four main fractions will be used frequently throughout this text, it is convenient that each be indicated by a separate symbol. Using S (substrate) as a generic symbol for organic material concentration (expressed as COD), the following parameters can be defined:

S_{ti} = (total) influent COD concentration
S_{bi} = biodegradable influent COD concentration
S_{ni} = non biodegradable influent COD concentration
S_{bpi} = biodegradable, particulate influent COD concentration
S_{bsi} = biodegradable, dissolved influent COD concentration
S_{npi} = non biodegradable, particulate influent COD concentration
S_{nsi} = non biodegradable, dissolved influent COD concentration

From the above definitions one has:

$$\begin{aligned} S_{ti} &= S_{bi} + S_{ni} \\ &= S_{bsi} + S_{bpi} + S_{nsi} + S_{npi} \end{aligned} \tag{3.1}$$

It will prove to be convenient to introduce the following fractions:

$$f_{ns} = S_{nsi}/S_{ti} \tag{3.1a}$$
$$= \text{non biodegradable, dissolved influent COD fraction}$$
$$f_{np} = S_{npi}/S_{ti} \tag{3.1b}$$
$$= \text{non biodegradable, particulate influent COD fraction}$$
$$f_{sb} = S_{sbi}/S_{bi} \tag{3.1c}$$
$$= \text{fraction of the biodegradable COD that is dissolved and easily biodegradable}$$

Applying these definitions one has:

$$S_{ni} = (f_{ns} + f_{np}) \cdot S_{ti} \quad \text{and} \tag{3.2}$$
$$S_{bi} = (1 - f_{ns} - f_{np}) \cdot S_{ti} \tag{3.3}$$

Figure 3.1 is a graphical representation of Eqs. (3.1 to 3.3). The methods to experimentally determine the value of the different fractions will be discussed later in this chapter. The numerical values of the fractions may vary significantly for different wastewaters, especially in the case of industrial wastewaters. Table 3.1 shows some examples.

The division of the influent organic material in four fractions is a simplification of a more complex reality, but it is adopted since a more complex model does not lead to a better simulation of reality and thus would be an unnecessary sophistication for the purpose of developing a general description of the activated sludge behaviour.

3.1.2 Activated sludge composition

The sludge concentration can be determined experimentally: the sludge is filtered and weighed after drying at 103°C, thus obtaining the total suspended solids concentration (TSS). The TSS can be divided into organic and inorganic solids. The mass of organic solids can be determined by means of ignition at

550°C, a temperature that results in the complete combustion of the organic solids. The organic solids concentration is calculated from the weight loss during the ignition. Due to the fact that the organic solids disappear during the ignition, these are also called volatile suspended solids (VSS), distinguishing them from the remaining, fixed suspended solids (FSS).

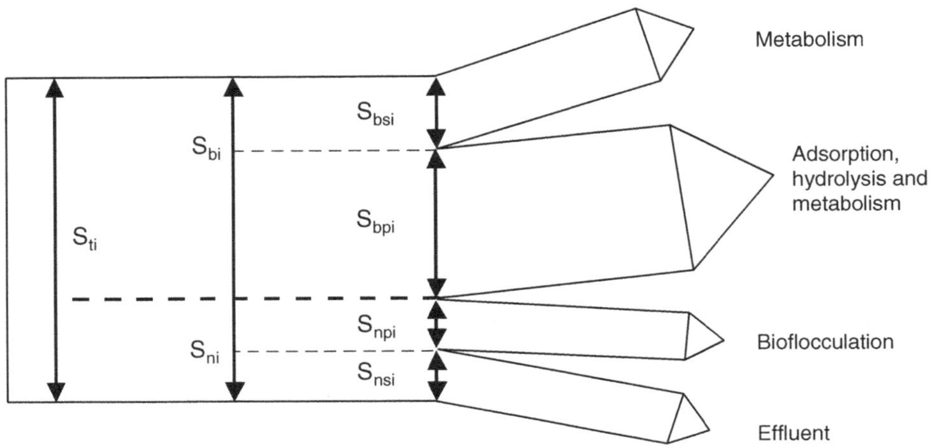

Figure 3.1 Characterisation of the influent COD in different fractions and their relation to the main processes in the activated sludge system

Table 3.1 Experimentally determined values of the influent organic material fractions for different types of wastewater

Type of wastewater	Fractions			Reference
	f_{ns}	f_{np}	f_{sb}	
Municipal sewage:				
– Campina Grande – Brazil (raw)	0.07	0.05	0.25	Coura Dias et al. (1983)
– Cape Town – South Africa:				
– Raw sewage	0.09	0.12	0.25	Marais and Ekama (1976)
– Pre-settled sewage	0.12	0.02	0.37	Marais and Ekama (1976)
– Burlington – Canada (raw)	0.12	0.25	–	Sutton et al. (1979)
Industrial wastewater:				
– Distillage (alcohol distillery)	0.02	0.02	0.80	(unpublished research)
– Black liquor (paper mill – pulp)	0.40	0.10	0.35	Macedo (1990)
– Petrochemical	0.20	0.06	0.25	Neto et al. (1994)

The inorganic sludge is generated by flocculation of inorganic influent material such as clay, silt and sand and by precipitation of salts during the biological treatment. In sludge from municipal wastewaters, the inorganic sludge fraction is in the order of 20 to 35% of the total sludge concentration.

In order to describe the activated sludge behaviour, Marais and Ekama (1976) suggested a subdivision of the volatile suspended solids (i.e. the organic sludge) in two basic fractions: (I) active sludge, composed of the living micro-organisms that act in the metabolism of the influent organic material and (II) inactive sludge composed of organic material that does not exhibit metabolic activity. It is important to stress, that this division is theoretical and that there is no test to directly determine the active or inactive sludge concentration: only the sum of the two can be determined experimentally. The division is justified by the fact that it leads to a rational model of the activated sludge system, capable of predicting the measurable parameters under strongly varying operational conditions.

3.1.2.1 Active sludge

The active sludge is generated from synthesis of influent organic material. The micro-organisms in the activated sludge system are composed of a large number of species of bacteria, fungi and protozoa. Depending on the operational conditions, more complex organisms like ciliates and rotifers may also be present. The composition of the active sludge may differ considerably from one system to the other, depending on the nature of the influent wastewater and the operational conditions. In spite of the complex nature of the active sludge mass, in this text it will be considered (for the purpose of modelling) as an equivalent bacterial suspension. To test the validity of this assumption, the predictions generated by the model will be compared to experimentally observed results. It must be stressed that although bacteria are predominant in the active sludge, its actual behaviour may be very different from a pure culture of bacteria.

3.1.2.2 Inactive sludge

The inactive sludge is composed of non biodegradable organic material and can be subdivided in two fractions in accordance with its origin: (I) the inert sludge and (II) the endogenous residue. The inert sludge fraction is generated from the accumulation of particulate non-biodegradable organic material present in the influent. This material is flocculated and becomes part of the solid phase, forming the inert fraction.

The endogenous residue has its origin in the decay of living bacteria cells, a process occurring continuously in the activated sludge system. During the decay process of the active sludge, part of the microbial mass is oxidised in a process called endogenous respiration. However, only part of the cellular mass is biodegradable: after decay a fraction of the decayed active sludge remains in the activated sludge as a non-biodegradable particulate fraction. The existence of the endogenous residue will be demonstrated and quantified in Chapter 12.

3.1.2.3 Inorganic sludge

Apart from the different organic fractions of the sludge, there is also an inorganic one. Inorganic solids may accumulate in the sludge from inert influent material such as silt and clay. An interesting issue is that the combustion of the organic material, containing not only carbon, hydrogen and oxygen but also other components such as phosphorus and metal ions will result in formation of phosphates, bicarbonates and metal-oxides. Due to combination of cell-internal metal ions (e.g. Na^+, K^+) with atmospheric oxygen, the mass of inert solids *after* combustion will be higher than the mass *before* combustion (e.g. $K^+ \rightarrow K_2O$). The mass increase of the inert fraction upon combustion is often not considered: when the calculation of the inorganic sludge mass fraction is only based on the accumulation of influent inorganic suspended solids, then the organic sludge mass fraction can be overestimated.

In experiments where a sludge batch was fed with only biodegradable COD (and nutrients), the formed inorganic residue of the organic sludge had a value of 10 – 15%. So the measured volatile sludge mass (active, inert and endogenous) is accompanied by a fixed solids mass fraction of $1/0.85 - 1 = 11 - 17\%$. Due to the presence of inorganic suspended material in the feed, in practice the ratio between volatile and total suspended solids tends to be less than 0.85 – 0.9. Depending on the origin of the wastewater and the operational conditions (pre-sedimentation, applied sludge age, quality of the sewer system), the ratio between volatile and total solids for domestic sewage will be in the order of 0.60 to 0.80. For industrial wastewaters containing a very low or even no inorganic material, this ratio will be close to the maximum value of 0.85 – 0.9.

3.1.2.4 Definition of sludge fractions

Having defined the different sludge fractions, it is convenient to introduce symbols for each. Using the letter "X" to generically indicate sludge concentration one has:

X_a = active sludge concentration (mg VSS $\cdot\, l^{-1}$)
X_e = endogenous sludge concentration (mg VSS $\cdot\, l^{-1}$)
X_i = inert sludge concentration (mg VSS $\cdot\, l^{-1}$)
X_v = organic or volatile sludge concentration (mg VSS $\cdot\, l^{-1}$)
X_m = mineral, fixed or inorganic sludge concentration (mg FSS $\cdot\, l^{-1}$)
X_t = sludge concentration (mg TSS $\cdot\, l^{-1}$)

From the definitions it follows that:

$$X_v = X_a + X_e + X_i \tag{3.4}$$

$$X_t = X_v + X_m = X_v/f_v \tag{3.5}$$

where f_v is the volatile sludge fraction:

- 0.65–0.75 for raw sewage;
- 0.70–0.80 for pre-settled sewage
- 0.80–0.90 for wastewaters without mineral suspended solids
- 0.50–0.70 for UASB plants treating raw sewage

Along with the three organic sludge fractions defined above, others may exist, depending on the operational conditions. If the sludge age is very short, the sludge wastage rate may be so high that there is not enough time for the metabolisation of all the influent biodegradable material, especially at low temperatures. In that case flocculation of the particulate biodegradable organic material in the influent will occur and this material will be adsorbed (stored) on the active sludge mass. Thus it is possible that part of the discharged organic sludge is actually flocculated influent organic material. The stored material fraction depends on the rate of metabolisation, the sludge age and on the composition of the influent organic material.

If nitrification takes place in the activated sludge system, a population of nitrifying bacteria (ammonium oxidizers and nitrite oxidizers) will develop. In the case of municipal sewage, the mass of nitrifying bacteria is very small compared to the total organic sludge mass, typically no more than a few percent. Finally, in the case of systems designed for biological phosphorus removal, a specific biomass will develop (phosphate accumulating organisms or bio-P organisms), with an increased phosphorus content of up to 38%. Naturally, in this case the ratio between volatile sludge and total sludge will be lower.

3.1.3 Mass balance of the organic material

When an activated sludge system receives a constant load of organic material, a sludge mass will develop that is quantitatively and qualitatively compatible with this load and the prevailing operational conditions. Under steady state conditions there is no accumulation of influent organic material, therefore it will either be discharged with the effluent, discharged with the excess sludge, or it will be transformed into inorganic products by oxidation.

Hence the daily applied mass or flux of influent organic material will be equal to the sum of the fluxes of (I) organic material in the effluent, (II) organic material contained in the excess sludge, and (III) the flux of oxidised material. There are basically only two transformations possible for the organic material in the activated sludge system:

- Transformation into organic sludge by biochemical (anabolism, decay) or physical processes (flocculation, adsorption);
- Oxidation into inorganic products.

Figure 3.2 shows a schematic representation of a basic activated sludge system. It can be observed that the ingoing mass flux of influent organic material can only leave the activated sludge system through three distinct routes, identified as follows:

- Part of the influent organic material is not removed from the liquid phase and leaves the activated sludge system together with the effluent (MS_{te} in Figure 3.2);
- A second fraction of the organic material is transformed into organic sludge and is discharged as excess sludge (MS_{xv});
- The third fraction of the organic material is oxidised (MS_o).

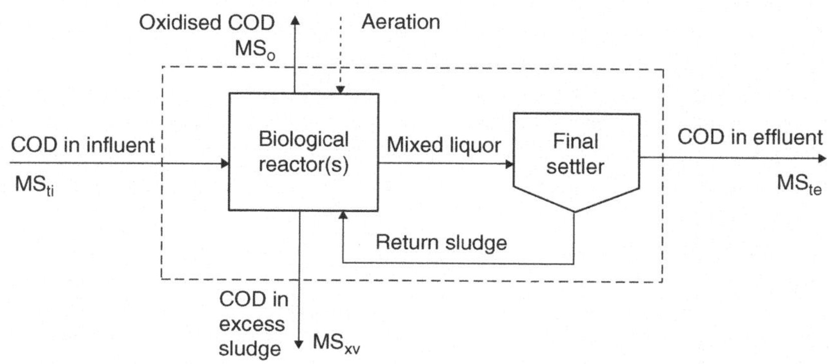

Figure 3.2 Flow diagram of the steady state activated sludge system and the associated COD fluxes

As all fractions are generated from the influent organic material (MS_{ti}), one has:

$$MS_{ti} = MS_{te} + MS_{xv} + MS_o \tag{3.6}$$

where:

MS_{ti} = daily applied COD mass ($kg\ COD \cdot d^{-1}$)
MS_{te} = daily COD mass in the effluent ($kg\ COD \cdot d^{-1}$)
MS_{xv} = daily COD mass in the excess sludge ($kg\ COD \cdot d^{-1}$)
MS_o = daily mass of oxidised COD ($kg\ O_2 \cdot d^{-1}$ or $kg\ COD \cdot d^{-1}$)

Eq. (3.6) expresses that in an activated sludge system under steady state conditions the flux of influent organic material is equal to the fluxes of organic material or its products that leave the activated sludge system. In order to verify the validity of Eq. (3.6) it is necessary to transform the fluxes MS_{ti}, MS_{te}, MS_{xv} and MS_o into experimentally measurable parameters. The COD fluxes in the influent and effluent can be transformed easily.

$$MS_{ti} = Q_i \cdot S_{ti} \tag{3.7}$$

$$MS_{te} = (Q_i - q) \cdot S_{te} \tag{3.8}$$

where:

Q_i = influent flow ($m^3 \cdot d^{-1}$)
q = excess sludge flow ($m^3 \cdot d^{-1}$)
S_{ti} = influent COD ($mg\ COD \cdot l^{-1}$)
S_{te} = effluent COD ($mg\ COD \cdot l^{-1}$)

The flux of organic material discharged as excess sludge can be determined from the volatile sludge concentration and the dissolved COD concentration in the excess sludge. Knowing that there is a proportionality between the volatile solids mass and its COD ($f_{cv} = 1.5\ mg\ COD \cdot mg\ VSS^{-1}$) one has:

$$MS_{xv} = q \cdot (f_{cv} \cdot X_v + S_{te}) \tag{3.9}$$

In Eq. (3.9) it is assumed that the dissolved COD concentration in the excess sludge is equal to the effluent COD concentration, a supposition that will prove to be justified (refer also to Appendix 2. The flux of oxidised organic material, MS_o, can be determined from the consumption of dissolved oxygen (DO) in the mixed liquor. By definition, in order to oxidise 1 kg of COD, the oxygen requirement will be 1 kg of O_2. Hence, the flux of oxidised organic material will be numerically equal to the flux of consumed oxygen. The latter flux is equal to the product of the reactor volume and the oxygen uptake rate (OUR). The OUR is the mass of oxygen consumed per unit of time in a unit volume of mixed liquor and can be determined experimentally.

The principle of the OUR test is the following: while the influent flow rate continues as normal, the aeration of the mixed liquor is interrupted. After the interruption the decrease of the DO concentration with time (due to consumption) is observed and -preferably- recorded. The decrease of the DO concentration is linear with time and the gradient of this linear function is equal to the OUR. A more detailed description of the OUR test and its limitations can be found in Appendix 1, Appendix 2 and Section 4.2.

The value of the OUR determined as described above equals the total oxygen uptake rate. However, part of the consumed oxygen may have been used for nitrification in the activated sludge system. It is possible to estimate the consumption rate for nitrification (O_n) from the increase of the nitrate concentration in the activated sludge system. Thus the OUR for the oxidation of organic material (O_c) can be determined indirectly, by subtracting the oxygen uptake rate for nitrification (O_n) from the total oxygen uptake rate (O_t):

$$O_c = O_t - O_n \tag{3.10}$$

where:

O_t = total OUR ($mg\ O_2 \cdot l^{-1} \cdot d^{-1}$)
O_n = OUR for nitrification ($mg\ O_2 \cdot l^{-1} \cdot d^{-1}$)
O_c = OUR for oxidation of organic material ($mg\ O_2 \cdot l^{-1} \cdot d^{-1}$)

Having established the value of O_c, the flux of oxidised organic material is determined as:

$$MS_o = O_c \cdot V_r \tag{3.11}$$

where V_r = reactor volume

Now, using the expressions of Eqs. (3.7 to 3.11) in Eq. (3.6), one has:

$$\begin{aligned} Q_i \cdot S_{ti} &= (Q_i - q) \cdot S_{te} + q \cdot (f_{cv} \cdot X_v + S_{te}) + O_c \cdot V_r \text{ or} \\ S_{ti} &= S_{te} + (q/Q_i) \cdot f_{cv} \cdot X_v + O_c \cdot R_h \end{aligned} \tag{3.12}$$

where R_h = liquid retention time = V_r/Q_i

In Eq. (3.12) all variables are measurable, so that the validity of the equation can be verified experimentally. However, in general it will be unlikely that an exact equality of the two sides of Eq. (3.12) is found. This is partly due to the fact that the tests are subject to experimental errors, but also because the activated sludge system usually is not operated under rigorously steady state conditions, which is a presupposition for the validity of Eq. (3.12). For this reason the recovery factor for organic material is defined as:

$$B_o = (MS_{te} + MS_{xv} + MS_o)/MS_{ti} = (S_{te} + (q/Q_i) \cdot f_{cv} \cdot X_v + O_c \cdot R_h)/S_{ti} \tag{3.13}$$

From Eq. (3.13) it can be concluded that the theoretical value of the recovery factor is identical to one. Due to analytical errors, the value of B_o will deviate from its theoretical value. However, when the average value of a series of steady state experiments over a period (for example a few weeks) is considered, the deviation between the theoretical and the experimental value of the recovery factor will typically be less than 10%. Stated differently, if there is a systematic difference between the theoretical and experimental value of B_o, there is good reason to suspect that one or more of the tests used to calculate B_o is not being carried out properly or that the activated sludge system is not yet operating under steady state conditions. On the other hand, a closing mass balance (i.e. an experimental B_o value between 0.9 and 1.1) is a clear indication that the system was operating under steady state conditions and that the tests to determine B_o were carried out correctly. Hence, the verification of a closing mass balance is a powerful indication that the experimental data are reliable.

When BOD is used (the alternative parameter for organic material), it is not possible to verify if the mass balance closes. In the previous section it was shown that in the activated sludge system a non-biodegradable sludge fraction, the endogenous residue, is generated from the decay of active sludge. Part of the biodegradable influent material (with associated BOD demand) is converted into non-biodegradable endogenous residue (without associated BOD demand) so that the mass balance cannot close: the activated sludge system is a "BOD sink" in which BOD disappears without corresponding oxidisation.

The value of the BOD flux in the effluent and in the excess sludge, together with the oxygen consumption for the oxidation of biodegradable organic material in the influent, will always be smaller than the BOD flux in the influent. The fact that it is not possible to verify whether the mass balance closes, when BOD is used as a quantitative parameter for organic material, is a very serious disadvantage for this test. In addition to the shortcomings of the BOD test discussed earlier, this is another reason that in the present text COD rather than BOD is used to quantify the concentration of organic material.

EXAMPLE 3.1

As an example of a mass balance calculation, the experimental data of Dias et al. (1981) in Table 3.2 will be analysed. In this experiment a bench scale activated sludge system treating raw sewage was operated at 5 different sludge ages in five sets of experiments (I to V). Table 3.2 shows the results of the daily analyses, reported per set in column 1. The OUR was determined for oxidation of carbonaceous material only (nitrification was inhibited by adding allyl-thio urea, a toxic compound for nitrifiers, but not for heterotrophs).

Table 3.2 Experimental results of an activated sludge system (steady state conditions)

Set	V_r (litre)	Q_i ($l \cdot d^{-1}$)	q ($l \cdot d^{-1}$)	S_{ti} ($mg \cdot l^{-1}$)	S_{te} ($mg \cdot l^{-1}$)	X_v ($mg \cdot l^{-1}$)	OUR_c ($mg \cdot l^{-1}$)	B_o (–)
I	10	16	3.33	730	127	1060	20.3	1.04
II	12	16	1.20	691	97	2235	19.6	1.02
III	15	16	0.75	780	91	2538	23.6	1.03
IV	12	16	0.60	785	155	3012	25.8	1.00
V	15	14	0.50	803	77	2686	21.5	0.97

Solution

Applying Eq. (3.13) for each of the five sets of experiments, the B_o values can be calculated. As an example, for set I one has:

$$B_o = (S_{te} + (q/Q) \cdot f_{cv} \cdot X_v + R_h \cdot O_c)/S_{ti}$$
$$= (127 + (3.33/16) \cdot 1.5 \cdot 1060 + (10/16) \cdot 20.3 \cdot 24)/730 = 1.04$$

The calculated values for B_o are in the last column of Table 3.2. It can be noted that in all experiments the B_o values tend towards the theoretical value of 1.00. The weighted average of all sets of experiments was $B_O = 1.02$, which means that there is a difference of 2% between the experimental and the theoretical value of B_o. As this difference is very small, it is concluded that the experimental data are reliable.

For the analysis of the behaviour of the activated sludge system, it is convenient to have explicit expressions for the different COD fractions (I) discharged with the effluent, (II) discharged as excess sludge, and (III) oxidised. To find these expressions Eq. (3.13) may be rewritten as note:

$$B_o = S_{te}/S_{ti} + (q/Q_i) \cdot f_{cv} \cdot X_v/S_{ti} + R_h \cdot O_c/S_{ti} \quad \text{or}$$
$$B_o = mS_{te} + mS_{xv} + mS_o \tag{3.14}$$

Note that this is a simplified equation as the values of mS_{te} and mS_{xv} are not compensated for $q \cdot S_{te}$, however the effect is very small. The values of mS_{te}, mS_{xv} and mS_o are defined as:

$mS_{te} = S_{te}/S_{ti}$
= fraction of the influent COD discharged in the effluent
$mS_{xv} = (q/Q_i) \cdot f_{cv} \cdot X_v/S_{ti}$
= fraction of the influent COD discharged with the excess sludge
$mS_o = R_h \cdot O_c/S_{ti}$
= fraction of the influent COD oxidised in the process.

The numerical values of these fractions are of very great importance for a description of the behaviour of the activated sludge system: the fraction mS_{te} is indicative for the effluent quality, the value of mS_{xv} is representative for the sludge production (and consequently for the design of the excess sludge treatment units) and the mS_o value is a measure for the oxygen demand in the process (and hence for the oxygenation capacity to be installed). As an example for the data of set I in Table 3.2 one can calculate the following values for the three fractions defined above:

$$mS_{te} = S_{te}/S_{ti} = 127/730 = 0.17$$
$$mS_{xv} = (q/Q_i) \cdot f_{cv} \cdot X_v/S_{ti} = 3.33/16 \cdot 1.5 \cdot 1060/730 = 0.45$$
$$mS_o = (V_r/Q_i) \cdot O_c/S_{ti} = 10/16 \cdot 20.3 \cdot 24/730 = 0.42$$

The sum of the three fractions is equal to the value of the recovery factor B_o.

$$B_o = 0.17 + 0.45 + 0.42 = 1.04$$

In Table 3.3 the values of the fractions mS_{te}, mS_{xv} and mS_o have been calculated for each of the sets of experiments in Table 3.2. The experiments show that the applied operational conditions have an influence on the values of the fractions, especially in the case of mS_{xv} and mS_o. The mass balance for organic material allows the determination of the values of the COD fraction in the effluent, in the excess sludge and oxidised in the reactor. However, in practice it is of more interest to be able to *predict* the division of the influent organic material over the three fractions, rather than to calculate their existing values. In order to be able to do so, it is necessary to develop a model to describe the behaviour of the activated sludge system in a quantitative manner, so that theoretical values for the fractions mS_{te}, mS_{xv} and mS_o can be calculated. In the next sections a model is developed that allows the three fractions to be estimated as a function of the concentration and composition of the influent organic material and the operational conditions of the activated sludge system.

Table 3.3 Values of the fractions mS_{te}, mS_{xv} and mS_o as determined from the 5 sets of experiments listed in Table 3.2

Exp.	mS_{te}	mS_{xv}	mS_o	B_n	R_s
I	0.17	0.45	0.42	1.04	3
II	0.13	0.37	0.52	1.02	10
III	0.12	0.23	0.68	1.03	20
IV	0.20	0.29	0.59	1.08	20
V	0.10	0.17	0.69	0.97	30

3.2 MODEL NOTATION

Several parameters have already been introduced in the previous sections. As discussed, most model parameters used in this book are constructed from:

- One or more *CAPITAL* letters identifying the main class;
- One or more $_{subscript}$ letters identifying the subclass or type.

As an example: X_v is composed of the capital letter (X), which stands for biomass or sludge concentration and the subscript letter ($_v$), which stands for volatile. So X_v means volatile sludge concentration. Furthermore, in this book a parameter can often be expressed in different formats, as for instance in the case of S_e – the effluent COD concentration ($mg\ COD \cdot l^{-1}$):

- MS_e for effluent COD load ($kg\ COD \cdot d^{-1}$);
- mS_e for mass of effluent COD per mass of applied COD ($mg\ COD \cdot mg^{-1}\ COD$).

This approach will prove to be very convenient, as (I) it significantly reduces the number of parameters that need to be defined and (II) it is a very logical approach to managing model parameters. It will however require some effort (and practice) to become familiarised with this method of model notation. In this section the fundamental logic will be briefly explained. In the steady state model, most parameters can be expressed as:

- Concentration;
- Total mass or total mass flow (preceded by a capital "M");
- Specific production or consumption or specific unit mass present in the system per unit mass daily applied COD (preceded by an "m" in normal font).

The different representations will be further detailed below.

(a) Concentration
The units of measure (UoM) are either $mg \cdot l^{-1}$ or $g \cdot l^{-1}$, the latter being equivalent to $kg \cdot m^{-3}$. Depending on the context, the concentration can be expressed as either:

- Per litre influent or -effluent;
- Per litre reactor volume.

(b) Total mass or total mass flow

In the case of total mass (kg), this only involves those parameters relating to the sludge mass present in the activated sludge system (kg VSS or kg TSS):

- MX_a = active sludge mass (kg VSS) $(= V_r \cdot X_a)$;
- MX_i = inert sludge mass (kg VSS) $(= V_r \cdot X_i)$;
- MX_e = endogenous sludge mass (kg VSS) $(= V_r \cdot X_e)$;
- MX_v = organic sludge mass (kg VSS) $(= MX_a + MX_i + MX_e)$;
- MX_t = total sludge mass (kg TSS) $(= MX_v/f_v)$.

In the case of total mass flows ($kg \cdot d^{-1}$), the following subclasses can be identified:

(1) Applied influent or -effluent load, for example:
- MS_{ti} = average COD load ($kg\,COD \cdot d^{-1}$) $(= Q_i \cdot S_{ti})$;
- MP_{ti} = average phosphorus load ($kg\,COD \cdot d^{-1}$) $(= Q_i \cdot P_{ti})$;

(2) Production (e.g. excess sludge, biogas, …):
- ME_v = organic excess sludge production ($kg\,VSS \cdot d^{-1}$) $(= MX_v/R_s)$;
- ME_t = excess sludge production ($kg\,TSS \cdot d^{-1}$) $(= MX_t/R_s$ or $ME_v/f_v)$;
- MS_{xv} = organic excess sludge production (in $kg\,COD \cdot d^{-1}$) $(= f_{cv} \cdot ME_v)$;

(3) Consumption (e.g. oxygen, nutrients):
- MO_c = oxygen demand for COD oxidation ($kg\,O_2 \cdot d^{-1}$) $(= V_{aer} \cdot O_c)$;
- MO_n = oxygen demand for nitrification ($kg\,O_2 \cdot d^{-1}$) $(= V_{aer} \cdot O_n)$;
- MN_l = nitrogen content of excess sludge ($kg\,N \cdot d^{-1}$) $(= f_n \cdot ME_v)$.

(c) Specific production/consumption or sludge mass per unit mass daily applied COD

In the case of unit sludge mass that will develop in the system per unit mass of daily applied COD, the same parameters exist as for total sludge mass. The total sludge mass is divided by the applied daily COD load, so the unit of measure is $kg/(kg\,COD \cdot d^{-1})$ or $kg \cdot d \cdot kg^{-1}$ COD.

- mX_a = active sludge mass per kg daily applied COD $(= MX_a/MS_{ti})$;
- mX_i = inert sludge mass per kg daily applied COD $(= MX_i/MS_{ti})$;
- mX_e = endogenous residue per kg daily applied COD $(= MX_e/MS_{ti})$;
- mX_v = organic sludge mass per kg daily applied COD $(= mX_a + mX_e + mX_i)$;
- mX_t = total sludge mass per kg daily applied COD $(= MX_t/MS_{ti} = mX_v/f_v)$.

In the case of load-, consumption- or production per unit mass of applied COD, these parameters are defined as daily load, production or consumption divided by the daily applied influent COD load MS_{ti}. Therefore the unit of measure is equal to $(kg \cdot d^{-1})/(kg\,COD \cdot d^{-1})$ or $kg \cdot kg^{-1}$ COD.

(1) Applied influent load or load in effluent:
- mS_{te} = fraction of influent COD discharged with the effluent $(= MS_{te}/MS_{ti})$.

(2) Production (e.g. excess sludge, biogas, …):
- mE_v = specific production of organic excess sludge ($kg\,VSS \cdot kg^{-1}$ COD);
  $= MX_v/R_s$ or ME_v/MS_{ti};
- mE_t = specific excess sludge production ($kg\,TSS \cdot kg^{-1}$ COD);
  $= MX_t/R_s$ or ME_t/MS_{ti}.

(3) Consumption (e.g. oxygen, nutrients):
- mO_c = mass of oxygen consumed per unit mass applied COD (= MO_c/MS_{ti});
- mN_l = mass of nitrogen required per unit mass applied COD (= MN_l/MS_{ti});

In the subsequent sections a steady state model will be developed, which will allow calculation of the values of the parameters above (fractions, specific consumption and -production), based on (I) the influent composition, (II) a limited number of kinetic-/stoichiometric parameters and (III) the applied sludge age. It will then be an easy matter to calculate for instance the total excess sludge production demand, simply by multiplying the calculated value of mE_t with the design daily influent COD load: $ME_t = mE_t \cdot MS_{ti}$.

3.3 STEADY-STATE MODEL OF THE ACTIVATED SLUDGE SYSTEM

3.3.1 Model development

The first important step towards modelling the activated sludge system is to simplify the system to the largest extent possible. First an ideal activated sludge system for COD removal will be considered with one completely mixed reactor, operating under constant flow and load conditions. The term ideal indicates that (I) all the biodegradable organic material is effectively metabolised in the process and (II) the settler is a perfect liquid-solid separator in the sense that there are no suspended solids in the effluent and that the sludge hold-up in the settler is negligible in relation to the sludge mass in the biological reactor.

The term constant flow and load implies that the excess sludge and the influent both have a constant flow rate and composition. As for the influent, it is important that the average daily COD loads are comparable. A fixed quantity of excess sludge discharge is necessary to establish a constant sludge mass in the process, characterised by the fact that the sludge growth rate is equal to the withdrawal rate due to excess sludge wastage. It is also assumed that the sludge is discharged directly from the reactor (hydraulic wasting) and that the composition of the excess sludge is equal to that of the mixed liquor in the reactor.

Later in this chapter a general model will be discussed that can also be applied when the above restrictions do not apply, resulting in a much more complex process description. In Figure 3.3 the processes that form the basis of the ideal steady state model for the activated sludge system are represented. When a wastewater containing organic material is placed in contact with an activated sludge mass under aerated conditions, the following processes will occur: metabolism, decay and bioflocculation.

(a) Metabolism

The biodegradable organic material in the influent is removed from the liquid phase and metabolised by the sludge. It was observed in Chapter 2 that this process leads to both sludge growth (anabolism) and oxygen consumption (catabolism).

(b) Decay

It is postulated that sludge decay is independent of metabolic processes and that part of the decayed active sludge is oxidised to inorganic compounds, whereas the remainder accumulates in the reactor as endogenous residue until it is discharged with the excess sludge. The oxygen consumption due to oxidation of active sludge is called endogenous respiration, to distinguish it from the oxidation of influent organic material, which is called exogenous respiration. The independence of endogenous and exogenous respiration will be demonstrated in Chapter 12.

(c) Bioflocculation

The particulate non-biodegradable organic material in the influent is not affected by the metabolic activity of the sludge, but is removed physically from the liquid phase by flocculation. The flocculated material constitutes the inert organic sludge fraction. In the model of Figure 3.3 the biodegradable fractions and the particulate non-biodegradable fractions are removed from the liquid phase, but the fourth fraction, dissolved non-biodegradable organic material is not affected in any way by the activated sludge system and is discharged without modifications into the effluent.

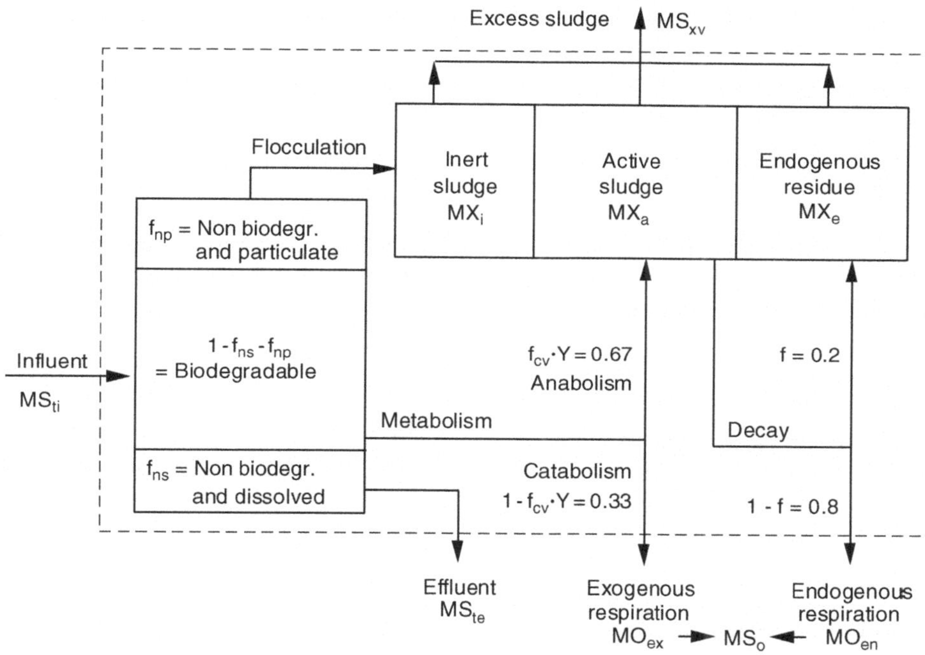

Figure 3.3 Overview of the processes that develop in an ideal activated sludge system

3.3.1.1 Definition of sludge age

Having defined the conditions to formulate the simplified model, the most important operational variable will now be defined: the sludge age R_s. This parameter indicates the average retention time of the sludge in the system and is defined as the ratio between the sludge mass present in the system and the daily sludge mass discharged from it. Using the model of Figure 3.3 and assuming hydraulic sludge wasting (i.e. excess sludge discharge directly from the aeration tank, which has many benefits that will be discussed in later sections), one has:

$$\begin{aligned} R_s &= MX_t/ME_t \\ &= V_r \cdot X_t/(q \cdot X_t) = V_r/q \end{aligned} \qquad (3.15)$$

where:

R_s = sludge age (d)
MX_t = sludge mass in the system (kg TSS)
ME_t = daily discharge of excess sludge (kg TSS $\cdot$ d^{-1})

Equation (3.15) can also be written in another way:

$$q = V_r/R_s \tag{3.16}$$

Equation (3.16) expresses that the flow of excess sludge, when discharged directly from the biological reactor, is a fraction $1/R_s$ of the reactor volume, i.e. over a period of R_s days the volume of wasted sludge is equal to the reactor volume.

The sludge age is independent of the liquid (or hydraulic) retention time R_h. This parameter is defined as the ratio between the reactor volume and the influent flow:

$$R_h = V_r/Q_i \tag{3.17}$$

Using the definitions for the sludge age and the liquid retention time, it is now possible to derive expressions to predict the values of the COD fractions mS_{te}, mS_{xv} and mS_o, which is the objective of the simplified model for the activated sludge system.

3.3.1.2 COD fraction discharged with the effluent

In the ideal activated sludge system, the effluent COD and the COD of the liquid phase of the mixed liquor are both equal to the concentration of non-biodegradable dissolved organic material in the influent, S_{nsi}. From the definition of S_{nsi} in Eq. (3.2), one has:

$$mS_{te} = S_{te}/S_{ti} = S_{nsi}/S_{ti} = f_{ns} \tag{3.18}$$

Hence the simplified model predicts a constant effluent COD, independent of the sludge age or the liquid retention time and equal to the non biodegradable, dissolved influent COD fraction.

3.3.1.3 COD fraction in the excess sludge

The determination of this fraction is more complicated and requires derivation of expressions for the three fractions that compose the organic sludge: inert sludge, active sludge and endogenous residue.

(a) The inert sludge X_i

The inert sludge concentration can be calculated easily from a simple mass balance using Figure 3.3. The inert sludge is generated by flocculation of the particulate and non-biodegradable material in the influent and is discharged in the excess sludge. Loss of X_i with the effluent is ignored as $q \cdot X_i >> (Q_i - q) \cdot X_{ie}$. Since the inert material is not affected by biochemical processes, the mass flow in the excess sludge must be equal to the influent mass flow, so that:

$$q \cdot X_i = Q_i \cdot X_{ii} \tag{3.19}$$

where

X_{ii} = concentration of non-biodegradable suspended solids in the influent (mg VSS.l^{-1})

The concentration X_{ii} can be correlated to the particulate and non-biodegradable COD fraction in the influent, by recognising the proportionality between COD and volatile suspended solids ($f_{cv} = 1.5$ mg $COD \cdot mg^{-1}$ VSS):

$$X_{ii} = S_{npi}/f_{cv} = (f_{np}/f_{cv}) \cdot S_{ti} \tag{3.20}$$

Now, using Eq. (3.20) in Eq. (3.19) and inserting the relationship $q = V_r/R_s$ leads to:

$$\begin{aligned} X_i &= (f_{np}/f_{cv}) \cdot (Q_i/q) \cdot S_{ti} \\ &= f_{np} \cdot R_s \cdot MS_{ti}/(f_{cv} \cdot V_r) \text{ or } f_{np} \cdot R_s \cdot S_{ti}/(f_{cv} \cdot R_h) \end{aligned} \tag{3.21}$$

(b) The active sludge X_a

As can be observed in Figure 3.3, the active sludge concentration is affected by three processes: (I) sludge growth due to synthesis, (II) decay and (III) sludge wastage. The variation of the active sludge concentration can be expressed as the sum of these three processes:

$$dX_a/dt = (dX_a/dt)_g + (dX_a/dt)_d + (dX_a/dt)_e \tag{3.22}$$

where:

X_a = active sludge concentration (mg $VSS.1^{-1}$)
dX_a/dt = rate of change of the active sludge concentration (mg $VSS \cdot l^{-1} \cdot d^{-1}$)
$(dX_a/dt)_g$ = growth rate due to synthesis (mg $VSS \cdot l^{-1} \cdot d^{-1}$)
$(dX_a/dt)_d$ = decay rate of active sludge (mg $VSS \cdot l^{-1} \cdot d^{-1}$)
$(dX_a/dt)_e$ = wastage rate of active sludge in excess sludge (mg $VSS \cdot l^{-1} \cdot d^{-1}$)

Under steady state conditions, the active sludge concentration does not change with time:

$$dX_a/dt = 0 = (dX_a/dt)_g + (dX_a/dt)_d + (dX_a/dt)_e \tag{3.23}$$

The active sludge growth rate is proportional to the utilisation rate of biodegradable material, with a yield of Y kg active sludge synthesised per kg utilised COD. In the ideal activated sludge system the utilisation rate of biodegradable material will be equal to the feed rate ($V_r \cdot r_u = Q_i \cdot S_{bi}$), so that the substrate utilization rate can be calculated as:

$$r_{us} = S_{bi} \cdot Q_i/V_r = S_{bi}/R_h \tag{3.24}$$

where $r_{us} = (dS_{bi}/dt)_u$ = utilisation rate of biodegradable material (mg $COD \cdot l^{-1} \cdot d^{-1}$)

Having defined r_{us}, the growth rate of active sludge can be calculated as:

$$r_g = (dX_a/dt)_g = Y \cdot r_{us} = Y \cdot S_{bi} \cdot Q_i/V_r = Y \cdot S_{bi}/R_h \tag{3.25}$$

where Y = yield coefficient for active sludge (mg $VSS \cdot mg^{-1}$ COD)

In Chapter 12 it will be shown that the decay rate of active sludge can be expressed as a first order process with respect to the active sludge concentration:

$$r_d = (dX_a/dt)_d = -b_h \cdot X_a \tag{3.26}$$

where b_h = decay constant for active sludge (d^{-1})

The rate at which the active sludge concentration decreases due to sludge wastage can by definition be expressed as:

$$\begin{aligned} R_s &= \text{(active sludge mass)/(wastage rate of active sludge)} \\ &= V_r \cdot X_a/[V_r \cdot (-dX_a/dt)_e] \end{aligned}$$

Hence:

$$r_e = (dX_a/dt)_e = -X_a/R_s \tag{3.27}$$

Substituting Eqs. (3.25 to 3.27) in Eq. (3.23), the following expression is obtained for the active sludge concentration:

$$Y \cdot Q_i \cdot S_{bi}/V_r - b_h \cdot X_a - X_a/R_s = 0 \quad \text{or}$$

$$X_a = [Y \cdot R_s/(l + b_h \cdot R_s)] \cdot Q_i \cdot S_{bi}/V_r \tag{3.28}$$

Now by using Eq. (3.3) to substitute for S_{bi} one has:

$$\begin{aligned} X_a &= [(1 - f_{ns} - f_{np}) \cdot Y \cdot R_s/(1 + b_h \cdot R_s)] \cdot Q_i \cdot S_{ti}/V_r \\ &= (1 - f_{ns} - f_{np}) \cdot C_r \cdot S_{ti}/R_h \end{aligned} \tag{3.29}$$

where:

$$C_r = Y \cdot R_s/(1 + b_h \cdot R_s) \tag{3.30}$$

C_r represents the *active* sludge mass present in the system per unit mass daily applied *biodegradable* organic material. The inverse of C_r is the COD utilisation rate per unit mass active sludge, also known as the specific utilisation rate of organic material, which will be discussed in Section 3.3.3.6

(c) The endogenous residue X_e

Once again, under steady state conditions the concentration of the endogenous residue does not change with time. Thus the concentration can be calculated from the fact that the production rate is equal to the withdrawal rate:

$$(dX_e/dt) = 0 = (dX_e/dt)_d + (dX_e/dt)_e \tag{3.31}$$

where (dX_e/dt) is equal to the rate of change of endogenous residue concentration. Indices "d" and "e" refer to active sludge decay and excess sludge wastage respectively.

Upon decay of active sludge, a constant fraction is transformed into endogenous residue, whereas the remainder is oxidised. Hence, the production rate of endogenous residue is proportional to the active sludge decay rate and the proportionality constant is equal to the fraction of decayed active sludge remaining as endogenous residue. Hence:

$$(dX_e/dt)_d = -f \cdot (dX_a/dt)_d = f \cdot b_h \cdot X_a \tag{3.32}$$

where f = fraction of decayed active sludge transformed into endogenous residue.

The rate of decrease of the endogenous residue concentration due to sludge wastage is calculated using Eq. (3.27):

$$(dX_e/dt)_e = -X_e/R_s \tag{3.33}$$

Substituting Eqs. (3.32 and 3.33) in Eq. (3.31) one has:

$$f \cdot b_h \cdot X_a - X_e/R_s = 0 \quad \text{or}$$

$$X_e = f \cdot b_h \cdot R_s \cdot X_a \tag{3.34}$$

(d) The organic sludge

The organic or volatile sludge concentration is equal to the sum of the three fractions: inert, active and endogenous residue. Hence, from Eqs. (3.21, 3.29 and 3.34) one has:

$$X_v = X_a + X_e + X_i = [(1 - f_{ns} - f_{np}) \cdot C_r \cdot (1 + f \cdot b_h \cdot R_s) + f_{np} \cdot R_s/f_{cv}] \cdot S_{ti}/R_h \tag{3.35}$$

The expression for the organic sludge concentration is particularly important because this parameter can be determined experimentally, allowing the possibility to verify if the calculated theoretical concentration is equal to the actual value. After having derived an expression for the organic sludge concentration, it becomes a simple matter to calculate the sludge mass in the reactor and the excess sludge production. The product of the volatile sludge concentration and the reactor volume V_r gives the sludge mass MX_v. For a particular sludge age R_s, the sludge production rate will be a fraction $1/R_s$ of the existing sludge mass.

$$MX_v = V_r \cdot X_v = [(1 - f_{ns} - f_{np}) \cdot (1 + f \cdot b_h \cdot R_s) \cdot C_r + f_{np} \cdot R_s/f_{cv}] \cdot MS_{ti} \quad \text{and} \tag{3.36}$$

$$ME_v = V_r \cdot X_v/R_s = [(1 - f_{ns} - f_{np}) \cdot (1 + f \cdot b_h \cdot R_s) \cdot C_r/R_s + f_{np}/f_{cv}] \cdot MS_{ti} \tag{3.37}$$

where:

MX_v = organic sludge mass in the system (kg VSS)
ME_v = daily organic sludge production (kg VSS $\cdot$ d^{-1})

Having established an expression for the sludge production rate and knowing that there is a proportionality between the organic sludge mass and its COD, it is now possible to calculate the fraction of the influent COD that is wasted as excess sludge:

$$\begin{aligned} mS_{xv} &= f_{cv} \cdot ME_v/MS_{ti} = f_{cv} \cdot (V_r \cdot X_v/R_s)/(Q_i \cdot S_{ti}) \\ &= f_{cv} \cdot (1 - f_{ns} - f_{np}) \cdot (1 + f \cdot b_h \cdot R_s) \cdot C_r/R_s + f_{np} \end{aligned} \tag{3.38}$$

EXAMPLE 3.2

An activated sludge system for secondary treatment is operated at a sludge age of 10 days, under the following conditions:

- MS_{ti} = 6000 kg COD $\cdot$ d^{-1}
- $f_{ns} = f_{np} = 0.1$
- V_r = 5000 m^3
- Y = 0.45 kg VSS $\cdot$ kg^{-1} COD
- $b_h = 0.24\ d^{-1}$
- f_v = 0.75 kg VSS $\cdot$ kg^{-1} TSS

Calculate the concentrations of the various sludge fractions and determine the organic excess sludge production per unit mass applied COD, i.e. both in terms of organic solids (mE_v) and in terms of COD (mS_{xv}).

Solution

$$X_i = (f_{np}/f_{cv}) \cdot R_s \cdot MS_{ti}/V_r \quad (3.21)$$
$$= (0.1/1.5) \cdot 10 \cdot 600/5000 = 0.8\,kg\,VSS \cdot m^{-3}$$
$$C_r = Y \cdot R_s/(1 + b_h \cdot R_s) \quad (3.30)$$
$$= 0.45 \cdot 10/(1 + 0.24 \cdot 10) = 1.32\,kg\,VSS \cdot d \cdot kg^{-1}\,COD$$
$$X_a = (1 - f_{ns} - f_{np}) \cdot C_r \cdot MS_{ti}/V_r \quad (3.29)$$
$$= 0.8 \cdot 1.32 \cdot 6000/5000 = 1.27\,kg\,VSS \cdot m^{-3}$$
$$X_e = f \cdot b_h \cdot R_s \cdot X_a \quad (3.34)$$
$$= 0.2 \cdot 0.24 \cdot 10 \cdot 1.27 = 0.61\,kg\,VSS \cdot m^{-3}$$
$$X_v = X_i + X_a + X_e = 2.68\,kg\,VSS \cdot m^{-3} \quad (3.35)$$

Now $mE_v = ME_v/MS_{ti} = (V_r \cdot X_v/R_s)/MS_{ti} = 1340/6000 = 0.22$ kg VSS $\cdot$ kg^{-1} COD and after multiplication with the proportionality constant f_{cv}:

$$mS_{xv} = f_{cv} \cdot mE_v = 1.5 \cdot 0.22 = 0.34\,kg\,COD \cdot kg^{-1}\,COD\ applied$$

Alternatively, you could also have calculated mS_{xv} directly with Eq. (3.38):

$$mS_{xv} = f_{cv} \cdot (1 - f_{ns} - f_{np}) \cdot (1 + f \cdot b_h \cdot R_s) \cdot C_r/R_s + f_{np}$$
$$= 1.5 \cdot 0.8 \cdot (1 + 0.2 \cdot 0.24 \cdot 10) \cdot 1.32/10 + 0.1 = 0.34\,kg\,COD \cdot kg^{-1}\,COD\ applied$$

3.3.1.4 COD fraction oxidised for respiration

Oxygen is consumed for both exogenous and endogenous respiration. The oxygen uptake rate (OUR) due to exogenous respiration O_{ex} is determined from Figure 3.3, where it is shown that upon metabolism of 1 gram of COD, there will be a production of active sludge equal to Y gram of VSS with a COD value of $f_{cv} \cdot Y$ gram COD. Hence the remaining fraction of $(1 - f_{cv} \cdot Y)$ gram COD will be oxidised and for that oxidation, by definition, an oxygen mass of $(1 - f_{cv} \cdot Y)$ gram O_2 is required. Hence the exogenous oxygen consumption rate can be expressed as:

$$O_{ex} = (1 - f_{cv} \cdot Y) \cdot r_u$$
$$= (1 - f_{cv} \cdot Y) \cdot Q_i \cdot S_{bi}/V_r \quad (3.39)$$

The OUR for endogenous respiration O_{en} is calculated from the oxidation rate of the decayed activated sludge, which is the difference between the decay rate and the production rate of the endogenous residue:

$$r_o = (dX_a/dt)_d - (dX_e/dt)_d$$
$$= b_h \cdot X_a - f \cdot b_h \cdot X_a = (1 - f) \cdot b_h \cdot X_a \quad (3.40)$$

where r_o = oxidation rate of the decayed active sludge

Again using the proportionality constant f_{cv} the endogenous respiration rate can be calculated:

$$O_{en} = f_{cv} \cdot r_o$$
$$= f_{cv} \cdot (1 - f) \cdot b_h \cdot X_a \quad (3.41)$$

The total OUR for the oxidation of organic material is equal to the sum of the values for exogenous and for endogenous respiration:

$$O_c = O_{ex} + O_{en}$$

Using Eqs. (3.3 and 3.29) to substitute for S_{bi} and X_a leads to:

$$O_c = (1 - f_{ns} - f_{np}) \cdot (1 - f_{cv} \cdot Y + f_{cv} \cdot (1 - f) \cdot b_h \cdot C_r) \cdot Q_i \cdot S_{ti}/V_r \quad (3.42)$$
$$= (1 - f_{cv} \cdot Y + f_{cv} \cdot (1 - f) \cdot b_h \cdot C_r) \cdot MS_{bi}/R_h$$

The influent COD fraction that is oxidised in the activated sludge system is now expressed as:

$$mS_o = MO_c/MS_{ti} = (V_r \cdot O_c)/(Q_i \cdot S_{ti})$$
$$= (1 - f_{ns} - f_{np}) \cdot [(1 - f_{cv} \cdot Y) + f_{cv} \cdot (1 - f) \cdot b_h \cdot C_r] \quad (3.43)$$

EXAMPLE 3.3

Continuing with the previous example, calculate the oxygen demand for exogenous and endogenous respiration in the activated sludge system.

Solution

$$O_{ex} = (1 - f_{cv} \cdot Y) \cdot MS_{bi}/V_r \quad (3.39)$$
$$= (1 - 1.5 \cdot 0.45) \cdot (1 - 0.2) \cdot 6000/5000 = 0.31\ \text{kg}\ O_2 \cdot m^{-3} \cdot d^{-1}$$
$$O_{en} = f_{cv} \cdot (1 - f) \cdot b_h \cdot X_a \quad (3.41)$$
$$= 1.5 \cdot (1 - 0.2) \cdot 0.24 \cdot 1.27 = 0.36\ \text{kg}\ O_2 \cdot m^{-3} \cdot d^{-1}$$
$$mS_o = MO_c/MS_{ti} = V_r \cdot (O_{ex} + O_{en})/MS_{ti} \quad (3.43)$$
$$= 5000 \cdot (0.31 + 0.36)/6000 = 0.56\ \text{kg COD} \cdot \text{kg}^{-1}\ \text{COD applied}$$

3.3.1.5 Model summary and evaluation

Equation (3.43) completes the construction of the simplified model in the sense that now expressions have been derived for the division of the influent COD into fractions in the effluent (mS_{te}), in the excess sludge (mS_{xv}) and oxidised into stable end products (mS_o). For convenience the expressions are repeated below:

(1) Fraction of influent COD remaining in the liquid phase:

$$mS_{te} = f_{ns} \quad (3.18)$$

(2) Fraction of influent COD discharged with the excess sludge:

$$mS_{xv} = f_{cv} \cdot (1 - f_{ns} - f_{np}) \cdot (1 + f \cdot b_h \cdot R_s) \cdot C_r/R_s + f_{np} \quad (3.38)$$

(3) Fraction of influent COD oxidised into stable end products:

$$mS_o = (1 - f_{ns} - f_{np}) \cdot [(1 - f_{cv} \cdot Y) + f_{cv} \cdot (1 - f) \cdot b_h \cdot C_r] \quad (3.43)$$

It is interesting to note that the sum of the three fractions is identical to unity over the whole range of sludge ages, as the (theoretical) model COD mass balance should always close:

$$B_o = mS_{te} + mS_{xv} + mS_o = 1.0 \quad (3.44)$$

From the model summary it can be concluded that in fact the hydraulic retention time R_h is not at all important for the definition of the main parameters in activated sludge system performance: i.e. excess sludge production and oxygen demand. On the other hand, the influence of the sludge age is crucial: remember that R_s is also present in $C_r = Y \cdot R_s/(1 + b_h \cdot R_s)$. This conclusion is exemplified in Figure 3.4, where it is shown that:

- The oxidized COD fraction (mS_o) increases at higher values of R_s;
- The COD fraction discharged with excess sludge (mS_{xv}) decreases at higher values of R_s;
- COD fraction discharged with effluent (mS_e) is not influenced.

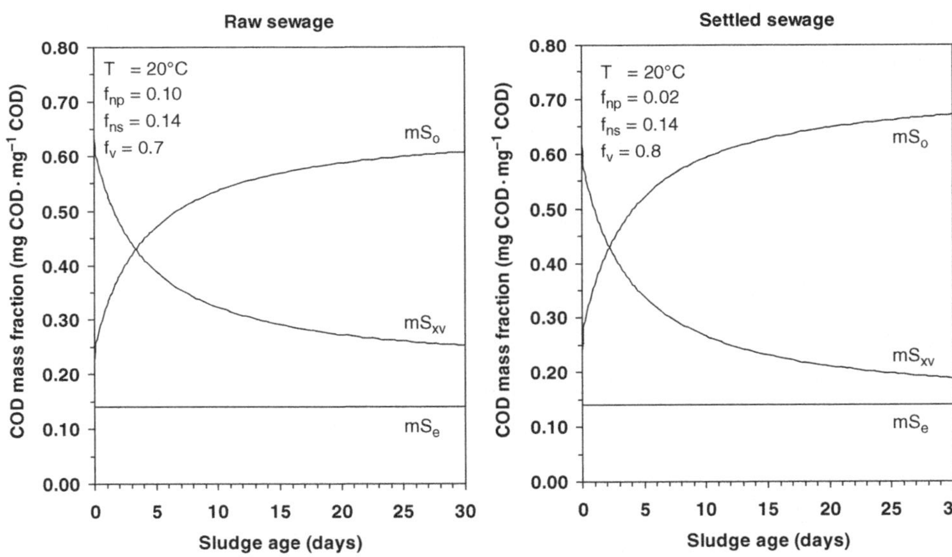

Figure 3.4 Model behaviour of the division of influent COD into the fractions mS_e, mS_{xv} and mS_o, as function of the sludge age

The basic equations forming the ideal steady state model for COD removal are summarized in Table 3.4 to Table 3.7.

Table 3.4 Mass-based equations for the COD mass balance

Par.	Equations	Eq. no.	Daily total
mS_{xv}	$= MS_{xv}/MS_{ti}$	(3.38)	$MS_{xv} = mS_{xv} \cdot MS_{ti}$
	$= f_{cv} \cdot mE_v$		
	$= f_{cv} \cdot (MX_v/R_s)/MS_{ti}$		
	$= f_{cv} \cdot (1 - f_{ns} - f_{np}) \cdot (1 + f \cdot b_h \cdot R_s) \cdot C_r/R_s + f_{np}$		
mS_o	$= MO_c/MS_{ti} = V_r \cdot O_c/(Q_i \cdot S_{ti})$	(3.43)	$MO_c = mS_o \cdot MS_{ti}$
	$= (1 - f_{np} - f_{ns}) \cdot (1 - f_{cv} \cdot Y + (1 - f) \cdot f_{cv} \cdot b_h \cdot C_r)$		
mS_e	$= MS_{te}/MS_{ti} = (Q_i \cdot S_{se})/(Q_i \cdot S_{ti})$	(3.18)	$MS_{te} = mS_e \cdot MS_{ti}$
	$= f_{ns}$		
B_o	$= mS_e + mS_o + mS_{xv} = 1.0$	(3.44)	

Table 3.5 Concentration-based equations of the activated sludge system

Par.	Equations	Eq. no.
C_r	$= Y \cdot R_s/(1 + b_h \cdot R_s)$	(3.30)
S_{bi}	$= (1 - f_{ns} - f_{np}) \cdot S_{ti}$	(3.3)
S_{te}	$= f_{ns} \cdot S_{ti}$	(3.1a)
X_i	$= f_{np} \cdot R_s \cdot S_{ti}/(f_{cv} \cdot R_h)$ or $f_{np} \cdot R_s \cdot MS_{ti}/(f_{cv} \cdot V_r)$	(3.21)
X_a	$= C_r \cdot S_{bi}/R_h$ or $(1 - f_{ns} - f_{np}) \cdot C_r \cdot S_{ti}/R_h$	(3.29)
X_e	$= f \cdot b_h \cdot R_s \cdot C_r \cdot S_{bi}/R_h$ or $(1 - f_{ns} - f_{np}) \cdot f \cdot b_h \cdot R_s \cdot C_r \cdot S_{ti}/R_h$	(3.34)
X_v	$= [(1 + f \cdot b_h \cdot R_s) \cdot C_r + f_{np} \cdot R_s/f_{cv}] \cdot S_{bi}/R_h$	(3.35)
	$= [(1 - f_{ns} - f_{np}) \cdot (1 + f \cdot b_h \cdot R_s) \cdot C_r + f_{np} \cdot R_s/f_{cv}] \cdot S_{ti}/R_h$	
O_c	$= (1 - f_{cv} \cdot Y + (1 - f) \cdot f_{cv} \cdot b_h \cdot C_r) \cdot S_{bi}/R_h$	(3.42)

As discussed above, it can be seen that in all concentration based equations the hydraulic retention time is present, which may give the (erroneous) impression that this parameter is of fundamental importance to model basic activated sludge process behaviour. As will be demonstrated, in all cases the above equations can be rewritten in the form of mass equations, from which the hydraulic retention time is deleted (Table 3.6 and Table 3.7). In these equations, the masses rather than the concentrations are considered as variables, so for example:

$$mX_i = MX_i/MS_{ti} = V_r \cdot X_i/(Q_i \cdot S_{ti}) = R_h \cdot X_i/S_{ti}$$

Now, inserting Eq. (3.21) for X_i, mX_i can be written explicitly as:

$$mX_i = f_{np} \cdot R_s/f_{cv} \qquad (3.45)$$

where mX_i = mass of inert organic sludge in the system per unit mass daily applied COD

Table 3.6 Mass-based equations for sludge fractions in the activated sludge system

Par.	Equations	Eq. no.	Total mass
mX_i	$= MX_i/MS_{ti} = (V_r \cdot X_i)/(Q_i \cdot S_{ti}) = f_{np} \cdot R_s/f_{cv}$	(3.45)	$MX_i = mX_i \cdot MS_{ti}$
mX_a	$= MX_a/MS_{ti} = (V_r \cdot X_a)/(Q_i \cdot S_{ti}) = (1 - f_{ns} - f_{np}) \cdot C_r$	(3.46)	$MX_a = mX_a \cdot MS_{ti}$
mX_e	$= MX_e/MS_{ti} = (V_r \cdot X_e)/(Q_i \cdot S_{ti}) = (1 - f_{ns} - f_{np}) \cdot C_r \cdot f \cdot b_h \cdot R_s$	(3.47)	$MX_e = mX_e \cdot MS_{ti}$
mX_v	$= MX_v/MS_{ti} = V_r \cdot (X_e + X_a + X_i)/(Q_i \cdot S_{ti}) = mX_i + mX_a + mX_e$	(3.48)	$MX_v = mX_v \cdot MS_{ti}$
	$= (1 - f_{ns} - f_{np}) \cdot (1 + f \cdot b_h \cdot R_s) \cdot C_r + f_{np} \cdot R_s/f_{cv}$		
mX_t	$= MX_t/MS_{ti} = (V_r \cdot X_t)/(Q_i \cdot S_{ti}) = mX_v/f_v$	(3.49)	$MX_t = mX_t \cdot MS_{ti}$

Table 3.7 Mass-based equations for excess sludge production

Par.	Equations	Eq. no.	Daily production
mE_v	$= ME_v/MS_{ti}$	(3.50)	$ME_v = mE_v \cdot MS_{ti}$
	$= mX_v/R_s$		$ME_v = MX_v/R_s$
	$= (1 - f_{ns} - f_{np}) \cdot (1 + f \cdot b_h \cdot R_s) \cdot C_r/R_s + f_{np}/f_{cv}$		
mE_t	$= ME_v/MS_{ti}$	(3.51)	$ME_t = mE_t \cdot MS_{ti}$
	$= mX_t/R_s$		$ME_t = MX_t/R_s$
	$= mE_v/f_v$		
	$= [(1 - f_{ns} - f_{np}) \cdot (1 + f \cdot b_h \cdot R_s) \cdot C_r/R_s + f_{np}/f_{cv}]/f_v$		

EXAMPLE 3.4

An activated sludge system is operated at a sludge age of 10 days under the following conditions:

- $Q_i = 10{,}000\ m^3 \cdot d^{-1}$
- $S_{ti} = 800$ mg COD $\cdot\ l^{-1}$
- $f_{ns} = 0.05$
- $f_{np} = 0.15$
- $f_v = 0.8$ mg VSS $\cdot$ mg^{-1} TSS
- $Y = 0.45$ mg VSS $\cdot$ mg^{-1} COD
- $T = 14°C$
- $b_{h14} = 0.24 \cdot 1.04^{(20-14)} = 0.19\ d^{-1}$

Characterise the system performance by calculating:

- The sludge composition and -quantity.
- The division of influent COD over the COD mass fractions;

Solution

As a first step, determine the sludge mass and -composition that will develop in the activated sludge system. C_r is equal to 1.55 kg VSS $\cdot$ d $\cdot$ kg^{-1} COD. Using the mass-based equations from Table 3.6, the biomass composition and quantity in $kg \cdot d \cdot kg^{-1}$ COD can be calculated. The total sludge masses present in the system are obtained by multiplication with MS_{ti} ($8000\ kg\ COD \cdot d^{-1}$):

$$
\begin{aligned}
mX_i &= f_{np} \cdot R_s/f_{cv} = 0.15 \cdot 10/1.5 = 1.00 && \rightarrow MX_i = 8000\,kg\,VSS \\
mX_a &= (1 - f_{ns} - f_{np}) \cdot C_r = (1 - 0.2) \cdot 1.55 = 1.24 && \rightarrow MX_a = 9942\,kg\,VSS \\
mX_e &= (1 - f_{ns} - f_{np}) \cdot C_r \cdot f \cdot b_h \cdot R_s && \\
&= (1 - 0.2) \cdot 1.55 \cdot 0.2 \cdot 0.19 \cdot 10 = 0.47 && \rightarrow MX_e = 3772\,kg\,VSS \\
mX_v &= mX_i + mX_a + mX_e = 1.00 + 1.24 + 0.47 = 2.71 && \rightarrow MX_v = 21{,}714\,kg\,VSS \\
\text{Finally } mX_t &= mX_v/f_v = 2.71/0.8 = 3.39 && \rightarrow MX_t = 27{,}142\,kg\,VSS
\end{aligned}
$$

Having defined the total sludge mass that will develop in this activated sludge system, it is easy to calculate the excess sludge production as $ME_t = MX_t/R_s = 27{,}142/10 = 2714\ kg\ TSS \cdot d^{-1}$.

Note that indeed the hydraulic retention time does not influence the mass of sludge that will develop (but only the concentration). The division of the influent COD over the different COD mass fraction can be calculated as:

$$mS_e = f_{ns} = 0.05$$

$$mS_{xv} = f_v \cdot f_{cv} \cdot mE_t = 0.8 \cdot 1.5 \cdot 2714/8000 = 0.41$$

As the theoretical COD mass balance always closes, this determines the value of mS_o as $1.00 - 0.05 - 0.41 = 0.54$. Alternatively (and to check on your calculation), mS_o can also be calculated directly with Eq. (3.43):

$$
\begin{aligned}
mS_o &= (1 - f_{np} - f_{ns}) \cdot (1 - f_{cv} \cdot Y + (1 - f) \cdot f_{cv} \cdot b_h \cdot C_r) \\
&= (1 - 0.2) \cdot (1 - 1.5 \cdot 0.45 + (1 - 0.2) \cdot 1.5 \cdot 0.19 \cdot 1.55) = 0.54
\end{aligned}
$$

3.3.2 Model calibration

Equations (3.18, 3.38 and 3.43) show that the fractions mS_{te}, mS_{xv} and mS_o depend on several parameters. Table 3.8 summarises the eight factors that influence the steady state model of the activated sludge system for COD removal and attributes typical values when this is possible.

The sludge mass parameters (Y, f and f_{cv}) have constant values and the decay constant b_h is affected only by temperature. The values of these constants were determined by extensive experimental research, which is described in Chapter 12. As the sludge age is an operational variable that must be selected by the designer, this leaves only three unknown factors in Table 3.8: the temperature and the non-biodegradable COD fractions of dissolved (f_{ns}) and particulate (f_{np}) material in the influent.

In the case of sewage treatment, the temperature may be estimated taking into consideration the climate in the region where the activated sludge system is to be constructed, while for industrial wastewaters it may be estimated from the temperature at which the effluent is produced.

The value of the non-biodegradable influent COD fractions can only be determined experimentally, requiring an activated sludge system to be operated under steady state conditions for various sludge ages. Alternatively, the use of respirometrics has proven to be a powerful tool for model calibration, as will be discussed in Appendix 1 and 2. The steady state based determination of f_{ns} and f_{np} proceeds with the following steps:

(1) For at least one but preferably more values of the sludge age, the fractions mS_{te}, mS_{xv} and mS_o are determined experimentally when steady state conditions have been established;
(2) Check with Eq. (3.14) if the mass balance closes: i.e. if the sum of the three fractions deviates less than 10% from unity ($|B_o - 1| < 0.1$);
(3) With the aid of the measured values for mS_{te}, select the value of f_{ns} that leads to the best correlation between experimental data and theoretical prediction, i.e. equate the value of f_{ns} to the average ratio of the effluent and influent;
(4) Having established the f_{ns} value and using experimental values for mS_{xv} and mS_o, select the f_{np} value that gives the closest correlation between the experimental results and the theoretical predictions for mS_{xv} and mS_o.

Table 3.8 Factors that influence the ideal steady state model for COD removal and their typical values

Parameter	Symbol	Typical value
Yield coefficient (heterotrophs)	Y	0.45 mg COD · mg^{-1} VSS
Fraction of decayed active sludge remaining as endogenous residue	f	0.2 mg VSS · mg^{-1} VSS
COD/VSS ratio for organic sludge	f_{cv}	1.5 mg COD · mg^{-1} VSS
Decay rate constant for active sludge	b_h	$0.24 \cdot 1.04^{(T-20)}$ d^{-1}
Soluble, non biodegradable influent COD fraction	f_{ns}	Variable – influent parameter (mg COD · mg^{-1} COD)
Particulate, non biodegradable influent COD fraction	f_{np}	Variable – influent parameter (mg COD · mg^{-1} COD)
Sewage temperature (minimum)	T	Variable – local parameter (°C)
Sludge age	R_s	Variable – design parameter (d)

Naturally, the procedure presented above is only valid when the behaviour of the activated sludge system approaches ideality: i.e. when the concentration of suspended solids in the effluent is very low. An example of the determination of the f_{ns} and f_{np} values is presented in Figure 3.5. The collected data refer to an experiment conducted with raw sewage from the city of Campina Grande by Dias et al. (1981), which was discussed in Example 3.1. This data set was complemented by Van Haandel and Catunda (1985 and 1989), while Table 3.2 deals specifically with the data presented by Dias et al. (1981). In Table 3.2 it can be seen that the value of the recovery factor B_o deviates less than ten percent from the theoretical value of one, so that the data is considered acceptable. The experimental values of mS_{te}, mS_{xv} and mS_o were calculated using Eq. (3.14), as indicated in Table 3.3 while Eq. (3.15) was used to calculate the sludge age R_s. In Figure 3.5 the measured values of mS_{te}, mS_{xv} and mS_o are shown as a function of the sludge age R_s. In so far as the non biodegradable and dissolved influent COD fraction is concerned, Figure 3.5a shows that the ratio of effluent and influent soluble COD oscillates around 0.14

so that this value is accepted as the "best" value for f_{ns}. Once the f_{ns} value has been established, the f_{np} value is determined as follows:

(1) With the aid of Eq. (3.38), calculate as a function of the sludge age theoretical values of mS_{xv} for different f_{np} values;
(2) Plot the theoretical mS_{xv} curves as a function of R_s for the chosen f_{np} values;
(3) Similarly, using Eq. (3.43), calculate and plot theoretical curves of mS_o as a function of R_s for the same series of f_{np} values;
(4) By comparing the theoretical curves of mS_{xv} and mS_o and the experimental results, the f_{np} value that gives the closest correlation between experimental and theoretical results is selected as the "best" value for the sewage under consideration.

In the case of Figure 3.5, theoretical curves were generated for values of f_{np} ranging from 0.00 to 0.12. Figure 3.5b and c show clearly that the value $f_{np} = 0.06$ results in the closest correlation between the theoretical and experimental values. In Figure 3.5 there is a close correspondence between theory and practice over the entire sludge age range from 2 to 30 days.

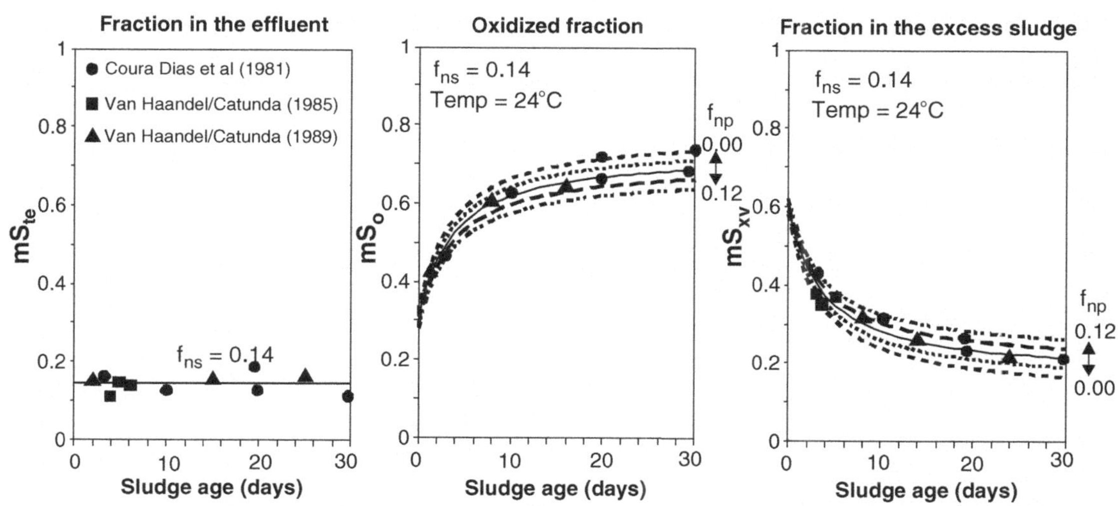

Figure 3.5 Model calibration: experimental and theoretical values of COD fractions mS_{te}, mS_{xv} and mS_o for different values of f_{ns} and f_{np}

In practice the sludge age will typically be longer than 2 days and shorter than 30 days. Therefore the steady state model for the activated sludge system can be used for most full scale plants, when temperature is not very much lower than the one prevailing during the investigation: $T = 24 \pm 2°C$ (at low temperatures, in combination with a short sludge age, the utilisation of organic material may be incomplete, see Section 1.4 and Appendix 3). This conclusion is of great practical importance, because the parameters that the simplified model predicts are exactly those parameters that are of most interest in practice: (I) the COD fraction remaining in the effluent (or in other words, the COD removal efficiency),

(II) the fraction of the influent COD discharged as excess sludge (or the sludge production), and (III) the fraction of the influent COD oxidised in the process (determining how much oxygenation capacity must be installed).

In practice it is often difficult or even impossible to carry out the experimental investigations required to determine the fractions f_{ns} and f_{np}. In such cases the only alternative may be to estimate the values of these fractions, based on the available information about the nature of the wastewater and other parameters like the presence of pre-treatment systems and social-economic habits.

Pre-treatment systems like septic tanks tend to lead to a decrease of the biodegradable organic material (due to anaerobic digestion in the tank) and of the suspended solids concentration (due to settling). Hence pre-treated sewage tends to have a high f_{ns} value and a low f_{np} value. The use of garbage grinders and the habit of scouring of pots with sand are examples of social economic habits influencing the composition of sewage: the garbage grinders lead to the presence of a high concentration of particulates (both biodegradable and non biodegradable) and the use of sand tends to increase the mineral sludge fraction. In Figure 3.6 the influence of f_{ns} on the activated sludge system is analysed. The values of mS_{te}, mS_{xv} and mS_o are plotted as function of the sludge age for $f_{np} = 0.1$ and different f_{ns} values.

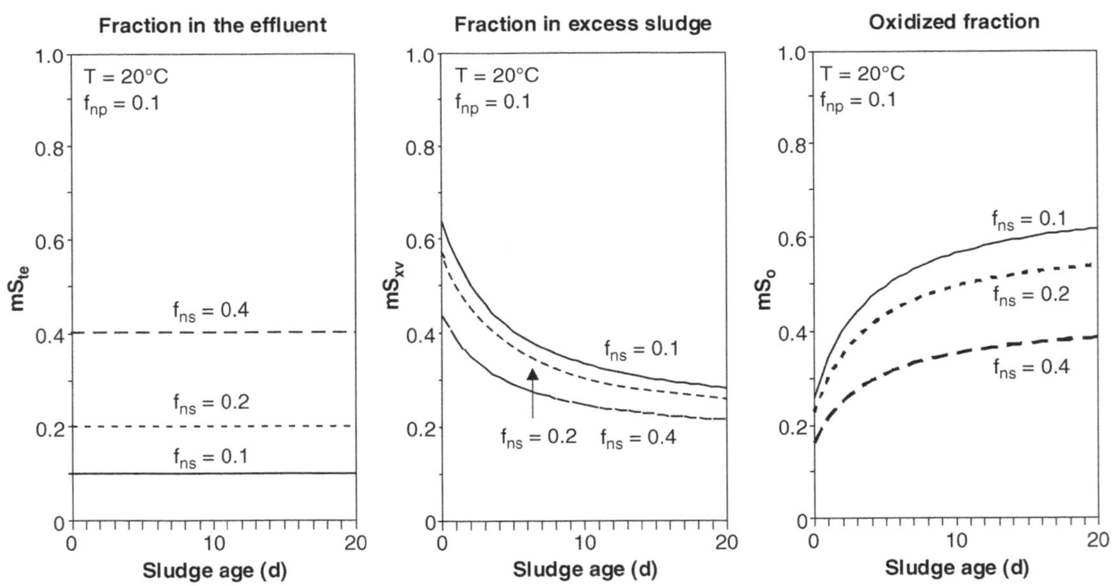

Figure 3.6 Evaluation of the influence of the value of the f_{ns} fraction on the values of mS_{te}, mS_{xv} and mS_o

Municipal sewage usually has a f_{ns} value in the range of 0.1 (raw sewage) to 0.2 (pre-treated sewage). Larger values are encountered in some industrial wastes: for example black liquor from paper mills (especially when pulp is used instead of recycled paper) contains a high concentration of non-biodegradable lignin. In Figure 3.6 it can be observed that a 100% increase from $f_{ns} = 0.1$ to $f_{ns} = 0.2$ has little influence on sludge production and a modest influence on oxygen consumption.

In Figure 3.7 the influence of the value of f_{np} on activated sludge behaviour is evaluated. An f_{ns} value of 0.1 was adopted and the values of mS_{xv} and mS_o are shown as function of the sludge age for different f_{np} values: $f_{np} = 0.02$ (sewage after efficient primary sedimentation or dissolved industrial waste), $f_{np} = 0.10$ (raw municipal sewage) and $f_{np} = 0.25$. The latter value was found from the data presented by Sutton et al. (1979) using sewage at Burlington, Canada. The high value possibly can be attributed to the North-American habit of using garbage grinders.

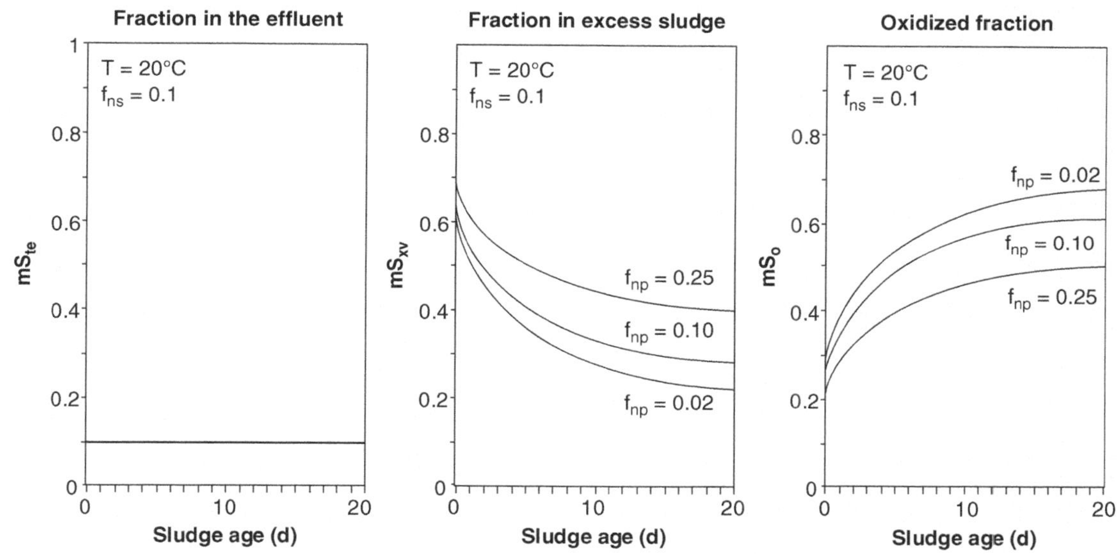

Figure 3.7 Evaluation of the influence of the value of the f_{np} fraction on the values of mS_{te}, mS_{xv} and mS_o

It can be observed from Figure 3.7 that variations of the f_{np} value lead to very significant changes in the basic behaviour of the activated sludge system, especially at long sludge ages. For example, an increase from f_{np} from 0.02 to 0.25 causes an increase of mS_{xv} from 0.20 to 0.40 when the sludge age is 20 days. At the same time the mS_o value decreases from 0.70 to 0.50. When it is impossible to determine the values of f_{ns} and f_{np}, an estimate must be made. In the case of municipal sewage, the following approach may be used for design purposes: when the sludge production is estimated, a low f_{ns} value (for example 0.05) and a high f_{np} value (for example 0.15) are adopted. When oxygen consumption is estimated low values for both are adopted (for example $f_{ns} = f_{np} = 0.05$). Thus, the estimates for both sludge production and oxygen consumption are conservative and probably a little above the actual values, so that both sludge handling and aeration capacity will be adequate for the demand.

3.3.3 Model applications

3.3.3.1 Sludge mass and composition

The mass equations listed in Table 3.6 can be used to calculate the masses of the different fractions that compose the sludge as a function of the sludge age, when the daily organic load is known. In Figure 3.8, the masses of inert, active, endogenous, organic, mineral and total sludge per unit mass of daily applied COD (mX_i, mX_a, mX_e, mX_v, mX_m and mX_t) are plotted as functions of the sludge age for raw sewage ($f_{np} = 0.10$ and $f_{ns} = 0.14$) and settled sewage ($f_{np} = 0.02$ and $f_{ns} = 0.14$).

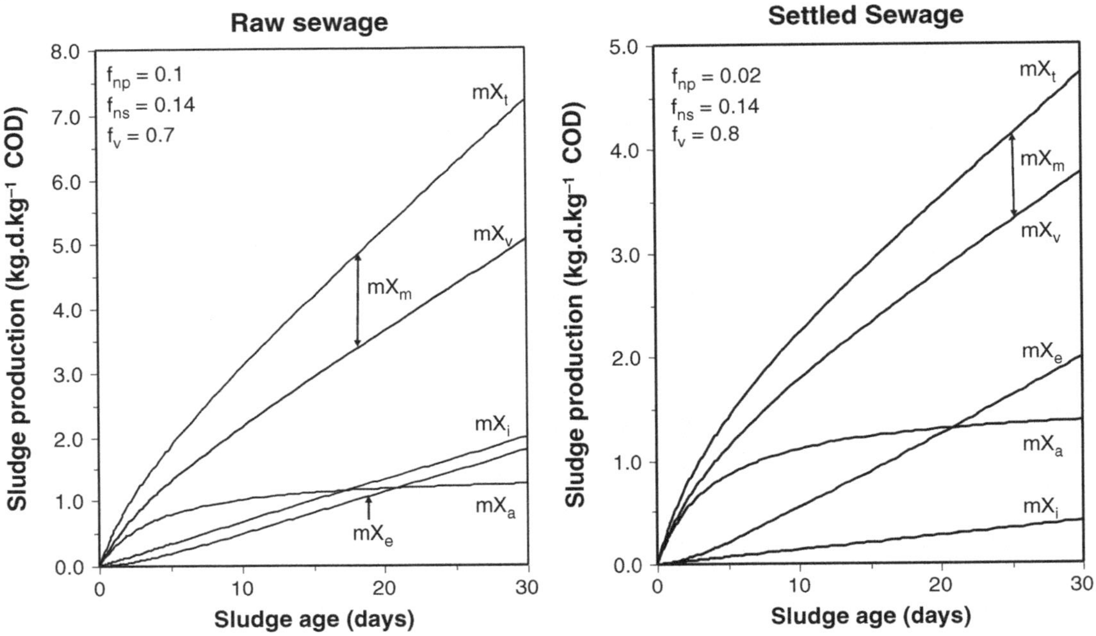

Figure 3.8 Sludge mass per unit mass daily applied COD for the different sludge fractions for raw and settled sewage

The figures show two important aspects: (I) the sludge mass present in the system depends heavily on the characteristics of the influent organic material and (II) the active sludge fraction decreases with increasing sludge age. Since the active sludge fraction is an important parameter, it is interesting to derive an expression for it. The active sludge fraction can either be defined as a fraction of the organic or of the total sludge concentration:

$$f_{av} = mX_a/mX_v = (1 - f_{ns} - f_{np}) \cdot C_r/[(1 - f_{ns} - f_{np}) \cdot (1 + f \cdot b_h \cdot R_s) \cdot C_r + f_{np} \cdot R_s/f_{cv}] \qquad (3.52)$$

$$f_{at} = mX_a/mX_t = (1 - f_{ns} - f_{np}) \cdot C_r/ \\ [(1 - f_{ns} - f_{np} \cdot (1 + f \cdot b_h \cdot R_s) \cdot C_r + f_{np} \cdot R_s/f_{cv}] \cdot f_v = f_{av} \cdot f_v \qquad (3.53)$$

where:

f_{av} = ratio between active and volatile sludge mass
f_{at} = ratio between active and total sludge mass
f_v = ratio between volatile and total sludge mass (organic sludge fraction)

Figure 3.9 shows values of f_{av} and f_{at} as functions of the sludge age for raw and settled sewage. It can be noted in Figure 3.9 that the active sludge fraction depends heavily on the composition of the influent organic material. For example, for raw sewage the active sludge fraction $f_{av} = 0.47$ at a sludge age of 10 days. In the case of settled sewage, for the same sludge age the active fraction is much higher: $f_{av} = 0.63$. In the case of settled sewage an active fraction $f_{av} = 0.47$ is only possible for a sludge age of more than 20 days.

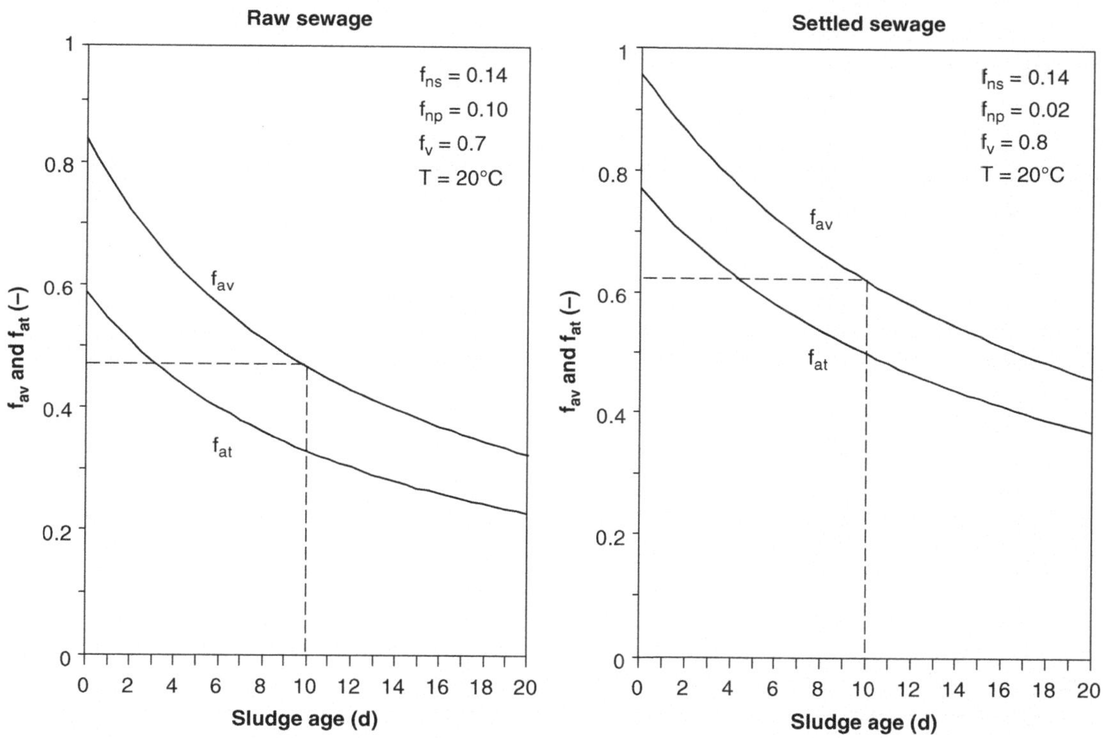

Figure 3.9 Active sludge fraction as a function of the sludge age for raw and settled sewage

EXAMPLE 3.5

An activated sludge system treats raw sewage ($f_{ns} = 0.14$ and $f_{np} = 0.10$) and is operated at a sludge age of 20 days. It is required to increase the organic load by 50% from 10 to 15 ton COD per day without increasing the sludge mass in the system. Answer the following questions:

- What will be the new maximum sludge age?
- How much does the fraction of influent COD wasted as excess sludge change?
- How much will the oxygen consumption increase?

Solution

From Figure 3.8a it is determined that mX_v (the organic sludge mass per unit mass of daily applied influent COD) for the given sludge age of 20 days is equal to 3.6 mg VSS · d · mg^{-1} COD. Hence the maximum sludge mass in the system is:

$$MX_v = mX_v \cdot MS_{ti} = 3.6 \cdot 10 = 36 \text{ ton VSS}$$

After the load increase, the total sludge mass is not to increase, so consequently the sludge age must be reduced. For the new sludge age the mX_v value is given as:

$$mX_v = MX_v/MS_{ti} = 36/15 = 2.4 \text{ mg VSS} \cdot d \cdot mg^{-1} \text{ COD}$$

Again using Figure 3.8, it is noted that for $mX_v = 2.4$ the sludge age is 12 days. Hence it is concluded that the sludge age must be reduced from 20 to 12 days, due to the increase of the organic load. The change of the active sludge fraction can be evaluated with the aid of Figure 3.9: for $R_s = 20$ days one has $f_{av} = 0.33$ while for $R_s = 12$ days the value of $f_{av} = 0.45$.

To evaluate the influence of the load increase and the consequential sludge age reduction on the oxygen consumption, first the fractions mS_{xv} and mS_o are calculated for the original load and for $R_s = 20$ days:

$$mS_{xv} = f_{cv} \cdot mX_v/R_s = 1.5 \cdot 3.6/20 = 0.27$$
$$mS_o = 1 - 0.14 - 0.27 = 0.59$$

After the reduction of the sludge age to $R_s = 12$ days one has:

$$mS_o = 1 - 0.14 - 1.5 \cdot 2.4/12 = 0.56$$

Hence, the oxygen demand increases from $0.59 \cdot 10 = 5.9$ ton $O_2 \cdot d^{-1}$ *before* the load increase to $0.56 \cdot 15 = 8.4$ ton $O_2 \cdot d^{-1}$ *after* the load increase, i.e. the increase of the load by 5 ton COD · d^{-1} results in an increase of the oxygen consumption of $8.4 - 5.9 = 2.5$ ton $O_2 \cdot d^{-1}$.

At the same time, there is an increase of the effluent load from $0.14 \cdot 10 = 1.4$ ton COD · d^{-1} to $0.14 \cdot 15 = 2.1$ ton COD · d^{-1}. The COD mass discharged as excess sludge increases from its initial value of $MS_{xv} = MS_{ti} - MS_o - MS_{te} = 10 - 5.9 - 1.4 = 2.6$ ton COD · d^{-1} to $MS_{xv} = 15 - 8.4 - 2.1 = 4.5$ ton COD · d^{-1}. The excess sludge production is $4.5/f_{cv} = 3.0$ ton VSS · d^{-1}, an increase of 67% compared to its initial value of $2.6/1.5 = 1.7$ ton VSS · d^{-1}.

3.3.3.2 Biological reactor volume

In the previous section it was established that in a steady state activated sludge system a sludge mass will develop that is compatible with the daily applied COD load. When the sludge mass of the system is known, the reactor volume can be calculated after defining the sludge concentration (X_v or X_t) that is to be maintained:

$$\begin{aligned} V_r &= MX_v/X_v \\ &= [(1 - f_{ns} - f_{np}) \cdot (1 + f \cdot b_h \cdot R_s) \cdot C_r + f_{np} \cdot R_s/f_{cv}] \cdot MS_{ti}/X_v \end{aligned} \tag{3.54}$$

The volume per unit mass daily applied COD can be expressed as:

$$v_r = V_r/MS_{ti} = mX_v/X_v$$
$$= [(1 - f_{ns} - f_{np}) \cdot (1 + f \cdot b_h \cdot R_s) \cdot C_r + f_{np} \cdot R_s/f_{cv}]/X_v \quad (3.55)$$

Figure 3.10 shows the biological reactor volume per unit mass daily applied COD as a function of the sludge age for different sludge concentrations for typical values of both raw and settled sewage. Equation (3.55) shows that the volume per unit mass daily applied COD depends on the following factors:

- Sludge concentration;
- Sludge age;
- Composition of organic material (f_{ns} and f_{np});
- Temperature (influences b_h).

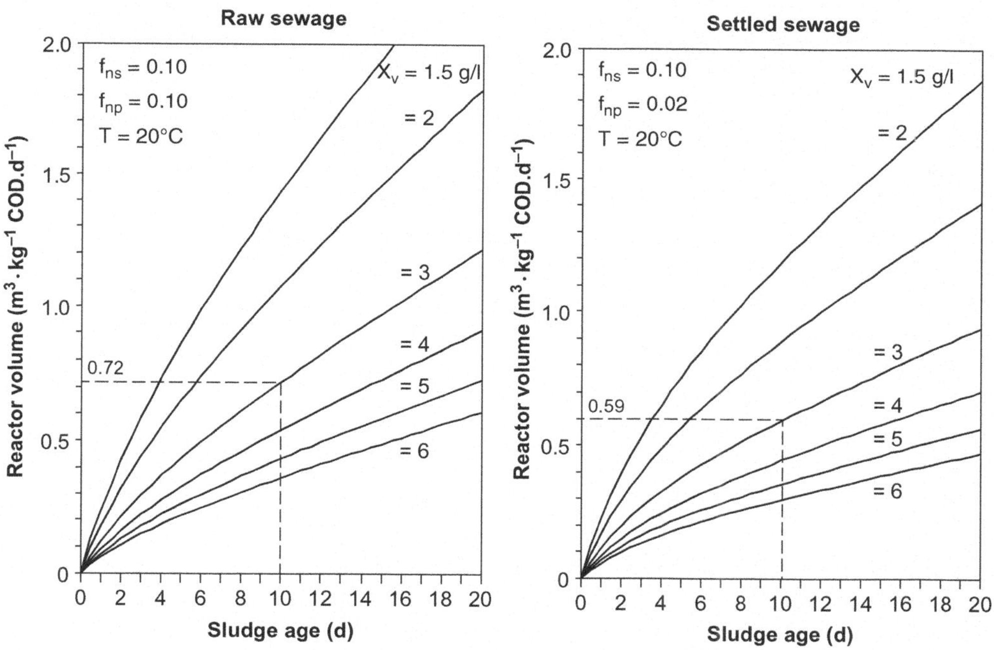

Figure 3.10 Volume of the biological reactor of an activated sludge system per unit mass daily applied COD as a function of the sludge age for different sludge concentrations for raw and settled sewage

In the case of municipal sewage it is possible to calculate the volume per capita, if the COD contribution per inhabitant is known:

$$V_{hab} = S_{hab} \cdot v_r \quad (3.56)$$

where:

V_{hab} = required reactor volume per inhabitant
S_{hab} = daily COD contribution per inhabitant

EXAMPLE 3.6

What is the value of the per capita reactor volume for an activated sludge system operating at a sludge age of 10 days and a sludge concentration of $X_v = 3$ g VSS · l^{-1}, if a per capita contribution of $S_{hab} = 75$ g COD · $inh^{-1} \cdot d^{-1}$ is assumed? Evaluate this for both raw ($f_{ns} = 0.10$; $f_{np} = 0.10$) and settled sewage ($f_{ns} = 0.10$; $f_{np} = 0.01$), using Figure 3.10.

Solution

In the case of raw sewage, the left-hand graph in Figure 3.10 is used to determine the reactor volume. For $R_s = 10$ days and a volatile sludge concentration of 3 g VSS · l^{-1}, one has $v_r = 0.72$ $m^3 \cdot kg^{-1}$ COD · d^{-1}. For the per capita contribution of 75 g COD · $inh^{-1} \cdot d^{-1}$ or 1000/75 = 13.3 inh · kg^{-1} COD · d^{-1}, the per capita volume $V_{hab} = 720/13.3 = 54$ l · inh^{-1}.

Similarly, in the case of settled sludge one has in the right-hand graph in Figure 3.10b for $R_s = 10$ and $X_v = 3$ g VSS · l^{-1} a volume of $v_r = 0.59$ $m^3 \cdot kg^{-1}$ COD · d^{-1}. Hence for the same per capita contribution (i.e. 13.3 inh · kg^{-1} COD · d^{-1}) the per capita volume $V_{hab} = 590/13.3 = 44$ l · inh^{-1}.

In practice the COD contribution per capita is of the order of 35 g (slums) to 100 g COD · $inh^{-1} \cdot d^{-1}$ (middle class). It can be noted that the reactor volume is independent of the sewage concentration.

3.3.3.3 *Excess sludge production and nutrient demand*

The excess or surplus sludge production can be calculated directly from the sludge mass in the activated sludge system. Knowing that the excess sludge production is a fraction $1/R_s$ of the existing sludge mass, one has:

$$mE_v = mX_v/R_s = (1 - f_{ns} - f_{np}) \cdot (1 + f \cdot b_h \cdot R_s) \cdot C_r/R_s + f_{np}/f_{cv} \quad (3.50)$$

where:

mE_v = volatile sludge mass produced per unit mass applied COD (mg VSS · mg^{-1} COD)

Figure 3.11 shows the excess sludge production as a function of the sludge age for $f_{ns} = f_{np} = 0.10$ (raw sewage) as well as $f_{ns} = 0.10$ and $f_{np} = 0.02$ (settled sewage). Along with carbon, volatile sludge is composed of several elements, of which nitrogen and phosphorus are the most important ones. The nitrogen fraction of volatile sludge is typically around 10% of the organic sludge mass.

The phosphorus mass fraction is about 2.5% both for active and inactive organic sludge in completely aerobic systems. When systems are designed for biological phosphorus removal, this fraction increases to typical values of 6 – 8% as will be discussed in Chapter 7.

To compensate for nutrient losses in the excess sludge, the wastewater must supply the activated sludge with new nutrients. If insufficient nutrients are present in the influent, the activated sludge system will not function properly: e.g. problems with bulking sludge may appear. The minimum mass of nutrients required in the influent can be calculated from the excess sludge production. For nitrogen one has:

$$mN_l = f_n \cdot mE_v \quad (3.57)$$

where:

mN_l = mass of nitrogen needed for sludge production per unit mass applied COD
f_n = mass fraction of nitrogen in organic sludge = 0.1 g N · g^{-1} VSS

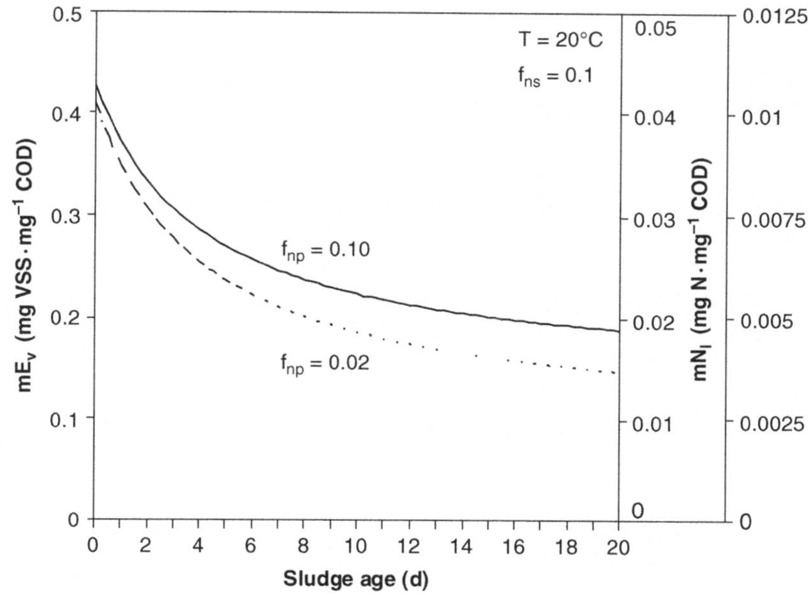

Figure 3.11 Typical profile of organic sludge production and nutrient demand as function of the sludge age for raw ($f_{np} = 0.10$) and settled sewage ($f_{np} = 0.02$)

For phosphorus the corresponding expression is:

$$mP_l = f_p \cdot mE_v \tag{3.58}$$

where:

mP_l = phosphorus mass required for sludge production per unit mass applied COD
f_p = fraction of phosphorus in organic sludge = 0.025 g P · g^{-1} VSS.

As an example in Figure 3.11 the values of mN_l and mP_l have been plotted as function of the sludge age. This figure can be used to estimate the minimum COD:N:P ratio in the wastewater required for healthy biomass growth, as will be demonstrated in Example 3.7.

Once the values of mN_l and mP_l have been established, it is a simple matter to calculate the corresponding minimum required nutrient concentrations in the influent. For nitrogen one has:

$$mN_l = MN_l/MS_{ti} = (Q_i \cdot N_l)/(Q_i \cdot S_{ti}) = N_l/S_{ti} \quad \text{or}$$
$$N_l = mN_l \cdot S_{ti} = f_n \cdot mE_v \cdot S_{ti} = f_n \cdot [(1 - f_{ns} - f_{np}) \cdot [(1 + f \cdot b_h \cdot R_s) \cdot C_r/R_s + f_{np}/f_{cv}] \cdot S_{ti} \tag{3.59}$$

Similarly, for phosphorus one calculates:

$$P_l = mP_l \cdot S_{ti} = f_p \cdot mE_v \cdot S_{ti} = f_p \cdot [(1 - f_{ns} - f_{np}) \cdot [(1 + f \cdot b_h \cdot R_s) \cdot C_r/R_s + f_{np}/f_{cv}] \cdot S_{ti} \tag{3.60}$$

where N_l and P_l are the influent nitrogen and phosphorus concentrations required to compensate for nutrient demand for excess sludge production.

In the case of domestic wastewater, the concentrations of the nutrients will be much higher than the minimum requirements for sludge production. However, in industrial wastewaters, especially those of vegetable origin, the nutrient concentrations are low and additional nutrients may have to be added to the influent to have a properly functioning system.

The above equations are valid for activated sludge systems without an anaerobic- or aerobic digester. If these are installed, then the recycle of the liquid phase of the digested sludge (reject water) to the biological reactor will reduce the nutrient demand, as in the sludge digester part of the organic material is converted into biogas and the corresponding nutrients will be released as ammonium and phosphate. In the case of nitrogen removal systems this release of nitrogen constitutes a significant additional nitrogen load to the plant. The subject of nutrient release during digestion will be further discussed in Chapter 12. Table 3.9 summarizes the different mass-based equations for nutrient demand.

Table 3.9 Mass equations for nutrient demand

Par.	Mass equations for nutrient demand	No.	Total demand
mN_l	$= MN_l/MS_{ti}$		
	$= f_n \cdot ME_v/MS_{ti}$	(3.59)	$MN_l = mN_l \cdot MS_{ti}$
	$= f_n \cdot mE_v = f_n \cdot mX_v/R_s$		
mP_l	$= MP_l/MS_{ti}$		
	$= f_p \cdot ME_v/MS_{ti}$	(3.60)	$MP_l = mP_l \cdot MS_{ti}$
	$= f_p \cdot mE_v = f_p \cdot mX_v/R_s$		

EXAMPLE 3.7

Raw sewage is treated in an activated sludge system operating at a sludge age of 8 days. The wastewater has the following composition: COD = 660 mg · l^{-1} ($f_{ns} = f_{np} = 0.10$); TKN = 50 mg N · l^{-1} and total phosphorus = 3 mg P · l^{-1}. Furthermore T = 26°C, so $b_h = 0.3\ d^{-1}$. Estimate the sludge production and the effluent nitrogen and phosphorus concentrations, using Figure 3.11.

Solution

From Figure 3.11 the sludge production and the demand for N and P for a sludge age of 8 days are determined, assuming the typical nutrient mass fractions of $f_n = 0.1$ mg N · mg^{-1} VSS and $f_p = 0.025$ mg P · mg^{-1} VSS:

- $mE_v = 0.23$ mg VSS · mg^{-1} COD;
- $mN_l = 0.023$ mg N · mg^{-1} COD (Eq. 3.57);
- $mP_l = 0.0058$ mg P · mg^{-1} COD (Eq. 3.58).

Hence the sludge production will be 23% of the applied COD mass. The required minimum influent nitrogen concentration for sludge production is:

$$N_l = mN_l \cdot S_{ti} = 0.023 \cdot 660 = 15\,mg\,N \cdot l^{-1}$$

Hence, if no denitrification takes place, the effluent nitrogen concentration will be:

$$N_{te} = N_{ti} - N_l = 50 - 15 = 35\,mg\,N \cdot l^{-1}$$

Similarly for phosphorus the required minimum concentration for sludge production is:

$$P_l = mP_i \cdot S_{ti} = 0.0058 \cdot 660 = 3.8\,mg\,P \cdot l^{-1}$$

Hence, the concentration of phosphorus in the influent ($3.0\ mg\ P \cdot l^{-1}$) will be insufficient for sludge production and the difference between the demand and the available concentration ($3.8 - 3.0 = 0.8$ $mg\ P \cdot l^{-1}$) must be added. Without this addition, problems may arise with poorly settling sludge. However, the phosphorus demand may be satisfied by supernatant return from the sludge digestion unit (see Chapter 12).

EXAMPLE 3.8

An activated sludge system treats $2000\ m^3 \cdot d^{-1}$ of an industrial wastewater with a COD : TKN : P ratio of 1000 : 2 : 2. Under normal operational conditions the sludge production (mE_v) is equal to 0.3 $mg\ VSS \cdot mg^{-1}\ COD$. DAP [di-ammonium phosphate or $(NH_4)_2HPO_4$] and urea (NH_2CNH_2) are used to add the deficient nutrients. What will be the minimum concentrations of these compounds per litre influent when an influent COD concentration of $1000\ mg \cdot l^{-1}$ is assumed? It is known that urea is much cheaper than DAP.

Solution

As $mE_v = 0.3\ mg\ VSS \cdot mg^{-1}\ COD$ one has:

$$mN_l = f_n \cdot mE_v = 0.1 \cdot 0.3 = 0.03\ mg\ N \cdot mg^{-1}\ COD \quad \text{and}$$
$$mP_l = f_p \cdot mEv = 0.025 \cdot 0.3 = 0.0075\ mg\ P \cdot mg^{-1}\ COD$$

For the influent concentration of $S_{ti} = 1000\ mg\ COD \cdot l^{-1}$ the minimum required concentrations of nitrogen and phosphorus are calculated as:

$$N_l = 0.03 \cdot S_{ti} = 30\ mg\ N \cdot l^{-1} \text{ and } P_l = 0.0075 \cdot S_{ti} = 7.5\ mg\ P \cdot l^{-1}$$

For the given nitrogen and phosphorus influent concentrations ($2\ mg \cdot l^{-1}$ of both N and P), the deficits are $30 - 2 = 28\ mg\ N \cdot l^{-1}$ and $7.5 - 2 = 5.5\ mg\ P \cdot l^{-1}$. The minimum dosage of DAP is calculated by considering that 1 mol DAP (128 g) has 2 moles of N (28 g) and 1 mol of P (31 g).

Thus the addition of 1 mg · l^{-1} of DAP results in an increase of 28/128 = 0.21 mg N · l^{-1} and 31/128 = 0.23 mg P · l^{-1}. Hence for the required 5.5 mg P · l^{-1} there is a demand of 5.5/0.23 = 24 mg DAP · l^{-1}. For the demand of 28 mg N · l^{-1} the required addition would be 28/0.21 = 133 mg DAP · l^{-1}. However, this would lead to excess phosphorus in the effluent.

It is better to add just enough DAP to cover the phosphorus demand (adding 24 mg · l^{-1} for 5.5 mg P · l^{-1} and 0.21 · 24 = 5.0 mg N · l^{-1}) and then to supply urea to make up for the remaining nitrogen demand. Knowing that urea has a nitrogen fraction of 28/60 = 0.47 mg N · mg^{-1} urea, then for the residual N demand of 28 − 5 = 23 mg N · l^{-1} an addition of 23/0.47 = 49 mg · l^{-1} urea is required. For the flow of 2000 m^3 · d^{-1}, the nutrient demand is 2000 · 0.024 = 48 kg · d^{-1} of DAP and 2000 · 0.049 = 98 kg · d^{-1} of urea. The residual nutrient concentration would then be zero.

3.3.3.4 Temperature effect

Temperature influences the ideal steady state model for COD removal through its effect on the active sludge decay rate. In Table 3.8 the decay constant is given as $b_h = 0.24 \cdot 1.04^{(T-20)}$. Hence at increased temperature, the decay rate will rise and with it the oxygen consumption for endogenous respiration. Consequently the sludge production rate will decrease. The influence of temperature on OUR and the sludge production rate and the active fraction is presented graphically in Figure 3.12, where these parameters have been plotted as functions of the sludge age for temperatures 14°C and 28°C, which may be considered to be respectively the minimum and maximum temperatures of sewage in subtropical and tropical regions. The curve for 20°C has also been indicated.

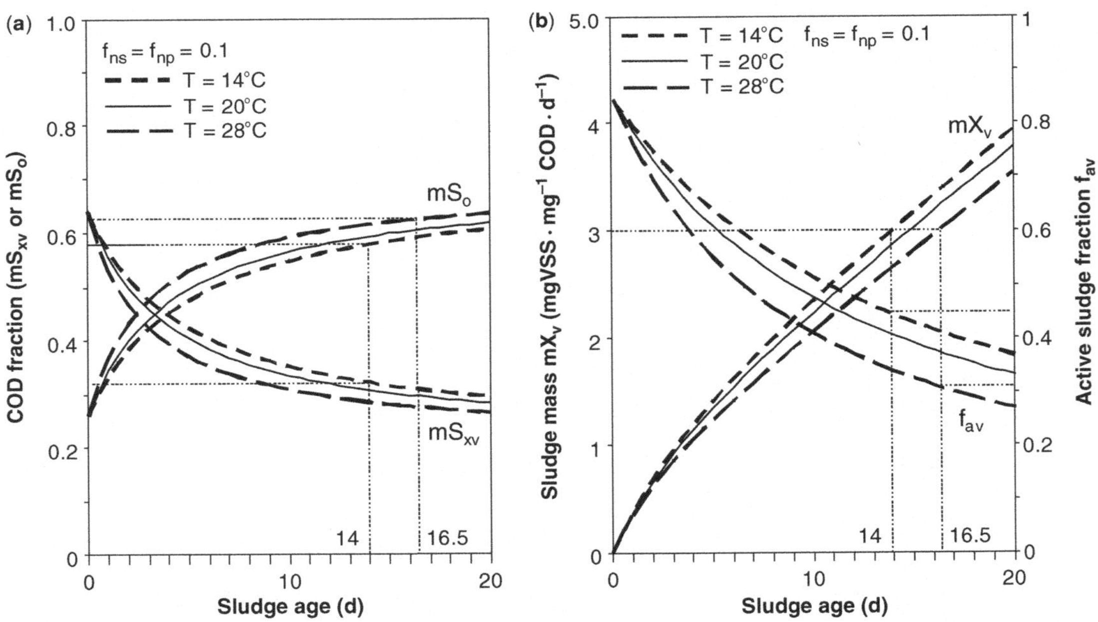

Figure 3.12 Effect of the temperature on the production of sludge, the oxygen consumption (Fig. 3.12a) and the active fraction (Fig. 3.12b)

EXAMPLE 3.9

An activated sludge system is designed to contain a maximum sludge mass of 3.0 mg VSS · d · mg^{-1} COD. If the non-biodegradable fractions are $f_{ns} = f_{np} = 0.10$, calculate for 14°C and for 28°C the following parameters:

- The maximum sludge age that can be applied;
- The oxygen consumption;
- The active sludge fraction for this maximum sludge age.

Solution

The maximum sludge age for $mX_v = 3.0$ mg VSS · d · mg^{-1} COD can be calculated from Eq. (3.48) or determined from the graph in Figure 3.12b (for the specified f_{ns} and f_{np} values). For $mX_v = 3.0$ and $T = 14°C$ one has $R_s = 14$ days and for $T = 28°C$ one has $R_s = 16.5$ days.

In Figure 3.12a it can be seen that the oxygen consumption per unit mass influent COD $mS_o = 0.62$ for $T = 28°C$ and $R_s = 16.5$ days and $mS_o = 0.58$ for $T = 14°C$ and $R_s = 14$ days. It is concluded that in the same system and maintaining an equal volatile sludge concentration an increase of temperature from 14 to 28°C results in an increase of the oxidised COD fraction from 58 to 62%, due to a higher endogenous respiration rate; an increase of $(62 - 58)/58 = 7\%$ or 0.5% per °C.

The oxygenation capacity of an aerator tends to decrease with increasing temperature (due to the decrease of oxygen solubility, although this is partly compensated by an increase in the oxygen transfer rate), so that the oxygenation capacity must be designed for the highest temperature to be expected. Alternatively the process can also be operated at a shorter sludge age when the temperature increases.

The active sludge fraction is calculated from Eq. (3.52): $f_{av} = 0.45$ for $R_s = 14$ days and $T = 14°C$ and $f_{av} = 0.32$ for $R_s = 16.5$ days and $T = 28°C$. Hence there is a considerable reduction of the active sludge fraction, due to the higher decay rate that occurs when the temperature increases. However, since the metabolic capacity of the sludge increases as well, the treatment capacity of the system may actually increase.

3.3.3.5 True yield versus apparent yield

A fundamental parameter in any theoretical model of the activated sludge system is the yield coefficient. At this point it might be interesting to elaborate on the difference between the "true or biochemical" yield (Y) and the "apparent or observed" yield (Y_{ap}), the latter being equal to the specific excess sludge production. In practice the concepts of true yield and apparent yield are often confused with each other and hence erroneously applied.

The true or biochemical yield is defined by biochemical considerations: i.e. its value is based on the amount of chemical energy that can be released from organic matter upon oxidation and subsequently used for growth. Naturally the type of micro-organisms involved, the type of compounds degraded and the nature of the available oxidant might influence this value. However, if these factors are maintained constant, it may be expected that the value of Y is constant as well. This is in fact the case for municipal sewage treatment, with its cocktail of organic compounds and different bacterial species, as the experimental values reported for the true yield are all remarkably constant at around 0.45 mg VSS · mg^{-1} COD.

Furthermore, the true yield only considers instantaneous growth on externally supplied and biodegradable COD, or stated differently, the associated oxygen demand is only due to exogenous respiration. The value of the true yield is thus equal to that of mS_{xv} (or mE_v) at a (theoretical) sludge age of zero days, when endogenous respiration is absent, as can be demonstrated when considering Eq. (3.38):

$$mS_{xv} = [(f_{cv} \cdot (1 - f_{ns} - f_{np}) \cdot (1 + f \cdot b_h \cdot R_s) \cdot Y \cdot R_s/(1 + b_h \cdot R_s)]/R_s + f_{np}$$

This equation can be simplified when a completely biodegradable material is considered (i.e. $f_{ns} = f_{np} = 0$) and by removing R_s from both the numerator and the denominator of the first term:

$$mS_{xv} = f_{cv} \cdot (1 + f \cdot b_h \cdot R_s) \cdot Y/(1 + b_h \cdot R_s)$$

Now it becomes evident that for $R_s = 0$ days mS_{xv} will indeed be equal to Y, as the above equation then further simplifies into $mS_{xv} = f_{cv} \cdot Y = 1.5 \cdot 0.45 = 0.67$ mg COD · mg^{-1} COD. This is graphically illustrated in Figure 3.13a, where it can be observed that for $R_s = 0$ days (and $S_{bi} = S_{ti}$) the value of mS_{xv} (and mE_v) is indeed identical to the value of the theoretical yield Y.

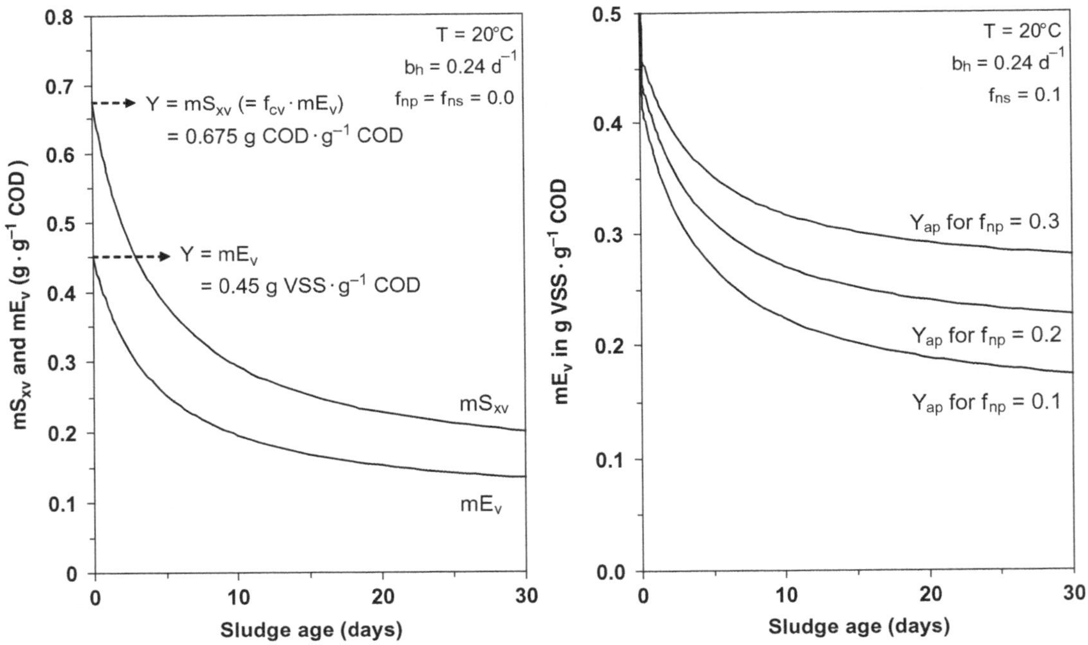

Figure 3.13 Apparent yield approaching true yield for $R_s = 0$ days and $f_{ns} = f_{np} = 0$ (left), and influence of f_{np} on the value of apparent yield $Y_{ap} = mE_v$ (right)

Whereas the true yield is essentially constant, this certainly is not the case for the apparent sludge yield, defined as the actual excess sludge production divided by the applied COD load. As the value of the apparent yield is numerically identical to mE_v (or mS_{xv}), it includes the effect of accumulation of inert organic particulate material and those of endogenous respiration: i.e. the decrease of active biomass and the production of endogenous residue. Clearly the value of the apparent yield will depend on the applied sludge age, the temperature and on the influent composition (f_{ns} and f_{np}). Therefore it is concluded that the use of literature values for the apparent yield should only be considered if it is certain that the design conditions (temperature, sludge age, influent composition) are similar to those under which the literature value was determined. Figure 3.13b shows the influence of particulate non-biodegradable COD fraction (f_{np}) on the value of the apparent yield ($Y_{ap} = mE_v$).

EXAMPLE 3.10

For a certain activated sludge system, operated at T = 20°C, the following experimental data have been determined:

- Sludge production = 1500 kg VSS · d^{-1};
- Applied COD load = 5000 kg COD · d^{-1};
- Biodegradable COD fraction = 70% of the influent COD;
- Soluble non biodegradable COD in effluent = 10% of the influent COD.

Determine the apparent yield and estimate the operational sludge age. Use the default values of the kinetic- and stoichiometric parameters.

Solution

It is easy to determine the apparent yield, as $Y_{ap} = mE_v$:

$$Y_{ap} = ME_v/MS_{ti} = 1500/5000 = 0.3\,\text{mg VSS}\cdot\text{mg}^{-1}\,\text{COD}$$

The value of the sludge age can be estimated from Eq. (3.50), as all other parameters are known:

$$mE_v = (1 - f_{ns} - f_{np}) \cdot (1 + f \cdot b_h \cdot R_s) \cdot C_r/R_s + f_{np}/f_{cv}$$
$$C_r = 0.45 \cdot R_s/(1 + 0.24 \cdot R_s)$$
$$f_{np} = 1 - 0.7 - 0.1 = 0.2$$
$$mE_v = 0.8 \cdot (1 + 0.2 \cdot 0.24 \cdot R_s) \cdot C_r/R_s + 0.2/1.5 = 0.3$$

With trial and error this equation can be solved for R_s = 6 days.

3.3.3.6 *F/M ratio*

In the development of the simplified model, the sludge age evolved as the fundamental process variable. In practice a different parameter is amply used in design and analysis of activated sludge systems: the F/M ratio. This parameter seeks to express the ratio between the influent organic material (F for "food") and the bacterial mass available to metabolise it (M for "micro-organism mass"). Usually the parameter F is taken as the influent COD mass, whereas M is taken to be equal to the volatile sludge mass, so that the F/M ratio is expressed as kg COD · kg^{-1} VSS · d^{-1}. In the terminology of the simplified model one has:

$$F/M = MS_{ti}/MX_v = 1/mX_v \tag{3.61}$$

Hence, the F/M ratio can also be expressed as a function of the sludge age:

$$F/M = 1/mX_v = 1/[(1 - f_{ns} - f_{np}) \cdot (1 + f \cdot b_h \cdot R_s) \cdot C_r + f_{np} \cdot R_s/f_{cv}] \quad (3.62)$$

In Figure 3.14 the F/M ratio is plotted as a function of the sludge age for raw sewage and settled sewage. Figure 3.14a has been calculated for long sludge ages (R_s from 4 to 20 days) and Figure 3.14b for short sludge ages (0 to 4 days). When Figure 3.14 is analysed, it can be noted that the F/M ratio is an ambiguous parameter: for the same sludge age it exhibits very different values for different f_{np} values.

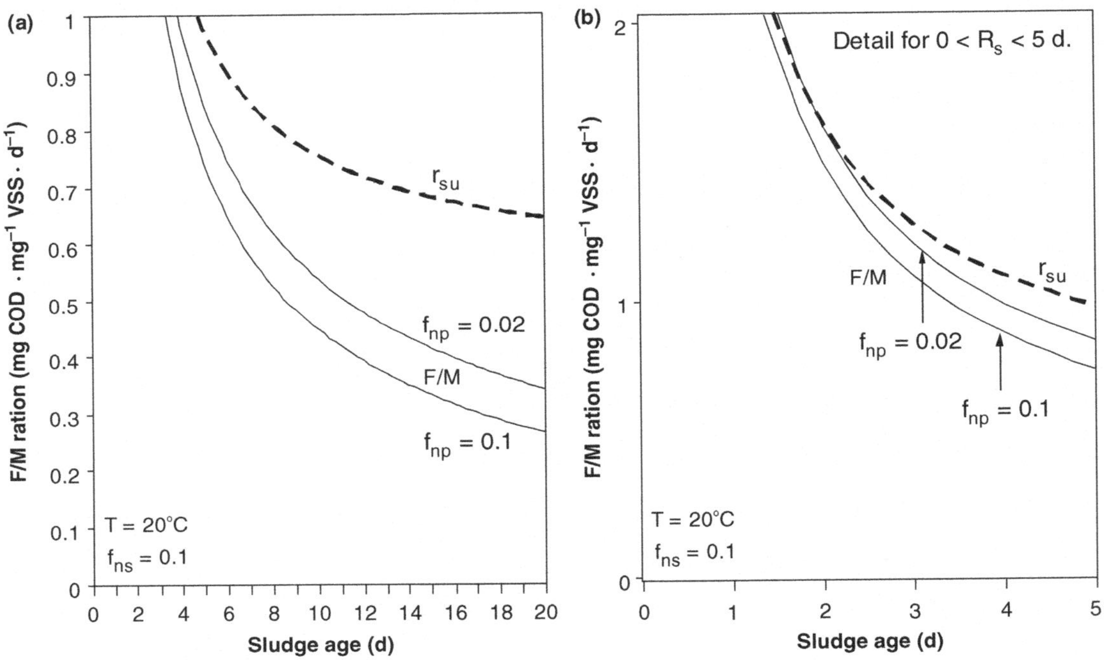

Figure 3.14 F/M ratio and specific utilisation rate r_{su} as function of the sludge age

Consider for instance the two activated sludge systems shown in Figure 3.15, with equal volume and -sludge mass and receiving the same COD load. The F/M ratio applied to system A and system B is identical, i.e. $500/400 \cdot 3 = 0.4$ kg COD $\cdot$ kg^{-1} VSS $\cdot$ d^{-1}. However, it can be observed that the system response is significantly different: system A produces much more excess sludge and exerts considerably less oxygen demand than system B.

The F/M ratio, being equal for both systems, obviously is incapable of explaining the observed system behaviour. On the other hand, when the applied sludge age is calculated, the reason for the dissimilar behaviour is obvious. For system A the sludge age is equal to $1200/127 = 9.5$ days, while for system B it is 15 days. As has been explained in earlier sections, operation at a higher sludge age will result in an increase in oxygen demand and a decrease in excess sludge production. Refer also to the theoretical curves shown in Figure 3.4.

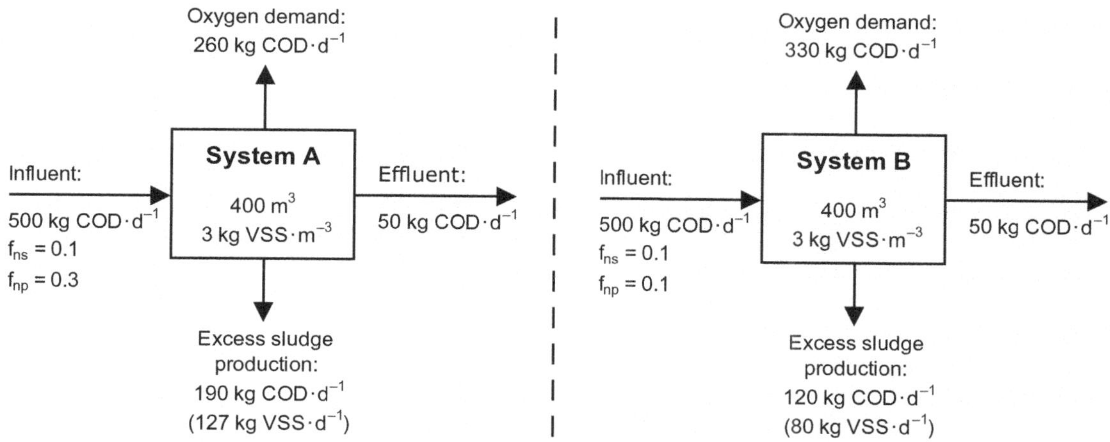

Figure 3.15 Comparison of behaviour of two activated sludge systems operated at *equal* F/M ratio of 0.4 kg COD · kg^{-1} VSS · d^{-1} but at different sludge ages

Hence, the fact that different systems are operated at the same F/M ratio does not mean that the sludge ages are equal or that these processes are otherwise comparable. In this context it would be more meaningful to define an alternative parameter, indicating the ratio between the mass of daily applied *biodegradable* material and the available *active* sludge mass. This parameter represents the specific utilisation rate of biodegradable organic material by the sludge, or in other words, the metabolised COD mass per unit mass active sludge per day:

$$r_{su} = MS_{bi}/MX_a = (1 + b_h \cdot R_s)/(Y \cdot R_s) = 1/C_r \tag{3.63}$$

where:

r_{su} = specific utilisation rate of biodegradable influent organic material by the active sludge (mg COD · mg^{-1} X_a · d^{-1}).

The r_{su} value has been plotted as a function of the sludge age in Figure 3.14. Note that this parameter is *independent* of the composition of the influent in terms of f_{ns} and f_{np}. Another important aspect that can be observed especially in Figure 3.14b, is that both the F/M ratio and r_{su} increase as the sludge age decreases. In reality the values of F/M and r_{su} will have an upper limit, because of limitations to the capacity of bacteria to metabolise organic material. Hence, there is a minimum sludge age below which it will not be possible for the bacteria to remove all the biodegradable organic material in the influent. The steady state model for COD removal only has validity for sludge ages above this minimum, where ideal behaviour is approached. In Section 3.4 the non-ideal active sludge process will be discussed, where a kinetic model is presented that allows the determination of this minimum sludge age.

3.3.4 Selection and control of the sludge age

In the previous sections it has been established that the most important operational parameter of the activated sludge system is the sludge age. Therefore, attributing the correct value to this parameter is of great importance.

A short sludge age as used in the so called high-rate processes ($F/M>1$ to 2 g $COD \cdot g^{-1}$ $VSS \cdot d^{-1}$, i.e. $R_s<1.5$ to 2 days) may allow almost complete utilisation of biodegradable material at higher temperatures, but the solids retention time is too short for extensive decay and the associated endogenous respiration. Hence, the oxygen consumption in these processes will be low, whereas the sludge production is high (Figure 3.5) and the fraction of active (biodegradable) sludge is also high (Figure 3.9). For this reason, in high-rate processes the units for sludge treatment are large, whereas the reactor itself is relatively small (see also Chapter 14). A disadvantage of a very short sludge age is that the predators of free bacteria (those not aggregated to flocs) do not have sufficient residence time to develop, so that the effluent quality is reduced: part of the active sludge will be discharged as free bacteria in the effluent. For that reason effluent BOD and VSS concentrations will be relatively high. At longer sludge ages (above 5 to 8 days), predators of free bacteria will develop and BOD and VSS concentrations can be very small (<5 to 10 $mg \cdot l^{-1}$), if the final settler works properly.

In Europe and the United States initially many activated sludge systems were designed for operation at a very short sludge age, even though the final effluent had a somewhat inferior quality. In regions with a warm climate, if removal of organic material is the only or principal objective of an activated sludge system, there is a very solid argument for using a short sludge age. The main disadvantage of a short sludge age is the high excess sludge production, but in tropical regions this can be used as an advantage. Using anaerobic digestion (which can be applied at ambient temperatures in regions with a warm climate), the large and highly biodegradable excess sludge mass can be converted into methane, which in turn may be used for power generation. This energy can then be used to cover the energy needs of the aeration process. Hence, it becomes possible to use the chemical energy of the organic material in the wastewater in the treatment process. In principle the activated sludge process may even become independent of external energy sources. The quantitative aspects of this configuration are discussed in Chapter 12.

However, often the applied sludge age is not determined alone by considerations concerning the removal of organic material. Due to stricter legislation, in general the activated sludge system will also have to remove nutrients (nitrogen and phosphorus) and the removal of these constituents requires a certain minimum operational sludge age for the activated sludge system. Chapter 5 presents the theory to determine the minimum sludge age required for nitrogen removal. Once the sludge age has been selected, it is important to maintain the chosen value by an adequate discharge of excess sludge. This discharge can be directly from the reactor (hydraulic control of the sludge age) or from the return sludge flow. The latter option is much applied in practice, because the return sludge is always more concentrated than the mixed liquor in the reactor. Hence a smaller flow needs to be discharged to withdraw the same mass of solids. However, the potential advantage of withdrawing sludge from the return sludge flow is non-existent when the sludge is thickened before being introduced into the sludge treatment unit, as in practice will be very often the case. The thickened sludge concentration is practically independent from the influent suspended solids concentration, so that the same concentration will be obtained after thickening from both mixed liquor and return sludge (Section 12.2). Consider that return sludge typically has a concentration between 6 – 12 g $TSS \cdot l^{-1}$, whereas thickened sludge usually is in the range of 25 – 60 g $TSS \cdot l^{-1}$. Anyone who has ever performed a SVI experiment can testify that the initial increase in sludge concentration, corresponding to the decrease in sludge blanket level, is relatively rapid. Very quickly gravity compaction will be the limiting process instead of zone settling, causing the rate of concentration increase to be drastically reduced. So in effect, by increasing the solids concentration in the feed to the thickener, only a fraction of the rapid zone settling phase is eliminated.

On the other hand, hydraulic sludge age control has important advantages over control by discharging from the return sludge flow. For example, due to variations in the influent flow, the flow of mixed liquor to the settler and the sludge mass in the settler vary considerably. In Figure 3.16 typical profiles of the sludge concentration in the mixed liquor and in the return sludge are presented as a function of time (WRC, 1984). As can be observed, the mixed liquor concentration oscillates around an average of $4.5\ g \cdot l^{-1}$, whereas the maximum value of the return sludge concentration is more than twice as high as the minimum value. Hence the sludge mass in a unit volume of return sludge is highly variable and consequently precise sludge age control is difficult, even if the excess sludge concentration is analysed regularly (which often is not the case). The sludge concentration in the reactor is much less variable and largely independent of influent flow fluctuations. Thus sludge age control by direct discharge of excess sludge from the reactor is much more reliable. In fact, when hydraulic sludge age control is applied, i.e. every day a fraction $1/R_s$ of the aeration tank volume is discharged as excess sludge, then the control of the sludge age is by definition perfect.

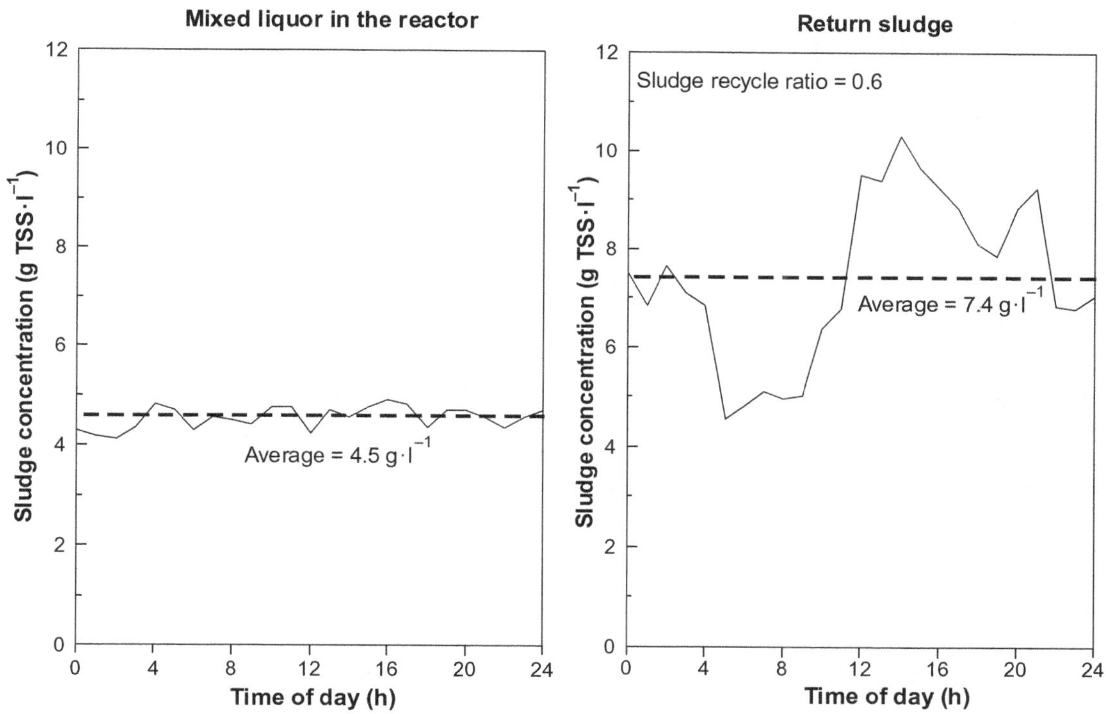

Figure 3.16 Typical daily profiles of variation of the mixed liquor- and the return sludge concentrations in an activated sludge system

Strictly speaking, for considerations of sludge age control alone, this would even dispense with the notoriously time-consuming and inaccurate measurement of the reactor sludge concentration. Having said all of this, it should be stressed that in the above analysis the discharge of suspended solids in the effluent is supposed to be negligible compared to that with the effluent, i.e. an ideal final settler is assumed. In case of diluted wastewaters, e.g. after anaerobic pre-treatment and especially when combined with a high value of the sludge age, this assumption might not be justified. Should hydraulic sludge wasting be applied in this case, then the actual sludge age might be considerably less than the design sludge age. However, even in this case direct discharge of excess sludge from the reactor is recommended, due to its simplicity, but the volume discharged should now be adapted for the anticipated loss of sludge with the effluent.

An alternative for sludge age control is control of the reactor sludge concentration to a setpoint value, for instance around $4\ g \cdot l^{-1}$. Although this method is frequently applied due to its apparent simplicity, it is principally wrong and in actuality not all that easy. First of all, it will result in a difference between operational- and design sludge age (sometimes significantly), depending on the deviation between the actual COD load and the design COD load that is applied. Performance will thus be different from what is expected, and often will be inferior. Furthermore, errors of more than ten percent during the analysis of suspended solids are not uncommon. Operators know this and therefore will wait several days to confirm a decreasing or increasing trend in sludge concentration, before adjustments to the excess sludge flow are made. Inexperienced operators tend to overcompensate, causing large fluctuations in reactor sludge concentration and load to the excess sludge treatment units. Therefore it is counter productive to discharge sludge based on the result of the latest sludge concentration test. Sludge discharge should be pre-emptive and not corrective.

3.4 GENERAL MODEL OF THE ACTIVATED SLUDGE SYSTEM

In the preceding sections, an ideal steady state model for COD removal was presented, allowing a description of the activated sludge system in terms of the removal efficiency of organic material, sludge production and oxygen consumption. To develop the steady state model, it was assumed that the activated sludge system is operated under steady state conditions, with complete mixing in a single reactor and with complete utilisation of the biodegradable material. Furthermore, the settler was assumed to be an ideal and instantaneous phase separator. It was shown that it is possible to operate the activated sludge system under conditions very similar to those assumed for the model. In Figure 3.5 an excellent correlation was observed between actual and theoretically predicted activated sludge behaviour. Although it is possible to operate the system under near ideal conditions, in practice an activated sludge system usually will not comply with all the imposed conditions. The following factors may cause non-ideal behaviour:

- In practice, activated sludge systems almost never operate under steady state conditions: the normal situation is that both flow and load exhibit a strong daily variation. This is true for municipal sewage, but also (and sometimes even more so) for industrial wastewater. In the case of municipal sewage, approximately half of the organic load is produced in only 4 hours (from 08:00 to 12:00), leaving 20 hours for the production of the other half. Even when some flow and load equalisation occurs in the sewer system, the wastewater treatment plant will still experience cyclic organic and hydraulic load variations. It is possible that during the periods of maximum loading, the bacterial mass is incapable of metabolising all the incoming organic material. In such cases there is non-ideal behaviour due to the cyclic variations of the influent flow and load. At any rate, the OUR due to exogenous respiration will vary with time;
- Many activated sludge systems are composed of several reactors instead of one completely mixed unit. Since the entire influent flow is usually discharged in the first of the series of reactors, this reactor will tend to be overloaded and metabolism of the influent organic material will be incomplete. Hence there will be a concentration gradient of biodegradable material in the reactor series, with consequential differences in the OUR in the different reactors, which characterises non-ideal behaviour;

- The sludge age applied in the activated sludge system may be so short that the existing sludge mass is unable to metabolise all the influent biodegradable material, so that it might be present in the wasted sludge and in the effluent. As we will see in Chapter 4, this will never be the case in nutrient removal systems due to the prolonged sludge age;
- In practice the settler will be non ideal in two aspects: (I) some suspended solids may escape together with the effluent and (II) the sludge mass in the settler may constitute a considerable fraction of the total mass present in the system, particularly when the influent flow (and hence the mixed liquor flow to the settler) is maximum, as it will be during peak flow conditions, for instance due to heavy rainfall.

In order to develop a model that describes the removal of organic material and the consequential processes of sludge growth and oxygen consumption under non-ideal operational conditions, it is necessary to take into consideration the rate at which the different processes develop in the system. In the simplified model, kinetics are of no importance because it is supposed that the utilisation of biodegradable material is immediate and complete. The best-known kinetic model for metabolism of organic material by micro-organisms was developed by Monod in a study about sugar fermentation by yeasts. The essence of the kinetic model by Monod can be summarised in two items:

(1) The growth rate of micro-organisms is proportional to the rate of substrate metabolism. This point has been used already to define the yield coefficient and may be formulated as follows:

$$r_g = (dX/dt)_g = Y \cdot r_u = -Y \cdot (dS/dt)_u \tag{3.64}$$

where:

r_g = $(dX/dt)_g$ = growth rate of micro-organisms ($mg\ VSS \cdot l^{-1} \cdot d^{-1}$)
X = micro-organism concentration ($mg\ VSS \cdot l^{-1}$)
$(dS/dt)_u$ = substrate utilisation rate ($= r_u$) in $mg\ COD \cdot l^{-1} \cdot d^{-1}$
Y = yield coefficient ($mg\ VSS \cdot mg^{-1}\ COD$)

(2) The substrate utilisation rate depends on the substrate concentration:

$$r_u = K_m \cdot S/(S + K_s) \cdot X \tag{3.65}$$

where:

r_u = substrate utilisation rate ($mg\ COD \cdot l^{-1} \cdot d^{-1}$)
K_m = specific utilisation rate constant of the substrate ($mg\ COD \cdot mg^{-1}\ X_a \cdot d^{-1}$)
K_s = half saturation concentration constant ($mg\ COD \cdot l^{-1}$)

Combining Eqs. (3.64 and 3.65) one has:

$$\begin{aligned}(dX/dt)_g &= Y \cdot K_m \cdot S/(S + K_s) \cdot X \\ &= \mu_m \cdot S/(S + K_s) \cdot X \end{aligned} \tag{3.66}$$

where μ_m = maximum specific growth rate constant for the micro-organisms

In the decades following the publication of the work of Monod (1948), several researchers tried to apply the model to the activated sludge system. The greatest difficulties were to define the parameters S and X. In Monod's investigation these parameters were well defined, as single substrates (substrates consisting of only one chemical substance) were fermented by pure culture yeasts. Therefore the concentration of both was unmistakably defined. In contrast, in an activated sludge system there are many suspended, colloidal and soluble compounds that are all metabolised simultaneously by a highly diversified mass of micro-organisms. Therefore the definition of the parameters "X" and "S" in the case of sewage treatment in the activated sludge system is problematic.

Initially the BOD_5 concentration of the liquid phase of the mixed liquor was taken as the substrate concentration and the micro-organisms concentration was equated to the volatile solids concentration (Garret and Sawyer, 1952, Lawrence and McCarty, 1970). While these models were important for improving understanding of the basic mechanisms of the activated sludge system, they were unable to predict its quantitative behaviour under dynamic flow and load conditions. The most important failures of these models were:

- The concentration of living micro-organisms (active sludge) is only a fraction of the volatile sludge mass and this fraction depends heavily on the composition of the influent organic material and the operational conditions in the activated sludge system (notably the sludge age, see Figure 3.9). It is therefore concluded that there is no obvious relationship between the measurable volatile solids concentration (X_v) and the relevant parameter, which is the micro-organism or active sludge concentration (X_a).
- The BOD in the liquid phase is not necessarily indicative for the concentration of substrate available for metabolism. In most wastewaters the influent organic material is composed partially of suspended solids. If the BOD concentration of the liquid phase of the mixed liquor is to be determined, it is necessary to effect the separation of the solid phase (sludge) from the liquid phase and in the process some of the particulate biodegradable influent material will unavoidably also be removed. Hence the BOD concentration in the liquid phase of the mixed liquor under actual process conditions may be higher than the measured BOD concentration.

Another problem related to the particulate nature of part of the influent organic material is that this part cannot be metabolised directly by the bacteria. In order to describe the utilisation of particulate biodegradable material, several authors have suggested the mechanism of adsorption of the particulate material on the active sludge, followed by external hydrolysis of the adsorbed material, resulting in the production of easily biodegradable organic material that can be metabolised by the bacteria (Katz and Rohlich, 1956, Blackwell 1971, Andrews and Busby, 1973; Dold, Ekama and Marais, 1980).

The concept of adsorption of organic material is very important for the development of a kinetic model for the activated sludge system: through the adsorption mechanism, the biodegradable material may be removed from the liquid phase without metabolisation taking place. Hence, there is no direct relationship between the concentration of biodegradable material in the liquid phase (having the BOD concentration as a quantitative parameter) and the concentration of biodegradable material available for the micro-organisms. It is concluded that even if it were possible to determine the BOD in the liquid phase of the mixed liquor, this parameter would not be adequate to describe the biodegradable material concentration available to the micro-organisms.

Marais and Ekama (1976) made an important contribution to the development of a kinetic model for the activated sludge system, suggesting that the rate of substrate utilisation can be related to the oxygen consumption rate. In Eq. (3.39) it was shown that there is a proportional relationship between the utilisation of organic material and the corresponding oxygen consumption:

$$O_{ex} = (1 - f_{cv} \cdot Y_h) \cdot r_u \tag{3.39}$$

Since the growth rate of biomass is related directly to the substrate utilisation rate one has:

$$r_g = Y \cdot r_u = Y_h/(1 - f_{cv} \cdot Y_h) \cdot O_{ex} \tag{3.67}$$

3.4.1 Model development

At the University of Cape Town (UCT) a general model for the activated sludge system has been developed. This model describes the quantitative variation of the most important parameters of the process: the fractions mS_{te}, mS_{xv} and mS_o, as well as measurable parameters such as the effluent COD concentration S_{te}, the volatile sludge concentration X_v and the oxygen uptake rate O_c. The model has been tested at bench scale and pilot scale units as well as at large full scale activated sludge plants (Johannesburg and Cape Town) under the most widely varying operational conditions. It was possible to correctly predict the activated sludge behaviour under varying flow and load conditions in reactors in series for a wide range of sludge ages (2 to 20 days) and for temperatures between 14 and 24°C. In all cases the correlation between the theoretically predicted and the experimentally determined values of the parameters was excellent.

Figure 3.17 is a representation of the processes that are related to organic material in the activated sludge system as described by the general model. As in this model the utilisation of biodegradable material is not necessarily complete, the kinetics of this utilisation are of fundamental importance. Since the small molecules of dissolved biodegradable material can be used directly by the bacteria, its utilisation rate will be higher than that of the particulate material that requires adsorption and hydrolysis before it can be utilised by the bacteria. Thus a subdivision between these two influent fractions is made. In accordance with Figure 3.17 the following processes relative to the utilisation of biodegradable material and sludge activity can be distinguished:

- Metabolism of the dissolved biodegradable or easily biodegradable material;
- Removal and solubilisation of the particulate biodegradable or slowly biodegradable material (adsorption and hydrolysis);
- Active sludge growth and decay;
- Consumption of oxygen.

(1) Utilisation of the easily biodegradable material.
The metabolism of the easily biodegradable material is described by means of the conventional Monod equation:

$$r_{us} = (dS_{bs}/dt)_{us} = K_{ms} \cdot X_a \cdot S_{bs}/(S_{bs} + K_{ss}) \tag{3.68}$$

where:

r_{us} = utilisation rate of easily biodegradable material ($mg\ COD \cdot l^{-1} \cdot d^{-1}$)
S_{bs} = COD concentration of the easily biodegradable material ($mg\ COD \cdot l^{-1}$)
K_{ms} = specific utilisation rate of easily biodegradable organic material ($mg\ COD \cdot mg^{-1}\ X_a \cdot d^{-1}$)
K_{ss} = half rate (Monod) constant (growth on S_{bs}) in $mg\ COD \cdot l^{-1}$

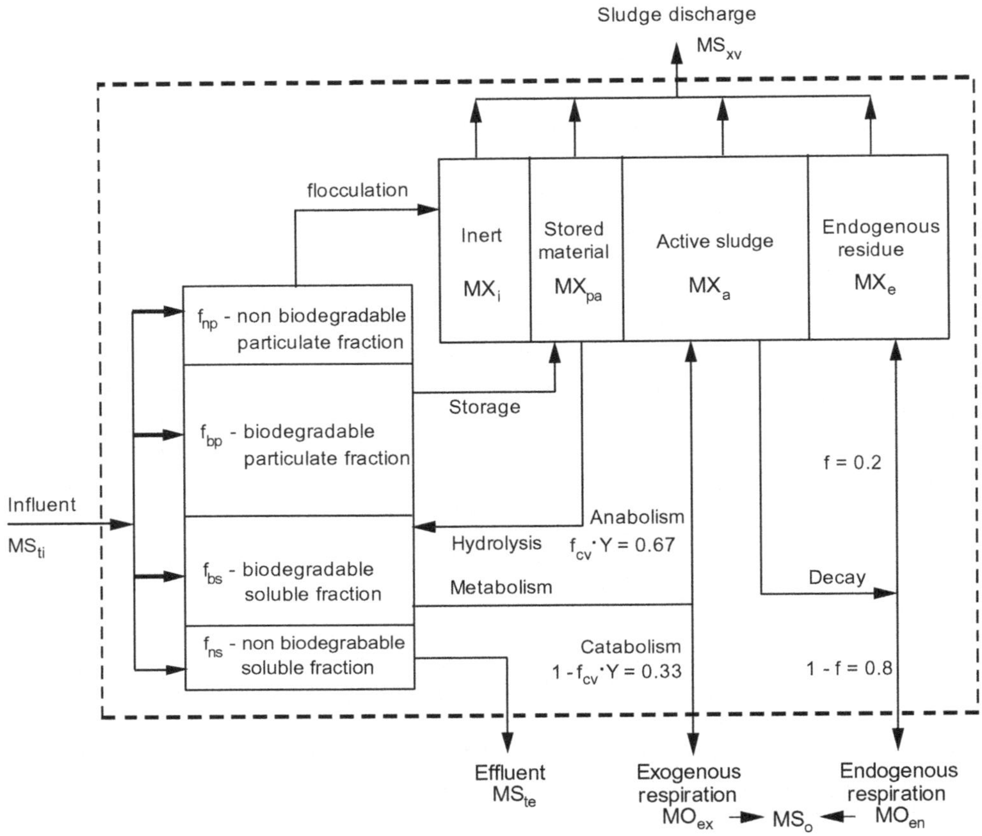

Figure 3.17 Schematic representation of the processes of organic material removal in the activated sludge system under non-ideal conditions

(2) Utilisation of the slowly biodegradable material

The first step of the utilisation of the slowly biodegradable material is adsorption to the active sludge, but it is only metabolised after it has been transformed into easily biodegradable material by the process of hydrolysis. The rate of the adsorption process was expressed by Dold, Ekama and Marais (1980) as:

$$r_a = -(dS_{bp}/dt)_a = K_a \cdot X_a \cdot S_{bp} \cdot (K_{ap} - S_{pa}/X_a) \tag{3.69}$$

where:

r_a = adsorption rate ($mg\ COD \cdot l^{-1} \cdot d^{-1}$)
S_{bp} = concentration of slowly biodegradable material ($mg\ COD \cdot l^{-1}$)
S_{pa} = concentration of adsorbed material ($mg\ COD \cdot l^{-1}$)
K_a = adsorption rate constant ($litre \cdot mg^{-1}\ X_a \cdot d^{-1}$)
K_{ap} = adsorption saturation constant ($mg\ COD.\ mg^{-1}\ X_a$)

The S_{pa}/X_a ratio indicates the mass of stored COD per unit mass of active sludge. The adsorption rate is proportional to the concentration of slowly biodegradable material in the mixed liquor, S_{bp}, and to the factor (K_{sp} - S_{pa}/X_a). Hence the adsorption rate approaches zero when the S_{pa}/X_a ratio approaches the value K_{sp}. The value of K_{ap} is thus indicative for the saturation of X_a with stored material. The experiments by Dold et al. (1980) indicate a value of 1.5 mg COD $\cdot$ mg^{-1} X_a.

Hydrolysis is a slow process and limits the utilisation rate of particulate material. Dold et al. (1980) suggested a modified Monod equation to describe hydrolysis. As the material is stored on the surface of the active sludge, the relevant expression to describe the concentration is not the mass per unit of volume but rather the mass per unit active sludge mass, so that:

$$\begin{aligned} r_{hi} &= (dS_{pa}/dt)_{hi} = K_{mp} \cdot X_a \cdot (S_{pa}/X_a)/(S_{pa}/X_a + K_{sp}) \\ &= K_{mp} \cdot X_a \cdot S_{pa}/(S_{pa} + K_{sp} \cdot X_a) \end{aligned} \tag{3.70}$$

where:

r_{hi} = hydrolysis rate of stored material (mg COD $\cdot$ l^{-1} $\cdot$ d^{-1})
K_{mp} = specific utilisation rate of slowly bio-degradable (adsorbed) organic material in mg COD $\cdot$ mg^{-1} $X_a \cdot d^{-1}$
K_{sp} = half rate (Monod) constant (growth on S_{pa}) in mg COD $\cdot$ mg^{-1} X_a

(3) Active sludge growth

The rate of active sludge growth can easily be expressed after having derived equations for the utilisation rate of the easily biodegradable material. From Figure 3.17 and Eq. (3.25) one has:

$$r_g = (dX_a/dt)_g = Y \cdot r_{us} = Y \cdot K_{ms} \cdot X_a \cdot S_{bs}/(S_{bs} + K_{ss}) \tag{3.71}$$

(4) Decay of the active sludge

Parallel to and independent of the sludge growth, decay of the active sludge occurs as a first order process, together with the associated appearance of an endogenous residue:

$$r_d = -(dX_a/dt)_d = -b_h \cdot X_a \text{ and} \tag{3.72}$$

$$r_{xe} = (dX_e/dt)_d = f \cdot r_d = f \cdot b_h \cdot X_a \tag{3.73}$$

(5) Consumption of oxygen

The OUR for oxidation of the organic material is the sum of the rates of endogenous and exogenous respiration. With the aid of Eqs. (3.39 and 3.41) one has:

$$\begin{aligned} O_c &= O_{ex} + O_{en} \\ &= (1 - f_{cv} \cdot Y) \cdot r_{us} + f_{cv} \cdot (1 - f) \cdot r_d \end{aligned} \tag{3.74}$$

Once the rates of the basic processes have been defined, it becomes a simple matter to describe the reaction rates of the different parameters involved in the activated sludge metabolism. For this one has to consider the processes that influence the concentration of a particular component. For example, the concentration of the easily biodegradable material decreases due its utilisation by the active sludge (rate r_{ns}), but it increases due to hydrolysis of stored material (rate r_{hi}). Hence the rate of change of the easily biodegradable material due to reactions can be written as:

$$r_{sbs} = r_{hi} - r_{us} \tag{3.75}$$

In Table 3.10 the expressions for the kinetics of processes in the activated sludge system and the reaction rates of the different concentrations affected by it have been brought together. Table 3.10 shows that the general activated sludge model for the removal of organic material is a rather complex set of differential equations. As several of these have no analytical solution, the solution of the set of equations must be obtained by numerical methods. In order to be able to use the general model, first the kinetic constants in the model must be determined. Either "typical" values can be used, or more preferably, the values are determined by means of experimental investigation.

Table 3.10 Process kinetics and production rates in the general activated sludge model

Par.	Equation	Description	No.
r_{us}	$= K_{ms} \cdot X_a \cdot S_{bs}/(S_{bs} + K_{ss})$	Utilisation of easily biodegr. material	(3.68)
r_a	$= K_a \cdot X_a \cdot S_{bp} \cdot (K_{ap} - S_{pa}/X_a)$	Adsorption of slowly biodegr. material	(3.69)
r_{hi}	$= K_{mp} \cdot X_a \cdot S_{pa}/(S_{pa} + K_{sp} \cdot X_a)$	Hydrolysis of stored material	(3.70)
r_g	$= Y \cdot K_{ms} \cdot X_a \cdot S_{bs}/(S_{bs} + K_{ss})$	Growth of active sludge	(3.71)
r_d	$= - b_h \cdot X_a$	Decay of active sludge	(3.72)
r_{xe}	$= f \cdot b_h \cdot X_a$	Production rate of endogenous residue	(3.73)
O_c	$= (1 - f_{cv} \cdot Y) \cdot r_{us} + f_{cv} \cdot (1 - f) \cdot r_d$	Oxygen consumption rate	(3.74)
r_{sbs}	$= r_{hi} - r_{us}$	Net production of easily biodegr. material	(3.75)
r_{sbp}	$= - r_a$	Net production of slowly biodegr. material	(3.76)
r_{spa}	$= r_a - r_{hi}$	Net production of adsorbed biodegr. mater.	(3.77)
r_{xa}	$= r_g - r_d = Y \cdot r_{us} - r_d$	Net production of active sludge	(3.78)

3.4.2 Model calibration

The calibration of the general model consists essentially of attributing values to the model constants in the differential equations that were developed above. The procedure is to overload the system continually or periodically with biodegradable material and to determine the following measurable parameters as a function of time: (I) COD of the liquid phase, (II) concentration of the volatile sludge concentration and (III) the oxygen uptake rate O_c.

The general model is then used to generate theoretical profiles of the measurable parameters, for different sets of values of the constants. Those values that result in the closest correlation between the experimental and theoretical values are accepted. As an example of the calibration procedure for the general model two calibration methods are discussed in Appendix 2: i.e. the application of cyclic loads and the application of batch loads.

3.4.3 Application of the general model

The most important practical applications of the general activated sludge model are:

- The determination of the sludge age required to obtain an substantially complete removal of the biodegradable organic material, i.e. approaching an ideal activated sludge system;
- The determination of O_c as a function of time and space (and hence the oxygenation capacity to be installed) in systems with non-ideal behaviour, which may for instance be caused by a plug-flow configuration and/or a variation in input flow and -load.

Table 3.11 summarizes the values of the kinetic constants, together with the temperature dependencies as determined by Dold, Ekama and Marais (1980). Computer simulations using these values show that the utilisation of organic material in an activated sludge system is nearly complete, even at very short sludge ages. For temperatures >18°C the required minimum sludge age is only 1.5 to 2 days (Van Haandel and Marais, 1981), refer also to Appendix 3. In practice, certainly when an activated sludge process is designed for nutrient removal, the sludge age will be much higher than the minimum value required for organic material removal.

Table 3.11 Values of kinetic constants and temperature dependencies (Dold et al., 1980)

Symbol	Description	Value at 20°C	Temp. dependency
K_{ms}	Specific utilisation rate of easily biodegradable organic material	20 mg COD · mg^{-1} X_a · d^{-1}	$1.2^{(T-20)}$
K_{mp}	Specific utilisation rate of slowly bio-degradable (adsorbed) organic material	3.0 mg COD · mg^{-1} X_a · d^{-1}	$1.1^{(T-20)}$
K_{ss}	Half rate (Monod) constant (S_{bs})	5.0 mg COD · l^{-1}	–
K_{sp}	Half rate (Monod) constant (S_{pa})	0.04 mg COD · mg^{-1} X_a	$1.1^{(T-20)}$
K_a	Adsorption rate constant	0.25 litre · mg^{-1} X_a · d^{-1}	$1.1^{(T-20)}$
K_{ap}	Adsorption saturation constant	1.5 mg COD · mg^{-1} X_a	–
b_h	Decay rate of heterotrophs	0.24 mg X_a · mg^{-1} X_a · d^{-1}	$1.04^{(T-20)}$

In general, at such a short sludge age the concentration of organic material in the effluent of an activated sludge system will be higher than the calculated value from the general model. This is not due to a failure in the kinetic expressions, but to the inability of the settler to separate the free bacteria (i.e. not attached to the sludge flocs), which are abundant at short sludge ages.

In regions with a low temperature, the sludge age required for complete removal of organic material is considerably higher and when the process is operated at a short sludge age, part of the wasted sludge will actually be stored or adsorbed organic material that has not yet been metabolised. In this case the activated sludge process functions also as a bioflocculation process. Some activated sludge processes are explicitly designed for this and part of the removal of the influent organic material is achieved in the anaerobic sludge digester.

For nutrient removal processes the operating sludge age is in general much higher than the minimum required for metabolisation of the organic material. In Appendix 3 a method is discussed to estimate the metabolised organic material fraction as a function of the values of the kinetic constants and the influent biodegradable material composition.

With regard to the O_c in systems with more than one reactor and/or variable flow and load conditions, profiles can be generated with the aid of the general model to predict the required oxygenation capacity. A computer simulation is the only alternative to experimental determination in order to be able to estimate the O_c under non-ideal conditions.

3.5 CONFIGURATIONS OF THE ACTIVATED SLUDGE SYSTEM

The different variants of the suspended growth version of the activated sludge system have two things in common: (I) the biomass is present in the form of macroscopic sludge flocs suspended in the mixed liquor, which can be separated from the liquid phase by quiescent settling, and (II) the oxygen demand is satisfied by mechanical aeration using air or pure oxygen. Since the early conception of the activated sludge process by Lockett and Ardern in 1914 several variants of the system have been developed to effect the removal of organic material and suspended solids from wastewaters. Presently a large number of commercial names exist for different configurations. In the following sections the principles of the main variants are discussed.

3.5.1 Conventional activated sludge systems

As was shown in Figure 3.2, the conventional activated sludge system is composed of one or more biological reactors (aeration tanks), in which the sludge is kept in uniform suspension due to mechanical aeration. The mixed liquor is directed to the settler where phase separation of the solid and the liquid phase takes place, after which the latter is discharged as the final effluent. Activated sludge systems are almost always constructed in concrete or steel (the latter is often used for smaller systems), but sometimes in developing countries the aeration tank consists only of a simple excavation with a reinforced floor and -sidewalls to avoid erosion. An important aspect of conventional activated sludge systems is the hydraulic regime in the aeration tank.

There are two extremes: the completely mixed reactor, in which mixing is immediate and complete and the plug-flow regime, in which no longitudinal mixing takes place at all and therefore will contain concentration gradients of both substrate and oxygen. The completely mixed reactor has the following advantages:

- Uniform oxygen demand in the reactor, which makes control of oxygen concentration easier and allows the aeration devices (aerators or diffusers) to be distributed uniformly as well;
- Maximum resistance to toxic discharges or sudden overloads, as the influent is distributed over the entire reactor volume, resulting in instant dilution of the toxic material.

However, currently activated sludge systems are often constructed as rectangular instead of square units, which favours the plug-flow regime rather than the completely mixed regime. The reasons for this are:

- The effluent quality of plug-flow reactors is somewhat better than that of completely mixed reactors, as a substrate concentration gradient exists throughout the length of the plug-flow reactor and because short circuiting from influent to effluent is not possible;

– It is believed that the plug-flow regime produces a sludge with better settling characteristics. However, today several mechanisms exist to avoid the development of poorly settling sludge, which will be discussed in Chapter 8.

As for the effluent quality, both experiments and theory show that the removal of organic material is substantially complete under either hydraulic regime as long as the sludge age is longer than the required minimum.

If axial (longitudinal) mixing is incomplete in the aeration tank, the OUR will be larger in the influent feed zone than in the end of the reactor where mixed liquor is discharged to the settler. For that reason a variable (step) aeration intensity is applied, which is higher near the feed and lower near the discharge zone (Figure 3.18a). On the other hand it is also possible to avoid a non-uniform OUR in a plug flow reactor, by introducing the influent in several points along the reactor length (a configuration denominated step loading, which is displayed in Figure 3.18b). Several activated sludge configurations have been developed to effect the removal of nitrogen and phosphorus in addition to that of organic material and suspended solids. These configurations require more than one reactor and are discussed in detail in the next two chapters.

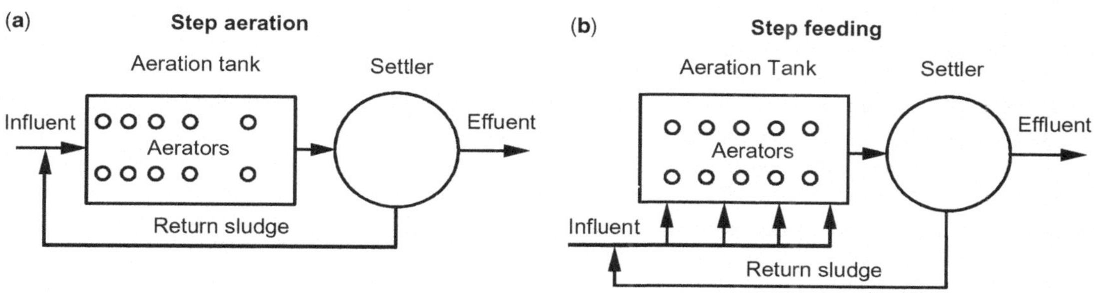

Figure 3.18 Plug flow type systems with step aeration (a) and step feeding (b)

3.5.2 Sequential batch systems

The first activated sludge systems were composed of a single reactor that processed sequential batches of wastewater for a certain period while aeration was applied. This was followed by a period in which the aeration was switched off, which transformed the reactor into a settler, from which the effluent was discharged and a new batch could be taken in. Hence in batch-wise operation, four different phases plus an optional one can be distinguished (see also Figure 3.19):

(1) Fill: a wastewater batch is fed to the sludge mass already present in the tank. During this phase the aerator may or may not be switched on;
(2) React: treatment of the wastewater (removal of the organic material and suspended solids with the reactor filled and the aerators on or off);
(3) Settle: sludge settling in the reactor in a quiescent environment (aeration and/or mixers off);
(4) Discharge: the clarified supernatant (treated effluent) is discharged and, if required, excess sludge is withdrawn as well;
(5) Pause: optional phase which is applied if the wastewater quantity to be treated is much smaller than the design flow, which will reduce aeration costs.

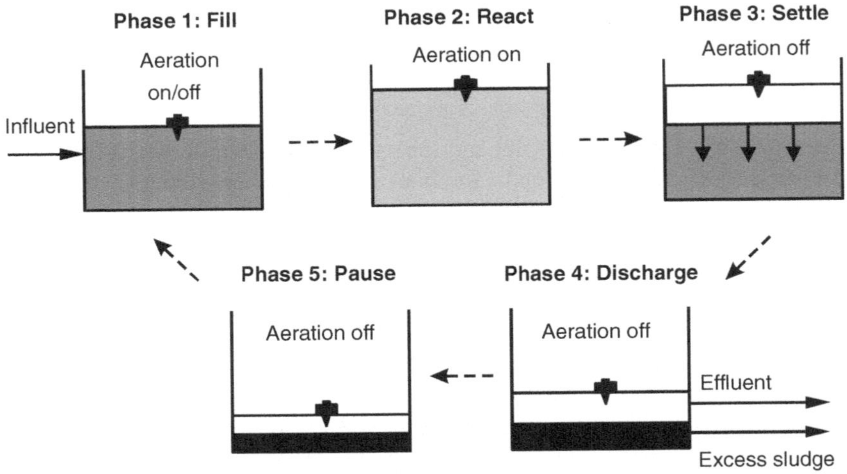

Figure 3.19 Typical operational cycle of a sequential batch reactor (SBR)

The duration of the phases depends on the nature and composition of the influent wastewater as well as on the concentration and composition of the sludge in the reactor.

The sequential batch reactor almost became obsolete when systems were developed with one or more continuous biological reactors, complemented with a separate settler. However, recently there has been a renewed interest in sequential batch reactors, especially when smaller industrial wastewater plants are being considered.

Advantages that are attributed to it are a better effluent quality, simplicity of operation and lower investment costs, due to the absence of a final settler. On the other hand it must be taken into consideration that the installed aeration capacity of a sequential batch system is considerably larger than that of a comparable conventional activated sludge system to compensate for the idle time required for decanting and settling, which can be up to 20 – 30% depending on the number of cycles per day and the duration of the different phases. This is further aggravated if anaerobic or anoxic periods are required, as will be the case in nutrient removal processes. For example, the aeration time in SBR systems for nutrient removal is only 20 to 25% of the total cycle time, the remainder being occupied with denitrification and settling. As the total oxygen consumption in the SBR is necessarily equal to that of a conventional system with continuous aeration, the required oxygen transfer rate and therefore also the installed aeration capacity in an SBR will be about 3 to 4 times larger.

Another disadvantage that is often attributed to SBR systems is the inflexibility in dealing with flow variations, as the SBR only receives influent during a minor part of the total cycle time. This can be resolved in several ways: for small applications a buffer feed tank can be constructed (as tank volume is much cheaper than reactor volume) while for larger treatment plants a series of parallel SBR reactors can be constructed, which are operated out of sequence with each other. So at all times one SBR will be available to receive wastewater. Furthermore at high flow rates a reduction of the cycle time may be temporarily implemented.

A recent modification of the SBR is the (aerobic) granulated sludge bed reactor (Appendix A9). It has been demonstrated that under specific conditions a granular activated sludge may develop in an SBR type of reactor. As the granular sludge has excellent settling characteristics, the sludge concentration can be increased and the settling time reduced. Both result in reduced reactor volume as compared to conventional activated sludge treatment. To cultivate aerobic sludge granules, the preferential growth of substrate accumulating organisms is required, such as bio-P organisms. Aerobic granulated sludge will be discussed in Appendix A9.

3.5.3 Carrousels

The carrousel system is also known as circulation- or circuit system, to indicate the main difference from a completely mixed or plug-flow system: in a carrousel the mixed liquor is recirculated at relative high speed (e.g. 20 $cm \cdot s^{-1}$) through a long curved reactor of limited width. The layout of a modern carrousel often resembles a car racetrack. To remain in this analogy, during the hydraulic residence time a specific volume of mixed liquor will cover many “laps”. The first of these systems was the so-called oxidation ditch, which was developed by Pasveer in Holland as a treatment unit to be used by small communities (500 to 1000 inhabitants). In the original and simplest version, all treatment processes were carried out in a single physical unit. The biological reactor had an ellipsoidal form and consisted of an excavated ditch in which the sludge circulated. This circulation was induced by a surface aerator, which also introduced the oxygen needed for the treatment of the wastewater (Figure 3.20a). Using a time controller, the aerator was switched off at regular intervals (normally at night, when the sewage flow is small) and after settling the clarified supernatant was discharged. Hence the operation was similar to a sequencing batch reactor. Any incoming sewage during the settling period was accumulated in the sewage network. As the sludge age was very long (25 to 50 days), the sludge in the system was already well stabilised and could be applied directly on sludge drying beds without any further treatment.

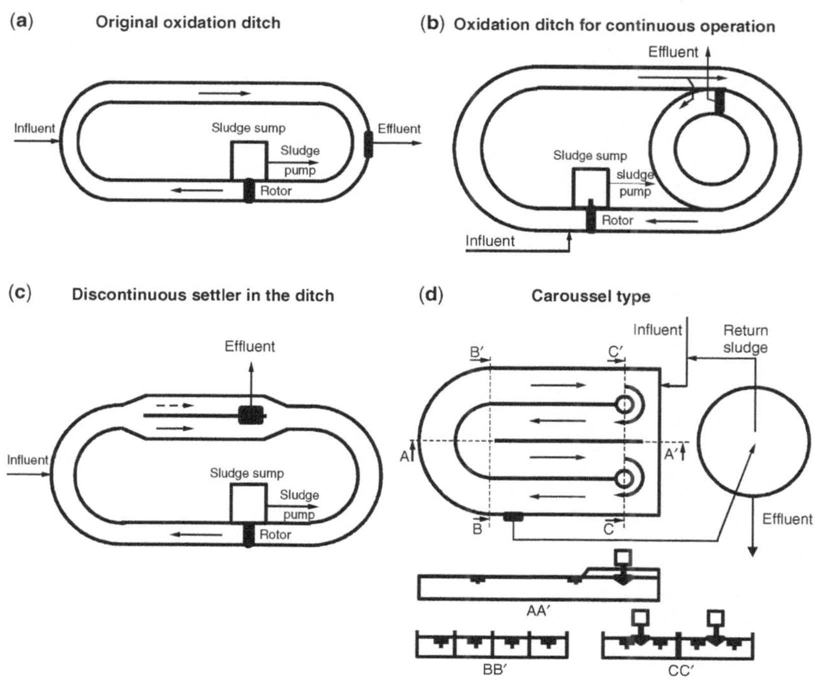

Figure 3.20 Different configurations of carrousel type systems

After the successful introduction of this system, several new versions of the oxidation ditch were developed to handle larger wastewater flows. The first modification was the introduction of an auxiliary ditch where the sludge was settled before discharge, so that the system could be operated continuously (Figure 3.20b). Another alternative was to have divisions in the ditch, so that parts of it could be used as a settler (Figure 3.20c). However, the real breakthrough of the circulation system came when the oxidation ditch was converted in the 1970s into a system with a dedicated settler: the Carrousel system (Figure 3.20d), with many units in operation throughout the world. While the first carrousels were all equipped with surface aerators, for the dual purpose of circulating the mixed liquor through the reactor and for providing aeration, newer carrousels may be equipped with dedicated propulsors and diffused aeration systems, which allows them to operate at increased liquid depth (5 – 6 m).

A main advantage attributed to the carrousel is that this reactor type combines features of both CSTR and plug-flow reactors: the rapid circulation promotes intense mixing, which ensures that influent concentration peaks and/or toxic loads are quickly diluted.

On the other hand, the large length of the reactor makes it very easy to establish an oxygen gradient by turning aerators on or off. This improves flexibility when dealing with highly variable wastewater flows and -compositions and is very useful in nutrient removal processes. However, the existence of rapid fluctuations in oxygen level as experienced by the micro-organisms seems to make these systems more susceptible to problems with poor sludge settleability and/or sludge bulking (refer also to Chapter 9).

3.5.4 Aerated lagoons

Only completely mixed lagoons will be discussed here, i.e. lagoons in which the aeration intensity is sufficient to avoid sludge settling and to maintain a uniform sludge suspension. The aerated lagoons are distinguished from other activated sludge variants by the fact that they do not have a final settler or another mechanism to retain the activated sludge. Therefore in an aerated lagoon the sludge age is always equal to the liquid retention time: $R_s = R_h$. Although the absence of the final settler is an operational- and cost advantage, the price in terms of effluent quality is high:

- The aerated lagoon is large compared to a conventional activated sludge system treating the same organic load. In aerated lagoons in general, liquid retention times (and hence sludge ages) in the order of 1 to 4 days are applied. In contrast, for sewage treatment in conventional activated sludge systems, a typical value of the liquid retention time is around 8 hours (reactor + settler) for a comparable sludge age. Hence the aerated lagoon is 3 to 10 times larger than a comparable conventional activated sludge system. On the other hand the cost per unit volume of lagoons is lower, because a lagoon is normally only an excavated hole with rudimentary protection against erosion, so that the total cost may actually be smaller. An advantage of the large volume is that occasional toxic loads may be diluted and hence their effect will be reduced. Similarly, sudden organic and hydraulic overloads can be accommodated more easily.
- A second, very serious, disadvantage of the aerated lagoon is that in the absence of a final settler the effluent in principle has the same composition as the mixed liquor so that biodegradable material and suspended solids will be discharged. As a consequence, the effluent quality of aerated lagoons is poor in terms of BOD, COD and TSS concentration.

For these reasons aerated lagoons are often only applied as pre-treatment units, with some form of complementary treatment of the effluent. In practice the effluent is often discharged into a second, non-aerated lagoon, where the sludge settles out and accumulates and from which the effluent is discharged. The settled sludge will be digested at the bottom of the lagoon, but the non-biodegradable fraction will accumulate. Occasionally the accumulated sludge has to be removed.

The lagoons may be utilised as single units or in a series configuration. The effluent quality in terms of COD and BOD (due to endogenous respiration) and suspended solids (VSS and TSS) can be calculated by applying the simplified model to each lagoon consecutively. It is important to note that the simplified model is only valid if two conditions are satisfied: (I) the retention time in the first lagoon must be sufficient for substantially complete removal of the biodegradable material, and (II) the sludge must be kept in suspension by the aeration. The validity of the second presupposition depends on the energy dissipation of the aeration units in the mixed liquor, as well as on the geometry of the lagoons. With regard to geometry, experience has shown that the energy dissipation required per cubic meter to maintain the sludge in suspension is higher in smaller lagoons than in larger ones. Von der Emde (1969) has provided the following empirical equation to calculate the required energy dissipation in an aerated lagoon on the basis of results obtained in a large number of full scale lagoons:

$$P_{min} = 450/(V_r)^{1/2} \tag{3.79}$$

where P_{min} = minimum energy dissipation to maintain the sludge in suspension ($W \cdot m^{-3}$)

EXAMPLE 3.11

Assuming a transfer efficiency of 0.75 kg $O_2 \cdot kWh^{-1}$, calculate the power requirement of an aerated lagoon treating raw sewage ($f_{ns} = f_{np} = 0.1$) at an average temperature of 26°C and a retention time of 3 days. Also determine the BOD, COD, VSS and TSS concentrations in the effluent, as well as the oxygen demand and the required power input. The influent COD concentration is 660 mg $COD \cdot l^{-1}$ and the influent flow is 5000 $m^3 \cdot d^{-1}$.

Solution

To determine the volatile suspended solids concentration Eq. (3.35) can be used, which can be simplified considering that in this particular case $R_s = R_h$:

$$\begin{aligned} X_v &= [(1 - f_{ns} - f_{np}) \cdot (1 + f \cdot b_h \cdot R_s) \cdot Y/(1 + b_h \cdot R_s) + f_{np}/p] \cdot S_{ti} \\ &= \{(1 - 0.1 - 0.1) \cdot [1 + (0.2 \cdot 0.30 \cdot 3) \cdot 0.45/(1 + 0.30 \cdot 3)] + 0.1/1.5\} \cdot 660 \\ &= 193 \text{ mg VSS} \cdot l^{-1} \end{aligned}$$

The COD concentration in the effluent will be the same as in the mixed liquor and can be calculated as the sum of the non-biodegradable dissolved material ($S_{ns} = f_{ns} \cdot S_{ti}$) and the COD that can be attributed to the VSS concentration. Hence:

$$\begin{aligned} S_{te} &= f_{ns} \cdot S_{ti} + f_{cv} \cdot X_v \\ &= 0.1 \cdot 660 + 1.5 \cdot 193 = 356 \text{ mg COD} \cdot l^{-1} \end{aligned}$$

It is concluded that the COD removal efficiency is only 1 – 356/660 = 46%. In an activated sludge system the efficiency would be $1 - f_{ns} = 90\%$. The BOD concentration is caused by oxygen consumption by the active sludge due to endogenous respiration. In the BOD bottle exponential decay of the active sludge will take place and after 5 days (at 20°C) the concentration will be:

$$X_{a5} = X_{a0} \cdot \exp(-b_h \cdot t) = X_{ao} \cdot \exp(-0.24 \cdot 5) = 0.3 \cdot X_{a0}$$

where:

X_{a5} = active sludge concentration at the end of the incubation period of 5 days
X_{a0} = initial active sludge concentration (t = 0 days)

Hence during the BOD test there is a decay of 70% of the active sludge. Knowing that only a fraction $(1 - f) = 0.8$ of the decayed active sludge is oxidised and that the consumption is equal to $f_{cv} = 1.5$ mg $O_2 \cdot$ mg VSS^{-1}, the oxygen consumption corresponding to endogenous respiration is calculated as:

$$BOD = f_{cv} \cdot (1 - f) \cdot (X_{a0} - X_{a5}) = f_{cv} \cdot (1 - f) \cdot 0.7 \cdot X_{a0} = 0.84 \cdot X_{a0}$$

Now, using Eq. (3.29) to calculate Xa and considering that the f_{ns} fraction does not contribute to BOD demand, one has:

$$\begin{aligned} BOD &= [0.84 \cdot (1 - f_{ns} - f_{np}) \cdot Y/(1 + b_h \cdot R_s)] \cdot S_{ti} \\ &= [0.84 \cdot (1 - 0.1 - 0.1) \cdot 0.45/(1 + 0.30 \cdot 3)] \cdot 660 = 107\,\text{mg BOD} \cdot l^{-1} \end{aligned}$$

If the influent BOD concentration is estimated at half the COD influent concentration $(0.5 \cdot S_{ti})$, the removal efficiency can be calculated as $1 - 107/(0.5 \cdot 660) = 68\%$. Theoretically in an activated sludge system the effluent would not have any BOD at all, although in practice between 5–15 mg $VSS \cdot l^{-1}$ will be present in the effluent, of which part consists of X_a.

The oxygen demand in the lagoon can be calculated as the difference between the influent and effluent COD load:

$$\begin{aligned} MS_o &= MS_{ti} - MS_{te} = 5000 \cdot (0.66 - 0.356) \\ &= 1520\,\text{kg}\,O_2 \cdot d^{-1} = 63\,\text{kg}\,O_2 \cdot h^{-1} \end{aligned}$$

As the (given) oxygen transfer efficiency is $0.75\,\text{kg}\,O_2 \cdot \text{kWh}^{-1}$, the required aeration power is 63/0.75 = 85 kW. Using the Von der Emde equation, the minimum power required to keep the sludge in suspension in a lagoon with a size of $3 \cdot 5000 = 15{,}000\ m^3$ is $450/(15{,}000)^{1/2} = 3.7\ W \cdot m^{-3}$, much less than the power necessary to transfer the oxygen into the lagoon, estimated as $85{,}000/15{,}000 = 5.6\ W \cdot m^{-3}$. It is concluded that the required power for aeration will likely be sufficient to maintain the sludge in the lagoon in suspension.

Chapter 4
Aeration

4.0 INTRODUCTION

Aeration in the activated sludge system is applied primarily to effect the transfer of atmospheric oxygen to the mixed liquor, where it is consumed to oxidise organic material and, if applicable, ammonium and H_2S. At the same time the turbulence resulting from agitation of the mixed liquor by the aerators needs to be sufficient to keep the sludge flocs in suspension. For most activated sludge systems, the oxygen demand per unit reactor volume is so high, that the introduced turbulence is more than sufficient to keep a homogenous suspension in the mixed liquor. Aerators can be classified into two main types: (I) diffused air systems, where air bubbles are introduced in the bottom of the reactor and oxygen transfer takes place during the upflow path of these bubbles and (II) mechanical or surface aerators where air bubbles are introduced in the liquid phase and simultaneously drops of mixed liquor are suspended into the air. Diffused aeration systems rely on positive displacement (e.g. rotary lobe) or centrifugal blowers to supply air to a submerged aeration grid. Figure 4.1 shows schematic representations of several common aerator types. The main types of diffused aeration systems are:

- Fine bubble aeration, often composed of porous ceramic domes or discs mounted on the bottom of the aeration tank (requiring a higher differential pressure), or alternatively plate or tubular membranes (operating at slightly lower differential pressure). The oxygen transfer efficiency is high. An example of a high efficiency fine bubble aeration system can be found in Figure 4.2, which shows a plate membrane;
- Coarse bubble aeration, often composed of non-porous domes, discs or tubes that produce larger air bubbles. The required differential pressure is lower than that of fine bubble aeration, but the aeration efficiency is also lower. On the other hand, these systems are less vulnerable to fouling and scaling;
- Jet aeration: a liquid stream is recirculated through a Venturi type ejector. This creates a pressure drop causing ambient air (or air from a compressor) to be sucked in and discharged into the reactor together with the recirculation stream. High turbulence in the ejector ensures the formation of small to medium sized air bubbles.

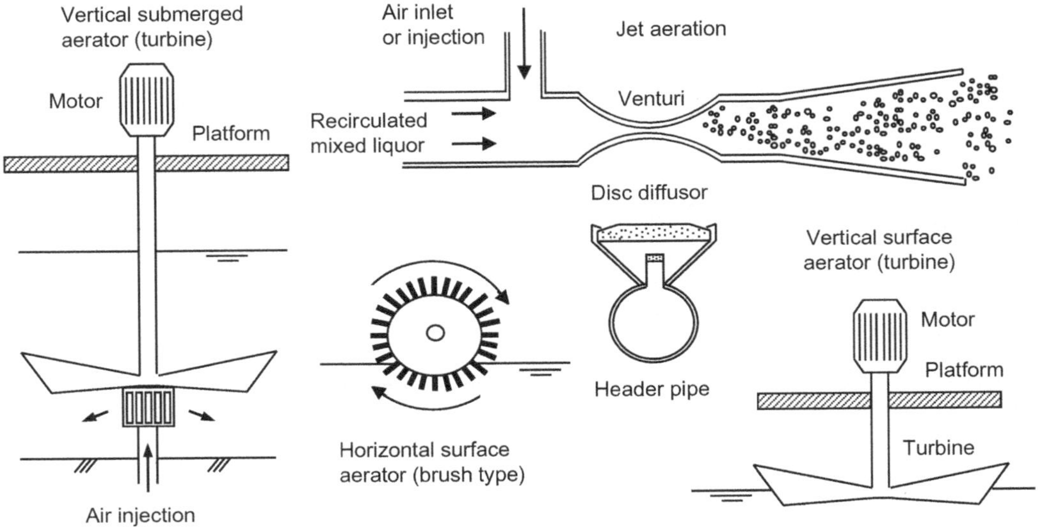

Figure 4.1 Schematic representation of several types of aeration systems

Figure 4.2 Modern high efficiency plate aerators used for fine bubble diffused aeration, courtesy of DHV BV

As for the mechanical surface aerators, two main types are in use: vertical and horizontal:

- Horizontal shaft surface aerators, operating at a low rotating speed of 20 to 60 rpm. They are mounted on fixed platforms and each surface aerator has its own motor and transmission. Examples are brush aerators and disk aerators;
- Vertical shaft aerators. A propeller or rotor violently agitates the water, introducing air bubbles into the mixed liquor and suspending liquid droplets in the air. The propeller zone is often covered to prevent excessive aerosol formation. As surface aerators, they can be mounted on fixed or floating platforms. The motor may be directly coupled to the propeller or rotor, in which case the rotation speed is high, or there may be a gear-box to reduce the rotation speed of the propeller or rotor. Even though the cost are higher, experience suggests that the low speed propellers have a lower incidence of breakdowns;

- The submerged vertical shaft aerator can be considered a hybrid system: it combines the functions of mixing and aeration. A turbine or hyperbolic mixer is mounted on the bottom of the aeration tank. Air is supplied by a compressor and injected below the mixer, where the shear stress produces small air bubbles. As the pressure drop over the air injection element is very low (as no pores are required), the required aeration energy is lower than that of a diffused aeration system. However, this effect is reduced due to the power demand of the turbine mixer.

An advantage of the hybrid system is that it allows flexible operation: opening or closing the air supply will turn the reactor aerobic, anoxic or anaerobic, but regardless the sludge will be maintained in suspension by the mixer. This system is particularly suited for small or compartmentalised systems.

Surface aerators are significantly less expensive than diffused aeration systems while they are also less vulnerable to fouling and scaling. On the other hand, the oxygen transfer efficiency expressed in kg $O_2 \cdot kWh^{-1}$ is lower. A second disadvantage of surface aerators is that achieving proper aeration in reactors with a depth of 4 metres or more is difficult. Some suppliers equip their surface aerators with draft tubes, extending from below the turbine blades to near the bottom of the reactor, which induces a vertical flow circulation. This makes it possible to operate surface aerators up to depths of 6 metres or more. Surface aerators are particularly suited for circuit systems such as carrousels, as a horizontal circulation over the reactor may be induced. Thus the need for additional equipment to circulate the mixed liquor may be dispensed with. Another option is the combination of diffused aeration with specific mixers ('propulsors') to induce a circulation flow, as can be observed in Figure 4.3.

Figure 4.3 When diffused aeration is used instead of surface aeration, propulsor mixers are installed in order to induce the required recirculation flow in a Carrousel® system - STP Veenendaal, The Netherlands. Courtesy of DHV BV

4.1 AERATION THEORY

Aeration theory is based on Henry's law: at equilibrium, the partial pressure of a component in the gas phase is proportional to the concentration of this component dissolved in the liquid phase. In the case of aeration, the liquid phase is the mixed liquor and the gas phase is air, whereas the component in question is oxygen. Hence, equilibrium exists if:

$$DO_s = k_H \cdot p_{O2} \tag{4.1}$$

where:

DO_s = saturation concentration of dissolved oxygen in the mixed liquor ($mol \cdot l^{-1}$ or $mg \cdot l^{-1}$)
k_H = Henry's constant ($mg \cdot l^{-1} \cdot atm^{-1}$)
p_{O2} = partial pressure of oxygen in air = 0.21 atm at atmospheric pressure

Note that the value of k_H is dependent on the temperature.

In principle, by using Eq. (4.1) one can calculate the equilibrium dissolved oxygen concentration in the mixed liquor for the existing atmospheric pressure and the temperature at hand. Fortunately, standard tables are available where the equilibrium dissolved oxygen concentration is listed as a function of temperature, pressure and salinity. In biological treatment systems, the dissolved oxygen concentration in the mixed liquor will be less than the saturation value, because oxygen is being consumed by the micro-organisms. Under those conditions there is a natural tendency of atmospheric oxygen to be transferred to the mixed liquor. According to Fick's law, the transfer rate is proportional to the difference between the saturation concentration and the actual dissolved oxygen concentration in the mixed liquor:

$$(dDO_l/dt) = k_{la} \cdot (DO_s - DO_l) \tag{4.2}$$

where:

dDO_l/dt = transfer rate of atmospheric oxygen ($mg\ O_2 \cdot l^{-1} \cdot h^{-1}$)
DO_s = saturation dissolved oxygen concentration in the mixed liquor ($mg\ O_2 \cdot l^{-1}$)
DO_l = dissolved oxygen concentration in the mixed liquor ($mg\ O_2 \cdot l^{-1}$)
k_{la} = oxygen transfer coefficient (h^{-1})

The value of the transfer constant depends on the type of aeration system, the geometry of the reactor, operational conditions (temperature, atmospheric pressure) and the presence of impurities in the mixed liquor. In practice, the concept of oxygen transfer efficiency (oxygenation capacity) is often used.

The oxygen transfer efficiency of an aerator is the maximum oxygen transfer rate under standard operational conditions. The oxygen transfer rate is measured in pure water without oxygen ($DO_l = 0$) under atmospheric pressure (1.013 bar or 760 mm Hg) and at a temperature of 20°C. The oxygen transfer efficiency of an aerator (in $mg\ O_2 \cdot l^{-1} \cdot h^{-1}$ or $kg\ O_2 \cdot h^{-1}$) is given as:

$$OT = (dDO_l/dt)_{max} = k_{la} \cdot DO_s \tag{4.3a}$$

$$OT = (dDO_l/dt)_{max} = k_{la} \cdot DO_s \cdot V_r \tag{4.3b}$$

In fact what is really important is not the oxygen transfer efficiency under standard conditions, but under the actual process conditions of the mixed liquor in the activated sludge system. In order to calculate the oxygen transfer efficiency under process conditions several corrections to the value of the k_{la} factor and the saturation DO concentration must be made. The influence of these factors will now be discussed.

4.1.1 Factors affecting k_{la} and DO_s

The presence of impurities in the wastewater (notably surface active substances, like soaps and detergents) reduces the transfer rate of oxygen and the solubility of oxygen. Furthermore mixing intensity, tank geometry and the type of aeration system all have an effect on the k_{la} value. Several factors are used to compensate for above effects. The first one is the α-factor, which expresses the ratio of the transfer rate in mixed liquor and in pure water under otherwise identical conditions. The effects of tank geometry and mixing intensity on the transfer rate are included in this factor as well. As to be expected, the α-factor is highly variable at values between 0.4 and 1.2. Typical values for domestic sewage are $\alpha = 0.4 - 0.8$ for diffused aeration systems and 0.6 to 1.2 for surface aeration systems (Metcalf et al., 2003). The α-factor in industrial wastewaters can deviate considerably from these "normal" values and ideally should be determined experimentally, as will be demonstrated in Section 4.2.

The effect of the suspended solids concentration on the α-factor is not very large when the sludge concentration is less than 8 g $TSS \cdot l^{-1}$, as will almost invariably be the case in conventional activated sludge systems. However, higher sludge concentrations are applied in more recent reactor configurations such as the membrane bioreactor (MBR - Chapter 10). The nature of the sludge and its interaction with the liquid phase (e.g. formation of extra-cellular polymers) also seems to play an important role. For MBR systems, experimental α-factors of 0.5 – 0.6 or even lower have been measured at sludge concentrations between 10 – 15 g $TSS \cdot l^{-1}$. Excessive aeration costs are one reason why MBR suppliers tend to move away from operation at high sludge concentrations. Very high sludge concentrations can also be found in aerobic sludge digesters. Baker, Loehr and Anthonisen (1975) showed that at a sludge concentration of 30 $g \cdot l^{-1}$, the α-factor was only two-thirds of the value measured at a concentration of 10 $g \cdot l^{-1}$.

An important aspect to consider in the design of diffused aeration systems is that oxygen transfer efficiency tends to decrease in time, due to biofouling and scaling effects. Due to the small pore size fine bubble systems are more vulnerable than coarse bubble systems. To account for diffuser fouling a second factor F is introduced. According to Metcalf et al. (2003), for domestic wastewater this factor ranges between 0.65 and 0.9, depending on the degree of fouling, which is a function of the wastewater characteristics and the duration of the period in which the diffusers have been in service. It is not uncommon that the product of $\alpha \cdot F$ is lumped into a single value for α, which explains why lower values for α are reported for fine bubble systems than for coarse bubble systems.

Theimpurities in the mixed liquor not only affect the transfer coefficient but also the saturation concentration DO_s itself. Thus a correction factor β is introduced, which is the ratio of the saturation concentration of DO in mixed liquor and in pure water under otherwise identical conditions. An important factor in the β value is salinity. For example, under atmospheric conditions the DO_s value of sweet water at 20°C is 9.08 mg $O_2 \cdot l^{-1}$, while for seawater (40 g $TDS \cdot l^{-1}$) it is only 7.17 mg $O_2 \cdot l^{-1}$. The β value is often reported to be 0.90 to 0.98 in the case of domestic sewage, with 0.95 as a typical average.

4.1.2 Effect of local pressure on DO_s

When the local atmospheric pressure differs from the standard pressure at sea level of 1 atm (1.013 bar or 760 mm Hg), the saturation concentration of dissolved oxygen in water can be related to the actual atmospheric pressure and the water vapour pressure:

$$DO_s = DO_{sp} \cdot (p - p_w)/(p_s - p_w) \tag{4.4a}$$

where:

DO_s = saturation concentration of dissolved oxygen at a pressure p
DO_{sp} = saturation concentration of dissolved oxygen at standard pressure
p = actual atmospheric pressure (bar or mm Hg)
p_w = water vapour pressure (bar or mm Hg)
p_s = standard pressure = 1 atm = 1.013 bar or 760 mm Hg.

The atmospheric pressure depends on the altitude above sea level. Table 4.1 shows values for different altitudes, as well values of the water vapour pressure as function of the temperature. When diffusers are submerged, air will be introduced beneath the water level (often diffusers are located at about 0.3 m from the bottom of the reactor) and the resulting pressure of the water column will result in a higher value of the oxygen saturation concentration DO_s.

Table 4.1 Relationship between atmospheric pressure and altitude (left) and between temperature and water vapour pressure (right)

Altitude (m)	Pressure		Temperature (°C)	Vapour pressure (p_w)	
	(mbar)	(mm Hg)		(mbar)	(mm Hg)
0	1013	760	0	6	4.6
500	953	715	5	9	6.5
1000	897	673	10	12	9.2
1500	844	633	15	17	12.8
2000	793	595	20	23	17.5
2500	746	560	25	32	23.7
3000	700	525	30	42	31.7
(1013 mbar = 760 mmHg)			35	56	42.0

In practice this results in an oxygen saturation concentration gradient from the diffusers up to the liquid-air interface. This can be approximated using the *average* submerged depth of the diffusers. Knowing that one bar is equal to a water column with a height of 10 m, Eq. (4.4a) can be adapted to:

$$DO_s = DO_{sp} \cdot [p - p_w + (H_{liq} - H_{dif})/20]/(p_s - p_w) \tag{4.4b}$$

where:

DO_s = saturation concentration of dissolved oxygen at a pressure P
DO_{sp} = saturation concentration of dissolved oxygen at standard pressure
p = actual atmospheric pressure (bar)
p_w = water vapour pressure (bar)
p_s = standard pressure = 1.013 bar

H_{liq} = liquid level in the reactor
H_{dif} = height of diffusers above reactor floor

Note that this effect is for a large part absent in surface aeration systems, as most of the oxygen transfer takes place at the surface area of suspended droplets, i.e. at atmospheric pressure.

4.1.3 Effect of temperature on k_{la} and DO_s

Temperature influences the transfer of oxygen: not only because it affects the transfer coefficient k_{la}, but also because of the influence on the saturation concentration DO_s. In the range of 0 to 50°C, the value of DO_s can be approximated as:

$$DO_{sT}/DO_{s20} = 51.6/(31.6 + T) \quad (4.5)$$

where:

DO_{s20} = saturation concentration of dissolved oxygen at 20°C = 9.1 mg $O_2 \cdot l^{-1}$
DO_{sT} = saturation concentration of dissolved oxygen at temperature T (in °C)

The influence of temperature on the oxygen transfer constant has been described with an Arrhenius equation (Eckenfelder and Ford, 1968):

$$k_{laT} = k_{las} \cdot \theta^{(T-20)} \quad (4.6)$$

where:

k_{laT} = oxygen transfer constant at T°C
k_{las} = oxygen transfer constant at 20°C
θ = temperature dependency factor of the transfer coefficient

Values for θ have been estimated between 1.020 – 1.028 for diffused air systems (Eckenfelder and Ford, 1968) and around 1.012 for surface aeration (Landberg et al., 1969).

EXAMPLE 4.1

An activated sludge system is located at 1250 m altitude where the average liquid temperature is 10°C. Answer the following questions:

- What will be the expected equilibrium dissolved oxygen concentration (DO_s)? The standard equilibrium concentration (DO_{ss}) at 1 atm and 20°C is 9.1 mg $O_2 \cdot l^{-1}$;
- If diffused aeration is used, how will this effect DO_s concentration? Assume that H_{liq} = 5 m and H_{dif} = 0.3 m;
- What will be the effect of an increase in reactor temperature to 18°C?

Solution

(1) Calculate the equilibrium do-concentration for the given conditions

Use Eq. (4.5) to calculate the value of DO_s at 10°C

$$DO_{sT} = 51.6/(31.6 + T) \cdot DO_{ss}$$
$$= 51.6/(31.6 + 10) \cdot 9.1 = 11.3\,mg \cdot l^{-1}$$

Adapt for the altitude with Eq. (4.4a). Use the data in Table 4.1: $p = (897 + 844)/2 = 870$ mbar and $p_w = 12$ mbar.

$$DO_s = DO_{sT} \cdot (p - p_w)/(p_s - p_w)$$
$$= 11.3 \cdot (870 - 12)/(1013 - 12) = 9.7\,mg \cdot l^{-1}$$

(2) Incorporate the effect of diffused aeration on the DO_s concentration

Now use Eq. (4.4b) to include the effect that the introduction of oxygen below the liquid surface area will have, as will be the case for diffused aeration:

$$DO_s = DO_{sT} \cdot [p - p_w + 1000 \cdot (H_{liq} - H_{dif})/(2 \cdot 10)]/(p_s - p_w)$$
$$= 11.3 \cdot [870 - 12 + 1000 \cdot (5 - 0.3)/(2 \cdot 10)]/(1013 - 12) = 12.3\,mg \cdot l^{-1}$$

(3) Determine the effect of a temperature increase to 18°C in the reactor

The value of DO_s will decrease: DO_s at 18°C = 9.47 $mg \cdot l^{-1}$ and $DO_{s18}/DO_{s10} = 9.47/11.3 = 0.84$. This ratio will also apply to DO values adjusted for altitude and water height above the diffusers in step (2). So $DO_s = 0.84 \cdot 12.3 = 10.3\,mg \cdot l^{-1}$.

On the other hand, the value of k_{la} will increase. Use Eq. (4.6) to calculate the effect of the temperature increase on k_{la} by comparing the value of k_{la} at 20°C with that at 10°C and 18°C. The value of the temperature coefficient θ is $1.024^{(10-20)} = 0.79$ at 10°C and $1.024^{(18-20)} = 0.95$ at 18°C. The ratio between k_{la} at 18°C and 10°C = 0.95/0.79 = 1.21. So the overall effect on oxygen transfer rate = $0.84 \cdot 1.21 = 1.01$ or a very small increase

4.1.4 Oxygen transfer efficiency for surface aerators

Taking into account the factors discussed above, which have an effect on either the value of the oxygen saturation concentration or the k_{la} value, the oxygen transfer efficiency under non-standard conditions can be related to the oxygen transfer efficiency under standard conditions as:

$$OT_a/OT_s = k_{laa} \cdot (DO_{sa} - DO_l)/(k_{las} \cdot DO_{ss}) \quad (4.7)$$

where:

OT_a = actual oxygen transfer efficiency ($kg\ O_2 \cdot kWh^{-1}$)
OT_s = oxygen transfer efficiency under standard conditions ($kg\ O_2 \cdot kWh^{-1}$)
DO_{ss} = saturation concentration of dissolved oxygen at 20°C and 1 atm (= $9.1\ mg \cdot l^{-1}$)
DO_{sa} = saturation concentration of dissolved oxygen under actual conditions
DO_l = actual dissolved oxygen (normally equal to the target DO setpoint value)

The value of k_{laa}, the oxygen transfer coefficient under actual conditions, is equal to:

$$k_{laa} = \alpha \cdot k_{las} \cdot \theta^{(T-20)} \tag{4.8}$$

This allows Eq. (4.7) to be rewritten to yield the actual oxygen transfer efficiency:

$$OT_a = [\alpha \cdot \theta^{(T-20)} \cdot (DO_{sa} - DO_l)/DO_{ss}] \cdot OT_s \tag{4.9}$$

DO_{sa} can also be written explicitly:

$$DO_{sa} = \frac{(p - p_w) \cdot 51.6 \cdot \beta \cdot DO_{ss}}{(p_s - p_w) \cdot (31.6 + T)} \tag{4.10}$$

Note that for surface aeration DO_{sa} is not corrected to compensate for diffuser submergence. Now, introducing Eq. (4.10) in Eq. (4.9), a general expression for the relationship between the oxygen transfer efficiency under actual and standard conditions can be derived:

$$OT_a = OT_s \cdot \alpha \cdot \theta^{(T-20)} \cdot \left[\frac{(p - p_w) \cdot 51.6 \cdot \beta \cdot DO_{ss}}{(p_s - p_w) \cdot (31.6 + T)} - DO_l\right]/DO_{ss} \tag{4.11}$$

For surface aerators, suppliers often quote a standard oxygen transfer efficiency OT_s in terms of $kg\ O_2 \cdot kWh^{-1}$. Using this standard oxygen transfer efficiency and the relationship between actual and standard oxygen transfer efficiency developed above, the required aeration power can be calculated as:

$$P_{aer} = MO_t/(24 \cdot OT_a) \tag{4.12}$$

where:

P_{aer} = required average aeration (motor) power (kW)
OT_a = actual oxygen transfer efficiency ($kg\ O_2 \cdot kWh^{-1}$)
MO_t = oxygen consumption ($kg\ O_2 \cdot d^{-1}$)

To account for daily fluctuations in oxygen demand and peak loads, a peak factor is commonly used, typically with a value around 1.5 – 2: i.e. to install a larger aerator.

EXAMPLE 4.2

A mechanical surface aerator is rated for an oxygen transfer efficiency of 2.0 kg $O_2 \cdot kWh^{-1}$ under standard conditions. What will the oxygen transfer efficiency be in an activated sludge system ($DO_l = 2$ mg $O_2 \cdot l^{-1}$) at an altitude of 800 m in winter (T = 18°C) and summer (T = 28°C).

Calculate the required aeration capacity (i.e. the required aeration power), for an average daily oxygen consumption (MO_t) of 2500 kg $O_2 \cdot d^{-1}$, when no peak factor is applied. Assume typical values for α, β and θ.

Solution

At 800 m the atmospheric pressure is 919 mbar or 690 mm Hg (interpolation of Table 4.1). Furthermore the vapour pressure is 16 mm Hg at 18°C and 30 mm Hg at 28°C. Using Eq. (4.11) the ratio between actual and standard oxygen transfer efficiency is calculated as:

$$OT_a/OT_s = 0.8 \cdot 1.012^{(18-20)} \cdot \left[\frac{(690-16) \cdot 51.6 \cdot 0.9 \cdot 9.1}{(760-16) \cdot (31.6+18)} - 2\right]/9.1 = 0.49 \text{ at } 18°C$$

$$OT_a = 0.49 \cdot 2.0 = 0.98 \text{ kg } O_2 \cdot kWh^{-1}$$

$$P_{aer} = MO_t/(24 \cdot OT_a) = 2500/(24 \cdot 0.98) = 106 \text{ kW}$$

$$OT_a/OT_s = 0.8 \cdot 1.012^{(28-20)} \cdot \left[\frac{(690-30) \cdot 51.6 \cdot 0.9 \cdot 9.1}{(760-30) \cdot (31.6+28)} - 2\right]/9.1 = 0.43 \text{ at } 28°C$$

$$OT_a = 0.43 \cdot 2 = 0.88 \text{ kg } O_2 \cdot kWh^{-1}$$

$$P_{aer} = MO_t/(24 \cdot OT_a) = 2500/(24 \cdot 0.86) = 121 \text{ kW}$$

In both cases the oxygen transfer efficiency under process conditions is well below that under standard conditions. Furthermore the oxygen transfer efficiency decreases with increasing temperature for the chosen θ value of 1.012, which is assumed to be representative for mechanical aerators. It is concluded that when the temperature rises, the increase in the value of the k_{la} constant does not compensate for the negative effect of the reduction of the oxygen saturation concentration. As the oxygen demand tends be higher with increasing temperature as well (due to a higher decay rate of active sludge), one would design the aeration capacity for the highest liquid temperature expected.

4.1.5 Power requirement for diffused aeration

In diffused aeration the term standard oxygen transfer efficiency (OT_s) is also used, now indicating the percentage of oxygen transferred from the pressurised air bubble to the mixed liquor upon leaving the diffuser until reaching the liquid surface. In Table 4.2 typical OT_s values for common diffuser types are listed (adapted from Metcalf & Eddy, 2003). Note that the OT_s values in this table have been standardised to 4.5 m depth. To adapt for a different liquid level above the diffusers, the following equation can be used:

$$OT_s = OT_{4.5} \cdot [(H_{liq} - H_{dif})/4.5]^{0.8} \tag{4.13}$$

Table 4.2 Standardised oxygen transfer efficiency and typical air flow per diffuser element for several diffuser types

Diffuser type	Air flow rate ($Nm^3 \cdot h^{-1}$)	OT_s (%) at 4.5 m submergence
Ceramic discs	0.5–6	25–35
Ceramic domes	1–4	27–37
Ceramic plates	3.5–8.5	26–33
Rigid porous plastic tubes	4–7	28–32
Non–rigid porous plastic tubes	1.5–12	26–36
Perforated membrane tubes	1.5–7	26–36
Jet aeration	100–500	15–24

Similar to surface aeration, for diffused aeration the OT_a/OT_s ratio can be defined as:

$$OT_a/OT_s = k_{laa} \cdot (DO_{sa} - DO_l)/(k_{las} \cdot DO_{ss}) \tag{4.14}$$

where:

OT_a = actual oxygen transfer efficiency (%)
OT_s = oxygen transfer efficiency under standard conditions (%)

The value of k_{laa} is given by Eq. (4.8), while DO_{sa} can be expressed as:

$$DO_{sa} = \frac{(p - p_w + (H_{liq} - H_{dif})/20) \cdot 51.6 \cdot \beta \cdot DO_{as}}{(p_s - p_w) \cdot (31.6 + T)} \tag{4.15}$$

This leads to:

$$OT_a = OT_s \cdot \alpha \cdot F \cdot \theta^{(T-20)} \cdot \left[\frac{(p - p_w + (H_{liq} - H_{dif})/20) \cdot 51.6 \cdot \beta \cdot DO_{as}}{(p_s - p_w) \cdot (31.6 + T)} - DO_l\right]/DO_{ss} \tag{4.16}$$

In diffused aeration systems, the air is supplied by rotary lobe blowers (for smaller capacities) or centrifugal blowers (for large capacities, typically 5000 $m^3 \cdot h^{-1}$ or larger). To determine the required blower power, two additional parameters need to be defined: the air mass flow rate and the blower discharge pressure. Knowing that the oxygen mass fraction in air is 20.9% and the molar weights of air and oxygen are 29 respectively 32 $g \cdot mol^{-1}$, the air mass flow rate in $kg \cdot h^{-1}$ can be calculated as:

$$Q_{air} = \frac{MO_t \cdot 29}{24 \cdot 32 \cdot 0.209 \cdot OT_a} \tag{4.17}$$

To calculate the volumetric flow rate in $Nm^3 \cdot h^{-1}$, divide the mass flow rate with the density of air, which at standard conditions is equal to 1.29 $kg \cdot m^{-3}$. The discharge pressure of the blower is calculated by taking into account the submergence level of the diffusers and the pressure drop over pipeline and valves:

$$p_{dis} = p + (H_{liq} - H_{dif})/10 + \Delta p \quad (4.18)$$

where:

p_{dis} = discharge pressure of the blower (bar)
Δp = pressure drop over pipeline and air diffuser elements: default values are 0.05 bar for coarse bubble aeration and 0.15 bar for fine bubble aeration

Now, the average power requirement for aeration can be calculated as:

$$P_{aer} = \frac{Q_{air} \cdot R \cdot T_{in} \cdot [(p_{dis}/p)^{0.283} - 1]}{3600 \cdot 29.7 \cdot 0.283 \cdot \eta_{aer}} \quad (4.19)$$

where:

P_{aer} = blower power requirement (kW)
Q_{air} = air flow ($kg \cdot h^{-1}$)
R = gas constant = 8.314 $kJ \cdot mol^{-1} \cdot K^{-1}$
T_{in} = blower inlet temperature (°K): this is usually not equal to the liquid temperature!
η_{aer} = blower efficiency (usually around 70–80%)

To calculate the *installed* motor power, P_{aer} should be calculated for worst case conditions, i.e. for the expected peak oxygen demand and for the maximum ambient (air inlet) temperature.

EXAMPLE 4.3

Determine the average power consumption and the installed aeration capacity of a diffused aeration system to be installed in a new activated sludge system. Use the following data:

- $V_r = 10{,}000\ m^3$; $MO_t = 8000\ kg\ O_2 \cdot d^{-1}$ or $333\ kg\ O_2 \cdot h^{-1}$
- Local elevation = 600 m; T_{air} (avg/max) = 15/35°C; T_r (avg) = 15°C;
- $H_{liq} = 4$ m; $H_{dif} = 0.3$ m; $\Delta p = 0.15$ barg;
- $OTE_{4.5} = 24\%$; $DO_{ss} = 9.1$ and $DO_l = 2\ mg\ O_2 \cdot l^{-1}$;
- $\theta = 1.024$; $\alpha \cdot F = 0.7$; $\beta = 0.95$; peak factor = 2; $\eta_{aer} = 64\%$.

Solution

From Table 4.1 the atmospheric pressure at the local altitude of 600 m is determined as 942 mbar or 0.942 bar. From the same table, for T = 15°C the water vapour pressure is estimated as 0.017 bar. The standard OTE at 3.7 m submergence is calculated with Eq. (4.13):

$$OT_s = OT_{4.5} \cdot [(H_{liq} - H_{dif})/4.5]^{0.8} = 24\% \cdot [(4.0 - 0.3)/4.5]^{0.8} = 21\%$$

The ratio between OTE under standard- and actual conditions is calculated with Eq. (4.16):

$$OT_a/OT_s = \alpha \cdot F \cdot \theta^{(T-20)} \cdot \left[\frac{(p - p_w + (H_{liq} - H_{dif})/20) \cdot 51.6 \cdot \beta \cdot DO_{ss}}{(p_s - p_w) \cdot (31.6 + T)} - DO_l\right]/DO_{ss}$$

$$= 0.7 \cdot 1.024^{(-7)} \cdot \left[\frac{(0.942 - 0.017 + 3.7/20) \cdot 51.6 \cdot 0.95 \cdot 9.1}{(1.013 - 0.017) \cdot (31.6 + 13)} - 2\right]/9.1 = 0.60$$

So $OT_a = 0.60 \cdot 21\% = 12\%$. The required air flow in $kg \cdot h^{-1}$ is calculated with Eq. (4.17):

$$Q_{air} = \frac{MO_t \cdot 29}{24 \cdot 32 \cdot 0.209 \cdot OT_a} = \frac{8000 \cdot 29}{24 \cdot 32 \cdot 0.209 \cdot 12\%} = 11,821\,kg \cdot h^{-1}$$

Eq. (4.18) is used to calculate the blower discharge pressure:

$$p_{dis} = p + (H_{liq} - H_{dif})/10 + \Delta p$$
$$= 0.942 + 3.7/10.3 + 0.15 = 1.45\,bar$$

Now the required average blower power can be calculated with Eq. (4.19):

$$P_{aer} = \frac{Q_{air} \cdot R \cdot T_{in} \cdot [(p_{dis}/p)^{0.283} - 1]}{3600 \cdot 29.7 \cdot 0.283 \cdot \eta_{aer}} = \frac{11,821 \cdot 8.314 \cdot 288 \cdot [(1.45/0.942)^{0.283} - 1]}{3600 \cdot 29.7 \cdot 0.283 \cdot 64\%} = 190\,kW$$

The procedure is repeated for the peak oxygen demand of $2 \cdot 333 = 666\ kg \cdot h^{-1}$ in order to calculate the installed aeration capacity as 414 kW. Make sure to calculate the installed blower power for the maximum (most unfavourable) ambient temperature of T = 35°C.

4.2 METHODS TO DETERMINE THE OXYGEN TRANSFER EFFICIENCY

In practice it may be very useful to determine or to verify the actual oxygen transfer efficiency of an aerator in an activated sludge system. When the process is not yet operational, this can be done directly in the biological reactor by aerating clean water free of dissolved oxygen. If the system is already in operation, the possibility exists to evaluate the oxygen transfer efficiency by using the steady state technique, presented later in this section.

4.2.1 Determination of the standard oxygen transfer efficiency

To carry out the determination of the standard oxygen transfer efficiency, the reactor in which the aerator is placed is filled with clean water. The dissolved oxygen is removed chemically by adding sodium sulphite (which is oxidised to sulphate), using cobaltous chloride as a catalyst. For water saturated with dissolved oxygen about 100 mg $Na_2SO_3 \cdot l^{-1}$ is used (the stoichiometric demand for 9 mg O_2 is 71 mg Na_2SO_3).

In order to determine the oxygen transfer efficiency, it is necessary to correlate the increase of the dissolved oxygen concentration to the duration of the aeration period. This correlation can be established by integrating Eq. (4.2):

$$dDO/dt = k_{la} \cdot (DO_s - DO_l) \quad \text{or}$$
$$\ln[(DO_s - DO_0)/(DO_s - DO_l] = k_{la} \cdot t \text{ or } \ln(DO_s - DO_l) = k_{la} \cdot t + \ln(DO_s - DO_0) \qquad (4.20)$$

where:

DO_0 = initial dissolved oxygen concentration (at t=0)
t = aeration time

Equation (4.20) shows that the relationship between the natural log of the oxygen deficit and the aeration time is linear and that the gradient of the corresponding straight line is equal to the transfer constant k_{la}. The following procedure to determine k_{la} is given:

(1) The aerator is installed under normal operational conditions (immersion depth, velocity). Water is added to the reactor and aeration is applied until saturation is attained. This oxygen saturation value is determined
(2) The dissolved oxygen is chemically removed by adding $CoCl_2$ ($0.5\ mg \cdot l^{-1}$) and Na_2SO_3 (40 to 50% in excess of the calculated stoichiometric quantity).
(3) Aeration is continued;
(4) As soon as dissolved oxygen is detected again, the increase in concentration is recorded as a function of time;
(5) The power consumption during the duration of the test is determined;
(6) From a semi log plot of the of dissolved oxygen deficit as a function of time, the k_{la} value is determined as the gradient of the "best-fit" straight line;
(7) The oxygenation capacity of the aerator is given (Eq. 4.3b) as:

$$OT_a = k_{la} \cdot DO_s \cdot V_r\ (kg\ O_2 \cdot h^{-1});$$

(8) The actual oxygen transfer efficiency can also be expressed as the ratio between the calculated value of the oxygenation capacity and the (actual) power consumption:

$$OT_a = OT/P_{aer} \qquad (4.21)$$

EXAMPLE 4.4

In a new activated sludge system, 10 aerators of 15 kW each are placed in a 5000 m^3 reactor. A test is carried out to verify if the standard oxygen transfer efficiency is indeed 2.2 kg $O_2 \cdot kWh^{-1}$ or higher at 20°C, as specified by the supplier. The test is carried out at sea level and at 22°C. Table 4.3 gives the dissolved oxygen concentration as a function of the aeration time. The power consumption recorded during the test was $10 \cdot 15 = 150$ kW.

Table 4.3 Experimental results from a test to determine the oxygen transfer efficiency

Time (minutes)	DO_l ($mg\ O_2 \cdot l^{-1}$)	DO_s-DO_l ($mg\ O_2 \cdot l^{-1}$)	$\ln(DO_s-DO_l)$ (–)	k_{la} ($min.^{-1}$)
0	1.0	7.7	2.04	(–)
1	2.2	6.5	1.87	0.17
2	3.0	5.7	1.74	0.15
4	4.2	4.5	1.50	0.13
6	5.2	3.5	1.25	0.13
8	6.0	2.7	0.99	0.13
12	7.0	1.7	0.50	0.12
16	7.7	1.0	0.00	0.15
32	8.6	0.1	−2.30	0.13

Solution

At 22°C the dissolved oxygen saturation concentration is 8.7 $mg\ O_2 \cdot l^{-1}$. In Table 4.3, column 3 lists the deficit ($DO_s - DO_l$), column 4 the natural log of the deficit and column 5 the values of the transfer constant k_{la}, calculated as $[\ln(DO_s - DO_l)]/t$ using Eq. (4.20). The experimental results suggest a value of 0.13 min^{-1} for k_{la} at 22°C. Hence, the oxygen transfer efficiency is $k_{la} \cdot DO_s = 0.13 \cdot 8.7 = 1.13\ mg\ O_2 \cdot l^{-1} \cdot min^{-1}$ at 22°C or $1.13/(1.012)^2 \cdot 9.1 = 1.16\ mg\ O_2 \cdot l^{-1} \cdot min^{-1}$ at 20°C.

Therefore, in the 5000 m^3 reactor the maximum transfer rate of oxygen at 20°C would be $5000 \cdot 1.16$ g $O_2 \cdot min^{-1}$ or 347 kg $O_2 \cdot h^{-1}$. As the measured power consumption during the test was 150 kWh, the observed maximum oxygen transfer efficiency is $347/150 = 2.31$ kg $O_2 \cdot kWh^{-1}$, even more than specified by the supplier (2.2 kg $O_2 \cdot kWh^{-1}$).

4.2.2 Determination of the actual oxygen transfer efficiency

If an activated sludge system is already in operation, the procedure outlined above to determine the oxygen transfer coefficient cannot be applied, because of the oxygen consumption by the micro-organisms. However, if the system operates under steady-state conditions the k_{la} value can still be determined, as the dissolved oxygen concentration in the liquid phase does not change because the oxygen transfer rate is equal to the oxygen consumption rate. Hence:

$$(dDO_l/dt) = 0 = (dDO_l/dt)_a - O_t = k_{la} \cdot (DO_s - DO_l) - O_t \quad (4.22)$$

where:

dDO_l/dt = rate of change of the dissolved oxygen concentration
$(dDO_l/dt)_a$ = aeration rate
O_t = total oxygen uptake rate

Rearranging Eq. (4.22) leads to

$$k_{la} = O_t/(DO_s - DO_l) \quad (4.23)$$

EXAMPLE 4.5

When the activated sludge system of the previous example is taken in operation, it is determined that the dissolved oxygen concentration stabilises at 1.5 mg $O_2 \cdot l^{-1}$ for an oxygen uptake rate of 30 mg $O_2 \cdot l^{-1} \cdot h^{-1}$ (20°C). The saturation oxygen concentration in the mixed liquor is 8.8 mg $O_2 \cdot l^{-1}$ at 20°C. Calculate the value of constant k_{la}.

Solution

From Eq. (4.23) one has:

$$k_{la} = O_t/(DO_s - DO_l) = 30/(8.8 - 1.5) = 4.1\,h^{-1} = 0.068\,min^{-1}$$

Comparing the data in Example 4.4 and Example 4.5, it is concluded that the efficiency under process conditions can be expressed as:

$$OT_a/OT_s = k_{laa}/k_{las} = 0.068/0.13 = 0.52$$

In practice Eq. (4.23) is not very useful, because for the calculation of the k_{la} value it is necessary to determine the O_t (OUR) and DO_l values, while the system is in normal operation. In theory, one can determine the OUR in the aeration tank by switching off the aerators and observing the decrease of the dissolved oxygen concentration in time. However, this procedure is not feasible due to a number of practical constraints: when the aerators are switched off, the agitation of the mixed liquor will also cease. Not only will the influent no longer be distributed over the reactor, but there will also be a tendency for the sludge to settle. As the value of OUR depends on both the substrate- and the sludge concentration, without agitation its value will deviate significantly from its normal value, depending on the position of the oxygen probe in the reactor. It is concluded that the OUR cannot be measured in the reactor if there is no stirring device independent of the aerators. Normally full-scale reactor aerators have the double function of oxygen transfer and of mixing of the mixed liquor.

An often applied method to overcome these difficulties is to continue normal operation of the plant and take samples of the mixed liquor to determine the OUR. This procedure, although widely used, is *fundamentally wrong* and may lead to results that underestimate the real OUR value by 30 to 50%, depending on the influent composition. This large error is due to the rapid rate of utilisation of the easily biodegradable part of the organic material. The associated oxygen demand will also be sustained only for a very short period. Hence, if a sample is withdrawn from the mixed liquor to determine the OUR, the readily biodegradable material will be rapidly depleted and as a consequence the associated oxygen consumption will not be detected. In nitrifying systems this difference will be even more accentuated.

The preferred alternative is the use of a respirometer: basically a very small reactor where the in- and outgoing oxygen concentrations of a mixed liquor stream taken from the reactor are continuously measured (or measured at very short intervals). This technique is explained in detail in Appendix 1. However, a respirometer of good quality is quite costly and skilled operators or process engineers are required for interpretation of the results. This is perhaps why respirometers have still not found wide application, despite their many advantages.

A practical alternative is to operate an aeration tank without introducing influent and to determine the OUR from samples withdrawn from the tank. In this case the aeration tank is operated as a batch reactor and under such conditions it is perfectly valid to withdraw samples to carry out OUR tests: the results obtained with the samples will closely reflect the value of the OUR inside the tank as neither in the sample nor in the reactor will there be oxygen uptake associated with the oxidation of readily biodegradable material. However, if the influent flow to the reactor is simply interrupted in order to carry out OUR tests, the oxygen demand in it will decrease rapidly and as a consequence the dissolved oxygen concentration increases and may approach the saturation value. Under such conditions it is difficult to determine the transfer constant k_{la} accurately, because its value depends on the difference between saturation and actual DO concentration.

For this reason it is important to load the reactor heavily before the tests are carried out, so that at least the initial dissolved oxygen concentration is low and the difference with the saturation value is large. If there is more than one aeration tank, accomplishing the overloading of a single reactor can be accomplished simply by diverting the entire influent flow to one reactor. If there is only one aeration tank, it may be possible to temporarily interrupt the influent flow and accumulate wastewater in the sewerage network. After a sufficient waiting period, the influent line is opened, causing the accumulated wastewater to be discharged at high rate into the reactor. Due to the applied overload, the OUR in the reactor will increase and when the aeration capacity is maintained constant, the dissolved oxygen concentration will decrease. When the dissolved oxygen concentration is very low (for example 1 mg $O_2 \cdot l^{-1}$), the feed to the reactor is completely interrupted while aeration is continued as normal. OUR tests are then carried out with samples withdrawn from the tank at regular intervals (for example every half hour), while the dissolved oxygen concentration in the reactor is recorded as a function of time. Now with the aid of the OUR and the dissolved oxygen concentration values determined at the different intervals, and knowing the saturation concentration (which should be determined in the effluent), the corresponding values of the k_{la} value can be calculated from Eq. (4.23).

In the case of surface aerators, the calculated k_{la} values will tend to vary with time because the immersion depth of the aerator decreases in time during the test. This is caused by the initial increase of the water level due to the applied overload prior to the test itself: the large influent flow required to reduce the dissolved oxygen concentration will also increases the water level in the aeration tank. During the subsequent period of testing, mixed liquor will continue to be discharged and the water level will decrease gradually until discharge of mixed liquor equals the sludge recycle flow. The fluctuation of the water level during the OUR test offers a possibility to calculate the k_{la} value of the aerators as a function of the immersion depth and hence determine the optimum depth for maximum aeration efficiency.

EXAMPLE 4.6

A 30,000 m^3 aeration tank is subjected to maximum loading between 8:00 and 10:00 h, after which the influent flow is interrupted. During the subsequent period of aeration the following parameters are determined at intervals of 30 minutes:

- The OUR (O_c) in mixed liquor batches withdrawn from the reactor;
- The dissolved oxygen concentration of the mixed liquor;
- The water level on the effluent weir of the reactor (indicating the water level in the reactor).

The experimental data are summarised in Table 4.4. The saturation concentration was determined in the effluent (having the same temperature as the aeration tank), yielding a value of 7.6 mg $O_2 \cdot l^{-1}$.

Determine the optimal immersion depth of the aerators in order to maximise the value of the transfer constant k_{la}.

Solution

The data in Table 4.4 show how the O_c and the DO_l values vary with time in the reactor after the influent flow was interrupted. The value of the dissolved oxygen deficit and the water depth on the effluent weir are also indicated.

Table 4.4 Determination of the oxygen transfer constant k_{la} according to Example 4.6

Time	O_t ($mg\ O_2 \cdot l^{-1} \cdot h^{-1}$)	DO_l ($mg\ O_2 \cdot l^{-1}$)	$DO_s - DO_l$ ($mg\ O_2 \cdot l^{-1}$)	H (cm)	k_{la} (h^{-1})
8:00	–	–	–	7.5	–
10:00	51	0.8	6.8	15	7.5
10:30	42	1.5	6.1	12	6.9
11:00	38	2.0	5.6	8.5	6.8
11:30	35	2.2	5.4	7.5	6.5
12:00	28.5	3.2	4.4	7.0	6.5
12:30	23	4.0	3.6	7.0	6.4
13:00	20	4.2	3.4	6.5	5.9

Now by applying Eq. (4.23) for the different intervals that OUR tests were carried out, the constant k_{la} is calculated as a function of the water depth. The values in Table 4.4 show that the maximum value of the constant was obtained at 10:00 hours, when the water level was 15 cm ($k_{la} = 7.5\ h^{-1}$). The DO_l value reflects the oxygen concentration measured near the aerators.

Once the value of the transfer constant has been established, it is also possible to determine the efficiency of the aerator if the power consumption is monitored at the same time. The oxygen transfer efficiency is expressed in terms of transferred oxygen mass per unit of consumed power:

$$OT_a = k_{la} \cdot (DO_s - DO_l) \cdot V_r / P_{aer} \tag{4.24}$$

where:

OT_a = oxygen transfer efficiency ($kg\ O_2 \cdot kWh^{-1}$)
P = consumed power (kW)
V_r = volume of the aeration tank (m^3)

The maximum efficiency is obtained when there is no dissolved oxygen present in the liquid:

$$OT_m = k_{la} \cdot DO_s \cdot V_r / P_{aer} \tag{4.25}$$

The value of OT_m represents the maximum oxygen mass that can be transferred to the mixed liquor per unit of consumed power by the aerator. Note that OT_m equals OT_s if it is specified at $T = 20°C$. It is important to verify if the OT_m value specified by the manufacturer can really be obtained in the reactor. Under normal operational conditions the efficiency of the aerators in the reactor will always be lower than OT_m for two reasons:

(1) It is necessary to maintain a certain minimum dissolved concentration in the mixed liquor to maintain the performance of the activated sludge process (for example 1 to 2 mg $O_2 \cdot l^{-1}$);
(2) The dissolved oxygen concentration in the aeration tank is always stratified: near the aeration units (the point of introduction of the oxygen into the mixed liquor), the concentration will always be higher than in the bulk of the liquid.

It is necessary to maintain the minimum concentration in the bulk of the mixed liquor (where the biological reactions take place). The steeper the stratification profile of dissolved oxygen is in the reactor, the higher will be the required dissolved oxygen concentration at the point of introduction, where the k_{la} value is determined. Stratification to some extent may be attributable to the design of the aerators, but also operational conditions (principally the value of the OUR) are important. The ratio between the effective or actual efficiency and the maximum value (for $DO_l = 0$ mg $O_2 \cdot l^{-1}$) is obtained by dividing Eq. (4.24) by Eq. (4.25):

$$OT_a/OT_m = (DO_s - DO_{min})/DO_s \qquad (4.26)$$

where:

OT_a = oxygen transfer efficiency under actual operational conditions
DO_{min} = dissolved oxygen concentration in the transfer zone of the aerator required to maintain a certain minimum dissolved oxygen concentration in the bulk of the mixed liquor

Naturally a reduction of the oxygen transfer efficiency will require a larger power consumption to affect the same oxygen transfer, and hence will lead to an increase in operational costs. Minimisation of the dissolved oxygen stratification in the aeration tank is therefore of great importance.

EXAMPLE 4.7

The power consumption by the surface aerators was determined simultaneously with the OUR and dissolved oxygen (DO) concentration in the previous example and is listed in Table 4.5. In order to maintain a minimum bulk mixed liquor concentration of 1 mg $\cdot l^{-1}$, the DO concentration in the zone near the aerators had to be maintained at 2.5 mg $\cdot l^{-1}$, i.e. the DO stratification was 1.5 mg $\cdot l^{-1}$. Determine the aeration efficiency under actual operational conditions.

Solution

With the aid of the data for dissolved oxygen and k_{la} as function of time, Eq. (4.22) is used to calculate the flux of transferred oxygen at the different times. The maximum transfer for $DO_l = 0$ mg $\cdot l^{-1}$ is also calculated. For example at 10.00 hrs, it can be observed in Table 4.5 that $DO_l = 0.8$ mg $O_2 \cdot l^{-1}$ and $k_{la} = 7.5$ h^{-1}. With Eq. (4.22) the oxygen transfer rate at 10:00 can be calculated as $(dDO/dt) = 0$ and $O_t = k_{la} \cdot (DO_s - DO_l) = 51$ mg $O_2 \cdot l^{-1} \cdot h^{-1} = 0.051$ kg $O_2 \cdot m^{-3} \cdot h^{-1}$

Table 4.5 Consumed power during the test for oxygen transfer efficiency determination

Time	Power (kW)	Transfer coefficient (h^{-1})	Actual transfer ($kg\ O_2 \cdot h^{-1}$)	Maximum transfer ($kg\ O_2 \cdot h^{-1}$)	Transfer efficiency ($kg\ O_2 \cdot kWh^{-1}$)	
					Maximum (at $DO_l = 0$)	Actual (at $DO_l = 2.5$)
10:00	938	7.5	1734	1938	2.07	1.38
10:30	779	6.9	1428	1779	2.28	1.52
11:00	737	6.8	1292	1753	2.34	1.57
11:30	721	6.5	1190	1674	2.32	1.55
12:00	737	6.5	969	1673	2.27	1.52
12:30	729	6.4	782	1650	2.26	1.51
13:00	713	5.9	680	1520	2.13	1.43

The above calculation is not entirely correct because it presupposes that the dissolved oxygen concentration is constant at the time of determination. In fact there is a very small rate of increase: in Table 4.5 the dissolved oxygen concentration increases from 0.8 mg $O_2 \cdot l^{-1}$ at 10:00 hrs to 1.5 mg $O_2 \cdot l^{-1}$ at 10:30 hrs. This is an increase of $1.5 - 0.8 = 0.7$ mg $O_2 \cdot l^{-1}$ in 0.5 hr, which means that the rate of dissolved oxygen increase ($\Delta DO/\Delta t$) was $0.7/0.5 = 1.4$ mg $O_2 \cdot l^{-1} \cdot h^{-1}$. If necessary, Eq. (4.22) above can be corrected to account for this effect: $O_t = k_{la} \cdot (DO_s - DO_l) - (\Delta DO/\Delta t)$. For the given volume of the reactor of 30,000 m^3, the mass of transferred oxygen per hour is now calculated as: $0.051 \cdot 30{,}000 = 1734$ kg $O_2 \cdot h^{-1}$. Table 4.5 shows the values of the actual and maximum oxygen transfer in columns 4 and 5. Knowing the power consumption, the aeration efficiency is calculated by dividing the flux of transferred oxygen by the power consumption. For example at 10:00 hrs, the maximum transfer is 1938 kg $O_2 \cdot h^{-1}$ and the power consumption is 938 kW, so that the maximum aeration efficiency with these aerators is:

$$OT_m = 1938/938 = 2.07\,kg\,O_2 \cdot kWh^{-1}$$

For DO_l is 2.5 mg $O_2 \cdot l^{-1}$ in the aeration zone, the actual oxygen transfer efficiency is:

$$OT_a = OT_{max} \cdot (DO_s - DO_{min})/DO_s = 2.07 \cdot (7.6 - 2.5)/7.6 = 1.38\,kg\,O_2 \cdot kWh^{-1}$$

The calculated values for the maximum efficiency ($DO_l = 0$ mg $O_2 \cdot l^{-1}$) and the actual efficiency ($DO_l = 2.5$ mg $O_2 \cdot l^{-1}$) are shown in the last two columns of Table 4.5. The data in Table 4.4 and Table 4.5 reveal an interesting fact: the liquid level in the reactor, as indicated by the liquid height (H) on the effluent weir, affects both the value of the transfer constant k_{la} and the power consumption. The data show that the level with the maximum k_{la} value and hence the highest oxygen transfer (at H = 15 cm at 10:00 hrs), does not coincide with the level where the lowest power consumption per kg oxygen transferred was measured (at H = 8.5 cm at 11:00 hrs). At 11:00 the actual oxygen transfer efficiency is 1.57 kg $O_2 \cdot kWh^{-1}$, which is larger than the 1.38 kg $O_2 \cdot kWh^{-1}$ measured at 10:00 hrs. It is concluded that the additional oxygen transfer obtained when the liquid level rises from 8.5 to 15 cm does not compensate for the extra power consumption that the aerator requires. A liquid level of 8.5 cm is more advantageous from the point of view of minimisation of operational costs.

The data of the example were obtained at the CETREL plant at the Camaçari Petrochemical Complex in Brazil and show that the method described in the previous sections not only serves to determine the actual oxygen transfer efficiency of aerators under operating conditions, but also supplies a method to optimise the immersion depth of the aerators in order to decrease energy use (Van Haandel et al., 1997).

Chapter 5

Nitrogen removal

5.0 INTRODUCTION

In an activated sludge plant designed for tertiary treatment, the objective is to remove nutrients, suspended solids and organic matter. During the last few decades, the importance of nutrient removal has increased as a result of the necessity to avoid eutrophication of water bodies receiving untreated wastewater and the effluent of wastewater treatment plants. For this reason, many new wastewater treatment plants are now designed for tertiary treatment.

Apart from the important repercussions on effluent quality, tertiary treatment also has a beneficial influence on the performance of the wastewater treatment process itself. This is particularly noticeable in the case of nitrogen removal. The development of nitrification in an activated sludge process is practically unavoidable when the sewage reaches temperatures of 22 to 24°C, which will be the case for at least part of the year in tropical and subtropical regions. The formed nitrate can be used by most micro-organisms in the activated sludge as a substitute to dissolved oxygen. In an anoxic environment, characterised by the presence of nitrate and the absence of dissolved oxygen, the nitrate ion can be reduced by organic matter to nitrogen gas: this process is called denitrification.

If the biological reactor is kept completely aerobic, the nitrified mixed liquor will flow to the final settler, where an adequate environment for denitrification is established as soon as the oxygen is consumed; this will take only a few minutes. Microscopic nitrogen gas bubbles will appear, predominantly inside the sludge flocs, causing them to rise to the liquid surface of the final settler, where a layer of floating sludge will be formed that will eventually be discharged with the effluent. Of course, the effluent quality will be very poor due to the presence of suspended solids. Another negative aspect of floating layers of denitrifying sludge and the loss of sludge with the effluent refers to the operational stability of the activated sludge process. The loss of biomass reduces the sludge age and the remaining sludge mass may be insufficient to metabolise the influent organic matter, resulting in a reduction of secondary treatment efficiency. Furthermore the reduction of the sludge age may lead to wash out of nitrifiers and hence interrupt the nitrification process. As nitrate is no longer produced, denitrification ceases as well and the problem of floating sludge layers will disappear. Thus favourable conditions are established for a recuperation of the secondary treatment efficiency. First, the fast growing heterotrophs will be re-established and efficient organic matter removal will resume. Subsequently the slower growing nitrifiers will also return in the sludge mass in sufficient quantities to nitrify the applied ammonium load. Nitrate will be formed and a new cycle of operational instability will be initiated.

In contrast, when nitrification and denitrification are controlled and occur as planned in the activated sludge process itself, before the mixed liquor reaches the final settler, this unit will behave as a normal liquid-solid separator. This allows an effluent containing very low concentrations of suspended solids, organic matter and nitrogen to be produced.

Nitrogen removal also has important economic consequences. In the nitrification process both oxygen and alkalinity are consumed, while in the denitrification process part of this consumption may be recovered. In the case of municipal sewage, the oxygen demand for nitrification is about one-third of the total demand. From the stoichiometrics of the reactions involving nitrogenous matter, it can be calculated that $\frac{5}{8}^{th}$ or 63% of the oxygen demand for nitrification may be recovered in the denitrification process. Thus in activated sludge processes with complete biological nitrogen removal (nitrification + denitrification), oxygen consumption will be $\frac{5}{8} \cdot \frac{1}{3} = \frac{5}{24}^{th}$ or about 21% lower than in comparable processes with nitrification only. Since aeration is the main part of the operational costs in an activated sludge process, a 21% reduction of the oxygen demand is very significant in economic terms.

Another effect of reactions with nitrogenous matter is on alkalinity. The alkalinity consumption by nitrification may result in a reduction of pH. The magnitude of this reduction depends on the initial alkalinity and the oxidised ammonium concentration. In many cases the pH tends to become unacceptably low and alkalinity addition, for instance in the form of lime ($Ca(OH)_2$), will be necessary. During denitrification, half of the alkalinity consumption for nitrification is recovered. Thus the alkalinity demand will be smaller and in many cases after the introduction of denitrification, addition of alkalinity is no longer necessary.

5.1 FUNDAMENTALS OF NITROGEN REMOVAL

5.1.1 Forms and reactions of nitrogenous matter

Nitrogenous matter in wastewaters is mainly composed of inorganic ammonium nitrogen, which can be present in gaseous (NH_3) and ionic form (NH_4^+), and organic nitrogen (urea, amino acids and other organic compounds with an amino group). Sometimes wastewaters contain traces of oxidised forms of nitrogen, mainly nitrite (NO_2^-) and nitrate (NO_3^-). Different to organic matter, nitrogenous matter can be defined quantitatively and unequivocally by one parameter: the nitrogen concentration in its different forms. In practice, spectrophotometric tests and specific ion electrodes are used to determine the concentrations of ammonium, nitrate and nitrite. Organic nitrogen can be determined after its conversion to ammonium nitrogen by chemical digestion. The sum of the organic and ammonium concentrations is called Total Kjeldahl Nitrogen, TKN.

In the activated sludge process several reactions may occur that change the form of nitrogenous matter. Figure 5.1 shows the different possibilities: (a) ammonification or the inverse: ammonium assimilation by the organisms, (b) nitrification and (c) denitrification.

(a) Ammonification/assimilation

Ammonification is the conversion of organic nitrogen into ammonium, whereas the inverse process, the conversion of ammonium into organic nitrogen, is called bacterial anabolism or assimilation. Considering that the pH in mixed liquor is typically near the neutral point (pH = 7), ammonium will be present predominantly in its ionic form (NH_4^+) and the following reaction equation may be written:

$$RNH_2 + H_2O + H^+ \longleftrightarrow ROH + NH_4^+ \quad (5.1)$$

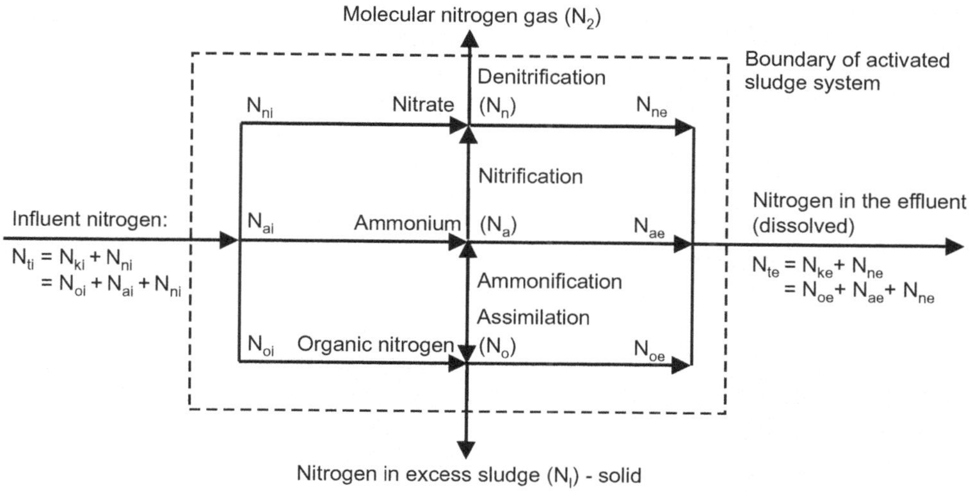

Figure 5.1 Schematic representation of the different forms of nitrogenous material present in wastewater and the main transformation reactions that occur in the activated sludge process

(b) Nitrification

Nitrification is the biological oxidation of ammonium, with nitrate as the end product. The reaction is a two-step process, mediated by specific bacteria: in the first step ammonium is oxidised to nitrite and in the second step nitrite is oxidized to nitrate. It has been assumed for a long time that the ammonium oxidation was only mediated by the bacterial species *Nitrosomonas spp.* However, recent research indicates that in fact other bacterial species might also be involved or even dominant (such as *Nitrosococcus spp.*). Likewise, the complementary step of nitrite oxidation, is no longer only mediated by species such as *Nitrobacter spp.* Therefore in this text the general terms ammonium oxidizers and nitrite oxidizers will be used. Both ammonium- and nitrite oxidizers can only develop biochemical activity in an environment containing dissolved oxygen. The two reactions (excluding nitrifier biomass growth) can be written as:

$$NH_4^+ + \tfrac{3}{2}O_2 \longrightarrow NO_2^- + H_2O + 2H^+ \quad (5.2a)$$

$$NO_2^- + \tfrac{1}{2}O_2 \longrightarrow NO_3^- \quad (5.2b)$$

$$NH_4^+ + 2\,O_2 \longrightarrow NO_3^- + H_2O + 2H^+ \quad (5.2)$$

(c) Denitrification

Denitrification is the biological reduction of nitrate to molecular nitrogen, with organic matter acting as a reductor. For organic matter with a general structural formula $C_xH_yO_z$, the half reactions of this redox process can be expressed as:

(1) Oxidation reaction:

$$C_xH_yO_z + (2x - z)\,H_2O \longrightarrow x\,CO_2 + (4x + y - 2z)\,H^+ + (4x + y - 2z)\,e^- \quad (5.3a)$$

(2) Reduction reaction:

$$e^- + \tfrac{6}{5}H^+ + \tfrac{1}{5}NO_3^- \longrightarrow \tfrac{1}{10}N_2 + \tfrac{3}{5}H_2O \tag{5.3b}$$

(3) Overall redox reaction (excluding growth of bacterial cell-mass):

$$\begin{aligned} &C_xH_yO_z + (4x + y - 2z)/5H^+ + (4x + y - 2z)/5NO_3^- \\ &\qquad \longrightarrow x\,CO_2 + (2x + 3y - z)/5H_2O + (4x + y - 2z)/10N_2 \end{aligned} \tag{5.3}$$

The TKN concentration in municipal sewage typically is in the range of 40 to 60 $mg\,N \cdot l^{-1}$, i.e. a fraction in the range of 0.06 to 0.12 of the influent COD. Furthermore, generally about 75% of the total TKN concentration will be in the form of ammonium nitrogen while the remaining 25% is predominantly made up of organic nitrogen. In the activated sludge process, organic nitrogen is converted rapidly and almost quantitatively to ammonium nitrogen (ammonification). If nitrification occurs and the oxygenation capacity is sufficient, the oxidation of ammonium nitrogen will be almost complete. If after nitrification the formed nitrate is removed by denitrification, the total nitrogen concentration in the effluent is in general smaller than 5 to $10\,mg\,N \cdot l^{-1}$. It can be concluded that excellent biological nitrogen removal is possible from municipal sewage, with a removal efficiency of 90% or more.

5.1.2 Mass balance of nitrogenous matter

Figure 5.1 shows nitrogen leaves the activated sludge process in one of the following forms:

- As solid matter in the excess sludge (N_l);
- As dissolved matter in the effluent: ammonium (N_{ae}), nitrate/nitrite (N_{ne}) and soluble organic nitrogen (N_{oe});
- As gaseous material (in the form of molecular nitrogen) to the atmosphere (N_2).

In Figure 5.1 the possibility of ammonium volatilisation is not considered because this process only has importance when the pH approaches a value of 9 or more. A significant fraction of the ammonium will then be present in the unionised form. In practice such a situation can only develop under very special conditions.

Depending on the liquid-solid separation efficiency of the final settler, a certain fraction of the suspended solids present in the mixed liquor will not be retained. Naturally, these solids will contain organic nitrogen (N_{ope}). Thus part of the nitrogen in the produced excess sludge (N_l) will end up in the effluent and not in the excess sludge flow. Stated otherwise, part of N_l leaves as N_{ope} in the effluent. However, as $N_{ope} \ll N_l$, this can generally be ignored when the nitrogen mass balance is calculated. On the other hand, when strict nitrogen effluent limits apply, the presence of organic nitrogen in the effluent should be considered as the contribution of N_{ope} to N_{te} may be significant. As an indication, the volatile suspended solids concentration in the effluent of a well designed final settler is typically between 5–10 $mg\,VSS \cdot l^{-1}$, with an associated nitrogen content between 0.5 and $1.0\,mg\,N \cdot l^{-1}$. The presence of particulate organic nitrogen in the effluent and its implications on the calculation of nitrogen removal performance is discussed in more detail in Appendix 5;

Using the concepts developed for mass balance calculations of organic material, the nitrogen recovery factor can be defined as the ratio of the nitrogen mass fluxes leaving and entering the activated sludge process:

$$B_n = (MN_l + MN_{te} + MN_d)/MN_{ti} \quad (5.4)$$

where:

B_n = mass balance recovery factor for nitrogenous material (–)
MN_l = flux of nitrogenous matter in the excess sludge ($kg\ N \cdot d^{-1}$)
MN_{te} = flux of nitrogenous matter in the effluent ($kg\ N \cdot d^{-1}$)
MN_d = flux of denitrified nitrogen ($kg\ N \cdot d^{-1}$)
MN_{ti} = flux of nitrogenous matter in the influent ($kg\ N \cdot d^{-1}$)

Eq. (5.4) is only useful when the different fluxes are formulated in terms of measurable parameters, so that the B_n value can be determined experimentally and compared to its theoretical value of one. For the nitrogen flux leaving the activated sludge system in the excess sludge, an expression was already derived in the previous chapter:

$$MN_l = f_n \cdot MX_v/R_s \quad (5.5)$$

The fluxes in the influent and the effluent are easily calculated as:

$$MN_{ti} = Q_i \cdot (N_{oi} + N_{ai} + N_{ni}) = Q_i \cdot N_{ti} \quad (5.6)$$

$$MN_{te} = Q_i \cdot (N_{oe} + N_{ae} + N_{ne}) = Q_i \cdot N_{te} \quad (5.7)$$

where:

N_t = total nitrogen concentration ($mg\ N \cdot l^{-1}$)
N_a = ammonium nitrogen concentration ($mg\ N \cdot l^{-1}$)
N_o = organic nitrogen concentration ($mg\ N \cdot l^{-1}$)
N_n = nitrate nitrogen concentration ($mg\ N \cdot l^{-1}$)

The indices "i" and "e" refer to influent and effluent respectively. In Eqs. (5.6 and 5.7) the nitrite concentration is assumed to be insignificant, which in practice is usually justified. If this is not the case, then it indicates a process disturbance that should be remedied.

In order to calculate the denitrified nitrogen flux, the process configuration must be taken into consideration. When the objective of the process is nitrogen removal, there will be anoxic zones where denitrification takes place. The flux of removed nitrogen is calculated as the product of the flow passing through the anoxic reactor and the decrease of the nitrate nitrogen concentration in it. Hence:

$$MN_{dk} = Q_k \cdot \Delta N_{nk} \quad (5.8)$$

where:

MN_{dk} = flux of denitrified nitrogen in anoxic reactor "k" ($kg\ N \cdot d^{-1}$)
Q_k = flow rate to reactor "k" ($m^3 \cdot d^{-1}$)
= influent- and return sludge flow plus possibly other recycle streams
ΔN_{nk} = NO_3–N concentration difference between inlet and outlet in anoxic reactor "k"

For a system consisting of "k" anoxic reactors, the total nitrogen flux that is denitrified can be expressed as:

$$MN_d = \sum_{k=1}^{K} MN_{dk} = \sum_{k=1}^{K} (Q_k \cdot \Delta N_{nk}) \tag{5.9}$$

Now, using the expressions of Eqs. (5.5, 5.6, 5.7 and 5.9):

$$B_n = \left(f_n \cdot MX_v/R_s + Q_i \cdot N_{te} + \sum_{k=1}^{K} Q_k \cdot \Delta N_{nk} \right) / (Q_i \cdot N_{ti}) \tag{5.10}$$

In Eq. (5.10) all parameters on the right hand side are measurable, so that it is possible to calculate the nitrogen recovery factor based on experimental data.

EXAMPLE 5.1

As an example of the application of mass balance recovery concept for nitrogen removal, the experimental data obtained during the operation of a nitrogen removal pilot plant are discussed (Ekama et al., 1976). The process was composed of five reactors with 5 m^3 volume each and a final settler. The first reactor (receiving all the influent) was unaerated, whereas the other four were aerated. The average temperature was 21.6°C. Figure 5.2 shows the flow scheme of the process.

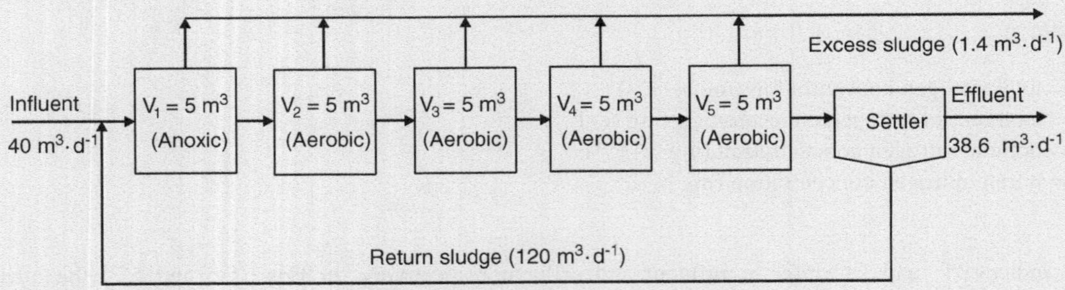

Figure 5.2 Flow scheme of the pilot process from Example 5.1

Table 5.1 shows the average values of the analytical results from a 18 day period, when a steady state performance had already been established. Figure 5.2 and Table 5.1 contain all the information required to perform the mass balance calculations as shown below.

Solution

The following calculation procedure is followed to determine the nitrogen recovery factor B_n:

(1) Calculate MN_{ti} with Eq. (5.6)

$$N_{ti} = Q_i \cdot (N_{oi} + N_{ai} + N_{ni}) = 40 \cdot (45.1 + 0.3) = 1816 \, gN \cdot d^{-1}$$

(2) Calculate MN_l

(a) Calculate the sludge age with Eq. (3.15)

$$R_s = V_r/q = 25/1.4 = 18\,\text{days}$$

(b) Calculate MN_l

Use Eq. (5.5), assuming the waste sludge concentration is equal to the average VSS concentration of 2469 mg $VSS \cdot l^{-1}$.

$$MN_l = f_n \cdot V_r \cdot X_v/R_s = 0.1 \cdot 25 \cdot 2469/18 = 343\,gN \cdot d^{-1}$$

(3) Calculate MN_{te} with Eq. (5.7)

$$MN_{te} = Q_i \cdot (N_{oe} + N_{ae} + N_{ne}) = 40 \cdot (1.9 + 8.7) = 424\,g\,N \cdot d^{-1}$$

(4) Calculate MN_d

The data in Table 5.1 indicate that denitrification occurred in the first reactor and in the final settler, as in these two units the nitrate concentration decreased. The nitrate concentration entering into the first reactor N_{n0} is calculated as the weighted average of the concentrations in the influent- and recycle flows. Assuming the nitrate concentration in the return sludge flow is equal to the effluent nitrate concentration, one has:

$$N_{n0} = (Q_i \cdot N_{ni} + Q_r \cdot N_{ne})/(Q_i + Q_r) = (40 \cdot 0.3 + 120 \cdot 8.7)/(40 + 120) = 6.6\,mg\,N \cdot l^{-1}$$

As the nitrate concentration in the flow leaving the first reactor was 1.2 mg $N \cdot l^{-1}$ (Table 5.1), the nitrate decrease equals $\Delta N_{n1} = 6.6 - 1.2 = 5.4$ mg $N \cdot l^{-1}$. Hence the flux of removed nitrogen in the first reactor was:

$$MN_{d1} = (Q_i + Q_r) \cdot \Delta N_{n1} = (40 + 120) \cdot 5.4 = 864\,g\,N \cdot d^{-1}$$

Similarly the flux of nitrogen removed in the final settler is calculated as:

$$MN_{dd} = (Q_i + Q_r) \cdot \Delta N_{nd} = (40 + 120) \cdot (9.8 - 8.7) = 176\,g\,N \cdot d^{-1}$$

Table 5.1 Average values of process parameters in the pilot experiment of Example 5.1

Parameter		Influent	Reactor					Effluent
			1	2	3	4	5	
COD	$mg \cdot l^{-1}$	477	25	19	18	18	18	18
TKN	$mg\,N \cdot l^{-1}$	45.1	9.2	4.8	3.3	2.6	2.0	1.9
NH_4^+	$mg\,N \cdot l^{-1}$	32.8	9.4	3.4	0.6	0.3	0.2	0.0
NO_3^-	$mg\,N \cdot l^{-1}$	<0.3	1.2	5.8	9.2	9.7	9.8	8.7
VSS	$mg\,VSS \cdot l^{-1}$	–	2550	2447	2466	2406	2477	–
OUR	$mg\,O_2 \cdot l^{-1} \cdot h^{-1}$	–	–	59.3	36.5	23.4	19.3	–

where:

MN_{dd} = denitrified nitrogen in the final settler ($g N \cdot d^{-1}$)

ΔN_{nd} = decrease of the nitrate concentration in the final settler ($mg N \cdot l^{-1}$)

Now the total flux of nitrogen removed by denitrification can be calculated as:

$$MN_d = MN_{dl} + MN_{dd} = 864 + 176 = 1040\, g N \cdot l^{-1}$$

Having calculated all the relevant nitrogen fluxes, the nitrogen recovery factor can be determined with the aid of Eq. (5.4):

$$B_n = (MN_l + MN_{te} + MN_d)/MN_{ti} = (343 + 424 + 1040)/1816 = 1807/1816 = 0.995$$

In the example there is only a 0.5% difference between the sum of the experimental values of the nitrogen fluxes to and from the pilot plant. This indicates that the analytical procedures were correct and for this reason the data can be attributed a high degree of reliability.

It is interesting to note that once one has established that the nitrogen mass balance closes, it is also possible to determine the recovery factor for the organic material. To do this, first the three fractions mS_e, mS_{xv} and mS_o must be calculated.

In the case of the above example one has:

$$mS_e = S_e/S_{ti} = 18/477 = 0.038$$
$$mS_{xv} = f_{cv} \cdot mE_v = f_{cv} \cdot X_v \cdot V_r/(R_s \cdot MS_{ti}) = 1.5 \cdot 2469 \cdot 25/(18 \cdot 40 \cdot 477) = 0.270$$

The value of mS_o is calculated as the sum of the oxygen consumption for organic matter and the equivalent oxygen recovered in the denitrification process:

$$mS_o = MS_o/MS_{ti} = (MO_c + MO_{eq})/MS_{ti}$$

In the above expression, the oxygen consumption for the oxidation of organic matter is the difference between the total consumption and the consumption for nitrification. The total consumption in the four aerobic reactors is:

$$MO_t = V_1 \cdot (O_{t2} + O_{t3} + O_{t4} + O_{t5})$$
$$= 5 \cdot (59.3 + 36.5 + 23.4 + 19.3) \cdot 24 = 16{,}620\, g O_2 \cdot d^{-1}$$

In order to calculate the oxygen consumption for nitrification, the flux of nitrified ammonium is determined as the difference between the TKN flux in the influent and the fluxes leaving the system in the effluent or the excess sludge. Knowing there is an oxygen consumption of 4.57 mg O_2 per mg N nitrified (refer to Section 5.1.3.1), one has in the case of the example:

$$MO_n = 4.57 \cdot (MN_{ti} - MN_{ni} - MN_l - MN_{oe} - MN_{ae})$$
$$= 4.57 \cdot (1816 - 40 \cdot 0.3 - 343 - 40 \cdot 1.9) = 6329\, g O_2 \cdot d^{-1}$$

Hence the oxygen consumption for oxidation of organic matter is:

$$MO_c = MO_t - MO_n = 16,620 - 6329 = 10,291\ g\ O_2 \cdot d^{-1}$$

The equivalent oxygen recovery is equal to 2.86 mg O_2 per mg N denitrified (refer to Section 5.1.3.1). Thus, the total mass of equivalent oxygen recovered in the denitrification process can be determined as:

$$MO_{eq} = 2.86 \cdot MN_d = 2.86 \cdot 1040 = 2974\ g\ O_2 \cdot d^{-1}$$

Now, the fraction of influent COD that is oxidised in the activated sludge is determined as:

$$mS_o = (MO_c + MO_{eq})/MS_{ti} = (10,291 + 2974)/(40 \cdot 477) = 13,210/19,080 = 0.695$$

Finally, the recovery factor for organic matter can be calculated as:

$$B_o = mS_e + mS_{xv} + mS_o = 0.038 + 0.270 + 0.695 = 1.003$$

It can be concluded that the mass balance for organic material also closes: the experimentally determined recovery factor is practically equal to the theoretical value of 1.0. In practice it can be expected that the recovery factors B_o and B_n deviate more from the theoretical value than in the above example. The main reason is that most activated sludge processes are not operated completely under steady state conditions.

5.1.3 Stoichiometrics of reactions with nitrogenous matter

5.1.3.1 Oxygen consumption

Only nitrification and denitrification are of interest when calculating the oxygen consumption of reactions with nitrogenous matter. Figure 5.3 schematically shows the electron transfer that will occur in the nitrification- and the denitrification processes.

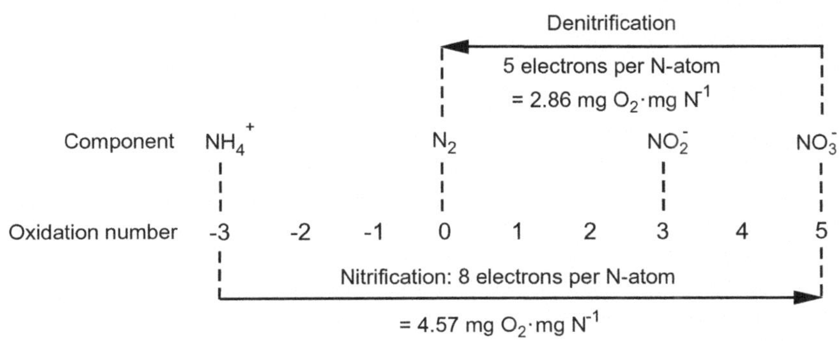

Figure 5.3 Variation of the oxidation number of the nitrogen atom in the processes of full nitrification and -denitrification

In the nitrification process, the oxidation number of the nitrogen atom in ammonium increases from −3 to +5 by the transfer of 8 electrons to the electron acceptor (oxidant): i.e. oxygen. These electrons are accepted by two molecules (four atoms) of oxygen (thereby changing its oxidation number from 0 to −2). Hence, for the nitrification of l mol of ammonium nitrogen (14 g N), there is a demand for two moles (64 g) of oxygen, so that the stoichiometric oxygen consumption can be calculated as 64/14 or 4.57 mg $O_2 \cdot mg\ N^{-1}$.

In the denitrification process, nitrate (oxidation number +5) is reduced by organic matter to molecular nitrogen (oxidation number 0), so that 5 electrons are transferred per nitrogen atom. Hence, of the 8 electrons released by nitrogen in the nitrification process, 5 electrons are recovered when nitrate is reduced to nitrogen. Thus, in oxidimetric terms, the nitrate has an oxidation capacity of $\frac{5}{8}^{th}$ of the oxygen used in the production of the nitrate by nitrification. In other words, a fraction equal to $\frac{5}{8}^{th}$ or 62.5% of the oxygen consumption in the nitrification process can be recovered as "equivalent oxygen" in the process, i.e. $0.625 \cdot 4.57 = 2.86$ mg $O_2 \cdot mg\ N^{-1}$. It can be concluded that there is a net oxygen consumption of 4.57−2.86 = 1.71 mg $O_2 \cdot mg\ N^{-1}$ during complete biological removal of nitrogen.

As shown in Example 5.1, in a process with nitrogen removal the following equations can be derived to express the oxygen demand for nitrification and the oxygen recovery from denitrification:

$$MO_n = 4.57 \cdot MN_c \tag{5.11}$$

$$MO_{eq} = 2.86 \cdot MN_d \tag{5.12}$$

So the total oxygen demand in an activated sludge process with nitrogen removal is equal to:

$$MO_t = MO_c + MO_n - MO_{eq} \tag{5.13}$$

EXAMPLE 5.2

A wastewater contains 600 mg $COD \cdot l^{-1}$ and 60 mg $TKN \cdot l^{-1}$. It has been established that 10% of the influent COD is discharged with the effluent while 30% leaves the system in the excess sludge. The effluent TKN concentration is 3 mg $N \cdot l^{-1}$. Denitrification is complete.

Determine the fraction of the oxygen consumption necessary for oxidation of nitrogenous matter in the cases of (a) nitrification only and (b) nitrification plus denitrification.

Solution

(1) Calculate the oxygen consumption for the removal of organic matter per litre influent:

$$O_c = (1 - mS_e - mS_{xv}) \cdot S_{ti} = (1 - 0.1 - 0.3) \cdot 600 = 360\ mg\ O_2 \cdot l^{-1}$$

(2) Calculate the nitrogen concentration (expressed as mg $N \cdot l^{-1}$ of influent), leaving the system together with the excess sludge:

$$N_l = f_n \cdot mE_v \cdot S_{ti} = 0.1 \cdot (0.3/1.5) \cdot 600 = 12\ mg\ N \cdot l^{-1}$$

(3) Calculate the nitrified TKN concentration:

$$N_c = N_{ai} + N_{oi} - N_{ae} - N_{oe} - N_l = 60 + 0 - 3 - 12 = 45\ mg\ N \cdot l^{-1}$$

(4) Calculate the oxygen consumption for nitrification:

$$O_n = 4.57 \cdot N_c = 206\,mg\,O_2 \cdot l^{-1}$$

(5) Calculate the equivalent oxygen recovered in the denitrification process:

$$O_{eq} = 2.86 \cdot N_c = 129\,mg\,O_2 \cdot l^{-1}$$

When only nitrification is considered, the total oxygen consumption expressed per litre influent $O_t = O_c + O_n = 360 + 206 = 566$ mg N $\cdot$ l^{-1} of which a fraction $O_n/O_t = 206/566 = 36\%$ is consumed for the oxidation of ammonium.

In the case of nitrification followed by denitrification, the total oxygen consumption decreases to $O_t = O_c + O_n - O_{eq} = 360 + 206 - 129 = 437$ mg $O_2 \cdot l^{-1}$ and the fraction of the oxygen consumed by the nitrogenous material is reduced to (437–360)/437 = 18%. It can be concluded that in this example the inclusion of denitrification in the process configuration reduces oxygen consumption from 566 to 437 mg $O_2 \cdot l^{-1}$, a reduction of 23%.

5.1.3.2 Effects on alkalinity

The processes of ammonification, nitrification and denitrification influence the (carbonate) alkalinity of mixed liquor and hence the pH in an activated sludge system. In this section it will be demonstrated that the effect on alkalinity can be calculated from simple stoichiometric relationships using the reaction equations of the three processes (Eqs. 5.1 to 5.3). Then, in the next section the relationship between alkalinity and pH will be explored.

It can be observed that in all of the Eqs. (5.1 to 5.3) the hydrogen ion is involved: in the ammonification process and the denitrification process there is a consumption of 1 mol H^+ per mol N, whereas during nitrification there is a release of 2 moles H^+ per mol N.

Knowing that the formation of 1 mol of H^+ (mineral acidity) is equivalent to the consumption of 1 mol of alkalinity or $\frac{1}{2}$ mol of $CaCO_3$ (50 g $CaCO_3$), the following alkalinity changes are calculated:

- Ammonification process: production of 1 meq or 50 g $CaCO_3$ per mol N;
- Nitrification process: consumption of 2 meq or $2 \cdot 50 = 100$ g $CaCO_3$ per mol N;
- Denitrification process: production of 1 meq or 50 g $CaCO_3$ per mol N.

The alkalinity changes resulting from above processes are summarized in Table 5.2. In the case of municipal sewage, the alkalinity effect of ammonification is usually very small, as the following analysis will show. The concentration of ammonified nitrogen in the activated sludge process is given by the difference between the organic nitrogen present in the influent and the sum of the organic nitrogen fractions contained in the effluent and the excess sludge (see Figure 5.1) so that:

$$\Delta N_{am} = N_{oi} - N_{oe} - N_l \tag{5.14}$$

where:

ΔN_{am} = ammonified nitrogen concentration in the activated sludge process (mg N $\cdot$ l^{-1} influent)

Table 5.2 Alkalinity change resulting from reactions with nitrogenous matter

Reaction	Alkalinity change	
	$(meq \cdot mg^{-1}\ N)$	$(mg\ CaCO_3 \cdot mg^{-1}\ N)$
Ammonification	$+\frac{1}{14}$	+3.57
Nitrification	$-\frac{1}{7}$	−7.14
Denitrification	$+\frac{1}{14}$	+3.57

Normally in the case of municipal sewage, the organic nitrogen concentration in the effluent is small, only 1 or 2 $mg\ N \cdot l^{-1}$, whereas the values of N_{oi} and N_l are both approximately equal to 25% of the influent TKN concentration. Hence the variation of the organic nitrogen concentration in the activated sludge process will be very small. Consequently the associated alkalinity change will also be limited and can be expressed as:

$$\Delta Alk_{am} = 3.57 \cdot \Delta N_{am} = 3.57 \cdot (N_{oi} - N_{oe} - N_l) \tag{5.15}$$

The alkalinity change due to nitrification is calculated from the concentration of nitrified ammonium. This concentration is equal to the difference of the influent TKN concentration (N_{ki}) and the sum of the TKN concentrations in the effluent (N_{ke}) and the excess sludge (N_l):

$$N_c = N_{ki} - N_{ke} - N_l \tag{5.16}$$

where N_c = influent ammonium concentration, nitrified in the system ($mg\ N \cdot l^{-1}$)

The effect of nitrification on alkalinity can be expressed as:

$$\Delta Alk_n = -7.14 \cdot N_c = -7.14 \cdot (N_{ki} - N_{ke} - N_l) \tag{5.17}$$

The alkalinity change due to denitrification depends on the removed nitrate concentration. This concentration can be calculated as:

$$N_d = N_{ni} + N_c - N_{ne} \tag{5.18}$$

Hence the alkalinity change resulting from denitrification can be expressed as:

$$\Delta Alk_d = 3.57 \cdot \Delta N_d = 3.57 \cdot (N_{ni} + (N_{ki} - N_{ke} - N_l) - N_{ne}) \tag{5.19}$$

The total alkalinity change in the activated sludge process from the reactions of nitrogenous matter will be equal to the sum of the alkalinity changes calculated for ammonification, nitrification and denitrification.

$$\begin{aligned}\Delta Alk_t &= \Delta Alk_{am} + \Delta Alk_n + \Delta Alk_d \\ &= 3.57 \cdot (N_{oi} - N_{oe} - N_l) - 7.14 \cdot (N_{ki} - N_{ke} - N_l) + 3.57 \cdot (N_{ni} + N_{ki} - N_{ke} - N_l - N_{ne})\end{aligned} \tag{5.20}$$

Knowing that the TKN concentration (N_k) is equal to the sum of the concentrations of organic (N_o) and ammonium nitrogen (N_a), Eq. (5.20) can be simplified to:

$$\Delta Alk_t = -3.57 \cdot (N_{ai} - N_{ae} - N_{ni} + N_{ne}) = +3.57 \cdot (\Delta N_a - \Delta N_n) \tag{5.21}$$

where:

$\Delta N_a = N_{ae} - N_{ai}$

$= $ variation of the ammonium concentration ($mg\,N \cdot l^{-1}$)

$\Delta N_n = N_{ne} - N_{ni}$

$= $ variation of the nitrate concentration ($mg\,N \cdot l^{-1}$)

All parameters on the right hand side of Eq. (5.21) can be measured experimentally by standard tests. Hence it is possible to calculate the stoichiometric alkalinity change due to the combined effect of ammonification, nitrification and denitrification in the activated sludge process. Furthermore, it is also possible to measure the alkalinity change directly.

In Figure 5.4 the calculated (according to Eq. 5.21) and the observed alkalinity change in different activated sludge processes have been compared. The data in Figure 5.4 refers to very diverse systems: (I) without nitrification, (II) only nitrification and (III) both nitrification and denitrification. In all cases there is an excellent correlation between the calculated and the observed alkalinity change, for a very large range of changes (ΔAlk between -600 and $+100$ $mg \cdot l^{-1}$ $CaCO_3$) and for very diverse operational conditions. Thus, the conclusion is justified that the alkalinity change in an activated sludge process is predominantly due to the stoichiometric effects of the reactions with nitrogenous material: ammonification, nitrification and denitrification.

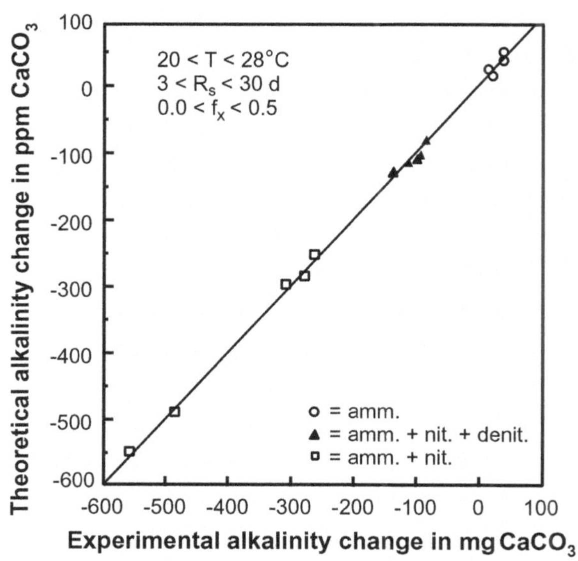

Figure 5.4 Calculated versus experimentally observed alkalinity change in a number of activated sludge processes

EXAMPLE 5.3

Consider again the activated sludge process represented in Example 5.1:

- Calculate the alkalinity change predicted by the model;
- Estimate the alkalinity change in the process if denitrification would not occur.

Solution

With the aid of Eq. (5.21) and Table 5.1 the total alkalinity change is calculated as:

$$\begin{aligned}\Delta Alk_t &= 3.57 \cdot (\Delta N_a - \Delta N_n) = -3.57 \cdot (N_{ai} - N_{ae} - N_{ni} + N_{ne})\\ &= -3.57 \cdot (32.8 - 0.0 - 0.3 + 8.7) = -147\,mg\,CaCO_3 \cdot l^{-1}\end{aligned}$$

Without denitrification the alkalinity changes only due to ammonification and nitrification. The nitrogen concentration in the excess sludge is estimated as $N_l = MN_l/Q_i = f_n \cdot V_r \cdot X_v/(Q_i \cdot R_s) = 8.6\,mg\,N \cdot l^{-1}$, so using Table 5.1 one can calculate ΔN_{am}:

$$\begin{aligned}\Delta N_{am} &= N_{oi} - N_{oe} - N_l\\ &= (45.1 - 32.8) - (1.9 - 0.0) - 8.6 = 12.3 - 1.9 - 8.6 = 1.8\,mg\,N \cdot l^{-1}\end{aligned}$$

The effect of ammonification on the alkalinity change is now calculated as:

$$\begin{aligned}\Delta Alk_a &= 3.57 \cdot \Delta N_{am}\\ &= 3.57 \cdot 1.8 = 6\,mg\,CaCO_3 \cdot l^{-1}\end{aligned}$$

The nitrified ammonium concentration is calculated as:

$$\begin{aligned}N_c &= N_{ki} - N_{ke} - N_l\\ &= 45.1 - 1.9 - 8.6 = 34.6\,mg\,N \cdot l^{-1}\end{aligned}$$

$$\Delta Alk_n = -7.14 \cdot N_c = -247\,mg\,CaCO_3 \cdot l^{-1}$$

Hence, without denitrification the alkalinity change would amount to:

$$\Delta Alk_t = \Delta Alk_{am} + \Delta Alk_n = 6 - 247 = -241\,mg\,CaCO_3 \cdot l^{-1}$$

5.1.3.3 Effects on pH

Having established the relationship between the reactions of nitrogenous matter and the alkalinity change in an activated sludge process, it is now possible to evaluate the effect of these reactions on the pH of the mixed liquor.

First it must be recognised that the pH in activated sludge processes is set mainly by the carbonic system $CO_2 \longleftrightarrow HCO_3^- \longleftrightarrow CO_3^{2-}$, because this system is present at much higher concentrations than other acid-base systems. The equilibrium of the weak acid and associated base $NH_4^+ \longleftrightarrow NH_3$ is not important when the pH is in the neutral range as in the case of mixed liquor: almost all ammonium will be present in the ionised form. Other equilibriums with a pK value (negative logarithm of the dissociation constant) in the neutral pH range, for example $H_2PO_4^- \longleftrightarrow HPO_4^{2-}$ (pK = 7.2) and $H_2S \longleftrightarrow HS^-$ (pK = 7.0) are not important because the concentrations of phosphate and sulphide in mixed liquor are much lower than the concentrations of the carbonic system, as demonstrated by Van Haandel et al. (1994).

For the carbonic system, the relationship between alkalinity and pH can be derived from the model developed by Loewenthal and Marais (1976). This model describes the interrelationship between alkalinity, acidity and pH in aqueous solutions. For the carbonic system the alkalinity is defined as:

$$Alk = [HCO_3^-] + 2 \cdot [CO_3^{2-}] + [OH^-] - [H^+] \tag{5.22}$$

where $[X]$ = concentration of X in $mol \cdot l^{-1}$

In order to correlate pH and alkalinity, it is necessary to eliminate the concentrations $[HCO_3^-]$, $[CO_3^{2-}]$ and $[OH^-]$ from Eq. (5.22), using the relevant dissociation equations:

$$\text{(a) } CO_2 + H_2O \overset{k_1}{\longleftrightarrow} HCO_3^- + H^+ \tag{5.23}$$

$$\text{(b) } HCO_3^- \overset{k_2}{\longleftrightarrow} CO_3^{2-} + H^+ \tag{5.24}$$

$$\text{(c) } H_2O \overset{k_w}{\longleftrightarrow} H^+ + OH^- \tag{5.25}$$

From Eq. (5.23), the chemical equilibrium can be written as:

$$k_1 = [HCO_3^-] \cdot [H^+]/[CO_2] \text{ or } [HCO_3^-] = k_1/f_m \cdot [CO_2]/[H^+] = k_1^* \cdot [CO_2]/(H^+) \tag{5.26}$$

where:

$[X]$ = activity of X in $mol \cdot l^{-1}$
k_1 = equilibrium constant of the CO_2 dissociation = $4.45 \cdot 10^{-7}$ (at 20°C)
k_1^* = "real" equilibrium constant of the CO_2 dissociation (on molar base)
f_m = activity coefficient for a monovalent ion in the mixed liquor

Similarly one has:

$$k_2 = [CO_3^{2-}] \cdot [H^+]/[HCO_3^-] \text{ or}$$

$$[CO_3^{2-}] = k_2 \cdot (f_m/f_d) \cdot [HCO_3^{2-}]/(H^+) = k_1^* \cdot k_2^* \cdot [CO_2]/(H^+)^2 \tag{5.27}$$

$$k_w = [OH^-] \cdot [H^+] \text{ or } [OH^-] = (k_w/f_m)/[H^+] = k_w^*/[H^+] \tag{5.28}$$

where:

k_2 = equilibrium constant for bicarbonate dissociation = $4.69 \cdot 10^{-11}$ at 20°C
k_2^* = "real" equilibrium constant of the bicarbonate dissociation
k_w = equilibrium constant for the dissociation of water = 10^{-14} at 20°C
k_w^* = "real" equilibrium constant for the dissociation of water
f_d = activity factor for a bivalent ion

Finally by substituting Eqs. (5.26, 5.27 and 5.28) in Eq. (5.22) an expression linking $[H^+]$ and alkalinity is obtained:

$$Alk = [CO_2] \cdot (k_1^*/[H^+] + 2 \cdot k_1^* \cdot k_2^*/[H^+]^2) + k_w^*/[H^+] - [H^+] \quad (5.29)$$

Knowing that $pH = -\log[H^+]$ one has $[H^+] = 10^{-pH}$ and

$$Alk = [CO_2] \cdot 10^{(pH-pk1^*)} \cdot [(1 + 2 \cdot 10^{(pH-pk2^*)}) + 10^{(pH-pkw^*)} - 10^{-pH}] \quad (5.30)$$

From Eq. (5.30), the pH can be calculated for any alkalinity value if the dissolved carbon dioxide concentration is known. This concentration depends on the production rate of this gas from the oxidation of organic matter and the removal efficiency from the liquid phase due to the stripping effect of the aeration system.

In Figure 5.5 several pH curves as a function of alkalinity have been drawn, for CO_2 concentrations ranging from 0.5 mg $CO_2 \cdot l^{-1}$ (the saturation concentration at 20°C) to 10 mg $CO_2 \cdot l^{-1}$ (i.e. 20 times super-saturated). To construct the diagram, a temperature of 20°C and activity coefficients $f_m = 0.90$ and $f_d = 0.67$ were assumed. These values correspond to a ionic force of 0.01 as calculated from the Debye-Hückel theory, as shown by Loewenthal et al. (1976) and are fairly typical for sewage. Figure 5.5 shows that for alkalinities greater than 35 mg $\cdot l^{-1}$ $CaCO_3$, the pH does not respond significantly to alkalinity changes. For example, an alkalinity increase from 35 to 500 mg $\cdot l^{-1}$ results in an increase of the pH value of less than one unit. In contrast, for alkalinities smaller than 35 mg $\cdot l^{-1}$, the pH value depends strongly on the alkalinity value. An alkalinity decrease from 35 to 0 mg $\cdot l^{-1}$ causes the pH to drop from the neutral range to a value of approximately 4.2.

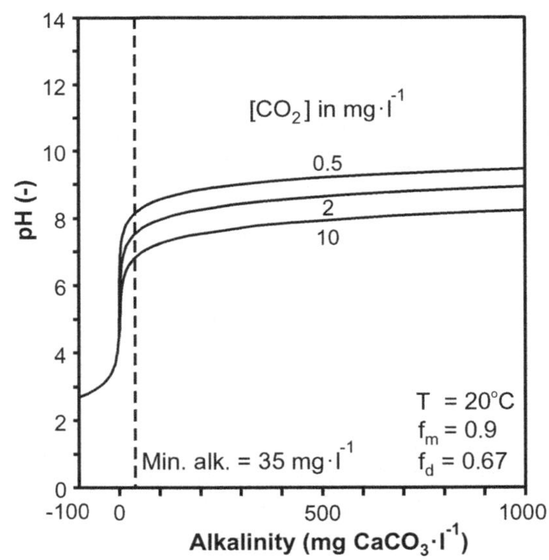

Figure 5.5 pH value as function of the alkalinity concentration in mixed liquor

A low pH value affects the activity of micro-organisms. In particular the activity of nitrifying bacteria has been shown to decrease at low pH values: i.e. below a pH of 6 nitrification virtually ceases. Hence to ensure stable and efficient nitrification, it is necessary that the alkalinity is maintained at a value higher than 35 $mg \cdot l^{-1}$ $CaCO_3$, so that approximately neutral pH is guaranteed. It is interesting to note that Haug and McCarty (1971), on the basis of an experimental investigation, established the same minimum alkalinity value as the one calculated from theory above.

Now it is possible to estimate the minimum influent alkalinity required to ensure a stable and neutral pH value in an activated sludge process:

$$Alk_i > 35 + \Delta Alk_t \text{ or } Alk_i > 35 + 3.57 \cdot (\Delta N_a - \Delta N_n) \tag{5.31a}$$

$$\text{Or approximated: } Alk_i > 3.57 \cdot (10 + \Delta N_n - \Delta N_a) \tag{5.31b}$$

where:

Alk_i= influent alkalinity ($mg\ CaCO_3 \cdot l^{-1}$)
Alk_e= effluent alkalinity ($mg\ CaCO_3 \cdot l^{-1}$)

EXAMPLE 5.4

What would be the minimum alkalinity of the sewage in the activated sludge system of Example 5.1 required to ensure a stable and neutral pH value?

Solution

In Table 5.1 it can be observed that $N_{ai} = 32.8$; $N_{ae} = 0.0$; $N_{ni} = 0.3$ and $N_{ne} = 8.7\ mg\ N \cdot l^{-1}$. Hence by using Eq. (5.31b) the minimum required influent alkalinity is calculated as:

$$Alk_i > 3.57 \cdot [10 + (8.7 - 0.3) - (0.0 - 32.8)] = 183\ mg \cdot l^{-1}\ CaCO_3$$

In practice the alkalinity present in the influent may be less than the minimum value required to maintain a stable pH in the activated sludge process. This is a particular risk when nitrification without subsequent denitrification occurs in the process. In such cases it is necessary to increase the influent alkalinity, which is usually done by addition of lime or caustic. Without the addition of alkalinity, the behaviour of the activated sludge process will be irregular; there will be periods with nitrification and the consequential decrease of alkalinity and pH, until a pH value is established that is inhibitory for nitrification. When nitrification ceases, alkalinity automatically increases and pH rises, so that once again favourable conditions for nitrification are established and a new cycle of instability is initiated. If denitrification is included as a treatment step, the decrease of alkalinity will be smaller and often there will be no need for lime addition at all.

5.2 NITRIFICATION

Nitrification is a two-step biological process, but only the first step – oxidation of ammonium to nitrite – is normally of importance for the nitrification kinetics in an activated sludge system. When the nitrifying population is well established in the activated sludge process the second step, oxidation of nitrite to nitrate, is so fast that it can be considered as instantaneous for all practical purposes. Consequently, the nitrite concentration in the effluent of activated sludge systems is in general very small. In the following text the general term "nitrifiers" will be used to describe both ammonium- and nitrite oxidisers.

5.2.1 Nitrification kinetics

Downing et al. (1964) were the first to show that the growth of nitrifiers in the oxidation process of ammonium can be described by Monod kinetics:

$$(dX_n/dt) = (dX_n/dt)_g + (dX_n/dt)_d \tag{5.32a}$$

$$(dX_n/dt)_g = \mu \cdot X_n = \mu_m \cdot X_n \cdot N_a/(N_a + K_n) \tag{5.32b}$$

$$(dX_n/dt)_d = -b_n \cdot X_n \tag{5.32c}$$

where:

(dX_n/dt) = net rate of change in nitrifier concentration (mg VSS $\cdot l^{-1} \cdot d^{-1}$)
$(dX_n/dt)_g$ = net rate of change in nitrifier concentration due to growth (mg VSS $\cdot l^{-1} \cdot d^{-1}$)
$(dX_n/dt)_d$ = net rate of change in nitrifier concentration due to decay (mg VSS $\cdot l^{-1} \cdot d^{-1}$)
X_n = nitrifier concentration (mg VSS $\cdot l^{-1}$)
μ = specific growth rate of nitrifiers (d^{-1})
μ_m = maximum specific growth rate of nitrifiers (d^{-1})
b_n = decay rate of nitrifiers (d^{-1})
K_n = Monod half saturation constant (mg N $\cdot l^{-1}$)

In the Monod equation, the parameter μ represents the growth rate of the micro-organisms per time unit. For example, a value of $\mu = 0.6\ d^{-1}$ means that the daily rate of micro-organism synthesis is equal to 60% of the mass initially present. Equation (5.32b) shows that the μ value depends on the ammonia concentration N_a. At high N_a concentration (saturation) the maximum growth rate μ_m is attained. The constant K_n is equal to the substrate concentration for which $\mu_m = \frac{1}{2}\mu$, and for that reason is called the "half" saturation constant. The basic equation of Downing et al. (1964) can be used to calculate the residual ammonium concentration in a completely mixed, steady state activated sludge process. Under these conditions, the mass of nitrifiers in the system will not change: the net growth rate (defined as the growth rate minus the decay rate) is equal to the discharge rate due to abstraction of excess sludge. Hence:

$$(dX_n/dt) = 0 = (dX_n/dt)_g + (dX_n/dt)_d + (dX_n/dt)_e \tag{5.33}$$

The rate of change of the nitrifier concentration due to the discharge of excess sludge $(dX_n/dt)_e$, can be expressed as:

$$(dX_n/dt)_e = -X_n/R_s \tag{5.34}$$

Now, using Eqs. (5.32b and c and 5.34) in Eq. (5.33) one has:

$$(dX_n/dt) = 0 = \mu_m \cdot X_n \cdot N_a/(N_a + K_n) - b_n \cdot X_n - X_n/R_s \tag{5.35}$$

X_n can be deleted from Eq. (5.35) and after some rearranging, the ammonium concentration in the mixed liquor of a completely mixed activated sludge process is given as:

$$N_a = K_n \cdot (b_n + 1/R_s)/[\mu_m - (b_n + 1/R_s)](= N_{ae}) \tag{5.36}$$

For a completely mixed process, by definition this ammonium concentration is equal to the residual ammonia effluent concentration. This residual ammonium concentration, which is indicative of the efficiency of the nitrification process, depends on the values of the three kinetic parameters (μ_m, K_n and b_n) and the value of one operational variable: the sludge age R_s. It is interesting to note that the *residual* ammonium concentration does not depend on the *initial* concentration, as under steady state conditions a nitrifying sludge mass develops that will be compatible with the applied nitrogen load.

Equation (5.36) can be rewritten to yield the value of the sludge age as function of the residual ammonium concentration, i.e. the sludge age required to reduce the ammonium concentration to a value N_a:

$$R_s = (1 + K_n/N_a)/[\mu_m - b_n \cdot (1 + K_n/N_a)] \tag{5.37}$$

When the activated sludge system is operated at the *minimum* sludge age for nitrification, this implies that for this sludge age the nitrification capacity will be very small. The residual ammonium concentration will thus always be much higher than the value of the half saturation value K_n.

In that case the ratio K_n/N_a will be $\ll 1$ and Eq. (5.37) is simplified to:

$$R_{sn} = 1/(\mu_m - b_n) \tag{5.38}$$

where R_{sn} = minimum sludge age required for nitrification

Equation (5.38) expresses that nitrification will not develop if the sludge age is shorter than a minimum value of $R_{sn} = 1/(\mu_m - b_n)$, because the rate of nitrifier discharge in the excess sludge will then exceed the net growth rate. However, when the sludge age R_s is higher than the minimum value, nitrification will develop and its efficiency will depend on the sludge age and the kinetic constants K_n, μ_m and b_n. Figure 5.6 shows a typical profile of the effluent ammonium concentration in a completely mixed nitrification reactor, calculated with Eq. (5.36) for T = 20°C. Nitrification does not develop for sludge ages lower than the minimum sludge age $R_{sn} = 1/(0.4 - 0.04) = 2.8$ days. The ammonium concentration rapidly decreases at sludge ages higher than R_{sn}, until at a certain sludge age R_{sm} the effluent ammonium is equal to the *specified* residual ammonia concentration N_{ad}. The value of the R_{sm} is of great practical importance, as operation at a sludge age higher than R_{sm} will allow an anoxic zone to be included, while at the same time N_{ae} remains equal to N_{ad}. The value of R_{sm} can be explicitly calculated when Eq. (5.37) is slightly reworked to:

$$R_{sm} = 1/[\mu_m/(1 + K_n/N_{ad}) - b_n] \tag{5.39}$$

In Figure 5.6 the value of R_{sm} is indicated as well. For the specified conditions it can be calculated with Eq. (5.39) as $R_{sm} = 1/[0.4/(1+1.0/1.0) - 0.04] = 6.3$ days.

When Monod kinetics are assumed to be representative for nitrification, this also implies that there is a trade-off between nitrification rate and residual ammonia concentration. When the Monod equation for nitrifier growth rate (Eq. 5.32a) is analyzed, then the impact of the ammonium concentration in the reactor on the nitrification rate is apparent. This is shown in Figure 5.7, where the relative nitrification rate, equal to the Monod factor $N_a/(K_n + N_a)$, is plotted for different temperatures. When the data for the curve of 20°C is analyzed, it can be observed that for typical effluent ammonium concentrations (1 to 2 mg NH_4-N · l^{-1}), the actual nitrification rate is only 50 to 67% of the maximum nitrification rate.

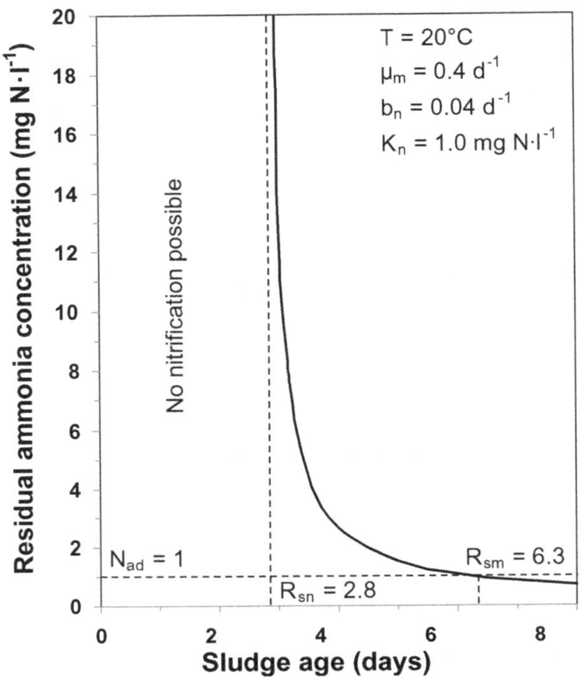

Figure 5.6 Typical profile of the residual ammonium concentration as function of the sludge age

Figure 5.7 also shows that an increase in ammonium conversion capacity might often be possible, but only at the expense of a higher residual ammonium concentration. For instance, when the ammonium concentration increases from 1 to 2 mg N · l^{-1} (at 20°C), the ammonium conversion capacity increases with 33%.

When a nitrogen peak load is applied to an activated sludge system, the ammonium concentration will increase. This in turn increases the rate of nitrification and hence a new (but higher) equilibrium ammonium concentration will be established. So in the design of the nitrification process, it is important to consider a temporary increase in the ammonium effluent concentration during peak load conditions. This will enable (part of) the additional nitrogen load to be removed. In other words, the specified residual ammonium concentration should then be less than the effluent limit minus the expected ammonium increase during peak loading. However, there are more issues to consider, as will be discussed in Section 5.2.4 and Example 5.6.

In general it is advantageous to use a plug flow reactor for nitrification, as it allows an ammonium concentration gradient to develop over the length of the reactor. The front end of the nitrification reactor will operate at a higher ammonium concentration and hence at increased nitrification rate than the back end of the reactor. The *average* nitrification rate will be higher than that in a completely mixed system of the same size. When the nitrification reactor is designed for completely mixed conditions but constructed as a plug-flow reactor, then the effluent ammonium concentration will always be somewhat lower than the specified residual concentration. Thus the plug-flow system will have some "spare" capacity available to handle peak nitrogen loads without exceeding the ammonium effluent limit.

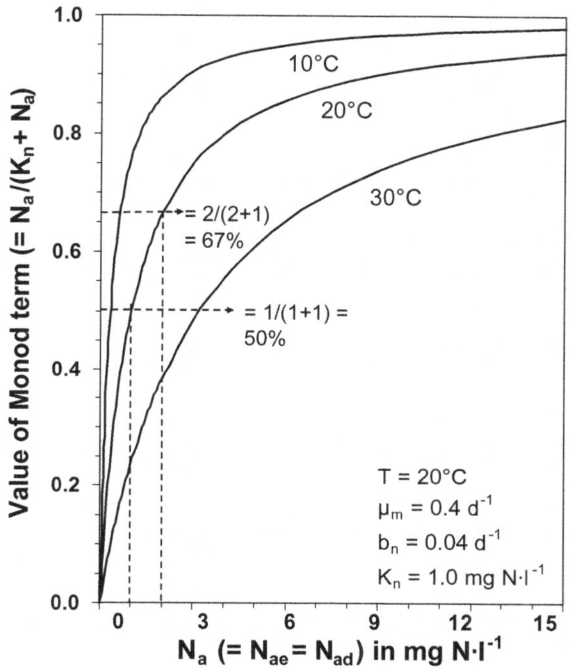

Figure 5.7 Relative nitrification rate as function of the ammonium concentration for different temperatures

EXAMPLE 5.5

An activated sludge process is designed for nitrification. Assuming a completely mixed reactor, calculate for the minimum and maximum expected temperature:

- The minimum sludge age for nitrification ($=R_{sn}$);
- The minimum sludge age where the residual ammonium concentration equals the specified one, i.e. where $N_{ae}=N_{ad}$ ($=R_{sm}$).

Furthermore, evaluate the ammonium removal performance for the selected R_s value at minimum and maximum temperature.

Use the following data:

- $N_{ad}=1$ mg $N \cdot l^{-1}$;
- At $T_{max}=20°C$: $\mu_m=0.40\ d^{-1}$; $b_n=0.04\ d^{-1}$ and $K_n=1$ mg $N \cdot l^{-1}$;
- At $T_{min}=10°C$: $\mu_m=0.13\ d^{-1}$; $b_n=0.03\ d^{-1}$ and $K_n=0.31$ mg $N \cdot l^{-1}$.

Solution

The values of R_{sn} and R_{sm} at $T = 20°C$ have been determined earlier in this section as 2.8 and 6.3 days. Using Eqs. (5.38 and 5.39), it can be calculated that at the minimum temperature of 10°C they are equal to:

$$R_{sn} = 1/(\mu_m - b_n) = 1/(0.13 - 0.03) = 10.2\,\text{days}$$

$$R_{sm} = 1/[\mu_m/(1 + K_n/N_{ad}) - b_n] = 1/[0.13/(1 + 0.31/1.0) - 0.03] = 14.6\,\text{days}$$

Design should always be based on the worst case, or in this case the lowest temperature. So by definition the value of $N_{ae} = N_{ad} = 1.0$ at 10°C. Use Eq. (5.36) to calculate N_{ae} at 20°C:

$$N_a = K_n \cdot (b_n + 1/R_s)/[\mu_m - (b_n + 1/R_s)] = 1 \cdot (0.04 + 1/14.6)/[0.4 - (0.04 + 1/14.6)] = 0.37\,\text{mg N} \cdot l^{-1}$$

The effluent ammonia concentration will thus be lower than the specified effluent limit during a large part of the year. It is interesting to evaluate the additional nitrogen load that can be handled without exceeding the ammonium limit. At 10°C the answer is simple: as $N_{ae} = N_{ad} = 1.0$ mg $N \cdot l^{-1}$, any increase in nitrogen load will immediately result in $N_{ae} > N_{ad}$. So there is no margin. However, there may still exist some flexibility if the ammonium effluent limit is based on a flow proportional 24 hrs sample. Considering that the TKN load typically varies over the day, this allows periods with excess ammonium load in the effluent to be compensated with periods of lower than average load (when $N_{ae} > N_{ad}$). At 20°C the value of the Monod constant is equal to $0.37/(1 + 0.37) = 0.27$. So theoretically, the nitrogen load can be increased at least $1/0.27 = 3.7$ times before the effluent ammonium concentration increases to 1.0 mg $N \cdot l^{-1}$.

After Downing's work, many researchers have carried out experimental investigations to determine the kinetic parameters for nitrification in the activated sludge process. Table 5.3 to Table 5.5 show experimental values of μ_m, b_n and K_n. It can be observed that the data obtained by the different authors have a very large spread. This may partially be attributed to differences in the experimental methods, for instance the oxygen concentration used during the test. The influence of the bulk oxygen concentration on the measured value of μ_m will be discussed later in this section and also in Appendix A4. However, certainly the fact that different wastewaters have been used must have had an influence. Thus it can be concluded that the value of the kinetic parameters of the nitrifiers depends on the origin of the wastewater. Ideally these values should be determined for each specific design case.

In order to be able to compare the data collected at different temperatures, all values have been corrected to a standard value at 20°C, using the temperature dependencies as determined by Ekama and Marais (1976):

- $\mu_{mT} = \mu_{m20} \cdot 1.123^{(T-20)}$;
- $b_{nT} = b_{n20} \cdot 1.04^{(T-20)}$;
- $K_{nT} = K_{n20} \cdot 1.123^{(T-20)}$.

Table 5.3 Values of the maximum nitrifier growth rate μ_m according to various authors

μ_{mT} (d^{-1})	T (°C)	μ_{m20} (d^{-1})	Reference
0.33	15	0.66	Barnard (1991)
0.47	15	0.45	Kayser (1991)
0.33	20	0.33	Downing et al. (1964)
0.33–0.65	20	0.33–0.65	Ekama et al. (1976)
0.34–0.40	12	0.86–1.01	Gujer et al. (1974)
0.45	15	0.73	Eckenfelder (1991)
0.40–0.50	14	0.80–1.00	Gujer (1977)
0.50	20	0.50	Lawrence et al. (1973)
0.53	25	0.26	Sutton et al. (1979)
0.57	16	0.76	Gujer et al. (1974)
0.94	29	0.33	Lijklema (1973)
1.08–1.44	23	0.76–1.02	Poduska et al. (1974)

Table 5.4 Values of the nitrifier decay rate b_n as determined by various authors

b_{nT} (d^{-1})	T (°C)	b_{n20} (d^{-1})	Reference
0.0	20	0.0	Downing et al. (1964)
0.0	15	0.0	Downing et al. 1964)
0.0	10	0.0	Gujer (1979)
0.04	20	0.04	Ekama et al. (1976)
0.12	29	0.09	Lijklema (1973)
0.12	23	0.11	Poduska et al. (1974)

Table 5.5 Values of nitrifier Monod constant K_n according to various authors

K_{nT} ($mg \cdot l^{-1}$)	T (°C)	K_{n20} ($mg \cdot l^{-1}$)	Reference
0.0	23	0.04	Poduska et al. (1974)
0.2	15	0.1	Downing et al. (1964)
0.2	20	0.2	Downing et al. (1964)
0.2	10	0.6	Gujer (1977)
0.5	14	1.0	Ekama et al. (1976)
1.0	20	1.0	Ekama et al. (1976)
1.0	20	1.0	Lijklema (1973)

To evaluate the influence of the values of the kinetic parameters for nitrification on the efficiency of the process, the following procedure has been followed. Table 5.3 to Table 5.5 suggest average values at 20°C of $\mu_m = 0.4\ d^{-1}$; $b_n = 0.04\ d^{-1}$ and $K_n = 0.5\ mg\ N \cdot l^{-1}$. The influence of the values of these parameters on the residual ammonium concentration is shown in Figure 5.8.

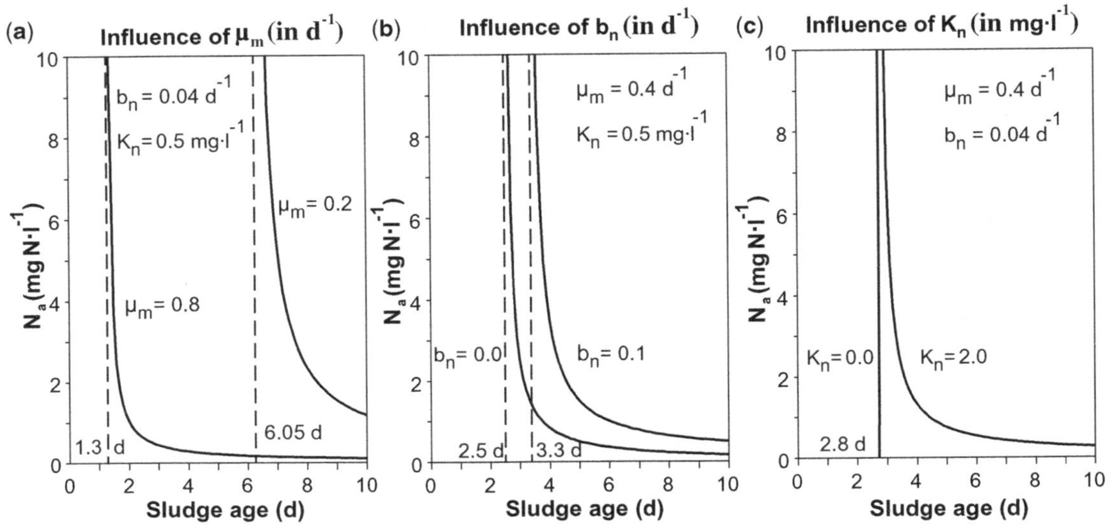

Figure 5.8 Residual ammonium concentration as a function of different values of the kinetic parameters μ_m, b_n and K_n

Figure 5.8a shows the residual ammonium concentration N_a as a function of the sludge age for average b_n and K_n values ($b_n = 0.04\ d^{-1}$ and $K_n = 0.5$ mg $N \cdot l^{-1}$) and for two values of μ_m, one extremely high ($\mu_m = 0.8\ d^{-1}$) and the other extremely low ($\mu_m = 0.2\ d^{-1}$). Hence, the difference between the curves for the residual ammonium concentration in Figure 5.8a reflects the influence of the different μ_m values (the curves were calculated using Eq. 3.33).

Similarly, in Figure 5.8b, the influence of the value of the decay rate b_n is analysed for average values of the other kinetic parameters: $\mu_m = 0.4\ d^{-1}$ and $K_n = 0.5$ mg $N \cdot l^{-1}$. The residual ammonium concentration N_a is calculated as a function of the sludge age for a very high value of the decay rate ($b_n = 0.1\ d^{-1}$) and without decay rate at all ($b_n = 0.0\ d^{-1}$). The difference between the two curves is due exclusively to the variation of the b_n value.

Finally, in Figure 5.8c the influence of the K_n value on the residual ammonium concentration is evaluated. For average values of the other two parameters ($\mu_m = 0.4\ d^{-1}$ and $b_n = 0.04\ d^{-1}$) curves were drawn for N_a as a function of R_s for $K_n = 2$ mg N. l^{-1} (very high value) and $K_n = 0.0$ mg N. l^{-1} (very low value). From Figure 5.8 the following conclusions can be drawn:

- The influence of μ_m on the residual ammonium concentration - and hence on nitrification efficiency - is much more pronounced than that of the other two parameters b_n and K_n;
- For sludge ages of more than 50% beyond the minimum sludge age for nitrification R_{sn}, the residual ammonium concentration is so low that for practical purposes nitrification may be considered to be complete.

As the minimum sludge age for nitrification depends mainly on the value of μ_m, it is necessary to analyse why such large differences in the values of μ_m are reported in Table 5.3. The values of the parameters b_n and K_n are of minor importance.

The factors influencing the μ_m value can be divided in two categories: (I) factors related to the origin of the wastewater and (II) factors related to the operational conditions in the activated sludge process.

In so far as the origin of the wastewater is concerned, there are several compounds that are known to inhibit nitrification. There are clear indications that the μ_m value depends on the fraction of industrial waste in municipal wastewater. In the case of a small industrial contribution, the μ_m value is determined in the range of 0.5 to 0.7 d^{-1} at 20°C, but this value decreases to 0.25 to 0.3 d^{-1} or even lower when the proportion of industrial wastewater in the total influent is higher. Wilson and Marais (1976) measured an μ_m value of 0.17 d^{-1} for a predominantly industrial waste.

In the case of purely industrial wastewaters, the μ_m may be very small: a research project at CETREL in Brazil, where petrochemical wastes are processed, showed a μ_m value of less than 0.1 d^{-1} at a temperature of 26°C, which is equivalent to $\mu_m < 0.05\ d^{-1}$ at 20°C. The dominant influence of the origin of the wastewater on the μ_m value indicates that this parameter should be seen as a sewage characteristic rather than a kinetic constant. In so far as operational conditions are concerned, the following factors have been shown to influence nitrification and particularly the μ_m value: temperature, DO concentration and pH.

(a) Temperature

The temperature has a strong influence on the μ_m value as the research results obtained by several authors show. Often a simplified Arrhenius equation is used to describe the influence of temperature, i.e.:

$$\mu_{mT} = \mu_{m20}\theta^{(T-20)} \tag{5.40}$$

where θ = Arrhenius temperature dependency coefficient

Table 5.6 shows the experimental values of the temperature dependency determined by several authors. The θ value ranges from 1.11 to 1.13; which means that the μ_m value increases by 11 to 13% per degree Celsius of temperature increase. Hence, the μ_m value doubles for every 6 to 7 degrees Celsius of temperature increase. The influence of the temperature on the growth rate of the nitrifiers has an important repercussion on the activated sludge process. In regions with a moderate climate, wastewater temperatures in winter are in the range of 8 to 14°C, resulting in a low value of μ_m. For a medium value of μ_m of 0.4 d^{-1} at 20°C, one would expect values 0.2 d^{-1} at 14°C and 0.1 d^{-1} at 8°C. From Eq. (5.38), it is calculated that the minimum sludge age for nitrification in this case will be in the range of 6 to 14 days. Therefore in Europe, it is common that activated sludge processes for nutrient removal are operated at a sludge age of more than 15 days. In contrast, in tropical regions water and sewage temperatures are much higher. For example, in Campina Grande in North East Brazil (a.k.a. the Queen of the Borborema Heights), the average temperature is 26°C during summer.

Table 5.6 Temperature dependency of the maximum specific growth rate of nitrifiers

Temp. factor (θ)	Temperature interval (°C)	Reference
1.116	19–21	Gujer (1977)
1.123	15–20	Downing et al. (1964)
1.123	14–20	Ekama et al. (1976)
1.130	20–30	Lijklema (1973)

If again it is assumed that $\mu_m = 0.4\ d^{-1}$ at 20°C, then the μ_m value at sewage temperature is calculated as $\mu_m = 0.8\ d^{-1}$ at 26°C, so that the minimum sludge age for nitrification is now only $R_{sn} = 1.25$ days. In practice, the activated sludge process will be almost invariably operated at higher sludge age, so that nitrification will develop if enough oxygenation capacity is available.

(b) Dissolved oxygen concentration

The influence of the dissolved oxygen concentration on nitrification kinetics has been the object of several studies. Several authors have proposed a Monod type equation to incorporate the influence of the dissolved oxygen concentration (Stenstrom and Poduska, 1980). In the IWA activated sludge models no. 1 and 2, this approach has also been followed. Both ammonium and dissolved oxygen are considered substrates and the maximum specific growth rate is expressed as:

$$\mu = \mu_m \cdot N_a/(N_a + K_n) \cdot DO/(DO + K_o) \quad (5.41)$$

where:

DO = dissolved oxygen concentration ($mg\ O_2 \cdot l^{-1}$)
K_o = half saturation constant ($mg\ O_2 \cdot l^{-1}$)

The value attributed to K_o varies considerably between different authors and values ranging from 0.3 to 2.0 $mg\ O_2 \cdot l^{-1}$ have been published. This wide range may be due to the fact that it is only possible to determine the dissolved oxygen concentration in the bulk of the liquid phase. In the sludge flocs, where consumption occurs, the dissolved oxygen concentration is lower than in the bulk. The oxygen consumption creates a concentration gradient from the floc surface (where the dissolved oxygen concentration is considered to be equal to the bulk concentration) to the centre. Figure 5.9 schematically shows the dissolved oxygen concentration profile in a sludge floc as a function of the distance to its centre (a spherical floc is assumed).

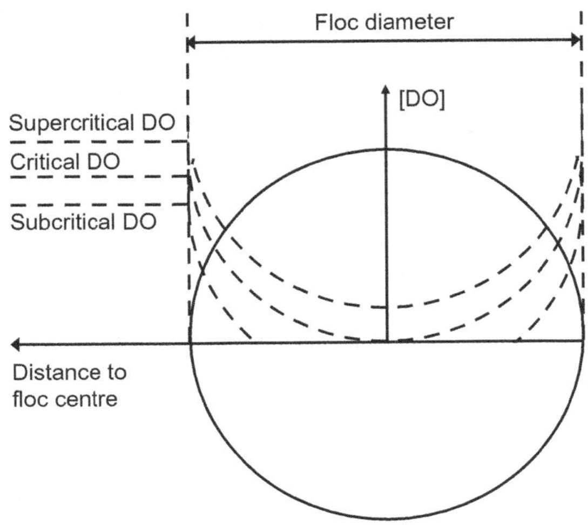

Figure 5.9 Dissolved oxygen concentration gradient as function of distance from the floc surface

Depending on the existing bulk dissolved oxygen concentration and the rates of dissolved oxygen transport and -consumption within the floc, anoxic micro regions may develop in the floc centre, where no dissolved oxygen is present and where, as a consequence, no nitrification will take place. Instead denitrification may develop.

This phenomenon is called simultaneous denitrification and is often observed in circulation systems such as the carrousel, which essentially is a completely mixed system (for all components except oxygen) in which the mixed liquor is subjected to an oxygen gradient over the length of the reactor. The minimum bulk dissolved oxygen concentration that is required to maintain the centre of the flocs in an aerobic state depends on several factors such as floc size, stirring intensity, temperature and the oxygen uptake rate. As these factors may differ significantly between different active sludge processes, the required minimum dissolved oxygen concentration will vary as well. In general a bulk dissolved oxygen concentration of $2\ mg\ O_2 \cdot l^{-1}$ is sufficient to prevent oxygen limitation in the nitrification process.

(c) Mixed liquor pH

Several authors have found approximately constant μ_m values over the pH range from 7 to 8.5. For pH values below or beyond this range, the value of μ_m decreases rapidly, as shown in Figure 5.10. In practice, many wastewaters (e.g. municipal sewage) have a pH value between 7 and 8. In the activated sludge process the pH tends to decrease, because of the consumption of alkalinity resulting from nitrification and an increase of acidity due to the production of CO_2 from the oxidation of organic matter. For this reason, unless the influent alkalinity is high, as can be the case after anaerobic pre-treatment, the mixed liquor pH will be less than 8. Hence, generally only the lower pH limit of mixed liquor is of practical importance. As discussed in Section 5.1.3, a pH value below 7 can be avoided when the mixed liquor alkalinity is maintained above a minimum value of $35\ mg \cdot l^{-1}\ CaCO_3$.

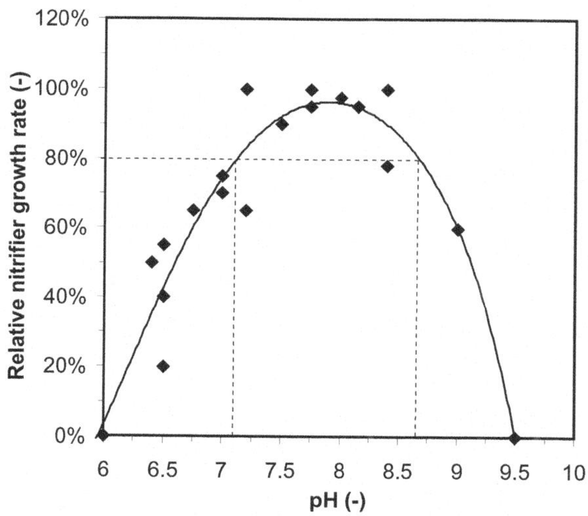

Figure 5.10 Influence of the pH on the nitrification rate. Summary graph based on the data collected by Ekama et al. (1975), Malan et al. (1966), Downing et al. (1966), Sawyer et al. (1973) and Antoniou et al. (1990)

5.2.2 Nitrification in systems with non aerated zones

In activated sludge systems designed for biological nitrogen removal, part of the reactor volume is not aerated, in order to allow for denitrification. The presence of these anoxic zones influences the nitrification efficiency, because the nitrifiers can only grow in an aerobic environment. If it is assumed that the decay of the nitrifiers is not affected by the presence or absence of dissolved oxygen, the effect of anoxic zones on nitrification can be evaluated as follows: in a steady state system the total nitrifier mass MX_n is constant and can be expressed as:

$$dMX_n/dt = 0 = (dMX_n/dt)_g + (dMX_n/dt)_d + (dMX_n/dt)_e \tag{5.42}$$

Indices g, d and e refer to growth, decay and discharge with the excess sludge respectively. Since the nitrifiers only grow in an aerobic environment one has:

$$dMX_n = (1 - f_x) \cdot V_r \cdot (dX_n/dt)_c = (1 - f_x) \cdot V_r \cdot \mu_m \cdot X_n \tag{5.43}$$

f_x = anoxic sludge mass fraction

V_r = biological reactor volume (aerobic plus anoxic zones)

By substituting Eq. (5.43) in Eq. (5.42) and using Eqs. (5.32 a to c and 5.33) one has:

$$N_a = K_n \cdot (b_n + 1/R_s)/[(1 - f_x) \cdot \mu_m - b_n - 1/R_s] \tag{5.44}$$

The expression to calculate the residual ammonium concentration in a process containing anoxic and aerobic zones (Eq. 5.44) is very similar to the one derived by Downing for the completely aerobic process (Eq. 5.36). When the two equations are compared, it can be noted that the presence of the anoxic sludge mass fraction f_x has the effect of a reduction of the μ_m value by a factor $(1 - f_x)$ i.e.:

$$\mu'_m = (1 - f_x) \cdot \mu_m \tag{5.45}$$

where μ'_m = apparent maximum nitrifier growth rate in systems with non aerated zones (d^{-1})

Figure 5.11 shows the residual ammonium concentration as a function of the anoxic sludge mass fraction for three different μ_m values: 0.2 d^{-1} (low), 0.4 d^{-1} (normal) and 0.8 d^{-1} (high). It can be noted that for each of these cases there is a maximum anoxic sludge mass fraction above which nitrification does not occur. Equation (5.44) can also be written explicitly in terms of the anoxic sludge mass fraction:

$$f_x = 1 - (1 + K_n/N_a) \cdot (b_n + 1/R_s)/\mu_m \tag{5.46}$$

When a certain nitrification efficiency is to be maintained and therefore a maximum residual ammonium concentration is specified, there is a consequential maximum to the sludge mass fraction that can be placed in an anoxic environment. This maximum anoxic mass fraction f_m can be calculated from Eq. (5.46) by substituting N_a with the specified effluent residual ammonium concentration N_{ad}:

$$f_m = 1 - (1 + K_n/N_{ad}) \cdot (b_n + 1/R_s)/\mu_m \tag{5.47}$$

The maximum anoxic sludge mass fraction not only depends on the specified residual ammonium concentration, but also on the sludge age and the kinetic constants for nitrification.

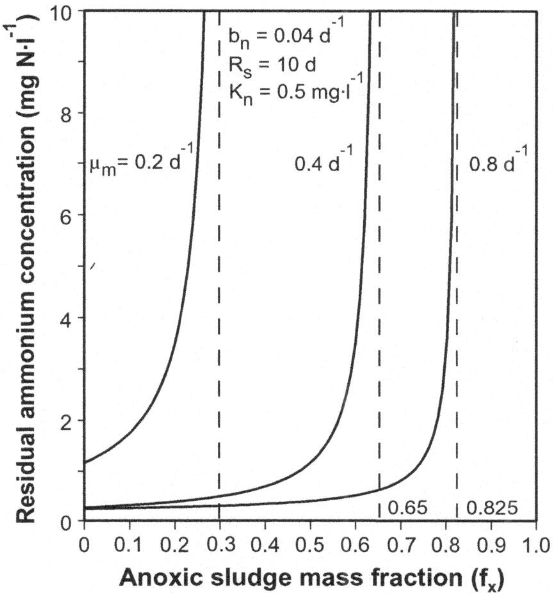

Figure 5.11 Residual ammonium concentration as a function of the anoxic sludge mass fraction for different values of μ_m

The values of K_n and b_n have relatively little influence on the value of f_m and when no information is available default values may adopted, such as:

- $K_n = 0.5 \cdot 1.123^{(T-20)}$
- $b_n = 0.04 \cdot 1.03^{(T-20)}$

In contrast, the influence of μ_m on the maximum anoxic sludge mass fraction is considerable. In Figure 5.12, f_m values are shown plotted as a function of the sludge age for μ_m values between 0.2 and 0.8 d^{-1}.

The numeric value of f_m is of great practical importance: the extent of denitrification that is possible increases as the anoxic sludge mass fraction is enlarged. Hence in principle, to maximise the nitrogen removal capacity of a system, the largest possible anoxic sludge mass fraction should be selected. However, apart from the maximum set by the need for efficient nitrification, there are other factors that may influence the value of f_m: the removal efficiency of organic matter and the sludge settleability.

When the anoxic sludge mass fraction is very large, there is the possibility that the metabolism of organic matter in the process becomes incomplete because the rate of metabolism in an anoxic environment is lower than in an aerobic environment. In such a case, the organic matter may still be removed efficiently from the liquid phase, but the sludge production will increase, because part of the stored organic matter will not be metabolised but will instead be discharged as excess sludge.

Furthermore in processes with a high anoxic sludge mass fraction, sludge settleability may be poor and development of filamentous or bulking sludge may be frequent (refer to Chapter 9), possibly because of the presence of non-metabolised organic matter in the sludge. Thus there is an upper limit to the anoxic sludge mass fraction, independent of the maximum value set by the requirements for efficient nitrification.

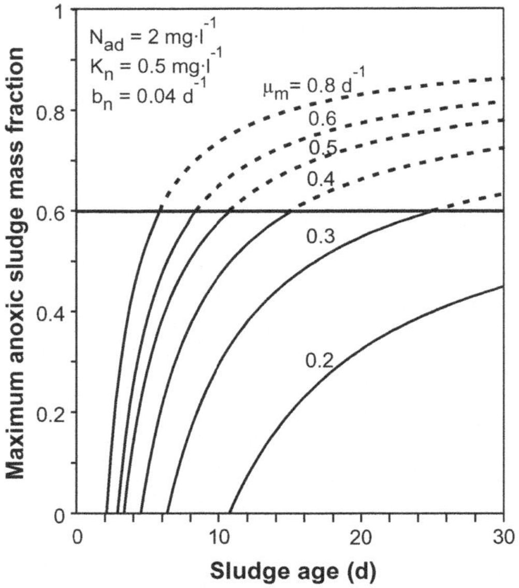

Figure 5.12 Maximum anoxic sludge mass fraction f_m as a function of the sludge age for different values of μ_m

Presently, there are full-scale plants with an anoxic sludge mass fraction of fifty percent that operate satisfactorily, but there is little information about the possibility to increase the anoxic sludge mass fraction beyond this point. In the Netherlands for example, the anoxic mass fraction in activated sludge system designed for nitrogen removal seldom exceeds forty percent. Based on the results of a pilot plant study by Arkley et al. (1982), the water research commission of South Africa (1984) suggests a maximum value of $f_m = f_{max} = 0.6$. This value is indicated in Figure 5.12 as well.

There may yet be another limitation to the value of the anoxic sludge mass fraction. As f_m increases, the volume of the aerobic reactors decreases and consequently the OUR increases. Hence, to maintain the flocs in an aerobic environment (i.e. to prevent anoxic conditions *within* the sludge floc), operation at higher bulk dissolved oxygen concentration is required. The higher dissolved oxygen concentration in turn leads to an increased energy requirement for aeration. Due to the increase of aeration costs, an increase of f_m may become unattractive from the point of view of economics.

5.2.3 Nitrification potential and nitrification capacity

The nitrification potential is defined as the TKN concentration in the influent that *can* be nitrified, i.e. is available for nitrification. This concentration can be expressed as:

$$N_p = N_{ki} - N_{oe} - N_l \tag{5.48}$$

where:

N_p = nitrification potential ($mg N \cdot l^{-1}$)
N_{ki} = influent TKN concentration ($mg N \cdot l^{-1}$)
N_l = nitrogen concentration required for sludge production ($mg N \cdot l^{-1}$)
N_{oe} = organic nitrogen in the effluent ($mg N \cdot l^{-1}$)

To reduce model complexity, the N_{oe} fraction is assumed to contain both the soluble and the particulate organic nitrogen. In reality, the particulate organic nitrogen fraction is part of the nitrogen present in the produced excess sludge N_l, in this case leaving with the effluent because of imperfect solid-liquid separation in the final settler. The consequences of this modelling decision are small but will be discussed nonetheless in Appendix 5. The value of N_l has been determined previously with Eq. (3.59). Using this expression in Eq. (5.48) one has:

$$N_p = N_{ki} - N_{oe} - f_n \cdot [(1 - f_{ns} - f_{np}) \cdot (1 + f \cdot b_h \cdot R_s) \cdot C_r/R_s + f_{np}/f_{cv}] \cdot S_{ti} \quad (5.49)$$

The nitrification capacity is defined as the influent TKN concentration that ***is*** effectively nitrified in the activated sludge process. Hence the nitrification capacity is the difference between the nitrification potential and the effluent ammonium concentration.

$$N_c = N_p - N_{ae} = N_{ki} - N_{oe} - N_l - N_{ae} \quad (5.50)$$

where N_c = nitrification capacity ($mg\ N \cdot l^{-1}$)

Using Eq. (5.44) for N_a and Eq. (3.59) for N_l, the following equation is derived:

$$\begin{aligned} N_c = {} & N_{ki} - N_{oe} - f_n \cdot [(1 - f_{ns} - f_{np}) \cdot (1 + f \cdot b_h \cdot R_s) \cdot C_r/R_s + f_{np}/f_{cv}] \cdot S_{ti} \\ & - K_n \cdot (b_n + 1/R_s)/[(1 - f_x) \cdot \mu_m - b_n - 1/R_s] \end{aligned} \quad (5.51)$$

Figure 5.13 shows the values of N_p, N_c and f_m as function of the sludge age for the following conditions:

- Composition and concentration of the influent organic matter (used to calculate N_l):
  - $f_{ns} = f_{np} = 0.1$;
  - $S_{ti} = 500\ mg\ COD \cdot l^{-1}$.
- Nitrification kinetic parameters:
  - $\mu_m = 0.3\ d^{-1}$ (Figure 5.13a) and $0.6\ d^{-1}$ (Figure 5.13b);
  - $K_n = 1.0\ mg\ N \cdot l^{-1}$ and $b_n = 0.04\ d^{-1}$;
  - $N_{ad} = 2\ mg\ N \cdot l^{-1}$ and $f_{max} = 0.6$.
- Influent nitrogen concentration:
  - $N_{ki} = N_{ti} = 50\ mg\ N \cdot l^{-1}$.

5.2.4 Design procedure for nitrification

When an activated sludge system is designed for both COD removal and nitrification, it is sized based on the requirements for nitrification, as this process will be rate-limiting. The following design procedure is recommeded:

(1) Attribute values to the kinetic parameters (μ_m, b_n and K_n)
This is done for worst case conditions, i.e. for $T = T_{min}$, where T_{min} is the lowest expected average reactor temperature, which is often equal to the lowest recorded wastewater temperature (monthly average). If the temperature dependencies of the kinetic parameters are not known, then the default ones specified in Section 5.2.1. can be used. When a conservative design is required, select a low value for the specific nitrifier growth rate μ_m, which will result in a higher value of the design sludge age.

(2) Specify the influent TKN load
The design of the activated sludge system should be based on the average daily TKN- and COD loads and not on the maximum loads. Remember that eventually a sludge mass will develop that is compatible with

average, not *maximum* COD- and TKN loads. This does not mean that these maximum loads should be ignored in the design process, as they will definitely have an impact on oxygen demand and effluent quality. In the case of nitrification systems, the correct method to create margin to handle TKN peak loads is to select conservative values for μ_m and/or N_{ad}. Alternatively the use of buffer volume could be considered, especially when the ratio between peak- and average TKN load is high. Part of the daily TKN peak load (e.g. the morning peak) is stored and treated at a later time, when the load to the system is much less. As buffer volume is much cheaper than reactor- or settler volume, the reduction in flow and load will significantly reduce the construction costs of all downstream treatment units.

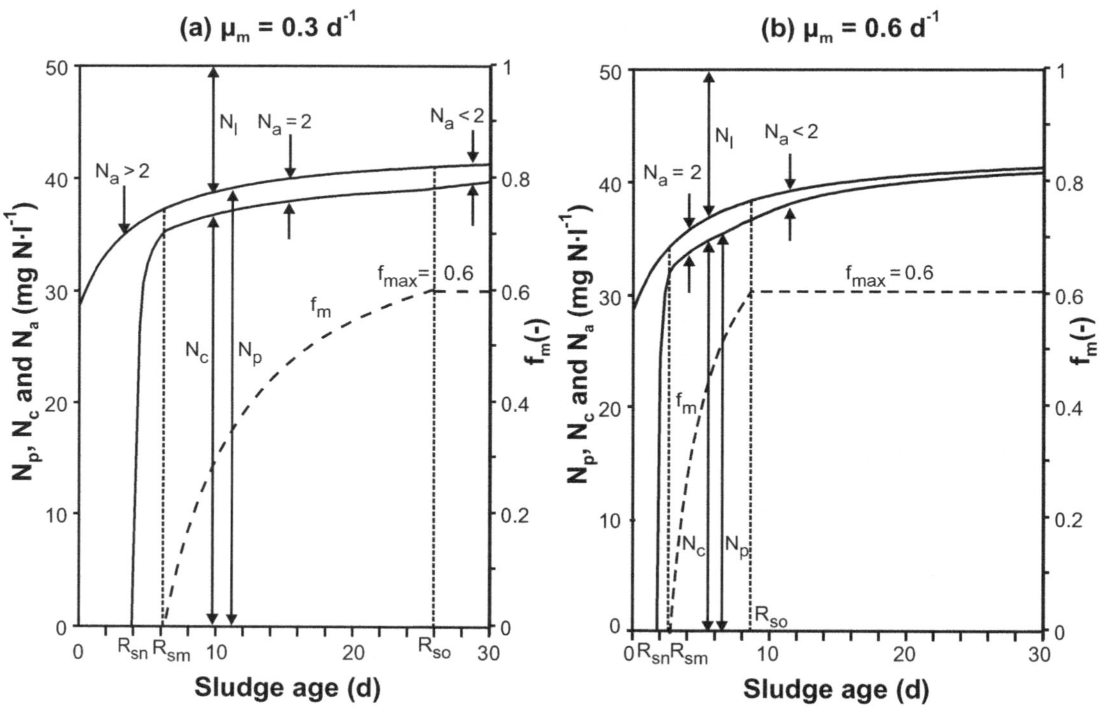

Figure 5.13 Values of N_c, N_p, N_l and N_a as function of R_s for different values of μ_m. The values of R_{sn} ($f_m = 0$), R_{sm} ($N_a = N_{ad}$) and R_{so} ($f_m = f_{max}$) are also indicated

(3) Select a reactor configuration (plug-flow or completely mixed)

The treatment performance of a plug-flow reactor is superior to that of a completely mixed reactor of equal volume. The concentration gradient that develops over the length of the reactor allows higher conversion rates at the head of the reactor while the effluent limits will still be met at the back end. For example, when a completely mixed reactor is operated at $N_{ae} = 1.0\ mg\,N \cdot l^{-1}$, then in a compartmentalized system of equal total volume (where $n = 3$, i.e. approaching plug-flow conditions) the concentation in the first reactor might be around $3\ mg\,N \cdot l^{-1}$, which decreases to $0.7\ mg\,N \cdot l^{-1}$ in the last reactor.

Effluent NH_4–N limits are typically in the same range as the K_n value, which means that even a small increase in N_a, for instance during peak loading, will already result in a significant increase on the nitrification rate. Therefore it is recommended that especially the nitrification reactor is constructed in plug-flow configuration. The beneficial effect of a plug-flow configuration on nitrification performance is more pronounced at higher temperatures and at lower values of N_{ad}. As a general indication, when the nitrification design is made for a completely mixed system, then a plugflow reactor will in general be able to handle short term increases (2–3 hrs) of TKN load of up to 40–100% without compromising effluent quality.

(4) Specify the desired residual ammonia concentratition (N_{ad})

This need not always be equal to the effluent limit. As mentioned in step (2), selection of a conservative value for N_{ad} is recommended in order to create margin to handle for peak TKN loads, as it allows the nitrification rate to be increased without directly violating the effluent ammonium discharge limit. This will be explained in Example 5.6. Also consider the effect that flow or time proportional sampling will have on the average NH_4–N concentration in the effluent: i.e. temporary peak values will be compensated by lower values later in the day.

(5) Calculate the required sludge age (R_s)

Calculate with Eq. (5.39) the minimum required *aerobic* sludge age R_{sm} (for which $N_{ae} = N_{ad}$). In principle this is the sludge age that should be selected, as design margin is already created in the selection of μ_m and N_{ad}, and also because design is based on lowest expected reactor temperature (winter). So for a large part of the time, actual system performance will be much better than the design performance.

(6) Calculate all other system parameters with the theory presented in Chapter 3

EXAMPLE 5.6

Make an indicative, conservative design for a nitrifying activated sludge system capable of meeting an effluent nitrogen limit of 1.0 mg $N \cdot l^{-1}$, based on a 24 hrs composite sample. During the morning peak flow, the TKN load will increase significantly: detailed flow- and load data as provided by the client are shown in Table 5.7. Use the following additional data:

- At the design temperature $T_{min} = 14°C$, the values of the kinetic parameters are $\mu_m = 0.2\ d^{-1}$; $b_n = 0.03\ d^{-1}$ and $K_n = 0.5$ mg $N \cdot l^{-1}$;
- Assume $MN_{le} = 15\%$ of MN_{ti};
- During the peak load period the effluent nitrogen limit may be temporary exceeded, as long as this is compensated for during the periods of lower loading;

Table 5.7 Flow and load data of Example 5.6. Peak load duration is 4 hours

Time period	Flow (m^3)	COD load (kg COD)	[COD] (mg COD $\cdot l^{-1}$)	TKN load (kg N)	[TKN] (mg N $\cdot l^{-1}$)
Daily total	24,000	12,000	500	1200	50
Daily avg hourly	1000	500	500	50	50
Peak load period	1250	ND	ND	87.5	70
Avg rest of day	950	ND	ND	42.5	45

Note: ND = not determined

Solution

In this example we have opted for a conservative design, based on a completely mixed configuration (where a plug-flow would be constructed) and the application of margin in the selection of N_{ad}, i.e. $N_{ad} = 0.5$ mg N·l^{-1} instead of 1.0 mg N·l^{-1}. For the minimum design temperature of 14°C, the value of R_{sm} is calculated with Eq. (5.39) as:

$$R_{sm} = 1/[\mu_m/(1 + K_n/N_{ad}) - b_n]$$
$$= 1/[0.3/(1 + 0.5/0.5) - 0.03] = 11.8 \text{ days}$$

At 14°C and for the average nitrogen load, the average nitrification rate MN_c is equal to $MN_{ti} - MN_l - MN_{te} = (1200 - 0.15 \cdot 1200 - 12) = 1008$ kg N·d^{-1} or 42 kg N·h^{-1} at the target ammonium effluent concentration of 0.5 mg N·l^{-1}. Based on the daily flow rate of 24,000 m^3·d^{-1} and $N_{ad} = 1$ mg N·l^{-1}, the allowable discharge of ammonium with the effluent is 24 kg N·d^{-1}. Assuming for simplicity that, when the treatment plant is receiving less than the average load, the value of N_{ae} will be equal to 0.5 mg N·l^{-1} (in practice it will be slightly less), then the maximum ammonium nitrogen discharge with the effluent that may be allowed during the hours of peak flow is equal to $24 - 20 \cdot 950 \cdot 0.0005 = 14.5$ kg N·d^{-1}, or 3.625 kg N·h^{-1}. This corresponds to a maximum (peak load) effluent ammonium concentration of $14.5 \cdot 1000/(4 \cdot 1250) = 2.9$ mg N·l^{-1}.

Table 5.8 shows the value of the Monod factor and the increase in nitrification rate resulting from operation at effluent ammonium concentrations higher than 0.5 mg N·l^{-1}. During the peak flow period the nitrogen load is increased to 87.5 kg N·h^{-1}, an increase of 75% compared to the average nitrogen load. The required nitrification capacity during peak flow is $87.5 - 0.15 \cdot 87.5 - 3.625 = 70.75$ kg N·h^{-1}.

Table 5.8 Monod factor and nitrification rate for different values of N_{ae}

N_{ae} (mg N·l^{-1})	Monod factor (–)	Increase of monod factor compared to $N_{ae} = 0.5$ mg N·l^{-1}	Nitrification capacity (kg N·h^{-1})
0.5	0.50	0%	42.0
1.0	0.67	33%	56.0
1.5	0.75	37%	57.7
2.0	0.80	60%	67.1
2.5	0.83	67%	69.9
3.0	0.86	71%	71.9

As long as the *required* nitrification capacity does not exceed the theoretical *maximum* value of $42/0.5 = 84$ kg N·h^{-1}, the effluent ammonium concentration during peak flow can be calculated from the Monod equation. The value of the Monod factor is equal to $70.75/42.0 \cdot 0.5 = 0.843 = N_a/(K_n + N_a)$. Rewriting the equation yields $N_a = N_{ae} = 0.843 \cdot 0.5/(1{-}0.843) = 2.7$ mg N·l^{-1}.

This is slightly less than the maximum allowable value of $N_{ae} = 2.9$ mg N·l^{-1}, so the solution is acceptable. However, the nitrification capacity at 2.7 mg N·l^{-1} will be slightly lower than at 2.9 mg N·l^{-1}, while simultaneously less ammonium will be discharged with the effluent. Therefore, the calculation has to be iterated and finally only a slight increase in the ammonium effluent concentration is required to reach an equilibrium at $N_{ae} = 2.74$ mg $N \cdot l^{-1}$, where $MN_{ti} - MN_c - MN_l - MN_{ae} = 87.5 - 71.0 - 13.1 - 3.4 = 0.0$ kg $N \cdot h^{-1}$.

Note that the evaluation was performed at the lowest expected reactor temperature, so in fact during the rest of the year the nitrification performance will be much better. Furthermore, in practice the flow regime in the nitrification reactor will never be completely mixed, which will reduce the expected effluent nitrogen concentration. Finally, when comparing the peak flow rate and -duration ($4 \cdot 1250$ $m^3 \cdot h^{-1} = 5000$ m^3) with the reactor size, typically 8000 to 12,000 m^3 for the design conditions of this example, it is obvious that a significant buffering effect will occur.

5.3 DENITRIFICATION

The necessary conditions for the denitrification process to develop in an activated sludge process can be summarised as:

(1) Presence of a facultative bacterial mass;
(2) Presence of nitrate and absence of DO in the mixed liquor (i.e. an anoxic environment);
(3) Suitable environmental conditions for bacterial growth;
(4) Presence of an electron donor (nitrate reductor).

(1) Presence of a facultative bacterial mass

Facultative bacteria are characterised by the fact that they can use both oxygen and nitrate as an oxidant for organic matter. A large fraction of the bacterial mass that develops in an activated sludge process is facultative. It has been established experimentally that activated sludge generated under aerobic conditions will use nitrate immediately when it is placed in an anoxic environment (Heidman, 1979). The rate of nitrate utilisation continues without change, as long as the anoxic condition and the availability of organic matter persist. However, anoxic oxidation of organic matter occurs at a lower rate than aerobic oxidation under otherwise comparable conditions.

(2) Presence of nitrate and absence of dissolved oxygen in the mixed liquor

In general, nitrogen in wastewater is present in the form of ammonium or organic nitrogen. Thus, the necessity to have nitrate present in an anoxic environment normally implies the need for nitrification as a prerequisite for denitrification. The magnitude of the nitrate concentration has little influence on the denitrification rate: when the nitrate concentration is higher than 0.5 mg $N \cdot l^{-1}$, the denitrification rate will be independent of the nitrate concentration.

The presence of dissolved oxygen in mixed liquor inhibits the development of denitrification. It is difficult to quantify this influence because concentration gradients of dissolved oxygen will develop in the flocs so that the micro-environment in a floc may be very different from the bulk of the liquid phase (see also Figure 5.9). In effect, efficient (though irregular) nitrate removal has been observed in aerobic activated sludge processes (Pasveer, 1965 and Maatsche, 1971), mainly those of the carrousel type. This can be explained only if it is accepted that anoxic micro regions are formed within the flocs. In general it has been observed that a dissolved oxygen concentration of more than 0.2 to 0.5 mg $O_2 \cdot l^{-1}$ reduces the rate of denitrification significantly.

(3) Suitable conditions for bacterial growth

Temperature and mixed liquor pH are among the most important environmental conditions for bacterial growth. The denitrification rate increases with temperature until an optimum is reached at 40°C. At temperatures above 40°C, the denitrification rate is quickly reduced due to decay of biomass. The influence of temperature on denitrification kinetics is discussed in more detail in Section 5.3.2.2.

Concerning the influence of pH, it has been established that the denitrification rate has a maximum value for the pH range between 7 and 8.5, whereas for pH values lower than 6 and higher than 8.5 there is a sharp decrease in denitrification activity. It is very unlikely that a pH $>$ 8.5 is established in an activated sludge process. On the other hand, a low pH value, e.g. pH $<$ 6 is not only inhibitory for denitrification, but also for nitrification, rendering nitrogen removal practically impossible.

For municipal wastewater, it was shown earlier in Section 5.1.3.3 that in order to maintain the pH in the optimal range of $7 < pH < 8$, a minimum alkalinity of 35 $mg \cdot l^{-1}$ $CaCO_3$ is required in the mixed liquor.

Another environmental requirement for efficient denitrification is that toxic compounds must be either absent or present at a low concentration. There is little information about the influence of specific compounds on the denitrification rate, except from the influence of the hydrogen ion mentioned above (pH). However, nitrifiers are often much more sensitive to the presence of toxic materials than the heterotrophic bacteria. Hence, in general, if nitrification is possible in an activated sludge process, so is denitrification.

(4) Presence of an electron donor

The presence of an electron donor is essential for the reduction of nitrate. The electron donor in the denitrification process is biodegradable organic matter. In accordance with the nature of organic matter two different types of denitrifying systems can be defined:

- Systems with an external carbon source. In these systems the organic matter is added to the mixed liquor after nitrification is complete. Methanol is among the most frequently used organic compounds for denitrification but other materials (ethanol, acetone and acetic acid) have been used as well;
- Systems with an internal carbon source. In this case the influent organic matter is used for the reduction of nitrate. Alternatively, the bacterial mass generated in the activated sludge process may also be used (endogenous respiration).

The choice of the type of organic matter to be used is of fundamental importance for the configuration of the denitrification system. The relationship between the source of organic matter and the system configuration will be discussed in the next section.

5.3.1 System configurations for denitrification

5.3.1.1 Denitrification with an external carbon source

Denitrification using an external source of organic matter was first implemented by Barth, Bremmer and Lewis (1969). They developed the process that is schematically represented in Figure 5.14. The system is composed of three biological reactors in series, each one having a dedicated settler. The result is the development of a different sludge in each of the reactors, hence its name: the three sludge system.

In the first reactor, which is a conventional aerobic activated sludge process operated at a short sludge age, the influent organic matter will be removed. The effluent from the first settler flows into the second reactor, also aerobic, where nitrification takes place. The sludge in this reactor is composed mainly of nitrifying bacteria. The nitrified effluent is discharged into the third reactor, operated under anoxic conditions for denitrification to take place. As the nitrified effluent is substantially free of biodegradable organic matter, this must be added to effect the reduction of nitrate. Often methanol is used because of its relatively low price and its easy handling.

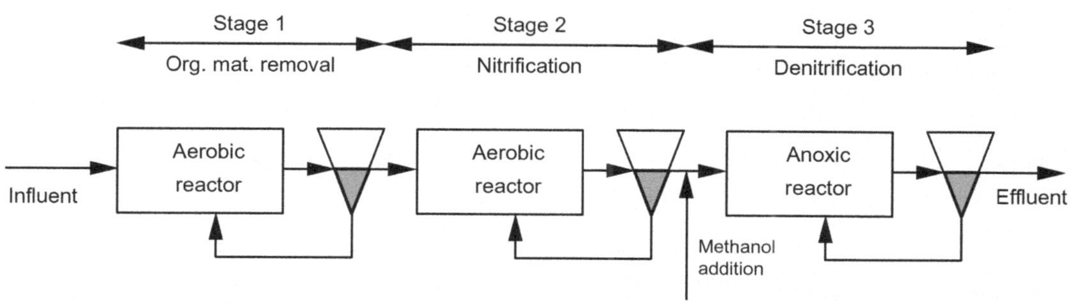

Figure 5.14 Denitrification with an external source of carbon (three-sludge system)

Three-sludge systems have been constructed and operated successfully at full scale. However, the construction and operational costs of this system is very high, not only due to the fact that three different systems must be constructed, but also because of the need to add the external electron donor. Christensen et al. (1977) calculated from full-scale data a consumption of 2.2 to 2.5 mg CH_3OH per mg denitrified nitrogen.

5.3.1.2 Denitrification with an internal carbon source

(1) Early designs

In the so-called single sludge systems, the influent organic matter is used for the biological reduction of nitrate. In these systems the same sludge is placed alternately in an aerobic environment (for nitrification) and in an anoxic environment (for denitrification). The alternation can be realised by periodically interrupting the aeration in a single reactor, as for example is done in sequencing batch reactors (SBR's). Alternatively, the reactor volume can be divided into a continuously aerated reactor and a permanently anoxic reactor, with sludge recirculating between both reactors. The latter option is more practical and has found more application in large full-scale plants. SBR reactors are often used when smaller or relatively simple systems are required (due to the fact that no final settler is required).

Wurhmann (1964) operated the first single sludge system. The Wurhmann system or post-denitrification system (Figure 5.15b) is composed of two reactors, the first one aerobic and the second anoxic. The influent enters into the first reactor, where nitrification develops, together with removal of almost all biodegradable organic material. The nitrified mixed liquor passes to the second reactor, where the sludge is kept in suspension by moderate stirring, but no aeration is applied. In this anoxic reactor – also called the post denitrification (post-D) reactor – reduction of nitrate takes place. The organic material available for the nitrate reduction is non-metabolised influent material and organic material released during the decay of active sludge in the anoxic reactor. The mixed liquor leaving the second reactor passes through a settler and is recirculated to the aerobic reactor.

The denitrification rate in the Wurhmann system is low, due to the low concentration of biodegradable organic material in the post-D reactor. If denitrification of a considerable nitrate concentration is required, it is necessary that a large fraction of the sludge is located in the anoxic reactor. However, the size of the anoxic sludge mass fraction is limited because of the requirement that nitrification (a prerequisite for denitrification) must be efficient (refer to Figure 5.12).

(2) Present designs

In the pre-D system proposed by Ludzack and Ettinger (1964) and improved by Barnard (1970), the influent organic material is the main electron donor source for denitrification. In this system, there are two reactors in series, the first one anoxic and the second aerobic. The nitrate formed in the second reactor is returned to the anoxic reactor through direct recirculation of mixed liquor from the second to the first reactor and together with the return sludge flow from the final settler (refer to Figure 5.15a).

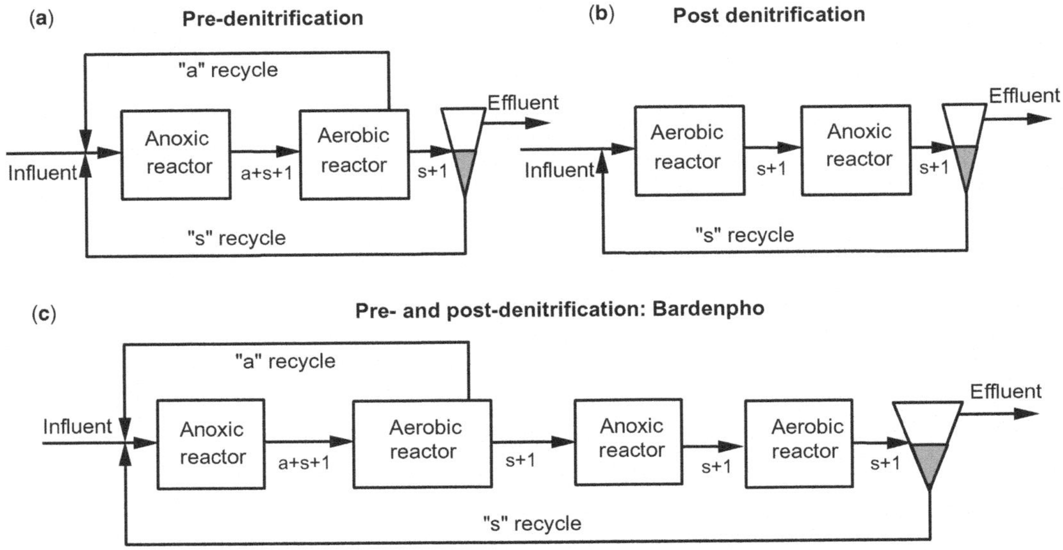

Figure 5.15 Configuration of three widely used designs for biological nitrogen removal: pre-D (a), post-D (b) and Bardenpho (c)

This system is called a pre-denitrification (pre-D) system, because the anoxic reactor is placed before the aerobic reactor. Under otherwise comparable conditions, the pre-D system has a higher denitrification rate than the post-D system, because the concentration of biodegradable organic material is much higher. However, the pre-D system has one important disadvantage: complete nitrate removal is not possible. A fraction of the nitrate generated in the aerobic reactor is discharged directly from the settler without passing through an anoxic reactor. The maximum nitrate removal efficiency of the pre-D system depends on the recirculation rates from the aerobic reactor and from the settler to the anoxic reactor. However, pre-denitrification designs are still applied, mainly when the following conditions apply:

- Complete nitrogen removal is not required;
- The COD concentration in the influent is insufficient to remove all the nitrate, i.e. the (N_{ti}/S_{ti}) ratio is unfavourable. If so, a pre-D system might in fact be the optimal configuration, assuming the addition of an external carbon source is not an option.

Barnard (1973) proposed the Bardenpho system, thus combining the advantage of the post-D system (feasibility of complete denitrification) with that of the pre-D system (high-rate denitrification).

Figure 5.15c shows the Bardenpho system. It is composed of four reactors, the second and the fourth being aerobic and the first and the third anoxic. Nitrification takes place in the second reactor.

In the Bardenpho process both pre- and post denitrification are applied. In the first reactor a large part of the nitrate is removed. The remaining nitrate is reduced in the third reactor and a mixed liquor, substantially free of nitrate, passes to a (optional) fourth reactor, from where it flows to the final settler. The function of the fourth reactor is to provide a short period of re-aeration (the fourth reactor is much smaller than the other ones).

This ensures that the sludge does not remain excessively long in an anoxic environment: without the re-aeration reactor, the sludge would be continuously in an anoxic environment from the third reactor through the settler and back to the first reactor. Re-aeration also removes nitrogen bubbles formed in the post-D reactor, which might otherwise cause problems in the final settler due to aggregation to sludge flocs, resulting in flotation of the sludge blanket. As an alternative for the fourth aerobic reactor, a cascade can be placed between the post-D reactor and the final settler, if the hydraulic profile permits this.

The feasibility to produce an effluent with a very low total nitrogen concentration has made the Bardenpho configuration a very popular design. When the single sludge system (and particularly the Bardenpho system) is compared to the three sludge system several important advantages of the former become apparent:

- In the single sludge system there is no cost for the addition of organic material. In contrast, the costs of adding organic material to the three sludge system are considerable as the following evaluation shows. For an assumed per capita contribution of nitrogen in the sewage of $10\ g\ N \cdot hab^{-1} \cdot d^{-1}$ and an estimated requirement for sludge production of $2\ g\ N \cdot hab^{-1} \cdot d^{-1}$ (i.e. twenty percent of the influent TKN), the nitrification potential is $8\ g\ N \cdot hab^{-1} \cdot d^{-1}$. If the consumption of external organic material is $2.5\ g\ CH_3OH \cdot g\ N^{-1}$ (Christensen et al., 1977), the daily per capita methanol consumption for denitrification is $2.5 \cdot 8 = 20$ g. This quantity amounts to about $10\ litre \cdot hab^{-1} \cdot year^{-1}$ with a cost comparable to that of aeration: US$ 3 to 5 per capita and per annum;
- In the single sludge system part of the oxygen used for nitrification can be recovered as "equivalent" oxygen for the oxidation of organic material. In Section 5.1.3.1, it was shown that the use of nitrate for the oxidation of organic material reduces oxygen consumption by some twenty percent. For complete denitrification, the nitrate mass to be denitrified equals $8\ g\ N \cdot hab^{-1} \cdot d^{-1}$. Knowing that 1 mg N is equivalent to 2.86 mg O_2, it can be calculated that denitrification reduces the oxygen demand by $8 \cdot 2.86 = 23\ g\ O_2 \cdot hab^{-1} \cdot d^{-1}$. If it is further assumed that the energy consumption of the aerators is $1\ Wh \cdot g^{-1}\ O_2$, the application of denitrification reduces the required power by $23\ Wh \cdot hab^{-1} \cdot d^{-1}$ or $23/24 = 1\ W \cdot hab^{-1}$ The reduction of $1\ W \cdot hab^{-1}$ in power consumption is very significant in economic terms, because aeration is the largest item of the operational costs for wastewater treatment plants. On an annual basis the reduction of energy consumption amounts to $8.7\ kW \cdot hab^{-1}$, which at an assumed price of $0.10\ US\$ \cdot kWh^{-1}$ results in a cost reduction of almost US$ 1 per capita · $year^{-1}$;
- In the single sludge system, the alkalinity produced during denitrification can be used in the process. In Section 5.1.3.2, it was demonstrated that in the activated sludge process there is an alkalinity consumption of $7.14\ mg\ CaCO_3 \cdot mg\ N^{-1}$ in the nitrification process and a production of 3.57 mg $CaCO_3 \cdot mg\ N^{-1}$ during the denitrification process. Hence in single sludge systems half of the alkalinity consumed during nitrification can be recovered when denitrification is complete.
- In the three-sludge system, nitrification and denitrification develop sequentially in the second and the third part of the system respectively. Thus, the recovery of alkalinity by denitrification in the last part of the system cannot be used to balance the consumption of alkalinity due to nitrification in the second part. For this reason, in the three-sludge system there is usually a need for alkalinity addition (e.g.

lime), whereas the alkalinity of most municipal wastewaters is high enough to operate a nitrogen removing single sludge system without alkalinity addition;

- In the three-sludge process, it is very difficult to match the dosage of organic material with the nitrate concentration so that neither organic material nor nitrate are present in the final effluent. In practice it will be required that a small aerobic reactor is added after the third reactor, where excess organic material is removed biologically, thereby further complicating the already complex configuration of the three sludge system;
- For the biological excess removal of phosphorus it is necessary to create a truly anaerobic zone, characterised by the absence of both dissolved oxygen and nitrate. Such an anaerobic reactor is only feasible in a single sludge system with a pre-D reactor. Thus in the three sludge system biological phosphorus removal is not possible, which reduces its applicability in practice.

There is one advantage that the three-sludge process may have compared with the single sludge system: in a single sludge system nitrification occurs in the aerobic part of the system. In a system with a large anoxic sludge fraction (which in practice will usually be required), the sludge age needs to be relatively high and hence a large treatment system is required. Thus it is possible that the reactor volume of the single sludge process is larger than the volume of the three reactors of the three sludge system together. However, this possible advantage will certainly not compensate for the very serious disadvantages inherent to the three sludge system as discussed above. For that reason, only the single sludge system will be considered further.

5.3.2 Denitrification kinetics

Marais and his group of research workers at the University of Cape Town developed an empirical model for the kinetics of denitrification. This model is an extension to the model for the removal of organic material presented in the previous chapter.

5.3.2.1 Sludge production in anoxic/aerobic systems

When the data published on sludge production in aerobic/anoxic systems is compared, it can be concluded that sludge production is not affected by the presence of anoxic zones and is equivalent to that of a purely aerobic system. In this context the experimental data collected by Sutton et al. (1979), presented in Figure 5.16, are possibly the most illustrative.

The organic sludge mass per unit mass of daily applied COD (mX_v) is plotted as function of the sludge age for different anoxic sludge mass fractions, temperatures and sludge ages (both pre-D and post-D systems). There is a close correlation between the experimental data and the theoretical curves of mX_v, which have been drawn using Eq. (3.48), derived in Chapter 3. for completely aerobic systems:

$$mX_v = (1 - f_{ns} - f_{np}) \cdot (1 + f \cdot b_h \cdot R_s) \cdot C_r + f_{np} \cdot R_s/f_{cv} \tag{3.48}$$

In the example presented in Figure 5.16, the closest correlation between the data of Sutton et al. (1979) and theory is obtained for $f_{ns} = 0.11$ and $f_{np} = 0.25$. The correlation between experimental data and theory is close over a wide range of temperatures (7 to 26°C), sludge ages (3 to 35 days) and anoxic sludge mass fractions ($0.00 < f_x < 0.82$). On the basis of the data by Sutton et al. (1979) and others, it is concluded that all the parameters and constants that determine the sludge production in an aerobic activated sludge process can be applied unchanged in processes with anoxic zones, i.e.: $Y = 0.45$ mg VSS $\cdot$ mg^{-1} COD; $f = 0.2$; $f_{cv} = 1.5$ mg COD $\cdot$ mg^{-1} VSS and $b_h = 0.24 \cdot 1.04^{(T-20)}$ d^{-1}.

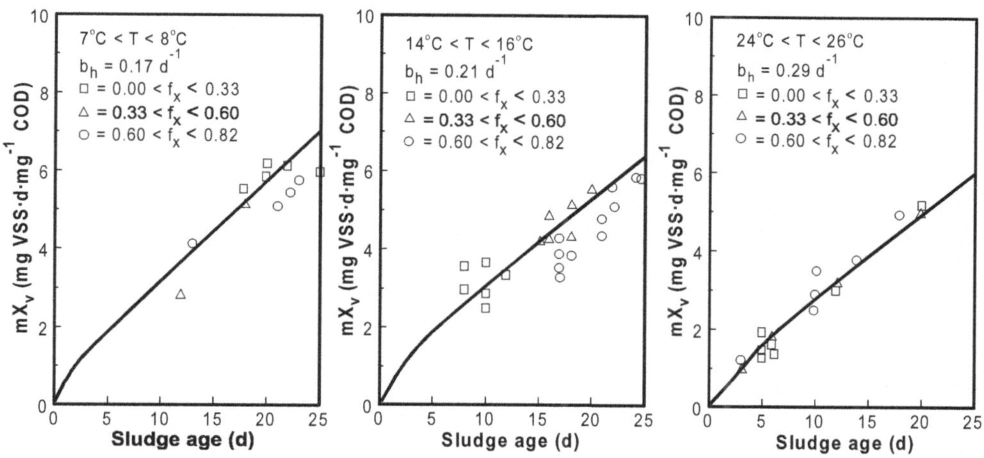

Figure 5.16 Theoretical and experimental values of the organic sludge mass production per unit mass of daily applied COD (mX_v)

5.3.2.2 Denitrification rates

Denitrification rates can be conveniently determined in an anoxic plug flow reactor. A true plug-flow reactor is characterised by the fact that no back-mixing occurs: the mixed liquor flows as a "piston" from the inlet to the outlet of the reactor. In Fig. 4.15 the experimental set up of a system with an anoxic plug flow reactor is shown. The retention time in the anoxic reactor increases proportionally with its volume (length). Hence by withdrawing samples at different points, it is possible to obtain a nitrate concentration profile in the anoxic reactor as a function of the contact time. The denitrification rate at any moment is given by the gradient of the nitrate concentration profile. Typical nitrate concentration profiles as observed in pre-D and post-D reactors are presented in Figure 5.18. The decrease of the nitrate concentration tends to be linear with time. This indicates that nitrate removal is a zero order process with respect to the concentration of nitrate. The nitrate concentration profile in a pre-D reactor indicates that two phases can be distinguished:

- A primary phase with a short duration (a few minutes) with a high denitrification rate;
- A secondary phase during the remaining anoxic retention time, with a constant but lower denitrification rate.

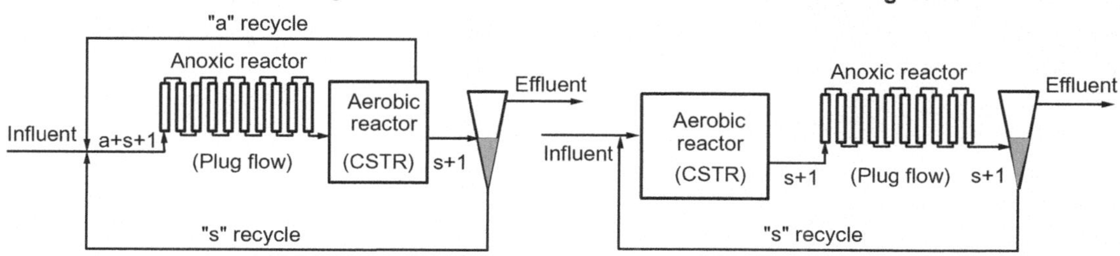

Figure 5.17 Schematic representation of the experimental set-up for the determination of the denitrification kinetics

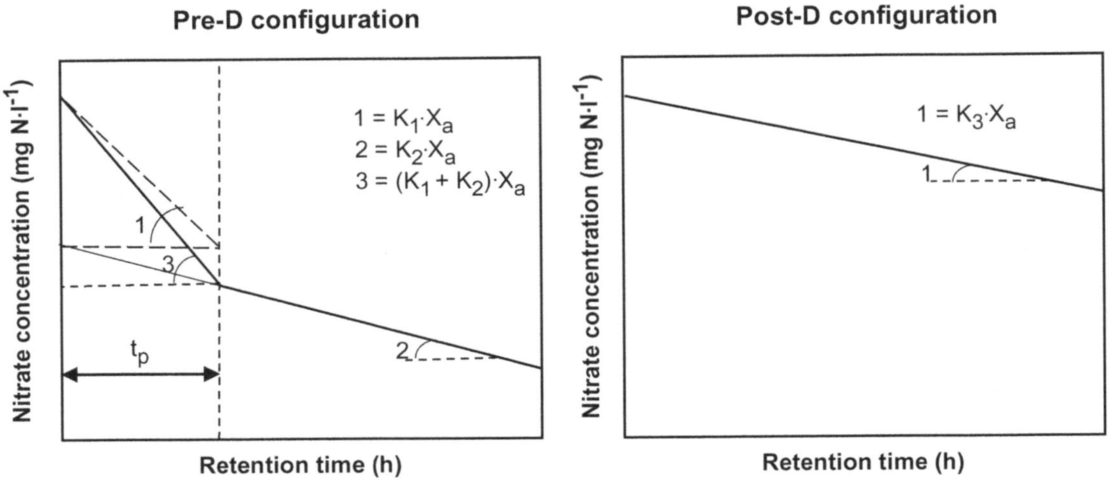

Figure 5.18 Nitrate concentration profiles observed in anoxic plug-flow reactors for pre-D and post-D configurations

In the post-D reactor there is also a linear profile of the nitrate concentration as a function of retention time, but the denitrification rate in the post-D reactor is always smaller than in the secondary phase of a pre-D reactor. From the data obtained with plug flow reactors by Stern et al. (1974), Wilson et al. (1976) and Marsden et al. (1974), it has been established that the denitrification rate is proportional to the *active* sludge concentration and can be expressed as:

$$r_d = (dN/dt) = -K \cdot X_a \tag{5.52}$$

$K =$ denitrification constant ($mg\ N \cdot mg^{-1}\ X_a \cdot d^{-1}$)

The denitrification behaviour in the pre-D reactor can be described using two constants: K_1 for the primary phase and K_2 for the secondary phase. It can be imagined that during the primary phase two denitrification processes develop simultaneously and that only one of these two continues during the secondary phase as indicated by the interrupted lines in Figure 5.18. In that case one would have: $K = K_1 + K_2$ in the primary phase and $K = K_2$ in the secondary phase. Van Haandel et al. (1981) showed that the high value of the denitrification rate during the primary phase is associated with the simultaneous utilisation of both easily and slowly biodegradable material. In the secondary phase the easily biodegradable material is depleted and the denitrification rate is only due to the utilisation of slowly biodegradable material. The denitrification rate can be written as:

$$r_d = dN/dt = r_{ds} + r_{dp} = (K_1 + K_2) \cdot X_a \quad (t < t_p) \quad \text{and} \tag{5.53a}$$

$$r_d = dN/dt = r_{dp} = K_2 \cdot X_a \quad (t > t_p) \tag{5.53b}$$

where:

r_{ds} = denitrification rate associated with the utilisation of easily biodegradable material
r_{dp} = denitrification rate associated with the utilisation of slowly biodegradable material

K_1 = denitrification constant for easily biodegradable organic material ($mg\ N \cdot mg^{-1}\ X_a \cdot d^{-1}$)
K_2 = denitrification constant for slowly biodegradable organic material ($mg\ N \cdot mg^{-1}\ X_a \cdot d^{-1}$)
t_p = duration of the primary phase (d)

In the post-D reactor, denitrification is only associated with the utilisation of slowly biodegradable material and endogenous respiration. Since the concentration of this slowly biodegradable material will be smaller in a post-D reactor than in a corresponding pre-D reactor, the denitrification rate will also be lower. The rate of nitrate removal in a post-D reactor can be expressed as:

$$r_d = dN/dt = K_3 \cdot X_a \tag{5.54}$$

The kinetic expressions for denitrification in Eqs. (5.53 and 5.54) are all zero order equations: in a steady state situation, the denitrification rate does not change with time as the active biomass can be considered to be constant. Therefore the above expressions can be used to calculate nitrate removal in anoxic reactors, independent of its hydraulic regime, and may also be applied to completely and partially mixed reactors.
Van Haandel et al. (1981) calculated the values of the denitrification rate constants K_1, K_2 and K_3 from the experimental results obtained by several authors, all using municipal wastewater as influent. From the data obtained by Stern et al. (1974), Wilson et al. (1976), Marsden et al. (1974), Van Haandel et al. (1981), Nichols (1981) in South Africa; Sutton et al. (1969) in Canada; Heide (1975) in the Netherlands and Heidman (1979) in the United States, the following average values were calculated for the range of temperatures from 12 to 26°C:

$$K_1 = 0.72 \cdot 1.2^{(T-20)} \tag{5.55a}$$

$$K_2 = 0.10 \cdot 1.08^{(T-20)} \tag{5.55b}$$

$$K_3 = 0.08 \cdot 1.03^{(T-20)} \tag{5.55c}$$

Unpublished research, using municipal wastewater from Campina Grande (Brazil), shows that the formulas in Eq. (5.55) remain valid for temperatures up to 28°C. In all cases the data were obtained with wastewaters containing only minor industrial contributions. Ekama et al. (2008) demonstrated that the denitrification rates determined above for municipal wastewater are indeed comparable with the kinetic expressions for the anoxic growth of heterotrophic bacteria as used in the Activated Sludge Models No.1 to 3 (Henze et al., 1994 to 1998). However, it is quite possible that in wastewaters with a significant or predominant industrial contribution the constants have different values due to a different composition of the influent organic material or the presence of toxic materials.

5.3.2.3 Minimum anoxic mass fraction in the pre-D reactor

In the previous section it was shown that the denitrification rate in the pre-D reactor is high, as long as easily biodegradable organic material is present. As the objective of the anoxic reactors is to remove nitrate, it is important that the denitrification rate is kept as high as possible. Therefore it is necessary that the retention time in the pre-D reactor is sufficiently long to guarantee complete utilisation of the easily biodegradable material. To determine the minimum retention time, the removal rate of the easily biodegradable material is compared with the feeding rate to the pre-D reactor. The feeding rate of easily biodegradable material can be expressed as:

$$r_{sbs} = Q_i \cdot S_{bsi}/V_1 = S_{bsi}/R_1 = f_{sb} \cdot S_{bi}/R_1 \tag{5.56}$$

where:

r_{sbs} = feeding rate of easily biodegradable material to the pre-D reactor ($mg\ N \cdot l^{-1} \cdot d^{-1}$)
R_{h1} = hydraulic retention time in the pre-D reactor = V_1/Q_i (d)
V_1 = volume of the pre-D reactor

The utilisation rate of easily biodegradable material is proportional to the associated denitrification rate $r_{ds} = K_1 \cdot X_a$ (Eq. 5.53). In the process of utilisation, a fraction of $(1 - f_{cv} \cdot Y)$ is oxidised. As stoichiometrically 1 mg NO_3-N equals 2.86 mg O_2, the utilisation rate of easily biodegradable material can be expressed as:

$$r_{ds} = (1 - f_{cv} \cdot Y)/2.86 \cdot r_{us} = f_{dn} \cdot r_{us} \tag{5.57}$$

where:

$f_{dn} = (1 - f_{cv} \cdot Y)/2.86$ = denitrification constant, which has a value of 0.114 if the default values of f_{cv} and Y are accepted
r_{ds} = denitrification rate due to the utilisation of easily biodegradable material
r_{us} = utilisation rate of easily biodegradable material

Now the minimum required retention time in the pre-D reactor to remove the easily biodegradable organic material can be calculated by the following condition:

$$r_{sbs} = r_{us} \tag{5.58}$$

Using Eqs. (5.56 and 5.57) in Eq. (5.58) one has:

$$f_{sb} \cdot S_{bi}/R_{min} = r_{ds}/f_{dn} = K_1 \cdot X_a/f_{dn} \tag{5.59}$$

R_{min} = minimum retention time required for complete utilisation of the easily biodegradable material in the pre-D reactor (d)
Substituting for X_a from Eq. (3.29) and rearranging:

$$R_{min}/R_h = f_{dn} \cdot f_{sb}/(K_1 \cdot C_r) \tag{5.60}$$

The minimum retention time R_{min} is associated to a minimum sludge mass fraction in the pre-D reactor. Since $R_{min} = V_{min}/Q_i$ one has:

$$R_{min}/R_h = (V_{min}/Q_i)/(V_r/Q_i) = V_{min}/V_r = f_{min} \quad \text{or} \quad f_{min} = f_{dn} \cdot f_{sb}/(K_1 \cdot C_r) \tag{5.61}$$

where:

V_{min} = minimum pre-D reactor volume required for complete utilisation of easily biodegradable material
f_{min} = minimum anoxic sludge mass fraction in the pre-D reactor

As can be observed in Figure 5.19, for "normal" values of f_{sb}, K_1 and b_h, the minimum fraction f_{min} is always very small. The value of f_{min} decreases at increasing temperature and sludge age:

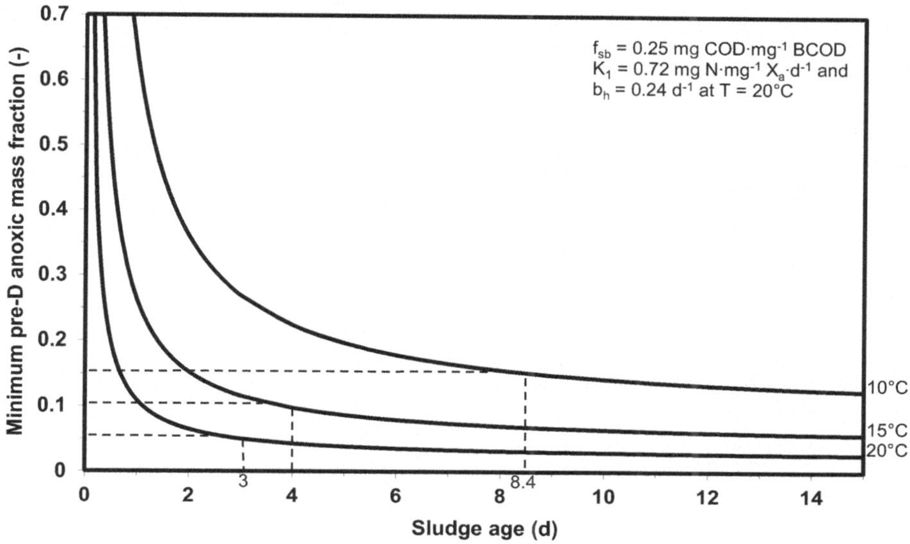

Figure 5.19 Minimum anoxic sludge mass fraction (pre-D zone) required for full utilization of easily biodegradable COD for denitrification, at different temperatures

- For T = 10°C and R_s = 8.4 days, f_{min} = 0.15;
- For T = 15°C and R_s = 4.0 days, f_{min} = 0.10;
- For T = 20°C and R_s = 3.0 days, f_{min} = 0.05.

Typically, at the temperatures indicated, the sludge age would have to be much higher to allow for nitrogen removal. Furthermore, the anoxic sludge mass fraction in the pre-D zone of a full-scale activated sludge system will invariably be much larger than f_{min}. Therefore it can be concluded that the utilisation of easily biodegradable material can be considered complete in the pre-D reactor, provided that sufficient nitrate is available.

5.3.3 Denitrification capacity

In practice, the most important parameter in a nitrogen removing activated sludge system is the amount of nitrate that can be removed per litre of influent. This parameter is called the denitrification capacity and is determined from Eqs. (5.53, 5.54 and 5.61) as shown below.

5.3.3.1 Denitrification capacity in a pre-D reactor

If the volume of a pre-D reactor is insufficient for complete removal of the easily biodegradable material, the removed nitrate mass can be expressed as:

$$\begin{aligned} MN_d &= r_d \cdot V_1 \\ &= (K_1 + K_2) \cdot X_a \cdot V_1 \quad (V_1 < V_{min}) \end{aligned} \tag{5.62}$$

where:

MN_d = mass of removed nitrate per time unit
V_1 = pre-D reactor volume

Knowing that the volume of influent entering into the pre-D reactor per time unit is equal to the influent flow Q_i, the removed nitrate concentration per litre of influent is given as:

$$\begin{aligned} D_{c1} &= MN_d/Q_i \\ &= (K_1 + K_2) \cdot X_a \cdot V_1/Q_i \end{aligned} \tag{5.63}$$

where D_{c1} = denitrification capacity in the pre-D reactor ($V_1 < V_{min}$)
Substituting for X_a from Eq. (3.29) one has:

$$\begin{aligned} D_{c1} &= (K_1 + K_2) \cdot C_r \cdot S_{bi} \cdot V_1/V_r \\ &= (K_1 + K_2) \cdot C_r \cdot f_{x1} \cdot S_{bi} (f_{x1} < f_{min}) \end{aligned} \tag{5.64}$$

where f_{x1} = sludge mass fraction in the pre-D reactor

If the retention time in the pre-D reactor is sufficient for complete removal of the easily biodegradable material and if enough nitrate is available, the denitrification capacity can be calculated by considering separately the denitrification due to both easily biodegradable and slowly biodegradable material. In so far as the easily biodegradable material is concerned, the stoichiometric relationship from Eq. (5.57) can be used.

$$\begin{aligned} MN_{ds} &= (1 - f_{cv} \cdot Y)/2.86 \cdot MS_{bsi} = f_{dn} \cdot f_{sb} \cdot Q_i \cdot S_{bi} \text{ or} \\ N_{ds} &= D_{c1s} = MN_{ds}/Q_i = f_{dn} \cdot f_{sb} \cdot S_{bi} \end{aligned} \tag{5.65}$$

where:

MN_{ds} = removed nitrate mass per time unit, associated to the utilisation of easily biodegradable material (MS_{bsi})
N_{ds} = nitrate removal in mg N per litre of influent through utilisation of S_{bsi}
D_{c1s} = denitrification capacity in the pre-D reactor per litre of influent using S_{bsi}

The mass of removed nitrate per unit time due to the utilisation of slowly biodegradable material in a pre-D reactor can be calculated as:

$$MN_{dp} = K_2 \cdot X_a \cdot V_1 \tag{5.66}$$

Now, using the same procedure as above, the removed nitrate concentration due to the utilisation of slowly biodegradable material in mg $N \cdot l^{-1}$ influent (N_{dp}) is:

$$N_{dp} = D_{c1p} = K_2 \cdot C_r \cdot f_{x1} \cdot S_{bi} \tag{5.67}$$

The denitrification capacity of the pre-D reactor is the sum of the values of D_{c1s} and D_{c1p}. From Eqs. (5.65 and 5.67) one has:

$$D_{c1} = D_{c1s} + D_{c1p}(=N_{ds} + N_{dp}) \\ = (f_{dn} \cdot f_{sb} + K_2 \cdot C_r \cdot f_{x1}) \cdot S_{bi} \text{ for } f_{x1} > f_{min} \quad (5.68)$$

5.3.3.2 Denitrification capacity in a post-D reactor

$$D_{c3} = K_3 \cdot C_r \cdot f_{x3} \cdot S_{bi} \quad (5.69)$$

Where:

D_{c3} = denitrification capacity of a post-D reactor (mg N · l^{-1} influent)
f_{x3} = sludge mass fraction in the post-D reactor

In Figure 5.20 the denitrification capacities of a pre-D and a post-D reactor (D_{c1} and D_{c3}) are plotted as a function of the anoxic sludge mass fraction for a sludge age of 10 days and under the following conditions: S_{bi} = 400 mg COD · l^{-1}; T = 20°C; f_{sb} = 0.24. The ratio D_c/S_{bi} is also indicated (on the right hand scale).

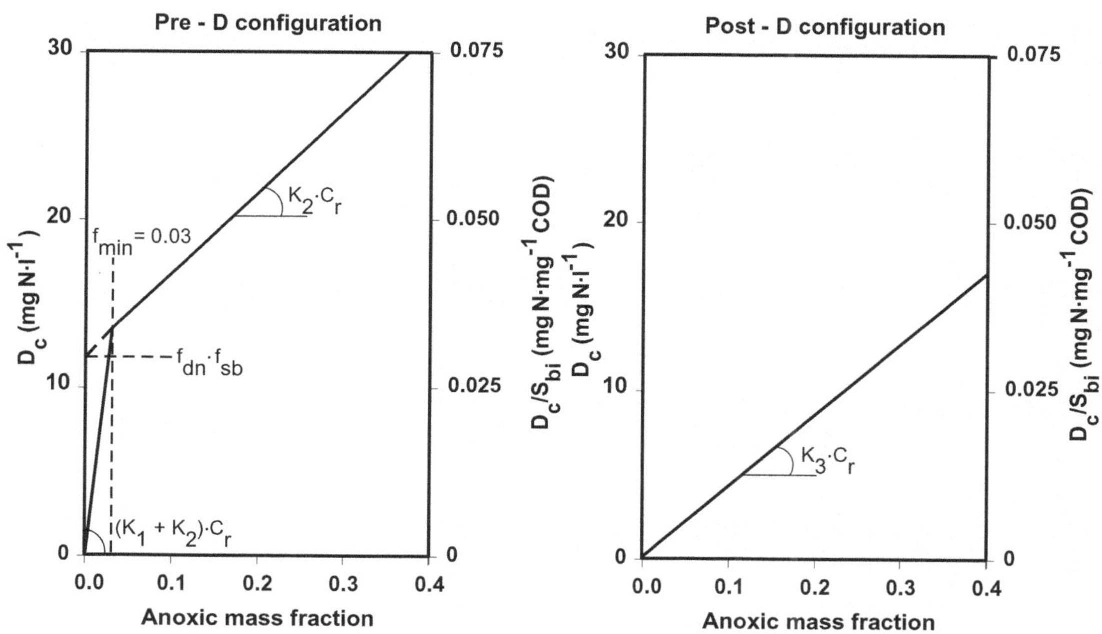

Figure 5.20 Denitrification capacity as a function of the anoxic sludge mass fraction for a sludge age of 10 days in a pre-D and a post-D anoxic reactor

It can be observed that the denitrification capacity depends on the following factors:

- Concentration- and composition of the influent organic material, i.e. S_{ti} and the values of the fractions f_{ns}, f_{np} and f_{sb};
- Sludge age: the value of $C_r = Y \cdot R_s/(1 + b_h \cdot R_s)$ increases at higher sludge age and thus the value of D_c will be higher as well;
- Temperature: the values of the denitrification rate constants K_2 and K_3 increase at higher temperatures, resulting in an increase of D_c. On the other hand, the value of the decay constant b_h will be higher as well, which reduces the overall temperature effect;
- Size of the anoxic sludge mass fractions: when f_{x1} and f_{x3} increase in size, so do the denitrification capacities D_{c1} and D_{c3}. In practice, the values of f_{x1} and f_{x3} are limited by the requirement to maintain efficient nitrification and good sludge settleability.

EXAMPLE 5.7

Determine the denitrification capacity of the activated sludge process of Example 5.1, assuming $f_{sb} = 0.20$. Verify if the experimentally observed nitrate removal corresponds to the calculated denitrification capacity.

Solution

The composition of the organic material can be calculated from the influent and effluent COD concentrations and the concentration of volatile sludge X_v in Table 5.1. With $S_{ti} = 477$ mg COD $\cdot$ l^{-1}, $S_{te} = 18$ mg COD $\cdot$ l^{-1} and $X_v = 2469$ mg VSS $\cdot$ l^{-1}, the following values are calculated:

$$f_{ns} = S_{te}/S_{ti} = 18/477 = 0.04$$

$$mX_v = MX_v/MS_{ti} = V_r \cdot X_v/(Q_i \cdot S_{ti})$$

$$= 25 \cdot 2469/(40 \cdot 477) = 3.24 \text{ mg VSS} \cdot \text{d} \cdot \text{mg}^{-1} \text{ COD}$$

For the applied sludge age $R_s = 18$ days and a temperature of 21.6°C, the values of C_r and b_h are calculated as:

$$b_h = 0.24 \cdot 1.04^{(21.6-20)} = 0.26 \text{ d}^{-1}$$

$$C_r = Y \cdot R_s/(1 + b_h \cdot R_s) = 0.45 \cdot 18/(1 + 0.26 \cdot 18) = 1.45 \text{ mg VSS} \cdot \text{d} \cdot \text{mg}^{-1} \text{ COD}$$

Now, equating the previously calculated value of mX_v to Eq. (3.48), the value of f_{np} can be calculated (as it is the only unknown parameter):

$$mX_v = (1 - f_{ns} - f_{np}) \cdot (1 + f \cdot b_h \cdot R_s) \cdot C_r + f_{np} \cdot R_s/f_{cv}$$

$$= (1 - 0.04 - f_{np}) \cdot (1 + 0.2 \cdot 0.26 \cdot 18) \cdot 1.45 + f_{np} \cdot 18/1.5)$$

$$= (0.96 - f_{np}) \cdot 2.78 + f_{np} \cdot 0.12, \text{ or } f_{np} = 0.062$$

As the total non-biodegradable COD fraction is now known, the biodegradable COD concentration is calculated as:

$$S_{bs} = (1 - f_{ns} - f_{np}) \cdot S_{ti} = (1 - 0.04 - 0.062) \cdot 477 = 429\,\text{mg COD} \cdot l^{-1}$$

$$S_{bsi} = f_{sb} \cdot S_{bi} = 86\text{mg COD} \cdot l^{-1}$$

The values of the denitrification rate constants in the pre-D zone are calculated with Eq. (5.55).

$$K_1 = 0.72 \cdot 1.2^{(T-20)} = 0.72 \cdot 1.2^{(1.6)} = 0.964\,\text{mg N} \cdot \text{mg}^{-1}\,X_a.d^{-1}$$

$$K_2 = 0.1 \cdot 1.08^{(T-20)} = 0.1 \cdot 1.08^{(1.6)} = 0.113\,\text{mg N} \cdot \text{mg}^{-1}\,X_a.d^{-1}$$

The anoxic sludge mass fraction $f_{x1} = V_1/V_r = 5/25 = 0.2$. This is much larger than the minimum anoxic sludge mass fraction required for the removal of easily biodegradable organic material:

$$f_{min} = f_{dn} \cdot f_{sb}/(K_1 \cdot C_r) = 0.114 \cdot 0.25/(0.964 \cdot 1.45) = 0.016$$

Hence, as $f_{x1} > f_{min}$, Eq. (5.68) can be applied:

$$D_{c1} = (f_{dn} \cdot f_{sb} + K_2 \cdot C_r \cdot f_{x1}) \cdot S_{bi}$$
$$= (0.114 \cdot 0.20 + 0.113 \cdot 1.45 \cdot 0.2) \cdot (1 - 0.04 - 0.062) \cdot 477 = 23.8\,\text{mg N} \cdot l^{-1}$$

In Example 5.1 the daily removed nitrate mass in the pre-D zone was calculated as 864 $g\,N \cdot d^{-1}$. As the influent flow is 40 $m^3 \cdot d^{-1}$, the experimentally observed nitrate removal is $864/40 = 21.6$ mg $N \cdot l^{-1}$. This value corresponds to 91% of the model calculated value of $D_{c1} = 23.8$ mg $N \cdot l^{-1}$.

EXAMPLE 5.8

Continuing with Example 5.7, estimate the denitrification capacity for the following two cases:

- The last two reactors are anoxic (post-D configuration);
- The first and fourth reactor are anoxic (Bardenpho configuration).

Solution

Calculate the denitrification rate constant for post-denitrification:

$$K_3 = 0.08 \cdot 1.03^{(T-20)} = 0.08 \cdot 1.03^{(1.6)} = 0.084\,\text{mg N} \cdot \text{mg}^{-1} X_a \cdot d^{-1}$$

The denitrification capacity in the post-D configuration is equal to:

$$D_{c3} = K_3 \cdot C_r \cdot f_{x3} \cdot S_{bi} = 0.084 \cdot 1.45 \cdot 0.4 \cdot 429 = 20.8\,\text{mg N} \cdot l^{-1}$$

For the Bardenpho configuration ($f_{x1} = f_{x3} = 0.2$):

$$D_c = D_{c1} + D_{c3} = 23.8 + 20.8/2 = 34.2 \text{ mg N} \cdot l^{-1}$$

When the denitrification capacity in the post-D reactor ($20.8/2 = 10.4$ mg N $\cdot$ l^{-1} per reactor) is compared with the value calculated in the pre-D reactor in Example 5.7 (23.8 mg N $\cdot$ l^{-1}), it is concluded that under the specified conditions the pre-D reactor removes more than twice the amount of nitrate of the post-D reactor. It is interesting to compare the denitrification capacity of the Bardenpho configuration with the nitrification capacity. N_c is calculated with Eq. (5.50):

$$N_c = N_{ti} + N_{ni} - N_l - N_{te} = 45.1 + 0.3 - 343/40 - 1.9 = 34.9 \text{ mg N} \cdot l^{-1}$$

In the Bardenpho configuration, the denitrification capacity (34.2 mg N $\cdot$ l^{-1}) is marginally smaller than the nitrification capacity (34.9 mg N $\cdot$ l^{-1}). Thus in principle it is possible to produce an effluent with a very low nitrate concentration. However, to do so, it will be required to introduce a recirculation flow from the aerobic- to the pre-D reactor.

5.3.4 Available nitrate

In the previous sections two important parameters defining the nitrogen removal capacity of an activated sludge system have been introduced: i.e. the nitrification- and denitrification capacities. Complete nitrogen removal is only feasible when the denitrification capacity is larger or at least equal to the nitrification capacity. However, a second condition is that the supply of nitrate to the pre-D zone is matched with the nitrate removal rate, i.e. nitrate should be supplied only where sufficient denitrification capacity is available to remove it. To have a large pre-D denitrification capacity without any nitrate fed to it does not bring any advantages and likewise it does not make much sense to have a large nitrate recycle when the pre-D zone is already overloaded. To optimise the design of the nitrogen removal process, it is convenient to introduce a new parameter: available nitrate (N_{av}).

(a) Available nitrate in a pre-D configuration

As could be observed in Figure 5.16, in a pre-D configuration complete denitrification is impossible because it is impossible to return all the nitrate formed in the nitrification zone to the pre-D zone. Hence, assuming that the extent of denitrification occurring in the final settler is limited and can be ignored, the available nitrate in the pre-D zone is equal to:

$$N_{av1} = (a + s)/(a + s + 1) \cdot N_c \tag{5.70}$$

where factors "a" and "s" are defined as in Figure 5.16. The effluent nitrate concentration depends on the fact whether or not the pre-D zone is under- or overloaded with nitrate:

$$N_{ne} = 1/(a + s + 1) \cdot N_c \text{ for } D_{c1} \geq N_{av1} \text{ (under loaded pre-D zone)} \tag{5.71}$$

$$N_{ne} = N_c - D_{c1} \text{ for } D_{c1} \leq N_{av1} \text{ (overloaded pre-D zone)} \tag{5.72}$$

(b) Available nitrate in a Bardenpho configuration

In the Bardenpho configuration a new parameter is introduced: available nitrate in the post-D zone (N_{av3}). Furthermore, the return of nitrate to the pre-D zone is reduced in comparison to the pre-D configuration, as the nitrate that otherwise would have been present in the return sludge stream is now partially or even completely removed in the post-D zone. So, the value of N_{av1} is now defined as:

$$N_{av1} = a/(a + s + 1) \cdot N_c \text{ (complete denitrification)} \quad (5.73)$$

$$N_{av1} = a/(a + s + 1) \cdot N_c + s \cdot N_{ne} \text{ (incomplete denitrification)} \quad (5.74)$$

The value of N_{av3}, the available nitrate in the post-D zone, depends on whether or not the pre-D zone is overloaded with nitrate:

$$N_{av3} = N_c - N_{av1} \text{ for } D_{c1} \geq N_{av1} \text{ (under loaded pre-D zone)} \quad (5.75)$$

$$N_{av3} = N_c - D_{c1} \text{ for } N_{av1} \geq D_{c1} \text{ (overloaded pre-D zone)} \quad (5.76)$$

Once the values of D_{c1}, D_{c3}, N_{av1} and N_{av3} are known, the effluent nitrate concentration can be calculated as:

$$N_{ne} = N_c - N_{av1} - D_{c3} \quad \text{or}$$

$$= N_c/(a + s + 1) - D_{c3}/(s + 1) \text{ for } D_{c1} \geq N_{av1} \text{ (under loaded pre-D zone)} \quad (5.77)$$

$$N_{ne} = N_c - D_{c1} - D_{c3} \text{ for } N_{av1} \geq D_{c1} \text{(overloaded pre-D zone)} \quad (5.78)$$

As the highest rate of denitrification occurs in the pre-D zone, it makes sense to maximize the recirculation flow rate "a" and to recycle as much nitrate to the pre-D zone as possible. However, as can be observed in Figure 5.21, N_{av1} increases only marginally at higher values of the recirculation factor "a". Due to the low concentration (or even absence) of nitrate in the "s" recycle stream, this effects BDP systems even more.

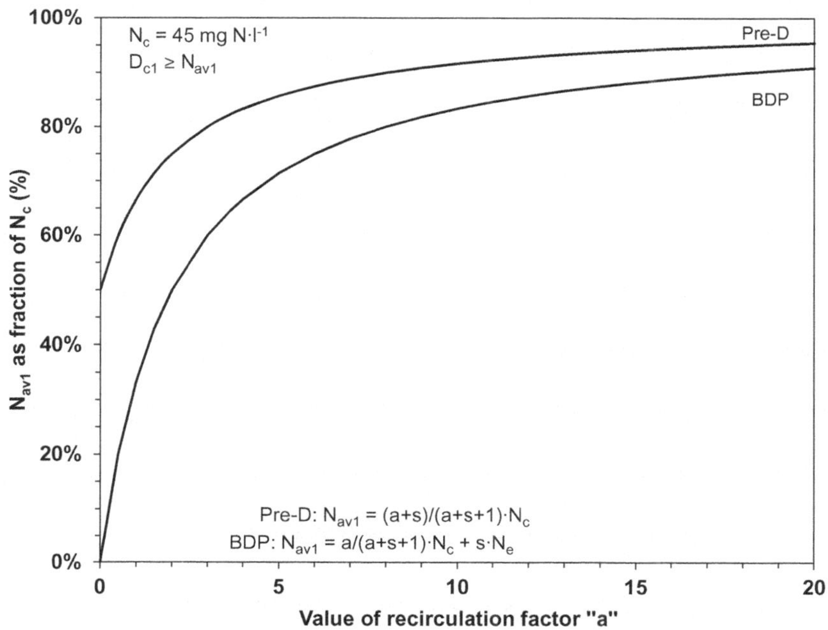

Figure 5.21 Ratio between N_{av1} and N_c as function of recirculation factor "a" for a pre-D and a BDP configuration

Consider the graphs shown in Figure 5.22, constructed for $N_c = 45$ mg N · l^{-1} and assuming that the pre-D denitrification capacity is not limiting: i.e. all nitrate returned will be removed. When it is required to reduce N_{ne} to ≤ 8 mg N · l^{-1}, then for the pre-D configuration the value of recirculation factor "a" needs to be $\geq$ 3.6. To reduce N_{ne} further, the required value of "a" increases rapidly. For example, a reduction of N_{ne} from 8 to 5 mg N · l^{-1} requires an increase of the recirculation rate from 3.6 to 7 times the influent flow rate, almost twice the original value. The application of high "a" recirculation factors will result in increased energy requirements for pumping (although these are small), but the main disadvantage is the increased return of dissolved oxygen to the pre-D zone. The oxygen reduces the available pre-D denitrification capacity due to the competition with nitrate for the use of easily biodegradable COD. The use of a high recirculation factor "a" is therefore not recommended, refer also to Section 5.4.2.3.

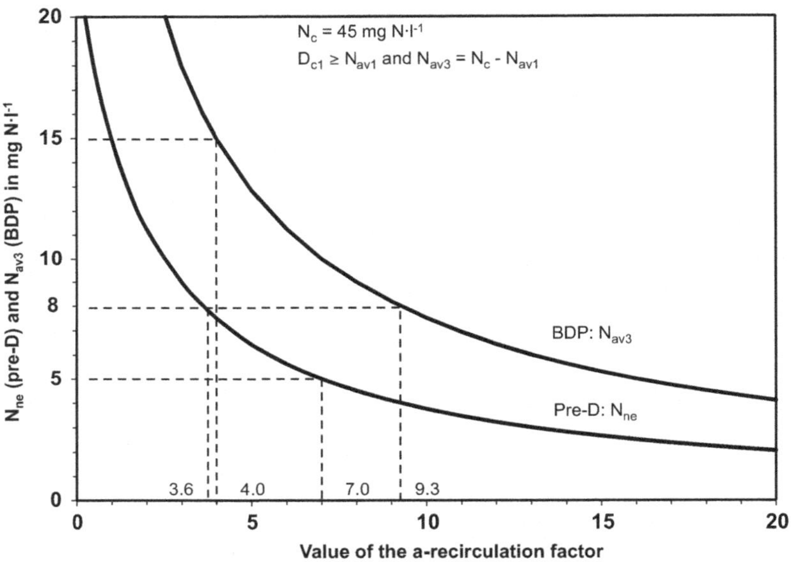

Figure 5.22 Nitrate available as in the effluent or in the post-D zone a function of the recirculation factor "a": i.e. N_{ne} for the pre-D configuration and N_{av3} for the BDP configuration

Now consider the BDP configuration. In Figure 5.22 it is assumed that the nitrate in the return sludge flow is removed in the pre-D reactor. It can be observed that $a = 9.3$ reduces $N_{av3} = N_c - N_{av1}$ to 8 mg N · l^{-1}. However, as part of the nitrate load will be removed in the post-D reactor, there is usually no need to reduce N_{av3} to such a low value. For example, supposing that $D_{c3} = 7$ mg N · l^{-1}, then a N_{av3} value of 15 mg N · l^{-1} would be sufficient to meet the effluent nitrate limit of $N_{ne} \leq 8$ mg N · l^{-1}. From $N_{ne} = N_c - N_{av1} - D_{c3} = 45 - N_{av1} - 7 = 8$ mg N · l^{-1}, the value of N_{av1} is calculated as 30 mg N · l^{-1}. According to Figure 5.23, $N_{av3} = 15$ mg N · l^{-1} corresponds to $a = 4$.

5.4 DESIGNING AND OPTIMISING NITROGEN REMOVAL

The model for nitrogen removal presented in the previous sections is based on experimental observations of denitrification in single sludge activated sludge processes. The only way to verify the validity of the model is to compare experimental values with the theoretical model values. Unfortunately, most literature data cannot be used for this purpose, because one or more parameters required to determine the nitrification- and denitrification capacity are not reported, such as the sludge age, the temperature or the anoxic sludge mass fraction. Furthermore, the anoxic reactors were often under loaded, so the availability of nitrate in the anoxic reactor was restricted and more nitrate could have been removed. In that case, the observed nitrate removal will always be inferior to the denitrification capacity.

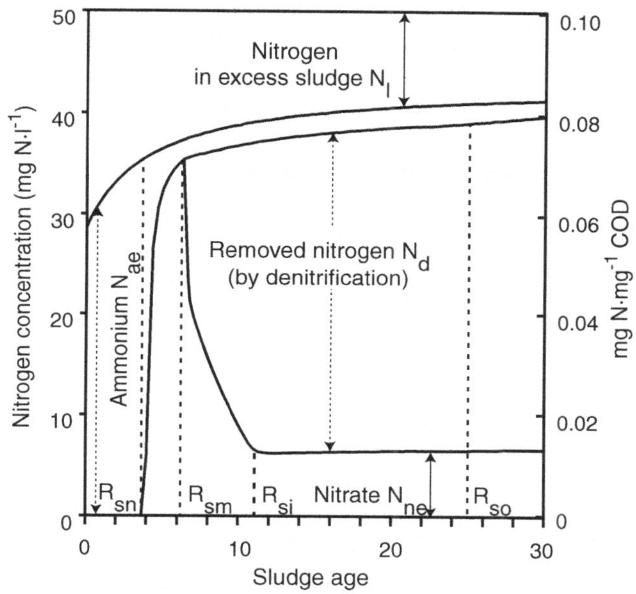

Figure 5.23 Calculation example: nitrification- and denitrification capacity in a pre-D configuration as a function of the sludge age for the maximum allowable anoxic sludge mass fraction

However, in all cases where it was possible to verify the validity of the model, a close correlation was found between the predicted values of removal and the observed values. This was demonstrated in pre-D and post-D reactors of nitrogen removal systems operating under the most diverse conditions:

- Size of the activated sludge system: up to 60,000 m^3;
- Applied sludge age from 3 to 35 days;
- Temperature from 8 to 28°C;
- Anoxic sludge mass fractions from 10 to 82%;
- Pre-D, post-D and Bardenpho configurations;
- Influent COD values between 220 and 850 mg COD $\cdot l^{-1}$;
- a- and s- factors of 0.2 to 6 times the size of the influent flow;
- Municipal sewage from South Africa, United States, Canada, the Netherlands and Brazil.

The data show that the model adequately describes nitrogen removal in single sludge activated sludge system. On the other hand, there are also limits to the model validity, as for instance the denitrification capacity depends on factors that vary from one wastewater to another:

- The concentration and composition of the influent organic material;
- The denitrification rate constant K_2 (in the case of a large proportion of industrial wastewater being present in the influent).

Because of the variability of several factors determining the model for nitrogen removal, ideally the values of the model parameters should be determined experimentally for each wastewater, prior to the start of the design. Hence it is important to have a simple and reliable calibration method to determine these factors. Chapter 3 and Appendix 2 present a procedure to determine the parameters defining the composition of the organic material. Furthermore, in Appendix 4 experimental methods will be presented to determine the value of the kinetic parameters for nitrification (μ_m, b_n and K_n) and the denitrification constants K_2 and K_3.

5.4.1 Calculation of nitrogen removal capacity

The concepts of nitrification capacity, denitrification capacity and available nitrate are very convenient to describe nitrogen removal in the activated sludge process, as demonstrated in the following example. Consider the nitrogen removal in an activated sludge process characterised by the following parameters:

– $N_{ti} = 50\ mg\ N \cdot l^{-1}$	– $f_{ns} = 0.10$	– $\mu_m = 0.3\ d^{-1}$
– $S_{ti} = 500\ mg\ COD \cdot l^{-1}$	– $f_{np} = 0.06$	– $b_n = 0.04\ d^{-1}$
– $N_{ad} = 2\ mg\ N \cdot l^{-1}$	– $f_{sb} = 0.25$	– $K_n = 1\ mg\ N \cdot l^{-1}$
– $T = 20\ °C$	– $a = 4$ and $s = 1$	– $K_2 = 0.1\ mg\ N \cdot mg^{-1}\ X_a \cdot d^{-1}$

The nitrification- and denitrification capacity can be calculated as function of the sludge age using Eq. (5.51) for N_c and Eqs. (5.54 and 5.68) for D_{c1}. To calculate D_{c1}, it is necessary to first determine the maximum allowable anoxic sludge mass fraction f_m as a function of sludge age, using Eq. (5.47). In Figure 5.23 the curves of N_c, D_{c1} and f_m are shown.

The value of N_{av1} is indicated in Figure 5.24 as a function of the sludge age for recirculation factors $a = 4$ and $s = 1$, i.e. for $N_{av} = (4 + 1)/(4 + 1 + 1) \cdot N_c = \frac{5}{6} \cdot N_c$. The value of N_{av1} represents the maximum nitrogen concentration that can be removed in a pre-D activated sludge process. Figure 5.24 is a useful illustration that demonstrates the utility of the concepts of nitrification- and denitrification capacity. With increasing sludge age the following situations can be observed:

(1) When $R_s < R_{sn}$, nitrification is impossible. The minimum sludge age for nitrification R_{sn} is given by Eq. (5.38):

$$R_{sn} = 1/(\mu_m - b_n) = 1/(0.3 - 0.04) = 3.85\ \text{days}$$

(2) For $R_s > R_{sn}$, nitrification is possible. However, it is not yet possible to comply with the condition that $N_{ae} \leq N_{ad}$, the specified effluent ammonium concentration. The reduction of N_{ae} to a value $\leq N_{ad}$ is only possible when the applied sludge age is higher than R_{sm}, which can be calculated from the condition that $f_m = 0$ i.e.:

$$f_m = 0 = 1 - (1 + K_n/N_{ad}) \cdot (1/R_{sm} + b_n)\mu_m \quad (5.47)$$

$$R_{sm} = 1/[\mu_m/(1 + K_n/N_{ad}) - b_n] = 1/[0.3/(1 + 1/2) - 0.04] = 6.25\ \text{days} \quad (5.39)$$

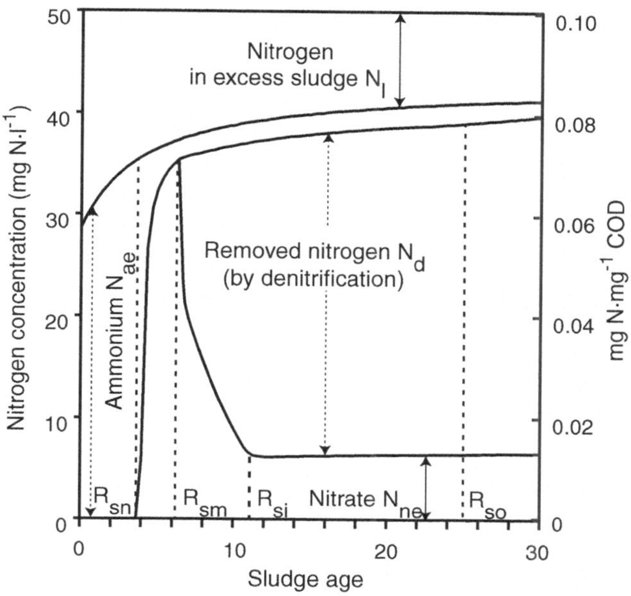

Figure 5.24 Calculation example: division of the nitrogen present in the influent over the different nitrogen fractions as a function of the sludge age

(3) For sludge ages beyond R_{sm}, it is possible to meet the specified residual ammonium concentration N_{ad} *and* to include an anoxic zone in the system as well. Using Eq. (5.47) to determine the anoxic sludge mass fraction, the denitrification capacity can be calculated from Eqs. (5.64 or 5.68). The nitrification capacity is calculated with the aid of Eq. (5.50):

$$f_m = 1 - (1 + K_n/N_{ad}) \cdot (b_n + 1/R_s)/\mu_m \tag{5.47}$$

$$D_{c1} = (K_1 + K_2) \cdot C_r \cdot S_{bi} \cdot V_1/V_r$$

$$= (K_1 + K_2) \cdot C_r \cdot f_{x1} \cdot S_{bi}(f_{x1} < f_{min}) \tag{5.64}$$

$$D_{c1} = N_{ds} + N_{dp}$$

$$= (f_{dn} \cdot f_{sb} + K_2 \cdot C_r \cdot f_{x1}) \cdot S_{bi}(f_{x1} > f_{min}) \tag{5.68}$$

$$N_c = N_{ti} - N_l - N_{ae} - N_{oe} \tag{5.50}$$

(4) For a particular sludge age R_{so}, the maximum anoxic sludge mass fraction f_m will be equal to the maximum allowable value f_{max}. For the given operating conditions and for $f_{max} = 0.6$ the value of R_{so} is calculated as:

$$f_m = f_{max} = 0.6 = 1 - (1 + K_n/N_{ad}) \cdot (1/R_{so} + b_n)/\mu_m \tag{5.47}$$

$$R_{so} = 1/[\mu_m \cdot (1 - f_{max})/(1 + K_n/N_{ad}) - b_n] \tag{5.79}$$

(5) For the example considered:

$$R_{so} = 1/[0.3 \cdot (1 - 0.6)/(1 + 1/2) - 0.04] = 25 \text{ days}$$

(6) When $R_s > R_{so}$, then both nitrification capacity and the denitrification capacity increase marginally with the sludge age. D_{c1} will increase slightly more than N_c.

Using the values of N_c and D_{c1}, the effluent nitrogen concentration can be calculated as a function of the sludge age. The presence of organic nitrogen in the effluent is ignored.

(a) $R_s < R_{sn}$
Below this sludge age nitrification is not possible. Hence, the ammonium concentration is equal to the nitrification potential. Obviously it doesn't make sense to include an anoxic zone as no nitrate will be formed (it is assumed that nitrate is not present in the influent). Biological nitrogen removal will not take place.

(b) $R_{sn} < R_s < R_{sm}$
In this range of sludge ages nitrification will develop. The effluent ammonium concentration is given by Eq. (5.36). An anoxic zone cannot yet be included without compromising ammonia effluent quality. The nitrate concentration will be equal to the nitrification capacity. Again, biological nitrogen removal will not take place.

(c) $R_s > R_{sm}$
Now it becomes possible to include an anoxic reactor. At increasing sludge age, the maximum allowable anoxic sludge mass fraction will increase as well and so will the denitrification capacity. For a particular sludge age $R_s = R_{si}$, the value of D_{c1} will be equal to N_{av1} so that:

$$D_{c1} = N_{av1} \text{ or } (f_{dn} \cdot f_{sb} + K_2 \cdot C_r \cdot f_m) \cdot S_{bi} = N_c \cdot (a + s)/(a + s + 1) \quad (5.80)$$

The value of R_{si} can be graphically determined from Figure 5.23 and is equal to 11 days. Alternatively, this value can also be calculated by trial and error with Eq. (5.80). In the range $R_{sm} < R_s < R_{si}$, the nitrate returned to the pre-D zone N_{av1} exceeds the available denitrification capacity D_{c1}. It can be concluded that the anoxic reactor is overloaded with nitrate. The nitrate load in excess of the denitrification capacity will be returned to the aerobic reactor. It is therefore possible to reduce the recirculation factors "a" and "s" and thus the value of N_{av1} until N_{av1} is equal to D_{c1}, without reducing the degree of nitrogen removal.

For example, when $R_s = 10$ days and for the conditions specified in this example, one can calculate N_{av1} as 37.2 $mg\,N \cdot l^{-1}$ using Eq. (5.70) and D_{c1} as 28.9 $mg\,N \cdot l^{-1}$ using Eq. (5.68). It can be concluded that it is indeed possible to reduce the recirculation to the pre-D reactor. Assuming that $s = 1$, the value of a is calculated from $D_{c1} = 28.9 = N_{av1} = N_c \cdot (a + s)/(a + s + 1)$. This equation can be solved for $(a + s) = 3.4$, so $s = 1$ and $a = 2.4$.

(d) $R_{sm} < R_s < R_{si}$
In this range of sludge ages, the ammonium effluent concentration will be constant: $N_{ad} = 2\ mg\,N \cdot l^{-1}$ ($f_{x1} = f_{max}$). The nitrate concentration in the effluent will be equal to the difference between the nitrification capacity and the denitrification capacity: $N_{ne} = N_c - D_{c1}$.

(e) $R_{si} < R_s < R_{so}$

In this case, $D_{c1} > N_{av1}$ and the anoxic reactor is under loaded, even when maximum recirculation ($a + s = 5$) is applied. All nitrate recirculated to the anoxic reactor will be removed. The effluent nitrate concentration will be equal to the fraction of the nitrification capacity that is discharged directly from the aerobic reactor to the effluent, without passing through the anoxic reactor: $N_{ne} = N_c/(a + s + 1)$. It is assumed here that no denitrification will take place in the settler. The ammonium concentration will be constant at $N_{ad} = 2\ mg\ N \cdot l^{-1}$, as f_{x1} is equal to f_m. In this range of sludge ages, the nitrogen removal efficiency could be increased by taking part of the pre-D reactor and using it to create a post-D reactor.

(f) $R_s > R_{so}$

Now the anoxic sludge mass fraction is limited by the condition that it may not exceed a maximum value: $f_x < f_{max}$ or $f_x < 0.6$. In this range of sludge ages, the residual ammonium concentration will be smaller than the specified value N_{ad}. The value of N_{ae} can be calculated with the aid of Eq. (5.44). As $D_{c1} > N_{av1}$, the effluent nitrate concentration is given as $N_{ne} = N_c/(a + s + 1)$.

In Figure 5.24 the division of the influent nitrogen concentration over the different nitrogen fractions N_{ae}, N_{ne}, N_l and N_d is shown as function of the sludge age, for the conditions specified in this calculation example. It can be observed in Figure 5.24 that for a sludge age of 11 days almost all of the influent nitrogen concentration of $50\ mg\ N \cdot l^{-1}$ is removed. The effluent nitrogen concentration N_{te} is equal to $N_{ad} + N_{ne}$. The value of $N_{ne} = N_c/(a + s + 1)$ is $38.0/6 = 6.3\ mg\ N \cdot l^{-1}$, so $N_{te} = 2.0 + 6.3 = 8.3\ mg\ N \cdot l^{-1}$.

At the selected sludge age of 11 days the nitrogen concentration that is removed with the excess sludge N_l equals $10.0\ mg\ N \cdot l^{-1}$. Hence, the denitrified nitrogen concentration $N_d = N_{ki} - N_{ad} - N_{ne} = 50 - 2 - 6.3 - 10.0 = 31.7\ mg\ N \cdot l^{-1}$, which is equal to the denitrification capacity for $R_s = 11$ days. If it is desired to reduce the effluent nitrogen concentration any further, it will be necessary to increase the sludge age and modify the reactor configuration of the process, transforming it from a pre-D system to a Bardenpho system. The optimisation of the Bardenpho system will be discussed in the next section.

EXAMPLE 5.9

For the calculation example of Section 5.4.1, demonstrate that the pre-D zone is indeed overloaded for $R_{sm} < R_s < R_{si}$, i.e. that the nitrate recirculation to the pre-D zone can be decreased without reducing nitrogen removal efficiency. The following data are given:

- $R_s = 9$days and $f_{x1} = f_m = 0.24$;
- $b_h = 0.24\ d^{-1}$, $K_1 = 0.72$, $K_2 = 0.10$ and $C_r = 1.28\ mg\ VSS \cdot d \cdot mg^{-1}\ COD$;
- $S_{bi} = 420\ mg\ COD \cdot l^{-1}$and $f_{sb} = 0.25$;
- $N_c = 35.4\ mg\ N \cdot l^{-1}$, $a = 4$ and $s = 1$

Calculate the lowest value of the a-factor that can be applied without reducing nitrate removal.

Solution

Check whether Eq. (5.68) can be used to calculate the value of D_{c1}:

$$f_{min} = f_{dn} \cdot f_{sb}/(K_1 \cdot C_r)$$
$$= 0.11 \cdot 0.25/(0.72 \cdot 1.28) = 0.03$$

For $R_s = 9$ days the value of $f_x = f_{x1} = f_m \gg f_{min}$. Thus D_{c1} can be calculated with Eq. (5.68) as:

$$\begin{aligned} D_{c1} &= (f_{dn} \cdot f_{sb} + K_2 \cdot C_r \cdot f_{x1}) \cdot S_{bi} \\ &= (0.11 \cdot 0.25 + 0.10 \cdot 1.28 \cdot 0.24) \cdot 420 \\ &= 25.1\ \text{mg N} \cdot l^{-1}\text{influent} \end{aligned}$$

For the current values of "a" and "s", the available nitrate in the pre-D reactor is equal to:

$$\begin{aligned} N_{av1} &= (a + s)/(a + s + 1) \cdot N_c \\ &= (4 + 1)/(4 + 1 + 1) \cdot 35.4 = 29.5\ \text{mg N} \cdot l^{-1} \end{aligned}$$

As $N_{av1} > D_{c1}$, the pre-D reactor is indeed overloaded with nitrate. The minimum value of the "a" recirculation required to maintain the same nitrate removal performance can be calculated from:

$$N_{av1} = D_{c1} \text{ or } (a + s)/(a + s + 1) \cdot N_c = D_{c1}$$

After rearranging:

$$(a + 1)/(a + 2) \cdot 35.4 = 25.1\ \text{mg N} \cdot l^{-1}$$

$$(a + 1) = 0.71 \cdot (a + 2) \longrightarrow 0.29 \cdot a = 0.42 \longrightarrow a = 1.43$$

It can be checked that N_{av1} is indeed equal to D_{c1} for $a = 1.43$

$$\begin{aligned} N_{av1} &= (a + s)/(a + s + 1) \cdot N_c \\ &= (1.43 + 1)/(1.43 + 2) \cdot 35.4 = 25.1\ \text{mg N} \cdot l^{-1} \end{aligned}$$

Thus it is possible to reduce "a" to 1.43 without decreasing nitrate removal in the pre-D reactor.

EXAMPLE 5.10

Again for the example in Section 5.4.1, demonstrate that for $R_s > R_{si}$ it is advantageous to take part of the pre-D zone and allocate it to a post-D zone. Estimate the decrease in the effluent nitrate concentration if the pre-D configuration is converted to a BDP configuration. The following additional data are given:

- $R_s = 12$ days and $f_{x1} = f_m = 0.38$;
- $C_r = 1.39$ mg VSS · d · mg^{-1} COD;
- $N_c = 36.3$ mg N · l^{-1};
- $a = 4$ and $s = 1$;
- $K_3 = 0.08$ mg N · mg^{-1} VSS · d^{-1}

Solution

As a first step calculate the available nitrate in the pre-D zone for the specified conditions:

$$N_{av1} = (5/6) \cdot N_c = 30.3 \text{ mg N} \cdot l^{-1}$$

When N_{av1} is equated to D_{c1}, it can be verified that for $f_{x1} = 0.31$, D_{c1} is equal to N_{av1}:

$$D_{c1} = (f_{dn} \cdot f_{sb} + K_2 \cdot C_r \cdot f_{x1}) \cdot S_{bi}$$
$$= (0.11 \cdot 0.25 + 0.10 \cdot 1.39 \cdot 0.31) \cdot 420 = 30.3 \text{ mg N} \cdot l^{-1}$$

For the pre-D configuration (with $D_{c1} \geq N_{av1}$), N_{ne} can be calculated as:

$$N_{ne} = 1/(a + s + 1) \cdot N_c$$
$$= 1/6 \cdot 36.3 = 6.1 \text{ mg N} \cdot l^{-1}$$

In a BDP configuration, as $D_{c1} \approx N_{av1}$, N_{av3} can be calculated as:

$$N_c - D_{c1} = 36.3 - 30.3 = 6.1 \text{ mg N} \cdot l^{-1}$$

The maximum value of $f_{x3} = f_m - f_{x1} = 0.38 - 0.31 = 0.07$
D_{c3} is given by Eq. (5.69):

$$D_{c3} = K_3 \cdot C_r \cdot f_{x3} \cdot S_{bi}$$
$$= 0.08 \cdot 1.39 \cdot 0.07 \cdot 420 = 3.3 \text{ mg N} \cdot l^{-1}$$

So the effluent nitrate concentration in the BDP configuration can be calculated as

$$N_{ne} = N_c - D_{c1} - D_{c3} = 36.3 - 30.3 - 3.3 = 2.8 \text{ mg N} \cdot l^{-1}$$

Converting the pre-D into a BDP configuration will thus reduce the effluent nitrate concentration from 6.1 to 2.8 mg $N \cdot l^{-1}$, without requiring additional reactor volume.

5.4.2 Optimised design of nitrogen removal

The objectives of design optimisation for nitrogen removal activated sludge processes are:

- To produce an effluent with a minimum total nitrogen concentration;
- To carry out this nitrogen removal at minimum construction- and operational costs.

Before starting with the optimisation procedure, it is necessary to remember that there are several constraints for single sludge activated sludge processes designed for nitrogen removal:

(1) The anoxic sludge mass fraction is limited by two independent criteria:
   - The nitrification efficiency must be high, which implies a certain minimum for the sludge mass fraction in the aerobic zones and a corresponding maximum for the anoxic sludge fraction;
   - The settling properties of the sludge may be affected by an excessive anoxic sludge mass fraction. It is suggested that the anoxic sludge mass fraction should not be larger than sixty percent.

(2) The value of the nitrate recirculation factor "a" and that of the return sludge factor "s" have upper constraints from a viewpoint of energy efficiency and denitrification efficiency.

(3) As the headloss in the "a" recirculation is always smaller and since the nitrate concentration in the aerobic reactor is at least as high as in the settler, the "a"-recycle is always more cost-efficient than the "s" recycle to introduce nitrate into the pre-D reactor. Furthermore, the value of the sludge recycle factor "s" should be dictated by the requirements for efficient settling only. In practice this often results in an "s"-recycle with a value of $0.5 < s < 1.5$ (refer also to Chapter 8).

(4) The value of the "a"-recycle is more difficult to assess. Ideally, the size of the mixed liquor recirculation flow must be such that the available nitrate in the pre-D reactor is exactly equal to its denitrification capacity. Considering that the denitrification rate in the pre-D reactor is always larger than in the post-D reactor (Figure 5.19 and Figure 5.21) it is, at least in principle, advantageous to have a large pre-D reactor and hence a high value of the "a"-recycle would be required. On the other hand, even if the head loss is low, the high recycle flow will lead to increased operational costs. Furthermore when a high "a"-recirculation flow is imposed, the mass of oxygen recycled to the anoxic zone can be considerable. Later in this section the optimal value of the "a"-recycle is evaluated.

A variable of great importance that has not yet been discussed is the proportion between nitrogenous and organic material in the wastewater: N_{ti}/S_{ti}. Note that this book does not use the more common COD/N (S_{ti}/N_{ti}) ratio, but this is purely for practical purposes: as S_{ti} is typically larger than N_{ti}, it means that N_{ti}/S_{ti} can be normalized to a value between 0 and 1.

The value of this ratio is heavily dependent on the origin of the wastewater. Low values (< 0.02 mg $N \cdot mg^{-1}$ COD) are found for wastewater from agricultural industries, such as distillate from alcohol plants, black liquor from cellulose production and effluent from breweries. High values (up to 0.16 mg $N \cdot mg^{-1}$ COD) are typical for wastewater from industries processing animal products like tanneries, slaughterhouses and dairy factories.

In the case of municipal sewage, the N_{ti}/S_{ti} ratio is closely associated with the protein consumption of the population. For example, in the United States (where meat consumption per capita is high), the ratio N_{ti}/S_{ti} is about 0.12 mg $N \cdot mg^{-1}$ COD, whereas in the cities with a predominance of vegetarians (India, certain regions in Africa), the ratio is found to be only 0.04 to 0.06 mg $N \cdot mg^{-1}$ COD.

In general there is a linear correlation between the wealth of a contributing population and the N_{ti}/S_{ti} ratio in the sewage. The equations that define the nitrification and denitrification capacities show that N_c is proportional with the influent TKN concentration, whereas D_c is proportional to the influent COD concentration. When the N_{ti}/S_{ti} (or N_c/S_{bi}) ratio is low, it is easy to create a denitrification capacity large enough to completely remove the nitrate formed in the system. However, even then for complete nitrogen removal a Bardenpho system is required.

5.4.2.1 Complete nitrogen removal

If it is assumed that denitrification is complete in both anoxic reactors, it can be observed from Figure 5.16c that a fraction $a/(a + s + 1)$ of the nitrification capacity will be denitrified in the pre-D reactor, while the remaining fraction (s + l)/(a + s + l) will be removed in the post-D reactor. Hence, for complete denitrification it is necessary that:

$$D_{c1} = (f_{dn} \cdot f_{sb} + K_2 \cdot C_r \cdot f_{x1}) \cdot S_{bi} = a/(a + s + 1) \cdot N_c \quad (5.81)$$

$$D_{c3} = K_3 \cdot C_r \cdot f_{x3} \cdot S_{bi} = (s + 1)/(a + s + 1) \cdot N_c \quad (5.82)$$

Writing the sludge mass fractions explicitly in Eqs. (5.81 and 5.82) one has:

$$f_{x1} = \frac{(N_c/S_{bi}) \cdot a/(a + s + l) - f_{dn} \cdot f_{sb}}{K_2 \cdot C_r} \quad (5.83)$$

$$f_{x3} = \frac{(N_c/S_{bi}) \cdot (s + l)/(a + s + l)}{K_3 \cdot C_r} \quad (5.84)$$

The largest N_c/S_{bi} ratio for which complete denitrification is possible, indicated as $(N_c/S_{bi})_o$, can be calculated knowing that for this N_c/S_{bi} ratio the anoxic sludge mass fraction will be maximum. Hence, with the aid of Eqs. (5.83 and 5.84) for f_{x1} and f_{x3}:

$$f_m = f_{x1} + f_{x3} \quad \text{or}$$

$$f_m = \frac{(N_c/S_{bi})_o \cdot [a/(a + s + 1)] - (f_{dn} \cdot f_{sb})}{K_2 \cdot C_r + \left[(N_c/S_{bi})_o \cdot (S + 1)(a + s + l)\right]/(K_3 \cdot C_r)} \quad (5.85)$$

After rearranging, the ratio $(N_c/S_{bi})_o$ can be written explicitly as:

$$(N_c/S_{bi})_o = \frac{(a + s + l) \cdot (f_{dn} \cdot f_{sb} + K_2 \cdot C_r \cdot f_m)}{a + (K_2/K_3) \cdot (s + 1)} \quad (5.6)$$

Once the value of the $(N_c/S_{bi})_o$ ratio has been determined, the corresponding ratio $(N_{ti}/S_{ti})_o$ can be calculated using Eqs. (3.3 and 5.50).

$$S_{bi} = (1 - f_{ns} - f_{np}) \cdot S_{ti} \quad (3.3)$$

$$N_c = N_{ti} - N_l - N_{ad} - N_{oe} \quad (5.50)$$

which yields:

$$(N_{ti}/S_{ti})_o = (1 - f_{ns} - f_{np}) \cdot (N_c/S_{bi})_o + (N_l + N_{ad} + N_{oe})/S_{ti} \quad (5.87)$$

Finally, when $(N_c/S_{bi})_o$ is written explicitly with Eq. (5.86), one can derive Eq. (5.88) :

$$(N_{ti}/S_{ti})_o = \frac{(1 - f_{ns} - f_{np}) \cdot (f_{dn} \cdot f_{sb} + K_2 \cdot C_r \cdot f_m) \cdot (a + s + 1)}{a + (K_2/K_3) \cdot (s + 1)} + (N_l + N_{ad} + N_{oe})/S_{ti} \quad (5.88)$$

Obviously Eq. (5.88) is only valid for $R_s \geq R_{sm}$. Equation (5.88) shows that several factors influence the value of the largest TKN/COD ratio allowing complete denitrification:

(1) Composition of the influent organic material (f_{ns}, f_{np} and f_{sb});
(2) Kinetic parameters for denitrification (K_2 and K_3);
(3) Kinetic parameters for nitrification (μ_m, K_n and b_n);
(4) Temperature, which influences the values of kinetic constants (K_2, K_3, K_n, b_n, b_h and μ_m);
(5) Organic nitrogen in the effluent (N_{oe});
(6) Specified residual ammonium concentration (N_{ad});
(7) Recirculation factors (a and s);
(8) Sludge age (R_s).

The values of factors 1 to 5 listed above cannot be randomly selected in a particular design, but should rather be considered as "given" design values. In principle, the specified residual ammonium concentration N_{ad} (factor 6) can be specified by the designer, but in general the activated sludge system will have to comply to a strict effluent ammonium limit anyway. Hence, in fact only the recirculation factors a and s and the sludge age (factors 7 and 8) may be considered to be design variables. As will be shown in Chapter 8, the value of the s-recycle factor should be determined by the requirements for efficient liquid-solid separation in the final settler.

Therefore, the factors to be defined in optimising nitrogen removal are the a-recycle and the sludge age. One of the methods to determine a suitable a-recirculation factor is to plot the $(N_{ti}/S_{ti})_o$ or $(N_c/S_{bi})_o$ ratio as a function of the sludge age for different "a" values. This is shown in Figure 5.25 for values of the "a" recirculation factor ranging from 1 to 10. The value of f_m is indicated as well. The following parameter values were used to construct Figure 5.25:

$- T = 20°$	$- f_{sb} = 0.25$	$- K_n = 1.0\ mg\ N \cdot l^{-1}$
$- S_{ti} = 500\ mg \cdot l^{-1}$	$- \mu_m = 0.3\ d^{-1}$	$- K_2 = 0.10\ mg\ N \cdot mg^{-1}\ X_a \cdot d^{-1}$
$- f_{ns} = 0.10$	$- b_n = 0.04\ d^{-1}$	$- K_3 = 0.08\ mg\ N \cdot mg^{-1}\ X_a \cdot d^{-1}$
$- f_{np} = 0.15$	$- s = 1$	

It can be observed in Figure 5.26 that the required sludge age for the complete removal of nitrogen tends to decrease when the recirculation factor "a" increases. The choice of the optimal recirculation factor then would seem to become a question of economics, where the costs for pumping of nitrified mixed liquor to the pre-D reactor (energy and pumps) are compared to the construction and operational costs as a function of the sludge age. However, note that the reduction of the denitrification capacity resulting from oxygen recycle to the pre-D zone has not been considered in Figure 5.25 (refer also to Section 5.4.2.3). Furthermore, as discussed in Section 5.3.4, the law of diminishing returns applies here as well: beyond a certain point a further increase of the a-factor will yield only a very limited increase in the value of N_{av1}.

At high recirculation rates, the effect of the additional return of nitrate to the pre-D zone will thus be (more than) compensated by the mass of oxygen recycled. In practice, depending on the value of N_c, the optimal value of the a-recirculation factor will therefore almost invariably be less than a = 4–6 for a pre-D configuration and less than a = 6–10 for a BDP configuration. There are several exceptions, for example:

- Recirculation systems such as carrousels do not have dedicated nitrification and denitrification reactors, but consist of one or several large "racetrack" loops. They are designed with surface aerators that act also as propulsors in order to induce a recirculation flow. The value of recirculation factor "a" is often as high as 10–20;

- Industrial systems treating wastewater with a high N_{ti}/S_{ti} ratio, requiring the addition of an external carbon source. Often it is preferred to add this external carbon source to the pre-D zone, in order to prevent accidental overdosing in the post-D zone with the consequential risk of exceeding the effluent COD limit. In this case the effect of DO recycle to the pre-D zone can be compensated through the addition of excess external carbon source.

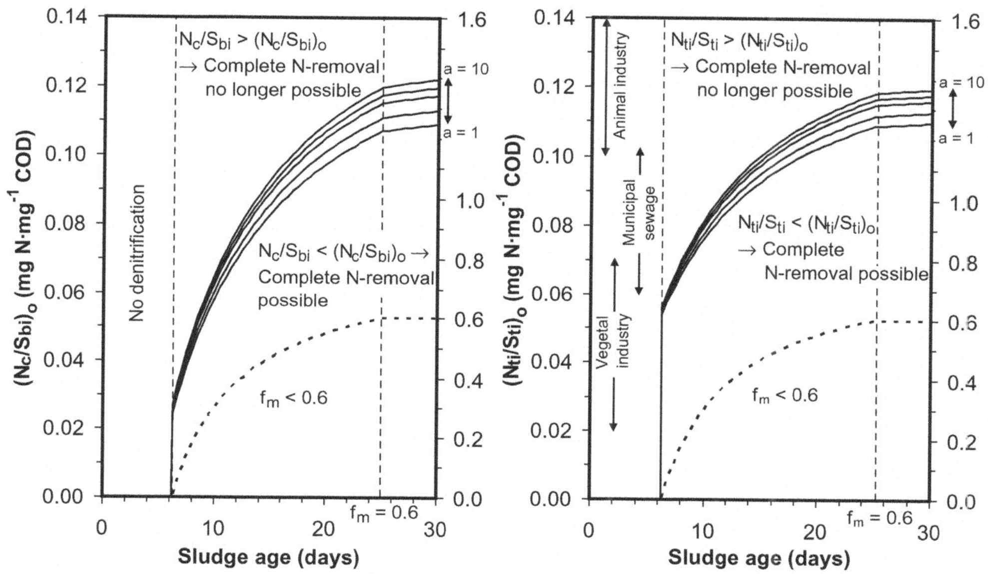

Figure 5.25 Maximum ratio between nitrification capacity and biodegradable COD (left) and TKN and total COD (right) in the influent allowing complete nitrate removal, as a function of R_s and for different values of "a"

5.4.2.2 Incomplete nitrogen removal

For high N_{ti}/S_{ti} ratios or other unfavourable conditions, for example a low temperature, the presence of toxic compounds or a low easily biodegradable COD concentration, complete denitrification might not be possible, unless the applied sludge age is so high that the system will become unacceptably large or alternatively, the addition of an external carbon source is required. If the addition of an external carbon source is not considered a viable option due to the associated costs, the maximum nitrogen removal efficiency may be determined as follows: in the pre-D reactor of the Bardenpho system, nitrate is being introduced with the "a" recycle from the nitrification zone and with the "s" recycle from the final settler so that:

$$D_{c1} = (f_{dn} \cdot f_{sb} + K_2 \cdot C_r \cdot f_m) \cdot S_{bi} = a/(a + s + 1) \cdot N_c + s \cdot N_{ne} \tag{5.89}$$

The effluent nitrate concentration is given by the difference between the nitrate concentration in the aerobic reactor and the nitrate removal in the post-D reactor. When Eq. (5.89) is valid, the nitrate concentration in the pre-D reactor will be zero and the concentration in the aerobic reactor will be a factor $1/(a + s + 1)$ of N_c, as the influent is diluted $(1 + a + s)$ times before it reaches the aerobic reactor. The decrease of nitrate concentration in the post-D reactor is equal to $D_{c3}/(s + 1)$, so that:

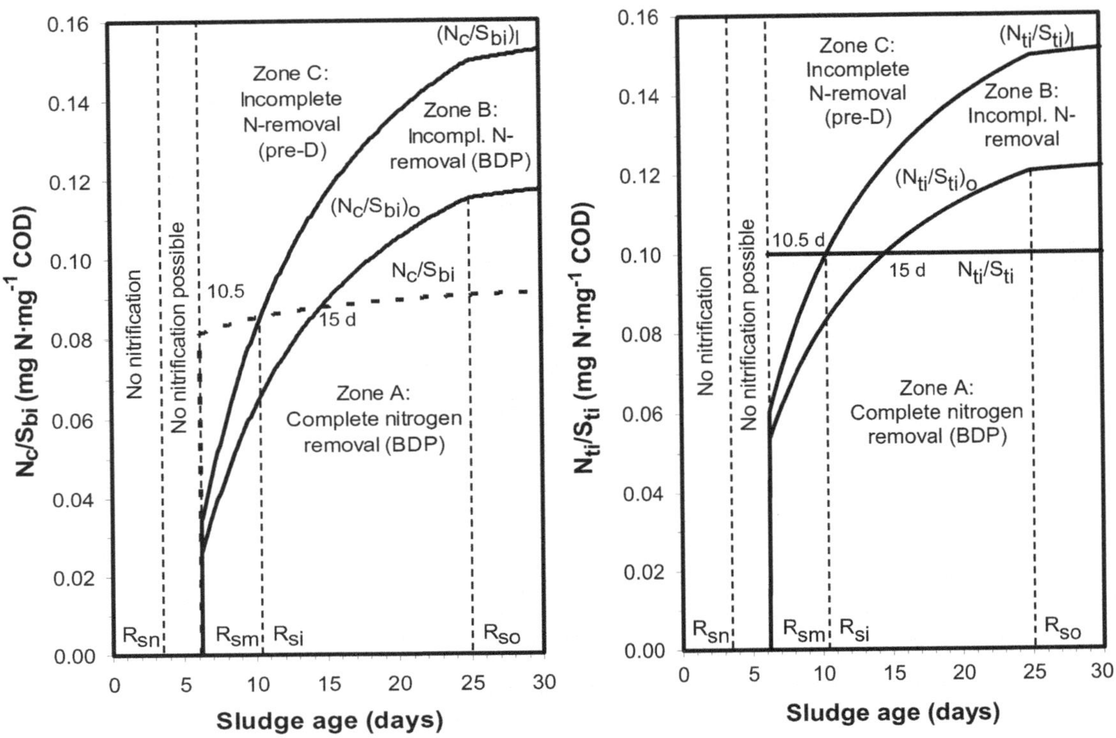

Figure 5.26 Value of the ratio $(N_c/S_{bi})_o$ and (N_c/S_{bi}) (left) and $(N_{ti}/S_{ti})_o$ and (N_{ti}/S_{ti}) (right) as function of the sludge age for the conditions of the example discussed in Section 5.4.2.1

$$N_{ne} = N_c/(a + s + 1) - D_{c3}/(s + 1) \tag{5.90}$$

Now, by substituting for D_{c3} from Eq. (5.69) and knowing that $f_{x3} = f_m - f_{x1}$ one has:

$$N_{ne} = N_c/(a + s + 1) - k_3 \cdot C_r \cdot (f_m - f_{x1}) \cdot S_{bi}/(s + 1) \tag{5.91}$$

By using Eq. (5.91) in Eq. (5.89) one calculates that:

$$f_{x1} = \frac{(N_c/S_{bi}) \cdot (a + s)/(a + s + 1) - f_{dn} \cdot f_{sb} - K_3 \cdot C_r \cdot f_m \cdot s/(s + 1)}{C_r \cdot [K_2 - K_3 \cdot s/(s + 1)]} \tag{5.92}$$

In Eq. (5.92) the calculated value of f_{x1} is the one that leads to the highest nitrate removal in the Bardenpho system, if complete denitrification is not possible. The residual nitrate concentration is given by Eq. (5.90).

In Eq. (5.92), as the N_c/S_{bi} ratio increases, so does the value of f_{x1} (while at the same time the value for f_{x3} decreases), until $f_{x1} = f_m$ (and $f_{x3} = 0$). For this limiting value of N_c/S_{bi}, the entire anoxic sludge mass fraction is placed in the pre-D reactor. In other words, the Bardenpho configuration has ceased to be advantageous and it is therefore changed into a pre-D system. The value of the ratio N_c/S_{bi} for which this occurs is obtained by substituting $f_{x1} = f_m$ (not f_{max}!) in Eq. (5.92):

$$(N_c/S_{bi})_l = (f_{dn} \cdot f_{sb} + K_2 \cdot C_r \cdot f_m) \cdot (a + s + 1)/(a + s) \tag{5.93}$$

where $(N_c/S_{bi})_l$ = limiting ratio for the applicability of the Bardenpho process

The $(N_c/S_{bi})_l$ ratio can also be expressed as $(N_{ti}/S_{ti})_l$:

$$(N_{ti}/S_{ti})_l = \frac{(1 - f_{ns} - f_{np}) \cdot (f_{dn} \cdot f_{sb} + K_2 \cdot C_r \cdot f_m) \cdot (a + s + 1)}{(a + s)} + (N_l + N_{ad} + N_{oe})/S_{ti} \tag{5.94}$$

Equations (5.93 and 5.94) are only valid for $R_s > R_{sm}$. In Figure 5.26 (left) the curves of $(N_c/S_{bi})_l$ and $(N_c/S_{bi})_o$ have been plotted as a function of sludge age for the same conditions used to construct Figure 5.26, however the "a" recycle is now fixed at a value of 4. In the right-hand side of Figure 5.26 the corresponding curves for N_{ti}/S_{ti} are shown. Now, basically three different situations can be distinguished:

(1) Zone A: low N_{ti}/S_{ti} ratio i.e. $N_{ti}/S_{ti} < (N_{ti}/S_{ti})_o$. In this case the proportion between nitrogenous material and biodegradable organic material is favourable and complete nitrate removal is possible;
(2) Zone B: average N_{ti}/S_{ti} ratio, i.e., $(N_{ti}/S_{ti})_o < (N_{ti}/S_{ti}) < (N_{ti}/S_{ti})_l$. In this case, complete denitrification is not possible, but the lowest possible effluent nitrogen concentration is still obtained in a Bardenpho configuration;
(3) Zone C: high N_{ti}/S_{ti} ratio, i.e. $(N_{ti}/S_{ti}) > (N_{ti}/S_{ti})_l$. In this case, the proportion between nitrogenous and biodegradable organic material is unfavourable for nitrate removal. The lowest nitrogen concentration in the effluent is obtained in a pre-D system.

The value of N_c/S_{bi} and N_{ti}/S_{ti} have also been plotted in Figure 5.26. So, for the example it can be observed that full nitrogen removal is only possible when $R_s > 15$ days. For $10.5 < R_s < 15$ days, complete denitrification is no longer possible, but a BDP configuration will still result in maximum nitrogen removal. For $R_s < 10.5$ days, it is better to change to a pre-D configuration.

In general, for regions with a warm climate the ratio between the TKN and the COD concentration in raw sewage is such that complete nitrogen removal is feasible, even at relatively short sludge ages (5 to 10 days). For sewage with a large industrial wastewater fraction may lead to a low μ_m value and the consequential need to increase the sludge age. On the other hand some industrial wastewaters (especially those of vegetable origin) have a low N_{ti}/S_{ti} ratio so that complete nitrogen removal is relatively easy. Primary and/or anaerobic pre-treatment of the raw sewage has a negative effect on the nitrogen removal process, as in such systems more organic than nitrogen material is removed. The N_{ti}/S_{ti} ratio will increase, requiring in a longer sludge age or resulting in incomplete nitrogen removal.

An analysis of the factors that affect the required sludge age for complete nitrogen removal reveals that the maximum specific nitrifier growth rate μ_m is the most important one. As it is known that this value tends to vary considerably depending on the origin of the wastewater, it is important to measure its value experimentally whenever possible. In Appendix 4 the method used to determine this parameter is explained.

5.4.2.3 Effect of recirculation of oxygen on denitrification capacity

In the previous sections the detrimental effect of the recirculation of oxygen to the pre-D zone on the denitrification capacity has already been indicated. It will be demonstrated in this section that this effect should not be underestimated, especially for high values of the a-recirculation factor. Due to the recirculation of oxygen to the pre-D zone, oxygen instead of nitrate is consumed for the oxidation of COD. Hence the denitrification capacity will be reduced. Taking into account the oxygen equivalence of nitrate (2.86 mg O_2 per mg NO_3–N), the reduction of D_{c1} can be calculated as:

$$\Delta D_{c1} = a \cdot DO_l/2.86 \tag{5.95}$$

where DO_l = dissolved oxygen concentration in the recirculation stream, generally equal to the DO setpoint in the nitrification zone

The magnitude of this effect is indicated in Figure 5.27, where it can be observed that the combination of a high recirculation rate and a high DO_l concentration (e.g. due to overdesign of the aeration system) can be very detrimental to nitrogen removal indeed. But even at lower values (for instance for a = 4 and $DO_l = 2\ mg \cdot l^{-1}$) the effect is already considerable: 2.8 mg $N \cdot l^{-1}$ or between 5 and 10% of the denitrification capacity typically required for a municipal activated sludge system.

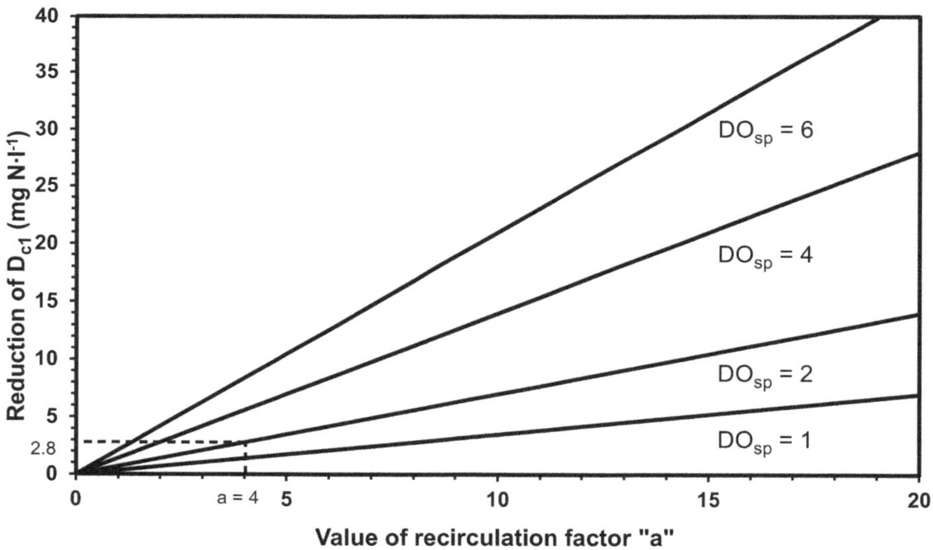

Figure 5.27 Reduction of D_{c1} as function of recirculation factor "a" for different DO concentrations in the end of the nitrification zone

Another example is given in Figure 5.28, which shows for a pre-D system and for two design cases (i.e. a low- and a high influent nitrogen concentration) the following parameters as function of the recirculation factor "a":

– Available nitrate N_{av1} as fraction of N_c;

– Calculated effluent nitrate concentration N_{ne} (note: without considering ΔD_{c1});
– The value of ΔD_{c1} for $DO_l = 2\ mg\ O_2 \cdot l^{-1}$

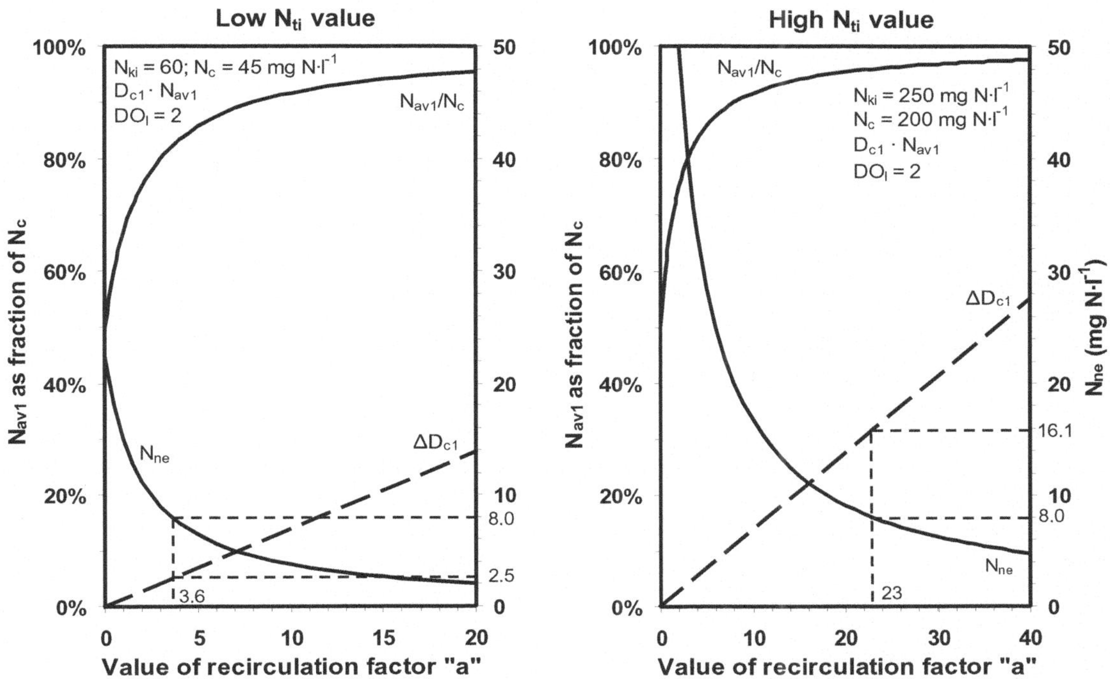

Figure 5.28 Value of N_{av1} and N_{ne} as function of the recirculation factor "a" in a pre-D system for low- and high values of N_c, calculated *excluding* the effect of ΔD_{c1}

It is interesting to evaluate for both cases (i.e. for a low- and high value of N_{ti}) what the value of the a-recirculation factor will have to be in a pre-D system in order to reduce N_{ne} to a value of $8\ mg\ N \cdot l^{-1}$ or less:

(a) The case of a low nitrogen influent concentration ($N_{ti} = 60\ mg\ N \cdot l^{-1}$)

It is assumed that $15\ mg\ N \cdot l^{-1}$ will end up either in the excess sludge (N_l) or as ammonium or organic nitrogen in the effluent ($N_{ae} + N_{oe}$). Thus $N_c = 60–15 = 45\ mg\ N \cdot l^{-1}$ and in order to meet the effluent N_{ne} limit of $8\ mg\ N \cdot l^{-1}$, the required pre-D denitrification capacity D_{c1} is equal to $45–8 = 37\ mg\ N \cdot l^{-1}$. Assuming that this pre-D denitrification capacity is available, the value of N_{av1} shoud be at least $37\ mg\ N \cdot l^{-1}$ as well. The value of N_{av1} is equal to $(a + s)/(a + s + 1) \cdot N_c = 37$, which can be solved for $a = 3.7$ (for $s = 1$).

(b) The case of a high nitrogen influent concentration ($N_{ti} = 250\ mg\ N \cdot l^{-1}$)

It is assumed that $N_c = 200\ mg\ N \cdot l^{-1}$. In order to meet the effluent N_{ne} limit, $200 - 8 = 192\ mg\ N \cdot l^{-1}$ needs to be denitrified. Thus $N_{av1} = (a + s)/(a + s + 1) \cdot N_c = 192$, which can be solved for $a = 23$ (for $s = 1$).

Now it is interesting to calculate the reduction in pre-D denitrification capacity from oxygen recirculation for the a-recirculation factors calculated above (for $DO_l = 2$ mg $O_2 \cdot l^{-1}$).

- $N_{ti} = 60\,mg\,N \cdot l^{-1} \longrightarrow$ for $a = 3.6$, $\Delta D_{c1} = a \cdot DO_l/2.86 = 7.2/2.86 = 2.5\,mg\,N \cdot l^{-1}$;
- $N_{ti} = 250\,mg\,N \cdot l^{-1} \longrightarrow$ for $a = 23$, $\Delta D_{c1} = 46/2.86 = 16\,mg\,N \cdot l^{-1}$.

Thus on top of the denitrification capacity required for nitrate removal, additional "denitrification capacity" is required to remove the recycled oxygen. In the case of a low N_{ti} value, the required D_{c1} increases to $37 + 2.5 = 39.5$ mg $N \cdot l^{-1}$, while in the case of high N_{ti} value, the required D_{c1} will be $192 + 16 = 208$ mg $N \cdot l^{-1}$.

From Figure 5.28 it can be observed that in the example of a low influent nitrogen concentration (representative for municipal sewage), already at $a = 6$ the increase of ΔD_{c1} cancels out the anticipated reduction of N_{ne} resulting from the increase of N_{av1}. Alternatively phrased: contrary to what might be expected, the effluent nitrate concentration will probably *not* decrease when the a-factor is increased from 5 to 6, *unless* excess denitrification capacity is still available in the pre-D zone. However, even then the alternative of creating a post-D zone will be much more effective. The detrimental effect on nitrate removal from an increase of the a-factor is even more adverse for $a > 10$.

In the example of a high influent nitrogen concentration, a higher value of the a-recirculation can be accepted, as the rate of the increase of N_{av} will initially be higher than the rate of increase of ΔD_{c1}. In the example from Figure 5.28, for $a = 15$ the reduction of N_{ne} is canceled out by the increase of ΔD_{c1}.

The remarks made above regarding the reycle of oxygen to the pre-D zone apply also to the post-D zone. Similar to ΔD_{c1} the presence of oxygen in a post-D zone will result in a reduction of D_{c3}. The value of ΔD_{c3} can be calculated as:

$$\Delta D_{c3} = (s + 1) \cdot DO_l/2.86 \qquad (5.96)$$

As the sum of (s + 1) is generally less than 2, the effect of ΔD_{c3} on the post-D denitrification capacity will not be large. Reducing the DO_l concentration at the end of the nitrification zone (if reactor geometry permits this) will reduce both ΔD_{c1} and ΔD_{c3}, but at the expense of a reduced nitrification capacity (refer also to Figure 5.9). However, if the activated sludge system is overdesigned, a certain reduction in nitrification capacity can be tolerated.

EXAMPLE 5.11

An activated sludge system treats a wastewater with a TKN concentration of $153\,mg\,N \cdot l^{-1}$. The following system characteristics relevant to nitrogen removal may be assumed:

- $s = 1$;
- $N_{ad} = 1$; $N_l = 30$; $N_{oe} = 2$; $N_c = N_{ti} - N_l - N_{oe} - N_{ad} = 120\,mg\,N \cdot l^{-1}$;
- Pre-D system: $D_{c1} = 125\,mg\,N \cdot l^{-1}$;
- BDP system: $D_{c1} = 95$ and $D_{c3} = 30\,mg\,N \cdot l^{-1}$;
- $DO_l = 2\,mg\,O_2 . l^{-1}$;
- $N_{ted} = 10\,mg\,N \cdot l^{-1} \longrightarrow N_{ne} = N_{ted} - N_{oe} - N_{ad} = 7\,mg\,N \cdot l^{-1}$.

Calculate the expected effluent nitrate concentration as a function of the a-recirculation factor for both the pre-D and BDP configuration, taking into account the reduction of denitrification capacity in the pre- and post-D zones due to the effect of oxygen recirculation over the system (ΔD_{c1} and ΔD_{c3}).

Solution

The curves of N_{ne} for pre-D and BDP configuration are shown in Figure 5.29. As an example the N_{ne} values for $a = 6$ will be calculated.

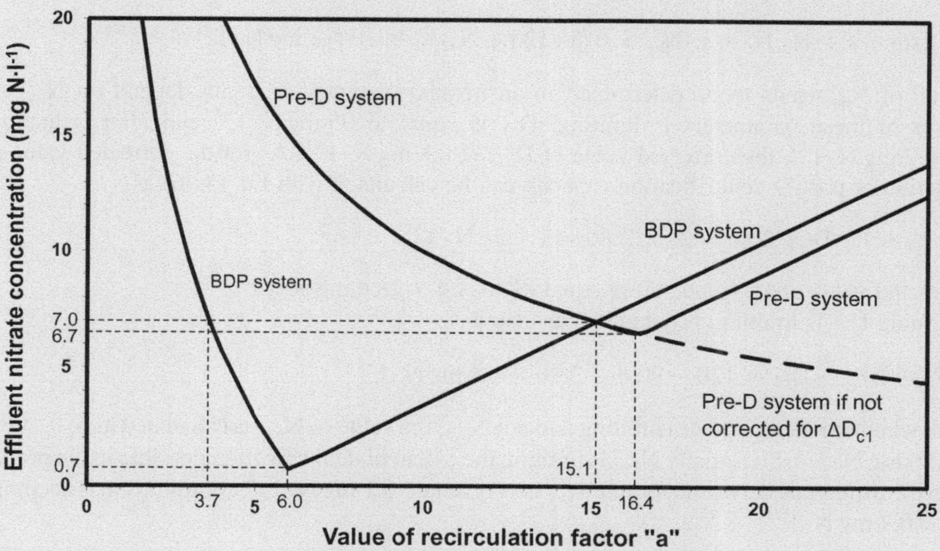

Figure 5.29 Lowest possible effluent nitrate concentration for the BDP and pre-D systems of Example 5.11 as function of the recirculation factor "a", when the effect of oxygen recirculation on the pre-D and post-D denitrification capacities is included

(1) Pre-D configuration

For a pre-D configuration, Eq. (5.70) can be used:

$$N_{av1} = (a + s)/(a + s + 1) \cdot N_c = 7/8 \cdot 120 = 105 \text{ mg N} \cdot l^{-1}$$

The reduction of the pre-D denitrification capacity can be calculated with Eq. (5.95) as:

$$\Delta D_{c1} = a \cdot DO_l/2.86 = 6 \cdot 2/2.86 = 4.2 \text{ mg N} \cdot l^{-1}$$

The pre-D denitrification capacity corrected for ΔD_{c1} is therefore equal to $D_{c1} = 125 - 4.2 = 120.8$ mg $N \cdot l^{-1}$. As the corrected value of D_{c1} is still larger than N_{av1}, the pre-D zone remains underloaded and $N_{ne} = N_c - N_{av1} = 120 - 105 = 15 \text{ mg N} \cdot l^{-1}$.

In this particular example (high value of N_{ti}), sufficient denitrification capacity is available in the pre-D reactor to compensate for the recycle of oxygen, at least when the a-recirculation factor has a value of 6. As can be observed in Figure 5.29, for a = 16.4 this is no longer the case, resulting in a rapid increase of the effluent nitrate concentration for a-values beyond this value. The minimum value of N_{ne} that can be obtained in the pre-D configuration of this example is 6.7 mg $N \cdot l^{-1}$ (for a = 16.4), slightly below the specified effluent limit. Note that theoretically, when ΔD_{c1} is ignored, N_{ne} can be reduced to 1.5 mg $N \cdot l^{-1}$ or less, when a > 100.

(2) Bardenpho configuration
If incomplete nitrogen removal is assumed, then Eq. (5.74) can be used

$$N_{av1} = a/(a + s + 1) \cdot N_c + s \cdot N_{ne} = 6/8 \cdot 120 + N_{ne} = 90 + N_{ne}\ \text{mg N} \cdot l^{-1}$$

The value of N_{ne} needs to be determined in an iterative manner, as it can depend on N_{av1} or D_{c1}, whichever of these parameters is limiting. D_{c1} is equal to 95 mg $N \cdot l^{-1}$ and after reduction with $\Delta D_{c1} = 4.2$ mg $N \cdot l^{-1}$, the corrected value of $D_{c1} = 90.8$ mg $N \cdot l^{-1}$. As for the corrected value of D_{c3}, the reduction of post-D denitrification capacity can be calculated with Eq. (5.96) as:

$$\Delta D_{c3} = (s + 1) \cdot DO_l/2.86 = 2 \cdot 2/2.86 = 1.4\ \text{mg N} \cdot l^{-1}$$

The corrected value of D_{c3} is therefore equal to 30–1.4 = 28.6 mg $N \cdot l^{-1}$

Assuming D_{c1} is limiting, N_{ne} can be calculated as:

$$N_{ne} = N_c - D_{c1} - D_{c3} = 120 - 90.8 - 28.6 = 0.6\ \text{mg N} \cdot l^{-1}$$

To check whether indeed D_{c1} was limiting and not N_{av1}, the value of N_{av1} is calculated as $6/8 \cdot 120 + 1 \cdot 0.6 = 90.6$ mg $N \cdot l^{-1}$. So actually N_{av1} is limiting the extent of denitrification possible in the pre-D zone, although the difference between D_{c1} and N_{av1} is very small (0.1 mg $N \cdot l^{-1}$). Equilibrium is accomplished for $N_{ne} = 0.7$ mg $N \cdot l^{-1}$:

$$N_{av1} = 90 + 0.7 = 90.7\ \text{mg N} \cdot l^{-1} \quad \text{and}$$

$$N_{ne} = N_c - N_{av1} - D_{c3} = 120 - 90.7 - 28.6 = 0.7\ \text{mg N} \cdot l^{-1}$$

It was assumed that D_{c3} was limiting and could be used to calculate N_{ne}, as denitrification was incomplete. To check this assumption, N_{av3} is calculated as $N_c - N_{av1} = 29.3$ mg $N \cdot l^{-1}$, which is indeed slightly larger than D_{c3}.

As can be observed in Figure 5.30, the calculated value of $N_{ne} = 0.7$ mg $N \cdot l^{-1}$ corresponds to the minimum concentration that can be obtained in a BDP configuration. As expected the BDP configuration is able to deliver a much better nitrate effluent concentration for the range of a-values of interest. When the a-recirculation factor is increased beyond a value of 6, then the pre-D zone of the BDP configuration is no longer capable to absorb the mass of oxygen recycled, reducing the value of $D_{c1} - \Delta D_{c1}$ below the value of N_{av1}. As a result, the effluent nitrate concentration will quickly increase.

It can be concluded from this example that selection of an appropriate a-recirculation factor is indeed of crucial importance. On the other hand, as long as the BDP system of Example 5.11 is operated with an a-recirculation factor between 3.5 to 15, the effluent nitrate concentration will comply to the limit of 7.0 mg $N \cdot l^{-1}$. It can be observed that, as already indicated previously, in general the highest degree of nitrate removal is obtained for relatively low values of the a-recirculation factor (4 < a < 8).

5.4.2.4 Design procedure for optimized nitrogen removal

In the previous sections the theory required to design an activated sludge system for nitrogen removal has been discussed. To conclude this topic, the general procedure for optimized design will be summarized here:

(1) Assume default values for the recirculation factors "a" and "s"
For instance $a = 4$ for a pre-D configuration and $a = 6$ for a BDP configuration. In both cases, a sludge recycle factor of $s = 1$ is recommended. This value should be validated during the optimised design of the system consisting of an aeration tank and a final settler (Chapter 8). For these values of "a" and "s" sufficient nitrate will be returned to the pre-D zone while the decrease in denitrification capacity resulting from the recycle of oxygen to the anoxic zones will be limited. Furthermore, this reduction will be partly compensated by the denitrification in the final settler (for incomplete nitrogen removal only).

Depending on the temperature of the mixed liquor, typically about 6–8 mg of denitrified nitrogen per litre of return sludge can be accepted before the produced nitrogen gas will result in problems with rising sludge (Henze et al., 1992), refer also to Appendix A8. For the high sludge age typically applied to nitrogen removal systems, the extent of denitrification in the final settler is generally less than $6 - 8$ mg $N \cdot l^{-1}$, due to the low active fraction in the sludge fraction, which will result in a low rate of endogenous denitrification.

(2) Specify the required effluent nitrogen quality
This means attributing values to N_{ad}, N_{oe} and N_{te}. This by default determines the maximum allowed effluent nitrate concentration, as $N_{ne} = N_{te} - N_{ad} - N_{oe}$. Refer to Section 5.2.4 for more information on the proper selection of N_{ad} for the design of the nitrification process.

(3) Plot N_c/S_{bi}, $(N_c/S_{bi})_o$ and $(N_c/S_{bi})_l$ as function of R_s
Use Eqs. (5.86 and 5.93) to plot the graphs. Alternatively N_{ti}/S_{ti}, $(N_{ti}/S_{ti})_o$ and $(N_{ti}/S_{ti})_l$ can also be used. If this is desired, use Eqs. (5.88 and 5.94). Check if complete nitrogen removal is possible at reasonable values of R_s (zone A in Figure 5.27). If so, the minimum sludge age for complete nitrogen removal can be determined from the intersection of N_c/S_{bi} with $(N_c/S_{bi})_o$.

- Whenever possible, design for complete nitrogen removal. Use Eqs. (5.83 and 5.84) to determine the values of f_{x1} and f_{x3}. As for the effluent quality: $N_{ae} = N_{ad}$; $N_{ne} = 0$; $N_{te} = N_{ke} = N_{ae} + N_{oe}$. Compensate for oxygen recirculation if needed;
- If complete nitrogen removal is not possible, one should determine whether a BDP or a pre-D configuration is most advantageous. A BDP configuration should be selected when the ratio N_c/S_{bi} is located in Zone B, while a pre-D configuration is better when N_c/S_{bi} is located in Zone C of Figure 5.27.

(4) In case of incomplete nitrogen removal, check if the effluent nitrogen limits are met
To do so, only the effluent nitrate concentration is calculated, as N_{ae} and N_{oe} have already been specified. For the pre-D configuration use Eqs. (5.71 or 5.72):

- $N_{ne} = N_c - D_{c1}$ for $R_{sm} < R_s < R_{si}$ (overloaded pre-D zone);
- $N_{ne} = N_c/(a + s + 1)$ for $R_s > R_{si}$ (underloaded pre-D zone → this is in fact an incorrect choice as a BDP configuration should have been selected instead of a pre-D system;
- The value of f_{x1} is equal to f_m.

For the BDP configuration use Eqs. (5.77 or 5.78):

- This is not applicable for $R_{sm} < R_s < R_{si}$ (overloaded pre-D zone), as in this case a pre-D configuration will result in better nitrogen removal;
- $N_{ne} = N_c/(a + s + 1) - D_{c3}/(s + 1)$ for $R_s > R_{si}$ (fully loaded pre-D zone i.e. $N_{av1} = D_{c1}$);
- Calculate the value of f_{x1} with Eq. (5.92), which defines f_{x3} as $f_m - f_{x1}$.

(5) If a suitable solution cannot be obtained, consider the following actions

- Increase the sludge age;
- Increase the value of nitrate recirculation factor "a". However, remember that an increase of a to values higher than 8 will yield only *very* limited benefits. Furthermore, beware of the reduction of D_{c1} due to return of oxygen: $\Delta D_{c1} = a \cdot DO_l/2.86$, where DO_l typically is equal to $DO_{sp} = 2\ mg\ O_2 \cdot l^{-1}$
- Decrease the N_{ti}/S_{bi} ratio. For instance bypass flow around the primary settler or anaerobic pre-treatment (if applicable) or consider external carbon source dosing (e.g. methanol);

(6) Finalise the design with the theory presented in this book

Among other things, this includes the calculation of:

- Volume and total sludge mass;
- Excess sludge production and aeration demand;
- Final settler-, sludge thickener- and sludge digester volume.

EXAMPLE 5.12

For the design case detailed below, calculate the sludge age for which complete nitrogen removal is possible and the sludge age for which complete nitrogen removal is no longer possible and pre-D and BDP systems yield comparable results. Characterize the nitrogen removal performance for both cases, using the following data:

- $S_{ti} = 650\ mg\ COD \cdot l^{-1}$; $T = 12°C$;
- $f_{ns} = 0.1$; $f_{np} = 0.12$ and $f_{sb} = 0.25$;
- $N_{ti} = 50$; $N_{ad} = 1$ and $N_{oe} = 2\ mg\ N \cdot l^{-1}$;
- $b_h = 0.18d^{-1}$; $K_2 = 0.065$ and $K_3 = 0.063\ mg\ N \cdot mg^{-1}\ VSS \cdot d^{-1}$;
- $\mu_m = 0.16d^{-1}$; $K_n = 0.40\ mg\ N \cdot l^{-1}$ and $b_n = 0.03d^{-1}$;
- $a = 4$ and $s = 1$; $f_{max} = 50\%$.

Ignore the effect of oxygen being introduced to the pre-D and post-D zones.

Solution

Use Eqs. (5.38 and 5.39) to calculate the minimum sludge age for nitrification (R_{sn}) and the minimum sludge age for which inclusion of an anoxic zone becomes possible (R_{sm}), i.e. when $N_{ae} = N_{ad}$ (while $f_m = 0$):

$$R_{sn} = 1/(\mu_m - b_n)$$
$$= 1/(0.16 - 0.03) = 7.9\ days$$

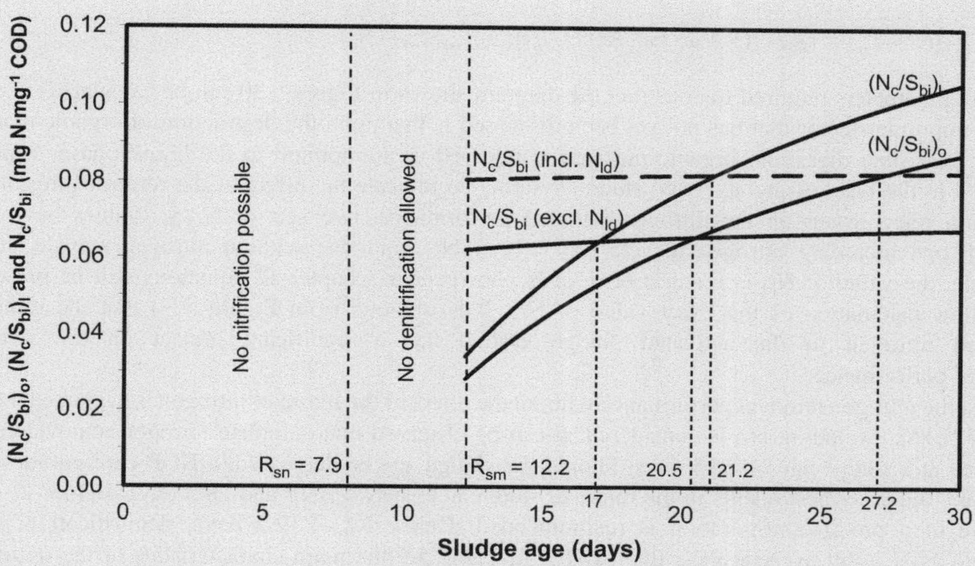

Figure 5.30 Graphical determination of intersection of N_c/S_{bi} with $(N_c/S_{bi})_o$ and $(N_c/S_{bi})_l$ with and without considering release of nitrogen during digestion (N_{ld})

$$R_{sm} = 1/[\mu_m/(1 + K_n/N_{ad}) - b_n]$$
$$= 1/[0.16/(1 + 0.40/1.0) - 0.03] = 12.2 \text{ days}$$

Now, use Eqs. (5.86 and 5.93) to construct plots of N_c/S_{bi}, $(N_c/S_{bi})_o$ and $(N_c/S_{bi})_l$ as function of the sludge age, which can be used to determine:

- The sludge age that allows full nitrogen removal: intersection of N_c/S_{bi} and $(N_c/S_{bi})_o$;
- The sludge age where BDP ceases to advantageous: intersection of N_c/S_{bi} and $(N_c/S_{bi})_l$.

$$(N_c/S_{bi})_o = \frac{(a + s + 1) \cdot (f_{dn} \cdot f_{sb} + K_2 \cdot C_r \cdot f_m)}{a + (K_2/K_3) \cdot (s + 1)} \tag{5.86}$$

$$(N_c/S_{bi})_l = (f_{dn} \cdot f_{sb} + K_2 \cdot C_r \cdot f_m) \cdot (a + s + 1)/(a + s) \tag{5.93}$$

Most of the parameters required to calculate $(N_c/S_{bi})_o$ and $(N_c/S_{bi})_o$ have already been specified, with the exception of C_r, f_m and N_c.

$$C_r = Y \cdot R_s/(1 + b_h \cdot R_s) \tag{3.30}$$

$$f_m = 1 - (1 + K_n/N_{ad}) \cdot (b_n + 1/R_s)/\mu_m \tag{5.47}$$

$$N_c = N_{ti} - N_l - N_{ad} - N_{oe} \tag{5.50}$$

Finally, to calculate N_c, the value of N_l is required

$$N_l = f_n \cdot [(1 - f_{ns} - f_{np}) \cdot [(1 + f \cdot b_h \cdot R_s) \cdot C_r/R_s + f_{np}/f_{cv}] \cdot S_{ti} \quad (3.59)$$

Now all parameters required to construct the diagram shown in Figure 5.30 can be calculated.

An important factor that has not yet been discussed is that upon the destruction of organic material during anaerobic digestion, organic nitrogen is released as ammonium to the liquid phase, which is returned to the head of the activated sludge system. To indicate the effect of the return of this sludge digestion reject water on the nitrogen removal performance, two sets of N_c/S_{bi} values have been plotted: one including nitrogen recycle ($N_c' = N_c + N_{ld}$) and one without nitrogen recycle. In the example, the value of N_{ld} is fixed at 60% of N_l, however in Chapter 12 equations will be presented that allow calculation of the exact value of N_{ld}. It is obvious from Figure 5.30 that the return of digested nitrogen to the activated sludge system has a significant impact on the nitrogen removal performance.

First the nitrogen removal performance without the effect of the return of nitrogen is evaluated, i.e. the line of N_c/S_{bi} excluding N_{ld} is considered. It can be observed that complete nitrogen removal can be obtained at a sludge age of 20.5 days. Should the sludge age be decreased, a BDP configuration will continue to deliver best results in the range of sludge ages between 17 and 20.5 days. Below 17 days, the use of a pre-D configuration is recommended. Below $R_{sm} = 12.2$ days, denitrification is not possible as N_{ae} still exceeds the value of N_{ad}. In Table 5.9 the main characteristics of the optimized solutions are listed:

5.9 System characteristics of the optimised solutions of Example 5.12 (excl. N_{ld})

Parameter	Incomplete N removal (pre-D)	Complete N-removal (BDP)	Eq. no.
R_s	17.0	20.5	–
f_{x1}	0.20	0.12	5.83 / 5.92
f_{x3}	–	0.17	5.84
N_c	32.7	33.3	5.51
D_{c1} ($= N_{av1}$)	27.2	22.2	5.68
D_{c3} ($= N_{av3}$)	–	11.1	5.69
N_{ne}	5.4	0.0	5.77/5.71

When the return of nitrogen in the reject water is considered, this has the following effects:

- The sludge age when a Pre-D system ceases to be advantageous over a BDP configuration, i.e. the intersection of N_c/S_{bi} and $(N_c/S_{bi})_l$: R_s increases from 17.0 to 21.2 days;
- The sludge age when complete nitrogen removal becomes feasible, i.e. the intersection of N_c/S_{bi} and $(N_c/S_{bi})_o$: R_s increases from 20.5 to 27.2 days.

Chapter 6

Innovative systems for nitrogen removal

6.0 INTRODUCTION

Tertiary treatment systems for nitrogen rem oval have now been around for several decades and in many cases excellent nitrogen removal has been demonstrated. However, under unfavourable conditions it may be difficult to obtain the desired level of nitrogen removal efficiency, for instance because:

- When nitrogen systems are overloaded, priority is given to nitrification. The anoxic sludge mass fraction is then often reduced to a level that insufficient denitrification capacity remains for proper denitrification;
- The ratio between TKN and COD (or actually N_c/S_{bi}) in the influent is high, which makes nitrogen removal more difficult, as the nitrate production is directly related to the TKN concentration in the influent, whereas the denitrification capacity is directly linked to the presence of (biodegradable) COD;
- Anaerobic sludge digestion is now commonly applied, either on site or at a central sludge treatment facility. During the solids digestion process, a large quantity of nitrogen is released to the liquid phase ($\pm$ 10% of the digested sludge mass), which will be returned to the activated sludge system, where it will increase the TKN/COD ratio in the influent;
- When the tertiary treatment plant combines biological phosphorus removal with nitrogen removal, part of the unaerated sludge mass will be allocated to the anaerobic zone instead of the anoxic zone (often 10–15% of the total sludge mass);
- In bio-P removal systems, the mechanism responsible for phosphorus removal from the wastewater is through discharge with the excess sludge. Therefore, a low sludge age enhances bio-P removal at the expense of nitrogen removal, whereas the opposite is true for a high sludge age;
- Application of primary settlers or anaerobic pre-treatment units will increase the ratio between TKN and COD in the pre-treated wastewater. In the case of primary settling only suspended solids will be removed, whereas most of the nitrogen will be present in soluble form as NH_4-N. When anaerobic pre-treatment is applied, this will be further aggravated as now a large fraction of the soluble biodegradable COD will be removed as well, while again the soluble NH_4-N will not be affected.

Because of the high rate of population growth, especially in developing countries, many wastewater treatment plants are becoming overloaded. Furthermore, effluent discharge limits tend to become more strict worldwide. Concerning nitrogen removal, in the European Union current typical effluent discharge limits are <1–2 mg NH_4-N $\cdot l^{-1}$ and <10 mg $\cdot l^{-1}$ total nitrogen. The limits for discharge into vulnerable water bodies are even stricter, requiring the application of novel reactor concepts such as the membrane bioreactor or installation of an effluent polishing step, for example chemically- or biologically enhanced sand filtration.

For all of these reasons, in the last two decades significant research effort has been directed towards developing technologies that increase the nitrogen removal performance of existing wastewater treatment plants, while avoiding the need for a costly expansion of the activated sludge system volume. A focal point has been the separate treatment of the reject water that originates from the dewatering of (anaerobically) digested sludge and from the sludge drying units, both of which are very rich in ammonium.

As discussed in Chapter 3, there is a significant nitrogen demand for excess sludge production. In the case of municipal sewage, this typically amounts to 15 to 40% of the nitrogen load, depending on the applied sludge age and the influent COD/N ratio.

Unless a very high sludge age is applied, the excess sludge must be stabilised by anaerobic digestion, before it is dewatered. In the digestion process, the nitrogen associated with the mineralised sludge is released to the water phase as ammonium. After phase separation, the liquid phase of the digester effluent is returned to the activated sludge process, thus increasing the nitrogen load to be removed. During sludge stabilisation typically some 30 to 40% of the volatile suspended solids are digested. Therefore, the nitrogen content of the digested sludge represents 5 to 15% of the nitrogen load in municipal sewage. When a primary settler is present, or on larger plants with centralized sludge digestion facilities, this fraction will be even higher. The subject of anaerobic digestion, including calculation of the quantity of nitrogen released during the digestion process, will be discussed in Chapter 12. The soluble nitrogen concentration in the digester is typically in the range of 500–1500 mg N $\cdot l^{-1}$ while, because of the need to heat the digester, the reject water temperature is relatively high as well at 30–35°C. Due to the ammonification of organic nitrogen to ammonium in the digester, the generated alkalinity is approximately equivalent to the molar ammonium concentration. Therefore the effluent of an anaerobic digester contains about half of the alkalinity required to compensate for the alkalinity demand for full nitrification.

Due to the high nitrogen concentration it can be advantageous to treat this reject water in a separate side stream process, especially if the main activated sludge process is overloaded and cannot meet the desired effluent limits. Research into this area started in the 1990s. For example, in the Netherlands the foundation of applied water research (STOWA) funded several studies in the period 1994–1998 with the specific objective to develop and test new biological and physical-chemical methods for the removal of nitrogen from sludge digestion reject water. Since then, several of the most promising processes have been upscaled and delivered to the market. The following processes will be discussed here:

- Nitrogen removal over nitrite instead of over nitrate (nitritation – denitritation);
- Anaerobic ammonium oxidation;
- Combined nitritation and anaerobic ammonium oxidation;
- Bioaugmentation.

The nitritation–denitritation process is a modification of the traditional systems for biological nitrogen removal where ammonium is oxidised by autotrophic organisms in a two step process (nitrification) and the oxidised form of nitrogen is reduced to molecular nitrogen by heterotrophic bacteria in an anoxic environment (denitrification). However, in this case the process conditions (e.g. sludge age, pH, DO concentration and temperature) are controlled to inhibit the second step in the nitrification process, i.e. the conversion of nitrite to nitrate. Hence the oxidised form of nitrogen in this process is nitrite and not nitrate. This reduces oxygen consumption while simultaneously less organic material is required for denitrification.

The process of anaerobic ammonium oxidation involves the application of a recently discovered micro-organism that is able to use nitrite for the oxidation of ammonium, with molecular nitrogen as the main end product. This process is actually anoxic, due to the presence of both nitrite and nitrate, but as the term anaerobic ammonium oxidation is now widely used we have not changed it in this book. This process has the fundamental advantage that there is no need for organic material to reduce the oxidised nitrogen. However, it can only be applied if both nitrite and ammonium are present in approximately equimolar concentrations in the wastewater, which seldom is the case. Therefore application of anaerobic ammonium oxidation needs to be combined with the nitritation process, acting as a source of nitrite. Finally, bioaugmentation is the most conventional of the new developments. The reject water is treated conventionally in a side stream process, where after the biomass rich in nitrifiers is returned to the main activated sludge system. This seeding effect increases the nitrification capacity or alternatively, allows operation at reduced aerobic sludge age.

6.1 NITROGEN REMOVAL OVER NITRITE

As can be observed from Figure 6.1, biological nitrogen removal in the activated sludge process can follow two pathways: (I) ammonium oxidation to nitrate and subsequent denitrification (reduction) of nitrate to molecular nitrogen (N_2) or (II) ammonium oxidation to nitrite and denitritation of nitrite to N_2. The nitrification reactions can be written as:

$$NH_4^+ + 1\tfrac{1}{2}O_2 \longrightarrow NO_2^- + H_2O + 2\,H^+ \quad \text{(ammonium oxidation or nitritation)} \tag{5.2a}$$

$$NO_2^- + \tfrac{1}{2}O_2 \longrightarrow NO_3^- \quad \text{(nitrite oxidation)} \tag{5.2b}$$

From these equations it can be observed that nitrogen removal over nitrite can result in a considerable reduction in oxygen demand, as oxidation to nitrite requires only 75% of the oxygen demand of oxidation to nitrate.

Similar to the overall redox reaction for the denitrification process, the redox reaction for denitritation can be derived as follows:

(1) Oxidation reaction of organic material:

$$C_xH_yO_z + (2x - z)H_2O \longrightarrow xCO_2 + (4x + y - 2z)H^+ + (4x + y - 2z)e^- \tag{2.1a}$$

(2) Nitrite reduction:

$$e + \tfrac{4}{3}H^+ + \tfrac{1}{3}NO_2^- \longrightarrow \tfrac{1}{6}N_2 + \tfrac{2}{3}H_2O \tag{2.9c}$$

(3) Overall redox reaction:

$$C_xH_yO_z + \tfrac{1}{3}(4\cdot x + y - 2\cdot z)H^+ + \tfrac{1}{3}(4\cdot x + y - 2\cdot z)/3\,NO_2 \longrightarrow xCO_2 + \tfrac{1}{6}\cdot(4\cdot x + y - 2\cdot z)N_2 + \tfrac{1}{3}(2\cdot x + 2\cdot y - z)H_2O \tag{6.1}$$

Sludge reject water does not contain sufficient biodegradable COD to remove the nitrate or nitrite produced from nitrification or nitritation. Therefore addition of an external carbon source is required if nitrogen removal is desired. When it is assumed that methanol is used (one of the cheaper commercial carbon sources available), the following (catabolic) redox reaction equations can be written:

$$NO_2^- + \tfrac{1}{2}CH_3OH + H^+ \longrightarrow \tfrac{1}{2}N_2 + \tfrac{1}{2}CO_2 + 1\tfrac{1}{2}H_2O \tag{6.2}$$
$$NO_3^- + \tfrac{5}{6}CH_3OH + H^+ \longrightarrow \tfrac{1}{2}N_2 + \tfrac{5}{6}CO_2 + 2\tfrac{1}{6}H_2O \tag{6.3}$$

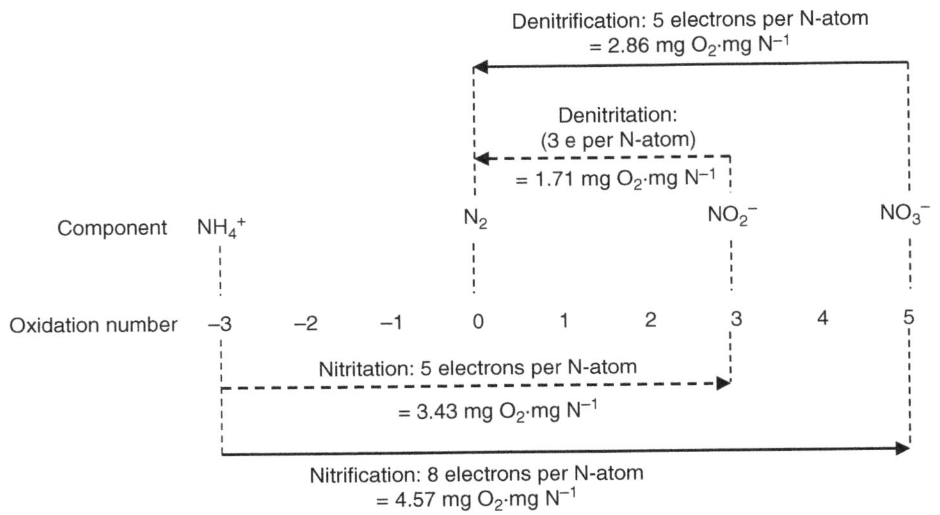

Figure 6.1 Variation of the oxidation number of the nitrogen atom in the processes of full- and partial nitrification and -denitrification

From these equations it can be concluded that removal of nitrite requires only $0.5/0.83 = 60\%$ of the COD required for removal of nitrate. However, note that Eqs. (6.2 and 6.3) only consider the effect of the *catabolic* reactions, as the anabolic reactions are ignored (cell mass growth followed by decay/endogenous respiration). Therefore, depending on the applied sludge age, the actual COD consumption will be significantly higher. The theory presented earlier in Chapter 5 can be used to predict the COD consumption for each specific case.

6.1.1 Basic principles of nitritation

There are basically two approaches that can be used to force the biological nitrogen removal process to use the nitrite route instead of the nitrate route (Van Loosdrecht, 2008):

- Selection based on specific growth rate, resulting in removal of the nitrite oxidisers from the system (selective wash-out);
- Applying suboptimal conditions, for instance a low dissolved oxygen concentration, a high nitrite- or ammonium concentration or an unfavourable pH. In this case nitrite oxidation will only be partly

inhibited. Therefore a second selection factor is required: for example removal of the produced nitrite which will deprive the nitrite oxidisers of their substrate. This application will be discussed in Section 6.3.3.

At temperatures below 20°C, oxidation of nitrite generally proceeds at a higher rate than oxidation of ammonium, whereas the opposite is true for temperatures above 20°C. At higher temperatures the difference in oxidation rate becomes more accentuated and under those circumstances it is possible to limit the two-step nitrification process to the first step (nitritation) only and thus to prevent the generation of nitrate. In Figure 6.2a, typical net growth rates (μ_m–b_n) for both ammonium- and nitrite oxidisers are plotted as a function of the temperature. The corresponding minimum aerobic sludge age can be calculated from Eq. (5.38) as $R_{sn} = 1/(\mu_m - b_n)$ and is indicated in Figure 6.2b. In the temperature range of practical interest to most activated sludge processes (10–25°C), the growth rate of the ammonium oxidisers is either lower than or practically equal to the growth rate of the nitrite oxidisers, which makes it very difficult to limit the nitrification process to the generation of nitrite. However, above 25°C the difference in maximum growth rate and hence in required minimum aerobic sludge age becomes more significant. As heated anaerobic sludge digesters typically operate at 30 to 37° C, in the case of reject water treatment it now becomes possible to use the sludge age as a selection parameter to induce nitrogen removal to nitrite.

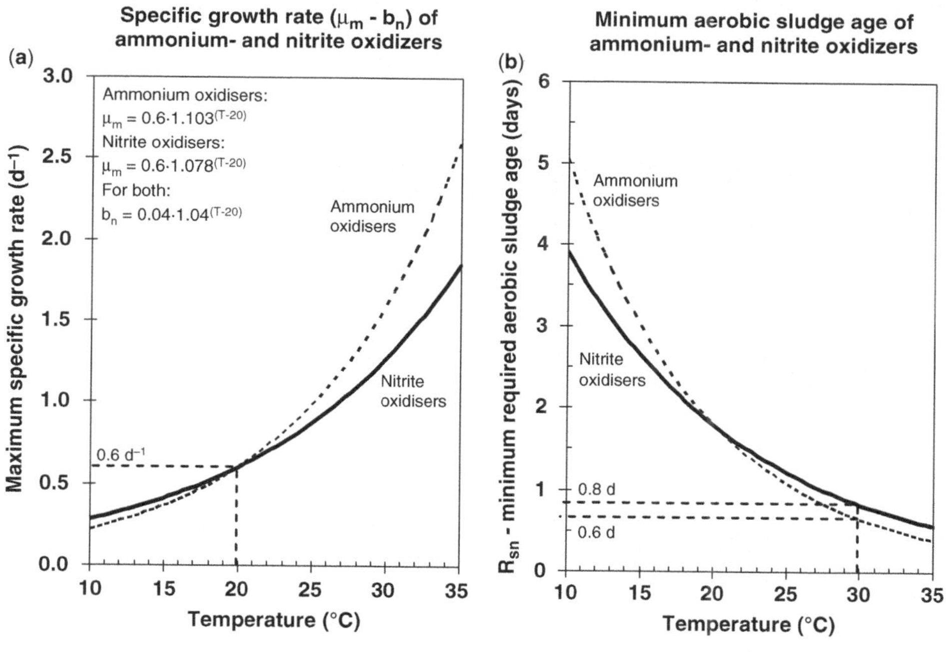

Figure 6.2 Typical profiles of net growth rate (Fig. a) and minimum required aerobic sludge age (Fig. b) for ammonium- and nitrite oxidisers, as function of the temperature: adapted from Veldhuizen et al. (1997) and Jetten et al. (2000)

This is the philosophy behind the SHARON process (Single reactor for High activity Ammonium Removal Over Nitrite, recently renamed Stable High Ammonium Removal Over Nitrite), the first nitritation-denitritation system to be implemented on full-scale. The applied aerobic sludge age depends on the operational temperature, as shown in Figure 6.2b. Because the design sludge age is so low and considering that the volume of reject water to be treated is small, it becomes feasible to operate the reactor as a chemostat: i.e. a reactor without sludge retention, in which the sludge age is equal to the hydraulic retention time.

In principle it is also possible to use a system with biomass retention, for instance an SBR or MBBR. This allows the volume of the reactor to be reduced, but only within certain limits as oxygen transfer will soon become a limiting factor (as will excessive foaming). In general, for ammonium concentrations higher than 400–500 mg $N \cdot l^{-1}$ there is no advantage to the use of a reactor configuration with sludge retention (Van Loosdrecht, 2008).

Figure 6.3 shows a simplified flow scheme of a sewage treatment plant that includes a single reactor nitritation-denitritation process for sidestream nitrogen removal. The nitritation- denitritation process consists of a single completely mixed reactor that is operated at a temperature between 30–35°C and is subjected to alternating aerobic and anoxic conditions. The aerobic sludge age is controlled at a value that is low enough to prevent growth of the nitrite oxidisers. All excess sludge is discharged with the effluent. As a result, the suspended solids concentration in the effluent of the nitritation-denitritation reactor will be relatively high and consequently the effluent will contain some organic nitrogen as well.

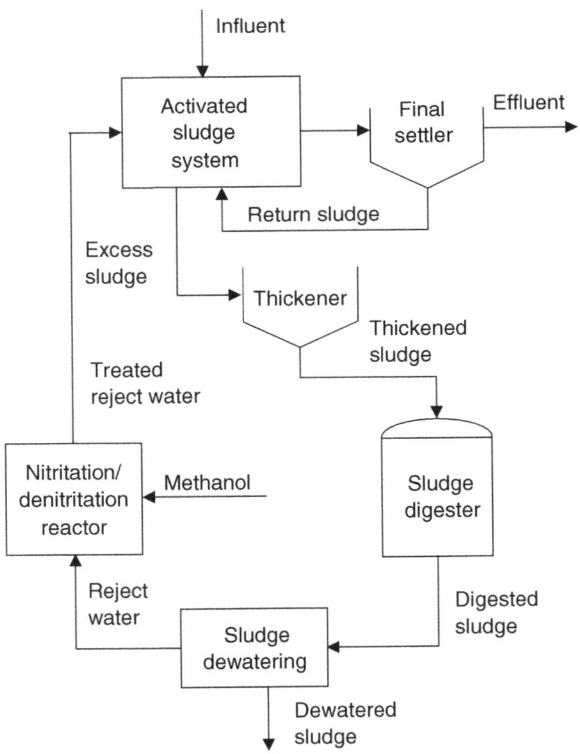

Figure 6.3 Flow scheme of a sewage treatment plant including nitrogen removal over nitrite

Micro-organisms will predominantly be present as (clusters of) free bacteria rather than conglomerated into sludge flocs. The presence of suspended solids in the effluent is generally not a problem, as the effluent will be sent to the main activated sludge system, where the suspended solids and bacteria will be rapidly flocculated onto the activated sludge flocs.

The combination of a short hydraulic residence time with a high ammonium influent concentration allows high volumetric nitrogen loading rates to be applied: full-scale SHARON reactors have demonstrated ammonium conversion rates between 0.4 to 0.8 kg $N \cdot m^{-3} \cdot d^{-1}$ with conversion percentages of 80 to 95%. Recent (undisclosed) pilot research by Biothane Systems International demonstrated loading rates between 0.8–1.0 kg $NH_4\text{-}N \cdot m^{-3} \cdot d^{-1}$ with similar conversion efficiencies for an MBBR reactor. At higher loading rates other factors may become limiting, such as substrate inhibition and/or -toxicity, oxygen transfer- and diffusion rates and an excessive reactor temperature because of the release of reaction heat.

6.1.2 Kinetics of high rate ammonium oxidation

The value of the maximum growth rate of the high-rate ammonium oxidiser variant (μ_m) is around 1.5–2.5 d^{-1} in the temperature range of interest (30–35°C). As for the decay rate, this can be estimated from the data reported by Jetten et al. (2000) on the loss in nitritation rate observed after interruption of the feed to a lab-scale nitritation reactor for a prolonged period of time and at different temperatures. When it is assumed that the decrease in both maximum ammonium removal rate and -nitrite production rate is directly proportional to the decrease in the active ammonium oxidising biomass, then the value of b_n can be calculated from $(dX_n/dt)_d = -b_n \cdot X_n$, or $X_{n,t} = X_{n0} \cdot \exp(-b_n \cdot t)$. From Jetten's data, the value of b_n can be estimated as 0.23 d^{-1} at 35°C, much larger than the value typically found for nitrifiers: $b_n = 0.04 \cdot 1.04^{(35-20)} = 0.07\ d^{-1}$. The difference is possibly due to increased predation, as the bacteria are dispersed in suspension instead of concentrated (and less accessible) in sludge flocs.

A disadvantage of selecting on fast-growing or "feast" types of micro-organisms, used to an abundant supply of substrate, is the lower substrate affinity compared to slow-growing or "famine" type of organisms that are normally encountered in the activated sludge process. Jetten et al. (2000) established that the K_n value of fast-growing ammonium oxidisers (identified as Nitrosomonas eutropha) ranged from 20 to 60 mg $NH_4\text{-}N \cdot l^{-1}$ at temperatures between 30 to 35°C. For K_{n20} values between 0.5 and 1.0 mg $N \cdot l^{-1}$ and using the temperature dependency relationship $K_{nT} = K_{n20} \cdot 1.123^{(T-20)}$, the K_n value of conventional nitrifiers is estimated at 1.5–5.5 mg $N \cdot l^{-1}$ for the same temperature range.

Due to the combination of a high value of K_n and a short aerobic sludge age, it will not be possible to obtain a low effluent ammonium concentration. In practice, full-scale nitritation–denitrification systems report effluent ammonium concentrations between 10–50 mg $N \cdot l^{-1}$. As can be observed in Figure 6.4, this corresponds well with the range of N_{ae}-values predicted by Eq. (5.36), calculated for different values of K_n (20, 40 and 60 mg $N \cdot l^{-1}$). For comparison the curve for conventional nitrification is indicated as well (for $K_n = 3$ mg $N \cdot l^{-1}$). As digestion reject water contains sulphide, a conservative value of the maximum growth rate has been selected: $\mu_m = 0.2\ d^{-1}$ at 20°C. A further decrease of the ammonium effluent concentration of a nitritation reactor would require an increase in the applied aerobic sludge age. However, this would invalidate the whole concept of the nitritation-denitritation process, as at higher values of the sludge age nitrite oxidizers will again become established and conversion of nitrite to nitrate can no longer be prevented, at the expense of additional oxygen- and COD demand. Furthermore, the main treatment objective of a side-stream nitrogen removal process is *not* maximum ammonium removal. When sludge digestion reject water is treated, it is sufficient to remove the bulk of the nitrogen load it contains: the effluent of the nitritation-denitritation reactor will be returned to the activated sludge process where residual ammonium and nitrite will be removed.

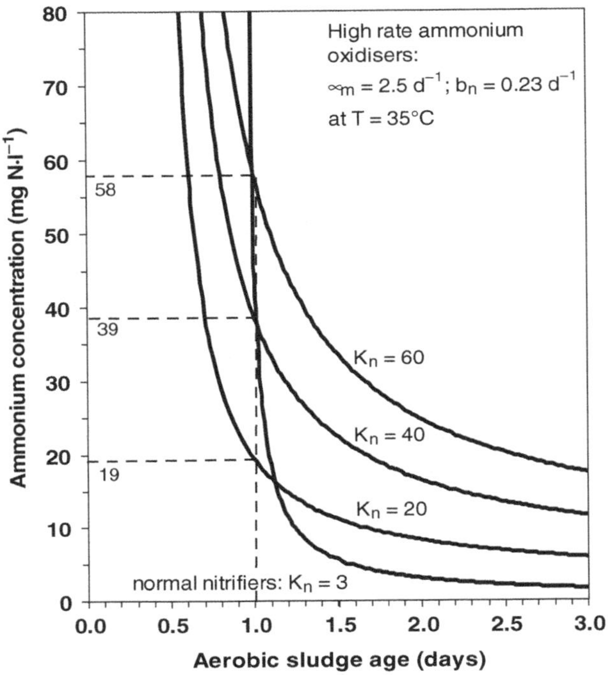

Figure 6.4 Influence of the K_n value on the effluent ammonium concentration

The value of the overall nitrifier yield Y_n is reported in literature as 0.10 to 0.15 g VSS · g^{-1} N oxidised, with a typical value of 0.12 g VSS · g^{-1} N, i.e. for the complete oxidation reaction from ammonium to nitrate (Metcalf & Eddy, 2003). Currently no values have been reported for the yield of the two individual groups of nitrifiers. However, an estimate can be made based on the following reasoning: the complete oxidation of one molecule of NH_4^+ to NO_3^- will deliver 8 electrons to the micro-organisms. The first step, oxidation of NH_4^+ to NO_2^- supplies 6 electrons, while the second step, oxidation of NO_2^- to NO_3^-, supplies only 2 electrons. Assuming the value of the yield coefficient is directly proportional to the number of electrons transferred, then $Y_{ao} = 0.09$ g VSS · g^{-1} N and $Y_{no} = 0.03$ g VSS · g^{-1} N.

6.1.3 Reactor configuration and operation

In the nitrification process, regardless of whether oxidation proceeds to nitrite or nitrate, the removal of 1 mole of ammonium results in an acid production of 2 moles of protons (H^+) or the equivalent consumption of 2 moles of bicarbonate (HCO_3^-). A pH value outside the range between 6.5 and 8.5 results in severe inhibition of the nitrification process, due to the presence of free (undissociated) nitrous acid or free ammonia. Not surprisingly, the same type of pH dependency applies to the nitritation-denitritation process, where Jetten et al. (2000) observed that the conversion efficiency of ammonium into nitrite is maximum for pH values between 7.1 and 7.8. At a pH value below 7.1 the conversion efficiency quickly decreases: while at a pH value of 6.8 about 50% of the influent ammonium is still converted, this ceases completely when the pH is reduced to a value of 6.4 or less. During anaerobic sludge digestion one mole of bicarbonate will be produced per mole of mineralised organic nitrogen, while two moles are required for the subsequent nitritation to NO_2^-. As a result, the alkalinity in the reject water will be consumed by the time that fifty percent of the ammonium has been converted.

To compensate for the decrease in alkalinity and pH, an inorganic base can be added, such as $Ca(OH)_2$ or NaOH. Alternatively the produced nitrite or nitrate can be denitrified, which will also generate alkalinity. In practice methanol is often added as a carbon source for denitrification, as this is cheaper than the addition of an inorganic base and because it will simultaneously reduce the demand for COD in the activated sludge system to which the treated reject water is returned. Depending on the alkalinity initially present in the reject water, partial- or full denitrification may be required for pH control. Based on a review of several full-scale nitritation-denitritation reactors, Van Betuw et al. (2008) observed that in the case of sludge digestion reject water treatment, a denitrification efficiency of 70% is sufficient to maintain a suitable pH value. Overall ammonium removal efficiencies of 80–98% have been observed in full-scale nitritation reactors, operated at aerobic retention times of 1.3 to 1.8 days.

In general, the nitritation-denitritation system can be either constructed as a single tank with alternating aerobic- and anoxic periods or as a series of two tanks, the first tank aerobic and the second one anoxic. The main advantage of the single tank concept is simplicity of construction. Furthermore the alkalinity produced during denitrification is immediately available to compensate for the acid produced in the aerobic phase. On the other hand, if the reactor is fed in the anoxic phase, part of the ammonium in the feed will be discharged directly with the effluent. However, the effect of this "short-circuiting" on the effluent ammonium concentration is limited, due to the dilution of the influent in the tank. A second advantage of the single reactor configuration is that the size of the anoxic and aerobic sludge mass fractions can easily be adjusted by manipulation of the duration of the aerated and anoxic periods. On the other hand, the installed aeration capacity will be significantly larger, as during the periods of aeration the oxygen transfer rate will necessarily have to be higher to compensate for the unaerated period. For the configuration with two tanks, a recirculation pump will be required to return the produced alkalinity in the denitrification tank to the nitrification tank. The advantage of this configuration is that the influent will always be nitrified first and therefore the ammonium effluent concentration will be lower than in the other configuration. On the other hand, methanol consumption will be slightly higher as the easily biodegradable influent COD in the reject water will be metabolised in the aerobic zone. A typical process cycle of the single tank nitritation-denitritation reactor consists of an aerobic period of 80 minutes followed by an anoxic period of 40 minutes (Ten Have, 2004). Therefore in the two tank configuration the aerobic tank will be twice the size of the anoxic tank.

6.1.4 Required model enhancements

Nitrogen removal over nitrite can be incorporated in the ideal steady state model, providing the following modifications are introduced:

(1) The nitrifiers are included as a biomass fraction: this subject is discussed in Appendix A6.2

(2) The oxygen demand for nitritation is reduced by 25% compared to that for nitrification

$$MO_n = 0.75 \cdot 4.57 \cdot MN_c = 3.42 \cdot MN_c \tag{6.4}$$

(3) The equivalent oxygen recovery for denitritation is reduced by 40% as well

$$MO_{eq} = 0.6 \cdot 2.86 = 1.72 \cdot MD_c (\text{or actually } 1.72 \cdot MN_d) \tag{6.5}$$

(4) The value of f_{dn} is increased to reflect the decreased COD demand for reduction of nitrite

$$f_{dn} = (1 - f_{cv} \cdot Y_h)/(0.6 \cdot 2.86) = 0.189 \tag{6.6}$$

6.2 ANAEROBIC AMMONIUM OXIDATION

Mulder et al. (1995) were the first to observe the removal of ammonium in an anoxic fluidised bed reactor, under simultaneous conversion of nitrate to nitrogen gas. The stoichiometrics of the reaction in this reactor were experimentally determined as:

$$5\,NH_4^+ + 3\,NO_3^- \longrightarrow 4\,N_2 + 9\,H_2O + 2\,H^+ \tag{6.7}$$

Van de Graaf et al. (1996) later demonstrated that it involved a biological process, as it could be inactivated by heating, antibiotics and gamma radiation. The bacteria involved have since been identified by Strous et al. (1999) as belonging to a group of micro-organisms called Planctomycetes. The anaerobic ammonium oxidation process has been further developed at the University of Delft in the Netherlands and it was patented under the name Anammox. Similar to other anaerobic micro-organisms, provided suitable process conditions are applied, the Anammox bacteria can be cultivated in the form of granules, as shown in Figure 6.5.

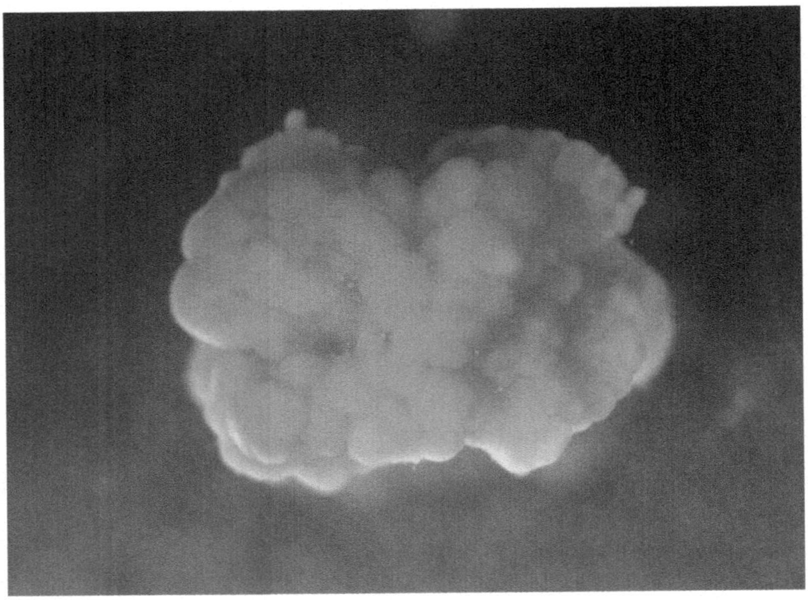

Figure 6.5 Microscopic picture of granule containing Anammox bacteria. Courtesy of Paques BV

Based on thermodynamic considerations, the existence of autotrophic bacteria capable of ammonium oxidation under anoxic conditions had been predicted as early as 1977 by Broda, as the free energy release from the oxidation of ammonium with nitrite or nitrate per electron equivalent ($-361\ kJ \cdot mol^{-1}$)

does not differ substantially from the energy release from oxygen ($-315\ kJ \cdot mol^{-1}$). Later research by Van de Graaf (1995) and Strous et al. (1998) showed that it was nitrite rather than nitrate that was being used as substrate. According to Jetten et al. (1996), the metabolic reactions involved are:

(a) Catabolic reaction (dissimilation):

$$NH_4^+ + NO_2^- \longrightarrow N_2 + 2\,H_2O \tag{6.8a}$$

(b) Anabolic reaction (assimilation):

$$HCO_3^- + 0.2\,NH_4^+ + 2.1\,NO_2^- + 0.8\,H^+ \longrightarrow CH_{1.8}O_{0.5}N_{0.2} + 2.1\,NO_3^- + 0.4\,H_2O \tag{6.8b}$$

In a pilot research project, Jetten et al. (2000) established the following overall equation, which has been confirmed by the treatment results of the first full-scale Anammox reactor in Rotterdam, The Netherlands:

$$NH_4^+ + 1.32\,NO_2 + 0.066\,HCO_3^- + 0.13\,H^+ \longrightarrow$$
$$1.02\,N_2 + 0.066\,CH_2O_{0.5}N_{0.15} + 0.26\,NO_3^- + 2.03\,H_2O \tag{6.8c}$$

Using the Anammox process it is in principle possible to remove ammonium from wastewater under anoxic conditions without the requirement of an organic carbon source. However, there are some issues to consider:

- For complete ammonium removal, the wastewater requires ammonium and nitrite in a molar ratio varying between 1:1.18 (Heijnen, 1996) and 1:1.32 (Jetten et al., 2000). Such a ratio is rarely encountered in wastewater and certainly not in domestic sewage. However, as will be discussed later, this can be circumvented when the Anammox process is combined with nitritation, either in single or double stage configuration. At a significantly different molar nitrite/ammonia ratio, either ammonium or nitrite will remain. Post-treatment will be required or alternatively the limiting component could be added;
- Nitrate will be formed at a stoichiometrical ratio of 0.2–0.3 mg NO_3-N per mg NH_4-N removed. To remove the nitrate, some form of post-treatment will be required, but in general the main activated sludge system can be used for this. If insufficient COD is available, an external source of carbon is required;

6.2.1 Anammox process characteristics

Growth rate and yield of Anammox bacteria are low, as is typical for anaerobic processes. The net growth rate ($\mu = \mu_{man} - b_{an}$) is estimated as 0.04 to 0.06 d^{-1} at 35°C, while the yield is 0.11 g VSS $\cdot\ g^{-1}$ NH_4-N removed. While the low growth rate and yield normally are considered as an advantage, as it reduces the production of excess sludge, the downside in this case is that it results in a slow reactor start-up when insufficient seed sludge is available and a slow recovery after process upsets;

Substrate toxicity for both ammonium and nitrite has been observed by Jetten et al. (1996). While the inhibition by ammonium is limited and only occurs at high ammonium concentrations (larger than several hundreds mg NH_4-N $\cdot\ l^{-1}$), nitrite inhibition reduces the growth rate by 50 to 60% at nitrite concentrations above 70 mg NO_2-N $\cdot\ l^{-1}$. The nitrite concentration in the Anammox reactor should therefore be controlled at values below 50 mg NO_2-N $\cdot\ l^{-1}$. Nitrite is irreversibly toxic at concentrations higher than 70 mg $NO_2\text{-}N \cdot l^{-1}$ at longer exposure times, which might constitute a problem for a process where nitrite is a substrate and present in high concentrations in the influent. For example, when the Anammox reactor receives partially nitrified reject water with a high nitrite concentration (250–750 mg $NO_2\text{-}N \cdot l^{-1}$), incomplete metabolisation of the nitrite in the Anammox reactor might easily result in toxic nitrite levels. To eliminate this risk, the Anammox should be operated under conditions of nitrite limitation. On the other hand, short time exposure (less than one day) to higher levels of nitrite (up to 100 mg $NO_2\text{-}N \cdot l^{1}$) can be tolerated. The Anammox bacteria have demonstrated tolerance to nitrate, which is important as it is a by-product of their own metabolic process. Oxygen is not irreversibly toxic as it is to some other types of anaerobic organisms: after removal of the oxygen, the anaerobic ammonium oxidation process will resume. However, sulphides and alcohols (especially methanol) are toxic at low concentrations. Sulphate also might represent a problem, as under anaerobic conditions it is converted into sulphide by sulphate reducing bacteria. However, under anoxic conditions the sulphate reducers are inhibited. Therefore, if sulphate is present in the feed then nitrate should be added during the start-up period until the Anammox bacteria are firmly established and the production of nitrate is sufficient to suppress sulphate reduction.

In Figure 6.6 the temperature and pH dependency of Anammox bacteria are shown. The optimum pH range is located between 7.0 and 8.5, which means that for a Anammox system receiving partially nitritied reject water, no provisions have to be made for pH adjustment. The pH of the nitritation reactor effluent will have a value between 6.5 and 6.8 and according to the reaction equation of Eq. (6.8) there is some consumption of protons in the Anammox process, resulting in a slight increase in the pH value.

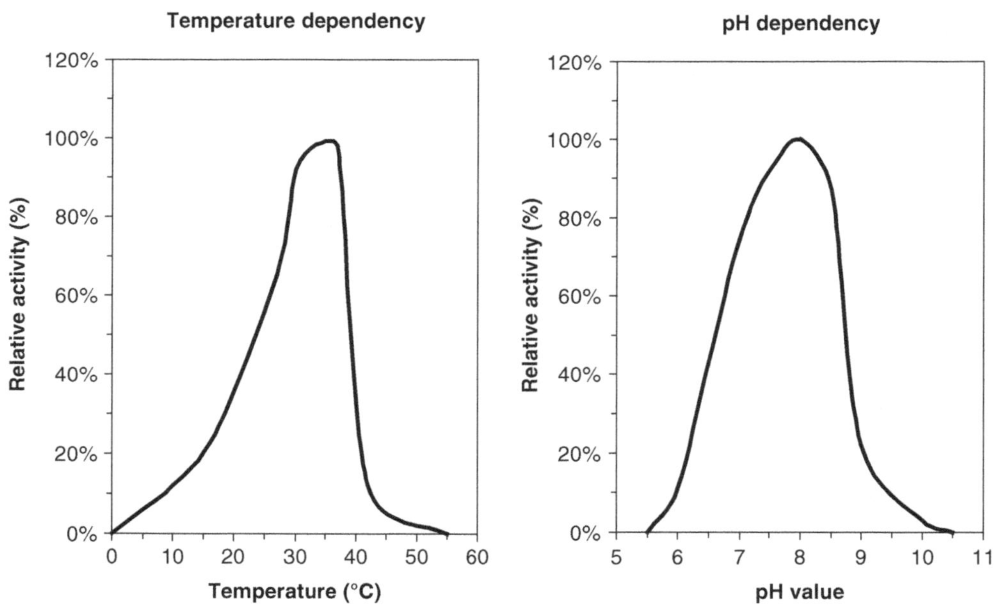

Figure 6.6 Temperature and pH dependency of Anammox bacteria (based on data from Jetten et al., 1996)

The optimum process temperature is within the range of 30 to 37°C. At temperatures above 37°C the ammonium conversion rate rapidly decreases (Jetten et al., 1996), while the conversion of nitrite continues to increase up to a temperature of 42°C, followed by a rapid decrease. Either different bacteria or a different metabolic pathway might be involved. A possible explanation might be the utilization of released COD from increased biomass decay at higher temperatures.

6.2.2 Reactor design and configuration

On pilot- and lab scale, the Anammox process has been operated predominantly in fluidised-bed and SBR configurations. However, other configurations might actually be more suitable. According to Jetten et al. (2000), the following criteria are important in the selection of a reactor type for the Anammox process:

- Suspended solids retention: while the amount of Anammox biomass retained in the system should be maximised as both growth rate and yield are very low, incoming suspended solids should preferably not be retained as this would significantly lower the volumetric nitrogen removal capacity. The reduction in sludge age resulting from the accumulation of suspended solids could potentially result in loss of Anammox bacteria from the reactor;
- Mixing intensity: as the reactor feed contains a high nitrite concentration which is inhibitory to Anammox organisms, it is important to dilute the influent quickly. Reactors with a true plug-flow regime should therefore be avoided;
- High volumetric nitrogen conversion rates: i.e. requiring a high biomass concentration for suspended systems or a high specific surface area for biofilm systems.

Based on these criteria the most suitable reactor types are:

- Granulated sludge bed systems (such as the Expanded Granular Sludge Bed (EGSB) which will be discussed in Chapter 13 and the Internal Circulation (IC) reactor, see Figure 6.6), both with a biomass concentration between 30 and 80 kg TSS $\cdot$ m^{-3} in the lower reactor section, equivalent to a specific biofilm surface area of 1000–2000 $m^2 \cdot m^{-3}$ reactor volume;
- Membrane bioreactors (see also Chapter 7): this configuration should only be selected if the concentration of non biodegradable suspended solids in the influent is very small, as these solids will be retained by the membranes. Accumulation of inert material in the reactor will reduce the sludge age, possibly below the minimum required value;
- Moving bed biofilm reactors (MBBR), filled with support material with an average biofilm surface of 350 $m^2 \cdot m^{-3}$ reactor volume.

Figure 6.7 shows a schematic representation of the first full-scale Anammox reactor built at Rotterdam Dokhaven, The Netherlands, which is an IC type reactor that is also used for high rate anaerobic treatment of industrial wastewaters.

The reactor is divided into a lower and upper compartment, each with a dedicated gas-liquid-solid separator. The lower section contains the expanded granular sludge bed and may be considered as completely mixed. The biogas (mainly nitrogen in this case) that is produced in the bottom compartment is collected in the bottom separator and induces an upward flow of liquid/gas to the top section through the riser. In the top section, gas is separated from the liquid and part of the liquid is returned to the bottom of the first compartment through the downer. Thus a recirculation flow over the reactor is induced (gas-lift principle). Additionally, nitrogen gas is collected at the top of the reactor, pressurised and injected at the bottom of the lower compartment. Both actions induce an expansion of the sludge bed while the return flow also serves to dilute the incoming nitrite concentration.

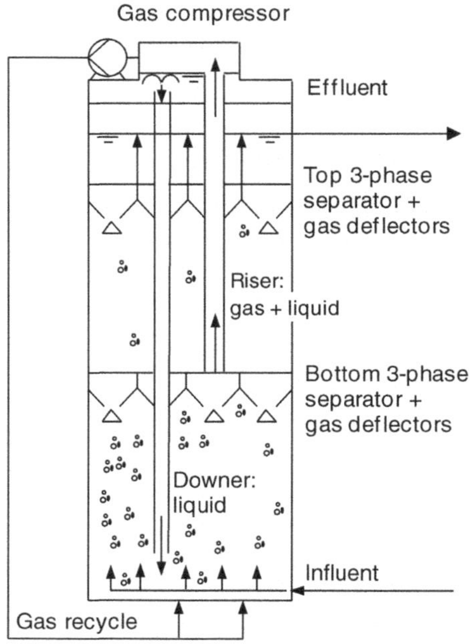

Figure 6.7 Schematic representation of an Anammox reactor in IC configuration (two-stage nitritation-Anammox process)

The first separator retains most of the sludge granules contained in the upward liquid flow leaving the lower compartment. In the top compartment the flow regime is tranquil and in plug-flow mode, as most of the gas has been removed, allowing the biomass carried over from the lower compartment to settle and be returned to the lower compartment. The top separator polishes the reactor effluent from most of the remaining sludge granules while flocculent suspended solids are not (or only partly) retained. This two-compartmental approach allows the application of high nitrogen loads to the lower compartment, while achieving a relatively low nitrite- and suspended solids concentration in the effluent. The lower compartment is operated at a bulk nitrite concentration between 10–30 mg NO_2-N $\cdot l^{-1}$, while part of this nitrite is removed in the top section. The ammonium effluent concentration is higher and depends on the ammonium/nitrite ratio in the influent to the Anammox reactor.

As the diffusion coefficient of ammonium is in the same order of magnitude as the diffusion coefficient of nitrite, ammonium will never be the limiting factor. Both components are able to penetrate into a significant depth (or all) of the biofilm or sludge granule, which maximizes the amount of biomass available for the anaerobic ammonium oxidation process. Penetration of nitrite and ammonium is essentially complete for sludge granules with a (typical) average diameter of 500 µm. For preliminary sizing of an IC or EGSB type Anammox reactor the following design guidelines can be used:

- $r_{no2} = 0.3 - 0.4\,g\,N \cdot kg^{-1} VSS \cdot d^{-1}$ (specific nitrite removal rate);
- $X_t = 50 - 75\,kg\,TSS \cdot m^{-3}$ (average sludge concentration in sludge bed section);
- $X_v = 25 - 5\,kg\,VSS \cdot m^{-3}$ (average organic sludge concentration in sludge bed section).

In practice, sustained total nitrogen removal rates of 5 to 8 kg $N \cdot m^{-3} \cdot d^{-1}$ have been observed at full-scale installations.

6.3 COMBINATION OF NITRITATION WITH ANAMMOX

It is obvious from the previous section that the Anammox process may offer a significant reduction in the requirements for energy and COD. However, as a standalone process it is not very useful as wastewaters with a suitable influent composition are hard to find. On the other hand, the combination of Anammox with the nitritation process is quite feasible and in fact very attractive for treatment of sludge digestion reject water. Two configurations are currently applied:

- A two stage configuration consisting of a nitritation reactor followed by an Anammox reactor. An example is the SHARON-Anammox process operated at Rotterdam Dokhaven, basically a completely mixed reactor without sludge retention, followed by an IC reactor. Alternatives to the above reactor types are also possible: for example the completely mixed reactor can be replaced by a MBBR operated under oxygen limitation while the IC reactor can be replaced by an EGSB reactor;
- A single reactor system using either biofilm on carriers, sludge granules or even suspended growth processes. The EGSB or IC, the MBBR or even an SBR are the most appropriate reactor types. The basic idea is to generate a dissolved oxygen gradient in the biofilm or granule, where the outer part is aerobic (nitritation) and the inner part anaerobic (anaerobic ammonium oxidation). The three main configurations are CANON (completely autotrophic nitrogen removal over nitrite), OLAND (oxygen limited autotrophic nitrification–denitrification) and DEMON (de-ammonification). Due to the possibility to treat more types of wastewater than only sludge reject water and because of the reduction in construction costs, currently the one-reactor concept is favoured.

6.3.1 Two stage configuration (nitritation reactor–Anammox)

When the molar NH_4^+/HCO_3^- ratio of a wastewater is between 1.0–1.2, which is fairly typical for reject water from anaerobic sludge digestion, it is relatively easy to use a nitritation reactor to produce an effluent with the appropriate ammonium/nitrite mixture for Anammox treatment. Therefore the combination of a nitritation reactor with the Anammox process can be very attractive. As denitritation is not applied in the nitritation reactor, the production of alkalinity will be insufficient to allow for full conversion of the ammonium into nitrite.

Jetten et al. (2000) observed that when a nitritation reactor was continuously aerated without pH control, a steady state situation developed in which nitritation proceeded for approximately 50 - 60 percent. At this point most of the bicarbonate was consumed and the resulting decrease in pH precluded further nitritation. As feeding of the reactor with reject water was continued, an equilibrium was established with the pH value oscillating between 6.5 and 6.8 and an average ammonium removal of 53%, resulting in an effluent with a NO_2-N/NH_4-N ratio of around 1.1. In the absence of an anoxic period, the nitrite concentration in the reactor exceeded 500 mg $N \cdot l^{-1}$. However, nitrite inhibition was not observed on the ammonium oxidisers in the nitritation reactor.

Figure 6.8 can be used to evaluate the potential benefits of the combination of nitritation and anaerobic ammonium oxidation. For the calculation optimum conditions have been assumed: i.e. all ammonium is converted and the effluent of the nitritation reactor contains a mixture with a molar NO_2/NH_4 ratio of 1.32 (57% nitrite and 43% ammonium), so that according to Eq. (6.8) both nitrite and ammonium will be completely removed in the Anammox reactor. This allows the *maximum* theoretical savings in methanol- and oxygen consumption to be calculated. It should be emphasised that the anabolic reactions have not been included in this analysis. For full nitrification to nitrate, the oxygen consumption is 2 moles O_2 per mole of NH_4-N. For partial nitrification to nitrite only 1.5 moles O_2 are required. When only 53% of the influent ammonium is oxidised to nitrite, the stoichiometric requirement can thus be calculated as $0.53 \cdot 1.5 = 0.86$ mol $O_2 \cdot mol^{-1}$ NH_4-N removed. The theoretical methanol demand can be calculated from stoichiometrics as well: 0.5 mole of methanol is required for the reduction of 1 mole of nitrite while 0.83 mole of methanol is needed for the reduction of 1 mole of nitrate. Although in the combined nitritation-Anammox process itself the addition of methanol is not required, the nitrate produced in the Anammox reactor will have to be removed in a subsequent treatment step. The stoichiometric methanol consumption is $0.83 \cdot 0.43 \cdot 0.26 = 0.093$ mol methanol $\cdot$ mol^{-1} NH_4-N removed.

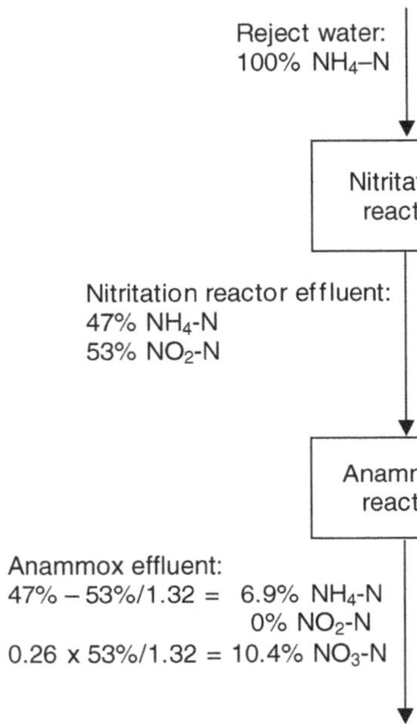

Figure 6.8 Overall efficiency of two-stage nitritation-Anammox treatment, adapted from Jetten et al. (2000)

The theoretical savings compared to conventional biological nitrogen removal and to a stand-alone nitritation/denitritation process are summarized in Table 6.1. In practice, the reduction in both oxygen- and methanol demand will be less as the anabolic reactions have not been included (i.e. methanol consumption for biomass growth and respiration) and because the composition of the nitritation reactor effluent in terms of nitrite and ammonium is rarely ideal.

Jetten et al. (2000) operated two lab-scale Anammox SBR reactors for a period of 100 days on effluent from a lab-scale nitritation reactor. Both Anammox reactors were operated at 35°C and the hydraulic residence time in each reactor was one day. The Anammox reactors were operated under conditions of nitrite limitation, resulting in complete removal of all nitrite and the presence of residual ammonium in the Anammox effluent. The treatment results are shown in Table 6.2. During the experiment, the average volumetric nitrogen loading rate of the nitritation reactor was $1.2\ kg\ N \cdot m^{-3} \cdot d^{-1}$, with an average conversion efficiency of ammonium to nitrite of 53%. The nitrogen loading rate in the Anammox reactor averaged $0.3\ kg\ N \cdot kg^{-1}\ VSS \cdot d^{-1}$. The maximum observed nitrogen conversion rate in the Anammox reactor was between $0.6–0.8\ kg\ N \cdot kg^{-1}\ VSS \cdot d^{-1}$, with excellent nitrogen conversion efficiency. The overall ammonium removal efficiency wass 94%.

Table 6.1 Theoretical maximum reduction in oxygen- and methanol demand of a nitritation-denitritation system and a combined nitritation-Anammox system compared to conventional biological nitrogen removal

Characteristic	Conventional system	Nitritation-denitritation	Nitritation/Anammox
Pathway	100% to NO_3^-	100% to NO_2^-	57% to NO_2^-
Molar ratio (mol $O_2 \cdot mol^{-1}$ N)	2	1.5	0.86
Reduction in oxygen demand compared to:			
– Conventional system	N.A.	25%	60%
– Nitritation-denitritation	N.A.	N.A.	43%
Molar ratio (mol $CH_3OH \cdot mol^{-1}$ N)	0.83	0.50	0.093
Reduction in methanol demand compared to:			
– Conventional activated sludge	N.A.	40%	89%
– Nitritation-denitritation	N.A.	N.A.	81%

Table 6.2 Performance of the lab-scale nitritation – Anammox system treating reject water from the sludge digestion plant Sluisjesdijk – Rotterdam, The Netherlands, based on data from Jetten et al. (2000)

Par. (mg $N \cdot l^{-1}$)	Nitritation reactor			Anammox reactor		
	Influent	Effluent	Removal	Influent	Effluent	Removal
NH_4^+-N	1180	600	580	600	70	530
NO_2^--N	0	550	−550	550	0	550
NO_3^--N	0	0	0	0	12	−12
N_l	0	30	−30	30	30	0
N_t	1180	1180	0	1180	112	1068

Organic nitrogen (N_l) is present in the Anammox effluent, as a result of the biomass production in the nitritation reactor, but this amount will be small compared to a stand-alone nitritation-denitritation reactor as no methanol is consumed, i.e. no growth of heterotrophs.

The observed molar nitrite/ammonium ratio in the effluent of the nitritation reactor was equal to 550/600 or 0.92. As this value is much lower than the stoichiometric value of 1.32 according to Eq. (6.8), it was not possible to remove all the ammonium present in the effluent of the nitritation reactor. The maximum theoretical removal of ammonium in the Anammox reactor is equal to 550/1.32 or 417 mg NH_4-N $\cdot l^{-1}$, with complete consumption of all nitrite. However, the *measured* removal of ammonium was 530 mg NH_4-N $\cdot l^{-1}$, more than 113 mg NH_4-N $\cdot l^{-1}$ higher than stoichiometrically expected. Thus in the experiment from Table 6.2, the actual ratio between NO_2-N and NH_4-N removed was 1.04.

It is interesting to note that the measured effluent nitrate concentration (12 mg N $\cdot l^{-1}$) also deviates considerably from the stoichiometrical production according to Eq. (6.8c): i.e. $530 \cdot 0.26 = 138$ mg N $\cdot l^{-1}$. One explanation could be that the "missing" nitrate has been denitrified back to nitrite by heterotrophic organisms, using the organic material present in the sludge reject water. Considering the additional amount of nitrite made available this way, the Anammox bacteria can oxidise more ammonium: i.e. $(138 - 12)/1.32 = 95\ mg\ N \cdot l^{-1}$. Now the mass balance almost closes according to Eq. (6.8): i.e. the ratio between nitrite- and ammonium removed is equal to $(550 + 126)/530 = 1.28$. So the above hypothesis does seem plausible. In practice, the observed ratio between nitrite- and ammonium removed in the Anammox process might thus be lower than 1.32, depending on the extent of the denitrification of nitrate that will occur.

6.3.2 Case study: full scale SHARON - Anammox treatment

Van Betuw et al. (2008) evaluated the performance of the first full-scale implementation of combined nitritation-Anammox treatment. Up to 2006 Paques have constructed two additional full-scale two-step systems, but from that time onwards the one-step configuration is preferred (refer also to the next section).

The Sluisjesdijk sludge treatment plant receives the excess sludge produced at the Dokhaven sewage treatment plant, Rotterdam (The Netherlands) plus that of two other wastewater treatment plants, with a combined capacity of 460,000 P.E. The reject water from the anaerobic digesters contains 10 to 16% of the total nitrogen load to the Dokhaven plant.

In 1999 a decommissioned sludge thickener with a volume of 1800 m^3 was converted into a SHARON reactor, with the objective to remove a substantial part of the nitrogen load in the reject water and thereby to improve the nitrogen removal capacity of the Dokhaven plant. The design flow and load were 770 $m^3 \cdot d^{-1}$ and 827 kg $NH_4\text{-}N \cdot d^{-1}$ respectively. The applied aerobic sludge age varied between 1.0 – 1.5 days. Methanol was dosed for pH control by denitritation. The SHARON reactor was operated in a sequenced mode with a cycle time of 80 minutes aeration followed by 40 minutes denitrification. The operational temperature was maintained at 30–35°C.

The sidestream treatment was extended in 2002 with the construction of an Anammox reactor, in order to further reduce the costs of aeration and external carbon source dosing. In order to provide the nitrite/ammonium mixture required for Anammox treatment, pH control in the SHARON reactor by methanol addition was no longer required. As a result, the reactor pH decreased to a value of 5.9–6.2. The aerated-non aerated cycle times were not adapted: i.e. the aerobic sludge age was maintained at 1 to 1.5 days. The average conversion of ammonium to nitrite is 53%, very close to the optimum value of 57%. The nitritation reactor effluent contains on average 500 mg $NH_4\text{-}N \cdot l^{-1}$ and 650 mg $NO_2^-\text{-}N \cdot l^{-1}$. Only low concentrations of nitrate are measured in the SHARON effluent (on average 8 mg $NO_3\text{-}N \cdot l^{-1}$), indicating that nitrite oxidizers are indeed almost absent in the reactor.

The Anammox reactor is an IC type (Figure 6.6) with a volume of 70 m^3 and a height of 16 m. The design nitrogen load is 500 kg $N \cdot d^{-1}$, or about 7 kg $N \cdot m^{-3} \cdot d^{-1}$. To promote mixing, the produced off-gas is recycled with a 4 kW compressor at a rate of 70 $Nm^3 \cdot h^{-1}$. The reactor is operated at a temperature of 30–36°C.

A plate pack (lamella) separator is installed upstream the Anammox reactor in order to remove the suspended solids present in the SHARON effluent, which would otherwise decrease the sludge age. Possibly this treatment step is no longer required once the Anammox process is firmly established. However, this cannot yet be confirmed, as to date the flow has been bypassed around the separator only once and then only for a limited duration of time (several weeks). Although no negative effects were observed on the performance of the Anammox reactor, a longer test period is required before a firm conclusion can be made.

The start-up of the Anammox reactor required much more time than anticipated, as initially almost no Anammox seed sludge was available. During the start-up period in total 9 m^3 of seed sludge from a pilot scale reactor was added. Several process upsets delayed the start-up considerably: toxification by methanol, nitrite and biocides. It took almost 3 years before significant Anammox granule formation was observed. The increase in Anammox treatment capacity allowed the operation of the SHARON reactor to be changed from nitritation-denitritation to nitritation only. After 3.5 years the design nitrogen removal capacity was reached. Similar to anaerobic granulated sludge systems, it is expected that the start-up period can be significantly shortened once sufficient quantities of granular Anammox excess sludge are available to seed new reactors.

The nitrogen removal performance of the Anammox reactor in the period from mid 2007 to mid 2008 is summarized in Table 6.3 and confirms the data from lab-scale experiments. During this period the average nitrogen loading was 390 kg $N \cdot d^{-1}$, with peak loads reaching up to 684 kg $N \cdot d^{-1}$. Scaling was a significant problem in the Anammox reactor, caused by precipitation of magnesium- and calcium phosphates, resulting in the accumulation of a heavy gritty sludge in the reactor. To reduce the precipitation rate, CO_2 is injected at a ratio of 0.2 to 0.8 kg CO_2 per kg N applied, with the objective to reduce the pH in the reactor to a value between 6.9 – 7.0. Even with this control measure in place every two months one cubic metre of (mainly inorganic) bottom sludge has to be discharged. This might be a specific problem for the two reactor configuration, as a large quantity of CO_2 is stripped from the reject water in the nitritation reactor due to the intense aeration.

Table 6.3 Nitrogen removal performance of the full-scale Anammox reactor located at the sludge treatment plant Sluisjesdijk - Rotterdam, The Netherlands (Van Betuw et al., 2008)

Parameter ($mg\ N \cdot l^{-1}$)	Influent		Effluent		Removal %
	Range	Average	Range	Average	
NH_4-N	300–550	400	20–50	35	92%
NO_2-N	500–700	582	2–20	10	98%
NO_3-N	2–15	8	N.D.	112	N.A.
Total-N	800–1250	990	N.D.	157	84%

6.3.3 Single reactor configurations

A different approach to the sequential application of nitritation and anaerobic ammonium oxidation is the combination of these processes in a single reactor. An example is the CANON process developed by Paques: an acronym for Completely Autotrophic Nitrogen removal Over Nitrite. The CANON process uses a granular sludge bed reactor in which the two processes are combined through application of the appropriate combination of film thickness and oxygen concentration in the bulk liquid (Hiao et al., 2002).

The reactor is continuously aerated, controlled by an on-line nitrite measurement. Due to the need for aeration, the height of the reactor is restricted to 6 meter. Control of the oxygen concentration by itself may not be enough to remove the nitrite oxidisers from the system, although it seems that nitrite oxidisers have a lower affinity for oxygen than ammonium oxidisers. However, in the case of combined nitritation and anaerobic ammonium oxidation, the rapid removal of nitrite by the Anammox bacteria acts as a supporting selection factor, which will deprive the nitrite oxidisers of their substrate. Selecting and maintaining an appropriate oxygen concentration in the reactor is a balancing act, as the optimal value depends on the film thickness, the oxygen consumption and the nitrogen removal rate.

Several advantages can be attributed to one reactor configurations. One of them is the reduction in investment costs due to a decrease in both volume and complexity. On the other hand, due to the conflicting environmental conditions required by the two processes (nitritation and Anammox), both will be operated under suboptimal conditions. Full-scale nitrogen conversion rates between 0.6–2.0 kg $N \cdot m^{-3} \cdot d^{-1}$ have been reported, which is significantly less than the 5–10 kg $N \cdot m^{-3} \cdot d^{-1}$ that can be achieved in a dedicated Anammox reactor. On the other hand, the overall reactor volume of the two reactor configuration will be significantly larger, due to the large volume required by the nitritation reactor, which is operated at 1–1.5 days hydraulic retention time.

An important disadvantage of a two reactor system is that application of nitritation to a waste stream will only result in an appropriate mixture of ammonium and nitrite when the molar concentrations of alkalinity and ammonium are approximately equal. Only in this case, when the alkalinity is exhausted, which will be the case when approximately 50% of the ammonium has been converted to nitrite, the pH will rapidly drop, thereby stopping the conversion process. In contrast, in the single-reactor process the main selection criterion is oxygen limitation, not the combination of selection on growth rate and application of a low pH. Therefore it might also be used for wastewater streams with a different molar ratio between alkalinity and ammonium. In fact, as wash-out of nitrite oxidisers by means of a reduction of the sludge age at a high temperature is no longer the selection mechanism, there seems to be no fundamental reason why these one reactor systems cannot be applied at lower temperatures, although nitrogen removal rates would probably be significantly lower.

Another advantage attributed to single reactor systems is that the nitrite, which is produced in the aerobic part of the biofilm, will be immediately removed: low nitrite concentrations will reduce the potential for gaseous NO and N_2O emissions, known contributors to the greenhouse effect. On the other hand, the low oxygen concentration applied might actually induce formation of NO and N_2O. The present knowledge on the dynamics of NO and N_2O production does not yet allow a firm conclusion as to which process configuration is advantageous in this respect.

At the time of writing (2011), five full-scale CANON processes (or one-step Anammox systems, which is the new trade name used by Paques) have been constructed: the first one in the Netherlands in 2006, which will be discussed in the case study below, the second in Switzerland treating reject water from sludge digestion and three large ones in China at yeast and glutamate factories, with a total design nitrogen removal capacity of 21,000 kg $N \cdot d^{-1}$.

The first one-step Anammox reference was constructed for the wastewater treatment company "Waterstromen" at the sewage treatment plant (STP) in Olburgen, The Netherlands. This STP treats process wastewater from the potato industry and centrate from the sludge dewatering unit of the neighboring municipal sewage treatment plant. The wastewater from the potato processing plant contains proteins, starch and phosphate and the load of pollutants is equal to 160,000 population equivalents. Until 2003 this stream was subjected only to anaerobic pre-treatment in an UASB reactor, followed by discharge to the local STP, resulting in a significant contribution to the total nitrogen and phosphate loads. After Waterstromen was requested to reduce these loads, in 2006 the one-step Anammox®-process was implemented, precided by a struvite precipitation reactor (Paques, 2011b).

The effluent from the UASB reactors, combined with a small reject water stream, is introduced first in the struvite precipitation reactor (PHOSPAQ). This reactor is an aerated crystallization reactor where phosphorus- and residual COD removal are combined. Under addition of MgO, phosphate is removed by precipitation as struvite, also known as Magnesium-Ammonium-Phosphate or MAP, with a structural formula $MgNH_4PO_4 \cdot 6H_2O$. The aeration of the reactor allows the biological conversion of residual COD present in the anaerobic effluent, but also provides for the mixing energy required to obtain a good struvite quality. In addition, CO_2 stripping raises the pH, which stimulates the struvite formation process.

The produced struvite is harvested from the bottom of the reactor. As the heavy metal content is less than 5% of the allowed value, the product complies with the EU standards for fertilizers and can be used as slow-release fertilizer (Paques, 2011b).

The system consists of two PHOSPAQ reactors of each 300 m^3, operated in parallel, followed by a single 600 m^3 one stage Anammox system. The design- and actual performance and loading rates are listed in Table 6.4, while the influent and effluent qualities are summarised in Table 6.5.

Table 6.4 Design and actual loading- and removal rates of the full scale 2 × 300 m^3 PHOSPAQ + 600 m^3 one-step Anammox® system located at STP Olburgen (based on data provided by Paques, 2011a)

Parameter	UoM	Design average	Design maximum	Actual
Flow	$m^3 \cdot d^{-1}$	2750	3600	2600
COD load				
– daily load	$kg \cdot d^{-1}$	1700	2550	1690
– daily removal	$kg \cdot d^{-1}$	900	1350	1040
Phosphate:				
– daily load	$kg\ P \cdot d^{-1}$	200	250	208
– daily removal	$kg\ P \cdot d^{-1}$	160	180	169
– loading rate	$kg\ P \cdot m^{-3} \cdot d^{-1}$	0.33	0.42	0.35
– removal rate	$kg\ P \cdot m^{-3} \cdot d^{-1}$	0.27	0.30	0.28
Ammonium:				
– daily load	$kg\ N \cdot d^{-1}$	1000	1350	780
– daily removal	$kg\ N \cdot d^{-1}$	920	1190	728
– loading rate	$kg\ N \cdot m^{-3} \cdot d^{-1}$	1.67	2.25	1.30
– removal rate	$kg\ N \cdot m^{-3} \cdot d^{-1}$	1.53	1.98	1.21
Total nitrogen:				
– daily removal(1)	$kg\ N \cdot d^{-1}$	780	950	611
– removal rate(1)	$kg\ N \cdot m^{-3} \cdot d^{-1}$	1.30	1.58	1.02

Note: (1) Based on the difference between ammonium load and the sum of ammonium, nitrite and nitrate loads in the effluent of the one-step Anammox. It is assumed that little or no particulate organic nitrogen is present in the PHOSPAQ effluent.

Table 6.5 Performance of the full-scale 2 × 300 m^3 PHOSPAQ + 600 m^3 one-step Anammox® system located at STP Olburgen (based on data provided by Paques, 2011a)

Parameter	UoM	Influent		Effluent		Removal %
		Avg	Range	Avg	Range	
COD	$mg \cdot l^{-1}$	650	300–900	250	150–400	62%
Phosphate	$mg\ P \cdot l^{-1}$	80	60–100	15	15–20	81%
Ammonium	$mg\ N \cdot l^{-1}$	300	200–400	20	10–30	93%
Nitrite	$mg\ N \cdot l^{-1}$	–	–	10	5–20	
Nitrate	$mg\ N \cdot l^{-1}$	–	–	35	20–50	

According to the data reported by Paques (2011a), 750 tons of MgO or consumed per year while the (recovered) struvite production is equal to 400 ton · yr^{-1}. Assuming no magnesium is present in the feed, then the stoichiometrical dosing requirement according to the structural formula of struvite is 1.0 mol Mg · mol^{-1} P. For the observed average daily PO_4-P removal of 169 kg · P · d^{-1} (5.5 kmol · d^{-1}) and the average daily consumption of MgO (685 kg · d^{-1} or 17.0 kmol · d^{-1}), it can be calculated that the applied molar MgO dosing rate of 3.1 mol Mg per mol P exceeds the stoichiometrical dosing rate considerably. Based on the observed phosphate removal, the expected struvite production is 1340 kg · d^{-1} or 488 ton · yr^{-1}, which corresponds reasonably well with the reported struvite production of 400 ton · yr^{-1}. The difference might be due to shutdown periods and the presence of residual struvite in the effluent.

A second configuration of combined nitritation-Anammox treatment is OLAND, which stands for Oxygen Limited Autotrophic Nitrification – Denitrification (Kuai et al., 1998). Similar to the CANON process, oxygen limitation is the key parameter for selection on ammonium oxidizers. A supplementary selection criterion is maintenance of a pH value of 7.9, as nitrite oxidisers are more vulnerable to a high concentration of unionized ammonia than ammonium oxidizers.

Initially it was postulated that regular (aerobic) ammonium oxidizers were responsible for the observed oxidation of ammonium with nitrite, as it had been demonstrated previously that ammonium oxidisers are much more versatile than originally considered. Under micro-aerophilic conditions these organisms can combine hydroxylamine (NH_2OH) with NO_2^- to give N_2O gas (Bock, 1995), whereas under anoxic conditions they can convert ammonium with NO_2 to NO gas (Hippen et al., 1997). Although both reactions result in removal of ammonium from the water phase, they are highly undesirable as the end products are potent greenhouse gases.

Pynaert et al. (2003) showed that in the case of the OLAND process, the coexistence of ammonium oxidisers and Anammox bacteria in a single biofilm was responsible for the observed autotrophic nitrogen removal. Several configurations have been examined since then, for instance a single reactor SBR, rotating discs (Pynaert et al., 2003) and two membrane reactors in series (Wyffels et al., 2004). However, the latest research efforts have been directed at developing a single reactor SBR configuration in which the formation of granulated sludge is promoted. Removal rates of up to 0.45 g N · l^{-1} · d^{-1} have been reported (Vlaeminck et al., 2009). However, to date no full-scale OLAND system has been constructed.

The last single reactor system to be discussed is the DEMON process (Wett, 2006). The acronym DEMON refers to de-ammonification, which again comprises of the processes of nitritation and anaerobic ammonium oxidation. The main feature of this system is that the aeration is controlled by the pH value. The system consists of a SBR reactor that is operated with a cycle time of 8 hrs (6 hrs reaction +2 hrs settling and decanting). Aeration is intermittent and start and stop are controlled by the pH value. When the lower pH setpoint of 7.04 is reached, aeration is interrupted. In the absence of oxygen the nitritation process stops while the anaerobic ammonium oxidation process continues. Thus some alkalinity is produced, while simultaneously alkalinity is introduced with the influent (reject water). When the upper pH setpoint of 7.06 is reached, aeration is resumed. The oxygen setpoint of the aeration control is only 0.3 mg O_2 · l^{-1} in order to prevent rapid nitrite accumulation, which can be toxic to the Anammox biomass. Furthermore this prevents further oxidation of nitrite to nitrate.

Growth of the Anammox bacteria takes place predominantly in granular form and there are positive results reported on the use of a hydrocyclone to selectively waste the less heavy (non-Anammox) sludge fraction (De Mooij et al., 2010).

From 2004 onwards a DEMON reactor has been used for the treatment of reject water from the 200,000 P.E. sewage treatment plant located at the city of Strass, Austria (Wett, 2007). The SBR with a volume of 500 m^3 is operated with an average biomass concentration of 4.3 g TSS · l^{-1} (3 g VSS · l^{-1}). The DSVI of the biomass is 74 ml · g^{-1}, which corresponds to a sludge with medium settleability. The average operating temperature was 27.8°C. No heating was applied, but the SBR is located inside a building. The average daily nitrogen load in 2005 was 215 kg N · d^{-1} (a reject water flow of 117 $m^3 \cdot d^{-1}$ with an average ammonium concentration of 1845 mg N · l^{-1}. The observed nitrogen removal was excellent, with average effluent ammonium and nitrite concentrations of 180 mg N · l^{-1} and 4.4 mg N · l^{-1} respectively. The ammonia removal was 90.3%, while overall nitrogen removal was slightly less at 85.8%, due to the production of nitrate. However, as observed with the SHARON–Anammox process, part of the produced nitrate was denitrified with the biodegradable COD in the reject water. The volumetric nitrogen removal capacity (including production of nitrate) was 0.37 kg N · $m^{-3} \cdot d^{-1}$. This is lower than the 0.8–2.0 kg N · $m^{-3} \cdot d^{-1}$ reported for CANON, but on the other hand the system is much simpler to build and operate. In 2006 a second DEMON has been started up in the sewage treatment plant of Glarnerland, Switzerland (Nyhuis et al., 2006). At the time of writing (2010) there have been nine systems installed with another six under construction.

6.4 BIOAUGMENTATION

For an activated sludge system designed for nitrogen removal, the main design criteria are the aerobic sludge age required to meet the specified effluent ammonium limit (R_{sm}) and the size of the anoxic zones required to meet the effluent nitrate limit. By adding nitrifying bacteria to the activated sludge system the nitrification capacity is increased. In principle it becomes possible to operate at sub-optimal aerobic sludge age and to increase the sludge mass fraction allocated to denitrification. This may not be an easy task when reactor volumes are fixed, but will be much easier in the case of carrousels. Several configurations have been proposed and at least one of these has been implemented at full scale: the BABE configuration or Bio-Augmentation Batch Enhanced (Berends et al., 2002). The bio-augmentation reactor is basically a small completely mixed reactor, where part of the return sludge from the final settler of the main activated sludge system is mixed with sludge reject water as shown in Figure 6.9. The objective is to remove the ammonium in the concentrated nitrogen stream and to increase the nitrification capacity of the main activated sludge system by "seeding" it with the nitrifiers produced in the bioaugmentation reactor. As the return sludge flow is small compared to the reject water flow, the bioaugmentation reactor can be heated, thereby increasing the nitrogen removal rate. Similar to nitritation systems, methanol is added in order to control pH, as it is cheaper than the addition of caustic and because it simultaneously reduces the demand for COD in the activated sludge system.

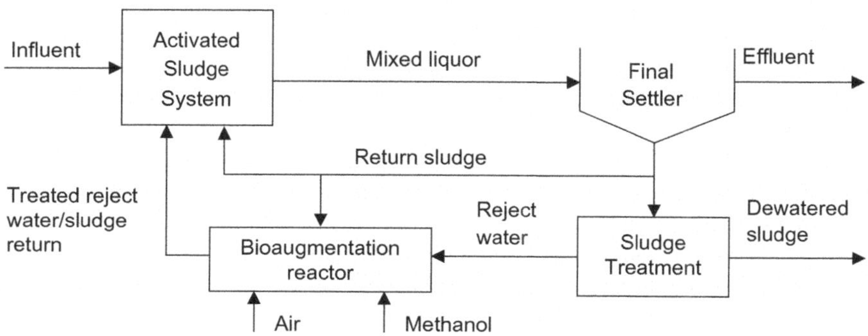

Figure 6.9 Bio-augmentation configuration for digestion reject water nitrogen removal

When the reactor is operated in SBR mode, there will be no need for excess sludge discharge, as an equilibrium will form in which, when total sludge volume after settling is too large to be contained in the reactor, some excess sludge will be discharged together with the effluent. This allows operation of the reactor at maximum sludge concentration and -nitrogen removal capacity, while simultaneously seeding the main activated sludge process with ammonium oxidisers. There is no specific selection on ammonium oxidisers, as the biomass in the bio-augmentation reactor is continuously supplied with fresh return sludge. As nitrite oxidisers will be present in the return sludge, there will be oxidation of nitrite to nitrate regardless of the sludge age. Consequently methanol demand will be higher than in an equivalent nitritation reactor.

Apart from reducing the ammonium load to the main activated sludge system by treating the reject water stream, the second beneficial effect of a bioaugmentation reactor is the return of the produced nitrifiers to the activated sludge system (seeding), thereby increasing the nitrification capacity. The nitrifiers in the bioaugmentation reactor grow as an integrated part of the sludge flocs present in the return sludge, in contrast to nitritation reactors operated without sludge retention, where growth of the micro-organisms is dispersed. Jetten et al. (2000) conducted a pilot scale study in which an activated sludge system was continuously inoculated with ammonium oxidisers cultivated in suspension in a nitritation reactor. The increase of the nitrification rate was lower than could be expected from the mass of ammonium oxidisers fed to the activated sludge system. This was not due to lack of retention of the ammonium oxidisers in the aeration tank, but to selective grazing by protozoa. The ammonium oxidisers fed from the nitritation reactor may be subjected to higher predation since they are present outside of the relatively sheltered micro-environment of a sludge floc when introduced into the aeration tank. Therefore they are subsequently more likely to be present in the outer layers of the sludge flocs and more accessible to protozoans. In contrast, the ammonium oxidisers from a bioaugmentation reactor already form an integral part of the sludge floc before introduction into the activated sludge system. Therefore, the influx of nitrifiers from the bioaugmentation reactor might have a larger effect on the nitrification capacity of the main activated sludge system than that from a comparable nitritation reactor.

The existence of protozoan predation was demonstrated in a research study by van Loosdrecht et al. (1997), where the fraction of ammonium oxidisers removed by predation was estimated to be as high as 75%. On the other hand, the observed decrease in effluent ammonium concentration in full-scale wastewater treatment plants where nitritation reactors are operational suggests that a lower degree of predation might occur in practice.

6.5 SIDE STREAM NITROGEN REMOVAL: EVALUATION AND POTENTIAL

The side stream techniques discussed in this chapter, with the exception of OLAND, have all been implemented at full-scale and may now be considered as proven technologies. However, the decision to implement side stream nitrogen removal should only be made after careful analysis of the actual situation at hand and of the available alternatives.

For instance, as long as the main activated sludge system has sufficient capacity to treat the ammonium nitrogen load released from anaerobic sludge digestion, a sidestream nitrogen removal system is not really advantageous. If the total effluent nitrogen effluent limit is complied with, but the ammonium- or Kjeldahl nitrogen limit is not, then a reduction in anoxic volume could be considered. If both limits are exceeded, then it should first be attempted to optimise or improve the performance of the activated sludge system.

One of the measures that could be considered is the addition of methanol to the anoxic zone. This allows the volume of the anoxic zone to be reduced, as the denitrification rate and hence the denitrification capacity will increase due to the availability of more easily biodegradable COD. The reduction in the value of f_x

allows for operation at a higher aerobic sludge age, which will decrease the effluent ammonium concentration. Although the methanol consumption will be higher compared to that in a side stream nitrogen removal system using the nitrite pathway, this has to be balanced against the additional investment costs of new reactors with associated equipment, piping and instrumentation.

Another option that could be explored is operation at a higher sludge age: this will reduce the ammonium effluent concentration and increase the denitrification capacity as well. However, both the installed oxygenation capacity and the allowable solids loading rate of the final settler can become limiting. Increasing oxygenation capacity may be relatively simple, especially if there is room to add aeration elements. If the final settler capacity is limiting, this is not so easily remedied. Improving the settling characteristics of the sludge by optimising the operation of the activated sludge process or through addition of chemicals (metal salts or PE) may create some capacity here. The alternative, extension of the final settler surface area, may be prohibitively expensive.

Another popular measure is to modify the operation of the activated sludge system by the introduction of floating support material in the aeration tank, to which a biomass layer will adhere. Retrofitting existing activated sludge systems into a moving bed bioreactor or MBBR configuration in order to increase treatment capacity of overloaded systems has gained significant popularity in the last decade. The transformation of the aeration tank into a MBBR will have the following effects:

- The sludge mass that can be maintained in the system will increase and so will the sludge age and the nitrogen removal capacity;
- The sludge mass that is attached to the support material will be retained in the biological reactors. Therefore, only the biomass that detaches from the support material because of shear stress will be added to the mixed liquor load to the final settler. The increase in solids load to the final settler will thus be limited.

Should the result of the system evaluation be that side stream nitrogen removal is indeed the preferred solution, then the discussion turns to the selection of the most appropriate side stream treatment process.

The one-stage nitritation-Anammox combination is very attractive considering the reduction in overall resource utilisation, as a large part of the ammonium will be removed without oxygen and COD demand. From a sustainability viewpoint (e.g. the emission of CO_2 and the use of electricity), this combination is superior to the other techniques. However, these advantages should be balanced against the added investment costs: although small in volume, the Anammox reactor will not be cheap, as it is equipped with expensive internal three-phase separators. In the case of an MBBR, expensive carriers media will have to be purchased. The DEMON process is cheaper as it is a simple SBR process. On the other hand the reaction rates are lower. In general the start-up process (and recovery) of Anammox reactors is very slow, though this situation will improve when more full-scale reactors are available and seed sludge can be purchased.

Due to the development of one-reactor systems, applicability of the nitritation/Anammox process to the treatment of high strength nitrogen wastewater streams other than digestion reject water is facilitated, as the ratio between NO_2/NH_4 required by the Anammox bacteria can be produced by other methods than control of the sludge age and pH alone.

As for standalone nitritation and bioaugmentation reactors: these systems are particularly suited for application in situations where a simple and robust system is required that requires removal of the bulk of the nitrogen. Aeration- and methanol demand will be higher in the bio-augmentation reactor as the ammonium will be removed over the nitrate pathway.

On the other hand, the seeding efficiency of ammonium oxidisers to the main activated sludge system will be higher for the bioaugmentation reactor, as the nitrifiers will grow integrated in the (return) sludge flocs and are thus less susceptible to grazing by protozoa.

The application of nitritation/Anammox systems to the effluent of anaerobic reactors might be a very interesting option in regions with a warm climate. In these regions, combined anaerobic-aerobic treatment offers many advantages compared to aerobic treatment (refer to Chapter 13), but during anaerobic pre-treatment the TKN/COD ratio increases to the point where conventional nitrogen removal will no longer be possible. At least in principle it should be possible to subject the anaerobic effluent to treatment in a one step nitritation-Anammox system. This would allow for nitrogen removal almost without the need for organic material, although some form of post-treatment might be necessary to meet the effluent limits (see also Section 13.5.2.3). At present, the feasibility of combining anaerobic pre-treatment with anaerobic ammonium oxidation has yet to be proven in practice. However, considering the fact that Anammox bacteria are abundant in nature, where a low substrate concentration is the rule rather than the exception, there seem to be no fundamental barriers that prevent application to low strength wastewater streams.

Chapter 7

Phosphorus removal

7.0 INTRODUCTION

Phosphorus in sewage is present predominantly in the form of ortho-phosphates, with a minor fraction of organic phosphate, incorporated in proteins. In the activated sludge process, most of the organic phosphorus is mineralised and consequently (ortho) phosphate will dominate in the effluent. The main problem associated with the presence of phosphorus in water is that, being a nutrient required for growth, it is often responsible for the excessive growth of aquatic life, also called eutrophication. This reduces the quality of the water and thus the suitability for reuse. For this reason in many countries effluent standards have been implemented with regard to phosphorus.

Phosphorus in surface waters originates from two main sources: run-off water from cultivated lands where chemical fertiliser has been used and (2) discharges from untreated wastewater (mainly municipal sewage). The phosphorus concentration in sewage depends strongly on the social economic profile of the contributing population. An important source of phosphorus used to be polyphosphate-based washing powders. However in many countries these have now been replaced by products without phosphate. For this reason the phosphorus content in municipal sewage has been reduced considerably in recent years. Another source of phosphorus in sewage is the consumption of proteins (meat, fish and dairy products). In many developing countries the consumption of proteins is still limited compared to that of the richer regions, but in general it is increasing worldwide. In developed countries the ratio between phosphorus and COD in the sewage ranges between 0.02 and 0.03 mg $P \cdot mg^{-1}$ COD, while in Brazil for example this is typically between 0.01 and 0.02 mg $P \cdot mg^{-1}$ COD. For sewage with a COD value of 500 $mg \cdot l^{-1}$, this results in a phosphorus concentration between 5 and 10 mg $P \cdot l^{-1}$. For industrial wastewaters, the nature of the industry determines the ratio between phosphorus and COD. For industries processing animal products, the ratio is around 0.03 mg $P \cdot mg^{-1}$COD or even more, while for industries processing vegetable products the concentration of phosphorus in many cases is not even sufficient to cover the demands for the production of excess sludge. Addition of phosphate will then be necessary to prevent operational problems such as sludge bulking.

7.1 BIOLOGICAL PHOSPHORUS REMOVAL

7.1.1 Mechanisms involved in biological phosphorus removal

As the phosphorus mass fraction in volatile sludge is about 2.5% of the VSS concentration, the phosphorus content in wastewater is partially removed in a conventional activated sludge system together with the produced excess sludge. For a nitrogen removal activated sludge system treating municipal sewage and operating at an average sludge age of 8 to 12 days, the excess sludge production is around 0.25 mg $VSS \cdot mg^{-1}$ COD. The phosphorus removal will then be approximately $0.25 \cdot 0.025 = 0.006$ mg $P \cdot mg^{-1}$ COD. Thus for sewage with a P/COD ratio between 0.01 and 0.02 mg $P \cdot mg^{-1}$ COD, an effluent concentration between 0.004 and 0.014 mg $P \cdot mg^{-1}$ COD can be expected. This amounts to 2 to 7 mg $P \cdot l^{-1}$ for municipal sewage with a COD concentration of 500 $mg \cdot l^{-1}$. In Chapter 3 an expression was developed to determine the concentration of phosphorus in the influent that is removed as part of the excess sludge:

$$P_l = mP_l \cdot S_{ti} = f_p \cdot mE_v \cdot S_{ti} = f_p \cdot (1 - f_{np} - f_{np}) \cdot [(1 + f \cdot b_h \cdot R_s) \cdot C_r/R_s + f_{np}/f_{cv}] \cdot S_{ti} \quad (3.60)$$

Generally, it will be required to reduce the effluent phosphorus concentration to a value lower than 1 mg $P \cdot l^{-1}$. However, when discharge of organic phosphorus in the excess sludge is the only means of phosphorus removal, this is only possible under favourable conditions: a low P/COD ratio combined with a short sludge age. In wastewaters with a higher level of nutrients and/or activated sludge systems operating at a higher sludge age, additional methods of phosphorus removal will be necessary.

Initially the methods used for phosphorus removal were all based on physical-chemical processes, especially the addition of metal salts ($FeCl_3$, $FeSO_4$) or lime. This results in the precipitation of metal-phosphorus complexes such as ferric phosphate ($FePO_4$), calcium phosphate ($Ca_3(PO_4)_2$), apatite ($Ca_5(OH)(PO_4)_3$) and struvite (NH_4MgPO_4). There are two important disadvantages associated to this strategy: (I) a certain overdosing of metal salts is necessary to obtain the required low effluent phosphorus value, resulting in high costs of chemicals and a significant increase of excess sludge production and (II) the accumulation of ions (increased salt content) may seriously restrict the reuse possibilities of the effluent.

For these reasons, research on the subject of phosphorus removal became focused on biological removal methods. In the last two decades, knowledge of the biological phosphorus removal processes and the feasibility and optimisation of this process has increased enormously. Nowadays, in modern municipal wastewater treatment plants for which phosphorus removal is to be achieved, physical-chemical treatment methods will only be auxiliary to biological methods. For nutrient removal plants (where removal of both nitrogen and phosphorus is required), the main reason to apply physical-chemical treatment is the restricted availability of easily biodegradable COD, required to achieve simultaneously the objectives of nitrogen- and phosphorus removal. In Section 7.3, the use of chemical precipitation as the principal method to remove phosphorus will be discussed.

The removal mechanism involved in biological excess phosphorus removal (or bio-P removal) is so-called luxury phosphorus uptake. Under appropriate operational conditions a sludge mass will develop that contains a significantly increased phosphorus content, compared to the 2.5% normally found in conventional activated sludge systems. Using artificial substrate (i.e. acetate), phosphorus mass fractions of up to 38% weight have been obtained. In systems designed for bio-P removal, a mixed population will develop consisting of the "normal" biomass with a phosphorus content of 2.5% and an "enriched" bio-P sludge mass containing 38% phosphorus. An average phosphorus mass fraction between 80 to 100 mg $P \cdot g^{-1}$ VSS can be expected, depending on the concentration and composition of the organic material in the influent, the operational conditions and the configuration of the sewage treatment plant. Although the mechanisms involved in the bio-P removal process are complex, the following conditions are fundamental in order for luxury phosphorus uptake to occur:

(1) The inclusion of an anaerobic zone in the process configuration (i.e. a zone without oxygen or nitrate present). This resulted in an increase in the mass fraction of phosphorus in the biomass. Many researchers established that the biomass in this anaerobic environment released phosphate to the liquid phase of the sludge, e.g. Barnard (1975), Comeau et al. (1985), and Wentzel et al. (1988). In the subsequent anoxic- and aerobic zones, the uptake of phosphorus by the biomass was such that the phosphorus fraction in the biomass in a system containing an anaerobic zone was much higher than in conventional activated sludge systems operated at the same sludge age. This process is called luxury phosphorus uptake. The increased phosphorus fraction in the biomass results in a higher degree of phosphorus removal (by means of disposal with the excess sludge).

(2) To create an anaerobic environment in activated sludge systems, Barnard (1975) suggested modifying the Bardenpho system through the installation of an anaerobic zone upstream of the pre-D reactor. In this zone an anaerobic environment will be established when the nitrate removal in the Bardenpho system is complete or virtually complete. If not, then first all nitrate recycled to the anaerobic zone will be removed by denitrification.

(3) Siebritz and Marais (1982) demonstrated that exposing the activated sludge to an anaerobic environment was necessary, but this condition alone was not sufficient to induce the process of phosphate release and absorption. The phosphorus fraction in the sludge was dependent on the concentration of easily biodegradable material in the anaerobic zone. A minimum concentration of 25 mg $COD \cdot l^{-1}$ is required to induce phosphate release, which explains why phosphate release in an anaerobic zone is not always observed. Thus the nature and concentration of the organic material in the anaerobic zone play an important part as well.

(4) The presence of nitrate in the anaerobic zone will result in partial or complete removal of the easily biodegradable material. To protect the anaerobic zone against contamination with nitrate, at the University of Cape Town the UCT configuration was developed: an activated sludge system consisting of three reactors (anaerobic/anoxic/aerobic). The a- and s-recycles are returned to the anoxic zone, where a low concentration of nitrate is maintained by manipulation of the a-recycle. An additional r-recycle is introduced, returning mixed liquor from the anoxic zone to the anaerobic zone. This ensures minimal introduction of nitrate in the anaerobic zone. It should be noted that the sludge concentration in the anaerobic zone will only be a fraction $r/(1+r)$ of the sludge concentration in the other reactors.

(5) Wentzel et al. (1986) demonstrated that the organic material required for triggering the release of phosphate in the anaerobic zone was largely composed of volatile fatty acids (predominantly acetate). Using a synthetic feed with acetate, Wentzel et al. (1986) managed to cultivate a sludge mass with a phosphorus mass fraction of up to 38%. These organisms are called phosphate accumulating organisms (PAO) or bio-P organisms;

(6) A scientific model was developed to explain the observed empirical phenomenon of excess phosphorus accumulation:

    (a) The presence of sufficient volatile fatty acids in the anaerobic reactor upstream of the anoxic/aerobic reactors triggers the development of a bacterial population that normally is not encountered in an activated sludge system: the bio-P organisms;

(b) In this anaerobic environment, the non bio-P biomass is not able to use the available substrate, as it lacks a suitable electron acceptor such as dissolved oxygen or nitrate. However, the bio-P organisms have the option to absorb the volatile fatty acids in the form of internal cell-polymers such as polyhydroxy-butyrate (PHB). To supply the energy required for this process, the bio-P organisms use the previously stored polyphosphate, which is split into orthophosphate (PO_4^{3-}) and then released from the cell. In the anoxic- and oxic conditions encountered in the subsequent reactors, the PHB is used by the bio-P organisms as a source of energy supply and bacterial growth. Part of the released energy is used to regenerate the polyphosphate released in the anaerobic reactor. In this process, phosphate is absorbed from the liquid phase by the bio-P organisms;

(c) Due to the capability of the bio-P organisms to "harvest" organic material in the anaerobic phase and store it internally, a competitive advantage is gained over the other heterotrophic micro-organisms present in an activated sludge system. Thus it is possible to establish a sludge rich in bio-P organisms, with a phosphorus level much higher than the fraction of $0.025 \text{ mg P} \cdot \text{mg}^{-1}$ VSS found in conventional activated sludge systems.

(d) The maximum phosphorus level is dependent on the composition of the wastewater (phosphorus and volatile fatty acids content), but will not exceed the maximum value of $0.38 \text{ mg P} \cdot \text{mg}^{-1}$ VSS as measured in enhanced cultures;

(e) As it is vital for the growth of bio-P organisms that volatile fatty acids are present in the anaerobic reactor, it is of crucial importance that the return of nitrate to the anaerobic reactor is avoided, as the volatile fatty acids will then be used for denitrification by non bio-P organisms.

After extensive research by the research group of van Loosdrecht and Heijnen in the Netherlands, Smolders et al. (1994) presented a detailed metabolic model explaining the phenomena observed above on the level of cell microbiology:

- Bio-P organisms contain three internal cell storage products relevant for excess phosphorus removal: polyphosphate, polyhydroxy-alkanoates (mainly present as PHB) and glycogen;
- Under anaerobic conditions, volatile fatty acids are taken up from the liquid phase and stored as PHB. An important intermediate in this process is $NADH_2$, an energy carrier released during the formation of PHB from glycogen. The energy required comes from the hydrolysis of polyphosphate and the subsequent formation of ATP;
- Under anoxic or aerobic conditions, the stored PHB will be oxidised to CO_2, releasing energy in the form of $NADH_2$. This will be used to create ATP, which in turn will allow the bio-P organisms to grow and restock with polyphosphate and glycogen. This process is graphically displayed in Figure 7.1;
- The main difference between the metabolism of bio-P organisms under anoxic and aerobic conditions is the ratio between ATP formed/$NADH_2$ used: this ratio is about 40% lower under anoxic conditions. This explains the lower growth rate observed under anoxic conditions and also applies to "normal" heterotrophic organisms.

There are other groups of micro-organisms with the capacity to store low molecular organic compounds such as VFA under anaerobic conditions, with subsequent growth under aerobic conditions. A particular group of interest are the Glycogen Accumulating Organisms or GAO. These bacteria rely only on intracellular glycogen as a source of energy and carbon storage, as demonstrated by Filipe et al. (2001). Therefore the glycogen accumulating organisms do not exhibit the behaviour that is so typical for bio-P organisms, such as phosphorus release under anaerobic conditions and take up of phosphorus under aerobic conditions. The proposed metabolism is indicated in Figure 7.2. Several strains have been identified, with different affinities for propionate and acetate but often with a similar VFA uptake rate as the bio-P bacteria.

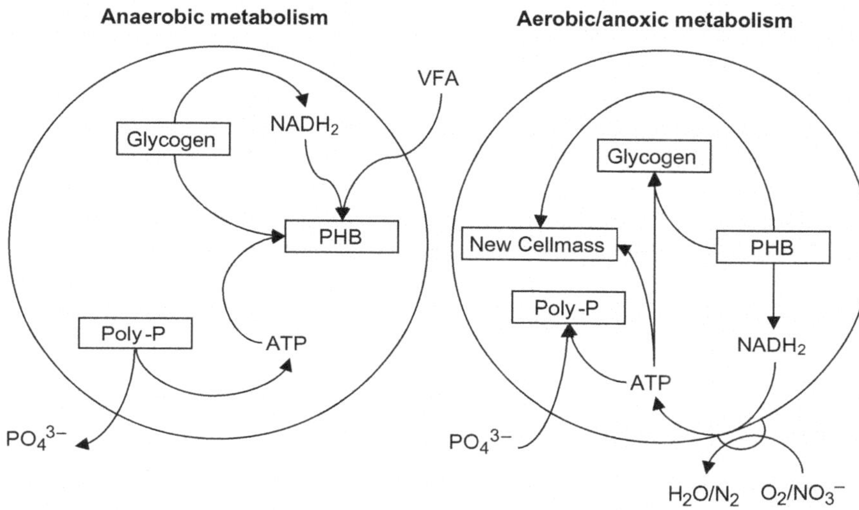

Figure 7.1 Metabolism of bio-P organisms (PAO) under anaerobic and -oxic conditions, according to Smolders et al. (1994)

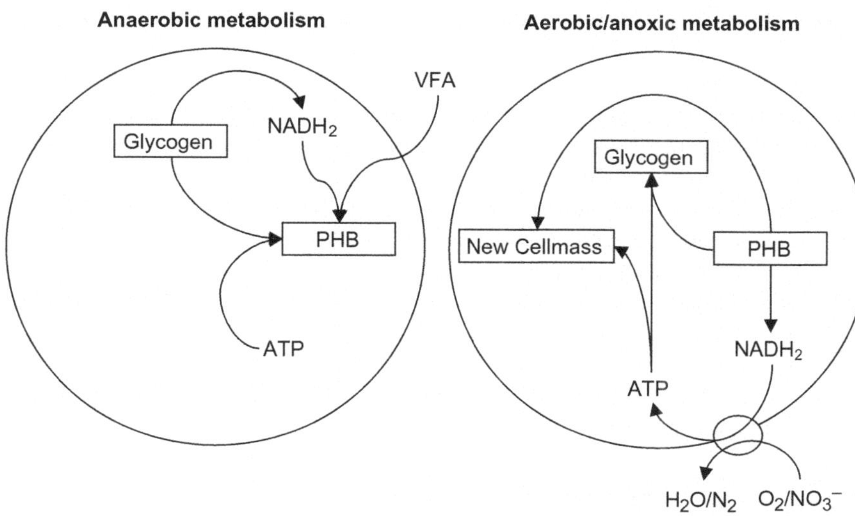

Figure 7.2 Metabolism of glycogen accumulating organisms (GAO) under anaerobic and -oxic conditions, according to Filipe et al. (2001)

Due to the direct competition for VFA between bio-P organisms and glycogen accumulating organisms, this makes the presence of the latter group in an activated sludge system very undesirable when bio-P removal is envisaged. Therefore it is important to assess how the dominance of bio-P organisms over glycogen accumulating organisms can be assured. It seems that the main factors are:

- The pH value, which influences the amount of energy required for transport of components through the cell membrane. Under otherwise comparable conditions, it has been found that bio-P organisms tend to dominate for pH values higher than 7.25;
- The temperature, as it is reported by Lopez Vazques (2008) that at a temperature lower than 20°C bio-P organisms will dominate while higher temperatures favour glycogen accumulating organisms. However, the effect of pH seems to be more important, so a higher pH value will allow bio-P organisms to dominate, even at temperatures higher than 20°C;
- The influent P/VFA ratio. When phosphorus is absent from the influent for extended periods of time, bio-P organisms lose their internal poly-P mass and will be unable to compete. When enriched cultures are studied, typically a P/COD ratio of 0.006 $g\ P \cdot g^{-1}$ COD is used to cultivate glycogen accumulating organisms while a much higher P/COD ratio of 0.04 $g\ P \cdot g^{-1}$ COD is used when bio-P organism are grown.

7.1.2 Bio-P removal system configurations

Various system configurations have been developed for biological phosphorus removal, all of which have been extensively applied in practice. The main difference between these systems is the way in which an anaerobic zone is maintained and protected against the introduction of nitrate. In the following sections several system configurations are discussed.

(a) Phoredox and A/O configuration

The Phoredox system (Figure 7.3a) proposed by Barnard (1976) is composed of two reactors in series, of which the first (receiving the influent) is anaerobic and the second aerobic. The return sludge flow is recirculated from the final settler to the anaerobic reactor. There are no other recirculation streams between the reactors. The Phoredox system should only be used when nitrogen removal is not required, as it requires operation at low sludge age. Consequently, an anoxic zone is not needed and the system volume is rather compact.

The A/O system (Timmerman, 1976) has the same configuration as the Phoredox system, but due to a compartmentalisation of the anaerobic zone a plug-flow regime is induced, which promotes the conversion of easily biodegradable material to acetate and increases the phosphorus removal capacity. Currently the Phoredox process has only found application in regions with a cold climate, principally in Europe and the US. In regions with temperate and hot climates the applicability is limited, as nitrification cannot be prevented completely, even at low sludge ages. The introduction of nitrate in the anaerobic zone will then be unavoidable and results in a reduction of the phosphorus removal capacity of the system. Burke et al. (1990) demonstrated that it was impossible to prevent partial nitrification in a pilot scale Phoredox system operated at a sludge age of only three days at 20°C.

(b) The modified pre-D, A^2/O and Bardenpho configurations

In the modified pre-D and Bardenpho system (Figure 7.3b and c), an anaerobic zone is added upstream of the pre-D anoxic reactor. The anaerobic reactor receives the influent and the return sludge flow. However, if the removal of nitrate is not complete, then nitrate will be introduced in the anaerobic zone. This reduces the availability of easily biodegradable material to the bio-P organisms and thus decreases the phosphorus removal capacity of the system. The modified pre-D system is equivalent to the plug-flow A^2/O system. The modified Bardenpho configuration has been widely applied, although it has been replaced in popularity by the UCT and modified UCT configurations.

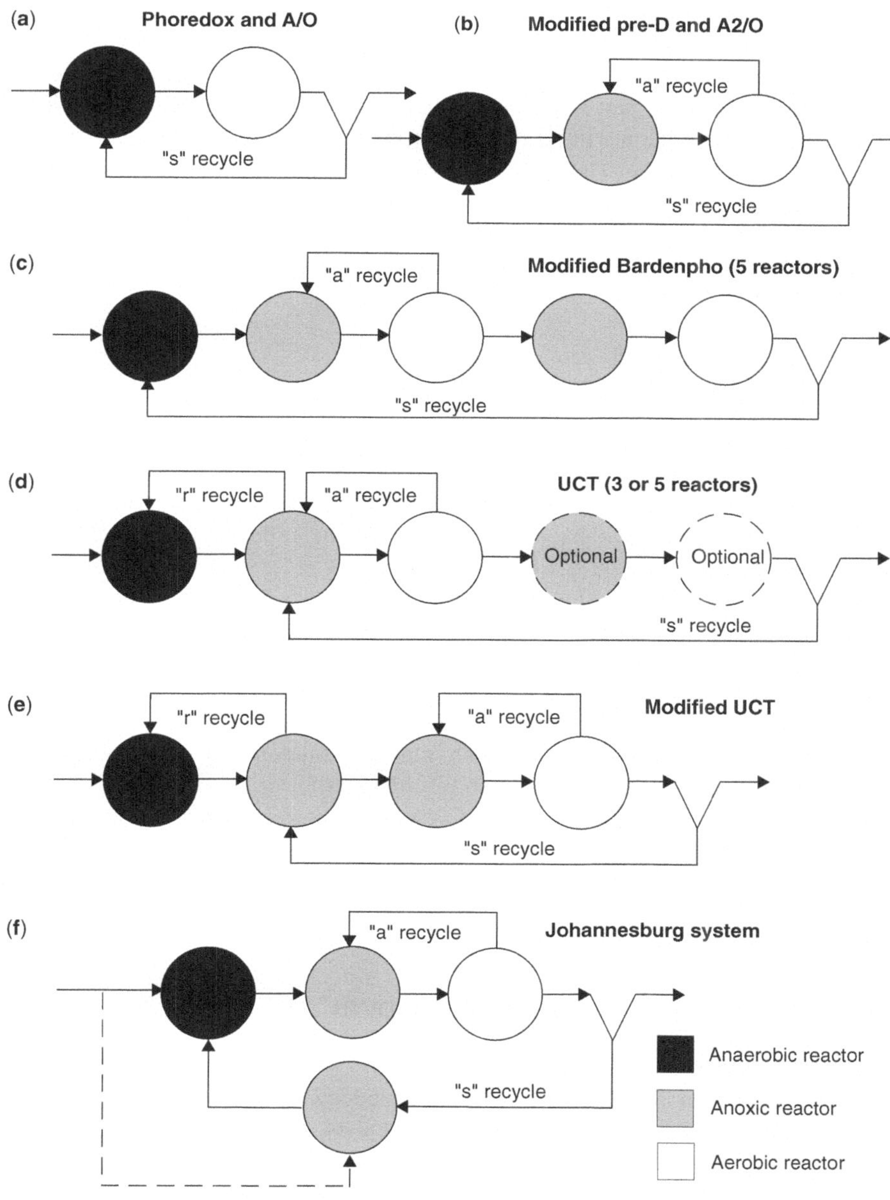

Figure 7.3 Common system configurations for biological phosphorus removal

(c) UCT system and modified UCT configurations

In the UCT system proposed by Rabinowitz and Marais (1980) and represented in Figure 7.3d, the introduction of nitrate in the anaerobic zone is avoided, because the recycle stream is taken from the anoxic instead of the aerobic zone. In the anoxic zone the concentration of nitrate is controlled at a low level by manipulation of the recirculation factor "a", in such a way that the nitrate available for denitrification in the pre-D zone is always smaller than the available denitrification capacity.

The modified UCT system (Figure 7.3e) was designed to ensure that the introduction of nitrate in the anaerobic zone is impossible, even with a variable nitrate concentration in excess of the denitrification capacity. The anoxic zone is split into two parts, introducing the return sludge in the first (upstream) part and using the second (downstream) part for denitrification of the nitrate recycled with recirculation "a". Under these conditions, denitrification will be complete in the first part of the anoxic zone and no nitrate will be returned to the anaerobic zone. The disadvantage of this configuration is that the anoxic zone as a whole is under-loaded with nitrate. Therefore a larger total anoxic volume is required compared to the modified Bardenpho configuration. If this is ignored in design, then the nitrate concentration in the effluent will be higher than expected, which might cause problems in the final settler, such as the formation of a layer of floating sludge.

(d) The Johannesburg configuration

Figure 7.3f shows the Johannesburg system (Osborn and Nicholls, 1978), where the mixed liquor from the aerobic zone passes through the final settler, while the return sludge is directed to an anoxic zone. As the sludge concentration in the final settler is a factor $(s + 1)/s$ larger than the mixed liquor entering the final settler, the denitrification rate in the post-D zone will also be proportionally increased. This means that it is possible to produce a mixed liquor without nitrate in the discharge of the post-D reactor to the anaerobic zone, even while there will be nitrate present in the effluent. This configuration can be advantageous if effluent nitrate limits are not very strict.

Table 7.1 compares the various configurations discussed above. It can be observed that the main difference is the way in which the anoxic zone is used. In the systems with the highest degree of protection against nitrate recycle to the anaerobic zone, the pre-D anoxic zone is relatively under-loaded and the removal of nitrate will be smaller than the denitrification capacity. On the other hand, the higher one exploits the denitrification capacity in the pre-D anoxic zone, the lower the protection of the anaerobic zone will be against contamination with nitrate. Consequently the process of biological phosphorus removal will become more vulnerable to disturbances.

It is possible to design a flexible activated sludge system for nutrient removal that allows the system configuration to be modified relatively easy as the conditions change (for instance the ratio TKN/COD, temperature, sludge age, f_{sb}, μ_m or presence of toxic materials). Figure 7.4 shows a configuration that permits operation in all of the alternative configurations discussed above, through manipulation of the recirculation flows and the relocation of aerators.

7.1.3 Model of biological phosphorus removal

7.1.3.1 Enhanced cultures

Based on the concepts presented in the previous section, a model was developed at the university of Cape Town (UCT) to describe the processes involved in biological phosphorus removal, including the release of phosphorus in the anaerobic zone and the excess phosphorus uptake in the subsequent aerobic zone. This model is an extension to the ideal steady state activated sludge model and includes the presence of bio-P organisms.

Table 7.1 Comparison of different configurations for biological phosphorus removal

Configuration	Advantages	Disadvantages
Phoredox and A/O	– Small and simple system – Short residence time	– No nitrogen removal – In hot or moderate climates the system will not be reliable
Modified Pre-D and A^2/O	– High denitrification rate – Short sludge age	– Might not function properly (due to recirculation of nitrate) – Incomplete denitrification – Tendency to induce sludge bulking
Modified BDP (3 or 5 reactors)	– Excellent configuration for nitrogen removal	– If denitrification is incomplete then nitrate will be recycled to the anaerobic zone, adversely affecting P-removal
UCT	– Prevents recirculation of nitrate	– Utilisation of denitrification capacity is inefficient
Modified UCT	– Ensures absence of nitrate in the anaerobic zone	– Utilisation of denitrification capacity is inefficient (even more so than in the UCT system)
Johannesburg	– Efficient use of denitrification zone	– Incomplete denitrification

To investigate the behaviour of bio-P organisms, Wentzel et al. (1986) operated a number of activated sludge systems using acetate as the only source of COD in the influent, resulting in a culture enhanced with bio-P organisms. Based on experimental observations it was concluded that:

- In the anaerobic zone there is a proportional relationship between the absorbed acetate concentration and the concentration of released phosphate. This constant f_{pr} has a value of 0.5 mg $P \cdot mg^{-1}$ COD absorbed;
- The PHB generated in the anaerobic zone will be completely utilised in the subsequent aerobic zone;
- The absorption of phosphorus in the anoxic- and aerobic zones by the bio-P organisms produces polyphosphate, which is stored internally. Depending on the mass of acetate present in the influent (and thus on the mass of PHB formed), this can result in a maximum phosphorus content in the active biomass of 38%.

When the bio-P organisms are compared with the micro-organisms normally present in activated sludge systems, the following differences can be observed (apart from the increased phosphorus content):

(a) Release of phosphate in the anaerobic zone

Under anaerobic conditions and in the presence of an adequate substrate (VFA, such as acetate), the bio-P organisms transform internally stored polyphosphate into phosphate, a process that releases the energy required for the absorption of VFA. The release of phosphate is described as:

$$P_r = f_{pr} \cdot S_{VFA} \tag{7.1}$$

where:

P_r = phosphate concentration released to the liquid phase ($mg\ P \cdot l^{-1}$)
S_{VFA} = concentration of volatile fatty acids ($mg\ COD \cdot l^{-1}$)
f_{pr} = phosphorus release constant = 0.5 $mg\ P \cdot mg^{-1}$ COD)

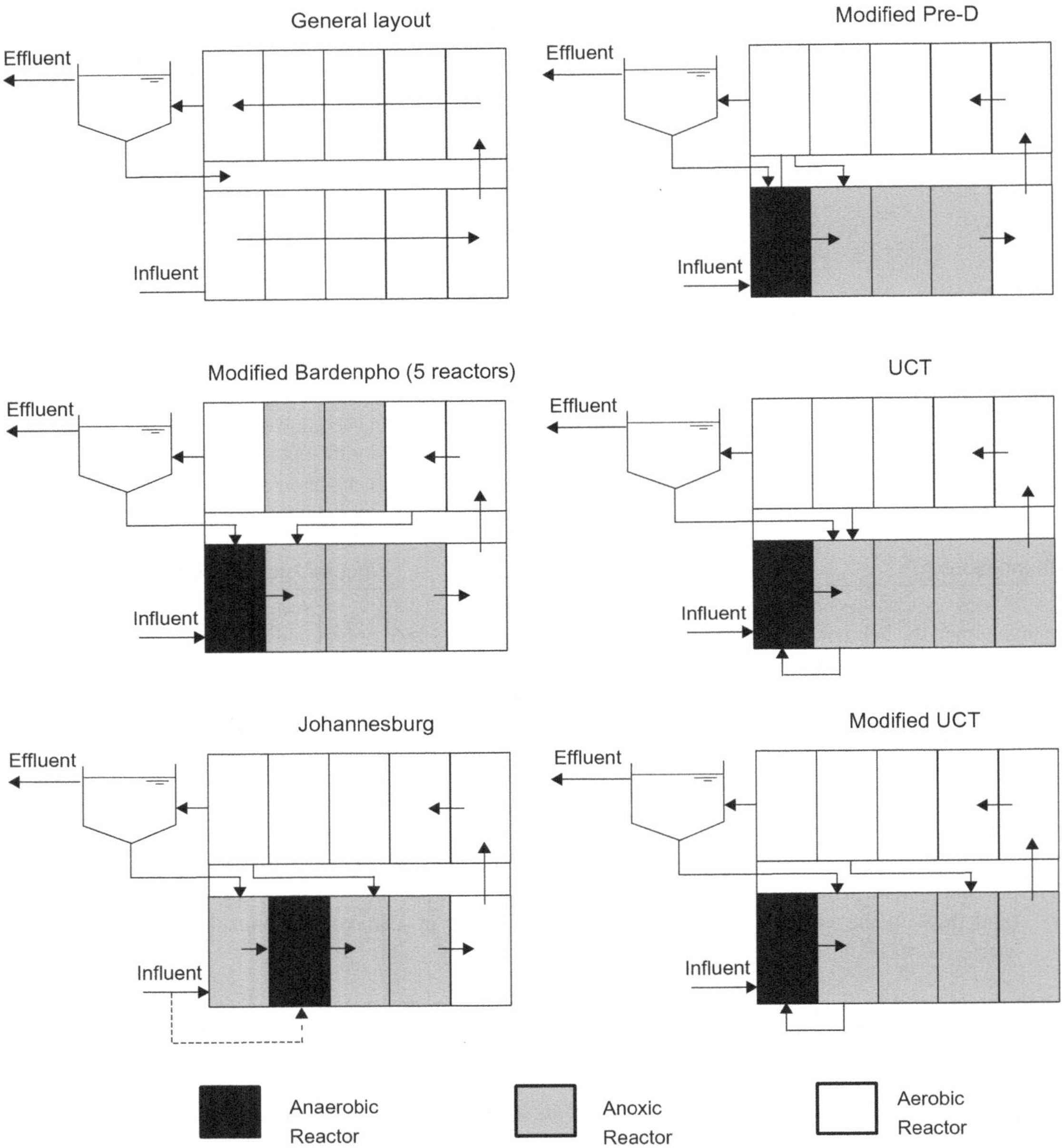

Figure 7.4 Flexible system layout that allows a wastewater treatment plant to be operated in different bio-P removal configurations with only small modifications required

(b) Decay rate of bio-P organisms

The decay rate of bio-P organisms is significantly lower than that of the other (heterotrophic) bacteria in activated sludge. The value for the decay rate constant b_p was determined experimentally as 0.04 d^{-1} at 20°C.

(c) Quantity and composition of activated sludge and the endogenous residue

It was determined that 25% of the bacterial mass remained as endogenous residue after decay: $f_{ep} = 0.25$. However, the endogenous residue does not contain the high poly-P content of the active bio-P biomass. Instead the phosphorus content was equal to that of normal biomass: i.e. 2.5%. Thus, when bio-P organisms decay, the stored polyphosphate is released into the liquid phase.

(d) Ratio VSS/TSS

Due to the large inorganic fraction in bio-P organisms (mainly internally stored polyphosphate), the ratio between VSS and TSS (f_{vp}) is as low as 0.46 mg VSS $\cdot$ mg^{-1} TSS. This is significantly smaller than the f_v value of normal activated sludge, which typically is between 0.70 to 0.85 mg VSS $\cdot$ mg^{-1} TSS. The excess sludge production is therefore much higher in systems with bio-P sludge than in conventional systems.

(e) Denitrification

Wentzel et al. (1986) observed that the denitrification rate of the bio-P organisms in an anoxic environment was very small and for all practical purposes could be ignored. Therefore in the first version of the Activated Sludge Model no. II (Henze et al., 1994), which included bio-P removal for the first time, bio-P organisms were therefore modelled as being incapable of denitrification. However, as in practice significant phosphate uptake is observed in the anoxic zones of full-scale bio-P removal plants, it must be concluded that this is incorrect.

A possible explanation to account for the observation of Wentzel et al. (1986) on the absence of denitrifying bio-P organisms may be found in the data reported by Kuba et al. (1995). The cytochrome oxidation enzyme, required for aerobic oxidation, is always present in the heterotrophic bio-P organisms, even when the organisms have been cultivated under strictly anaerobic-anoxic conditions. However, this is not the case for the equivalent enzyme required for anoxic oxidation (nitrate reductase). So, when denitrifying bio-P organisms are cultivated under anaerobic-aerobic conditions, a large part of the nitrate reductase is deactivated and the denitrification capacity decreases accordingly. Once anoxic conditions are established, resynthesis of the enzyme is not immediate and it takes a long period for the denitrification capacity to recover. As the experimental work of Wentzel et al. (1986) was done with strictly anaerobic-aerobic systems, the absence of denitrification can thus be explained.

When the aerobic- and anoxic P-uptake of sludge from two full-scale wastewater treatment plants operating in UCT configuration was compared, Kuba et al. (1994) estimated the fraction of bio-P organisms capable of denitrification at 40 to 50% of the total bio-P biomass. Janssen et al. (2004) also determined this fraction at 11 full-scale bio-P removal systems in the Netherlands: six with dedicated anoxic zones and five aeration circuits (carrousels). The average ratio between anoxic- and aerobic P-uptake was 0.54 for the systems with separate anoxic zones and 0.63 for the carrousels. Perhaps coincidentally, this ratio is very close to the value of 0.6 observed between the growth rate of heterotrophic bacteria under anoxic and aerobic conditions, resulting from the decreased ratio between ATP formed/$NADH_2$ used under anoxic conditions. Therefore it might very well be possible that in an activated sludge process specifically designed for combined nitrogen- and biological phosphorus removal in fact all bio-P organisms are capable of denitrification.

In practice this fraction may be lower due to adverse conditions or sub-optimal design, resulting in loss of denitrification capacity, for example due to inactivation of part of the enzymes used in the denitrification process as explained above. Therefore for design purposes a more conservative estimate of this fraction might be used. For instance, from the values reported by Kuba et al. (1994), the anoxic bio-P biomass fraction of the two sewage treatment plants that were investigated can be estimated as $0.4/0.6 = 0.67$

and 0.5/0.6 = 0.83 respectively. In Table 7.2 the values of a number of key parameters of bio-P organisms are compared to those of the biomass normally present in conventional activated sludge systems.

Figure 7.5 compares, as a function of the sludge age, the specific sludge mass per unit mass daily applied COD of the different sludge fractions in a conventional activated sludge system, in an enhanced culture of bio-P organisms and in a mixed activated sludge system. The mixed culture is based on a division of influent COD in which 25% of the influent COD is available for bio-P organisms and 75% for the conventional

Table 7.2 Parameters of bio-P organisms compared to regular heterotrophic organisms (determined at 20°C)

Parameter	Symbol	Bio-P Organisms	Regular Heterotrophs	UoM
Phosphorus content	f_{pp}/f_p	0.38	0.025	$mg\ P \cdot mg^{-1}\ X_a$
Decay rate	b_p/b_h	$0.04 \cdot 1.06^{T-20}$	0.24	d^{-1}
Endogenous residue	f_{ep}/f	0.25	0.20	(−)
P-fraction end. residue	f_p	0.025	0.025	$mg\ P \cdot mg^{-1}\ X_e$
Ratio VSS/TSS	$f_{vp}^{(1)}/f_v$	0.46	0.80–0.85	$mg\ VSS \cdot mg^{-1}\ TSS$
Denitrifying fraction	f_{pd}	0.6–1.0	1.0	(−)
Denitrification rate	K_2/K_3	0.10/0.08	0.10/0.08	$mg\ N \cdot mg^{-1}\ X_a \cdot d^{-1}$
Anaerobic P- release	f_{pr}	0.5	–	$mg\ P \cdot mg^{-1}\ COD$

Note (1): Lower f_v value only applies to active part of bio-P biomass, due to presence of polyphosphates

EXAMPLE 7.1

Determine the maximum phosphorus concentration that can be removed from the influent in an activated sludge system equipped with an anaerobic zone, when the influent substrate concentration of 500 mg $COD \cdot l^{-1}$ is completely in the form of acetate. Assume a temperature of 20°C and a sludge age of 10 days. Compare the total sludge mass that will develop in the system with that of a conventional system.

Solution

In the calculation presented below, S_{bi} is equal to S_{ti} as all organic material is biodegradable. Assuming that a bio-P biomass will develop and that the utilisation of the influent organic material is complete, one has:

$$mX_{ap} = Y \cdot R_s/(1 + b_p \cdot R_s)$$

$$= 0.45 \cdot 10/(1 + 0.04 \cdot 10) = 3.2\ mg\ VSS \cdot d \cdot mg^{-1}\ COD$$

$$mX_{ep} = f_{ep} \cdot b_p \cdot R_s \cdot mX_{ap}$$

$$= 0.25 \cdot 0.04 \cdot 10 \cdot 3.2 = 0.32\ mg\ VSS \cdot d \cdot mg^{-1}\ COD$$

The discharge of phosphorus with the excess sludge per unit mass of daily applied COD is:

$$mP_l = (f_{pp} \cdot mX_{ap} + f_p \cdot mX_{ep})/R_s$$
$$= (0.38 \cdot 3.2 + 0.025 \cdot 0.32)/10 = 0.12 \, mg\,P \cdot mg^{-1}\,COD$$

For the influent COD concentration of 500 mg $\cdot$ l^{-1}, the value of P_l, the concentration of phosphorus that theoretically can be removed from the influent with the excess sludge is $0.12 \cdot 500 = 61.5$ mg P $\cdot$ l^{-1}. In comparison, in a comparable conventional activated sludge system (receiving only biodegradable COD) the value of P_l would be much lower:

$$P_l = f_p \cdot (1 + f \cdot b_h \cdot R_s) \cdot C_r/R_s \cdot S_{ti}$$
$$= 0.025 \cdot (1 = 0.2 \cdot 0.24 \cdot 10) \cdot 0.45 \cdot 10/(1 = 0.24 \cdot 10)/10 \cdot 500$$
$$= 0.0049 \cdot 500 = 2.45 \, mgP.l^{-1}$$

The ratio between P_l in the enhanced system (61.5 mg P $\cdot$ l^{-1}) and in the conventional system (2.45 mg P $\cdot$ l^{-1}), i.e. $61.5/2.45 = 25$, is even more than could be expected based on the difference in phosphorus content of the different sludges: i.e. $0.38/0.025 = 15$. This is due to the lower decay rate of the bio-P organisms compared to that of the other heterotrophs. For this reason, the active fraction in the excess sludge is much higher for bio-P sludge than for conventional sludge: $mX_{ap} = 3.2$ against $mX_a = 1.3$ mg VSS $\cdot$ d $\cdot$ mg^{-1} COD.

In both cases it should be noted that due to imperfect solid-liquid separation in the final settler, part of P_l will be present in the effluent instead of in the excess sludge. The consequences of this model simplification will be discussed in Section 7.1.3.4. The volatile excess sludge production of the enhanced bio-P biomass culture can be calculated as:

$$mE_{vp} = (mX_{ap} + mX_{ep})/R_s = (3.2 + 0.32)/10 = 0.35 \, mgVSS \cdot mg^{-1}\,COD$$

The total excess sludge production is given by:

$$mE_{tp} = (mX_{ap}/f_{vp} + mX_{ep}/f_v)/R_s$$
$$= (3.2/0.46 + 0.32/0.8)/10$$
$$= 0.74 \, mg\,TSS \cdot mg^{-1}COD$$

In the conventional activated sludge system the volatile and total excess sludge production can be calculated as:

$$mE_v = (1 + f \cdot b_h \cdot R_s) \cdot C_r/R_s$$
$$= (1 + 0.2 \cdot 0.24 \cdot 10) \cdot 1.32/10$$
$$= 0.20 \, mg\,VSS \cdot mg^{-1}COD$$

$$mE_t = mE_v/f_v = 0.20/0.80 = 0.24 \, g\,TSS \cdot mg^{-1}COD$$

It is concluded that under the specified conditions, the excess sludge production in a system with biological excess phosphorus removal $= 0.74/0.24 = 3$ times higher than in a conventional activated sludge system.

organisms. In the next section it will be demonstrated that this is approximately the expected ratio for domestic sewage. In Figure 7.5 it can be observed that in systems with bio-P biomass:

- The active sludge mass fraction is much higher than in conventional systems, due to the slow decay rate of bio-P biomass;
- On the other hand, the total sludge production will be much higher as well due to the high inorganic mass fraction (stored poly-P) of the bio-P organisms.

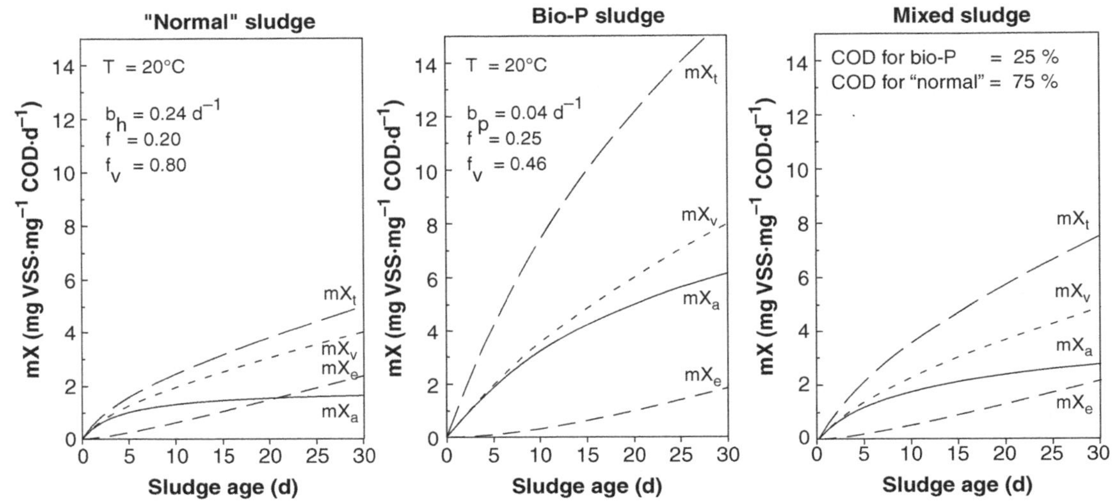

Figure 7.5 Comparison of the sludge mass and -composition in a conventional activated sludge system, an enhanced culture of bio-P organisms and a mixed culture typical for a municipal activated sludge system designed for bio-P removal

7.1.3.2 Mixed cultures

In many municipal wastewater treatment plants, the organic fraction in the influent is not in the form required by the bio-P organisms, i.e. present as volatile fatty acids. In general the fraction of VFA in domestic wastewaters is less than 10% of the total COD concentration, even when the residence time in the sewer system is long and some fermentation has occurred (which produces VFA). Wentzel (1985) demonstrated that in an anaerobic environment, the conventional bacteria present in activated systems are able to convert the easily biodegradable material into VFA, which then can be absorbed by the bio-P organisms.

The bio-P organisms themselves are not capable of converting the easily biodegradable material into VFA. So in those cases where the wastewater does not contain sufficient VFA, the presence of conventional bacteria is a necessity to generate sufficient substrate for the bio-P organisms. Therefore, in bio-P removal systems there will always be a mixed culture of conventional micro-organisms and bio-P organisms. Wentzel et al. (1990) developed a model to describe the behaviour of an activated sludge system with such a mixed culture, based on the UCT configuration.

(1) If nitrate is introduced into the anaerobic zone, the concentration of easily biodegradable material is reduced according to the following expression:

$$S'_{bsi} = S_{bsi} - r \cdot N_{ne} \cdot 2.86/(1 - f_{cv} \cdot Y) \cdot K_1/(K_1 + K_2) \tag{7.2}$$

Where S'_{sbi} = influent concentration of easily biodegradable material after correction for denitrification in the anaerobic zone

The factor $K_1/(K_1 + K_2)$ reflects the proportion of easily- versus slowly biodegradable organic material that is used for nitrate reduction.

(2) In an anaerobic environment the easily biodegradable material will be fermented into VFA. Assuming there is no VFA present in the influent, Wentzel et al. (1990) proposed the following expression:

$$dS_{VFA}/dt = K_c \cdot X_{ah,an} \cdot S_{bs} \tag{7.3}$$

where:

K_c = fermentation constant = 0.06 litre · mg^{-1} VSS · d^{-1} or 60 $m^3 \cdot kg^{-1}$ VSS · d^{-1}
$X_{ah,an}$ = concentration of active (*non bio-P*) sludge in the anaerobic zone. Note that the subscript "h" is added here to differentiate between bio-P biomass and normal heterotrophic biomass

In an UCT configuration, the value of $X_{ah,an}$ will be less than that of X_{ah} in the other reactors, as the thickened sludge from the final settler is not returned to the anaerobic reactor. The dilution factor equals $r/(r + 1)$. The following expression can be derived for the residual concentration of the easily biodegradable material in the effluent of a completely mixed anaerobic reactor:

$$S_{bs} \frac{S'_{bsi}/(1 + r)}{1 + f_{an} \cdot K_c \cdot MX_{ah}/(Q_i \cdot (r + 1))} \tag{7.4}$$

And for a series of N equally sized anaerobic reactors:

$$S_{bsN} = \frac{S'_{bsi}/(1 + r)}{[1 + f_{an} \cdot K_c \cdot MX_{ah}/(Q_i \cdot N \cdot (r + 1))]^N} \tag{7.5}$$

where:

f_{an} = anaerobic sludge mass fraction
MX_{ah} = total active (heterotrophic) sludge mass in the system.

(3) All fermented organic material (plus any VFA present in the influent) will be taken up by the bio-P organisms and stored as PHB (this is a relatively rapid process). Therefore the concentration of organic material sequestered by the bio-P organisms is given as:

$$MS_{seq} = Q_i \cdot S'_{bsi} - (1 + r) \cdot Q_i \cdot S_{bsN} \tag{7.6}$$

(4) In the aerobic zone, the sequestered material is used by the bio-P organisms for growth and for the absorption of phosphate from the liquid phase in order to re-synthesize the depleted storage of intracellular polyphosphate. The residual organic material in the liquid phase of the aerobic reactor will only be utilised by the non bio-P organisms. Therefore the total active sludge mass in the system can be expressed as $MX_a = MX_{ah} + MX_{ap}$, where:

$$MX_{ap} = C_{rp} \cdot MS_{seq} = Y \cdot R_s/(1 + b_p \cdot R_s) \cdot MS_{seq} \tag{7.7}$$

$$MX_{ah} = C_{rh} \cdot (MS_{bi} - MS_{seq}) = Y \cdot R_s/(1 + b_h \cdot R_s) \cdot (MS_{bi} - MS_{seq}) \tag{7.8}$$

The formulas presented above are not yet sufficient for the calculation of the performance of an activated sludge system with biological phosphorus removal. For this it will be necessary to determine the residual concentration of easily biodegradable material in the anaerobic zone. This concentration can be calculated using the iterative procedure outlined below:

(1) Assume that the conversion of easily biodegradable material in the anaerobic zone is complete and calculate the active non bio-P biomass that will develop:

$$S_{bsN} = 0 \rightarrow MS_{seq} = MS'_{bsi} - 0 = MS'_{bsi}$$

$$MX_{ah} = Y \cdot R_s/(1 + b_h \cdot R_s) \cdot (MS_{bi} - MS_{seq})$$

(2) With the value calculated above for MX_{ah}, calculate the new value for S_{bsN} (Eq. 7.5);
(3) Then use the value of S_{bsN} from step (2) to recalculate MX_{ah} with Eq. (7.8);
(4) Repeat step (2) and (3) until the values of S_{bsN} and MX_{ah} are stable.

Once the concentration of the influent biodegradable organic material that is sequestered by the bio-P organisms is known, all other important system parameters can be calculated.

$$MX_{ep} = (f_{ep} \cdot b_p \cdot R_s) \cdot MX_{ap} = f_{ep} \cdot b_p \cdot R_s \cdot C_{rp} \cdot MS_{seq} \tag{7.9}$$

$$MX_{eh} = (f \cdot b_h \cdot R_s) \cdot MX_{ah} = f \cdot b_h \cdot R_s \cdot C_{rh} \cdot (MS_{bi} - MS_{seq}) \tag{7.10}$$

Where MX_{ah} and MX_{ap} have been calculated with Eqs. (7.7 and 7.8). The inert organic sludge is calculated as usual with Eq. (3.45):

$$MX_i = f_{np} \cdot R_s/f_{cv} \cdot MS_{ti} \tag{3.45}$$

The total sludge mass in the mixed system can be calculated as:

$$MX_v = MX_{ah} + MX_{eh} + MX_i + MX_{ap} + MX_{ep} \tag{7.11}$$

$$= (1 + f \cdot b_h \cdot R_s) \cdot C_{rh} \cdot (MS_{bi} - MS_{seq}) + f_{np} \cdot R_s/f_{cv} \cdot MS_{ti} + (1 + f_{ep} \cdot b_p \cdot R_s) \cdot C_{rp} \cdot MS_{seq}$$

$$MX_t = (MX_a + MX_e + MX_i + MX_{ep})/f_v + MX_{ap}/f_{vp} \tag{7.12}$$

Finally the phosphorus removal can be calculated as:

$$MP_l = f_p \cdot (MX_{ah} + MX_{eh} + MX_i + MX_{ep})/R_s + f_{pp} \cdot MX_{ap}/R_s \tag{7.13}$$

EXAMPLE 7.2

Determine the potential removal of phosphorus from the influent, as well as the main other system parameters, of an UCT system treating municipal wastewater. Assume the following characteristics and conditions: $Q_i = 1000\ m^3 \cdot d^{-1}$; $S_{ti} = 500$ mg $COD \cdot l^{-1}$; $f_{ns} = f_{np} = 0.1$; $f_{sb} = 0.25$; $f_{an} = 0.15$; $f_{x1} = 0.35$, $R_s = 10$ d; $T = 20°C$; $N = 1$ and $r = 1$. Furthermore, assume that nitrate is not present in the recycle stream to the anaerobic zone.

Solution

(1) Determine the amount of VFA formed in the anaerobic zone

When it is initially assumed that all easily biodegradable material in the influent is converted into VFA then:

$$S_{bs} = 0 \quad \text{and} \quad S_{seq} = S'_{bsi} = S_{bsi} = f_{sb} \cdot S_{bi} = 0.25 \cdot 400 = 100\,\text{mg COD} \cdot l^{-1}$$
$$MX_{ah} = Q_i \cdot C_{rh} \cdot (S_{bi} - S_{seq})$$
$$MX_{ah} = 1000 \cdot 1.32 \cdot (400 - 100)/1000 = 397\,\text{kg VSS}$$

Now the residual concentration of easily biodegradable material in the effluent of the anaerobic zone is calculated with Eq. (7.4):

$$S_{bs} = \frac{S'_{bsi}/(1+r)}{1 + f_{an} \cdot K_c \cdot MX_{ah}/(Q_i \cdot (r+1))} = \frac{100/2}{1 + 0.15 \cdot 60 \cdot 397/(1000 \cdot 2)} = 17.9\,\text{mg COD} \cdot l^{-1}$$

The value of S_{bs} calculated above is used to recalculate MX_a as 445 kg VSS which in turn results in a new value for $S_{bs} = 16.7$ mg $COD \cdot l^{-1}$. The third iteration results in $MX_a = 441$ kg VSS and $S_{bs} = 16.7$ mg $COD \cdot l^{-1}$, which are accepted as the final values.

(2) Determine the concentration of the different sludge fractions

(A) Phosphate accumulating organisms

Use the concentration S_{bsi} at the inlet- and S_{bs} at the outlet of the anaerobic zone, to calculate the daily amount of organic material sequestered by the bio-P organisms:

$$MS_{seq} = Q_i \cdot (S_{bsi} - (r+1) \cdot S_{bs}) = 1000 \cdot (100 - 2 \cdot 16.7)/1000 = 67\,\text{kg COD.d}^{-1} \tag{7.6}$$

This allows the active mass of bio-P organisms to be calculated:

$$C_{rp} = Y \cdot R_s/(1 + b_p \cdot R_s) = 0.45 \cdot 10/(1 + 0.04 \cdot 10) = 3.21$$
$$MX_{ap} = C_{rp} \cdot MS_{seq} = 3.21 \cdot 67 = 214\,\text{kgVSS} \tag{7.7}$$

The endogenous residue generated during the decay of the bio-P organisms is defined as:

$$MX_{ep} = f_{ep} \cdot b_p \cdot R_s \cdot MX_{ap} = 0.25 \cdot 0.04 \cdot 10 \cdot 214 = 21.4\,\text{kg VSS} \tag{7.9}$$

(B) Normal heterotrophic sludge

The active sludge mass of the normal heterotrophic activated sludge has already been calculated above: $MX_{ah} = 441$ kg VSS. So the endogenous residue generated during the decay of the normal sludge is:

$$MX_{eh} = f \cdot b_h \cdot R_s \cdot MX_{ah} = 0.2 \cdot 0.24 \cdot 10 \cdot 441 = 212\,kgVSS \quad (7.10)$$

(C) Total and volatile sludge mass and production

The amount of inert organic sludge is calculated from the particulate, non biodegradable organic fraction in the influent:

$$MX_i = f_{np} \cdot MS_{ti} \cdot R_s/f_{cv} = 0.1 \cdot 500 \cdot 10/1.5 = 333\,kgVSS \quad (3.45)$$

The total mass of volatile sludge is calculated as:

$$MX_v = MX_{ah} + MX_{eh} + MX_i + MX_{ap} + MX_{ep} = 441 + 212 + 333 + 214 + 21 = 1222\,kgVSS \quad (7.11)$$

The total mass of sludge is given as:

$$MX_t = (MX_{ah} + MX_{eh} + MX_i + MX_{ep})/0.8 + MX_{ap}/0.46 = 1725\,kg\,TSS \quad (7.12)$$

The production of excess sludge is a fraction $1/R_s$ of the total sludge mass:

$$ME_v = MX_v/R_s = 1222/10 = 122\,kgVSS \cdot d^{-1}$$

$$ME_t = MX_t/R_s = 1725/10 = 173\,kgTSS \cdot d^{-1}$$

(3) Phosphorus removal

The removal of phosphorus from the influent with the excess sludge is equal to 38% of the mass of the active bio-P organisms discharged from the system and 2.5% of the other volatile sludge mass fractions:

$$MP_l = f_p \cdot (MX_{ah} + MX_{eh} + MX_i + MX_{ep})/R_s + f_{pp} \cdot MX_{ap}/R_s = 10.7\,kg\,P \cdot d^{-1} \quad (7.13)$$

It is interesting to compare to compare the results calculated above with those of a conventional activated sludge system without an anaerobic zone:

$$MX_a = C_r \cdot MS_{bi} = 1.32 \cdot 400 = 528\,kgVSS$$

$$MX_e = f \cdot b_h \cdot R_s \cdot MX_a = 0.2 \cdot 0.24 \cdot 10 \cdot 528 = 253\,kgVSS$$

$$MX_i = f_{ns} \cdot R_s \cdot MS_{ti}/f_{cv} = 0.1 \cdot 10 \cdot 500/1.5 = 333\,kgVSS$$

$$MX_v = MX_a + MX_e + MX_i = 528 + 253 + 333 = 1114\,kgVSS$$

$MX_t = MX_v/0.8 = 1392\,kg\,TSS$

$ME_v = MX_v/R_s = 111\,kg\,VSS \cdot d^{-1}$ and $ME_t = MX_t/R_s = 139\,kg\,TSS \cdot d^{-1}$

$MP_l = f_p \cdot ME_v = 0.025 \cdot 111 = 2.8\,kg\,P \cdot d^{-1}$

It can be concluded that in the conventional system both the volatile sludge mass (1114/1222 = 90%) and the total sludge mass (1392/1725 = 80%) are smaller than those in the mixed system with bio-P organisms. On the other hand, the removal of phosphorus in the conventional system is only a fraction 2.8/10.7 = 23% of that in the bio-P system.

The model of Wentzel et al. (1990) described above has been validated extensively in a series of experimental studies where all the important factors where varied:

- Type of bio-P system: Phoredox, modified Bardenpho, UCT, modified UCT and Johannesburg configuration;
- Operational conditions: different values of sludge age, recirculation factors and anaerobic-, anoxic- and aerobic sludge mass fractions;
- Wastewater characteristics: temperature, concentration and composition of the organic material, ratio TKN/COD and ratio P/COD.

It was verified that all measured parameters closely correlated with the simulated model values. Therefore the model may be considered to be a reliable instrument to describe and predict the biological removal of phosphorus in activated sludge systems.

7.1.3.3 Denitrification of bio-P organisms

As discussed before, an aspect that was not clear from the model by Wentzel (1990) is the denitrification rate that occurs in the anoxic zone following the anaerobic zone. While the pure culture of bio-P organisms did not display a significant denitrification capacity, in the mixed culture denitrification did exist, in fact even at a higher rate than in the conventional system designed for nitrogen removal. Clayton (1989, 1991) investigated this issue and presented the following findings:

- In activated sludge systems with an anaerobic zone, the denitrification rate can still be described with the following generic equation: $r_D = k \cdot X_a$;
- The primary denitrification phase in the pre-D reactors is much reduced or does not exist at all, which implies that the concentration of easily biodegradable organic material in the effluent of the anaerobic zone is low. This can be explained as a result of the processes observed in the anaerobic zone, such as the absorption of VFA and the release of phosphorus, which remove a large part of the available easily biodegradable organic material;
- The value of denitrification constant K_2, corresponding to the utilisation of slowly biodegradable material in the pre-D reactor, is about 2.5 times higher than the value in conventional nitrogen removal systems. The value of denitrification constant K_3 is about 1.5 times higher;
- Clayton hypothesised that a possible reason for the increase of the denitrification constants was an increase in the hydrolysis rate of the slowly biodegradable organic material, as a result of the inclusion of an anaerobic zone.

The existence of denitrifying bio-P organisms was not considered in this hypothesis. Therefore the calculation of the denitrification rates was based only on the active non bio-P biomass. However, as will be demonstrated in Example 7.3, the observed increase in denitrification rate per unit mass X_a can be very well explained when the denitrification by the bio-P biomass is taken into consideration.

This means that the values of the denitrification constants K_2 and K_3 in bio-P removal systems are comparable to those in conventional biological nitrogen removal systems. The denitrification capacity in a bio-P removal system can be modelled when Eqs. (5.68 and 5.69) are adapted, differentiating into nitrate removal of bio-P biomass and of non bio-P biomass.

$$D_{c1} = [f_{dn} \cdot (f_{bsp} \cdot f_{pd} + f_{bsh}) \cdot f_{sb} + K_2 \cdot f_{x1} \cdot (C_{rh} \cdot f_{bh} + C_{rp} \cdot f_{bp} \cdot f_{pd})] \cdot S_{bi} \quad (7.14)$$

$$= (f_{dn} \cdot f_{sb} + K_2 \cdot f_{x1} \cdot C_r) \cdot S_{bi} \quad \text{for} \quad f_{pd} = 1 \quad (5.68)$$

$$D_{c3} = K_3 \cdot f_{x3} \cdot (C_{rh} \cdot f_{bh} + C_{rp} \cdot f_{bp} \cdot f_{pd}) \cdot S_{bi} \quad \text{and} \quad (7.15)$$

$$= K_3 \cdot f_{x3} \cdot C_r \cdot S_{bi} \quad \text{for} \quad f_{pd} = 1 \quad (5.69)$$

where the following fractions are defined:

f_{pd} = fraction of bio-P organisms capable of denitrification

$f_{bsp} = S_{seq}/S_{bsi}$
= fraction of S_{bsi} sequestered by bio-P organisms (7.16)

$f_{bsh} = (S_{bsi} - S_{seq})/S_{bsi}$
= fraction of S_{bsi} consumed by normal heterotrophs (7.17)

$f_{bp} = S_{seq}/S_{bi}$
= fraction of S_{bi} sequestered by bio-P organisms (7.18)

$f_{bh} = (S_{bi} - S_{seq})/S_{bi}$
= fraction of S_{bi} consumed by normal heterotrophs (7.19)

EXAMPLE 7.3

Assuming that the value of the non-aerated sludge mass fraction f_m is 50% in Example 7.2, estimate the maximum denitrification in a pre-D system and in a Bardenpho system with two denitrification zones of equal size. Calculate this for two different values of f_{pd}, the fraction of bio-P organisms capable of denitrification: 80% and 100%. Use $K_2 = 0.10$ and $K_3 = 0.08$ mg N · mg^{-1} VSS · d^{-1}.

Solution

The anaerobic mass fraction is 0.15 and $f_m = 0.5$, therefore the anoxic mass fraction is equal to 0.35. The influent composition can be calculated from the data of the previous example:

$$S_{bsp} = S_{seq} = 100 - 2 \cdot 16.7 = 66.6 \text{ mg COD} \cdot l^{-1} \rightarrow f_{bsp} = 66.7/100 = 0.67$$

$$S_{bsh} = 2 \cdot 16.7 = 33.7 \text{ mgCOD} \cdot l^{-1} \qquad \rightarrow f_{bsh} = 1 - 0.67 = 0.33$$

$S_{bp} = S_{seq} = 66.6 \text{ mg COD} \cdot l^{-1} \qquad \rightarrow f_{bp} = 67.4/400 = 0.17$

$S_{bh} = 400 - 66.6 = 333.4 \text{ mg COD} \cdot l^{-1} \qquad \rightarrow f_{bh} = 1 - 0.17 = 0.83$

(1) Pre-D configuration

First the denitrification capacity for $f_{pd} = 0.8$ is calculated, using Eq. (7.14):

$$D_{c1} = [f_{dn} \cdot (f_{bsp} \cdot f_{pd} + f_{bsh}) \cdot f_{sb} + K_2 \cdot f_{x1} \cdot (C_{rh} \cdot f_{bh} + C_{rp} \cdot f_{bp} \cdot f_{pd})] \cdot S_{bi}$$
$$= [0.11 \cdot (0.67 \cdot 0.8 + 0.33) \cdot 0.25 + 0.10 \cdot 0.35 \cdot (1.32 \cdot 0.83 + 3.21 \cdot 0.17 \cdot 0.8)] \cdot 400$$
$$= 31.3 \text{ mg N} \cdot l^{-1}$$

For $f_{pd} = 1.0$, the value of D_{c1} is slightly higher at 34.3 mg $N \cdot l^{-1}$. In comparison, when the denitrification capacity is calculated according to the hypothesis of Clayton, the bio-P organisms do not exhibit anoxic activity. Instead the value of K_2 increases from 0.10 to 0.25 mg $N \cdot mg^{-1}$ VSS $\cdot$ d^{-1} and the value of $D_{c1} = 0.25 \cdot 1.32 \cdot 0.35 \cdot 0.83 \cdot 400 = 38.6$ mg $N \cdot l^{-1}$.

(2) Bardenpho configuration

In the case where the anoxic sludge mass is equally divided over the pre-D and post-D reactors, both will have an anoxic mass fraction of $0.35/2 = 0.175$.

For $f_{pd} = 0.8$ the denitrification capacity will be equal to:

$$D_{c1} = [0.11 \cdot (0.67 \cdot 0.8 + 0.33) \cdot 0.25 + 0.10 \cdot 0.175 \cdot (1.32 \cdot 0.83 + 3.21 \cdot 0.17 \cdot 0.8)] \cdot 400$$
$$= 20.6 \text{ mg N} \cdot l^{-1}$$

And using Eq. (5.11):

$$D_{c3} = K_3 \cdot f_{x3} \cdot (C_{rh} \cdot f_{bh} + C_{rp} \cdot f_{bp} \cdot f_{pd}) \cdot S_{bi}$$
$$= 0.08 \cdot 0.175 \cdot (1.32 \cdot 0.83 + 3.21 \cdot 0.17 \cdot 0.8) \cdot 400 = 8.6 \text{ mg N} \cdot l^{-1}$$

The combined denitrification capacity $D_{c1} + D_{c3} = 29.1$ mg $N \cdot l^{-1}$. For $f_{pd} = 1.0$ the combined denitrification capacity $D_{c1} + D_{c3} = 22.8 + 9.2 = 32.0$ mg $N \cdot l^{-1}$. Under Clayton's hypothesis that bio-P organisms do not exhibit anoxic activity, the denitrification capacity will be:

$$D_{c1} = 0.25 \cdot 1.32 \cdot 0.175 \cdot 0.83 \cdot 400 = 19.3 \text{ mg N} \cdot l^{-1}$$

$$D_{c3} = 0.12 \cdot 1.32 \cdot 0.175 \cdot 0.83 \cdot 400 = 9.3 \text{ mg N} \cdot l^{-1}$$

$$D_{c1} + D_{c3} = 19.3 + 9.3 = 28.6 \text{ mg N} \cdot l^{-1}$$

It can be observed from Example 7.3 that the expected denitrification capacity for the bio-P removal system, with 80 to 100% of the bio-P organisms capable of denitrification, corresponds very well with the denitrification capacity calculated under the assumption that only the normal heterotrophic biomass is able to denitrify, i.e. with an increased K_2 and K_3 value. Furthermore it can be concluded that the inclusion of anoxic bio-P organisms into the simplified steady state model leads to a very good description of the two observed phenomena: i.e. phosphate uptake in the anoxic zone and the (perceived) increase of the denitrification rate when based on the active non bio-P organisms only. It also suggests that, when conditions are favourable, the fraction of bio-P organisms capable of denitrification will be high.

A typical domestic wastewater with a COD concentration of 500 mg $\cdot$ l^{-1} will have a TKN concentration between 40 and 50 mg N $\cdot$ l^{-1}, of which about 10–15 mg N $\cdot$ l^{-1} will be used for the production of excess sludge (N_l). So the expected nitrification capacity is between 25 to 40 mg N $\cdot$ l^{-1}. It is concluded that the denitrification capacity that can be created in a bio-P removal system is approximately equal to the expected nitrification capacity. Therefore the degree of nitrogen removal can be high and the production of an effluent with a low level of both nitrogen and phosphorus is possible.

7.1.3.4 DISCHARGE OF ORGANIC PHOSPHORUS WITH THE EFFLUENT

The total phosphorus concentration in the effluent (P_{te}) is composed of two fractions: soluble inorganic phosphate (P_{pe}) and organic phosphorus (P_{oe}). The organic phosphorus fraction P_{oe} consists of a soluble (P_{ose}) and a particulate fraction (P_{ope}), which forms part of the volatile suspended solids in the effluent. The value of P_{ose} is often low (typically between 0.1 and 0.2 mg P $\cdot$ l^{-1}) and is unlikely to be influenced by the applied process conditions.

Depending on the liquid-solid separation efficiency of the final settler, a certain fraction of the suspended solids present in the mixed liquor will not be retained. Naturally, these solids will contain organic phosphorus (P_{ope}). Thus part of the phosphorus in the produced excess sludge (P_l) will end up in the effluent and not in the excess sludge flow. Stated otherwise, part of P_l leaves as P_{ope} in the effluent:

$$P_{ope} = f_p \cdot X_{ve} = f_p \cdot f_v \cdot X_{te} \tag{7.20}$$

For activated sludge systems without bio-P removal ($f_p = 0.025$ g P $\cdot$ g^{-1} VSS), a well performing final settler will produce an effluent with 5–15 mg TSS $\cdot$ l^{-1}, containing 0.1 to 0.3 mg P $\cdot$ l^{-1} of organic particulate phosphorus. This is a significant contribution to the total phosphorus concentration in the effluent, especially considering the trend towards stricter phosphorus effluent limits.

When bio-P removal is applied, the contribution to P_{ope} to P_{te} will be significantly larger, as the phosphorus content of the bio-P organisms may reach a maximum of 0.38 mg P $\cdot$ mg^{-1} VSS. For a bio-P removal system, the average f_p value of the combined volatile biomass will typically be between 0.04 to 0.08 g P $\cdot$ g^{-1} VSS. This results in typical P_{ope} values of 0.4 to 0.8 mg P $\cdot$ l^{-1} for bio-P removal systems. For the design of municipal sewage treatment systems, if more specific data is not available, the following default values are suggested: 0.2 mg P $\cdot$ l^{-1} for conventional activated sludge systems and 0.6 mg P $\cdot$ l^{-1} for bio-P removal systems.

In the steady state model, the value of P_{oe} includes both P_{ose} and P_{ope}. For those cases where phosphorus removal is not required, P_l is not corrected for the loss of organic phosphorus with the effluent and is therefore slightly overestimated. This results in an equal underestimate of the concentration of phosphate that will be present in the effluent, as this is calculated according to:

$$P_{pe} = P_{ti} - P_l - P_{oe} = P_{ti} - P_l - P_{ose} - P_{ope} \tag{7.21}$$

When phosphorus removal is required, this might result in a design where the mass of phosphorus to be removed is underestimated and the effluent phosphorus limit is not met. Therefore in the case of biological- or chemical phosphorus removal, it is recommended to use the exact value of P_l (i.e. corrected for loss of particulate organic phosphorus with the effluent). Refer also to a similar discussion about the presence of particulate organic nitrogen in the effluent (Appendix 5). The "exact" phosphorus concentration discharged with the excess sludge is equal to:

$$P_{lx} = P_l - P_{ope} \tag{7.22}$$

7.2 OPTIMISATION OF BIOLOGICAL NUTRIENT REMOVAL

7.2.1 Influence of wastewater characteristics

The substrate used by the bio-P organisms is generated in the anaerobic zone from the fermentation of easily biodegradable organic material S_{bsi} in the influent into volatile fatty acids. Therefore the value of S_{bsi} is of crucial importance.

$$S_{bsi} = f_{sb} \cdot (1 - f_{ns} - f_{np}) \cdot S_{ti} \tag{3.3}$$

The effect of the S_{bsi} concentration on the performance of a bio-P removal process is twofold:

- It influences the quantity of phosphorus that can be released in the anaerobic zone;
- It has an effect on the rate of phosphorus removal.

In Figure 7.6a the relationship between phosphorus removal and the concentration of easily biodegradable COD in the influent is shown as a function of the fraction f_{sb}. Phosphorus removal was evaluated for an anaerobic mass fraction (f_{an}) of 0.15, operated in two modes: a single completely mixed reactor and two completely mixed reactors in series. The second option allows a higher degree of phosphorus removal, as the conversion of S_{bsi} to VFA is more complete. As expected, an increase in f_{sb} results in higher phosphorus removal.

In Figure 7.6b the ratio between the COD concentration in the influent and the quantity of phosphorus removed is given. Figure 7.6b clearly shows that the extent of phosphorus removal is significantly higher when the concentration of COD in the influent is increased. This is explained by the fact that the fermentation process (in which VFA is generated from S_{bsi}) is a first order process and therefore proceeds more rapidly at higher values of S_{bsi}. Therefore it is important to maintain a high COD concentration in the influent. Infiltration of rainwater into the sewer system should be minimised as this will lower the COD concentration (but it will *not* lower the COD/P ratio!). Phosphorus removal in combined sewer systems may therefore be problematic, especially if nitrogen removal is also required in the same treatment plant. In such cases, biological phosphorus removal is often supplemented by chemical precipitation of phosphorus with metal salts.

When the performance was evaluated of several municipal sewage treatment plants designed for bio-P removal, nitrogen removal or both, the Water Research Commission of South Africa made the following recommendations (WRC, 1984):

- When the concentration of easily biodegradable COD in the influent (S_{bsi}) is less than 60 mg COD $\cdot$ l^{-1}, it is very unlikely that (significant) bio-P removal will develop. When $S_{bsi} > 60$ mg COD $\cdot$ l^{-1}, bio-P removal is possible as long as recirculation of nitrate to the anaerobic zone is prevented. The extent of bio-P removal that can be obtained increases proportionally to the increase of easily biodegradable COD in the influent;

- Whether recirculation of nitrate to the anaerobic zone can be prevented depends on the extent of nitrogen removal that can be achieved (this depends on the ratio COD/TKN in the sewage) and on the type of bio-P removal configuration that is adopted. For the South-African situation (minimum wastewater temperature >14°C), the following recommendations were made:

  - COD/TKN >13 mg COD · mg^{-1} TKN. As complete nitrate removal is possible, the modified Bardenpho configuration is the most appropriate configuration, although part of the nitrate will be removed in the anaerobic zone. This high COD/TKN ratio is not very common and in generally only found in countries with a high proportion of vegetarians in the population, such as India;
  - COD/TKN in the range of 9–13 mg COD · mg^{-1} TKN. Complete nitrate removal is no longer possible. However, it is still possible to prevent nitrate recirculation to the anaerobic zone, as long as a modified UCT configuration is adopted.
  - COD/TKN in the range of 7–9 mg COD · mg^{-1} TKN. Now the modified UCT process is no longer capable to prevent nitrate recirculation to the anaerobic zone. To maximize nitrogen removal, a UCT configuration is recommended, which uses the available denitrification capacity more effectively. It is very important to control the nitrate recirculation flow ("a") adequately in order to maintain a low nitrate concentration in the pre-D reactor;
  - COD/TKN <7 mg COD · mg^{-1} TKN. It becomes very unlikely that significant bio-P removal can be obtained in activated sludge systems with nitrification.

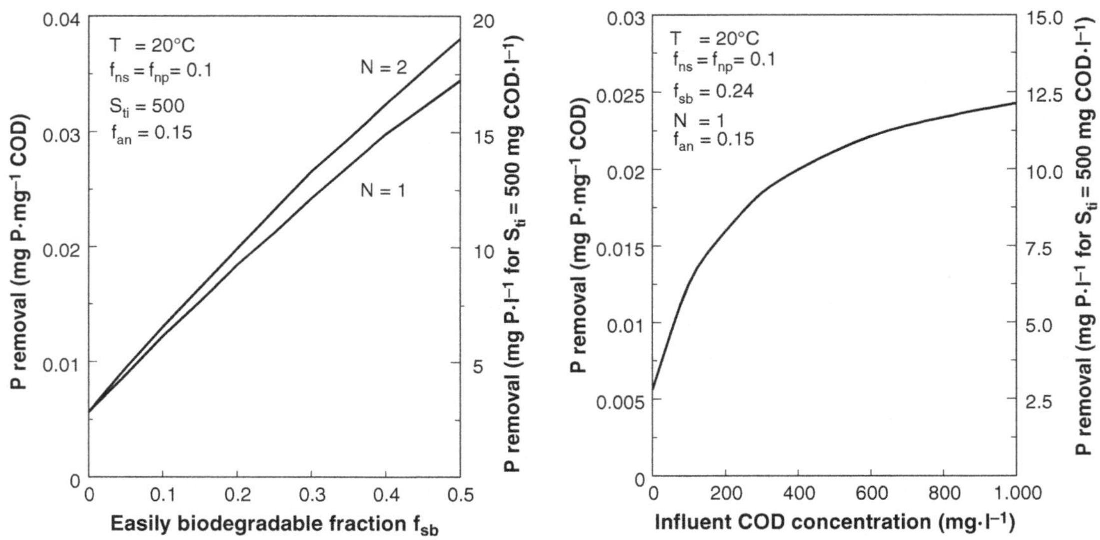

Figure 7.6 Influence of the influent COD composition (a) and concentration (b) on the degree and rate of biological phosphorus removal

7.2.2 Improving substrate availability for nutrient removal

As discussed earlier, both biological phosphorus removal and biological nitrogen removal are constrained by the availability of easily biodegradable material and as such may be competing processes. However, if the activated sludge system has been configured to promote the establishment of a large fraction of denitrifying bio-P organisms, the extent of competition will be limited and mainly due to the fact that the inclusion of an anaerobic zone will reduce the anoxic mass fraction.

The availability of biodegradable COD will be further restricted when primary sedimentation or anaerobic pre-treatment are applied. The removal efficiency of COD from the wastewater is higher than that of nitrogen and phosphorus, making extensive nitrogen- and phosphorus removal more difficult.

However, most activated sludge systems treating municipal wastewater may be compliant with the current effluent limits that are applied (such as total nitrogen ≤ 10 mg $N \cdot l^{-1}$ and total phosphorus ≤ 1–2 mg $P \cdot l^{-1}$), providing that they are properly designed and operated. Depending on the situation, supplementary phosphorus removal using chemical precipitation of phosphorus might be required. This scenario will be demonstrated in Example 14.14.

The following design approach can be applied to treatment of municipal sewage:

- Design the activated sludge system for both biological nitrogen- and phosphorus removal;
- Priority is given to biological nitrogen removal, i.e. the aerobic sludge age should be high enough to allow for nitrification and a sufficient quantity of easily degradable organics should remain available for denitrification in order to comply with the nitrogen effluent discharge limits;
- The system is further optimised to maximise biological excess phosphorus removal;
- The biological phosphorus removal process is supplemented by additional chemical methods as required, e.g. simultaneous precipitation with metal salts.

In many cases the availability of S_{bsi}, the easily biodegradable COD concentration in the influent, is a limiting factor. In theory it is attractive to increase S_{bsi} using biological processes, as an alternative to the addition of an external carbon source or to chemical phosphorus removal. The two main alternatives that will be discussed here are anaerobic pre-treatment and hydrolysis of primary sludge. In both cases the anaerobic treatment is only partial, i.e. the anaerobic process is constrained to the production of VFA and is not allowed to proceed to the production of methane.

(a) Anaerobic pre-treatment

The two main concepts that have been developed for partial anaerobic pre-treatment of the influent flow are:

- The hydrolysis upflow sludge blanket reactor or HUSB (Wang, 1994);
- The activated primary tank or APT (Rössle et al., 2001).

The main drawback of both concepts is that at low temperatures anaerobic hydrolysis proceeds at a very low rate. Heating of the wastewater flow is clearly not cost-effective and prohibits application of these concepts in regions with a cold to moderate climate. The HUSB is basically a UASB system, operating at a reduced sludge age so that only hydrolysis and acid fermentation develop. The result is that volatile fatty acids are produced instead of methane. As an additional benefit, the removal efficiency of suspended solids is very high, as suspended solids are entrapped in the sludge blanket. In the Netherlands the HUSB was extensively researched between 1992 and 1996. However, the results indicated that this process was not attractive in countries with a temperate climate as:

- During the winter the municipal sewage temperature ranges from 6 to 10°C. At these temperatures the hydrolysis process is very slow, requiring either a very high sludge age (and large reactor volume) or the installation of heat exchangers to increase the reactor temperature;
- At a typical HRT of 2 hours, the amount of additional VFA produced from the raw influent was on average only about 40 mg $COD \cdot l^{-1}$, approximately sufficient for the removal of 2–4 mg $P \cdot l^{-1}$;
- Hydrolysis of suspended solids will be the limiting process, not methanogenesis. It will therefore be difficult to prevent methane production, especially at higher temperatures.

In the activated primary tank, the sludge blanket in the primary settler is allowed to increase, thus increasing solids retention time and allowing for the development of sludge hydrolysis. Application of an APT might be advantageous if a primary settler is already constructed.

Primary settler effluent is recycled through the sludge bed to transfer the produced VFA to the influent. The main disadvantages of the APT concept are:

- At high rainwater flows the high level of the sludge bed may lead to primary settler failure and subsequent increased solids washout to the activated sludge system;
- VFA production is low, especially at low temperatures.

(b) Hydrolysis of primary sludge

The difference between this method and those discussed under (a) to increase the influent VFA concentration is that it is not the wastewater flow but the excess sludge flow that is subjected to hydrolysis, which allows the application of heating to speed up the reaction rates. Two configurations have been proposed for hydrolysis of primary sludge:

- The use of the primary sludge thickener. In moderate climates, a solids retention time of 2.5–4 days is typically applied to primary sludge thickeners and this is sufficient for some hydrolysis to occur. However, VFA production will only be low to moderate. Sludge recirculation is required to transfer the produced VFA to the supernatant, which will interfere with the primary function of the thickener, i.e. to increase the solids content in the primary sludge;
- The use of a dedicated (completely mixed) sludge hydrolysis reactor. This reactor can be heated to the optimum temperature of 30 to 35°C (mesophilic digestion) or even 50–55°C (thermophilic digestion), as the volumetric flow to be treated is much smaller compared to the wastewater flow. The hydraulic residence time is 5–10 hours. The hydrolysed sludge is separated from the liquid using gravity- or mechanical thickening and the reject water is directed to the anaerobic zone of a bio-P system.

The latter configuration is considered to be the most attractive. In Denmark, pre-precipitation of phosphorus in the primary settler using metal salts, followed by sludge hydrolysis, has been the subject of extensive research (i.e. the HYPRO process by Harremoes et al., 1991). As COD removal in the primary settler increases when pre-precipitation is applied, so does the VFA yield in the hydrolysis reactor. Some full-scale installations have been built on this principle in Scandinavia and they seem to be performing well. A study by De Jong et al. (1996) summarizes some operational and process data regarding sludge hydrolysis reactors:

- It is recommended to mix the fresh primary sludge with the hydrolysed sludge;
- The solids concentration in the reactor should not exceed 20 kg TSS $\cdot$ m^{-3};
- The pH should not be allowed to decrease below a value of 5.5–6.0;
- Nitrogen released from digestion is returned to the activated sludge system, together with the produced VFA, as with conventional anaerobic sludge digestion;
- The VFA yield that can be obtained from the primary sludge is significantly influenced by the nature of the sewer system. A long anaerobic residence time (as in pressure lines) might result in a partly hydrolysed influent, where primary sludge hydrolysis will not increase VFA production significantly anymore. On the other hand, the presence of oxygen in a gravity collection sewer system might reduce the concentration of available easily biodegradable COD;
- The yield of soluble COD from the primary sludge VSS averaged between 90–120 mg COD $\cdot$ g^{-1} VSS (at 25–30°C), but values as far apart as 40 to 400 mg COD $\cdot$ g^{-1} VSS have been reported as well;
- The recovery of produced VFA can be increased by thickening the hydrolysed sludge.

In the same study by De Jong et al. (1996), the break-even point for the cost of primary sludge hydrolysis compared to that of acetate addition was found for a VFA production of 160 mg COD $\cdot$ g^{-1} VSS (for an acetate cost of 500 US$ per ton). However, it is difficult to consistently achieve this VFA production in regions with a cold to temperate climate. Influent- and/or primary sludge hydrolysis has not found wide application for the following reasons:

- The investment costs of a system for the hydrolysis of influent or -sludge are much higher than those of an external carbon source dosing system, although the operational costs are much lower;
- The operational uncertainties regarding the VFA yield of the hydrolysis process are such that often a back-up external carbon source dosing system needs to be installed anyway;

7.2.3 Optimisation of operational conditions

A wastewater treatment plant for biological nitrogen and phosphorus removal is designed for the production of an effluent free of organic matter, suspended solids and macronutrients. As mentioned above, the simultaneous removal of nitrogen and phosphorus can be difficult as the conditions for maximum removal of these nutrients are conflicting: for optimum phosphorus removal a large anaerobic zone is indispensable, but this will limit the size of the anoxic zone and consequently reduces the denitrification capacity. On the other hand, if both anaerobic- and anoxic zones are large, then the aerobic zone will be small and the nitrification process is less efficient (and possibly also the removal efficiency of organic material), apart from the risk of developing bulking sludge, which will be discussed in Chapter 9. Furthermore phosphorus removal efficiency will increase at lower sludge ages, whereas a high degree of nitrogen removal requires higher sludge ages.

In general the design of any wastewater treatment plant is subject to a set of requirements, which will all impose constraints to the design and might even be contradictory. Such priorities depend, among other things, on the effluent limits set by the authorities, but could also be (in descending order of importance):

- Extensive removal of organic biodegradable material and suspended solids;
- Almost complete removal of ammonia and biodegradable organic nitrogen, i.e. to concentrations between $N_{ad} < 1$ to 2 mg N $\cdot$ l^{-1} (this demand in general will ensure that organic material and suspended solids removal will also be efficient);

- Low total effluent phosphorus concentration ($P_{pe} < 1$ mg P · l^{-1});
- Low total effluent nitrogen concentration, for example $N_{te} < 10$ mg N · l^{-1}.

The numerical values of the maximum allowable effluent concentrations depend on legislation, available options for reuse and the nature of the receiving water. In the design of a nutrient removal plant the values of the following three parameters have to be defined: the sludge age, the size of anaerobic-, anoxic- and aerobic sludge mass fractions and the value of the different recirculation factors "a", "s" and "r".

(a) Sludge age

To minimise the construction- and operational costs, the activated sludge system should be designed and operated at the lowest possible sludge age permitting the production of effluent with the desired quality. Figure 7.7 shows the biological phosphorus removal in UCT systems as a function of the sludge age and for different mass fractions of the anaerobic zone ($f_{an} = 0.1$; 0.2 and 0.3).

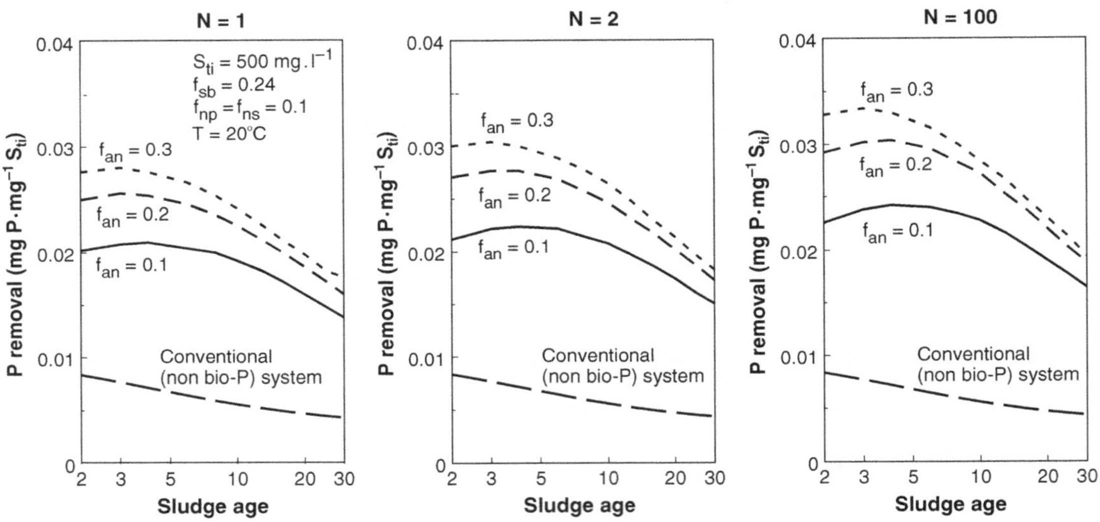

Figure 7.7 Phosphorus removal (in UCT configuration) as a function of the sludge age for different values of f_{an} and N, compared to that of a conventional (non bio-P) system

For each case three different configurations of the anaerobic zone where considered, i.e. a single completely mixed reactor (N = 1), two completely mixed reactors in series (N = 2) and a long series of completely mixed reactors approximating plug flow conditions (N = 100). Furthermore it has been assumed in Figure 7.7 that "r", the recirculation factor from the anoxic to the anaerobic zone has a value of 1 and that no nitrate is recirculated to the anaerobic zone. Figure 7.7 shows that for all considered design cases the phosphorus removal reaches a maximum at a rather short sludge age: $R_s = 3$ to 5 days. For shorter sludge ages, due to the low active sludge concentration, the conversion of easily biodegradable material to VFA will be incomplete, resulting in a reduced availability of substrate for the bio-P organisms. On the other hand, for sludge ages longer than 3 to 5 days almost all easily biodegradable material S_{bsi} will have been converted into VFA, but as the discharge of excess sludge will decrease at higher sludge ages, the removal of phosphorus will be lower as well. It can also be observed that the subdivision of the anaerobic zone increases the phosphorus removal capacity, as more organic material will be converted into VFA. However, increasing the number of subdivisions to more than two will only result in a minor increase in phosphorus removal capacity, but it will add to the investment costs. In those cases when both biological phosphorus removal and biological nitrogen removal are desired, it is best to prioritize the sizing criteria for nitrogen removal. In practice this means that the sludge age will be (much) higher than the optimum for bio-P removal. If needed, chemical (simultaneous) phosphorus removal can be used to supplement the biological phosphorus removal, as will be discussed in Section 7.3.2.2. Furthermore, the effect of an increase of the sludge age on the residual phosphorus concentration might be smaller than expected, as there are several compensating effects:

- The active heterotrophic sludge mass will increase, resulting in improved substrate availability to the bio-P bacteria according to Eqs. (7.4 and 7.5);
- The bio-P bacteria exhibit a lower die-off rate than the heterotrophic biomass. Hence the active bio-P fraction, containing the poly-P mass, will decrease only slowly with increasing sludge age;
- The mass of phosphorus discharged with the excess sludge ($f_p \cdot MX_{vh}/R_s$ and $f_{pp} \cdot MX_{ap}/R_s$) decreases asymptotically as function of the sludge age (i.e. slower at higher sludge age).

EXAMPLE 7.4

Indicate the effect of the value of the selected sludge age on the residual phosphorus concentration of an UCT system treating municipal wastewater. Assume the following characteristics and conditions: $Q_i = 1000\ m^3 \cdot d^{-1}$; $S_{ti} = 600$ mg $COD \cdot l^{-1}$; $P_{oe} = 0.25$ mg $P \cdot l^{-1}$; $f_{ns} = f_{np} = 0.1$; $f_{sb} = 0.25$; $f_{an} = 0.15$; $T = 20°C$; $N = 2$ and $r = 1$. Assume that nitrate is not present in the recycle stream "r" to the anaerobic zone.

Solution

Using the theory presented in Section 7.1.3.2, the residual biodegradable COD concentration in the effluent of the anaerobic zone (S_{bsN}) can be calculated. Use the general equation Eq. (7.5), as the anaerobic zone is now divided into two parts. Once the value of S_{bsn} is known, the biomass composition and -quantity can be calculated as demonstrated in Example 7.2.

Figure 7.8 shows the increase of the residual phosphorus concentration over the range of sludge ages from 10 to 20 days. Perhaps contrary to expectation, the increase is quite modest at about 2.5 mg $P \cdot l^{-1}$ for a 10 days increase in sludge age, or approximately 0.25 mg P per day.

This can be explained by taking into account the active bio-P mass that develops and the mass of phosphorus discharged as part of the bio-P excess sludge. Whereas for instance the active heterotrophic biomass MX_{ah} increases with only 14% from 511 to 593 kg VSS, the active bio-P biomass increases at a much higher rate (+ 38%) from 302 to 490 kg VSS. The mass of phosphorus discharged from the system with the active bio-P excess sludge will thus be a smaller fraction of a larger whole. For example, at $R_s = 10$ days the discharge of phosphorus is equal to $0.38 \cdot 302/10 = 11.5$ kg $P \cdot d^{-1}$ while at $R_s = 20$ days it decreases to $0.38 \cdot 490/20 = 9.2$ kg $P \cdot d - 1$, a decrease of only 23%.

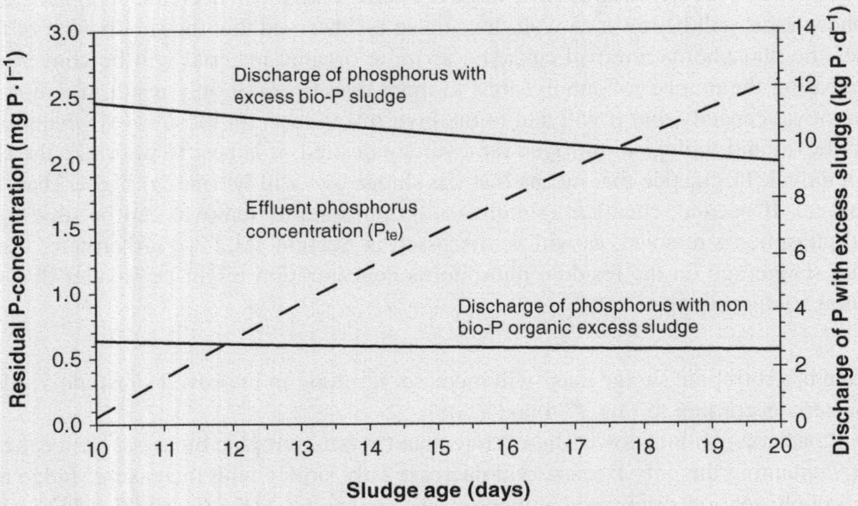

Figure 7.8 Increase in residual effluent phosphorus concentration and decrease in the mass of phosphorus removed with the bio-P biomass and with the other organic excess sludge fractions as function of the sludge age, according to the conditions of Example 7.4

(b) Allocation of the sludge mass fractions to anaerobic and anoxic zones

For each sludge age, a certain minimum aerobic sludge mass fraction is required to maintain the efficiency of the nitrification process (Eq. 5.44). Thus a maximum value of $f_m = f_{max}$ is defined for the sum of the anaerobic- and anoxic mass fractions. In Figure 7.7 the influence of the size of the anaerobic zone (f_{an} = 0.1, 0.2 and 0.3) on the phosphorus removal efficiency was already indicated. It can be observed that an increase of the anaerobic fraction results in an increased phosphorus removal capacity. However, the increase of f_{an} from 0.2 to 0.3 has only a minor effect. Furthermore, a large anaerobic zone will reduce the volume available for the anoxic zone and consequently also the denitrification capacity.

Apart from a high nitrate concentration in the effluent, this might result in serious operational problems such as uncontrolled denitrification in the final settler and propensity for sludge bulking. For this reason the size of the anaerobic mass fraction f_{an} is in practice limited to a value between 0.1 and 0.2. The anoxic zone is often subdivided into two or more parts. This subdivision has several objectives: (I) protection of the anaerobic zone against introduction of nitrate, which is in effect the intention of the subdivision of the pre-D reactor as used in the UCT- and modified UCT systems and (II) optimisation of nitrogen removal efficiency using pre-D and post-D reactors. The denitrification capacity is always larger in the pre-D reactor than in a post-D reactor of comparable size, but complete denitrification can only be obtained if a post-D reactor is installed. Figure 7.9 shows the effect of the size of the anaerobic mass fraction on the denitrification capacity in a bio-P removal system for a modified pre-D and Bardenpho configuration. As can be expected, the denitrification capacity decreases rapidly when the anaerobic mass fraction increases as the anoxic mass fraction will have to decrease ($f_x \leq f_m \leq f_{max} = 0.6$). The denitrification capacity in the pre-D configuration is higher than that in the Bardenpho configuration, due to the higher rate of denitrification in the pre-D versus the post-D reactor. However, complete denitrification can only be obtained in a Bardenpho configuration. Otherwise part of the nitrate load generated in the aerobic reactor, i.e. the part that is not recycled to the pre-D reactor, will leave with the effluent.

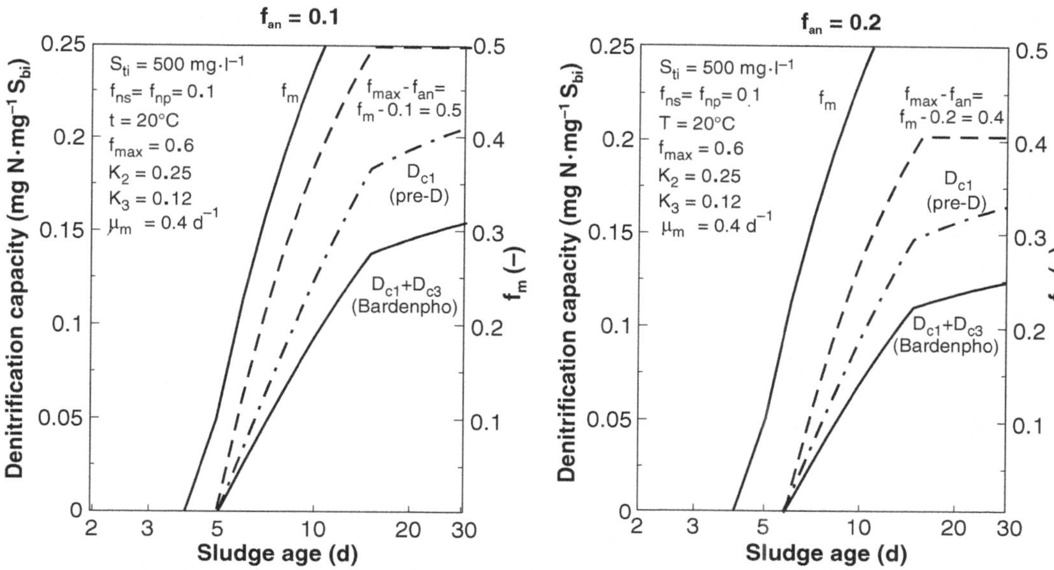

Figure 7.9 Denitrification capacity in tertiary systems for bio-P and nitrogen removal as function of the sludge age for different anaerobic mass fractions (modified pre-D and modified BDP configuration)

Therefore, for complete nitrogen removal in a bio-P removal configuration the following requirements should be met: (I) operation in modified Bardenpho or UCT configuration, (II) a combined denitrification capacity $D_{c1} + D_{c3}$ either exceeding or equal to the nitrification capacity N_c and (III) a post-D denitrification capacity D_{c3} at least equal to the fraction of N_c not returned to the pre-D zone: $D_{c3} \geq (s+1)/(a+s+1) \cdot N_c$.

(c) Recirculation factors

The UCT and modified UCT configuration have three internal recirculation flows:

(1) The return sludge flow "s";
(2) The nitrified mixed liquor flow from the aerated zone to the pre-D reactor "a";
(3) The flow of denitrified mixed liquor flow from the pre-D anoxic reactor to the anaerobic zone "r".

The value of "s", the return sludge- or sludge recycle factor, is set by the requirements for the proper operation and design of the final settler as will be discussed in Chapter 8. The value of the "a" factor is limited by the condition that the nitrate concentration in the pre-D reactor will have to be low, not only to avoid recirculation of nitrate to the anaerobic zone, but also to reduce the risk of sludge bulking.

The value of the "r" factor is in practice always equal to about one. A smaller value permits a high concentration of easily biodegradable organic material in the anaerobic zone, as there is little dilution of the influent with the recirculated mixed liquor. However, at the same time the sludge concentration in the anaerobic reactor will be reduced, because in the UCT configuration the return sludge is recycled to the anoxic zone instead of to the anaerobic zone. This means that a large anaerobic volume is required in order to obtain a proper bio-P removal performance. On the other hand, when the value of the "r" factor is large, the concentration of easily biodegradable organic material S_{bs} in the anaerobic zone will be low and this will result in a decrease in phosphate release. Furthermore, if the anaerobic zone is small, part of the easily biodegradable organic material may be carried over to the anoxic zone and will then not be available for bio-P removal. The value of "r" is one is therefore a compromise between the two unfavourable extremes.

7.2.4 Resolving operational problems

Tertiary treatment systems designed for biological nitrogen- and phosphorus removal are among the most sophisticated technologies available in the field of wastewater treatment. However, in practice there are several factors that may reduce the efficiency and reliability of the treatment system.

The first factor refers to operational stability: it should be realised that the bio-P organisms in the activated sludge system are in fact a very large reservoir of phosphorus. Under adverse conditions this phosphate can be released to the liquid phase. Such a situation might occur, for instance, when aeration is interrupted for several hours. In this case, phosphate release will continue only in the anaerobic zone and will not occur in the aerobic zone as it requires the availability of VFA. However, the absorption of the released phosphate will cease in the normally aerated zone, as the uptake of phosphorus is linked to utilisation of stored PHB and such a process requires either anoxic- or aerobic conditions. While the decrease in oxygen concentration in the aerobic reactor in the event of the interruption of aeration is obvious, the nitrate concentration will also decrease in time as nitrification will cease as well.

In the anaerobic zone the concentration of released phosphorus will be equal to fifty percent of the mass of VFA taken up by the bio-P organisms. Thus for a typical domestic sewage containing 100 $mg \cdot l^1$ of easily biodegradable COD, an interruption of the air supply for a couple of hours could result in several tens of milligrams per litre of phosphorus in the effluent. Under these conditions it may be preferable to temporarily by-pass the influent over the anaerobic zone, so that no VFA becomes available for the bio-P organisms.

A similar problem might occur when toxic shock loads are introduced into the system, which might reduce the OUR and thus also the phosphorus absorption rate. In the same context, it is important to notice that an inadequate aeration capacity will also jeopardise the efficiency of the phosphorus removal process. If the utilisation of PHB in the aerobic zone is incomplete due to lack of an oxidant, the energy generated by the bio-P organisms will be insufficient to completely regenerate the polyphosphate released in the anaerobic zone, resulting in the discharge of phosphorus in the effluent. However, the problem most frequently encountered in bio-P systems is that the capacity for biological phosphorus removal is insufficient to produce an effluent with a low phosphorus level. In the earlier sections it was explained that it is important to protect the anaerobic zone against introduction of nitrate. In other cases oxygen is introduced, for example due to some form of pre-treatment involving aeration (e.g. aerated sand traps or dissolved air flotation for removal of oil and fats). This oxygen will then be used to oxidise part of the easily biodegradable COD in the wastewater, reducing the fraction of bio-P organisms in the system and thus also the phosphorus removal capacity.

However, in most cases when biological phosphorus removal capacity is insufficient, this is due to an unfavourable ratio between phosphorus in the influent and VFA present or generated in the anaerobic zone. In this case it should be attempted to increase the easily biodegradable material in the influent. The three main alternatives, which are all technically feasible, but will result in additional operations and/or costs, are:

- Addition of an easily biodegradable carbon source (such as acetate or methanol);
- Pre-treatment of the influent using acid fermentation, generating additional VFA from the proteins, fats and carbohydrates present in the wastewater;

- Treatment of the excess primary sludge with acid fermentation (without subsequent methanogenesis) and recirculation of the generated VFA to the anaerobic reactor.

Another problem in the operation of systems with bio-P removal is the handling of the biological excess sludge. During anaerobic sludge digestion, a large part of the bio-P organisms will be hydrolysed. As a consequence, the internally stored polyphosphate is also released from the cell and this can result in a very high phosphate concentration in the digester effluent (up to 200 mg $P \cdot l^{-1}$). Part of the phosphorus precipitates as metal-phosphate complex and another (small) part will be present in the stabilised anaerobic excess sludge. The remaining soluble phosphorus in the digester effluent can amount to more than half of the phosphorus load in the influent to the activated sludge system.

Direct recirculation of the anaerobic digestion water to the activated sludge system is therefore not practical, as this would overload the biological system with phosphorus: prior to recirculation to the activated sludge process the phosphorus will have to be removed. In general physical-chemical treatment methods are used to eliminate the phosphate from the liquid phase of the digested sludge. An alternative that is might be applied is to dewater the excess sludge from bio-P systems directly, i.e. without prior anaerobic digestion, with a combination of mechanical thickeners (e.g. band filters) and dewatering equipment such as decanter centrifuges or belt filter presses.

7.3 CHEMICAL PHOSPHORUS REMOVAL

Chemical phosphorus removal is based on the formation- and subsequent precipitation of insoluble metal-phosphate salts. Metal salts are added to either the influent, the mixed liquor or the effluent. The metal ions will form insoluble metal-phosphate complexes with free ortho-phosphate ions (PO_4^{3-}, hereafter referred to as "phosphate"). The resulting inorganic sludge is removed from the system together with the excess sludge. However, even when chemical phosphorus removal is applied, still a significant part of the phosphorus present in the influent (between 10 and 40%) will be removed as part of the organic excess sludge (P_l). The value of P_l depends on the applied sludge age.

Chemical phosphorus removal used to be a very common method for phosphorus removal prior to the development of bio-P systems. Nowadays, in municipal sewage treatment it is being replaced by biological methods. However, it can still be useful, for instance if the activated sludge system does not have enough capacity for complete biological phosphorus removal, when the influent composition (P/COD ratio) is unfavourable or when the concentration of phosphorus to be removed is relatively small, for instance in industrial wastewater.

7.3.1 Stoichiometrics of chemical phosphorus removal

The main chemicals used in chemical phosphorus removal are metal salts and lime.

7.3.1.1 Addition of metal salts

When trivalent metal salts, i.e. metal (III) salts, are added to water, then after dissociation the metal ions will react as:

$$Me^{3+}(aq) + PO_4^{3-}(aq) \longrightarrow MePO_4(s) \text{ or rather}$$
$$Me^{3+}(aq) + H_2PO_4^-(aq) \longrightarrow MePO_4(s) + 2H^+ (\text{in the pH range of interest}) \quad (7.23)$$

The precipitation reaction of phosphate requires metal ions with a valence of (+ 3). If metal (II) salts such as $FeSO_4$ are used, then as a preliminary step oxidation of the metal ion from Me^{2+} to Me^{3+} is required. In the case of $FeSO_4$:

$$Fe^{2+}(aq) + H^+ + \frac{1}{2}O_2 \longrightarrow Fe^{3+}(aq) + \frac{1}{2}H_2O \tag{7.24}$$

When sulphide is present in the influent, metal sulphides (MeS or Me_2S_3) will be formed as well, which are poorly soluble and will precipitate as well:

$$Me^{2+} + S^{2-} \longrightarrow MeS\,(s) \tag{7.25}$$

$$2\,Me^{3+} + 3\,S^{2-} \longrightarrow Me_2S_3\,(s) \tag{7.26}$$

Although the theoretical molar Me(III)/P dosing ratio according to Eq. (7.23) is equal to one, it will always be necessary to apply the metal salt in excess of the stoichiometric requirements if a low effluent phosphate concentration is to be achieved. One reason is that the pH in an activated sludge system is above the optimal range for precipitation of most metal-phosphate complexes, which would require a pH between 5 and 6. Another reason is that the metal ions also react with water to form hydrated complexes such as Me $(H_2O)_6^{3+}$.

These complexes tend to accumulate into a positively charged superstructure under simultaneous release of H^+ and H_2O. The formation of metal hydrates can be simplified as:

$$Me^{3+}(aq) + 3\,H_2O \longrightarrow Me(OH)_3\,(s) + 3\,H^+ \tag{7.27}$$

Apart from precipitation as $MePO_4$, phosphate is also removed by adsorption to the positively charged hydrated metal complexes. In general the required stoichiometric molar Me(III)/P dosing ratio increases when lower PO_4^{3-} concentrations are desired, because an increasing fraction of the Me^{3+} ions will complex with water, due to the limited availability of PO_4^3. From the reaction equations above, it is obvious that the addition of metal salts will lead to an increase in solids production. Furthermore, the Me $(OH)_3$ formed in Eq. (7.27) is subjected to a further series of reactions, resulting in a mixture of metal oxides, -hydroxides and hydrate water. This mixture can be approximated with the formula $Me_2O_3 \cdot n\,H_2O$.

When the suspended solids concentration of the chemical sludge is determined (i.e. by drying at 103°C), the free water will evaporate but the hydrate water will not, as it is chemically bound to the metal complex. This will lead to an overestimate of the suspended solids content of the wastewater or mixed liquor. Furthermore, after ignition at 600°C the hydrate water will evaporate, resulting in an overestimate of the volatile mass fraction.

In an experimental investigation by Voors et al. (1993), it was determined that for a molar Me/P dosing ratio of 2 mol $Me \cdot mol^{-1}$ P, the hydrate water content amounted to approximately 4% of the suspended solids and to 6.5% of the observed weight loss at 600°C. In this same research it was established that the average number of hydrate water molecules per molecule Fe_2O_3 was equal to n = 2.7. Coincidentally the chemical composition of $Me(OH)_3$ equals ½ $Me_2O_3 \cdot 3H_2O$. Therefore it is possible to use the simplified composition $Me(OH)_3$ in Eq. (7.27) to estimate the production of chemical sludge resulting from the dosing of metal salts.

Table 7.3 Metal salts used for chemical phosphorus removal

Metal salt	MW metal ($g \cdot mol^{-1}$)	MW metal salt ($g \cdot mol^{-1}$)	Density ($kg \cdot m^{-3}$)	Metal content ($kg\ Me \cdot m^{-3}$)
40% wt Ferric chloride - $FeCl_3$	55.8	162.2	1400	193
– May contain traces of heavy metals that might preclude reuse of the excess sludge.				
40% wt Ferric chloride sulphate–$FeClSO_4$	55.8	187.4	1500	180
– Reduces chloride load to wastewater by 67%.				
Ferrous sulphate (powder)–$FeSO_4 \cdot 7H_2O$	55.8	277.9	1900	382
– Cheapest chemical, but requires dissolution before dosing. – Fe^{2+} requires oxidation to Fe^{3+} to be effective.				
Aluminium sulphate (powder)–$Al_2(SO_4)_3 \cdot 16H_2O$	27.0	630.4	2170	186
– Requires dissolution before dosing.				
30% wt Aluminium chloride - $AlCl_3$	27.0	133.5	1300	79
– Effective but more expensive than ferric chloride.				
PAC - Polymerised Aluminium Chloride $[Al_2(OH)_{(6\text{-}x)}Cl_x \cdot yH_2O]_n$ with $y \approx 2 \cdot x$ and $n \approx 14$	27.0	MW depends on x-value: 3710 for $x = 2$	1250–1300	255 (for $x = 2$)
– Effect on pH is reduced as less H^+ is produced. – More expensive than other products.				

In Table 7.3 several common metal salts are listed that are used for chemical phosphorus removal. Sometimes waste sludge from municipal drinking water treatment plants is used as an alternative. This precipitated sludge consists primarily of metal oxides and hydroxides (principally iron). The removal mechanism for phosphorus is therefore adsorption to the positively charged metal-hydroxide complex. It may be necessary to lower the pH to increase the activity of the drinking water sludge. Although drinking water sludge is a cheap source of metal salts, it may be polluted with heavy metals, which restricts the usage for mainstream precipitation, due to possible emission of heavy metals with the effluent to the environment. An alternative is to use the drinking water sludge for sidestream precipitation, for example to precipitate the phosphate released during sludge digestion, as the stabilised sludge in general will be either land-filled or incinerated.

7.3.1.2 Addition of lime

When lime $Ca(OH)_2$ is used for chemical phosphorus removal, the main removal mechanism is the precipitation of phosphate as calcium hydroxy apatite: $Ca_5(PO_4)_3OH$. As calcium is often already present in sufficient quantities in the wastewater and the removal of phosphorus is more efficient at higher pH, the main function of the lime addition is to increase the pH to a value above 8. The following reactions are of interest:

$$Ca(OH)_2\,(aq) + H_2PO_4^- \longrightarrow CaHPO_4\,(s) + H_2O + OH^- \tag{7.28}$$

$$5\,Ca^{2+}(aq) + 3\,PO_4^{3-} + OH^- \longrightarrow Ca_5(PO_4)_3OH\,(s) \tag{7.29}$$

$$Ca^{2+}(aq) + CO_3^{2-} \longrightarrow CaCO_3\ (s)\ (\text{at pH} > 9.5) \tag{7.30}$$

Table 7.4 lists a number of commercially available products used in lime precipitation for chemical phosphorus removal. The use of lime for phosphorus removal is declining in popularity due to the production of a large quantity of inert sludge and because of the difficulties involved in the handling, storage and addition of lime. For large wastewater treatment plants, a lime recovery system might be installed in which the calcium carbonate in the sludge is reconverted into lime. In regions with acid soils, the alkaline sludge may be used for pH correction in agricultural applications.

Table 7.4 Different varieties of lime and caustic used for chemical phosphorus removal

Metal salt	MW ($g \cdot mol^{-1}$)	Density ($kg \cdot m^{-3}$)	Remarks
Pure lime (powder) – CaO	56.1	3300	– Requires pre-treatment steps to convert into $Ca(OH)_2$
Hydrated lime (powder) – $Ca(OH)_2$	74.1	2200	– Requires mixing with water (suspension)
15 – 40% wt lime milk (suspension) – $Ca(OH)_2$	74.1	1100–1250	– Can be directly be applied, more expensive
40% wt sodium hydroxide – NaOH	40	1300	– Less chemical sludge produced – More expensive

As the dosing of lime is not stoichiometrically linked to the phosphorus content, but is only intended to increase the pH value, lime dosing is mainly interesting at higher phosphorus concentrations: e.g. in side stream precipitation processes. Optionally caustic soda NaOH is used as an alternative, although this is more expensive. As less $CaCO_3$ will be formed, the chemical sludge production will be lower compared to that resulting from the addition of lime.

7.3.1.3 Effects on pH

The use of metal salts has an effect on the pH, because in the precipitation process alkalinity is consumed and acidity is released. If it is assumed that about half of the metal ions combine with phosphate and the other half precipitate as hydroxide, then using Eqs. (7.23 to 7.27) the overall reaction equation can be written as:

$$2\,MeCl_3 + H_2PO_4^- + 3H_2O \longrightarrow Me(OH)_3 + MePO_4 + 5\,H^+ + 6\,Cl^- \tag{7.31}$$

However, the effect on the alkalinity is not very large, as can be deducted from the following calculation: if for example 6.2 mg $P \cdot l^{-1}$ (0.2 mmol $P \cdot l^{-1}$) are to be removed through addition of metal(III) salts in a 2 : 1 molar Me/P ratio, the alkalinity effect can be estimated from Eq. (7.31) as $0.2 \cdot 5 = 1$ meq $\cdot\, l^{-1}$ or 0.5 mmol $CaCO_3$ (50 mg $CaCO_3 \cdot l^{-1}$). If a metal(II)salt is added, 1 mol of H^+ is consumed in the oxidation of Me^{2+} to Me^{3+}. Thus for metal(II)salts, the alkalinity consumption is only $0.2 \cdot (5 - 2) = 0.6$ meq $\cdot\, l^{-1}$ per mmol of phosphorus removed.

To estimate whether the influent alkalinity is sufficient to maintain an appropriate pH for the biological processes, both the consumption of alkalinity by chemical phosphorus removal and that of nitrogen removal should be considered (refer to Section 5.1.3.2). Along with a pH effect, there is also an increase in the salt concentration in the effluent, which can be easily calculated from the stoichiometry of the reactions in Eqs. (7.23 to 7.31).

7.3.2 Chemical phosphorus removal configurations

Figure 7.10 schematically shows the four main process configurations used for chemical phosphorus removal:

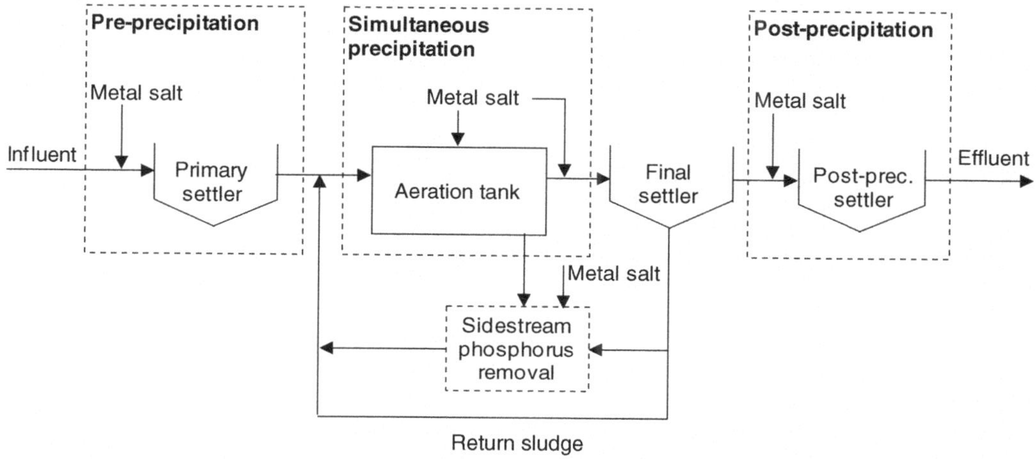

Figure 7.10 Main process configurations applied in chemical phosphorus removal

(1) Pre-precipitation:
Chemicals are added to the raw influent or in the primary clarifier. The metal-phosphate complex precipitates and is removed together with the primary sludge;

(2) Simultaneous precipitation:
Chemicals are added to the mixed liquor. The metal-phosphate complex is removed together with the excess biological sludge. A distinction can be made between those activated sludge processes where chemical precipitation is the main mechanism for phosphorus removal and those where it is used to supplement bio-P removal;

(3) Post-precipitation:
Chemicals are added to the effluent of the clarifier. A dedicated treatment unit removes the metal salts;

(4) Side-stream precipitation:
Phosphorus is concentrated into a low-volume sidestream with a high phosphorus content. This sidestream is then subjected to chemical phosphorus removal. Thickened- or digested sludge streams are also rich in phosphorus. Sidestream precipitation can be used as a supplementary process to bio-P removal.

In Table 7.5 the main characteristics (advantages and disadvantages) of these configurations are compared. The design data presented in the subsequent sections is based on an extensive set of data collected from full-scale municipal wastewater treatment plants in the Netherlands (De Jong et al., 1993 and Janssen et al., 2002). Most sewage systems in the Netherlands are socalled common sewers in which both rainfall and sewage are collected, resulting in a rather diluted sewage with the following average influent composition:

- Total COD between 300 and 700 mg $COD \cdot l^{-1}$;

Table 7.5 Comparison of the main configurations for chemical phosphorus removal

Configuration	Advantages / disadvantages
Pre-precipitation	
Advantages:	– Increased removal of suspended solids and COD/BOD in primary settler, although this may actually be a disadvantage for activated sludge systems designed for nitrogen removal. – Lowest risk on discharge of metals with the effluent. – Can be used to reduce organic- and nitrogen loading to the biological treatment step of overloaded systems. – Biological treatment capacity is not decreased.
Disadvantages:	– Only an alternative if a primary settler already exists, otherwise very expensive. – Primary sludge may be more difficult to dewater. – Highest use of chemicals. – High production of chemical sludge. – Efficient phosphorus removal may lead to shortage of phosphorus for bacterial growth in subsequent biological processes. – Polymer dosage may be necessary to enhance settling.
Simultaneous precipitation (main process for phosphorus removal)	
Advantages:	– Lowest cost option. – Improved settling characteristics of secondary sludge.
Disadvantages:	– Loss of biological treatment capacity as inorganic fraction of sludge will increase. – High chemical sludge production. – If sludge age decreases, nitrification may suffer. – pH control may be necessary to compensate for lower pH. – Lime addition is only possible if pH in return sludge stream is adjusted.
Post-precipitation	
Advantages:	– Lowest effluent phosphorus concentration. – No impact on biological processes. – Can be combined with other effluent post-treatment methods.
Disadvantages:	– Highest risk of discharge of metal salts with the effluent. – Additional treatment units. – High investment costs. – Sludge difficult to dewater.
Simultaneous precipitation (supplementary process)	
Advantages:	– Lowest chemical use. – Lowest operational costs. – Most environmentally friendly solution.
Disadvantages:	– Interferes in complex biological system with many interactions. – Investment costs to create additional volume→if plant is under-loaded part of the anoxic reactor could be used. – To some extent competition with nitrogen removal for available COD and unaerated volume. – Suitable excess sludge treatment required.
Sidestream precipitation	
Advantages:	– Lower phosphorus content in the organic sludge→lower particulate organic phosphorus effluent concentration. – Reduced usage of chemicals.
Disadvantages:	– Additional treatment units required. – High investment costs.

- Total nitrogen between 30 and 80 mg $N \cdot l^{-1}$;
- Total phosphorus between 6 and 15 mg $P \cdot l^{-1}$.

7.3.2.1 Pre-precipitation

The simplified flow diagram of the pre-precipitation process is displayed in Figure 7.11. This process is often implemented for objectives other than chemical phosphorus removal alone:

- Reduction of the organic load to the biological system The formed $Me(OH)_3$ acts as a flocculant and increases removal of suspended solids and colloids in the influent. Average BOD removal in primary settlers in Dutch municipal sewage treatment plants increased from 25–40% to 50–75% after the addition of Me(III)salts (De Jong et al., 1993). If denitrification is required, pre-precipitation is not recommended as it will significantly increase the N/COD ratio in the pre-settled influent;
- Odour prevention if sulphides are present, through the precipitation of metal sulphides.

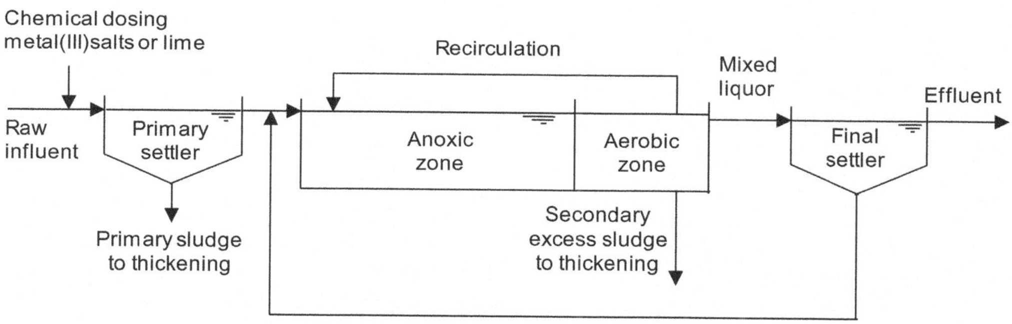

Figure 7.11 Simplified flow scheme of phosphorus removal by pre-precipitation

When pre-precipitation is applied, biological excess phosphorus removal will not develop, as the bulk of the phosphorus will be removed before the biological unit operations. Stoichiometrically, one mole of metal salt is sufficient to precipitate one mole of phosphate. However, much more metal ions are required to compensate for the formation of metal oxides, -hydroxides and sulphides, which actually enhance the removal of suspended solids in the influent. The particulate phosphorus fraction in the influent will precipitate together with the suspended solids as primary sludge and therefore will not exert any Me(III) demand. The sludge age of the system under consideration should be considered when the Me/P_{ti} dosing ratio is selected, as sufficient phosphorus should remain present in the effluent of the primary settler to satisfy the nutrient demand of the produced excess sludge (Eq. 3.60). Molar Me/P dosing ratios of 1 to 5 moles Me(III) per mole influent phosphorus have been reported by full-scale treatment plants with pre-precipitation of phosphorus. The applied dosing ratio depends on the required degree of phosphorus removal, as can be observed in Figure 7.12. Sometimes polymer is added as well, to act as flocculant and to improve settling.

All metal salts listed in Table 7.3 can be used for pre-precipitation, except ferrous sulphate, as the Fe^{2+} ion requires oxidation to Fe^{3+} before it is effective. Application of lime is not common as the pre-treated influent will have to be neutralised prior to biological treatment.

Assuming the biological reactors and final settler of the activated sludge system perform well (i.e. a suspended solids concentrations in the effluent $<$10–20 mg $TSS \cdot l^{-1}$), the effluent total phosphorus

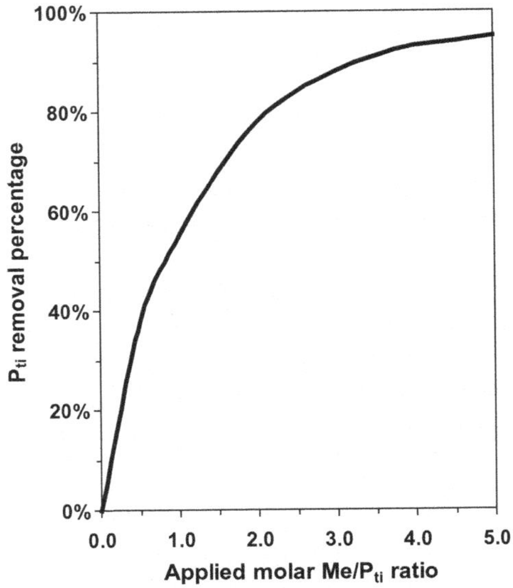

Figure 7.12 Percentage removal of the phosphorus concentration present in the influent in the primary settler as function of the molar Me/P_{ti} dosing ratio. Based on a review by De Jong et al., 1993

concentration can be reduced to a value lower than 1.0 mg $P \cdot l^{-1}$. The concentration of phosphorus in the effluent of the primary clarifier should remain high enough to sustain the requirements for biomass growth (i.e. P_l). If the activated sludge system is designed for nitrogen removal, pre-precipitation is often not a suitable technique as the N/COD ratio will increase: most of the nitrogen in the influent will be present in the form of soluble NH_4^+. The effect of the change in influent COD load and -composition to the activated sludge system can be estimated with the theory that will be presented in Section 7.3.3 (refer also to Example 14.1 in Section 14.5.1).

EXAMPLE 7.5

An existing activated sludge system consisting of a primary settler followed by conventional secondary treatment is adapted to remove phosphorus. As a primary settler is already present, a pre-precipitation configuration is considered. Calculate the daily consumption of 40% wt $FeCl_3$ and the primary excess sludge production that will result, when it is required to reduce P_{te} to a value ≤ 1 mg $P \cdot l^{-1}$, for the following conditions:

- Influent composition: $S_{ti} = 500$ mg $COD \cdot l^{-1}$ and $P_{ti} = 15$ mg $P \cdot l^{-1}$;
- Assume that the solids removal efficiency is equal to the COD removal efficiency: $\eta_1 = 50\%$ (when $FeCl_3$ is added);
- Organic fraction of the primary excess sludge: $f_v = 0.6$;
- Secondary excess sludge production: $mE_{v2} = 0.4$ mg $VSS \cdot mg^{-1}$ COD;
- Density of 40% $FeCl_3 = 1400$ $kg \cdot m^{-3}$.

Solution

Fist calculate the P-removal in the activated sludge system. The influent COD concentration will be reduced after primary settling:

$$S'_{ti} = (1 - 0.5) \cdot 500 = 250\,\text{mg COD} \cdot l^{-1}$$

The quantity of phosphorus that will be removed with the secondary excess sludge (in $mg\,P \cdot l^{-1}$) can be calculated as:

$$P_l = f_p \cdot mEv_2 \cdot S'_{ti} = 0.025 \cdot 0.4 \cdot 250 = 2.5\,\text{mg}\,P \cdot l^{-1}$$

The concentration of phosphorus to be removed in the primary settler is equal to:

$$P_{chem} = P_{ti} - P_{te} - P_l = 15 - 2.5 - 1.0 = 11.5\,\text{mg}\,P \cdot l^{-1}$$

The fraction of the influent phosphorus concentration that has to be removed is 11.5/15–77%. Use Figure 7.12 to determine the required molar Fe^{3+}/P_{ti} dosing ratio as approximately 2.0 for 77% removal. Both the influent phosphorus concentration and the concentration of phosphorus that has to be removed by pre-precipitation can be expressed on a molar basis: $P_{chem} = 11.5/31 = 0.37\,\text{mmol} \cdot l^{-1}$ and $P_{ti} = 15/31 = 0.48\,\text{mmol} \cdot l^{-1}$.

The daily consumption of Fe^{3+} is thus equal to $2 \cdot 0.48 = 0.97\,\text{mmol}\,Fe^{3+} \cdot l^{-1}$ or $0.97 \cdot 55.8 = 54$ mg $Fe^{3+} \cdot l^{-1}$. At 40% wt and $1400\,\text{kg} \cdot m^{-3}$ density, the iron content of the $FeCl_3$ solution is 192.5 g $Fe^{3+} \cdot l^{-1}$ or mg $Fe^{3+} \cdot ml^{-1}$, so the daily consumption of 40% wt $FeCl_3$ is $54/192.5 = 0.28\,ml \cdot l^{-1}$ influent. Knowing the percentage solids removal in the primary settler and the quantity of $FeCl_3$ added, the primary excess sludge production can be calculated as:

- Formation of $FePO_4 = 0.37 \cdot 155.8 = 55.9$ mg $TSS \cdot l^{-1}$
- Formation of $Fe(OH)_3 = (0.97–0.37) \cdot 106.8 = 63.7$ mg $TSS \cdot l^{-1}$
- Primary organic sludge: $\eta_l \cdot S_{ti}/(f_{cv} \cdot f_v) = 50\% \cdot 500/(1.5 \cdot 0.6) = 277.8$ mg $TSS \cdot l^{-1}$

The total primary sludge production is equal to $55.9 + 63.7 + 277.8 = 397.5$ mg $TSS \cdot l^{-1}$ influent.

7.3.2.2 Simultaneous precipitation

The basic process configuration of simultaneous precipitation is shown in Figure 7.13. Simultaneous precipitation can be used either as the principal mechanism to remove phosphorus or only as a supplementary process to support biological phosphorus removal. In the latter case the chemical dosing requirements are of course significantly reduced. When simultaneous precipitation is the main mechanism for phosphorus removal, the biological processes in the activated sludge system will be only designed for removal of nitrogen and organic material. Some phosphorus will be removed together with the produced organic excess sludge, but the mechanism of phosphorus release and -luxury uptake will not develop. Chemical dosing is used to remove the surplus phosphorus not used for biological growth, which will otherwise end up in the effluent. In general, dosing with a molar Me/P ratio of one mole Fe per mole influent P is sufficient to reduce the effluent PO_4-P concentration to a value below

2 mg $P \cdot l^{-1}$. A further increase of the Me/P ratio to 1.7 mol $Fe \cdot mol^{-1}$ P will be sufficient to reduce the effluent PO_4-P concentration to values lower than 1 mg $P \cdot l^{-1}$ (De Jong et al., 1993). Results from full-scale installations in the Netherlands regarding the effectiveness of dosing metal (III) salts are indicated in Figure 7.14. All metal salts listed in Table 7.3 can be used for simultaneous precipitation. When Fe(II) is used, it should be added in the aerated zone of the activated sludge tank, where Fe(II) will be oxidised to Fe(III). The oxygen required for oxidation of Fe^{2+} to Fe^{3+} can be calculated from stoichiometric considerations and will be small: in general between 0.5–2% of the total oxygen demand of the wastewater.

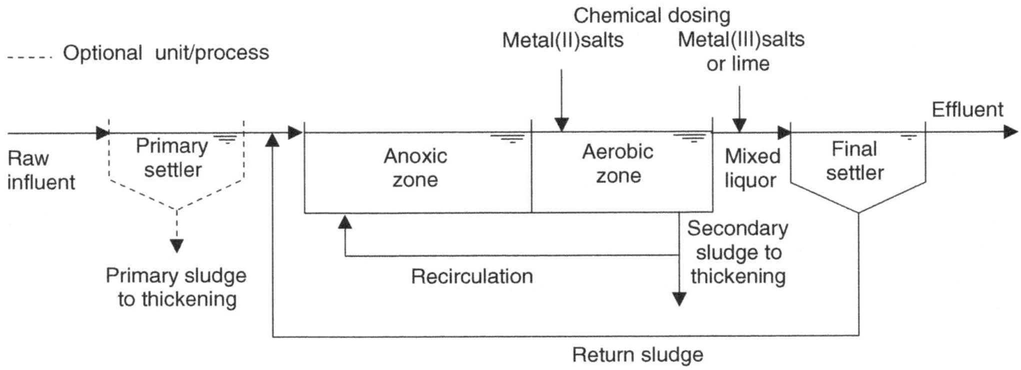

Figure 7.13 Simultaneous precipitation as the main process for phosphorus removal

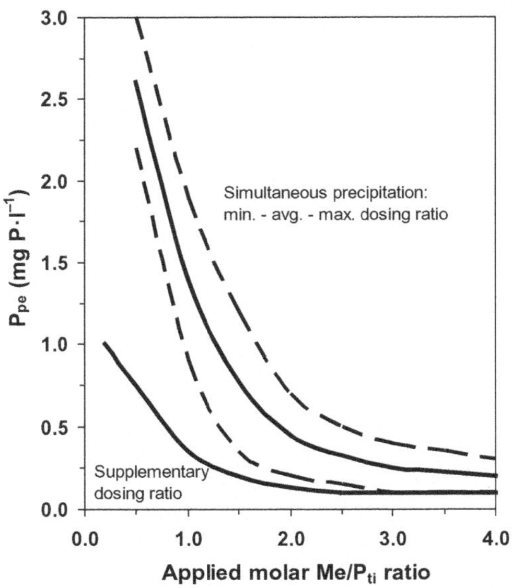

Figure 7.14 Phosphate effluent concentration as function of the applied molar Me/P_{ti} ratio for simultaneous phosphorus removal. Based on a review by De Jong et al., 1993

Metal (III) salts are added in the overflow of the aeration tank to the final settler, in order not to restrict phosphorus availability to the biomass. The chemical sludge production can be calculated with Eqs. (7.23 to 7.26). As a rough indication: in Dutch sewage treatment plants, excess sludge production increased by 10% at a Me/P_{ti} molar dosing ratio of 0.5 and by 50% at a Me/P_{ti} molar dosing ratio of 3.0.

For optimal performance of the phosphorus removal process, the pH of the reactor should be lower than 7, as the solubility of $FePO_4$ is minimum at pH = 5.3 and that of $AlPO_4$ at pH = 6.3. However, for optimum nitrogen removal the pH value should be between 7 and 7.5.

When lime is used, the effluent phosphorus concentration is not dependent on the influent phosphorus concentration, but on the pH in the reactor. Lime addition is therefore mainly interesting at high phosphorus concentrations. De Jong et al. (1993) indicated that a pH increase to at least 8.5–8.7 is required to reduce effluent total phosphorus concentration to a value below $2.0\ mg\ P \cdot l^{-1}$. For an effluent total phosphorus concentration lower than $1.0\ mg\ P \cdot l^{-1}$, a pH value above 9.0 is required. The lime dosing requirements for municipal sewage are between 150–350 g $Ca(OH)_2$ per m^3 influent. At these dosing rates, the production of excess sludge increases by 20 to 50%. The dosing location is situated in the overflow from the aeration tank to the final settler and not in the aeration tank itself, as the resulting high pH would severely inhibit the nitrification- and denitrification processes. The return sludge flow might require pH correction in order to maintain the pH in the aeration tank at the desired value.

EXAMPLE 7.6

It is considered to use simultaneous precipitation for phosphorus removal in an existing activated sludge system operated at a sludge age of 10 days. Calculate the daily consumption of 40% wt $FeCl_3$ required to reduce P_{te} to a value $\leq 1\ mg\ P \cdot l^{-1}$ for the following conditions:

- $Q_i = 2000\ m^3 \cdot d^{-1}$ and $S_{ti} = 500\ mg\ COD \cdot l^{-1}$;
- $P_{ti} = 15$ and $P_{oe} = 0.25\ mg\ P \cdot l^{-1}$;
- $V_r = 1000\ m^3$; $mE_v = 0.3\ mg\ VSS \cdot mg^{-1}\ COD$; $f_v = 0.7\ mg\ VSS \cdot mg^{-1}\ TSS$;
- $T = 20°C$

How much will the sludge concentration increase, assuming the sludge age is not changed?

Solution

Determine the current sludge concentration in the activated sludge system:

$$MS_{ti} = Q_i \cdot S_{ti} = 2000 \cdot 0.5 = 1000\ kg\ COD \cdot d^{-1}$$

$$X_t = MX_t/V_r = R_s \cdot ME_t/V_r = R_s \cdot mE_v/f_v \cdot MS_{ti}/V_r$$

$$= 10 \cdot 0.3/0.7 \cdot 1000/1000 = 4.3\ g\ TSS \cdot l^{-1}$$

The quantity of phosphorus to be removed by simultaneous precipitation is equal to:

$$
\begin{aligned}
MP_{chem} &= MP_{ti} - MP_{te} - MP_{l} \\
&= Q_i \cdot (P_{ti} - P_{te}) - f_p \cdot ME_v \\
&= 2000 \cdot (15 - 1)/1000 - 0.025 \cdot 0.3 \cdot 1000 \\
&= 20.5\,\text{kg P} \cdot d^{-1} \quad \text{or} \quad 10.3\,\text{mg P} \cdot l^{-1}
\end{aligned}
$$

On a molar basis $MP_{chem} = 20.5/31 = 0.66$ kmol $\cdot\, d^{-1}$ and $MP_{ti} = 0.97$ kmol $\cdot\, d^{-1}$. The maximum allowed effluent phosphate concentration can be calculated as:

$$P_{pe} = P_{te} - P_{oe} = 1 - 0.25 = 0.75\,\text{mg P} \cdot l^{-1}$$

The average molar Fe^{3+}/P_{ti} dosing ratio required to meet the maximum allowable effluent phosphate concentration of $P_{pe} = 0.75$ mg $P \cdot l^{-1}$ is equal to 1.5 mol $Fe^{3+} \cdot mol^{-1}$ P (Figure 7.14). The $FeCl_3$ consumption rate is calculated as:

$$
\begin{aligned}
M_{ME} &= 1.5 \cdot 0.97 = 1.5\,\text{kmol}\,Fe^{3+} \cdot d^{-1} \\
&= 1.5 \cdot (55.8 + 3 \cdot 35.5) = 236\,\text{kg}\,FeCl_3 \cdot d^{-1}\ \text{or}\ 81\,\text{kg}\,Fe^{3+} \cdot d^{-1}
\end{aligned}
$$

The Fe-content of 40% wt $FeCl_3$ is 192.5 g $Fe^{3+} \cdot l^{-1}$ So the consumption of 40% wt $FeCl_3 = 81/192.5 = 0.42\ m^3 \cdot d^{-1}$. Finally the chemical sludge production is calculated as:

$$
\begin{aligned}
ME_{mep} &= 0.66 \cdot 155.8 = 103\,\text{kg}\,FePO_4 \cdot d^{-1} \\
ME_{meoh} &= (0.97 - 0.66) \cdot 106.8 = 32.7\,\text{kg}\,Fe(OH)_3 \cdot d^{-1}
\end{aligned}
$$

The mass of chemical sludge that will be present in the activated sludge system is:

$$MX_{chem} = R_s \cdot ME_{chem} = 10 \cdot (103 + 32.7) = 1358\,\text{kg TSS}$$

So the chemical sludge concentration $X_{chem} = MX_{chem}/V_r = 1358/1000 = 1.36$ g TSS $\cdot\, l^{-1}$. The sludge concentration will increase to $X_{t,new} = X_t + X_{chem} = 4.3 + 1.36 = 5.64\ g \cdot l^{-1}$ as a result of simultaneous precipitation, or an increase of 24%.

When simultaneous chemical phosphorus removal is used as a supplementary process, as shown in Figure 7.15, the activated sludge system will be designed for both biological phosphorus and nitrogen removal. Chemical dosing is used to remove the excess phosphorus not taken up by the bio-P organisms or the normal heterotrophic biomass. Care should be taken not to overdose, as this will restrict the availability of phosphate for the bio-P organisms and might lead to a downward spiral in which the fraction of bio-P organisms in the sludge decreases, less phosphate is biologically removed, the effluent phosphorus limits are not met, the chemical dosing rate is increased and so on. Chemical dosing requirements for supplementary precipitation are significantly smaller than those for simultaneous precipitation as the main process. For municipal wastewater, a molar Me/P ratio between 0.2 to 0.5 mol

Me per mol influent P is generally sufficient to reduce the effluent PO_4-P to values below 1 mg $P \cdot l^{-1}$ (Janssen et al., 2002). The effluent phosphate concentration for different molar Me/P dosing ratios is indicated in Figure 7.14.

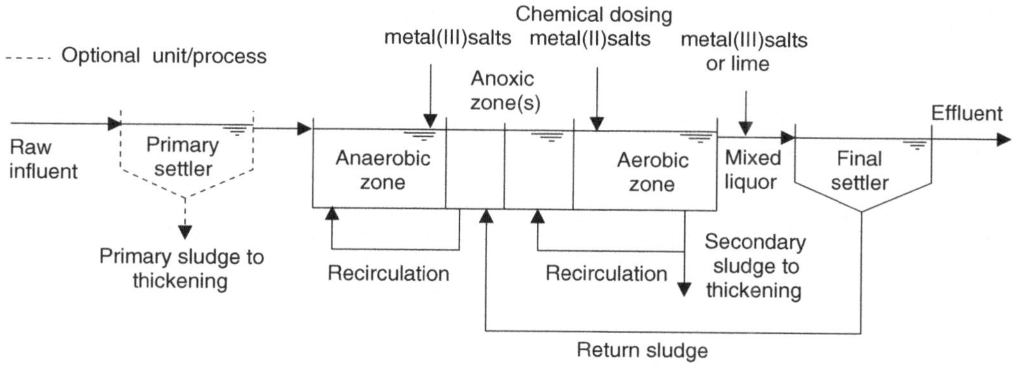

Figure 7.15 Simultaneous precipitation used to supplement the biological phosphorus removal capacity

All metal salts listed in Table 7.3 can be used for simultaneous (supplementary) precipitation. Several dosing locations for metal (III) salts can be considered: at the end of the aerobic zone in the overflow of the aeration tank to the final settler or at the end of the anaerobic zone where the phosphate concentration is highest and stoichiometric dosing requirements lowest, but the risk of phosphate limitation for the bio-P organisms is highest as well. Metal (II) salts can be added at the head of the aerobic zone. The typical increase in excess (chemical) sludge production reported for Dutch municipal sewage treatment plants is between 5 and 10%, but this can be calculated from stoichiometric considerations as well. The pH value depends on the requirements for nitrification and is thus in general between 7 and 8.

Table 7.6 Effects of simultaneous precipitation on sludge characteristics

Characteristic	Effect on sludge characteristics
Settling	– Improved settleability: DSVI value generally decreases. – For metal salts the decrease is modest, while Al-salts have a better effect than Fe-salts. A reduction from 120–140 ml $\cdot$ g^{-1} TSS to values $<$100 ml $\cdot$ g^{-1} TSS has been observed in Dutch sewage treatment plants. – Lime additions results in a SVI decrease up to 50%.
Thickening	– Metal salts: no significant effect. – Lime: thickened concentration can be increased significantly (up to 5–7% dry solids).
Dewatering	– Metal salts: no significant effect. – Lime: dewaterability increases: a 2 to 7% increase in dry solids content can be expected.

7.3.2.3 *Post-precipitation*

Post precipitation is only applied when very low effluent phosphorus values are required. Capital costs are significantly higher than for the other methods, as an additional unit operation has to be included. Therefore post-precipitation has so far not found very wide application, though this may change in the future if effluent limits become stricter and effluent polishing might be required anyway. All metal (III) salts and lime can be used in post-precipitation. Figure 7.16 shows the simplified process configuration for post-precipitation.

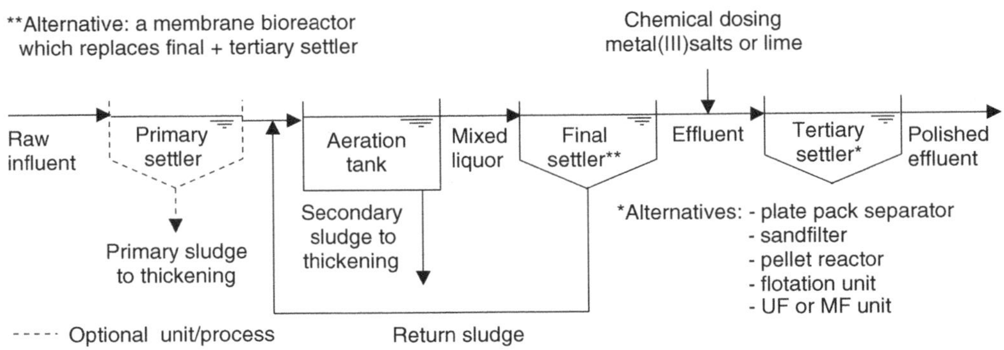

Figure 7.16 Post-precipitation as the main process for phosphorus removal

When polyelectrolyte is used, effluent total phosphorus concentrations <1.0 mg $P \cdot l^{-1}$ are feasible. The applied Me/P dosing ratio is similar to that for pre-precipitation. A chemical sludge will be formed: this tertiary sludge often does not settle well, at DSVI values >300 ml $\cdot$ g^{-1}. It can be separated in a dedicated settler, a plate pack separator or a dissolved air flotation unit. As at values <1 mg $P \cdot l^{-1}$ the main contribution to the total phosphorus concentration in the effluent will come from the organic particulate phosphorus fraction, efficient removal of suspended solids from the effluent is crucial to performance. The dry solids content of the chemical sludge is between 2 and 5 kg TSS $\cdot$ m^{-3} and can be increased by thickening to 5–15 kg TSS $\cdot$ m^{-3}, somewhat higher when lime addition has been used. The thickened tertiary sludge can be dewatered together with the thickened primary and secondary excess sludge. When very low effluent phosphorus values are to be achieved (<0.2 mg $P \cdot l^{-1}$) the following techniques may be considered:

(1) Pellet reactors: an upflow fluidised bed reactor with a bed consisting of granules of fine sand. The effluent of the final settler, to which lime is added, flows upward through the fluidised bed while struvite ($MgNH_4PO_4$) precipitates on the granules. As the granule grows, the apparent density decreases and at a certain point the granule will be flushed out of the reactor with the effluent. The granules are separated from the effluent and the end product can be sold commercially as fertiliser;
(2) Chemical precipitation in combination with sand filters;
(3) Ultra-filtration;
(4) Membrane bioreactors, i.e. activated sludge systems in which the final settler has been replaced by an ultra-filtration membrane unit, refer to Chapter 10.

Method (2) to (4) will retain all organic particulate phosphorus contained in the effluent of the final settler. Soluble phosphate is for a large part removed by methods (1) and (2), for the other methods removal in the upstream processes is required, either biologically or chemically.

7.3.2.4 Sidestream precipitation

The two main configurations in use for sidestream phosphorus precipitation are:

- Precipitation of phosphorus in return streams from sludge treatment;
- A variation on the regular biological phosphorus removal process in which, following bio-P removal with the excess sludge, phosphate is concentrated into a smaller sidestream that is then subjected to chemical precipitation.

(1) Phosphate release from sludge treatment processes

Although biological phosphorus removal has many advantages over chemical phosphorus removal but one disadvantage is the potential release of the accumulated poly-P phosphorus from the bio-P biomass during sludge treatment and subsequent recycle of PO_4^{3-} to the activated sludge system. Contrary, release of phosphate will not occur for chemically precipitated phosphate.

During sludge treatment the following processes may release phosphate to the liquid phase: thickening, anaerobic digestion and dewatering. During gravitational thickening of excess bio-P sludge, a hydraulic retention time of one day is already sufficient to release about 50% of all polyphosphate. After two days, release is almost complete. Thickening of secondary excess sludge together with primary sludge will lead to increased phosphate release, due to the presence of easily biodegradable COD in the primary sludge.

Phosphate release as a result of biomass decay will in general not be significant in a thickener. When the thickener is properly designed and operated, i.e. the amount of turbulence and short-circuiting in the thickener is limited, the released phosphate will remain in the liquid phase of the thickened sludge and will not be returned to the activated sludge system with the thickener supernatant. As an alternative to gravitational thickening of excess bio-P sludge, direct dewatering or mechanical thickening might be considered, as due to the short processing time phosphate release will be minimal, except when:

- The excess sludge has been stored in a sludge buffer tank for a prolonged period of time. If the function of the sludge buffer tank is to level out fluctuations in the solids content of the excess sludge, it may be considered to discharge the excess sludge directly from the aeration tanks, apart from other benefits of this arrangement, discussed in Chapter 3;
- The excess sludge is taken from the return sludge line, while retention time in the final settler is excessive, resulting in anaerobic conditions and the subsequent release of stored polyphosphate into the liquid phase.

During sludge digestion all polyphosphate in the digested sludge is degraded to ortho-phosphate, as well as a large part of the organic phosphorus contained in the organic sludge. Chemically precipitated phosphate is not released into the liquid phase during sludge digestion. Part of the released phosphate may complex with metal ions. Bio-P organisms in general contain a significant amount of Mg^{2+} (absorbed by the bio-P organisms to compensate for the negative electrical charge of the PO_4^{3-} ions). During sludge digestion, the released Mg^{2+} will precipitate with PO_4^{3-} as $Mg_2(PO_4)_3$. If NH_4^+ is present (as it normally will be), $MgNH_4PO_4$ (struvite) is formed and finally, PO_4^{3-} might form complexes with zeolithes (the aluminium silicates present in detergents). Furthermore, it might be considered to enhance the chemical precipitation process by increasing the pH value. When for instance lime is added to the sludge, preferably a mixture of CaO and MgO, the liquid phase of the sludge will contain the three components that are required for struvite precipitation (Mg^{2+}, NH_4^+ and PO_4^{3-}), while also the pH will be sufficiently high (pH > 9).

When thickened or digested sludge is dewatered, all phosphate present in the liquid phase will be returned to the activated sludge process. For conventional activated sludge systems without chemical phosphorus removal, this may be up to 10–20% of the influent phosphorus load; for bio-P systems this will be even higher. Depending on the biological phosphorus removal capacity of the activated sludge process and the COD/P and COD/N ratio of the influent, it may be necessary to remove this additional phosphorus load by dosing metal (III) salts, either in the digester or in the dewatering operation itself. The latter option has the advantage that a metal (III) salt will improve the dewaterability of the sludge and will also reduce odor problems due to precipitation with sulphide. Sometimes chemical sludge from drinking water preparation is used for this purpose.

(2) Sidestream phosphorus removal (as a supplementary process)

Sidestream phosphorus removal requires the application of biological excess phosphorus removal in the mainstream process. The phosphorus biologically removed from the main process is concentrated into a much smaller side stream and is then removed by means of chemical precipitation. For sidestream removal, all techniques used for post-precipitation can be used: full-scale installations have been built with gravity sedimentation and with pellet reactors. Sidestream phosphorus removal can be classified into two basic configurations, which differ in the location from which the sidestream is taken, i.e. from the anaerobic zone, e.g. the BCFS process or from the aerobic zone or from the clarifier, e.g. the Phostrip process.

(a) Sidestream taken from the anaerobic zone

The BCFS process (a Dutch acronym for biological-chemical phosphorus and nitrogen removal) has been developed at the University of Delft and has since been implemented in the Netherlands in a number of full-scale installations. Figure 7.17 gives a schematic representation of the configuration. Mixed liquor is taken from the end of the anaerobic zone, where the phosphate concentration is highest. In a dedicated settler the phosphate-rich supernatant is separated from the sludge, which is then returned to the anoxic zone.

The supernatant is either sent to the thickener, or directly to the sludge digester together with the excess sludge. Metal (III) salts or lime are added to precipitate the free phosphate ions. As the phosphorus is released using the easily biodegradable COD present in the influent, the addition of a carbon source is not required. However, the quantity of phosphorus removed should be carefully controlled. Excessive phosphorus removal might restrict the regeneration of polyphosphates in the aerobic zone, required by the bio-P organisms to retain their comparative advantage over normal heterotrophic biomass: i.e. it will limit the sequestration of VFA by the bio-P organisms in the anaerobic zone.

(b) Sidestream taken from the aerobic zone or clarifier

In this process configuration, shown in Figure 7.18, mixed liquor is taken from the aerobic zone and directed to an anaerobic phosphate stripper. An easily biodegradable carbon source (e.g. acetate) is added to release the stored polyphosphate. After solids/liquid separation, the sludge is returned to the first anoxic compartment or directly to the anaerobic zone as the nitrate concentration of the sludge is low due to the previous addition of a carbon source. The supernatant is either sent to a dedicated phosphate removal unit, for example a pellet reactor, or to the sludge thickener. In both cases, metal (III) salts or lime are added to precipitate the free phosphate. An advantage of sidestream phosphorus removal from the aerobic zone is that there is no risk of disturbing the biological phosphorus removal process. On the other hand, it involves additional operational costs as an external carbon source is required.

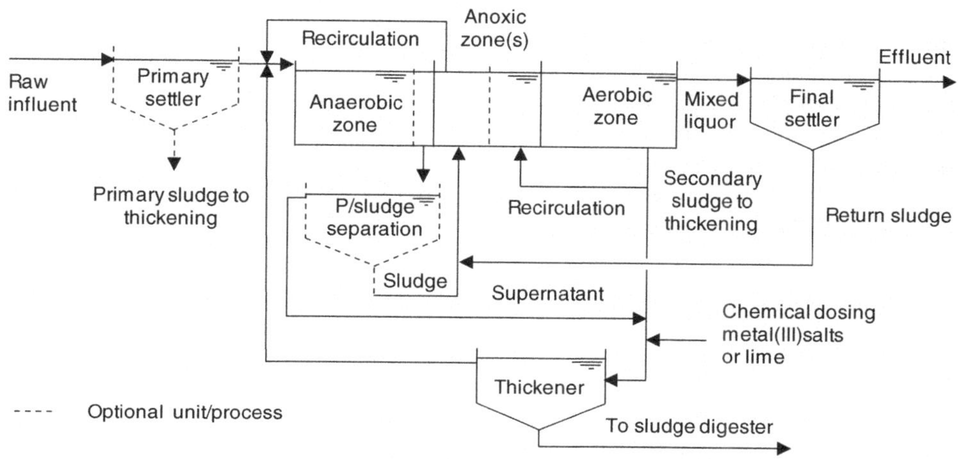

Figure 7.17 Side-stream precipitation: BCFS configuration

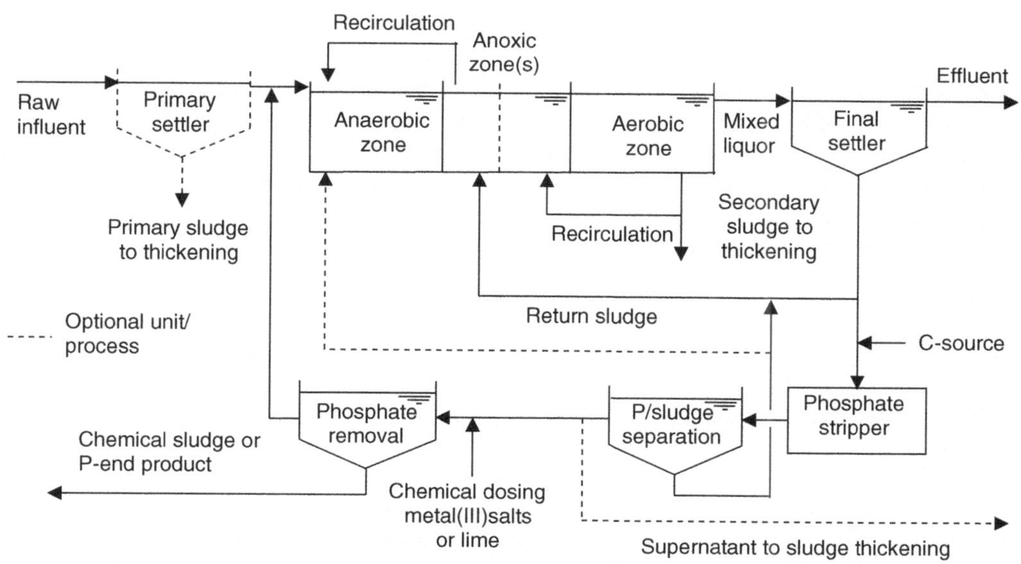

Figure 7.18 Side-stream precipitation: sidestream taken from the end of the aerobic zone or from the clarifier

7.3.3 Design procedure for chemical phosphorus removal

In this section a general procedure is presented for the design of chemical phosphorus removal with either pre-precipitation or simultaneous precipitation. The following important system parameters are calculated:

(I) MP_{chem} – the phosphorus mass to be removed by chemical precipitation, (II) M_{ME} – the amount of metal salt required and (III) ME_{chem}–the amount of chemical sludge produced. This procedure is approximately similar for both configurations and calculates sequentially:

(1) The influent phosphorus load – MP_{ti};
(2) The effluent phosphorus load – MP_{te};
(3) The removal of phosphorus with the secondary excess sludge – MP_{lx2};
(4) The removal of phosphorus required by means of chemical precipitation – MP_{chem};
(5) The consumption of metal salt–M_{ME};
(6) The removal of phosphorus with the primary excess sludge–MP_{l1} (pre-precipiation only);
(7) The chemical sludge production–ME_{chem}.

(1) Calculate MP_{ti} – the influent phosphorus load ($kg\ P \cdot d^{-1}$)

$$MP_{ti} = Q_i \cdot P_{ti} \qquad (7.32)$$

(2) Calculate MP_{te} – the effluent phosphorus load ($kg\ P \cdot d^{-1}$)

$$MP_{te} = Q_i \cdot (P_{ose} + P_{ope} + P_{pe}) \leq Q_i \cdot P_{td} \qquad (7.33)$$

Where P_{td} = the effluent discharge limit for total phosphorus

(3) Calculate MP_{lx2} – removal of phosphorus with the secondary excess sludge ($kg\ P \cdot d^{-1}$)
It depends on the selected method of chemical phosphorus removal whether phosphorus will be removed only with the secondary excess sludge or also (or predominantly) together with the primary excess sludge.

As discussed in Section 7.1.3.4, the value of MP_{l2} should be corrected for the value of MP_{ope} in order not to underestimate the quantity of phosphorus that is to be removed. The net quantity of phosphorus disposed with the secondary excess sludge equals:

$$MP_{l2x} = f_p \cdot ME_{v2} - Q_i \cdot P_{ope} \qquad (7.34)$$

In the case of pre-precipitation it is not necessary to differentiate between the phosphorus load removed as part of organic suspended solids and the part which precipitates as inorganic metal-phosphate salts. The reason is simple: the applied molar Me/P ratio is based on required P_{ti} removal only and does not make this differentiation either.

When pre-precipitation is applied, the influent COD load to the activated system will be reduced as a result of the removal of suspended solids. Furthermore, the pre-settled wastewater COD composition changes as well. This means that the secondary excess sludge production will be significantly smaller. Assuming there is no difference in removal between the biodegradable and non biodegradable solids fractions, the influent COD after primary settling can be characterised as:

$$f'_{ns} = f_{ns}/(1 - \eta_{x1} \cdot (f_{np} + f_{bp})) \qquad (7.35)$$

$$f'_{np} = (1 - \eta_{x1}) \cdot f_{np}/(1 - \eta_{x1} \cdot (f_{np} + f_{bp})) \qquad (7.36)$$

where:

η_{x1} = *solids* removal efficiency (typically between 40–70% for primary settlers with pre-precipitation). Note: do not confuse η_{x1} with η_{l}, which is defined as *COD* removal efficiency and will discussed in Section 12.1): basically a simplified approach to define primary settler performance

f_{bp} = slowly biodegradable (mainly particulate)COD fraction in the raw wastewater

$$= (1 - f_{sb}) \cdot (1 - f_{np} - f_{ns}) \quad (7.37)$$

f_{bs} = easily biodegradable (mainly soluble)COD fraction in the raw wastewater

$$= f_{sb} \cdot (1 - f_{np} - f_{ns}) \quad (7.38)$$

The value of f_{sb}, the easily degradable fraction of the biodegradable COD, will increase as a result of the solids removal. If it is assumed that no hydrolysis takes place in the primary settler, one has:

$$f'_{sb} = f'_{bs}/(f'_{bs} + f'_{bp}) \quad (7.39)$$

where:

$$f'_{bs} = f_{bs}/(1 - \eta_{x1} \cdot (f_{np} + f_{bp})) \quad (7.40)$$

$$f'_{bp} = (1 - \eta_{x1}) \cdot f_{bp}/(1 - \eta_{x1} \cdot (f_{np} + f_{bp})) \quad (7.41)$$

(4) Calculate MP_{chem} – removal of phosphorus by means of chemical precipitation ($kg\ P \cdot d^{-1}$)
The mass of phosphorus to be removed by chemical precipitation can be calculated as:

$$MP_{chem} = MP_{ti} - MP_{l2x} - MP_{pe} - MP_{ope} - MP_{ose} \quad (7.42)$$

(5) Calculate M_{ME} – consumption of metal salt ($kg \cdot d^{-1}$)
The Me/P dosing ratio is expressed in mole Me per mole P. In order to calculate the consumption of metal salt, it will be necessary to convert the mass flow rate of phosphorus from $kg\ P \cdot d^{-1}$ to $kmoles\ P \cdot d^{-1}$, using the molar weight of phosphorus ($31\ kg \cdot kmol^{-1}$). Then, after multiplication with the appropriate molar Me/P ratio, the equivalent number of kmoles of metal salt can be converted to kg metal salts using the molar weights listed in Table 7.3.

(a) Consumption of metal salts for pre-precipitation
In case of chemical pre-precipitation of phosphorus, the recommended molar Me(III)/P dosing ratio is expressed in Figure 7.12 as a function of the required removal efficiency based on the *influent* total phosphorus concentration P_{ti}. The required removal efficiency is equal to $MP_{chem.}/MP_{ti}$. When MP_{chem} (in $kmol\ P \cdot d^{-1}$) is multiplied with the selected molar Me^{3+}/P ratio, this results in the consumption of metal salt in $kmol \cdot d^{-1}$.

(b) Consumption of metal salts for simultaneous precipitation
For simultaneous precipitation (both mainstream and supplementary), the molar Me/P dosing ratio depends on the allowed effluent phosphate concentration P_{pe}, which can be calculated by subtracting the concentration of organic phosphorus ($P_{oe} = P_{ose} + P_{ope}$) from the total phosphorus limit P_{td} in the effluent.

For this P_{pe} concentration, one can use Figure 7.14 to determine the required molar Me/P dosing ratio. This ratio is multiplied with MP_{ti} (in kmol $P \cdot d^{-1}$), which in turn yields the quantity (in kmol $\cdot d^{-1}$) of metal salt required.

(6) Calculate MP_{l1} – removal of phosphorus with the primary excess sludge (kg $P \cdot d^{-1}$)
If applicable, the daily mass of organic phosphorus that is removed in the primary settler (MP_{l1}) can be calculated as:

$$MP_{l1} = \eta_{x1} \cdot f_p \cdot Q_i \cdot S_{pi}/f_{cv} \tag{7.43}$$

where:

η_{x1} = removal efficiency of suspended solids in primary settler (%)

$$S_{pi} = \text{particulate COD concentration in the influent} \\ = f_{np} \cdot S_{ti} + (1 - f_{np} - f_{ns}) \cdot (1 - f_{sb}) \cdot S_{ti} \tag{7.44}$$

The phosphorus mass fraction in primary sludge is typically lower than the value used for biological secondary excess sludge, i.e. $f_p = 0.025$ mg $P \cdot mg^{-1}$ VSS. A typical value of $f_p = 0.015$ may be used for primary sludge generated from municipal sewage.

(7) Calculate ME_{chem} – total chemical sludge production
Assuming that sulphide is not present in the wastewater in significant quantities and that all metal added will precipitate, the chemical sludge will contain the following two main components: $MePO_4$ and $Me(OH)_3$. For simultaneous precipitation, the quantity of $MePO_4$ (in kmol $\cdot d^{-1}$) produced is equal to MP_{chem} (in kmol $P \cdot d^{-1}$) and the production of $Me(OH)_3$ will be equal to $M_{ME} - MP_{chem}$.

$$ME_{mp} = MP_{chem}/31 \cdot mw_{mp} \tag{7.45}$$

$$ME_{mo} = (M_{ME}/mw_{ms} - MP_{chem}/31) \cdot mw_{mo} \tag{7.46}$$

where:

ME_{mp} = chemical excess sludge production in the form of $MePO_4$ (kg TSS $\cdot d^{-1}$)

ME_{mo} = chemical excess sludge production in the form of $Me(OH)_3$ (kg TSS $\cdot d^{-1}$)

mw_{ms} = molar weight of the metal salt (kg $\cdot kmol^{-1}$)

mw_{mp} = molar weight of the metal phosphate (kg $\cdot kmol^{-1}$)

mw_{mo} = molar weight of the metal hydroxide (kg $\cdot kmol^{-1}$)

For pre-precipitation, the calculation is similar with one exception: for the calculation of the mass of $MePO_4$ formed, it is required to reduce the value of MP_{chem} by MP_{l1}, the mass of organic phosphorus removed with the primary excess sludge, as this does not exert any demand of metal(III) ions:

$$ME_{mp} = (MP_{chem} - MP_{l1})/31 \cdot mw_{mp} \tag{7.47}$$

$$ME_{mo} = (M_{ME}/mw_{ms} - (MP_{chem} - MP_{l1})/31) \cdot mw_{mo} \tag{7.48}$$

Refer also to the extensive calculation examples in Sections 14.2.5 and 14.5, where the designs of different configurations for tertiary nutrient removal (both nitrogen and phosphorus) are compared.

Chapter 8

Sludge settling

8.0 INTRODUCTION

Activated sludge settlers are applied to effect the separation of the solid (sludge) and the liquid phase of mixed liquor. They are operated as continuous units, discharging an effluent that is substantially free of solids, while the settled solids are recycled to the aeration tank. Activated sludge settlers are also called final- or secondary settlers or -clarifiers to distinguish them from primary settlers that are used for settling of raw wastewater.

Due to the high suspended solids concentration, settling in final settlers is of an entirely different nature from the one observed in more dilute suspensions like most wastewaters. In dilute suspensions, the interactions between individual flocs or particles are weak or non existent and therefore each particle acquires its own settling velocity as a result of frictional and gravitational forces. In more concentrated suspensions like mixed liquor, a matrix of interlinked particles is formed, all settling with the same rate: the zone settling velocity. This type of settling is not dictated by the forces acting on each particle alone, but rather by the forces acting on the whole interlinked sludge mass. For zone settling to apply, the minimum sludge concentration is in the range of 0.5 to $1 \cdot 0\ g \cdot l^{-1}$. At lower concentrations the flocs are too dispersed and tend to settle as individual particles. As the suspended solids concentration of mixed liquor is normally within the range of 2 to $5\ g \cdot l^{-1}$, zone settling is unavoidable in final settlers of the activated sludge process. In contrast, in most wastewaters the suspended solids concentration is less than $0.5\ g \cdot l^{-1}$, so that particle settling occurs.

Final settlers perform two functions simultaneously: (I) clarification, i.e. the liquid-solid separation necessary to produce an effluent free of suspended solids, and (II) thickening, i.e. the increase of the suspended solids concentration in the return sludge flow during its passage through the final settler. Depending on the settling characteristics of the sludge and the operational conditions of the final settler (i.e. the solids load of incoming mixed liquor and the concentration of the outgoing return sludge), either one or the other function is the limiting factor for the solids loading rate that may be applied. This maximum solids loading rate in turn defines the minimum surface area of the final settler, and hence its minimum volume.

In this chapter some factors influencing activated sludge settleability will be identified, while methods to determine the settling characteristics are discussed. Equations are derived to design and optimise final settlers as a function of the settling characteristics and operational conditions. The same expressions can also be used for the design and optimisation of sludge thickeners, which are used to concentrate sludge prior to its discharge to dewatering- or sludge processing units such as digesters. This will be discussed in Chapter 12.

In practice, two common problems may reduce the performance of sludge settlers: the development of filamentous sludge with poor settling characteristics and scum formation. The reasons that lead to the development of these problems, as well as preventative and remedial measures are also discussed in this chapter.

8.1 METHODS TO DETERMINE SLUDGE SETTLEABILITY

8.1.1 Zone settling rate test

Zone settling may be observed in the batch settler described by White (1975) and shown schematically in Figure 8.1. The apparatus consists of a transparent vertical cylinder, in which a batch of sludge is placed. A stirrer is connected to the central axis, which in turn is driven by a low rotation motor, and this gently stirs the sludge.

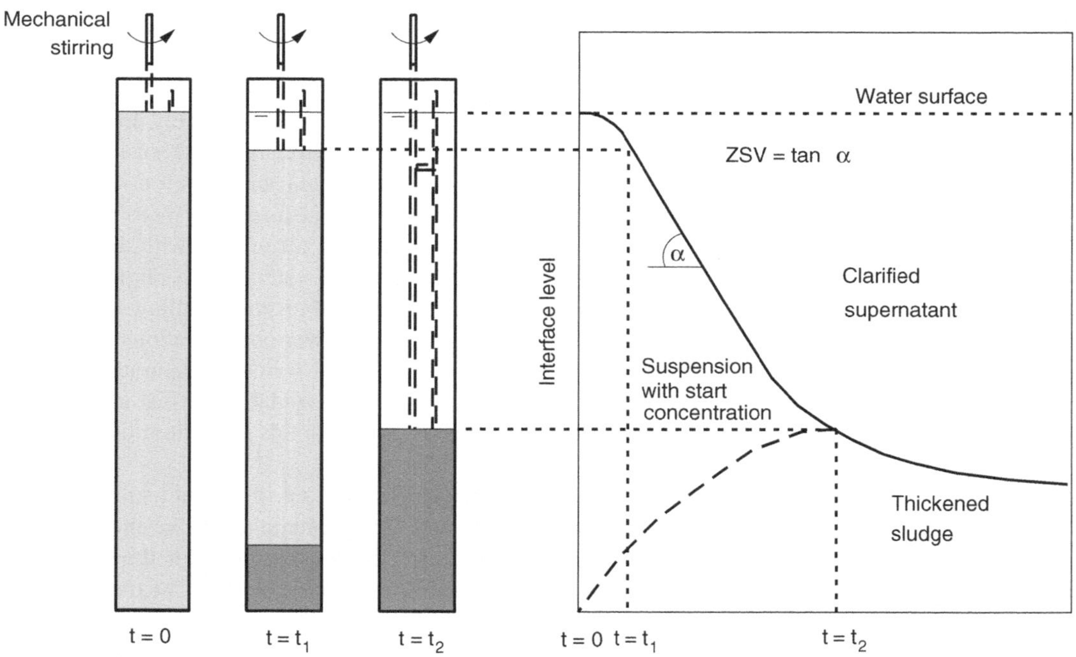

Figure 8.1 Experimental set-up to determine the zone settling velocity (left) and a typical curve of the interface displacement in time (right)

After placing the sludge batch in the cylinder, the following behaviour can be observed:

- A short time (a few minutes) after placing the sludge in the cylinder, a sharp interface is formed, separating the clear supernatant not containing suspended solids in the upper part and settling sludge in the lower part;
- In the region below the interface all sludge particles settle at the same rate, so that the interface is also displaced at the same rate;
- Simultaneously at the bottom of the cylinder, sludge with a higher concentration accumulates. With time, an ever larger fraction of the sludge particles become part of this concentrated sludge;
- After some time, the sludge-supernatant interface approaches the region of concentrated sludge and its rate of displacement starts to decrease gradually.

Figure 8.1 also shows a typical curve of the interface displacement with time. The zone settling velocity is defined as the gradient of the linear (or linearised) part of this curve. Several cylinders may be used in parallel, in order to determine the zone settling velocity simultaneously for different sludge concentrations. The use of different concentrations allows the experimental determination of the relationship between the sludge concentration and the zone settling velocity. Several research workers have investigated the relationship between the zone settling velocity and the activated sludge concentration. The best known models are those proposed by Vesilind (1968) and by Dick (1972). The models describe the relationship between zone settling velocity and sludge concentration as follows:

(1) Vesilind's equation: $ZSV = v_0 \cdot \exp(-k \cdot X_t)$ (8.1)

(2) Dick's equation: $ZSV = V_0 \cdot (X_t)^K$ (8.2)

where:

ZSV = zone settling velocity
X = activated sludge concentration
V_0, K, v_0 and k = sludge settleability constants

In order to evaluate which of the two equations better describes the settling behaviour of activated sludge, the following method may be used:

- Obtain experimental data of the zone settling velocity as a function of the sludge concentration;
- Plot this data in a suitable diagram: semi log (natural) for Vesilind's equation and log-log paper for Dick's equation;
- Draw the "best-fit straight line" through the experimental points. The gradient of this straight line will be the constant k (Vesilind) or K (Dick) and the linear coefficient is log v_0 or log V_0.

Smollen and Ekama (1984) analysed their own data from systems in South Africa, as well as that of other activated sludge processes (Pitman, 1980 and 1984; Ibama 1984; Tuntoolavest and Grady, 1980 and Rachwall et al., 1981) and concluded that in all cases Vesilind's equation led to a better description of the actual settling behaviour of activated sludge. These results are corroborated by Catunda et al. (1992) and for this reason Vesilind's equation will be adopted in this text as the basis for the description of activated sludge settling.

EXAMPLE 8.1

Using stirred batch cylinders it is attempted to determine the Vesilind settleability constants for a certain sludge. The height of the solid-liquid interface has been measured for 6 different sludge concentrations as a function of time for 30 minutes. The results are shown in Table 8.1 below. Determine the values of k and v_0.

Table 8.1 Level of solids-liquid interface (in cm under the liquid surface) as a function of the elapsed time and for different sludge concentrations

Time	Sludge concentration (g TSS · l^{-1})					
(min)	$X_t = 1.2$	$X_t = 2.0$	$X_t = 3.1$	$X_t = 3.9$	$X_t = 4.9$	$X_t = 6.2$
0	−0.0	−0.0	−0.0	−0.0	−0.0	−0.0
1	−4.5	−3.5	−2.5	−2.0	−1.5	−0.5
2	**−9.5**	**−9.5**	−6.0	−5.0	−3.0	−2.0
3	**−20.0**	**−19.0**	**−11.5**	**−8.5**	**−5.5**	**−3.5**
4	**−31.5**	**−27.0**	**−17.0**	**−12.0**	**−8.0**	**−5.0**
6	**−54.0**	**−42.5**	**−27.5**	**−19.5**	**−13.5**	**−7.5**
8	−63.5	**−59.5**	**−38.0**	**−26.5**	**−19.0**	**−10.0**
10	−71.0	−68.0	**−49.5**	**−33.5**	**−23.0**	**−12.0**
12	−76.0	−71.0	**−57.5**	**−43.5**	**−28.5**	**−14.0**
15	−82.0	−75.5	−65.5	**−54.0**	**−35.0**	**−18.0**
20	−85.0	−79.0	−71.0	−64.5	**−48.0**	**−25.5**
25	−86.5	−81.0	−73.5	−69.0	**−57.5**	**−30.5**
30	−87.5	−83.5	−76.0	−72.5	−61.0	**−37.0**

Solution

To determine the values of the settleability constants, follow the procedure as described above:

(1) Plot the position of the interface as a function of time at different concentrations. Figure 8.2 shows the data points and associated curves (dotted lines);
(2) Decide which part of the curve may be considered as linear. In Table 8.1 these data points are marked in "bold";
(3) Draw the best-fit straight line through the linear part of each curve. These straight lines are indicated in Figure 8.2;
(4) Determine the zone settling velocity (ZSV) as the gradient ($\Delta Y/\Delta X$) of the straight lines. The values are indicated in Figure 8.2;
(5) Plot the values of ln(ZSV) as a function of the sludge concentration as demonstrated in Figure 8.3 and draw the best-fit straight line through the data points;
(6) Use Figure 8.3 to determine the value of the constants: k is equal to the gradient of the straight line ($\Delta Y/\Delta X$) and v_0 is equal to the intersection of the straight line with the vertical axis. In this example $k = 0.44\ l \cdot g^{-1}$ and $v_0 = \exp(2.96) = 19.2\ cm \cdot min^{-1} = 11.7\ m \cdot h^{-1}$.

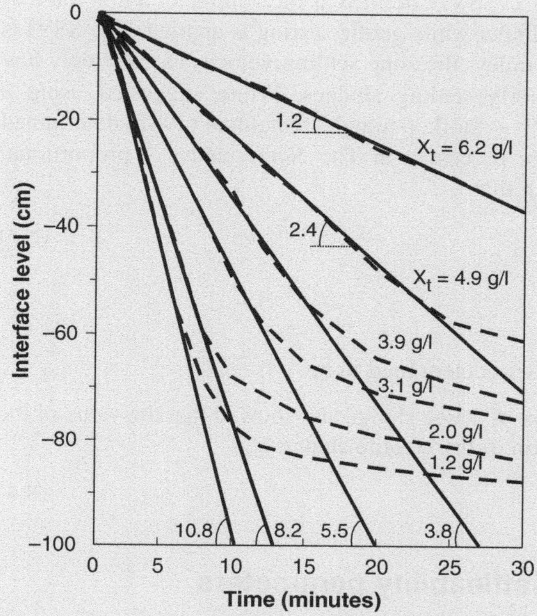

Figure 8.2 Graphical representation of the data in Table 8.1

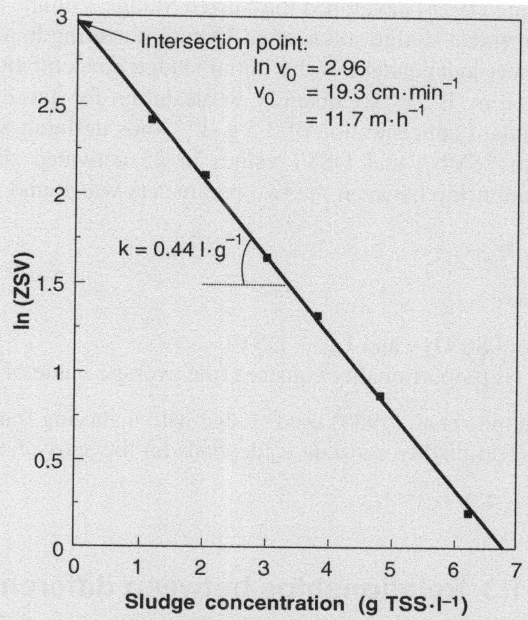

Figure 8.3 Semi log diagram of the zone settling velocity versus the sludge concentration based on the data in Table 8.1

8.1.2 Alternative parameters for sludge settleability

The zone settling velocity test as described in the previous section is not particularly suitable for routine use at wastewater treatment plants because it is very tedious and time consuming. For this reason many research workers have tried to find alternative ways to express sludge settleability in quantitative terms.

The Sludge Volume Index or SVI (Mohlman, 1934) is probably the best known and most widely applied test for sludge settling. In this test, a certain volume of mixed liquor (for example one litre) is placed in a calibrated cone or cylinder. After a set time (for example 30 minutes) the remaining sludge volume is read off and after determination of the initial sludge concentration the volume of settled sludge per gram of solids is calculated. This number expresses the sludge volume index (SVI or SVI_{30}). As this test is extremely simple to carry out it has found wide application. Unfortunately, it is not a very useful test for quantitative work. Its principal shortcoming is that the outcome of the test depends on the initial sludge concentration. A true parameter for sludge settleability should be independent of the tested sludge concentration.

In an attempt to eliminate the influence of sludge concentration, Stobbe (1964) developed the Diluted Sludge Volume Index or DSVI. This test is based on the observation that when the sludge volume after settling is less than about 25% of the initial volume, the calculated SVI value is practically constant and does not depend on the initial sludge concentration. Thus Stobbe (1964) suggested diluting sludge batches until the volume of the diluted suspension, after settling, has a volume of 200 ml or less per litre initial volume.

White (1975) developed the Stirred Sludge Volume Index (SSVI), defined as the volume of a unit mass of suspended sludge solids after 30 min of settling in a cylinder while gentle stirring is applied. The SSVI is almost independent of the initial sludge concentration, unless the zone settling velocity is extremely low ($< 1\ m \cdot h^{-1}$). To quantify settleability for these poorly settling sludges, White suggested using a standard concentration of $3.5\ g \cdot l^{-1}$, thus defining $SSVI_{3.5}$. Stofkoper and Trentelman (1982) determined both $SSVI_{3.5}$ and DSVI values in 25 activated sludge processes in The Netherlands. A proportional relationship between the two parameters was found such that:

$$I_{ssv} = c_p \cdot I_{dsv} \tag{8.3}$$

where

$I_{ssv} = SSVI_{3.5}$ and $I_{dsv} = DSVI$
c_p = proportionality constant (the average value of c_p was determined to be $^2/_3$)

Catunda et al. (1989) used sludge with a varying fraction of active sludge and showed that the value of the proportionality constant c_p depends on the active fraction of the volatile sludge f_{av}:

$$c_p = 1 - 0.35 \cdot f_{av} \tag{8.4}$$

8.1.3 Relationships between different settleability parameters

Pitman (1984) developed an empiric relationship between the constants of Vesilind's equation and I_{ssv}. Analysing his data obtained during six years of full-scale investigation, the following correlation was established:

$$v_0/k = 68 \cdot \exp(-0.016 \cdot I_{ssv}) \tag{8.5}$$

Ekama and Marais (1986) analysed their data and that of others (White 1975, Rachwall et al. 1981, Koopman and Cadee, 1983) and concluded that Pitman's empiric expression gave a good description for all. They also verified that there was a relationship between v_0/k and k:

$$k = 0.88 - 0.393 \cdot \log(v_0 \cdot k) \tag{8.6}$$

Knowing the value of k, v_0 can now be calculated with the aid of Eqs. (8.5 and 8.6):

$$v_0 = (v_0/k) \cdot k \tag{8.7}$$

Catunda et al. (1989) observed that by rearranging Eqs. (8.5 to 8.7), it is possible to express k and v_0 explicitly as a function of I_{ssv}. By substituting Eq. (8.5) in Eq. (8.6) one has:

$$k = 0.16 + 2.7.10^{-3} \cdot I_{ssv} \tag{8.8}$$

Substitution of Eqs. (8.7 and 8.8) in Eq. (8.5) results in:

$$v_0 = (10.9 + 0.18 \cdot I_{ssv}) \cdot \exp(-0.016 \cdot I_{ssv}) \tag{8.9a}$$

In the range of I_{ssv} values that are of practical interest for the activated sludge process, Eq. (8.9a) is almost linear and in good approximation can be expressed as:

$$v_0 = 11.2 - 0.06 \cdot I_{ssv} \tag{8.9b}$$

Equations (8.8 and 8.9) can be used to calculate the constants k and v_0 directly from the I_{ssv} value, without carrying out the zone settling velocity test. The constants can also be calculated from I_{dsv} data using Eq. (8.3 and 8.4). It must be remembered that the empiric relationships of this section are all based on experimental work with municipal sewage as wastewater. It is possible that the relationships do not hold for predominantly industrial wastewaters.

Catunda et al. (1989) carried out a research project to evaluate the influence of sludge concentration and composition of activated sludge generated from municipal wastewater on the values of the settleability constants: k, v_0, I_{dsv} and I_{ssv}. The investigation was carried out on pilot scale by operating an aerated lagoon ($R_s = 2$ d) and a series of 4 aerobic digesters digesting the excess sludge from the aerated lagoon. The active sludge fraction in the sludge varied between 83% (in the aerated lagoon) and 14% (in the last digester of the series). The main relationships derived from this investigation are listed below:

$$I_{ssv} = 25 + 25 \cdot f_{av} + 5 \cdot X_t \quad (8.10a)$$

$$I_{ssv} = (1 - 0.35 \cdot f_{av}) \cdot I_{dsv} \quad (8.10b)$$

$$k = 0.16 + 0.003 \cdot I_{ssv} \quad (8.10c)$$

$$v_0 = 16 - 0.1 \cdot I_{ssv} \quad (8.10d)$$

When the results by Catunda et al. (1989) and those from Ekama and Marais (1986) are compared, the following is concluded:

- The relationships between k and I_{ssv} (Eqs. 8.8 and 8.10c) and v_0 and I_{ssv} (Eqs. 8.9 and 8.10d) found by Catunda et al. (1989) do not differ significantly from those suggested by Ekama and Marais (1986);
- The relationship between I_{ssv} and I_{dsv} as described by Van Haandel and Catunda (1992) is comparable with the results from Stofkoper and Trentelman (1982), when the active sludge fraction is very high ($f_{av} \rightarrow 0.9$). This is not unexpected, as in the period between 1970 and 1985 nutrient removal was not yet required and most activated sludge processes in Holland were therefore operated at a short sludge age;
- The sludge concentration and -composition influence the values of the settleability constants. While the influence of the sludge concentration is relatively small, the value of the active fraction has a very marked effect on I_{ssv} and hence on the values of k and v_0: a higher active sludge fraction results in a decrease in sludge settleability.

EXAMPLE 8.2

For an activated sludge system operating at a short sludge age ($f_{av} = 0.9$), sludge settling volumes were determined for several concentrations.

- Experiment 1: $X_t = 8.4\ g \cdot l^{-1}$ and the sludge volume $= 512\ ml \cdot l^{-1}$;
- Experiment 2: $X_t = 4.2\ g \cdot l^{-1}$ and the sludge volume $= 364\ ml \cdot l^{-1}$;
- Experiment 3: $X_t = 2.1\ g \cdot l^{-1}$ and the sludge volume $= 192\ ml \cdot l^{-1}$.

From above data, estimate the values of the Vesilind constants.

Solution

The SVI values are calculated as:

- Experiment 1: SVI = 512/8.4 = 61 $ml \cdot g^{-1}$;
- Experiment 2: SVI = 364/4.2 = 87 $ml \cdot g^{-1}$;
- Experiment 3: SVI = 192/2.1 = 91 $ml \cdot g^{-1}$.

As the final sludge volume of experiment 3 after settling was less than 200 $ml \cdot l^{-1}$, this means that the calculated value of the SVI represents the value of the diluted sludge volume index I_{dsv} as well. Thus for this sludge $I_{dsv} = 91\ ml \cdot g^{-1}$. For the active fraction of the volatile sludge ($f_{av} = 0.9$), the value of I_{ssv} can now be calculated as:

$$I_{ssv} = I_{dsv} \cdot (1 - 0.35 \cdot f_{av}) = 91 \cdot (1 - 0.35 \cdot 0.9) = 62\,ml \cdot g^{-1}$$

Using Eq. (8.10c and d):

$$k = 0.16 + 0.003 \cdot I_{ssv} = 0.351 \cdot g^{-1}$$

$$v_0 = 16 - 0.1 \cdot I_{ssv} = 9.8\,m \cdot h^{-1}$$

8.2 MODEL FOR SETTLING IN A CONTINUOUS SETTLER

The term "clarifier" or "secondary clarifier" only represents one of the two main processes taking place in the settler. Thickening is the second important process. Therefore throughout this text the term "settler" has been used. Final settlers in activated sludge processes operate under conditions of continuous flow and load. Mixed liquor flows from the aeration tank to the final settler while return sludge, containing the settled and concentrated sludge, is pumped back to the aeration tank. The clarified effluent flow (equal to the wastewater influent flow when the excess sludge production is ignored) is discharged from the system.

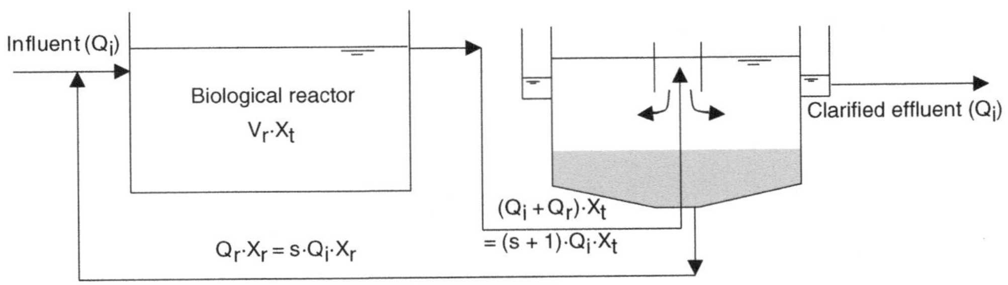

Figure 8.4 Schematic representation of an activated sludge process with a final settler

Figure 8.4 shows a schematic representation of a final settler with the incoming and outgoing flows. In order to describe settling in a continuous final settler, the following assumptions and approximations are made:

(1) The flow entering the final settler is the sum of the wastewater flow and the return sludge flow. The incoming flux of solids can be expressed as $(Q_i + Q_r) \cdot X_t$, while the outgoing flux is equal to

$Q_r \cdot X_r$. Assuming that no sludge accumulation takes place in the final settler, which is justified under steady state conditions, the incoming and outgoing solids flux are equal, so that:

$$(Q_i + Q_r) \cdot X_t = Q_r \cdot X_r \quad \text{or} \quad X_r = X_t \cdot (s + 1)/s \tag{8.11}$$

where:

X_r = return sludge concentration
s = recirculation factor = Q_r/Q_i

(2) The incoming flow $(Q_i + Q_r)$ is distributed uniformly over the cross sectional area at a certain inlet level. The flow direction in the final settler is vertical. In the supernatant region above the level of incoming solids, the liquid is free of solids and rises to the effluent discharge level at the top of the final settler. This upward velocity is called the hydraulic loading rate and can be expressed as:

$$T_s = Q_i/A_d \tag{8.12}$$

where:

T_s = hydraulic loading rate (upward velocity of the supernatant)
A_d = cross sectional area of the final settler

(3) Below the level of the incoming sludge a suspension is formed that flows in a downward direction to the return sludge discharge at the bottom of the final settler. The downward velocity of the liquid phase is given by:

$$u = Q_r/A_d = s \cdot Q_i/A_d \tag{8.13}$$

where u = downward velocity of the liquid phase due to the return sludge flow, or sludge abstraction rate

(4) In the lower part of the final settler, the solids have a settling velocity in addition to the downward velocity of the liquid phase, which means that they move downwards in the liquid phase. The settling velocity is given by Vesilind's equation (Eq. 8.1).

(5) The displacement velocity of the solids is given by the sum of the liquid velocity "u" and the settling velocity "v". Hence the solids flux, defined as the solids mass passing per unit area and per time unit at a certain level in the final settler can be expressed as:

$$\begin{aligned} F &= X \cdot (v + u) \\ &= F_v + F_u \\ &= X \cdot (v_0 \cdot \exp(-k \cdot X) + s \cdot Q_i/A_d) \end{aligned} \tag{8.14}$$

where:

F = solids flux passing at a particular level in the final settler
X = suspended solids concentration at a certain level in the final settler
F_v = solids flux due to settling
F_u = solids flux due to return sludge abstraction

(6) The solids loading rate is defined as the mass of suspended sludge solids entering the final settler per unit settler area and per unit time:

$$F_{sol} = X_t \cdot (s + 1) \cdot Q_i / A_d \tag{8.15}$$

where F_{sol} = solids loading rate

(7) An essential condition for the final settler to perform properly is that the sludge loading rate does not exceed the solids flux at any level in the settler. If this condition is obeyed, all solids are transported from the feeding point to the abstraction point and no accumulation of solids will occur in the final settler. However if at some level between the inlet and the abstraction point the solids loading rate exceeds the solids flux, then at that level solids will accumulate at a rate equal to the difference between the solids loading rate and the solids flux. Eventually the final settler will be completely filled with solids resulting in the discharge of sludge together with the effluent. The basic condition for a final settler to perform properly can therefore be expressed as:

$$F = F_v + F_u > F_{sol} \quad \text{for } X_t < X < X_r \tag{8.16}$$

Equation (8.16) forms the basis for final settler design. To evaluate the behaviour of the solids flux F, the components F_v and F_u are calculated. In Figure 8.5a the solids flux due to settling F_v is shown plotted as a function of the solids concentration X. It is assumed that Vesilind's equation applies.

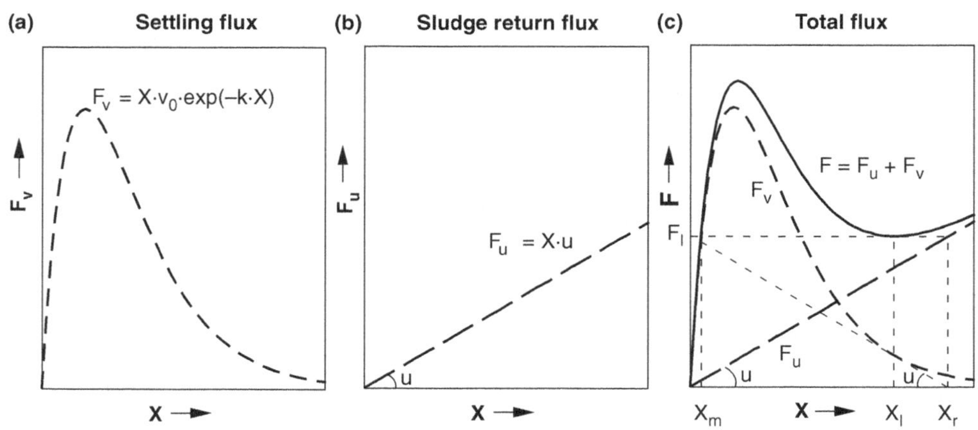

Figure 8.5 The solids flux due to settling (a), due to return sludge flow abstraction (b) and the resulting total flux (c) in a continuous settler as function of the sludge concentration

Figure 8.5b shows the solids flux due to return sludge abstraction F_u, also as a function of the sludge concentration. In Figure 8.5c the resulting total flux F is plotted. For the selected values of the sludge concentrations at the inlet point of the final settler and of the return sludge at the outlet (X_t and X_r respectively, Figure 8.5c), the curve F has a relative minimum F_l, for a sludge concentration X_l at some intermediate value between X_t and X_r. The flux F_l limits the maximum solids transport to the abstraction point in the settler and for that reason is called the limiting flux. The corresponding sludge concentration X_l is called the limiting concentration. Figure 8.5c also shows how the limiting sludge concentration can be determined on the basis of geometry when the return sludge concentration X_r and the batch settling curve F_v (Figure 8.5a) are known. Figure 8.5c presents the method developed by Yoshioka et al. (1957):

(1) Draw a straight line tangential to the batch settling curve F_v, passing through the point X_r at the horizontal axis
(2) The limiting flux corresponding to the chosen X_r value is found as the intersection of the straight line and the vertical axis.

From Figure 8.5c it is quite clear that the value of the limiting flux depends directly on the return sludge concentration. Therefore the limiting flux, i.e. the maximum flux that can be transported in the final settler, is determined by the thickening function of the settler through which the return sludge concentration is produced. Furthermore, Figure 8.5c shows that in the case of an inlet sludge concentration X_t greater than the limiting concentration X_l, the resulting flux curve F is a function that increases proportional to the increase in sludge concentration in the range from X_t to X_r. Hence in this case, the largest flux that can be transported through the final settler is equal to the inlet flux with concentration X_t and is independent on the outlet concentration X_r. When the inlet concentration is smaller than a particular minimum concentration X_m, the flux related to the inlet concentration will be smaller than the limiting flux and hence limits the solids transport in the final settler. It is concluded that when the inlet concentration X_t is greater than the limiting concentration X_l or smaller than the minimum concentration X_m, the flux related to the inlet sludge concentration is the maximum flux that can be transported through the settler. This maximum flux is determined by the clarification function of the final settler.

It can be observed in Figure 8.5c that Yoshioka's method to determine the limiting flux is only applicable if it is possible to draw a tangent line to the concave part of the batch settling curve F_v. There is a critical concentration X_c such that, for any return sludge concentration $X_r < X_c$, it is not possible to draw this tangent line and consequently the limiting flux and the limiting concentration do not exist. The tangential line passing through the critical concentration point at the horizontal axis intersects the curve F_v at the point where its gradient is maximum. This occurs at the inflection point of this curve (F_i, X_i), a situation that is shown in Figure 8.6. It can be noted that the limiting flux has its maximum value when the return sludge concentration is equal to the critical concentration $X_r = X_c$. In this situation, the downward liquid velocity is also maximum. The observations above on the solids flux curve F and its components F_v and F_u can be summarised as follows:

- The maximum flux that can be transported depends either on the inlet concentration (equal to the mixed liquor concentration) or on the outlet (return sludge) concentration;
- In the first case, clarification is the limiting function of the final settler and consequently will determine settler design;
- In the second case, sludge thickening is the limiting function and the criteria for thickening will determine final settler design;

- Thickening is limiting when the inlet concentration has a value between the minimum concentration X_m and the limiting concentration X_l and when the outlet concentration X_r is greater than the critical concentration X_c;
- In all other cases the limiting function of the settler will be clarification.

In order to establish if sludge settling will be determined by clarification or by thickening, it is necessary to derive expressions for the concentrations X_l, X_m and X_c. In the following sections these expressions will be derived with the aid of Vesilind's equation.

8.2.1 Determination of the limiting concentration X_l

In Figure 8.5c the straight line tangent to the curve F_v and passing through X_l can be written as:

$$F = m \cdot (X - X_r) \tag{8.17}$$

where m = gradient of the straight line $= (dF/dX)_{x=xi} = v_0 \cdot (1 - k \cdot X_l) \cdot \exp(-k \cdot X_l)$

At the tangential point, the value of curve F_v is equal to the value of the straight line so that:

$$F = (X_l - X_r) \cdot v_0 \cdot (1 - k \cdot X_l) \cdot \exp(-k \cdot X_l) = X_l \cdot v_0 \cdot \exp(-k \cdot X_l) \quad \text{or}$$
$$X_l = (X_r/2) \cdot [1 + (1 - 4/(k \cdot X_r))^{0.5}] \tag{8.18}$$

Now the limiting flux can be determined as the intersection of the vertical axis with the straight line of Eq. (8.17) for X = 0 (see Figure 8.5c):

$$F_l = m \cdot X_r = X_r \cdot v_0 \cdot (k \cdot X_l - 1) \cdot \exp(-k \cdot X_l) \tag{8.19}$$

where X_l is given by Eq. (8.18). Using Figure 8.5c, the downward velocity of the liquid phase in the lower part of the settler is given by:

$$u = F_l/X_r = v_0 \cdot (k \cdot X_l - 1) \cdot \exp(-k \cdot X_l) \tag{8.20}$$

8.2.2 Determination of the critical concentration X_c

The tangential straight line passing through the point X_c at the horizontal axis is also described by Eq. (8.17), but the gradient of the line is now maximum as in Figure 8.6. Hence:

$$(dm/dX) = (d^2F_v/dX^2)_{x=xi} = 0 \quad \text{or} \quad X_i = 2/k \text{ and } F_i = 2 \cdot v_0/(k \cdot e^2) \tag{8.21}$$

where:

- X_i = sludge concentration at the inflection point of curve F_v
- F_i = batch settling flux at the inflection point of F_v
- e = natural logarithm basis ≈ 2.71

The gradient of the straight line through (X_i, F_i) and $(X_c, 0)$ is equal to the derivative of F_v at the coordinate (X_i, F_i):

$$m = (dF_v/dX)_{x=2/k} = -v_0/e^2 \tag{8.22}$$

Hence, the straight line is given by:

$$F - 2 \cdot v_0/(k \cdot e^2) = -v_0/e^2 \cdot (X - 2/k) \quad \text{or} \quad F = -(v_0/e^2) \cdot (X - 4/k) \tag{8.23}$$

Now, the critical concentration can be determined, knowing that $F = 0$ when $X = X_c$:

$$X_c = 4/k \tag{8.24}$$

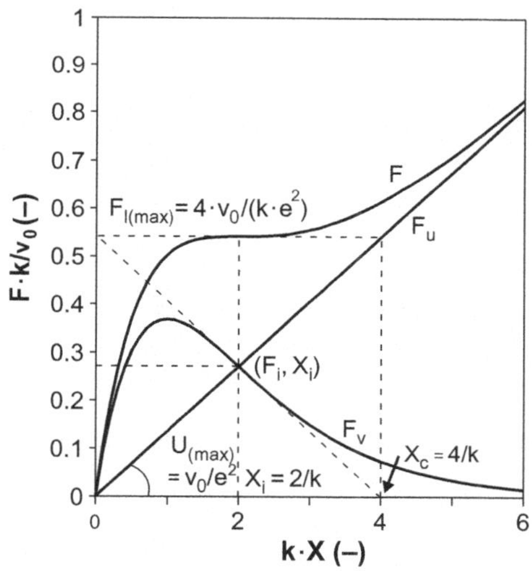

Figure 8.6 F_v and F_u curves as function of the sludge concentration for $X_r = X_c = 4/k$ (the coordinates are dimensionless)

8.2.3 Determination of the minimum concentration X_m

The minimum concentration X_m is determined by the condition that for this concentration the solids flux is equal to the limiting flux F_l (see Figure 8.5c). Hence:

$$F_{X=Xm} = (F_v + F_u)_{X=Xm} = F_l \tag{8.25}$$

After substituting Eqs. (8.14 and 8.19) in Eq. (8.25) and rearranging one has:

$$X_m \cdot \exp(-k \cdot X_m) = (X_r - X_m) \cdot (k \cdot X_l - 1) \cdot \exp(-k \cdot X_l) \tag{8.26}$$

Equation (8.26) does not have an analytical solution, but the values of X_m and X_l can be calculated as a function of X_r for any k-value using numerical methods. To represent the X_l and X_m graphically, it is convenient to construct an dimension less diagram, using $k \cdot X_r$ at the horizontal axis and $k \cdot X_l$ and $k \cdot X_m$ at the vertical axis. Figure 8.7 shows the $k \cdot X_l$ and $k \cdot X_m$ values as a function of $k \cdot X_r$, calculated with the aid of Eqs. (8.18 and 8.25) respectively. The value $k \cdot X_c = 4$ is also indicated on the horizontal axis. For $k \cdot X_c = 4$ one has $k \cdot X_l = k \cdot X_m = 2$, i.e. one has the critical sludge concentration displayed in Figure 8.6.

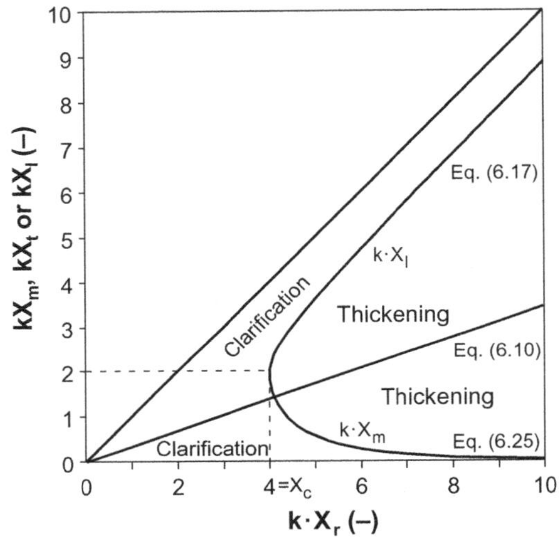

Figure 8.7 $k \cdot X_l$ and $k \cdot X_m$ values as function of $k \cdot X_r$ indicating whether clarification or thickening is the limiting process

Figure 8.7 has considerable practical utility: for any "pair" of inlet- and outlet concentrations X_t and X_r of a final settler, it can immediately be determined which of the two functions of the settler is limiting: clarification or thickening. For $k \cdot X_r > k \cdot X_c = 4$ and $k \cdot X_m < k \cdot X_t < k \cdot X_l$, the limiting process is thickening. For all other cases the limiting process will be clarification. In Section 8.3 it will prove convenient to relate the inlet and outlet sludge concentrations X_t and X_r. This relationship, expressed in Eq. (8.11), is also indicated in Figure 8.7.

EXAMPLE 8.3

A final settler with a surface area of 300 m^2 receives a mixed liquor flow of 300 $m^3 \cdot h^{-1}$ with a sludge concentration of $6\ g \cdot l^{-1}$. The Vesilind constants are $k = 0.4\ l \cdot g^{-1}$ and $v_0 = 8\ m \cdot h^{-1}$, while a recirculation factor $s = 1$ is being applied. Demonstrate that the final settler is under loaded.

Solution

For $X_t = 6\,g \cdot l^{-1}$ and $s = 1$ one has $X_r = X_t \cdot (s + 1)/s = 6 \cdot 2 = 12\,g\ TSS \cdot l^{-1}$

$$X_l = (X_r/2) \cdot [1 + (1 - 4/(k \cdot X_r))^{0.5}] = 12/2 \cdot [1 + (1 - 4/(0.4 \cdot 12))^{0.5}] = 8.45\,g\ TSS \cdot l^{-1}$$

The limiting flux is given by:

$$F_l = X_r \cdot v_o \cdot (k \cdot X_l - 1) \cdot \exp(-k \cdot X_l) = 12 \cdot 8 \cdot (0.4 \cdot 8.45 - 1) \cdot \exp(-0.4 \cdot 8.45) = 7.8\,kg\ TSS \cdot m^{-2} \cdot h^{-1}$$

The limiting flux can also be determined using a graphical analysis, using Yoshiaka's method. In Figure 8.8 the curve of $F_v = X \cdot v_o \cdot \exp(-k \cdot X)$ is plotted for the values given in the example. A line can be drawn tangential to the curve, starting from the value of $X_{rs} = 12\ g \cdot l^{-1}$. It can be observed that F_l is indeed equal to $7.8\ kg \cdot m^{-2} \cdot h^{-1}$. The value of $X_c = 4/k = 4/0.4 = 10\ g\ TSS \cdot l^{-1}$ is indicated as well. The applied solids loading rate is:

$$F_{sol} = (s + 1) \cdot X_t \cdot Q_i/A = 300 \cdot 6/300 = 6.0\,kg\ TSS \cdot m^{-2} \cdot h^{-1}.$$

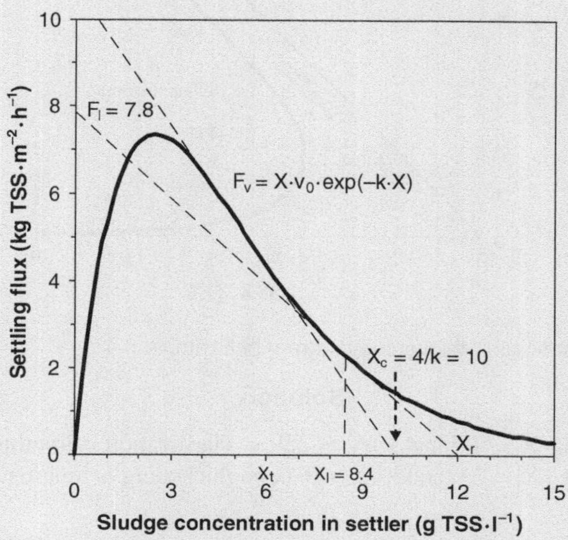

Figure 8.8 Determination of limiting solids flux according to Yoshioka's method in Example 8.3

It is clear that the applied solids loading rate ($6.0\ kg\ TSS \cdot m^{-2} \cdot h^{-1}$) is significantly less than the limiting flux ($7.8\ kg\ TSS \cdot m^{-2} \cdot h^{-1}$). Therefore the final settler is not receiving the maximum solids loading rate. Static point analysis (discussed in Section 8.5) can be used to determine how much the solids load can be increased without overloading the final settler.

EXAMPLE 8.4

Determine in Figure 8.9 whether clarification or thickening is limiting for the two cases listed below:

Case A:

- $X_t = 4$ and $X_r = 8\ g \cdot l^{-1}$ ($s = 1$);
- Fair settleability: $k = 0.365\ l \cdot g^{-1}$ and $v_0 = 9\ m \cdot h^{-1}$.

Case B:

- $X_t = 4.35$ and $X_r = 13.1\ g \cdot l^{-1}$ ($s = 0.5$);
- Poor settleability: $k = 0.46\ l \cdot g^{-1}$ and $v_0 = 6\ m \cdot h^{-1}$.

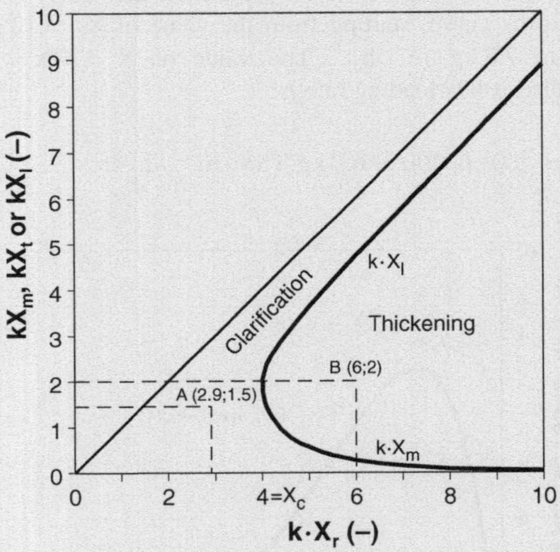

Figure 8.9 Use of the clarification-thickening diagram in Example 8.4

Solution

Case A: coordinates are $k \cdot X_t = 1.5$ and $k \cdot X_r = 2.9$, so clarification is limiting.
Case B: coordinates are $k \cdot X_t = 2.0$ and $k \cdot X_r = 6.0$, so thickening is limiting.

8.3 DESIGN OF FINAL SETTLERS

8.3.1 Optimised design procedure for final settlers

Equation (8.16) forms the basis of secondary settler design. It expresses that the solids loading rate must never exceed the largest flux that can be transported through the final settler. This equation is valid for both clarification and thickening. Using Eq. (8.16) as a starting point, expressions will be derived for the maximum hydraulic loading rate that can be applied as function of the sludge concentration. This is the most important parameter for settler design, because it reflects the ratio between the influent flow rate and the cross-sectional area.

(a) Clarification

When clarification is the limiting process, Eq. (8.16) can be described as:

$$F = (F_v + F_u)_{X=Xt} = F_{sol} \qquad (8.27)$$

or using Eq. (8.14)

$$X_t \cdot (v_0 \cdot \exp(-k \cdot X_t) + s \cdot Q_i/A_{dmin}) = X_t \cdot (s+1) \cdot Q_i/A_{dmin}$$

Now, by applying the definition of hydraulic loading rate in Eq. (8.12) and rearranging, it is possible to write the maximum hydraulic rate explicitly as a function of X_t:

$$\ln(T_{sm}/v_0) = -k \cdot X_t \quad \text{or}$$
$$T_{sm} = Q_i/A_{dmin} = v_0 \cdot \exp(-k \cdot X_t) \quad \text{or} \quad A_{dmin} = Q_i/v_0 \cdot \exp(k \cdot X_t) \qquad (8.28)$$

where:

T_{sm} = maximum hydraulic loading rate
A_{dmin} = minimum cross sectional settler area

Equation (8.28) shows that the maximum hydraulic loading rate is proportional to the constant v_0 and has an inverse exponential relationship with the constant k and the suspended solids concentration in the inlet of the final settler X_t. It can also be noted that T_{sm} is independent of the recirculation factor "s" and the return sludge concentration X_r.

(b) Thickening

Applying Eq. (8.16) to the situation when thickening is the limiting process one has:

$$F_l = X_r \cdot v_0 \cdot (k \cdot X_l - 1) \cdot \exp(-k \cdot X_l) = F_{sol} = X_t \cdot (s+1) \cdot Q_i/A_{dmin} \qquad (8.29)$$

The corresponding maximum hydraulic loading rate is given by:

$$\ln(T_{sm}/v_0) = \ln((k \cdot X_l - 1)/s) - k \cdot X_l \quad \text{or}$$
$$T_{sm} = [v_0 \cdot (k \cdot X_l - 1)/s] \cdot \exp(-k \cdot X_l) \qquad (8.30)$$

Equation (8.30) shows that the maximum hydraulic loading rate in the case of thickening is proportional to the constant v_0 and a complex function of the constant k, the recirculation factor "s" and the return sludge concentration X_r. In this case the T_{sm} value does not depend on the inlet concentration X_t. With the aid of Eqs. (8.28 and 8.30) and Figure 8.7 it becomes a simple matter to calculate the maximum hydraulic loading rate of an activated sludge settler for any pair of inlet and outlet concentrations, X_t and X_r, provided the values of the constants k and v_0 are known.

- For the selected values of $k \cdot X_t$ and $k \cdot X_r$, determine in Figure 8.7 if the limiting criterion for design is clarification or thickening;
- Use Eq. (8.28) for clarification or Eq. (8.30) for thickening to determine the maximum hydraulic loading rate.

EXAMPLE 8.5

Considering again the cases of Example 8.4, answer the following questions:

- What will be the value of the maximum allowable hydraulic loading rate T_{sm} that can be applied in Case B?
- Determine the performance for Case B, when s is increased so that clarification becomes the limiting process;
- Check whether for this "critical" recirculation rate it is true that both thickening and clarification are now the limiting process.

Solution

As determined in Example 8.9, in Case B the limiting process is thickening. Use Eq. (8.16) to calculate the limiting concentration:

$$X_l = (X_r/2) \cdot [1 + (1 - 4/(k \cdot X_r))^{0.5}]$$
$$= (13/2) \cdot [1 + (1 - 4/(0.46 \cdot 13))^{0.5}] = 10.3\ g \cdot l^{-1}$$

Use Eq. (8.30) to calculate the maximum allowable hydraulic loading rate:

$$T_{sm} = [v_0 \cdot (k \cdot X_l - 1)/s] \cdot \exp(-k \cdot X_l)$$
$$= [6 \cdot (0.46 \cdot 10.3 - 1)/0.5] \cdot \exp(-0.46 \cdot 10.3) = 0.4\ m \cdot h^{-1}$$

From Figure 8.9 it can be determined that clarification becomes the limiting process when the value of $k \cdot X_r$ is reduced from 6 to 4. For $k = 0.46 \rightarrow X_r = 4/0.46 = 8.7\ g \cdot l^{-1}$ and $s = X_t/(X_r - X_t) = 4.35/(8.7 - 4.35) = 1$. Now calculate T_{sm} using the formula for clarification (Eq. 8.28):

$$T_{sm} = v_0 \cdot \exp(-k \cdot X_t)$$
$$= 6 \cdot \exp(-0.46 \cdot 4.35) = 0.8\ m \cdot h^{-1}$$

It can be observed that in this particular case an increase of recirculation factor s from 0.5 to 1.0 has resulted in an increase of the value of T_{sm} with a factor two. However, a further increase of s will have no effect as clarification is already the limiting process, in which case the value of T_{sm} is independent of the value of s.

If it is indeed true that for this "critical" recirculation rate $s_c = s = 1.0$ both clarification and settling are limiting, then the calculated value of T_{sm} should also be equal to $0.8\ m \cdot h^{-1}$ when the equations for thickening are used:

$$X_l = (X_r/2) \cdot [1 + (1 - 4/(k \cdot X_r))^{0.5}]$$
$$= (8.7/2) \cdot [1 + (1 - 4/(0.46 \cdot 8.7))^{0.5}] = 4.3\ g \cdot l^{-1}$$

$$T_{sm} = [v_0 \cdot (k \cdot X_l - 1)/s] \cdot \exp(-k \cdot X_l)$$
$$= [6 \cdot (0.46 \cdot 4.3 - 1)/1.0] \cdot \exp(-0.46 \cdot 4.3) = 0.8\ m \cdot h^{-1}$$

So indeed for $s = s_c$ both clarification and thickening are the limiting process.

(c) Design procedure for final settlers

The first step in the final settler design procedure is to define the appropriate value of the design influent flow rate Q_i. In the case of a design according to the solids flux theory the average flow rate is often selected. As the final settler will be high (4 m < H_d < 6 m), significant buffer capacity will be available for temporary storage of sludge under conditions of peak flow. In Section 8.5 the static point analysis will be presented, which can be used to evaluate the performance of the final settler under variable operational conditions. The estimated time to final settler failure, defined as a sludge blanket level of less than 0.3–0.5 m below the level of the overflow weir, is compared to the predicted peak flow duration. If the design is not considered adequate, then a larger design flow should be selected.

The next step is to attribute values to the Vesilind parameters k and v_0. If the Vesilind characteristics are known, these values can be adopted. If not, it is recommended to select those for poor settleability: $k = 0.46\ l \cdot g^{-1}$ and $v_0 = 6\ m \cdot h^{-1}$. Remember that settling characteristics of sludge are not a "fixed" quality and tend to vary in time as a result of operational conditions.

Select the target inlet (X_t) and return sludge concentration (X_r) or alternatively select X_t and s which will then define X_r according to Eq. (8.11): $X_r = X_t \cdot (s + 1)/s$. Now determine in Figure 8.7 whether clarification or thickening is the limiting process in the final settler, as demonstrated in Example 8.4. Calculate the maximum hydraulic loading rate T_{sm} that can be applied: use Eq. (8.28) when clarification is limiting and Eq. (8.30) when thickening is limiting. In the latter case, use Eq. (8.18) to calculate the limiting sludge concentration.

It is important to consider that the maximum hydraulic loading rate determined by Eqs. (8.28 or 8.30) is a *theoretical* value based on a mathematical model, which is based on a number of assumptions (refer to Section 8.3), which in practice may not always be realistic. Therefore the maximum hydraulic loading rate that can be applied in practice will always be less than the calculated T_{sm} value. For example, a correction must be made for the fact that not all of the settler volume is effectively utilised for liquid-solid separation: part of it is a stagnant zone, so that the hydraulic loading rate in the effective surface area is higher than the ratio Q_i/A_{dmin}. Therefore the surface area must be adjusted by a safety factor to be able to handle the hydraulic loading rate under actual operational conditions.

The value of the safety factor depends on the size of the stagnant volume fraction, resulting from the non-ideality of the settler. In practice the dead volume fraction is often in the range of 30 to 40%. Taking into consideration that there are other adverse conditions as well (wind effects, density currents due to development of temperature gradients and to the downward flux of settling solids), it is concluded that a safety factor s_{fd} of 1.5 to 2.0 should be considered in design.

The determination of the maximum allowable hydraulic loading rate of the final settler is the most important part of the design. Once the value of T_{sm} is established, the final settler design is completed as follows:

(1) Establish a suitable safety factor to guarantee proper operation of the final settler under adverse conditions. If it is not possible to determine the value of s_{fd} experimentally (which will be demonstrated in Example 8.7), then a default value is selected, for instance $s_{fd} = 2$;

(2) For the applied safety factor the required cross-sectional settler area is calculated;

$$A_{dmin} = s_{fd} \cdot Q_i/T_{sm} \tag{8.31}$$

(3) An adequate depth (in practice generally around 4–6 m) is selected and the final settler volume is calculated:

$$V_d = A_{dmin} \cdot H_d = s_{fd} \cdot H_d \cdot Q_i/T_{sm} \tag{8.32}$$

From Eq. (8.32) the final settler volume required per unit influent flow can be derived as well:

$$v_d = V_d/Q_i = s_{fd} \cdot H_d/T_{sm} = s_{fd} \cdot (H_d/v_O)/(T_{sm}/v_O) \tag{8.33}$$

Equation (8.33) shows that it is possible to calculate the required settler volume per unit influent flow (v_d) if the values of s_{fd}, H_d, v_O and T_{sm} are known. Equation (8.33) can be rearranged as:

$$\ln(v_d) = \ln(s_{fd} \cdot H_d/v_O) - \ln(T_{sm}/v_O) = \ln(s_{fd} \cdot H_d/v_O) + k \cdot X_t \text{ (clarification)} \quad \text{or} \tag{8.34}$$

$$\ln(v_d) = \ln(s_{fd} \cdot H_d/v_o) + k \cdot X_l - \ln((k \cdot X_l - 1)/s) \text{ (thickening)} \tag{8.35}$$

The v_d value is equal to the *hydraulic* retention time of the liquid in the settler. In practice, this retention time is subjected to an upper and a lower limit. The lower limit is imposed by the hydraulics of the settler: if the *actual* retention time (or contact time) is shorter than one hour, the efficiency of solid-liquid separation tends to be poor due to excessive turbulence. On the other hand, a very long retention time in the final settler may lead to denitrification with the consequential formation of a floating sludge layer. This may also induce growth of filamentous organisms, which are responsible for poor settling behaviour of the sludge. Thus the contact time in a final settler is usually not longer than about three hours. Taking these limits into consideration one has (when Q_i is expressed in $m^3 \cdot d^{-1}$):

$$1\text{h} < 24 \cdot V_d/[(s+1) \cdot Q_i] < 3\text{h} \quad \text{or} \quad (s+1) < 24 \cdot v_d < 3 \cdot (s+1) \tag{8.36}$$

If the v_d value found by Eqs. (8.33 to 8.35) is not within the range set by Eq. (8.36), other values must be selected for X_t and/or X_r.

8.3.2 Determination of the critical recirculation rate

Eqs. (8.28 or 8.30) show that the hydraulic loading rate T_{sm} decreases with increasing sludge concentration X_t. This increase is exponential in the case of clarification and even more accentuated in the case of thickening. As the increase of the required settler volume with increasing sludge concentration is so rapid, in principle it is not advantageous to have thickening as the limiting function of the final settler.

The minimum surface area of the final settler can be reduced if the value of the recirculation factor is increased until clarification becomes the limiting function of the final settler, as shown in Example 8.5. Furthermore, Eq. (8.28) shows that the hydraulic loading rate (and hence the final settler volume) is independent of the recirculation factor when clarification is the limiting process. Therefore, in principle one will choose the minimal recirculation factor required for clarification, as a further increase will no longer yield a decrease in required surface area. This minimum recirculation factor for clarification is called the critical recirculation factor s_c.

The value of the critical recirculation factor can be determined conveniently using Figure 8.7, where the straight line represents the inlet sludge concentration X_t as a function of the return sludge concentration X_r. In conformity with Eq. (8.11), the critical recirculation factor can now be calculated by intersecting the straight line with the curve for X_l or X_m as a function of X_r. It can also be observed in Figure 8.7 that for recirculation factor $s < 1$, the straight line of X_t intersects with X_m, whereas for $s > 1$ the intersection is with X_l. Hence two cases exist:

$$X_t = s_c/(s_c + 1) \cdot X_r = X_l = (X_r/2) \cdot [1 + (1 - 4/(k \cdot X_r))^{0.5}] \quad \text{for } s > 1 \qquad (8.37a)$$

$$X_t = s_c/(s_c + 1) \cdot X_r = X_m \quad \text{for } s < 1 \qquad (8.37b)$$

In Eq. (8.37b) the X_m value is given by Eq. (8.26). Equation (8.37a) can be solved analytically:

$$k \cdot X_r = (s_c + 1)^2/s_c \ (\text{for } s_c > 1) \qquad (8.38a)$$

$$k \cdot X_t = (s_c + 1) \ (\text{for } s_c > 1) \qquad (8.38b)$$

Equation (8.37b) does not have an analytical solution, but can be solved numerically. In Figure 8.10 the critical recirculation factor is shown as function of the adimensional unit $k \cdot X$ which is very useful when the values of $k \cdot X_t$ and $k \cdot X_r$ need to be determined for a particular s_c. For example, when $s_c = 0.5$ it can be determined in Figure 8.10 that $k \cdot X_t = 1.37$ and $k \cdot X_r = 4.11$ g $\cdot$ l^{-1}. It can be verified that effectively $k \cdot X_t = s_c/(s_c + 1) \cdot k \cdot X_r = 0.5/1.5 \cdot 4.11 = 1.37$.

kX_t	s_c	kX_t	s_c
2.0	1.00	1.0	0.30
1.9	0.90	0.9	0.25
1.8	0.81	0.8	0.21
1.7	0.73	0.7	0.18
1.6	0.66	0.6	0.14
1.5	0.59	0.5	0.11
1.4	0.52	0.4	0.08
1.3	0.46	0.3	0.06
1.2	0.40	0.2	0.03
1.1	0.35	0.1	0.01

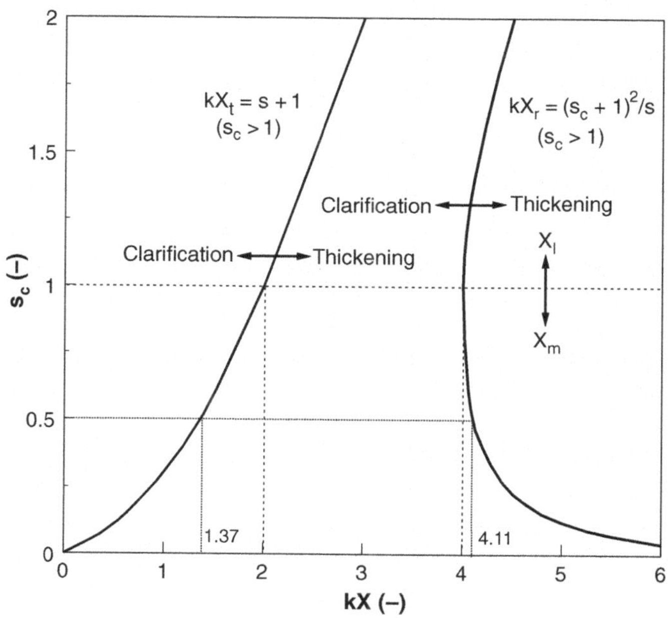

Figure 8.10 Relationship between $k \cdot X_t$ and $k \cdot X_r$ and the critical recirculation factor s_c

EXAMPLE 8.6

Design a final settler according to the solids flux theory presented in the previous sections, based on the following data:

- $Q_i = 500\,m^3 \cdot h^{-1}$; $f_{av} = 0.4$ and $X_t = 4\,kg\,TSS \cdot m^{-3}$;
- $s_{fd} = 2$ and $H_d = 4\,m$.

Solution

Estimate the values of the Vesilind constants with the empirical equations of Section 8.1.3:

$$\begin{aligned} I_{ssv} &= 25 + 25 \cdot f_{av} + 5 \cdot X_t \\ &= 25 + 25 \cdot 0.3 + 5 \cdot 4 = 55\,ml \cdot g^{-1}\,TSS \end{aligned} \tag{8.10a}$$

$$\begin{aligned} k &= 0.16 + 0.003 \cdot I_{ssv} \\ &= 0.16 + 0.003 \cdot 55 = 0.331 \cdot g^{-1} TSS \end{aligned} \tag{8.10c}$$

$$\begin{aligned} v_0 &= 16 - 0.1 \cdot I_{ssv} \\ &= 16 - 0.1 \cdot 55 = 10.5\,m \cdot h^{-1} \end{aligned} \tag{8.10d}$$

Calculate the volume of the final settler, assuming clarification is limiting ($s = s_c$):

$$\begin{aligned} T_{sm} &= v_0 \cdot \exp(-k \cdot X_t) \\ &= 10.5 \cdot \exp(-0.33 \cdot 4) = 2.86\,m \cdot h^{-1} \end{aligned} \tag{8.28}$$

The minimum required surface area of the final settler can be now calculated, taking into account the value of the safety factor s_{fd}:

$$\begin{aligned} A_{dmin} &= s_{fd} \cdot Q_i / T_{sm} \\ &= 2 \cdot 500/2.86 = 349\,m^2 \end{aligned} \tag{8.31}$$

$$V_d = 349 \cdot 4 = 1398\,m^3$$

Use Figure 8.10 to determine the minimum required value of the sludge recycle factor, i.e. the critical recirculation factor s_c: $k \cdot X_t = 0.33 \cdot 4 = 1.3 \rightarrow s_c = 0.46$. In practice a larger value of s may be adopted, but this will not influence the design of the final settler as clarification is already limiting. Now all that remains is to check whether the retention time in the final settler is acceptable. Either hydraulic or actual retention time can be used, using Eq. (8.36). In the case of the hydraulic retention time ($v_d = 1398/500 = 2.8$ hrs):

$$(s + 1) < v_d < 3 \cdot (s + 1) -> 1.46 < 2.8 < 4.38 (=OK)$$

In the case of the actual liquid retention time or contact time: $1\,hr < 24 \cdot V_d/[(s + 1) \cdot Q_i] < 3\,hr$. The actual retention time $= 1398/(1+0.46) \cdot 500 = 1.9$ hr. As the retention time in the final settler is between the recommended limits of 1 to 3 hr, the design is acceptable. Note that in practice probably more conservative settling characteristics will be used for the final settler design than the values used in this example.

8.3.3 Graphical optimization of final settler operation

Figure 8.11 shows the T_{sm} values as a function of the inlet suspended solids concentration for several values of the recirculation factor. The curves were calculated with the aid of Eqs. (8.28 and 8.30) for clarification and thickening respectively. For convenience an non dimensional presentation was used with $k \cdot X_t$ on the horizontal axis and the natural logarithm of the v_0/T_{sm} ratio on the vertical axis. If the constants k and v_0 are known, Figure 8.11 allows an immediate determination of T_{sm} for any value of the mixed liquor suspended solids concentration and for different values or the return sludge factor "s".

The diagram may be interpreted in the following way: the straight line corresponding to clarification (Eq. 8.28) divides Figure 8.11 in two parts:

- One part (the lower triangle) is characterised by the condition that $\ln(v_0/T_s) < k \cdot X_t$. The operational conditions are inadequate and the settler cannot function due to an excessive solids load and/or hydraulic load;
- In the other part of the diagram (the upper triangle), where $\ln(v_0/T_s) > k \cdot X_t$, the settler will function if the values of the operational variables are adequate.

Figure 8.11 also shows that there are three variables that influence final settler performance: the sludge concentration X_t, the hydraulic loading rate T_s and the sludge recycle factor s. In most existing plants the values of the operational variables X_t and T_s are not determined by considerations regarding optimisation of the final settler: the sludge mass (and hence its concentration X_t) is determined by the sludge age, whereas the hydraulic loading rate T_s is given by the ratio of the influent flow and the surface area of the final settler. Hence when the final settler is already constructed, in many cases only one variable can be selected by the operator: the sludge recycle factor.

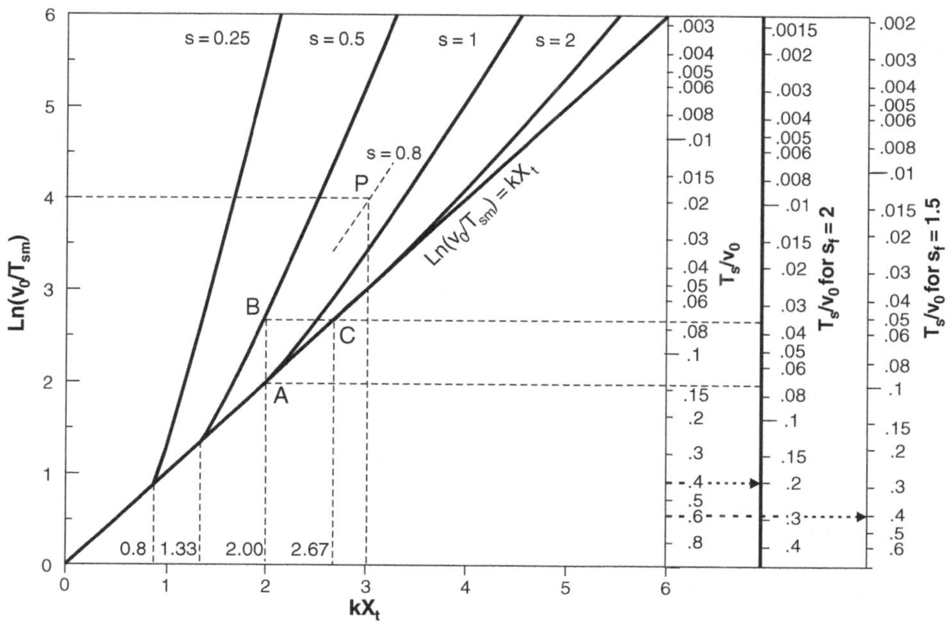

Figure 8.11 $\ln(v_0/T_{sm})$ ratio as a function of the adimensional product $k \cdot X_t$ for different values of the recirculation factor s

The appropriate value of this factor can easily be identified: for any given combination of values of the variables X_t and T_s calculate the corresponding values $k \cdot X_t$ and $\ln(v_0/T_s)$. These two values are the coordinates of a point in Figure 8.11. The necessary recycle factor can now be determined graphically by interpolation of the thickening curves for different values of "s", such that the curve passes through the intersection point P.

To account for dead volume, density currents etc., in Figure 8.11 the values of T_s/v_0 are also indicated with a safety factor of $s_{fd} = 2$ (right hand scale). The scale for $s_f = 2$ is produced by displacing the ordinate numeric values by a factor 2: if the scale value is 0.8 for $s_{fd} = 1$ (first value of the ordinate scale for $s_{fd} = 1$) then at the same level there will be a value of $0.8/2 = 0.4$ for $s_{fd} = 2$. Similarly the scale can be produced for any desired s_{fd} value, by sliding the ordinate scale downwards by a factor such that the numeric value N on the $s_{fd} = 1$ scale becomes N/s_{fd} on the scale with a safety factor s_{fd}. Figure 8.11 again demonstrates that the hydraulic loading rate that can be applied on a final settler depends of the following factors:

- Values of the settleability constants k and v_0;
- Sludge concentration of the mixed liquor in the inlet to the settler;
- Value of the sludge recycle factor s (if thickening is the limiting function);
- Value of the safety factor.

EXAMPLE 8.7

In a certain final settler, solid-liquid separation is satisfactory when it is operated under the following conditions: $X_t = 5\ g \cdot l^{-1}$, hydraulic loading rate $T_s = 0.5\ m \cdot h^{-1}$ and recycle factor $s = 1$. The values of the Vesilind constants have been determined previously as $k = 0.4\ l \cdot g^{-1}$ and $v_0 = 7\ m \cdot h^{-1}$. Answer the following questions:

- When the hydraulic loading rate is increased to $0.55\ m \cdot h^{-1}$ the final settler fails, independent of the value of the applied recycle factor "s". Estimate the value of the safety factor s_{fd}.
- When the sludge concentration X_t is maintained at $5\ g \cdot l^{-1}$ and the hydraulic loading rate T_s is reduced to $0.25\ m \cdot h^{-1}$, what is the minimum value of the sludge recycle factor "s" ?
- If a hydraulic loading rate T_s of $0.25\ m \cdot h^{-1}$ is applied, what will be the maximum possible concentration X_t and what will be the required minimum value of sludge recycle factor "s"?

Solution

For the data given, $k \cdot X_t = 0.4 \cdot 5 = 2$. The largest ratio of T_s/v_0 that can be applied for $T_s = 0.5\ m \cdot h^{-1}$ and $v_0 = 7\ m \cdot h^{-1} = T_s/v_0 = 0.5/7 = 0.07$. Using Figure 8.11, point A is identified as the intersection of the curve for $s = 1$ with $k \cdot X_t = 2$. The value of the ratio T_{sm}/v_0 for an ideal settler ($s_{fd} = 1$) would be 0.135. T_{sm} is $7 \cdot 0.135 = 0.94\ m \cdot h^{-1}$. As the applied T_s in reality is $0.5\ m \cdot h^{-1}$ the value of the safety factor can be calculated as $s_{fd} = T_{sm}/T_s = 0.94/0.5 = 1.88 \approx 2$.

The minimum value of s for $X_t = 5\ g \cdot l^{-1}$ is calculated as follows: $k \cdot X_t$ remains 2 and the value of $T_{sm}/v_0 = 0.25/7 = 0.036$. Again using Figure 8.11 and applying the safety factor of 2, then point B is identified. It can be observed that the corresponding s curve (going through B) is marginally higher than that for $s = 0.5$.

As for the last question: when clarification is limiting, for $s_{fd} = 2$ and $T_{sm}/v_0 = 0.25/7 = 0.035$ in Figure 8.11 the corresponding value of $k \cdot X_t$ can be determined as 2.67 or $X_t = 2.67/0.4 = 6.7\ g \cdot l^{-1}$ (point C). The required recycle rate is determined as $s = 2$. For lower values of the sludge recycle factor s, clarification ceases to be the limiting process. For thickening the required area will be larger.

8.3.4 Optimisation of the system of biological reactor and final settler

The optimisation procedure of the system comprising of a biological reactor and a final settler may be applicable to the following situations:

- When the activated sludge plant is designed, certain values are assumed for the settling constants k and v_0 and the design optimisation is carried out for these values. The problem is that the values of the constants tend to fluctuate considerably in time. Therefore, for conservative design, the chosen values must be such that liquid-solid separation will be efficient, even under adverse conditions, which will be discussed in Chapter 9. This design approach will be discussed in this section;
- Once the plant has been constructed on the basis of the optimised design is operational, the actual settling constants at any time may be different from the values adopted for design. Hence a different problem is posed, i.e. to carry out an operational optimisation of the plant, which means the selection of the optimal operational conditions for the actual values of k and v_0 (Section 8.5).

In the previous sections it was shown that it is possible to rationally design a final settler for specified values of the inlet and return sludge concentration if the settling constants k and v_0 are known. The objective of the design optimisation procedure is to determine values for X_r and X_t such that the activated sludge process is operationally stable and the efficiency of liquid-solid separation in the final settler is high, while total costs are minimum.

The total costs consist of construction- and operational costs. The former is by far the most important here and defined mainly by the volumes of the aeration tank and the final settler. The aeration tank volume is inversely proportional to the sludge concentration and the settler volume increases exponentially with this concentration, as long as the critical recirculation factor s_c is applied. The main factor that determines the operational costs is the value of the sludge recirculation factor "s": at larger return sludge flow rates, the pumping costs will increase.

The optimisation procedure for settler design involves the optimisation of two operational variables: X_t and s. These two then define a third variable X_r by Eq. (8.1). In principle, the chosen recirculation factor will be equal to s_c, unless there is a reason why this value cannot be applied. Thus the following optimisation procedure is suggested:

(1) Select values for sludge settleability (k and v_0), the safety factor s_{fd} and settler height H_d;

(2) Initially it is assumed that the critical recirculation factor may be used for design optimisation. This allows the clarification expression (Eq. 8.33) to be used for calculation of the volume of the final settler:

$$v_d = s_{fd} \cdot (H_d/v_0) \cdot \exp(k \cdot X_t) \quad \text{or} \quad V_d = Q_i \cdot s_{fd} \cdot (H_d/v_0) \cdot \exp(k \cdot X_t)$$

Note that the same settler volume is obtained when the thickening expression Eq. (8.34) is used, which is to be expected when the critical recirculation factor is applied and both thickening and clarification are limiting processes;

(3) Use Eq. (3.55) to calculate the aeration tank volume:

$$v_r = mX_t \cdot S_{ti}/X_t \quad \text{or} \quad V_r = mX_t \cdot MS_{ti}/X_t$$

(4) The v_r and v_d values as well as their sum $v_t = v_r + v_d$ are plotted as a function of X_t, the sludge concentration in the aeration tank, and the minimum value of v_t is determined. Alternatively, if it is considered more convenient, plot V_r, V_d and V_t. The corresponding sludge concentration is in

principle the optimal value, assuming construction costs per cubic metre of settler volume are equal to those of the biological reactor.

(5) If the costs per m^3 unit volume are known, then it is also possible to calculate the minimum total construction costs of the system reactor-settler (refer also to Chapter 14);

$$MC_{rd} = C_r \cdot V_r + C_d \cdot V_d \tag{8.39}$$

(6) For the optimal sludge concentration selected in step (4) or (5) determine the critical recirculation factor s_c and verify if the (actual) retention time in the final settler is within the desired range of approximately one to three hours (Eq. 8.36);

(7) If the retention time in the final settler is too long there are two options:
   - Increase the recirculation factor "s" to a value larger than the critical value;
   - Decrease the sludge concentration X_t below the optimal value, thus accepting less than optimal operation and/or construction costs, but designing a system with an adequate retention time in the final settler.

   In practice the operational value of s will often be larger than the design value of "s" (which is equal to s_c), in order to prevent accidental overloading of the final settler should the influent flow rate increase or the sludge concentration be somewhat higher than anticipated. The additional pumping costs will be small and the penalty associated with exceeding the effluent limits or overloading the final settler will be much larger.

(8) If the retention time in the final settler is too low, increase the design reactor biomass concentration X_t.

As an example, in Figure 8.12 the values of v_r, v_d and v_t are shown as function of the sludge concentration in the aeration tank X_t for the following conditions: $s_{fd} = 2$; $H_d = 4$ m; $S_{ti} = 0.5\ g \cdot l^{-1}$ and $mX_t = 2$ mg $TSS \cdot d \cdot mg^{-1}$ COD. This mX_t value corresponds to a sludge age of about 8 days in the case of raw sewage (Eq. 3.49). In Figure 8.12, three characteristic pairs of Vesilind constants were considered:

(a) Poor settleability: $k = 0.46\ l \cdot g^{-1}$ and $v_0 = 6\ m \cdot h^{-1}$ (Figure 8.12a);
(b) Medium settleability $k = 0.36\ l \cdot g^{-1}$ and $v_0 = 9\ m \cdot h^{-1}$ (Figure 8.12b);
(c) Good settleability: $k = 0.31\ l \cdot g^{-1}$ and $v_0 = 12\ m \cdot h^{-1}$ (Figure 8.12c).

For the specified conditions and in the case of medium settleability, it can be observed that the minimum volume v_t for the aeration tank-settler system is obtained for a sludge concentration $X_t = 4.15\ g \cdot l^{-1}$, with $v_d = 0.17$ and $v_r = 0.24\ m^3 \cdot m^{-3} \cdot d^{-1}$ so that $v_t = 0.41\ m^3 \cdot m^{-3} \cdot d^{-1}$. Using Figure 8.10 and Figure 8.12 it is possible to construct Figure 8.13, where the critical recirculation factor s_c and the actual settler retention time $v_d/(s_c + 1)$ are plotted as function of the sludge concentration X_t.

Again considering medium settleability (Figure 8.13b), it can be observed that the actual retention time in the settler is 2.5 hours for the optimum concentration of $4.15\ g \cdot l^{-1}$, which is within the required range of l to 3 hours. Hence, for the optimal sludge concentration and the critical recirculation factor, the retention time in the settler is adequate and for this reason these values can be accepted as the optimal values for design. Thus the optimal design for the aeration tank and final settler of Figure 8.13b can be summarised as:

- Sludge concentration in the aeration tank $X_t = 4.15\,g \cdot l^{-1}$;
- Recirculation factor $s_c = 0.56$ (critical, determined with Figure 8.10);
- Return sludge concentration $X_r = X_t \cdot (s_c + 1)/s_c = 4.15 \cdot 1.56/0.56 = 11.5\,g \cdot l^{-1}$.

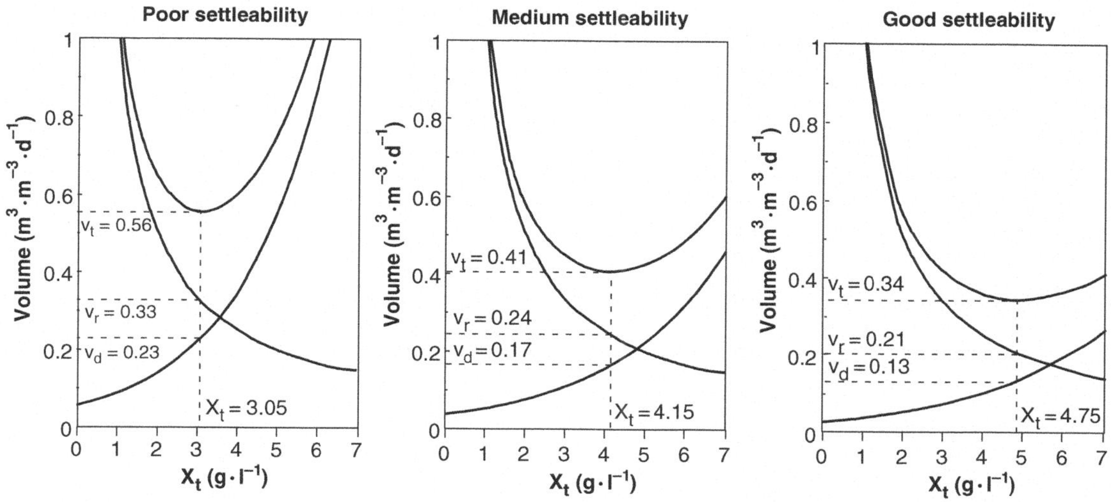

Figure 8.12 v_r, v_d and v_t as a function of the sludge concentration, when it is assumed that the critical recirculation factor is applied

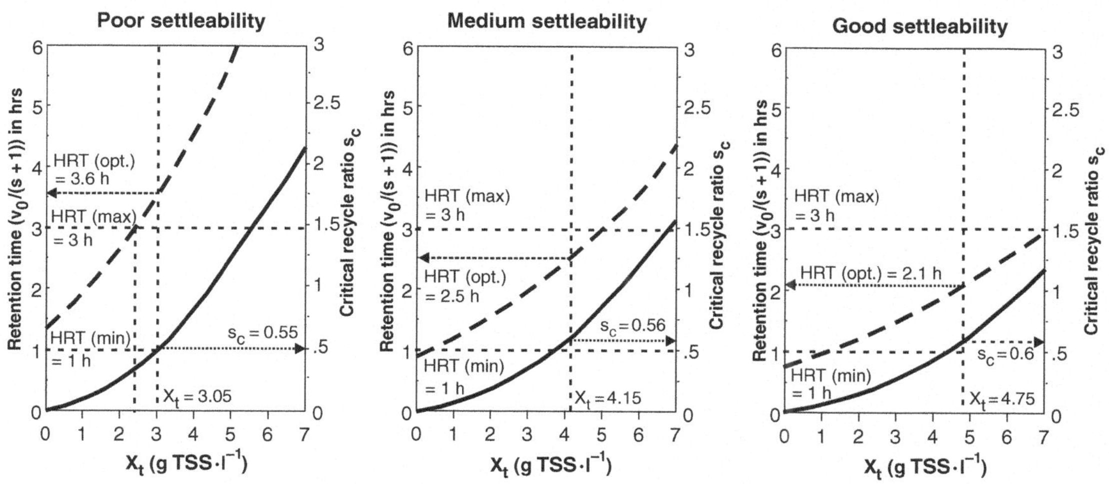

Figure 8.13 Critical recirculation factor and retention time in the settler as function of the sludge concentration (based on the data presented in Figure 8.12)

In the case of poor settleability ($k = 0.46\ l \cdot g^{-1}$; $v_0 = 6\ m \cdot h^{-1}$), the design optimisation leads to a retention time in the final settler that may be considered as excessively long. For the sludge concentration resulting in the minimum total volume ($X_t = 3.05\ g \cdot l^{-1}$) and the critical recirculation factor $s_c = 0.55$, the actual retention time is 3.6 hours. If this is considered too long, one possibility is to increase the recirculation factor, thereby reducing the retention time from 3.6 to 3 hours. The required recirculation factor can be calculated as $(1 + s)/(1 + s_c) = 3.6/3 = 1.2$. Hence $s = 1.2 \cdot 1.55 - 1 = 0.86$.

The second possibility is to apply a lower mixed liquor concentration by increasing the reactor volume. In Figure 8.13a it can be observed that the maximum retention time of 3 hours is obtained for the critical recirculation factor $s = 0.55$ for a sludge concentration of $2.4\ g \cdot l^{-1}$. When this concentration is adopted, the total volume v_t is equal to $0.60\ m^3 \cdot m^{-3} \cdot d^{-1}$, which is larger than the calculated minimum ($v_t = 0.56\ m^3 \cdot m^{-3} \cdot d^{-1}$).

Which of the options is preferred in the final design depends on a comparison between the value of the increase of the operational costs (sludge recirculation factor from 0.55 to 0.86 or 61%) and the increase of the construction costs (v_t from 0.54 to 0.60 or 8%).

8.3.5 Validation of the optimised settler design procedure

Several empirical guidelines for final settler design are used, based on one or more of the criteria listed below:

- Hydraulic loading rate ($m \cdot h^{-1}$);
- Solids loading rate ($kg\ TSS \cdot m^{-2} \cdot h^{-1}$);
- Sludge volume loading rate ($litre \cdot m^{-2} \cdot h^{-1}$);
- Weir overflow rate ($m^3 \cdot m^{-1} \cdot h^{-1}$).

Based on a empirical observations at full-scale plants, maximum values for one or more of the criteria listed above are recommended. In this section several of the most common methods are briefly reviewed and compared to the solids flux design method presented in this book.

8.3.5.1 US EPA design guidelines

The design guidelines as formulated by the US EPA (1975) can be summarised as:

- Average hydraulic loading rate T_s between 0.7 and $1.35\ m \cdot h^{-1}$;
- Peak hydraulic loading rate T_{sm} between 1.7 and $2.0\ m \cdot h^{-1}$;
- Average solids loading rate F_{sol} between 4.1 and $6.1\ kg\ TSS \cdot m^{-2} \cdot h^{-1}$;
- Peak solids loading rate F_{sol} is $10.2\ kg\ TSS \cdot m^{-2} \cdot h^{-1}$;
- Side wall depth H_d between 3.7 and 4.4 m;
- Weir loading rate $<10.4\ m^3 \cdot m^{-1} \cdot h^{-1}$ for small clarifiers and $<15.5\ m^3 \cdot m^{-1} \cdot h^{-1}$ for large clarifiers.

If compliance to all of the above criteria is not possible, the designer will have to decide on the priority. It is remarkable to observe that the settling characteristics of the sludge do not define the design of the final settler. However, although inadequate, these guidelines are still frequently applied as they are very simple to use.

8.3.5.2 WRC and modified WRC design guidelines

The WRC design method is based on the solids flux theory and adapted to the conditions and sludge characteristics prevailing in the UK. White (1975) correlated the $SSVI_{3.5}$ index to the values of the settling constants and obtained the following empirical formula to determine the maximum allowable solids loading rate:

$$\begin{aligned} F_m &= X_t \cdot (T_s + u) \\ &= 8.85 \cdot (100/I_{ssv})^{0.77} \cdot u^{0.68} \end{aligned} \tag{8.40a}$$

where:

F_m = maximum solids flux (kg TSS $\cdot$ $m^{-2} \cdot h^{-1}$)
T_s = hydraulic overflow rate (= Q_i/A_d in $m \cdot h^{-1}$)
u = underflow rate or downward velocity in the settler (= $s \cdot Q_i/A_d$ in $m \cdot h^{-1}$)
I_{ssv} = $SSVI_{3.5}$ ($ml \cdot g^{-1}$ TSS)

Ekama et al. modified the WRC design procedure (Ekama et al., 1986; Ekama et al., 1997) and noted that Eq. (8.40a) was only valid up to a certain critical value of the underflow rate u. Again using I_{ssv}, the following empirical formulas were defined to determine this critical underflow rate:

$$u = 1.612 - 0.00793 \cdot I_{ssv} \quad \text{for } I_{ssv} < 125 \text{ ml} \cdot g^{-1} \text{ TSS and} \tag{8.40b}$$

$$u = 1.612 - 0.00793 \cdot I_{ssv} + 0.0015 \cdot (I_{ssv} - 125) \quad \text{for } I_{ssv} > 125 \text{ ml} \cdot g^{-1} \text{ TSS} \tag{8.40c}$$

8.3.5.3 STORA/STOWA design guidelines

According to the original Dutch STORA guidelines developed in 1981 (STORA, 1981 and Stofkoper et al., 1982), design is based on the application of a maximum sludge volume loading rate. This sludge volume loading rate T_{vxm} is defined as $v_x/A_d = X_t \cdot I_{dsv}/A_d$ ($l \cdot m^{-2} \cdot h^{-1}$). The design procedure was based on a extensive evaluation of the performance of full-scale final settlers, where the mixed liquor flow to the settler was increased until failure was observed. The maximum solids loading rate T_{vxm} is a function of the sludge volume v_x. Using the appropriate value of T_{vxm}, an equivalent hydraulic overflow rate T_{sm} can be calculated:

$$T_{sm} = 300/v_x \quad \text{for } 200 < v_x < 300 \, l \cdot m^{-3} \tag{8.41a}$$

$$T_{sm} = \tfrac{1}{3} + 200/v_x \quad \text{for } 300 < v_x < 600 \, l \cdot m^{-3} \tag{8.41b}$$

$$T_{sm} = 400/v_x \quad \text{for } v_x > 600 \, l \cdot m^{-3} \tag{8.41c}$$

Eqs. (8.41a to c) are valid for $X > 2 \, g \cdot l^{-1}$ or $v_x > 200 \, ml \cdot l^{-1}$, whichever is limiting. An important difference to the solids flux procedure is that the design of the final settler is based on the maximum sustained *peak* influent flow rate (e.g. rainwater flow conditions) instead of the *average* influent flow rate and that during peak flow the reactor sludge concentration is assumed to decrease from its original value X_t as part of the sludge mass is transferred to the final settler. This approach originates from the Dutch situation where combined sewers are used and relatively large fluctuations between dry weather and rain weather flow are common. Furthermore, as a significant part of the country is situated below sea-level, due to the resulting high groundwater level the sidewall depth is limited to a maximum of 2–2.5 m.

In a recent research project (STOWA, 2002) eleven full-scale settlers were re-evaluated. It was observed that the value of T_{vxm} varied between 250 and 500 $l \cdot m^2 \cdot h^{-1}$. The original guidelines from 1981 are still commonly applied in the Netherlands. Refer to Appendix A7 for a detailed description of the original and revised guidelines

8.3.5.4 ATV design guidelines

The German ATV design procedure from 1976 precedes the STORA guideline but is similar to it in many aspects. Most importantly, the surface area of the final settler is defined by the maximum sludge volume loading rate T_{vxm}. In the case of the ATV the value of T_{vxm} is approximately $400\ l \cdot m^{-2} \cdot h^{-1}$ for $v_x = 200\ ml \cdot l^{-1}$ and decreases slowly to a value of $200\ l \cdot m^{-2} \cdot h^{-1}$ for $v_x = 1000\ ml \cdot l^{-1}$. The range of T_{vxm} values corresponds to a maximum concentration of $30\ mg\ TSS \cdot l^{-1}$ in the effluent as observed in full-scale settlers. Multiplying the sludge volume loading rate T_{vxm} with the sludge volume v_x, one calculates the allowed maximum hydraulic loading rate T_{sm} as:

$$T_{sm} = 2400/(v_x)^{1.34} \tag{8.42}$$

Similar to the STORA guideline, Eq. (8.42) is valid for $X > 2.0\ g \cdot l^{-1}$ or $v_x > 200\ ml \cdot l^{-1}$ and sizing is based on the peak sustained rainwater flow. The main distinguishing feature from the STORA guideline is that the depth of the settler is now an explicit design criterion: increasing the depth allows a higher proportion of the sludge mass to be stored in the final settler and thus reduces the sludge volume loading rate during peak flow. The ATV procedure therefore allows a trade-off to be made by the designer between required settler surface area and settler depth.

In 1991 the ATV guideline was revised. The main changes were that a higher sludge volume loading rate was allowed ($T_{vx} < 450\ l \cdot m^{-2} \cdot h^{-1}$) and that settler depth was increased. The latter resulted from application of stricter effluent limits ($X_{te} < 20\ mg\ TSS \cdot l^{-1}$). The overall result was a slight increase in calculated settler volume compared to the ATV 1976 guidelines. Refer to Appendix A7 for a detailed description of the original and revised guidelines.

8.3.5.5 Solids flux compared with other design methods

When the empirical relationships discussed in the previous sections are compared to the design model based on the solids flux theory presented in this chapter (Eqs. 8.28 and 8.30), it can be noted that the latter explicitly recognises the influence of:

- Sludge concentration;
- Sludge settleability, characterised by the constants k and v_0 (or I_{ssv});
- The recirculation factor when it is of relevance (i.e. in the case of thickening).

The criteria of ATV and STORA also seek to quantify the influence of sludge concentration and settleability, but not of the recirculation factor. On the other hand, the ATV explicitly recognises the influence of the settler depth as a design variable. Another important difference is that the design of ATV and STORA is based on sustained peak flow, while the design according to the flux theory is based on the average influent flow.

The sidewall depth of final settlers designed according to the solids flux theory is often considerably higher than those selected according to the ATV and especially the STORA design, while the recirculation rate will also be higher. The assumption that this depth is sufficient to handle a sustained peak flow situation should be checked using the static point procedure that will be outlined in Section 8.5. Finally, the EPA design criteria are surprisingly inadequate as they do not recognise any of the three basic factors that influence sludge settling.

One should keep in mind that the STORA and ATV guidelines are empirical measures and based on observations in a single country: in The Netherlands and Germany respectively. As a result, physical design- or site characteristics, which might be country specific, are implicitly included in the design procedure. For example the STORA guideline from 1981 was based on a set of 21 full-scale final settlers which shared the following characteristics:

- Settler diameter > 30 m, with a sidewall depth between 1.5–2.5 m and a bottom inclination of $0.08\ m.m^{-1}$;

- Circular, conical settler equipped with a bottom scraper. Mixed liquor enters in a centre flocculation well (no deflection baffle) and effluent is discharged over a single peripheral effluent weir;
- The ratio between rainwater and dry weather flow in The Netherlands is quite high at typical values between 2 and 3;
- No nitrogen removal in the activated sludge process;
- High I_{dsv} values (avg. 140–190 $ml \cdot g^{-1}$ TSS) as sludge bulking control measures (such as a selector) had not yet been implemented.

Thus a certain precaution is required when generalising these empiric guidelines. This disadvantage does not apply to the solids flux theory, which is based on sludge characteristics and in principle is independent of settler characteristics. However, in our design method it is assumed that Vesilind's equation is valid. The experimental results of many researchers justify this assumption, but the practical applicability of the method depends fundamentally on the values of the two Vesilind constants: k and v_0. The experimental results presented in Section 8.1 show that the values of the constants can be estimated from the stirred sludge volume index:

$$k = 0.16 + 0.003 \cdot I_{ssv} \quad \text{and} \quad v_0 = 16 - 0.1 \cdot I_{ssv} \tag{8.10c and 8.10d}$$

$$\text{where } I_{ssv} = (25 + 25 \cdot f_{av} + 5 \cdot X_t) \tag{8.10a}$$

These correlations were obtained using sludge generated from raw sewage so the values of the constants may be quite different for industrial wastewaters. Even in the case of sewage from one source, there were large fluctuations in the data. The I_{ssv} values had a standard deviation of 27% for sludge with a high active sludge fraction ($f_{av} = 0.76$) and 10% for sludge with a low active sludge fraction ($f_{av} < 0.16$). However, notwithstanding these limitations, starting from the observations above three situations can be distinguished in order to characterise sludge settleability:

(a) Good settleability
This situation is characterised by an I_{ssv} value corresponding to sludge with a low active fraction ($f_{av} < 0.3$). Using Eq. (8.10a) one has (note that by definition $X_t = 3.5\ g \cdot l^{-1}$ for I_{ssv}):

$$I_{ssv} = 25 + 25 \cdot 0.3 + 5 \cdot 3.5 = 50\,ml.g^{-1}, \text{ hence } k = 0.311 \cdot g^{-1} \quad \text{and} \quad v_0 = 11\,m \cdot h^{-1}$$

(b) Medium settleability
This situation is characterised by an I_{ssv} value corresponding to sludge with a high active fraction ($f_{av} = 0.9$). Using again Eq. (8.10a):

$$I_{ssv} = 25 + 25 \cdot 0.9 + 5 \cdot 3.5 = 65\,ml \cdot g^{-1}, \text{ hence } k = 0.36l \cdot g^{-1} \quad \text{and} \quad v_0 = 9.5\,m \cdot h^{-1}$$

(c) Poor settleability
To characterise this situation an I_{ssv} of 100 $ml \cdot g^{-1}$ is adopted. This value is justified by the following reasoning: the average I_{ssv} value for sludge with a high active sludge fraction is 65 $ml \cdot g^{1}$ and has a standard deviation of 27% or $0.27 \cdot 65 = 17.5\ ml \cdot g^{-1}$. Thus, statistically 95% of the sludge with medium settleability will have an I_{ssv} value below the average plus two times the standard deviation: $65 + 2 \cdot 17.5 = 100\ ml \cdot g^{-1}$. In only 5% of the cases the sludge will have an I_{ssv} above 100 $ml \cdot g^{-1}$, so that the qualification "poor" is justified. For $I_{ssv} = 100\ ml \cdot g^{-1}$, one has $k = 0.46\ l \cdot g^{-1}$ and

$v_0 = 6\ m \cdot h^{-1}$. It may be noted that several authors (Smollen and Ekama, 1984) consider an I_{ssv} of $100\ ml \cdot g^{-1}$ as the maximum value for "normal" sludge. Sludges with a larger I_{ssv} value are labelled filamentous, with atypical settling characteristics (bulking sludge).

For each of the situations characterising settling: (a) good, (b) medium, and (c) poor, the hydraulic loading rate can be determined as a function of the sludge concentration, using the solids flux theory explained in Section 8.3.1, especially Eqs. (8.28 and 8.30). Using the results obtained by Stofkoper et al. (1982) (Eq. 8.3) or by Catunda et al. (1993) for sludge with a high active fraction one has $I_{dsv} = 1.5 \cdot I_{ssv}$. Figure 8.14a to c show for different sludge settleabilities the hydraulic loading rate as a function of the sludge concentration. The applied safety factor s_{fd} has a value of 2. Furthermore it is assumed that the critical recirculation factor is applied in the design procedure (i.e. clarification is limiting).

The validity of the presented solids flux design method can now be evaluated by comparing the calculated maximum hydraulic overflow rate T_{sm} with the results obtained from the design criteria developed by several research institutions, discussed in the previous sections: i.e. the ATV guidelines (1976) and the STORA guidelines (1981).

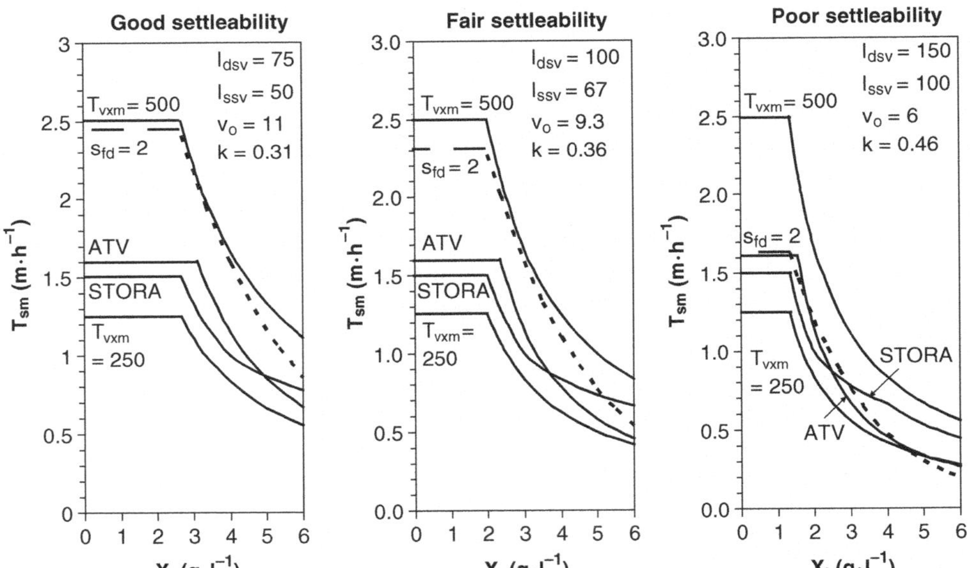

Figure 8.14 Theoretical T_{sm} values ($s_{fd} = 2$) as a function of the sludge concentration for good, medium and poor settleability, as compared to empirical values from the ATV (1976), STORA (1981) and the ranges indicated by the STOWA 2002 experimental results ($250 < T_{vxm} < 500\ l \cdot m^{-2} \cdot h^{-1}$)

Figure 8.14 also shows the findings from the STOWA project in 2002: the upper curve corresponds to $T_{vxm} = 500\ l \cdot m^{-2} \cdot h^{-1}$, while the lower curve corresponds to $T_{vxm} = 250\ l \cdot m^{-2} \cdot h^{-1}$. The T_{sm} curves shown in Figure 8.14 should be interpreted as the maximum hydraulic loading to the final settler at equilibrium: i.e. when the applied solids loading rate to the settler is equal to the solids removal capacity.

When Figure 8.14 is analysed, it can be observed that there is a close correlation between the theoretical values of T_{sm} derived in this chapter and the empirical values observed in full scale installations over the complete range of practical interest where the empiric curves are valid ($200 < v_x < 600\ l \cdot m^{-3}$). This close correlation is observed for poor, fair and well settling sludges. From Figure 8.14 it is also confirmed that $s_{fd} = 2$ leads to a good correlation between empirical and theoretical results.

Having established that Eqs. (8.28 and 8.30) form an adequate basis for final settler design and optimisation, it remains to be decided which values for I_{ssv}, v_0, k and s_{fd} are to be adopted. For conservative design, to ensure adequate final settler performance when the sludge characteristics are "normal" (i.e. not filamentous), the settling characteristics for poor settling sludge may be selected: $k = 0.46\ l \cdot g^{-1}$ and $v_0 = 6\ m \cdot h^{-1}$. This roughly corresponds to $I_{ssv} = 100\ ml \cdot g^{-1}$ and $I_{dsv} = 150\ ml \cdot g^{-1}$. As for the value of the safety factor, $s_{fd} = 2$ can be used. Of course a final settler that is designed on this basis will also have a satisfactory performance when the sludge settleability is fair or good.

The value of the safety factor that was adopted in order to obtain a good fit between theory and the empirical data is relatively high ($s_{fd} = 2$). However, it has to be considered that the conditions for the theoretical and experimental curves are not equal. A first difference is that the solids flux theory is derived for a constant flow rate and its expression indicates that the final settler will fail if a constant maximum flow is sustained. On the other hand, for the experimental curves the maximum influent flow could only be sustained for such a time as long as the (stored) water quantity lasted. Thus in many cases in the experimental Stora procedure the final settler would have failed, had it been possible to sustain the high influent flow for a longer time. For the empirical model this would be considered as satisfactory behaviour because heavy rains (and consequential maximum flows) normally only last for a relatively short time.

Another difference is that the theoretical curves are based on batch tests at constant sludge mass. In the experimental procedures the sludge concentration in the aeration tank tends to decrease as sludge accumulates in the final settler. This continues until a maximum of 30% of the total sludge mass has been transferred to the final settler. If it is considered that under normal conditions the sludge mass in the final settler would probably not exceed more than 5%, it is concluded that under maximum load the mixed liquor concentration can decrease by as much as 25%, and this reduction will of course allow application of a much higher flow rate to the settler. When the sewage flow returns to normal the sludge mass will gradually be returned to the aeration tank.

8.4 PHYSICAL DESIGN ASPECTS FOR FINAL SETTLERS

Figure 8.15 shows a schematic representation of a final settler. The cylindrical form is more common, although there are also rectangular units. The lateral depth is in the range of 2 to 6 m and the bottom has a slope of 2 to 8°, so that the settler is deeper at the centre. For economical reasons the settler diameter is limited to about 50 to 60 m. The following details can be observed in Figure 8.15:

(1) Inlet structure: in most settlers the mixed liquor inlet is at the centre, but there are settlers with peripheral feed. Figure 8.15 shows a common construction in which the sludge enters ascencionally (minimum velocity of $1\ m \cdot s^{-1}$). Near the liquid surface the inlet tube ends and the mixed liquor flows radially to the surface in an open cylinder, having a depth of about half the settler sidewall height and a diameter of 10% of the settler diameter (WPCF, 1977);

(2) Effluent outlet structure: normally the effluent is discharged into a peripheral gutter provided with triangular weirs (V-notches). Either one or two weirs can be used. Installing two weirs reduces weir overflow liquid speed by 50% and therefore theoretically reduces entrainment of solids. In practice increased suspended solids concentrations have been reported as well. Usually a foam retention baffle is placed to avoid carry-over and discharge of floating material (foam, sludge, fats). The effluent discharge rate should not exceed 3 to 5 $m^3 \cdot h^{-1}$ per metre of weir length in order to avoid

currents that might draw sludge particles into the effluent flow. Figure A7.6 shows a picture of a typical effluent gutter of a final settler;

(3) Sludge return device: this device is of fundamental importance for the performance of the settler and is composed of the following elements:
   a. A rotating bridge from the centre to the perimeter, moving with a rotation velocity of 2 to 5 rph;
   b. Fixed to the bridge, the bottom scraper causes light turbulence in the lower part of the settler and "sweeps" the settler bottom, helping to move settled sludge towards the centre;
(4) Central hopper for accumulation of settled sludge for return to the aeration tank;
(5) Skimming device for floating material. Floating material, principally fats and biological foam, is removed from the liquid surface by means of a skimming device, connected to the rotating bridge. The floating material is discharged into a special sump from where it is usually pumped to the sludge dewatering unit.

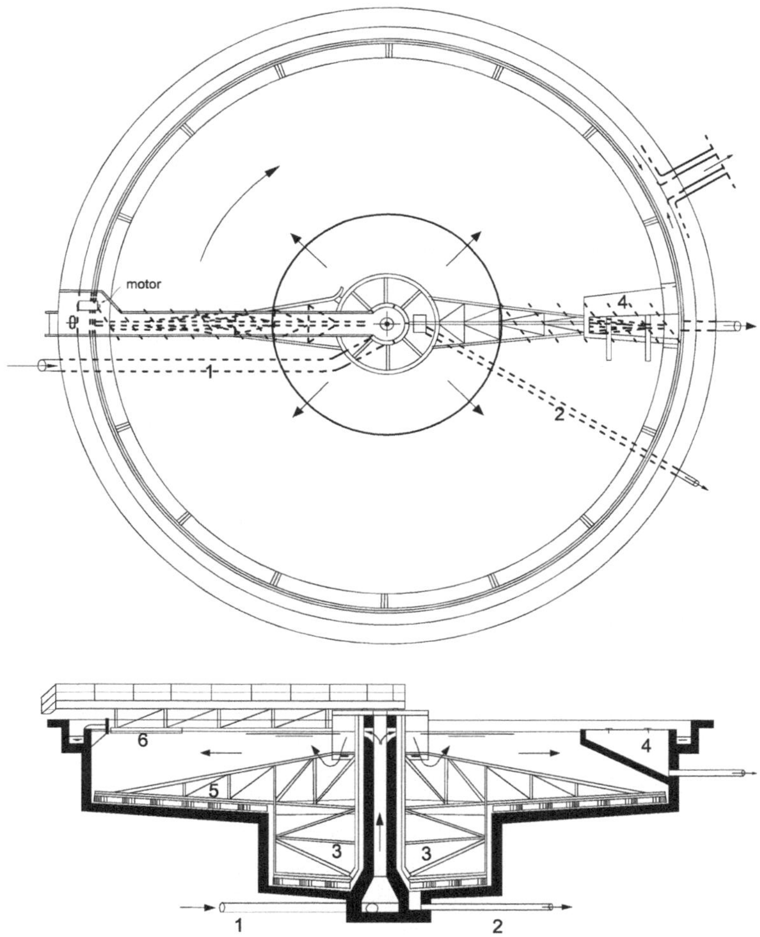

Figure 8.15 Schematic representation of construction details of an activated sludge settler: 1 - inlet structure; 2 - sludge outlet; 3 - sludge hopper; 4 - scum outlet; 5 - rotating bridge and 6 - skimming device

The construction material for settlers is almost invariably concrete, although small settlers may be constructed in steel. Steel is also used for the moving parts and the bridge, scraper and skimmer. Depending on the corrosivity of the effluent, the V-notches in the overflow weir may be constructed using corrosion resistant material such as PVC or polyester, otherwise steel is commonly used.

8.5 FINAL SETTLERS UNDER VARIABLE LOADING CONDITIONS

In the preceding sections it was tacitly assumed that the influent flow and hence the hydraulic loading rate and the solids loading rate were constant. However, usually the influent flow Q_i is subject to cyclic variations within a one-day period. In addition, there may be large fluctuations around the average flow due to rain events et cetera. Keinath (1980) introduced the concept of the static point to evaluate the influence of a varying hydraulic- and solids load on settler behaviour. In the diagram depicting F_v as a function of the sludge concentration, the static point P is defined in the point (F_t, X_t) characterised by:

$$F_t = X_t \cdot Q_i/A_d = X_t \cdot T_s \tag{8.43}$$

with F_t = actual solids flux applied to the settler

Figure 8.16 shows the static point P graphically as the intersection point of two straight lines: line I passes through the origin and line II through the value of the return sludge concentration on the horizontal axis. The gradients of the lines are defined as:

$$m_I = F_t/X_t = (Q_i \cdot X_t/A_d)/X_t = Q_i/A_d \quad \text{and} \tag{8.44}$$

$$m_{II} = -F_t/(X_r - X_t) = -s \cdot F_t/X_t = -(s \cdot Q_i \cdot /A_d)/X_t = -s \cdot Q_i/A_d = -u \tag{8.45}$$

The static point P, the associated straight lines I and II and the batch settling curve F_v define the operational state of the settler, as shown in Figure 8.16. Basically three different situations can be distinguished:

- The settler is critically loaded. This situation is characterised by the fact that line II is tangential to the curve F_v (line IIa with static point P_c in Figure 8.16);
- The settler is under loaded. In this case there is no intersection point between line II and curve F_v in the concentration range between X_t and X_r (line II with static point P_{min});
- The settler is overloaded. In this case there is an intersection point of line II with curve F_v in the concentration range between X_t and X_r (line IIb with static point P_{max}).

In reality, the influent flow Q_i and consequently the solids loading rate $F_{sol} = Q_i \cdot X_t/A_d$ will vary with time and correspondingly the static point P will be displaced vertically. If the size of the return sludge flow (and thus the value of "u", the gradient of line II) is kept constant it is possible that during a cycle of varying load, periods of overloading are followed by periods of under loading. When overloading occurs, sludge will accumulate in the settler and the interface that separates supernatant and settling sludge will rise. During the subsequent period of under loading the accumulated sludge will be withdrawn and recycled to the aeration tank. In practice the variations of the solids loading rate require the presence of an adequate buffer volume in the settler for sludge accumulation. This is the main reason for having a relatively deep final settler (4 to 6 metres).

It should be pointed out that above analysis should be based on the *net* available surface area: i.e. A_d/s_{fd}. Otherwise the response of the final settler behaviour and flow- and load variations will be significantly underestimated. On the other hand, the static point analysis does not consider the effect that the solids transfer from biological reactor to final settler will have on the biomass concentration in the reactor (i.e. equal to the mixed liquor feed to the final settler), which is significant.

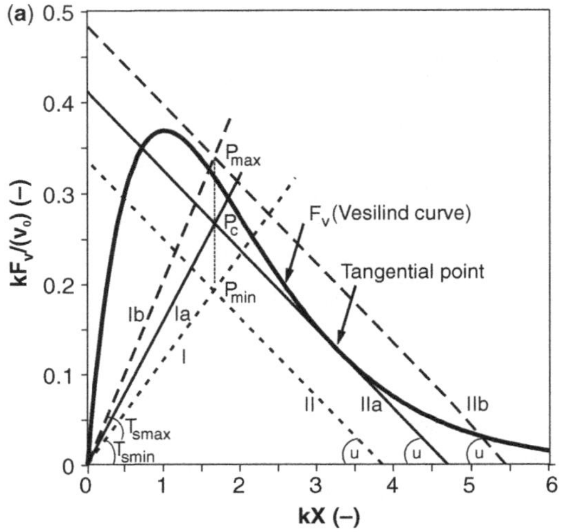

Figure 8.16 Graphical representation of the concept of the static point P and of situations of under- and overloading of the settler

EXAMPLE 8.8

A final settler is operated with an average hydraulic loading rate of $T_s = 1\ m \cdot h^{-1}$ and a recirculation factor $s = 0.5$. The mixed liquor concentration is $X_t = 3\ g \cdot l^{-1}$ and the Vesilind constants are $k = 0.5\ l \cdot g^{-1}$ and $v_0 = 10\ m \cdot h^{-1}$ (Figure 8.17).

(1) Show that the settler is under loaded for the average flow rate;
(2) How much can the influent flow be increased without accumulation of sludge in the settler (while maintaining the recycle rate s constant at $s = 0.5$)?;
(3) What is the maximum influent flow rate that can be applied when s is optimised?;
(4) If, for the maximum influent flow rate of item (3), the recirculation factor is kept at its original value of $s = 0.5$, what will be the rate of sludge accumulation in the settler and what will be the rising velocity of the interface between sludge and supernatant?

Solution

(1) Assess current performance of the settler. Figure 8.17 shows the situation defined by the example, line I and II apply. The static point P for average flow is at the intersection point of the straight lines I and II. Line I is given by $F = T_s \cdot X$ and passes through the static point for $X = X_t$, i.e. when $F = T_s \cdot X_t = 1 \cdot 3 = 3\ kg \cdot m^{-2} \cdot h^{-1}$. As $s = 0.5$ it follows that $X_r = (s + 1)/s \cdot X_t = 3 \cdot X_t = 9\ g \cdot l^{-1}$ and $u = s \cdot T_s = 0.5 \cdot 1 = 0.5\ m \cdot h^{-1}$. Hence, the straight line is defined by the equation $F = u \cdot (X - X_r) = -0.5 \cdot (X - 9)$. In Figure 8.17 it can be noted that line II does not have an intersection point

with the batch settling curve in the concentration range between $X_t = 3$ and $X_r = 9\ g \cdot l^{-1}$. It is concluded that the settler is under loaded. The average solids loading rate $F_{sol} = (s + 1) \cdot T_s \cdot X_t = 1.5 \cdot 1 \cdot 3 = 4.5$ $kg \cdot m^{-2} \cdot h^{-1}$.

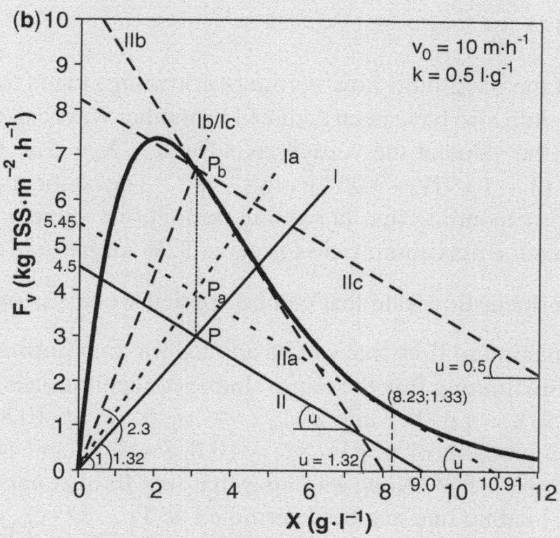

Figure 8.17 Application of the static point concept in Example 8.8

(2) How much can the influent flow be increased without accumulation of sludge in the settler?

Line Ia and IIa apply. To evaluate the maximum load that can be applied while maintaining $u = 0.5\ m \cdot h^{-1}$, the straight line II is displaced vertically, until it is tangential to the batch settling curve F_v (line IIa in Figure 8.17). The gradient of the line will still be $u = 0.5$ and will be equal to the derivative of the batch settling curve when $X = X_l$, so that:

$$u = 0.5 = (dF_v/dX)_{x=xl} = v_0 \cdot (1 - k \cdot X_l) \cdot \exp(-k \cdot X_l)$$

Using numerical methods one calculates the value of the limiting sludge concentration X_l as $8.23\ g \cdot l^{-1}$. With the aid of the value of the limiting concentration, the corresponding value of the batch settling curve for $X = X_l$ can be calculated: $F_v = 10 \cdot 8.23 \cdot \exp(-0.5 \cdot 8.23) = 1.33\ kg \cdot m^{-2} \cdot h^{-1}$. Now, the corresponding return sludge concentration can be calculated as:

$$X_r = X_l + (F_v)_{x=xl}/u = 8.23 + 1.33/0.5 = 10.91\ g \cdot l^{-1}$$

The limiting flux is given by: $F_l = u \cdot X_l = 0.5 \cdot 10.91 = 5.45\ kg \cdot m^{-2} \cdot h^{-1}$. For the recirculation factor $s = 0.5$ the maximum solids loading rate that can be transported in the settler equals $5.45\ kg \cdot m^{-3} \cdot h^{-1}$. To calculate the maximum flow that can be applied for $s = 0.5$ it is convenient to express the flow entering the settler as:

$$Q = (Q_r + z \cdot Q_{av}) = (0.5 + z) \cdot Q_{av}$$

where Q_{av} = average influent flow; Q_r = sludge recycle flow (= constant) and $z \cdot Q_{av}$ = influent flow at a particular moment.

For the maximum flow that can be applied before overload occurs, the solids loading rate will be equal to the limiting flux, i.e.:

$$F_{sol} = (0.5 + z) \cdot Q_{av} \cdot X_t/A_d = (0.5 + z) \cdot T_s \cdot X_t$$
$$F_{sol} = F_l \rightarrow (0.5 + z) \cdot 1 \cdot 3 = 5.45 \rightarrow z = 1.32$$

As $z = 1.32$, this means that the maximum flow before overloading (maintaining $s = 0.5$) is 32% larger than the average flow. This could also have been deduced in another way: calculate for the straight line IIa passing through $X_r = 10.91$ the value at the vertical axis for $X = X_t = 3\ g \cdot l^{-1}$. The value of IIa at the vertical axis equals $-0.5 \cdot (3 - 10.91) = 3.95\ kg \cdot m^{-2} \cdot h^{-1}$. The static point P_a is defined by the coordinates (3;3.95). The corresponding line Ia passing through the static point has a gradient of $T_s = 3.95/3 = 1.32\ m \cdot h^{-1}$. Hence the maximum ratio Q_i/A_d is 32% larger than the given average of 1.0.

(3) What is the maximum influent flow rate that can be applied when s is optimised?

Line Ib and IIb apply. The maximum flow that can be applied for an optimised s value can be calculated knowing that for the maximum limiting flux the return sludge concentration is given by $X_r = X_c = 4/k = 8\ g \cdot l^{-1}$. In this case $X_l = 2/k = 4\ g \cdot l^{-1}$ and $F_{lmax} = 4 \cdot v_0/(k \cdot e^2) = 10.8\ kg \cdot m^{-2} \cdot h^{-1}$ (Eq. 6.19). Line IIb tangent to F_v has a gradient of $u = F_l/X_c = 10.8/8 = 1.35$ and for $X = X_t = 3$ the value of $F_v = 1.35/(8 - 3) = 6.8\ kg \cdot m^{-2} \cdot h^{-1}$. Now, knowing that line Ib also passes through the static point P_b, the maximum hydraulic loading rate can be determined as $T_{sm} = z \cdot Q_{av}/A_d = 6.8/3 = 2.3\ m \cdot h^{-1}$.

It is concluded that the maximum flow that can be applied before overloading occurs is 130% larger than the average value. However, simultaneously it will be necessary to increase the recirculation factor to $s = X_t/(X_r - X_t) = 3/(8 - 3) = 0.6$. Hence, for the maximum influent flow of $2.3 \cdot Q_{av}$, the return sludge flow Q_r would have to be equal to $0.6 \cdot 2.3 = 1.38 \cdot Q_{av}$, which means an increase with a factor 2.76 compared to the original value of $Q_r = 0.5 \cdot Q_{av}$.

(4) If for the maximum influent flow rate of item (3) the sludge abstraction rate is maintained at the original value of $u = 0.5\ m \cdot h^{-1}$, what will be the rate of sludge accumulation in the settler and what will be the rising velocity of the interface level between sludge and supernatant?

Line Ic (equal to Ib) and IIc apply. When a hydraulic loading rate of $T_s = 2.3\ m \cdot h^{-1}$ and a return sludge abstraction rate u of $0.5\ m \cdot h^{-1}$ are applied, the settler is overloaded (line IIc in Figure 8.17). For these values the recirculation factor is determined as $s = u/T_s = 0.5/2.3 = 0.22$ and the solids loading rate and the limiting flux are calculated as:

$$F_{sol} = (s + 1) \cdot T_s \cdot X_t = (0.22 + 1) \cdot 2.3 \cdot 3 = 8.1\ kg \cdot m^{-2} \cdot h^{-1}$$
$$F_l = 5.45\ kg \cdot m^{-2} \cdot h^{-1} \text{(as calculated under item 2)}$$

Hence, there will be an accumulation rate of $F_{ac} = F_{sol} - F_l = 8.1 - 5.45 = 2.65\ kg \cdot m^2 \cdot h^{-1}$. The concentration of the accumulated sludge will be equal to the limiting sludge concentration $X_l = 8.23\ g \cdot l^{-1}$ (as calculated under item 2), so that the level of the sludge-supernatant interface will rise with a rate of $F_{ac}/X_l = 2.65/8.23 = 0.32\ m \cdot h^{-1}$. Since the sidewall depth of the final settler is usually in the range of 4 to 6 m, it is concluded that the overload may persist for several hours before solids will appear in the effluent.

Chapter 9

Sludge bulking and scum formation

9.0 INTRODUCTION

Serious problems may occur in the solids-liquid separation step of activated sludge systems. Overloading of the final settler has already been discussed and is basically the result of poor design or from increased flow- and load to the final settler. In the preceding chapter it was presupposed that the sludge maintains certain time-invariable settling characteristics. In practice the sludge settleability may vary considerably and at times the settleability can become so poor, that liquid-solid separation in the final settler is only partial, even when the settler was adequately designed and operated. When the reason for this behaviour is loss of compressibility of the sludge flocs, this phenomenon is called sludge bulking. Another problem that may occur is formation of a layer of scum or foam on top of the reactor or the final settler, resulting in smell problems and potentially loss of biomass with the effluent.

9.1 MICROBIAL ASPECTS OF SLUDGE BULKING

To explain the reasons for the occasional appearance of poorly settling sludge, it is necessary to first discuss the question why activated sludge forms macroscopic flocs that can be separated from the liquid phase by settling. According to Jenkins et al. (2004) the basic mechanism that drives activated sludge floc formation can be described as:

- Microbial adhesion, due to the presence of extracellular polymers that form bridges between the micro-organisms. In healthy biological sludge, these extracellular polymers typically make up between 15 to 20% of the dry sludge weight;
- At near neutral pH, these polymers carry a net negative charge, which allows divalent ions such as Ca^{2+} and Mg^{2+} to interact with the polymers and results in bridge formation between the micro-organisms.

Several researchers have shown that the macrostructure of sludge flocs is formed by filamentous organisms, i.e., micro-organisms that produce branches many times longer than the cell diameter. Szegin (1978) postulated that these filaments create a framework within the floc that gives it mechanical strength and to which the bacterial cells can attach. In absence of filamentous organisms, flocs will be perfectly spherical. When filaments are present, flocs will be irregular and can grow larger, as they are stronger. Therefore filaments are indispensable for development of strong, healthy sludge flocs. When filament growth is insufficient, small and weak flocs will develop with a low settling velocity.

The growth of micro-organisms with long and strong filaments allows the formation of macroscopic flocs that remain intact even in the turbulent environment of the aeration tank. However, when the filamentous organisms grow excessively, sludge settleability will be reduced due to two factors: (I) the floc becomes less dense, so that its apparent weight and hence its settling velocity tends to decrease and (II) the approximation of flocs becomes difficult because of the long filaments that form "bridges" between flocs. On the other hand, it is possible that the growth of filamentous organisms is insufficient. In that case the flocs are weak and tend to disintegrate to form small flocs that have a very low settling velocity (pin point flocs). So, basically three different situations can be distinguished relative to the filament content and structure of the sludge floc (see also Figure 9.1):

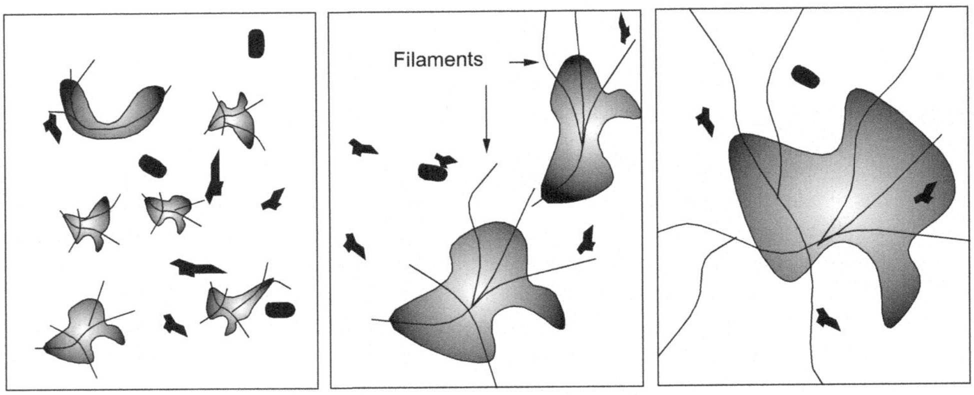

Figure 9.1 Schematic representation of floc structure of pin point flocs (left), ideal (healthy) sludge flocs (middle) and filamentous (bulking) sludge flocs (right)

(a) Ideal, non filamentous flocs:
  - Balance between filaments and floc formers resulting in big and strong flocs;
  - Filaments do not protrude much outside of the flocs and so do not hinder the sludge flocs to approach each other;
  - Zone settling velocity larger than $1.0\ m \cdot h^{-1}$ at $4.0\ g\ TSS \cdot l^{-1}$;
  - Clear supernatant (less than $15\ mg\ TSS \cdot l^{-1}$) and compact settling volume with a low to moderate DSVI value, typically less than $120\ ml \cdot g^{-1}$ TSS).

(b) Pin point flocs:
  - Few filamentous organisms present;
  - Small and weak flocs (typically less than 75 μm);
  - Due to the low filament content, the flocs are vulnerable to shear from the turbulence introduced by aeration and pumps, causing floc breakup;
  - Turbid supernatant, as smallest flocs and debris settle poorly;
  - Low DSVI values, as due to the absence of filaments extending out of the flocs the sludge has excellent thickening characteristics.

(c) Filamentous or bulking sludge:
- Filamentous organisms predominate, protrude outside the flocs and interfere in the settling process. The formation of diffuse, stretched out flocs with low apparent density might also occur;
- Big and strong flocs;
- Very clear supernatant as the flocs "sweep" the liquid phase as the interface travels downward;
- High to very high DSVI values, typically between 150–200 $ml \cdot g^{-1}$ TSS but up to 400 $ml \cdot g^{-1}$ TSS in extreme cases.

Two other types of microbial growth morphology are sometimes observed, not linked to the relative content of filaments in the bacterial population:

(d) Dispersed growth:
- Caused by absence or disruption of exo-polymer bridging;
- As a consequence, cells do not adhere to each other;
- Possible causes are:
  - Selective growth of non-flocculating bacteria due to an excessive F/M ratio;
  - High concentrations of monovalent- relative to divalent cations;
  - Deflocculation by toxic materials and surfactants.

(e) Viscous or zoogleal bulking:
- Due to an excessive quantity of extracellular polymers present in the biomass, which might reach up to 90% of the sludge mass;
- Dispersed flocculent cells are surrounded by large amounts of water retentive polymers;
- Often linked to shortage of nutrients;
- Very high DSVI values possible (up to 400–500 $ml \cdot g^{-1}$ TSS).

Several research workers have contributed to define what could be called an "excessive fraction" of filamentous organisms and to identify the main types of these organisms. Lee et al. (1983) tried to obtain a relationship between macro-parameters SVI or DSVI and the presence of filaments in the sludge. Sludge samples were microscopically examined and the total length of filaments per gram of sludge was determined. The resulting correlation is shown in Figure 9.2. After a more or less gradual increase of the DSVI value to 150 $ml \cdot g^{-1}$ up to a total filament length of 30 $km \cdot g^{-1}$, the DSVI rapidly increases when the total length of the filaments exceeds 30 $km \cdot g^{-1}$. Lee et al. (1983) concluded that above a DSVI value of 150 $ml \cdot g^{-1}$, the filamentous organisms dominated and formed a "bulking" sludge. This value is now generally accepted as the transition value between normal and bulking sludge.

Eikelboom and Van Buijsen (1981) completed a noteworthy investigation that led to the development of an identification system for filamentous organisms by microscopic observation of the activated sludge. Parameters such as sludge morphology, relations with other organisms and the form of the flocs where used to identify the filamentous organisms. Whenever it was not possible to identify the micro-organism as representing an existing species, they added a number for identification. This system of classification, though updated several times, is still widely used as the principal method for identifying filamentous micro-organisms.

Studies to identify the predominant filamentous organisms by Eikelboom and Van Buijsen's method were carried out in several countries. These studies showed that about ten types of filamentous bacteria are responsible for more than 90% of the problems related to poor settleability of activated sludge. However the relative proportion of these ten types varies considerable from one country to the other. Jenkins et al. (1986 and 2004) attributed these variations to differences in the raw sewage composition and operational conditions in the treatment plant: most notably the sludge age and applied F/M ratio.

Simultaneously to the identification of filamentous organisms, data concerning the conditions that stimulate their growth was obtained. Thus, it was possible to correlate the appearance of some types of filamentous organisms to particular operational conditions of the activated sludge process or to the characteristics of the influent wastewater. Table 9.1 shows the conditions that favour the growth of filamentous sludge and the type of bacteria that will typically develop.

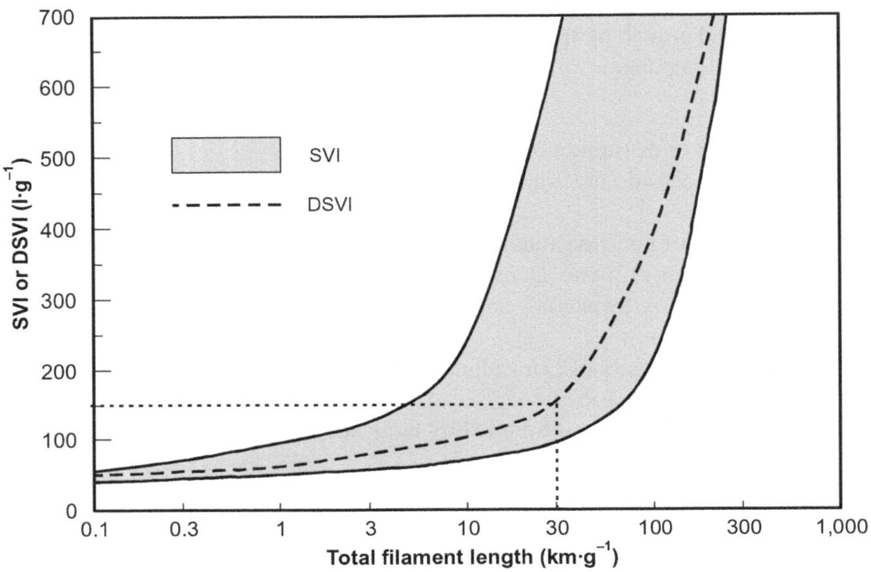

Figure 9.2 Relation between the DSVI and the filament length per unit mass of sludge

Microscopic examination is a very important tool in control of sludge bulking problems, as it enables identification of the bacteria causing problems. Furthermore, as changes in microbial composition often precede changes in activated system behaviour, it can be used to prevent bulking problems from materializing, provided appropriate measures are taken. It can also be used to monitor the effectiveness of the applied control measures, by observing the response in microbial population and composition. It is recommended to use microscopic examination in conjunction with the design and operational data of the activated sludge system under investigation. The interested reader might consider for instance Eikelboom (2000), who significantly updated and expanded on his earlier work from 1981.

According to Jenkins et al. (2004) and Tandoi et al. (2006), filamentous bacteria share the following typical characteristics:

- Most filamentous bacteria cannot denitrify and those who can only to NO_2^- and not to NO_3^-. The main types of denitrifying filaments are M. parvicella and Type 0092;
- Many filaments are capable of storing internal cell products such as PHB, glycogen, poly-phosphate etc. Some store sulphur particles, for instance *Thiotrix spp.*, *Beggiatoa spp.*, Types 021N and 0914;
- Filamentous bacteria often have a lower growth rate than most floc-forming bacteria, but a higher substrate affinity, as will be demonstrated in the next section

These characteristics will prove very useful when examining the causes of sludge bulking and for the development of remedial measures.

Table 9.1 Types of filamentous organisms in activated sludge as indicators of the operational conditions of sludge bulking

Cause of sludge bulking	Indicator type of filamentous bacteria
Low DO concentration	Type 1701; S. natans; H. hydrossis
Low F/M ratio or low substrate concentration	M. parvicella; *Nocardia spp.* Types 0041; 0675; 1851 and 0803
Nitrogen removal configuration (anoxic/aerobic zones)	M. parvicella; H. hydrossis; *Nocardia spp.*; Types 021N; 0042; 0675; 0092; 0581; 0961 and 0803
Septic sewage: presence of reduced sulphur compounds (such as S^{2-}) and VFA	Thiotrix I and II; *Beggiatoa spp.*; N. limicola II Types 021N; 0411; 0092; 0581; 0914 and 0961
High grease / oil content	*Nocardia spp.*; M. parcivella and Type 1863
Nutrient deficiency	Thiotrix I and II; S. natans; H. hydrossis ; N. Limicola III; Type 021N, 0041 and 0675
Low pH value	Fungi

Note: the numbers refer to the system developed by Eikelboom and Van Buijsen (1981)

9.2 CAUSES AND CONTROL OF SLUDGE BULKING

Even after several decades of research there is still considerable debate on the causes of sludge bulking and the remedial measures to be taken. In the subsequent sections the main causes and theories on development and prevention of filamentous sludge bulking will be discussed:

- Low substrate concentration;
- Incomplete denitrification (anoxic-aerobic bulking);
- Low oxygen concentration;
- Septic wastewater (i.e. wastewater containing reduced sulphur compounds and VFA);
- Low pH (fungal sludge bulking);
- Nutrient deficiency.

The reason why filamentous sludge develops can often be established by identifying the type of filamentous bacteria that are predominant in the sludge, taking into account the design and operational conditions of the wastewater treatment plant. By eliminating the cause, presumably the problem can be permanently solved.

9.2.1 Sludge bulking due to a low reactor substrate concentration

Chudoba (1973) considered that activated sludge contains two dominant types of organisms: floc formers and filamentous organisms. Depending on the operational conditions of a particular activated sludge process, one or the other will have the highest net growth rate and hence will dominate. When the filamentous organisms dominate this can result in sludge bulking.

In the 1970s Chudoba developed a hypothesis to explain why sludge bulking was observed in many activated sludge systems at that time: mostly aerobic systems with a short residence time designed for secondary treatment only. Chudoba suggested that the values of the parameters in the Monod kinetics for growth on easily biodegradable substrate (μ_{max} and K_s) were different for the two types of organisms. In Figure 9.3 the specific growth rate curves of the two types of organisms are displayed. It can be observed that Chudoba attributed lower values for μ_{max} and K_s to the filamentous organisms.

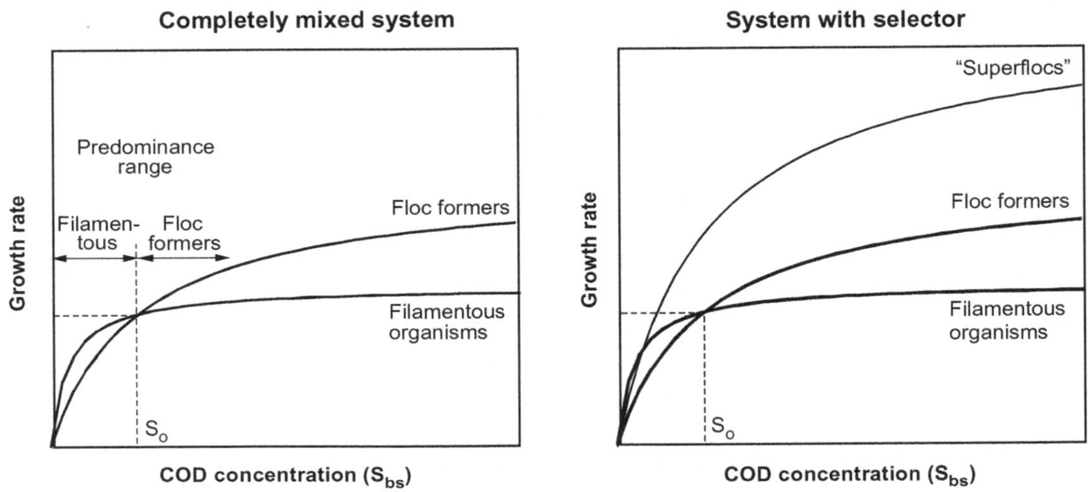

Figure 9.3 Growth rate of floc formers and filamentous organisms as function of the concentration in completely mixed systems and in systems with a selector

This means that below a certain minimum substrate concentration (S_o in Figure 9.3), the growth rate of the filamentous organisms will exceed that of the floc formers, resulting in sludge bulking. Above substrate concentration S_o, the floc formers will dominate and the sludge will settle well. As the objective of biological wastewater treatment is to produce an effluent with a low concentration of biodegradable material, in most activated sludge systems the substrate concentration will thus be lower than S_o and therefore favourable conditions for sludge bulking will exist.

The same hypothesis that explains the phenomenon of sludge bulking also offers a method to avoid it. Chudoba suggested the installation of a small aerobic reactor, which he called the selector, in front of the main aerobic system. In this selector the return sludge is brought into contact with the influent, resulting in a concentration of biodegradable material higher than S_o. It is assumed that the floc formers are able to outcompete the filamentous organisms by rapid absorption of- and subsequent growth on the easily biodegradable material. In the main reactor the remaining easily biodegradable solids concentration will be low ($< S_o$), but this will not compensate for the advantage that the floc formers have already had in the selector. In short, the selector stimulates selective growth of the floc formers, hence its name. Although the filamentous organisms are not eliminated from the system, the selector controls their presence and in general a sludge with good settling characteristics is expected. In any case, a certain amount of filamentous organisms is required as a "backbone" for floc formation.

An alternative hypothesis was offered by the group of Prof. Marais (Casey et al., 1994). Two activated sludge systems were operated under equal conditions, one with and the other without a selector. The oxygen uptake rate (OUR) was measured when influent was added to batches of mixed liquor of each system. In Figure 9.4 the resulting typical OUR profiles are displayed. Immediately after the influent was added, there is a high OUR peak associated with the utilisation of easily biodegradable material. In the second phase, the OUR decreases and is associated with the utilisation of slow biodegradable material. Finally, the OUR decreases to a base level that is associated with endogenous respiration.

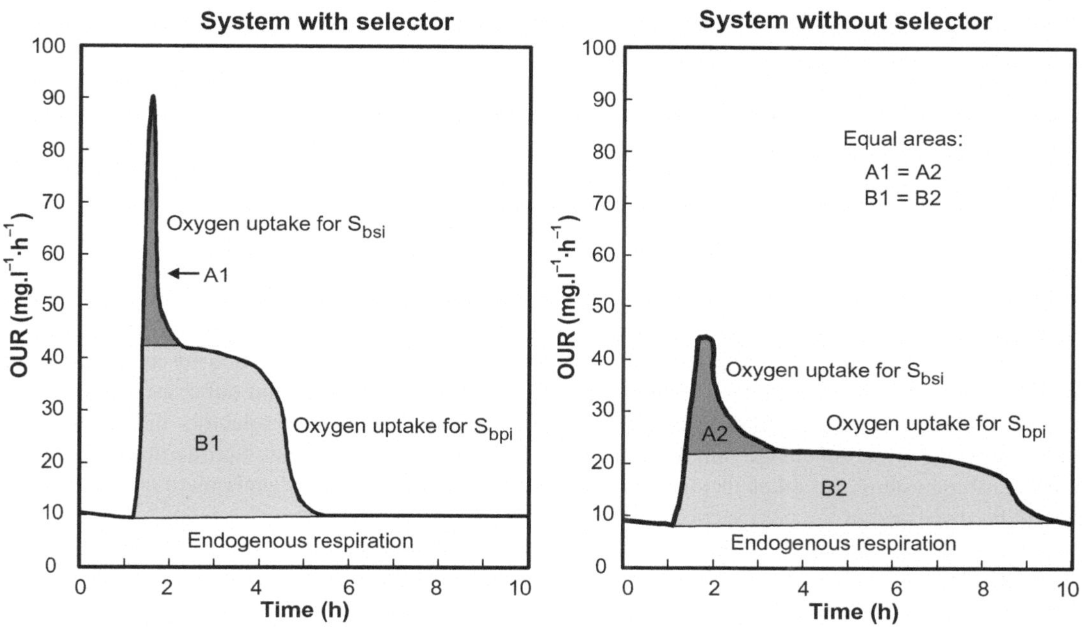

Figure 9.4 OUR profile of a mixed liquor batch after influent addition for a system with and without a selector

The peak OUR of the batch from the system with a selector was about 3 to 5 times higher than that of the system without a selector, while total exogenous oxygen demand was equal for both batches. This can be explained if in systems with a selector the floc formers have a 3 to 5 times higher growth rate than the floc formers in systems without a selector. These "super" floc formers also outgrow the filamentous organisms and will thus always dominate in the system. The growth curve of the super floc formers is indicated in Figure 9.3b.

Later research indicated that the selector stimulates the growth of micro-organisms that accumulate cell-internal polymers. In the subsequent (aerobic) reactor, diffusion through cell wall has already occurred, which allows for a rapid onset of respiration. Therefore it seems likely that these "super" floc formers are in fact organisms capable of storing internal cell polymers.

9.2.2 Guidelines for selector design

The common application of selectors in aerobic activated sludge systems in the US and Europe has generally produced satisfying results. However, the theories described above were derived at a time when most activated sludge systems were purely aerobic and designed for secondary treatment only. Afterwards, aerobic selectors have not been very successful in preventing sludge bulking in activated sludge processes designed for nutrient removal, as the causes for sludge bulking in this case turned out to be different. Initially it was tried to emulate the success of the aerobic selectors by installing anoxic and anaerobic selectors, but the results have been mixed at best.

A distinction can be made between two types of selectors:

- *Kinetic* selectors, like the ones described in the previous section. The selection mechanism is based on exploiting differences in the values of the kinetic parameters of the filamentous and floc-forming micro-organisms, such as growth rate and substrate affinity;
- *Metabolic* selectors, where the idea is to implement conditions that significantly stimulate the growth of organisms having a specific metabolic route available or alternatively to inhibit the growth of organisms that do not have it. For example, an anaerobic selector stimulates the growth of organisms with the capability of internal cell storage of polymers, such as the bio-P organism and reduces substrate availability to other micro-organisms.

As initially it was assumed that filamentous organisms were all strictly aerobic, it was presupposed that anaerobic or anoxic selectors would seriously disadvantage the filaments by reducing substrate availability. However, as discussed previously, there are filaments capable of (partial) denitrification, such as M. parvicella, which effectively ends the possibility of *metabolic* selection by anoxic selectors. This still does not prohibit kinetic selection, as differences in growth rate and substrate affinity might still exist between facultative filaments and floc-formers. However, anoxic selectors did not have the same suppressing effect on sludge bulking in systems for nitrogen removal that their aerobic counterparts had for aerobic systems, so other factors favouring growth of filamentous organisms must exist as well. This will be discussed in the next section.

As to anaerobic selectors, these units are very comparable to the anaerobic zone used in bio-P removal. Provided the design is such that the selector is protected against the return of nitrate, bio-P bacteria will develop, which are all floc-forming bacteria and are known to contribute to the formation of large, strong flocs. However, anaerobic selectors will not resolve the problems associated with the bulking sludge variant found in nitrogen removal systems.

The precondition to effective selector performance is that a significant proportion of the available substrate is utilized by floc-forming bacteria. According to a STORA report addressing the subject of selector design (Van Starkenburg et al., 1994), several mechanisms can be involved: (I) adsorption on active sludge flocs, (II) uptake and metabolization of soluble substrate under aerobic conditions and (II) uptake of soluble substrate under anaerobic or anoxic conditions. The adsorption of substrate on active sludge flocs is a rapid (physical-chemical) process that is not influenced by the presence or absence of oxygen and nitrate. This process removes predominantly particulate- and colloidal substrate.

The second process, uptake and (partial) metabolization of soluble substrate under aerobic conditions, represents the classical aerobic selector as intended by Chudoba. Sludge loading is high at 0.4–1.0 $kg \cdot COD \cdot kg^{-1} VSS \cdot d^{-1}$, while the contact time is (necessarily) short at only 10–15 minutes. The aeration capacity should be sufficient to sustain an oxygen respiration rate of 40–50 $g\ O_2 \cdot kg^{-1} VSS \cdot h^{-1}$.

Another important design consideration is that respiration rate of the return sludge should be at the endogenous level before mixing it with the influent in the selector, as otherwise insufficient metabolic capacity will be available for uptake and metabolization of the fresh influent substrate. Above a F/M ratio of 0.8 $kg\ COD \cdot kg^{-1} VSS \cdot d^{-1}$, it will be difficult for the return sludge to reach the endogenous respiration level. For a typical raw municipal sewage this is approximately equivalent to a sludge age of 5–6 days or less (refer also to Figure 3.14).

As for the third mechanism, the uptake of soluble substrate under anoxic or anaerobic conditions, this concerns *metabolic* selection against organisms that do not denitrify or do not have the capability of storing substrate anaerobically. Under anoxic conditions both uptake and usage can occur, but the rate is reduced compared to aerobic conditions. Under anaerobic conditions only uptake will occur. The design contact time in the selector is significantly longer: 45–60 minutes for anoxic selectors and up to 2 hrs for anaerobic selectors.

Although the effectivity of anoxic selectors is debatable, on the other hand the anaerobic selector will preselect for development of bio-P bacteria, capable of taking up VFA under release of phosphorus (refer to Chapter 7). It seems that only flocculent bacteria are capable of bio-P accumulation. Although glycogen accumulating organisms may develop in the absence of sufficient phosphate, these are also flocculent bacteria.

Several guidelines on the sizing of selectors are summarized in Table 9.2, based on the STORA research conducted by Van Starkenburg et al. (1994) and on the review data gathered by Jenkins et al. (2004).

Table 9.2 Main design criteria for different types of selectors: adapted from Van Starkenburg et al. (1994) and the review data by Jenkins et al. (2004)

Mechanism	Contact time (minutes)	Sludge loading rate ($g\ COD \cdot g^{-1}\ VSS \cdot d^{-1}$)	Respiration rate ($g\ O_2 \cdot g^{-1}\ VSS \cdot d^{-1}$)
Adsorption	5–10	1.0–1.5	0
Anoxic/anaerobic	60(1) 45–120(2)	No guideline given(1) 1.5 overall and 5–6 in 1st compartment(2)	0
Aerobic	10–15	0.5–1.5(1) 3–4 overall and 10–12 in 1st compartment(2)	50–60

Notes: (1) Van Starkenburg et al. (1994) and (2) Jenkins et al. (2004)

Concerning the physical layout of selectors, the following aspects should be considered:

- The effect of the selector significantly increases when the unit is constructed as 3–4 compartments in series;
- The design of the selector should allow for flexible operation, for instance a partial bypass of influent- and return sludge flows in order to achieve the desired loading rates;
- It is recommended that the selector volume can be modified as well, using movable baffles and exit gates. If the selector is too large, the sludge loading rate may be insufficient. On the other hand, if it is too small (excessive loading), then EPS may be formed leading to viscous bulking;
- The volume of an anoxic- or anaerobic selector should be considered as a part of respectively the pre-D and the anaerobic zone.

9.2.3 Control of bulking sludge in anoxic-aerobic systems

With the development of activated sludge systems designed for nutrient removal, other types of filamentous organisms started to appear. This phenomenon was first discovered in 1984 by Blackbeard et al. (1984) in

South Africa. Initially, Jenkins (1986) attributed the appearance of organisms like M. parvicella; H. hydrossis; Nocardia sp.; Type 021N; 0042; 0675; 0092; 0581; 0961 and 0803 to a low F/M ratio or equivalent to a high sludge residence time. Casey et al. (1994) demonstrated that the presence of these organisms only occurred in anoxic-aerobic systems designed for nitrogen removal and not in strictly aerobic systems. In aerobic systems operated at a high sludge age, the causes of sludge bulking were often determined as a low oxygen concentration, septic sewage or shortage of nutrients.

Once these causes were eliminated, the presence of filamentous organisms could be controlled with a selector. On the other hand, in systems with alternating anoxic/aerobic zones the implementation of a selector was shown to be ineffective. Especially systems with intermittent aeration, such as carrousels, are vulnerable to sludge bulking (Kruit et al., 2001 and Hartley et al., 2008). Casey et al. (1994 and 1999) established that the condition for filamentous organisms to appear in nitrogen removal systems is an anoxic sludge mass fraction larger than 30%. It is interesting to note that in completely anoxic systems, a sludge develops with good settleability characteristics, comparable to those of a completely aerobic system.

The findings that (I) the micro-organisms responsible for sludge bulking in denitrifying systems are different from those in aerobic systems, (II) the selector did not eliminate the filamentous organisms, and (III) there was a clear relationship between the relative size of the anoxic zone and the severity of the sludge bulking problem, led Casey et al. (1994) to conclude that the presence of filamentous organisms in nitrogen removal systems (called anoxic-aerobic or AA bulking) should be attributed to the metabolism of the heterotrophic organisms in alternating anoxic-aerobic environments. This could be explained by the following observations:

- In an aerobic environment, the consumption of oxygen by bacteria is mediated by two cytochromes denominated "O" and "aa3". Cytochrome "O" is constantly active while cytochrome "aa3" is dormant but can be activated;
- Under anoxic conditions, nitrate is reduced to molecular nitrogen, with several intermediate products that will be present in the micro-organisms:

$$NO_3^- \longrightarrow NO_2^- \longrightarrow NO\,(\text{nitric oxide}) \longrightarrow N_2O\,(\text{nitrous oxide}) \longrightarrow N_2$$

- During the denitrification process all intermediate products appear. Only when denitrification is complete then the intermediates will have disappeared;
- If nitric oxide is present in micro-organisms in an aerobic environment, it will inhibit the utilisation of oxygen by cytochrome "O", which normally would have been used. Instead "aa3" will be used, but the activation of this dormant cytochrome will require considerable time. During the period that "aa3" is not yet fully active, the organism will continue to denitrify at low rate, even when oxygen is present;
- If no nitric oxide is present in micro-organisms when the sludge is transferred from an anoxic- to an aerobic zone, then cytochrome "O" will be activated in the normal manner and the consumption of oxygen starts immediately.

The following experimental procedure can be used to demonstrate the above behaviour:

- Take two batches of sludge from the anoxic zone of a pre-D system and wait until denitrification is complete;
- Add NO_2^- to one of the batches so nitric oxide will be formed inside the micro-organisms;
- Add allylthiourea (ATU) to inhibit nitrification, which would remove NO_2^-;

– Before the nitrite removal is complete, add influent to both batches, start aeration and measure the resulting oxygen uptake rate. The resulting OUR profiles are shown in Figure 9.5.

As can be observed, the OUR in absence of nitric oxide follows the familiar pattern: an initial peak OUR due to the utilisation of readily biodegradable organic material, followed by a decrease in OUR when slowly biodegradable material is utilised, until a final baseline OUR is reached, which is equal to the endogenous respiration rate (for more information on the use of respirometrics refer to Appendix 1 and 2).

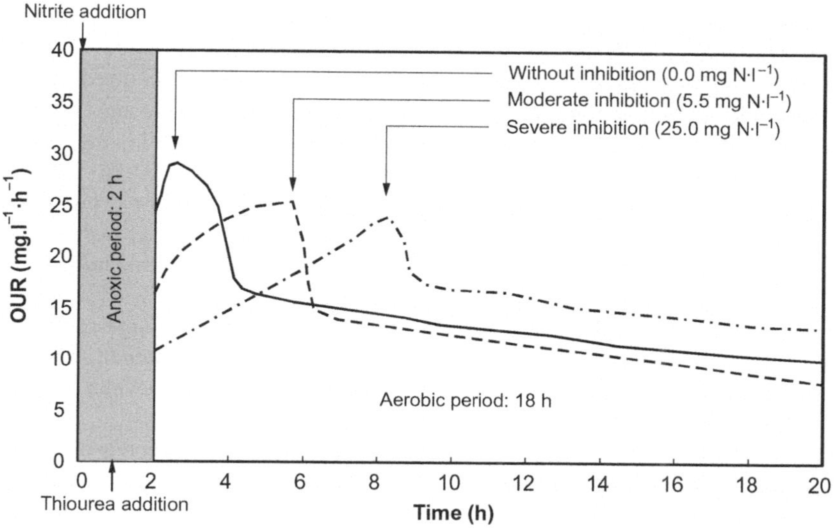

Figure 9.5 Typical OUR profiles in sludge batches taken from a pre-D system, with and without the addition of nitrite

In contrast, the sludge batches with nitric oxide $(NO_2)^-$ exhibit a prolonged period in which the OUR only gradually rises from its initial low value to a much lower peak value. From there on the OUR follows the same pattern as above. Another finding was that if sufficient easily biodegradable COD was present in the anoxic zone, sludge bulking did not develop, as the intracellular nitric oxide is then rapidly reduced.

It is important to consider that the behaviour of filamentous- and floc forming organisms is very different with respect to denitrification. The filamentous organisms that are capable of denitrification such as M. parvicella are nitrate reductors only, i.e. producing nitrite, while the floc formers in general are able to use all the intermediate components between nitrate and nitrogen for the oxidation of organic material. This also implies that nitric oxide will never be present in the cells of the filamentous organisms, while it might be present in the floc forming organisms. Therefore the latter might be inhibited by nitric oxide if denitrification is not complete. Thus if nitrite (and nitric oxide) is present at the end of the anoxic zone or period, the filamentous organisms have an advantage as the growth rate of the floc formers will be reduced. As nitric oxide inhibits the utilisation of oxygen and thus the substrate consumption rate, growth on oxygen will be inhibited and although denitrification will take place this process is slower than pure aerobic growth. The filamentous organisms may then become dominant and thus cause sludge bulking problems.

As the cause of bulking in systems with denitrification now appears to be clear, the strategy to combat sludge bulking in these systems can be developed. Basically, it is imperative that denitrification in the anoxic zone is complete. Thus the system has to be operated in such a way that the denitrification capacity of any anoxic zone that discharges into an aerobic zone is larger than the availability of nitrate in this anoxic zone. When denitrification is complete, the disadvantage of the floc formers compared to the filamentous organisms is removed and sludge with good settling characteristics will develop.

This strategy to combat sludge bulking in anoxic-aerobic systems will in general reduce nitrogen removal efficiency, as the availability of nitrate in the pre-D anoxic zone will be deliberately less than the pre-D denitrification capacity. If the nitrate concentration in the mixed liquor discharged to the settler is high, denitrification can develop in the settler, which might lead to problems with floating sludge.

Tsai et al. (2003) observed that the anoxic-aerobic (A-A) sludge bulking theory developed by Casey et al. (1999) in many cases was able to explain the reason why sludge bulking occurred in nutrient removal systems. However, under certain conditions even complete denitrification in the anoxic zone (i.e. ensuring the total absence of NO) could not prevent sludge bulking by M. parvicella, the main A-A bulking filament. Therefore the A-A bulking sludge theory was supplemented by the following additional hypothesis: M. parvicella cannot use nitrate or nitrite as a nutrient source for growth and requires the presence of free ammonia. Therefore, if nitrification is rapid and complete, growth of M. parvicella is inhibited. This hypothesis also explains several common observations on full-scale nutrient removal plants:

- The seasonal proliferation of M. parvicella with a peak in the end of winter or early spring and a minimum at the end of summer or early autumn. This can be explained by the increase and decrease of nitrification capacity (and hence effluent ammonium levels) resulting from the temperature fluctuations of the activated sludge system;
- The increase in DSVI with decreasing aerobic mass fraction, as this will decrease nitrification capacity and increase ammonia effluent concentration;
- The relationship between a low dissolved oxygen concentration at the end of the aerobic zone and sludge bulking problems, as nitrifiers have a relatively poor affinity for dissolved oxygen and once again reduced nitrification efficiency will result.

The modified A-A sludge bulking theory of Tsai et al. (2003) is supported by the findings of Kruit et al. (2001), who evaluated the performance of four full-scale carrousels in different nutrient removal configurations in terms of bulking sludge problems. The research project confirmed that a dedicated (separate) pre-D anoxic zone with full denitrification upstream of the circuit consistently decreased DSVI values to less than 150 $ml \cdot g^{-1}$ TSS. To obtain a further decrease of the DSVI to values less than 120 $ml \cdot g^{-1}$ TSS, a dedicated strictly aerobic zone was required at the end of the circuit, with $DO \geq 1.5\ mg \cdot l^{-1}$ and $NH_4\text{-}N \leq 1\ mg\ N \cdot l^{-1}$.

Finally, Hartley (2008) examined different control strategies to reduce sludge bulking in carrousel systems designed for nitrogen removal and concluded the following:

- Sludge settleability and effluent nitrogen concentration both are dependent on the size of the anoxic fraction and highest sludge settleability coincides with lowest total effluent nitrogen concentration;
- The effluent ammonia-N to nitrate-N ratio can be used as an alternative for the anoxic sludge mass fraction, which is in practice difficult to establish in a carrousel;
- Contrary to the findings by Tsai (2003), the absolute concentration of effluent ammonia did not seem to appear to affect sludge settleability. Rather, the ammonia to nitrate ratio (which represents the value of the anoxic sludge mass fraction) appears to be the dominant factor;

- The optimum ammonia to nitrate ratio in the effluent depends on the applied sludge age but is roughly equal to one. Therefore an operating rule of thumb is to operate with about equal effluent ammonia and nitrate nitrogen concentrations, through appropriate control of the DO setpoint.

9.2.4 Other causes of sludge bulking

The following causes of sludge bulking will be discussed in this section:

- Low DO bulking;
- Septic wastewater;
- Nutrient deficiency;
- Fungal bulking.

Several filamentous organisms are able to grow at low DO concentrations. These organisms tend to proliferate in completely aerobic systems. The most probable mechanism behind this phenomenon is insufficient availability of oxygen inside the sludge floc. As filamentous organisms extent outside the floc and into the bulk liquid, they do not suffer the oxygen limitation to the same extent as the floc formers located inside the sludge floc. There is no fixed lower threshold concentration below which this phenomenon occurs, as the value of the limiting bulk DO concentration depends on applied organic sludge loading rate. The penetration depth of oxygen into the sludge floc is a function of the respiration rate and the bulk DO concentration (refer also to Figure 5.9). Therefore low DO sludge bulking has even been observed at systems operated at a high bulk DO concentration. When this type of sludge bulking is diagnosed, the following measures can be taken:

- Increase the oxygen bulk concentration. For systems operated at an $F/M \leq 0.5$ kg COD $\cdot$ kg^{-1} VSS $\cdot$ d^{-1} (approximately equivalent to a sludge age of 6–10 days for municipal sewage), typically 2 mg $O_2 \cdot l^{-1}$ is considered sufficient, to be maintained at the location where the oxygen demand is highest;
- Reduce oxygen consumption through a decrease of the sludge age. This will decrease the total sludge mass present in the system and will therefore reduce the endogenous oxygen demand. However, it will also result in increased excess sludge production and might conflict with the requirements for nitrogen removal;

Septic wastewater is characterized by the presence of both VFA and reduced sulphur compounds (HS^- and H_2S). In general this wastewater exhibits a bad smell due to the presence of sulphides and has a dark colour, caused by precipitation of FeS. Examples of septic wastewater are for example sewage subjected to a long residence time in the sewer system, anaerobically pre-treated wastewater and reject water from sludge treatment processes. Several filamentous organisms have demonstrated a competitive advantage under these conditions as they can grow on reduced sulphur or on low weight VFA. Sludge bulking induced by septic wastewater can be remedied by refreshing the wastewater prior to entry in the main activated sludge system. The following measures may be considered:

- Pre-aeration, although this may release odours and thus might require off-gas treatment;
- Oxidation using chlorination, ozonisation and H_2O_2 addition;
- Chemical precipitation with metal salts as FeS.

Sludge bulking resulting from nutrient deficiency can be very severe and is often related to a shortage of either nitrogen and phosphorus. Whether one of these macro-nutrients is limiting can be easily established with Eqs. (3.58 and 3.60). If insufficient nitrogen or phosphorus is present, the deficit must be added. In general municipal sewage contains sufficient macronutrients and micronutrients (for instance Ca, K, S, Fe, Ni and Mo) to sustain healthy growth. However, for industrial wastewater this might be different.

While nitrogen and phosphorus deficiency is common in wastewaters from vegetal origin, chemical wastewaters and condensates may be lacking in many compounds. Industrial wastewater treatment companies often sell their own proprietary mixtures of micronutrients.

As to the last category, fungal bulking, this is a rare phenomenon as fungi are normally not dominant in an activated sludge system. However, they can dominate at low pH (< 6.5), in which case the remedy is simple: increase the pH value. In case of a nitrifying system the application of denitrification might already be sufficient to raise the pH value sufficiently, apart from the many other benefits associated to it: i.e. a reduction in oxygen demand, increased operational stability and prevention of a rising sludge blanket in the final settler. Otherwise lime or caustic can be used to increase pH.

9.3 NON-SPECIFIC MEASURES TO CONTROL SLUDGE BULKING

In the previous sections specific measures have been described to prevent sludge bulking when the cause of the problem is known. However, in practice situations may exist when the measures discussed above do not apply or do not (yet) have any effect. In this case, general non-specific methods are used to suppress the growth of filamentous organisms. The general strategy for control of filamentous sludge can thus be defined as follows:

1. Use microscopic examination and evaluate the operational conditions of the system and the characteristics of the influent to determine what might be the cause(s) for the development of filamentous sludge;
2. If the reason for the appearance of bulking sludge can be rectified immediately, the appropriate measures must be taken:
   (a) In case of septic wastewater: use pre-chlorination or pre-aeration;
   (b) If there is a shortage of nitrogen or phosphorus for growth: add these nutrients;
   (c) If the pH is low: add alkalinity to the influent;
   (d) If nitrite is entering the aerated zone: decrease the recirculation of nitrate to the pre-D zone (a-factor) or increase the size of the pre-D zone;
   (e) If ammonia is present in the effluent, increase the nitrification capacity, for instance by increasing the sludge age, aeration capacity or aerobic sludge mass fraction.

If the reason is one that cannot be remedied immediately (for example if the aeration capacity is insufficient), rapid non-specific methods may be used to alleviate the situation, without necessarily solving the underlying problem. These methods can be divided into the following categories:

- Manipulation of the sludge recirculation factor "s";
- Manipulation of the inlet location of the influent;
- Application of materials to improve sludge settleability;
- Use of toxic compounds for selective elimination of filamentous micro-organisms.

(1) Manipulation of the sludge recirculation factor

This measure only has an effect if thickening is the limiting process in the final settler. In Figure 8.5c it can be noted that the limiting flux increases with the recirculation factor until the critical value s_c, which is reached when clarification becomes the limiting function of the final settler. There is no point in increasing the recirculation ratio any further beyond this critical value, as this will not increase the liquid-solid separation capacity of the final settler. It may even have an adverse effect due to the increase in turbulence that will result.

(2) Manipulation of the inlet location of the influent

Figure 9.6 is presented to demonstrate how the location of the inlet can influence the operation of an activated sludge process:

(a) System A: standard configuration. It consists of two completely mixed aeration tanks in series followed by a final settler, with return sludge recirculated to the first reactor (which also receives the influent). Consequently the sludge concentration in the two reactors is virtually identical;

(b) System B: step feed configuration. The same aeration tanks in series but the influent flow is now distributed: a fraction "f" is discharged to the first tank (which also receives the return sludge) and the remaining fraction (1–f) is directed to the second tank. In this configuration the sludge concentration in the second tank is smaller than in the first one;

(c) System C: contact stabilization. Again the same two aeration tanks in series. The influent is discharged into the second tank together with the stabilised sludge from the first tank: i.e. after settling, the return sludge is directed to the first tank where it is aerated without feeding (stabilisation). This configuration is equivalent to system B for $f = 0$.

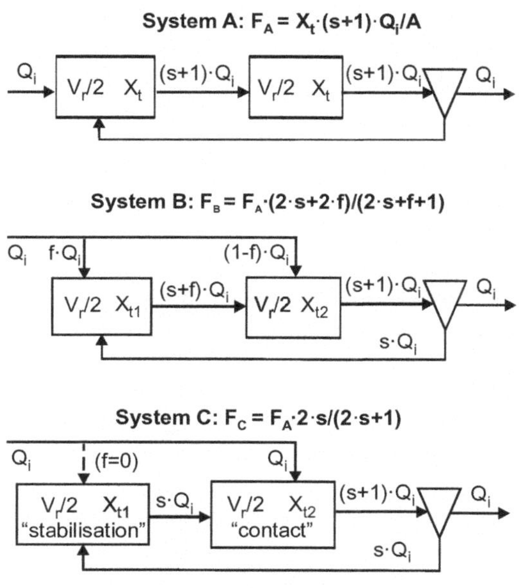

Figure 9.6 Similar system configurations, but with different solids loading rates to the settler

When the three systems shown above treat the same flow of wastewater and the operational conditions are equal, then all systems will develop the same sludge mass MX_t. However, the concentration of sludge in the reactors and the solids loading rate to the final settler are very different. In system A the sludge concentration can be expressed as $X_A = MX_t/V_t$ and the solids loading rate to the settler is given by: $F_{sol,A} = X_A \cdot (s+1) \cdot Q_i/A_d$.

In system B the solids flux entering the second reactor $F_{sol,B}$ is $(s+f) \cdot Q_i \cdot X_{t1}$, which equals the flux leaving this reactor: $(s+1) \cdot Q_i \cdot X_{t2}$. Hence:

$$(s+f) \cdot Q_i \cdot X_{t1} = (s+1) \cdot Q_i \cdot X_{t2} \text{ or } X_{t1} = X_{t2} \cdot (s+1)/(s+f) \tag{9.1}$$

Knowing that $MX_t = V_1 \cdot X_{t1} + V_2 \cdot X_{t2}$ and $V_1 = V_2$, one has:

$$\begin{aligned} X_{t2} &= (2 \cdot s + f + 1)/(2 \cdot s + 2 \cdot f) \cdot MX_t/V_r \\ &= (2 \cdot s + f + 1)/(2 \cdot s + 2 \cdot f) \cdot X_A \end{aligned} \tag{9.2}$$

Now the solids loading rate for configuration B is expressed as:

$$F_{sol,B} = (2 \cdot s + 2 \cdot f)/(2 \cdot s + f + 1) \cdot F_{sol,A} \tag{9.3}$$

For the special case of contact stabilisation (system C) one has $f = 0$ and:

$$F_{sol,C} = 2 \cdot s/(2 \cdot s + 1) \cdot F_{sol,A} \tag{9.4}$$

Hence the ratio of the solids loading rate to the settlers in systems A, B and C can be written as:

$$F_{sol,A}:F_{sol,B}:F_{sol,C} = 1:(2 \cdot s + 2 \cdot f)/(2 \cdot s + f + 1):2 \cdot s/(2 \cdot s + 1) \tag{9.5}$$

Equation (9.5) shows that although systems A, B and C are equal with respect to the reactor- and final settler volume, the influent flow and -load and the sludge mass and -composition, the solids loading rate to the final settler will be different. In Figure 9.7 the ratios $F_{sol,B}/F_{sol,A}$ and $F_{sol,C}/F_{sol,A}$ are shown plotted as function of the recirculation factors. It is very clear that in a system normally operated in accordance with the configuration of system A, the solids loading rate to the final settler can be significantly reduced if the location of the influent inlet point can be changed and the system is operated in the configuration of system B or C. The reduction in solids loading rate may lead to a better performance of the final settler. However, effluent quality will be slightly less.

(3) Application of materials to increase the sludge settleability

In practice the materials used to improve sludge settleability can be classified into two categories: coagulants such as metal salts and lime and flocculants such as poly-electrolytes. Coagulants like lime and ferric salts have been used to increase the settling velocity of filamentous flocs, by increasing the specific weight of sludge flocs and to form a voluminous precipitate that sweeps down the flocs. Typical dosing rates are between 10–25 mg Fe or Al per litre influent. It should be noted that the use of coagulants or lime results in a significant increase in the production of excess sludge. Typically dosing is applied either in the aeration basin or in the overflow to the final settler. Poly-electrolytes may be used to increase sludge density by reducing the bridge forming effect of the filaments, both between two flocs and within the flocs.

The appropriate type of polyelectrolyte and dosing rate can be determined in sludge batches using jar test equipment. Typical dosing rates are in the range of 2–10 mg active polyelectrolyte per litre mixed liquor flowing to the final settler. Poly-electrolytes are usually added into the mixed liquor overflow to the final settler, as they are biodegradable. Polyelectrolyte dosing does not result in increased excess sludge production, but application is quite expensive: typically 5–10 times more expensive than the use of chlorination.

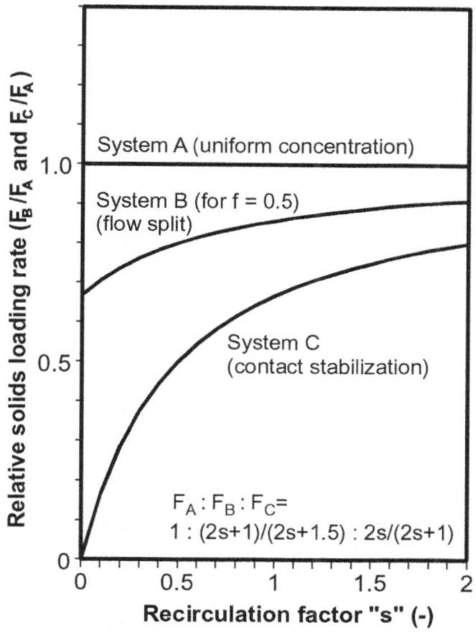

Figure 9.7 Ratio between the solids loading rates of the configurations in Figure 9.6

EXAMPLE 9.1

An activated sludge system is operated under the following conditions:

$Q_i = 5000\,m^3 \cdot d$; $s = 0.5$; $X_t = 4\,g \cdot l^{-1}$ and $V_r = 1000\,m^3$;

Answer the following questions:

- What is the solids loading rate to the final settler, assuming the system is operated as configuration "A" ?
- What will be the solids loading rate when operation is changed to a step feed configuration ($f = 0.5$) or to a contact stabilization configuration ($f = 0$);
- For the step feed configuration, confirm this loading rate explicitly by calculating the sludge concentration entering the final settler.

Solution

(1) Definition of the solids load to the final settler
The solids load is defined as:

$$F_{solA} = X_t \cdot (s + 1) \cdot Q_i = 4 \cdot (0.5 + 1) \cdot 5000 = 30{,}000\,kg \cdot d^{-1}$$

(2) Definition of the solids load in step feed and contact stabilization
In the case of step feeding, for f = 0.5 and s = 0.5 one has:

$$F_{solB} = (2 \cdot s + 2 \cdot f)/(2 \cdot s + f + 1) \cdot F_{solA}$$
$$= 0.8 \cdot 30{,}000 = 24{,}000\,kg \cdot d^{-1} \text{ or } 80\% \text{ of configuration "A"}$$

In the case of contact stabilization, for f = 0 and s = 0.5 one has:

$$F_{solC} = 2 \cdot s/(2 \cdot s + 1) \cdot F_{solA}$$
$$= 0.5 \cdot 30{,}000 = 15{,}000\,kg \cdot d^{-1} \text{ or } 50\% \text{ of configuration "A"}$$

(3) Confirmition of step feed solids load to the final settler
This requires the mass balances over reactor 1 and reactor 2 to be solved. The following equations apply (after Q_i has deleted from both sides of the equations):

Reactor 1: $s \cdot X_{rs} = (s + f) \cdot X_{t1}$ or $X_{t1} = 0.5 \cdot X_r$
Reactor 2: $(s + f) \cdot X_{t1} = (s + 1) \cdot X_{t2}$ or $X_{t2} = \frac{2}{3} \cdot X_{t1} = \frac{1}{3} \cdot X_r$
Furthermore $MX_t = V_1 \cdot X_{t1} + V_2 \cdot X_{t2} = 4000\,kg\,TSS$, so using the relations derived above:
$500 \cdot 0.5 \cdot X_r + 500 \cdot \frac{1}{3} \cdot X_r = 4000$

This can be solved for $X_r = 9.6\,g \cdot l^{-1}$ and $X_{t2} = {}^1/_3 \cdot X_r = {}^1/_3 \cdot 9.64 = 3.2\,g \cdot l^{-1}$. Now the solids loading rate can be calculated as $F_{sol} = 3.2 \cdot (1 + 0.5) \cdot 5000 = 24{,}000\,kg \cdot d^{-1}$. This is indeed 80% of the value for the conventional system configuration.

(4) Use of toxic compounds for selective elimination of filamentous micro-organisms
Among the components used to eliminate filamentous micro-organisms, the most important are chlorine in its different forms: i.e. liquid NaOCl or on-site generated as gaseous Cl_2 while sometimes hydrogen peroxide is used, although the efficiency of the latter chemical is much less. The toxic effect of these chemicals is based on a strong oxidizing action.

Filamentous bacteria are more sensitive to these toxic chemicals than floc formers, as the filaments often extend outside of the sludge floc into the bulk phase where the concentration of the toxic chemical is highest. The specific area of filamentous bacteria is much larger and so is their exposure. Therefore the addition of these chemicals to the sludge in a suitable concentration and frequency will lead to the selective destruction of the filamentous organisms. Often a rapid decrease in DSVI is observed, in the case of unsheathed filaments often within 5 days. If sheathed filaments are present, then more time is required because, although the filaments will die, their sheaths will remain in the sludge. As the only means of disposal from the system is in the excess sludge discharge, it will typically take at least 1–2 times the sludge age before the effect on DSVI becomes noticeable.

The selected dosing location should guarantee intensive mixing (and dilution) with the sludge. Furthermore, to minimise the chlorine demand it is important that the organic material concentration is as low as possible, so that the added chemical is used effectively to eliminate the filamentous organisms and not for the oxidation of organic material. Nitrite and sulphide are both oxidised by chlorine, while ammonia is converted into mono-chloramine, which is a far less potent disinfectant than chlorine or hypochlorite. Usually, the most appropriate dosing point is the return sludge channel, preferably before an elbow bend in the pipe to increase turbulence. Alternatives are directly into the final settler center feed well or into a dedicated side stream, if the return sludge flow is small compared to the reactor volume.

In so far as the required chlorine addition is concerned, several design parameters are used, such as the applied daily mass per unit mass of sludge in the process (2–10 g $Cl_2 \cdot kg^{-1}$ VSS $\cdot d^{-1}$) and the application rate at the addition point (kg $Cl_2 \cdot kg^{-1}$ TSS) and the initial chlorine concentration (mg $Cl_2 \cdot l^{-1}$). Another design variable is the frequency at which the sludge passes the dosing location. For a dosing point located in the return sludge channel this parameter can be expressed as:

$$f_r = s \cdot Q_i \cdot X_r/(V_r \cdot X_t) = s \cdot Q_i/V_r \tag{9.6}$$

where f_r = average daily frequency of exposure at the chlorine injection point.

Jenkins et al. (1986, 2004) suggested that the value of f_r should not be less than 3 d^{-1}. If the frequency of exposure is significantly lower than that, additional chlorine should be added in the aeration basin or (preferably) in a dedicated side-stream. It is important to monitor the effect of chlorine addition on the sludge settleability and effluent quality. The first effect can be determined quantitatively by tests like SVI or its improved versions $SSVI_{3.5}$ or DSVI. The applied dosing rate of the toxic compound must be carefully controlled so as not to overdose. An increase in the effluent turbidity is an indication that the chlorine addition is excessive and that the active sludge flocs are being destroyed.

9.4 CAUSES AND CONTROL OF SCUM FORMATION

The formation of a thick scum or foam layer at the surface is a common problem in many wastewater treatment plants. The scum may be present as a thick and viscous floating scum layer when the environment is tranquil, as in the anoxic reactors or the final settler, or as foam in the aerated zones. The scum may cause serious operational problems, such as a reduction in the oxygen transfer efficiency of surface aerators. Furthermore, it may spread over the entire treatment plant, making it slippery and creating unsafe situations. The formed scum layer may have such a large volume that it contains a considerable fraction of the total active sludge mass, which will then not take part in the biological conversion process. Finally, in regions with a hot climate, the scum layer may start to decompose anaerobically and cause odour problems.

Foam and scum can have either physical-chemical or microbial origins. Examples of physical-chemical causes are the presence of fats, grease or surfactants in the wastewater, or excessive energy input because of over-aeration. Microbial scum might result from very high F/M loading rates (e.g. during start-up), from nutrient deficiency, denitrification or from the development of hydrophobic filaments that attach to air bubbles and float to the surface.

In wastewater treatment plants a differentiation can be made between two types of foaming mechanisms (Jenkins et al., 2004): two phase and three phase dispersion. The two phase dispersion is formed by air bubbles in water and represents what most people would consider as "foam". The foam collapses when the water layer separating the air bubble from surrounding air becomes too thin to contain the pressure inside the bubble, causing it to burst. The liquid film thins out by draining and evaporation. Two-phase foams are stabilised by surfactants, which allows a thinner water layer before surface rupture occurs. The second mechanism that stabilizes foams is the presence of hydrophobic particles. The three phase foam contains gas, water and hydrophobic solids: i.e. filamentous organisms. If the solid particles are large enough, they may bridge the water film, preventing drainage of water from the foam and thus stabilizing it. The suspended solids fraction in (collapsed) foam can be substantial, up to 4–6% wt. This foam type represents what most people consider as "scum". Foam bubbles are less stable than scum, more fluffy and can easily collapse. Foam is often caused by:

- Excessive aeration;
- Application of high organic loading rates/very low sludge ages (e.g. during start-up). For instance Type 1863 foam appears at sludge ages lower than 2 days and forms a white-grey foam that easily collapses;
- Toxic- or temperature shocks;
- Presence of excessive content of fats & grease or detergents (surfactants) in the sewage.

Once the cause is addressed, foaming generally quickly ceases. The following measures can be applied for general non-specific control: spraying of water on the surface with jet nozzles and dosing of antifoam oil: 2–10 $mg \cdot l^{-1}$ based on influent flow.

The two main causes for scum formation in activated sludge systems are denitrification in the final settler and the presence of filamentous organisms, causing entrapment of gas bubbles by hydrophobic micro-organisms that will float on the surface area. At normal values of the sludge age denitrification in the final settler is practically unavoidable for temperatures around 20°C and for effluent nitrate concentrations larger than 6–8 $mg\ N \cdot l^{-1}$ (Henze et al., 1993). On the other hand, at a very high sludge age, or when anaerobic effluent is treated, endogenous denitrification rates are too low to cause problems (refer to Section A8).

Denitrification results in the production of micro bubbles of nitrogen gas once the liquid is saturated. These micro bubbles of nitrogen gas are very efficient at floating activated sludge because they are produced inside the sludge flocs, firmly attach to them and hence lower the apparent density of the sludge floc.

As to the second cause of scum formation, often this is caused by the same filamentous organisms that are responsible for sludge bulking. It has also been established that the appearance of scum is generally associated with the presence of bulking sludge: the scum appears before the bulking sludge problems become manifest and only disappears when a large part of the bulking sludge has already been removed. The most common scum forming micro-organisms are M. parvicella, often a nuisance in winter time and *Nocardia spp.*, which is often present in the summer. There are many possible causes for the growth of *Nocardia spp.* and M. parvicella, the most common being:

- High grease- and oil content in the sewage (inadequate pre-treatment);
- Low F/M or high sludge age applied;
- Anoxic-aerobic sludge bulking;
- Septic sewage;
- Low dissolved oxygen concentration;
- Presence of ammonium in the end of the aerobic zone.

It is easy to distinguish between scum caused by denitrification and filamentous organisms (Jenkins, 2004). Denitrification scum is often accompanied by the occurrence of small bubbles in the final settler. Furthermore there is no difference between the count of filamentous organisms in the scum and in the mixed liquor, whereas there is a much higher abundance of filaments in scum when filamentous organisms are involved. Finally during the DSVI test a floating layer will be formed if denitrification is the problem, whereas a greasy surface layer will be formed when filaments are the problem.

M. parvicella is often present in anoxic-aerobic systems as it is capable of partial denitrification to nitrite. Conventional selectors (both aerobic and anoxic types) are not effective against this organism. Possible control measures against M. parvicella have been extensively discussed in the section on anoxic-aerobic bulking and will be repeated only briefly here:

- Ensure complete denitrification in pre-D compartment;
- Prevent low dissolved oxygen concentrations in aerobic zones;
- Ensure that nitrification is complete (< 1 mg $N \cdot l^{-1}$) at the end of the aerobic zone;
- Intermittent aeration systems (carrousels) are much more susceptible to bulking/foaming problems with M. parvicella. A separate pre-D anoxic zone and equal NO_3-N/NH_4-N concentrations in the final effluent improve sludge settleability considerably.

Jenkins et al. (2004) recommend the following control measures against *Nocardia spp.*:

- Installation of aerobic- or anoxic selectors;
- Installation of anaerobic selectors, but only if bio-P bacteria develop;
- Selective wasting of foam from the reactor, e.g. using a flotation cell with fine bubble aeration;
- Regular chlorination methods are not effective as Nocardia is mainly present inside the sludge flocs. However, spraying a "mist" on the foam can be very effective;
- Application of cationic polymers, which will flocculate free Nocardia on the sludge flocs, hence reducing the foaming potential.

In general scum problems increase significantly if scum is retained on the surfaces of treatment units. So in order to prevent or reduce scum problems, the design of the wastewater treatment plant should be such that scum is not retained selectively and that scum removed from the system is not recirculated. The most important design measures are:

- The connection between two subsequent reactors and between the reactor and final settler should be designed as an overflow device, maintaining a difference in water level between the two reactors. This way it will be impossible for the scum to move upstream in the reactor chain while it also facilitates the discharge to the final settler where it can be removed. To avoid structural problems when filling up or draining a reactor, it will be necessary to install some form of gates or valves in the bottom part of the division wall between the reactors which can be opened as required;
- In case surface aerators are used with variable oxygen input capacity, the preferred method of control is either through adjustment of the immersion depth of the rotor blades or with frequency control on the motor (controlling the rotational speed). On/off control should be avoided;
- The surface skimmer should discharge into the grease trap, from where the scum is transported to thickening, stabilisation and final disposal. Recycling to the biological reactors should be avoided, as the concentration of filamentous organisms in the scum is much higher than in the mixed liquor (seeding effect);

- In the anoxic zone the mixers should have a rotational centrifugal movement, inducing the mixed liquor to be moved away from centre, so that the scum does not become trapped around the axis of the mixer;
- The dissipated power of the mixers should not exceed the requirements for effective mixing of the reactor contents;
- The suction lines of recirculation and sludge recycle pumps should be located well under the surface level, in order to prevent recirculation of scum;
- The inlet point of influent, sludge recycle and recirculation flows should be positioned to prevent formation of dead zones in the reactor where the scum can become trapped.

Finally, some non-specific control measures to prevent scum formation are the following:

- Reduce the operational sludge age: i.e. operate at minimum sludge age required to meet treatment objectives;
- Add anaerobic digestion supernatant to the aeration tank;
- Use anti scum agents;
- Spray water on the settler surface to break up the scum layer.

Reduction of the sludge age is only possible when the units for sludge treatment have enough capacity to handle the increase in sludge production. More importantly, in the case of nutrient removal processes, a reduction of the sludge age is in general not very attractive, as it might become impossible to achieve the required effluent standards. The scum formation is suppressed if the sludge age is below the minimum sludge age for growth of *Nocardia spp.* and M. parvicella, but this value is so low that in general nitrification will also be suppressed.

Lechevalier (1974) observed that the supernatant of anaerobic digesters contains material (possibly sulphide) that is toxic for *Nocardia spp.* and that addition of the supernatant to the mixed liquor may remove the scum. The experiences with anti-scum agents have so far not been very effective for scum removal. On the other hand foam suppression can be quite successful using anti-foam. The most successful method, spraying water on the scum surface in the settler, has been shown to be effective for control of scum formation. The sump for skimmed material must have sufficient capacity to receive and transport the removed material.

Chapter 10

Membrane bioreactors

10.0 INTRODUCTION

The traditional activated sludge system is currently the most popular and most widely implemented wastewater treatment system. The main reasons for this success are flexibility in design, good effluent quality at reasonable costs, high process stability (especially when compared to anaerobic systems) and relative ease of operation. In the last decades significant improvements have been made in the design and operation of the activated sludge system: e.g. extension with nutrient removal, improved effluent quality, reduced aeration cost and reduced sludge separation problems. The basic configuration of the activated sludge process has remained essentially the same in all these years: i.e. an aeration basin with optional anoxic and/or anaerobic zones followed by a final settler.

Assuming that the biological treatment capacity has been properly sized, the final settler may be considered as the "Achilles' heel" of the conventional activated sludge process. The performance of the final settler is crucial in order to meet a low effluent suspended solids concentration and to retain the biomass in the system. Influent flow variations and changes in microbial population (e.g. bulking sludge) may result in loss of solids with the effluent. During rainfall conditions a large part of the reactor biomass may be transferred to the final settler, reducing treatment capacity. Prolonged hydraulic overloading of the final settler will result in an increase in the effluent suspended solids concentration and possibly in the loss of a substantial part of the biomass. Therefore, final settlers are often sized for peak flow rates. Furthermore the biomass concentration in the aeration tank will be low, typically between 3 to $5\,gTSS \cdot l^{-1}$, in order to reduce the solids loading rate of the final settler. The disadvantages of the conventional activated sludge system thus can be summarised as:

- A low biomass concentration in the aeration tank, requiring a large treatment volume;
- The final settler requires a significant surface area;
- The effluent contains suspended solids, precluding re-use of effluent for high quality purposes without further treatment steps (effluent polishing);
- The effluent quality is vulnerable to upsets in final settler performance, due to the possibility of a high effluent suspended solids concentration, with the associated organic nitrogen and -phosphorus content.

Until the end of the 20th century, the only serious alternatives were the attached film reactor and the sequential batch reactor (SBR). The use of attached film reactors has been steadily declining although trickling filters are still used in developing countries as a low-tech aerobic (post-) treatment step. Disadvantages of this system are limited flexibility with respect to achieving nutrient removal and frequent operational problems, such as freezing, clogging of the packed bed and odour problems.

In a SBR all treatment steps are combined in a single reactor: i.e. feeding, aeration, mixing, settling and decanting. Recently the SBR has regained popularity, mostly for small-scale industrial applications. Disadvantages of the SBR are the need for influent buffering and that the installed aeration capacity is larger than in a comparable activated sludge system. This will partly offset the reduction in investment costs achieved by the removal of the final settler. The buffer tank stores all influent received during the process phases when the SBR cannot be fed (e.g. decanting and settling). Furthermore, in case of municipal sewage treatment, the buffer tank should also be sized for rainwater flows. It is possible to dispense with the buffer tank if several SBR's are operated in parallel or when a shorter cycle time can be applied during peak flow.

Recently several new activated sludge process configurations have been developed:

- The membrane bioreactor (MBR): a modification of the conventional activated sludge system in which the final settler is replaced by micro- or ultra-filtration membranes that retain all suspended solids, allowing only the clean effluent ("permeate") to pass;
- The aerobic granulated sludge bed reactor (GSBR): a modification of the conventional SBR system in which, through application of specific operational conditions, a granulated sludge is obtained that settles extremely well. This configuration will be further discussed in Appendix A9;
- The moving bed bioreactor (MBBR), refer to Chapter 11, a hybrid of the conventional activated sludge system and the packed bed reactor. The aeration tank is filled with a plastic support medium that is suspended in the mixed liquor and is retained in the reactor by a screen. A biofilm will form on the support medium, while depending on the specific configuration suspended biomass (sludge flocs) may be present as well. This allows a much higher sludge concentration to be maintained, while simultaneously the solids loading rate to the final settler (or flotation clarifier) is significantly reduced.

These configurations are basically all modifications to the conventional activated sludge system. When more sustainable solutions are required, for instance regarding the consumption of energy and the emission of greenhouse gases, other wastewater treatment technologies might be more appropriate. Sustainable nitrogen removal systems have already been presented in Section 6, while the application of anaerobic pre-treatment is discussed in Chapter 13.

10.1 MEMBRANE BIOREACTORS (MBR)

The main difference between a conventional activated sludge system and a membrane bioreactor is that the gravitational separation of solids/liquid in the final settler has been replaced by a separation process based on the filtration of mixed liquor through a membrane, as is displayed in Figure 10.1.

As discussed in Chapter 8, at higher sludge concentrations the gravity settling flux F_v decreases rapidly while the applied solids flux to the final settler F_{sol} increases rapidly. This is the reason why the biomass concentration in conventional activated sludge systems rarely exceeds a value of $6\ g\ TSS \cdot l^{-1}$ and many cases is even lower at $3-4\ g\ TSS \cdot l^{-1}$. When the gravity settler is replaced by membranes, the maximum sludge concentration in the biological reactor is set by filtration- instead of settleability characteristics. However, a second constraint will be the reduction in oxygen transfer efficiency resulting from operation at high suspended solids concentrations, combined with an increase in the volumetric oxygen uptake rate. In practice, depending on the wastewater treated and the membrane characteristics and -configuration, reactor biomass concentrations up to 50 kg TSS · m^{-3} have been observed, although values between 8 and 15 kg TSS · m^{-3} are much more common.

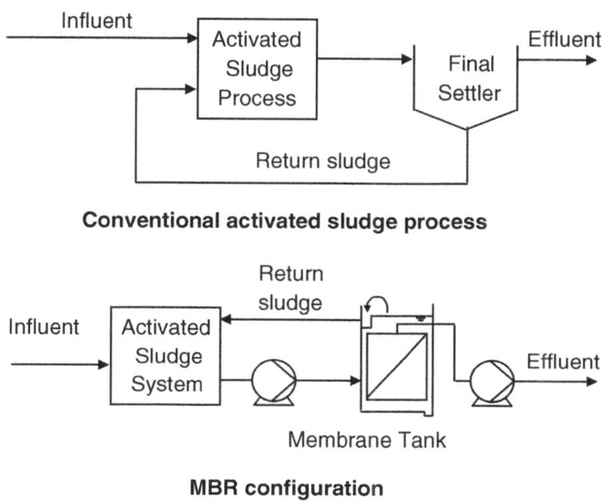

Figure 10.1 Comparison of the basic configuration of a conventional activated sludge system and an MBR equipped with submerged membranes

The membranes used for mixed liquor filtration fall in the micro- to ultra-filtration range, i.e. with pore diameters between 0.01 to 0.4 μm. Membranes effectively remove all suspended solids including free bacteria and colloidal material, while some viruses may pass (unless attached to suspended solids) as indicated in Figure 10.2.

The key drivers for MBR implementation are all related to the possibility of operating at a higher sludge concentration in the biological reactor and to produce an effluent that is free of suspended solids. Therefore, application of MBR might be advantageous when:

- The availability of space is limited. A MBR requires less volume and surface than a conventional activated sludge system, due to the operation at a higher biomass concentration and because the final settler is replaced by a much smaller membrane unit;
- Strict effluent quality limits apply and/or the effluent will be re-used for high quality purposes (e.g. process water). MBR treatment will remove all suspended solids including bacteria, the associated organic nitrogen and -phosphorus content and to some extent adsorbed heavy metals, viruses and endocrine disruptors. Conventional activated sludge systems and SBR will require an additional filtration step to polish the effluent;
- The capacity of existing wastewater treatment plants is to be enlarged (retrofitting);
- Saline wastewaters are treated, which often result in small and weak flocs;
- Difficult to degrade COD is to be removed. However, only when this COD either adsorbs to the sludge or is of a size (e.g. large colloids) that can be retained by the membranes. A very high sludge age is applied in this case. This is not impossible with conventional activated sludge systems, but problems with sludge fines escaping with the effluent are likely to occur, due to pinpoint floc formation.

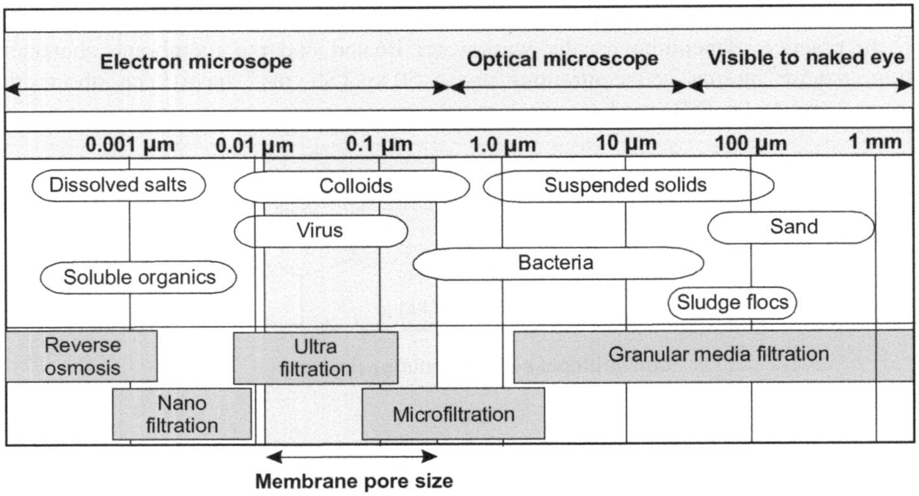

Figure 10.2 Typical range of separation diameters of several filtration techniques. The particle size of common wastewater constituents is indicated as well

The use of UF/MF membranes in the production of potable-, process- and boiler feed make-up water is already very common, dating as far back as the 1970's. While UF/MF membranes are also used for effluent polishing, e.g. when reuse of treated effluent is desired, this is not considered to be true MBR treatment as it involves an additional treatment step after an existing final settler, while in a true MBR the membranes *replace* the final settler. Furthermore the membranes used for effluent polishing are subjected to a low content of suspended solids ($<10-50\ \text{mg TSS} \cdot \text{l}^{-1}$), compared to concentrations of 10 to $50\ \text{g TSS} \cdot \text{l}^{-1}$ in an MBR. This means that membranes for potable water production cannot be used in an MBR installation as they are simply not robust enough.

Although the potential of using membranes as a replacement for a final settler is clear, it was only at the beginning of the 1990s that the first full-scale MBR's were implemented. The most important reasons for this were:

- High membrane investment- and replacement costs;
- High energy use compared to conventional activated sludge systems;
- Rapid drop in membrane flux and -performance as a result of organic-, biological- and chemical fouling;
- Poor membrane and module quality and integrity.

However, membrane technology has developed rapidly over the last 20 years and at present it is considered a mature technology.

10.2 MBR CONFIGURATIONS

All filtration processes operate on the same basic principle: a pressure is applied to force liquid through a semi-permeable barrier, retaining all particles larger than a certain cut-off pore size. In the case of MBR, the flow of the filtrate can be directed either inside-out (cross-flow filtration) or outside-in (submerged membranes). The two different filtration concepts are displayed schematically in Figure 10.3.

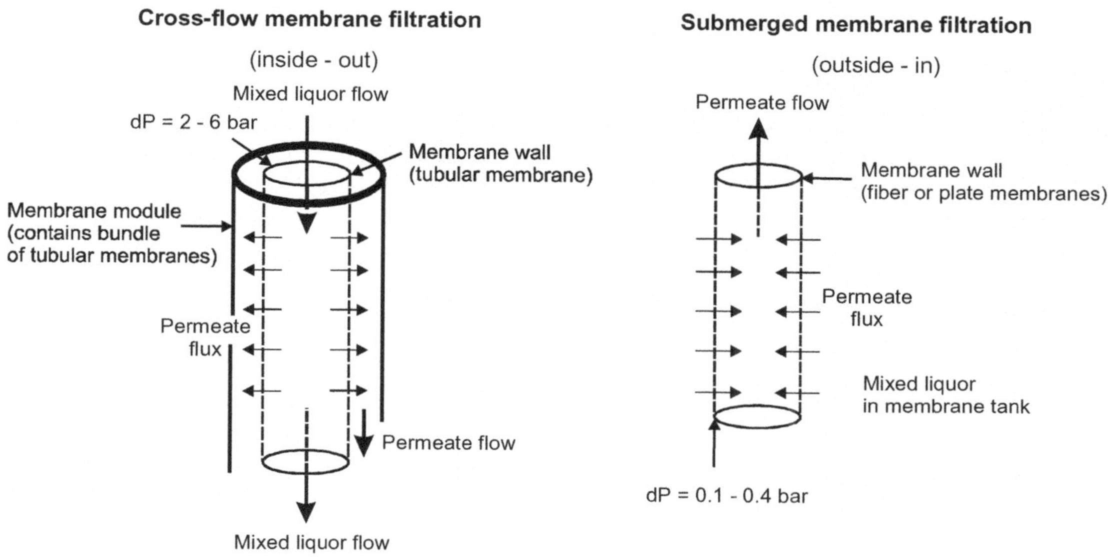

Figure 10.3 Basic working mechanism of cross-flow membranes (left) and submerged membrane filtration (right)

In the cross-flow membrane configuration a mixed liquor flow is pumped through a bundle of tubular membranes mounted together in a closed pipe, the membrane module. The differential pressure over the membrane surface forces the clean effluent (the permeate) through the membrane wall out of the tubes (inside-out principle). A later development are the submerged membranes, which are mounted in mixed liquor in a dedicated membrane tank. A slight vacuum is applied on the membranes by a permeate extraction pump and this induces a liquid flow through the membrane wall (outside-in principle).

The main differences between cross-flow and submerged membrane filtration are summarised below (for an explanation of the terminology refer to Section 10.3.1):

- The permeate flux (the flow of filtered effluent) is directed inside-out for cross-flow membranes (operating at overpressure) and outside-in for submerged membranes (operating at partial vacuum);
- Cross flow membranes are operated at a much higher differential pressure over the membranes ($\Delta p_{TM} = 2-6$ bar) than submerged membranes (0.1–0.4 bar);
- Cross-flow membranes may be operated at higher suspended solids concentration: $15-50\ \mathrm{kg\ TSS \cdot m^{-3}}$ compared to $12-16\ \mathrm{kg\ TSS \cdot m^{-3}}$ for submerged membranes (i.e. in the membrane tank). However, operation at a high biomass concentration will cause other problems: the oxygen transfer rate and -efficiency are significantly reduced and excessive foaming may become an issue. Sometimes pure oxygen is used for aeration when a MBR is operated at a very high biomass concentration;
- Submerged membranes are operated in a constant flux-variable pressure mode: i.e. the differential pressure over the membrane increases in time due to fouling of the membrane surface, while the

membrane flux remains constant (as it is set by the permeate pump capacity). For cross-flow membranes this is exactly the opposite: the applied pressure remains the same but the membrane flux decreases in time (due to fouling).

10.2.1 Submerged MBR

There are two types of submerged membranes: plate- and fibre membranes. Each type has specific advantages and disadvantages, which will be discussed later. However, the basic configuration is similar and is visualised in Figure 10.4. The membranes are placed in a dedicated membrane tank. In earlier MBR configurations the membranes were mounted directly in the aeration tank, but this resulted in serious operational and maintenance problems.

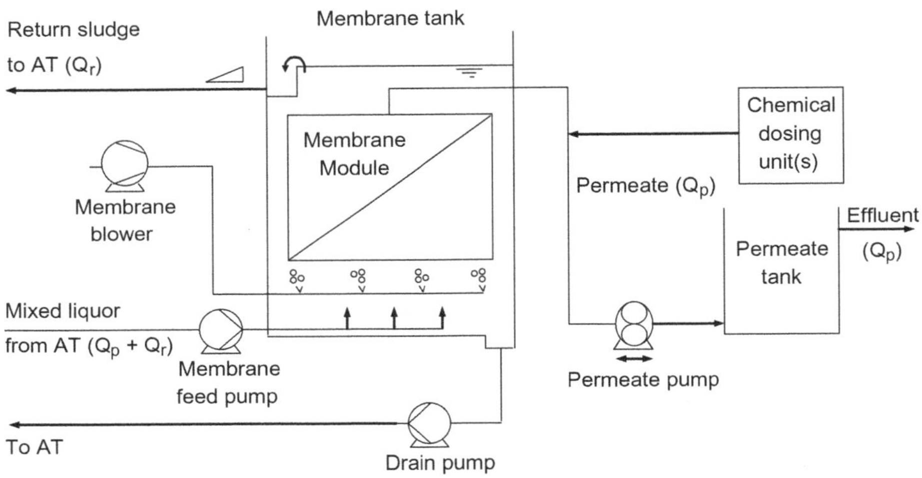

Figure 10.4 Schematic system layout of a submerged MBR system (membrane section)

The number of tanks depends on the size of the influent flow, but often a minimum of two membrane tanks is used in order to allow for cleaning or maintenance. The membrane tank is fitted with one or more membrane modules, each composed of several membrane elements. Module configuration will be discussed in Section 10.3.4. All modules in a membrane tank are connected to a single common permeate header, which is connected to the permeate pump. When the membrane tank is in the normal production mode, mixed liquor from the bioreactor is pumped to the membrane tank and distributed below the membrane modules. The alternative is sometimes also used: mixed liquor flows by gravity to the membrane tank and is pumped back to the aeration tank. The permeate pump applies a light vacuum of typically 0.1–0.2 bar to all membrane elements (plates or fibres), inducing the flow of effluent ("permeate") through the membrane surface and towards the permeate tank for final disposal.

As a result of the extraction of solids-free permeate, the suspended solids concentration in the membrane tank will increase. A sludge cake layer is formed on the membrane surface and if no fresh mixed liquor is supplied, at a certain point the pressure drop over the membranes will be so large that permeate extraction ceases completely. This situation should be avoided at all times, as extensive cleaning will be required to recover the membranes. To limit the increase of the suspended solids concentration in the membrane tank, the mixed liquor flow to the membrane tank, which is equal to the sum of permeate- and return sludge flow rates ($Q_p + Q_r$), is several times larger than the permeate flow rate. The return sludge flow rate is thus equal to the difference between the membrane feed flow rate and the permeate flow rate. The minimum value of the sludge recycle factor can easily be calculated from a mass balance over the membrane tank, if the maximum allowed sludge concentration X_{rmax} in the membrane tank is specified:

$$s_{min} = X_t/(X_{r\,max} - X_t) \tag{10.1}$$

For $X_t = 12$ g TSS $\cdot\, l^{-1}$ and $X_{rmax} = 15$ g TSS $\cdot\, l^{-1}$, the value of s_{min} is calculated as $12/(15 - 12) = 4$. Whereas the *average* value of the permeate flow rate Q_p will be equal to $(Q_i - q) \approx Q_i$, the *instantaneous* value of Q_p may differ from Q_i, as the liquid level in the aeration tank is maintained at the specified setpoint by means of frequency control of the permeate pump, resulting in a variation of Q_p over time.

This is an important difference to a conventional activated sludge system where the value of Q_e follows that of Q_i automatically, although with a time lag, as the flow though the different basins is induced by a small difference in hydraulic level.

When permeate is extracted, a sludge cake layer is formed on the outside of the membranes. While this increases the filtration effectivity as smaller particles are retained, it also increases filtration resistance and thus results in a higher membrane differential pressure (often called trans-membrane pressure or Δp_{TM}). In order to maintain a stable membrane performance, the thickness of the sludge layer must be controlled. One or several mechanisms can be employed:

- Continuous- or periodic aeration underneath or in between the membranes to promote mixing, which refreshes the sludge cake layer;
- Periodic backflushing of the membranes with clean permeate from the permeate buffer tank in order to remove the sludge layer and flush the membrane pores. This is only possible with fibre membranes. The permeate flow is reversed by either changing the pump rotational direction or by opening and closing the appropriate automatic valves;
- Application of a periodic idle time where permeate extraction is stopped while the membranes remain aerated. This strategy is called relaxation and is generally only applied to plate membranes as an alternative to backflushing.

Depending on the nature of the wastewater and the operational conditions, periodic cleaning of the membranes may be necessary (this will be discussed in Section 10.4). Several cleaning chemicals can be used. Depending on the selected cleaning method and the type of membranes used, the membrane tank might require previous draining to increase cleaning effectiveness.

10.2.2 Cross-flow MBR

Cross-flow MBR systems do not require a membrane tank. Instead, stacked racks or skids each typically containing 3 to 7 cross-flow modules are placed near the aeration tank. Each cross-flow module contains a bundle of tubular membranes: at present (2011) for mixed liquor filtration a diameter of 8 mm or 5.2 mm is typical. In general there is a trend towards using tubes with a smaller diameter as this increases the surface area per module and reduces the required recirculation flow rate: the result is an increase in permeate flux and a reduction in power consumption. An additional benefit is that the smaller tubes can be backflushed, should this be required. Membrane modules are available in several standard sizes: common sizes are the 3" and 8" modules. The Pentair (Norit) 8" module has a diameter of 206 mm, a length of 3 meter and a surface area of 27 m^2 (8 mm tubes) and 33 m^2 (5.2 mm). The Pentair 3" module has a diameter of 90 mm, a length of 3 m and a surface area of 4 m^2 (8 mm tubes) and 5.1 m^2 (5.2 mm tubes).

Mixed liquor is recirculated at high speed (3–5 $m \cdot s^{-1}$) through the membrane tubes at medium pressure (3–7 bar). The pressure that is required depends on the nature of the wastewater to be treated, the suspended solids concentration and the number, type and configuration of the membrane modules. The applied pressure forces part of the liquid through the membrane wall, where the produced effluent (permeate) is collected. The permeate flow is perpendicular to the direction of the main recirculation flow in the membrane tubes, hence the name cross-flow filtration. The liquid velocity in the membrane tubes is high in order to ensure that sufficient turbulence is induced to refresh the sludge cake layer and to remove fouling from the membrane surface. The recirculation flow through the membrane tubes is therefore much larger than the permeate flow. Cross-flow membranes can be operated at higher membrane fluxes than submerged membranes and are significantly less vulnerable to fouling. Two different configurations for cross-flow MBR are used (both are shown in Figure 10.5):

(a) Conventional cross-flow MBR, operated as a once-through system in which the complete membrane feed flow (minus the produced permeate) is returned to the activated sludge system;
(b) Feed & bleed cross-flow MBR in which a large part of the flow leaving the last module is recirculated (Q_{rec}) using a cross-flow recirculation pump. The return sludge flow Q_r is much smaller in this case.

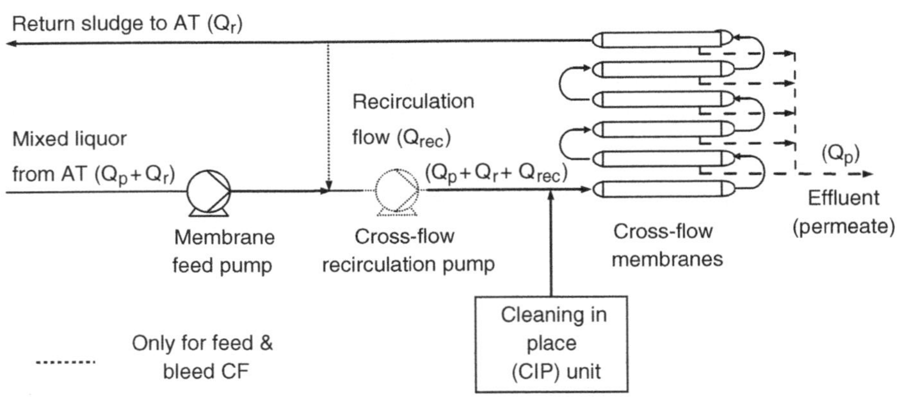

Figure 10.5 Schematic system layout of a cross-flow MBR: both conventional and feed & bleed configuration. Only the membrane section is shown.

(a) Conventional cross-flow MBR

In conventional cross-flow MBR, the membrane section consists of one or more parallel skids, each fitted with 3 to 7 horizontal membrane modules placed in series. The skids are grouped together in membrane units that are fed by a dedicated feed pump, forcing the mixed liquor ($Q_p + Q_r$) through the membrane modules. The pressure applied to the membrane tubes induces a flow of permeate (Q_p) through the membranes into the module, from where it is collected in a header and discharged as effluent. The return sludge flow Q_r is discharged back into the bioreactor. A cleaning in place (CIP) unit is available to periodically clean the membranes by recirculating a chemical solution over the modules (refer to Section 10.4.4).

The pressure at the inlet of the first membrane module is high, in order to compensate for the pressure loss in the system (0.6–0.8 bar per module at 3–5 $m \cdot s^{-1}$). For a 6 module rack with a minimal Δp_{TM} of 2 bar (i.e. the required pressure at the outlet of the last module should be at least 2 bar), the discharge pressure of the recirculation pump is equal to $6 \cdot 0.7 + 2 = 6.2$ bar. The average Δp_{TM} in this case is $(6.2 + 2)/2 = 4.1$ bar. The required flow rate ($Q_p + Q_r$) through the membrane module is determined by:

- The required flow velocity through the tubes;
- The diameter of the modules;
- The diameter of the membrane tubes.

For a standard 8" Pentair module with 5.2 mm tubes and a membrane surface area of 33 m^2, the required module flow is 212 $m^3 \cdot h^{-1}$ for a cross-flow velocity of 4 $m \cdot s^{-1}$. Assuming a membrane flux of 100 litre $\cdot m^2 \cdot h^{-1}$, the permeate production of this module is $33 \cdot 100 = 3.3$ $m^3 \cdot h^{-1}$. The ratio between feed- and permeate flow is dependent on the permeate production, which is a product of the number of modules and the membrane flux, as indicated in Table 10.1.

Table 10.1 Feed to permeate (F/P) ratio as function of the number of 8" Pentair modules (5.2 mm tubes) per skid for a 4 $m \cdot s^{-1}$ cross-flow velocity and a flux of 100 lmh

Number of modules	Inlet pressure (barg)	Feed flow ($m^3 \cdot h^{-1}$)	Permeate flow ($m^3 \cdot h^{-1}$)	F/P ratio (–)
3	4.1	212	9.9	21.4
4	4.8	212	13.2	16.1
5	5.5	212	16.5	12.8
6	6.2	212	19.8	10.7
7	6.9	212	23.1	9.2

A disadvantage of the conventional cross-flow configuration is the large power demand required to pressurise the feed stream. For a pump with water or (diluted) sludge service ($\rho \approx 1000$ $kg \cdot m^{-3}$), the electrical power consumption can be approximated with:

$$P_{el} = Q \cdot 100 \cdot \Delta p/(3600 \cdot \eta_{el}) \tag{10.2}$$

where

P_{el} = electrical power consumption (kW)
Q = pump flow rate ($m^3 \cdot h^{-1}$)
Δp = pump differential pressure or pump head (barg)
η_{el} = pump efficiency (typically 60–70% for centrifugal pumps)

Not only is this very costly (and not really sustainable), but it will also result in an increase in mixed liquor temperature. Not surprisingly this configuration is only used for small installations or for effluent polishing, where due to the much lower suspended solids concentration the required Δp_{TM} is significantly reduced.

EXAMPLE 10.1

A customer is considering treatment of its wastewater ($60\ m^3 \cdot h^{-1}$) in a conventional cross-flow MBR. The following information is obtained from the membrane supplier:

- Design flux $F_m = 100$ litre $\cdot\ m^2 \cdot h^{-1}$ at a minimum Δp_{TM} over the membranes of 2 barg;
- Membrane surface area per 8" module with 5.2 mm tubes: $A_{mod} = 33\ m^2$;
- Required module feed flow $Q_f = 212\ m^3 \cdot h^{-1}$;
- Pressure drop over one module $\Delta p_{mod} = 0.7$ barg;
- Pump efficiency $\eta_{el} = 65\%$.

Calculate the energy requirements for a MBR configuration with three respectively six 8" modules per membrane skid.

Solution

As a first step the permeate production per membrane skid is calculated. For the 3 module configuration, $Q_p = 3 \cdot A_{mod} \cdot F_m = 3 \cdot 33 \cdot 0.1 = 9.9\ m^3 \cdot h^{-1}$. To treat the wastewater flow $60/9.9 = 6.06$ skids will be required: it is assumed 6 skids will be sufficient. The required pump head can be calculated taking into account the minimum trans-membrane pressure Δp_{TM} and the pressure drop over the modules Δp_{mod}.

$$\Delta p = \Delta p_{TM} + 3 \cdot \Delta p_{mod} = 2 + 3 \cdot 0.7 = 4.1 \text{ barg}$$

The required membrane feed flow is equal to $6 \cdot Q_f = 6 \cdot 212 = 1272\ m^3 \cdot h^{-1}$. The power consumption from the membrane feed pumps can now be calculated as:

$$P_{el} = Q \cdot 100 \cdot \Delta p/(3600 \cdot \eta_{el}) = 1272 \cdot 100 \cdot 4.1/(3600 \cdot 0.65) = 223 \text{ kW} \qquad (10.2)$$

The procedure can be repeated for 6 modules per skid. Now only 3 skids are needed. $\Delta P = 6.2$ barg and the membrane feed flow is $3 \cdot 212 = 636\ m^3 \cdot h^{-1}$. This yields a power consumption of 169 kW. It can be concluded that the 6 module configuration is more energy efficient, though the power requirement is still very high at $169/60 = 2.8\ kWh \cdot m^{-3}$ wastewater treated.

(b) Feed & bleed cross flow configuration

This configuration set up is similar to the conventional cross-flow MBR described above and is shown in Figure 10.6. However, a large part of the cross-flow out of the last module is recirculated to the first module by means of the cross-flow recirculation pump. A smaller part, typically 2 to 6 times the permeate flow, is returned as socalled "bleed" to the bioreactor, in order to prevent the build-up of suspended solids in the cross-flow membrane system to unacceptable levels. Thus only the make-up or "feed" flow from the bioreactor needs full pressurization to the required pressure of 4 to 7 bar, while the recirculation flow only needs repressurization to compensate for the pressure loss over the series of membrane modules: as discussed above this is approximately 0.6–0.8 bar per module, including the losses in the connecting bends. Therefore this configuration has a lower energy consumption than the conventional cross-flow MBR.

Figure 10.6 Pentair X-Flow cross-flow skid under construction at the Istac leachate WWTP site in Istanbul – Turkey. Courtesy of Pentair X-Flow

However, the biomass that is recirculated over the membrane modules experiences continuous heavy shear stresses in which the sludge flocs are pulverised. It is still a subject of debate in the MBR community whether membrane performance benefits most from subjecting a small part of the biomass in the MBR to a very high shear stress or a large part of the biomass to limited shear stress.

In Figure 10.7, the energy consumption of a conventional MBR system with 6 modules in series per skid is compared to a feed & bleed MBR. It can be observed that in order to produce 100 $m^3 \cdot h^{-1}$ of permeate in the conventional cross-flow system $10.7 \cdot 100 = 1070\ m^3 \cdot h^{-1}$ of membrane feed flow requires pressurisation to 6.2 barg: the required electrical power is 284 kW. For the feed & bleed system only 500 $m^3 \cdot h^{-1}$ is pressured to 6.2 barg, while the 570 $m^3 \cdot h^{-1}$ of recirculated flow only requires repressurisation from 2 to 6.2 barg: the required electrical power is $132 + 102 = 240$ kW, a reduction of 17% compared to the conventional configuration.

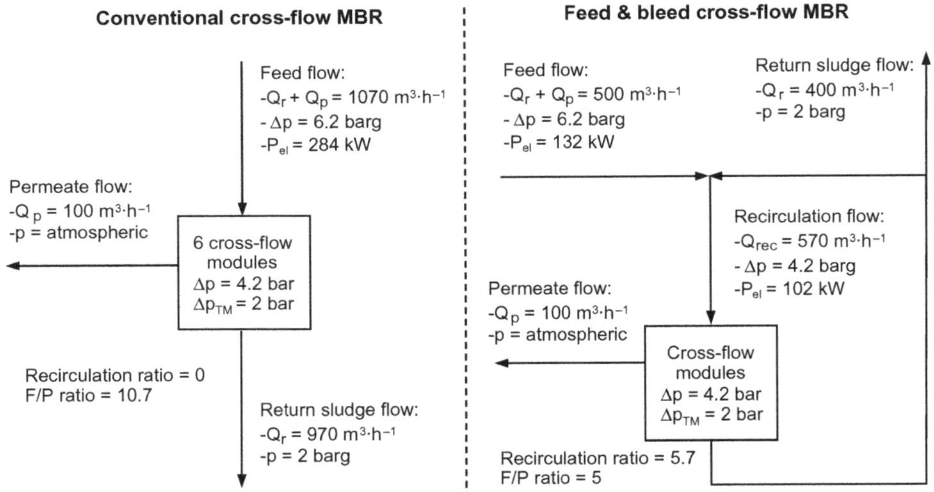

Figure 10.7 Typical mass balance of conventional and feed & bleed cross-flow MBR (100 $m^3 \cdot h^{-1}$ treatment capacity)

EXAMPLE 10.2

The energy consumption of the conventional cross-flow MBR of the previous example is higher than expected. Evaluate the implementation of feed & bleed cross-flow MBR for the 6 modules per skid configuration. Use the data from Example 10.1. The ratio between bleed and permeate flow should be at least 4 to limit the build-up of suspended solids in the membrane loop.

Solution

As in the previous example, 3 skids of 6 modules each will be needed, requiring a total membrane feed flow of 636 $m^3 \cdot h^{-1}$. However, the flow "bled" back to the aeration tank ($= Q_r$!) is now only $4 \cdot 60 = 240$ $m^3 \cdot h^{-1}$. The net feed flow to the system is calculated as $Q_p + Q_r = 60 + 240 = 300$ $m^3 \cdot h^{-1}$. This defines the value of the recirculation flow as:

$$Q_{rec} = Q_f - (Q_p + Q_{rs}) = 636 - 300 = 336\,m^3 \cdot h^{-1}$$

The total power consumption consists of two parts: the net feed flow that requires pressurization to 6.2 barg and the recirculation flow that only needs repressurization from 2 to 6.2 = 4.2 barg. Hence the power consumption can be calculated with Eq. (10.2) as:

$$P_{el} = (300 \cdot 100 \cdot 6.2 + 336 \cdot 100 \cdot 4.2)/(3600 \cdot 0.65) = 140\,kW$$

It can be concluded that modifying operation from conventional to feed & bleed cross flow reduces power consumption by $(169-140)/169 = 17\%$, at the expense of an additional pump. The power consumption remains very high at $140/60 = 2.3$ $kWh \cdot m^{-3}$ wastewater treated.

10.2.3 Comparison of submerged and cross-flow MBR

Initially only cross-flow membrane modules were used in MBR treatment but, as demonstrated above, the energy costs proved to be prohibitive for large scale application to low strength wastewater such as municipal sewage. On the other hand, cross-flow MBR installations have been very successful on the industrial market, and many small industrial systems have been constructed from the beginning of the 1990s.

The submerged membrane systems use less energy (refer to Example 10.3) and are currently replacing the cross-flow systems in popularity, especially since membrane prices have decreased considerably in the last couple of years. Industrial (and to a lesser extent municipal) application of submerged membranes dates back to the end of the 1990s. The investment costs of submerged membranes are higher than those of cross-flow membranes, but the energy consumption of cross-flow membranes is much higher, although this is partly compensated by the energy required for aeration of the submerged membranes. Compared to a submerged MBR, some advantages of a cross-flow MBR are:

- Superior operational reliability and significantly reduced vulnerability to membrane fouling, which makes application to difficult wastewaters easier;
- The performance of cross-flow membranes is less dependent on sludge characteristics than that of submerged membranes;
- Cross-flow membranes are also more robust than submerged membranes: this allows application at higher temperatures (up to 60°C) and also allows more intense cleaning;
- Cross-flow membranes are easily accessible, which facilitates maintenance.

Apart from the high energy consumption, a second disadvantage of cross-flow MBR is that this configuration is less suitable to handle large fluctuations in the feed flow rate. Submerged membranes can be operated (temporarily) above the normal operating flux by increasing the flow rate of the permeate pump (of course within certain limits), although this will result in an increase in the differential pressure over the membranes.

Figure 10.8 shows a typical example. When the membrane flux of a submerged membrane is increased from 20 to 60 litre $\cdot$ $m^{-2} \cdot h^{-1}$, this results in an increase of the Δp_{TM} with 0.2 bar. This peak membrane flux cannot be sustained for very long, as the increase in Δp_{TM} will increase the fouling rate, leading to a further increase in Δp_{TM} until finally cleaning is required. Therefore the maximum duration of peak flux operation is often restricted to less than 1 or 2 days. However, for cross-flow membranes, a similar increase in trans-membrane pressure will only yield a very small increase of the membrane flux, as the membranes are already operated at high pressure. In Figure 10.8 it can be observed that an increase in Δp_{TM} of 0.2 bar results in a very modest additional membrane flux of 3 litre $\cdot$ $m^{-2} \cdot h^{-1}$ or only 2%.

The operational flexibility of submerged membrane systems is very convenient when wastewater with a large ratio between peak and average flow is treated. An example is municipal sewage from combined sewer systems, often with a high ratio between rainy weather flow (RWF) and dry weather flow (DWF). As demonstrated in Figure 10.8, a peak flow of three times the average flow can be sustained by the submerged MBR, provided that the peak flow duration is not excessive. Thus as long as the expected RWF/DWF ratio is smaller than three, the membrane surface area can be sized based on the average flow rate. On the other hand, the cross-flow MBR system will have to be sized based on the peak flow rate, resulting in additional investment costs.

To demonstrate the effect this will have, the following example is given. Consider a municipal sewage treatment plant with an average dry weather flow of 100 $m^3 \cdot h^{-1}$ and a RWF/DWF ratio of 3. For submerged membranes, assume an average flux of 20 $l \cdot m^{-2} \cdot h^{-1}$ and a peak flux of 50 $l \cdot m^{-2} \cdot h^{-1}$. The required membrane surface is therefore equal to the minimum of the following two values: $100{,}000/20 = 5000\ m^2$ or $300{,}000/50 = 6000\ m^2$.

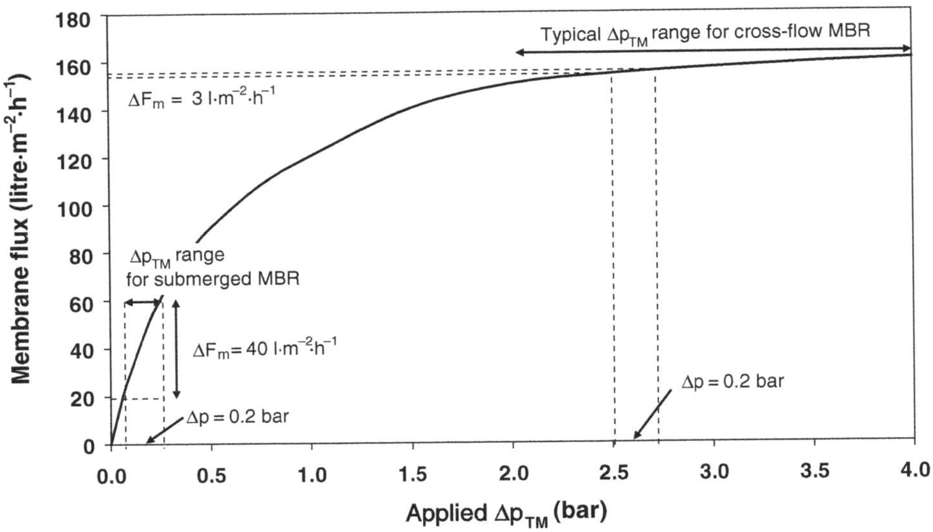

Figure 10.8 Typical response of the membrane flux of cross-flow and submerged membranes on an increase of the applied Δp_{TM}

In this case the membrane surface area for peak flow is limiting, which is 20% larger than the area required for the average flow rate. Now compare this to a cross-flow MBR installation where the membrane flux has a more or less fixed value of $150\ l \cdot m^{-2} \cdot h^{-1}$. The required area for average flow is only 100,000/150 or 667 m^2, but unfortunately the membrane unit will have to be sized for peak flow: $300{,}000/150 = 2000\ m^2$, i.e. three times the area required for average flow. Taking the above discussion into account, it can be concluded that in general submerged MBR is preferred over cross-flow MBR. However, cross-flow MBR can certainly be considered an attractive alternative to submerged MBR for small-scale applications (up to an influent flow of $20–30\ m^3 \cdot h^{-1}$) or when difficult wastewaters are treated.

EXAMPLE 10.3

Returning to the previous example, perhaps submerged membranes are the better choice after all. Evaluate the power demand of a submerged MBR system, using the following data:

Membranes:

- Average net membrane flux $F_m = 10\ l \cdot m^{-2} \cdot h^{-1}$ at 0.2 barg (this includes the back flush);
- Applied F/P ratio = 8;
- Membrane surface area per module: $A_{mod} = 1000\ m^2$;
- Process cycle: $T_{prod} = 400$ s and $T_{bf} = 20$ s.

Pumps:

- Permeate pump: $\Delta p = 2.0$ barg and $\eta = 65\%$;
- Feed pump: $\Delta p = 0.15$ barg and $\eta = 65\%$.

Membrane blower:

- Specific membrane aeration requirement: 0.4 $Nm^3 \cdot m^{-2} \cdot h^{-1}$ (no aeration is applied during backflushing);
- Overall blower efficiency $\eta = \eta_{el} \cdot \eta_{ad} = 65\%$;
- $\Delta p = 0.05$ barg, $H_{liq} - H_{dif} = 3.0$ m; T = 20°C.

Solution

The required membrane surface area can be calculated as $A_m = 60 \cdot 1000/10 = 6000\ m^2$ or 6 modules. The required aeration capacity $= 6000 \cdot 0.4 = 2400\ Nm^3 \cdot h^{-1}$ or $2540\ m^3 \cdot h^{-1}$ at 20°C. The required membrane blower power can be calculated as:

$$P_{aer} = 0.0981 \cdot Q_{air} \cdot p_{atm} \cdot X/\eta$$

Where:

$$X = [(p_{atm} + (H_{liq} - H_{dif})/10 + \Delta p)/p_{atm})\text{^}0.283] - 1$$
$$= [(1.013 + 3/10 + 0.05)/1.013)\text{^}0.283] - 1 = 0.086$$
$$P_{aer} = 0.0981 \cdot 2541 \cdot 1.013 \cdot 0.086/0.65 = 33.3\ \text{kW}$$

To calculate the consumed power for aeration, consider that the net aeration time = 400/420 = 95%. Taking into account the back flush time (when no aeration is required), $P_{aer} = 95\% \cdot 33.3 = 31.8$ kW. Alternatively, Eqs. (4.18 and 4.19) could also be used to calculate the blower power.

The membrane tank feed flow rate $Q_f = 8 \cdot Q_p = 8 \cdot 60 = 480\ m^3 \cdot h^{-1}$. The required pump power of the feed pump is $P_{el} = 480 \cdot 100 \cdot 0.15/(3600 \cdot 65\%) = 3.1$ kW and that of the permeate pump is $P_{el} = 60 \cdot 100 \cdot 2/(3600 \cdot 65\%) = 5.1$ kW. The total power demand of the submerged MBR is 31.8 + 3.1 + 5.1 = 40 kW. This amounts to 24% of the energy demand of the conventional cross-flow configuration and 29% of that of the feed & bleed CF configuration. The power consumption per m^3 wastewater treated is indeed significantly reduced at $40/60 = 0.67\ kWh \cdot m^{-3}$ wastewater treated.

A more recently developed configuration is the socalled "low pressure cross-flow" (LPCF) MBR, shown in Figures 10.9 to 10.12. In this configuration cross-flow membranes are used, but features of cross-flow and submerged membrane systems are combined. Energy consumption is reduced by limiting the liquid velocity in the tubular membranes to approximately $0.5-1\ m \cdot s^{-1}$. To induce sufficient shear stress to keep the membrane surface clean, air is injected in the feed line to the modules. Therefore the membrane modules require a vertical orientation. Furthermore, contrary to the cross-flow systems, the modules are installed parallel instead of in series. Similar to submerged membranes, the membranes are periodically back flushed with permeate and/or allowed to "relax". The applied membrane differential pressure is comparable to that of submerged membranes, i.e. between 0.1 to 0.4 bar.

As the membrane flux is comparable or even lower than that of submerged membranes, the main advantage is that the tubular membranes experience lower fouling rates and are more robust than their submerged counterparts, which makes cleaning easier. Furthermore there is the possibility to change operation of the membranes to cross-flow mode during peak flow situations, provided the cross-flow pump has been sized for this. This significantly reduces the required cross-flow membrane surface area for peak flow demand.

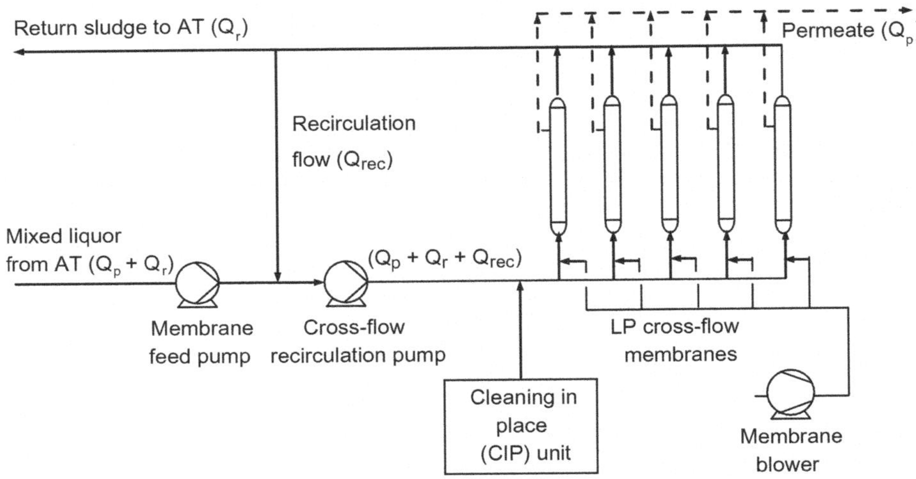

Figure 10.9 System configuration for low pressure cross-flow membranes

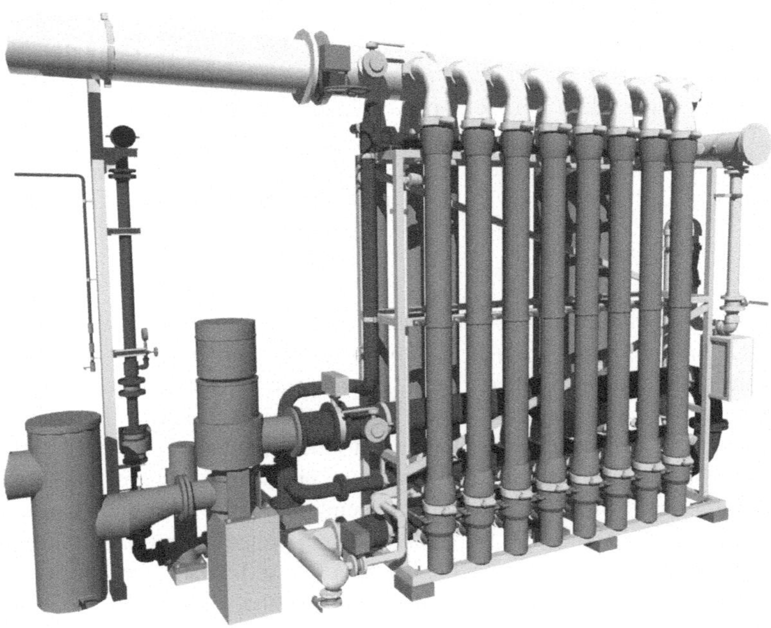

Figure 10.10 General arrangement drawing of a Pentair Airlift™ skid – parallel upflow LPCF configuration. Courtesy of Pentair X-Flow

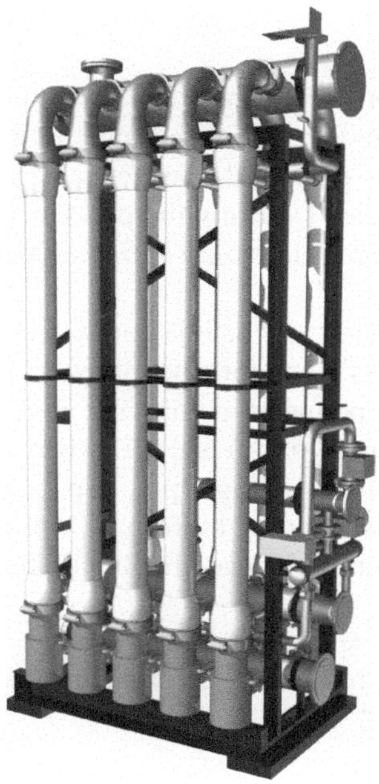

Figure 10.11 Detail of a 10 module parallel Pentair Airlift^{TM} skid. Courtesy of Pentair X-Flow

Figure 10.12 Pentair Airlift^{TM} modules installed at the 14,000 P. E. hybrid MBR (150 $m^3 \cdot h^{-1}$ MBR, 500 $m^3 \cdot h^{-1}$ total capacity) in Ootmarsum, the Netherlands. Courtesy of Pentair X-Flow

10.3 MBR DESIGN CONSIDERATIONS

10.3.1 Theoretical concepts in membrane filtration

The following parameters can be used to characterise membranes and to evaluate membrane performance:

F_m = membrane flux (litre $\cdot$ $m^{-2} \cdot h^{-1}$)
Δp_{TM} = trans-membrane pressure: i.e. differential pressure over the membranes (bar)
Φ = membrane permeability (litre $\cdot$ $m^{-2} \cdot h^{-1} \cdot bar^{-1}$)

The flux of permeate through a membrane is a function of the applied force (i.e. the trans-membrane pressure) and the resistance of the membrane. The membrane resistance is mainly influenced by the size of the membrane flux, the degree of membrane fouling and the liquid temperature. For MBR to be competitive in the wastewater treatment market, it is important to achieve and maintain a high membrane flux, while minimising energy consumption. This implies that the membranes will have to be operated at the lowest Δp_{TM} possible. Significant progress has been made in the last decade in the design of membrane materials that minimise resistance and in the development of module configurations and operational procedures that minimise fouling. An increase in Δp_{TM} will result in an increase in membrane flux if the membrane permeability remains constant. However, apart from increased energy consumption, the increased Δp_{TM} will also accelerate membrane fouling and increase membrane resistance.

The Δp_{TM} of a cross-flow membrane can be determined from the pressure difference between two pressure indicators: the first one mounted on the mixed liquor side and the second one on the permeate side of the membranes. There is a difference in pressure at the front and back end of a membrane module, so in fact the Δp_{TM} decreases over the length of the module. Typically the pressure is measured only before the first and after the last module: all other pressures are inferred assuming a linear decrease of the pressure per module.

For submerged membranes, the Δp_{TM} determination is more complicated as it is not possible to directly measure the pressure on the mixed liquor side (as this is an open tank). Therefore the following procedure is used: a pressure indicator is mounted on the suction side of the permeate pump. The pressure is then determined at two different moments: (I) when the pump is stopped (p_{static}) and (II) when the pump is running in process mode ($p_{dynamic}$) . The difference between the two readings ($p_{static}-p_{dynamic}$) is equal to the pressure drop over the membrane or the Δp_{TM}. It is necessary to measure p_{static} continuously in order to compensate for fluctuations in barometric pressure that influence the liquid height in the membrane tank (static head).

As the value of the membrane flux depends on the applied Δp_{TM}, a better parameter to evaluate membrane performance and status is the permeability ($\text{litre} \cdot m^{-2} \cdot h^{-1} \cdot bar^{-1}$), defined as the membrane flux per unit measure of applied Δp_{TM}:

$$\Phi = F_m/\Delta p_{TM} \tag{10.3}$$

As the relationship between the applied Δp_{TM} and the resulting membrane flux is not linear, it is important that permeability values are always measured at the same standardised membrane flux rate. The maximum permeability is dependent on the membrane type and the nature of the wastewater treated. Although a high permeability is better, it is not so much the absolute value of the permeability, but rather the rate of decline of permeability that is important.

A rapid decrease of membrane permeability indicates that that there are problems: for instance the operational membrane flux is too high or bacterial growth or slime formation is blocking the membrane pores. The permeability is influenced mainly by the degree of fouling and changes in temperature: at higher temperatures the permeability increases as the water viscosity decreases and the membrane pores tend to expand. Ideally the temperature effect is determined for the specific combination of wastewater and membrane, but often empiric correction methods are used. One method is based on the relationship between (sweet) water viscosity and temperature (reference temperature is 15°C):

$$\Phi_{T,ref} = \Phi_T \cdot (\nu_T/\nu_{T,ref}) \tag{10.4}$$

where ν_T; $\nu_{T,ref}$ = viscosity at process temperature and at reference temperature (cP)

$$\begin{aligned} \nu_{T,ref} &= 0.0006 \cdot T^2 - 0.0517 \cdot T + 1.9285 \\ &= 1.288 \text{ for } T_{ref} = 15^\circ C \end{aligned} \tag{10.5}$$

T = reactor temperature

Another correlation is (at a reference temperature of 20°C):

$$\Phi_{T,20} = \Phi_T \cdot \exp[-0.026 \cdot (T - 20)] \quad (10.6)$$

where Φ_T and $\Phi_{T,ref}$ = permeability at process and at reference temperature (litre · $m^{-2} \cdot h^{-1} \cdot bar^{-1}$)

Each membrane has a certain characteristic operational flux range, which depends among other things on the type of wastewater treated and the liquid temperature. Wastewater origin is a key determinant of the applicable flux. While an submerged MBR for municipal wastewater treatment is designed for average gross fluxes of 20–25 litre · $m^{-2} \cdot h^{-1}$ at the reference temperature, for industrial applications this may be significantly lower (5–15 litre · $m^{-2} \cdot h^{-1}$).

EXAMPLE 10.4

A submerged MBR system with 4000 m^2 installed membrane surface area is operated (t = 1) under the following conditions: T = 18°C and $Q_p = 76\ m^3 \cdot h^{-1}$. The operator checks the installation and finds that the Δp_{TM} over the membranes is 100 mbarg. The net membrane flux is calculated as 76,000/4000 or 19 $l \cdot m^2 \cdot h^{-1}$. The permeability Φ_T is calculated as 190 liter · $m^2 \cdot h^{-1} \cdot bar^{-1}$.

Two weeks later (t = 2), the operator checks again and finds that T = 24°C; $Q_p = 108\ m^3 \cdot h^{-1}$ and Δp_{TM} = 140 mbarg. Is performance better or worse than before? Has permeability declined and if so, is there reason to worry? The manufacturer recommends cleaning whenever normalized permeability $\Phi_{T,20} < 175$ liter · $m^2 \cdot h^{-1} \cdot bar^{-1}$.

Solution

To calculate the permeability at t = 2 first the membrane flux must be calculated

$$F_m = Q_p/A_m = 108,000/4000 = 27 \text{ liter} \cdot m^2 \cdot h^{-1}.$$

The permeability at 24°C can be calculated as

$$\Phi_T = F_m/\Delta p_{TM} = 27/0.14 = 193 \text{ liter} \cdot m^2 \cdot h^{-1} \cdot bar^{-1} \quad (10.3)$$

This is in fact a little higher than the permeability observed at t = 1 (190). However, one can only compare the readings once the permeability has been normalized to T = 20°C.

$$\Phi_{T,20} = \Phi_T \cdot \exp[-0.026 \cdot (T - 20)] \quad (10.6)$$

$$\text{At T} = 18°\text{C: } \Phi_{T,20} = 188 \cdot \exp[-0.026 \cdot (18 - 20)] = 200 \text{ liter} \cdot m^2 \cdot h^{-1} \cdot bar^{-1}$$

$$\text{At T} = 24°\text{C: } \Phi_{T,20} = 193 \cdot \exp[-0.026 \cdot (24 - 20)] = 174 \text{ liter} \cdot m^2 \cdot h^{-1} \cdot bar^{-1}$$

It can be concluded in fact permeability has declined by 26 liter · $m^2 \cdot h^{-1} \cdot bar^{-1}$ in two weeks time and has decreased below the value recommended by the manufacturer for cleaning. A regular cleaning cycle is due.

It is important to consider that in MBR design different definitions of flux are being used:

- The *gross* flux is the flux that can be sustained for a long period under the specified operational conditions. When membrane fluxes much higher than this gross flux are applied, a consequence will be rapid membrane fouling, requiring frequent cleaning;
- The *net* flux is calculated when the gross membrane flux is corrected with a factor η_m, in order to compensate for reduced permeate production from periodic back flushing (fibre membranes) or idle time/relaxation (plate membranes). Typically the value of η_m is between 0.8 and 0.95. Cross-flow membranes require neither back flushing nor relaxation, so $\eta_m = 1$ and the average gross and net membrane fluxes are thus equal.

As explained in the previous section, sufficient membrane surface area needs to be installed to treat both average- and peak flow. The membrane supplier will select a design (gross) membrane flux and a value for η_m, taking into account the nature of the wastewater. Often for peak flows a higher flux rate is selected, provided the peak flow duration is limited. Membrane sizing should always be done for the minimum liquid temperature expected, as the membrane permeability decreases rapidly at lower temperature. The required membrane surface area is evaluated for both average- and peak influent flow using the following formula.

$$A_m = Q_i \cdot 1000/(F_m \cdot \eta_m) \quad (10.7)$$

The larger of the two calculated surface areas is selected in design. For municipal applications with combined sewers this will often be the surface area required for peak flow treatment.

10.3.2 Impact on activated sludge system design

The main differences between MBR and conventional activated sludge systems are:

(a) The necessity of liquid level control in the MBR;
(b) An MBR is operated at a higher mixed liquor concentration;
(c) The return sludge flow of a (submerged) MBR is aerobic instead of anoxic while the flow rate is significantly larger as well. The return sludge is sent to the entry of the aerobic zone instead of to the pre-D zone. This will have an impact on nitrogen removal capacity, as the value of N_{avl} is reduced. Furthermore the biomass concentration in the pre-D zone will be diluted.

(a) Liquid level control
In a conventional activated sludge system the level of the hydraulic profile decreases moving from the influent inlet → activated sludge tank → final settler → effluent outlet. An increase in influent flow therefore results in a slight increase in the liquid level in the activated sludge tank, which will in turn increase the flow to the final settler, increase liquid level in the final settler and finally will lead to an increase in effluent flow. Thus, the effluent flow rate follows changes in the influent flow rate (with a time lag) and the variation in liquid level height in the individual units will be very limited. Therefore it is very unlikely that the activated sludge tanks will ever overflow. However, this self-regulating behaviour does not apply to the MBR as the effluent (permeate) is removed by the permeate pump instead of by gravity flow. A mismatch between influent- and permeate flow may very quickly lead to either an overflowing activated sludge tank or an empty one.

Therefore the permeate flow rate is controlled by the liquid level in the activated sludge tank. Membrane units will be taken in and out of operation as required to match the influent flow. If required, the membrane flux will be increased or decreased as well although in general it is better to operate the membranes at constant flux. If possible (for smaller influent flows), an upstream buffer volume might be used. This will also reduce the required membrane surface area, as peak flow rates are levelled off.

(b) Operation at a higher sludge concentration

The main benefit of operating at a higher sludge concentration is a reduction in the required treatment volume. However, an undesired side-effect is that peak influent flow rates may have a larger impact on effluent quality as the hydraulic residence time in the MBR is smaller as well. On the other hand, in conventional activated sludge systems a sustained peak flow may cause a substantial part of the biomass to be transferred from the biological reactor to the final settler, which will also reduce treatment capacity. This does not apply to MBR systems.

Another effect is that, assuming the applied sludge age remains unchanged, an threefold increase of the sludge concentration from 4 to 12 kg TSS $\cdot$ m^{-3} will result in an equal increase in the oxygen uptake rate. However, at the same time a high sludge concentration will decrease the oxygen transfer efficiency. For domestic wastewater the following α-factors are suggested:

- Between 0.6–0.8 for a mixed liquor concentration of 3–6 kg TSS $\cdot$ m^{-3};
- Between 0.5–0.6 for a mixed liquor concentration between 10–12 kg TSS $\cdot$ m^{-3};
- Around 0.4 for a mixed liquor concentration between 15–20 kg TSS $\cdot$ m^{-3}.

Apart from higher aeration costs and depending on the wastewater treated, the increased aeration intensity might cause foaming problems. Therefore provisions should be included in the MBR design to either prevent or control foaming (e.g. foam breaking by spraying nozzles or using antifoam oil).

(c) Aerobic sludge recycle

The return sludge flow from a conventional activated sludge system will be anoxic or anaerobic, as all oxygen present in the mixed liquor feed flow will have been consumed in the final settler. On the other hand, the return sludge flow from the (aerated) membrane tank will be highly aerobic (typically between 4–6 mg $O_2 \cdot l^{-1}$). For cross-flow MBR systems, the oxygen concentration will be approximately equal to that maintained at the end of the aerobic zone (1–2 mg $O_2 \cdot l^{-1}$), as the hydraulic residence time in the membrane modules will be very short. Furthermore, the return sludge flow rate will also be considerably larger, ranging from 4–6 times the influent flow for submerged MBR systems and feed & bleed cross-flow systems and up to 10–20 times for conventional cross-flow systems.

The return sludge flow from an MBR should therefore not be returned to the anaerobic- or anoxic zone if nutrient removal is to be achieved. Otherwise, the oxygen introduced by the sludge recycle stream will significantly reduce the mass of easily biodegradable COD available for denitrification and biological phosphorus removal (refer also to Section 5.4.2.3).

As an example, for $s = 5$ and $DO_{mt} = 4$ mg $O_2 \cdot l^{-1}$ (the oxygen concentration in the return sludge from the membrane tank), the consumption of S_{bs} is equal to $\Delta S_{bs} = s \cdot DO_{mt}/(1 - f_{cv} \cdot Y) = 20/0.33 = 60$ mg COD $\cdot l^{-1}$. This can also be expressed directly as a reduction of D_{c1}, i.e. $\Delta D_{c1} = s \cdot DO_{mt}/2.86 = 5 \cdot 4/2.86 = 7.0$ mg N $\cdot l^{-1}$ influent. Naturally, the ratio between ΔD_{c1} and ΔS_{bs} of $7/60 = 0.114$ is equal to f_{dn} $(= (1 - f_{cv} \cdot Y)/2.86)$. Therefore, in MBR systems the return sludge flow should always be directed to the head of the aerobic zone, where the aeration demand is highest. In Figure 10.13 and Figure 10.14 the Bardenpho and UCT configurations adapted for submerged MBR are shown.

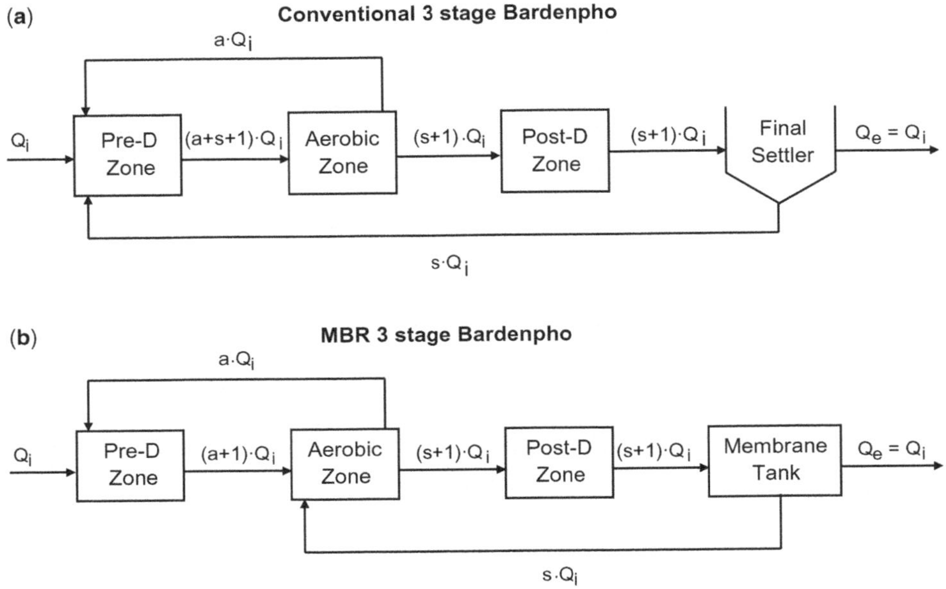

Figure 10.13 Comparison between the Bardenpho configurations for a conventional activated sludge system and a submerged MBR

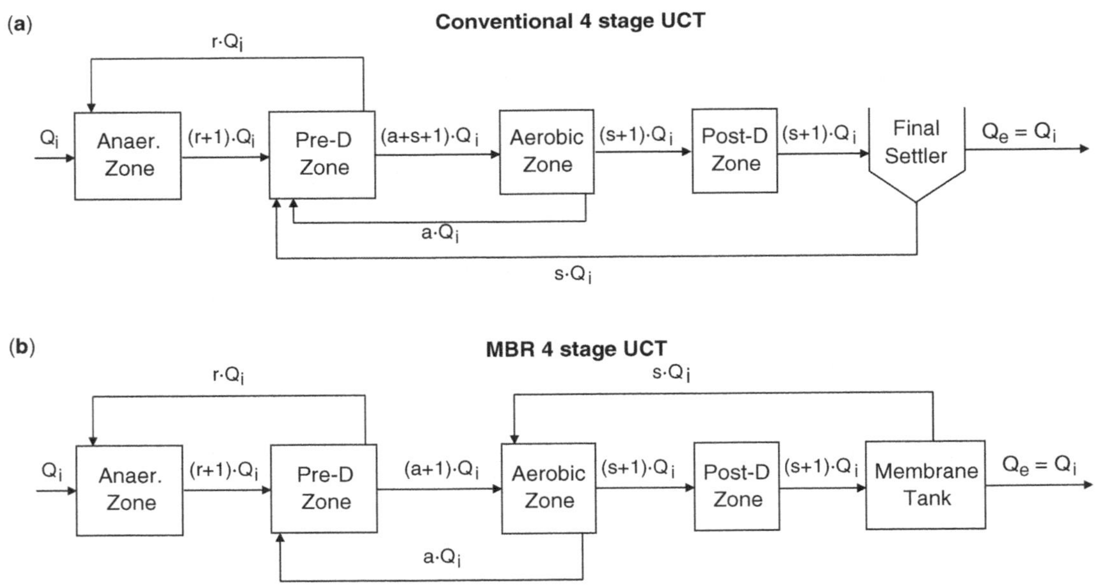

Figure 10.14 Comparison between the UCT configurations for a conventional activated sludge system and a submerged MBR

Compared to a conventional activated sludge system, the main effects of redirecting the (aerobic) return sludge flow in an MBR from the anoxic zone to the head of the aerobic zone are the following:

(a) Increased dilution of the sludge concentration in the pre-D and anaerobic zones;
(b) A decrease in the available nitrate concentration in the pre-D zone (N_{av1}), as the return sludge flow ("s") no longer returns nitrate to the pre-D zone;
(c) An increased oxygen load to the pre-D and post-D zones as the values of both a- and s-recycles will increase, while oxygen content in these streams will be higher as well;
(d) A reduction of required aeration capacity in the aeration tank due to the return of oxygen with the return sludge flow.

(a) Increased dilution of the sludge concentration in pre-D and anaerobic zones
As the sludge return flow is now aerobic, it will no longer be returned to the pre-D anoxic zone. Therefore the biomass concentration in the pre-D zone of an MBR will be a factor $a/(a+1)$ lower than in the aerobic zone. For example, for $a = 4$ the biomass concentration in the pre-D zone of an MBR will be 20% lower than that in the aerated zone. To compensate for the resulting loss of denitrification capacity, the volume of the MBR pre-D zone needs to be enlarged with a factor $(a+1)/a$. As a result, the reduction in MBR volume resulting from operation at a higher biomass concentration will be smaller than expected.

In the UCT system configuration for MBR the "r"- recycle is taken from the diluted pre-D zone, which means that the biomass concentration in the anaerobic zone of an MBR will be further diluted. Consequently for the UCT configuration the volumes of both the anaerobic- and pre-D zone of an MBR have to be increased with a factor $(a+1)/a$.

(b) Reduction of the N_{av1} value
The return sludge flow of an MBR is sent to nitrification zone instead of the pre-D zone, which will influence the value of N_{av1}. In Table 2.1 the formulas required to calculate N_{av1} are summarized, both for complete- and partial nitrogen removal. In general a higher "a"-recirculation value will be required in a MBR system to match the available nitrate with pre-D denitrification capacity. Only if denitrification is complete will there be no difference between the value of N_{av1} in a MBR and conventional configuration, as also in the conventional activated sludge system the s-recycle will then not contain any nitrate.

Table 10.2 Available nitrate in pre-D zone (N_{av1}) in conventional and MBR systems

Configuration	Conventional	MBR	Eq.
Pre-D configuration:	$(a+s)/(a+s+1) \cdot N_c$	$a/(a+1) \cdot N_c$	(10.8)
BPD complete N-removal	$a/(a+s+1) \cdot N_c$	$a/(a+s+1) \cdot N_c$	(10.9)
BPD incomplete N-removal	$a/(a+s+1) \cdot N_c + s \cdot N_{ne}$	$a/(a+s+1) \cdot (N_c + s \cdot N_{ne})$	(10.10)

(c) Increased oxygen load to the pre-D and post-D zones
The increase in the a-recirculation factor will result in an increased oxygen load to the pre-D zone. Likewise, the increase in the s-recycle factor will result in a higher oxygen load to the post-D zone. The equations to calculate the reduction in denitrification capacity due to the introduction of oxygen are summarized below:

$$\Delta D_{c1} = a \cdot DO_1/2.86 \tag{5.95}$$

$$\Delta D_{c3} = (s+1) \cdot DO_1/2.86 \tag{5.96}$$

As can be observed in Table 10.3, for typical values of a and s the reduction of the denitrification capacity in submerged MBR systems is almost twice the vale in a conventional system. Also in absolute terms this is significant, as ΔD_c represents about 20% of the denitrification capacity typically available in a municipal sewage treatment plant. If nitrogen removal is required, this is a very serious disadvantage for the application of MBR to wastewaters with a low COD/N ratio. The very high s-recycle factor applied for conventional cross-flow MBR effectively prohibits the use of a post-D zone, as it would almost certainly be aerobic, unless the volume is very large.

Table 10.3 Reduction of D_{c1} and D_{c3} resulting from oxygen recycle to the anoxic zones in conventional and submerged MBR systems for typical values of "a" and "s"

Configuration	Conventional	Submerged MBR
Typical recycle factors (municipal sewage)	a = 4 and s = 1	a = 6 and s = 4
ΔD_{c1}: pre-D capacity reduction ($mg\ N \cdot l^{-1}$)	$4 \cdot 2/2.86 = 2.8$	$6 \cdot 2/2.86 = 4.2$
ΔD_{c3}: post-D capacity reduction ($mg\ N \cdot l^{-1}$)	$(1+1) \cdot 2/2.86 = 1.4$	$(4+1) \cdot 2/2.86 = 3.5$
Total reduction: ΔD_c ($mg\ N \cdot l^{-1}$)	4.2	7.8

Figure 10.15 and Figure 10.16 highlight the important findings from the previous discussions. In Figure 10.15 the nitrogen removal performance in a pre-D MBR system is evaluated for a typical municipal sewage (case A: $N_c = 46\ N \cdot l^{-1}$) and an industrial wastewater (case B: $N_c = 200\ mg\ N \cdot l^{-1}$). For the sake of this example it is assumed that sufficient denitrification capacity is available in the pre-D zone, so that N_{ne} depends on the value of N_{av1}, whereas in reality often D_{c1} will be limiting. N_{av1} has been calculated with the equations in Table 10.2. In Figure 10.15 the effect of oxygen recirculation on the value of D_{c1} has been ignored in the calculation of N_{ne}, although to indicate the magnitude of this effect the value of ΔD_{c1} has been shown in the graph as well. It can be observed that especially for the industrial wastewater (case B) it will be very difficult to obtain a low effluent nitrate concentration, especially considering the rapid increase of ΔD_{c1} at higher values of "a".

In Figure 10.16 the BDP configuration is evaluated, where only municipal sewage is considered (case A). Again it is assumed that sufficient pre-D denitrification capacity is available (N_{av1} is limiting), while furthermore $D_{c3} = 10\ mg\ N \cdot l^{-1}$. It can be observed that for the same value of "a", N_{av1} will be higher (and hence N_{ne} lower) in a conventional BDP system than in an MBR system. Therefore the a-recirculation value in the MBR system will have to be higher to meet the same effluent nitrate limit. However, ΔD_{c1} will then increase as well. Furthermore, it can be observed that the value of $\Delta D_{c1} + \Delta D_{c3}$ is already significantly higher for the MBR configuration. Thus, as already indicated above, meeting low nitrate effluent limits may be hard in the MBR configuration. For an extensive MBR nutrient removal design case, refer to Example 14.15.

(d) Reduction of required aeration capacity in the aeration tank

MBR vendors often state as an advantage of their systems that the contribution of the oxygen recycle from the (aerated) MBR tank reduces the oxygen demand in the aeration tank considerably. This contribution can be calculated as $s \cdot Q_i \cdot DO_{mt}$. As will be shown in Example 14.15, the reduction in required oxygen transfer capacity is around 3–5 % only, while the energy requirements of membrane aeration are significant.

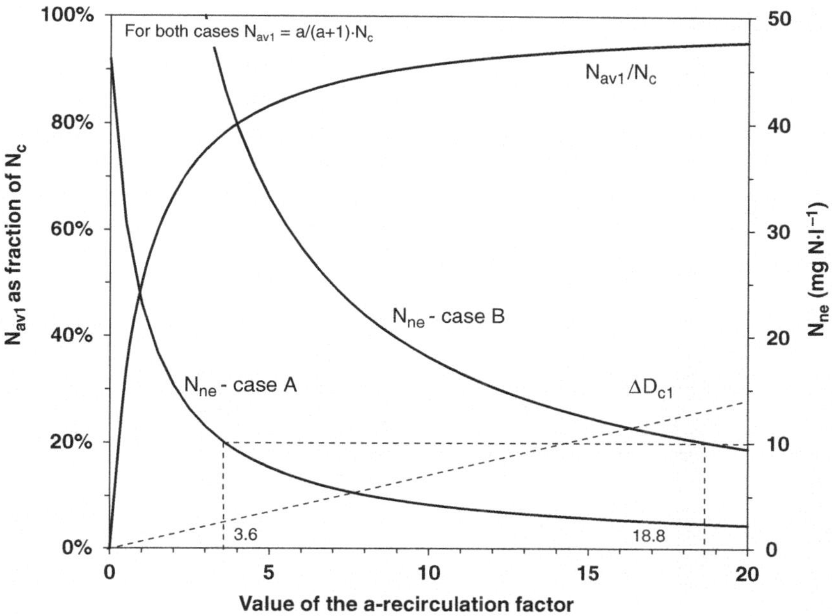

Figure 10.15 MBR in pre-D configuration: ratio N_{av1}/N_c and N_{ne} as function of the a-value for case A and B.

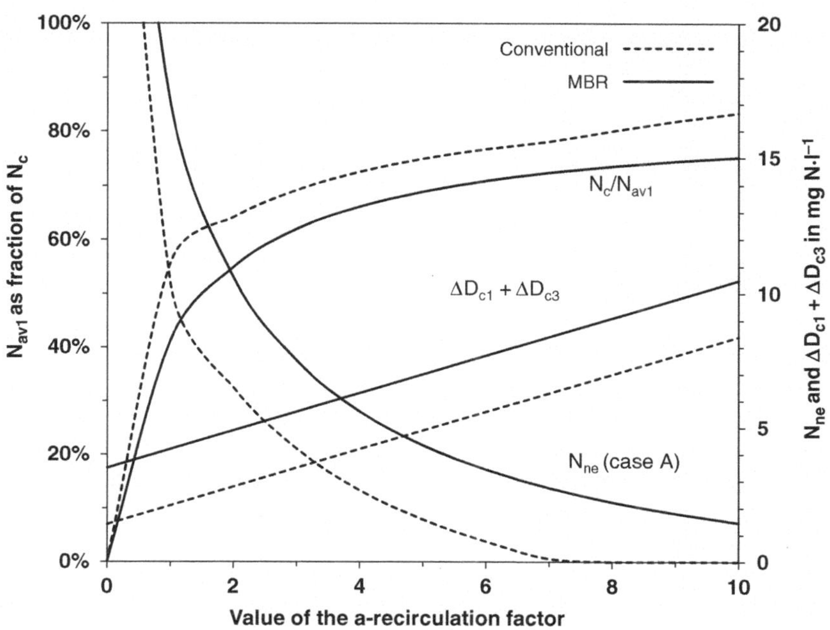

Figure 10.16 Comparison between N_{av1}/N_c and N_{ne} in a MBR and conventional BDP system, when N_{av1} is limiting (case A only).

10.3.3 Pre-treatment

Membranes act as an absolute barrier for suspended solids and all non-biodegradable particulate material can only leave with the excess sludge. Therefore, all membrane types are very susceptible to macro-fouling by plastic, hairs, etc. These materials accumulate in the mixed liquor and tend to block the entrance of cross-flow modules or end up in the connection between the fibres and the header of submerged membranes. This phenomenon is called clogging (see Figure 10.17). Apart from blocking part of the membrane surface, the accumulated debris exerts considerable strain on the connection between fibre and header and thus reduces membrane lifetime.

Figure 10.17 Serious clogging problems at the top header of hollow fibre membranes due to improper design of the pre-treatment unit

Manually cleaning a large number of membranes is very labour intensive, so prevention is much preferred. For raw wastewater, the recommended pre-treatment configuration consists of a 6–7 mm coarse screen (e.g. raked- or step screen), followed by a 0.6–1.0 mm fine screen (e.g. rotating brush raked screens, vibrating static screens, drum filters or rotary wedge-wire screens). If a primary settler is used the fine screen may be omitted, but then either a 2–3 mm basket filter has to be installed downstream of the primary settler (which might require significant cleaning effort) or a 3 mm coarse screen upstream of the settler.

Other undesired wastewater constituents are oil, fat and grease, as these materials are often slowly biodegradable and tend to clog the membrane pores, thereby significantly reducing the permeability. For municipal wastewater an aerated sand trap can be installed, with the added benefit of removing sand particles which have a scouring effect on the membranes if allowed to reach the membrane tanks. If primary settling is employed, a sand trap can be omitted. For industrial applications, gravity oil separation with plate pack separators followed by dissolved air flotation to remove free- and emulsified oil is frequently applied.

10.3.4 Module configuration – submerged MBR

Several permeate header configurations exist: modules with a top header, a top and a bottom header or a bottom header only. Early modules were predominantly designed with a top header. However, a disadvantage of connecting the membranes to a single header is that the vacuum near the end of the fibre will be lower than that near the header, which will result in imbalanced permeate flux rates and increases local fouling rates. The same problem appears when the membranes are backflushed: i.e. the cleaning efficiency is reduced near the end of the fibre. With two headers the pressure drop in the fibre is reduced by fifty percent and as a consequence the permeate withdrawal is much more balanced over the length of the fibre.

All top header module configurations are particularly vulnerable to clogging with floating debris (e.g. hair). To prevent this, modules with a single bottom header have been developed. Apart from reduced vulnerability to clogging, aeration equipment can be positioned in between the fibres instead of under the module bottom header. This results in improved turbulence at reduced aeration intensity. Simultaneously dead zones directly above the bottom header are avoided where sludge and debris may settle (a phenomenon called sludging), which causes problems similar to those resulting from clogging. In Figure 10.18 an example of a modern fibre membrane module in bottom header configuration is shown.

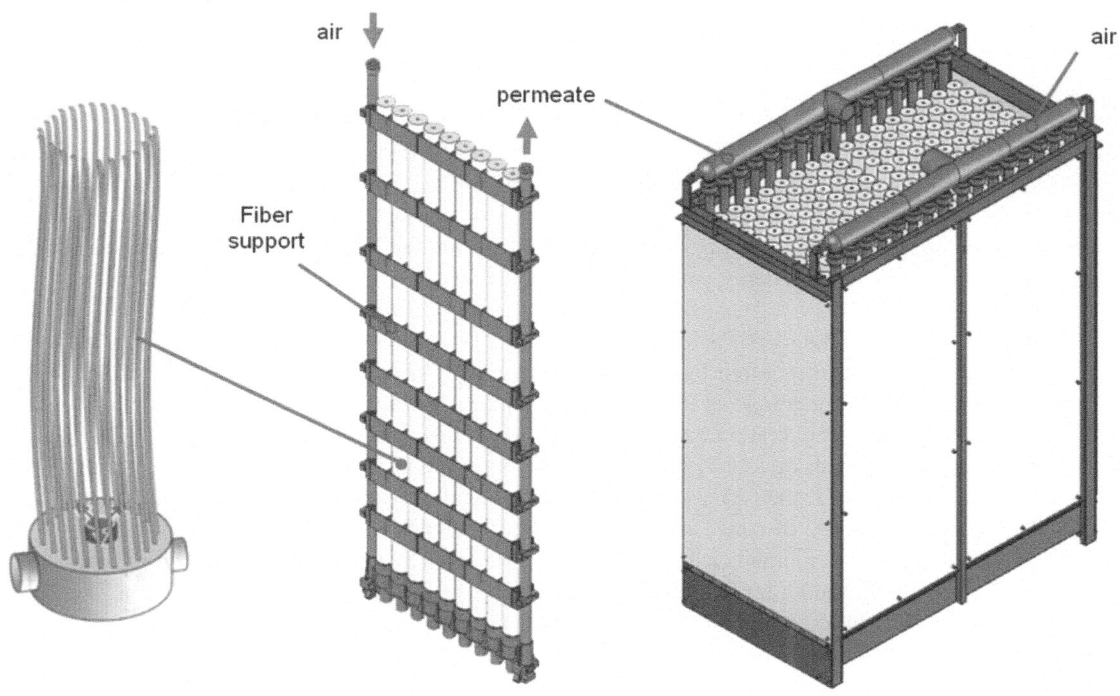

Figure 10.18 Schematic representation of a submerged fibre membrane module, equipped with bottom permeate header (Koch Membrane Systems). The modular design consists of bundles of membranes mounted in a pot, grouped together in rows. The number of rows per module can be varied. Each bundle is aerated with an aeration device mounted within the pot

There are two types of submerged membranes: plate- and fibre membranes. Plate membranes are available in top header configuration only. Both membrane types have been applied in full-scale installations with good results and at similar fluxes, and for both membrane types several suppliers are available.

Table 10.4 Comparison of advantages and disadvantages of plate- and fibre membrane module configurations

Type	Advantages	Disadvantages
Plates	– Robust – Less susceptible to clogging compared to the fibre membrane with top header or two header configuration – Simpler system & process control configuration – Manual cleaning possible – Low frequency of cleaning	– Less specific surface area per m^3 module volume – Back flushing not possible – Higher aeration requirements – More susceptible to channelling: the air speed between the two plates is high but at the plate surface itself it is low. This leads to solids build-up on the membrane surface – Automated cleaning is expensive
Fibres	– Back flushing possible – High specific surface area – Lower aeration requirements – Completely automated cleaning possible	– Susceptible to clogging, depending on module configuration – Manual cleaning non-practical – More complex system

10.3.5 Module aeration – submerged MBR

Submerged membrane systems require aeration in order to create sufficient turbulence around the membrane surface to refresh the sludge cake layer accumulating on this surface. Cross-flow membranes do not require aeration, with the exception of the low pressure cross-flow systems. For submerged membrane systems, aeration is the highest energy consumer and significant effort has been directed at ways to optimise aeration patterns and to reduce energy consumption, for instance:

- Intermittent aeration: the membranes are aerated only 50% of the time. This results in a reduction of aeration demand and in required blower capacity, as it doubles the membrane surface that a single blower can serve. An example is the air-cycling concept developed by Zenon, a supplier of fibre membranes. Intermittent aeration can be implemented in different ways. For example:
  - 15 sec. all modules in first tank, then 15 sec. all modules in second tank;
  - 15 sec. module 1 + 3, then 15 seconds module 2 + 4 (all modules in the same tank);
  - 15 sec. modules 1 + 3 of tank 1 and 2, then 15 sec. modules 2 + 4 of tank 1 and 2;
  - 15 sec. all "even" membrane elements (2, 4, …, 20), then 15 sec. all "uneven" membrane elements (1, 3, …, 19) in all modules in one or several tanks;
- Reduced aeration duration and -intensity whenever the membranes are in standby-mode, i.e. when there is no permeate production;
- Reduced aeration intensity at higher wastewater temperature, as the degree of fouling is often lower at higher wastewater temperatures;
- Control of the aeration intensity based on the applied membrane flux: i.e. a low membrane flux will result in a low aeration intensity and a high membrane flux will require a high aeration intensity. This method is used by Koch Membrane Systems for their fibre membrane module;

- Intermittent membrane operation: a period of maximum membrane flux and maximum aeration followed by an idle time in which no aeration and no flux is applied. This method is used by Kubota for their plate membrane modules.
- Many combinations of the above – with trade names such as AirWave, AirFlush, Air-cycling, continuous aeration, jet aeration, etc.

10.3.6 Key design data of different membrane types

In Table 10.5 various key design and operating parameters of the different membrane configurations and -types are summarised.

Table 10.5 Summary of design data of different membrane types

Parameter	UoM	Fibre	Plate	Cross-flow	LPCF
Gross flux at 15°C): (normal – peak flux)					
– municipal	$l/(m^2 \cdot h)$	15 – 50	15 – 45	50 – 150	15 – 30
– industrial		5 – 15	5 – 15	50 – 100	5 – 15
$\eta_m^{(1)}$ (net/gross flux)	(–)	0.85 – 0.95	0.8 – 0.9	1.0	0.85 – 0.95
Sludge concentration in membrane tank	$g\ TSS \cdot l^{-1}$	8 – 15	8 – 15	12 – 30	8 – 15
Module volume	m^3	2 – 12	3 – 10	0.1	0.1
Specific surface area	$m^2 \cdot m^{-3}$	100 – 150	40 – 70	200 – 330	200 – 330
Specific aeration requirement$^{(2)}$	$Nm^3 \cdot m^{-2}$	0.1 – 0.5	0.3 – 1.0$^{(3)}$ 0.2 – 0.6$^{(3)}$	N.A.	0.5 – 0.6
Operating Δp_{TM}	bar	0.1 – 0.2	0.1 – 0.2	2 – 4	0.2 – 0.5
Velocity in tubes	$m \cdot s^{-1}$	N.A.	N.A.	3 – 5	0.5 – 1.0
Recirculation ratio	(–)	4 – 8	4 – 8	10 – 20$^{(4)}$	30 – 40$^{(4)}$

Notes: (1) Defined as gross flux corrected by back flush and/or relaxation. Cleaning is not taken into account;
(2) During production mode: corrected for the percentage of time that the modules are aerated;
(3) Single decker resp. double decker configuration (two modules placed on top of each other);
(4) Recirculation ratio is defined by the required minimum velocity in the membrane tubes and the number of modules per train. High ratio for LPCF is due to parallel operation.

10.4 MBR OPERATION

10.4.1 Operation of submerged membranes

Submerged membranes are typically operated according to a standard process cycle containing the following phases:

- Production mode: the permeate pump is running in forward direction and permeate is produced. The membranes are aerated intermittently to refresh the sludge cake layer on the membrane surface, while the membrane feed pump is continuously recirculating mixed liquor over the membrane tank;
- Back flush mode: the permeate pump reverses direction (or this is done by opening/closing the appropriate automatic valves). During back flush, the pump flow rate is typically increased to 110

to 120% of the flow rate in production mode. Aeration and recirculation of mixed liquor over the membrane tank continue as normal. This process mode only applies to fibre membranes, plate membranes cannot be back flushed;

- Relaxation mode: the permeate pump is stopped. This process mode is used for plate membranes as an alternative to the backflush mode for fibre membranes: aeration and recirculation of mixed liquor over the membrane tank continue as normal. The process cycle of fibre membranes usually does not contain a relaxation step;
- Standby mode: this is actually not part of the process cycle. At low influent flow, one or more membrane tanks are temporarily taken out of operation and switched over to standby mode. The recirculation of mixed liquor over the membrane tank may be stopped and aeration frequency and intensity are reduced considerably. At regular intervals the active and stand-by membrane tanks switch duty;
- Cleaning: refer to Sections 10.4.3 and 10.4.4.

The duration of the different steps in the process cycle depends on the type of wastewater treated, the sludge characteristics and rate of fouling. Typical process cycles for submerged membranes treating domestic wastewater are shown in Table 10.6.

Table 10.6 Typical process cycles for a submerged MBR with fibre- or plate membranes treating domestic wastewater

Process cycle step	Fibre membranes	Plate membranes
Production	400 s	415 s
Back flush at 115%	20 s	—
Relaxation	—	45 s
Total cycle time	420 s	460 s
η (ratio net/gross flux)	$(400 \cdot 1.0 - 20 \cdot 1.15)/420 = 0.9$	$(415 \cdot 1.0)/460 = 0.9$

10.4.2 Operation of cross-flow membranes

Cross-flow membranes are continuously operated and therefore do not have different steps in their process cycle. At low influent flow, one or more membrane units are taken out of operation, so depending on the influent flow, a membrane skid is either in production- or in stand-by mode. In general, cross-flow membranes are allowed to run without cleaning until the flux decreases below a certain preset limit.

10.4.3 Membrane fouling

Membrane fouling results in declining membrane fluxes and -permeability and is considered as the largest operational problem in MBR treatment. Even well designed and operated MBR systems will experience some degree of membrane fouling. Fouling can be classified as either macro- or micro-fouling, depending on the size of the fouling component. Macro-fouling is the plugging of membranes and headers with gross material such as hair, plastic and other types of debris. It can be removed manually, but this is a very labour intensive job. Macro-fouling can and should be prevented to a large extent by installing the proper type of pre-treatment.

Micro-fouling results in blocking of membrane pores and can be of organic-, biological- and/or inorganic nature. It often involves ad- or absorption to the membrane surface and thus cannot be removed by back flushing and/or membrane aeration. When significant fouling is expected, membranes are designed for operation at low flux. At high fluxes (resulting in higher Δp_{TM}) the pollutants may be entrapped deep in the membrane pores, making removal by back flushing or even chemical cleaning much more difficult.

(a) Inorganic fouling

This type of fouling (also known as scaling) is caused by precipitation of inorganic solids on the membrane surface. Typical precipitates that may cause problems in wastewater are:

- $CaCO_3$: the solubility of this component depends on the pH. Due to the intensive aeration in the membrane tank, CO_2 will be stripped from the mixed liquor, resulting in a pH increase of typically less than 0.5–1.0. At higher pH values $CaCO_3$ has a lower solubility. This type of scaling can be removed by acid cleaning (e.g. HCl or H_2SO_4) at low pH;
- $FePO_4$ and $CaPO_4$: these metal salts can be a problem when chemical phosphorus removal is applied and the precipitation reaction has not yet been completed in the aerated zone of the biological reactor. It can be removed by acid cleaning at low pH;
- Iron oxides and hydroxides (Fe_2O_3, $Fe(OH)_3$ etc). Iron can be either present in the wastewater or added in case of chemical phosphorus removal. The complexes are characterised by their orange colour and can be removed by complex-forming organic acids such as citric- or oxalic acid. The pH is adjusted to a low value using a strong inorganic acid such as HCl.

(b) Organic fouling

This type of fouling can be caused by deposits of suspended- or colloidal organic material, such as oil, grease and fats or by organic molecules with either an affinity for the membrane material or with a high molecular weight. Both can be adsorbed to or absorbed in the membranes and thereby block membrane pores or reduce pore size. If oil, fat and grease are present in the influent in large quantities, they should be removed using proper pre-treatment. At lower concentrations they may be degraded by the biomass. Oil, fat and grease can be removed using NaOH and/or surface active compounds such as EDTA. Other organic components can be removed using strong oxidants such as NaOCl or H_2O_2 at a high pH value.

(c) Biofouling

Biofouling is the result of biological growth (bacteria, fungi etc.) on the surface of the membranes. This may result in the blocking of pores with biomass or the partial covering of the membrane surface with slime.

Biofouling can be prevented, or at least significantly reduced, if the MBR system is well-designed. It is important to prevent biological activity (growth) in the membrane tanks: therefore nitrification and COD removal should be essentially complete in the last biological reactor, before entering the membrane section. Should biological fouling occur, it can be removed with strong oxidants such as NaOCl or H_2O_2 at high pH or with NaOH.

10.4.4 Membrane cleaning

Almost all types of (micro)-fouling can be removed by chemical cleaning. However, as in general abrasive chemicals are used in the cleaning process (strong oxidants, acids and bases), cleaning reduces membrane lifetime and thus increases membrane replacement costs. Therefore the most important guideline to design and operating MBR systems is to prevent fouling by proper design of upstream treatment facilities and by proper operation of the biological process. While in other membrane applications (e.g. effluent polishing, RO and nano-filtration) special cleaning chemicals may be added to prevent fouling, this is not feasible in MBR treatment, given the high suspended solids concentration and the large recycle to the biological tank. One of the main findings of the Beverwijk MBR research project is that regular (but less intensive) cleaning is much better than sporadic (but intensive) cleaning (van de Roest et al., 2002). Not only is the membrane lifetime increased, but the average membrane permeability will be higher as well. This is exemplified in Figure 10.19, which shows membrane permeability as a function of operating time for two different cleaning regimes:

- A regime in which regularly (e.g. once every two weeks of once every month) a mild chemical cleaning is executed, supplemented with a more intensive cleaning whenever permeability decreases below a certain lower setpoint level;
- A regime where permeability is allowed to drop to a low level followed by an intensive cleaning to recover permeability.

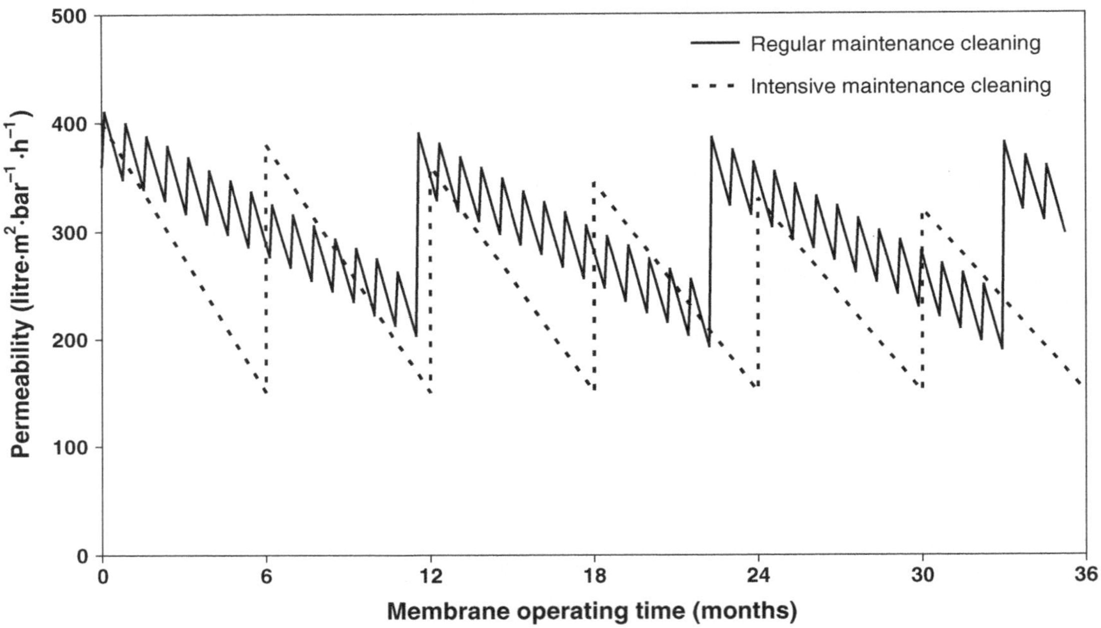

Figure 10.19 Effect of cleaning regime (regular maintenance versus intensive cleaning) on the development of membrane permeability in time

Depending on the nature of the micro-fouling, a single chemical or a combination of the chemicals listed in Section 10.4.3 can be selected and used in one (or more) cleaning steps. Stored permeate is mixed with the selected chemicals and backflushed through the membranes. Alternatively, the membranes can be soaked in a chemical solution. In the early stages of MBR development, the membranes were removed from the membrane tank and cleaned in a dedicated cleaning tank. As this was a very labour intensive and time consuming procedure, several other cleaning procedures have been developed since then:

- Regular maintenance cleaning in situ (fibre membranes and cross-flow membranes);
- Regular maintenance cleaning in air (fibre membranes);

- Intensive cleaning (plate membranes)
- Recovery cleaning (fibre membranes).

(a) Regular maintenance cleaning – in situ

In this cleaning method, the membranes remain submerged in the biomass in the membrane tank. Regular cleaning can be applied to fibre membranes of all header configurations. The membrane feed flow and membrane aeration are stopped and the membranes are backflushed several times with stored permeate from the permeate buffer tank.

Depending on the nature and degree of fouling, one or more chemicals are added, either together (e.g. NaOCl and NaOH) or in separate cleaning cycles. The chemicals are injected in the permeate pipeline near the permeate pump to promote mixing. In between the different cycles (with different chemicals), the membrane tank is flushed with mixed liquor using the membrane feed pump to prevent undesired chemical reactions, such as the formation of Cl_2 from the reaction of HCl and NaOCl. When the cleaning is ready, the membrane tank is flushed thoroughly with mixed liquor from the bioreactor to prevent loss of chemicals with the effluent, before it is put back into operation.

For top- and two header configurations, the backflush flow rate is 110–120% of the normal permeate flow rate. The applied pressure difference (maximum 0.5 bar) forces the chemicals through the membranes. The larger flow rate is required in order to supply enough pressure to the bottom part of the membranes, as at this location the static water pressure of the mixed liquor in the membrane tank (which counteracts the applied backflush pressure) is maximum. Therefore, chemical usage is relatively inefficient as more cleaning chemicals pass through the upper part of the membranes than required.

For bottom headers a different mechanism has been developed by Koch Membrane Systems: the back flush flow is only 10–15% of design permeate flow. This is sufficient to keep the membranes filled with cleaning liquid. The chemicals will diffuse through the membrane wall as a result of the concentration gradient. This concept saves on chemical consumption and may potentially be used for top- and two header configurations as well.

In general, cross-flow membranes require less cleaning than submerged membranes, as the high velocity in the membrane tubes keeps the surface relatively clean. The membranes are either cleaned when the flux decreases below a certain threshold value or, as is increasingly done, cleaning is part of a regular maintenance procedure and is executed when due, for instance 4–12 times per year. Cleaning of cross-flow membranes can be much more effective as the volume of the membrane modules to be cleaned is much smaller than that of the comparable membrane tank. Therefore it is possible (from a cost perspective) to use industrial cleaning liquids instead of a pH adjusted single chemical dissolved in water, as is the case for submerged membranes. These cleaning liquids are often a blend of oxidants, inorganic acids, surfactants (anionic, cationic or non-ionic), detergents, NaOH, EDTA etc. and are more effective in removing fouling.

For cross-flow membranes, the cleaning procedure is the following: a cleaning solution is prepared in the CIP tank and heated to the desired temperature. The biomass in the membrane rack to be cleaned is removed and replaced by water or permeate. The module to be cleaned is connected to the CIP tank, from which the cleaning solution is recirculated at low pressure over the membrane modules. If required the membranes can be backflushed as well.

(b) Regular maintenance cleaning in air

The main difference between cleaning in situ and cleaning in air is that the membrane tank is drained before cleaning. This has two advantages: (I) the use of chemicals is lower, as the chemicals will no longer react with the suspended solids in the mixed liquor present during in-situ cleaning and (II) the cleaning can be done at higher temperature, which improves cleaning results. In this case a dedicated CIP tank is required, in which the cleaning liquid, often stored permeate, is heated prior to cleaning. This type of cleaning cannot be applied to membranes with a bottom header configuration, as these modules require the static pressure from the liquid in the membrane tank in order to distribute the cleaning liquid over the full length of the fibres. In between cleanings the membrane tank is drained. After the final cleaning the membrane tank is flushed with biomass.

(c) Intensive maintenance cleaning

Plate membranes cannot be flushed back because at backflush pressures above 0.05–0.1 bar the plates may inflate and burst. Cleaning is therefore a manual activity. One at a time, a module is connected to a cleaning storage tank containing about 3 times the liquid volume contained in the plates. The cleaning liquid inside the storage vessel is emptied by gravity into the plate membrane module and slowly diffuses through the membrane. It is not recommended to clean all modules together, as it is then impossible to know if all modules received sufficient cleaning liquid. As a result this cleaning method is very labour intensive.

Automation of the cleaning procedure is costly, especially at larger implementations, as every plate module would require a dedicated open-close valve. Clearly, cleaning of all modules is not something one would like to do every other week. Therefore the frequency of this cleaning is low (1 or 2 cleanings a year), which is the reason why the regular cleaning philosophy is not used for plate membranes. As the consumption of chemicals is very high during an intensive cleaning, this will have repercussions on the performance of the biology (toxicity and excessive foaming) while effluent quality (chlorinated components) may (temporarily) suffer as well.

(d) Recovery cleaning

Recovery cleaning is applied to restore the permeability of fibre membranes to the initial operating value, or whenever regular maintenance cleaning does not have the required effect. It should be noted that permeability of fibre membranes cannot be restored to the initial value measured directly after the start of operation. It is normal that membrane permeability drops very quickly from the initial value to a lower, more or less stable value. For municipal wastewater treatment, a typical drop in initial (standardised) permeability can be observed from 800 to 400 litre $\cdot$ $m^{-2} \cdot h^{-1} \cdot bar^{-1}$ in a couple of days. Even recovery cleaning will not be able to restore permeability to a value higher than this 400 litre $\cdot$ $m^{-2} \cdot h^{-1} \cdot bar^{-1}$.

During recovery cleaning the membrane tank is completely drained and flushed with clean water or permeate. The tank is then filled with permeate or water and the required chemical is added at a relatively high concentration. Membrane aeration is resumed and the membranes are allowed to soak for several hours. In between- and after cleanings the membrane tank is drained to the bioreactor and flushed with water or permeate. Once again, as a significant quantity of chemicals is used, there may be negative effects on the biology (foaming, toxicity and effluent quality).

10.5 MBR TECHNOLOGY: EVALUATION AND POTENTIAL

At present MBR treatment may be considered a proven technology for wastewater treatment. In the last 25 years significant progress has been made in the design and operation of MBR systems. One of the reasons is the availability of government funding, which has made it possible to conduct extensive research on a practical scale. An example is the research project conducted at the Beverwijk WWTP in the Netherlands, where membrane suppliers, engineering firms and water boards have worked together to test several pilot MBR systems during a two year period (2001–2002), with additional pilot plants tested afterwards. The main results of this project were increased reliability of operation, increased membrane lifetime, reduced energy cost and a better understanding of the nature, prevention and removal of membrane fouling.

Several hundreds of small (5–50 $m^3 \cdot h^{-1}$) industrial installations have been built from 1990 onwards in Europe and North America. Also in Japan a large number (> 1000) of small industrial and municipal MBRs have been constructed. However, large-scale application of MBR to municipal wastewater lagged behind due to the diluted nature of this wastewater. From 1998 onwards, several municipal MBR's have been constructed, sometimes as a demonstration project with partial governmental funding, for example the 800 $m^3 \cdot h^{-1}$ (full-scale) demonstration plant constructed in Varsseveld, The Netherlands (Figure 10.20).

Figure 10.20 Aerial view of first municipal MBR constructed in The Netherlands, located in Varsseveld: left – old STP; right – from bottom to top: pre-treatment, Carrousel®, MBR tanks (submerged Zenon hollow fiber modules) and blower- and pump rooms. Courtesy DHV/Rhine and IJssel Waterboard, The Netherlands

With membrane prices decreasing, MBR is becoming more competitive compared to conventional treatment, but still the number of municipal installations in use is not very large. As mentioned in the introduction of Section 10.1, the main drivers to implement a MBR instead of conventional activated sludge system are:

- Limited availability of space;
- Strict effluent limits, i.e. due to discharge of effluent into a vulnerable water body or the possible reuse of effluent;
- Difficulties for solid/liquid separation in the final settler.

Disadvantages of the MBR process compared to a conventional system are:

- Increased investment and operational costs;
- Increased complexity of the system, requiring skilled operators and a considerable amount of automation/instrumentation;
- Ecological considerations such as the use of chemicals for cleaning and the increased consumption of energy, which also increases greenhouse gas emissions.

As long as the effluent from a municipal wastewater treatment plant has to comply with effluent limits similar or less stringent to those currently applicable in the Netherlands (i.e. $N_{te} < 10$ mg $N \cdot l^{-1}$, $P_{te} <$ 1 mg $P \cdot l^{-1}$ and $X_{te} < 20$ mg $TSS \cdot l^{-1}$), treatment by means of conventional activated sludge systems will in general be sufficient and in many cases much more competitive. For low strength municipal wastewater, annualised investment costs of an MBR system are still about fifty percent higher than those of a conventional activated sludge system. This is not only due to the cost of installing (and replacing) the membranes. Other factors are the need for more extensive pre-treatment and a much higher degree of automation. Furthermore energy requirements are higher. The difference in investment costs might be reduced in the future, as the continuing competition between membrane suppliers might cause membrane prices to decrease further, or if land prices go up. For difficult and/or high-strength industrial wastewaters, MBR reactors are already an attractive alternative, as in this case the fraction of investment costs related to the membrane unit will be small compared to those of the biological treatment volume.

Should stricter effluent limits be applied in the future, then MBR will certainly become more attractive. This may be the case in Europe as a result of the EU water framework directive. In the period from 2000 to 2015, the member states will have to increase the water quality of surface- and ground water: this may result in stricter effluent limits. However, MBR will have to compete with other systems capable of delivering the required effluent quality, for example conventional activated sludge treatment followed by a polishing step such as sand filtration. If the effluent of the wastewater treatment plant is to be re-used as a high quality water source (e.g. as process water, boiler feed make-up water, cooling make-up water or even potable water), once again the MBR is an attractive alternative, as an effluent free of solids is produced, which can be directly processed in downstream processing operations such as nano-filtration or reverse osmosis.

Finally, if available space is limited MBR might be a very interesting solution. This may be the case for many industrial locations, but certainly also for municipal wastewater treatment. In a large number of cities in developing countries, the rapid expansion of the population has two main effects on the local wastewater treatment situation: (I) the existing conventional wastewater treatment plants are overloaded and (II) there will often be no space available for expansion of the existing wastewater treatment plant. Retrofitting the old conventional treatment plant into an MBR system could increase treatment capacity by a factor of 3 without any additional space requirements.

If limited space is a decisive criterion, then the moving bed biofilm reactor (MBBR) is a very serious competitor, especially as these systems are relatively simple and robust, although not necessarily much cheaper. The effluent quality of a MBBR is comparable to a conventional activated sludge process: i.e. it will contain suspended solids. Several hundreds of MBBRs have already been constructed in the last decade.

The recently developed aerobic granulated sludge process (refer to Appendix A9) might also become a serious competitor for MBR (and the conventional activated sludge process), but at the time of writing (2011) there are only a limited number of full-scale installations, mostly for smaller capacities. However, it is a much simpler, low-tech system than MBR, which makes it considerably cheaper, also when compared to conventional activated sludge systems. The main driver will be a reduction in treatment volume and -costs, not better effluent quality.

Chapter 11

Moving bed biofilm reactors

11.0 INTRODUCTION

The authors would like to acknowledge the important contribution to this chapter of Hallvard Ødegaard, Professor Emeritus at the Department of Hydraulic & Environmental Engineering, Norwegian University of Science and Technology, Trondheim (hallvard.odegaard@ntnu.no).

As discussed in the introduction of Chapter 10, the conventional activated sludge (CAS) system is still a very popular process configuration, which is mainly due to its flexibility and the possibility to meet stringent effluent limits at reasonable costs. However, there are also several issues associated to this configuration, which are mainly related to the limitations imposed by the performance of the solid/liquid separation step, i.e. the final settler:

- Only a limited biomass concentration can be maintained in the reactor, thereby requiring a large biological treatment volume;
- The final settler occupies a large surface area;
- The final settler performance is vulnerable to process upsets, such as peak flow rates and sludge bulking, which can compromise effluent quality.

Hence for a long time there has been a quest for alternative system configurations that allow operation at higher biomass concentrations and/or to replace the final solid-liquid separation step. An example of the latter is the Membrane Bio Reactor (MBR), which has been extensively discussed in Chapter 10. An alternative means to accomplish an increase in biomass concentration is to change to a biofilm process, where biomass grows either as an attached film to a support medium or as an independent sludge granule. Apart from the benefits associated to operation at increased sludge concentration, other advantages are that:

- There is greater flexibility in the selection of the method for separation of suspended solids from the effluent (i.e. compact settling, flotation or filtration), as the biomass concentration to be separated is in general at least ten times lower than that in a conventional activated sludge system while, depending on the configuration, the flow rate might be lower as well if sludge recirculation is not required;

- The attached biomass may become more specialized and active, resulting in a higher concentration of relevant organisms at any location in the process train, because there is no biomass return and because the protected environment of the biofilm allows the development of organisms that require a high sludge age.

Several types of biofilm systems have been around for a long time, such as trickling filters, rotating biological contactors, fixed media submerged biofilters, granular media biofilters, fluidized bed reactors etc. All of these have their advantages, but also (significant) disadvantages, so that none of these technologies has ever become really popular:

- The trickling filter can be a cheap solution but has a low volumetric treatment capacity and is prone to blocking of the filter bed, channelling and odour problems;
- Rotating biological contactors often experience mechanical failures and are more suited to small-scale applications;
- In fixed media submerged biofilters it is difficult to get a balanced distribution of the load over the whole carrier surface and again blocking/channelling is a problem;
- The granular media biofilters have to be operated discontinuously because of the need for backwashing;
- Fluidized bed reactors show hydraulic instability and are notoriously hard to operate.

Hence the need for improved biofilm configurations. The configuration that will be discussed here is the Moving Bed Biofilm Reactor (MBBR), which was invented in Norway by prof. H. Ødegaard at the Norwegian University of Science and technology (NTNU) in the late 1980s and and developed further in the early 1990s (Ødegaard et al., 1994, Ødegaard et al., 1999). The concept was commercialized by Kaldnes Miljöteknologi and later AnoxKaldnes – currently part of Veolia Water. The MBBR dispenses with the problems of clogging, channelling and influent distribution as the carrier medium is suspended in a completely mixed reactor and as such is allowed to move freely.

At present there are more than six hundred municipal and industrial AnoxKaldnes™ MBBR treatment plants in operation or under construction, in more than fifty different countries all over the world. Although AnoxKaldnes may be considered to be the most dominant MBBR supplier, there are other suppliers in this market as well, such as Aqwise, Eimco, Brightwater, Siemens, Headworks and Degrémont.

Another innovative biofilm configuration is the Aerobic Granular Sludge technology (AGS), which relies on the cultivation of sludge granules with a high settling velocity and will be discussed in Appendix A9. This technology was developed in the late 1990s at the Delft University of Technology, the Netherlands. The concept of using sludge granules to increase the biomass concentration is not new: it has been very successfully applied from the early 1990s onwards in the high rate anaerobic EGSB systems for the treatment of industrial wastewater, refer to Section 13.6), while there have also been earlier attempts on cultivating aerobic or anoxic sludge granules. However, the mechanism driving the aerobic granulation process was never really understood. The aerobic granulated sludge technology is commercialized by DHV BV under the trade name Nereda™, with approximately ten full-scale implementations in the period between 2005 and 2011.

This chapter gives an overview of the main principles underlying the MBBR technology and provides an overview of the process configurations that may be selected for green field plants as well as for upgrading of existing activated sludge plants. Design guidelines and treatment performance data are discussed as well.

11.1 MBBR TECHNOLOGY AND REACTOR CONFIGURATION

The MBBR was originally developed and patented as a pure biofilm reactor, i.e. without sludge recycle and hence without a significant concentration of suspended biomass in the reactor. However, lately there has been a growing interest in the application of the MBBR technology to so-called hybrid or Integrated Fixed film Activated Sludge systems (IFAS). In these reactors, part of the biomass is attached to the carriers, while another part is present in the form of suspended growth biomass. This necessitates the separation of biomass from the effluent and its recycle to the biological reactors.

The MBBR system is based on the use of a biofilm that grows attached to the surface area of specifically designed carriers made of polyethylene or polypropylene with a density close to that of water. The carriers are designed to proide a large protected surface for bacterial growth. The reactor volume is filled with carriers up to a maximum value of 67%. Due to their density being close to that of water and the fact that only part of the reactor volume is filled with carriers, the packed bed is allowed to move freely in the reactor as can be observed in Figure 11.1, hence the name moving bed biofilm reactor. Medium bubble aeration or mixing by slow-speed mixers is applied to keep the carriers in suspension and also in order to control the biofilm thickness.

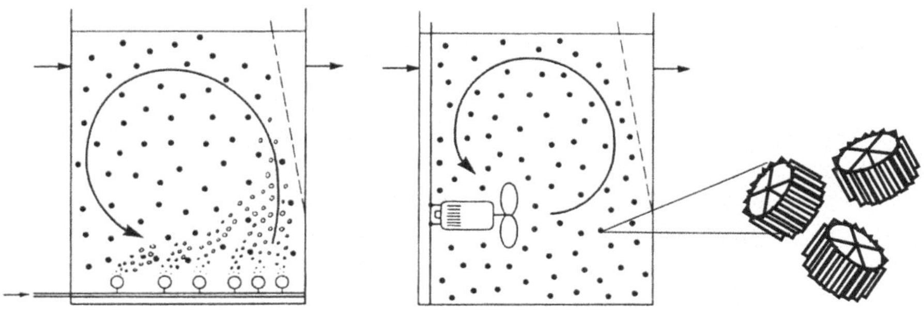

Figure 11.1 Operating principle of the MBBR process with aerobic reactors (left), anoxic and anaerobic reactors (middle) and the shape of the original biofilm carrier (Kaldnes K1). Courtesy of H. Ødegaard

Sieves or grids are used to retain the carriers in the reactor and, as mentioned earlier, there is no sludge recirculation from the final solids separation step to the biological reactors. Therefore, apart from the biofilm, there will be only a low concentration of suspended solids in the reactor, mainly originating from biofilm shearing or already present in the influent. Several reactors in series may be used to develop specialized bacteria in every stage, as depending on the requirements the reactor may be used for aerobic, anoxic or anaerobic processes.

It has been demonstrated (Ødegaard et al., 2000) that the available biofilm surface area is the key parameter in the design of biofilm processes, as it defines the treatment capacity of the reactor and therefore the organic surface loading rate that can be applied, expressed in g COD or g BOD $\cdot$ m^{-2} carrier area $\cdot$ d^{-1}. As in every biofilm process, diffusion of compounds in and out of the biofilm plays a key role. Because of the importance of diffusion, the thickness of the effective biofilm (the depth of the biofilm to which the substrates have penetrated) is significant. As the depth of full substrate penetration is normally less than 100 μm, the ideal biofilm is thin and evenly distributed over the surface of the carrier. To ensure that this condition is complied with, it is important that sufficient turbulence is introduced into the reactor, in order to transport the substrates to the biofilm and to maintain a low thickness of the biofilm by shearing forces. As can be observed in Figure 11.2, significantly less biomass is present on the outside of the carriers than on the (protected) inside. This is due to the fact that abrasion, resulting from carrier collisions, is limiting biofilm thickness.

Figure 11.2 Example of an aerobic MBBR reactor with biomass growth clearly visible on the carrier surface. Courtesy of Anglian Water

Various investigations have shown that the typical biomass concentration in MBBR reactors, defined as the product of specific surface area and average biofilm thickness, is in the order of 2–8 kg TSS $\cdot$ m^{-3} (Rusten et al., 1995a and 1998), which is about the same or higher as typically found in conventional suspended growth activated sludge systems. However, as the observed volumetric removal rate in the moving bed biofilm process can be several times higher (Rusten et al, 1995a), the biomass present in the MBBR must be much more viable, active and specific than in a comparable activated sludge process.

Figure 11.2 shows an example of an aerobic MBBR reactor. A picture of a carrier (taken under water) with attached biofilm is also shown. It can be observed that most of the biomass is growing on the protected inner surface of the carrier. In this specific example, the carrier contains more biomass than usual. The thickness of the biofilm is mainly dependent on the intensity of mixing and highly loaded reactors may therefore have a relatively thin biofilm, due to the high aeration intensity.

In a conventional activated sludge system, the development of an active biomass begins with the growth of zoogleal bacterial flocs. If the sludge age is high enough, these flocs are colonized by protozoa feeding on the free swimming bacteria and by doing so a clarified effluent is produced. In the moving bed process, the type of biofilm that develops depends on the applied organic loading rate (Mosey, 1996). High organic loading rates (30 g COD $\cdot$ m$^{-2}\cdot$ d^{-1}) produce compact bacterial biofilms, with the protozoan population either absent or limited to small free-swimming protozoa and Vorticella spp. Moderate loading rates (10–15 g COD $\cdot$ m$^{-2}\cdot$ d^{-1}) promote a more "fluffy" biofilm with a rich variety of ciliated protozoa while low loading rates ($<$5 g COD $\cdot$ m$^{-2}\cdot$ d^{-1}) produce a biofilm generally dominated by stalked ciliates. Contrary to most biofilm reactors, the MBBR utilizes the total tank volume for biomass growth, similar to the activated sludge system. However, contrary to the activated sludge system, it does not need a sludge recycle stream, as is also the case in other biofilm reactors. The biofilm carriers that move freely in the liquid phase of the reactor are retained within the reactor by a sieve arrangement at outlet.

As there is no sludge recirculation, only the surplus biomass produced in the MBBR reactor plus the particulate material that was present in the wastewater will have to be separated from the effluent. Furthermore, the hydraulic load to the separation step will be reduced as well, as the flow rate to this unit will be much smaller due to the absence of a sludge recycle flow (i.e. $Q_i < Q_i + Q_r$). As both the hydraulic- and solids load are reduced, a smaller separation unit can be selected, which is a considerable advantage over the conventional activated sludge system.

Figure 11.3 shows a typical design of an aerobic MBBR with the main functional elements that it is composed of:

- The reactor tank itself that may be of concrete, steel, glass-fibre reinforced plastic or another material;
- The sieves that are used to retain the carriers in the reactor – horizontal cylindrical sieves are shown which are most applicable to aerated MBBR reactors;
- The aeration system–medium bubble aerators are shown.

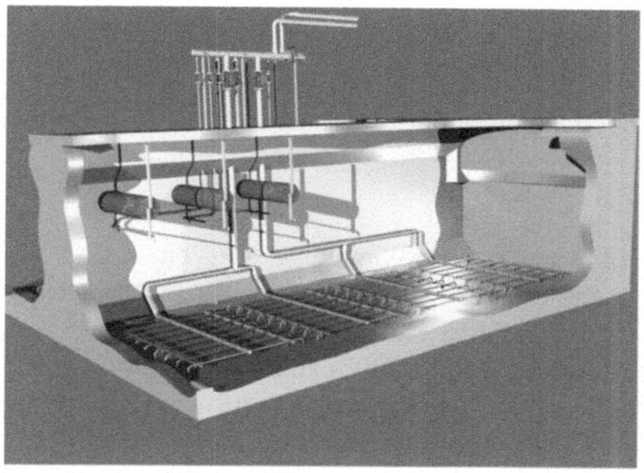

Figure 11.3 Typical design layout of an aerobic AnoxKaldnes MBBR reactor. Courtesy of KrügerKaldnes AS

11.1.1 Carriers used in MBBR processes

An important aspect in the selection of a suitable carrier type (also referred to as "media") for the MBBR process is that it should offer a high protected surface area per volume unit of carrier. Furthermore the carrier structure should be "open" enough to allow for a high mass transfer rate, both of substrate and oxygen and to avoid clogging of the carrier with biomass, which would eventually reduce the mass transfer rates significantly. Finally, the material used should be durable in order to sustain the wear and tear from constant carrier collisions.

The original biofilm carrier (Kaldnes K1) is made of high density polyethylene with a density ranging from $0.95 \pm 0.02\ g \cdot cm^{-3}$, which is shaped as a small cylinder with a cross on the inside of the cylinder and "fins" on the outside, see Figure 11.4. Various carriers of other sizes, shapes and materials have been introduced since then. Several commercially available MBBR carriers produced by AnoxKaldnes are shown in Figure 11.4. At present, the K1 carrier still dominates in terms of the number of applications, but the K3 and especially the K5 carriers are often preferred in new plants. In the case of the K5 carrier this is due to the increase in specific surface area, which allows the same biofilm mass to be maintained at significantly lower investment costs.

The Biofilm Chip M is specifically designed for slow growing organisms, such as nitrifiers and Anammox bacteria, as it offers an extremely high specific surface area. However, due to the very flat shape the mixing characteristics and oxygen transfer rate of this carrier type are poorer, thus requiring an increased energy input for mixing and higher air flow.

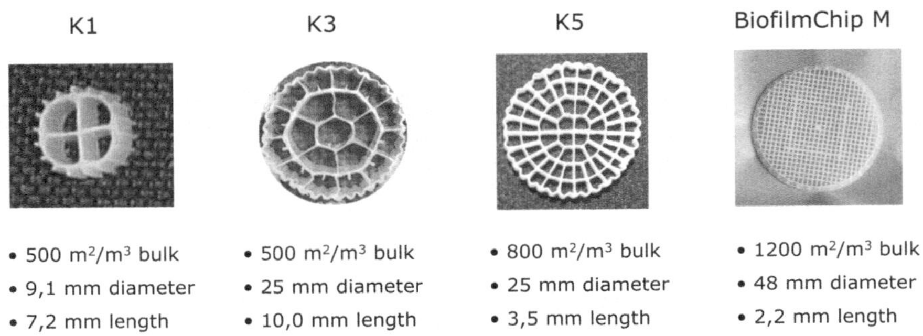

Figure 11.4 Several of the most popular carriers types used in the AnoxKaldnes MBBR – courtesy of AnoxKaldnes AB

Considering that the density of water is approximately $1.0\ g \cdot cm^{-3}$ and slightly less when aerated and depending on the temperature, the effect of the selection of a material with a density slightly less than water ($0.96\ kg \cdot m^{-3}$ in the case of HDPE carriers) is that the bed of carriers is easy to mix, as the carriers tend to "float". High fill rates of up to 70% can be used, although in general the fill fraction is in the range of 50–65%. In non-aerated reactors carriers with a slightly higher density are preferred, in general K1 or K3 heavy carriers with a density of $0.98\ g \cdot cm^{-3}$, in order to provide an optimal mixing regime under unaerated conditions.

One of the important advantages of the moving bed biofilm reactor is that the fill fraction of carrier in the reactor is flexible and only dependent on the required biofilm surface area, as long as the maximum recommended fill fraction is not exceeded. The maximum recommended fill fraction for the K1 and K3 carriers is 65%, each with a bulk specific, effective carrier area of $500\ m^2 \cdot m^{-3}$. Note that the *effective* specific surface area is only $500 \cdot 0.65 = 325\ m^2 \cdot m^{-3}$ at a fill fraction of 65%. For the disc type of carriers (K5 and Biofilm Chip) the recommended maximum fill fraction is slightly lower at 60% and 55% respectively. One may however select a lower fill fraction if this is possible. This is very convenient when the MBBR technology is used to upgrade activated sludge plants, often with a large volume, which means that the maximum fill ratio of 55–65% is not required.

One last issue to consider, especially for the IFAS type of plants, is that the plastic media itself also does occupy some reactor volume. This ranges from a volume fraction between 0.11 and $0.14\ m^3$ liquid reactor volume per m^3 media added for the K1, K3 and K5 carriers to $0.23\ m^3 \cdot m^{-3}$ for the Chip M.

11.1.2 Aeration system

Due to the presence of carriers in the reactor, it is important that the aeration system is robust and provides an even distribution of air over the reactor surface area. Sufficient energy should be introduced to maintain the carriers in suspension, which explains why generally medium bubble aeration is applied and not fine bubble aeration. However, even with a medium bubble diffuser system, it seems that the oxygen transfer in an MBBR is not inferior to that of a fine-bubble diffuser system in activated sludge. This is due to the fact that the air bubbles tend to attach for a period of time to the moving carriers, which increases the average retention time of the air bubbles in the water phase. Furthermore, air bubbles are also broken up into smaller parts when they contact the carriers. Both phenomena increase the oxygen transfer efficiency. A test carried out with K3 carriers at 50% reactor fill demonstrated that the oxygen transfer efficiency in clean water was 42% higher with carriers than without (Le Noir, 2009). The AOTR in a MBBR reactor depends on the type of aeration system selected, which is generally a medium bubble aeration system with 4 mm holes, the water depth, the fill fraction of the biomedia and finally the type of media used. Table 11.1 gives some indicative guidelines for different carrier types and liquid reactor depths. In general operation at a liquid depth of less than three meters is not recommended.

Table 11.1 Indicative AOTE values expressed in gram O_2 transferred per meter liquid depth at 20°C and 1 atm for several carrier types and at different reactor depths

Carrier type	Fill fraction	Liquid depth	AOTE
K1, K3	>25%	>4	12
K1, K3	>25%	<4	11
Biochip M	all	all	10

An MBBR for secondary treatment is generally operated at a dissolved oxygen concentration of 2 to 3 mg $O_2 \cdot l^{-1}$, depending on the reactor temperature. While at a temperature of 10–11°C the recommended DO concentration is 3 mg $\cdot l^{-1}$, at 12–15°C this decreases to 2.5 mg $O_2 \cdot l^{-1}$ and this is further reduced to 2 mg $O_2 \cdot l^{-1}$ at temperatures above 16°C.

11.1.3 Sieves and mixers

In general, the sieves used to retain the carriers in aerobic MBBR reactors are horizontally mounted cylindrical bar sieves, which have replaced the vertical setup in popularity. Rectangular, flat wedge wire sieves or perforated plates are often used in anoxic reactors. Both types are shown in Figure 11.5.

Figure 11.5 Several sieve configurations: aerobic MBBR with horizontal mounted cylindrical bar sieves (left) and anoxic MBBR with flat sieves and mixer mounted in the top left corner (right). Courtesy of KrügerKaldnes AS

In aerobic reactors, the movement of biofilm carriers originates from the agitation introduced by means of aeration, while in anoxic processes a mixer is required to keep the carriers in suspension. It is at present more common to use a vertical or inclined shaft mixer than a horizontal model. Refer for examples to Figure 11.5 and Figure 11.6.

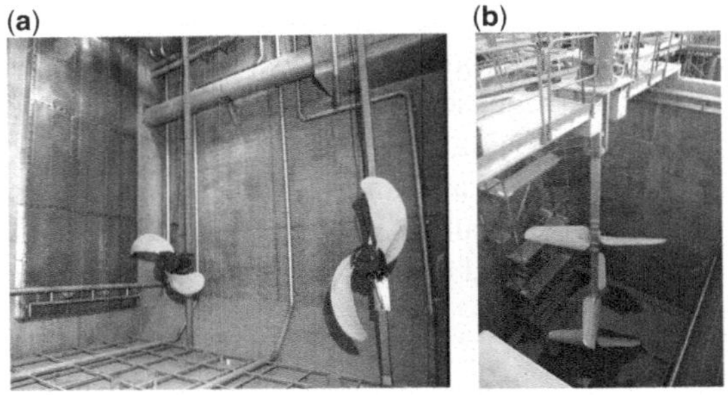

Figure 11.6 Several mixer arrangements: horizontal mixer (left) and vertical top entry mixer (right). Courtesy of AnoxKaldnes AB

11.2 FEATURES OF MBBR PROCESS

The most important advantage of the MBBR process is the compactness resulting from the high volumetric treatment capacity that can be achieved. This feature allows for easy upgrading of wastewater treatment plants as existing tanks may be converted into MBBR reactors. Due to the increase in volumetric treatment capacity an increased load can be applied without the need to enlarge the footprint of the treatment plant. Furthermore, often it is not necessary to fill the MBBR reactor up to the maximum fill fraction with carrier material. Thus, if at a later stage it is required to expand the treatment capacity, this can be done easily by adding more carriers. Of course the aeration system and the sieve arrangement should already be sized sufficiently large so that the future capacity expansion can be accommodated.

As the biomass is attached to the carrier material and is partly protected from the environment in the biofilm, the MBBR is relatively insensitive to temporary variations in pH and temperature or to toxic events. The biomass concentration that can be maintained in the MBBR is not dependent on the sludge settleability characteristics, but depends on the surface area available and thus on the amount of carriers added to the reactor. Due to the existence of substrate and oxygen (or nitrate) gradients, the inner part of the biofilm participates only at a low rate in the conversion process. As such it can be considered as a "reservoir" of biomass, to be called upon when the outer layers of biomass are overloaded. This allows utilisation when large variations in load can be expected, e.g. in tourist resorts.

As the biomass is attached to the carrier material, it is easier to maintain a nitrifying population in the reactor, also under cold conditions. So the typical seasonal variation in nitrification capacity observed in conventional activated sludge systems is reduced in a MBBR system. Furthermore, the use of a biofilm allows more specialized biomass to develop, for example to degrade recalcitrant organic compounds. In principle this is also possible in conventional suspended growth processes, but at the expense of a (very) high sludge age and concomitant system volume.

In pure MBBR systems no biomass recycle is required and both the solids and hydraulic loading rate to the solids separation step are reduced. Therefore, although the settleability of the smallest fraction of the suspended solids produced in an MBBR is typically not very good, the reduction in flow rate allows application of more expensive but also more efficient separation methods. In the IFAS systems there is a sludge recycle, but only part of the biomass (i.e. the suspended part) contributes to the solids load on the final settler. Thus also in this case a smaller separation unit will suffice. Temporary peak flows have the effect on conventional activated sludge systems that part of the biomass (typically up to 20–30%) may be transferred to the final settler, which reduces the biological treatment capacity. Provided that the outlet sieves have been properly sized, peak flows can be handled without a problem by MBBR reactors as the carriers are maintained in the reactor.

From the above list of advantages one might conclude that MBBR should always be the preferred solution, however this is, of course, not always the case, mainly due to the high costs of the carrier media (typically around 500–1500 US dollar per cubic meter of media volume, depending on the carrier type selected), which may very well more than off-set the decreased investment costs resulting from the reduction in treatment volume. Hence, if a new treatment plant is required and the availability of space is not an issue, then in many cases the conventional activated sludge process will require lower investment costs than an MBBR. However, the MBBR is certainly competitive if existing plants are to be upgraded due to more stringent requirements (e.g. nutrient removal) or to an increase in the applied load. This can usually be done without the need for new tanks.

As to construction of new treatment plants, MBBR can certainly be competitive if a small footprint is required, as often is the case for industrial applications or when there are large variations in the expected load in time, e.g. in touristic areas. Furthermore MBBR systems are relatively easy to operate, certainly when compared to the membrane bioreactors. Finally, due to the fact that treatment capacity can be sustained under low temperature conditions, a special niche market is that of municipal sewage treatment in countries with a cold climate, for instance the Nordic (Scandinavian) countries where the process was initially developed: several full scale municipal sewage treatment plants based on the MBBR process have been constructed in Norway and Sweden. However, full-scale municipal MBBR systems have also been successfully applied in regions with a warm climate, such as India and the Arab peninsula.

There are also some specific issues to consider with regards to MBBR, which might result in process upsets, especially if the contractor has insufficient process- and/or engineering knowledge. For example, the sieves are vulnerable to clogging and especially hairs, plastic rags and other fine debris should be removed upstream the MBBR reactors. If not, the sieve may become (partially) blocked which may result in flooding of the reactor and media carry-over. Coarse screening followed by 3 mm fine screens is generally sufficient.

Another aspect that should be considered it that the produced excess sludge is not stabilized, especially when operating in IFAS mode: in the case of nitrogen removal the sludge age of the suspended sludge fraction will typically be much lower than that of a comparable activated sludge system, as most of the nitrification capacity is provided by the biomas present in the biofilm. Thus in principle it will be necessary to separately stabilise the sludge. This might also be considered as a benefit, as it allows for increased biogas production from anaerobic sludge digestion.

The carriers should be distributed uniformly over the contents of the reactor in order to maximize the treatment capacity. In this respect the hydraulic flux is a factor to consider. The hydraulic flux is defined as the combined forward feed flow through the reactor, which is equal to the sum of the influent flow rate plus all recirculation flows, divided by the cross-sectional surface area perpendicular to the flow. If the MBBR is incorrectly designed or operated, the hydraulic flux may become too high, resulting in the accumulation of carriers at the far (downstream) end of the reactor, near or at the outlet sieves which may subsequently become blocked. An aeration scouring device can be provided to keep the sieves clean. In case of anoxic reactors, the mixers can be positioned in such a way that the flow is directed away from the sieves.

The maximum hydraulic flux also set limits to the nitrate recirculation flow rate. In general the value of the nitrate recirculation or "a" factor will have a value around 200–300% in MBBR systems and it should certainly not exceed a value of 500%. As this restricts the return of nitrate to and hence its availability in the pre-D reactor (N_{av1}), post-denitrification may be required in order to achieve a high degree of nitrogen removal. As especially in pure MBBR systems the rate of endogenous denitrification and denitrification on slowly biodegradable COD is low, the addition of an easily biodegradable external carbon source to the post-D reactor may be required. In fact even for IFAS systems this may be a necessity.

As nitrification mainly takes place in the biofilm, a relatively high bulk dissolved oxygen concentration is required to drive the transport of oxygen into the biofilm, typically between 3–7 mg $O_2 \cdot l^{-1}$ depending on the temperature. This increases the aeration costs as the difference between the setpoint- and the equilibrium dissolved oxyxgen concentrations is reduced, which is the driving force for oxygen transfer from air to liquid, while it also increases the load of oxygen that is returned to the anoxic zone.

Another more practical issue is that if maintenance is required inside a reactor filled with carriers, the carriers have to be removed first. This can be done by means of special air lift pumps, but if no empty reactor is available to receive the removed carriers, they need to be stored elsewhere, which might be a logistical problem.

The carriers are susceptible to scaling, for example with $CaCO_3$. The precipitate will not only reduce the surface area available for the biomass, but it also increases the apparent weight of the carrier, which makes it more difficult to keep it in suspension. In extreme cases all carriers might end up at the bottom of the reactor. For this reason application of MBBR to wastewater with a calcium concentration higher than 200 mg $\cdot l^{-1}$ is not recommended. Anaerobic effluent is especially known to create scaling problems, as there is generally a high bicarbonate concentration present, which will be partially stripped in the aeration tank, resulting in an increase of the pH value. On the other hand, if nitrification is applied then some alkalinity will be consumed, which reduces the pH and hence the scaling potential.

Foaming may occur especially during start-up of a plant and might result in media being carried over the reactor wall together with the foam. Apart from remediating the causes of foaming, general measures that can be taken are to ensure sufficient freeboard is present, to install spray nozzles and/or an anti-foam dosing system and to cover the top of the reactor with a net to retain the carriers.

The media may become damaged and it is recommended to store the big bags in which the carriers are delivered upright and side by side rather than in a single big pile, in order to prevent media being crushed. Furthermore the carriers should be handled with care during cold weather, as the media becomes brittle in freezing conditions. Storage in heat and direct sunshine should be avoided, as the media are not resistant to UV radiation.

11.3 MBBR PROCESS CONFIGURATIONS

The MBBR can be used in several ways for the treatment of wastewater. The most common process configurations are:

- Pure MBBR, where MBBR reactors in series are followed by a sludge separation unit;
- A high loaded pre-treatment MBBR upfront an existing activated sludge plant;
- Integrated fixed film processes (IFAS), a hybrid configuration where MBBR and CAS are combined in one or more reactors in a treatment plant;
- MBBR as post-treatment step after another type of pre-treatment process, such as an anaerobic reactor, an existing activated sludge system or a lagoon;

11.3.1 Pure MBBR

The MBBR was developed as a pure biofilm system in which MBBR reactors in series are followed by a biomass separation unit. There is no recycle of sludge because almost all the biomass is fixed on the carriers and only the non retained influent suspended solids plus the net growth of biomass (i.e. equal to the excess sludge production) will have to be separated from the MBBR effluent. Because of the absence of a sludge recycle, in each of the MBBR reactors a highly specialized biomass will develop, with an activity several times higher than that of an equivalent mass of "all-round" suspended biomass from an activated sludge system. Aerobic and anoxic reactors can be used in flexible arrangements, allowing for nutrient removal. The configuration shown in Figure 11.7 is a simple two stage configuration, with a high loaded stage for COD removal followed by a low loaded nitrification stage.

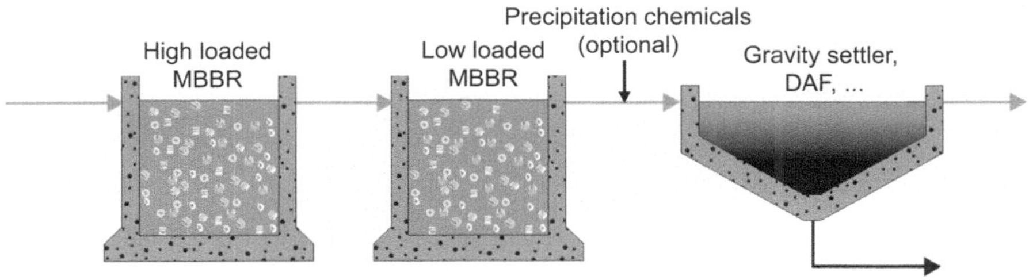

Figure 11.7 Typical pure MBBR configuration for BOD removal and nitrification. Addition of precipiation chemicals is optional and depends on the required effluent quality. Courtesy of AnoxKaldnes AB

A pure MBBR is typically selected when one or more of the following conditions or requirements exist:

- A very compact plant is needed;
- There are large variations in wastewater flow rate and composition;
- Robustness is required to deal with variations in temperature, load, pH or there might be some toxicity in the wastewater.

11.3.2 MBBR as pre-treatment

To an increasing extent the MBBR is also used for upgrading of activated sludge plants, either by using a highly loaded MBBR as a pre-treatment step before the activated sludge plant (an example is the Biofilm Activated Sludge or BAS^{TM} process from AnoxKaldnes) or by adding carriers to existing reactors (IFAS), which will be discussed in the next section.

In the case of MBBR for pre-treatment, a large part of the organic load in the wastewater will now be removed in the high loaded MBBR, allowing for the development of nitrification in the existing (hitherto overloaded) conventional reactor, as shown in Figure 11.8. In the MBBR typically 50–70% of the biodegradable COD load is removed at a high rate in a small volume. Therefore the load on the subsequent activated sludge system is decreased with a factor two to three. Furthermore, as most of the readily biodegradable organics are removed from the wastewater before it enters the activated sludge reactor, problems with the low F/M variant of bulking sludge (filamentous bacteria) are minimized. The low-loaded activated sludge system removes the remaining (slowly) biodegradable organic matter as well as a large part of the excess biomass from the MBBR, which reduces the overall sludge production of this system. This way a significant increase of the sludge age can be achieved, which increases the overall COD removal efficiency and might allow for nitrification to develop.

It is interesting to compare the microbial properties of the biomass present in a system consisting of a low-loaded activated sludge system plus selector with those that will develop in the combination of high-loaded MBBR plus low-loaded activated sludge system. In the first case the easily biodegradable organics are consumed by fast growing bacteria adapted to survive in the starvation conditions that characterise the environment created by a low-loaded activated sludge system. These fast growing bacteria constitute a large fraction of the biomass, while the plug-flow regime imposed in the selector ensures that the flocculent bacteria which settle best will dominate. On the other hand, the slow-growing bacteria needed to degrade difficult compounds will constitute only a small fraction of the total sludge mass and hence a high sludge age and a large reactor volume are needed to secure the required mass of slow-growers.

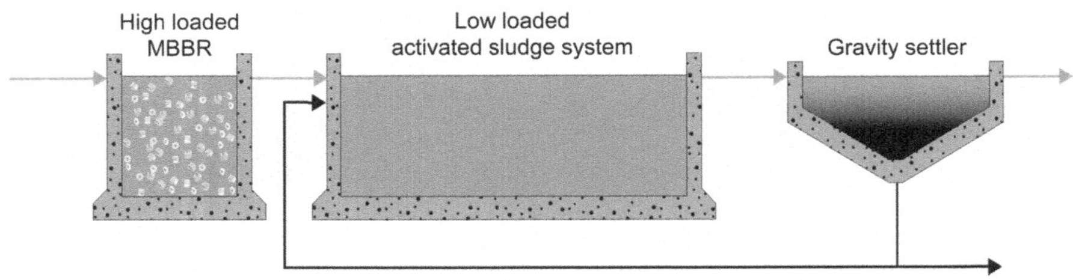

Figure 11.8 Typical BAS^{TM} configuration with a high loaded MBBR reactor to remove bulk of the COD load and a low-loaded AS system to polish the wastewater and/or for nitrification. Courtesy of AnoxKaldnes AB

In the BAS^{TM} configuration consisting of a high loaded MBBR followed by a CAS reactor, the readily biodegradable organics are consumed by fast growing biofilm bacteria that, when dislodged from the carrier, are not adapted to survive the harsh conditions in the subsequent low-loaded activated sludge system. Consequently, most of these fast growing bacteria decay and their (biodegradable) rest products are consumed. Due to the low concentration of easily biodegradable COD in the activated sludge reactor, the slow-growing bacteria that are required to degrade difficult compounds make up a large fraction of the biomass. Thus a relatively short sludge age with a concomitantly small reactor volume will suffice to ensure the presence of sufficient slow-growing micro-organisms.

The overall result is increased COD removal performance at equal sludge age or alternatively similar performance at lower sludge age. For example, the performance of a BAS^{TM} process operated at 6–10 days of apparent sludge age and a F/M ratio of 0.2–0.4 kg COD $\cdot$ kg^{-1} VSS $\cdot$ d^{-1} is comparable to that of a CAS system with a sludge age of 20 days and a F/M ratio of 0.2–0.25 kg COD $\cdot$ kg^{-1} VSS $\cdot$ d^{-1} or less. Therefore the treatment volume required by a BAS^{TM} system is only 30–50% of that of the low-rate CAS, while it exhibits considerably higher process stability and better tolerance against variations and disturbances. Due to the costs of the carriers, the equipment costs of a BAS^{TM} process will be higher, which is compensated by lower civil costs resulting from the reduction in volume.

An additional benefit is that the improved sludge settleability allows the downstream activated sludge system to be operated at increased biomass concentration, while simultaneously the effluent suspended solids concentration will be reduced. Due to the predation on the fast growing MBBR bacteria in the activated sludge system, the excess sludge production will typically be 30–50% lower, which will reduce nutrient demand.

11.3.3 MBBR as post-treatment

In this case an MBBR is placed after the existing wastewater treatment unit, as shown in Figure 11.9. The MBBR removes the residual ammonium load and provides for additional COD removal. In the biofilm specialized bacteria will develop that are able to remove part of the so-called "hard" or recalcitrant COD remaining after pre-treatment. This application is mostly used for municipal sewage treatment plants and for the pulp & paper industry.

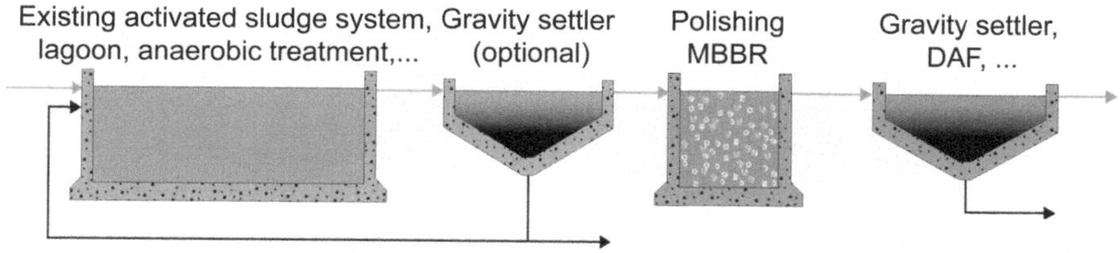

Figure 11.9 Polishing MBBR after pre-treatment. Courtesy of AnoxKaldnes AB

11.3.4 Integrated fixed film reactors

An alternative way to upgrade an activated sludge process using biofilm carriers is the so-called Integrated Fixed Film Activated Sludge (IFAS) process (an example is the AnoxKaldnes HYbrid Biofilm Activated Sludge or Hybas™ process), where the capacity of the activated sludge plant is enhanced by introducing carriers to part (or all) of the activated sludge volume and thus to establish two different types of biomass: one in suspension and one attached to the carriers.

This concept has become quite popular and is now frequently used to upgrade secondary treatment plants to achieve nitrification and/or nutrient removal. In many of these plants this objective has been achieved without the need for expansion of the volume of the existing activated sludge system.

In IFAS processes the carriers are generally only added to the nitrification zones and the post-denitrification zone, which significantly reduces the required investment costs. Suspended sludge is present in all reactors and is recirculated over the reactors with the return sludge. Even nitrification or nitrogen removal is required, the sludge age of the suspended sludge fraction is typically only a couple of days, much lower than required in a comparable conventional CAS configuration. This makes the IFAS configuration very suitable for bio-P removal, should this be desired. Figure 11.10 shows some examples of IFAS configurations.

Due to the continuous seeding with nitrifying biomass from the biofilm, the nitrification capacity in the suspended sludge can be substantial and will be higher than expected based on the applied sludge age alone.

11.4 PURE MBBR DESIGN AND PERFORMANCE

A pure MBBR process can be designed using empiric guidelines based on previous experience and/or based on mathematical models that may take into account all the various sub-processes (diffusion, hydrolysis, endogenous respiration etc.) that occur in a biofilm. The main suppliers of MBBR systems all have spreadsheets that are based on a combination of experience and mathematical modelling approaches, but for indicative sizing of reactor volumes etc., calculations based on empirical guidelines are generally sufficient. In this chapter only some empirical design guidelines are discussed, as well as typical treatment performances that can be expected.

11.4.1 Secondary treatment of municipal sewage

Because of the compactness of the process, the hydraulic residence time in (pure) MBBR reactors for carbonaceous matter removal from municipal sewage will typically be quite low (30–90 minutes), depending on the organic load, the strength of the wastewater and the temperature. Biodegradable, soluble organic matter is quickly degraded. Particulate organic matter is partly captured by the irregularities of the attached biomass, hydrolysed and degraded, while the rest leaves the reactor more or less unaffected.

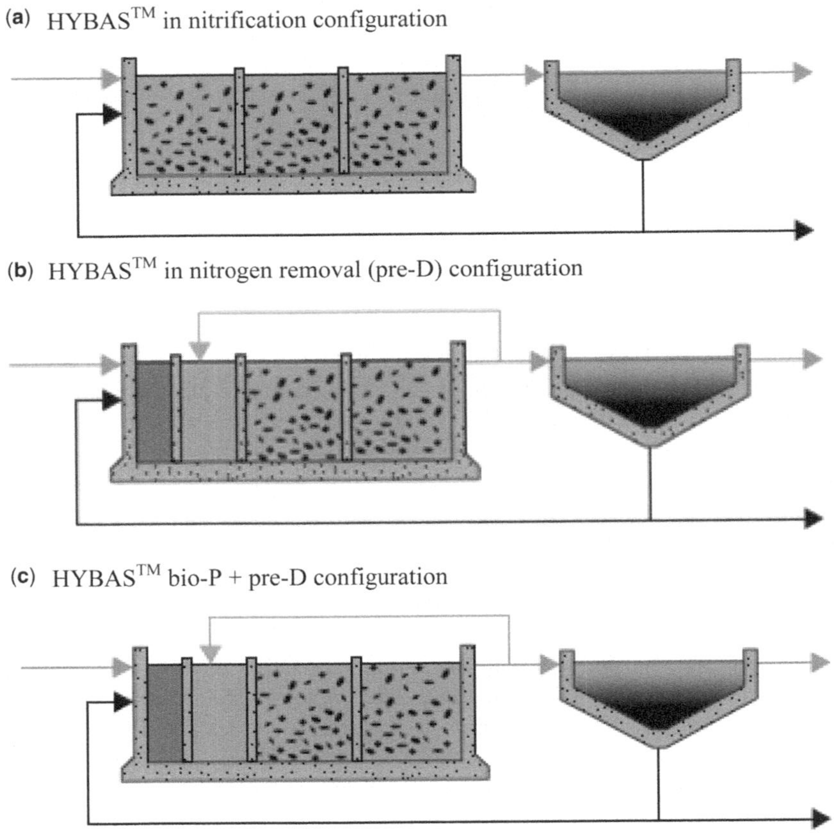

Figure 11.10 Several common Hybas™ (IFAS) configurations. Courtesy of AnoxKaldnes AB

In many countries, phosphate removal is required as well and this can be achieved by means of chemical precipitation with alum, iron chloride or pre-polymerized aluminium chloride (PAC) as coagulants. It is very common to install a coagulation/flocculation tank downstream the MBBR to precipitate ortho-phosphate before the effluent enters the solids separation step. The addition of coagulants/flocculants will significantly enhance TSS removal as well. For secondary treatment, it is recommended to use at least two MBBR stages in series: typically 25% of the total volume is allocated to the first stage and 75% to the second stage. This design ensures a high organic loading rate in the first stage of the bioreactor and a lower loading on the later stage(s). Thus one will take advantage of the fact that the MBBR biomass will operate at a higher removal rate at a high organic loading and will remove the easily biodegradable organic matter already in the first stage, before hydrolysing the remaining slowly biodegradable particulate organic matter predominantly in the later stage(s). This design should be combined with a relatively low hydraulic surface loading rate on the settling tank or preferably include the use of coagulants at a higher surface loading rate.

In order to evaluate the performance with regards to the degradation of organic matter independent of the final biomass separation step, one may look at the removal rate of soluble COD per area of biofilm as shown in Figure 11.11a. The straight line indicates the theoretical maximum value of 100% conversion of the applied load. It can be observed that the maximum SCOD removal rate is obtained at a SCOD loading rate around 30 g $SCOD \cdot m^{-2} \cdot d^{-1}$ (Ødegaard et al, 2000). Figure 11.11b shows the so-called "obtainable" total COD removal rate as function of the total COD loading rate. The obtainable TCOD loading rate is defined as the removal rate of total COD when 100% biomass separation is assumed. Figure 11.11b demonstrates that very high removal efficiencies may be obtained at loading rates of 50 to 100 g $COD \cdot m^{-2} \cdot d^{-1}$ or even higher, as long as good biomass separation downstream the MBBR can be ensured.

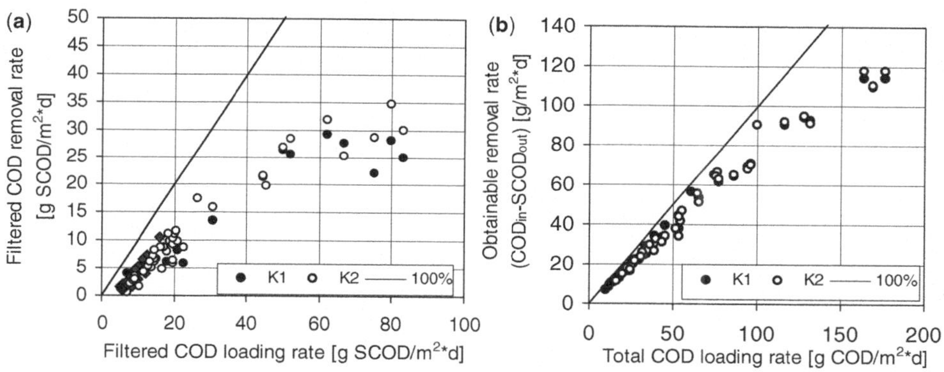

Figure 11.11 Soluble COD removal rate versus soluble COD loading rate (a) and "obtainable" COD removal rate versus total COD loading rate (b), determined at 10–15°C using pre-settled sewage from Ødegaard et al (2000)

When a pure MBBR system is designed for secondary treatment (<25 mg BOD_5 in effluent), the key design parameter is the organic surface area loading rate, expressed in g $BOD_5 \cdot m^{-2} \cdot d^{-1}$ or g $COD \cdot m^{-2} \cdot d^{-1}$. Based on the value selected, the required carrier surface area can be determined and hence the MBBR reactor volume, when the fill fraction is chosen and the specific surface area in $m^2 \cdot m^{-3}$ of the selected carrier is known (refer to Figure 11.4).

Another example of MBBR design rules for municipal sewage are the Norwegian design guidelines for wastewater treatment plants (Norsk Vann, 2009), in which the following design surface area loading rates are recommended for secondary treatment using pure MBBR:

- 8.0 g $BOD_5 \cdot m^{-2} \cdot d^{-1}$ at 10°C for two or more MBBR reactors in series – without any post-coagulation or flocculation applied;
- 11.5 $BOD_5 \cdot m^{-2} \cdot d^{-1}$ at 10°C for two or more MBBR reactors in series – with post-coagulation and flocculation.

According to the same guidelines, the organic surface area loading rate can be corrected to other design temperatures by the following Arrhenius equation:

$$A_T = A_{10} \cdot 1.08^{(T-10)} \tag{11.1}$$

where

A_T = organic area surface loading rate at design temperature T
A_{10} = organic area surface loading rate at 10°C as given above

To convert from BOD_5 to COD, the relationship of $COD = 2.15 \cdot BOD_5 + 50$ is considered typical for Norwegian raw sewage. According to the Norwegian guidelines, the design oxygen demand should be based on a consumption of $1 \text{ kg } O_2 \cdot kg^{-1}$ BOD_5 applied or higher, while a peak loading factor of 1.3 is suggested. Most of the Norwegian plants are designed for a typical organic surface area loading rate of 8–11.5 $g\ BOD_5 \cdot m^{-2} \cdot d^{-1}$ at a design temperature of 10°C with two reactors in series. In this case, the average effluent quality is in the order of $BOD_5 < 10\ mg \cdot l^{-1}$, $TSS < 10\ mg \cdot l^{-1}$ and total phosphorus $<0.5\ mg\ P \cdot l^{-1}$), while the plants are easy to operate (Ødegaard, 2006).

An interesting high-rate treatment concept for municipal sewage is based on the combination of a high rate MBBR preceded by fine sieves for pre-treatment followed by coagulation and DAF flotation. This results in an extremely compact process, as shown in Figure 11.12, with a total hydraulic residence time of less than one hour, which is able to meet the effluent limits for secondary treatment ($<25\ mg\ BOD_5 \cdot l^{-1}$).

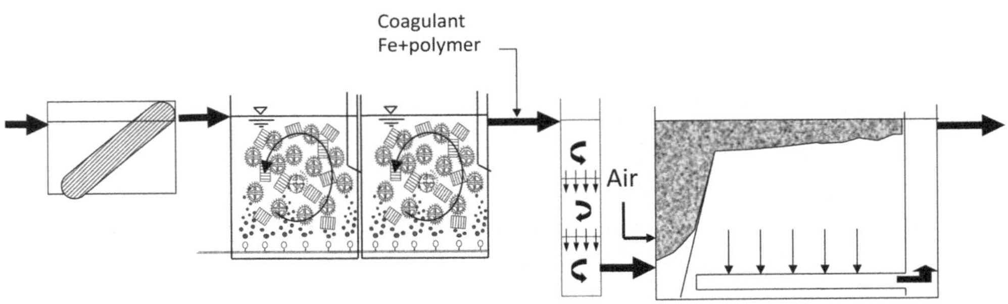

Figure 11.12 Treatment scheme of a high rate pure MBBR for secondary treatment, followed by coagulation-flocculation and flotation (Melin et al., 2004)

The theoretical concept behind the high rate MBBR process is that the biomass present in the biofilm only degrades the soluble, easily biodegradable organic matter, while the colloidal/particulate, slowly or non biodegradable organic matter passes through the reactor.

Then, after flocculation with polyelectrolyte and metal salts, this material is removed from the MBBR effluent in the DAF together with the biomass produced from the degradation of the soluble, organic matter. This requires an organic loading on the MBBR that is low enough for truly soluble organic matter to be converted and high enough to prevent significant hydrolysis and degradation of suspended- and colloidal organic matter. Due to the high putrescible (non-degraded) biodegradable fraction of the produced excess sludge, the combination with anaerobic sludge digestion is especially attractive.

A cationic polymer acts as the primary coagulant in order to reduce the dosage of iron to a very low value and thereby minimises hydroxide precipitation and consequently inert sludge production. Flotation is selected over a gravity settler because a highly loaded bioreactor may result in poorly settleable but easily floatable sludge and also because it is a low foot-print separation technique.

The performance of this treatment concept was evaluated at pilot scale by Melin et al., 2004. After passing the fine sieves (0.8 mm) the sewage was treated in two MBBR reactors in series, with a total

HRT ranging between 15–45 minutes at a temperature of 10 to 15°C. The applied COD loading rate was not allowed to exceed a value of 20–25 g SCOD $\cdot$ $m^{-2} \cdot d^{-1}$ and 65–85 g TCOD $\cdot$ $m^{-2} \cdot d^{-1}$, which is about 4–5 times higher than the normal design load for traditional biofilm reactors targeting secondary treatment effluent quality. The observed biomass production was 0.5 g TSS $\cdot$ g^{-1} SCOD removed. Subsequently coagulation/flocculation was applied in a dedicated coagulation/flocculation chamber with a HRT of 5–10 minutes, followed by DAF flotation with an hydraulic loading rate between 5–15 m $\cdot$ h-1 and a HRT of 20–25 minutes. The applied coagulant dosing rates were 5 mg poly-electrolyte plus 35 mg Fe per g TSS. In the pilot experiment this corresponded to about 1 mg $\cdot$ l^{-1} polymer and 7 mg $\cdot$ l^{-1} Fe for the 200 mg of suspended solids present in the MBBR effluent. At this level of chemical dosing, the final (separated) sludge production was very close to the observed suspended solids production in the MBBR, i.e. the precipitation resulted in very little additional inert sludge production.

11.4.2 Secondary treatment of industrial wastewater

For industrial wastewater treatment, the organic surface loading rate that can be applied and the COD removal efficiency that can be obtained are dependent on the type of wastewater to be treated. However, in Table 11.2 some indicative design guidelines are shown for pure industrial MBBR systems, based on wastewater with a high fraction of biodegradable COD.

Table 11.2 Indicative design guidelines for pure industrial MBBR systems based on the required $SBOD_5$ effluent value and SBCOD removal efficiency

Parameter	UoM	BOD effluent limits (mg $SBOD_5 \cdot l^{-1}$)		Moderate loading (incomplete removal)
		Very low <10 mg $\cdot$ l^{-1}	Low <15 mg $\cdot$ l^{-1}	
% SBCOD removal	%	>90%	70–80%	30–60%
Vol. loading rate[(1)]	g SCOD $\cdot$ $l^{-1} \cdot d^{-1}$	2–3	4–5	6–20
Surface loading rate	g SCOD $\cdot$ $m^{-2} \cdot d^{-1}$	20–30	30–50	80–150
Apparent yield[(2)]	g VSS $\cdot$ g^{-1} SCOD	0.20–0.25	0.25–0.3	0.3–0.45
HRT	hrs	>3	>3	>1.5

Notes: (1) Process temperature $\geq$*25°C*
(2) Expressed per gram of SCOD removed, *not applied*

The pure MBBR process has been applied to several industrial wastewaters, in general combined with coagulation in order to enhance biomass separation and/or achieve P-removal (Rusten et al., 1996 and 1998 and Marcolini et al., 2004). In the pulp and paper industry several MBBR plants have been operated at relatively low loadings (<2–3 kg COD $\cdot$ $m^{-3} \cdot d^{-1}$) in order to avoid the use of coagulants, as a low loading rate enhances the growth of filter-feeding micro-animals such as stalked ciliates and rotifers. As these animals filter the water and consume free-living bacteria, the latter concentration is kept on a low level, resulting in a clear liquid phase without the need for coagulant addition. Of special interest for the treatment of kraft mill wastewater is the use of a thermophilic MBBR process (Malmqvist et al, 2007a). Most forest industry waste- and process waters are discharged at high temperature and cooling to a temperature of around 35°C for mesophilic biological treatment may be costly. Furthermore thermophilic conditions prevent the growth of pathogenic bacteria. In spite of the fact that thermophilic treatment processes are generally considered to be more sensitive than conventional mesophilic processes, for example to variations in pH, long-term full-scale results from a Canadian and a Swedish plant (Malmqvist et al., 2007a) have demonstrated that the thermophilic MBBR process can be very stable and robust at COD removal rates of 8 kg $COD \cdot m^{-3} \cdot d^{-1}$ in both plants. An example of the application of MBBR for the petrochemical industry is for the Statoil/Hydro Ormen Lange facility at the coast of West-Norway. The Ormen Lange field is developed as a sub sea production facility at depths between 800 and 1100 meters and is linked to a processing plant on land. The process wastewater contains a variety of organic compounds, predominantly glycol. The treatment plant consists of two MBBR compartments with a total volume of 1500 m^3 and with a media fill fraction of 50% in two parallel lines, preceded by a small (160 m^3) aerated LSP (low sludge production) reactor. This is a reactor operated without sludge recycle. Biomass from the MBBR-reactors is separated by coagulation/flotation. The plant meets the following effluent requirements: 85 mg $TOC \cdot l^{-1}$, 125 mg $COD \cdot l^{-1}$, 15 mg $BOD_5 \cdot l^{-1}$, 0.4 mg phenols $\cdot l^{-1}$ and 1.5 mg oil $\cdot l^{-1}$.

11.4.3 Nitrification

To achieve nitrification, at least two MBBR reactors (sometimes three) in series are used, often constructed in a physical layout that promotes plug flow behaviour, for example by dividing a reactor into several compartments. Nitrification will be severely limited by the presence of organics in the first reactor, where the heterotrophic bacteria are dominating because they out-compete the nitrifiers for available oxygen in the biofilm. Therefore only in the second reactor nitrification will be established.

As long as the ammonium concentration in the MBBR is maintained above a certain critical minimum value, the nitrification rate will not be limited by ammonium but by oxygen instead. The value of this critical ammonium concentration ranges between 0.5–2.0 mg $NH_4\text{-}N \cdot l^{-1}$ and depends on the dissolved oxygen concentration applied in the liquid phase of the MBBR, as can be observed in Figure 11.13.

Figure 11.14 demonstrates that the organic surface loading rate applied to the nitrification reactor has a profound influence on the nitrification rate and should thus be maintained as low as possible. At 15°C and for organic surface loading rates above 5 g $BOD_5 \cdot m^{-2} \cdot d^{-1}$, nitrification will no longer occur at normal dissolved oxygen concentrations. Therefore it is recommended that the first MBBR reactor is designed to reduce the organic surface loading rate on the second reactor to 0.5 g $BOD_5 \cdot m^{-2} \cdot d^{-1}$ or less. Assuming the ammonium concentration is above the critical value, the subsequent reactor compartments can be designed for the nitrification rate that develops for the selected design oxygen concentration, using Figure 11.13.

If very low ammonium effluent values (<1 mg $NH_4\text{-}N \cdot l^{-1}$) are required, the last reactor compartment should be designed for an ammonium limited rate, in which case the oxygen concentration may be low as well, as it is no longer limiting the rate of nitrification.

Figure 11.13 clearly demonstrates that the nitrification rate is linearly dependent upon the oxygen concentration, up to more than 10 mg $O_2 \cdot l^{-1}$ (Ødegaard et al., 1994 and Æsøy et al., 1998). An advantage of the linear relationship between oxygen concentration and nitrification rate is that it is very easy to use for process control. An oxygen concentration of 2–3 mg $O_2 \cdot l^{-1}$ or less significantly reduces the nitrification rate, which is the reason that in general a nitrifying MBBR is operated between 3 to 7 mg $O_2 \cdot l^{-1}$. Note that this value is already close to the equilibrium concentration of 9.1 mg $O_2 \cdot l^{-1}$ at 20°C and 1 atm, although of course the equilibrium concentration will be higher at increased depth.

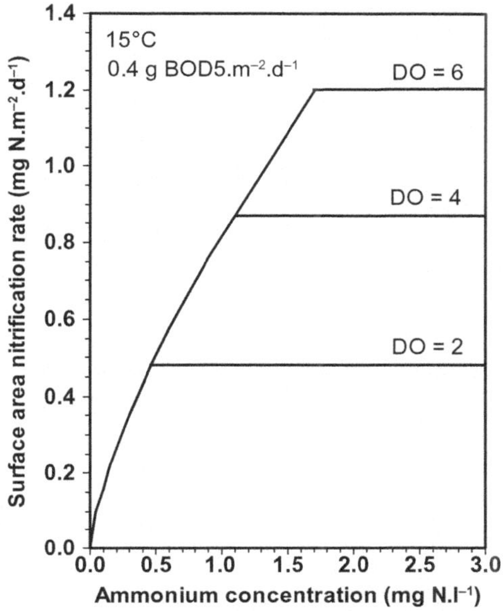

Figure 11.13 Influence of the ammonium concentration on the nitrification rate in a MBBR for several DO concentrations, at 15°C and at a surface organic loading rate of 0.4 g $BOD_5 \cdot m^{-2} \cdot d^{-1}$ (Rusten et al., 1995)

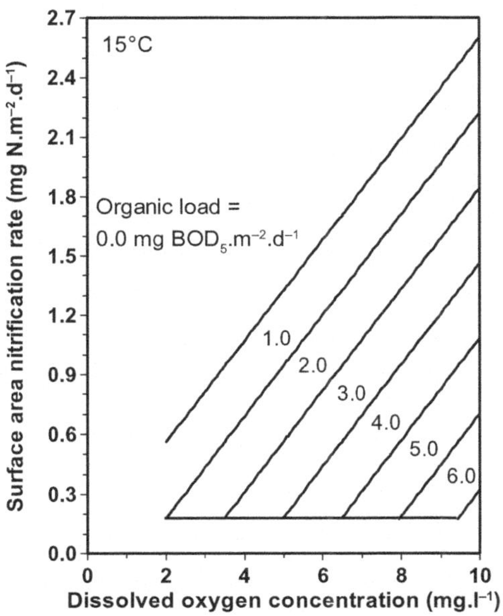

Figure 11.14 Effect of the DO concentration on the nitrification rate in a MBBR at different organic surface loading rates at 15°C and with ammonium present in excess (Hem et al., 1994)

The basic relationships shown in Figure 11.13 and Figure 11.14 can be used to design for MBBR based nitrification. For example, in the Norwegian design guidelines that were discussed earlier in Section 11.4.1, nitrifying MBBRs are designed for to comply with effluent limits of less than 15 mg $BOD_5 \cdot l^{-1}$ and less than 3 mg $NH_4\text{-}N \cdot l^{-1}$. Furthermore, the first MBBR reactor will be designed for BOD removal only at an organic surface loading rate of 4 g $BOD_5 \cdot m^{-2} \cdot d^{-1}$, in order to reduce the $SBOD_5$ concentration to 5 mg $\cdot l^{-1}$ or less, while the second reactor is designed for nitrification and is sized based on an ammonium surface loading rate of 0.8 mg $NH_4\text{-}N \cdot m^{-2} \cdot d^{-1}$. The design values given here are valid for an oxygen concentration >7 mg $O_2 \cdot l^{-1}$ and a temperature of 15°C. The required carrier surface area can be adjusted for the design temperature using the Arrhenius relationship shown in Eq. (11.1).

In some cases, when more than 90% ammonium removal is required or when the target effluent ammonium concentration is very low, for instance less than 1 mg $NH_4\text{-}N \cdot l^{-1}$, it is recommended to divide the nitrification volume into two parts, which should be designed based on the following ammonium surface loading rates:

- First nitrification compartment: 0.8 g $NH_4\text{-}N \cdot m^{-2} \cdot d^{-1}$ for $N_{ae} > 2$ mg $NH_4\text{-}N \cdot l^{-1}$;
- Second nitrification compartment: 0.3 g $NH_4\text{-}N \cdot m^{-2} \cdot d^{-1}$ for $N_{ae} < 2$ mg $NH_4\text{-}N \cdot l^{-1}$.

Hem et al (1994) proposed a more sophisticated method to determine the design nitrification rate, based on nitrification kinetics where the oxygen concentration is taken into account as well. After having first determined the organic matter removal volume as shown above, the nitrification rate (when NH_4-N is the limiting substrate) is determined by:

$$r_N = k \cdot (S_n)^n \tag{11.2}$$

where

r_N = surface nitrification rate (g $NH_4\text{-}N \cdot m^{-2} d^{-1}$)
S_N = NH_4-N concentration in the reactor (mg $N \cdot l^{-1}$)
n = reaction rate order–n is often set at 0.7 based on Hem et al (1994)
k = reactor rate constant, which is dependent on the organic loading rate, i.e. typical values are 0.4 without primary treatment; 0.5 with pre-denitrification and 0.6 with chemically enhanced primary treatment

NH_4-N is only rate limiting at low NH_4-concentrations (1–2 mg $NH_4\text{-}N \cdot l^{-1}$). At higher concentrations, r_N will be limited by the bulk liquid DO concentration and S_n should be replaced by S_{nDO}:

$$S_{nDO} = (DO_l - 0.5)/3.2 \tag{11.3}$$

where DO_l is the dissolved oxygen concentration in the reactor

11.4.4 Nitrogen removal

Nitrogen removal in a pure MBBR plant can, as with conventional activated sludge systems, be carried out by pre-denitrification, post-denitrification or by a combination of pre- and post-denitrification: refer also to the general discussion in Section 5.3 on the advantages and disadvantages of these nitrogen removal configurations.

Whereas the nitrification efficiency in an MBBR is typically much higher than that of a comparable activated sludge system, this is not the case for the denitrification efficiency. One reason is that the nitrification MBBR is operated at a DO concentration (3–7 mg $O_2 \cdot l^{-1}$), typically several times higher than that applied at the end of the conventional nitrification zone (1–2 mg $O_2 \cdot l^{-1}$). The effect of oxygen recirculation on denitrification capacity has been extensively discussed in Section 5.4.2.3. As a remedial measure it can be considered to install a dedicated de-oxygenation reactor downstream the nitrification reactor, operated at 1.5–2 mg $O_2 \cdot l^{-1}$, from which the nitrate recirculation stream to the pre-D reactor is taken.

Furthermore, as discussed in Section 11.2, there is a limit imposed on the nitrate recirculation flow rate that returns the nitrate to the pre-D stage, due to restrictions on the value of the maximum allowable forward flux in the MBBR reactors. Often the value of the nitrate recirculation factor "a" will be around 2–3. Hence, in the absence of a sludge recycle stream the fraction of the nitrification capacity returned as nitrate to the pre-D zone (i.e $N_{avl}/N_c = a/(a+1)$ for both pre-D and BDP configurations) will have a value between 0.67 and 0.75.

Finally, the MBBR system is less effective than the conventional activated sludge system in the utilisation of slowly biodegradable COD for denitrification. For example, even in cases where the pre-denitrification stage is limited by insufficient availability of easily biodegradable COD, it might still be necessary to have an aerobic COD oxidation stage upfront the nitrification stage in order to protect the nitrification reactor.

So if extensive nitrogen removal is required, then in situations where a CAS would be able to meet the effluent nitrogen limits using the COD present in the wastewater, in the case of pure MBBR an external carbon source may be required, which is dosed in the post-nitrification zone. A small aerobic MBBR is installed after the post-anoxic MBBR to remove any residual carbon source and to make sure the BOD and COD effluent limits are not exceeded.

Whenever possible the pre-denitrification MBBR configuration should be used because it ensures maximum utilization of the biodegradable COD in the influent and partial recovery of the alkalinity consumed in the nitrification process. Thus, pre-denitrification is the preferred configuration if the content of easily biodegradable COD in the raw water is high relative to the nitrogen concentration (i.e. a high SBCOD/N-ratio) and the required nitrogen removal efficiency is not so high, for example less than 75% total nitrogen removal. The denitrification rate in the MBBR can be limited by either the nitrate concentration or the biodegradable COD concentration. At nitrate concentrations above 3 mg $NO_3\text{-}N \cdot l^{-1}$, the denitrification rate will mostly be governed by the type and availability of easily biodegradable COD (Rusten et al., 1995b). If oxygen is supplied to the reactor with the inlet water or recirculated water, biodegradable organic matter will be consumed for oxygen respiration and thus reduce the available amount for denitrification. The limitation of the pre-denitrification process results from the fact that oxygen-rich water from the nitrification step will have to be returned to the pre-denitrification step if steps such as reducing the dissolved oxygen concentration in a de-oxygenation reactor are not taken to prevent this.

When a high degree of total nitrogen removal efficiency is needed (more than 75%), or if there is insufficient internal carbon source available, then a combination of pre- and post-denitrification is recommended.

If in addition the space available for the plant is very limited and the most compact solution has to used, then it could considered to combine post-denitrification with pre-precipitation. This gives the smallest overall reactor volume because most of the particulate and colloidal COD is removed in the pre-precipitation step. However, it also results in the highest requirement for external carbon source. In the post-denitrification reactor an easily biodegradable carbon source is added, so that the denitrification rate will be high (refer to Table 11.3). The post-denitrification mode has also other advantages over the pre-denitrification mode. It may lead to considerably lower total bioreactor volumes (40–50%) and it gives much better control of the nitrogen removal process. However, in order to minimise the consumption of external carbon source, combined pre- and post denitrification is generally preferred and used in all Norwegian nitrogen removal plants based on the MBBR process. A typical process configuration is shown in Figure 11.15.

Table 11.3 Surface area denitrification rates (g NO_3-N $\cdot$ m^{-2} $\cdot$ d^{-1}) and stoichiometrical COD demand in mg COD per mg NO_3-N removed for several sources of external carbon (Rusten et al, 1996)

Temperature	Denitrification rate/stoichiometrical ratio		
	Methanol	Ethanol/acetic acid	Mono-propylene glycol
≤5°C	0.8–1.2/5.0	1.5–2.3/5.5	0.9–1.1/5.0
10°C	1.2–1.8/4.5	2.7–3.5/5.0	1.3–1.7/4.5
≥15°C	1.5–2.3/4.0	3.2–4.0/4.5	1.6–2.2/4.0

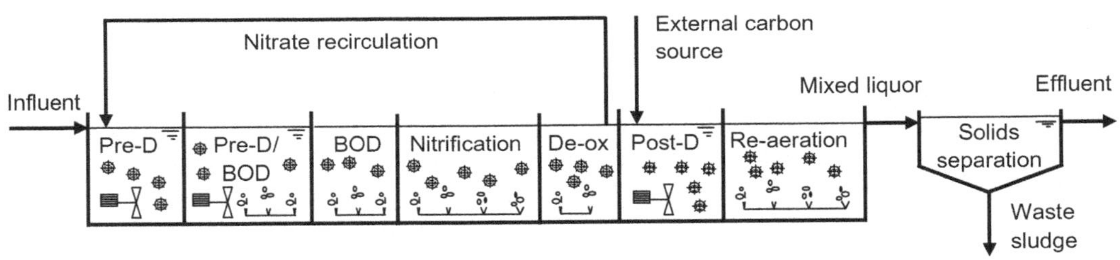

Figure 11.15 Typical process configuration of a pure MBBR system for combined pre- and post denitrification

The first compartment is anoxic (pre-denitrification) and receives recycled nitrate from the fifth (aerobic) compartment. If it is required to decrease the nitrate concentration in the pre-D zone to a value of 3 mg NO_3-N $\cdot$ l^{-1} or less, then it is recommended to divide this compartment into two reactors in series. The second compartment may be operated under anoxic or aerobic conditions, depending on the process conditions and the observed treatment performance, for example the extent of nitrate removal in the pre-D zone. Hereafter the wastewater passes two aerobic reactors (compartments three and four), for residual organic matter removal and for nitrification. The extent of nitrification may be controlled by the oxygen level in the nitrification reactor, as there is a linear relationship between the nitrification rate and the oxygen concentration at ammonium concentrations above 0.5–2 mg N $\cdot$ l^{-1}, as discussed in Section 11.4.3).

The fifth compartment serves as a de-oxygenation zone: the oxygen concentration is controlled at a value of 1.5 to 2.0 mg O_2 $\cdot$ l^{-1}. This compartment thus also contributes to the nitrification capacity, albeit at a significantly lower rate than in the preceding nitrification zone. The ammonium concentration in this compartment is generally low, typically less than 0.3 to 3 mg NH_4-N $\cdot$ l^{-1}, which means that the ammonium rather than the oxygen concentration is now limiting the nitrification rate. This compartment may be operated with or without aeration, depending on the oxygen demand for residual nitrification. The nitrification capacity in the fourth compartment should be controlled in order to ensure that sufficient ammonium is left in the effluent from this reactor to remove the residual oxygen in the de-ox zone to the desired level.

The sixth compartment is the post-denitrification zone, to which an external carbon source is added in order to remove the residual nitrate. To attain a high nitrate removal efficiency, typically a certain degree of excess carbon source is added, about $10\,mg\ COD \cdot l^{-1}$. Furthermore, if an effluent nitrate concentration of less than $2\,mg\ NO_3\text{-}N \cdot l^{-1}$ is required, it is recommended to divide the post-D compartment into two reactors, as the low nitrate concentration will otherwise reduce the denitrification rate in the entire post-D zone. In this case, the external carbon source is added to the first anoxic reactor to allow for high rate denitrification. In the second zone, denitrification will proceed at much lower rate. The specific excess sludge production in the post-D compartment is typically around 0.25 mg $TSS \cdot mg^{-1}$ COD consumed. Optionally a small aerobic compartment with a HRT of 15 minutes at peak flow is used for re-oxygenation as well as for removal of the residual biodegradable COD resulting from overdosing in the post-D zone. This zone could be omitted if the post-D reactor is divided into two compartments and is operated in such a way that removal of the added COD is ensured.

The process configuration shown in Figure 11.15 will ensure a very high degree of nitrogen removal and allows for flexibility of operation in order to maximise the extent of nitrification and minimise the consumption of external carbon source. For example, during summer operation less nitrification volume is needed because of the higher temperatures. Furthermore, the raw water sewage normally contains less oxygen than in the winter time. In this situation the second compartment may be used for denitrification (mixed and not aerated). The return of nitrate to the pre-D zone may be increased and the consumption of external carbon source in the post-D compartment will be reduced. Therefore, in the summer time a large fraction of the nitrogen load can be removed by pre-denitrification alone, reducing the consumption of external carbon source to the minimum required to meet the effluent nitrogen limits.

During winter operation, more nitrification capacity is needed and the second compartment is operated with aeration. The consumption of external carbon source in the post-D compartment will increase in order to compensate for the reduction in pre-D capacity. In a situation with cold wastewater, one may reduce the organic load by using pre-coagulation The residual organic material in the pre-treated wastewater will mainly consist of easily biodegradable organic matter, but as the denitrification capacity is reduced only a moderate nitrate recirculation flow rate is required, only 25% to 50% of the influent flow rate, thus minimising oxygen recirculation. The rest of the nitrate is removed in the post-denitrification step, where an online effluent nitrate analyzer controls the carbon source addition.

Rusten and Ødegaard (2007) have reported extensively on the performance of four Norwegian combined pre- and post-DN pure MBBR plants for the treatment of municipal sewage. The process configurations are shown in Figure 11.16, while selected design and operating data is presented in Table 11.4.

11.4.5 Phosphorus removal

In pure MBBR-processes, phosphorus removal is normally carried out through chemical precipitation/flocculation inbetween the MBBR reactor and the final solids separation step. The use of a dedicated coagulation/flocculation chamber is recommended, as it enhances the removal efficiencies of both phosphorus and suspended solids. In the Norwegian plants discussed in Table 11.4 chemical precipitation for phosphorus removal is applied. As can be observed, all plants demonstrate very low effluent total P-concentrations in the effluent, in general less than $0.2\,mg\ P \cdot l^{-1}$.

Biological phosphorus removal is not applied in pure MBBR systems, as this process requires the biomass to pass through sequential anaerobic/oxic conditions, but it is certainly a possibility for IFAS systems, where the phosphorus can be removed biologically using the suspended biomass.

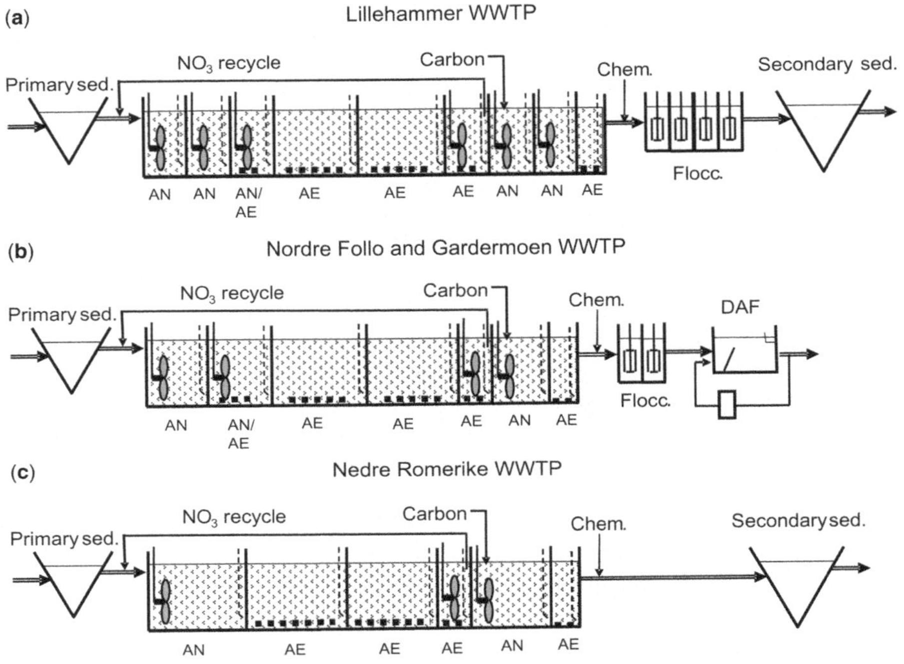

Figure 11.16 Process configurations of four full-scale Norwegian combined pre- and post-denitrification pure MBBR plants (Rusten and Ødegaard, 2007)

Table 11.4a Design and operating values of the four full-scale Norwegian sewage treatment plants shown in Figure 11.16 (Rusten and Ødegaard, 2007)

Parameter	UoM	Lille hammer	Nordre follo	Gardermoen	Nedre romerike
Avg. flow	$m^3 \cdot h^{-1}$	1200	750	920	2300
Max. flow	$m^3 \cdot h^{-1}$	1900	1125	1300	7200
Temperature	°C	3–14	6–14	4–14	7–14
MBBR vol.	m^3	3840	3710	5790	19,370
Carrier fill	%	65	66.2	58.5	42.7
Avg/max HRT		3.2/2.0	4.9/3.3	6.3/4.5	8.4/2.7
Carbon source	(–)	ethanol	methanol$^{(1)}$	glycol	methanol$^{(1)}$
Consumption	$g\ COD \cdot g^{-1}\ N$	3.3	2.2	2.4	no data

Note: (1) At present changed to glycol

Table 11.4b Design and operating values of the four full-scale Norwegian sewage treatment plants shown in Figure 11.16

Parameter	UoM	Lille hammer		Nordre follo		Gardermoen		Nedre romerike	
Effluent concentration/removal efficiency[1]									
BOD_5	$mg \cdot l^{-1}$	2.2	99%	2.8	98%	3.2	98%	4.0	95%
COD	$mg \cdot l^{-1}$	35	93%	39	91%	25	96%	27	93%
Total N	$mg\ N \cdot l^{-1}$	2.9	92%	9.7	73%	7.0	87%	5.0	83%
Total P	$mg\ P \cdot l^{-1}$	0.12	98%	0.2	96%	0.18	98%	0.05	99%

Note: (1) Based on average effluent concentrations

11.5 UPGRADING OF EXISTING ACTIVATED SLUDGE PLANTS

11.5.1 High rate pre-treatment MBBR for BOD/COD removal

The use of a high rate MBBR as a pre-treatment unit upstream of an (existing) activated sludge plant (for example the BAS™ process–refer for the basic flow scheme to Figure 11.8) is an interesting alternative to the pure MBBR process for applications when increased COD removal is required and conversion to a pure MBBR system is not economically attractive or not desired by the client.

The characteristics and main benefits of this process have been discussed in Section 11.3.2. In the design of the BAS™ process, it is especially important to know the distribution of soluble and particulate organic matter, as the design philosophy is to remove the easily biodegradable, soluble organic matter in the MBBR reactor and leave the particulate organic matter for the subsequent activated sludge system, which has much better flocculation characteristics anyway. Typically only 50–70% of the applied biodegradable SCOD load will be removed in the MBBR reactor.

For municipal applications, at soluble biodegradable COD (SBCOD) concentrations in the MBBR that are higher than 100–125 $mg \cdot l^{-1}$, the SCOD removal rate will reach a peak value of around 30 g $SCOD \cdot m^{-2} \cdot d^{-1}$.

At this loading rate only a very small fraction of the particulate COD will be converted into soluble COD, while at a much lower loading rate of around 5 g $SCOD \cdot m^{-2} \cdot d^{-1}$ as much as 60–70% of the particulate material may typically be hydrolysed into soluble compounds. The design of the activated sludge reactor should thus consider the non degraded soluble and particulate COD, the inert organic particulate COD from the influent and the biomass produced in the high rate MBBR system. Consequently, two variants of the high loaded MBBR configuration exist, i.e. with respect to whether the suspended solids are removed from the MBBR effluent. The benefit of suspended solids removal upstream of the activated sludge plant is that the organic loading rate of the activated sludge plant will be reduced. On the other hand, an additional unit is required and overall excess sludge production will be higher.

For industrial BAS™ it is not very common to have a solids separation step inbetween the MBBR and the activated sludge system. Some indicative design guidelines are provided in Table 11.5. Note that the BOD effluent limits in this table refer to the final effluent of the complete BAS™ system, i.e. *after* treatment in the subsequent activated sludge treatment step. Typical volumetric loading rates when applied to the whole BAS™ system are in the order of <2.5 kg $COD \cdot m^{-3} \cdot d^{-1}$ when maximum COD removal is required and up to 3–3.5 kg COD $m^{-3} \cdot d^{-1}$ when the required degree of COD removal is less.

Table 11.5 Indicative design guidelines for pre-treatment MBBR (BAS^{TM}) at 25°C based on the required *final* SBOD effluent value and the SBCOD removal efficiency

Parameter	UoM	BOD effluent limits ($mg\ SBOD_5 \cdot l^{-1}$)	
		Very low $< 10\ mg \cdot l^{-1}$	Low $< 15\ mg \cdot l^{-1}$
% SBCOD removal	%	>90%	70–80%
MBBR volumetric loading rate	$g\ SCOD \cdot l^{-1} \cdot d^{-1}$	8–15	8–15
MBBR surface loading rate	$g\ SCOD \cdot m^{-2} \cdot d^{-1}$	80–150	100–200
Filling degree	%	10–50	10–50
HRT	hrs	>1.5	>1.5

The BAS^{TM} process can be operated conventionally, allowing a slight excess of soluble nutrients (nitrogen and phosphorus) after the treatment, or under conditions of nutrient limitation: nutrient limited BAS or NLBAS (Malmqvist et al, 2007b). The main advantages of NLBAS operation are a considerable reduction in excess sludge production and therefore nutrient demand and often an improved sludge quality compared to conventional BAS^{TM} operation. On the other hand, nutrient limitation reduces the removal rate of organic material in the MBBR so that a larger reactor volume may be required. Furthermore, it is also important that the nutrient limitation is not too severe, else the organic removal efficiency of the whole BAS^{TM} system (MBBR + CAS) may suffer, while the biomass settling characteristics in the CAS may deteriorate as well (refer also to Chapter 9). The BAS^{TM} process, i.e. especially the nutrient limited version, has been successfully applied to the treatment of forest industry wastewaters. Other areas of application are the chemical industry including petrochemical wastewater, the pharmaceutical industry (Rosén et al., 1998 and Lexa et al., 2001), the textile industry and the food industry (dairy, vegetable processing).

In these cases, bench scale testing demonstrated that operation under nutrient limited conditions in the MBBR improved the settling characteristics in the activated sludge step. An example of a full-scale BAS^{TM} plant in the chemical industry is the SUT (Semcorp Utilities & Terminals) wastewater treatment plant, located on Jurong Island in Singapore. The wastewater to the plant originates from eight different chemical companies and contains mostly alcohols, aldehydes, organic salts, iso-phthalic acid, phenol and some minor quantities of other chemicals such as acetone, benzylaldehyde, m-toluic acid, benzene and xylene. In a pilot project for this plant (Wessman et al, 2004) the MBBR roughing stage achieved more than 85% removal of organic carbon and more than 90% removal of BOD_5, at a surface area loading rate of $24\ g\ COD \cdot m^{-2} \cdot d^{-1}$.

11.5.2 Upgrading of secondary CAS to nitrification

Often it is required that an existing activated sludge plant designed for secondary treatment needs to be upgraded in order to nitrify and reach low ammonium effluent values as well. One strategy would be to abandon the activated sludge process completely and convert the aeration volumes into pure MBBR reactors. As nitrification in a MBBR normally can be achieved within a hydraulic residence time of 3 hrs or less, the available reactor volume will in general be more than sufficient. However, the conversion to pure MBBR involves a large investment in biofilm carrier material and in many cases it is more cost-effective to enhance the operation of the activated sludge system using one of three alternative strategies:

1. Build a high-rate pre-treatment MBBR for BOD/COD removal upstream of the activated sludge reactor (which has been discussed in Sections 11.3.2 and 11.5.1), where two variants can be distinguished:
   a. With biomass separation after the MBBR ahead of the AS;
   b. Without biomass separation after the MBBR;
2. Convert the activated sludge system into an IFAS by means of the addition of moving bed biofilm carriers in (part of) the existing reactor volume (Section 11.3.4);
3. Combine a high rate pre-treatment MBBR with an IFAS.

Alternatives two and three will be discussed in the subsequent section with reference to the treatment results of pilot- and full scale plants.

11.5.3 Nitrification in IFAS processes

The IFAS process has become very popular for the upgrading of existing activated sludge plants in order to achieve nitrification. In this process all or part of the volume of an activated sludge system is retrofitted into an IFAS process. This way the content of nitrifying biomass of the system is increased and nitrification may be achieved in a volume that otherwise would not be sufficient to comply with the requested effluent ammonium limit.

The IFAS is not an easy process to design since the presence of nitrifiers shifts from the biofilm on the carriers to the suspended biomass and vice versa depending on the applied loads of organic material and ammonium. At a high sludge age (low organic load) then a larger fraction of the nitrification occurs in the suspended biomass, while at lower sludge ages a higher fraction of the nitrification takes place in the biofilm.

The IFAS process results in a higher degree of stability of the nitrification process than in an activated sludge system alone, because at increasing organic load and decreasing sludge age the subsequent decrease in nitrification capacity in the suspended biomass will be compensated by enhanced nitrification capacity in the biofilm.

Full nitrification in IFAS systems has been reported at aerobic sludge ages of less than three days, see Figure 11.17. Here it can also be observed that nitrification is achieved at a much lower aerobic sludge age than what would have been required in an conventional activated sludge system. The upper line in Figure 11.17 represents the German ATV design recommendations for the sludge age of nitrifying AS plants. The bottom line shows the minimum required aerobic sludge based on nitrifier kinetics. The figure indicates that the IFAS plants operate at a sludge age that on average is only 60% of what would be required in a conventional activated sludge system alone.

Trapani et al. (2009) conducted a series of pilot experiments to determine the nitrification rate of the suspended biomass in an IFAS system operated at an aerobic sludge age of 5.7 days and a temperature of 12°C. The reactor consisted of three equally sized compartments with a total HRT of 3.5–4.5 hours. The middle reactor compartment was filled for 60% with carriers and was operated at 3.5–4.5 mg $O_2 \cdot l^{-1}$. The observed nitrification rate of the suspended biomass was 3.17 g NH_4-N $\cdot$ kg^{-1} VSS $\cdot$ h^{-1}, much higher than to be expected for a nitrifying activated sludge system operating at this temperature and sludge age.

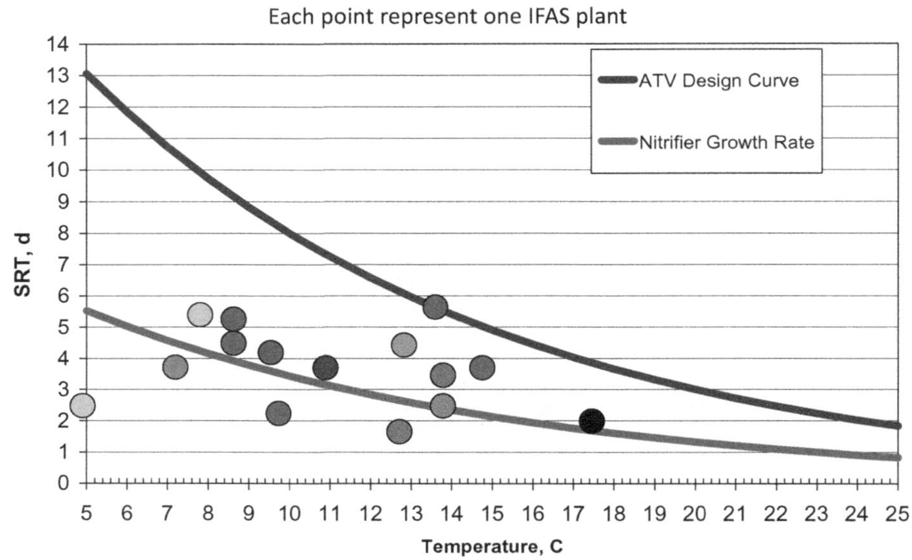

Figure 11.17 Operating SRT of several IFAS plants operating in the US, as function of the operating temperature, compared to the minimum aerobic sludge age for nitrification and the recommended sludge age for full nitrification according to the ATV-131 guidelines (Johnson, 2009)

The increase in nitrification capacity can be attributed to the "seeding effect" as attached biofilm with a high proportion of nitrifiers is continuously being sloughed-off from the carrier and subsequently becomes entrapped in the suspended biomass. Consequently, when the biofilm was analysed separately it was observed that the nitrification rate was even higher than that of the suspended biomass: 4.17 g NH_4-N $\cdot$ kg^{-1} VSS $\cdot$ h^{-1} at 12°C. This value corresponds to a surface area nitrification rate in the biofilm in the IFAS reactor compartment of 0.85 g NH_4-N $\cdot$ $m^{-2} \cdot d^{-1}$, which is close to the maximum value possible at the oxygen concentration applied. However, for the design of full-scale IFAS plants typically more conservative nitrification rates will be used, i.e. 1.0 to 1.5 g NH_4-N $\cdot$ kg^{-1} VSS $\cdot$ d^{-1} in the suspended biomass and between 0.5–0.75 g NH_4-N $\cdot$ $m^{-2} \cdot d^{-1}$ in the biofilm (at 15°C).

The design procedure for an IFAS system starts with an estimate of the aerobic sludge age of the suspended biomass that will develop. If it is higher than the minimum sludge age required for nitrification (i.e. around 2.5 days at 15°C), then the nitrification capacity of this biomass can be estimated based on the COD and nitrogen load and -composition in the raw sewage, using the design theory in this book (Chapter 5). When the nitrification capacity of the suspended biomass is subtracted from the TKN load in the raw sewage, this determines the biofilm nitrification capacity that is needed. Now the required biofilm carrier surface area can be calculated, which in turn defines the fill fraction of carriers. In general the available activated sludge volume is so large that only part of the reactor volume needs to be filled with carriers. If the selected carrier fill fraction is less than the maximum value allowed, this can be used as an advantage should the nitrogen load increase and more nitrification capacity be needed. This can easily be achieved by increasing the media fill fraction with additional carrier material.

An interesting application of a combination of a high rate MBBR followed by an IFAS process can be found in Sharjah, UAE. The objective was to convert the existing activated sludge system for secondary treatment into a nitrifying system for twice the existing load and without expanding the existing reactor volume of 13,000 m^3. The effluent quality after final settling needed to comply with the following limits:

- Less than 30 mg TSS $\cdot l^{-1}$;
- Less than 20 mg $BOD_5 \cdot l^{-1}$;
- Less than 3 mg NH_4-N $\cdot l^{-1}$.

The existing treatment plant consisted of a single aeration tank, which was divided in three compartments. The first 20% of the available reactor volume was converted into a high-rate MBBR pre-treatment reactor (BAS™) with a 50% fill rate of K3 carriers. This was followed by an IFAS in which carrier was used only in the last reactor compartment (30% of the available volume with a 50% fill rate of K3 carriers, while the remaining middle reactor compartment (50% of the available reactor volume) did not receive any carrier fill.

The high rate pre-treatment MBBR removes soluble COD at a rate of circa 20 g SCOD $\cdot m^{-2} \cdot d^{-1}$ and ensures that almost all the biodegradable soluble organics are removed ahead of the AS and IFAS compartments: typically between 30–60 mg SCOD $\cdot l^{-1}$ is present in the effluent from the high rate compartment of which a large part will be inert SCOD. Table 11.6 shows the results of two performance tests, demonstrating the capacity of this process. During the second performance test, the average daily flow to the plant was 51,700 $m^3 \cdot d^{-1}$, containing 13,750 kg $BOD_5 \cdot d^{-1}$ and 2785 kg TKN $\cdot d^{-1}$.

Table 11.6 Treatment efficiency of the high rate MBBR (BAS™) + IFAS during performance tests at the Sharjah WWTP – UAE (KrügerKaldnes, 2010)

Parameter	Nov. 2008 to Jan. 2009			April 2010 to May 2010		
	In	Out	% rem.	In	Out	% rem.
BOD_5	277.8	7.5	97.3	266.7	6.4	97.6
TKN	66.2	4.7	92.3	54.1	0.92	98.3

11.5.4 IFAS for nitrogen removal

In addition to upgrading of a secondary activated sludge plant to nitrification, it is also possible upgrade to full nitrogen removal. To reduce the required investment costs, in general carriers are only added to the nitrification, de-oxygenation and post-denitrification zones as indicated in Figure 11.18.

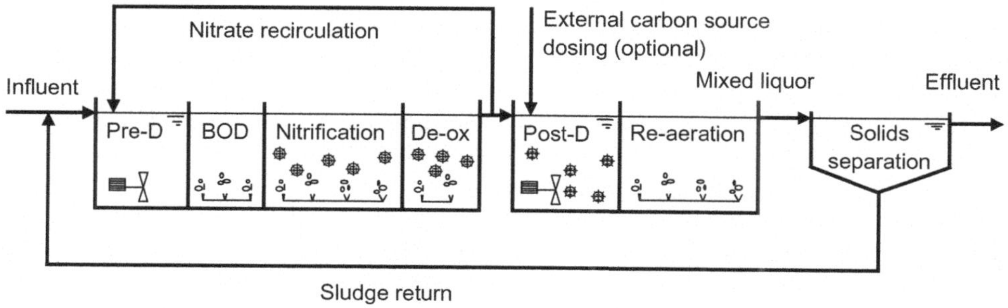

Figure 11.18 Typical configuration of a Hybas™ system for nitrogen removal

Biological P-removal may also be carried out in an IFAS-plant. In this concept the enhanced uptake of poly-P is carried out in the suspended biomass and design and operation of such a plant will be quite similar to that of an conventional activated sludge plant, the difference being that nitrification (and possibly denitrification) is carried out in a hybrid (IFAS) reactor.

A recent example of an industrial IFAS application is for a chemical industry complex in China (Wessman, 2008) where the Hybas^{TM} plant is treating anaerobically pre-treated chemical wastewater for the BOD and nitrogen removal. The IFAS volume of in total 11,250 m^3 is divided in three parts: a 2500 m^3 pre-anoxic reactor followed by a 5000 m^3 aerobic reactor (both containing only suspended biomass) and finally a 3750 m^3 hybrid aerobic reactor containing 1550 m^3 of carrier volume.

11.6 SOLIDS REMOVAL FROM MBBR EFFLUENT

Separation of biomass from pure MBBR systems is different from that in activated sludge- or IFAS systems as there is no sludge recycle flow. Therefore, the flow to the solids separation unit only contains the excess biomass produced in the MBBR plus any influent suspended solids that have not been retained or degraded. Typically, the suspended solids concentration in the feed of the solids separation unit is in the order of 150–250 mg $TSS \cdot l^{-1}$ when municipal sewage is treated, about twenty times less than in an activated sludge- or IFAS system. In addition to conventional methods for activated sludge separation, such as gravity settling and (to a much smaller extent) dissolved air flotation and membrane separation, other technologies can be applied as well, such as micro-sand ballasted lamella separation, micro-screening and media filtration. In the following sections the performance and design of different methods for suspended solids removal will be briefly discussed.

11.6.1 Gravity settling

Gravity settling is probably the most common method of biomass separation after an MBBR. However, the characteristics of MBBR biomass are not very favourable for settling without prior coagulation/flocculation. Most of the MBBR sludge will settle well, but especially in highly loaded MBBR systems the fine suspended solids in the effluent need to be coagulated in order to achieve a low suspended solids concentration. However, in most countries phosphate removal is required anyway which necessitates the addition of metal salts in the MBBR effluent before the final separation step.

Various types of coagulants may be considered. When phosphate removal is not required a cationic polymer may be used, in order to reduce the chemical excess sludge production. In addition, it has been demonstrated that a low dosage of a supplementary metal cation destabilizes the colloids and improves the suspended solids removal efficiency (Melin et al., 2004). As the cationic polymer replaces a large part of the metal cations that would have been required for destabilization, the combined coagulant dosing results in far less sludge production than the use of a high dose of metal coagulant alone. If it is decided to use only a cationic polymer, then a cationic poly-acryl-amide with a high molecular weight and a relatively low charge density is recommended. When combined metal/polymer coagulant dosing is selected, then poly-diallyl-dimethyl-ammonium chloride (poly-DADMAC) is recommended, which has a high charge density and a relatively low molecular weight.

When phosphate removal is required, a metal coagulant must be added in order to precipitate soluble phosphate, for example as $FePO_4$ or $AlPO_4$. An additional dosage of an anionic polymer as a flocculant enhances the settling velocity of the biomass-metal hydroxide-metal phosphate aggregates that will be formed and allows the application of higher surface overflow rates. The suspended solids concentration in the effluent after a pure MBBR process with settling/flotation in combination with coagulation-flocculation is often in the order of 10–15 mg TSS · l^{-1}. In the case of a well-designed BAS™ process this is in the range of 20–30 mg TSS · l^{-1}

It is important to take into account the variability of the flow in the design of the final settler. In the Norwegian design guidelines for treatment of municipal sewage, the following scenarios with regards to the influent flow rate are considered (Ødegaard, 2010):

- Q_{avg} (day) = average daily flow of all days in a year
- Q_{avg} (hr) = average hourly flow in a year
- Q_{des} (median day) = the median value of the maximum hourly flow for all days in a year
- Q_{des} (hr) = the maximum design flow that can be accepted in the final settler

In Table 11.7 recommended hydraulic overflow rates are given for the different design cases and different scenarios with respect to coagulation/flocculation.

Table 11.7 Proposed design values for gravity settling tanks ($H_d \geq 3$ m) for normally loaded MBBR reactors (less than 8 g $BOD_5 \cdot m^{-2} \cdot d^{-1}$)[6] treating pre-settled sewage[7] (Ødegaard, 2010)

Coagulation/flocculation	Hydraulic overflow rate ($m^3 \cdot m^{-2} \cdot h^{-1}$)			
	Q_{avg} (day)	Q_{avg} (hr)	Q_{des} (median day)	Q_{des} (hr)
No chemical addition or flocculation before settling[1]	0.50	0.60	0.75	1.10
Addition of cationic polymer coagulant before settling[2]	0.70	0.80	1.00	1.50
Addition of high dose of metal precipitant [3]	0.80	0.90	1.10	1.70
Addition of low dose of metal precipitant + cationic polymer coagulant before settling[4]	0.80	0.90	1.10	1.70
Addition of high of dose metal precipitant + anionic polymer coagulant before settling[5]	1.10	1.20	1.50	2.25

Notes: *(1) Settling tank directly after MBBR*
(2) Addition of polymer to the effluent of the MBBR before inlet to settling tank
(3) Addition of Al or Fe (i.e. for P- precipitation) followed by properly a designed flocculation tank before the settling tank
(4) Addition of cationic polymer to compensate for low metal dose followed by a properly designed flocculation tank before the settling tank
(5) Addition of Al or Fe (i.e. for P- precipitation) + addition of anionic polymer (as flocculant) followed by a properly designed flocculation tank before the settling tank
(6) When the organic loading rate is higher, all values are reduced by 10%. When the load is lower (i.e. when designing for nitrification), all values are increased by 10%
(7) When primary settling is not applied, all values are increased by 10%

11.6.2 Micro-sand ballasted lamella sedimentation

The MBBR is already a very compact bioreactor and when combined with a micro-sand ballasted lamella sedimentation tank (such as ACTIFLO®), the surface area of the total treatment plant becomes extremely small. It is possible to reach effluent suspended solids concentrations of less than 5 mg · l^{-1} at dosing rates of around 10–15 mg Fe $\cdot$ l^{-1} plus 1–2 mg $\cdot$ l^{-1} of anionic polymer and at hydraulic overflow rates in the settling tank of 50–100 m $\cdot$ h^{-1} (Ødegaard et al, 2010). At this time only a few full-scale combined MBBR-ACTIFLO® plants are in operation, most of them in Norway. Some of these plants are used in areas where an extreme variation in flow and load occurs over the year, for instance in tourist towns. The treatment plants are designed for secondary treatment and phosphorus removal.

11.6.3 Dissolved air flotation

Dissolved air flotation (DAF) is very efficient in separating biomass from MBBRs. Especially when combined with pre-coagulation, flotation achieves a very good separation efficiency in addition to resulting in a very compact treatment plant. It also produces a concentrated excess sludge with at least 40 g TSS $\cdot$ l^{-1}. In most cases it is used in combination with coagulation for chemical phosphorus removal, i.e. the plants are consisting of pre-treatment, MBBR, chemical addition (metal salt and polymer) followed by flotation. Dissolved air flotation reactors are usually designed based on surface overflow rate, even though this might not be the best parameter for design. Typical design values according to the Norwegian design guidelines are provided by Ødegaard et al. (2009):

- Tank depth: 2 to 3 m;
- Surface overflow rate: for Q_{max} (median day) 5 m $\cdot$ h^{-1} and for Q_{des} (hr) 10 m $\cdot$ h^{-1} (refer to Section 11.6.1 for the definitions);
- Saturation pressure : 400–600 kPa (4–6 bar), air saturation: 80–90%;
- Saturation water flow: 10–25% of Q_{des} (hr) depending on the content of suspended solids in the effluent and the degree of air saturation achieved.

The quantity of dissolved air provided to the DAF unit depends on the saturation water flow rate, the applied air pressure and the resulting air saturation percentage. When the lower limit for the saturation pressure is used, the higher limit for saturation water flow rate should be used and vice versa. The air saturation equipment should have a minimum capacity equivalent to 10% of Q_{des} (hr) at a pressure of 600 kPa.

11.6.4 Micro screening

Micro-screening can be used as a solids separation step directly downstream of a MBBR process, provided that the MBBR outlet TSS concentration does not exceed 150–200 mg $\cdot$ l^{-1}. Examples of suitable applications would be downstream of a tertiary nitrification MBBR or a post-denitrification MBBR that is placed after an existing activated sludge system. In other cases a flotation step upstream the micro-screening unit is recommended.

The solids separation effect of micro-screening, i.e. using drum- and disc filtration, is mainly based on sieving of particles and thus the particle size distribution is crucial for the separation result. Nominal pore sizes range from 10 µm to more than 100 µm depending upon the application. The principle of operation of most devices is relatively simple and straightforward, as can be observed in Figure 11.19, which shows a Hydrotech disc filter.

At the Rya sewage treatment plant in Göteborg–Sweden, thirty-two disc filters with a design filtration rate of 10 m $\cdot$ h^{-1} have been in operation since 2010 (Figure 11.20). The disc filters serve two objectives: biomass separation from the post-denitrifying MBBR and final filtration of the effluent from the secondary settlers. The design of this plant is further detailed in Mattson et al. (2009).

The performance of an Hydrotech disc filter was tested after a MBBR for post-denitrification at the Sjölunda wastewater treatment plant in Sweden (Persson et al., 2005). This wastewater treatment plant consisted of primary settling, a high-loaded activated sludge system for BOD removal, nitrification in trickling filters and finally post-denitrification in an MBBR utilising methanol as a carbon source. A DAF unit was used to remove the suspended solids from the MBBR effluent. Depending on the solids load in the MBBR effluent, the DAF could be operated with and without coagulation. The results from the pilot test, using a Hydrotech™ disc filter (model HSF 1702/1), carried out *directly* on the effluent of the post denitrification MBBR are shown in Table 11.8. Very low effluent TSS values are obtained. Furthermore it can be observed that operation at reduced pore size increased TSS removal efficiency from 81.8% to 88.5%, but at the cost of a significant reduction in filtration capacity.

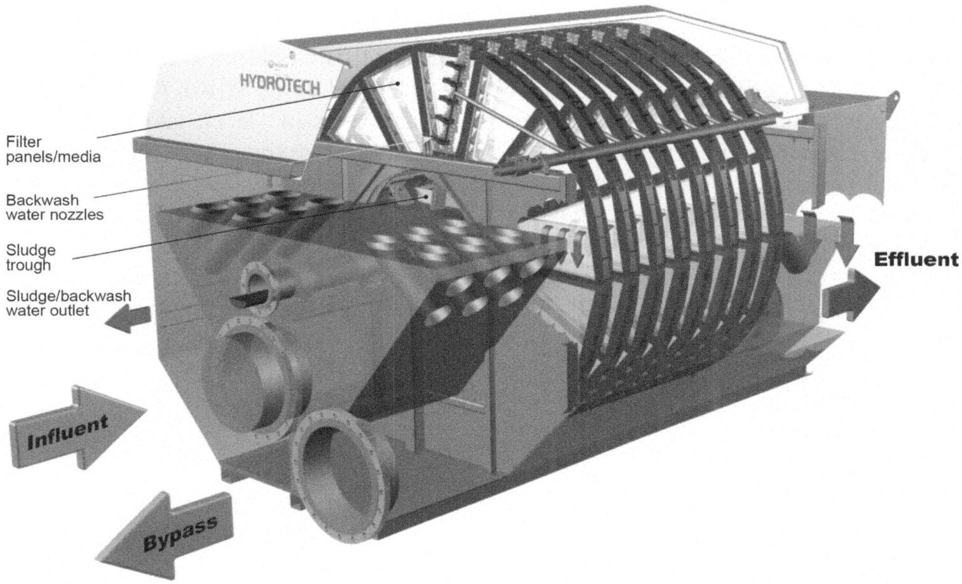

Figure 11.19 Physical layout and operating principles of a disc filter – courtesy of Hydrotech AB

Figure 11.20 Hydrotech disc filters installed at Rya WWTP, Gothenburg – Sweden, courtesy of Hydrotech AB

Table 11.8 Results of disc filter test on the effluent from a denitrifying MBBR at Sjölunda WWTP – Sweden. The suspended solids content in the feed water to the disc filters varied between 10 – 50 mg TSS · l^{-1} (Mattson et al., 2009)

Mesh pore size	Feed (mg TSS · l^{-1})		Effluent (mg TSS · l^{-1})		Filtration capacity ($m^3 \cdot m^{-2} \cdot h^{-1}$)	
	Avg.	**St. dev.**	**Avg.**	**St. dev.**	**Avg.**	**St. dev.**
18 µm	27.5	14.5	5.0	1.8	13.7	7.2
10 µm	30.5	10.8	3.5	1.3	4.8	2.2

Two separate pilot tests with disc filters were performed on MBBR effluent from the Gardermoen and Nordre Follo treatment plants, both located in Norway (Ødegaard et al., 2010). One important finding was that in order to maintain the filtration capacity at an acceptable level, taking into account the backwash water consumption, a mesh size of at least 20 µm was required. The results shown in Figure 11.21 and Figure 11.22 result from tests with 20 and 40 µm mesh sizes. In these experiments the applied filtration rates ranged between 2–6 m · h^{-1}, which in most cases was less than the maximum hydraulic capacity of the filter.

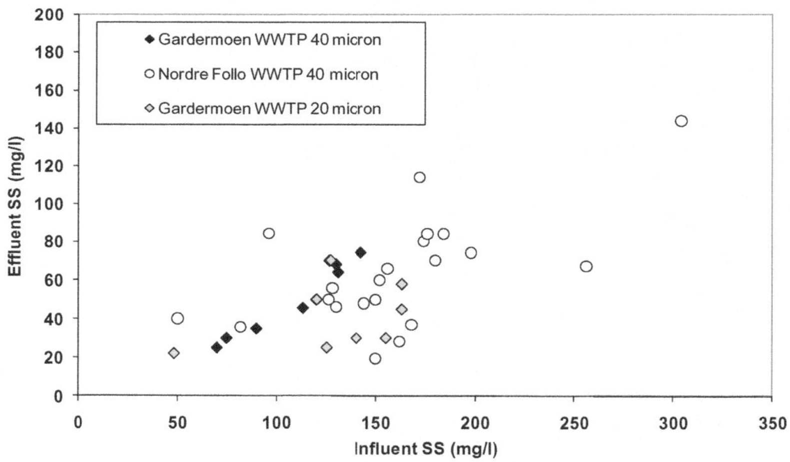

Figure 11.21 Effluent TSS as function of the influent TSS concentration with dosing of a cationic polymer but without the application of a separate pre-coagulation/flocculation unit (Ødegaard et al, 2010)

When disc filtration is applied to MBBR effluent without any previous separation steps, it is recommended to apply pre-coagulation/flocculation to remove the fine particle fraction. This was confirmed in the pilot tests when the effluent of disc filtration with only cationic polymer dosing was compared to that where a coagulation/flocculation unit was used. Figure 11.21 shows the effluent suspended solids concentration versus the applied polymer dose, for the case that a cationic polymer was dosed as the only coagulant. When compared to Figure 11.22, it is obvious that the effluent suspended solids concentration can be significantly reduced by the use of coagulation and flocculation ahead of the disc filter. An effluent suspended solids concentration well below 10 mg TSS · l^{-1} was achieved when a cationic polymer was used at a (relative high) dosing rate of 35 mg polymer · g^{-1} TSS. This corresponds to 5–7 mg cationic polymer · l^{-1} for a typical MBBR effluent suspended solids concentration of 150–200 mg TSS · l^{-1}. When the polymer dosing rate was increased furter, the hydraulic filter capacity could be further increased, but the effluent TSS concentration did not decrease further.

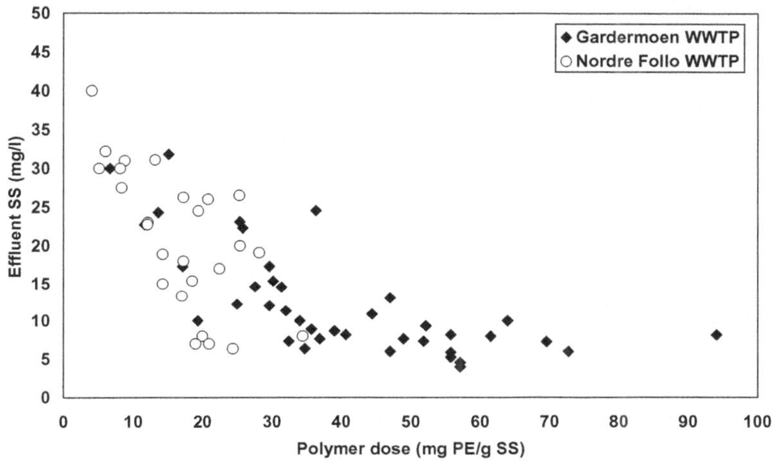

Figure 11.22 Effluent TSS after a 40 µm disc filter as function of applied polymer dosing rate, using a dedicated coagulation/flocculation chamber (Ødegaard et al, 2010)

Table 11.9 provides indicative design values for the use of disc filters for solids separation after a MBBR unit. As the table demonstrates, the energy consumption is low and mainly related to filter backwashing. Pumping to the filters is usually not required, as the filters can be placed in the gravity line due to the low differential pressure over the filter cloth. As discussed before, pre-coagulation/flocculation is normally required, at least when separating directly after a MBBR. The backwash water flow will be equal to approximately 2–4% of the treated flow rate.

Table 11.9 Indicative design values for hydraulic capacity and energy consumption of a Hydrotech Disc filter (HSF 2224) treating MBBR effluent (Ødegaard et al., 2010)

Feed TSS concentration	Filtration rate (m · h^{-1})		Energy consumption (Wh · m^{-3})	
	18 µm	40 µm	18 µm	40 µm
(1)40 mg TSS · l^{-1}	8	–	12	–
(2)100–200 mg TSS · l^{-1}	–	6–8	–	16

Notes: (1) MBBR for post-denitrification (after a final settler)
Coagulation/flocculation with 7–9 mg Al · l^{-1} and 3–4 mg · l^{-1} cationic polymer

11.6.5 Media filtration

Media filtration, in general using sand and/or anthracite, is mainly used as a final polishing step after settling or flotation. But in some cases sand filtration is also used directly after a MBBR, normally only when a nitrification- or post-denitrification MBBR is used as an additional treatment step, installed after the existing treatment system. For example in Klagshamn WWTP in Malmö, Sweden, five downflow dual media filters (anthracite + sand) are installed directly after a post-denitrification MBBR reactor. Each filter has a surface area of 44 m^2 and operates at a maximum filtration rate of 8.2 $m \cdot h^{-1}$. Ferric chloride is used only occasionally. The plant is consistently producing a final effluent with low TSS and total phosphorus values: <5 mg TSS $\cdot l^{-1}$ and total P < 0.2 mg $\cdot l^{-1}$.

Direct filtration might even be used directly after a high-rate aerobic MBBR without any other separation steps. If phosphate removal is required, a metal coagulant (Al or Fe) will be necessary, but if only SS- and BOD-removal is necessary, a cationic polymer alone or in combination with a low dose of metal salt may be sufficient. This would minimise sludge production and might make the use of direct filtration possible. A filter for such an application should have a high sludge retaining capacity in order to achieve acceptable filter run times. At present this concept has not yet been applied at full-scale.

11.6.6 Membrane filtration

Several combinations of MBBR and membrane (UF) separation are possible, as shown in Figure 11.23, both for submerged and cross-flow membranes. Several pilot studies on MBBR with UF separation have been reported so far, for example Leiknes et al. (2006), Ivanovic et al. (2006 and 2008) and Melin et al. (2005 and 2007). In most of these studies submerged Zenon hollow fibre membranes were used (Figure 11.23a). Recently, however, a comprehensive pilot study was carried out with separate, contained membranes after a high rate pre-separation process (Figure 11.23c and d.) by Ødegaard et al. (2010). At this time the (cautiously) recommended design fluxes are 25 $l \cdot m^{-2} \cdot h^{-1}$ for submerged membranes directly after MBBR treatment and 50 $l \cdot m^{-2} \cdot h^{-1}$ when membranes are used as a polishing step only.

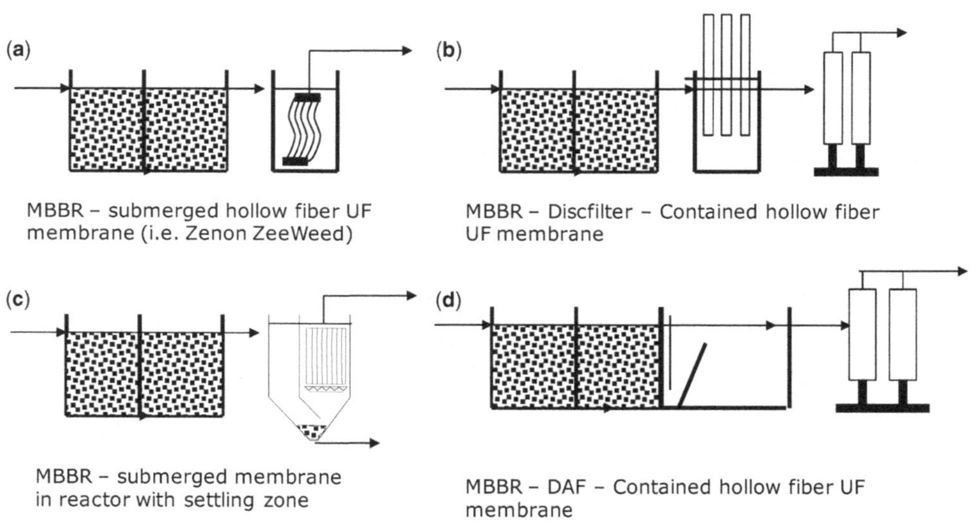

Figure 11.23 Different configurations for the combination of MBBR and UF filtration, courtesy of Ødegaard et al., 2010

Chapter 12

Sludge treatment and disposal

12.0 INTRODUCTION

The activated sludge process is very efficient at removing suspended solids, organic material and nutrients from the liquid phase, but at the same time the produced excess sludge constitutes a new problem. The treatment and final disposal of this sludge takes up a significant part of the material- and financial resources required for the wastewater treatment plant. The excess sludge from an activated sludge process has three undesirable aspects:

- Biological instability: depending on the sludge age, the excess sludge might contain a high fraction of biodegradable organic matter, consisting of the active biomass fraction and non degraded biodegradable material. If this is the case, the excess sludge will be putrescible and enters into decomposition within hours after the interruption of aeration;
- The hygienic quality of the excess sludge is very poor: a very large variety of viruses, bacteria and other pathogens (protozoa, amoebae, helminth eggs) are present;
- The suspended solids concentration in the excess sludge is low, in the range of 3 to $50\,g \cdot l^{-1}$ depending on the origin of the sludge and on the type of solid-liquid separation process used, requiring in a large excess sludge flow to be treated.

The main objective of sludge treatment is to reduce the fraction of biodegradable matter and the pathogen concentration, in order to obtain a stable and safe end product that does not constitute a public health risk. Two biological processes can be used: aerobic- and anaerobic sludge digestion, both having a positive effect on the hygienic quality of the sludge.

In addition, it is attempted to increase the solids concentration in order to reduce the excess sludge volume to be treated and disposed of. Upstream the sludge digesters the solids content is increased by excess sludge thickening, up to 2–5% wt solids for secondary sludge and 4–8% wt for primary sludge. Downstream the digesters the stabilised sludge is dewatered to 15–40% wt through the application of mechanical processes (filtration, centrifugation and flotation) or even higher using evaporation (up to 80–90% wt). Dewatering is often preceded by preparatory processes to accelerate or enhance liquid-solid separation, such as coagulation and flocculation with metal salts or poly-electrolytes. In this chapter the focus will be on the following subjects:

- Design of sludge thickeners using the solids flux theory and empirical guidelines;
- Aerobic sludge digestion;
- Anaerobic sludge digestion;
- The use of sludge drying beds, a very cost-effective dewatering method in regions with a warm climate.

12.1 EXCESS SLUDGE QUALITY AND QUANTITY

The quality and quantity of the excess sludge produced from an activated sludge process depends on the nature of the wastewater to be treated, the process configuration and the operational conditions of the process. When primary sedimentation is applied (this often implies that anaerobic sludge stabilisation will be used as well), the produced excess sludge mass will be larger than in the case of raw sewage treatment. This can be demonstrated as follows: if it is assumed that a fraction η_l of the influent COD is removed in the primary settler, the primary sludge production can be estimated as η_l/f_{cv} kg VSS $\cdot$ kg^{-1} COD applied.

If the organic sludge fraction in primary sludge is assumed to be equal to the organic fraction of the secondary sludge (i.e. an f_v value between 0.6 and 0.8), the total suspended solids production in the primary settler is equal to:

$$mE_{tl} = \eta_l/(f_{cv} \cdot f_v) \tag{12.1}$$

If it is further assumed that the primary sludge is thickened in the primary settler to a concentration X_{dl}, the flow of primary sludge per unit mass of applied COD is:

$$mq_l = mE_{tl}/X_{dl} = \eta_l/(f_{cv} \cdot f_v \cdot X_{dl}) \text{in} m^3 \cdot kg^{-1} COD \tag{12.2}$$

In the case of treatment of municipal sewage, the following values are found in practice:

- η_l = COD removal efficiency = 0.3 to 0.4, typically 0.33. If poly-electrolytes or metal salts; are used to enhance precipitation, the removed COD fraction can be as high as 0.5;
- f_v = 0.65 to 0.8 with a typical value of 0.75 g VSS $\cdot$ g^{-1} TSS;
- X_{dl} = 25–50 with a typical value of 40 g TSS $\cdot$ l^{-1}.

For the typical parameter values the specific primary excess sludge production and flow rate are calculated as $mE_{tl} = 0.33/(1.5 \cdot 0.75) = 0.29$ kg TSS $\cdot$ kg^{-1} COD and $mq_l = 0.29/40 = 7.3$ litre $\cdot$ kg^{-1} COD. For a per capita contribution of 100 g COD $\cdot$ d^{-1}, the estimated daily primary sludge production is 29 gram TSS in 0.73 litre $\cdot$ inh^{-1}. The mass of produced secondary sludge depends on the efficiency of primary settling, but also on the applied sludge age. With the aid of Eq. (3.51) and assuming a COD removal efficiency η_l in the primary settler, one has:

$$mE_{t2} = (1 - \eta_l) \cdot [(1 - f'_{ns} - f'_{np}) \cdot (1 + f \cdot b_h \cdot R_s) \cdot C_r/R_s + f'_{np}/f_{cv}]/f_v \tag{12.3}$$

where:

mE_{t2} = excess sludge production per unit mass applied COD in the activated sludge system
f'_{ns} = $f_{ns}/(1-\eta_l)$ = non-biodegradable dissolved COD fraction in the settled sewage: note that the *concentration* of this COD fraction is not affected by the settling process

f'_{np} = non-biodegradable particulate COD fraction in the settled sewage. In practice the fraction f'_{np} always has a low value (< 0.03)

In the absence of primary settling $\eta_1 = 0$, $f'_{ns} = f_{ns}$ and $f'_{np} = f_{np}$. In general, it is more difficult to thicken secondary excess sludge than primary sludge. In Section 12.2 it will be shown that in many cases the thickened secondary sludge concentration is not higher than 20 to 25 $g\,TSS \cdot l^{-1}$. The (thickened) secondary sludge flow per unit mass of daily applied COD is expressed as:

$$mq_2 = mE_{t2}/X_{th} \tag{12.4}$$

where:

mq_2 = secondary excess sludge flow per unit mass of daily applied COD in $m^3 \cdot kg^{-1}$ COD
X_{th} = secondary excess sludge concentration after thickening

Figure 12.1a shows the primary and secondary sludge production per unit mass of applied COD as a function of the sludge age in the activated sludge process. To construct the diagram the following values were assumed:

- COD removal efficiency in the primary settler: $\eta_1 = 0.33$;
- $f_{ns} = f_{np} = 0.10$ in the raw sewage;
- $f'_{ns} = f_{ns}/(1 - \eta_1) = 0.1/(1–0.33) = 0.15$ and $f'_{np} = 0.01$.

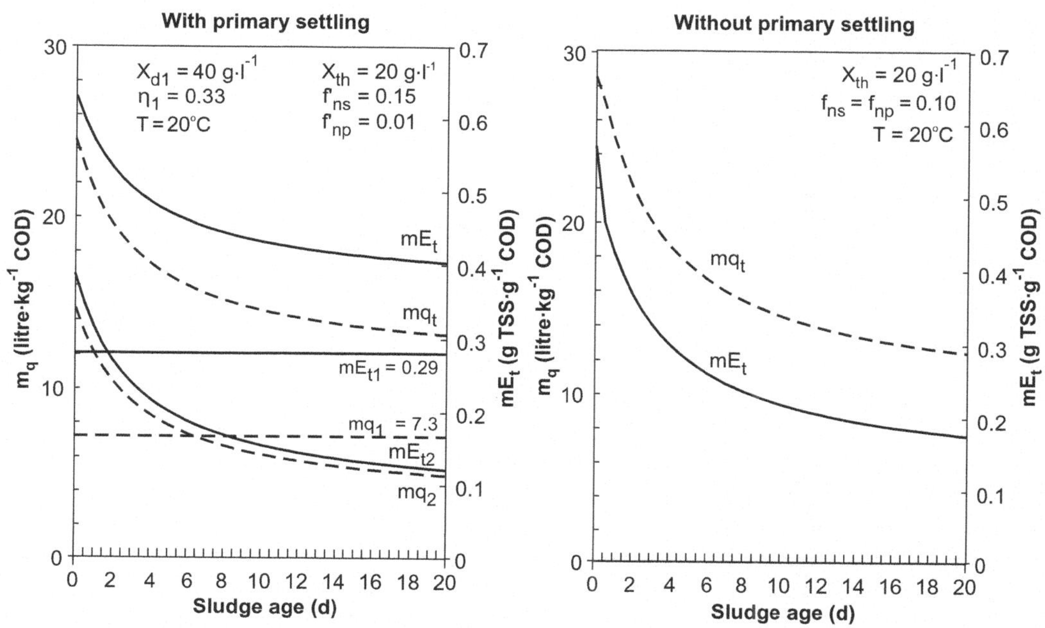

Figure 12.1 Typical profile of excess sludge production as a function of the sludge age for processes with (left) and without (right) primary settling

Figure 12.1b is similar to Figure 12.1a, but calculated for a configuration without primary settling. When the two diagrams are compared, it can be noted that primary settling tends to increase the production of excess sludge: for the range of sludge ages usually applied in practice, the total sludge production is 20 to 30% higher. In contrast, the volume of the excess sludge in the case of primary settling tends to be smaller, because primary sludge is easier to thicken and therefore will have a higher concentration than secondary sludge.

Sludge treatment (either aerobic or anaerobic) results in a reduction of 30 to 50% in the excess sludge solids content. After stabilisation the sludge usually has much better settling properties, so that a higher concentration may be obtained by thickening. Figure 12.2 shows the relationship between the sludge volume and the concentration of solids. In the same diagram the influence of the water content (humidity) on the physical characteristics of sludge is shown:

- Up to a solids concentration of 20–25%, the sludge behaves like a fluid;
- At higher concentrations of up to 30 to 35%, it is more of a "cake". Above this concentration, it is considered to be a solid;
- Granule formation begins when the solids concentration increases to 60–65%;
- Finally, above 80–90% solids content, the sludge is transformed into a fine powder.

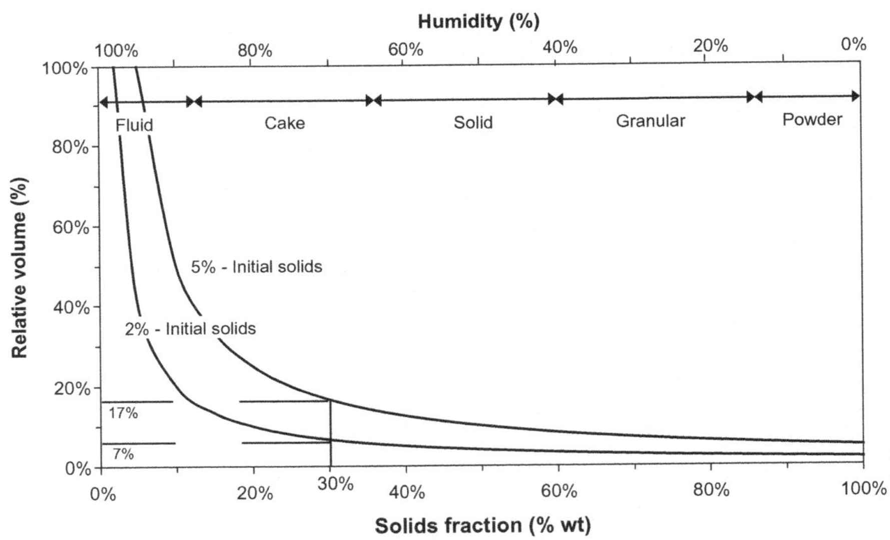

Figure 12.2 Relationship between solids fraction, humidity and relative sludge volume

It can be observed in Figure 12.2 that in order to transform stabilised excess sludge (with a solids content of 20 to 50 g $TSS \cdot l^{-1}$) into a cake (for example with 30% solids), the final volume will be only 7 to 17% of its initial value. This means that 83–93% of the water originally present will have to be removed. When a drier sludge is required, this percentage increases to more than 95%.

The composition of primary and secondary sludge is quite different. Whereas in the former, carbohydrates predominate and lipids have a significant concentration, in the latter the organic material is mostly of proteinic nature. Table 12.1 shows some experimental data on sludge composition from different authors.

Table 12.1 Composition of primary and secondary sludge as % dry weight of the organic sludge mass

Component	Primary sludge			Secondary sludge	
	(1)	(2)	(3)	(4)	(5)
Volatile fraction	79.7	73.5	75.0	59–75	79.0
Lipids	18.6	21.0	10.3	5–12	5.8
Cellulose	18.2	19.9	32.2	7	9.7
Hemicellulose			2.5		
Proteins	17.2	28.7	19.0	32–41	53.7

Sources: (1) = O'Rourke (1968), (2) = Eastman and Ferguson (1981), (3) = Higgins et al. (1982), (4) = US EPA (1979) and *(5) = Pavlostatis (1985)*

12.2 SLUDGE THICKENERS

A sludge thickener is a small but vitally important auxiliary unit in the activated sludge system with the specific objective to concentrate the excess sludge before it is sent to the sludge digester. The benefit of sludge thickening is a significant reduction in volumetric sludge production and thus also in the size of downstream sludge processing units, nowadays often an anaerobic digester followed by a sludge dewatering unit.

The primary sludge may or may not be thickened prior to digestion. The solids concentration is already quite high (30–50 g $TSS \cdot l^{-1}$) and especially in warmer climates there is a considerable risk that anaerobic digestion already develops in the thickener, which would significantly disturb the thickening process. On the other hand, the secondary excess sludge has a low concentration, especially when hydraulic wasting is implemented to control the sludge age, making a secondary sludge thickener a necessity.

The supernatant from the thickeners is returned to the aeration tank, so that its quality in terms of suspended solids concentration is of little importance. In general, no significant release of nitrogen and phosphorus is expected in the sludge thickener, as long as anaerobic decomposition does not develop. The one exception is bio-P sludge, where rapid release of stored poly-P into the liquid phase may be expected.

The excess sludge flow rate discharged into the thickener depends on the sludge age of the activated sludge process. By definition, when the sludge age has a value R_s, a fraction $1/R_s$ of the sludge mass in the process must be discharged daily as excess sludge. Hence, according to Eq. (3.51): $mE_t = mX_t/R_s$, where mE_t equals the suspended solids mass discharged as excess sludge per unit mass of applied COD. The inlet sludge concentration is either equal to the reactor sludge concentration X_t in the case of hydraulic sludge wasting, or (on average) equal to $X_r = (s + 1)/s \cdot X_t$ when sludge is discharged from the return sludge line. As discussed in Section 3.3.4, the use of hydraulic sludge wasting is always recommended.

In this chapter two methods for sludge thickener design are presented: (I) using the solids flux theory developed in Chapter 8 and (II) empirical guidelines relating the applied solids load and the DSVI value of the sludge to the attainable thickened sludge concentration.

12.2.1 Design of sludge thickeners using the solids flux theory

The theory for activated sludge settlers derived in Chapter 8 can also be applied to sludge thickeners. As the sludge concentration at the outlet of a thickener will be high, the limiting function will generally be thickening and the flux that can be transported is then given by the equations developed for the limiting solids flux, similar to Eqs. (8.18 and 8.19):

$$F_l = X_{th} \cdot v_0 \cdot (k \cdot X_l - 1) \cdot \exp(-k \cdot X_l) \tag{12.5a}$$

$$X_l = (X_{th}/2) \cdot [1 + (1 - 4/(k \cdot X_{th}))^{0.5}] \tag{12.5b}$$

The specific solids loading rate of a thickener i defined as

$$F_{sol} = mE_t/a_{th} = ME_t/A_{th} \tag{12.6}$$

Where a_{th} = cross sectional thickener area per unit mass daily applied COD in $m^2 \cdot d \cdot kg^{-1}$ COD

When the solids loading rate is equal to the limiting flux, the specific thickener area can be determined as a function of the thickened sludge concentration X_{th}:

$$\begin{aligned} a_{th} &= mE_t/F_{sol} = mE_t/F_l \\ &= [((1 - f_{ns} - f_{np}) \cdot (1 + f \cdot b_h \cdot R_s) \cdot C_r/R_s + f_{np}/f_{cv})/f_v]/ \\ &\quad [X_{th} \cdot v_0(k \cdot X_l - 1) \cdot \exp(-k \cdot X_l)] \end{aligned} \tag{12.7a}$$

where:

X_{th} = thickened sludge concentration at the point of abstraction of the sludge thickener (i.e. the design thickened sludge concentration)
X_l = limiting sludge concentration in the sludge thickener

For a sludge thickener with sidewall depth H_{th} and a safety factor s_{fth}, the volume of the thickener per unit mass of daily applied COD ($m^3 \cdot d \cdot kg^{-1}$ COD) can now be calculated as:

$$v_{th} = s_{fth} \cdot H_{th} \cdot a_{th} \tag{12.8a}$$

The sludge flow per unit mass of daily applied COD after thickening to a sludge concentration X_{th} is calculated as:

$$mq_{th} = mq \cdot (X_t/X_{th}) \text{ or more generally } mq_{th} = mq_w \cdot (X_w/X_{th}) \tag{12.9a}$$

where:

X_w = waste sludge concentration
mq_w = specific excess sludge flow in the thickener: mq_1 in the case of primary sludge and mq_2 in the case of secondary excess sludge, expressed in $m^3 \cdot kg^{-1}$ COD

The above equations can also be expressed as total values:

$$A_{th} = ME_t/F_l \tag{12.7b}$$

$$V_{th} = s_{fth} \cdot H_{th} \cdot A_{th} \tag{12.8b}$$

$$q_{th} = q \cdot (X_t/X_{th}) \text{or more generally } q_{th} = q_w \cdot (X_w/X_{th}) \tag{12.9b}$$

A much-applied sludge-wasting method is to discharge the excess sludge from the return sludge flow, as the suspended solids concentration is higher than that of the mixed liquor in the aeration tank. The assumption made to justify this procedure is that the higher inlet concentration will automatically lead to a higher outlet concentration in the thickener, so that a more concentrated sludge will be obtained. However, Eqs. (12.7a and b) show that this supposition has no theoretical ground as the thickened sludge concentration is independent of the inlet concentration. It can also be visually observed, for instance during DSVI experiments, that clarification and the first stage of thickening are relatively rapid processes. Likewise, it will take only a few hours out of the total thickener residence time of 1–2 days to increase the concentration from X_t to X_r. On the other hand, the increase from X_r to the thickened sludge concentration X_{th} is a much slower process.

It is therefore concluded that there is no clear advantage of using the return sludge flow instead of the mixed liquor in the aeration tank to feed the sludge thickener. On the contrary, there is a large disadvantage from discharging excess sludge from the return sludge line, as the return sludge concentration fluctuates considerably because of variations in the influent flow and/or sludge settleability. This makes it very difficult to control the mass of daily discharged excess sludge, which is of fundamental importance for adequate management and control of the sludge age.

For a more accurate control of the applied sludge age, it is preferable to discharge the excess sludge directly from the aeration tank, where the variations of the sludge concentration are much smaller. When excess sludge is discharged directly from the activated sludge process, a regime called "hydraulic sludge wasting" can be applied in which every day a volume of sludge is withdrawn equal to the total reactor volume V_r divided by the sludge age R_s. Refer also to Section 3.3.4.

Equation (12.7a) gives the required thickener area per unit mass of daily applied COD. It can be observed that this area is proportional to the organic load applied to the activated sludge process and in principle is independent of the hydraulic load. In this respect the design of the sludge thickener is different from that of the final settler: the area of the latter is proportional to the influent flow. Figure 12.3 shows the minimum thickener area per unit mass daily applied COD as a function of the desired outlet concentration for different values of specific excess sludge production (mE_t) and for different values of the settleability constants k and v_0, corresponding to poor, medium and excellent settleability.

In practice, the specific excess sludge production varies between 0.2 mg $TSS \cdot mg^{-1}$ COD (low rate systems – high sludge age) and 0.5 mg $TSS \cdot mg^{-1}$ COD (high rate systems – low sludge age). As can be observed in Figure 12.3, the relationship between the thickener area and the desired thickener outlet concentration approximates an exponential relationship for the values that are of practical interest. When sizing a sludge thickener, the following factors should be taken into consideration:

- A very large thickener area with a concomitant large volume might result in the onset of anaerobic decomposition of the excess sludge, which is undesirable (odour). Therefore the hydraulic retention time must not exceed 18 to 24 hours for system with a relatively high active sludge fraction (f_{av}>0.5–0.6), especially in regions with a warm climate;
- When the applied sludge age is high and/or temperatures are lower, the retention time might be higher, but for municipal sewage in general a maximum of 2 days is recommended. Anaerobic conditions may be prevented by recirculating fresh water over the thickener, for instance aerated effluent;

– The sludge thickener and digester should be designed for total minimum construction costs or, if costs per unit measure volume are not known, for minimum total volume. This minimum can only be evaluated if the retention time in the digester is known. This retention time, as well as the design optimisation of the thickener-digester system will be discussed in Sections 12.3.4 and 12.4.5.

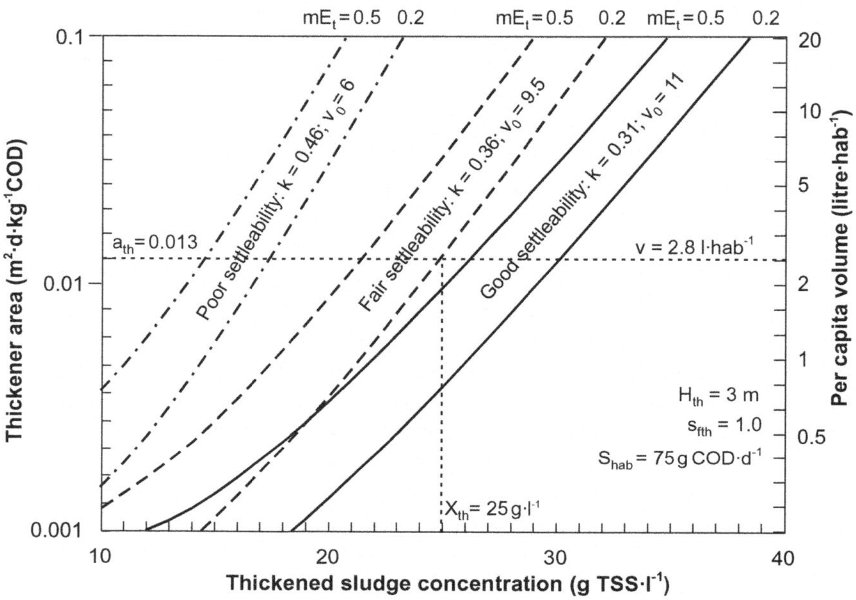

Figure 12.3 Thickener area per unit mass daily applied COD as a function of the target thickened sludge concentration

To complete the thickener design, it is necessary to specify the depth and the safety factor to be applied. The depth in a thickener is typically smaller than in a final settler, because there is no need to create a buffer capacity for sludge storage. When the thickener is overloaded, the excess load will simply be recycled to the aeration tank, together with the thickener effluent. The exception is when the dewatering unit is fed directly from the thickener and is operated intermittently, resulting in a high removal rate of thickened sludge in a relatively short period.

As a sludge thickener is much smaller than a final settler, the distribution of inlet flow over the surface area is easier and the dead volume fraction will be smaller as well. Therefore the safety factor to be applied may be less than the value of $s_{fd} = 2$ often adopted for final settlers, e.g. $s_{fth} = 1.5$.

EXAMPLE 12.1

What is the value of thickener volume per capita ($75\ g\ COD \cdot inh^{-1} \cdot d^{-1}$) in a low rate activated sludge process ($mE_t = 0.2\ g\ TSS \cdot g^{-1}\ COD$) for a required thickened sludge concentration of $25\ g\ TSS \cdot l^{-1}$, for medium sludge settleability ($k = 0.36\ l \cdot g^{-1}$ and $v_0 = 9.5\ m \cdot h^{-1}$).

What would be the concentrations of thickened sludge for good and poor settleability ($k = 0.31\ l \cdot g^{-1}$; $v_0 = 11\ m \cdot h^{-1}$ and $k = 0.46\ l \cdot g^{-1}$; $v_0 = 6\ m \cdot h^{-1}$ respectively)? Assume $s_{fth} = 1$, i.e. the thickener approaches ideality.

Solution

In Figure 12.3 it can be determined that for medium settleability and $mE_t = 0.2$ g TSS $\cdot$ g^{-1} COD, the required thickened sludge concentration of $X_{th} = 25\ g \cdot l^{-1}$ can be obtained for $a_{th} = 0.013\ m^2 \cdot d \cdot kg^{-1}$ COD. If a sidewall depth of 3 m is selected, the thickener volume is $0.013 \cdot 3 = 0.039\ m^3 . d \cdot kg^{-1}$ COD or, for a contribution of 75 COD $\cdot$ inh^{-1} . d^{-1}, the per capita volume is $0.075 \cdot 0.039 = 0.0029\ m^3 . inh^{-1}$ or 2.9 l . inh^{-1}. This can also be determined using Eqs. (12.5a and b):

$$X_l = (X_{th}/2) \cdot [1 + (1 - 4/(k \cdot X_{th}))^{0.5}]$$
$$= (25/2) \cdot [1 + (1 - 4/(0.36 \cdot 25))^{0.5}] = 21.8\ kg \cdot m^{-3}$$
$$F_l = X_{th} \cdot v_0 \cdot (k \cdot X_l - 1) \cdot \exp(-k \cdot X_l)$$
$$= 25 \cdot 9.5 \cdot (0.36 \cdot 21.8 - 1) \cdot \exp(-0.36 \cdot 21.8) = 0.63\ kg \cdot m^{-2} \cdot h^{-1}$$
$$a_{th} = mE_t/F_{sol} = mE_t/F_l$$
$$= 0.2/(0.63 \cdot 24) = 0.013 m^2 \cdot d \cdot kg^{-1} COD$$

For good settleability and $a_{th} = 0.013$, the thickened sludge concentration can be determined (again using Figure 12.3) as $X_{th} = 30.5\ g \cdot l^{-1}$ and for poor settleability as $X_{th} = 17.5\ g \cdot l^{-1}$.

It may be noted that, depending on the operational conditions of the activated sludge process, the thickener volume of 2.9 l . inh^{-1} may be excessive. For example, if the per capita aeration tank volume is 100 l and the sludge age is 10 days, the excess sludge flow per capita will be $q_{inh} = 100/10 = 10\ l \cdot d^{-1}$. In this case, the retention time in the sludge thickener is calculated as $R_{th} = V_{th}/q_{inh} = 2.9/10 = 0.29$ day or 7.0 hours. This value would be acceptable. However for a longer sludge age the retention time might be unacceptably long: for a sludge age of $R_s = 40$ days the excess sludge flow would be $q_{inh} = 100/40 = 2.5\ l \cdot d^{-1}$ and $R_{th} = 2.9/2.5 = 1.15$ days, a retention time that might well lead to operational problems due to (acid) fermentation of the sludge in the settler (at higher temperatures). On the other hand, the active sludge fraction in the sludge will be low in an activated sludge system operated at a high sludge age.

12.2.2 Design of sludge thickeners using empirical relationships

The solids flux theory considers an increase in the sludge concentration only as a result from the settling flux. As the calculated settling velocity decreases at higher sludge concentrations, the theoretical attainable thickened concentration is limited to 15–18 g $\cdot$ l^{-1} (poor settleability) to 25–30 g $\cdot$ l^{-1} (good settleability). However, the effect of gravity compression (compaction) is not considered. The weight of the accumulated sludge layer pushes downward, which will further thicken the bottom layer of sludge (refer also to Figure 8.1). Therefore, the maximum attainable thickened sludge concentration might be underestimated when the solids flux design is used. This effect is probably small when the hydraulic residence time in the thickener is short, as generally will be the case for sludge with a high active sludge fraction, originating from activated sludge systems operated at low sludge age. It may be more significant when higher hydraulic retention times are applied.

In general it might thus be better to design conservatively, as the additional construction costs for a larger thickener are small. Furthermore, if the design thickened sludge concentration is chosen too optimistic, the penalty in inadequate system performance may be large: i.e. insufficient solids degradation due to a reduced solids retention time in the anaerobic digester or insufficient dewatering capacity.

In Table 12.2 typical thickened sludge concentrations are given for different types of sludge. In general primary sludge can be thickened to a higher concentration than secondary sludge. Figure 12.4 shows the maximum attainable thickened secondary excess sludge concentration as function of the DSVI, determined for a solids loading rate of $50\ kg\ TSS \cdot m^{-2} . d^{-1}$. As can be observed, the degree of thickening can be quite high for low DSVI values. However, as a priori it will be difficult to predict the DSVI that will develop in the system and furthermore because sludge settleability may change in time, it would be unwise to base the design of a secondary sludge thickener on a very low DSVI value. In a more conservative range of $120 < DSVI < 150\ ml \cdot g^{-1}$ the maximum thickened sludge concentration varies roughly between 40 to $50\ g\ TSS \cdot l^{-1}$.

Table 12.2 Typical thickened sludge concentration for different sludge types without chemical conditioning (adapted from EPA, 1976 and Metcalf & Eddy, 2003)

Type of sludge	Solids content (% wt)		Solids loading rate ($kg \cdot m^{-2} \cdot d^{-1}$)	Max. hydraulic loading rate ($m \cdot d^{-1}$)
	Inlet	Outlet		
Primary sludge				
– $f_v > 0.65$	2–6	5–7	75–125	15–30
– $f_v < 0.65$	2–6	7–10	100–150	15–30
– digested	5–8	6–12	120	15–30
Secondary sludge	0.3–1.5	2–5	20–50	4–8
Mixture of primary and secondary sludge	0.5–4.0	4–7	25–80	6–12

According to Koot (1980), the following zones can be distinguished in a sludge thickener (Figure 12.5):

- $H_1 =$ inlet zone: depth of the sludge inlet below the surface area, which is typically between 0.4–0.6 m;
- $H_2 =$ clarification zone where solids-liquid separation occurs, typically $H_2 = 0.5$ m;
- $H_3 =$ compression zone where the actual thickening process takes place. Moving down the sludge concentration will increase. The height of this zone can be calculated with the following empirical expression:

$$H_3 = X_w \cdot q_w \cdot t_{comp}/(0.75 \cdot X_{th} \cdot A_{th}) \tag{12.10}$$

where $t_{comp} =$ retention time of the sludge in the compression zone (days);

- $H_4 =$ sludge removal zone (height of scraper), typically 0.3 m.

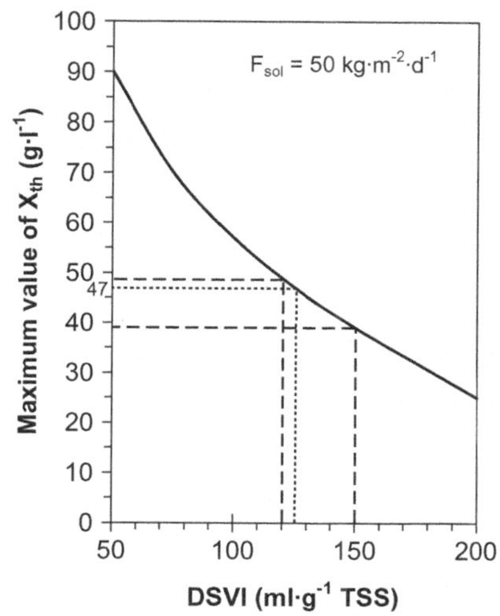

Figure 12.4 Maximum attainable thickened secondary excess sludge concentration as function of the DSVI (Kalbskopf, 1971)

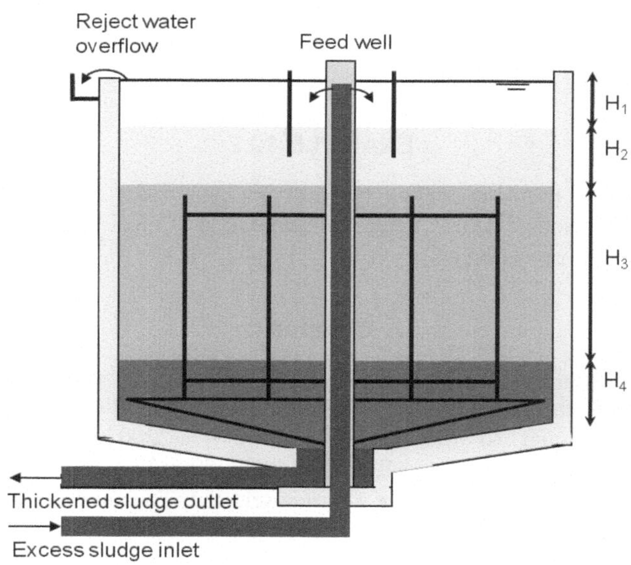

Figure 12.5 Schematic representation of a sludge thickener including thickening zones

In the design of a sludge thickener, values are attributed to the values of H_1 to H_4. The following steps are followed:

(1) Determine the solids loading rate F_{sol} that can be applied. For secondary sludge a default solids loading rate of 50 kg TSS · m $^{-2}$ · d^{-1} is suggested;

(2) Use the design DSVI value to estimate from Figure 12.4 the thickened sludge concentration that can be obtained;
(3) $q_w = ME_t/X_w = ME_t/X_t$ (in the case of hydraulic wasting);
(4) $A_{th} = ME_t/F_{sol}$;
(5) Select the allowable retention time t_{comp} in the compression zone (H_3-zone):
   - Maximum 2 days under normal conditions;
   - Maximum 1 day at high temperature, low R_s or a high nitrate concentration in the presence of biodegradable COD;
   - Calculate the height of the compression zone H_3 with Eq. (12.10);
(6) Select values for H_1, H_2 and H_4 or accept the following default values:
   - $H_1 = 0.5$ m; $H_2 = 0.5$ m and $H_4 = 0.3$ m or $H_1 + H_2 + H_4 = 1.3$ m;
(7) Calculate thickener height and volume:
   - $H_{th} = H_1 + H_2 + H_3 + H_4$;
   - The minimum value of $H_{th} = 3$ m;
   - $V_{th} = A_{th} \cdot H_{th}$
(8) Calculate thickened sludge flow
   - $q_{th} = ME_t/X_{th}$

In order to prevent odour problems and the onset of anaerobic acid digestion, it is recommended to supply at least 24–30 $m^3 \cdot m^{-2} \cdot d^{-1}$ of dilution water to the sludge thickener to maintain aerobic conditions, especially at temperatures higher than 20°C.

EXAMPLE 12.2

An activated sludge system produces 500 m^3 of waste sludge per day with a sludge concentration of 4 $g \cdot l^{-1}$. The long-term average DSVI value measured is 125 $ml \cdot l^{-1}$. Design a sludge thickener using the empirical design procedure.

Solution

Following the procedure outlined above, a solids loading rate of 50 $kg \cdot m^{-2} \cdot d^{-1}$ is selected. From Figure 12.4 the attainable thickened sludge concentration can be estimated as 47 $g \cdot l^{-1}$. The daily applied solids load is equal to $ME_t = q_w \cdot X_w = 500 \cdot 4 = 2000$ $kg \cdot d^{-1}$. At 50 $kg \cdot m^{-2} \cdot d^{-1}$ the required thickener surface area $A_{th} = ME_t/F_{sol} = 2000/50 = 40$ m^2, or a diameter of 7.2 m. For the sum of H_1, H_2 and H_4 the default value of 1.3 m is selected. To calculate the required depth of the compression zone Eq. (12.10) is used (for $t_{comp} = 2$ days):

$$H_3 = X_w \cdot q_w \cdot t_{comp}/(0.75 \cdot X_{th} \cdot A_{th})$$
$$= 4 \cdot 500 \cdot 2/(0.75 \cdot 47 \cdot 40) = 2.8 \text{ m}$$

So the total height of the sludge thickener $H_{th} = 2.8 + 1.3 = 4.1$ m. The thickener volume $V_{th} = 40 \cdot 4.1 = 164$ m^3. The retention time in the thickener is $164/500 = 8$ hrs, which is certainly acceptable.

12.3 AEROBIC DIGESTION

When active sludge is kept in an aerobic environment without feed, in time a reduction of the volatile solids concentration is observed, with a concurrent consumption of oxygen. These phenomena characterise aerobic sludge digestion and are attributed to the oxidation of microbial protoplasm, which releases the energy required to maintain vital cell functions. The oxidation of cellular matter is called endogenous respiration, in order to distinguish it from the oxidation of extra-cellular organic material, which is called exogenous respiration.

In the first model describing endogenous respiration and aerobic digestion (Lawrence and McCarthy, 1970), these processes were considered as first order processes with respect to the volatile solids concentration. Later the model was refined by differentiating between active or "viable" sludge and inactive or "unviable" sludge (Randall, 1975; Marais and Ekama, 1976 and Randall and Benefield, 1979), although the original model is still amply used in textbooks.

The difficulty with using the active sludge concentration as a parameter is that it cannot be measured directly. Thus, in order to evaluate the kinetics of aerobic sludge digestion, parameters must be identified that will change during aerobic sludge digestion, that can be measured easily, and that can be linked to the active sludge concentration. The following parameters can be used:

- Volatile sludge concentration;
- Mixed liquor alkalinity;
- Nitrate concentration;
- BOD of the digesting sludge;
- Oxygen uptake rate.

In the next section it will be shown that during aerobic sludge digestion these parameters change in accordance with a simple model as presented by Marais and Ekama (1976).

12.3.1 Kinetic model for aerobic sludge digestion

The model for aerobic sludge digestion is based on the following fundamental concepts:

(1) Only the active sludge concentration is susceptible to aerobic digestion and the decay rate of this active sludge is proportional to its concentration. The digestion process does not affect the inactive sludge fractions. The decay rate of the active sludge can be expressed as:

$$r_d = (dX_a/dt)_d = -b_h \cdot X_a \quad (12.11)$$

where r_d = active sludge decay rate

(2) Only part of the decayed active sludge is oxidised into inorganic products. The remaining fraction "f" accumulates as an inactive organic solid in the mixed liquor and is called the endogenous residue:

$$(dX_e/dt)_d = -f \cdot (dX_a/dt)_d = -f \cdot r_d \quad (12.12)$$

The existence of the endogenous residue has been demonstrated by Washington and Hetling (1962), who noted a linear increase in the sludge concentration in an activated sludge process, when this was fed with a soluble biodegradable substrate and without sludge discharge. It was assumed that the biodegradable influent material is metabolised and the active sludge thus generated decays to produce the endogenous residue, resulting in the increase of the sludge concentration. The observed increase of the sludge mass was 0.09 mg VSS · mg^{-1} COD applied. Considering that the active sludge production Y equals 0.45 mg active sludge per mg metabolised COD, it can be calculated that a fraction $f = 0.09/0.45 = 0.2$ of the decayed active sludge remains as the endogenous residue. Other research workers (Brodersen and McCarty, 1964; Marais and Ekama, 1976 and Dias et al., 1981) found similar values for "f", independent from operational parameters such as sludge age, temperature and stabilised sludge composition. Having established the value of the endogenous fraction "f", the constant "b_h" must be determined experimentally to define the kinetic model for aerobic digestion. This can be done conveniently by observing the aerobic digestion of activated sludge batches. In a batch reactor the decrease of the active sludge concentration in time can be calculated by integration of Eq. (12.11):

$$X_a = X_{ai} \cdot \exp(-b_h \cdot t) \tag{12.13}$$

where:

X_{ai} = initial active sludge concentration
t = aerobic digestion time

As the active sludge concentration cannot be determined experimentally, the validity of Eq. (12.13) cannot be established directly and must be confirmed by the behaviour of the parameters that are affected by aerobic digestion. The relationship between the active sludge concentration and the value of these parameters will now be derived.

12.3.1.1 Variation of the volatile sludge concentration

The variation of the volatile sludge concentration is equal to the decrease of the active sludge concentration and is partly compensated by the increase of the endogenous residue, which amounts to a fraction "f" of the decayed active sludge concentration. Hence:

$$X_{vi} - X_v = (X_{ai} - X_a) + X_e = (X_{ai} - X_a) + f \cdot (X_{ai} - X_a) \tag{12.14}$$

where:

X_{vi} = initial volatile sludge concentration
X_e = concentration of endogenous residue generated in the sludge batch

Substituting Eq. (12.13) in Eq. (12.14) one has:

$$X_v = X_{vi} - (1 - f) \cdot X_{ai} \cdot [1 - \exp(-b_h \cdot t)] \tag{12.15}$$

When all active sludge has decayed, the volatile sludge concentration has reached a stable minimum value, which can be calculated as:

$$X_{v\infty} = X_{vi} - (1 - f) \cdot X_{ai} \tag{12.16}$$

where $X_{v\infty}$ = volatile sludge concentration after completion of the active sludge decay

Now, the difference of the sludge concentration after a digestion time "t" and after complete digestion is expressed as:

$$X_v - X_{v\infty} = (1-f) \cdot X_{ai} \cdot \exp(-b_h \cdot t) \quad \text{or}$$
$$\log(X_v - X_{v\infty}) = \log[(1-f) \cdot X_{ai}] - 2.3 \cdot b_h \cdot t \tag{12.17}$$

Equation (12.17) can be used to determine b_h with the following experimental procedure:

(1) Determine the volatile sludge concentration in a digesting sludge batch as a function of time, until a constant value has been reached (this will take a few weeks);
(2) Plot the experimental values of X_v–$X_{v\infty}$ as function of the digestion time on semi log paper;
(3) The best-fit straight line through the experimental data has a gradient equal to $-2.3 \cdot b_h$.

12.3.1.2 Variation of the oxygen uptake rate

An alternative method to determine the value of b_h is by measuring the oxygen uptake rate (OUR) of a batch of digesting sludge. Since the oxidation of (1–f) gram of volatile solids (originating from the decay of l gram of active sludge) requires an oxygen mass of $f_{cv} \cdot (1\text{–}f)$ gram O_2, the oxygen uptake rate is directly linked to the decay rate:

$$O_c = f_{cv} \cdot (1-f) \cdot r_d = f_{cv} \cdot (1-f) \cdot b_h \cdot X_{ai} \cdot \exp(-b_h \cdot t) \quad \text{or}$$
$$\log O_c = \log[f_{cv} \cdot (1-f) \cdot b_h \cdot X_{ai}] - 2.3 \cdot b_h \cdot t \tag{12.18}$$

Equation (12.18) can be used to determine the constant b_h in the following way:

(1) Determine the O_c values as a function of time in a digesting active sludge batch;
(2) Plot the experimental data on semi log paper;
(3) The best-fit straight line through the experimental points has a gradient of $-2.3 \cdot b_h$.

Equation (12.18) does not take into consideration that nitrification of mineralised ammonia may occur, which increases the oxygen uptake rate. The oxygen requirement for nitrification can be included knowing that the production of ammonium will be a fraction f_n of the digested sludge concentration. Hence, per unit mass of decayed active sludge, there is a production of $f_n \cdot (1-f)$ g N, with an oxygen demand of 4.57 mg $O_2 \cdot mg^{-1}$ N. The OUR for nitrification can be expressed as:

$$\begin{aligned} O_n &= 4.57 \cdot f_n \cdot (1-f) \cdot r_d \\ &= 4.57 \cdot f_n \cdot (1-f) \cdot b_h \cdot X_{ai} \cdot \exp(-b_h \cdot t) \end{aligned} \tag{12.19}$$

Now the total OUR is given by:

$$\begin{aligned} O_t &= O_c + O_n \\ &= (f_{cv} + 4.57 \cdot f_n) \cdot (1-f) \cdot b_h \cdot X_{ai} \cdot \exp(-b_h \cdot t) \end{aligned} \tag{12.20}$$

Equation (12.20) shows that in the case of nitrification, the relationship between the logarithm of O_t and digestion time is also linear and has the same gradient as without nitrification. Therefore, the experimental procedure above for the determination of b_h is also valid when nitrification occurs. Alternatively, nitrification can also be inhibited through the addition of allyl-thio-urea (ATU).

12.3.1.3 Variation of the nitrate concentration

When nitrification occurs in the sludge batch, there is a production of f_n gram NO_3-N per gram of oxidised volatile sludge so that:

$$N_n - N_{ni} = f_n \cdot (X_{vi} - X_v) \quad (12.21)$$
$$N_{n\infty} - N_n = f_n \cdot (X_{vi} - X_{v,\infty}) \quad (12.22)$$

With $N_{ni} =$ initial nitrate concentration

Combining Eqs. (12.21 and 12.22) one has:

$$N_{n\infty} - N_n = f_n \cdot (X_v - X_{v,\infty}) = f_n \cdot (1 - f) \cdot X_{ai} \cdot \exp(-b_h \cdot t) \quad \text{or}$$
$$\log(N_{n\infty} - N_n) = \log[f_n \cdot (1 - f) \cdot X_{ai}] - 2.3 \cdot b_h \cdot t \quad (12.23)$$

As described above, Eq. (12.23) can be used to determine the value of b_h.

12.3.1.4 Variation of the alkalinity

The alkalinity of a sludge batch is influenced by the processes of ammonification and nitrification. In Section 5.1.3.2 it was established that the alkalinity production in the ammonification process is 3.57 mg $CaCO_3 \cdot mg^{-1}$ N, whereas there is a consumption of 7.14 mg $CaCO_3 \cdot mg^{-1}$ N during nitrification. Hence, there is a net alkalinity reduction of 3.57 mg $CaCO_3 \cdot mg^{-1}$ N when nitrification of the mineralised ammonia occurs, so that:

$$Alk_i - Alk_\infty = 3.57 \cdot (N_n - N_{ni}) \quad (12.24)$$

Using Eq. (12.23):

$$Alk - Alk_\infty = 3.57 \cdot f_n \cdot (1 - f) \cdot X_{ai} \cdot \exp(-b_h \cdot t) \quad \text{or}$$
$$\log(Alk - Alk_\infty) = \log[3.57 \cdot f_n \cdot (1 - f) \cdot X_{ai}] - 2.3 \cdot b_h \cdot t \quad (12.25)$$

where $Alk_\infty =$ alkalinity when digestion of the sludge batch is complete

Equation (12.25) is the fourth method that can be used to determine the decay rate constant b_h. The variation in nitrate concentration and alkalinity can only be used when nitrification of the mineralised ammonia occurs. In practice this will usually be the case in an aerobic digester. If nitrification does not occur, both the ammonium concentration and the alkalinity will increase and this will lead to an increase of the pH. Under those circumstances part of the ammonium may be stripped (as gaseous NH_3) by the aeration process, which reduces the alkalinity. Hence if nitrification does not occur, neither the concentrations of ammonium or nitrate nor the alkalinity can be used to determine the decay constant b_h.

EXAMPLE 12.3

A sludge batch is submitted to aerobic digestion. In Table 12.3 experimental results are presented for the variations in (I) OUR, (II) volatile suspended solids, (III) alkalinity and (IV) nitrate concentration, as a function of the digestion time. The sludge was taken from an activated sludge system operated at a sludge age of 4 days and the temperature was controlled at 22°C. Determine the decay rate constant b_h from these data according to the four methods presented above.

Table 12.3 Experimental results from the aerobic digestion of a batch of activated sludge, by Van Haandel et al. (1986)

Time (days)	O_t ($mg\ O_2 \cdot l^{-1} \cdot h^{-1}$)	Time (days)	X_v ($mg\ VSS \cdot l^{-1}$)	N_n ($mg\ N \cdot l^{-1}$)	Alkalinity ($mg\ CaCO_3 \cdot l^{-1}$)
0	43.6	0	4560	37	725
0.18	40.4	0.5	4100	69	585
0.82	33.5	1.0	3880	92	520
1.10	28.7	1.5	–	109	488
1.35	27.8	2.0	3840	110	435
1.87	25.1	2.5	3710	126	365
2.32	20.2	3.0	3490	138	355
2.87	18.9	3.5	3380	142	335
3.09	16.0	4.0	3320	158	295
3.33	16.5	4.5	3080	162	245
3.88	13.8	5.0	3080	178	222
4.17	12.4	5.5	–	185	175
4.41	13.0	6.0	2980	192	155
4.87	11.4				
5.26	10.1				
6.00	10.5				

Solution

The calculations required to determine the decay constant b_h can be summarised as follows:

(1) Plot the oxygen demand O_t as a function of the digestion time on semi log paper and determine the decay constant from the slope of the best-fit straight line through the experimental points as shown in Figure 12.6a: $b_h = [\ln(O_{t,0}) - \ln(O_{t,6})]/t = (3.65 - 2.08)/6 = 0.26\ d^{-1}$

(2) The intersection of the line with the vertical axis can be used to estimate the initial active sludge concentration. By extrapolating in Figure 12.6a to $t = 0$, the value of $\ln(O_t)$ can be estimated as 3.65 so that $O_{t,i} = 38.5\ mg\ O_2 \cdot l^{-1} \cdot h^{-1}$.

(3) With the aid of Eq. (12.20) the initial active sludge concentration can be calculated:

$$O_{t,i} = (f_{cv} + 4.57 \cdot f_n) \cdot (1 - f) \cdot b_h \cdot X_{ai} \cdot \exp(-b_h \cdot t)$$
$$= (1.5 + 0.457) \cdot (1 - 0.2) \cdot 0.26 \cdot X_{ai} = 38.5 \cdot 24 \text{ or } X_{ai} = 2255\ mg\ VSS \cdot l^{-1}$$

(4) Since the digestion was only applied for a period of 6 days there are no data available for complete digestion. However the value of the initial active sludge concentration can be used to estimate this value. For t = 0 the volatile sludge concentration $X_{vi} = 4560$ mg VSS $\cdot l^{-1}$. After complete decay of the active sludge, the reduction of the volatile sludge concentration will be: $X_{ai} \cdot (1-f) = 1800$ mg VSS $\cdot l^{-1}$ so that the final volatile sludge concentration is given by $X_{v\infty} = 4560 - 1800 = 2760$ mg VSS $\cdot l^{-1}$. Similarly the final values for nitrate and alkalinity can be calculated as: $N_{n\infty} = 240$ mg N $\cdot l^{-1}$ and $Alk_{\infty} = 10$ mg $\cdot l^{-1}$ $CaCO_3$.

(5) Plot the experimental data of $(X_v - X_{v\infty})$ as a function of the digestion time on semi log paper and determine the slope. If the data has a systematic tendency of deviating from a straight line (either convex or concave) the estimate for the final volatile solids concentration has been incorrect and must be reviewed.

(6) Plot the experimental data of $(N_{n\infty} - N_n)$ and $(Alk - Alk_{\infty})$ as a function of the digestion time on semi log paper and determine the corresponding decay constants.

In Figure 12.6 the experimental data of Table 12.3 are shown for each of the four methods, together with the best fit of the decay rate b_h. For all methods it can be observed that the straight lines predicted by theory correlate closely to the experimental data points. This allows determination of the value of the decay constant by no less than four independent methods. The results of the four determinations are practically equal while the small differences can be attributed to analytical errors. Hence the best estimate for the value of the decay constant is the average of the four obtained values: $b_{h22} = (0.262 + 0.276 + 0.252 + 0.240)/4 = 0.26\ d^{-1}$.

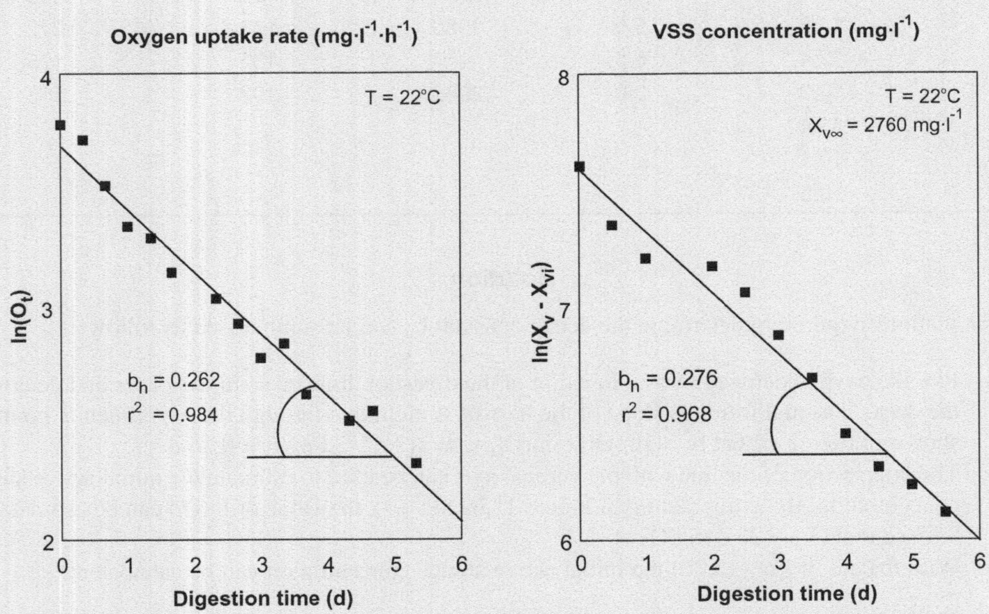

Figure 12.6a Determination of the decay rate b_h in Example 12.3 from the decrease of the oxygen uptake rate (O_t) and the volatile sludge concentration (X_v)

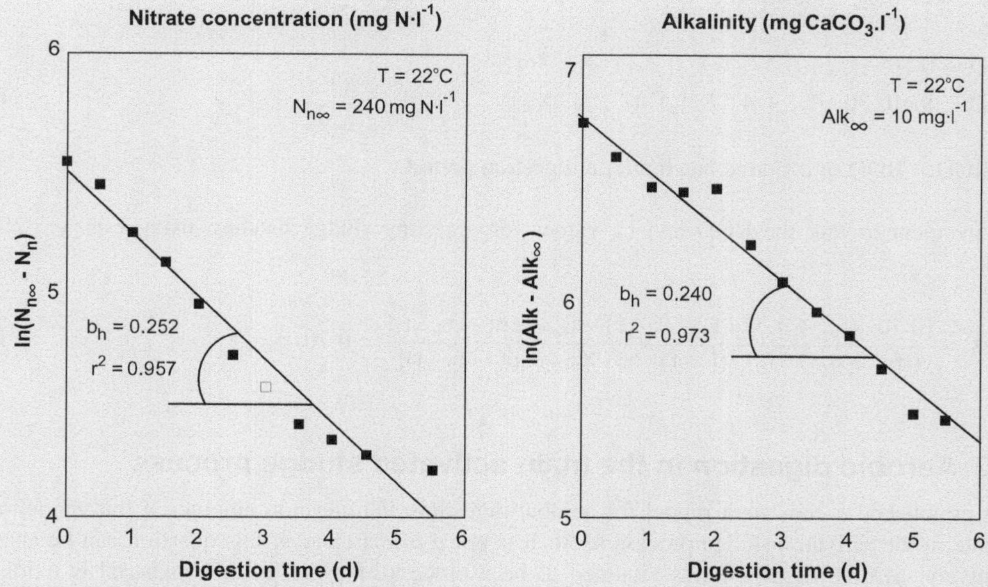

Figure 12.6b Determination of the decay rate b_h in Example 12.3 from the increase of the nitrate concentration (N_n) and the decrease of alkalinity (Alk)

Van Haandel et al (1985) applied the batch technique described above to determine the decay constant in the range of 20 to 32°C, observing the behaviour of the four parameters and obtained the following average result:

$$b_{hT} = 0.24 \cdot 1.04^{(T-20)} d^{-1} (20^{\circ}C < T < 32^{\circ}C) \tag{12.26a}$$

This result is very similar to the value obtained by Marais and Ekama (1976) for the range of temperatures from 12 to 20°C:

$$b_{hT} = 0.24 \cdot 1.029^{(T-20)} d^{-1} \text{for} 12^{\circ}C < T < 20^{\circ}C \tag{12.26b}$$

It must be emphasised that in both cases the value of the constant was independent of the initial active sludge fraction, i.e. it does *not* depend on the sludge age of the process from which the excess sludge is taken.

12.3.1.5 Variation of the BOD

The BOD value of digesting sludge may be estimated from the oxygen consumption during a incubation period of 5 days at 20°C.

$$\begin{aligned} BOD = \Delta DO &= (f_{cv} + 4.57 \cdot f_n) \cdot (1 - f) \cdot (X_{ai} - X_a5) \\ &= (f_{cv} + 4.57 \cdot f_n) \cdot (1 - f) \cdot X_{ai} \cdot [1 - \exp(- b_{h20} \cdot 5)] \end{aligned} \tag{12.27}$$

Because $\exp(-b_{h20} \cdot 5)$ is equal to $\exp(-0.24 \cdot 5) = 0.30$ one has:

$$\begin{aligned} BOD &= 0.70 \cdot (f_{cv} + 4.57 \cdot f_n) \cdot (1 - f) \cdot X_{ai} \quad \text{or} \\ \log BOD &= \log[0.70 \cdot (f_{cv} + 4.57 \cdot f_n) \cdot (1 - f) \cdot X_{ai}] \end{aligned} \tag{12.28}$$

where BOD = BOD of a sludge batch after a digestion period t

It is convenient to link the BOD and O_t values for digesting sludge batches, using Eqs. (12.20) and (12.28):

$$BOD/O_t = \frac{[0.70 \cdot (f_{cv} + 4.57 \cdot f_n) \cdot (1 - f) \cdot X_{ai} \cdot \exp(-b_h \cdot t)]}{[(f_{cv} + 4.57 \cdot f_n) \cdot (1 - f) \cdot b_h \cdot X_{ai} \cdot \exp(-b_h \cdot t)]} = 0.70/b_h \tag{12.29}$$

12.3.2 Aerobic digestion in the main activated sludge process

Having established a consistent model for aerobic digestion, the question emerges if this model is also applicable in the activated sludge process itself. It is not *a priori* clear, if this question can be answered affirmatively: aerobic digestion was assumed to be a process in which cellular material is oxidised to obtain the energy required to maintain vital functions of the micro-organisms. In an aerobic digester, the only source of organic matter for oxidation is the protoplasm of the active sludge. In contrast, in the activated sludge process there is also extra-cellular organic material present. It might be assumed that the bacteria would rather "save" their protoplasm and use preferentially the extra-cellular material, so that endogenous respiration would be substituted by exogenous respiration. However, it will now be shown that experimental results indeed indicate that the endogenous respiration rate is independent of the exogenous respiration rate.

Figure 3.5 shows the independence of endogenous and exogenous respiration. In this figure one can observe, plotted as a function of the sludge age, the experimentally determined fractions of the influent COD discharged in the effluent, oxidised and discharged as excess sludge. In the same figure the theoretical COD fractions are also indicated. In order to calculate these theoretical fractions, it was assumed that the value of the kinetic decay rate constant for aerobic digestion (b_h), as determined in the aerobic batch digester, could be used to describe the kinetics of endogenous respiration in the activated sludge process. As there is a very close correlation between the experimental and the theoretical values of the different COD fractions over the whole range of applied sludge ages, the assumed independence of exogenous and endogenous respiration seems perfectly justified.

The decay constant for active sludge determined in the previous section has a much larger value than the one commonly used in textbooks. This latter "constant" originates from a model in which the aerobic sludge digestion is described as a first order process in relation to the *volatile* (and not the active) sludge concentration:

$$r_v = (dX_v/dt)_d = -b_v \cdot X_v \tag{12.30}$$

where:

r_v = decay rate of volatile solids
b_v = proportionality constant = apparent aerobic decay constant

The numerical value attributed to b_v shows considerable variation, but generally is within the range of 0.04 to 0.10 d^{-1} for temperatures around 20°C, as compared to a value of $b_h = 0.24\ d^{-1}$ found by Marais and Ekama (1976) and by Van Haandel et al. (1985). The reasons for this large difference will be explained below. The following experimental method is often used to determine the b_v value (Ramalho, 1980): when an activated sludge process is operated under steady state conditions, a sludge mass develops that is compatible with the applied organic load. When this sludge mass is established, there is an equilibrium between the sludge growth due to synthesis and the sludge loss due to aerobic digestion and to sludge wastage.

$$(dX_v/dt) = 0 = r_c + r_v + r_e \qquad (12.31)$$

where:

(dX_v/dt) = rate of change of the volatile sludge concentration (= 0 at steady state)
r_c = growth rate of volatile solids = $Y_{ap} \cdot (S_{ti} - S_{te})/R_h$ with Y_{ap} as the apparent yield
r_v = decay rate of volatile solids = $- b_v \cdot X_v$
r_e = rate of change due to sludge discharge = $- X_v/R_s$

Substituting the expressions for r_c, r_v and r_e in Eq. (12.31):

$$Y_{ap} \cdot (S_{ti} - S_e)/R_h - X_v \cdot (b_v + 1/R_s) = 0 \quad \text{or}$$
$$1/R_s = Y_{ap} \cdot (S_{ti} - S_e)/(X_v \cdot R_h) - b_v \qquad (12.32)$$

Using $R_h = V_r/Q_i$, one has:

$$1/R_s = Y_{ap} \cdot (S_{ti} - S_{te}) \cdot Q_i/(X_v \cdot V_r) - b_v = Y \cdot F/M - b_v \qquad (12.33)$$

A slightly different definition of the F/M ratio is used here instead of the one defined previously in Section 3.3.3.5. Here "F" equals the removed COD load instead of the applied COD load. It is possible to determine b_v in the following manner (Figure 12.7):

(1) For different values of the sludge age the F/M ratio is determined. In the example of Figure 12.7, for $R_s = 2$, 3 and 4 days;
(2) The values of $1/R_s$ are plotted as a function of F/M;
(3) A straight line is drawn through the experimental points (line R_1 in Figure 12.7);
(4) With the aid of Eq. (12.33) the apparent yield coefficient Y_{ap} is determined graphically. For the straight line R_1 the gradient is $Y_{ap} = 0.40$ mg VSS · mg^{-1} COD;
(5) Extrapolate the line to $1/R_s = 0$ to find $(F/M)_{1/Rs=0}$ and the constant b_v can be calculated as $b_v = Y_{ap} \cdot (F/M)_{1/Rs=0}$

In the case of the example in Figure 12.7, line R_1 intercepts the horizontal axis ($1/R_s = 0$) at $(F/M)_{1/Rs=0} = 0.20$ mg COD · mg^{-1} VSS · d^{-1}. Hence in this case the value of the constant b_v would be $0.40 \cdot 0.20 = 0.08\ d^{-1}$.

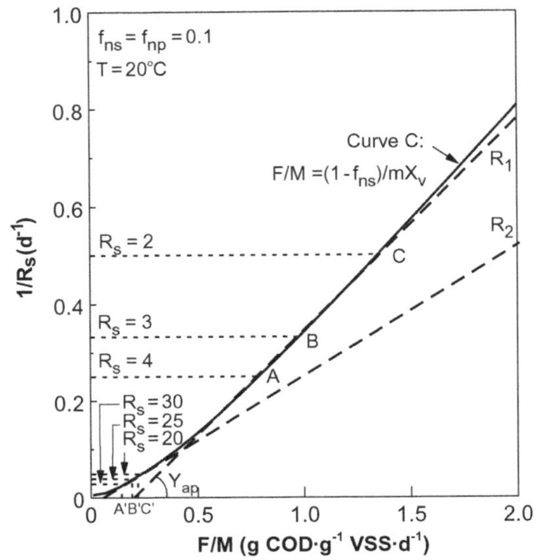

Figure 12.7 Graphical representation of the method to calculate the value of b_v

It must be emphasised that the above method, though much applied in practice, is *incorrect* as becomes apparent from the following analysis. The ratio F/M can be expressed as:

$$\begin{aligned} F/M &= (S_{ti} - S_t)/(X_v \cdot R_h) \\ &= (1 - f_{ns}) \cdot S_{ti} \cdot Q_i/(X_v \cdot V_r) = (1 - f_{ns})/mX_v \\ &= (1 - f_{ns})/[(1 - f_{ns} - f_{np}) \cdot (1 + f \cdot b_h \cdot R_s) \cdot C_r + f_{np} \cdot R_s/f_{cv}] \end{aligned} \quad (12.34)$$

From Eq. (12.34) it can be noted that the plot of $1/R_s$ as a function of F/M is in fact not a straight line as Eq. (12.33) suggests, but rather a curve that passes through the origin. In Figure 12.7 it can be clearly observed that the value of b_v is obtained from a linearization of the experimental relationship between $1/R_s$ and the F/M ratio. The obtained "straight" line depends heavily on the value of the sludge age used during the experiments. If these values are low (for example $R_s = 2$ to 4 days), then the corresponding b_v value will be relatively high: $b_v = 0.08\ d^{-1}$ for line R_1. For high sludge ages (for example $R_s = 20$ to 30 days) the value of b_v will be small ($0.03\ d^{-1}$ for line R_2).

It can be concluded that the value of the "constant" b_v in fact depends on the applied sludge age during the experiment. The reason is that at high sludge ages the active sludge fraction f_{av} will be small, resulting in a small apparent decay rate b_v for the volatile sludge mass as a whole. The opposite is true at low sludge ages: the X_a fraction is large compared to the total volatile sludge concentration resulting in a high apparent decay rate b_v. This also shows that the validity of the value of the b_v parameter is restricted to activated sludge processes that operate under very similar conditions to those used during the investigation (i.e. with the same f_{av} value). It is possible to calculate the theoretical value of b_v as a function of the sludge age. Figure 12.8 shows the relationship between b_v and R_s for $f_{np} = 0.02$ and for $f_{np} = 0.10$ (at T = 20°C). It can be observed that the value of b_v is influenced considerably by the values of R_s and f_{np} while in contrast the value of b_h does not depend on these variables. The range of values for b_v in Figure 12.8 covers the range usually found in the literature. Thus, Eq. (12.11) is capable of a correct prediction of the variations of the "constant" b_v as function of R_s.

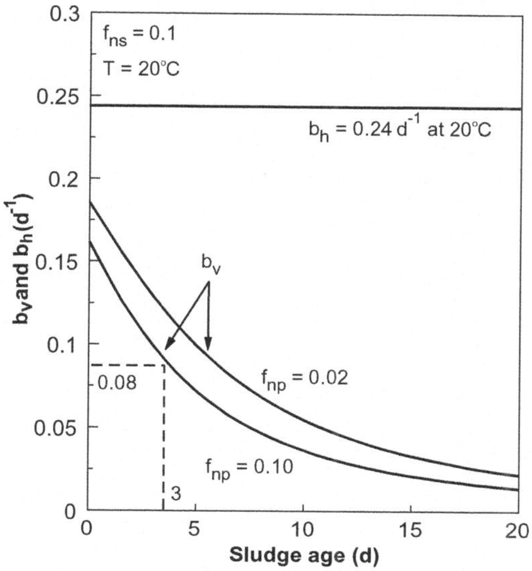

Figure 12.8 b_v and b_h values as function of the sludge age for different f_{np} values

Evidently, the model based on Eqs. (12.11 and 12.12) is much superior to the model suggested by Eq. (12.30): once the constant b_h has been determined, it can be used for any set of values of the parameters R_s and f_{np}. The value determined for b_v is only valid for similar values of R_s and f_{np} as were applied during the experimental investigation for its determination. For a different set of operational conditions a new determination of the value of b_v will be required.

12.3.3 Aerobic digester design

Aerobic digesters are usually constructed as completely mixed reactors. The reactor may be fed continuously or intermittently with excess sludge. The objective of the digestion is to reduce the fraction of biodegradable organic material to such a level that the digested sludge can be disposed of without problems. In practice this means that the active sludge fraction f_{av} should be reduced to 10–20 percent of the volatile sludge mass. For the design of the aerobic digester the following factors are important:

(1) Flow and composition of the sludge to be digested;
(2) Maximum allowable fraction of active sludge remaining after digestion;
(3) Digestion temperature;
(4) Configuration of the aerobic sludge digester: i.e. the number of digesters in series.

With respect to the sludge composition, a distinction must be made between the digestion of combined primary- and secondary excess sludge and that of secondary sludge only. In the former case, at first the primary sludge will be metabolised by the active sludge, which is an oxygen consuming process and leads to the production of more active sludge. Thus the advantages of having primary sedimentation (i.e. smaller aeration tank and less oxygen consumption) are lost to a large extent when aerobic sludge digestion is applied, because the digester will be large and will consume a considerable amount of oxygen.

Therefore in practice, primary settling is almost always combined with anaerobic sludge digestion. The composition of the excess sludge, the desired composition of the stabilised sludge and the temperature will normally be known, thus the only variables that the designer is able to optimise are the reactor configuration (i.e. the number of reactors) and the applied retention time.

The influence of the configuration on the digester performance can be evaluated using basic principles of chemical reactor engineering. Knowing that aerobic digestion is a first order process with respect to the active sludge concentration and that a digester operates as a completely mixed reactor, the decrease of the active sludge concentration in the case of continuous feeding can be expressed as (Levenspiel, 1972):

$$X_{ae} = X_{ai}/(1 + b_h \cdot R_d) \qquad (12.35)$$

where:

X_{ae} = active sludge concentration in the digester and its effluent
X_{ai} = active sludge concentration of the digester feed
R_d = retention time in the digester

In the case of a series of completely mixed digesters, the effluent from the first digester serves as the influent for the second and so on, until stabilised sludge is discharged from the last digester of the series. Equation (12.35) remains valid for each digester individually so that:

$$X_{aN} = X_{a(N-1)}/(1 + b_h \cdot R_N) \qquad (12.36)$$

where:

X_{aN} = active sludge concentration in digester n and its effluent.
$X_{a(N-1)}$ = active sludge concentration in digester (n−1)
R_N = retention time in the n-th digester

For a series of N digesters Eq. (12.36) leads to:

$$X_{aN}/X_{ai} = \sum_{n=1}^{N} 1/(1 + b_h \cdot R_n) \qquad (12.37)$$

X_{aN} = active sludge concentration in the discharge from the last digester

It is well known that for a first order process the highest efficiency is obtained when a series of completely mixed reactors all have the same retention time (Levenspiel, 1972) so that:

$$R_n = R_d/N \qquad (12.38)$$

For a very long series of reactors the behaviour of a plug flow reactor is approximated as can be observed from Eq. (12.39):

$$X_{ae}/X_{ai} = \lim_{N\to\infty} 1/(1 + b_h \cdot R_n)^N = \exp(-b_h \cdot R_d) \qquad (12.39)$$

Equation (12.39) shows that the relationship between the active sludge concentration and the retention time in a plug flow reactor is the same as in a batch reactor with a digestion period equal to R_d. Equation (12.37) permits an evaluation of the influence of the number of digesters on the digestion efficiency. To compare the performance of a single digester with a series of digesters, it is convenient to calculate the ratio X_{ae}/X_{ai} as a function of the total retention time R_d. In Figure 12.9 this relationship is presented for the single digester ($N = 1$), for two ($N = 2$) and four ($N = 4$) digesters in series as well as for the plug flow reactor ($N = \infty$) at temperatures of 20°C and 30°C. It can be observed that a higher temperature accelerates the digestion process significantly and that for any value of the retention time, the digester is more efficient when it has more reactors in series.

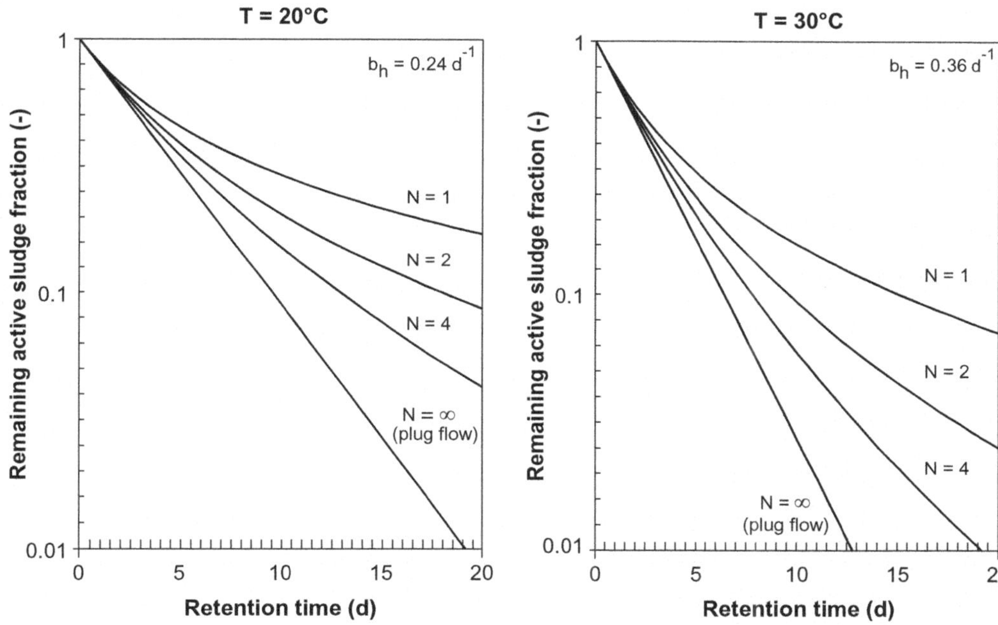

Figure 12.9 Residual active sludge fraction (f_{ae}) as function of the retention time in aerobic digesters for different numbers of reactors and at temperatures of 20 and 30°C

In practice, the objective of an aerobic digester is to reduce the active sludge fraction from a fraction f_{ai} in the excess sludge to a specified fraction f_{ae} in the stabilised sludge. The influence of the digester configuration on the performance can be evaluated by deriving an expression for the relationship between the retention time and the values of the incoming and outgoing active sludge fractions f_{ai} and f_{ae}. In the case of a single digester, the retention time to reduce the active sludge fraction from f_{ai} to f_{ae} is calculated as follows:

(1) The active sludge fraction in the influent of the digester is given by:

$$f_{ai} = X_{ai}/(X_{ai} + X_{nai}) = 1/(1 + X_{nai}/X_{ai}) \qquad (12.40)$$

where:

X_{ai} = incoming active sludge concentration
X_{nai} = incoming inactive sludge concentration ($X_i + X_e$)

(2) In the aerobic digester the active sludge concentration decreases to X_{ae}. Consequently, an endogenous residue mass will be formed with a concentration of:

$$X_{ee} = f \cdot (X_{ai} - X_{ae}) \tag{12.41}$$

where X_{ee} = endogenous residue concentration formed in the digester.

(3) Thus the active sludge fraction in the outgoing sludge is given by:

$$f_{ae} = X_{ae}/(X_{ae} + X_{nae} + X_{ee}) = 1/(1 + X_{nae}/X_{ae} + X_{ee}/X_{ae}) \tag{12.42}$$

where X_{nae} = inactive sludge concentration in the outgoing sludge

(4) Considering that the inactive sludge concentration remains constant in the digester one has:

$$X_{nae}/X_{ae} = X_{nai}/X_{ae} = (X_{nai}/X_{ai}) \cdot (1 + b_h \cdot R_d) \tag{12.43}$$

(5) Now using Eq. (12.40) in Eq. (12.43) gives:

$$1/f_{ae} = X_{nae}/X_{ae} = (1/f_{ai} - 1) \cdot (1 + b_h \cdot R_d) \tag{12.44}$$

(6) Finally, by substituting Eqs. (12.41 and 12.44) in Eq. (12.42) the following relationship is established:

$$R_d = 1/b_h \cdot \left[\frac{1/f_{ae} + f - 1}{1/f_{ai} + f - 1} - 1\right] = 1/b_h \cdot \frac{1/f_{ae} - 1/f_{ai}}{1/f_{ai} + f - 1} \tag{12.45}$$

In order to calculate the required retention time in a series of digesters (with equal volumes) to reduce the active sludge fraction from f_{ai} to f_{ae}, first the N-th digester of the series is considered. For this reactor Eq. (12.45) is valid so that:

$$R_N = 1/b_h \cdot \left[\frac{1/f_{aN} + f - 1}{1/f_{a(N-1)} + f - 1} - 1\right] \quad \text{or}$$
$$1/f_{aN} + f - 1 = [1/f_{a(N-1)} + f - 1] \cdot (1 + b_h \cdot R_n) \tag{12.46}$$

where:

f_{aN} = active sludge fraction in the sludge leaving the N-th digester
$f_{a(N-1)}$ = active sludge fraction in the sludge entering the N-th digester

Applying Eq. (12.46) for a series of N digesters one has:

$$1/f_{ae} + f - 1 = (1/f_{ai} + f - 1) \cdot (1 + b_h \cdot R_N)^N \quad \text{or}$$
$$R_N = 1/b_h \cdot \left\{\left[\frac{1/f_{ae} + f - 1}{1/f_{ai} + f - 1}\right]^{1/N} - 1\right\} \tag{12.47}$$

Now the total retention time can easily be calculated as:

$$R_d = N \cdot R_N = N/b_h \cdot \left\{ \left[\frac{1/f_{ae} + f - 1}{1/f_{ai} + f - 1} \right]^{1/N} - 1 \right\} \tag{12.48}$$

The (theoretical) minimum retention time for a large number of digesters (N→ ∞) is obtained when n appraches affinity and can be calculated from Eq. (12.48) as:

$$R_{dmin} = \lim_{N \to \infty} R_d = 1/b_h \cdot \ln \left[\frac{1/f_{ae} + f - 1}{1/f_{ai} + f - 1} \right] \tag{12.49}$$

Figure 12.10 shows the relationship between the retention time and the active sludge fraction in the excess sludge for different reactor configurations ($N = 1, 2, 4$ and ∞) and for residual active sludge concentrations of 10 and 20% in the stabilised sludge at 20°C. The required retention time is significantly influenced by the chosen digester configuration, especially when the active sludge fraction in the excess sludge is high and/or if a very low active sludge fraction in the stabilised sludge is to be obtained.

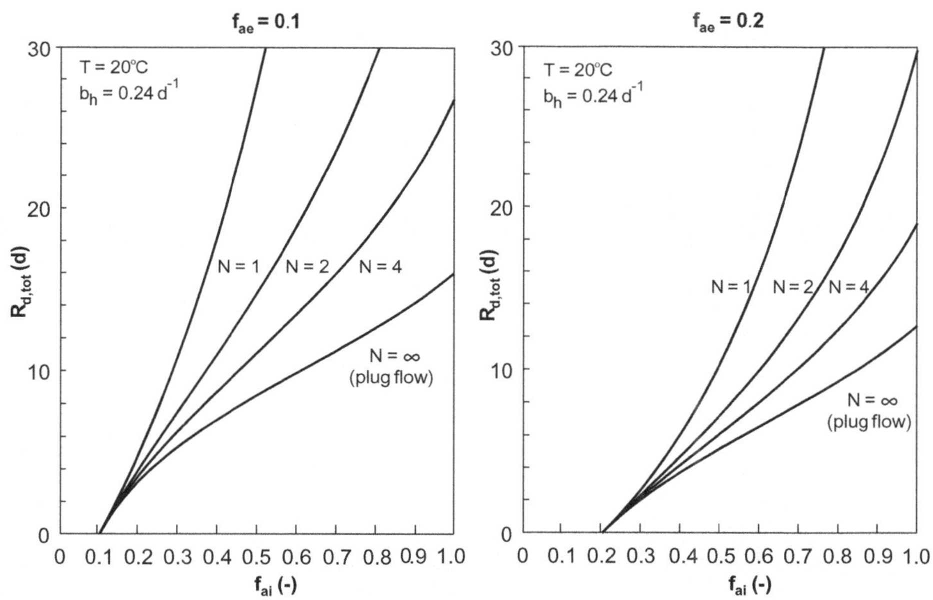

Figure 12.10 Minimum required retention time ($R_{d,tot}$) as function of the active sludge concentration in the excess sludge (f_{ai}) for different digester configurations ($N = 1, 2, 4$ and ∞) and for different stabilised sludge requirements ($f_{ae} = 0.1$ and $f_{ae} = 0.2$)

It can be observed that the required retention time for $f_{ae} = 0.10$ is always much higher than for $f_{ae} = 0.20$. Therefore it is important to establish the maximum permissible value of the active sludge fraction in the stabilised sludge. In practice the maximum f_{ae} value will depend on the method for final sludge disposal. The active sludge fraction affects the rheological and mechanical properties of the sludge as well as its biological and hygienic quality. The higher the degree of stabilisation, the more favourable mechanical properties (good settleability, low specific filter resistance) can be expected.

In so far as the biological properties are concerned, the most important parameters are the rate and extent of putrefaction (acid fermentation), when the stabilised sludge is placed in an anaerobic environment. The hygienic quality is of great importance when the stabilised sludge is used as a fertiliser in agriculture. Several researchers have shown that aerobic digestion is not a very efficient method to eliminate pathogens in sludge and often physical (drying) or chemical (lime treatment) are required to obtain an acceptable hygienic quality.

The active sludge fraction can be evaluated from OUR or BOD tests with stabilised sludge. The relationships between the OUR and the volatile sludge concentration can be derived from Eq. (12.20).

$$O_t = (f_{cv} + 4.57 \cdot f_n) \cdot (1 - f) \cdot b_h \cdot X_a = (f_{cv} + 4.57 \cdot f_n) \cdot (1 - f) \cdot b_h \cdot f_{av} \cdot X_v \quad \text{or}$$

$$O_t/X_v = (f_{cv} + 4.57 \cdot f_n) \cdot (1 - f) \cdot b_h \cdot f_{av} = f_{av} \cdot 15.7 \cdot 1.04^{(T-20)} (\text{mg } O_2 \cdot g^{-1} \text{ VSS} \cdot h^{-1}) \quad (12.50)$$

Using Eq. (12.29) in Eq. (12.50) the BOD per unit volatile sludge mass is calculated as:

$$BOD_{VSS} = BOD/X_v = (f_{cv} + 4.57 \cdot f_n) \cdot (1 - f) \cdot 0.7 \cdot f_{av} = 1.10 \cdot f_{av} \, (\text{mg BOD} \cdot \text{mg}^{-1} \text{ VSS}) \quad (12.51)$$

Equations (12.50 and 12.51) show that the active sludge mass fraction can be calculated easily if the OUR or the BOD per unit mass of volatile sludge are determined experimentally. Figure 12.11 shows the relationship between the BOD per unit mass of volatile sludge and the active sludge mass fraction. It can be observed that the BOD of stabilised sludge ($0.1 < f_{av} < 0.2$) is in the range of 0.10 to 0.25 mg O_2 . mg^{-1} VSS. In the same figure the relation between the OUR and the active sludge fraction is given for several temperatures. At 20°C the OUR per gram stabilised sludge is between 1.5 and 3.0 mg O_2 . g^{-1} VSS . h^{-1}.

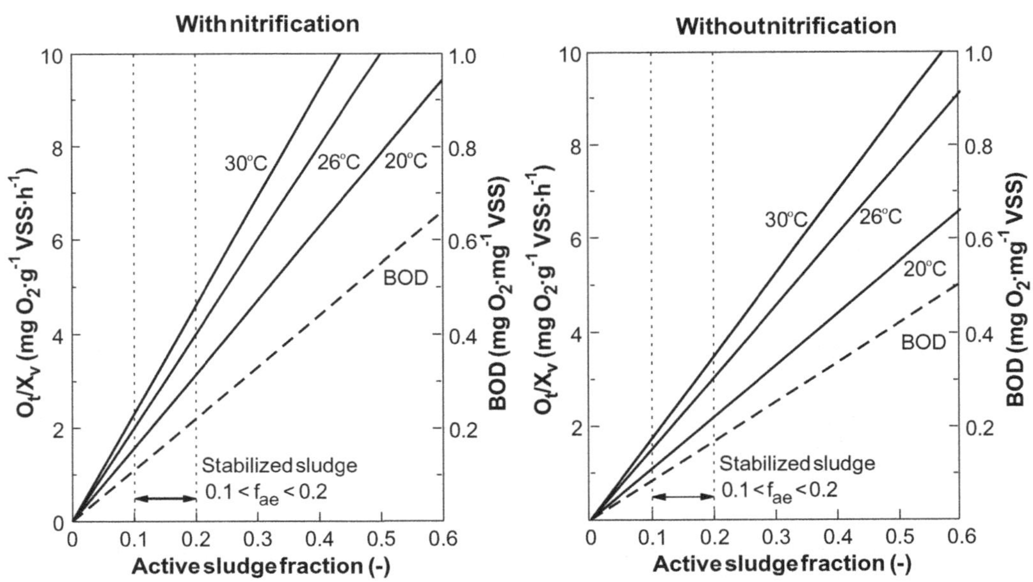

Figure 12.11 Theoretical OUR and BOD value per unit volatile sludge mass as function of the active sludge fraction, with and without taking into account nitrification of mineralised ammonia

12.3.4 Optimisation of aerobic sludge digestion

When aerobic sludge digestion is applied, in general a sludge thickener will precede the digester in order to increase the sludge concentration, with the objective to reduce the flow of excess sludge to the digester. As the required retention time in the digester to effect the reduction of the active sludge fraction from f_{ai} to f_{ae} is dependent on the excess sludge composition, the volume of the digester will be inversely proportional to the volume of the thickened sludge flow. The objective of the optimisation of the activated sludge system with aerobic digestion is to minimise construction cost of the biological reactor, thickener and digester while producing digested sludge with an active sludge fraction below a specified maximum value, for example $0.1 < f_{ae} < 0.2$. Two different situations can be considered:

- The sludge age in the activated sludge process is defined by factors unrelated to sludge stabilisation, for instance by the requirements for nutrient removal;
- The sludge age may be defined by the optimal value for sludge stabilisation.

In the first case the optimisation procedure is limited to a calculation for the minimum costs for construction of thickener and digester. In the second case, the construction costs of the aeration tank and the final settler must also be taken into consideration. Both cases will now be evaluated.

When the value of the sludge age is given, the composition and flow of the excess sludge to be digested are known. In this case, the volume of the thickener and the volume of the digester unit required to reduce the active sludge fraction to the desired value f_{ae} can be calculated. To calculate these volumes, the following factors are important: the sludge age, the settling characteristics of the sludge and the number of digesters in series. The sludge age in the activated sludge process determines the composition and flow of the excess sludge. By expressing the variables per unit daily applied COD mass ($kg\ SS \cdot d \cdot kg^{-1}$ COD), one has:

$$mX_a = (1 - f_{ns} - f_{np}) \cdot C_r \quad (3.46)$$

$$mX_v = (1 - f_{ns} - f_{np}) \cdot C_r \cdot (1 + f \cdot b_h \cdot R_s) + f_{np} \cdot R_s / f_{cv} \quad (3.48)$$

$$mX_t = mX_v / f_v \quad (3.49)$$

The mass of excess sludge per unit mass applied COD ($kg\ TSS \cdot kg^{-1}$ COD) is given by:

$$mE_t = mX_v / R_s \quad (3.51)$$

The active sludge fraction is equal to:

$$f_{av} = f_{ai} = mX_a / mX_v \quad (3.52)$$

If the thickener is designed using the solids flux theory, the settling characteristics determine the concentration of the excess sludge flow after thickening. The required thickener area per unit mass of daily applied COD ($m^2 \cdot d \cdot kg^{-1}$ COD) is given by:

$$a_{th} = mE_t / ([X_{th} \cdot v_0 \cdot (k \cdot X_l - 1) \cdot \exp(-k \cdot X_l))] \quad (12.7a)$$

For a thickener with depth H_{th} and a safety factor s_{fth}, the volume of the thickener per unit mass of daily applied COD ($m^3 \cdot d \cdot kg^{-1}$ COD) can now be calculated as:

$$v_{th} = s_{fth} \cdot H_{th} \cdot a_{th} \tag{12.8a}$$

The sludge flow per unit mass of daily applied COD after thickening to a sludge concentration X_{th} is calculated as (in $m^3 \cdot kg^{-1}$ COD):

$$mq_{th} = mE_t/X_{th} \tag{12.4}$$

The digester volume depends on the number of digesters in series, which in turn is dependent on the construction costs. While the combined volume of a series of digesters decreases when the number of digesters increases, the construction costs will not decrease correspondingly, because of the need for additional walls to separate the digesters and the additional equipment and instrumentation. There will be an optimal number of digesters with minimum construction costs. It is interesting to note that for any particular degree of desired stabilisation, the oxygen and alkalinity demands are the same, so that the operational costs are not influenced by the digester configuration. The total volume of the digestion unit per unit mass of daily applied COD ($m^3 \cdot d \cdot kg^{-1}$ COD) can be calculated with the aid of Eqs. (12.48 and 12.4):

$$v_{da} = q_{th} \cdot R_d = mE_t/X_{th} \cdot N/b_h \cdot \left\{ \left[\frac{1/f_{ae} + f - 1}{1/f_{ai} + f - 1} \right]^{1/N} - 1 \right\} \tag{12.52}$$

Now, the optimal values of the volumes of the thickener and the digester are determined by the minimum costs criterion. For a particular number of digesters in series N, the only variable is the thickened sludge outlet concentration. The higher this concentration, the larger will be the required thickener volume, but on the other hand the digester volume will be reduced. The objective function to be optimised can be formulated as:

$$(dC_t/dX_r) = 0 = C_{th} \cdot (dv_{th}/dX_{th}) + C_{da} \cdot (dv_{da}/dX_{th}) \tag{12.53}$$

where:

C_t = total construction costs per unit mass of daily applied COD
C_{da} = construction costs of the aerobic digester per unit mass of daily applied COD
C_{th} = construction costs of the thickener per unit mass of daily applied COD

As the variables v_{th} and v_{da} are complex functions of the thickened sludge concentration X_{th}, it is convenient to use numerical or graphical techniques to find the optimal solution of Eq. (12.53). In Figure 12.12 the optimisation procedure described above is shown. The volumes v_{da} of the digester unit and v_{th} of the thickener are shown plotted as functions of the thickened sludge concentration for the following conditions: $f_{ai} = 0.5$; $f_{ae} = 0.1$; $s_{fth} = 1$; $H_{th} = 4$ m; T = 20°C and $mE_t = 0.2$ mg TSS $\cdot$ mg^{-1} COD.
The calculations were made for poor settleability ($k = 0.46\ l \cdot g^{-1}$ and $v_0 = 144\ m \cdot d^{-1}$, top figures) and fair settleability ($k = 0.36\ l \cdot g^{-1}$ and $v_0 = 216\ m \cdot d^{-1}$, bottom figures). Furthermore, two configurations were

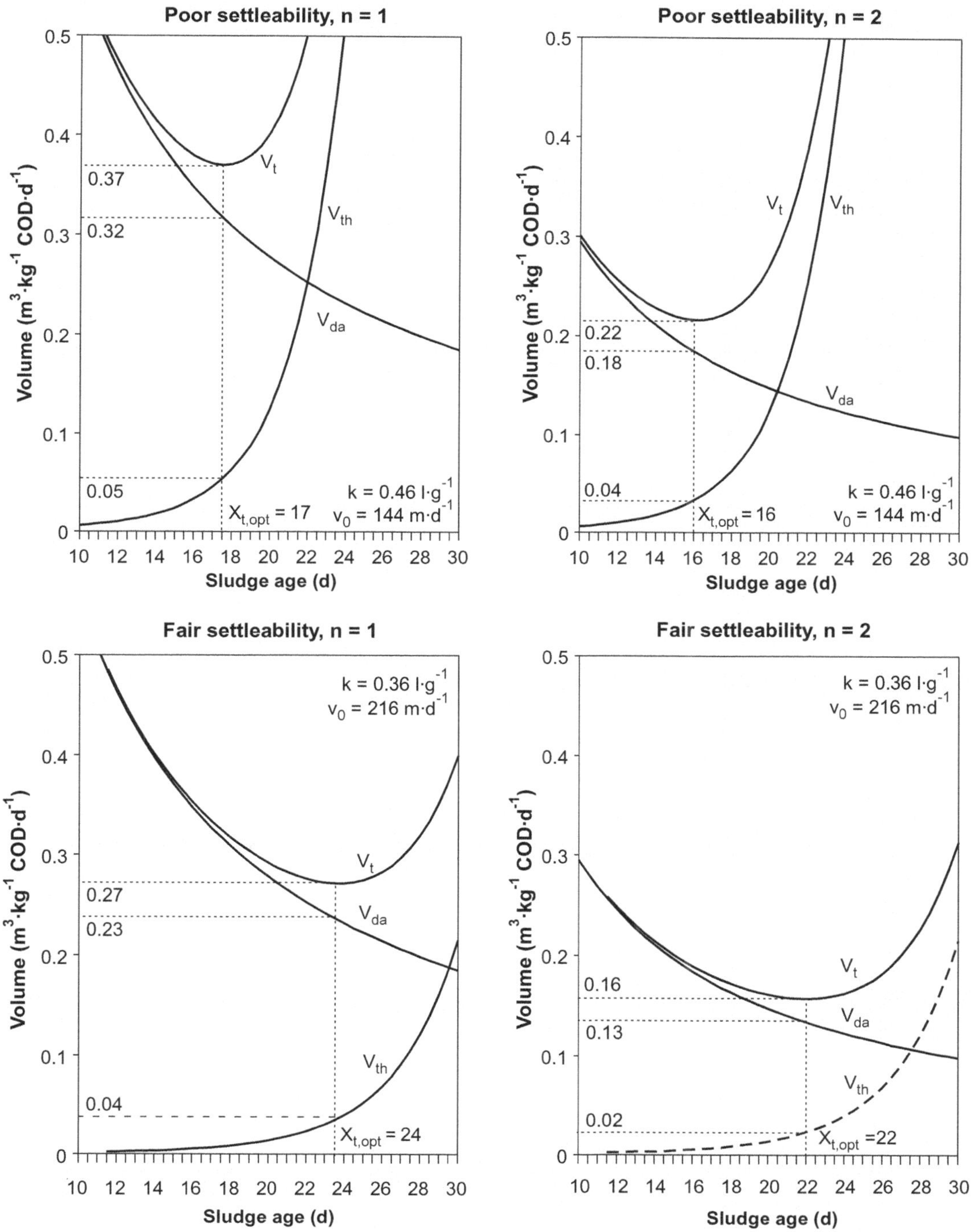

Figure 12.12 Graphical optimisation of an aerobic digester with a sludge thickener

considered: a completely mixed digester (N = 1, left) and two equally sized digesters in series (N = 2, right). From an analysis of Figure 12.12, the following may be concluded:

- In all cases the thickener volume is relatively small, but its presence is crucial in order to avoid a very large digester volume. Thus for example, for fair settleability and a series of two digesters (bottom-right diagram of Figure 12.12), the optimal volume of the thickener is 24 $\text{litre} \cdot \text{kg}^{-1}\,\text{COD} \cdot \text{d}^{-1}$, which produces a thickened sludge concentration of 22 $\text{g TSS} \cdot \text{l}^{-1}$. If it is assumed that the sludge concentration in the aeration tank was 4.4 $\text{g TSS} \cdot \text{l}^{-1}$, the sludge is thickened by a factor $24/4.4 = 5$, which means that the digester unit after thickening is 5 times smaller than without thickening. Thus if the thickener with a volume of 24 $\text{litre} \cdot \text{kg}^{-1}\,\text{COD} \cdot \text{d}^{-1}$ were omitted, the digester unit would have to be increased from 134 $\text{litre} \cdot \text{kg}^{-1}\,\text{COD} \cdot \text{d}^{-1}$ in Figure 12.12 to $5 \cdot 134 = 670$ $\text{litre} \cdot \text{kg}^{-1}\,\text{COD} \cdot \text{d}^{-1}$. The result would be that the digester is larger than the aeration tank!;
- The influence of the digester configuration on the required volume is considerable. For example, for poor settling (top figures in Figure 12.12) the optimal digester volume is 320 $\text{litre} \cdot \text{kg}^{-1}\,\text{COD} \cdot \text{d}^{-1}$ for the single digester, but only 190 $\text{litre} \cdot \text{kg}^{-1}\,\text{COD} \cdot \text{d}^{-1}$ for 2 digesters in series and 120 $\text{litre} \cdot \text{kg}^{-1}\,\text{COD} \cdot \text{d}^{-1}$ for a plug-flow digester. At the same time, the required thickener volume is also reduced when the number of digesters in series increases. When the number of digesters increases beyond two, the resulting reduction in digester volume will be relatively small and will probably not compensate for the extra construction costs and operational requirements;
- The settleability has a very marked influence on the required volume for sludge digestion. In Figure 12.12 the volumes for poor settleability (top) are about twice as large as the volumes in the case of fair settleability (bottom).

When the sludge age in the biological reactor can be selected without limitations imposed by other considerations such as nutrient removal, its optimal value will be determined by the minimal construction costs of the whole treatment system. In this case it will be necessary to carry out two optimisations at the same time: one for the system consisting of the aeration tank and the settler and the other for the system consisting of the thickener and digester. The procedure can be summarised as follows:

(1) For a particular sludge age the optimal sludge concentration in the aeration tank (using the method described in Section 8.3) is determined and the required volumes of the aeration tank and the final settler are calculated;

(2) For the same sludge age the optimal volumes of the thickener and the digester are calculated, using the method presented above.

(3) The total or specific construction costs are determined for the sludge age under consideration:

$$MC_t = C_r \cdot V_r + C_d \cdot V_d + C_{th} \cdot V_{th} + C_{da} \cdot V_{da} \tag{12.54a}$$

$$mC_t = C_r \cdot v_r + C_d \cdot v_d + C_{th} \cdot v_{th} + C_{da} \cdot v_{da} \tag{12.54b}$$

C = construction cost per unit volume
V = volume of unit
v = volume per unit daily applied COD mass

Indexes t, r, d, th and da refer to total, aeration tank, final settler, thickener and aerobic digester respectively;

(4) The procedure is repeated for different sludge ages and the total costs are plotted as a function of sludge age. The minimum total costs identify the optimal sludge age in the activated sludge process.

As an example of the procedure described above, the total volume for the treatment system with aerobic sludge digestion is calculated in Figure 12.13 for fair and poor sludge settleability respectively. To make the necessary calculations, it was assumed that the system has an optimally designed thickener and that the desired active sludge fraction in the stabilised sludge f_{ae} is 0.1. Furthermore the following data were used: non-biodegradable influent fractions f_{ns} and f_{np} are both equal to 0.1; the temperature is 20°C (i.e. $b_h = 0.24\ d^{-1}$); $H_d = 4$ and $s_{fd} = 2$ in the settler and $H_{th} = 4$ and $s_{fth} = 1.5$ in the thickener.

Figure 12.13 shows that, for the specified conditions, in both cases the optimal sludge age in the activated sludge plant is (coincidentally) 3.5 days. If it is assumed that the construction cost per unit of volume is equal for all treatment units, the minimum volume that is obtained for a sludge age of $R_s = 3.5$ days also indicates the minimum costs. In the case of fair settleability, the volumes would be: $v_r = 0.30$; $v_d = 0.20$; $v_{th} = 0.07$ and $v_{da} = 0.38\ m^3 \cdot kg^{-1}\ COD \cdot d^{-1}$, resulting in a total volume $v_t = 0.95\ m^3 \cdot kg^{-1}\ COD \cdot d^{-1}$.

For sludge ages beyond the optimal value, the total volume increases gradually and reaches a value of $v_t = 1.10\ m^3 \cdot kg^{-1}\ COD \cdot d^{-1}$ for $R_s = 10$ days. For sludge ages shorter than the optimal value there is a more rapid increase in the required volume, due to the large size of the aerobic digester. Although the same optimal sludge age of 3.5 days is calculated in the case of poor settleability, the reactor volumes are quite different: the total volume increases by 35% (from 0.95 to 1.30 $m^3 \cdot kg^{-1}\ COD \cdot d^{-1}$, due to the poor settling characteristics of the sludge.

12.3.5 Operational parameters of the aerobic digester

Once the optimal configuration of an aerobic sludge digester unit has been determined, it is a relatively simple matter to calculate the main operational parameters. These are:

- Reduction of the volatile sludge concentration;
- Oxygen uptake;
- Increase of the nitrate concentration;
- Alkalinity demand.

All these parameters are directly related to the oxidation of active sludge in the digester. The decrease of the active sludge concentration can be determined from the incoming and outgoing active sludge fractions (f_{ai} and f_{ae}) as follows:

$$f_{ae} = X_{ae}/X_{ve} = (X_{ai} - X_{ad})/[X_{vi} - (1 - f) \cdot X_{ad}] \qquad (12.55)$$

where X_{ad} = digested active sludge concentration (mg VSS $\cdot\ l^{-1}$)

By rearranging Eq. (12.55) one has:

$$X_{ad} = X_{ai} \cdot (1/f_{ai} - 1/f_{ae})/(1 - f - 1/f_{ae}) \qquad (12.56)$$

For any optimised system, the values of f_{ai}, f_{ae} and X_{ai} will be known and the value of X_{ad} can be calculated. The operational parameters can now be linked to X_{ad}. The decrease of the volatile sludge concentration is a fraction (1−f) of the value of X_{ad} (a fraction f remains as endogenous residue), so that:

$$X_{vd} = (1 - f) \cdot X_{ad} \qquad (12.57)$$

where X_{vd} = digested volatile sludge concentration (mg VSS $\cdot\ l^{-1}$)

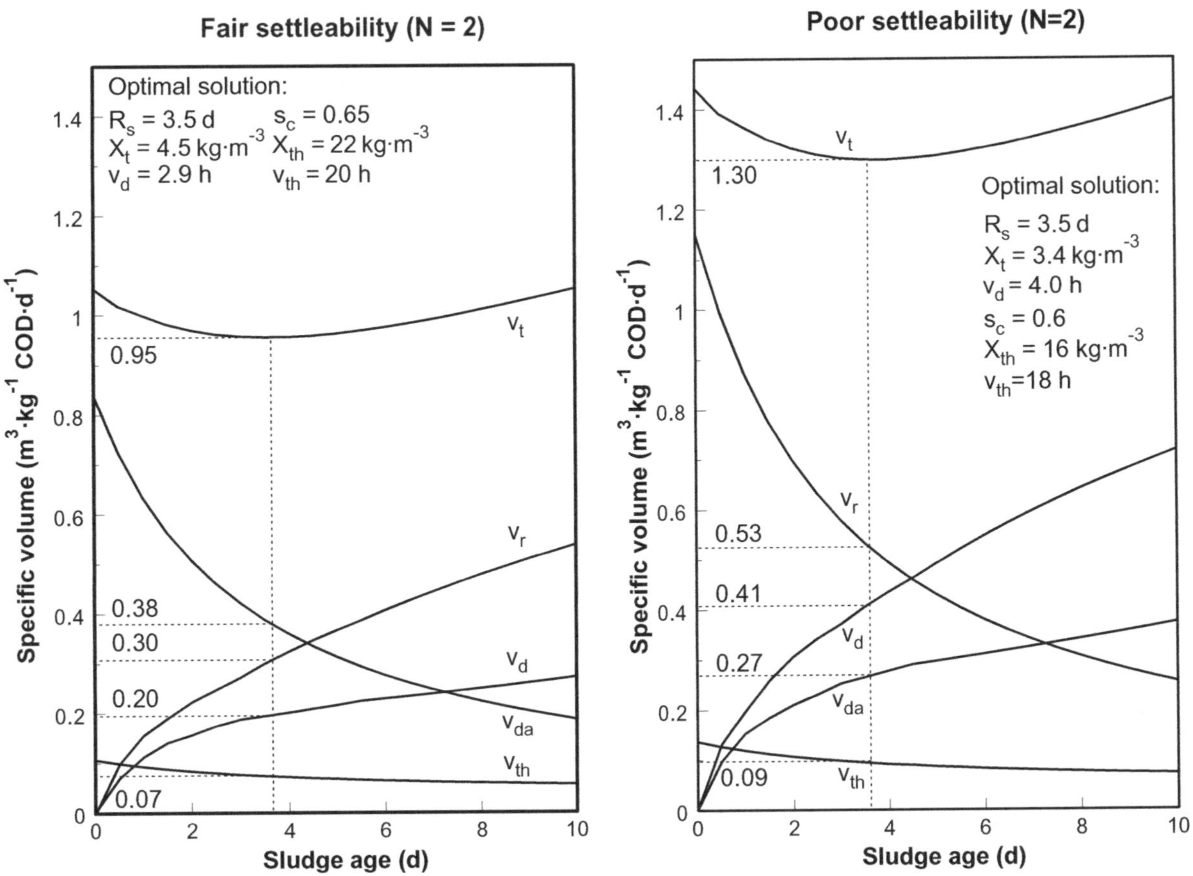

Figure 12.13 Optimisation of the activated sludge process with an aerobic digestion unit, based on total specific volume and without restrictions imposed to the applied sludge age

Similarly, the increase in the nitrate concentration and the alkalinity decrease are determined: knowing that the oxidation of 1 mg VSS results in the release of f_n mg N, which after nitrification is transformed into nitrate, consuming in these processes an alkalinity of 3.57 mg $CaCO_3.mg^{-1}$ N, one has:

$$N_{nd} = f_n \cdot X_{vd} = f_n \cdot (1 - f) \cdot X_{ad} \tag{12.58}$$

$$Alk_d = 3.57 \cdot N_{nd} = 3.57 \cdot f_n \cdot (1 - f) \cdot X_{ad} \tag{12.59}$$

where:

N_{nd} = nitrate production in the digester (mg $N \cdot l^{-1}$ of digested sludge)
Alk_d = alkalinity consumed in the digester (mg $CaCO_3 \cdot l^{-1}$ of digested sludge)

The oxygen uptake rate in the digester is calculated from the oxygen demand: the oxidation of one mg VSS requires f_{cv} mg O_2 for the organic matter and $4.57 \cdot f_n$ mg O_2 for nitrification of the released organic nitrogen. Hence:

$$V_{da} \cdot O_{td} = q \cdot (f_{cv} + 4.57 \cdot f_n) \cdot X_{vd} \quad \text{or} \quad O_{td} = (f_{cv} + 4.57 \cdot f_n) \cdot (1 - f) \cdot X_{ad}/R_d \tag{12.60}$$

EXAMPLE 12.4

Calculate the values of X_{vd}, N_{nd}, Alk_d and O_{td} in the two digesters in series of the previous example for fair sludge settleability (Figure 12.13), i.e. for a desired value of $f_{ae} = 0.1$:

Solution

(1) Determine the required retention time in the aerobic digesters

For the calculated optimal sludge age $R_s = 3.5$ d, one has $f_{ai} = X_a/X_v = 0.66$.

$$R_d = N/b_h \cdot \left\{ \left[\frac{1/f_{ae} + f - 1}{1/f_{ai} + f - 1} \right]^{1/N} - 1 \right\} \tag{12.48}$$

$R_d = (2/0.24) \cdot [(1/0.1 + 0.2 - 1)/(1/0.66 + 0.2 - 1)]^{0.5} - 1\} = 21.6\,d$, or
$R_{d1} = R_{d2} = 10.8\,d$

(2) Determine the concentration of digested active sludge X_{ad} and the active sludge concentration in the first and second digester X_{a1} and X_{a2}.

For the given thickened sludge concentration X_{th} of 22.0 g TSS $\cdot l^{-1}$ and a volatile sludge fraction f_v of 0.75, the active thickened sludge concentration X_{ai} can be calculated as $0.75 \cdot 0.66 \cdot X_t = 10.9$ g VSS $\cdot l^{-1}$. The digested active sludge concentration in the digesters is calculated as:

$$X_{ad} = X_{ai} \cdot (1/f_{ai} - 1/f_{ae})/(1 - f - 1/f_{ae})$$
$$= 10.9 \cdot (1/0.66 - 1/0.10)/(1 - 0.2 - 1/0.1) = 10.9 \cdot 0.92 = 10.1\,g\,VSS \cdot l^{-1}$$

$X_{a1} = X_{ai}/(1 + b_h \cdot R_1) = 10.9/(1 + 0.24 \cdot 10.8) = 3.0\,g\,VSS \cdot l^{-1}$

$X_{a2} = X_{a1}/(1 + b_h \cdot R_2) = 2.9/(1 + 0.24 \cdot 10.8) = 0.8\,g\,VSS \cdot l^{-1}$

It can now be verified that indeed $X_{ad} = 10.1 = X_{ai}–X_{a2} = 10.9–0.8 = 10.1$ g $VSS \cdot l^{-1}$

(3) Determine the values of X_{vd}, N_{nd}, Alk_d and O_{td}

With the values of X_{ai}, X_{a1} and X_{a2} calculated above, the operational parameters of the digesters can now easily be calculated. The results are shown in Table 12.4.

Table 12.4 Operational parameters of the optimised digester system of Figure 12.13

Parameter	UoM	Excess sludge	Digester 1	Digester 2
f_{ai}	(–)	0.66	0.29	0.1
X_{ad}	$g\ VSS \cdot l^{-1}$	–	7.9	2.2
X_{ai}	$g\ VSS \cdot l^{-1}$	10.9	3.0	0.8
X_{vi}	$g\ VSS \cdot l^{-1}$	16.5	10.1	8.4
X_{ti}	$g\ TSS \cdot l^{-1}$	22	15.7	12.9
NO_3^--N	$g\ N \cdot l^{-1}$	–	630*	824*
Alkalinity	$mg\ CaCO_3 \cdot l^{-1}$	–	−2250*	−2906*
OUR	$mg\ O_2 \cdot l^{-1} \cdot h^{-1}$	–	46.0	13.2
BOD	$g\ O_2 \cdot l^{-1}$	12.0	3.2	0.8
BOD	$g\ O_2 \cdot g\ VSS^{-1}$	0.73	0.32	0.1
Retention time	days	–	10.8	10.8

(*) = variation in the digesters is independent of the concentration in the excess sludge

It is interesting to calculate the ratio between the oxygen demand in the digester and the organic load to the activated sludge process. The volume of the two digesters is $0.38\ m^3 \cdot kg^{-1}\ COD \cdot d^{-1}$ (Figure 12.13a): i.e. $0.19\ m^3 \cdot kg^{-1}\ COD \cdot d^{-1}$ for each digester. The oxygen uptake rates are 46.0 and 13.2 $mg\ O_2 \cdot l^{-1} \cdot h^{-1}$ or 1.10 and 0.32 $g\ O_2 \cdot l^{-1} \cdot d^{-1}$ respectively (Table 12.4) so that:

$$mS_{od} = 0.19 \cdot (1.10 + 0.32) = 0.27\,kg\,O_2 \cdot kg^{-1}\,COD$$

where mS_{od} = fraction of influent COD oxidised in the digester

The calculated oxygen demand of 0.27 $kg\ O_2 \cdot kg^{-1}$ COD applied is very significant in economical terms. The oxygen consumption for the oxidation of organic material in the aeration tank is calculated from Eq. (3.43) for the optimal sludge age of $R_s = 3.5$ days:

$$mS_o = (1 - f_{ns} - f_{np}) \cdot [1 - f_{cv} \cdot Y + f_{cv} \cdot (1 - f) \cdot b_h \cdot C_r] = 0.46\,kg\,O_2 \cdot kg^{-1}\,COD$$

It is concluded that under the specified conditions the oxygen demand in the digesters of 0.27 $kg\ O_2 \cdot kg^{-1}$ COD (assuming nitrification of the liquefied nitrogen) is more than fifty percent of the oxygen consumption in the aeration tank of 0.46 $kg\ O_2 \cdot kg^{-1}$ COD (without nitrification).

EXAMPLE 12.5

An activated sludge system treats industrial effluents and produces 40 ton TSS $\cdot$ d^{-1} of sludge ($f_v = 0.75$) with an active fraction f_{av} of 0.6. The temperature is 30°C and the sludge exhibits good settling characteristics: $k = 0.25\ l \cdot g^{-1}$ and $v_0 = 12\ m \cdot h^{-1}$. The system consists of two aerobic digesters of 5000 m^3 each and two thickeners with a diameter of 18 m. Determine the optimal configuration to carry out sludge stabilisation. Calculate the required aeration power (assume that nitrification does not develop).

Solution

In Figure 12.14 several possible configurations for the sludge stabilisation are shown:

(a) Two parallel systems of thickener and digester;
(b) Two thickeners in parallel followed by two digesters in series;
(c) Two systems in series, each consisting of a thickener + digester;
(d) Thickener and two digesters plus a second thickener in series.

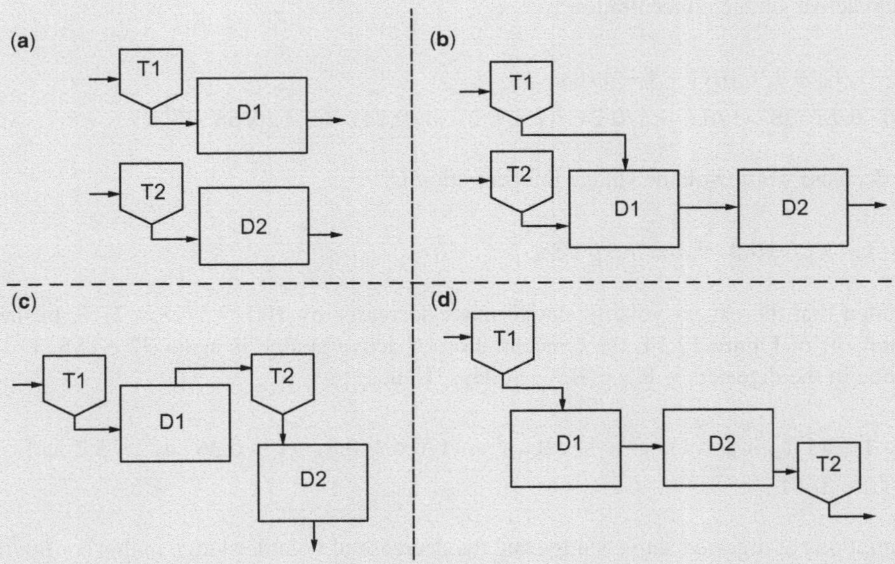

Figure 12.14 Schematic representation of different configurations for thickening and digestion of the sludge in Example 12.5

If all the sludge (40 ton $\cdot$ d^{-1}) is discharged into a single thickener with a diameter of 18 m (the thickener area is 314 m^2), the applied solids load is $40{,}000/314 = 123\ kg \cdot m^{-2} \cdot d^{-1}$ or $5.1\ kg \cdot m^{-2} \cdot h^{-1}$. By equating this value to the limiting flux and assuming thickener ideality (i.e. $s_{fth} = 1$), the concentration of the thickened sludge can be calculated from below equation:

$$F_l = F_{sol} = X_{th} \cdot (k \cdot X_l - 1) \cdot \exp(-k \cdot X_l) = 5.1\,kg\,TSS \cdot m^{-2} \cdot d^{-1}$$

For a single thickener, the trial and error solution is $28\,g\,TSS \cdot l^{-1}$. In case of having two parallel thickeners, the applied solids load to each is $2.55\,kg\,TSS \cdot m^{-2} \cdot d^{-1}$ and the thickened sludge concentration increases to $32\,g\,TSS \cdot l^{-1}$. As the applied sludge mass is $40\,ton \cdot d^{-1}$, the flow of thickened sludge is $40{,}000/28 = 1428\,m^3 \cdot d^{-1}$ for a single thickener and $40{,}000/32 = 1250\,m^3 \cdot d^{-1}$ for two thickeners in parallel.

In the case of configuration (a) of Figure 12.14, the thickened sludge concentration is $32\,g\,TSS \cdot l^{-1}$ and the flow of thickened sludge is divided over the two digesters so that each receives $625\,m^3 \cdot d^{-1}$. Hence the retention time is $R_d = 5000/625 = 8$ days and the active fraction in the digested sludge is given by Eq. (12.47):

$$\begin{aligned} 1/f_{ae} + f - 1 &= (1/f_{ai} + f - 1) \cdot (1 + b_h \cdot R_d) \\ &= (1/0.6 + 0.2 - 1) \cdot (1 + 0.36 \cdot 8) = 3.36 \quad \text{and} \quad f_{ae} = 0.24 \end{aligned}$$

The digested active sludge concentration is:

$$\begin{aligned} X_{ad} &= X_{ai} \cdot (1/f_{ai} - 1/f_{ae})/(1 - f - 1/f_{ae}) \\ &= 0.60 \cdot 0.75 \cdot 28 \cdot (1/0.6 - 1/0.24)/(1 - 0.2 - 1/0.24) = 9.4\,g\,VSS \cdot l^{-1} \end{aligned}$$

Hence the decrease of the volatile sludge concentration is:

$$X_{vd} = (1 - f) \cdot X_{ad} = 0.8 \cdot 9.4 = 7.5\,g\,VSS \cdot l^{-1}$$

It is concluded that the excess volatile sludge mass decreases by $100 \cdot 7.5/28 = 27\%$. In the case of configuration (b) of Figure 12.14, the concentration of active sludge is again $32\,g\,TSS \cdot l^{-1}$ and the retention time in the digesters is $R_{d1} = R_{d2} = 4$ days. Hence:

$$\begin{aligned} 1/f_{ae} + f - 1 &= (1/f_{ai} + f - 1) \cdot (1 + b_h \cdot R_{d1})^2 = (1/0.6 - 0.8) \cdot (1 + 0.36 \cdot 4)^2 = 5.2 \text{ and} \\ f_{ae} &= 0.17 \end{aligned}$$

The concentrations of digested active sludge and the decrease of volatile sludge in this configuration are $X_{ad} = 12.2$ g $VSS \cdot l^{-1}$ and $X_{vd} = 9.8$ g $VSS \cdot l^{-1}$. The decrease of volatile sludge in this configuration represents $100 \cdot 9.8/32 = 30\%$ of the excess volatile sludge mass.

In configuration (c) of Figure 12.14, in the first thickener the flux is $5.1\,kg \cdot m^2 \cdot d^{-1}$ and the concentration of thickened sludge is $28\,g\,TSS \cdot l^{-1}$ with $R_{d1} = 3.5$ d. Thus the fraction of active sludge leaving the first digester is:

$$1/f_{ael} + f - 1 = (1/f_{ai} + f - 1) \cdot (1 + b_h \cdot R_{d1}) = 1.96 \quad \text{and} \quad f_{ael} = 0.36$$

The concentration of digested active sludge is calculated from Eq. (12.56):

$$X_{ad1} = X_{ai} \cdot (1/f_{ai} - 1/f_{ae1})/(1 - f - 1/f_{ae1})$$
$$= 0.60 \cdot 0.75 \cdot 28 \cdot (1/0.6 - 1/0.36)/(1 - 0.2 - 1/0.36) = 7.0\,g\,VSS \cdot l^{-1}$$

Thus the decrease of the sludge concentration in the first digester is (Eq. 12.57)

$$X_{vd1} = (1 - f) \cdot X_{ad1} = 0.8 \cdot 7.0 = 5.6\,g\,VSS \cdot l^{-1}.$$

The concentration of the sludge leaving the first digester is 28−5.6 = 22.4 g TSS $\cdot l^{-1}$, resulting in a flux of 1428 · 22.4/324 = 99 kg $\cdot m^{-2} \cdot d^{-1}$ or 4.1 kg $\cdot m^{-2} \cdot h^{-1}$ in the second thickener. If it is assumed that the settleability does not change, the thickened sludge concentration in the second thickener can be calculated as 29 g TSS $\cdot l^{-1}$. The flow of sludge is then decreased from 1428 to 1428 · 22.4/29 = 1100 $m^3 \cdot d^{-1}$ and the retention time in the second digester becomes R_{d2} = 5000/1100 = 4.5 days. Hence the final active fraction will be:

$$1/f_{ae2} + f - 1 = (1/f_{ae1} + f - 1) \cdot (1 + b_h \cdot R_{d2}) = 5.20 \quad \text{and} \quad f_{ae2} = 0.17$$

The digested active sludge concentration and the decrease of volatile sludge are calculated as X_{ad} = 10.5 and X_{vd} = 8.4 g $\cdot l^{-1}$ respectively, i.e. a fraction of 100 · 8.4/28 = 30% of the volatile sludge is mineralised.

In configuration (d) of Figure 12.14 the situation in the first thickener is the same as in configuration (c). In the series of two digesters the final active sludge fraction is calculated as:

$$1/f_{ae} + f - 1 = (1/f_{ai} + f - 1) \cdot (1 + b_h \cdot R_{d1})^2$$
$$= (1/0.6 - 0.8) \cdot (1 + 0.36 \cdot 3.5)^2 = 4.42 \quad \text{and} \quad f_{ae} = 0.19$$

The digested active sludge concentration is now calculated as:

$$X_{ad} = X_{ai} \cdot (1/f_{ai} - 1/f_{ae})/(1 - f - 1/f_{ae})$$
$$= 0.60 \cdot 0.75 \cdot 28 \cdot (1/0.6 - 1/0.19)/(1 - 0.2 - 1/0.19) = 10.0\,g\,VSS \cdot l^{-1}$$

Hence:

$$X_{vd} = 0.8 \cdot 10.0 = 8.0\,g\,VSS \cdot l^{-1}$$

The solids flux to the second thickener will be 1428 · (28–8)/324 = 88 kg $\cdot m^{-2} \cdot d^{-1}$ or 3.7 kg $\cdot m^{-2} \cdot h^{-1}$. For this value, the concentration of thickened sludge is calculated as 29.8 g TSS $\cdot l^{-1}$: i.e. the flow of thickened digested sludge will be 20/29.8 · 1428 = 960 $m^3 \cdot d^{-1}$ (it is again assumed that the digestion does not affect the settleability of the sludge).

In Table 12.5 the results of the calculations are summarised. It can be concluded that option (a) is the most unattractive alternative: the stabilised sludge still has an active sludge fraction of 24%, which is more than the 20% generally adopted as an acceptable upper limit, while the flow of stabilised sludge is also larger than in the other options: 1428 $m^3 \cdot d^{-1}$. The other three options are more or less equivalent in terms of the residual active fraction. However, the digested sludge flow (and hence disposal costs) of option (d) is the smallest and for that reason might be adopted in practice.

Table 12.5 Summary of the performance of the different thickening and stabilisation configurations of Example 12.5

Parameter	UoM	Configuration (see Figure 12.14)			
		(a)	(b)	(c)	(d)
Residual active sludge fraction	–	0.24	0.19	0.17	0.19
Mass flow of stabilised sludge	ton TSS $\cdot d^{-1}$	30.6	28.6	28.0	28.6
Volumetric flow of stable sludge	$m^3 \cdot d^{-1}$	1428	1250	1100	960
Fraction mineralised volatile sludge	%	23	29	30	29

12.4 ANAEROBIC DIGESTION

When activated sludge is kept in an anaerobic environment, specialised bacteria will develop that use the excess sludge as a source of organic matter for fermentative metabolic processes. The end products of the fermentation are mainly methane and carbon dioxide. The overall conversion process of complex organic matter into methane and carbon dioxide can be divided into four steps (Gujer and Zehnder, 1983), as shown in Figure 12.15: hydrolysis, acidification, acetogenesis and methanogenesis.

In an anaerobic digester, the four processes occur simultaneously. When the anaerobic digester performs properly, the conversion of the intermediate products (i.e. the products of the first three steps) is virtually complete, so that the concentrations of these are low at any time. In the hydrolysis process, macro molecules like proteins, poly saccharides and fats that compose the cellular mass of the excess sludge are converted into molecules with a smaller molecular mass that are soluble in water: peptides, saccharides and fatty acids. The hydrolysis- or solubilisation process is carried out by exo-enzymes excreted by fermentative bacteria. Hydrolysis is a relatively slow process and generally limits the rate of the overall anaerobic digestion process.

The second step of the anaerobic digestion process is acidogenesis or acidification, a process that results in the conversion of the hydrolysed products into simple molecules with a low molecular weight, like volatile fatty acids (e.g. acetic-, propionic- and butyric acid), alcohols, aldehydes and gases like CO_2, H_2 and NH_3. Acidification is effected by a very diverse group of bacteria, the majority of which are strictly anaerobic, i.e. the presence of oxidants like oxygen or nitrate is toxic. Luckily for these strict anaerobes, there are always bacteria present that will scavenge the oxygen whenever it is available. The presence of these bacteria is important to remove all oxygen that might be introduced into the system, for instance together with the excess sludge. The acidogenic bacteria are able to metabolise organic material down to a very low pH of around 4.

In the third step, acetogenesis, the products of the acidification are converted into acetic acids, hydrogen and carbon dioxide by acetogenic bacteria. The first three steps of anaerobic digestion are often grouped together as acid fermentation. It is important to note that in the acid fermentation, no organic material is removed from the liquid phase: it is transformed into a form suitable as substrate for the subsequent process of methanogenesis.

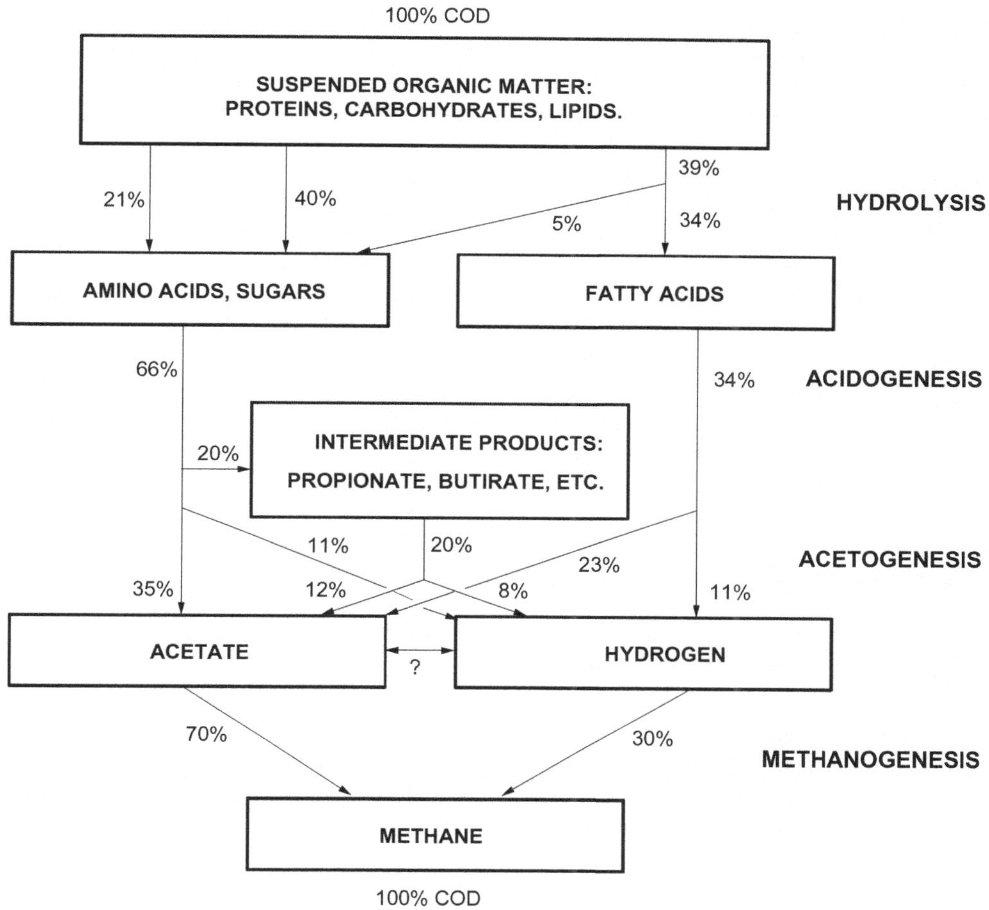

Figure 12.15 Schematic representation of the decomposition of excess activated sludge or other particulate organic material by means of anaerobic digestion

In the final step of the anaerobic digestion process: methanogenesis, the products of the acid fermentation (mainly acetic acid and hydrogen) are converted into CO_2 and CH_4. Only then will organic material be removed, as the produced methane gas will largely desorb from the liquid phase. In each of the four sequential steps, the catabolic reactions described above develop together with anabolic activity. The free energy released in the reactions is partially used for synthesis of the anaerobic bacterial populations. As the energy release from fermentative catabolism is relatively small (refer to Chapter 2), the yield coefficient is much lower than in aerobic processes. Therefore, a large fraction of the digested organic matter is converted into biogas (85 to 95%). In order to maintain an anaerobic sludge with a high metabolic activity, it is necessary to apply favourable environmental conditions. Among these factors the most important ones are temperature, pH, the absence of toxic materials and the availability of nutrients. The methanogens are very sensitive to adverse environmental conditions and for this reason it is always attempted to maintain optimal conditions for these bacteria, which will be discussed in Section 12.4.3.

The fundamental issue in the anaerobic digestion process is that equilibrium has to be maintained between acid- and methanogenic fermentation. As long as this equilibrium exists, the concentration of intermediate products from the conversion of organic material into biogas (many of which are acids) will be low and at any time the conversion of hydrolysed products to the final products is substantially complete. However, if for some reason the equilibrium is disrupted, there will be an accumulation of (acid) intermediates and consequently the pH of the digester will decrease. As methanogenesis requires a pH near the neutral value ($6.5 < pH < 7.5$), the decrease in pH might lead to a reduction of the methane production rate and a further accumulation of acids. As a consequence, the process of anaerobic digestion as a whole may fail due to "souring" of the reactor contents. The digester will only return to activity when the pH of the reactor is restored to a value near neutral pH, which can be effected through the addition of alkalinity.

When anaerobic digestion is compared with other methods of sludge stabilisation (particularly aerobic digestion), the following advantages and disadvantages apply:

(a) Advantages of anaerobic digestion:
  - Substantial improvement in the hygienic quality of the digested sludge because of the efficient removal of pathogens;
  - Production of a useful energy source in the form of biogas, which can be used for power generation (for instance to be used for aeration of the mixed liquor), or can be converted into liquid gas for car fuel, as is currently done by the wastewater companies in São Paulo (Sabesp) and Paraná (Sanepar);
  - Reduction of the mass of excess sludge and production of a stabilised sludge with excellent rheological properties for dewatering.

(b) Disadvantages of anaerobic digestion:
  - The construction costs of an anaerobic digester are considerable. It will be necessary to construct a relatively large unit which is closed to the atmosphere and equipped with complicated devices for feeding, mixing and (in case of low temperatures) heating of the digester contents;
  - The supernatant of the digester contains a high concentration of biodegradable material, principally ammonium. The return of the supernatant to the activated sludge process results in a significant increase of the nitrogen load;
  - Toxic material or operational errors may cause a disruption of the equilibrium between acid- and methanogenic fermentation. Correction of the operational problems is difficult and may require considerable time.

12.4.1 Stoichiometry of anaerobic digestion

There are two important aspects related to the stoichiometry of anaerobic digestion: (I) the effect of digestion on alkalinity and consequently on pH and (II) the potential biogas production and more specifically that of methane. If a structural formula of $C_5H_7O_2N$ is assumed to be representative for secondary sludge, then acid- and methanogenic fermentation can be expressed as:

(a) Acid fermentation

$$\begin{aligned} C_5H_7O_2N + 3H_2O &\rightarrow 2.5\,CH_3COOH + NH_3 \\ &\rightarrow 2.5\,CH_3COO^- + 1.5H^+ + NH_4^+ \end{aligned} \tag{12.61a}$$

(b) Methanogenic fermentation

$$2.5\,CH_3COO^- + 2.5\,H^+ \rightarrow 2.5\,CO_2 + 2.5\,CH_4 \tag{12.61b}$$

(c) Overall anaerobic digestion process

$$\begin{aligned} C_5H_7O_2N + 3\,H_2O + H^+ &\rightarrow 2.5\,CO_2 + 2.5\,CH_4 + NH_4^+ \quad \text{or} \\ C_5H_7O_2N + 4\,H_2O &\rightarrow HCO_3^- + 1.5\,CO_2 + 2.5\,CH_4 + NH_4^+ \end{aligned} \tag{12.61c}$$

Equation (12.61a) shows there is a production of 2.5 moles of acetic acid and l mol of ammonia per "mol" (113 grams) of digested biological sludge. After the dissociation of acetic acid (which is virtually complete at neutral pH) and the reaction of the hydrogen ion with ammonia, there is a net production of 1.5 mol H^+ per mol of sludge. Or, equivalently, there is a consumption of $1.5 \cdot 50 = 75$ g $CaCO_3$ per 113 g of acidified sludge. Hence, during acid fermentation there is an alkalinity consumption of $75/113 = 0.66$ g $CaCO_3$ per gram of acidified sludge.

During methanogenic fermentation, there is consumption of H^+ i.e. there is alkalinity production. With the aid of Eq. (12.61b), the alkalinity production is calculated as $2.5 \cdot 50 = 125$ g $CaCO_3$ per mol of digested sludge. Hence in the overall process (acid- plus methanogenic fermentation) there is an alkalinity production of 50 g $CaCO_3$ per 113 g of digested sludge or $50/113 = 0.44$ g $CaCO_3$ per gram of digested sludge. This increase can be attributed mainly to the mineralisation of organic nitrogen into NH_4^+ and is more than sufficient to maintain the pH in the suitable range for methanogenesis (Van Haandel, 1994).

It is known that the nitrogen content of primary sludge is lower than that of biological sludge. Probably a better approximation of the composition of primary sludge is to consider it as a mixture of proteins, carbohydrates and fats with the average structural formula $(CH_2O)_n$. The following reaction equations can be written:

(a) Acid fermentation

$$(CH_2O)_n \rightarrow n/2\,CH_3COOH \rightarrow n/2\,CH_3COO^- + n/2\,H^+ \tag{12.62a}$$

(b) Methanogenic fermentation

$$n/2\,CH_3COO^- + n/2\,H^+ \rightarrow n/2\,CH_4 + n/2\,CO_2 \tag{12.62b}$$

(c) Overall anaerobic digestion

$$(CH_2O)_n \rightarrow n/2\,CH_4 + n/2\,CO_2 \tag{12.62c}$$

Equations (12.62a and b) indicate that during anaerobic digestion of primary sludge there will be a consumption of 1 mol (50 g $CaCO_3$) of alkalinity per "mol" of primary sludge (60 grams) or $50/60 = 0.83$ g $CaCO_3 \cdot g^{-1}$ VSS. During the methanogenic fermentation, the consumed alkalinity will be recovered and the overall effect of complete anaerobic digestion of primary sludge is that alkalinity remains unchanged. In most situations when primary sludge is digested together with the biological excess sludge, there is always alkalinity production during anaerobic stabilisation of sludge. The value of the alkalinity increase will depend upon the TKN/VSS ratio in the mixed sludge, but it can be shown that for sewage the generation of alkalinity is always more than sufficient to maintain an optimal pH value in the digester (Van Haandel 1994). However, at lower TKN/VSS ratios both the alkalinity production and the buffer capacity in the digester will be small, so that souring may occur more easily. Even in the case of purely proteinic matter, souring is possible if for some reason methanogenic fermentation is inhibited.

The potential of methane generation in the digester can be calculated from stoichiometry by remembering that l gram of CH_4 (with a COD content of 4 grams) will be generated from the digestion of 4 grams organic matter expressed as COD. Thus the mass of produced methane is calculated from the total digested excess sludge production and the composition in terms of the mass fractions of primary and secondary sludge. Under the conditions prevailing in the digester (near to atmospheric pressure, temperature of 30°C), the volume of l mol of methane (16 gram) is about 25 litres or $25/16 = 1.6$ litres $\cdot g^{-1}$ CH_4. Eqs. (12.61 and 12.62) show that for both primary and secondary sludge, an equal number of moles of CO_2 and CH_4 are produced during anaerobic digestion. However, as CO_2 is more soluble in water where it will form bicarbonate, the released biogas is always richer in methane. In practice, the methane percentage in digester biogas is in the range of 55 to 70%.

EXAMPLE 12.6

In an anaerobic digester primary sludge $(CH_2O)_n$ and secondary sludge $(C_5H_7O_2N)$ are digested. The primary excess sludge production $mS_{xv1} = 0.3$ kg COD $\cdot$ kg^{-1} influent COD and the secondary excess sludge production $mS_{xv2} = 0.2$ kg COD $\cdot$ kg^{-1} influent COD. If the sludges have concentrations of 40 and 20 g TSS $\cdot$ l^{-1} respectively and the removal of the volatile solids in the digester is 40%, calculate per unit mass of influent COD:

(1) The fraction of influent COD that is digested in the system;
(2) The alkalinity production and alkalinity in the digester effluent;
(3) The methane and biogas production.

Assume that f_{cv} is equal to 1.5 g COD $\cdot$ g^{-1} VSS for both primary and secondary sludge.

Solution

(1) The excess sludge production is $0.4 \cdot 300 = 120$ g COD of primary sludge and $0.4 \cdot 200 = 80$ g COD of secondary sludge, so that in total $120 + 80 = 200$ g COD is sent to the digester per kg COD influent, or 20% of the influent COD;

(2) During the digestion of secondary sludge, the alkalinity production is 0.44 g $CaCO_3 \cdot g^{-1}$ VSS or 0.29 g $CaCO_3$ per gram digested COD. This results in a production of $0.29 \cdot 80 = 23.5$ g $CaCO_3 \cdot$ kg^{-1} influent COD. During the digestion of primary sludge, no variation in alkalinity is expected. For the assumed primary and secondary sludge concentrations of 40 and 20 g VSS $\cdot$ l^{-1}, the volume of the primary sludge is $120/(1.5 \cdot 40) = 2$ litre $\cdot$ kg^{-1} influent COD and that of the secondary sludge is $80/(1.5 \cdot 20) = 2.7$ litre $\cdot$ kg^{-1} influent COD. Thus, the sludge volume to be digested is $2 + 2.7 = 4.7$ litre $\cdot$ kg^{-1} influent COD. Since alkalinity production was estimated at a minimum of 23.5 g $CaCO_3 \cdot$ kg^{-1} COD, the alkalinity in the digester will increase by at least $23.5/4.7 = 5000$ mg $CaCO_3 \cdot$ l^{-1}, more than enough to establish an adequate pH for methanogenesis (Van Haandel et al., 1994);

(3) The methane production is equal to 25% of the digested COD mass i.e.: $0.25 \cdot 200 = 50$ g $CH_4 \cdot kg^{-1}$ COD or 3.12 mol $CH_4 \cdot kg^{-1}$ applied COD or $3.12 \cdot 25 = 78$ litre $CH_4 \cdot kg^{-1}$ applied COD. The theoretical molar CO_2 production is equal to the molar CH_4 production and will thus be 3.12 moles $CO_2 \cdot kg^{-1}$ applied COD. However, as there is an alkalinity production of 23.5 g $CaCO_3 \cdot kg^{-1}$ influent COD, it may be expected that a stoichiometric fraction (i.e. $23.5/50 = 0.47$ mol CO_2) will be absorbed and remain as bicarbonate in the liquid phase.

If it is assumed that the concentration of dissolved CO_2 is negligible compared to the bicarbonate concentration, then only $3.12-0.47 = 2.65$ mol CO_2 or $2.65 \cdot 25 = 66$ litres of CO_2 will desorb per kg of digested COD. Therefore, the total production of biogas may be expected to be $78 + 66 = 144$ litres per kg digested COD or $0.2 \cdot 174 = 55$ litres per kg influent COD, having a composition of $78/144 = 54\%$ of methane. In practice the biogas would probably be richer in methane because:

- Fats will be present so that more methane is generated than carbon dioxide;
- Alkalinity will be generated due to digestion of secondary sludge, resulting in an increase in the CO_2 conversion into bicarbonate;
- Depending on the pH, the dissolved CO_2 concentration may be a significant fraction of the bicarbonate concentration.

12.4.2 Configurations used for anaerobic digestion

The classical or low rate anaerobic digester is shown schematically in Figure 12.16. In this digester there is a vertical stratification and the following layers can be distinguished:

- Scum layer composed of non-biodegradable or slowly biodegradable material (like fat, oil and grease, surfactants, hair, rags, plastic etc.) floating on the liquid phase. When the temperature is very low, this layer may become very tough and obstruct the release of the produced biogas;
- Supernatant of a liquid phase with a relatively low solids concentration which forms as a result of sedimentation processes;
- Active digestion zone, the part of the anaerobic digester where the actual conversion of organic matter into biogas takes place;
- Stabilised sludge zone: the part of the digester in which the digested sludge accumulates and from where it is discharged for additional treatment or final disposal.

It can be observed that the digester performs two functions at the same time: sludge stabilisation by anaerobic digestion and separation of the digested solids from a supernatant substantially free of suspended solids. In practice the digester will not be very efficient, because the two functions are carried out in the same reactor while the optimal operational conditions for both are very different. For efficient digestion intense mixing is required, to ensure that good contact is established between the anaerobic biomass and the excess sludge. The mixing may be enhanced by recirculation of the produced biogas or by mechanical mixers. In contrast, a condition for efficient settling is a tranquil environment. As the digester is not very efficient in either digestion or phase separation, a large volume is required to obtain a properly stabilised sludge. The long retention time compensates for the inherent inefficiency of the low rate digestion concept.

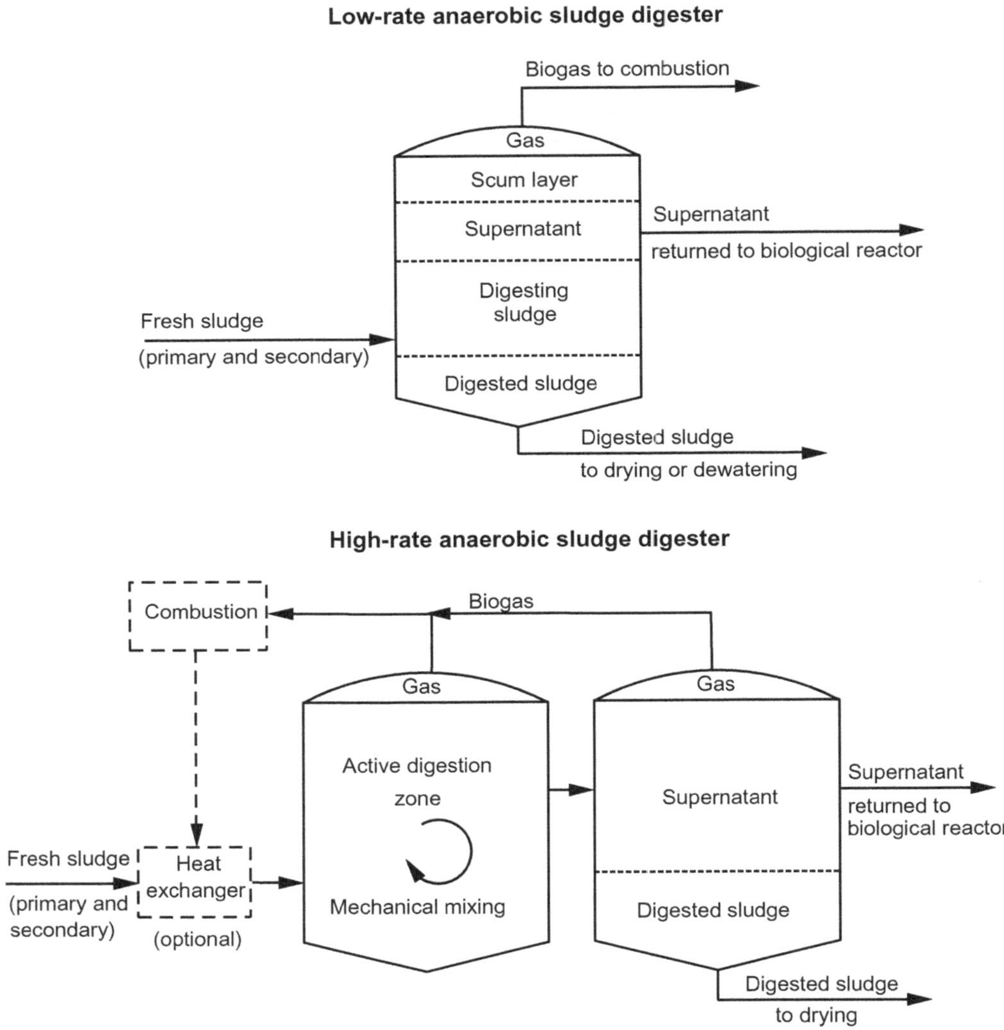

Figure 12.16 Basic configurations of low-rate (top figure) and high-rate (bottom figure) anaerobic sludge digesters. The second (degassing) digester is often much smaller than the main digester

In the 1950s, the high rate anaerobic digester was developed in which the entire volume of the digester is effectively used for digestion and phase separation is carried out in a separate unit specifically constructed for this purpose. An example is shown in Figure 12.33. Usually the second unit is constructed as a secondary digester similar to the two "digesters" in series concept, but this denomination is not very appropriate, because the two stages of anaerobic digestion (acid and methanogenic fermentation) will occur mainly in the first digester, with a possibility of some activity in the second one. Figure 12.33 shows an example of a full-scale anaerobic digester. Apart from phase separation, the secondary digester carries out a second function as a storage tank for digested sludge. The accumulated methanogenic sludge may be recirculated to the primary digester when this is convenient, for example, when there are signs of imminent souring. Furthermore, the secondary digester can be used as a single digester if the primary digester needs to be taken out of operation for maintenance. Thus the presence of the secondary digester improves the flexibility and operational stability of the anaerobic digestion process of excess sludge. However, even in the secondary digester, the liquid-solid separation efficiency may not be satisfactory as the digested sludge tends to float due to the adsorption of biogas bubbles to flocs that rise to the liquid surface and because often a considerable part of the solids in the digested sludge may have very poor settling properties. Recycling of the supernatant with a high concentration of suspended solids and other materials (nutrients) represents an extra load to the aeration tank. For this reason, there is a tendency to add a specific dewatering unit after the secondary digester. Since it may be assumed that the phase separation in such a unit will be better than in the secondary digester, it is to be expected that the recycling of the supernatant will cause less problems. The performance of the high rate digester may be stimulated by several measures.

(a) Continuous feeding

Continuous or semi-continuous introduction of the excess sludge favours stable performance of the digester. Intermittent feeding with a frequency of once per day or less leads to large fluctuations in the composition and concentration of the substrate and may result in a tendency for souring.

(b) Mixing of the primary reactor contents

Mixing favours a homogeneous composition of the mixed liquor in the digester and improves the contact between the anaerobic biomass and the excess sludge to be digested. In addition, possible toxic compounds are quickly diluted over the whole reactor volume, reducing the risk of disruption of the fermentation equilibrium. Furthermore the formation of a scum layer is avoided, thus preventing the danger of serious operational problems. Common methods for mixing are (I) pumping of mixed liquor with an external pump, often combined with external heat exchange, (II) internal mechanical mixers and (III) mixing by recycling of the produced biogas.

(c) Thickening and recycling of digested sludge

Tarp and Melbinger (1967) showed the advantages of recycling digested sludge and mixing it with excess sludge. The mixture can be concentrated to a much higher solids content than would be possible for the excess sludge alone. There is an upper limit to the suspended solids concentration in the digester: above 8 to 10% of suspended solids, the mixing of the digester contents becomes difficult due to the high viscosity of the mixed liquor. Furthermore the concentration of mineralised materials (ammonium, alkalinity) may reach such high values that toxicity becomes a problem, although methanogenic bacteria are able to adapt to very high ammonium concentrations: up to $2500\,mg \cdot l^{-1}$ as demonstrated by Rinzema (1989).

(d) Heating

The metabolic activity of the bacteria in the anaerobic digestion process increases up to the optimal temperature of 35 to 37°C. When heating of the anaerobic digester is applied, the produced methane is usually used as a fuel. Internal or external heat exchangers may be used. The decision as to whether heating of the digester is attractive, depends predominantly on the minimum environmental temperature. If it has a low value, a considerable increase in sludge activity can be expected when heating is applied and consequently the required digester volume will be much smaller. On the other hand, the equipment required to maintain a high and constant temperature is expensive and skilled labour is required to operate the digester. Hence heating is only attractive at low temperatures if proper performance without it is impossible. In tropical and subtropical regions, the environmental temperature in general is sufficiently high to avoid the need for heating of the digester.

12.4.3 Influence of operational parameters

Among the various operational parameters that influence the anaerobic digestion process, the most important ones are the temperature in the digester, pH value and stability, the presence of nutrients in the digester and the presence- and concentration of toxic materials in the digester.

(a) Temperature

In Figure 12.17 the experimental values of the anaerobic digestion rate determined by several researchers are shown plotted as a function of the temperature in the range of 1 to 45°C. The rate increases until a maximum is reached at 35 to 37°C. An increase of temperature beyond 39–40°C leads to a decrease in the rate, while above 42°C the mesophilic biomass decays. However, if the temperature is maintained consistently above 42°C, the digestion rate increases again and reaches an absolute maximum at 53 to 55°C (not shown in Figure 12.17). Thus a mesophilic digestion range to 42°C and a thermophilic range above this value can be distinguished.

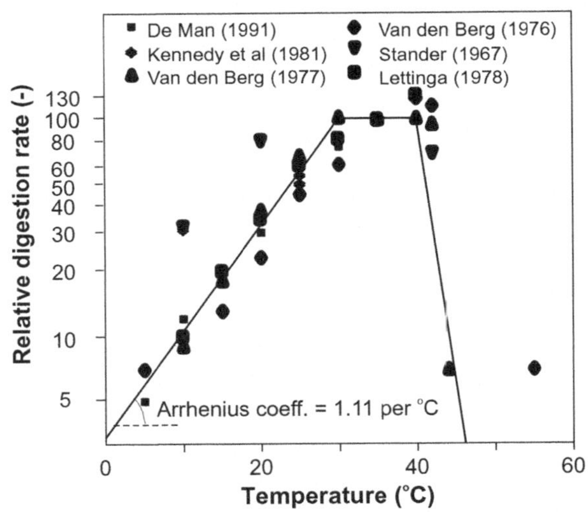

Figure 12.17 Digestion rate as a function of the temperature as determined by different authors

Although the maximum rate of thermophilic digestion is superior to the maximum for mesophilic digestion, up to the present only a few full-scale thermophilic sludge digesters have been built. Apparently the operational difficulties and costs involved in order to maintain the temperature at 53 to 55°C are not compensated by its advantages. In the future, this situation may change because thermophilic digestion has another important advantage: it is very efficient in the reduction of pathogens and produces a stabilised sludge with a good hygienic quality. Another application for thermophilic digestion might high solids wastewater streams, for instance thin stillage from bio-ethanol production, although the TKN concentration is sometimes very high. Ammonia toxicity will increase at higher temperatures (and pH) as the equilibrium between NH_4 and NH_3 will shift towards the toxic undissociated fraction (NH_3).

Temperature does not only influence the rate but also the extent of anaerobic digestion. The classical work by O'Rourke (1968) showed the influence of temperature on the decrease of the volatile solids concentration of primary sludge in an anaerobic digester. In Figure 12.18 the converted volatile solids fraction of primary sludge in a high rate digester is shown as a function of the retention time for different temperatures. It can be noted that the extent of anaerobic digestion tends to increase until the optimal temperature range of 35 to 37°C is reached. Figure 12.18 reveals another important point: the time required to obtain the maximum conversion efficiency increases when the temperature decreases. On the basis of O'Rourkes data presented in Figure 12.18, a diagram may be constructed, linking the minimum retention time for maximum solids conversion to the digestion temperature. Figure 12.19a shows the relationship for the data by O'Rourke as well as other research workers (McCarty, 1964 and Lin et al., 1987). McCarty suggested adopting a safety factor of 2.5 for the retention time in full-scale units.

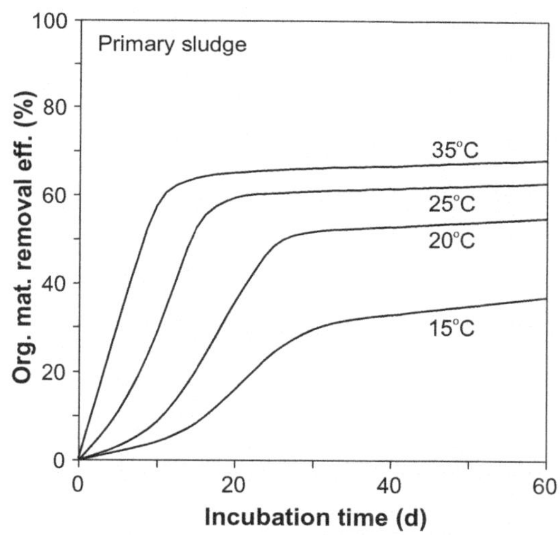

Figure 12.18 Solids conversion efficiency as a function of retention time at different temperatures, O'Rourke (1968)

In Figure 12.19b the values suggested by McCarty (1964) and the WPCF (1979) for the retention time in full-scale high rate digesters are shown plotted as a function of the digestion temperature. Taking into consideration the experimental data and design criteria presented above, the following empiric expression is suggested for the retention time in a high rate anaerobic digester:

$$R_{di} = 20 \cdot 1.1^{(20.T)} + 5(15^\circ C < T < 35^\circ C) \quad (12.63)$$

where:

R_{di} = retention time in the high rate digester (d)
T = temperature (°C) in the mesophilic range

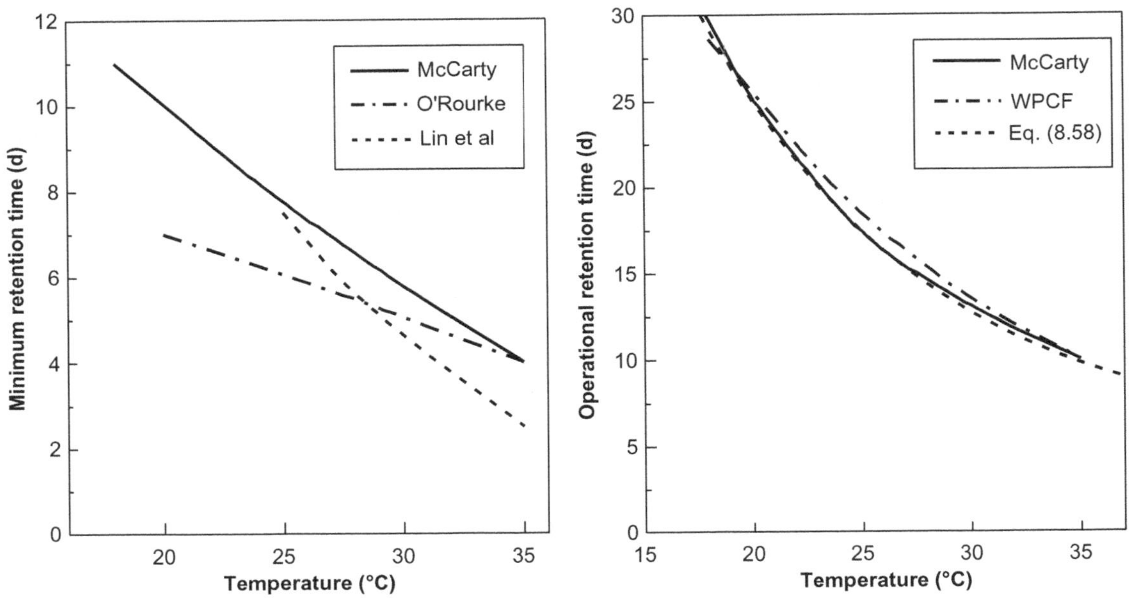

Figure 12.19 Minimum retention time (left) and practical values for full-scale design (right) as function of the operating temperature

The empiric relations of McCarty are valid for both primary and secondary sludge. Equation (12.63) is also indicated in Figure 12.19b.

(b) pH value

Maintaining the pH up or near the neutral value of 7 is a prerequisite for stable operation of the anaerobic digester. Methanogenesis occurs at a very low rate when the pH is outside a narrow pH range from about 6.5 to 7.5. It has been demonstrated in Section 12.4.1 that alkalinity is produced during anaerobic digestion and that bicarbonate is formed from the generated carbon dioxide. Therefore, the carbonic system is always the predominant buffer system in anaerobic digesters and its presence automatically ensures a pH in the appropriate range. However, if for some reason acid fermentation develops more rapidly than methanogenic fermentation, the accumulation of acetic acid (acting as a strong acid in the neutral pH range) will consume the alkalinity and as a consequence, the pH in the mixed liquor may drop to values as low as 4.5 to 5.0. Once a low pH value has been established, the methanogenesis rate will remain very low and the reactor can only return to its normal performance by introducing external alkalinity.

(c) Presence of nutrients

The presence of macronutrients in excess sludge is normally guaranteed in the case of secondary sludge or a mixture of primary and secondary sludge. The biological material to be digested has a high fraction of nitrogen and phosphorus (10 and 2.5% respectively) and a large part of it is mineralised during the anaerobic digestion process and will be available for the anaerobic bacteria in the form of ammonium and phosphate. However, Speece (1985) showed that the availability of micronutrients (especially iron) may be problematic and that sometimes the performance of digesters can be improved considerably when the appropriate salt is added.

(d) Presence of toxic materials

The methanogenic bacteria are very sensitive to the presence of toxic materials. In the case of sewage treatment, the presence of toxic materials is unlikely in primary sludge and almost impossible in secondary sludge. In the case of industrial wastes, the presence of toxins is a real danger especially in primary sludge. The toxic components can be divided into three classes according to their nature:

- Toxicity related to pH;
- Immediate toxicity even at low concentrations;
- Gradually increasing toxicity with increasing concentration of the material.

The toxicity of some weak acids and bases is related to pH, which determines the degree of dissociation. The components most likely to be found in sludge digesters are:

- Sulphides originating from the mineralisation of sulphur containing proteins and from the reduction of sulphate in the anaerobic digester, a process that develops parallel to anaerobic digestion. Normally the sulphide concentration will remain below 50 to 100 $mg \cdot l^{-1}$ and in that case it will not inhibit methanogenic activity seriously. Possible temporary problems may be overcome by the addition of ferrous salts (but not $FeSO_4$!), resulting in precipitation of FeS;
- Volatile fatty acids, which only have a significant concentration if there is a predominance of acid fermentation due to equilibrium disruption of the anaerobic digestion processes. The ionised form is not toxic. The unionised acids are toxic due to their diffusion through the cell membrane and subsequent dissociation in the bacterial cell. The unionised acids may be converted into ions by adding alkalinity to increase pH;
- Ammonium. Only the unionised form is toxic, but the anaerobic populations can adapt to very high concentrations. Ammonia mainly originates from the mineralisation of proteinic material and amounts to about 10% of the mass of digested secondary sludge. It is always possible to adapt the anaerobic sludge to the ammonia concentrations normally found in excess sludge digesters, at least when municipal excess sludge is treated. However, at high temperatures and/or high pH, the fraction of ammonium present in the undissociated and toxic form (NH_3) will rapidly increase. Hence ammonia toxicity is a significant problem in thermophilic treatment, especially for waste streams with a high content of TKN, such as thin stillage from ethanol production.

The components that have an immediate toxic effect are industrial products like biocides and chlorinated organics. These products may be introduced in the digester together with primary sludge. A distinction can be made between material with an irreversible action (toxic materials) and those that only have an effect as long as they are present in the reactor (inhibitory materials). It is possible that after the removal of an inhibitory material, the sludge requires a long time to recover its original activity. A special case is the presence of dissolved oxygen, which is toxic for the methanogens. Normally the introduced dissolved oxygen concentration is low and will be removed rapidly by facultative acidifying bacteria present in the digester.

Heavy metals such as Hg, Cd, Zn, Cr, Ni etc. form another special group. Their ions are very toxic for methanogens, but in the anaerobic digester their effect is often neutralised by sulphide that forms insoluble salts with these metal ions. In the class of concentration depending toxic materials, calcium and sodium may be important, especially if these are added to the reactor together with alkalinity (e.g. in the form of lime or caustic soda). Methanogenic bacteria can adapt very well to high concentrations of alkali and earth alkali metal ions, provided the increase of the concentration is gradual.

12.4.4 Performance of the high rate anaerobic digester

When the performance of the high rate anaerobic digester is discussed, the following aspects are of interest:

- Removal efficiency of the volatile suspended solids present in the excess sludge;
- Biogas production;
- Stabilised excess sludge production and composition;
- Nutrient release and recycling to the aeration tank.

12.4.4.1 Removal efficiency of volatile suspended solids

The removal efficiency of the volatile solids concentration in an anaerobic digester depends basically on three factors: the digestion temperature, the retention time in the digester and the nature of the excess sludge to be digested. In Figure 12.18 it can be observed that anaerobic digestion is more complete as the digestion temperature approaches the optimal range for mesophilic bacteria of 35 to 37°C. However, even at the optimal temperature, anaerobic digestion will not be complete: the maximum solids removal efficiency does not exceed 55 to 60% in the case of primary sludge and is even lower for secondary sludge.

Araújo and Van Haandel (1998) operated anaerobic digesters with secondary sludges of different compositions. The active sludge concentration was varied between 17 and 83% of the volatile sludge concentration. It was observed that in completely mixed anaerobic digesters, operated in steady state at 20 days retention time and a temperature of 25°C, the removal efficiency of the volatile solids depended on the active sludge fraction. The removal efficiency of the active sludge fraction was found to be equal to the efficiency found by O'Rourke (1969) for primary sludge. However, the reduction of the inactive sludge was much smaller, only 15% at 25°C. Based on these experimental results, the following empiric relationship is suggested to estimate the maximum removal efficiency of volatile solids as a function of temperature and sludge composition:

(a) For primary sludge and the active fraction in secondary sludge:

$$\eta_{dp} = (0.67 \cdot T + 36)/100 \tag{12.64}$$

(b) For inactive secondary sludge:

$$\eta_{dn} = (0.19 \cdot T + 10)/100 \tag{12.65}$$

where:

η_{dp} = removal efficiency for digestion of the primary or active sludge mass (%)
η_{dn} = removal efficiency for digestion of the inactive sludge mass (%)
T = temperature in °C (< 37°C)

12.4.4.2 Biogas production

The biogas production is directly related to the removal of volatile solids. Knowing that the average COD/VSS ratio of excess sludge (f_{cv}) is equal to 1.5 kg COd $\cdot$ kg^{-1} VSS and that the COD content of methane is 4 kg COd $\cdot$ kg^{-1} CH_4, the production of methane is calculated as $1.5/4 = 0.375$ kg $CH_4 \cdot kg^{-1}$ VSS.

Hence for a primary sludge production of mE_{v1} kg VSS $\cdot$ kg^{-1} COD and a secondary sludge production of mE_{v2} kg VSS $\cdot$ kg^{-1} COD, with an active fraction f_{av} and an inactive fraction ($1-f_{av}$), the digested influent COD fraction can be expressed as:

$$mS_d = f_{cv} \cdot [\eta_{dp} \cdot (mE_{v1} + f_{av} \cdot mE_{v2}) + \eta_{dn} \cdot (1 - f_{av}) \cdot mE_{v2}] \qquad (12.66)$$

Using the theoretical COD content of methane (4 g COD $\cdot$ g^{-1} CH_4), the specific methane production can be calculated as:

$$mM_d = mS_d/4 = f_{cv}/4 \cdot [\eta_{dp} \cdot (mE_{v1} + f_{av} \cdot mE_{v2}) + \eta_{dn} \cdot (1 - f_{av} \cdot mE_{v2})] \qquad (12.67)$$

where:

mS_d = digested influent COD fraction
mM_d = mass of methane produced per unit mass applied COD (mg $CH_4 \cdot mg^{-1}$ COD)

The volume of methane produced depends on the temperature and pressure of the biogas. Under normalized conditions (i.e. 0°C or 273°K and 1 atm), the gas volume of l mol of methane is 22.4 litres. Therefore, the volume of methane produced per kg digested sludge (assuming $f_{cv} = 1.5$ kg COD) is $1.5/4 \cdot 22.4/16 = 0.525$ Nm^3 $CH_4 \cdot kg^{-1}$ VSS. Or equivalently, per kg COD converted $0.525/1.5 = 0.35$ Nm^3 of CH_4 is produced. The (normalized) methane gas volume produced per unit of influent COD mass can be calculated directly as:

$$mQ_{ch4} = 0.525 \cdot [\eta_{dp} \cdot (mE_{v1} + f_{av} \cdot mE_{v2}) + \eta_{dn} \cdot (1 - f_{av}) \cdot mE_{v2}] \qquad (12.68)$$

Using the ideal gas law, the normalized methane gas volume production can be converted into actual gas volume. The molar gas volume under actual conditions can be calculated as:

$$V_{tp} = [1013/(1013 + p) \cdot (273.15 + T)/273.15] \cdot 22.4 \qquad (12.69)$$

V_{tp} = molar gas volume at actual pressure and temperature
p = actual gas pressure (typically between 20 to 60 mbarg)
T = actual gas temperature (equal to reactor temperature)

Table 12.6 conveniently lists the values of the conversion factors relating normal cubic metres to actual cubic metres, for typical values of digester pressure and temperature. As can be observed, the actual volumetric methane production is typically between 1 to 10% higher than the normalized methane production.

Table 12.6 Conversion factors from Nm^3 to actual m^3 of gas (at P, T)

Gas pressure (mbarg)	Temperature			
	20°C	25°C	30°C	35°C
20	1.05	1.07	1.09	1.11
30	1.04	1.06	1.08	1.10
40	1.03	1.05	1.07	1.09
50	1.02	1.04	1.06	1.08
60	1.01	1.03	1.05	1.07

The produced biogas volume depends on the methane fraction of the gas, which in the case of sludge digesters typically ranges from 55 to 70%. The biogas composition is dependent on the incoming excess sludge composition, the presence of sulphates, which will be converted into H_2S and hence reduce the mass of COD available to methanogic bacteria, and the pH, which will influence the CO_2 content of the biogas. Typically the CO_2 content varies between 20–35%. Other gas constituents are H_2S ($<$ 0.1 to 3%), water (3–4%) and some nitrogen ($<$ 1–2%). The latter component may be produced from denitrification. For a biogas containing a methane fraction η_{ch4}, the volume of produced biogas per unit of influent COD mass will be:

$$\begin{aligned} mQ_{bg} &= mQ_{ch4}/\eta_{ch4} \\ &= 0.525/\eta_{ch4} \cdot [\eta_{dp} \cdot (mE_{v1} + f_{av} \cdot mE_{v2}) + \eta_{dn} \cdot (1 - f_{av}) \cdot mE_{v2}] \end{aligned} \qquad (12.70)$$

where mQ_{ch4} and mQ_{bg} are the volume of methane respectively biogas produced per unit mass of applied COD ($Nm^3 \cdot$ kg COD)

12.4.4.3 Energy generation in anaerobic sludge digesters

The potential for energy production in the form of methane is of significant practical importance. It is interesting to compare the potential energy production to the energy demand for aeration. For most aerators operating under process conditions, the required energy for aeration is in the order of 1.0 to 1.6 $kWh \cdot kg^{-1}$ O_2 transferred (refer also to Chapter 4). The oxygen demand depends on the influent composition (COD concentration and -composition, nitrogen concentration) and the sludge age. These factors determine the fraction of influent COD that will be oxidised in the activated sludge process and the extent of nitrification that will take place. For an activated sludge process with nitrogen removal one has:

$$MO_t = MO_c + MO_n - MO_{eq} \qquad (5.13)$$

If nitrification and (therefore) denitrification does not take place, the energy demand can be linked directly to the oxidised COD fraction:

$$mS_o = (1 - f_{ns} - f_{np}) \cdot [1 - f_{cv} \cdot Y + f_{cv} \cdot (1 - f) \cdot C_r] \qquad (3.43)$$

The chemical energy in the produced methane can be determined from the combustion heat: 12,000 kCal or 50,400 $kJ \cdot kg^{-1}$ CH_4 (equivalent to 50,400/3600 = 14 $kWh \cdot kg^{-1}$ CH_4). However, when methane is used for power generation not all the available chemical energy can be harvested. When the gas is used in a boiler, the thermal efficiency can reach values up to 80–90%. If electrical power production is considered, then typically the energy conversion efficiency for conventional gas motors ranges from 35 to 42%. Note that this gross figure should be corrected for the power consumption of the unit itself (2–4%), for example from ventilators, air coolers etc. If combined heat and power (CHP) units are used, a large part of the released heat (40–50%) can be recovered in the form of steam or hot water. Figure 12.34 shows an example of a full-scale biogas motor, in this case without CHP unit.

Assuming a net energy conversion efficiency of η_{el}, the electrical power production potential from the production of methane can be estimated as:

$$mP_{el} = \eta_{el} \cdot 14 \cdot mM_d \quad \text{and} \quad P_{el} = \eta_{el} \cdot 14 \cdot MM_d/24 \tag{12.71}$$

Knowing that 1 kg CH_4 originates from the digestion of 4 kg of COD or $4/1.5 = 2.677$ kg VSS, the power production potential from anaerobic sludge digestion can also be expressed as:

$$mP_{el} = \eta_{el} \cdot 14/2.677 \cdot mE_d = \eta_{el} \cdot 5.25 \cdot mE_d \quad \text{and} \quad P_{el} = \eta_{el} \cdot 5.25 \cdot ME_d/24 \tag{12.72}$$

where:

P_{el} = electrical power production potential (kW)
η_{el} = energy conversion efficiency (30–35%)
mP_{el} = energy production potential in kWh per kg applied COD
mE_d = digested sludge mass per unit mass applied COD
ME_d = digested sludge mass (kg VSS $\cdot$ d^{-1})
MM_d = methane production (kg $CH_4 \cdot d^{-1}$)

12.4.4.4 Solids destruction and stabilised excess sludge production

The production of stabilised sludge can be calculated from the difference between the total mass of excess sludge and the digested sludge mass:

$$mE_d = \eta_{dp} \cdot (mE_{v1} + f_{av} \cdot mE_{v2}) + \eta_{dn} \cdot (1 - f_{av}) \cdot mE_{v2} \tag{12.73}$$

$$\begin{aligned} mE_{ve} &= mE_{v1} + mE_{v2} - mE_d \\ &= mE_{v1} + mE_{v2} - (\eta_{dp} \cdot (mE_{v1} + f_{av} \cdot mE_{v2} + \eta_{dn} \cdot (1 - f_{av}) \cdot mE_{v2}) \\ &= (1 - \eta_{dp}) \cdot (mE_{v1} + f_{av} \cdot mE_{v2}) + (1 - \eta_{dn}) \cdot (1 - f_{av}) \cdot mE_{v2} \end{aligned} \tag{12.74}$$

$$\begin{aligned} mE_{te} &= mE_{t1} + mE_{t2} - mE_d \\ &= (1 - \eta_{dp}) \cdot (mE_{v1} + f_{av} \cdot mE_{v2})/f_v + (1 - \eta_{dn}) \cdot (1 - f_{av}) \cdot mE_{v2}/f_v \end{aligned} \tag{12.75}$$

where:

mE_d = digested sludge mass per unit mass of applied COD ($= mS_d/f_{cv}$)
mE_{ve} = volatile stabilised sludge mass production per unit mass of applied COD
mE_{te} = stabilised sludge mass production per unit mass of applied COD

In Eq. (12.75) it is tacitly assumed that there is no net solubilisation in the digester. In reality, some solubilisation will occur as the soluble COD concentration of the digester effluent is higher than that of the liquid phase of the raw sludge. However, under normal conditions most of the liquefied material will be digested.

In general, the settleability of anaerobic digested sludge is much better compared to that of fresh (undigested) sludge. The values of the settling constants k and v_0 of digested sludge are in the order of 0.20 to 0.25 $l \cdot g^{-1}$ and 200 to 300 $m \cdot d^{-1}$ respectively. However, a problem is that during digestion a large number of microscopic sludge flocs are formed which practically do not settle. Therefore, the COD value of the liquid effluent is relatively high, between 500 and 1500 $mg \cdot l^{-1}$, of which a large fraction is not biodegradable.

12.4.4.5 Nutrient balance in the anaerobic digester

The most important characteristic of the effluent of the anaerobic digester is the high nitrogen concentration, predominantly in the form of ammonium. This ammonium originates from the ammonification of nitrogen released upon upon destruction of organic suspended solids. While phosphorus is also released, predominantly in the form of PO_4^{3-}, this generally precipitates together with the counter ions released from the bacterial cells (Ca^{2+}, Mg^{2+}, Fe^{2+}, Fe^{3+} and Al^{3+}). The precipitation process is enhanced because of the high pH value and the high carbonate concentration in the digestion water. If it is assumed that the release of ammonium from digested secondary sludge amounts to a fraction f_n per digested unit mass of excess sludge and that no nitrogen is released during the digestion of primary sludge, the release of ammonium during digestion can be estimated as:

$$mN_{ld} = f_n \cdot mE_d = f_n \cdot [\eta_{dp} \cdot f_{av} + \eta_{dn} \cdot (1 - f_{av})] \cdot mE_{v2} \quad (12.76)$$

mN_{ld} = nitrogen production in the digester per unit mass of applied COD

This will constitute an additional nitrogen load to the activated sludge system, which can also be expressed in $mg\ N \cdot l^{-1}$ influent:

$$N_{ld} = mN_{ld}/Q_i = f_n \cdot mE_d/Q_i \quad (12.77)$$

Finally, the nutrient discharge with the stabilised excess sludge is given as:

$$mN_{le} = f_n \cdot mE_{ve} \quad \text{or} \quad N_{le} = f_n \cdot ME_{ve}/Q_i \quad (12.78)$$

$$mP_{le} = f_p \cdot mE_{ve} \quad \text{or} \quad P_{le} = f_p \cdot ME_{ve}/Q_i \quad (12.79)$$

In nutrient deficient systems this is the minimal concentration of nitrogen and phosphorus that needs to be present in the influent to prevent nutrient deficiency. It can be concluded that there is a very large advantage for the anaerobic digestion process when compared with aerobic digestion. This is particularly so if there is no need to heat the reactor contents of the anaerobic reactor, as will be the case in tropical and subtropical regions. Under those circumstances, the energy consumption in a process with aerobic digestion can easily be more than twice the demand when anaerobic digestion is applied. It is concluded that anaerobic digestion will reduce the operational costs of activated sludge processes and should always be applied, unless it is technically impossible, as may be the case when industrial wastewaters with toxic components are treated.

EXAMPLE 12.7

For an activated sludge process operating at a sludge age of 6 days and a temperature of 20°C, determine the following parameters for a system configuration with and without primary settling:

(a) The COD fractions: (I) digested, (II) in the effluent, (III) oxidised and (IV) in the stabilised sludge;
(b) The methane production per unit mass of applied COD;
(c) Calculate the required energy demand for the oxidation of organic matter and the potential for energy generation from sludge digestion;
(d) The ammonia concentration in the digester effluent.

To make the necessary calculations the following data are given:

- $f_{ns} = f_{np} = 0.1$ for the raw sewage;
- In the case of primary settling, it is assumed that 80% of the particulate non-biodegradable material and 30% of the total influent COD will be removed;
- Oxygen transfer efficiency $OT_a = 1\ kWh \cdot kg^{-1}\ O_2$;
- Efficiency of energy generation $\eta_{el} = 33\%$.

Solution

(A) Without primary settling

Calculate the active (mE_{vxa}) and inactive excess sludge production (mE_{vnxa}). The active excess sludge production is given by:

$$mE_{vxa} = (1 - f_{ns} - f_{np}) \cdot C_r/R_s = 0.8 \cdot 0.45/(1 + 6 \cdot 0.24) = 0.15\,mg\,VSS \cdot mg^{-1}COD$$

The volatile excess sludge production (comprising both the active and inactive fractions) is calculated as:

$$mE_v = mE_{vxa} \cdot (1 + f \cdot b_h \cdot R_s) + f_{np}/f_{cv} = 0.26\,mg\,VSS \cdot mg^{-1}COD$$

Hence the production of inactive sludge $mE_{vnxa} = mE_v - mE_{vxa} = 0.11$ mg VSS · mg $^{-1}$ COD.

The fraction of the influent COD digested in the process is calculated from Eq. (12.66):

$$\begin{aligned} mS_d &= f_{cv} \cdot (\eta_{dp} \cdot mE_{xa} + \eta_{dn} \cdot mE_{nxa}) \\ &= 1.5 \cdot ((0.67 \cdot 20 + 36)/100 \cdot 0.15 + (0.19 \cdot 20 + 10)/100 \cdot 0.11) \\ &= 0.093\,mg\,COD \cdot mg^{-1}\,COD \end{aligned}$$

The influent COD fraction discharged into the effluent is equal to the non-biodegradable and soluble influent fraction, which is given as $f_{ns} = 0.10$. The oxidised influent COD fraction is calculated as:

$$mS_o = (1 - f_{ns} - f_{np}) \cdot [1 - f_{cv} \cdot Y + f_{cv} \cdot b_h \cdot (1 - f) \cdot C_r] = 0.8 \cdot (0.33 + 0.32) = 0.52$$

Finally the influent COD fraction discharged as stabilised sludge can be calculated as:

$$mS_{xve} = 1.0 - mS_{te} - mS_o - mS_d = 1.0 - 0.1 - 0.52 - 0.093 = 0.29$$

The fraction of solids that is converted in the digester is given by:

$$\eta_{xv} = (mS_d/f_{cv})/mE_{xv} = (0.093/1.5)/0.26 = 0.23$$

This percentage can also be calculated when it is considered that the COD fraction discharged as biological excess sludge from the aeration tank is given by $mS_{xv} = 1-0.1-0.52 = 0.38$ so that the converted fraction is $mS_d/mS_{xv} = 0.093/0.38 = 0.23$. The methane production per unit mass of influent COD is calculated with the aid of Eq. (12.67):

$$mM_d = mS_d/4 = 0.093/4 = 23\,g\,CH_4 \cdot kg^{-1}\,COD$$
$$mS_o = 0.52 \quad \rightarrow \quad energydemand = 0.52\,kWh \cdot kg^{-1}\,COD\ oxidised$$
$$mS_d = 0.093 \quad \rightarrow \quad mE_d = 0.093/1.5 = 0.062\ g\,VSS \cdot g^{-1}\ COD$$
$$mP_{el} = \eta_{el} \cdot 5.25 \cdot mE_d = 0.33 \cdot 5.25 \cdot 0.062 = 0.107\,kWh \cdot kg^{-1}\,COD$$

The ammonium production in the anaerobic digester can be estimated by Eq. (12.76):

$$mN_{ld} = f_n \cdot (\eta_{dp} \cdot f_{av} \cdot mE_v + \eta_{dn} \cdot (1 - f_{av}) \cdot mE_v)$$
$$= 0.1 \cdot (0.52 \cdot 0.09 + 0.15 \cdot 0.11) + 6.3\,g\,N \cdot kg^{-1}\,COD$$

The ammonia release from anaerobic digestion represents 5 to 10% of the influent TKN mass to the activated sludge system and must be taken into consideration when designing the activated sludge process for nitrogen removal.

(B) With primary settling

If a removal efficiency of 30% of the influent COD is assumed in the primary settler, the primary sludge production can be estimated as:

$$mE_{v1} = 0.3/f_{cv} = 0.2\,g\,VSS \cdot g^{-1}\,COD$$

In the raw sewage, the fractions of soluble and particulate organic material f_{ns} and f_{np} are equal to 0.10. As S_{nsi} is not settleable (dissolved matter) and 80% of S_{npi} and 30% of S_{ti} are removed during primary settling, the non-biodegradable fractions in the settled sewage are adapted after settling to:

$$f'_{ns} = 0.1/0.7 = 0.143 \quad \text{and} \quad f'_{np} = 0.02/0.7 = 0.029$$

The active sludge production is calculated as:

$$mE_{vxa} = 0.7 \cdot (1 - 0.143 - 0.029) \cdot C_r/R_s = 0.7 \cdot 0.83 \cdot 0.9375/6 = 0.11\,mg\,X_a \cdot mg^{-1}\,COD$$

The volatile biological sludge production is given by:

$$mE_v = mE_{xa} \cdot (1 + f \cdot b_h \cdot R_s) + f_{np}/f_{cv} = 0.14 + 0.029/1.5 = 0.17$$

Therefore the inactive secondary sludge production is expressed as:

$$mE_{vnxa} = mE_v - mE_{vxa} = 0.17 - 0.11 = 0.06$$

The active fraction f_{av} is equal to $(0.17 - 0.06)/0.17 = 0.65$

Now by using Eq. (12.66) one has:

$$mS_d = 1.5 \cdot [0.52 \cdot (mE_{v1} + f_{av} \cdot mE_{v2}) + 0.15 \cdot (1 - f_{av}) \cdot mE_{v2}]$$
$$= 1.5 \cdot (0.52 \cdot 0.31 + 0.15 \cdot 0.05) = 0.25$$

The COD fraction in the effluent is not affected by primary settling so that:

$$mS_{te} = f_{ns} = 0.1$$

The oxidised fraction after primary settling is expressed as:

$$mS_o = 0.7 \cdot (1 - f_{ns} - f_{np}) \cdot [1 - f_{cv} \cdot Y + f_{cv} \cdot b_h \cdot (1 - f) \cdot C_r]$$
$$= 0.7 \cdot 0.83 \cdot (0.33 + 0.32) = 0.36$$

The COD fraction discharged as stabilised sludge is:

$$mS_{xve} = 1 - mS_{te} - mS_o - S_d = 1 - 0.1 - 0.36 - 0.25 = 0.29$$

The methane production is determined as in Eq. (12.67):

$$mM_d = mS_d/4 = 0.25/4 = 62\,g\,CH_4 \cdot kg^{-1}\,COD$$
$$mS_o = 0.36 \rightarrow \text{energy demand} = 0.360\,kWh \cdot kg^{-1}\,COD\,\text{oxidised}$$

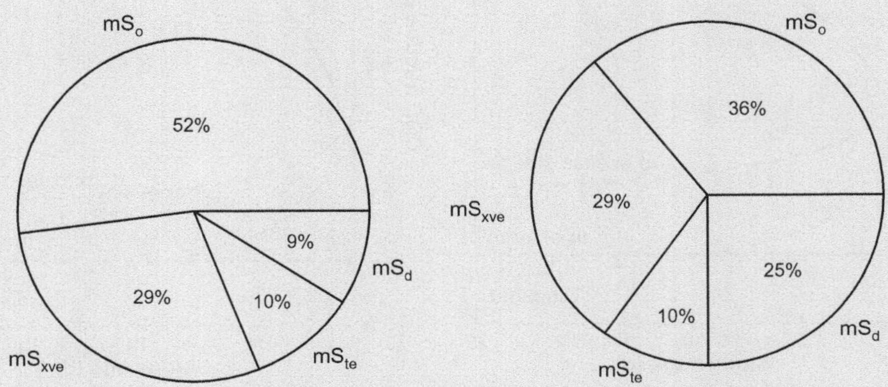

Figure 12.20 Division of the influent COD over fractions mS_e, mS_o, mS_{xv} and mS_d in Example 12.7, without (left) and with primary settling (right)

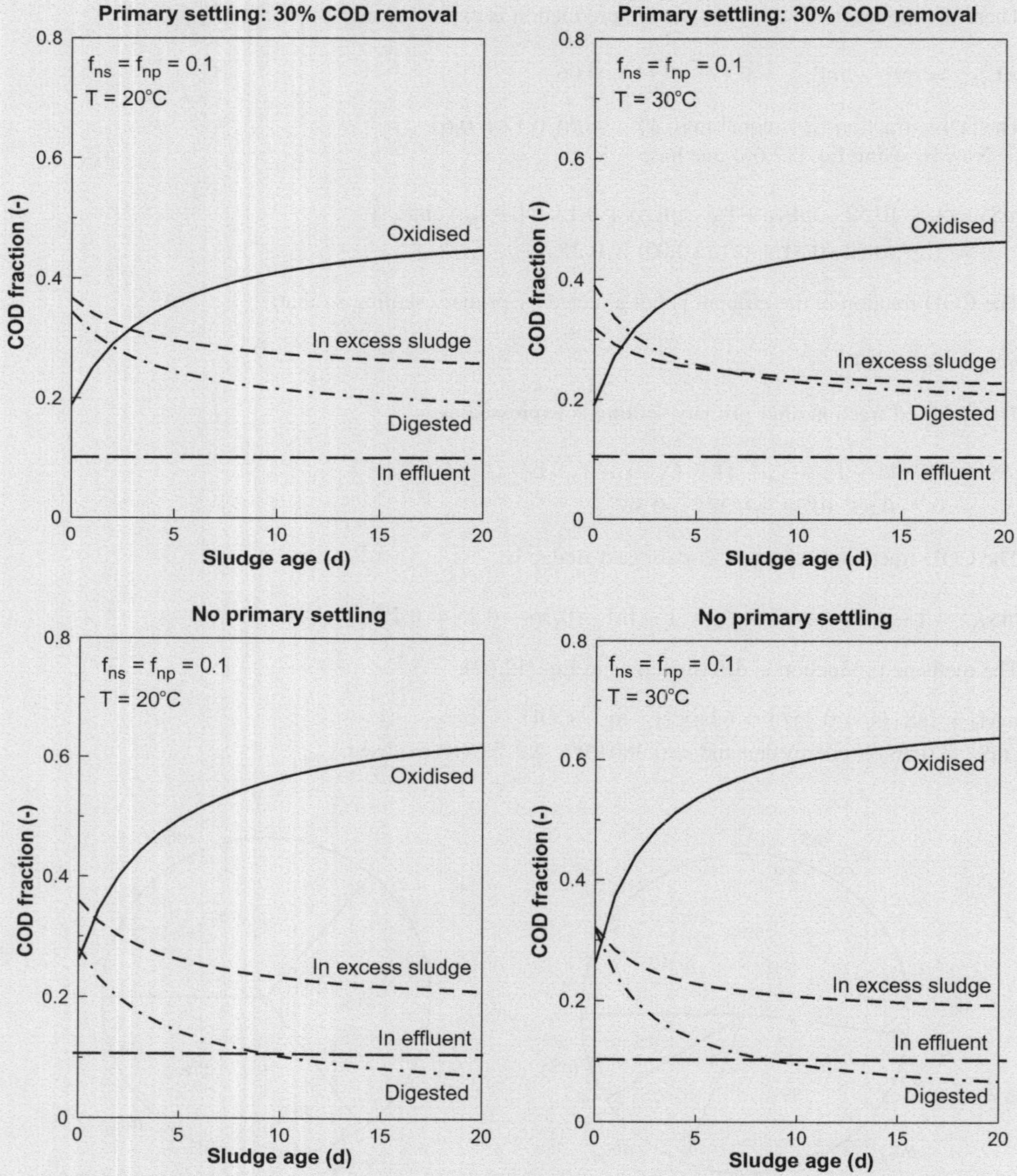

Figure 12.21 Division of the influent COD in Example 12.7 over fractions mS_e, mS_o, mS_{xve} and mS_d as function of the sludge age

$$mS_d = 0.25 \quad \rightarrow \quad mE_d = 0.25/1.5 = 0.167\,g\,VSS \cdot g^{-1}\,COD$$
$$mP_{el} = 0.33 \cdot 5.25 \cdot 0.167 = 0.289\,kWh \cdot kg^{-1}\,COD.$$

The ammonia production in the digester is estimated as in Eq. (12.76):

$$\begin{aligned} mN_{ld} &= f_n \cdot (\eta_{dp} \cdot mE_{vxa} + \eta_{dn} \cdot E_{vxna}) \\ &= 0.1 \cdot (0.52 \cdot 0.11 + 0.15 \cdot 0.05) = 6.5\,g\,N \cdot kg^{-1}\,COD \end{aligned}$$

The fraction of solids removal in the digester is now given by:

$$\eta_{xv} = (mS_d/f_{cv})/(mE_{v1} + mE_{v2}) = (0.25/1.5)/(0.16 + 0.2) = 0.46$$

Figure 12.20 graphically shows the division of the four COD fractions over (1) the effluent, (2) oxidised sludge, (3) digested sludge, and (4) the stabilised sludge, both for the case of primary settling and for raw sewage treatment. It can be noted that a considerable part of the removed organic material is degraded via the anaerobic pathway, especially if primary settling is applied.

The fraction of anaerobic degraded organic material increases at lower values of the sludge age, because the excess sludge will have a higher active fraction. In Figure 12.21 the COD fractions in the effluent and in the stabilised, oxidised and digested sludge are shown plotted as functions of the sludge age for temperatures of 20°C (left) and 30°C (right), maintaining the same assumptions as in the preceding example. Figure 12.21 shows that when primary settling is applied and the process is operated at a short sludge age, the influent COD fraction digested in the process may exceed the oxidised fraction, so that the process as a whole actually becomes predominantly anaerobic.

From the results of Example 12.7, it is clear that the potential for energy production represents a significant fraction of the energy demand at an activated sludge plant. In the case of Example 12.7, the energy that could be produced varies between 107/520 = 21% (no primary settling) and 289/360 = 80% (with primary settling) of the energy demand. If a shorter sludge age is applied and more efficient energy conversion techniques (gas turbines) are used, it is possible to operate the activated sludge process without external energy consumption at all: the chemical energy of the influent organic matter would then be sufficient to supply the energy requirements of the system.

12.4.5 Design and optimisation of anaerobic digesters

The most important design parameter of an anaerobic digester is the operational temperature. In accordance with Eq. (12.63), the retention time in the digester depends on this parameter, whereas Eqs. (12.64 and 12.65) show that the volatile solids removal efficiency is also affected. For design of unheated digesters, the average temperature of the coldest month may be taken. Once the operational temperature is known, the retention time can be calculated. With the daily (thickened) excess sludge flow rate, the required digester volume is determined. Alternatively, if the costs per m^3 unit volume are known, then it is also possible to calculate the minimum total construction costs of the system thickener – digester (refer to Chapter 14);

$$MC_{thdi} = C_{th} \cdot V_{th} + C_{di} \cdot V_{di} \tag{12.80}$$

In the above example, in the case of primary settling the total volume equals $0.23 + 0.51 = 0.74\ m^3 \cdot kg^{-1}\ COD \cdot d^{-1}$, which is slightly smaller than in the configuration without primary settling: $0.51 + 0.35 = 0.86\ m^3 \cdot kg^{-1}\ COD \cdot d^{-1}$. However, it was assumed that the sludge age in the aeration tank would be the same for both cases: i.e. $R_s = 5$ d.

EXAMPLE 12.8

Design an anaerobic digester to process the excess sludge produced from an activated sludge process treating sewage ($f_{ns} = f_{np} = 0.1$ and $f_v = 0.75$) at a sludge age of 5 days (with and without primary settling). Thickening to 25 g TSS $\cdot\ l^{-1}$ is assumed, while the operational temperature is 20°C, so that the retention time is calculated using Eq. (12.63):

$$R_{di} = 20 \cdot 1.1^{(20-T)} + 5 = 20 + 5 = 25 \text{ days}$$

Solution

(A) Without primary settling

It can be calculated that in the case of raw sewage treatment ($R_s = 5$ days), the excess sludge production is 0.26 kg VSS $\cdot\ kg^{-1}$ COD and 0.35 kg TSS $\cdot\ kg^{-1}$ COD. Hence for the thickened sludge concentration of 25 g TSS $\cdot\ l^{-1}$ the excess sludge flow is given by:

$$mq_{th} = 0.35/25 = 0.014\ m^3 \cdot kg^{-1}\ COD$$

Therefore for the calculated retention time of 25 days the volume of the digester will be:

$$v_{di} = R_{di} \cdot mq_{th} = 0.014 \cdot 25 = 0.35\ m^3 \cdot kg^{-1}\ COD \cdot d$$

In comparison, for a typical sludge concentration in the aeration tank of 3.5 g TSS $\cdot\ l^{-1}$, the required reactor volume is:

$$v_r = mX_t/X_t = R_s \cdot mE_t/X_t = 0.35 \cdot 5/3.5 = 0.5\ m^3 \cdot kg^{-1}\ COD \cdot d^{-1}$$

It is concluded that under the given circumstances the digester volume is 0.35/0.50 = 70% of the aeration tank volume. For activated sludge systems designed for nutrient removal, this ratio would be much lower as the operational sludge age in the activated sludge process will be higher, which reduces excess sludge production.

(B) With primary settling

In this case, a production of 0.22 g VSS $\cdot\ g^{-1}$ COD of primary sludge and 0.16 g VSS $\cdot\ g^{-1}$ COD of secondary sludge are calculated, resulting in a total excess sludge production of:

$$mE_t = (mE_{v1} + mE_{v2})/f_v = (0.22 + 0.16)/0.75 = 0.51\ g\,TSS \cdot g^{-1}$$

Hence after thickening the flow of sludge will be:

$$mq_{th} = 0.51/25 = 0.020\,m^3{\cdot}kg^{-1}\,COD \cdot d^{-1}$$

For the calculated retention time of 25 days this results in a digester volume $v_{di} = 0.51\ m^3 \cdot kg^{-1}$ COD. This value is significantly larger than in the configuration without primary settling. On the other hand, the volume of the aerobic reactor will be smaller. If again a mixed liquor concentration of 3.5 g TSS $\cdot\, l^{-1}$ is assumed, the aeration tank volume after primary settling is calculated as:

$$v_r = mX_t/X_t = 0.16 \cdot 5/3.5 = 0.23\,m^3.kg^{-1}\,COD \cdot d^{-1}$$

In practice this assumption may not always be justified. As in most wastewaters, organic matter is predominantly of a particulate nature while nitrogen is mostly dissolved (ammonium), and during the primary settling process a large proportion of organic matter will be removed. This will result in an increase in the TKN/COD ratio of the pre-settled influent. Thus, if nitrogen is to be removed in the activated sludge system, in general the required sludge age in a system treating settled wastewater will be longer than in a system treating raw wastewater. This increase in sludge age leads to a larger aeration tank volume. The effect of the increase of the TKN/COD ratio can be assessed quantitatively using the nitrification and denitrification models as presented in Chapter 5.

Once the minimum sludge age of the activated sludge process has been established, it is a relatively simple matter to optimise the entire treatment system. For the optimisation of the activated sludge system with anaerobic sludge stabilisation, the construction and the operational costs must be taken into consideration. The construction costs are mainly determined by the volume of the main treatment units: primary settler, aeration tank, final settler, thickener, digester and liquid-solid separation unit for the digested sludge. Furthermore, the required aeration capacity and, if applicable, the equipment for power generation are important items in the construction costs.

The operational costs mainly depend on the power consumption for aeration, the temperature in the digester (heated or not) and the concentration- and mass of stabilised sludge, which partially determine the cost for final disposal. On the other hand, methane production will be a cost reducing factor when used for power generation and/or heat generation.

The procedure for optimisation of the activated sludge system with stabilisation of activated sludge can be summarised as follows:

(a) Including primary sedimentation

(1) For the TKN/COD and P/COD ratio of the wastewater after settling, determine the minimum sludge age necessary to achieve the desired level of biological nutrient removal;

(2) Determine the mass and production of sludge for the calculated sludge age;

(3) Determine the required retention time in the digester for the operational temperature with Eq. 12.63;

(4) Determine the optimal volume for the sludge thickener and the digester for the known or assumed settling characteristics of the excess sludge;

(5) Use the settling characteristics to determine the optimal volumes of the aeration tank and the final settler;

(6) Determine the volume of the primary settler (normally a residence time of about two hours is adopted).

(b) For raw sewage treatment: Repeat step (1) to (5) for raw wastewater. Finally, taking into consideration the construction and operational costs, it will be decided which of the two options is the most attractive from a total cost point of view. Refer also to the extensive design examples in Chapter 14.

12.5 STABILISED SLUDGE DRYING AND DISPOSAL

After sludge digestion, the solids concentration in the digested sludge is not very high, usually in the order of 2 to 5% (20 to 50 g $TSS \cdot l^{-1}$). There are isolated cases where this thin slurry has been applied directly on land, but it is more usual to apply some form of liquid-solid separation to increase the solids content of the sludge and hence to reduce its volume before it is disposed of. The required degree of dewatering depends on the distance to the final destination of the dried sludge (transport costs) and the possible use of the dried sludge. Figure 12.22 shows several dewatering methods and some options for the final disposal of dried sludge.

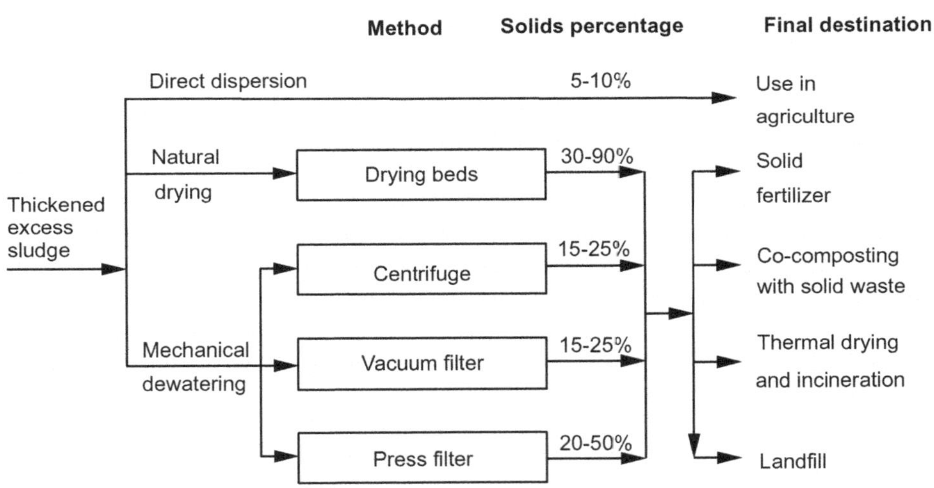

Figure 12.22 Methods for dewatering of stabilised sludge and for final sludge disposal

The liquid phase of the stabilised sludge consists of four fractions:

- Free water that can be separated from the solids by means of gravitational differences (gravitation or flotation). This fraction contains about 70% of the water in the sludge;
- Water adsorbed to the suspended- and colloidal solids. Part of this water can be removed by mechanical force (filtration, centrifugation) or through the addition of a flocculant;
- Water bound to the suspended solids through capillary forces. The difference between this fraction and the previous one is a subtle one and resides basically in the fact that the applied force to separate the water is larger. The two fractions combined constitute about 20% of the total mass of water;
- Cellular water (about 10%), which forms an integral part of the suspended solids and can only be removed when the cellular membrane is ruptured. This cannot be achieved through mechanical

means but is possible using biological methods or by altering the physical state of the water, either by evaporation or freezing.

The most important criteria for the selection of the liquid-solid separation method are the cost of the process and its reliability. It is interesting to note that the energy requirements for sludge dewatering are in the order of 1 Wh · m^{-3} for processes using natural gravitation like thickening, 1 kWh · m^{-3} for mechanical processes like filtration or centrifugation and 1000 kWh · m^{-3} for evaporation. Therefore it is concluded that the inclusion of thickening, at least as a pre-treatment process, should always be considered due to the low operational costs of this process. Of course, a higher degree of water removal can be achieved using the more energy intensive methods.

12.5.1 Natural sludge drying

Natural sludge drying is carried out in constructed beds that resemble (slow) sand filters: in a concrete or brickwork box a layer of sand rests on stratified stones, having a drainage system at the bottom. This is schematically shown in Figure 12.23 and Figure 12.24.

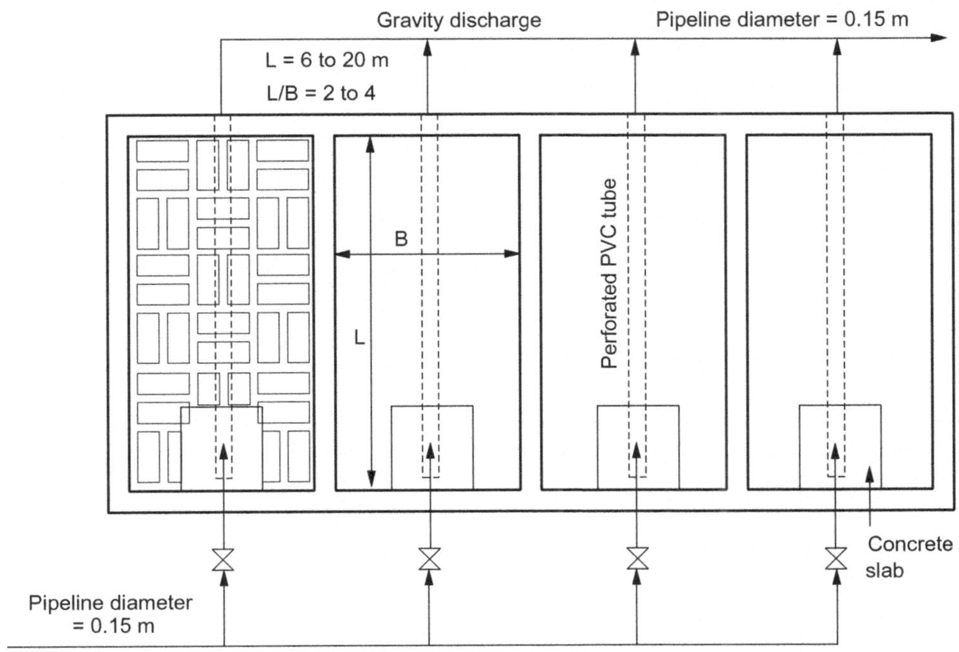

Figure 12.23 Typical layout of sludge drying beds (top view)

The sludge to be dried is placed on top of the sand and liquid-solid separation is achieved by percolation and evaporation. In regions with a hot climate, in a relative short time (1 to 3 days) a large part of the sludge water (80%) percolates and a semi solid cake is obtained with a concentration of 150 to 200 g TSS · l^{-1}, depending on the degree of stabilisation of the sludge. From then on, evaporation is the only mechanism that can effect a further reduction in the water content of the sludge. A "final" product with very high dry solids content is obtained in the form of a granular solid. In this respect, the sludge drying bed is different from mechanised systems for liquid-solid separation, as these produce a semi-solid cake with a much lower dry solids percentage, i.e. in the range of 15 to 30%.

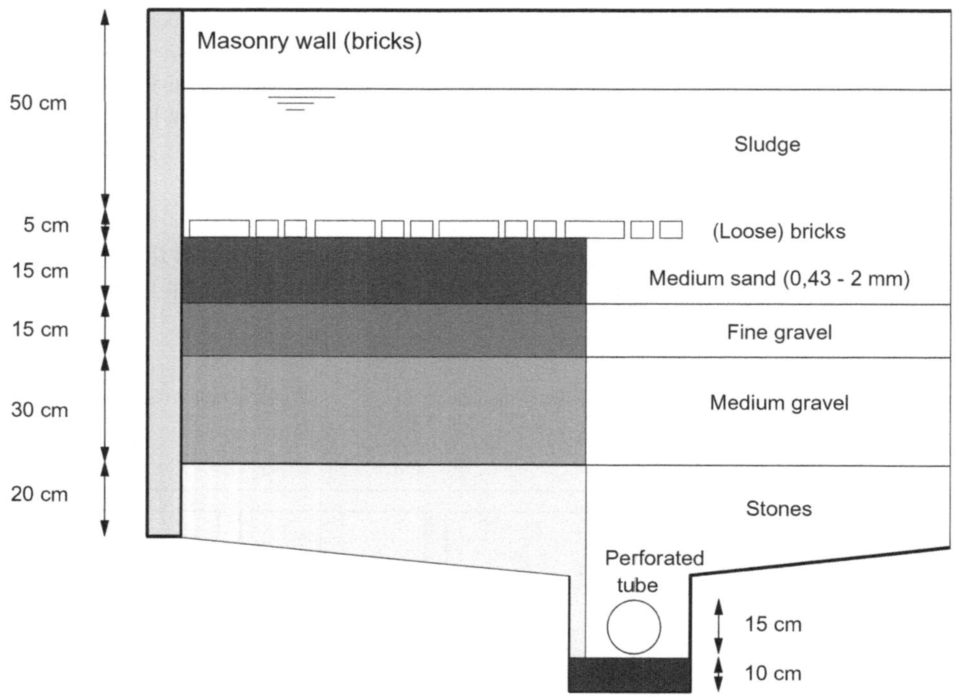

Figure 12.24 Cross section of the drainage layer of a sludge drying bed

The main design parameter of sludge drying beds is the productivity, defined as the maximum sludge mass to be applied per unit bed area per day ($kg\ TSS \cdot m^{-2} \cdot d^{-1}$). The value of the sludge bed productivity depends on various factors:

- Nature of the sludge and desired humidity of the end product;
- Climatic conditions;
- Operation & maintenance requirements of the sludge drying beds.

With respect to the nature of the sludge, it is important to stabilise the sludge before drying, also to prevent odour problems. Climatic conditions that favour natural sludge bed drying are a low humidity of the air, a low precipitation rate, a lot of wind and a high temperature.

Van Haandel and Lettinga (1994) have determined that the maximum productivity in tropical regions is obtained for a solids load of 20 to 40 $kg\ TSS \cdot m^{-2}$. Under those conditions, in 10 to 20 days time, an end product with a dry solids content of 50% is produced, independent of whether the sludge stabilisation was anaerobic or aerobic. Taking into account the time required for cleaning the beds, a net productivity of 1 kg TSS · m^{-2} · d^{-1} can be achieved. In practice a more conservative number will be applied due to adverse conditions (rain) and the fact that the beds need maintenance as well, so that not all are available at all times.

EXAMPLE 12.9

An activated sludge system produces a stabilised excess sludge at a rate of 0.25 kg TSS · kg^{-1} COD and the per capita contribution is 100 g COD · inh^{-1} · d^{-1}. The sludge bed productivity is determined at 0.8 kg TSS · m^{-2} · d^{-1} for a final solids fraction of 60%. Calculate the required sludge drying bed area per inhabitant mass daily applied COD.

Solution

For a bed productivity of 0.8 kg TSS · m^{-2} · d^{-1} and an excess sludge production rate of 0.25 kg TSS · kg^{-1} COD, the required bed area can be calculated as $0.25/0.8 = 0.31$ m^2 · d · kg^{-1} COD. For a per capita contribution of 0.1 kg COD · d^{-1}, the required per capita sludge bed area is 0.031 m^2, i.e. 1 m^2 of bed is sufficient to serve $1/0.031 = 32$ inhabitants.

It is important to note that the sludge drying beds do not necessarily need to be constructed at the same location as the activated sludge plant. As the digested sludge flow rate is small compared to the sewage flow rate, the sludge can be pumped at low cost to a location convenient for construction of the beds. For a stabilised sludge production of 0.25 kg TSS · kg^{-1} COD and a concentration of 25 g TSS · l^{-1} (these are values that are found in practice), the sludge volume is $0.25/25 \cdot 1000 = 10$ litre · kg^{-1} COD. For a typical raw sewage COD concentration of 0.5 to 1.0 g COD · l^{-1}, the sewage volume is 1000 to 2000 litres · kg^{-1} COD, meaning that the sludge flow is only 10/(1000 to 2000) = 0.5 to 1.0% of the sewage flow.

EXAMPLE 12.10

Sludge with a solids concentration of 50 g · l^{-1} is to be dried by percolation and evaporation on a sludge drying bed to produce a final product with less than 50% humidity. How much water is removed per cubic meter of sludge? It is assumed that 1 kg of solids has a volume of 1 litre.

Solution

The original suspension has a solids content of 50 g · l^{-1} and a water content of 950 ml · l^{-1}, as can also be seen from Figure 12.25. If percolation and evaporation are considered as sequential and independent processes, then the percolation will proceed until a solids fraction of about 0.2 is reached. Hence, after percolation the original 1 litre of sludge contains 50 g solids (20%) and 200 ml water (80%), i.e. $950 - 200 = 750$ ml · l^{-1} has percolated.

During the process of evaporation that follows, the solids fraction increases to 50%, i.e. after evaporation the dried sludge contains 50 g solids and 50 ml water, so that $200 - 50 = 150$ ml has evaporated.

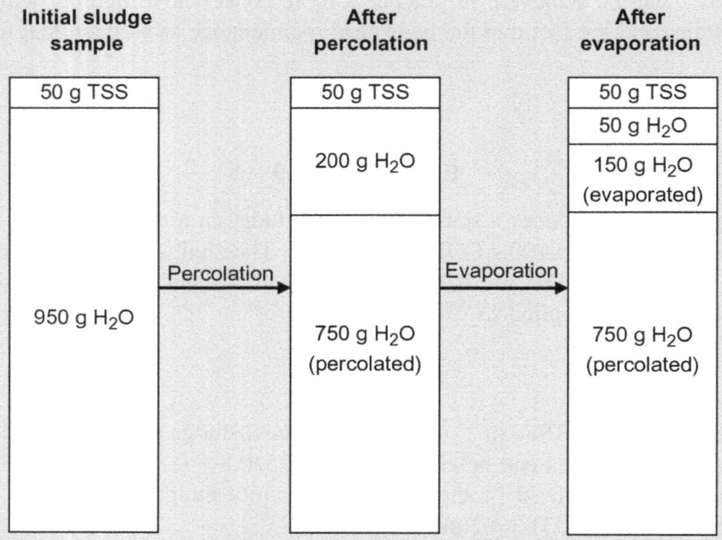

Figure 12.25 Schematic indication of percolated and evaporated water fractions of sludge during the drying process on sludge beds

From the example it is clear that in general much more water percolates than evaporates, but also that the evaporation step is essential in order to obtain a satisfactory end product. In the example, the sludge mass after percolation is still $250\,g \cdot l^{-1}$, which is 2.5 times more than the mass after evaporation ($100\,g \cdot l^{-1}$). Furthermore, the sludge mass after percolation is much more difficult to manipulate, because it is a sticky cake instead of a solid. The conclusion that percolation is important to remove a large fraction of the water leaves little room for the applicability of sludge lagoons as units for liquid-solid separation. In these lagoons percolation does not take place and all the sludge water will have to evaporate, which will take a very long time.

In terms of product quality, the sludge drying bed is superior when compared to other methods of liquid removal. Not only does the end product have a very high solids content (often more than 50%), the hygienic quality of the dried sludge is also significantly better. This is due to the combination of prolonged exposure time (two to three weeks in tropical regions) and high temperature (due to absorption of solar radiation by the black sludge). The hygienic quality can be increased even more by covering the sludge drying beds, e.g. with transparent plastic foil or glass. In this way, in tropic regions the temperature in the sludge bed can easily reach values up to 60°C or more in the top section of the sludge, eliminating all pathogenic bacteria and helminth eggs. However, the bottom section is not much affected by the absorption of solar energy, unless there is a combination of favourable factors: intense solar energy, a low solids loading rate and a low initial sludge humidity.

Maintenance of the sludge drying beds is simple and does not required skilled labour. The only disadvantage of the process is that the required area for the process is much larger than that of other sludge drying methods (although in the above example it was demonstrated that the required area is not excessive (0.02 to $0.025\ m^2 \cdot hab^{-1} \cdot d^{-1}$).

For these reasons, in many cases the application of sludge drying beds is the most attractive alternative for the removal of water from sludge, especially in regions with a warm climate. Drying beds can also be applied for additional drying of sludge cakes from mechanical drying system, increasing the solids concentration from 15 to 20% to 50 to 80%. This may lead to a significant reduction in the final sludge mass and hence of the costs for disposal, at least when the disposal costs are per ton sludge rather than per ton dry solids.

12.5.2 Design and optimisation of natural sludge drying beds

To allow rational design of sludge drying beds, the predicted sludge bed productivity should be known, i.e. the mass of solids that can be processed per unit area of bed per day, in order to obtain a certain final humidity. This productivity is defined as the ratio between the applied solids load and the time required to complete one drying cycle:

$$\eta_{sb} = F_s/t_c \tag{12.81}$$

where:

η_{sb} = sludge drying bed productivity ($kg\ TSS \cdot m^{-2} \cdot d^{-1}$)
F_s = applied solids load ($kg\ TSS \cdot m^{-2}$)
t_c = total drying cycle time (days)

A drying cycle consists of four sequential time periods:

t_1 = preparation of the sludge bed and application of the sludge to be dried
t_2 = percolation
t_3 = evaporation
t_4 = removal of dried sludge and cleaning of the bed for the next batch

Hence

$$t_c = t_1 + t_2 + t_3 + t_4 \tag{12.82}$$

The duration of periods t_1 and t_4 will mainly depend on factors related to the degree of mechanisation of the operations, but it will be much shorter than t_2 and certainly much shorter than t_3. The periods t_2 and t_3 depend on factors that are beyond control of operations: the nature of the sludge to be dried and the climate (air humidity, temperature, wind). The main operational variable is the solids loading rate, which is the mass of applied solids per unit area of sludge bed. As the process of sludge drying on beds depends on several uncontrollable factors, it is not possible to present a general equation for the required bed area. An empirical solution may be found by operating a small sludge drying bed pilot unit and experimentally determining the periods of percolation and evaporation (t_2 and t_3) for a particular value of the desired final sludge humidity as a function of the applied solids load under actual climatic conditions. The experimental procedure is outlined below.

12.5.2.1 Determination of the percolation time (t_2)

When sludge is applied to a drying bed, the processes of percolation (filtration) and evaporation start immediately, although the former at a much higher rate than the latter. The percolation water is collected and returned to the sewage treatment plant. The percolation rate depends on the degree of sludge stabilisation and the applied solids loading rate, but is independent of the grain size of the sand. This is in accordance with filtration theory, which postulates that the resistance of a filter is due to the cake layer that will build up and not due to the filter medium (sand). For the same reason there is no reduction in the percolation rate when a layer of loose bricks is placed on the sand to facilitate the removal of the dried sludge.

Using small pilot scale drying bed units like those represented in Figure 12.26, Van Haandel and Lettinga (1994) determined the percolation time (i.e. the time until percolation ends) as a function of the solids loading rate for different values of the initial solids concentration. They established an empiric equation that allows estimation of the percolation time (in days) as a function of the solids loading rate:

$$t_2 = F_S^2/220 + 1.5 \tag{12.83}$$

where F_s = solids loading rate, for $15 < F_s < 50\ kg\ TSS \cdot m^{-2}$

Figure 12.26 Pilot experiments with sludge drying beds to investigate the influence of rain on the dewatering of sludge

Thus for example, for a loading rate of $30\ kg\ TSS \cdot m^{-2}$ the percolation time t_2 can be estimated as $30^2/220 + 1.5 = 5.6$ days. Later research showed that there is a considerable variation in the required percolation time, even under apparently identical conditions. However, usually for anaerobic or aerobically digested sludges the value of t_2 will be between 5 and 15 days, for solids loads between 15 to $50\ kg\ TSS \cdot m^{-2}$. The percolation time is therefore much smaller than the time required for evaporation, as will be demonstrated in the next section.

12.5.2.2 Determination of the evaporation time (t_4)

After percolation, the evaporation rate can be determined by exposing the percolated sludge to sun and wind and observing the rate of weight loss of the sludge sample. Dividing the daily weight loss by the area of the pilot sludge bed, the evaporation rate can be determined. Pedroza et al. (2005) determined the evaporation rate in pilot sludge drying beds and compared it to the evaporation rate of a pure water sample placed in a similar sludge bed. The relative evaporation rate was defined as the ratio between the evaporation rate of sludge in a sludge drying bed and the evaporation rate of water under the same conditions.

A batch of 5.08 kg sludge with an initial humidity of 76.2% was placed in the sun and at regular intervals the extent of evaporation was determined from the weight loss. The solids mass of the sample was equal to $(1-0.762) \cdot 5.08 = 1.21$ kg TSS. Simultaneously the weight loss of a water sample with the same surface area placed in a bed with a black bottom (for solar energy absorption) was determined and the corresponding evaporation rate was calculated by dividing the average daily volume loss of this sample.

Table 12.7 Determination of the evaporation rate of a sludge- and water sample in pilot drying beds ($0.075\ m^2$) for an applied solids load F_s of 16.1 kg $TSS \cdot m^{-2} \cdot d^{-1}$

Evaporation period (days)	Sample weight (g)	Humidity (%)	Accumulated evaporation rate		Relative evaporation rate (%)
			In sludge	In water	
0	5080	76.2			
1	4865	75.1	5.1	4.5	113
7	4255	71.5	19.4	40.6	51
14	3475	65.2	37.8	78.9	51
21	3165	61.7	45.1	117.1	41
28	2840	57.4	52.7	162.6	34
35	2340	48.2	64.5	201.0	34
49	1780	32.0	77.6	292.2	28
56	1460	17.0	85.2	340.1	27
63	1320	7.6	88.5	384.3	24

Table 12.7 shows the experimental data used for the calculation of the relative evaporation rate. The calculation procedure can be summarised as:

(1) At different time intervals the mass of the sludge sample (column 2) is determined;

(2) Based on the weight of the sludge sample, the humidity of this sample is calculated as a function of time (column 3):

$$U = M_w/(M_s + M_w) = (M_t - M_s)/M_t \qquad (11.84)$$

where:

U = humidity of the evaporating sludge sample (%)
M_t = total sample mass (g)
M_w = water mass in the sample (g)
M_s = suspended solids mass in the sample (g)

(3) The evaporation rate is calculated from the rate of weight loss and the bed area (it is assumed that the weight loss is only due to evaporation), for both the sludge sample (column 3) and the water sample (column 4): note that only the weight loss of the sludge sample is presented in Table 12.7;

(4) The relative evaporation rate as function of time is determined from the ratio between the evaporation rates in the sludge and water samples (column 6) ;
(5) Now the relative evaporation rate (column 6) is linked to the humidity in the sludge sample (column 3). Figure 12.27 shows the resulting diagram for the data in Table 12.7.

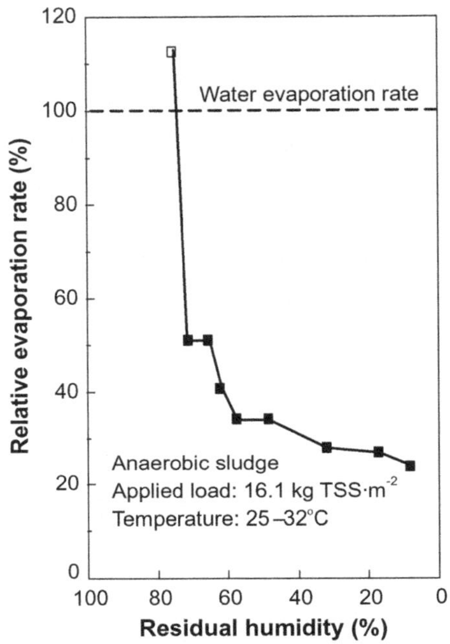

Figure 12.27 Relative evaporation rate R_{rel} as a function of residual humidity U_e (based on the data from Table 12.7)

The experimental results of Table 12.7 demonstrate the following important points:

(1) Initially the evaporation rate in the sludge sample exceeds the evaporation rate of the water sample (the relative evaporation rate is larger than 1.0 in Figure 12.27), but when the humidity of the sludge sample decreases so does the relative evaporation rate;
(2) At low residual humidity (between 10 to 30%), the sludge water evaporation rate is only 15 to 25% of the water evaporation rate;
(3) The relative evaporation rate tends to decrease at higher solids loading rates. This is also demonstrated in Figure 12.28, where experimental values of the relative evaporation rate are shown as a function of the residual humidity for different solids loading rates.

The low rate of evaporation in sludge batches is due to an inherent inefficiency associated with sludge drying beds: in the water samples the absorbed solar energy is distributed uniformly over the total water sample which facilitates the evaporation process, while the temperature increases only marginally.

In contrast, in the sludge samples there is a steep temperature gradient due to the absence of thermal convection as a mechanism for heat distribution and because of the low thermal conductivity of the sludge. As a result the top section may be very hot, resulting in heat being irradiated back to the atmosphere. For the sludge sample, this heat is lost for the purpose of water evaporation. This problem is further accentuated when a high solids loading rate is applied, as the water to be evaporated will mainly be present in the lower part of the sludge layer. As a result, the evaporation rate in sludge batches on drying beds is low, especially when the solids loading rate is high and the desired residual humidity is low.

Data like those presented in Table 12.7 can be used to calculate the required evaporation time for any combination of initial- and final humidity as a function of the solids loading rate, as long as the average water evaporation rate in the region where the beds will be placed is known. The thickness of the water layer to be evaporated is then divided by the average evaporation rate of water in the sludge batch. The height of the water layer in a sludge batch applied on a sludge drying bed depends on both the solids loading rate and the humidity [defined as $L = U \cdot X/(1–U)$], so that:

$$t_3 = [U_i \cdot F_s/(1 - U_i) - U_e \cdot F_s/(1 - U_e)]/(R_{ref} \cdot R_w) \tag{12.85}$$

where:

F_s = applied solids load
U_i = initial humidity after percolation (usually $\approx 80\%$)
U_e = desired final humidity after drying
R_{rel} = relative evaporation rate of water in the exposed sludge batch
R_w = water evaporation rate in the region under consideration

For a residual humidity in the range of 10 to 60% (in practice the desired final residual humidity will almost invariably be in this range), the average value of the relative evaporation rate in the exposed sludge batch R_{rel} can be estimated from Figure 12.28 as a function of the solids loading rate per drying cycle:

$$R_{rel} = 0.4 + 0.33 \cdot U_e/100 \quad \text{for } F_s = 10\,\text{kg TSS} \cdot m^{-2} \tag{12.86a}$$

$$R_{rel} = 0.33 + 0.33 \cdot U_e/100 \quad \text{for } F_s = 20\,\text{kg TSS} \cdot m^{-2} \tag{12.86b}$$

$$R_{rel} = 0.2 + 0.33 \cdot U_e/100 \quad \text{forF}_s = 30\text{to}50\,\text{kg TSS} \cdot m^{-2} \tag{12.86c}$$

Once the time required for evaporation has been determined, the total drying time can be estimated as the sum of the four cycle times (Eq. 12.82):

$$\begin{aligned} t_c &= t_1 + t_4 + t_2 + t_3 \\ &= (t_1 + t_4) + F_s^2/220 + 1.5 + \{F_s \cdot [1/(1 - U_i) - 1/(1 - U_e)]/(R_{rel} \cdot R_w)\} \end{aligned}$$

For $t_1 + t_4$, the time required to fill and empty the sludge bed, often a default value is taken, for example 2 days. With the aid of Eqs. (12.85 and 12.86) the evaporation time t_3 can be calculated, while the bed productivity can be determined with Eqs. (12.81 and 12.82). Figure 12.29 shows the productivity of sludge drying beds as a function of the applied solids load for different values of the final humidity for several water evaporation rates. The adopted duration for filling and emptying the beds and for percolation are indicated in Figure 12.29 as well. Figure 12.29 demonstrates the following points:

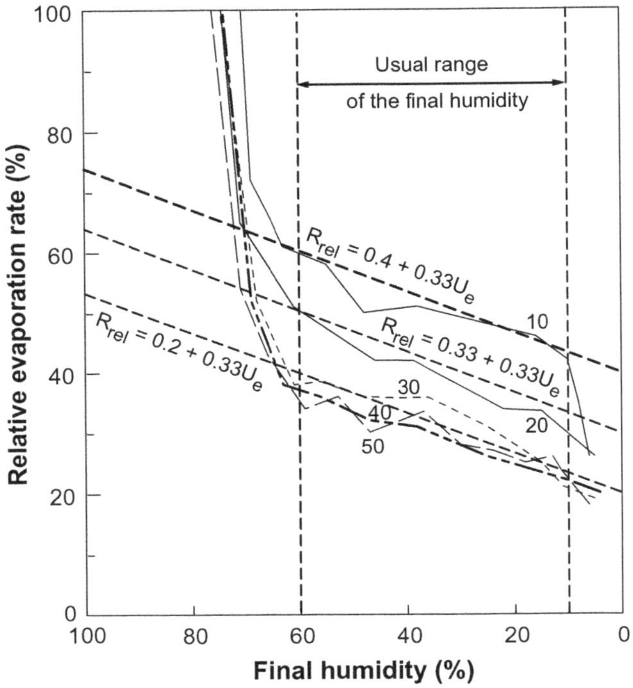

Figure 12.28 Relative evaporation rate R_{rel} as a function of residual humidity U_e for different solids loading rates (10 to 50 kg TSS · m^{-2})

EXAMPLE 12.11

Estimate the required time for evaporation (t_3) in a sludge bed when a solids load of 30 kg TSS · m^{-2} is applied and the humidity of the sludge is to be reduced from its initial value of 76% to a final value of 30%. The water evaporation rate R_w in the region is given as 5 mm · d^{-1}.

Solution

The relative evaporation rate can be estimated from Figure 12.28 or using Eq. 12.86c as $0.2 + 0.33 \cdot 60/100 = 0.4$ at $U_e = 60\%$ and 0.3 at $U_e = 30\%$, or an average value of $R_{el} = 0.35$. Although the relative evaporation rate will be higher in the range between $60\% < U_e < 76\%$, the average value calculated above is accepted as a conservative estimate for R_{rel}. The average evaporation rate of water in sludge drying beds will thus be $R_{el} \cdot R_w = 0.35 \cdot 5 = 1.75$ mm · d^{-1}.

For the given values of the applied solids load and the initial humidity, the water layer in sludge is initially equal to $0.76 \cdot 30/(1 - 0.76) = 95$ l · m^{-2} or 95 mm and for the stipulated residual humidity of 30% the final water layer will be $0.30 \cdot 30/(1 - 0.30) = 13$ mm. Thus the required amount of evaporation equals $95 - 13 = 82$ mm. For the estimated average evaporation rate of 1.75 mm · d^{-1}, this will take $82/1.75 = 47$ days.

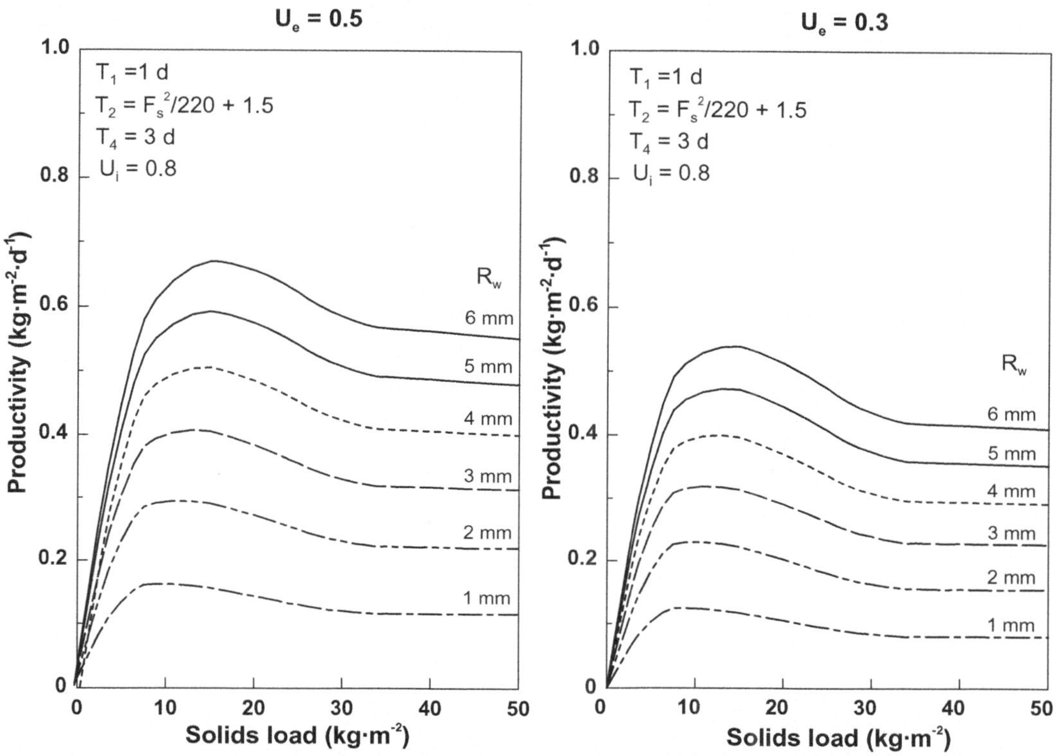

Figure 12.29 Sludge bed productivity as a function of the applied solids load for different values of R_w and for two values of U_e: 50% (left) and 30% (right)

(1) The productivity curve as function of the solids load exhibits a maximum in the range of 10 to 15 kg TSS $\cdot$ m^{-2}, which indicates the optimal value of the solids load. The optimal solids loading rate tends to increase at higher values of the water evaporation rate, whereas the desired residual humidity of the dry sludge does not have an influence;

(2) The drying bed productivity depends heavily on the selected final humidity. The bed productivity for a final humidity of 50% is more than 25% higher than the value for a final humidity of 30%.

According to the EPA (1985), the production of a stabilised and dewatered excess sludge with a low final humidity (i.e. between 10% – 20%) has several important advantages:

- A concentrated organic fertiliser for application in agriculture can be obtained;
- Transport costs will be reduced;
- The pathogen concentration (i.e. worm eggs) will be reduced.

Once the productivity of the sludge drying bed has been established, the design procedure can be summarised as follows:

(1) Establish the production rate of stabilised sludge;
(2) Determine the minimum required area for the sludge drying bed, defined as the ratio between the stabilised excess sludge production rate and the bed productivity;
(3) Determine the size of the maximum sludge batch that can be discharged during one day of operation. In the case of anaerobic digesters this value is usually around 30% of the digester volume;
(4) From the sludge production rate and the maximum stabilised sludge batch size, determine the minimum required discharge frequency;
(5) Determine the number of sludge drying beds.

EXAMPLE 12.12

Sludge drying beds are operated in a region where the average water evaporation rate is 5 $mm \cdot d^{-1}$ during the summer season and 3 $mm \cdot d^{-1}$ during winter time. The desired final humidity of the dried sludge is 50%. If the beds are operated at the optimal solids load, estimate the difference in required area resulting from climatic variations.

Solution

Figure 12.29 shows that for an evaporation rate of 5 $mm \cdot d^{-1}$ and a desired final residual humidity of 50%, the optimal solids load is about 15 $kg\,TSS \cdot m^{-2}$ and the corresponding bed productivity is 0.58 $kg\,TSS \cdot m^{-2} \cdot d^{-1}$. In winter time the water evaporation rate is reduced to 3 $mm \cdot d^{-1}$, so that the productivity is now only 0.39 $kg\,TSS \cdot m^{-2} \cdot d^{-1}$. Therefore the bed area has to be increased by a factor $0.58/0.39 = 1.5$ to be able to handle the same sludge mass as in summer time.

If during the summer period the same area is used as in the winter, the produced dry sludge will have a final humidity of less than 30%, as can be deducted from Figure 12.29 (right): for the optimal solids load and a water evaporation rate of 5 $mm \cdot d^{-1}$, the maximum productivity of the bed is 0.48 $kg\,TSS \cdot m^{-2} \cdot d^{-1}$, more than the required capacity of 0.39 $kg\,TSS \cdot m^{-2} \cdot d^{-1}$.

EXAMPLE 12.13

A wastewater treatment plant receives a COD load of 4 $ton \cdot d^{-1}$ and produces 1 $ton\,TSS \cdot d^{-1}$ of stabilised excess sludge, which is discharged on sludge drying beds with the objective to produce a sludge with a humidity of less than 50%.

It has been established in field experiments that a productivity of 0.55 $kg\,TSS \cdot m^{-2} \cdot d^{-1}$ can be attained for a solids loading rate of 20 $kg \cdot m^{-2} \cdot d^{-1}$. The maximum sludge batch discharge from the anaerobic digester (with a retention time of 30 days) is $1/3^{rd}$ of the liquid volume. The digested sludge concentration is 40 $g\,TSS \cdot l^{-1}$. Design the optimal sludge drying bed configuration. What is the per capita area needed for the sludge drying beds?

Solution

For a sludge drying bed productivity $\eta_{sb} = 0.55$ kg TSS · m^{-2} · d^{-1} and a sludge production rate of 1 ton TSS · d^{-1}, the required area of the sludge drying beds is $1000/0.55 \approx 1800$ m^2. If a solids load of 20 kg TSS · m^{-2} is applied, the total cycle time for drying is estimated as: $t_c = F_s/\eta_{sb} = 20/0.55 \approx$ 36 days. The thickness of the sludge layer on the sludge bed, directly after receiving a solids load of 20 kg TSS · m^{-2} with a solids concentration of 40 g TSS · l^{-1}, is equal to $20/40 = 0.5$ m.

At the end of the percolation process the solids fraction will have increased to 20% (or 200 kg TSS · m^{-3}), resulting in a reduction of the water layer thickness to $20/200 = 0.1$ m. At the end of the drying process (50% solids or 500 kg TSS · m^{-3}), the water layer thickness is only $20/500 =$ 0.04 m (it is assumed that the density of the dried sludge ≈ 1 k g · l^{-1}).

The minimum sludge discharge frequency will be 1/3rd of the retention time of 30 days (10 days), corresponding to a maximum sludge discharge batch of 10 · 1 ton TSS · d^{-1} or 10 ton TSS. The sludge can also be discharged at a higher frequency, which consequently will result in a smaller batch and–sludge bed size. However, the number of beds required will then increase.

Table 12.8 (column 1) shows the required sludge bed configurations for discharge intervals in the range from 1 (daily sludge discharge) to 10 days (minimum discharge frequency). When sludge is discharged daily, then the bed area is equal to the ratio between the daily stabilised sludge production and the design solids load or $1000/20 = 50$ m^2 (column 2). As the drying cycle has a duration of 36 days, at least 36 sludge drying beds will be required as well (column 3). Therefore, in practice this alternative would probably not be selected, because of the large effort involved in handling such a high number of drying beds.

Table 12.8 Various sludge drying bed configurations for different discharge intervals

Discharge interval (d)	Stabilised sludge discharge (ton TSS)	Area of single bed (m^2)	Minimum number of beds (-)	30% reserve capacity (-)
1	1	50	36	
2	2	100	18	24 × 100
3	3	150	12	16 × 150
4	4	200	9	12 × 200
5	5	250	7.2	
6	6	300	6.0	8 × 300
7	7	350	5.1	
8	8	400	4.5	6 × 400
9	9	450	4.0	
10	10	500	3.6	

The second option (one discharge every two days) would require 18 beds. In practice it might be considered to increase the number of beds by 1/3rd to allow for contingencies such as maintenance, excessive rain etc. Thus 24 beds would be required with an area of 100 m^2 each, resulting in a total area of 2400 m^2. Other (equivalent) alternatives are 12 × 200; 8 × 300 or 6 × 400 m^2 for discharge periods of 4, 6 and 8 days respectively.

If it is assumed that the per capita contribution is 100 g COD · d^{-1}, the contributing population for 4 ton COD · d^{-1} is equal to 40,000 and the required bed area is 2400/40.000 = 0.06 m^2 per inhabitant, or alternatively, per m^2 of bed area the stabilised sludge produced by 1/0.06 = 17 inhabitants can be dried.

12.5.2.3 Influence of rain on sludge drying bed productivity

Before evaluating the influence of rainfall on the performance of sludge drying beds, it is interesting to evaluate the height of the water layer that is removed per unit time and per unit area. For an initial and final humidity U_i and U_e and a bed productivity η_{sb}, the height of the water layer applied to the sludge bed is expressed as:

$$L_i = \eta_{sb} \cdot U_i/(1 - U_i) \quad (12.87)$$

At the desired final residual humidity, the height of the remaining water layer is:

$$L_e = \eta_{sb} \cdot U_e/(1 - U_e) \quad (12.88)$$

Therefore the layer of water removed from the bed (by percolation or evaporation) is:

$$\Delta L = L_i - L_e = \eta_{sb} \cdot [U_i/(1 - U_i) - U_e/(1 - U_e)] \quad (12.89)$$

For example, for $U_i = 95\%$, $U_e = 50\%$ and a bed productivity of 0.6 kg TSS · $m^{-2} \cdot d^{-1}$, the water removal rate is $\Delta L = 0.6 \cdot (0.95/0.05 - 0.5/0.5) = 10.8\ l \cdot m^{-2} \cdot d^{-1}$ or 11 mm · d^{-1}, which amounts to about 4000 mm · $year^{-1}$. This value is significantly higher than the yearly rainfall in many areas. One method to compensate for the addition of water to the sludge bed due to rainfall would be to increase the area of the sludge drying beds in the same proportion as the ratio between rainfall and the water throughput rate. If in the example above the annual rainfall equals 800 mm, the area of the beds could be increased by a factor of 800/4000 = 0.2 or 20% to compensate for the yearly precipitation.

In practice the required compensation may be much smaller, as established in a recent research project by Moreira et al. (2005). It can be observed visually that rainwater readily percolates through a sludge drying bed when the exposed sludge is already quite dry (i.e. when most of the sludge water has already been removed by percolation). On the other hand, if the sludge is still mainly a fluid, then the precipitated water is simply added to the water already present in the sludge batch.

To experimentally determine the influence of the sludge humidity on the fraction of rainfall water percolating through the sludge bed, sludge batches with different levels of humidity were placed in pilot sludge drying beds and exposed to the atmosphere (Figure 12.26). One series of batches was subjected to a solids load of 20 kg TSS · m^{-2} and the other one to 40 kg TSS · m^2. The values of the initial humidity were 80, 74, 70, 65 and 60%. Rainfall was simulated by daily sprinkling a 10 mm layer of water on the beds. The percolation was determined by weighing the sludge beds directly after and 24 hours after the "rainfall" event. The rain simulation was continued for 4 consecutive days. It can be observed from Figure 12.30 that a very large fraction (80 to 90%) of the rain water percolates through the bed when the humidity is less than about 65%. In contrast, at humidities above 70% the rainwater tends to accumulate on the bed, especially if the solids load is high, disrupting the dewatering process.

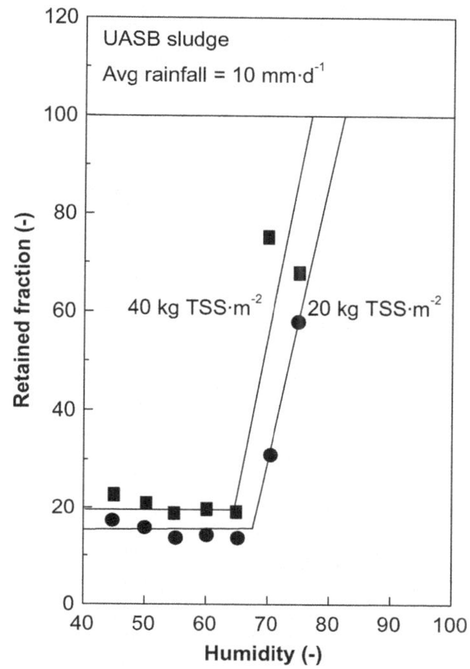

Figure 12.30 Retained water fraction on sludge beds after 10 mm · d^{-1} of artificial rain as function of the initial humidity U_i

In practice, the time required for a freshly applied sludge batch to attain a humidity of 65% is relatively short compared to the time to attain the final humidity. Thus statistically, the majority of the times when it rains the water will percolate because the humidity will already be lower than 65%. In these cases only a small fraction of the rainwater will actually be retained in the sludge to be dried. On the other hand, if persistent rainfall occurs on samples with a high humidity, it may be difficult to establish favourable conditions for natural sludge drying if the relative evaporation rate is in the same order of magnitude as the precipitation rate. In these cases it may be advisable to cover the sludge drying beds.

12.5.3 Accelerated sludge drying with external energy

When the required area for conventional sludge drying beds is unavailable or too expensive, it may be considered to reduce the area demand by applying external energy sources to accelerate the evaporation process. The sludge should be dewatered before the drying process, so that a cake instead of a fluid is processed.

Depending on the maximum temperature applied to the dried sludge, there may be the additional advantage of thermal removal of pathogens like helminth eggs. The required evaporation heat of water from a sludge bed is much larger than the heat required to increase the temperature of the dried sludge mass. This is shown in Example 12.14.

Two energy sources may be considered for use in sludge drying beds: solar energy and the combustion heat generated from the biogas produced from anaerobic treatment. If direct solar radiation is not the main source of energy, it is no longer required to construct the sludge drying beds in the traditional manner: a large area where a relatively shallow layer of sludge is applied.

EXAMPLE 12.14

After percolation, a sludge batch is applied to drying beds with a solids loading rate of 20 kg TSS · m^{-2}, 80% humidity and a temperature of 20°C. The sludge is dried and heated to a final product with a residual humidity of 30% and a temperature of 60°C.

What is the required amount of energy to effect the evaporation of water and to increase the temperature? Adopt an evaporation heat of 590 kCal · l^{-1} and a specific heat of 1 and 0.25 cal · g^{-1}.°C^{-1} for water and dry sludge respectively.

Solution

The initial water layer for $F_s = 20$ kg TSS · m^{-2} and $U_i = 80\%$ is equal to $L_i = 20 \cdot 0.8/(1 - 0.8) =$ 80 mm and the final layer $L_e = 20 \cdot 0.3/(1 - 0.7) = 9$ mm. Thus a water layer of 80 − 9 = 71 mm or 71 l · m^{-2} will have to be evaporated, requiring a heat of 590 · 71 = 41,900 kCal · m^{-2}. In contrast, the energy required to increase the temperature of 20 kg of solids and 20 l of water from 20 to 60°C at the end of the drying process will only require:

$$H = (C_w \cdot M_w + C_l \cdot M_l \cdot \Delta T = (1 \cdot 20 + 0.25 \cdot 20) \cdot 40 + 1000 \, \text{kCal} \cdot m^{-2}$$

where:
H = required energy per m^2 of bed
C_w = specific heat of water
M_w = mass of water in dried sludge to be heated
C_l = specific heat of sludge solids
M_w = mass of solids in dried sludge to be heated
ΔT = desired temperature increase

Hence, for the given conditions the amount of heat required to evaporate the water is more than 40 times larger than the amount of heat required to increase the temperature.

Instead, the use of an external energy source would make it feasible to construct a sludge drying tower with a height of several metres. The height of the sludge layer would then be determined by economic and technical considerations. Figure 12.31 shows a schematic diagram of a drying tower where solar energy and combustion heat are used. In addition, air flow could be induced to enhance the distribution of heat and to accelerate the evaporation process. The operation of the tower is semi continuous: daily discharges of excess sludge cake are introduced in the top section, while at the bottom the dried sludge is removed by a milling device or transport screw. This screw might be coupled to a temperature analyser so that the sludge is only discharged at a certain predetermined minimum temperature. This ensures that a dry and hygienically safe sludge is produced.

12.5.3.1 Use of solar energy

In the sludge drying process, solar energy can be used in three ways: (I) application of solar radiation directly on to the sludge bed, as in conventional sludge drying beds, (II) application of solar radiation directly on a sludge bed that is covered in order to reduce the loss of absorbed energy to the atmosphere, or (II) indirectly by absorption of solar radiation in a solar energy collector from where it is transferred using another medium (often water) to a sludge drying bed or tower.

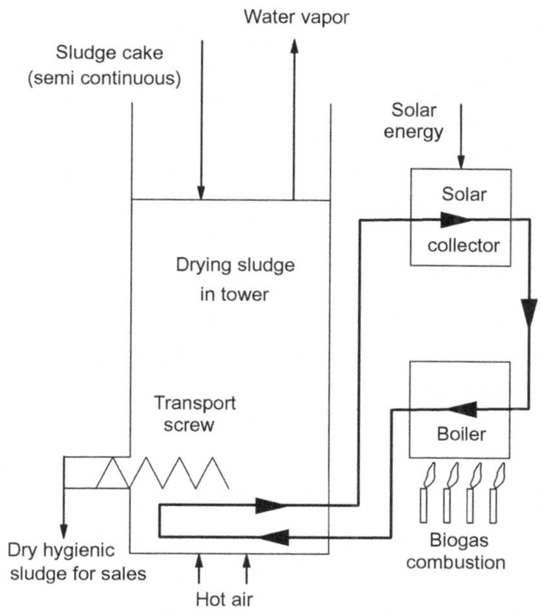

Figure 12.31 Schematic representation of a sludge drying tower with use of solar energy and heat from biogas combustion

In so far as the conventional drying beds are concerned, it was shown (Figure 12.27) that the energy use efficiency is low: 20 to 30% (compared to a pure water sample). Pedroza et al. (2006) demonstrated that covering the sludge drying bed has only a limited beneficial effect. During the daytime period the solar energy results in an increase of the sludge temperature, however during the hours with little or no solar intensity the accumulated energy is lost almost entirely to the atmosphere, even when material with a low thermal conductivity is used (e.g. glass with a 6 mm thickness).

The quantity of solar energy that can be collected from a specific area in a single day is determined by geographical considerations. However, even in regions with intense solar radiation, the maximum solar energy flux does not exceed 1200 $W \cdot m^{-2}$, so that even on the sunniest days (assuming six hours of sunshine with an average intensity of 800 $W \cdot m^{-2}$), the total quantity of harvested energy will not be more than 5 $kWh \cdot m^{-2}$.

In reality, the temperature increase of sludge batches on drying beds is much smaller than the maximum value estimated in the example above because:

- The average daily absorption of solar energy will usually be significantly lower than 5 $kWh \cdot m^{-2} \cdot d^{-1}$;

- The temperature increase will cause part of the accumulated solar energy to be irradiated back to the environment, even if the bed is covered with insulating material of low thermal conductivity;
- In the sludge layer, a thermal stratification profile will develop with the top section having a much higher temperature than the bottom section.

EXAMPLE 12.15

What is the maximum temperature increase of a sludge layer on a covered bed when the daily absorption of solar energy is 5 $kWh \cdot m^{-2}$, the solids load is 20 $kg\ TSS \cdot m^{-2}$ and the humidity is 70%? Disregard the thermal capacity of the solids.

Solution

The absorbed energy equals 5 $kWh \cdot d^{-1}$ or $5 \cdot 3600/4.2 = 4285$ $kCal \cdot m^{-2} \cdot d^{-1}$. The height of the water layer of the sludge batch $L_i = \eta_{sp} \cdot U_i/(1{-}U_i) = 20 \cdot 0.7/0.3 = 47\ l \cdot m^{-2}$. Since 1 litre of water has a thermal capacity of 1 $kCal \cdot °C^{-1}$, the absorbed heat can be used to increase the temperature by $4285/47 = 92°C$. It is concluded that under the given conditions there is a potential for a very significant increase of the temperature.

The occurrence of thermal stratification increases the irradiation of absorbed energy to the environment. Figure 12.32 shows measured temperature profiles in the top- and bottom sections of covered sludge drying beds (4 mm glass cover) for four different situations:

(1) High solids load and high humidity (Figure 12.32a): $F_s = 51$ $kg\ TSS \cdot m^{-2}$ and $U_i = 73\%$;
(2) High solids load and low humidity (Figure 12.32b): $F_s = 52$ $kg\ TSS \cdot m^{-2}$ and $U_i = 10\%$;
(3) Low solids load and high humidity (Figure 12.32c): $F_s = 18$ $kg\ TSS \cdot m^{-2}$ and $U_i = 68\%$;
(4) Low solids load and low humidity (Figure 12.32d): $F_s = 18$ $kg\ TSS \cdot m^{-2}$ and $U_i = 11\%$.

From the data in Figure 12.32, it is apparent that only the combination of a low solids load with a low humidity will result in a significant temperature increase in the bottom section of the sludge batch. Under these conditions, improvement in the hygienic quality of the sludge using solar energy would be feasible. As it has been established earlier in this section that the application of a cover above the sludge drying bed does not lead to a significant increase in the evaporation rate, the application of covers to increase bed productivity does not seem feasible.

On the other hand, thermal deactivation of pathogens is only possible when both load and humidity are low, i.e. after drying has already occurred. Taking into consideration the cost involved in covering a sludge bed, it might be considered to construct a mobile bed cover to sequentially cover the beds. The mobile cover is only used when a particular sludge bed is already dry, allowing for thermal removal of the pathogens. When the desired temperature has been maintained sufficiently long to effect the elimination of pathogens, the mobile cover is moved to the next bed, whereafter the dried and hygienically safe sludge is collected for use in agriculture. The use of a solar energy collector with heat exchange to the water phase in order to transfer energy to a sludge drying bed as indicated in Figure 12.31, is only advantageous if the total energy absorption by the solar collector is significantly larger than the efficiency of conventional beds, which is in the range of 15 to 30%. Given the cost and the complexity of operating a solar energy collector plus heat exchanger, this option is probably not economical.

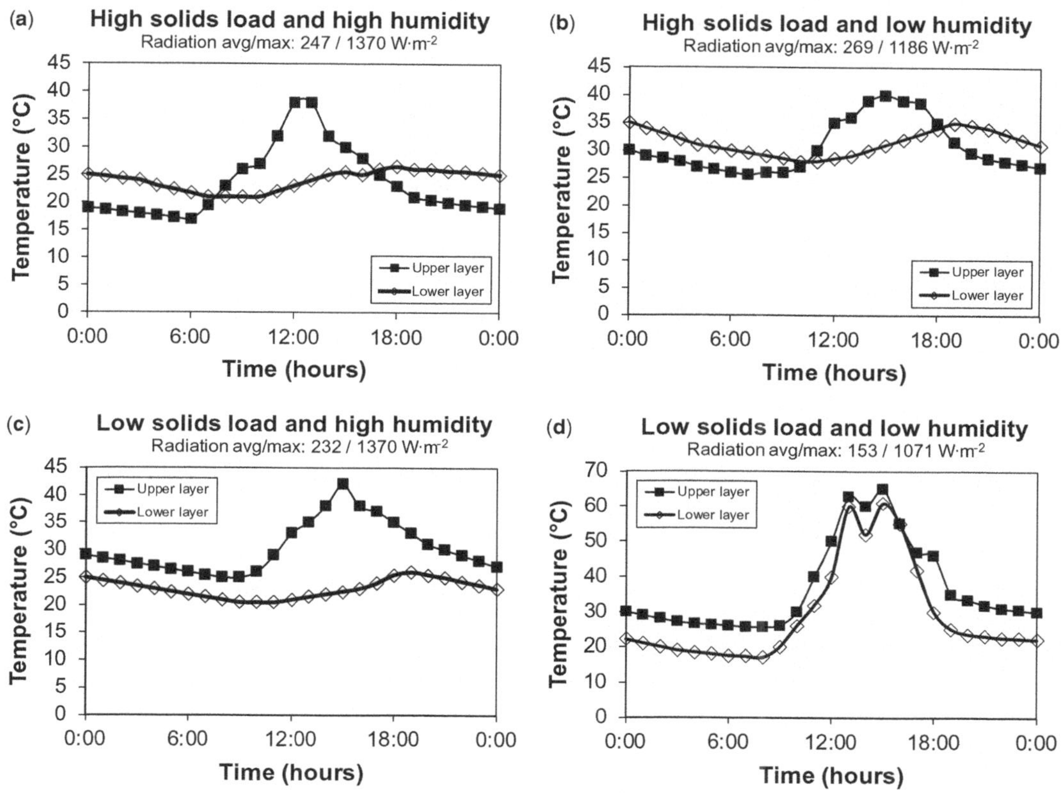

Figure 12.32 Temperature profiles in upper layer (2 cm depth) and lower layer (20 cm depth) in covered sludge drying beds operated with different solids loadings and with different initial humidities

12.5.3.2 Use of combustion heat from biogas

If anaerobic digestion is applied for pre-treatment of wastewater or for sludge stabilisation, the generated biogas can be converted into combustion heat and/or electricity. The combustion heat may then be used as an energy source in a sludge drying tower. To evaluate the potential feasibility of this alternative, the heat generated from the combustion of the biogas must be compared to the heat required for evaporation.

Due to strict limits on the concentration of undesirable compounds in excess sludge (e.g. heavy metals, chlorinated organics etc.), in many countries it has become impossible to use excess biological sludge as an organic fertiliser. Even disposal in a landfill in many cases is no longer a feasible option. Under these conditions sludge combustion may be the only alternative. This may be done in facilities where the chemical energy of combustion is used in a productive manner, such as in cement production where the sludge helps to maintain the very high temperature required in the production process and where the heavy metals present in the sludge are incorporated in the end product (cement).

More frequently, the excess sludge is combusted in a specific unit although the cost of constructing an incinerator with associated equipment is very high. Thus in many cases a large incinerator will be constructed which will serve several wastewater treatment plants. In such cases the distance from thewastewater treatment plant to the incinerator may be considerable, so that it is important to transport the sludge with a low humidity. A low humidity is also important because the cost of incineration is often based on the total weight of the sludge and not on the solids mass. A simple calculation shows that the combustion heat is more than sufficient to heat the wet sludge and to effect the evaporation of the water fraction. It has been demonstrated before that the evaporation heat is much larger than the heat to increase the temperature of the oxidised product.

The combustion heat of dried sludge can be estimated, knowing that for an organic material the value is approximately 13.7 $kJ \cdot g^{-1}$ COD or 20.6 $kJ \cdot g^{-1}$ VSS. Thus for an initial humidity of U_i the combustion heat of 1 g wet sludge is approximately equal to $H = 20.6 \cdot f_v \cdot (1 - U_i)$. The combustion heat is larger than the evaporation heat when the humidity is low (<0.8) and the volatile fraction is high (>0.5 – 0.6). In practice the dried sludge to be processed will be in the form of cakes from mechanical drying processes (centrifuges of filter presses) with an initial humidity in the range of 70 to 80%, which, depending on the value of f_v, may not be enough for auto sufficiency. If the sludge is well dried, the process is more than auto sufficient and electric energy may be generated.

EXAMPLE 12.16

In Example 12.7 (Figure 12.20) it was estimated that in an activated sludge system with primary settling and anaerobic digestion of the primary- and secondary sludge, the fraction of influent COD transformed into sludge is 29% whereas a fraction of 25% is digested. Under these conditions evaluate the feasibility of using the combustion heat of the produced methane to dry the digested sludge after percolation (80% humidity). Assume $f_v = 0.75$.

Solution

The sludge production rate is estimated as:

$$mE_t = 0.29/(f_v \cdot f_{cv}) = 0.29/(0.75 \cdot 1.5) = 0.26\,kg\,TSS \cdot kg$$

If after percolation the humidity is 80%, then the percolated sludge production will be equal to 0.26 kg TSS and $0.26 \cdot 0.8/(1 - 0.8) = 1.05$ kg water per kg COD. The heat required to evaporate 1.05 litres of water is $1.05 \cdot 590 = 608$ $kCal \cdot kg^{-1}$COD. The anaerobic sludge digestion process produces methane at a rate of $0.25/4 = 0.06$ $kg\ CH_4 \cdot kg^{-1}$ COD with a combustion heat of $12{,}000 \cdot 0.06 = 750$ $kCal \cdot kg^{-1}$ COD. It is concluded that the potential heat generation is marginally larger than the required evaporation rate, so that theoretically thermal drying would be feasible. In reality, the generated heat will probably not be sufficient to effect the total evaporation of water due to inefficiencies in harvesting the produced methane, the combustion process and the heat exchanger in the drying tower. If anaerobic pre-treatment is applied instead of primary settling more methane will be generated and less excess sludge produced. Under those conditions the use of methane for sludge drying might be feasible.

Figure 12.33 Two anaerobic sludge digesters at the municipal WWTP Leewarden in the Netherlands. Courtesy of Friesland Water Authority

Figure 12.34 Example of a 350 kW biogas combustion engine. Upfront the engine H_2S is removed from the biogas. Courtesy of B. Heffernan

Chapter 13
Anaerobic pretreatment

13.0 INTRODUCTION

In the previous chapters, it has been shown that the activated sludge process efficiently removes suspended solids, organic material and nutrients from wastewater. The high effluent quality is the main positive feature of the activated sludge process and the principal reason for its widespread application. On the other hand, from the viewpoint of economics, the activated sludge process is not attractive at all: both investment- and operational costs are higher than those of most other biological treatment systems. This is even worse for modern developments such as the MBR, that are much more energy intensive. In this chapter it will be shown that under suitable conditions, anaerobic pre-treatment of wastewater may result in considerable cost reductions, while a high effluent quality is maintained.

Modern anaerobic processes can remove organic material at a much higher rate than the activated sludge system. In the case of sewage digestion, a large fraction of the influent organic material can be removed in systems with a retention time of only a few hours (Van Haandel and Lettinga, 1994). On the other hand, anaerobic processes cannot produce the very low residual effluent concentration of suspended solids and organic material that is feasible in a well-tuned activated sludge process. However, the residual organic load in anaerobic treated wastewater may be polished in an activated sludge process operating at a very short hydraulic retention time. It will be shown that in the case of municipal sewage and under favourable climate conditions, the total retention time (and thus volume) of a combined anaerobic-aerobic system will be significantly less than that of a conventional activated sludge process.

Due the removal of a large fraction of the influent organic material in the anaerobic pre-treatment reactor, the oxygen demand for the removal of the residual organic load in the subsequent activated sludge system is much smaller than in the case of raw or settled sewage treatment. Furthermore, the anaerobic pre-treatment unit will act as an influent equalisation tank (with respect to concentration), which will reduce diurnal variations of the oxygen demand and hence lead to a further reduction of the required maximum aeration capacity.

Another option to reduce investment costs is to stabilize the aerobic excess sludge in the anaerobic pre-treatment reactor, provided that this unit is sized to handle the additional load of suspended solids. Thus, no dedicated unit for sludge stabilisation is required. In the previous chapter it was shown that for activated sludge systems operating at a short sludge age (i.e. the activated sludge system is not designed for nutrient removal), the volume of the (unheated) sludge digester (either aerobic or anaerobic) may be in the same order of magnitude as the reactor volume. The advantage of not having to construct the digester will thus lead to a very significant reduction in investment costs.

The fact that less oxygen is required for the removal of the residual organic material also leads to a reduction in the most important factor in the operational costs: electric power. Aeration costs may be even lower if electric power is generated from the combustion of methane in an appropriate generator (explosion motor or gas turbine). It can be shown that even for diluted wastes like municipal sewage, the potential for energy production is more than sufficient to cover the demand for aeration in the post treatment unit. Another important factor contributing to operational costs in a conventional activated sludge plant is the stabilisation and final disposal of the produced excess sludge. If anaerobic pre-treatment is applied, not only can the pre-treatment unit simultaneously be used as a sludge digester, but also the sludge production will be lower as anaerobic systems tend to produce less excess sludge than aerobic systems. On the other hand, when conventional activated sludge systems are equipped with anaerobic digesters, the production of stabilised excess sludge will be reduced as well.

More important than the reduction in the mass of excess sludge produced is the advantage that the stabilised anaerobic excess sludge concentration is much higher, ranging from 40 to 80 g $TSS \cdot l^{-1}$ at the bottom to 15–35 g $TSS \cdot l^{-1}$ at the top of the digestion zone. Thus, no thickener will be required and the excess sludge production is not only smaller in mass but also significantly smaller in volume, so consequently smaller flows have to be handled and final disposal of the sludge will be easier.

The advantages of anaerobic pre-treatment as described above are especially applicable if the wastewater temperature is sufficiently high to allow application of the high-rate anaerobic digestion process: both the extent and the rate of anaerobic digestion decrease when temperature is below the optimum range of 30 to 37°C (mesophilic digestion). For temperatures below 15°C, the application of anaerobic digestion for wastewater treatment is often found to be problematic. In the case of concentrated wastes, the chemical energy of the organic material is recovered in the form of methane and can be used to increase the wastewater temperature by heating. However, in the case of low strength wastes like municipal sewage, the available energy is insufficient for a significant increase in temperature. Hence, if the temperature of a dilute wastewater like sewage is low (below 15 to 17°C), the efficiency of anaerobic pre-treatment will decrease significantly. Nevertheless, full scale experiences obtained with UASB reactors in for example the state of Paraná have shown that in the winter still a significant fraction of the influent COD (60–70%) can be removed from raw sewage, at a temperature of 14–17°C and at a hydraulic retention time of 12 hours. Furthermore, in a large part of the world the average sewage temperature in the coldest month will be higher than 18°C, which makes anaerobic pre-treatment very attractive, as it allows the hydraulic retention time to be significantly reduced.

13.1 ANAEROBIC TREATMENT OF MUNICIPAL SEWAGE

The performance of anaerobic wastewater treatment processes has improved dramatically during the last few decades, due to a better understanding of the nature of anaerobic digestion. While classical reactors like the septic tank and anaerobic ponds have a rather low removal efficiency of suspended solids and organic material and require a very long retention time, more recent designs obtain a higher efficiency at a much shorter retention time. Modern anaerobic processes have two main characteristics that are responsible for this improvement: (l) they are equipped with devices to retain a large sludge mass in the process and (2) the flow of the wastewater through the systems is such that intense contact between the incoming organic material and the sludge is ensured. To ensure the presence of a large sludge mass, modern anaerobic treatment systems have dedicated solid-liquid separators to retain the sludge in the process.

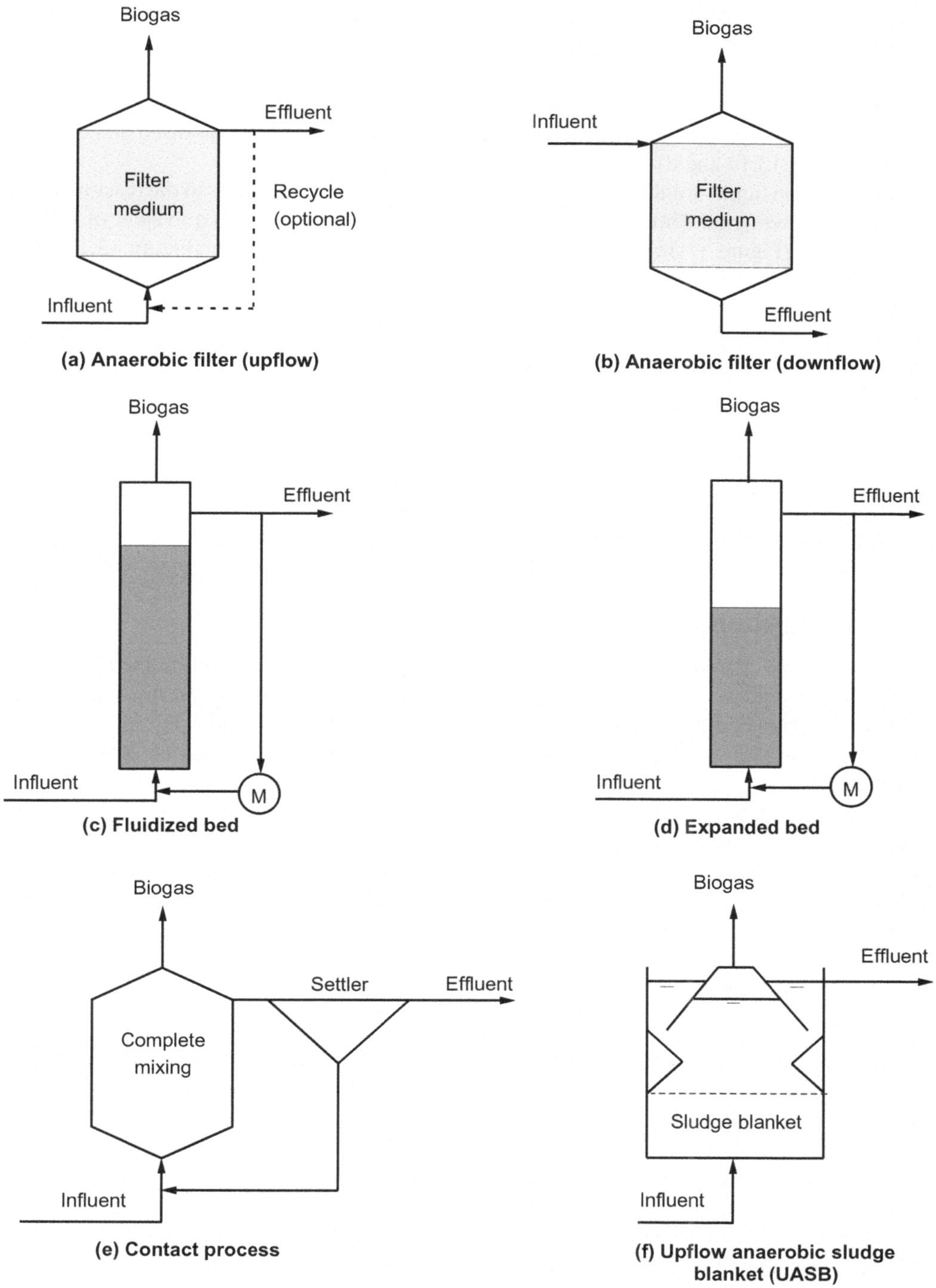

Figure 13.1 Basic configurations of modern high rate anaerobic systems

Figure 13.1 shows schematic reactor configurations of several modern anaerobic treatment systems. Basically two mechanisms of sludge retention can be distinguished:

- Systems based on sludge immobilisation, e.g. by attachment of the sludge to a solid carrier material. The upflow and downflow anaerobic filter (Figure 13.1a and b) and the fluidised and expanded bed reactors (Figure 13.1c and d) all belong to this category;
- Systems based on liquid-solid separation, with return of the separated solids to the reactor. The contact process uses an external settler and can be considered as the anaerobic equivalent of the activated sludge process (Figure 13.1e). The UASB reactor uses an internal settler (Figure 13.1f). A special case is the anaerobic SBR, where the reactor also functions as a settler i.e. there is no special separation device;
- The industrial UASB and EGSB (Expanded Granular Sludge Bed) systems that use granular sludge without support material may be considered as hybrid designs, as they rely on both biomass immobilization and liquid-solid separation (refer to Section 13.6).

Although the different types of modern anaerobic treatment systems have been applied to a great variety of industrial wastes, so far anaerobic sewage treatment has been restricted mainly to classic systems like the septic tank (for localised treatment) and anaerobic ponds. Modern anaerobic systems for sewage treatment are at present mostly used in countries with a warm climate, while application to sewage treatment in a moderate to cold climate is still very limited, although (heated) anaerobic sludge digesters are often used.

13.1.1 Configurations for anaerobic sewage treatment

In order to compare the different anaerobic treatment systems, one must first define the objectives of the anaerobic treatment system. These will depend upon the particular situation at the treatment site and the use of the effluent. However, invariably the system should offer the highest possible removal efficiency of organic material in the shortest possible hydraulic retention time (i.e. the volume of the system must be as small as possible). For this reason, it is useful to assess the relationship between the removal efficiency of organic matter and the hydraulic retention time for different anaerobic treatment systems. As the kinetics of the anaerobic digestion systems are not yet sufficiently developed to allow a satisfactory prediction of this relationship, it is necessary to establish empirical relationships based on experimental results. Experimental results obtained in full scale or pilot plants treating sewage will now be evaluated for different anaerobic treatment systems.

13.1.1.1 Anaerobic filter

The anaerobic filter (AF) process is mainly used for industrial wastewater treatment, though at a rather limited scale. It has been shown that organic loads of up to 10–20 kg $COD \cdot m^{-3} \cdot d^{-1}$ can be applied, when the concentration and nature of the organic matter are favourable. An important disadvantage of the AF-system concerns the high price of many carrier materials. Full-scale AF-systems have been implemented for the treatment of various types of industrial wastewaters, but for sewage the system is not used at large scale. When applied to sewage treatment, a septic tank is often used as a pre-treatment unit to reduce the risk of clogging in the filter bed.

Performance data of several pilot and bench scale anaerobic filters treating sewage are combined in Figure 13.2a, where the COD removal efficiency is plotted against the hydraulic retention time in a log-log diagram. The empirical equations proposed by Young (1990), which are also indicated in Figure 13.2a, are based on observations of anaerobic upflow filters with both loose and modular filling media. There is a considerable spread in the experimental data, which may be attributed partially to the differences in sewage characteristics (temperature and origin of sewage) and partially on the specific surface area of the carrier material in the filter. However, from the data in Figure 13.2a it is clear that there is a trend towards the following empiric relationship:

$$\log S_e/S_i = -c_1 \cdot \log(R_h) + c_2 \quad \text{or} \tag{13.1}$$

$$\eta_{COD} = 1 - S_e/S_i = 1 - c_2 \cdot (R_h)^{-c1} \tag{13.2}$$

where:

S = substrate concentration (COD), indices i and e refer to influent and effluent
η_{COD} = COD removal efficiency
c_1, c_2 = empiric constants

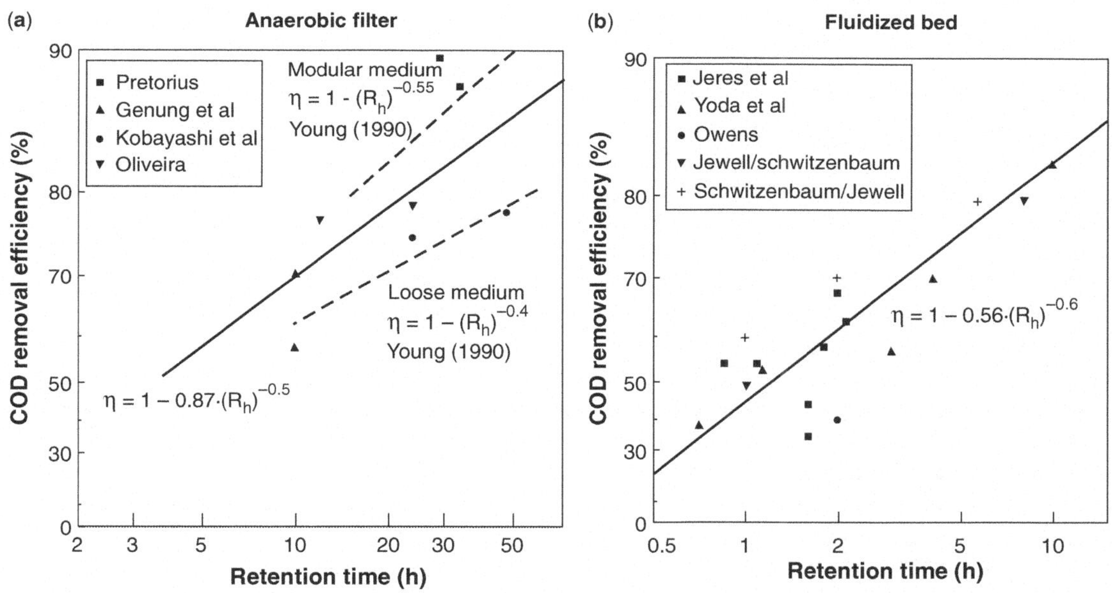

Figure 13.2 COD removal efficiency of raw sewage treatment as a function of retention time for anaerobic filters and fluidised bed reactors (on log-log scale)

From the experimental data for the anaerobic filter in Figure 13.2a, it follows that $c_1 = 0.50$ and $c_2 = 0.87$ (with R_h in hours) and hence Eq. (13.2) becomes:

$$\eta_{COD} = 1 - 0.87 \cdot (R_h)^{-0.50} \tag{13.3}$$

13.1.1.2 Fluidised and expanded bed systems

In the fluidised bed (FB) system, introduced by Jeris (1982), the carrier material consists of a granular medium, which is kept fluidised as a result of the frictional resistance of the upward wastewater flow.

Sand was initially used for the granular medium, to be replaced later by media with a lower density like anthracite and plastic materials, in order to reduce the required upflow liquid velocity and consequently also the pumping costs. In practice, considerable difficulties have been experienced in controlling the particle size and the density of the flocs. In fact, stable process performance often proved impossible.

The fixed film expanded bed (AAFEB) reactor differs from the fluidised bed concept because of the much lower upflow velocities applied; the sludge bed is only expanded by 10 to 20% (Jewell, 1982). Currently there is not much data available on full-scale installations for sewage treatment using the fluidised or the expanded bed systems, but several pilot and bench scale studies have been carried out.

Figure 13.2b shows experimental results of fluidised and expanded bed reactors in terms of organic matter removal efficiency as a function of the retention time. As in the case of anaerobic filters, there is a considerable spread in the results, but also a trend to a linear relationship on log-log scale. This relationship can be expressed by the following equation:

$$\eta_{COD} = 1 - 0.56 \cdot (R_h)^{-0.60} \tag{13.4}$$

13.1.1.3 Upflow Anaerobic Sludge Blanket (UASB) reactor

The Upflow Anaerobic Sludge Blanket (UASB) reactor was developed in the seventies by Professor Lettinga and his group at the University of Wageningen in the Netherlands. The UASB reactor is by far the most widely used high rate anaerobic system for anaerobic sewage treatment. Many full-scale plants, with capacities up to 350,000 $m^3 \cdot d^{-1}$, have been started up in recent years and many more are under construction. All full-scale UASB sewage treatment plants operate under tropical or subtropical conditions in countries like India and Brazil, while industrial UASB and EGSB treatment (high-strength wastewater) is also applied frequently in moderate to cold climates (refer to Section 13.6), as the produced biogas can be used to heat the reactor.

Figure 13.3 contains a schematic representation of the UASB system, while Figure 13.4 shows an artist impression of the recently developed Biothane municipal UASB (Upthane^{TM}). The most characteristic device of a UASB reactor is the three-phase separator, which is located on top of the reactor and divides it into a lower part, the digestion zone and an upper part, the settling zone. The wastewater is introduced as uniformly as possible on the reactor bottom, passes through the sludge bed and enters into the settling zone via the apertures between the phase separator elements. Due to the inclined walls of the phase separator, the area available to the liquid flow in the settling zone increases as it approaches the water surface, so that the upflow velocity of the liquid decreases as the liquid flows towards the discharge point. As a result of the decrease in liquid velocity, sludge flocs drawn into the settling zone are allowed to flocculate and/or settle out. Eventually, the weight of the accumulated sludge deposited on the phase separator will exceed the frictional force that keeps it on the inclined surface, causing it to slide back into the digestion zone. Thus, the presence of a settler on top of the digestion zone enables a high sludge concentration in the UASB reactor, while an effluent with a low concentration of suspended solids is discharged. The biogas bubbles rise up to the liquid-gas interface under the phase separator. This interface may be at the same level as the water-air interface in the settler or at some lower level if the gas hoods are pressurized, for example with an hydraulic seal (see Figure 13.3). Sludge flocs with adhering gas bubbles may rise up to the interface in the gas collector, but will settle when the gas bubbles are released to the gas phase. Baffles, placed beneath the apertures of the gas collector units, operate as gas deflectors and prevent biogas bubbles from entering the settling zone, where they would create turbulence and consequently interfere with the settling of sludge particles.

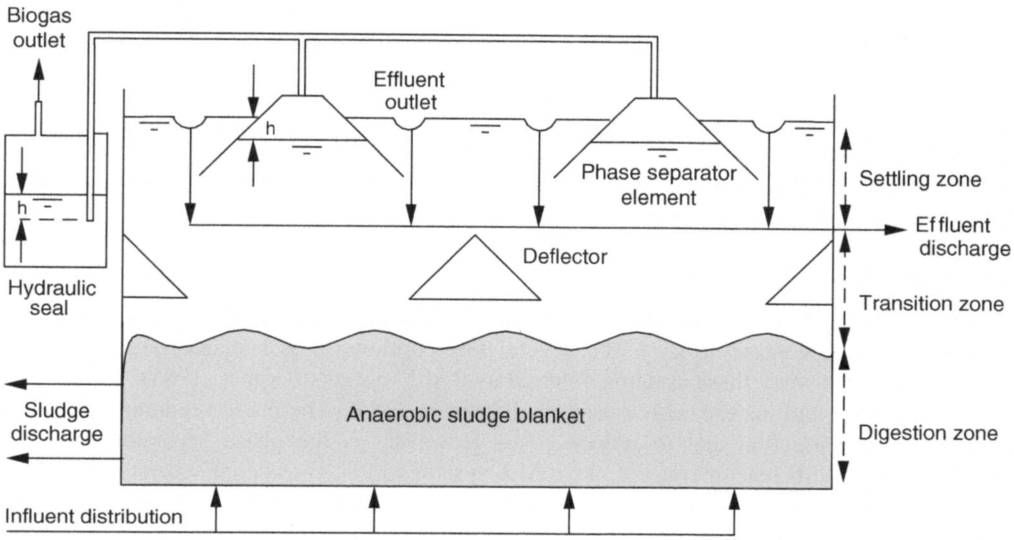

Figure 13.3 Schematic layout of the design of a typical municipal UASB reactor

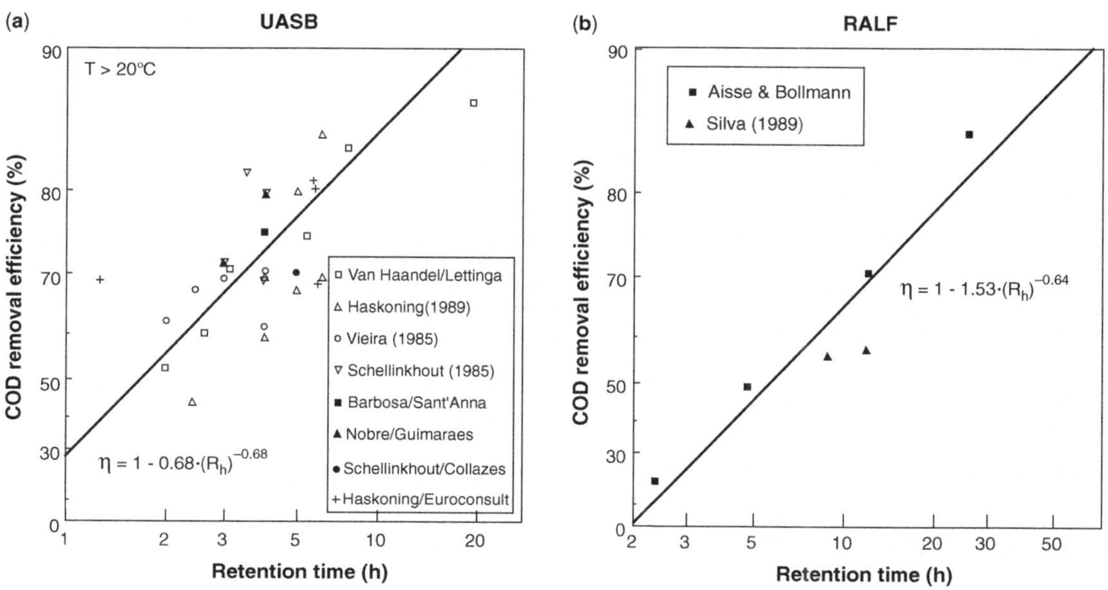

Figure 13.4 COD removal efficiency of raw sewage treatment as a function of retention time for UASB and RALF reactors (log-log scale – results up to 1990)

From the results obtained by the different authors in Figure 13.4a the following empiric equation may be derived:

$$\eta_{COD} = 1 - 0.68 \cdot (R_h)^{-0.68} \tag{13.5}$$

Later research by Cavalcanti et al. (2003) has yielded detailed empiric equations which can be used to predict the division of the influent COD over the effluent, the digested fraction and the excess sludge as a function of the anaerobic sludge age (Section 13.3).

13.1.1.4 The RALF system

In the Brazilian state of Paraná (predominantly subtropical, although the average sewage temperature in the coldest month typically decreases to 15–17°C), several dozen upflow sludge bed anaerobic treatment plants have been installed. However these systems, referred as RALF-systems (Gomes, 1985), are not equipped with a phase separator, but instead with a very small built-in settler. The phase separator was omitted in order to simplify construction and to reduce costs. However, as the phase separator constitutes an essential part of the UASB reactor concept, the RALF system will be considered separately.

Figure 13.5 is a schematic representation of a large RALF unit; while Figure 13.4b shows experimental results obtained with full-scale and pilot RALF installations. From the experimental data the following expression can be derived:

$$\eta_{COD} = 1 - 1.53 \cdot (R_h)^{-0.64} \tag{13.6}$$

It seems the concept of the RALF, at least in the form as shown in Figure 13.5, has been abandoned as the most recent large anaerobic reactors constructed in Paraná are now also of the UASB type, although with a modified GLS design (Heffernan et al., 2010).

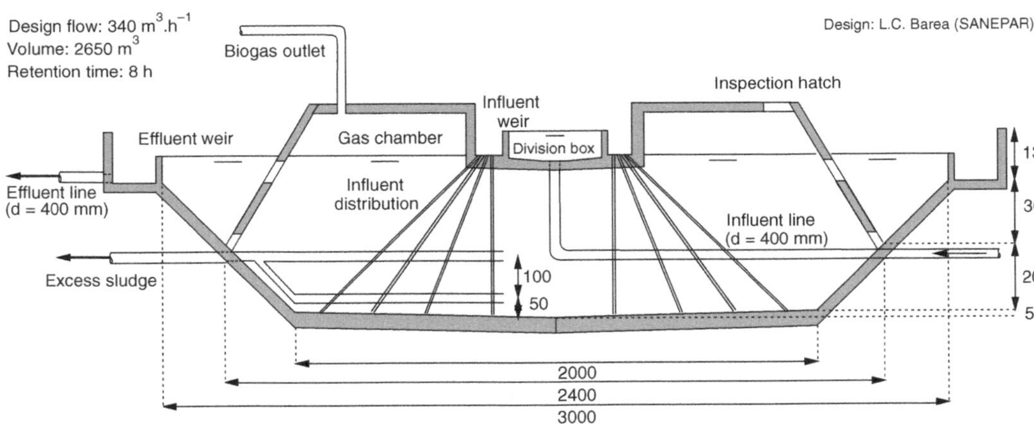

Figure 13.5 Schematic representation of a RALF unit built in Londrina–Brazil

13.1.2 Evaluation of different anaerobic configurations

In order to compare the different anaerobic treatment systems, their performance in terms of treatment efficiency is evaluated as a function of the sewage retention time. Figure 13.6 provides a comparison of the data derived from those shown in Figures 13.2 and 13.4. As only limited experimental data were available at the time these figures were created and operational and environmental conditions were diverse, the actual efficiency in the different treatment systems may deviate significantly from the predicted value. Nevertheless Figure 13.6 shows some interesting tendencies:

- For sewage temperatures higher than 20°C, obtaining a COD removal efficiency of 70–80% is possible for all of the systems considered, but the required retention times differ considerably;
- In the range of retention time values of practical interest, the performance of a UASB reactor and a fluidised/expanded bed reactor tends to be similar;
- The performance of a well designed UASB system is definitely superior to that of the sludge bed reactor not equipped with a settler (RALF) and the anaerobic filter.

In practice, the suitability of a treatment system is not determined by the required reactor volume alone. Other advantages and disadvantages of the treatment options should also be considered. Some of the processes presented above have such serious drawbacks, that they hardly offer any serious prospect for large scale application. The main negative aspects of the anaerobic filter are the high construction costs and particularly the probability of operational difficulties due to blockages, especially when treating wastewater with a high concentration of suspended solids such as raw sewage.

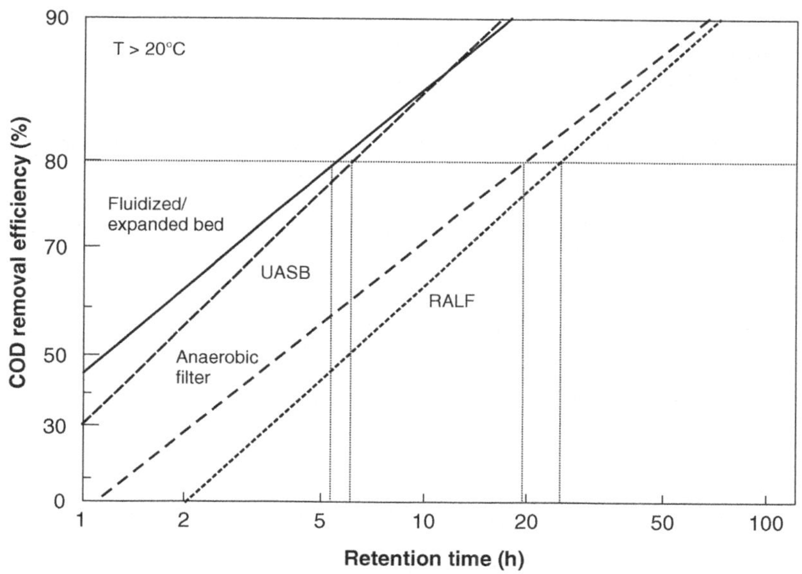

Figure 13.6 COD removal efficiency as function of R_h in several anaerobic systems (based on early pilot and full-scale results obtained up to 1994)

From Figure 13.6 it is concluded that for the same removal efficiency, the RALF reactor requires a retention time exceeding that of a UASB reactor by a factor of 4 to 5. Therefore in most cases, even though the three phase separator of the UASB can be omitted from design, the RALF reactor will be more expensive than a comparable UASB. The costs of the separator will be amply compensated by the reduction of the required reactor volume. When comparing the UASB reactor with the fluidised- and expanded bed reactors, it is clear that latter two systems have important disadvantages because of their operational instability and their need of additional pumping to maintain the carrier material in suspension.

The UASB reactor dispenses with the need for any pumping, providing that sufficient differential head is available at the inlet of the distribution system. The fluidised bed does not seem to be very suitable for sewage treatment because of the difficulties of retaining influent suspended solids in the reactor and of maintaining a large sludge mass in the reactor. Consequently the UASB concept is currently the most attractive option for anaerobic wastewater treatment and by far the most popular choice for high rate anaerobic systems in new implementations. Anaerobic filters are sometimes used as a secondary (polishing) treatment step (mainly for small installations in Brazil), while anaerobic ponds are still applied in those areas where the required technical skills are not available.

13.2 FACTORS AFFECTING MUNICIPAL UASB PERFORMANCE

The potential for anaerobic treatment in regions with a warm climate is obvious, considering the attainable COD removal efficiency of 70–80% at relatively modest hydraulic retention time, as reported in many of the early pilot and full-scale UASB studies (Figure 13.4). However, when the performance of some of the recently constructed large full-scale UASBs is considered, the results are less favourable (Chernicharo, 2007 and Oliveira et al., 2009). The latter evaluated UASB performance based on a review of recorded plant operating data, i.e. without visiting the plants involved. In Figure 13.7 the COD removal efficiency as predicted by Eq. (13.5) is compared to the findings of Chernicharo (2007) and to the predicted values according to the theoretical model that will be presented in Section 13.3. The latter graphs were calculated based on the following data: $S_{ti} = 550$; $f_{ns} = f_{np} = 0.1$; $SO_4^{2-} = 50\ mg \cdot l^{-1}$, of which 75% is assumed to be converted to H_2S; $X_{tu} = 17\ g \cdot l^{-1}$ and $X_{te} = 80\ mg\ TSS \cdot l^{-1}$, which is fairly conservative for a well designed GLS. The use of a small pre-aeration reactor for (partial) sulphide oxidation increases COD removal with an additional 2–5%. Finally, Figure 13.7 also indicates the results from an extensive field survey conducted in 2009 (Heffernan et al., 2010), including several large full-scale UASBs located in Brazil, India and the United Arab Emirates), with installed capacities ranging from 30,000 to 164,000 $m^3 \cdot d^{-1}$.

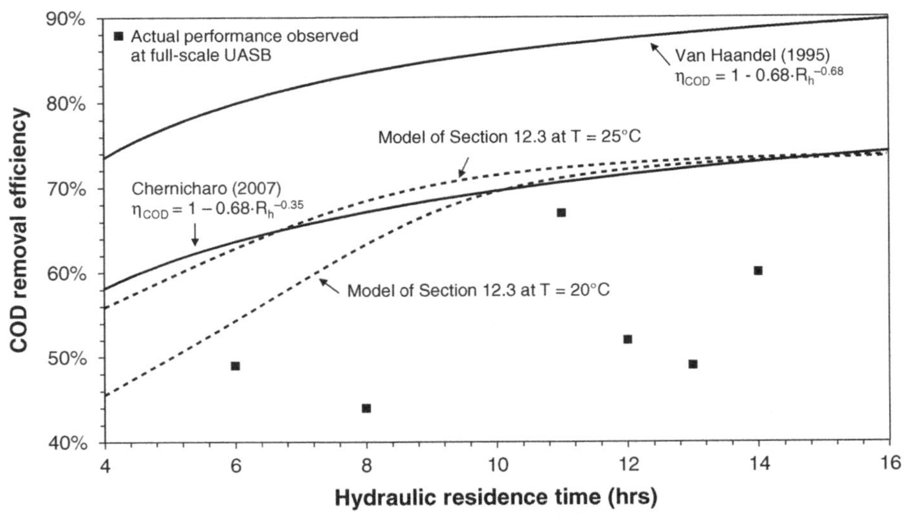

Figure 13.7 Observed COD removal efficiency in large UASB reactors by Heffernan et al. (2010)–black squares, compared to the predicted values according to Van Haandel (1995), Chernicharo (2007) and the theory in Section 13.3

An extensive report of the findings with respect to plant design, operation and performance can be found in Heffernan et al. (2010); Van der Lubbe et al. (2010) and in the chapter written together with Van Lier (2010). The authors concluded that the disappointing treatment results were not caused by any inherent flaw in the concept of UASB treatment. On the contrary, for each of the UASBs visited, the lower than expected COD removal efficiency could be fully explained by errors in design- and/or engineering, lack of qualified supervision and insufficient maintenance. Some of these findings will be detailed in the subsequent sections.

Another finding from the review study is that in the last two decades the progress regarding the design of municipal anaerobic sewage treatment plants has been rather slow. Notable exceptions are the work of Chernicharo (2005 and 2007), who provided valuable contributions on reactor design and engineering, construction and operational procedures and the research from the team of Professor Van Haandel on the subjects of process design and system behaviour (Cavalcanti et al., 2003 and Silva Filho et al., 2005), which will be discussed extensively in Section 13.3. Notwithstanding these individual endeavours, the basic design of the UASB is still pretty much the same as it was in the early 1990s, summarized by Van Haandel et al. (1994). However, to quote the famous Dutch soccer player Johan Cruyff "Elk nadeel heb z'n voordeel", there seems to be ample room for improvement in UASB design and engineering. A possible reason for the slow development of the UASB technology might have been its "hippie" origin, as originally the UASB was meant as a low-cost and low-tech technology to provide at least partial sewage treatment in developing countries. To minimize power consumption, maintenance requirements and taking into account the unavailability of skilled labour in the poorer countries, the design was based on one time pumping at the head of the STP with gravity flow through the rest of the installation. As a consequence the current design of the UASB reactor tends to be rather complicated. For example, the influent distribution system is mounted on the top of the reactor while flow is upward, which results in many conflicts during design and construction.

It should also be considered that without a suitable infrastructure in terms of technical and institutional capabilities, even the low-tech UASB will ultimately fail. Reactors will fill up, effluent weirs are misaligned and influent systems will block. Implementation of UASB reactors in countries without an appropriate infrastructure is therefore not recommended. On the other hand, the UASB will be continue to provide at least partial treatment (typically around 50% COD removal) for a long time, even when abused. Fortunately a significant part of the formerly underdeveloped world is now experiencing rapid development, for example countries like Brazil, India and China, resulting in increased technical and financial capabilities. Simultaneously stricter effluent limits often require application of more complex aerobic post-treatment. So there is no need to maintain the rock-bottom low cost approach towards the UASB and significant improvements in municipal UASB design (or perhaps more generally in anaerobic sewage treatment) are possible. Lessons can be learnt from the more advanced designs that are commonplace in the anaerobic treatment of industrial wastewater.

13.2.1 Design and engineering issues

In this section a number of key issues observed in the review of existing full-scale UASBs are summarized (Heffernan et al., 2010 and Van der Lubbe et al., 2010):

- Inadequate process design;
- Improper pre-treatment;
- Incorrect flow division and distribution;
- Poorly constructed effluent collection system;
- No provisions for scum removal and/or poor access for cleaning;

- Poor means of sludge sampling and excess sludge withdrawal;
- Poor quality of gas collection and utilization devices;
- Improper material selection and poor quality of construction.

(a) Inadequate process design

The first requirement for a well-performing sewage treatment plant is that the process design is adequate. The following issues were often observed:

- Failure to take into account the detrimental effect of sulphate on anaerobic process performance (Section 13.2.4). Prior to starting a new project the sulphate concentration in the raw sewage needs to be determined or estimated (for instance from data of drinking water production companies);
- Use of the UASB for aerobic excess sludge stabilization has often been cited as an important advantage. In our view for several reasons this is not recommended (Section 13.5.1.2). However, the effect is made much worse when the additional solids load is not explicitly considered during design. The aerobic sludge is only partially degradable under the conditions prevailing in the UASB and the undegraded solids will accumulate as an inert organic fraction in the anaerobic sludge. A significant reduction in anaerobic sludge age will result, which in turn reduces anaerobic treatment efficiency, increases the organic load to the aerobic post-treatment system, which ultimately results in more aerobic excess sludge being sent to the UASB and so on;
- Return of backflush water from filter units (e.g. sandfilters or submerged aerated filters) to the UASB creates a similar problem. Apart from the presence of poorly degradable suspended solids which decrease the anaerobic sludge age, the hydraulic load to the UASB is increased as well.

(b) Improper pre-treatment

Three constituents need to be removed from the raw sewage prior to treatment in an UASB:

- Sand and grit, which otherwise accumulate in the digestion zone of the UASB, reducing the volume available for anaerobic biomass and potentially blocking the inlet pipes. In general a square grit chamber with bottom scraper is sufficient;
- Large and fine debris such as fibres and hairs, that cause blockages throughout the whole influent distribution system, most notably in the inlet pipes and also at the overflow weirs, while they also tend to promote scum formation. In general a 40 mm bar screen followed by a 6 mm step screen is considered the most cost-effective combination to remove most of the debris. Raking mechanisms for screen cleaning should be avoided, as the rakes tend to push the debris through the bars. Improper construction of the screens require high operator attendance for cleaning and unblocking of the distribution system and if blockages left unchecked, severe short-circuiting in the UASB may result;
- Free oil, grease and fat (OGF). Given enough time and provided that the mixing intensity is sufficient, these components are perfectly biodegradable. However, in the UASB the conditions for degradation of this material are less favourable and formation of a scum layer is almost unavoidable. Apart from a nuisance, scum can block the overflow weirs and fill up the gas hoods. An aerated sand trap can significantly reduce the rate of scum formation as at least part of the free OGF will be removed. However, these systems are currently seldom installed in the pre-treatment section of an UASB.

(c) Incorrect flow division and distribution

Proper distribution of the inlet flow over the reactor surface area is important for proper UASB performance. As each inlet pipe typically covers 2–4 m^2 of reactor surface area, the inlet flow has to be divided into many equal parts. Typically a series of distribution boxes is used, which divide the flow over the individual reactors. From there the final division over the individual inlet pipes is achieved through one or more final distribution boxes. At every distribution box, it is important that a free fall overflow is ensured, even at peak flow. Assuming that the overflows are correctly aligned (often they are not!), then equal flow division is ensured.

If the height of the free fall overflow is insufficient and flooding of one or several outlets occurs, then differences in the pressure drop over the outlets will determine the flow through each pipe and imbalances may occur. To facilitate even distribution, it is also recommended to construct the distribution box as symmetrical as possible. Although all of this seems obvious, in many plants the influent distribution system was poorly designed and constructed, as can be observed in Figure 13.8. Examples of a well-designed final distribution box layout are shown in Figure 13.9.

Figure 13.8 A poorly designed final distribution box (3 out of 6 overflows not visible) resulting in uneven flow distribution (left). Covering these boxes resulted in the formation of scum, with a further impact on distribution (right) Courtesy of B. Heffernan

Another design issue is the installation of pipes with multiple outlets (Figure 13.10). While this saves on material costs, blockages of individual openings will be difficult to detect, as to a certain extent the flow will be divided over the other openings. Even if eventually the blockage is detected, cleaning of the plugged opening is by no means guaranteed. In contrast, in the one pipe – one outlet configuration the blockage of a pipe will be immediately detected as the liquid level in the corresponding inlet box will increase. Unblocking the pipe means unblocking the opening: the success of cleaning is indicated by the resulting decrease in liquid level in the inlet box.

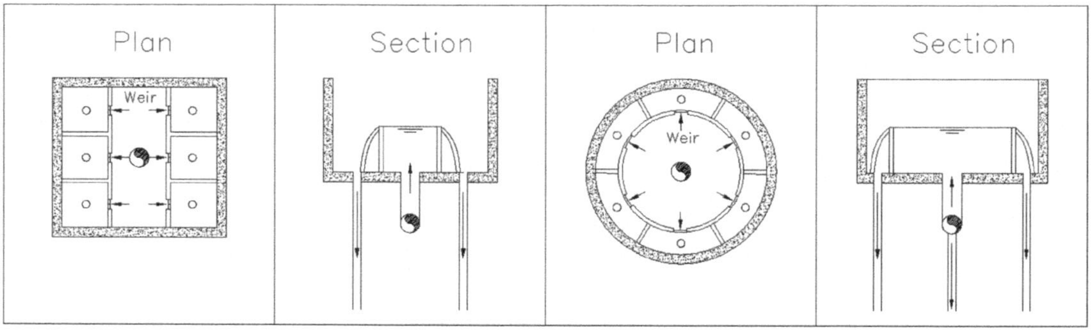

Figure 13.9 A schematic representation (plan and section) of a properly designed rectangular (square) and circular final distribution box. Courtesy of B. Heffernan

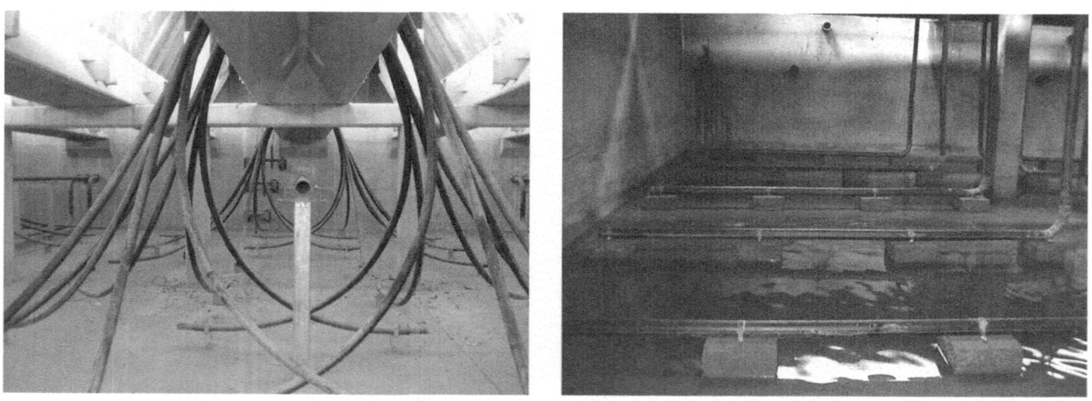

Figure 13.10 Recommended inlet distribution system, where each inlet pipe discharges at a single outlet point (left) and the alternative configuration where each inlet pipe has several discharge points (right). Courtesy of B. Heffernan

(d) Poorly constructed effluent collection system

Effluent withdrawal should be balanced over the surface of the reactor, in order to prevent short-circuiting between influent and effluent, resulting in localized high overflow velocities and the occurrence of dead (stagnant) zones in the reactor. In practice, two collection systems are used: effluent gutters fitted with overflow weirs and submerged perforated pipes.

If effluent gutters are installed, it is recommended to equip the overflow weirs with V-notches (triangular overflows) as this allows for easier alignment. However, frequently too many V-notches are used, i.e. more than the 1–2 per m^2 reactor area as recommended by Van Haandel et al. (1994). The resulting flow rate per

V-notch will then be too low and the liquid level above the base of the V-notch will be insufficient to facilitate weir alignment. Another issue to consider is that at the low liquid height resulting from an excessive number of V-notches, a small misalignment has proportionally a much larger effect on the overflow rate than at a higher liquid height. This can be demonstrated when one considers the following equation, which calculates the flow rate in $m^3 \cdot s^{-1}$ as function of the liquid height above the base of the V-notch (in meters) and the base angle α:

$$Q = 1.465 \cdot H^{2.5} \cdot \tan(0.5 \cdot \alpha) \quad (13.7a)$$

Consider the effect of a misalignment error of only 2 mm for a 90° V-notch. At low liquid height, the flow per V-notch increases for example from 0.053 $m^3 \cdot h^{-1}$ at 10 mm to 0.083 $m^3 \cdot h^{-1}$ at 12 mm, an increase of 58%. The same misalignment error of 2 mm, but now at increased liquid height above the base of the V-notch has a much smaller effect: for example 0.52 $m^3 \cdot h^{-1}$ at 25 mm to 0.63 $m^3 \cdot h^{-1}$ at 27 mm or an increase of only 21%.

Yet another detrimental effect of an excessive number of V-notches is that it will make the weir susceptible to blockages by scum. Due to the low flow rate the V-notches are no longer self-cleaning: the scouring intensity is insufficient to remove the blockage. This even applies when scum baffles are used. Once one V-notch is blocked, the floating scum layer starts to accumulate and soon adjacent V-notches will be blocked as well.

At several full-scale plants the quality of construction of the effluent weir was very poor, in one case with up to 3–5 cm difference in weir level over the length of the reactor. Obviously the difference in local overflow rates must have been enormous. The effluent gutter of another reactor did not have any V-notches at all and consequently whole sections of the effluent collection channel received no flow at all.

If submerged pipes are used then alignment is easier, while the effect of a remaining misalignment on reactor performance will be significantly reduced. The flow rate through a circular perforation is much less sensitive to variations in liquid height than the flow over a V-notch. The flow rate through a submerged perforation can be described as:

$$Q = \omega \cdot A \cdot (2 \cdot g \cdot H)^{0.5} \quad (13.7b)$$

where ω = contraction coefficient (0.66 for a circular opening), g is the gravitational contant (9.81 $m \cdot s^{-2}$) and H is the liquid level above the perforation (m). In this case, an increase from 10 to 12 mm yields an increase in flow rate from 0.57 to 0.63 $m^3 \cdot h^{-1}$ (+12%), while an increase from 25 to 27 mm results in an increase from 0.85 to 0.89 $m^3 \cdot h^{-1}$ (+5%). Thus, it is obvious that the submerged pipe is indeed less vulnerable to misalignment.

Another advantage of submerged pipes is that less turbulence is created on the surface area, thus minimizing the emissions of CH_4 and H_2S to the atmosphere, as corroborated by the results of Souza et al. (2010b), discussed in Section 13.2.5. This offers the possibility of localized vent gas collection in the effluent collection system. Finally, the risk that a perforation will be blocked by scum is low, provided that the velocity through the opening is sufficiently high.

(e) No provisions for scum removal or poor access for cleaning

Scum problems are a real nuisance for UASB reactors at two locations: (I) the surface of the settling compartment and (II) the gas-liquid interface under the gas hoods. As discussed previously, scum formation is almost inevitable due to the presence in the raw sewage of fats, oil and grease as well as surfactants and larger floating debris. This is partly a dietary issue, as the problems observed are much more serious in Brazil where meat consumption per capita is high, compared to India with predominantly vegetarians. Proper pre-treatment will significantly reduce the rate of scum formation and has been discussed previously. In principle scum formation on the reactor surface does not have to be a problem, provided scum baffles are not used, as the scum is then discharged together with the effluent over the effluent weir. However, as discussed above, the flow rate over the V-notches is often so low that they are easily blocked, as demonstrated in Figure 13.11. Once a V-notch is blocked, a scum layer will start to accumulate and rapidly spreads out over the surface. Cleaning V-notches is easy, but the scum problem is compounded by the recent trend to cover UASB reactors in order to prevent odour problems. Apart from the question whether covering the entire reactor is really necessary, as most of the odour is released at the overflows, the use of fixed concrete covers makes inspection and cleaning of the V-notch weirs very difficult, even if inspection hatches are installed. Should covers be required, then light-weight removable covers are recommended.

(a) (b)

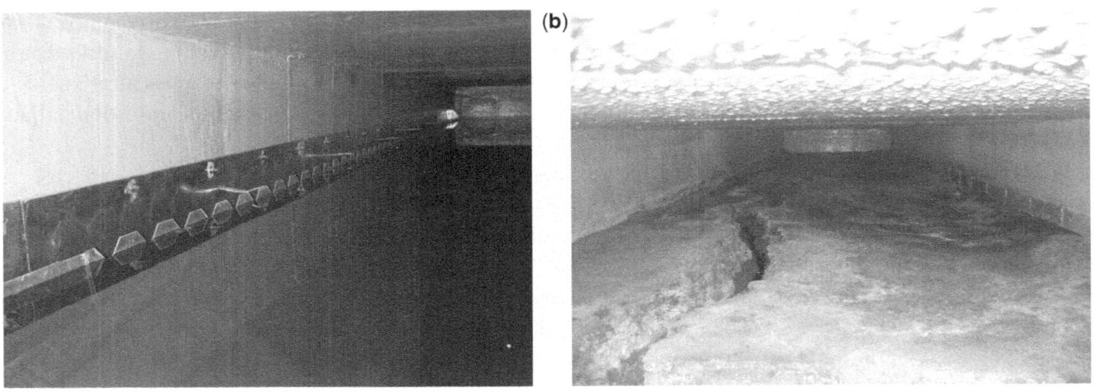

Figure 13.11 Scum accumulation in the settler section (no scum baffle installed). The effluent gutter is clearly visible during commissioning (a) but after 1 year a scum layer of approximately 30 cm in thickness has developed (b). Courtesy of B. Heffernan

Under the gas hoods scum formation is a more serious problem, as access is difficult. Release of gas from the liquid phase will be much more difficult when a hard and thick scum layer is present. When release of gas is erratic and violent, solids (scum) can be entrained in the gas piping and create obstructions. The resulting increase in gas pressure might damage the gas hoods. Furthermore the gas hoods may fill up completely with scum at which time erratic and uncontrolled discharges of gas into the settling zone may occur over the length of the gas hood. At this point there may be no alternative but to drain the reactor and clean it.

Removal of scum from the gas hood is difficult. Some plants have resorted to the installation of access hatches on top of the gas box. However, the construction of an access hatch that remains gas tight after repeated opening and closing has proven difficult, even when the size of the hatch is minimised. Insufficient gas-tightness of the gas hoods may again result in odour problems while it also prejudicates operation at the gas pressure required to transport the biogas to its point of treatment and utilization. Furthermore, as gas boxes are long and narrow, cleaning it from the restricted access provided by the hatch is a daunting task indeed.

An alternative approach was followed by Chernicharo et al. (2009), who developed a scum removal system to be installed inside the gas hood. It is based on the observation that development of a scum layer progresses through several stages, from a thin and liquid layer to a more viscous fluid and then to the final stage of a thick and crusted (solid) layer. In its liquid phase, the scum can still be withdrawn from the gas hood by means of gravity flow. Thus, periodically the contents of the gas hood are removed through a sequence of steps, for example as shown in Figure 13.12. Note that this procedure can only work if the gas boxes are gas tight. As an alternative to increasing the liquid level in the gas box through a reduction of the gas pressure, it is also possible to close the effluent line. An emergency overflow should be provided to prevent uncontrolled effluent spillage during the discharge procedure.

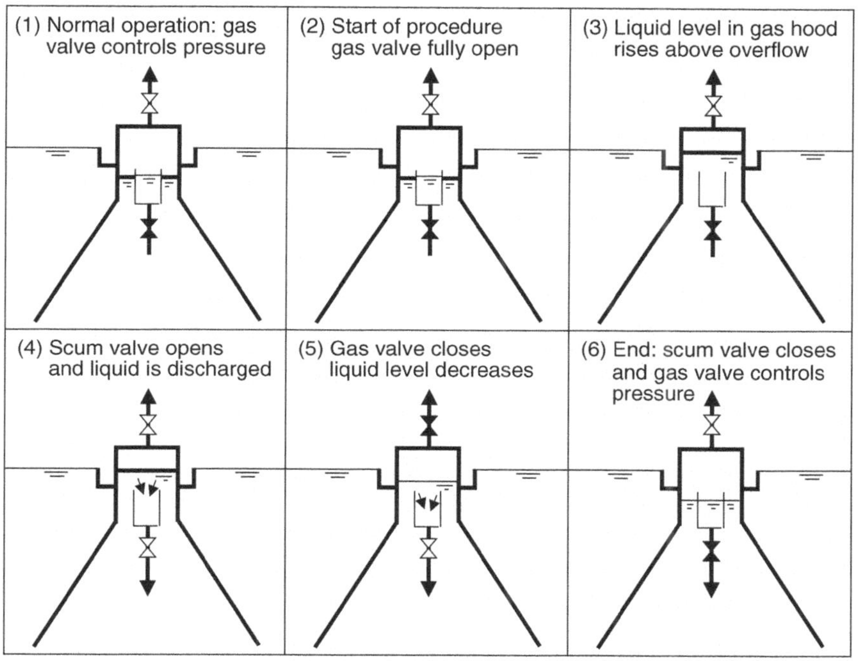

Figure 13.12 Scum removal from the gas hoods. Adapted from Chernicharo et al. (2009)

The key factor to sustained and successful operation of the scum removal system is to prevent the scum layer from turning into a hard or viscous layer…ever. As an example, in one of the plants visited a scum removal system was installed, but it had not been used in the first year after start-up due to commissioning problems in the sludge dewatering unit. As a result, the whole scum discharge system had become completely blocked by scum, meaning there is no way of rehabilitating the system except a complete reactor drain. Thus, scum disposal should take place on a regular basis. It could even be considered to automate the procedure.

Note that the blockage of the scum removal system could have been prevented if the scum would have been discharged into the effluent line and not to the sludge treatment units. As long as the scum discharge flow is high enough, the scum will be very diluted so that the COD content will be low. The decrease in UASB effluent quality will be very small while the scum will be removed in the aerobic post-treatment system anyway. Furthermore, this set-up also decreases the hydraulic load to the sludge treatment units.

Therefore, in our view, notwithstanding the problems associated to installing and operating a scum removal system in the gas hoods, it is strongly recommended to do so, as the alternative of emptying and cleaning a UASB reactor is a very labour intensive task.

(f) Poor means of sludge sampling and excess sludge withdrawal

In principle the treatment objectives of an UASB will be to maximize COD removal and/or to produce a stabilised sludge. Therefore the UASB should be operated at the highest sludge age possible, i.e. with the highest biomass inventory possible for the given reactor volume. This requires operation at the highest sludge blanket level possible and at maximum biomass concentration. A first condition for proper UASB solids inventory management is that it should be possible to determine the height of the sludge blanket accurately. Generally sludge sampling connections are installed in the reactor wall for this purpose. However, the number of sample connections is often inadequate. Take for instance an UASB with only three sampling points installed over the height of a 2.5 m digestion zone (at 0.3, 0.9 and 2.0 m). When the top of the sludge blanket is higher than 2 m from the reactor floor it can no longer be measured. The only tell-tale sign that the maximum sludge blanket height is exceeded is when the concentration of suspended solids in the effluent starts to increase.

A second problem with this method of sludge sampling is that the sampling pipes are often mounted close the reactor wall (refer also to Figure A3.3). It is not altogether certain if the concentration measured at the wall represents the true average reactor sludge concentration at that height, as mixing tends to be less intense near the walls. A much better (and simpler) solution is to install a sampling connection through the top of the gas hood and to sample the sludge blanket from above. This not only allows exact determination of the sludge blanket height but also allows sample collection from any desired height.

A proper system of sludge withdrawal is also important. In general it is recommended to be able to withdraw sludge from at least two levels in the sludge blanket. The reason for this is that a sludge profile will develop in the digestion zone with the more dense and better settling sludge (50–80 g $TSS \cdot l^{-1}$) at the bottom of the reactor and the less dense sludge (15–30 g $TSS \cdot l^{-1}$) at the top of the sludge blanket. Thus the active biomass concentration will be higher in the bottom sludge. Unless there is a specific reason to do so (e.g. excessive accumulation of inert material in the bottom sludge) it is recommended to waste from the top of the sludge blanket. This way the denser, better settling and more active sludge is maintained in the reactor with preference. Several UASBs with only a bottom withdrawal system in place exhibited poor biomass retention, characterised by a shallow sludge blanket and a low sludge concentration.

Finally, the sludge abstraction points should be distributed evenly over the reactor surface area. In general one abstraction point per 50–60 m^2 is considered as a good compromise between optimum performance and required investment costs.

(g) Improper material selection and poor quality of construction

Due to the presence of sulphides in the anaerobic effluent the potential for corrosion in an UASB is high, especially at those locations where a gas-liquid interface exists. This is especially true for covered reactors where the presence of condensed water, oxygen and H_2S accelerate the formation of H_2SO_4, which is very corrosive towards concrete. It is therefore recommended to apply lining (or coating, provided it is applied well, which is a difficult task) on all concrete parts upwards from 0.5 m below the waterline.

Use of lining inside the gas box has a second advantage in that it is impermeable to gas and therefore allows the gas boxes to be pressurised. The biogas can then be transported to the gas treatment equipment without the use of a gas compressor. Furthermore, all piping and valves on top of the reactors should be made of corrosion resistant (and solar radiation resistant) material, or alternatively a suitable coating is required. This also applies to other metal parts such as handrailing and ladders.

13.2.2 Operational- and maintenance issues

The main issues observed are the following (Van der Lubbe et al., 2010):

- Insufficient operator attendance and maintenance;
- Insufficient process knowledge.

Many examples of insufficient operator attendance were observed during the site visits. Most frequently this included blocked inlet pipes and accumulated scum on the surface of the settler blocking off whole sections of the effluent weir, which obviously should have been cleaned. In defence of the operators it should be pointed out that often no adequate means of unplugging the inlet pipes were provided. However, the solution can be simple, as observed at a Brazilian plant where a cleaning pipe with rubber seal is used. The pipe is placed on top of a blocked inlet pipe and flushed with pressurized service water (effluent). This solution is shown in Figure 13.13. Although in this UASB the inlet pipes were constructed with many bends and connections, problems with blocked inlet piping were nearly absent.

Figure 13.13 Effective method for unblocking clogged inlet pipes with pressurized water using a simple cleaning tube that seals off the inlet of the pipe. Courtesy of B. Heffernan

As to the issue of insufficient maintenance, one area where this was particularly obvious was in gas collection and treatment. Not one of the plants visited had a working gas flow meter and even worse, some of the flares were out of operation as well, resulting in continuous venting of methane to the atmosphere. Fine screens were sometimes also found to be in a poor condition, especially the raking mechanisms. The same applied to the grit classifiers that are used to transport the deposited sand/grit out of the sand trap.

Finally the issue of insufficient process knowledge. One might not expect an operator to know all finesses of biological treatment, but at least the plant or operations manager should have some grasp of the basic concepts involved. For instance that if a UASB is fed with wastewater, the result will be that sludge is produced. This sludge needs to be wasted as otherwise it will inevitably end up in the effluent. Several UASBs discussed in the study by Van der Lubbe et al. (2010) either did not withdraw excess sludge at all or the solids control strategy consisted simply of "opening a valve for a couple of turns for one minute per day". A simple yet effective sludge management strategy is to discharge a sludge batch with a fixed volume whenever the height of the sludge blanket exceeds a certain upper setpoint value. However, this presupposes that the operator has the means to determine the sludge blanket level with a certain degree of accuracy, which often is not the case as discussed in the previous section. Figure 13.14 shows an artist impression of the recently developed Veolia/Biothane municipal UASB reactor (UpthaneTM), in which the issues previously discussed have been properly addressed.

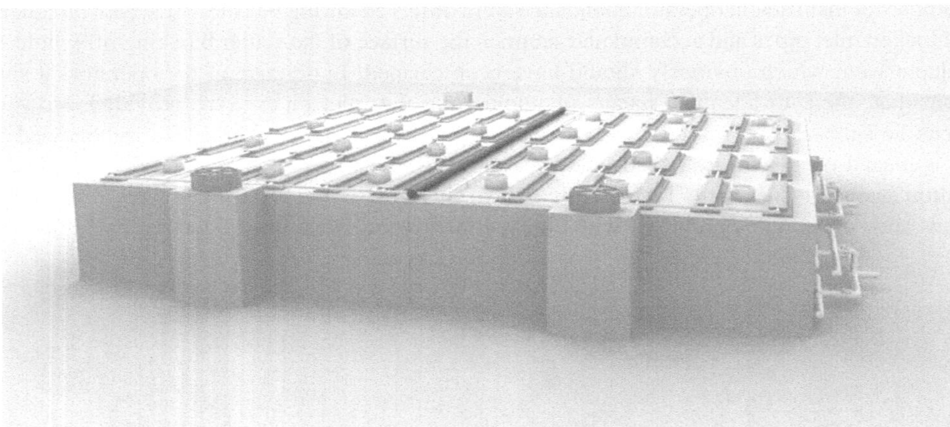

Figure 13.14 Artist impression of the recently developed Veolia Upthane^{TM} municipal UASB reactor. Courtesy of Veolia Water Solutions and Technologies (VWS) and Biothane Systems International

13.2.3 Inappropriate expectations of UASB performance

The initial publicity surrounding the municipal UASB might at least partly be to blame for this. Whereas there are doubtless many advantages associated to anaerobic treatment, these benefits have sometimes been exaggerated. Consequently, when performance is less than expected, the UASB technology is blamed for it. Some common examples of exaggerated expectations of UASB performance are presented below.

(a) Overestimated methane production

This is perhaps one of the most important items where appropriate management of expectations is due. In Section 12.4.4.2 it was demonstrated that the theoretical methane production is equal to 0.35 $Nm^3 \cdot kg^{-1}$ COD *converted*. However, some enthusiasts take this as per kg COD *applied*. Whereas for the industrial application of anaerobic treatment, often receiving wastewater with a high soluble biodegradable COD fraction, this approximation may be justified, for municipal sewage it is certainly not.

As will be demonstrated in Section 13.3, the conversion efficiency of influent COD into methane for municipal sewage is often in the range of 40–60%, depending on the wastewater composition, the applied sludge age and the temperature. The rest of the COD will end up either in the excess sludge or in the effluent. A second factor that reduces the methane yield in municipal sewage treatment is the diluted nature of the wastewater, resulting in poor methane recovery. Even though methane is not very soluble in water (14–18 mg · l^{-1} at 70% methane content in the biogas and at atmospheric pressure for temperatures between 15–30°C), the associated COD demand of 55–70 mg COD · l^{-1} represents up to 20 to 30% of the digested influent COD fraction. Finally there might be a significant concentration of sulphate present in the raw sewage, resulting in a competition between sulphate reducing- and methanogenic bacteria for available biodegradable COD. This subject will be discussed in more detail in Section 13.2.4. As a result of the factors discussed here, the actual methane production is typically much lower than the theoretical maximum of 0.35 $Nm^3 \cdot kg^{-1}$ COD, i.e. in the order of 0.1 to 0.15 $Nm^3 \cdot kg^{-1}$ COD applied.

(b) Less reduction in sludge production than expected

The sludge production in an anaerobic system will generally be lower than that of a comparable activated sludge system for secondary treatment, even if an aerobic post-treatment step is added. However, one should not compare the *true* yield, which for the anaerobic bacteria is indeed much lower than that of the aerobic bacteria, i.e. 0.05 versus 0.45 kg VSS · kg^{-1} COD removed. The observed solids production or *apparent* yield (mE_t) in both systems, when fed with municipal sewage, will be much more comparable. Rrefer to Section 3.3.3.5 for more information on the difference between true and apparent yield.

The accumulation of non-biodegradable or non-degraded particulate material constitutes the main contribution to excess solids production in a municipal UASB. The situation is different when a mainly soluble industrial wastewater is considered, when very low apparent yields are indeed possible, as there will be no accumulation of particulate material in the reactor. For a municipal sewage UASB an apparent yield of 0.10–0.15 mg TSS · mg^{-1} COD may be considered as a typical value. However, this figure reflects only the suspended solids discharge with the excess sludge and does not consider the suspended solids content in the UASB effluent. At a concentration of 40–100 mg TSS · l^{-1} this is considerable higher than in activated sludge systems. This "effluent excess sludge" is only partly degradable and accumulates in the aerobic post-treatment system, where it will contribute to the secondary excess sludge production.

The total apparent sludge yield (i.e. excess sludge plus effluent suspended solids) for a municipal UASB depends on the applied sludge age and the efficiency of the GLS separator, but will in general be around 0.2 to 0.25 mg TSS · mg^{-1} COD, of which approximately 0.10 mg TSS · mg^{-1} COD ends up in the effluent. If an aerobic post-treatment step is added for effluent polishing, then the total *discharged* excess sludge production (mE_{tu}+mE_{t2}) will be around 0.2–0.25 mg TSS · mg^{-1} COD. In comparison, the organic sludge production in a conventional activated sludge system treating raw sewage at a sludge age of 5 days and at 20°C typically has a value around 0.4–0.45 kg TSS · kg^{-1} COD applied and slightly higher (up to 0.50–0.55 kg TSS · kg^{-1} COD) if a primary settler is installed. It can be concluded that, when combined anaerobic-aerobic treatment is applied, a more realistic estimate of the reduction in excess sludge production will be between 30 and 50%. As a final note, it should also be considered that a significant number of conventional activated sludge systems are equipped with an anaerobic sludge digester. So in the end, both in the UASB configuration and in the activated sludge plant, all excess sludge produced will be anaerobically stabilised. As the solids load to the digester from a conventional activated sludge system is higher to begin with, the final stabilised sludge production will be higher as well, but assuming the biodegradable organics are largely removed, then the difference in stabilised excess sludge production will be reduced.

13.2.4 Presence of sulphate in municipal sewage

Until recently the presence of sulphate in municipal sewage was considered as a nuisance only, due to the smell and corrosion issues caused by the sulphides (H_2S, HS^-) produced under anaerobic conditions.

Whereas sulphate concentrations in industrial wastewater, e.g. from the pulp and paper industry, may be very high (up to several hundred milligrams), typically in municipal wastewater the concentration is much lower. Values between 40 to 70 mg $SO_4 \cdot l^{-1}$ were reported in some of the earlier work on anaerobic sewage treatment, which took place in the Netherlands and in South America (Brazil, Colombia), where high concentrations of sulphate are not common, perhaps due to the moderate to high precipition rates in these countries. However, in other parts of the world several recently constructed municipal UASB systems have encountered serious problems due to the presence of (very) high sulphate levels in the raw sewage. For instance, sulphate concentrations of 100–150 mg $\cdot l^{-1}$ are found in for instance Egypt and India and up to 400–600 mg $SO_4^{2-} \cdot l^{-1}$ in a particular location on the Arab peninsula. When the source of domestic drinking water contains a high level of sulphate, this eventually might end up in the municipal sewage as well. An exception is the use of seawater for potable water production, as the RO brine can be discharged back into the sea. Another factor that may contribute to a high sulphate concentration in the sewage is infiltration of salty or brackish groundwater into the sewer system, a problem that might occur in coastal areas.

When sulphate is present in the feed to an anaerobic reactor then sulphide will be formed by sulphate reducing bacteria. These sulphate reducing bacteria compete with methanogenic bacteria for the conversion of available easily biodegradable COD. In particular the hydrogen oxidising sulphate reducers have kinetic advantages (higher growth rate and higher substrate affinity) over the equivalent methanogenic bacteria. Stoichiometrically only 0.67 mg of *biodegradable* COD per mg of sulphate is required, so in many cases potentially all sulphate present in the wastewater can be converted into sulphide. Fortunately the results obtained at several full scale UASBs indicate otherwise (Van der Lubbe et al., 2010), which might be explained when it is considered that sulphate reduction and methane production occur simultaneously and the outcome of the competition for substrate is determined by differences in growth rate and substrate affinity. There is currently little data available on the relationship between the COD/SO_4^{2-} ratio in the raw sewage and the fraction of sulphate that will be converted into sulphide.

It can be attempted to theoretically quantify the expected SO_4^{2-} conversion through an estimate of the COD available to the (more competitive) hydrogen oxidising sulphate reducers. For a complex substrate as the organic material present in raw sewage, under anaerobic conditions typically 30% of the suspended and colloidal material is eventually converted into methane using hydrogen gas as an intermediate component (refer also to Figure 12.15). Furthermore, there will be acetate present that will be directly converted into methane. For example, let us consider a typical raw sewage that contains 600 mg COD $\cdot l^{-1}$, of which 50 mg $\cdot l^{-1}$ consists of acetate. When it is assumed that 50% of the influent COD is available for digestion, then approximately $0.3 \cdot (0.50 \cdot 600–50) = 75$ mg COD will be degraded via the hydrogen pathway. Assuming 67% of the hydrogen COD is used by the sulphate reducers due to their competitive advantage, this means that $67\% \cdot 75 = 50$ mg $\cdot l^{-1}$ COD is converted into inorganic H_2S-COD, equivalent to 25 mg H_2S-S $\cdot l^{-1}$. As an practical alternative, based on the treatment results from several full-scale UASBs the following indicative guidelines might be used for design:

- Ratio $COD/SO_4^{2-} = 12 \rightarrow$ 100% sulphate conversion;
- Ratio $COD/SO_4^{2-} = 9 \rightarrow$ 75% sulphate conversion;
- Ratio $COD/SO_4^{2-} = 6 \rightarrow$ 65% sulphate conversion;
- Ratio $COD/SO_4^{2-} = 3 \rightarrow$ 50% sulphate conversion.

Thus for example, for raw sewage containing 600 mg $\cdot l^{-1}$ COD and 50 mg $\cdot l^{-1}$ SO_4^{2-} (i.e. COD/SO_4^{2-} = 12), full conversion of sulphate to sulphide is assumed. In practice the conversion will be somewhat lower.

The detrimental effects of sulphate conversion, or to be more exact the production of sulphides are threefold:

- Health and safety issues, as a H_2S gas concentration of 800 ppm is lethal for 50% of the exposed population after 5 minutes exposure (LC_{50}), while concentrations over 1000 ppm can cause immediate collapse and loss of breathing capacity;
- Corrosion of concrete and metals;
- Negative impact on process performance;

H_2S is highly soluble in water, as it dissociates into HS^- or even S^{2-}, depending on the pH. As the pK_a of the equilibrium dissociation reaction from H_2S to HS^- is 6.96, it can be concluded that in the pH range of interest (7.0–8.5) most of the H_2S will be present in dissociated form: ranging from 50% HS^- at pH = 7 to 90% HS^- at pH = 8.

Part of the generated H_2S will be present in the biogas, depending on the COD/SO_4^{2-} ratio in the sewage. For municipal sewage the H_2S concentration in the biogas will be generally below 0.5%. Not all of the produced biogas will be collected and thus traces of H_2S gas may be emitted to the atmosphere at the UASB liquid surface and cause odour problems, mainly at points of high turbulence, for instance at the effluent weirs. The use of submerged effluent pipes instead of overflow weirs may significantly reduce H_2S emissions.

Corrosion is a second adverse affect of H_2S generation. While the produced H_2S itself is corrosive, it is principally the biological production of sulphuric acid that causes most of the problems. Whenever H_2S (or HS^-) is exposed to atmospheric oxygen, sulphur oxidizing bacteria (e.g. Thiobacillus spp.) growing on the surfaces of walls, pipes et cetera will rapidly oxidize the H_2S gas into H_2SO_4, which is a strong inorganic acid. Especially in semi-confined spaces (for example influent lifting stations and the headspace of covered UASB reactors, even when ventilated) the effect of corrosion on unprotected concrete and metals is very severe. Selection of appropriate materials or application of lining/coating is required in these cases.

The main detrimental effect of the presence of sulphate in the raw sewage on anaerobic process performance is that part of the biodegradable organic COD will be used to convert sulphate into H_2S, thereby forming inorganic soluble COD instead of methane. Obviously this will significantly reduce the methane production. Furthermore, as part of the biodegradable COD is not removed but merely converted from an organic into an inorganic form, the COD removal efficiency in the anaerobic treatment step will suffer. In full-scale municipal UASB reactors treating sulphate-rich wastewater, COD removal efficiencies can be reduced to values as low as 50–55% (Heffernan et al., 2010). In the aerobic post-treatment the presence of dissolved sulphides will result in increased oxygen demand. Furthermore, H_2S is known to be very toxic to nitrifying bacteria, significantly reducing their growth rate and requiring an increase in sludge age in order to meet ammonium effluent limits. Therefore, in general it is recommended to install a pre-aeration step upstream of the main aerobic system in order to reduce sulphide toxicity. Such a pre-aeration reactor is operated without biomass retention at a hydraulic residence time of 20–60 minutes. In this reactor a biologically mediated chemical oxidation of H_2S/HS^- will occur at a very fast rate, contrary to the slow reaction of the chemical oxidation process, even in the presence of a catalyst such as $FeCl_3$.

The end product of the sulphide oxidation in the pre-aeration reactor can be elemental sulphur (S^0), thiosulphate ($S_2O_3^{2-}$) or sulphate (SO_4^{2-}). However, the oxidation process is far from complete at the relatively short retention times of a pre-aeration reactor. Therefore mainly elemental sulphur (present in colloidal form) is formed. The composition of the end products of sulphide oxidation seems to depend on the applied molar O_2/S ratio, although much is still unclear on this subject. From the perspective of effluent toxicity this does not matter, as it is only the removal of H_2S that is important. In the subsequent aerobic reactors the intermediate products are then further oxidized, eventually to SO_4^{2-}. The main oxidation reactions of hydrogen sulphide with oxygen are summarized as:

(1) Complete oxidation of hydrogen sulphide (HS^-) to sulphate

$$HS^- + 2O_2 \longrightarrow SO_4^{2-} + H^+ \text{or } H_2S + 2O_2 \longrightarrow SO_4^{2-} + 2H^+ \quad (13.8a)$$

The stoichiometrical oxygen demand for complete oxidation of HS^- is equal to $4 \cdot 16/32 = 2.0$ mg O_2 per mg S oxidized.

(2) Two step oxidation of hydrogen sulphide to elemental sulphur and sulphate:

$$HS^- + 0.5O_2 \longrightarrow S^0 + OH^- \text{or } H_2S + 0.5O_2 \longrightarrow S^0 + H_2O \quad (13.8b)$$

$$S^0 + H_2O + 1.5O_2 \longrightarrow H_2SO_4 \longrightarrow SO_4^{2-} + 2H^+ \quad (13.8c)$$

From the reaction stoichiometrics it can be determined that the oxygen demand is equal to 0.5 mg O_2 per mg S for the first step and 1.5 mg O_2 for the second step.

(3) Two step oxidation of hydrogen sulphide to thiosulphate and sulphate:

$$2HS^- + 2O_2 \longrightarrow S_2O_3^{2-} + H_2O \quad \text{or} \quad 2H_2S + 2O_2 \longrightarrow S_2O_3^{2-} + H_2O + 2H^+ \quad (13.8d)$$

$$S_2O_3^{2-} + H_2O + 2O_2 \longrightarrow 2SO_4^{2-} + 2H^+ \quad (13.8e)$$

In this case, each step in the oxidation process requires 1 mg of O_2 per mg S oxidized. It can be observed that, apart from the oxygen demand, a secondary effect is the conversion of a weak acid (H_2S or HS^-) into a strong acid (H_2SO_4).

In addition to sulphate, sulphide may also be present in the raw sewage if the residence time in the sewer system is very long. Under anaerobic conditions a significant portion of the SO_4^{2-} present in the sewer will be reduced to H_2S, which subsequently reacts with iron to form ferric sulphide (hence the black colour of the sewage). This FeS represents a fraction of the COD demand in both anaerobic influent and -effluent and will not be removed in the UASB.

EXAMPLE 13.1

A municipal UASB treats a wastewater containing 600 mg COD $\cdot$ l^{-1} and 50 mg sulphate $\cdot$ l^{-1}. The COD removal efficiency is 65%. Estimate the fraction of the COD demand in the anaerobic effluent that is due to the presence of sulphides. Furthermore, indicate the effect on COD removal efficiency should the influent sulphate concentration increase to 100 mg $\cdot$ l^{-1}.

Solution

The initial COD/SO_4 ratio is $600/50 = 12$ and it may thus be expected that sufficient biodegradable COD is available for complete reduction of the sulphate. Therefore, 100% sulphate reduction is assumed, although in practice the conversion may be less. Knowing that the molecular weight of sulphate is 96 $mg \cdot l^{-1}$, with a sulphur content of $32/96 = 0.33$ g $S \cdot g^{-1}$ SO_4^{2-}, then the effluent sulphide concentration is equal to $0.33 \cdot 50 = 16.67$ mg $S \cdot l^{-1}$.

From Eq. (13.8a) it can be observed that by definition the COD of 1 mg of hydrogen sulphide-S is equal to two mg of O_2. Thus the effluent sulphide COD concentration $= 16.67 \cdot 2 = 33.3$ mg COD $\cdot l^{-1}$. This represents a fraction of $33.3/((1-0.65) \cdot 600) = 33.3/210 = 15.9\%$ of the COD in the anaerobic effluent.

If the sulphate concentration increases to 100 $mg \cdot l^{-1}$, while the inlet COD concentration remains unchanged, the ratio COD/SO_4 will decrease to a value of 6. In this case 65% conversion is assumed, or $0.65 \cdot 100 \cdot 0.33 = 21.7$ mg $S \cdot l^{-1}$, corresponding to a COD value of 43.3 $mg \cdot l^{-1}$. The effluent COD concentration increases by $43.3-33.3 = 10$ $mg \cdot l^{-1}$ from 210 to 220 mg COD $\cdot l^{-1}$, of which the sulphide fraction comprises $43.3/220 = 19.7\%$. As a result of the increase of the inorganic H_2S-COD concentration in the effluent, the COD removal in the anaerobic reactor decreases to from 65% to 63.3%.

13.2.5 Energy production and greenhouse gas emissions

The main driver for the implementation of anaerobic treatment is the reduction of the costs of wastewater treatment. This is mainly due to the significant decrease in energy demand compared to conventional activated sludge systems, as the oxygen consumption for carbon oxidation in the aerobic post-treatment system is typically reduced with 65–80% for a well designed and operated UASB. Furthermore, energy can be generated from the methane that is produced in the UASB. Both the reduction in energy demand and the potential for energy production reduce the carbon footprint of wastewater treatment plants. For municipal sewage, at 15 to 30°C and at an anaerobic sludge age of 30 to 60 days, typically about 40–60% of the influent COD is converted into methane. The methane can be used in a variety of ways:

- It can be burned in gas boilers (typically at 80–85% thermal efficiency) for the generation of heat: e.g. for steam production, for heating of (sludge) digesters or for sludge drying;
- Electricity generation with combustion motors (typical 35–38% net electrical efficiency). The overall energy efficiency can be significantly increased when the hot exhaust gases are used to heat steam or water by means of combined heat and power generation (CHP);
- Use the methane as cooking fuel, by connecting to a gas distribution grid in a nearby residential area. Suitable gas burners should be used, i.e. capable of handling a variable methane content in the cooking gas.

While electricity and/or heat generation are proven technologies, the safety issues related to the construction of a biogas distribution grid, at least when western safety standards are applied, are considerable. This means this option is mainly attractive for small-scale community based UASBs near to residential areas or even on the level of individual households, when the manure from livestock can be codigested.

13.2.5.1 Carbon footprint

It is interesting to evaluate and compare the carbon footprint of different wastewater treatment processes. First of all, consider the biological processes. The organic matter in wastewater may be considered as a renewable source of energy. As will be demonstrated below, there is little difference between anaerobic- and aerobic treatment in this aspect, provided the effluent quality is the same, which will be the case when aerobic systems are compared to anaerobic-aerobic systems and when the formed methane is oxidized (with or without energy generation). The catabolic reaction (energy production) for aerobic treatment is:

$$C_xH_yO_z + \frac{1}{4}\cdot(4x + y - 2z)O_2 \longrightarrow x\ CO_2 + y/2\ H_2O \tag{2.1}$$

It does not matter if nitrate or another oxidant is used instead of oxygen, as it is not oxygen consumption but carbon dioxide production that is of interest in these metabolic reactions. For anaerobic treatment the catabolic reaction is:

$$C_xH_yO_z + \frac{1}{4}\cdot(4\cdot x - y - 2\cdot z)\,H_2O \longrightarrow \frac{1}{8}\cdot(4\cdot x - y + 2\cdot z)\,CO_2 + \frac{1}{8}\cdot(4\cdot x + y - 2\cdot z)\,CH_4 \tag{2.10}$$

From the anaerobic reaction equation it can be observed that 50% of the carbon atoms are used for methane production instead of carbon dioxide production. However, in the subsequent oxidation of methane (combustion), this fraction will also be fully converted into carbon dioxide. Apart from the catabolic reactions discussed above, the anabolic reaction (biomass generation) should be considered as well. Often it is assumed that the excess sludge production of anaerobic systems is much smaller than that of aerobic systems, but for municipal sewage this is actually only true when secondary treatment (COD removal) is considered.

At the high sludge age required for nutrient removal, the increase in endogenous respiration will decrease the sludge production of an aerobic system significantly. Furthermore anaerobic sludge digestion is nowadays often applied, which further reduces the sludge production from aerobic systems. In the end, all the sludge eventually ends up as CO_2, either directly (combustion) or indirectly by biological and/or chemical degradation process (when disposed on land or in a landfill). Thus in the end there is no real difference in carbon dioxide production between modern aerobic- and anaerobic-aerobic systems when the overall biological degradation of the organic load is concerned. Therefore it may be concluded that the main difference in carbon footprint between aerobic- and anaerobic processes is due to differences in the energy consumption (aeration, dewatering) and the production of energy.

There are two important exceptions: (I) if the methane produced from digested sludge in a landfill is not collected and flared, then the contribution to greenhouse gas emissions will be *very* large, which benefits the anaerobic process due to its lower production of excess sludge and (II) the loss of dissolved methane with the anaerobic effluent, as will be discussed in Section 13.2.5.2.

Methane is considered to be a roughly twenty-one times more potent greenhouse than carbon dioxide, or in other words, the CO_2 equivalence of methane is equal to 21. In developed countries the disposal of excess sludge at landfills is no longer permitted, while generally the biogas of existing landfills is collected and either utilized or flared. For example, the most common route for sludge disposal in the Netherlands consists of anaerobic digestion, dewatering and sludge drying followed by incineration. In many developing countries the situation is different and the application of poorly stabilized excess sludge on agricultural land or in (uncontrolled) landfills is often the largest contributor to the overall emission of greenhouse gases in the wastewater treatment process.

To evaluate the impact on the carbon footprint of uncontrolled digestion of stabilized excess sludge following disposal in a landfill, then in principle it is not unrealistic to assume that, given sufficient time, the stabilized sludge is almost completely mineralized under anaerobic conditions. When a composition of the stabilized organic sludge of $C_7H_{10}O_3N$ is assumed ($f_{cv} = 1.48$ mg COD $\cdot$ mg^{-1} COD, refer to Table 2.5), the carbon mass fraction in the sludge is equal to 54%. In the anaerobic digestion of this sludge, equal molar quantities of methane and carbon dioxide are produced. Hence, the maximum methane production per g VSS of stabilized sludge disposed is equal to $0.54 \cdot 0.5 \cdot 16/12 = 0.36$ g $CH_4 \cdot g^{-1}$ VSS or $21 \cdot 0.36 = 7.56$ g $CO_2 \cdot g^{-1}$ VSS when the carbon equivalence of methane is considered. However, the guidelines of the Integrated Pollution Prevention and Control committee (IPPC, 2007) suggest a default value of 50% anaerobic degradation in a landfill. Furthermore, it is assumed that part of the produced methane (40%) is oxidised to carbon dioxide by aerobic bacteria present in the upper layers of the landfill. When the IPPC values are applied, then the expected methane emission is $0.5 \cdot (1-0.4) \cdot 0.36 = 0.11$ g $CH_4 \cdot g^{-1}$ VSS or 2.27 g $CO_2 \cdot g^{-1}$ VSS.

As to the wastewater treatment process itself, every kWh of consumed electricity results in an emission of CO_2, which depends on the means of electricity production: coal-fired plants, nuclear plants, gas-powered plants etc. Often country-specific emission factors are used that correspond to the mix of energy production methods in use. For example, for the US the CO_2 emission factor used in 2009 is equal to 0.718 kg CO_2 per kWh consumed (EPA, 2009). In comparison, if methane is used as a fuel source and a 35% combustion efficiency is assumed, then per kWh generated (= 3600 kJ) about 3600/35% = 10,300 kJ is required which is equivalent to 10,300/50,400 = 0.204 kg CH_4. Upon combustion, 1 kg of CH_4 produces 44/16 = 2.75 kg of CO_2. Thus the CO_2 emission factor per kWh produced is $0.204 \cdot 2.75 = 0.56$ kg CO_2. The average US number is higher because it is based on a mix of different fuel sources and it includes the energy losses from the transport of electricity from power plant to final consumer.

EXAMPLE 13.2

Compare the energy balance and carbon footprint of a conventional activated sludge plant designed for secondary treatment (including a heated digester) with that of a combined anaerobic-aerobic sewage treatment plant. The STP receives 50,000 $m^3 \cdot d^{-1}$ of raw sewage containing 548 mg COD $\cdot l^{-1}$ (i.e. 10,000 ton COD $\cdot yr^{-1}$) at a temperature of 25°C.

Use the system performance specified in Table 13.1, which is fairly typical for the system configurations involved and is based on a COD removal efficiency of 70% in the anaerobic reactor and recovery of all of the methane produced: an assumption that will be relaxed in Example 13.4. Assume an oxygen transfer efficiency of 1.2 kg $O_2 \cdot kWh^{-1}$ and an electrical efficiency of the gas motor of 35%. To estimate the greenhouse gas emissions resulting from the energy demand for aeration, use the typical US figure of 0.718 kg $CO_2 \cdot kWh^{-1}$.

Table 13.1 Performance of the conventional aerobic and the combined anaerobic – aerobic system for secondary treatment of Example 13.2

COD fraction	Symbol	Aerobic system	Anaerobic–aerobic
Oxidized	mS_o	50%	15%
In effluent	mS_{te}	10%	10%
In stabilised sludge	mS_{xve}	25%	15%
Digested	mS_d	15%	60%

Solution

Consider the conventional activated sludge system first. On a yearly basis, the total oxygen demand will be $0.5 \cdot 10,000 = 5000$ ton O_2, which requires 5000/1.2 = 4167 MWh. Using the US emission factor of

0.718, the yearly CO_2 emission is equal to 2990 ton. On the other hand, methane is produced from the digested sludge: $0.15 \cdot 10{,}000/4 = 375$ ton $CH_4 \cdot yr^{-1}$, which can be used for the production of energy: $35\% \cdot 375 \cdot 50{,}400/3600 = 1840$ MWh $\cdot yr^{-1}$, equivalent to an averted CO_2 emission of $1840 \cdot 0.718 =$ 1320 ton $\cdot yr^{-1}$. Thus the net CO_2 emission of the aerobic system is $2990-1320 = 1670$ ton $CO_2 \cdot yr^{-1}$.

As to the performance of the combined anaerobic-aerobic treatment system, the oxygen consumption is much lower (1500 ton $O_2 \cdot yr^{-1}$), corresponding to an electrical power consumption of 1250 MWh $\cdot yr^{-1}$ and resulting in a CO_2 emission of 900 ton $\cdot yr^{-1}$. Furthermore, due to the presence of the anaerobic reactor the methane production is increased to 1500 ton $CH_4 \cdot yr^{-1}$, from which 7350 MWh $\cdot yr^{-1}$ of electricity can be produced. This reduces the CO_2 emission with 5280 ton $CO_2 \cdot yr^{-1}$. The net CO_2 emission is therefore equal to $900-5280 = -4380$ ton $CO_2 \cdot yr^{-1}$, i.e. the anaerobic-aerobic system actually *reduces* (or averts) the emission of CO_2!

Clearly the anaerobic-aerobic system is much more sustainable than a conventional activated sludge system. However, there are several additional factors to consider such as (I) the potential emission of methane following disposal of the stabilized sludge, which has been discussed earlier in this section and will be evaluated below, (II) the impact of nitrogen removal on overall power consumption and carbon emissions, which will be discussed in Example 13.3 and (III) the loss of dissolved methane with the anaerobic effluent, discussed in Section 13.2.5.2 and Example 13.4. Based on the conditions of a specific design case, the performance with regards to the carbon footprint might even be inverted.

To evaluate the impact of uncontrolled digestion of stabilized excess sludge following disposal in a landfill on the carbon footprint, one can use the values previously calculated: 2.27 g $CO_2 \cdot g^{-1}$ VSS when the IPPC guidelines are used and 7.56 g $CO_2 \cdot g^{-1}$ VSS when full anaerobic degradation is assumed (and no aerobic CH_4 oxidation). Table 13.2 presents the CO_2 emissions for the two treatment configurations for the following three cases: (I) not taking into account potential emissions of methane from the landfill, (II) calculating the emissions according to the IPPC guidelines and (III) assuming full conversion of the organic material in the stabilised sludge to methane.

Table 13.2 Effect of methane emissions from (stabilized) excess sludge disposal at uncontrolled landfills on the carbon footprint of sewage treatment plants

Parameter	UoM	Aerobic	Anaerobic–aerobic
Case I – Reference case–disregarding CH_4 emissions from landfill			
– Stabilized sludge production	ton VSS $\cdot yr^{-1}$	1690	1010
– CO_2 emission excl $\cdot$ landfill	ton $CO_2 \cdot yr^{-1}$	1670	–4380
Case II – IPPC guidelines			
– CH_4 emission from landfill	ton $CH_4 \cdot yr^{-1}$	182	109
– Equivalent CO_2 emission	ton $CO_2 \cdot yr^{-1}$	3830	2300
– Total CO_2 emission	ton $CO_2 \cdot yr^{-1}$	5500	–2080
Case III – Complete anaerobic degradation–no aerobic CH_4 oxidation			
– CH_4 emission from landfill	ton $CH_4 \cdot yr^{-1}$	605	365
– Equivalent CO_2 emission	ton $CO_2 \cdot yr^{-1}$	12,730	7640
– Total CO_2 emission	ton $CO_2 \cdot yr^{-1}$	14,410	3260

From the results presented in Table 13.2 it is clear that, regardless of the the assumptions used, the contribution of uncontrolled methane emissions from the disposal of stabilized sludge at landfills to total CO_2 emissions is very significant. For instance, when the IPPC guidelines are considered, then in the case of the aerobic system the CO_2 emission from sludge disposal is 3830 ton · yr^{-1}, or more than two times the direct CO_2 emissions from the aerobic plant.

In the case of the anaerobic-aerobic system, as the stabilized sludge production is 40% lower so will be the CO_2 emission (i.e 2300 ton $CO_2 \cdot yr^{-1}$). The carbon footprint performance of the aerobic-anaerobic sewage treatment plant will decrease from 4380 ton · yr^{-1} to 2080 ton · yr^{-1} of averted CO_2 emissions, but on the other hand the difference with the aerobic system will be further accentuated. In any case, it is clearly demonstrated that the selected method of stabilized sludge disposal is the single most important factor in the carbon footprint of a sewage treatment plant.

EXAMPLE 13.3

For the data of the previous example, evaluate the impact on energy consumption and carbon footprint when nitrification is considered as well. Assume that the nitrification capacity is 34 mg N · l^{-1} for all system configurations.

For the aerobic system, evaluate two cases: (I) only nitrification and (II) nitrification + denitrification to less than 8 mg NO_3-N · l^{-1}. In this case, denitrification is primarily applied in order to reduce the potential for a rising sludge blanket in the final settler. The application of denitrification will increase the sludge age and hence the volume of the activated sludge system, which will not be considered here. However, it will also decrease the oxygen demand, although this is partly of-set because of the increase in sludge age. A second benefit is that due to the increase in sludge age the excess sludge production will be reduced.

In the anaerobic-aerobic configuration, the effluent of the anaerobic reactor will be almost depleted of (easily) biodegradable COD so that the extent of denitrification will be limited, irrespective of the applied sludge age. Thus, as will be demonstrated in Appendix A8, this eliminates the potential for development of a rising sludge blanket in the final settler. Therefore in the case of the anaerobic–aerobic configuration only nitrification will be considered. If nitrogen removal is desired, it is possible to bypass raw sewage directly to the anoxic zone of the aerobic post-treatment system, but as will be demonstrated in Section 13.5.2.1 this is generally only attractive at COD/TKN ratios higher than 14. Table 13.3 summarizes the performance of the three system configurations evaluated in Example 13.3.

Table 13.3 Performance of the conventional aerobic and the combined anaerobic – aerobic system for nitrogen removal of Example 13.3

COD fraction	Symbol	N	N–DN	UASB–N
Oxidized	mS_o	55%	60%	18%
In effluent	mS_{te}	10%	10%	10%
In stabilised sludge	mS_{xve}	23%	21%	12%
Digested	mS_d	12%	9%	60%

Note: N = nitrification; DN = denitrification

Solution

For all system configurations the nitrification capacity is equal, i.e. $365 \cdot 50{,}000 \cdot 34 \cdot 10^{-6} = 621$ ton $N \cdot yr^{-1}$, which requires $621 \cdot 4.57 = 2836$ tons of oxygen. The denitrification capacity required to reduce the nitrate concentration to 8 mg $N \cdot l^{-1}$ is $365 \cdot 50{,}000 \cdot (34–8) \cdot 10^{-6} = 475$ ton $N \cdot yr^{-1}$. The equivalent oxygen recovered is equal to $475 \cdot 2.86 = 1357$ ton $O_2 \cdot yr^{-1}$. The power consumption and -production and the associated carbon footprint can be calculated as demonstrated in the previous example. The results are shown in Table 13.4.

Table 13.4 Comparison of energy demand and carbon footprint for a conventional aerobic and an anaerobic-aerobic system for nitrogen removal

Parameter	UoM	N	N + DN	UASB + N
Oxygen demand	ton $O_2 \cdot yr^{-1}$	8340	7480	4640
Power consumption	$MWh \cdot yr^{-1}$	6950	6230	3860
Methane production	ton $CH_4 \cdot yr^{-1}$	300	225	1500
Power production	$MWh \cdot yr^{-1}$	1470	1100	7350
Net power consumption	$MWh \cdot yr^{-1}$	5480	5130	−3490
CO_2 emission (excl. sludge disposal)	ton $CO_2 \cdot yr^{-1}$	3930	3680	−2500
Stabilized sludge production	ton $VSS \cdot yr^{-1}$	1550	1420	810
CO_2 emission (incl. sludge disposal)	ton $CO_2 \cdot yr^{-1}$	9220	8510	255

Note: effect of CO_2 emission from sludge disposal based on an uncontrolled landfill (i.e. no biogas collection) according to the IPPC guidelines (IPPC, 2007)

The effect of stabilized sludge disposal to an uncontrolled landfill on the carbon footprint of the overall treatment process is indicated as well: in this example only the IPPC guidelines have been evaluated. It can be observed that both from a sustainability viewpoint (carbon footprint) and an economical viewpoint (power consumption), the performance of the anaerobic-aerobic system is superior to that of the conventional aerobic system.

13.2.5.2 Biogas utilization

Perhaps surprisingly given the large benefits as indicated in the previous two examples, at present it is still by no means common practice to utilize the biogas produced from anaerobic sewage treatment. For example, of the ten large scale UASBs visited by Van der Lubbe et al. (2010), with design capacities ranging from 30,000 to 165,000 $m^3 \cdot d^{-1}$, not a single one was equipped with equipment for biogas utilization, although in one plant a gas motor was being installed at the time of visit. The most commonly observed method of biogas disposal was by means of flaring, although several of the installations were in such a poor state that methane was emitted directly to the atmosphere. This should be prevented at all costs as the impact of methane emissions to the greenhouse effect is much higher than that of carbon dioxide.

These disappointing findings are very different from the practices observed in industrial anaerobic wastewater treatment plants or in municipal sludge digesters in developed countries, where energy production from biogas is much more common. It may be partly attributed to a lack of knowledge or the unavailability of appropriate equipment in some countries, while in other countries the costs of fossil energy are (still) so low that it is simply not economically viable to use the methane for energy generation. It can be concluded that significant improvement can and should be made in the area of energy recovery.

Unfortunately, not all of the methane produced in the UASB can be collected in the gas hoods, as part of the methane will remain dissolved in the anaerobic effluent. In anaerobic reactors treating high strength wastewater or slurries/waste sludge, this is not a problem as the wastewater flow rate will be low and therefore the loss of methane with the effluent will only be a small fraction of the methane produced per litre wastewater. On the other hand, for diluted wastewaters such as municipal sewage the methane fraction lost with the effluent can be as high as 20–40%. Apart from the obvious reduction in energy production, methane is also released into the atmosphere, resulting in a considerable contribution to the greenhouse gas emissions.

To address this issue, it is important to known what the methane concentration in the UASB effluent at the level of the overflow weir will be, i.e. corresponding to the non-collected methane production. A theoretical approach can be made using Henry's law to calculate the theoretical equilibrium dissolved methane concentration as a function of the temperature and the partial methane pressure, using the temperature coefficients A = 675.74 and B = 6.88 (Metcalf & Eddy, 2003) and knowing that one litre of water contains 55.6 moles:

$$K_H = 10^{(-A/(273+T)+B)} \tag{13.9a}$$

$$X_{ch4} = p_{ch4}/K_H \tag{13.9b}$$

$$[CH_4]_{eq} = 1000 \cdot X_{ch4} \cdot 55.6 \cdot 16 \tag{13.9c}$$

where

K_H = Henry coefficient (atm)
x_{ch4} = mol fraction of dissolved methane gas in water ($mol \cdot mol^{-1}$)
T = temperature (°C)
p_{CH4} = partial methane pressure (atm): the product of actual pressure (including the effect of liquid depth) and the molar (volume) fraction of the methane in the gas phase
$[CH_4]_{eq}$ = equilibrium dissolved methane concentration ($mg\ CH_4 \cdot l^{-1}$)

As an example, the equilibrium methane concentration at 1.0 atm, 70% methane content and 25°C can be calculated as:

$$H = 10^{(-675.74/(273+25)+6.88)} = 40{,}965\ \text{atm} \tag{13.9a}$$

$$x_{ch4} = 0.7/40{,}695 = 1.71 \cdot 10^{-5}\ mol \cdot mol^{-1} \tag{13.9b}$$

$$[CH_4]_{eq} = 1000 \cdot 1.71 \cdot 10^{-5} \cdot 55.6 \cdot 16 = 15.2\ mg\ CH_4 \cdot l^{-1} \tag{13.9c}$$

However, the theoretical approach by itself yields insufficient information to be able to predict the expected methane concentration in the UASB effluent, first of all because the value of the Henry coefficient is based on solubility of methane in pure water, which might very well be different from that in anaerobic effluent. Furthermore, it is likely that a certain degree of oversaturation exists. Methane is produced in the sludge blanket, which for most municipal UASB reactors will be located at 2.5 to 5.0 meter liquid depth. The formation of biogas bubbles is normally not observed in the settler section, even though the effluent slowly travels upward to the overflow weirs, resulting in a gradual release of the liquid pressure. This fact alone already shows that oversaturation of methane in the anaerobic effluent will exist.

In this book the assumption is made that the methane concentration in the top liquid layer (i.e. the effluent) of the UASB is *at least* equal to the equilibrium methane concentration at 2.5 m submergence, which is a typical depth of the deflector. Above this level, i.e. when the liquid enters the settler section, no more biogas will be produced nor collected. Table 13.5 shows the calculated effluent methane concentration at this depth as function of the liquid temperature.

Table 13.5 Calculated methane concentration in the effluent of an UASB reactor as function of the temperature and for different fractions of methane in the biogas. Based on the *equilibrium* methane concentration at 2.5 m liquid depth

Temperature (°C)	60% CH_4 ($mg\ CH_4 \cdot l^{-1}$)	70% CH_4 ($mg\ CH_4 \cdot l^{-1}$)	80% CH_4 ($mg\ CH_4 \cdot l^{-1}$)
15	19.2	22.4	25.6
20	17.5	20.4	23.3
25	16.0	18.7	21.3
30	14.7	17.1	19.6
35	13.5	15.7	18.0

Recently some interesting data on actual (i.e. measured) effluent methane concentrations was published by Souza et al. (2010a). A pilot-scale UASB reactor (0.35 m^3 with 4.0 m liquid height) and a demo-scale UASB (14 m^3 with 4.5 m liquid height), i.e. both with a liquid height comparable with full-scale reactors, were operated at the test site of the Federal University of Minas Gerais (UFMG) and the state water authority (COPASA), located at the Arrudas sewage treatment plant. In the experiment the methane concentration was measured just below the effluent overflow level. The temperature during the experiments was around 25°C, while the methane content in the biogas varied between 63 and 74%, depending on the applied hydraulic retention time. The results are shown in Table 13.6, where it can be observed that the measured effluent methane concentration ranged between 19.2 to 22 $mg\ CH_4 \cdot l^{-1}$. This coincides reasonably well with the values predicted by our theoretical approach (15.6 to 18.3 $mg\ CH_4 \cdot l^{-1}$), although our values tend to be slightly but consistently lower. This is indeed a strong indication that some degree of oversaturation occurs.

Table 13.6 Comparison between measured and theoretical methane concentration, based on the data from Souza et al. (2010a), for $T = 25°C$ and $p = 0.91$ atm

Parameter	UoM	Case 1	Case 2	Case 3
HRT	h	5.0	7.0	12.0
Reactor	(–)	pilot	pilot	demo
Methane content	%	63%	70%	74%
Measured methane concentration	$mg \cdot l^{-1}$	19.6	22.0	19.2
Theoretical methane concentration$^{(1)}$	$mg \cdot l^{-1}$	15.6	17.4	18.3
Degree of oversaturation	%	125%	127%	105%

Note: (1) Based on the theory presented in Section 13.2.5.2 and assuming a liquid depth of the deflector of 2.5 m

An interesting finding from Table 13.6, albeit based on a limited amount of data, is that the degree of oversaturation at a lower hydraulic retention time is considerably higher (around 125% at 5–7 hours) than at a higher hydraulic retention time (105% at 12 hours). A possible explanation might be that the increase in hydraulic retention time allows for more time to reach the equilibrium methane concentration in the liquid phase, but this remains to be validated.

Furthermore, it can be observed that there is a clear trend toward higher methane fractions in the biogas at increasing hydraulic residence time. An increase in the retention time (and thus in sludge age) results in increased conversion of organic material into methane. As pointed out by Cakir et al. (2005), this will reduce the nitrogen concentration in the biogas, as the mass of dissolved nitrogen gas entering the reactor does not change. It is interesting to observe that the nitrogen fraction in the biogas can be quite substantial in the case of anaerobic treatment of low-strength wastewater such as municipal sewage.

EXAMPLE 13.4

Obviously the potential loss of soluble methane from the effluent to the atmosphere will have repercussions on both energy production and carbon footprint of an anaerobic wastewater treatment plant. Continuing with the previous examples, evaluate the effect of the methane loss to the atmosphere on the performance of the configurations discussed in Example 13.2 and Example 13.3 with respect to carbon footprint. Assume that the dissolved methane concentration eventually lost to the atmosphere is equal to the equilibrium CH_4 concentration in the anaerobic effluent at 2.5 m liquid depth, at 20°C and 65% methane content in the biogas.

Solution

From Table 13.5 the value of the equilibrium dissolved methane concentration at 20°C and 65% methane content in the biogas can be interpolated as 19 mg $CH_4 \cdot l^{-1}$. In the anaerobic–aerobic system, the presence of dissolved methane in the UASB effluent (50,000 $m^3 \cdot d^{-1}$) results in a loss of 346 tons of CH_4 per year to the atmosphere, or 7260 tons of equivalent CO_2. However, apart from the direct effect of this methane emission on the carbon footprint, there is also an indirect effect as the power production from the collected methane will decrease with 1695 $MWh \cdot yr^{-1}$ and thus the associated averted CO_2 emission as well (-1220 ton $CO_2 \cdot yr^{-1}$). Figure 13.15 shows the performance of the systems for secondary treatment as function of the methane concentration in the anaerobic effluent, which is assumed to be emitted to the atmosphere. The effect of sludge degradation in landfills is indicated as well.

It can be observed that higher effluent methane concentrations will result in a steep increase in CO_2 emissions. When anaerobic sludge degradation in (uncontrolled) landfills is excluded from the evaluation, then up to a methane concentration of 13.5 $mg \cdot l^{-1}$ the carbon footprint of the anaerobic-aerobic system is still smaller than that of the conventional activated sludge system. However, at the expected effluent methane concentration of 19 mg $CH_4 \cdot l^{-1}$, the situation is inverted: 1670 versus 4100 tons of $CO_2 \cdot yr^{-1}$.

Now, when the effects of disposal of stabilized sludge to the landfill are included, the picture changes again. It has already been demonstrated that the CO_2 emissions from uncontrolled landfills are considerable. For example, when full degradation of the stabilized sludge is assumed, then the anaerobic–aerobic system has a smaller carbon footprint over the full range of effluent methane concentrations evaluated. On the other hand, when the IPPC guidelines are followed, then for the expected effluent dissolved methane concentration of 19 mg $CH_4 \cdot l^{-1}$, performance of the aerobic system is slightly better (6400 versus 5500 ton $CO_2 \cdot yr^{-1}$). However, it should be considered that not all major electrical consumers have been included in the analysis (e.g. influent lifting stations, return sludge lifting stations and sludge dewatering units) and in any case that the performance of the anaerobic - aerobic system in terms of power consumption and excess sludge production is still superior.

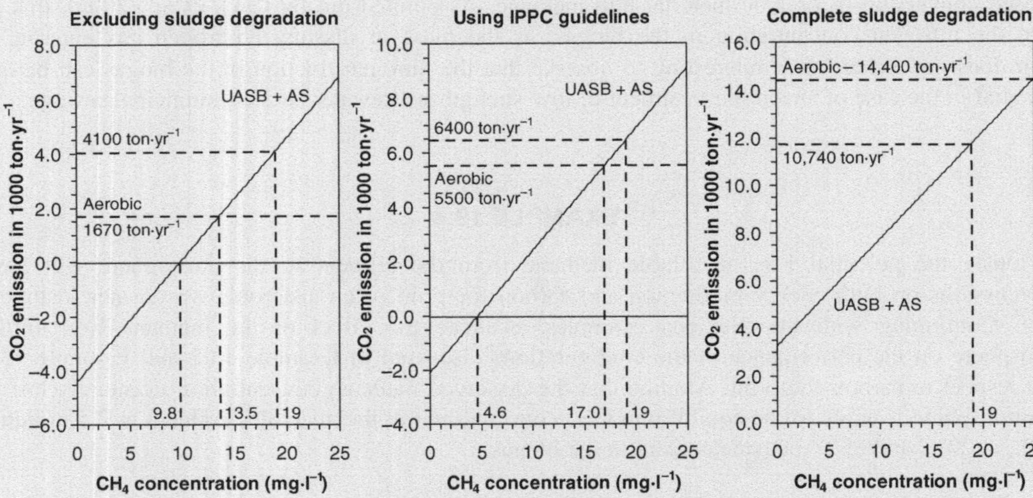

Figure 13.15 Effect of methane emissions from the anaerobic effluent and landfills on the carbon footprint of systems for secondary treatment (COD removal) for the conditions in Example 13.4

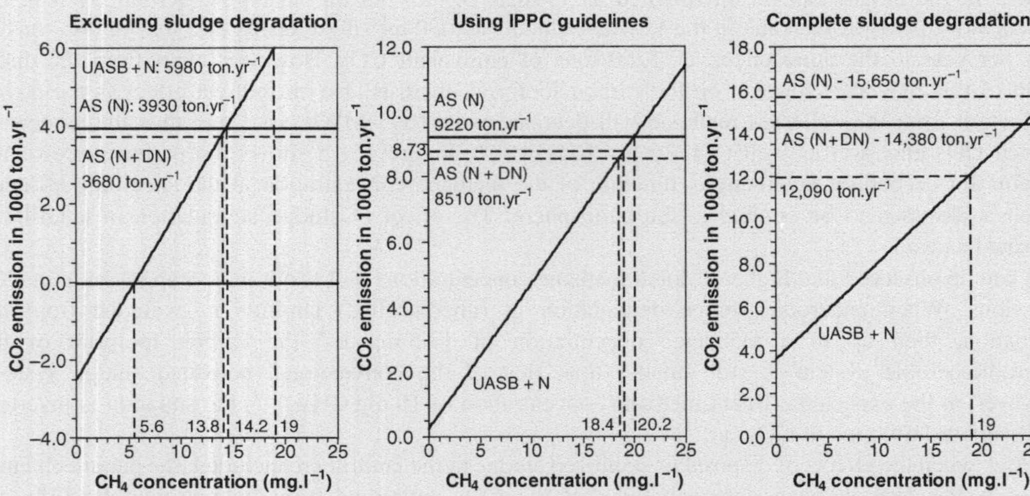

Figure 13.16 Effect of methane emissions from the anaerobic effluent and landfills on the carbon footprint of treatment systems for COD and (partial) nitrogen removal for the conditions in Example 13.4

When nitrification and denitrification are included in the analysis (Figure 13.16), the overall results are similar. When the IPPC guidelines are considered, it can be observed that the CO_2 emissions of the anaerobic system (8730 ton $CO_2 \cdot yr^{-1}$) are quite comparable to those of the aerobic system. The overall net power demand and net CO_2 emission of the systems for secondary and tertiary treatment (nitrogen removal) for the different cases is summarized in Table 13.7.

Table 13.7 Energy demand and carbon footprint for different configurations of aerobic and anaerobic - aerobic systems for the conditions in Example 13.4, including the effects of methane emission from the effluent (a value of 19 mg $CH_4 \cdot l^{-1}$) has been assumed and landfill (IPPC guidelines)

Parameter	UoM	Secondary		Tertiary	
		AS	UASB + AS	N+DN	UASB + N
Case I – Disregarding methane emissions from effluent and landfill:					
Power consumption	$MWh \cdot yr^{-1}$	2330	−6100	5130	−3490
CO_2 emission	ton $CO_2 \cdot yr^{-1}$	1670	−4380	3680	−2500
Case II – Disregarding methane emissions from landfill, including those from the effluent:					
Power consumption	$MWh \cdot yr^{-1}$	2330	−4400	5130	−1790
CO_2 emission	ton $CO_2 \cdot yr^{-1}$	1670	4100	3680	5980
Case III – Including methane emissions from both anaerobic effluent and landfill (IPPC):					
Power consumption	$MWh \cdot yr^{-1}$	2330	−4400	5130	−1790
CO_2 emission	ton $CO_2 \cdot yr^{-1}$	5500	6400	8510	8730

In general it can be concluded that the installation of a system for removal (and preferably utilization) of methane from the anaerobic effluent will have a decisive impact on the carbon footprint of the anaerobic-aerobic systems, and allows the combined anaerobic-aerobic system to be converted into a truly "green" wastewater treatment plant.

Disposal of stabilized sludge on uncontrolled landfills (or on agricultural land) is the most common practice for most developing countries, especially in regions with a warm climate, where application of UASB reactors is most appropriate anyway. Hence it can be concluded that if the overall wastewater treatment scheme *including* the disposal of stabilized excess sludge is considered, then combined anaerobic–aerobic treatment is much more sustainable (i.e. comparable carbon footprint but reduced power consumption) than conventional activated sludge treatment, even when the losses of dissolved methane in the anaerobic effluent are taken into account.

Notwithstanding this important conclusion, it is also clear that removing or preferably recovering and utilizing the methane present in the anaerobic effluent will have a further beneficial impact on the overall system performance. The recovery of methane from the anaerobic effluent increases the potential for power generation by 25 to 40%, while the impact on the carbon footprint is even more significant. On the other hand, in countries like Brazil and India greenhouse gas emissions are presently not an issue, although this might change in the future.

For any methane recovery scheme, it is important to establish what will happen to the dissolved methane contained in the anaerobic effluent. Most UASB reactors are equipped with V-notch type overflow weirs discharging into effluent gutters which in turn overflow into a common effluent channel. Especially the last step is often a rather turbulent process and there will be intensive contact between air and effluent, as can also be observed in Figure A6.1. This is sometimes done deliberately, in order to "refresh" the effluent. Therefore, a large part of the methane present in the UASB effluent will be emitted to the atmosphere at one of the various overflows. Note that odour problems due to H_2S, also present in the UASB effluent, are indeed often most noticeable at the overflows into the common effluent channel and not so much at the settler surface. The use of submerged pipes for effluent withdrawal and a closed effluent collection system could reduce these undesired gas emissions considerably.

Table 13.8 Comparison between measured and theoretical methane concentration, based on the data from Souza et al. (2010b), for T = 25°C and p = 0.91 atm

Parameter	UoM	Case 1	Case 2	Case 3
HRT	h	5.0	7.0	12.0
Reactor	(–)	pilot	pilot	demo
Type of effluent collection	(–)	V-notched gutter		subm. pipe
CH_4 conc. below overflow level	$mg \cdot l^{-1}$	19.6	22.0	19.2
CH_4 conc. in effluent pipe/gutter	$mg \cdot l^{-1}$	18.1	20.3	18.5
CH_4 conc. in reactor effluent pipe	$mg \cdot l^{-1}$	N.D.	N.D.	7.0

Souza et al. (2010b) presented data from an additional experiment with the reactors discussed previously, which reinforce the above assumptions. This data is summarized in Table 13.8. In the experiment, the methane concentration was measured just before the overflow level, just after the overflow and after a free fall of 3 m into a vertical reactor effluent pipe. The pilot reactor was fitted with V-notched effluent gutters while the demo-scale reactor was equipped with submerged effluent pipes. It can be observed that the decrease in methane concentration is small at the level of the effluent collection device, as the flow rate per individual V-notch or perforation is low, resulting in limited turbulence and hence a low rate of methane transfer. Furthermore, the methane emission at the V-notch overflow weir is higher than at the submerged perforated pipe, on average 1.6 (−7.5%) versus 0.7 mg $CH_4 \cdot l^{-1}$ (−3.6%). However, the decrease after the vertical 3 m drop in the reactor effluent pipe is much more significant: 11.5 mg $CH_4 \cdot l^{-1}$ or -60%. Thus, it may be assumed that most of the methane contained in the anaerobic effluent is lost to the atmosphere at some point in the effluent collection system. This can be prevented by covering the reactor, the effluent weirs and the effluent channels or by using submerged pipes for effluent collection. The collected vent gas contains hydrogen sulphide, ammonia and methane and thus requires treatment, together with the vent gas from other locations (e.g. influent lifting station, sludge dewatering unit). Several common methods of vent gas treatment include:

- Injection of the vent gas into the aerated zone of an activated sludge system. The injection depth will necessarily be small in order to limit the power demand of the ventilator and the bubble size will be coarse. Therefore it might be assumed that the contact between gas and liquid is not sufficient for complete removal of the contaminants;
- Treatment in a dedicated vent gas treatment system, such as:
  - A caustic scrubber (removes H_2S) or an acid scrubber (removes NH_3). However, methane is not removed;
  - Chemical scrubbers where strong oxidants such as NaOCl are used to oxidize the contaminants;

- A biological filter or scrubber (lava filter, packed bed) where both NH_3 and H_2S are removed and possibly (part of) the CH_4 as well;
- Activated carbon filters, which can remove all of the above compounds except methane but are generally more expensive to operate, as the activated carbon packing needs to be replaced or regenerated periodically;

The methane contained in the vent gas should preferably be put to useful means. An interesting concept would be to use the ventilation air as a source of oxygen supply for either energy generation equipment (gas motor or turbine) or alternatively the flare.

Apart from reducing methane emissions by combustion of this component to carbon dioxide, this would increase the power production from the collected methane by 20–40%. Example 13.5 discusses the impact on plant performance of such a scheme in more detail. Removal of the H_2S from the vent gas prior to combustion in gas motors will be required.

Another approach is to reduce the transfer of methane to the gas phase at the surface area and in the effluent collection system and recover the methane in a concentrated form from the effluent, allowing injection in the biogas line. One such method is through the use of a vacuum stripper. An interesting example of this type of application can be found in the Netherlands, albeit not in the field of wastewater treatment. Vitens, a potable water company, uses 25 million cubic metres of groundwater from peat lands as a water source in its production location in Tjerkgaast. This water is characterised by a high methane concentration (30–45 mg $CH_4 \cdot l^{-1}$), due to the long retention time of up to 10,000 yrs under anaerobic conditions. The conventional method for methane removal from ground water involves the use of perforated plates where a thin film of water is contacted with air. Typically 40–50 m^3 of air is applied per m^3 of groundwater. All of the stripped methane (i.e. 1150 tons of $CH_4 \cdot yr^{-1}$) is currently emitted to the atmosphere, corresponding to 0.16% of the total yearly Dutch greenhouse gas emissions. However, in 2012 a methane recovery system will be taken into operation, consisting of a vacuum degassing installation where 95% of the methane present in the groundwater will be collected (Visscher, 2010). The removal of the residual methane still requires aeration, but the aeration demand is reduced by approximately 50%. Conditioning of the methane by means of CO_2 removal and dewatering, followed by pressurization allows direct injection of the recovered methane gas into to the existing natural gas distribution grid.

Another method to recover methane from effluent is described by Cookney et al. (2010). The effluent passes through a hydrophobic membrane of a type that is already commercially used for degassing. The membrane is due to its hydrophobic nature not permeable for water but only for gases. To enhance gas transfer, the gas side of the membranes is either sweeped with a vent gas (like ambient air) or a vacuum is applied. As an alternative to combustion, there are several other possible uses for the methane present in the anaerobic effluent. These methods are all based on the high COD value of methane (4 g $COD \cdot g^{-1}$ CH_4), which therefore in principle could be used to enhance the denitrification capacity:

- A first concept relies on the microbial catalyzed oxidation of methane to methanol, mediated by methanotrophic bacteria (Werner et al., 1991):

  $$CH_4 + O_2 + 2H^+ \longrightarrow H_2O + CH_3OH \longrightarrow CO_2 + H_2O \qquad (13.10a)$$

  It is important to maintain the dissolved oxygen concentration below 1 mg $O_2 \cdot l^{-1}$, else the reaction according to Eq. (13.10a) will proceed to full conversion of methane to carbon dioxide (Rajapakse et al., 1999).

Unfortunately, the production of methanol can only occur in the aerobic zone where it will be consumed at least partly as well. This will result in increased aeration demand, partly off-setting the energy savings by anaerobic pre-treatment. On the other hand, if the oxygen concentration is controlled at a low value, simultaneous denitrification can occur,

$$CH_3OH + \frac{6}{5}\ NO_3^- \longrightarrow \frac{3}{5}\ N_2 + \frac{6}{5}\ OH^- + \frac{7}{5}\ H_2O + CO_2 \tag{13.10b}$$

However, operation at a low oxygen concentration reduces the reaction rates of both the nitrification and the denitrification process considerable. Furthermore, the nitrifiers present in the mixed liquor would be placed in direct contact with the sulphides in the anaerobic effluent, which is a known contributing factor to inhibited nitrifier growth. This will be discussed in more detail in Section 13.5.2;

- A variation on this concept is to use the methane and H_2S present in the biogas as a source of COD for denitrification. Pantoja Filho et al. (2010) operated an aerobic-anoxic fixed bed pilot reactor and achieved significant denitrification once the biogas was injected into the anoxic (lower) chamber. However, if the methane is not converted completely in the aerobic zone, it will end up in the aeration off-gas and recreates a significant problem;
- Finally, research is ongoing on the development of reactors where methane and or sulphide present in the anaerobic effluent are used directly for autotrophic denitrification. An example is a recent project at the University of Wageningen, which focuses on the use of a specific type of biomass capable of using methane directly for denitrification (using nitrite), but this project is at present (2010) still at the laboratory stage. The reaction equation involved is:

  $$3\ CH_4 + 8\ NO_2^- + 8\ H^+ \longrightarrow 3\ CO_2 + 4\ N_2 + 10\ H_2O \tag{13.10c}$$

  It remains to be seen whether this will be an economically viable concept, as the maximum extent of nitrate removal will probably be limited and the added complexity and costs of a dedicated reactor will be high.

As discussed above, an undesirable side-effect of anaerobic pre-treatment of municipal sewage is that the sulphides produced are very odorous, while they are also toxic to nitrifiers and reduce their growth rate significantly. For this reason, as discussed in Section 13.2.4, the anaerobic effluent is often subjected to a pre-aeration step, with the objective of (partial) oxidation of the sulphide to non toxic elemental sulphur, thiosulphate and sulphate. To prevent odour issues, the off-gas should be collected and subjected to vent gas treatment.

On the other hand, assuming the pre-aeration off-gas strips methane from the anaerobic effluent, it could also be used as combustion air for the gas motor or flare, as the oxygen content in the off-gas will still be relatively high. For example, when an OT_a of 14% is assumed, then the oxygen content in pre-aeration reactor off-gas will be reduced from 21 to 18%.

EXAMPLE 13.5

Evaluate the use of pre-aeration off-gas as combustion air for the 50,000 $m^3 \cdot d^{-1}$ municipal UASB discussed previously in Example 13.2 to Example 13.4, in terms of methane recovery. The sewage temperature is 25°C. Assume that 15 mg H_2S-S and 18.7 mg $CH_4 \cdot l^{-1}$ are present in the anaerobic effluent. The maximum allowable effluent methane concentration is ≤ 2.5 mg $CH_4 \cdot l^{-1}$. The following additional data is given for the pre-aeration reactor:

- Actual oxygen transfer efficiency $OT_a = 12\%$;
- Oxygen demand for H_2S oxidation to $S^0 = 0.5$ mg $O_2 \cdot mg^{-1}$ S;
- Density of air is 1.29 kg $\cdot$ Nm^{-3};
- Value of the Henry constant for methane $K_H = 40{,}965$ at 25°C.

Solution

Calculate the oxygen demand for partial H_2S oxidation to S^0 and the associated air input:

$$MO_t = 0.5 \cdot Q_i \cdot [H_2S\text{-}S]/1000 = 0.5 \cdot 50{,}000 \cdot 0.015 = 375 \text{ kg } O_2 \cdot d^{-1}$$

$$Q_{air} = 375/(24 \cdot 0.12 \cdot 0.21 \cdot 1.29) = 481 Nm^3 \cdot h^{-1}$$

Now compare this to the quantity of methane to be recovered from the effluent, which is equal to $Q_i \cdot (S_{ch4in} - S_{ch4out}) = 50{,}000 \cdot (0.0189–0.0025) = 820$ kg $CH_4 \cdot d^{-1}$

$$Q_{ch4} = 810 \cdot 4 \cdot 0.35 = 1148 \text{ Nm}^3 \text{ CH}_4 \cdot d^{-1} \quad \text{or} \quad 48 \text{ Nm}^3 \cdot h^{-1}$$

Disregarding the consumption of oxygen and stripping of CO_2 on the composition of the pre-aeration off-gas, the methane content of the gas escaping from the water phase is equal to $48/(48 + 481) = 9.1\%$. The equilibrium methane concentration in the pre-aeration effluent is:

$$X_{ch4} = 0.091/40{,}965 = 2.2 \cdot 10^{-6} \text{mol} \cdot \text{mol}^{-1} \tag{13.9b}$$

$$[CH_4]_{eq} = 1000 \cdot 2.2 \cdot 10^{-6} \cdot 55.6 \cdot 16 = 2.0 \text{ mg } CH_4 \cdot l^{-1} \tag{13.9c}$$

This is indeed below the limit of 2.5 mg $CH_4 \cdot l^{-1}$. The use of pre-aeration off-gas as combustion air increases methane recovery from 70% to 96.7%. It can be observed that in this example the methane content in the off-gas of 9.1% is within the explosion limits for methane (5–15%). However, this depends for a large part on the value assumed for the CH_4 concentration present in the UASB effluent at the inlet to the pre-aeration reactor. In this example it was assumed that all methane present in the UASB effluent at the overflow weir level was received in the pre-aeration reactor. In reality a large part of it will escape to the atmosphere somewhere in the effluent collection system.

If it is desired to reduce the methane fraction to a value below the explosion limit (intrinsic safety), a solution could be to increase the pre-aeration air flow. This will not only reduce the methane content in the off-gas, but it will also result in a more complete oxidation of H_2S to SO_4^{2-}, provided that the retention time in the pre-aeration reactor is sufficient. The air flow rate required for stoichiometric oxidation to SO_4^{2-} is equal to 1923 $Nm^3 \cdot h^{-1}$. At this flow rate, the methane fraction in the off-gas decreases to 2.4%, well below the explosion limit. As a secondary effect, the decrease in partial methane pressure will reduce the methane concentration in the effluent of the pre-aeration reactor approximately 0.5 mg $CH_4 \cdot l^{-1}$.

13.3 DESIGN MODEL FOR ANAEROBIC SEWAGE TREATMENT

In Section 13.1, the COD removal efficiency of several anaerobic reactor types was given as a function of the hydraulic retention time, thus giving the impression that this is the key parameter in the performance of anaerobic systems. However, it was demonstrated in Chapter 3 for aerobic treatment systems that it is not the hydraulic retention time but rather the sludge age that determines system performance in terms of organic material removal efficiency and excess sludge production. The hydraulic retention time is only important to the extent that it determines the sludge concentration as it defines the reactor volume for a given sewage flow. Thus the sludge age is the key design parameter for aerobic treatment.

In anaerobic treatment systems it is much harder to measure and control the sludge concentration, hence the sludge age is usually difficult to control. Furthermore, contrary to aerobic systems, anaerobic systems in general are operated at the maximum possible sludge age for the available reactor volume, as the objective is to maximise the digestion efficiency of biodegradable organic material. Operation at a high sludge age does not incur a "penalty" like excessive aeration costs for aerobic systems, although the required reactor volume will increase at higher values of the sludge age. Instead, as the sludge age is difficult to control in anaerobic treatment systems, the liquid retention time is often selected as the design parameter.

13.3.1 Sludge age as the key design parameter

In this section it will be demonstrated that similar to aerobic systems, the sludge age is the key design parameter in anaerobic systems. Cavalcanti et al. (2003) operated two UASB reactors with a different GLS separator design on raw sewage. One reactor was equipped with a conventional separator similar to the one presented in Figure 13.3, while the other was additionally equipped with parallel plates placed above the conventional separator, effectively transforming it into a high rate settler. Both systems were operated without intentional excess sludge discharge, so that each reactor was operated with the maximum sludge mass that could be retained for the applied organic- and hydraulic load. After steady state conditions had been established, the two systems, although operated at the same hydraulic retention time, showed a very different performance as the retained sludge mass, and thus also the operational sludge age in the systems, was not the same. The maximum sludge mass that could be obtained was influenced by the efficiency of the phase separation.

This is shown in Figure 13.17a, where the division of the influent organic material fed to the anaerobic reactor into three fractions is shown: (I) digested organic material present as methane in the biogas (II) excess sludge present as settleable organic material in the effluent and (III) non degraded organic material present in the settled effluent (mainly soluble and fine colloidal material). The experimental values of the sludge- and effluent COD fractions of the two UASB reactors have been plotted as a function of the hydraulic retention time. It can be observed that the reactor with the improved separator demonstrated a much better performance, as at all retention times the fraction of digested COD was much larger than in the conventional reactor. This is a clear indication that the hydraulic retention time alone is not a suitable indicator of the expected performance of a UASB reactor: the quality of the phase separator design, which influences sludge retention and hence the sludge age, is important as well.

In the same experiment, the sludge age of the two systems was determined as the ratio of the sludge mass present in the UASB reactor and the daily mass of sludge (settleable solids) discharged with the effluent. When the COD removal efficiency and sludge production are plotted as function of the sludge age, the data for the two GLS configurations now turns out to be very comparable as can be observed in Figure 13.17b. Thus it is clearly shown that it is in fact the anaerobic sludge age and not the hydraulic retention time that is the fundamental design parameter of anaerobic treatment systems. Hence it can be concluded that the performance of different types of anaerobic reactors treating the same wastewater will be comparable as long as the operational sludge ages are equal (and the effluent suspended solids concentration is comparable). In other words, the division of the influent COD into fractions that end up in the solid-, liquid- and gaseous phase tends to be identical in different anaerobic treatment systems treating similar wastewater, as long as these are operated at the same sludge age. In this respect, anaerobic treatment systems are thus similar to aerobic systems.

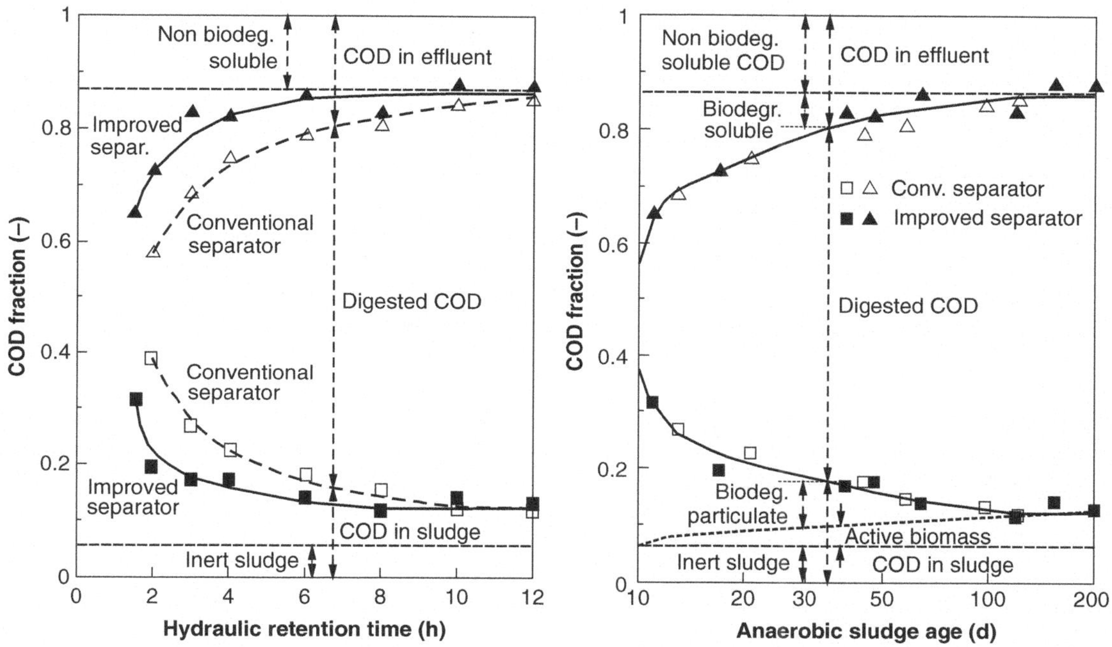

Figure 13.17 Influent COD fractions leaving in the effluent, the excess sludge and in the form of methane as function of the hydraulic retention time (left) and the anaerobic sludge age (right) for a conventional UASB and for a UASB with improved phase separator, both operating at 25°C

However, in practice it is difficult to predict the sludge age of an anaerobic system, which makes it difficult to use as a design parameter. Therefore, the hydraulic retention time is often used instead (using Eqs. 13.3 to 13.6), although as explained above different systems with the same hydraulic retention time may exhibit significantly different behaviour. Therefore, in this section empiric equations will be presented that allow an estimate of the sludge mass and sludge composition that will develop as function of the influent COD load and -composition and the applied anaerobic sludge age. This in turn facilitates the use of the anaerobic sludge age for the process design of UASB reactors. Apart from highlighting the influence of the sludge age as the key determinant to the performance of an anaerobic treatment system, the experimental curves of Figure 13.17 reveal some additional important characteristics:

(1) In addition to anaerobic biomass, there will also be an inert organic fraction present in the sludge, due to flocculation of non biodegradable particulate organic material present in the influent;

(2) Depending on the operational conditions of the anaerobic reactor (i.e. sludge age and temperature), there may be a significant fraction of biodegradable particulate organic material present in the sludge mass, resulting from the incomplete hydrolysis of organic material;
(3) As a consequence, the active biomass fraction in the organic excess sludge from an anaerobic reactor will be small;
(4) At low sludge ages, the particulate biodegradable organic material will constitute the largest fraction in the volatile sludge mass, so the excess sludge will be unstable and may be putrescible. In fact in many cases the minimum required sludge age in UASB reactors is set by the requirement to obtain a stabilised sludge rather than the permissible reduction in the COD removal efficiency;
(5) The division of influent COD into effluent-, excess sludge- and digested fractions will be different for systems treating sewage and for systems treating soluble wastewater: in the former, the hydrolysis of the particulate biodegradable material proceeds at low rate at low sludge ages and due to the flocculation of this material it will become part of the anaerobic sludge, while this process is absent in the latter;
(6) As the sludge age decreases, the value of the COD fraction that ends up in the effluent will increase due to a reduction in the removal of soluble biodegradable material in the anaerobic reactor;
(7) There is a minimum sludge age below which the anaerobic reactor fails altogether, i.e. the biological activity ceases because the bacterial populations are not able to develop. The reactor will then act as a primary settler.

From experimental data obtained in tropical regions (at a temperature of 25°C) empirical expressions have been derived to predict the fraction of *soluble* (i.e. non settleable) organic material in the UASB effluent (mS_{eu}) as function of the anaerobic sludge age R_{su} in the case of raw sewage treatment:

$$mS_{eu} = f_{ns} + f_{h2s} + 0.27 \cdot \exp[-0.04 \cdot (R_{su} - 4)] \tag{13.11}$$

The fraction f_{h2S} represents the COD concentration in the anaerobic effluent due to the conversion of biodegradable *organic* influent COD into biodegradable *inorganic* sulphide COD as a result of sulphate reduction (as discussed in Section 13.2.4). This fraction can be defined in a similar manner to f_{ns}:

$$f_{h2s} = S_{h2se}/S_{ti} \tag{13.12}$$

where S_{h2se} is the H_2S concentration in the anaerobic effluent expressed in mg $COD \cdot l^{-1}$

Note that Eq. (13.11) only considers non-settleable material, i.e. soluble and some colloidal COD, as all settleable material is assumed to be retained by the GLS separator and removed with the excess sludge. However in practice, due to the non-ideal performance of the GLS separator, a certain suspended solids concentration will remain in the effluent. As a consequence, the total COD concentration in the UASB effluent will be higher than that predicted by Eq. (13.11). The TSS concentration in the effluent of a well designed and operated UASB reactor generally varies between 40–100 mg $TSS \cdot l^{-1}$, which at typical values of the organic fraction (0.6–0.7) roughly equates to the same COD concentration.

The influent COD fraction converted into anaerobic sludge (mS_{xvu}), i.e. considering both the discharge from the reactor with the excess sludge and the loss of suspended solids with the effluent, is given by:

$$mS_{xvu} = f_{np} + f_{cv} \cdot Y_{an} \cdot (1 - f_{ns} - f_{np} - f_{h2s}) + 0.25 \cdot \exp[(-0.04 \cdot (R_{su} - 4)] \tag{13.13}$$

Hence the digested fraction mS_{du} can be expressed as:

$$\begin{aligned} mS_{du} &= 1 - mS_{eu} - mS_{xvu} \\ &= (1 - f_{cv} \cdot Y_{an}) \cdot (1 - f_{ns} - f_{np} - f_{h2s}) - 0.52 \cdot \exp[(-0.04 \cdot (R_{su} - 4)] \end{aligned} \quad (13.14)$$

where:

R_{su} = operating sludge age in the UASB reactor
Y_{an} = yield coefficient in an anaerobic environment (0.05 mg VSS $\cdot$ mg^{-1} COD) as determined by Cavalcanti et al. (2003)

Equations (13.11 to 13.14) are empirical expressions derived for municipal sewage and therefore only valid for conditions comparable to those under which they were generated. For industrial wastewaters, the relationships may be very different, due to the difference in organic material composition. In particular, the value of the particulate biodegradable fraction is important, because the limiting factor in anaerobic digestion is usually the rate of hydrolysis of this material. Furthermore, the sulphate concentration in the raw sewage might differ from the one encountered in Campina Grande, which was fairly typical for raw sewage (50–70 mg $SO_4^{2-} \cdot l^{-1}$). The reduction of sulphate reduces the conversion of COD into methane, whereas the produced sulphides will end up as inorganic COD in the effluent. Thus COD removal efficiency is reduced as well.

The guidelines to estimate the extent of sulphate conversion in the anaerobic reactor have been discussed in Section 13.2.4 and can be used to predict the sulphide COD concentration in the UASB effluent. Thus it is possible to compensate for a significantly deviating sulphate content in the raw sewage. However, when the sulphate concentration is very high, e.g. above 150 mg $SO_4 \cdot l^{-1}$, the decision to implement an anaerobic system should perhaps be reconsidered, although this also depends on the influent COD concentration. A high COD concentration in the raw sewage allows a higher sulphate concentration to be accepted, as the impact on COD removal efficiency will be smaller.

The empiric equations (13.11 to 13.14) indicate that for efficient treatment the required sludge age will be much higher than in the case of aerobic treatment. On the other hand, it is usually possible to accommodate several times more sludge per unit of volume in anaerobic systems, which allows the application of a high sludge age in a relatively compact system.

EXAMPLE 13.6

Determine the performance of an UASB reactor treating raw sewage at 50,000 $m^3 \cdot d^{-1}$ with a COD concentration of 600 mg $\cdot l^{-1}$ (MS_{ti} = 30,000 kg COD $\cdot d^{-1}$, $f_{ns} = f_{np} = 0.1$). Ignore the presence of sulphate in the wastewater. The UASB is operated at a sludge age of 35 days at a temperature of 25°C. Indicate the effect of the following factors on UASB performance:

- Loss of methane to the atmosphere, with and without system for methane recovery in place. (methane recovery reduces the emission to the atmosphere to a value $\leq$ 2.5 mg $CH_4 \cdot l^{-1}$);
- Effluent TSS concentration of 80 mg $\cdot l^{-1}$, with an organic fraction f_{vu} of 0.63.

Solution

The (ideal) division of the raw sewage COD into fractions in the effluent, discharged as excess sludge and digested can be calculated with Eqs. (13.11 to 13.14), where f_{h2s} is ignored:

$$
\begin{aligned}
mS_{eu} &= f_{ns} + 0.27 \cdot \exp[-0.04 \cdot (R_{su} - 4)] \\
&= 0.10 + 0.27 \cdot \exp[-0.04 \cdot (35 - 4)] = 0.18 \qquad (13.11) \\
mS_{xvu} &= f_{np} + f_{cv} \cdot Y_{an} \cdot (1 - f_{ns} - f_{np}) + 0.25 \cdot \exp[(-0.04 \cdot (R_{su} - 4)] \\
&= 0.10 + 1.5 \cdot 0.05 \cdot (1 - 0.2) + 0.25 \cdot \exp[(-0.04 \cdot (35 - 4)] = 0.23 \qquad (13.13) \\
mS_{du} &= 1 - mS_{eu} - mS_{xvu} = 1 - 0.18 - 0.23 = 0.59 \qquad (13.14)
\end{aligned}
$$

Assuming an ideal situation, i.e. when sludge- and methane losses with the effluent are ignored, then for the influent COD load of 30,000 $kg \cdot d^{-1}$ the daily mass flows are:

$$
\begin{aligned}
MS_{eu} &= 5340 \text{ kg COD} \cdot d^{-1} \\
MS_{xvu} &= 6970 \text{ kg COD} \cdot d^{-1} (= 4650 \text{ kg VSS} \cdot d^{-1}) \\
MS_{du} &= 17,690 \text{ kg COD} \cdot d^{-1} (= 4420 \text{ kg } CH_4 \cdot d^{-1})
\end{aligned}
$$

The calculated COD removal efficiency in the UASB is 1–(5340/30,000) = 82%. However, in practice this value needs to be corrected for the presence of suspended solids in the effluent, which in this case is equal to $50,000 \cdot 0.63 \cdot 0.08 = 2510$ kg VSS $\cdot d^{-1}$ (= 3770 kg COD $\cdot d^{-1}$). Note that this is more than fifty percent of the predicted excess sludge production.

As to the evaluation of the effect of the presence of methane in the effluent, when it is assumed that the methane concentration in the effluent is 18.7 mg $CH_4 \cdot l^{-1}$ (Table 13.5, for a methane content in the biogas of 70%), then the daily load of methane in the effluent is 935 kg $CH_4 \cdot d^{-1}$. If a system of methane recovery is in place, then less than $2.5/18.7 \cdot 935 = 125$ kg $CH_4 \cdot d^{-1}$ is lost to the atmosphere, whereas the remainder (810 kg $CH_4 \cdot d^{-1}$) is collected in the reactor or recovered from the effluent. The division of COD loads (when methane recovery is applied) will thus be:

$$
\begin{aligned}
MS_{eu} &= 5340 + 3770 = 9110 \text{ kg COD} \cdot d^{-1} \\
MS_{xvu} &= 6970 - 3770 = 3200 \text{ kg COD} \cdot d^{-1} (= 2135 \text{ kg VSS} \cdot d^{-1}) \\
MS_{du} &= 17,690 - 4 \cdot 125 = 17,190 \text{ kg COD} \cdot d^{-1} (= 4300 \text{ kg } CH_4 \cdot d^{-1}) \\
MS_{da} &= 500 \text{ kg COD} \cdot d^{-1} (= 125 \text{ kg } CH_4 \cdot d^{-1}) \longrightarrow \text{(to atmosphere)}
\end{aligned}
$$

The COD removal efficiency decreases from 82% to 70% due to the presence of suspended solids in the effluent. Note that the presence of dissolved methane in the effluent is ignored in the calculation of COD removal efficiency.

Now, without methane recovery, it is assumed that all methane in the UASB effluent will (eventually) be lost to the atmosphere, so the emission of methane is equal to 935 kg $CH_4 \cdot d^{-1}$. In this case, the value of MS_{du} will be reduced to 13,950 kg COD $\cdot d^{-1}$ while that of MS_{da} increases to 3740 kg COD $\cdot d^{-1}$. Methane recovery decreases to 81%, compared to 97% when a system of methane recovery is in place. As the residual methane in the effluent will not be detected, the observed COD removal efficiency is not affected.

13.3.2 Influence of the temperature

The wastewater temperature has a large impact on the performance of anaerobic treatment systems. In anaerobic systems treating complex wastewaters like sewage, the rate limiting step of the overall digestion process is the rate of hydrolysis of particulate matter. From interpolation of the data obtained by O'Rourke (1969) for the digestion of primary sludge (Figure 12.17), it can be observed that the anaerobic sludge age required to obtain significant removal of organic material by anaerobic digestion may be expressed as:

$$R_{su} = 15 \cdot 1.067^{(T-25)} (15°C < T < 35°C) \quad (13.15)$$

where R_{su} = anaerobic sludge age in the UASB reactor

The same temperature dependency was also reported by Gujer et al. (1993). Thus the COD fractions in the effluent and in the excess sludge of a UASB reactor treating raw sewage and operating at a temperature between 15 and 35°C can be expressed as:

$$mS_{eu} = f_{ns} + f_{h2s} + 0.27 \cdot \exp[-0.04 \cdot (R_{su} - 4)]/1.067^{(T-25)} \quad (13.16)$$

$$mS_{xvu} = f_{np} + f_{cv} \cdot Y_{an} \cdot (1 - f_{ns} - f_{np} - f_{h2s}) + 0.25 \cdot \exp[(-0.04 \cdot (R_{su} - 4)]/1.067^{(T-25)} \quad (13.17)$$

The corresponding COD fraction that is converted into methane in the UASB is given as:

$$mS_{du} = 1 - mS_{eu} - mS_{xvu} \quad (13.18)$$

In practice the design temperature of a UASB will rarely be higher than 25°C, although the operational temperature may well be higher in the summer time. Figure 13.18 shows the division of the influent organic material over the three fractions as function of the anaerobic sludge age for temperatures of 15 and 20°C. The same data was presented previously in Figure 13.17b for a temperature of 25°C. It can be observed from Figure 13.18 as well as from Eqs. (13.16 and 13.17) that the effluent quality of the UASB becomes poorer and the sludge production will be higher at lower sludge ages.

In Figure 13.17b the influent COD fractions that are transformed into active and inert sludge (mS_{xa} and mS_{xi}) are indicated, as well as the fraction of COD that is wasted as non degraded particulate material (mS_{xbp}). The endogenous residue fraction has not been indicated as a separate fraction and is included with the active biomass, because the decay rate of the anaerobic bacteria is very low and therefore the value of the endogenous residue fraction will be small, even at high values of the sludge.

In Figure 13.18 the value of a new parameter f_{pu} is indicated as well: this parameter represents the "putrescible fraction" in the anaerobic sludge and is composed of residual, non degraded biodegradable particulate COD. As such it is an indication of the biological stability of the anaerobic excess sludge (refer also to Section 13.3.3).

At lower sludge ages, the fraction of non digested biodegradable particulate material in the UASB excess sludge will increase and this may eventually lead to difficulties in sludge handling due to production of an unstable sludge. As an alternative to an increase in the anaerobic sludge age in the UASB, it could also be considered to stabilize the anaerobic excess sludge in a dedicated heated anaerobic digester. As the digestion rate will accelerate due to the operation at higher temperature and at higher mixing intensity, a relatively small digestion unit will suffice. The overall treatment volume will be significantly reduced, as will be demonstrated in Example 13.8.

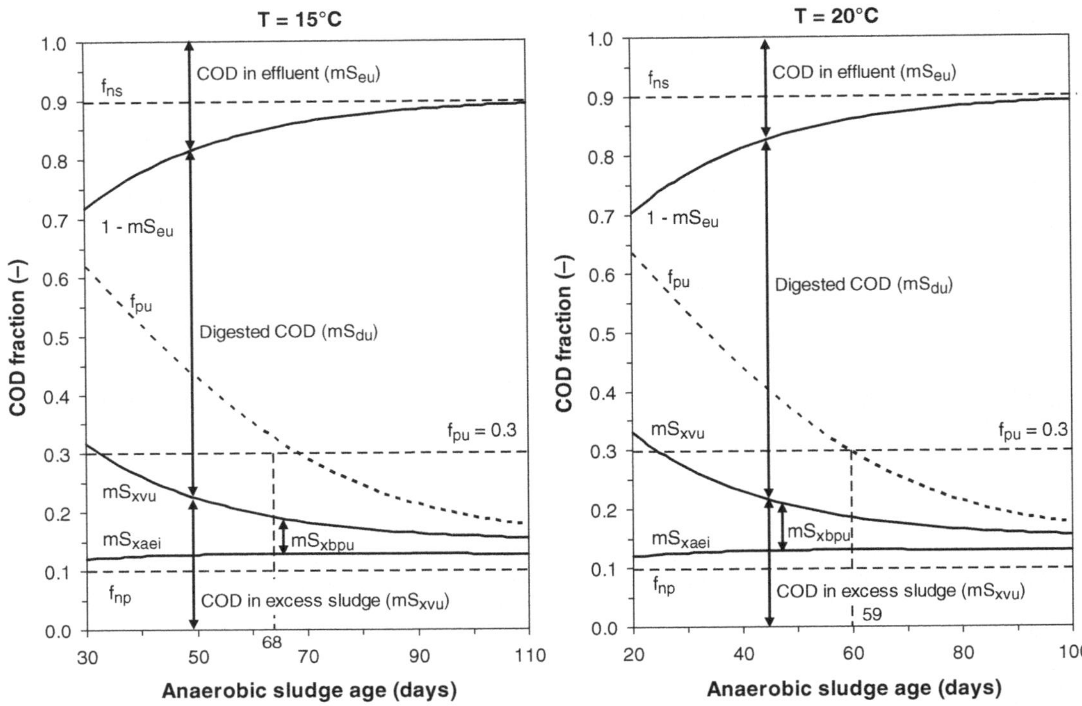

Figure 13.18 Division of the influent organic material into COD fractions (a) in the effluent (soluble), (b) in the excess sludge, and (c) transformed into methane, as function of the sludge age at 15 and 20°C ($f_{ns} = f_{np} = 0.10$)

13.3.3 Characterisation of anaerobic biomass

The organic anaerobic biomass can be characterised in terms of the sludge mass fractions (mX_{au}, mX_{eu}, mX_{iu} and mX_{bpu}), considering that when an influent COD fraction mS_{du} is catabolised (i.e. transformed into methane), the equivalent metabolised COD fraction will be:

$$mS_{mb} = mS_{du}/(1 - f_{cv} \cdot Y_{an}) \tag{13.19}$$

Using the above equation it is very easy to characterize the anaerobic biomass. The calculation of the active, endogenous and inert sludge mass that develops (in mg VSS · d · mg^{-1} COD) proceeds similar to that of the aerobic biomass fractions:

$$mX_{au} = Y_{an} \cdot R_{su}/(1 + b_{an} \cdot R_{su}) \cdot mS_{mb} \tag{13.20}$$

where b_{an} = decay constant for anaerobic sludge, which is very small ($\approx 0.01\ d^{-1}$)

$$mX_{eu} = f \cdot b_{an} \cdot R_{su} \cdot mX_{au} \tag{13.21}$$

$$mX_{iu} = f_{np} \cdot R_{su}/f_{cv} \tag{13.22}$$

From the difference between mS_{xvu} and the sum of the sludge mass fractions mX_{au} to mX_{iu} determined above, mX_{bpu} can be calculated:

$$mX_{vu} = mS_{xvu} \cdot R_{su}/f_{cv} = mX_{au} + mX_{eu} + mX_{iu} + mX_{bpu} \tag{13.23}$$

$$mX_{bpu} = mX_{vu} - mX_{au} - mX_{eu} - mX_{iu} \tag{13.24}$$

To finalize the characterization of the anaerobic biomass, it is necessary to estimate the value of f_{vu}. Due to the upflow mode of operation and the absence of aeration, the inorganic fraction of the anaerobic sludge tends to be higher than in aerobic systems, as heavy inert sludge will settle and because the value of the mineral fraction mX_{mu} tends to increase at higher values of the sludge age, similar to the fraction mX_{iu} as defined above. If the fraction of inert suspended solids in the influent ($f_{mi} = X_{mi}/S_{ti}$) is known, then X_{mu} (and hence f_{vu}) can be calculated from a simple mass balance:

$$mX_{mu} = f_{mi} \cdot R_{su} \tag{13.25}$$

$$mX_{tu} = mX_{vu} + mX_{mu} \tag{13.26}$$

$$f_{vu} = 1 - mX_{mu}/mX_{tu} \tag{13.27}$$

If the value of X_{mi} is not known, a default value for f_{mi} can be taken, typically between 0.08 to 0.10. It can be observed in Figure 13.17 and Figure 13.18 that, depending on the sludge age, the fraction of the influent COD that accumulates as biodegradable particulate solids in the UASB reactor (mS_{xbp}) may be significant and even predominant, especially when the temperature is low. Therefore it may be considered to establish a minimum limit for the anaerobic sludge age of a UASB reactor treating sewage, linked to the sludge composition and more specifically to the fraction of biodegradable material. This fraction may be expressed in terms of the putrescible biodegradable particulate material present in the anaerobic excess sludge ($f_{pu} = mX_{bpu}/mX_{xvu}$). Note that due to the low decay rate of the anaerobic sludge, the active biomass fraction is not included in the definition of f_{pu}.

At present no explicit guidelines are available on the maximum value of f_{pu} that avoids the generation of an unstable anaerobic excess sludge. In this book it is proposed that the maximum value of f_{pu} is equal to 0.3 mg VSS $\cdot$ mg^{-1} VSS. This is higher than the recommended value of 0.1–0.2 for the comparable parameter f_{ae} as used in aerobic digestion, the active sludge fraction remaining after aerobic digestion, as discussed in Section 12.3. However, it should be considered that the organic fraction of the UASB excess sludge will be lower than that of the excess sludge of a comparable activated sludge system for secondary treatment. The value of f_{pu} as a function of the anaerobic sludge age is indicated in Figure 13.18 as well.

Once the required sludge age of an anaerobic treatment unit has been established, based on the requirements for a certain minimum value of f_{pu} or a minimum required COD removal efficiency, the corresponding hydraulic retention time R_{hu} can also be estimated. From the calculated anaerobic sludge mass, the reactor volume and hence the value of R_{hu} can be calculated using the expected average sludge concentration in the UASB reactor. This can be demonstrated as follows: the sludge mass in a UASB reactor can be expressed either as the product of average sludge concentration and reactor volume or as the product of sludge age and daily excess sludge production. The daily excess sludge production is the product of the apparent yield coefficient or specific excess sludge production ($Y_{ap} = mE_{tu}$) and the organic load (MS_{ti}):

$$MX_{tu} = R_{su} \cdot ME_{tu} = R_{su} \cdot Y_{ap} \cdot Q_i \cdot S_{ti} = V_u \cdot X_{tu} \quad \text{or, after rearranging:}$$

$$R_{hu} = V_u/Q_i = R_s \cdot Y_{ap} \cdot S_{ti}/X_{tu} \tag{13.28}$$

In Eq. (13.28) the apparent yield coefficient Y_{ap} is the ratio between the *total* TSS production of the UASB reactor and the applied organic load:

$$Y_{ap} = mE_{tu} = (mX_{xau} + mX_{xiu} + mX_{xeu} + mX_{xbpu})/(f_{vu} \cdot R_{su}) \tag{13.29}$$

Note that in the above definition of the apparent yield no discrimination is made between excess sludge discharge and the loss of solids with the effluent, which can be considered as the second process of excess sludge discharge. In the case of UASB reactors the loss of sludge with the effluent cannot be ignored, as contrary to aerobic systems it will constitute a significant part of the total excess sludge discharge (typically between 40 to 60%, refer also to Example 13.7). The suspended solids concentration in the UASB effluent depends on the quality of the GLS design, the applied upflow velocity and the applied sludge age. As discussed before, the apparent yield expressed in mg VSS $\cdot$ mg^{-1} COD applied is usually several times larger than the anaerobic yield coefficient Y_{an}.

Equation (13.28) shows that R_{hu} is inversely proportional to the average sludge concentration in the treatment system. In practice, the sludge concentration will depend on the efficiency of the phase separator: an efficient separator will allow a higher sludge concentration to be maintained, which increases the treatment capacity of the reactor. It has been established for municipal sewage treatment, where a flocculent rather than a granular sludge develops, that the *average* sludge concentration X_{tu} is usually in the range of 15 to 18 g TSS $\cdot$ l^{-1} for a well designed and operated conventional UASB system. When enhanced GLS separators are used (e.g. when plate pack settlers are installed in the settling compartment), an average sludge concentration up to 25–30 g TSS $\cdot$ l^{-1} seems feasible, as demonstrated by the pilot results of Cavalcanti et al. (2003). However, at present this concept has not yet been implemented at full-scale for municipal UASB reactors, contrary to industrial installations.

For an influent concentration of 600 mg COD $\cdot$ l^{-1} and a typical value of the apparent yield Y_{ap} of 0.25 g TSS $\cdot$ g^{-1} COD for $R_{su} = 35$ days at 25°C, it can be calculated that the hydraulic retention time R_{hu} ranges from $35 \cdot 0.25 \cdot 0.6/(24 \cdot 15) = 8.4$ hours for the minimum sludge concentration of 15 g TSS $\cdot$ l^{-1} to 4.2 hours for the maximum concentration of 30 g TSS $\cdot$ l^{-1}. It is concluded that the calculated R_{hu} range is compatible with the typical range of values found in full-scale units, in accordance with the experimental data by Van Haandel et al. (1994) and those from Chernicharo (2007).

EXAMPLE 13.7

Finalise the design of the anaerobic pre-treatment system discussed in Example 13.6. The UASB is operated at a sludge age of 35 days and at 25°C. The following additional data are given:

- Decay rate of the anaerobic sludge $b_{an} = 0.01$ d^{-1};
- Concentration of inert suspended solids in raw sewage $X_i = 55$ mg ISS $\cdot$ l^{-1};
- Design upflow velocity $v_l = 0.65$ m $\cdot$ h^{-1};
- Average sludge concentration is 16 kg TSS $\cdot$ m^{-3};

Calculate the main system parameters (volume, surface area, excess sludge production and methane production). Is the anaerobic excess sludge sufficiently stabilised?

Solution

Using the value of mS_{du} derived in Example 13.6, the metabolized COD fraction is calculated as:

$$\begin{aligned} mS_{mb} &= mS_{du}/(1 - f_{cv} \cdot Y_{an}) \\ &= 0.59/(1 - 1.5 \cdot 0.45) = 0.64\,\text{mg VSS} \cdot \text{d} \cdot \text{mg}^{-1}\,\text{COD} \end{aligned} \tag{13.19}$$

$$\begin{aligned} mX_{au} &= Y_{an} \cdot R_{su}/(1 + b_{an} \cdot R_{su}) \cdot mS_{mb} \\ &= 0.05 \cdot 35/(1 + 0.01 \cdot 35) \cdot 0.64 = 0.83\,\text{mg VSS} \cdot \text{d} \cdot \text{mg}^{-1}\,\text{COD} \end{aligned} \tag{13.20}$$

$$\begin{aligned} mX_{eu} &= f \cdot b_{an} \cdot R_{su} \cdot mX_{au} \\ &= 0.2 \cdot 0.01 \cdot 35 \cdot 0.83 = 0.06\,\text{mg VSS} \cdot \text{d} \cdot \text{mg}^{-1}\,\text{COD} \end{aligned} \tag{13.21}$$

$$\begin{aligned} mX_{iu} &= f_{np} \cdot R_{su}/f_{cv} \\ &= 0.1 \cdot 35/1.5 = 2.33\,\text{mg VSS} \cdot \text{d} \cdot \text{mg}^{-1}\,\text{COD} \end{aligned} \tag{13.22}$$

$$\begin{aligned} mX_{bpu} &= mS_{xvu} \cdot R_{su}/f_{cv} - mX_{au} - mX_{eu} - mX_{iu} \\ &= 0.23 \cdot 35/1.5 - 0.83 - 0.06 - 2.33 = 2.2\,\text{mg VSS} \cdot \text{d} \cdot \text{mg}^{-1}\,\text{COD} \end{aligned} \tag{13.24}$$

$$mX_{vu} = mX_{au} + mX_{eu} + mX_{iu} + mX_{bpu} = 5.42\,\text{mg VSS} \cdot \text{d} \cdot \text{mg}^{-1}\,\text{COD} \tag{13.23}$$

$$mX_{mu} = f_{mi} \cdot R_{su} = 55/600 \cdot 35 = 3.21\,\text{mg ISS} \cdot \text{d} \cdot \text{mg}^{-1}\,\text{COD} \tag{13.26}$$

$$mX_t = mX_{vu} + mX_{mu} = 5.42 + 3.21 = 8.63\,\text{mg TSS} \cdot \text{d} \cdot \text{mg}^{-1}\,\text{COD} \tag{13.27}$$

From above data f_{vu} can be calculated as $5.42/8.63 = 0.63$ mg VSS $\cdot$ mg^{-1} TSS, while the value of the putrescible fraction f_{pu} is equal to $mX_{bpu}/mX_{vu} = 2.2/5.42 = 0.41$ mg VSS $\cdot$ mg^{-1} VSS, which is above the recommended limit of 0.3. However, in practice wastewater treatment plants are often under loaded compared to the design case, allowing operation at a sludge age that is higher than the design value. It can be calculated that an increase in sludge age of 50 days is sufficient to decrease f_{pu} to 0.3. The total sludge mass $MX_t = mX_{tu} \cdot MS_{ti} = 8.63 \cdot 30{,}000 = 259{,}000$ kg TSS, which at 16 kg TSS $\cdot$ m^{-3} results in a required liquid volume of 16,180 m^3. Note that in order to reduce the value of f_{pu} to 0.3 through an increase of R_{su} to 50 days ($mX_{tu} = 11.24$), the required UASB volume would be increased to 21,000 m^3 for the design load and flow.

The surface area is set by the ratio of influent flow rate and design upflow velocity as $50{,}000/(24 \cdot 0.65) = 3200$ m^2, for example divided over 8 units of each 20 x 20 m = 400 m^2. The liquid height is $16{,}180/3200 = 5.0$ m. Refer to Section 13.4 and Example 13.9 for a detailed design procedure for the UASB reactor itself.

The total daily excess production is $259{,}000/35 = 7380$ kg TSS $\cdot$ d^{-1} or 4650 kg VSS $\cdot$ d^{-1}. It is interesting to note that the value of $Y_{ap} = mE_v = 4650/30{,}000 = 0.15$ mg VSS $\cdot$ mg^{-1} COD is three times higher than the value of the true yield Y_{an} of 0.05 mg VSS $\cdot$ mg^{-1} COD. Expressed as total suspended solids, $Y_{ap} = mE_t = 7380/30{,}000 = 0.25$ mg TSS $\cdot$ mg^{-1} COD. A significant part of the produced excess sludge will leave with the effluent, equal to $50{,}000 \cdot 0.08 = 4000$ kg TSS $\cdot$ d^{-1}, or $4000/7380 = 54\%$ of the total excess sludge production.

The COD removal efficiency can be calculated from the soluble COD concentration in the effluent (S_{eu}) plus the COD content of the suspended solids. $S_{eu} = mS_{eu} \cdot S_{ti} = 0.18 \cdot 600 = 108$ mg COD $\cdot l^{-1}$ and $S_{tu} = 108 + 1.5 \cdot 0.63 \cdot 80 = 184$ mg COD $\cdot l^{-1}$. The overall COD removal efficiency (excluding residual methane) is $1 - 184/600 = 70\%$. The total methane production is $mS_{du} \cdot MS_{ti}/4 = 0.59 \cdot 30{,}000 = 4420$ kg $CH_4 \cdot d^{-1}$. Assuming a system for methane recovery is in place that reduces the effluent methane concentration to less than 2.5 mg $CH_4 \cdot l^{-1}$, the percentage methane recovery will be $1 - 125/4420 = 97\%$.

With respect to the selection of the optimum sludge age, anaerobic and aerobic treatment systems behave quite differently. Aerobic treatment systems have an optimal R_s value, which is determined by the need to develop some key feature of the treatment, for example organic material- or nutrient removal. Ideally the aerobic system is then designed for the lowest possible value of the sludge age. From Figure 13.17 and Eqs. (13.11 to 13.18), it can be observed that in anaerobic treatment systems maximum digestion (and substantially complete biodegradable organic material removal) is only achieved at a very high sludge age (> 100 days at 25°C). At lower sludge ages the digested fraction decreases and more biodegradable COD will end up in the effluent and in the excess sludge. Thus in order to maximise the mass of digested COD, the anaerobic reactor requires operation at the maximum sludge age possible. However, the resulting reactor size would then also be (very) large. In practice it is often not the maximisation of the digestion efficiency, but the production of a sufficiently stabilised anaerobic excess sludge, that is of importance: i.e. a volatile sludge where the ratio $f_{pu} = X_{bpu}/X_{vu}$ is below a certain specified value, for example $f_{pu} = 0.3$ mg VSS $\cdot$ mg^{-1} VSS. As could be observed in Example 13.7, at 25°C an anaerobic sludge age of at least 50 days is required. An interesting alternative, which will be discussed in Example 13.8, is to treat the partially stabilised anaerobic sludge in a small heated digester. Both the digestion rate and the degree of digestion increase at higher temperatures. Furthermore, the mixing intensity in a sludge digester will be much higher than in an UASB, creating improved conditions for the degradation of particulate organic material. Finally, as the excess sludge originates from an anaerobic system, there will be a "seed" of anaerobic bacteria to the digester, which might increase the metabolic capacity considerably.

EXAMPLE 13.8

Estimate the reduction in f_{pu} and the additional methane production when the excess sludge from the UASB system of Example 13.7 operated at $R_{su} = 35$ days is further stabilised in a heated digester at 35°C. Assume that Eqs. (12.63 to 12.65) can be used to characterize the digestion process and that the "active" fraction of the anaerobic excess sludge consists only of the non-degraded biodegradable organic material. Compare this to the alternative of operating the UASB at $R_{su} = 50$ days, for which the following additional data are given:

- $mS_{du} = 66\%$
- $mS_{eu} = 14\%$
- $mS_{xvu} = 20\%$
- $f_{vu} = 0.59$

Solution

Let us first consider the effect of using a heated digester. At T = 35°C, the required retention time is calculated with Eq. (12.63):

$$R_{di} = 20 \cdot 1.1^{(20-T)} + 5 = 20 \cdot 1.1^{(20-35)} + 5 = 9.8\text{days} \tag{12.63}$$

The digester volume will not be very large, as the anaerobic excess sludge has a high concentration. If the excess sludge is discharged from the top layer of the sludge blanket, a concentration of 20 g $TSS \cdot l^{-1}$ may be assumed. For the excess sludge production of $(1 - 0.54) \cdot 7380 = 3380$ kg $TSS \cdot d^{-1}$, this requires a digester volume V_{di} of $9.8 \cdot 3380/20 = 1650\ m^3$. If required, the digester volume can be further reduced using a mechanical sludge thickener, which allows the UASB excess sludge concentration to be increased to 40 – 60 g $TSS \cdot l^{-1}$. The degree of solids destruction in the heated digester can be estimated with Eqs. (12.64 and 12.65):

$$\eta_{dp} = (0.67 \cdot T + 36)/100 = 59\% \tag{12.64}$$

$$\eta_{dn} = (0.19 \cdot T + 10)/100 = 17\% \tag{12.65}$$

The volatile sludge production discharged with the excess sludge is calculated from Example 13.7 as $0.63 \cdot 3380 = 2130$ kg $VSS \cdot d^{-1}$, of which a fraction $f_{pu} = 0.41$ (865 kg $VSS \cdot d^{-1}$) may be considered as "active" (i.e. $f_{avu} = f_{pu}$). The digested sludge mass is equal to:

$$\begin{aligned} ME_d &= [\eta_{dp} \cdot f_{av} + \eta_{dn} \cdot (1 - f_{av})] \cdot ME_{vu} \\ &= [0.59 \cdot 0.41 + 0.17 \cdot (1 - 0.41)] \cdot 2130 \\ &= 725\ \text{kg VSS} \cdot d^{-1} \end{aligned} \tag{12.66}$$

This is equivalent to an additional methane production of $725 \cdot 1.5/4 = 270$ kg $CH_4 \cdot d^{-1}$, so total recovered methane production is $0.97 \cdot 4400 + 270 = 4570$ kg $CH_4 \cdot d^{-1}$. The putrescible fraction in the stabilised sludge is reduced to $f_{pu} = (1–0.59) \cdot 865/(2130–725) = 0.25$, well below the recommended maximum value of $f_{pu} = 0.3$. The total stabilised excess sludge production is equal to 3380–725 = 2655 kg $TSS \cdot d^{-1}$.

As to the system operated at $R_{su} = 50$ days, the excess sludge production can be calculated as $0.20 \cdot 30{,}000/(0.59 \cdot 1.5)–50{,}000 \cdot 0.08 = 2745$ kg $TSS \cdot d^{-1}$ (with a putrescible fraction $f_{pu} = 0.3$), while the recovered methane production is $0.97 \cdot 0.66 \cdot 30{,}000 = 4805$ kg $CH_4 \cdot d^{-1}$.

It can be concluded that the performance the system comprising of a UASB + heated digester is comparable in terms of methane production and better in terms of sludge stability than that of the single UASB. However, the real savings will originate from the reduction in required UASB volume. Operation at 50 days sludge age requires a total UASB reactor volume of $50 \cdot 0.23 \cdot 30{,}000/(0.59 \cdot 1.5 \cdot 16) = 21{,}075\ m^3$, which is an increase of about 5000 m^3 compared to the additional volume of 1650 m^3 required for the heated digester, or even much less if the UASB sludge is thickened.

13.4 UASB REACTOR DESIGN GUIDELINES

In the previous sections it was demonstrated that the performance of the municipal UASB can be determined as a function of the anaerobic sludge age, sewage flow rate and -composition and sewage temperature. Based on the mass of anaerobic sludge that will develop and the expected average biomass concentration in the reactor, the reactor volume can be calculated. In this section guidelines will be presented that can be used for indicative design of the UASB reactor with regards to the following physical aspects:

(a) Required surface area;
(b) Design of the GLS separator and gas box;
(c) Selection of the reactor height;
(d) Sizing of the influent distribution system;
(e) Effluent withdrawal;
(f) Sludge sampling and withdrawal.

(a) Required surface area

The reactor surface area is determined by the ratio between the sewage flow rate and the selected upflow velocity. In fact there are several upflow velocities to be considered: (I) in the digestion zone (v_l), (II) in the apertures between the GLS separators (v_a) and (III) at the overflow level (v_o). The reduction in area available for liquid upflow, especially at the apertures, will result in a significant increase in liquid upflow velocity. Refer also to Figure 13.3. In Table 13.9 recommended velocities for each of these zones are given, slightly adapted from the ones proposed by Chernicharo (2007).

Table 13.9 Recommended values for the upflow velocity ($m \cdot h^{-1}$) at different levels in the UASB reactor for municipal sewage, adapted from Chernicharo (2007)

Parameter	Average	Maximum(1)	Peak(2)
v_l - upflow velocity in the digestion zone	0.5–0.7	< 1.0	< 1.5
v_a - upflow velocity in the apertures	< 2.5	< 3.5	< 5.5
v_o - upflow velocity at the overflow level	< 0.8	< 1.2	< 1.8

Notes: (1) Maximum expected daily flow rate
(2) Allowable peak flow duration: maximum 2–4 hours

As a first step in the design process the minimum reactor surface area is determined, which is equal to the largest surface area required to comply with the recommended upflow velocity under design, maximum daily and peak flow conditions. At this stage in the design procedure only the upflow velocity in the digestion zone is considered:

$$A_{umin} = Q_i/(24 \cdot v_l) \tag{13.30}$$

When the ratio between peak and average flow rate is very high, or when the duration of the peak flow is larger than four hours, it is recommended to bypass part of the peak flow if it concerns diluted wastewater or to store the excess flow in a buffer tank for treatment at a later time. The alternative is to base the design of the UASB reactor on this high peak flow rate, which would be very expensive. If necessary, the bypassed flow can be subjected to some form of partial treatment prior to disposal, for example in high rate primary (lamella) settlers.

Whereas for small flow-rates ($< 250\ m^3 \cdot h^{-1}$) the reactor size might be customized, for larger applications often a standard modular design is used. Generally the reactors are square, though a rectangular shape can also be used. However, circular designs are not recommended when more than one reactor is constructed, as sharing of reactor walls is not possible, which would save on the use of concrete. Typically the length (L_u) and width (W_u) of a modular UASB reactor vary between 15 to 30 m. Once the surface area is determined and the values of L_u and W_u have been specified, this will yield the number of UASB reactors required (N).

(b) Design of the GLS separator

The GLS separator is constructed of three main parts (refer also to Figure 13.22):

- The deflector, which prevents the biogas from entering into the settling compartment;
- The gas collection plates, which guide the produced biogas into the gas box and simultaneously create a quiescent top settling compartment;
- The gas box, in which the biogas produced in the reactor is collected.

Starting with the gas box, in principle the surface area required for gas desorption from the liquid phase, i.e. the cumulative surface area of the gas boxes, is determined by the maximum allowable gas desorption rate expressed in $m^3 \cdot m^2 \cdot h^{-1}$ (or $m \cdot h^{-1}$). Chernicharo (2007) recommends a minimum/maximum gas evolution rate of respectively 1 and 3–5 $m \cdot h^{-1}$. However, due to the low rate of gas production in municipal UASBs, this will in practice never be a limiting design factor and the size of the gas boxes is determined by practical considerations. For example, the width of a gas box (W_{gb}) is usually determined by the requirements for providing safe access to operators, as it doubles as a walkway. The minimum recommended (outer) gas box width is therefore typically 0.5–0.7 m, and there is no need to go above this value as it will reduce the area available for settling and thus increases the overflow rate. When a minimum wall thickness of 150 mm is assumed (when concrete is used), this is in good agreement with the minimum internal gas box width of 250 mm, as recommended by Chernicharo (2007).

Gas boxes usually span the full length of the reactor, but Chernicharo (2007) provided an interesting alternative configuration in which two gas boxes are placed in series. Although more complicated to construct, this maximizes the overflow area for a given reactor surface area, while it also allows pre-fabrication off-site. The two arrangements are shown schematically in Figure 13.19, where the continuous lines indicate the gas box and the dashed lines indicate the gas collection plates. However, the remainder of this section will discuss the conventional configuration with a single gas box over the full length of the reactor.

The number of gas boxes (n) per reactor depends on the width of the reactor and on the objective to minimize the height of the GLS separator (gas box, collection plates and deflector). For example, if a single gas box is selected, then the projected height of the gas collection plates will be very large, which reduces the volume of the digestion zone. On the other hand, a large number of gas boxes will result in a low value of the projected height of the gas collection plates and thus a large digestion zone, but with little surface area available for settling. Therefore a sensible compromise should be selected, ideally by calculation of the GLS layout for different numbers of gas boxes installed. The liquid height of the gas box (H_{gb}) is typically at least 0.3 m (Chernicharo, 2007), to allow for installation of effluent collection channels with overflow weirs or submerged pipes.

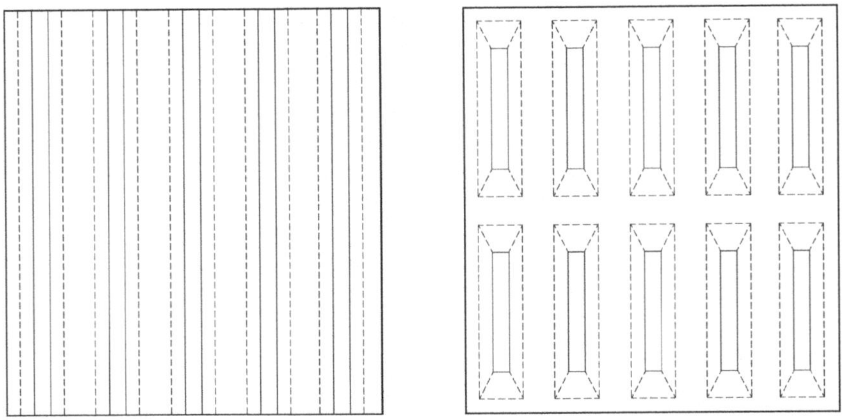

Figure 13.19 UASB gas box arrangements (top view): a single gas box over the full length of the reactor (left) and two gas boxes in series (right). Adapted from Chernicharo (2007)

The main factors that define the height of the GLS separator and hence that of the UASB reactor are the number of gas boxes and the inclination of the plates. Inclinations ranging from 45 to 60° have been applied. Higher plate inclinations result in increased interception of sludge particles, but also in an increased GLS height. According to Van Haandel et al. (1995), an inclination of 45° is the minimum value required to ensure that the settled solids can slide back into the digestion zone, but a higher value is recommended, although this will increase the height of the GLS separator. Often the gas deflector mounted below the aperture has the same inclination as the gas collection plate, though this is not strictly necessary. The deflector should overlap at least 0.1–0.15 m with both gas collection plates, in order to prevent the gas bubbles from entering into the settling zone. Several deflector designs have been applied in practice, as indicated in Figure 13.20.

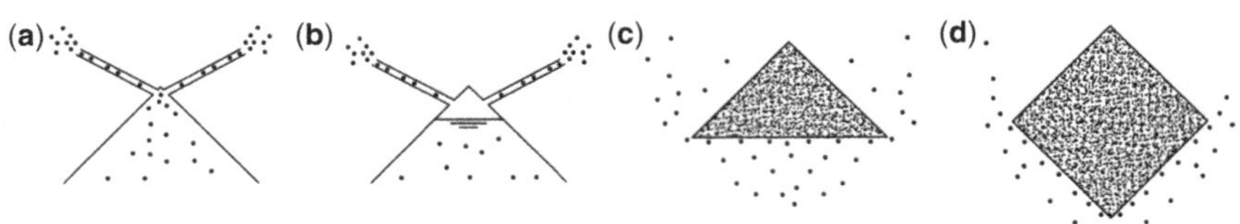

Figure 13.20 Several UASB deflector designs (courtesy of B. Heffernan)

The first design (Figure 13.20a) is an inverted V-shape, allowing the upflowing biogas-liquid-solids mixture to enter into the deflector. At regular intervals over the length of the deflector, pipes are installed, approximately 20 cm from the top of the deflector, directing the collected biogas-sludge mixture into the gas hood. However, because of the position of the pipes *below* the deflector top, a gas-liquid interface will be established in the top part of the deflector, which makes the formation of a scum layer inevitable. Scum build-up may eventually result in clogging of the pipes, which can lead to uncontrolled discharge of large gas bubbles over the bottom of the deflector and through the aperture into the settling compartment. Furthermore, the use of pipes to direct the biogas into the gas hood may result in locally increased turbulence near the aperture in those sections of the deflector where the pipes are installed. The second deflector design, shown in Figure 13.20b, is very similar to the first design but in this case the pipes are located at the very top of the deflector. This at least prevents development of a gas-liquid interface and therefore avoids the problems associated with scum formation.

The third design (Figure 13.20c) differs from the previous two designs in that the bottom of the deflector is now closed. Hence there is no gas-liquid interface with consequential risk of scum formation. However, as the bottom of the deflector is closed, but not inclined, there is a distinct possibility that biogas will be entrained with the effluent flow into the settler section, especially since the deflector is never perfectly constructed, which might result in the formation of gas pockets resulting in localised gas escape.

The final deflector design (Figure 13.20d) is closed as well, but additionaly equipped with an inclined bottom section. The biogas is therefore deflected into the gas collection compartment, as it has already a velocity in the direction of the gas box when reaching the deflector midpoint. On the other hand, the addition of a bottom section to the deflector increases the height compared to the first three designs. However, it should be considered that this has only a minor effect on the volume available for the digestion zone, as it is the widest point of the deflector, not the bottom, that sets the maximum height of the sludge blanket. The top of the sludge blanket should be maintained at least 0.2–0.3 m below this point. This strategy prevents biogas production above the deflector, where it might enter directly into the settling compartment. It is therefore concluded that the additional height of the deflector with an inclined bottom section does in fact not reduce the volume of digestion zone, whereas it has significant benefits.

Once the reactor surface area has been determined, as discussed in the previous section, the design of the GLS separator proceeds through the series of steps described below. Note that to facilitate the equations the flow rate has been expressed in cubic meter per reactor per hour ($m^3 \cdot h^{-1}$).

(1) For the selected reactor area and as function of the number of gas boxes, calculate the overflow velocity that will result and check whether it is in the recommended range. If not, discard that configuration. In principle the largest possible number of gas boxes is selected, as this will minimize the reactor height.

$$v_o = Q_i/(A_u - n \cdot W_{gb} \cdot L_u) \tag{13.31}$$

(2) For the selected aperture velocity, calculate the required aperture area and the aperture width. Take into account that there is half an aperture width on both extremes of the reactor, so that the number of apertures is equal to the number of gas boxes "n":

$$A_a = Q_i/v_a \tag{13.32}$$

$$W_a = Q_i/(L_u \cdot v_a \cdot n) \tag{13.33}$$

(3) Calculate the projected width of the collection plate. The projected width of all gas collection plates combined is equal to the width of the reactor not occupied by either the apertures or the gas boxes. Therefore the projected width of a single gas collection plate "a" is given by (remember each gas box has two collection plates):

$$a = (W_u - \Sigma W_{gb} + \Sigma W_a)/(2 \cdot n) \tag{13.34}$$

(4) Select the collection plate inclination α. For this inclination, calculate the projected plate height "b":

$$b = a \cdot \tan(\alpha) \tag{13.35}$$

(5) For the calculated aperture width, calculate the deflector height H_{df}. Unless there is a specific reason not to do so, accept the inclination used for the gas collection plates. Take into account that the width of the deflector should overlap at least 0.1 m with each of the gas collection plates:

$$H_{df} = (W_a + 2 \cdot 0.1)/2 \cdot \tan(\alpha) \tag{13.36}$$

(6) The total liquid height of the GLS separator is equal to the sum of the projected heights of gas collection plate and deflector plus the liquid height of the gas box.

$$H_{gls} = b + H_{df} + H_{gb} \tag{13.37}$$

(c) Determination of the required reactor height

Once the procedure outlined above has been followed, only one last element is needed to calculate the liquid height of the UASB reactor: the height of the digestion zone (H_{dig}).

$$H_{liq} = H_{dig} + H_{gls} \tag{13.38}$$

The digestion zone height follows from the required digestion volume, which is equal to the anaerobic sludge mass that has to be accommodated divided by the average biomass concentration in the digestion zone. This concentration typically varies between 50–60 $g \cdot l^{-1}$ at the bottom of the reactor to 15–25 $g \cdot l^{-1}$ at the top of the sludge blanket (Van der Lubbe et al., 2010). An average sludge blanket concentration of 30–40 $g \cdot l^{-1}$ can be obtained in well designed and operated UASB reactors. In practice the top of the sludge blanket will have to be maintained at some distance (typically at least 0.25 m) from the widest point of the deflector, in order to prevent sludge and biogas from spilling over into the settling zone. The total height of the digestion zone is thus calculated as:

$$H_{dig} = MX_{tu}/(N \cdot X_{tud} \cdot A_u) + 0.25 \tag{13.39}$$

where X_{tud} = average biomass concentration in the sludge blanket

Finally, to obtain the total reactor height, the height of the freeboard (H_{fb}) should be included, which is typically between 0.3 and 0.5 m.

$$H_u = H_{liq} + H_{fb} \tag{13.40}$$

In general, operation at a liquid height of less than 4 m is not recommended. In practice, most municipal UASB reactors have liquid heights between 4 and 5 m, but there is no fundamental reason not to go higher, for example to 6 or 7 m, which is a typical height of industrial UASB reactors. In fact, operation at higher liquid reactor height is very beneficial as there is a strong positive correlation between the reactor height and the average sludge concentration in the reactor. The height occupied by the GLS is constant for a given reactor design, so when the liquid height of the reactor is increased (i.e. the GLS separator is moved upwards), the volume available for the digestion zone increases, also in proportion to the total reactor volume. It is interesting to compare the average sludge concentration of reactors of equal volume but different liquid heights. Consider two reactors with a liquid height of 4 and 5 m respectively and equipped with a GLS device of 2 m. For an assumed sludge concentration *in the digestion zone* of $30\ g \cdot l^{-1}$, then the average sludge concentration in the 4 m high reactor is $(4\text{–}2)/4 \cdot 30 = 15\ g \cdot l^{-1}$, whereas that of the 5 m reactor is equal to $(5\text{–}2)/5 \cdot 30 = 18\ g \cdot l^{-1}$. Thus, for equal volume, a higher reactor can hold significantly more biomass. Naturally, there is a limit to the maximum allowable liquid height for a given reactor volume, which is determined by the minimum surface area required to comply with the velocity limitations discussed previously.

(d) Influent distribution

Influent distribution can be evaluated from the perspective of distributing flow *within* a reactor and *to* a reactor. In the first case the total flow to a reactor needs to be distributed as evenly as possible over the reactor surface area, in order to maximize the contact between sewage and anaerobic biomass and to minimize the risk of localized high velocities and short circuiting between inlet and outlet, which would reduce the effluent quality. In the second case the sewage flow to the treatment plant needs to be divided several times up to the level of the individual reactor. This is a common phenomenon in every sewage treatment plant with parallel units and will not be discussed here further, as it belongs to the realm of civil engineering.

As to the influent distribution within the reactor, so far, at least to the author's knowledge, the influence of the inlet point density on reactor performance has never been properly investigated at a full-scale UASB reactor. In practice, inlet point densities ranging from a maximum of 1 inlet per 2 m^2 to a minimum of 1 per 4.5 m^2 have been applied (Heffernan et al., 2010), without any clear and distinguishing effect on treatment performance. Thus, it seems other factors may be more important. The guidelines suggested by Chernicharo (2007) are listed in Table 13.10.

Table 13.10 Suggested inlet distribution point densities for municipal sewage UASB reactors–adapted from Chernicharo (2007)

Sludge concentration in digestion zone ($kg\ TSS \cdot m^{-3}$)	Volumetric loading rate ($kg\ COD \cdot m^{-3} \cdot d^{-1}$)	Suggested density (m^2 per inlet point)
> 40	< 1.0	0.5–1.0
	1.0–2.0	1.0–2.0
	> 2.0	2.0–3.0
20–40	< 2.0	1.0–2.0
	> 2.0	2.0–5.0

At lower volumetric loading rates the recommended inlet point density increases, due to the reduction in mixing intensity resulting from the decrease in both influent flow rate and biogas production. When a high inlet point density is selected, this has an effect on the construction costs, bearing in mind that for a typical 20 × 20 m UASB reactor (400 m^2) and for a density of 1 pipe per 2 m^2 surface area no less than 200 pipes are required, assuming that each inlet pipe will have only one outlet. The result will be a veritable "spaghetti" of pipes that all have to be led down from the distribution box through the limited space available between collection plates and deflector and from there to the designated area on the bottom of the reactor.

It is important that the design of the inlet pipes is such that air bubbles can escape back up the pipe. Otherwise the entrained air bubbles might conglomerate into larger air pockets that (temporary) block the inlet flow, until the upward force becomes so large that the air pocket is released back into the distribution box. This has a negative impact on the influent distribution. According to Van Haandel et al. (1994), a downflow velocity of $\leq 0.2\ m \cdot s^{-1}$ at the maximum flow rate will be sufficient to prevent air bubbles with a diameter of 2 mm or more to be entrained. On the other hand, a high velocity is required at the outlet of the pipe, typically mounted at 100–200 mm from the bottom of the reactor, in order to prevent deposition around the pipe outlet, which might result in blockages. At the maximum flow rate an outlet velocity of around $0.6\ m \cdot s^{-1}$, which is twice the design settling velocity for a grit channel, is considered to be sufficient (Van Haandel et al., 1994).

As to the design of the final distribution box, the number of boxes installed per reactor seems to be a matter of personal preference of the designer. If only one box is selected, the benefit is a simplified influent distribution system in which the reactor distribution box that divides the flow to a single UASB reactor over the final distribution boxes can be omitted. The downside is that the diameter of the final distribution box will necessarily be very large, which will conflict with the construction of the gas hoods and the effluent collection system. On the other hand, a high number of distribution boxes, for instance 12 or more, might be optimal from the viewpoint of balanced influent distribution, but will result in increased construction costs. As always, the optimum will be located somewhere in between. More importantly, it is crucial that the distribution box is symmetrical in design (refer also to Section 13.2.1) and that the overflow to the individual inlet pipes is over a V-notched free-fall weir. Sufficient head should be available to prevent the occurrence of obstructions. It is recommended that the minimum difference in height between the liquid level in the reactor and that in the distribution box (i.e. at the base of the V-notch) is at least 300 mm.

(e) Effluent withdrawal

This subject has already been extensively discussed in Section 13.2.1 and only the main points will be summarized here. It is recommended to use submerged effluent pipes instead of effluent gutters as the effect of misalignment on the flow distribution in the reactor is less severe and, perhaps more importantly, as emissions of sulphide and methane to the atmosphere are minimized. The effluent pipes, especially when operated semi-flooded, can in principle be integrated in an effluent methane recovery scheme.

Should effluent gutters be preferred, then is it is important that the flow over an individual V-notch is sufficiently large to prevent blockages by scum (the same applies to the circular perforations in submerged pipes). To reduce the effects of misalignment, at least 25 mm of head above the base of the V-notch should be available under conditions of average flow. For most municipal applications, this corresponds to 1 or 2 V-notches per m^2 of surface area, as calculated by van Haandel et al. (1994).

(f) Sludge sampling and withdrawal

Once the UASB reactor has been constructed, optimal performance will be achieved only when the digestion zone is operated as full as possible (while unintentional discharge of suspended solids with the effluent is avoided), with the highest concentration of active biomass. In general the bottom part of the digestion zone will contain the densest sludge (up to 40–80 g $TSS \cdot l^{-1}$), while this gradually decreases to 15–25 g $TSS \cdot l^{-1}$ at the top of the sludge blanket. Thus in principle two systems for excess sludge discharge should be provided: a top one from which the less dense sludge is discharged with preference and a bottom one that is only used to control the inorganic fraction in the biomass. According to Chernicharo, the diameter of the excess sludge discharge pipes should be at least 150 mm.

Often the excess sludge is withdrawn batch-wise from the reactor. To determine when a new sludge batch discharge is due, it is important to be able to establish the exact height of the sludge blanket. Often sampling collections are provided at the wall of the reactor, as shown in Figure 13.21. If this method of sludge sampling is selected, then it is recommended to install a minimum of five sampling connections over the height of the digestion zone, at intervals not exceeding 50 cm. However, even then it is not possible to use the reactor volume between the top sampling point and the widest point of the deflector for digestion, as there is no information available about the sludge concentration inbetween these points. Furthermore, it is important that the sampling points extent at least 50 cm into the reactor as hydraulic effects can result in locally increased sludge concentrations at the wall. The diameter of the sampling pipes should be at least 50 mm.

As an alternative, an opening can be provided in the gas box, connected to a pipe that extends below the gas/liquid interface in order to act as a hydraulic seal. A sampler can be introduced through the opening, which allows samples to be taken at any desired level. If desired, a pump can be used to facilitate sampling. The advantage of this sampling method is that is allows exact determination of the height of the sludge blanket.

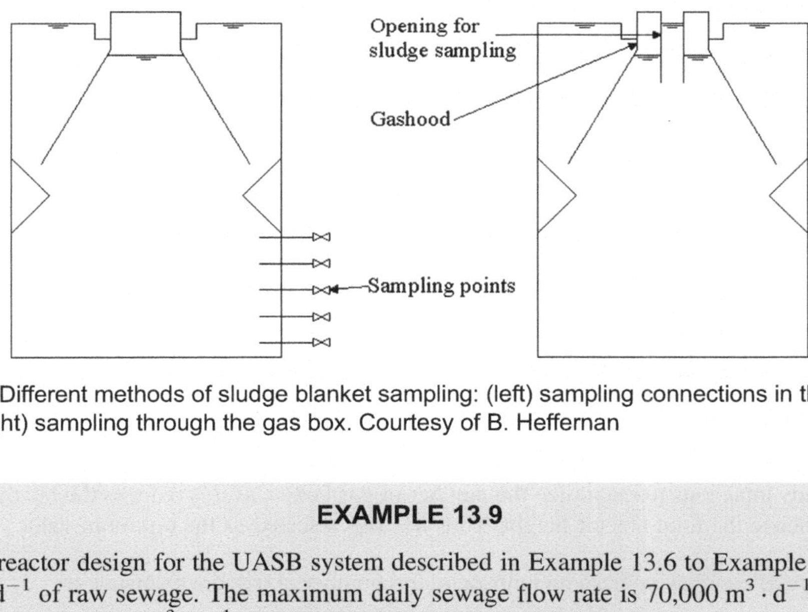

Figure 13.21 Different methods of sludge blanket sampling: (left) sampling connections in the wall of the UASB and (right) sampling through the gas box. Courtesy of B. Heffernan

EXAMPLE 13.9

Finalize the reactor design for the UASB system described in Example 13.6 to Example 13.8, treating 50,000 $m^3 \cdot d^{-1}$ of raw sewage. The maximum daily sewage flow rate is 70,000 $m^3 \cdot d^{-1}$ and the peak (hourly) flow rate is 4600 $m^3 \cdot h^{-1}$. The following additional data are given:

- Design upflow velocity for average flow $v_l = 0.65\ m \cdot h^{-1}$;
- Outer width of the gas box $W_{gb} = 0.6$ m;
- Liquid height of the gas box $H_{gb} = 0.4$ m;
- The inclination of the gas collection plates and the deflectors $\alpha = 52°$;
- A standard triangular deflector is used, with 0.1 m overlap on each collection plate;
- The gas box runs over the full length of the reactor;
- Average concentration in the sludge blanket $X_{tud} = 33\ g\ TSS \cdot l^{-1}$;
- Density of inlet pipes is 1 per 4 m^2.

Solution

First determine the minimum required UASB reactor surface area:

$$A_{umin} = Q_i/(24 \cdot v_l) = 50{,}000/(24 \cdot 0.65) = 3205\,m^2 \tag{13.30}$$

When a reactor of 20 x 20 m size is selected, then N = 8 reactors are required, each with a surface area A_u of 400 m^2. To facilitate the calculations, it is convenient to express the sewage flow rates per reactor and per hour:

- The average flow rate is $50{,}000/(24 \cdot 8) = 260\ m^3 \cdot h^{-1}$;
- The maximum daily flow rate is $70{,}000/(24 \cdot 8) = 365\ m^3 \cdot h^{-1}$;
- The peak flow rate is $4600/8 = 575\ m^3 \cdot h^{-1}$.

The selected UASB reactor size is large enough to ensure that the reactor upflow velocity limits in Table 13.9 are not exceeded for the other two cases: i.e. during maximum daily flow $v_l = 365/400 = 0.91 \leq 1.0\ m \cdot h^{-1}$ and during peak flow $v_l = 575/400 = 1.44 \leq 1.5\ m \cdot h^{-1}$.

The required overflow area A_o can be calculated in a fashion similar to A_u:

$$A_o = Q_i/v_o = 260/0.8 = 326\,m^2$$

However, the available overflow surface area depends on the area occupied by the gas boxes. The overflow velocity is therefore calculated as function of the number of gas boxes per reactor. Below this calculation is demonstrated for four gas boxes and for $W_{gb} = 0.6$ m.

$$v_o = Q_i/(A_u - n \cdot W_{gb} \cdot L_u) = 260/(400 - 4 \cdot 0.6 \cdot 20) = 0.74\,m \cdot h^{-1} \tag{13.31}$$

Table 13.11 shows the calculated overflow velocity for different numbers of gas boxes per reactor. It can be observed that for n > 6 the recommended overflow velocity limits from Table 13.9 are exceeded. In principle it is advantageous to maximize the number of gas boxes, as this reduces the height of the GLS separator and hence the total reactor height. Thus n = 6 is selected as the optimum value.

Table 13.11 UASB overflow velocity as function of the number of gas boxes installed

Parameter	UoM	Number of gas boxes (n)				
		4	**5**	**6**	**7**	**8**
Gas box width per reactor	m	2.4	3	3.6	4.2	4.8
Gas box area per reactor	m^2	48	60	72	84	96
Overflow area	m^2	352	340	328	316	304
Average overflow velocity	$m \cdot h^{-1}$	0.74	0.77	0.79	*0.82*	*0.86*
Maximum overflow velocity	$m \cdot h^{-1}$	1.04	1.07	1.11	1.15	1.20
Peak overflow velocity	$m \cdot h^{-1}$	1.63	1.69	1.75	*1.82*	*1.89*

In the next step, the required aperture area is calculated for the recommended average aperture velocity of $2.5\ m \cdot h^{-1}$ (Table 13.9).

$$A_a = Q_i/v_a = 260/2.5 = 104\,m^2 \tag{13.32}$$

Again it should be verified that the velocity limits, in this case of the aperture, are not exceeded during maximum daily or peak flow, which indeed is the case. Now the aperture width can be calculated as well:

$$W_a = Q_i/(L_u \cdot v_a \cdot n) = 260/(20 \cdot 2.5 \cdot 6) = 0.87\,m \tag{13.33}$$

As the width of the reactor occupied by the apertures and the gas boxes is now known, the projected width and height of the collection plates can be calculated:

$$\begin{aligned} a &= (W_u - \Sigma W_{gb} + \Sigma W_{ap})/(2 \cdot n) \\ &= (20 - 6 \cdot 0.6 - 6 \cdot 0.87)/(2 \cdot 6) = 0.93\,m\ \text{(projected width)} \end{aligned} \tag{13.34}$$

$$\begin{aligned} b &= a \cdot \tan(\alpha) \\ &= 0.93 \cdot \tan(52^\circ) = 1.19\,m\ \text{(projected height)} \end{aligned} \tag{13.35}$$

The projected height of the triangular deflector can be calculated once the aperture width is known:

$$\begin{aligned} H_{df} &= (W_a + 2 \cdot 0.1)/2 \cdot \tan(\alpha) \\ &= (0.87 + 0.2)/2 \cdot \tan(52^\circ) = 0.68\,m \end{aligned} \tag{13.36}$$

The liquid GLS height is equal to the sum of the projected heights of the gas collection plate and the deflector, plus the liquid height of the gas box:

$$H_{lgls} = b + H_{df} + H_{gb} = 1.19 + 0.68 + 0.4 = 2.28\,m \tag{13.37}$$

In order to calculate the liquid reactor height, all that remains is to determine the height of the digestion zone. In Example 13.7 the anaerobic sludge mass that develops was calculated as 259,000 kg TSS, which at $33\ g\ TSS \cdot l^{-1}$ and for 8 reactors amounts to a required digestion zone volume of $980\ m^3$ per reactor or 2.45 m of digestion zone height.

Taking into account that the top of the sludge blanket should be maintained at a minimum distance of 0.25 m below the widest point of the deflector, the liquid reactor height can be calculated as:

$$H_{liq} = H_{dig} + H_{gls} + 0.25 = 2.45 + 2.28 + 0.25 = 4.98\,m \tag{13.38}$$

Figure 13.22 shows the GLS design developed in Example 13.9. For the selected influent distribution density of $4\ m^2$ reactor surface per inlet pipe, 100 pipes will be required. Therefore for example four final distribution boxes with each 25 overflows can be selected.

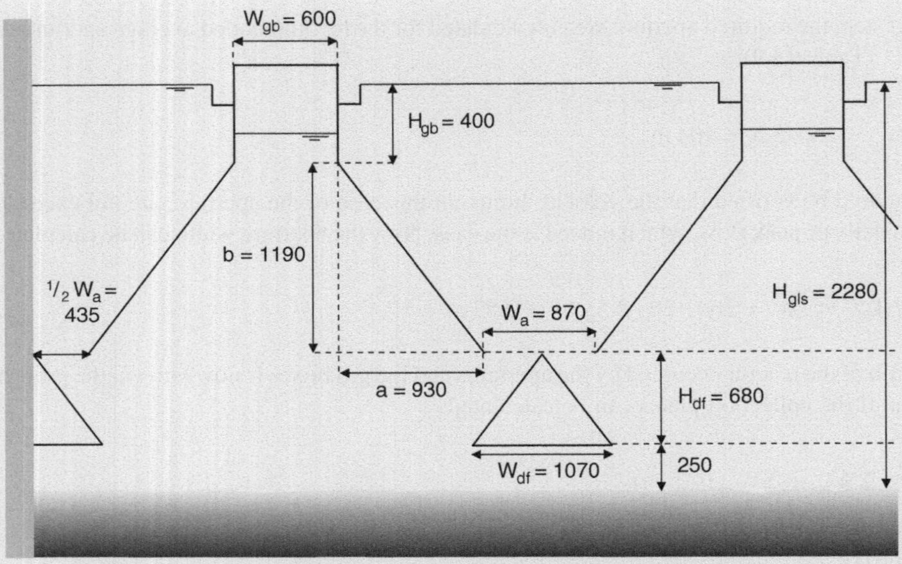

Figure 13.22 Schematic representation of the GLS design developed in Example 13.9

13.5 POST-TREATMENT OF ANAEROBIC EFFLUENT

It depends on the effluent discharge limits, ideally determined by the final destination of the treated effluent, whether a post-treatment step is required and if so, of which type. Typically as a minimum a polishing step is included in order to "refresh" the anaerobic effluent prior to discharge (i.e. mainly to remove H_2S and residual CH_4). Often it is also desired to remove part of the residual organic material in the anaerobic effluent, which for a significant part consists of suspended solids. As to the final destination of the treated effluent, several options are:

- Discharge to sea: in this case it might be decided to apply no further treatment to the UASB effluent, provided the point of discharge is located sufficiently far away from the coast, as the remaining contaminants are instantly diluted. However, the cost of construction of a subsea outfall can be considerable;
- Discharge to inland water bodies such as rivers, lakes etc. Here it depends on the specific legislation in place. Often COD, BOD and TSS concentrations in the anaerobic effluent will have to be further reduced. Phosphorus removal, nitrification or even nitrogen removal might be required as well;
- Reuse as irrigation water. In this case, the nitrogen and phosphorus left in the anaerobic effluent may actually be considered as beneficial, as they will constitute a source of nutrient supply to the crops. Nitrification will often be required, although in fact crops take up ammonium much easier than nitrate, which for a significant part is lost to the groundwater. Pathogen removal may be required: helminth eggs are the main health risk, but often some means of disinfection for pathogenic bacteria may be required as well. Micro-screening and filtration can be applied for the removal of helminth eggs, whereas chlorination and UV radiation are used for disinfection. These technologies will not be discussed in the book, the reader is referred to for instance Metcalf et al. (2003).

Anaerobic treatment processes are notoriously ineffective for the removal of nitrogen, phosphorus and pathogens, so if effluent limits are to be met for any of these components then suitable post-treatment is required. Some nitrogen and phosphorus will be removed with the anaerobic excess sludge: the nitrogen content of the organic UASB excess sludge ranges typically between 6 and 10% and that of phosphorus between 1.5–2.5%.

While phosphorus removal can be relatively easy achieved by means of chemical precipitation, biological nitrogen removal is much more difficult as the COD required for denitrification has been removed to a large extent in the anaerobic pre-treatment step. It is possible to bypass part of the raw sewage to the anoxic zone of the aerobic post-treatment in order to supply the required COD, but if the fraction of the raw sewage flow bypassed to the anoxic reactor becomes larger than 30% then anaerobic pre-treatment quickly ceases to be attractive (refer also to Section 13.5.2.1).

At present, the following systems have been applied at full-scale for the post-treatment of anaerobic effluent:

- Pre-, post- or flash aeration, where the anaerobic effluent is aerated for 20–60 minutes in order to remove the noxious and toxic sulphides by oxidation of sulphides to thiosulphate, sulphur and sulphate. Volatile compounds such as CH_4 are stripped as well;
- A series of ponds. It depends on the residual oxygen demand in the anaerobic effluent whether the first pond is facultative or aerobic. Sometimes aerators are installed to provide additional aeration;
- Trickling filters, basically consisting of a packed bed containing a cheap medium (e.g. lava rock) to which bacteria can attach. The anaerobic effluent is sprayed on top and percolates downward through the filter. Aeration is by natural draft and is sometimes enhanced by the use of ventilators;
- Dissolved air flotation units. A coagulant such as $FeCl_3$ is used to enhance the flocculation of the remaining suspended solids. Phosphate can be removed through precipitation as $FePO_4$. Some colloidal COD is removed as well and a clear effluent can be produced;
- One of the many variants of the activated sludge system. Conventional activated sludge systems, SBRs, MBBRs, MBRs and submerged aerated filters have all been applied regularly on anaerobic effluent.

Table 13.12 and Table 13.13 summarize the performance of several types of post-treatment in terms of the removal efficiency of different components from the anaerobic effluent. It can be observed that the activated sludge system is most flexible in meeting different effluent requirements. Therefore, in the remainder of this chapter, only the activated sludge system will be considered (often preceded by a flash aeration unit to prevent sulphide toxicity and odour issues).

13.5.1 Secondary treatment of anaerobic effluent

The basic configuration of a combined anaerobic-aerobic process for secondary treatment and nitrification (optional) is shown in Figure 13.23.

Along with the performance of the anaerobic pre-treatment system, discussed in the previous sections, the key questions to be answered in order to evaluate the feasibility of this configuration can be summarised as follows:

(1) Can the steady state activated sludge model for COD removal be applied to an activated sludge system receiving the effluent from an anaerobic pre-treatment unit?

(2) Does the anaerobic pre-treatment have an effect on the metabolic characteristics of the biomass (both autotrophic and heterotrophic bacteria) in the aerobic post-treatment unit?

(3) Does the anaerobic pre-treatment influence the settleability of the aerobic sludge in the post-treatment unit?
(4) What will be the effect of the recycle of aerobic excess sludge to the anaerobic pre-treatment unit?
(5) As the extent of denitrification is limited after anaerobic pre-treatment, is there a risk that denitrification in the final settler might result in problems with a floating sludge blanket?

Table 13.12 Performance of several post-treatment systems in terms of removal of organic material and nutrients

Configuration	Removal efficiency			
	COD/BOD/TSS	TKN/NH_4^+	N	P
Post-aeration	Soluble COD only	None	None	None
Ponds	Reasonable to good	Partly	Some	Some
Trickling filter	Reasonable	Partly	None	None
Dissolved air flotation	Good	Partly[(1)]	None	Good
Activated sludge system	Excellent	Good	Some[(2)]	Good
Submerged aerated filter	Good	Variable	None	None

Notes: (1) Organic nitrogen present in suspended solids is removed
(2) Good if bypass of raw sewage to the anoxic zone is applied

Table 13.13 Key characteristics of several post-treatment systems for the treatment of anaerobic effluent

Configuration	Characteristics
Post-aeration	- Removal of sulphides and other volatile compounds such as CH_4 - COD removal due to oxidation of H_2S - Cheap and simple system, but provides only partial treatment
Ponds	- Ammonia removal mainly due to to ammonia stripping as a result of the high pH that will develop - TSS in effluent might increase due to growth of algae - Cheap if land area is available and inexpensive
Trickling filter	- Vulnerable to clogging and short circuiting - Odour issues/fly problems have been reported - Effluent contains suspended solids which need to be removed - Ventilation off gas from the UASB reactor could be used to supply oxygen while simultaneously the off-gas is treated
Dissolved air flotation	- Small footprint - Rather advanced & complex system - Needs qualified maintenance and operating staff
Actived sludge system	- Nitrogen removal requires bypass of a significant part of the raw sewage to anoxic zone - Phosphorus removal by means of simultaneous precipitation
Submerged aerated filter	- Small footprint - Vulnerable to clogging with colloidal sulphur - Effluent contains suspended solids which need to be removed

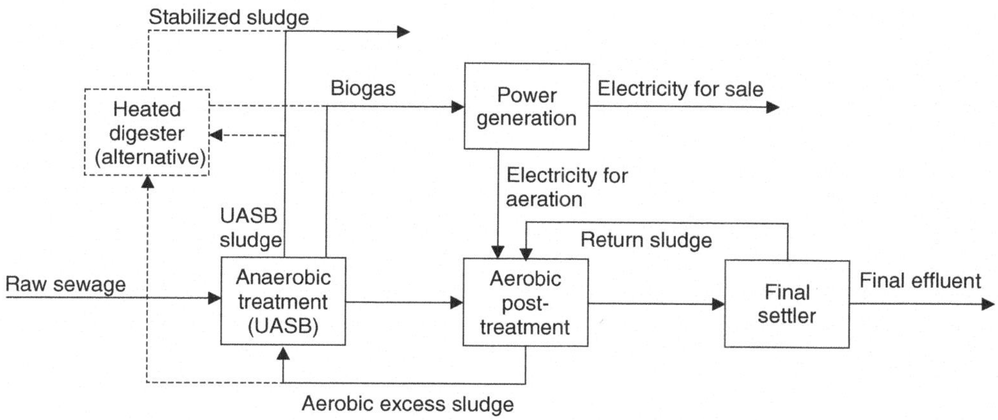

Figure 13.23 Basic configuration of an anaerobic-aerobic treatment system with digestion of the aerobic excess sludge in the anaerobic pre-treatment unit or in a heated digester

These questions will be addressed in the subsequent sections. However, first the attractiveness of combined anaerobic-aerobic treatment compared to a conventional aerobic system for secondary treatment is demonstrated in Example 13.10.

EXAMPLE 13.10

A petrochemical complex generates 4320 $m^3 \cdot h^{-1}$ of industrial wastewater with a COD concentration of 1500 $mg \cdot l^{-1}$ and with non biodegradable fractions $f_{ns} = 0.15$ and $f_{np} = 0.05$. The wastewater is treated in an activated sludge system with a liquid retention time of 1 day and at an operating sludge age of 20 days (T = 30°C). Due to the toxic nature of the wastewater, nitrification does not occur. The aerators have an efficiency of 1 kg $O_2 \cdot kWh^{-1}$.

If anaerobic pre-treatment is applied, what will be the required anaerobic efficiency to attain a situation of energetic self sufficiency of the entire wastewater treatment system (assume that the efficiency of the generator is 33% and the combustion heat is 50,400 $kJ \cdot kg^{-1}$ CH_4). Ignore the potential presence of sulphate in the wastewater and assume full methane recovery.

Solution

The biodegradable COD load is $(1 - f_{ns} - f_{np}) \cdot MS_{ti} = (1 - 0.15 - 0.05) \cdot 4320 \cdot 1.5 \cdot 24 = 124$ ton COD per day. For the specified temperature, the value of the decay constant $b_h = 0.36\ d^{-1}$. The fraction of the influent COD that will be oxidised in the conventional activated sludge system (mS_o) will be equal to 0.79. Hence for an oxygen transfer capacity of 1 kg $O_2 \cdot kWh^{-1}$, the power demand in the conventional activated sludge system will be $0.79 \cdot 124{,}000/24 = 4140$ kW.

If anaerobic digestion is applied and a fraction X of the biodegradable organic load is digested, the produced methane mass is $X \cdot 124/4 = 31 \cdot X$ ton $CH_4 \cdot d^{-1}$. From this mass of produced methane, the power generation will be $0.33 \cdot 31 \cdot X \cdot 50{,}400/(24 \cdot 3600) = X \cdot 8260$ kW for the assumed electric power conversion efficiency of 33%.

When it is assumed that the aerobic post-treatment unit will be operated at the same sludge age of 20 days as the existing treatment system, the resulting oxygen demand will be reduced to $(1-X) \cdot 124 \cdot 0.79 = (1-X) \cdot 99$ ton $O_2 \cdot d^{-1}$ or $(1-X) \cdot 4140$ kg $O_2 \cdot h^{-1}$. Equating power consumption to power generation yields a solution for $X = 33\%$. Thus it can be concluded that an anaerobic pre-treatment (digestion) efficiency of 33% is sufficient to compensate for the power demand of the aerobic post-treatment system.

In the above analysis, the COD mass incorporated in the anaerobic sludge and the methane generated from the anaerobic digestion of the aerobic excess sludge have not been taken into account, as it is assumed that the net effect of these two processes is very small compared to the total methane production potential. In principle, these factors will even increase the feasibility of obtaining self sufficiency from external power. The production of anaerobic sludge reduces the load of biodegradable organic material (and hence the oxygen demand) to the aerobic post treatment system, while the anaerobic digestion of the aerobic excess sludge in the pre-treatment unit increases the potential for power generation even further.

13.5.1.1 Applicability of the ideal steady state model for COD removal

As most of the influent organic material is removed in the anaerobic pre-treatment unit, the availability of biodegradable COD in the aerobic post-treatment unit will be restricted. However, the experience gathered so far from the operation of anaerobic-aerobic treatment systems clearly indicates that the steady state model for organic material removal in activated sludge systems remains valid.

The biodegradable COD concentration in the anaerobic effluent can be estimated from Eq. (13.16), which is valid for $15 < T < 35°C$. For a very high value of the anaerobic sludge age ($R_{su} \rightarrow \infty$), the soluble effluent COD concentration in the anaerobic pre-treatment unit reaches its minimum value, which is equal to $f_{ns} + f_{h2s}$. At lower values of the anaerobic sludge age the non-settleable effluent COD concentration will be larger, because it will then also contain soluble or colloidal biodegradable material. The value of this biodegradable fraction can be estimated as:

$$mS_{bu} = mS_{eu} - f_{ns} - f_{h2s} = 0.27 \cdot \exp[-0.04 \cdot (R_{su} - 4)]/1.067^{(T-25)} \quad (12.41)$$

Note that in Eq. (12.41) mS_{bu} is a fraction of the total COD in the *raw sewage* and not of that in the *anaerobic effluent*. The residual soluble biodegradable organic COD mainly consists of easily biodegradable COD (e. g. some residual VFA). The total COD concentration in the anaerobic effluent will be higher than the value predicted by Eq. (13.16), due to the presence of suspended solids that have not been retained by the GLS device. The effluent TSS concentration depends on the design and quality of the GLS separator, the applied upflow velocity in the reactor and on the quality of operation. In practice, effluent values are typically in the range between 40–120 mg $TSS \cdot l^{-1}$, where values between 40–80 mg $TSS \cdot l^{-1}$ represent a well designed and operated system, wheras higher values often indicate that the effluent weirs are misaligned and/or that insufficient excess sludge is withdrawn from the reactor.

Thus, in order to assess the COD load to the aerobic post-treatment system, the COD content of the effluent suspended solids should be considered as well. Part of the organic suspended solids will be inert, while another part is non-degraded slowly biodegradable material. It is assumed that the composition of the particulate organic material in the anaerobic effluent is identical to that of the biomass in the anaerobic reactor, which can be calculated with Eqs. (13.20 to 13.27). Only the non-degraded slowly biodegradable particulate material and a part (1–f) of the active anaerobic biomass are considered to be (slowly) degradable in the aerobic post-treatment system. When the COD removal efficiency in the anaerobic reactor η_{COD} is considered as well, then the different COD fractions as function of the total COD concentration in the anaerobic effluent are defined as:

$$f_{nsu} = f_{ns}/(1 - \eta_{COD}) \tag{13.42}$$
$$f_{bsu} = (mS_{eu} - f_{h2s} - f_{nsu})/(1 - \eta_{COD}) \tag{13.43}$$
$$f_{bpu} = (mX_{bpu} + (1 - f) \cdot mX_{au})/mX_{vu} \cdot S_{pu}/S_{tu} \tag{13.44}$$
$$f_{npu} = S_{pu}/S_{tu} - f_{pbu} \tag{13.45}$$
$$f_{h2su} = S_{h2su}/S_{tu} = f_{h2s}/(1 - \eta_{cod}) \tag{13.46}$$

The characterisation of the COD present in the UASB effluent as discussed above might be best explained using an example.

EXAMPLE 13.11

For the UASB system described in Example 13.6 to Example 13.8, estimate the composition of the anaerobic effluent, assuming an effluent TSS concentration of $80\ mg \cdot l^{-1}$ and an effluent H_2S concentration of $12\ mg\ S \cdot l^{-1}$ (originating from $36\ mg\ SO_4^{2-} \cdot l^{-1}$ in the raw sewage).

Solution

The soluble COD fraction in the UASB effluent mS_{eu} is equal to 18% of the total COD in the raw sewage or $108\ mg \cdot l^{-1}$ COD. Part of the soluble effluent COD fraction consists of inorganic COD associated to the H_2S in the anaerobic effluent. The value of $f_{h2s} = 2 \cdot 12/600 = 0.04$. For the applied anaerobic sludge age of 35 days, the value of the soluble (or more exactly, non settleable) biodegradable effluent fraction can be calculated with Eq. (12.41):

$$mS_{bu} = mS_{eu} - f_{ns} - f_{h2s} = 0.27 \cdot \exp[-0.04 \cdot (R_{su} - 4)]/1.067^{(T-25)}$$
$$= 0.18 - 0.10 - 0.04 = 0.04 \tag{12.41}$$

So $S_{bsu} = 0.04 \cdot 600 = 24\ mg\ COD \cdot l^{-1}$. As to the $80\ mg\ TSS \cdot l^{-1}$ in the anaerobic effluent, only a fraction $f_v = 0.63$ is organic so $X_{ve} = 0.63 \cdot 80 = 50\ mg\ VSS \cdot l^{-1}$ and $S_{pu} = 1.5 \cdot 50.3 = 75\ mg\ COD \cdot l^{-1}$. When it is assumed that the composition of the effluent suspended solids is identical to that in the sludge blanket, then the particulate COD concentration can be characterized as:

$$S_{bpu} = (mX_{bp} + (1 - f) \cdot mX_{au})/mX_{vu} \cdot S_{pu}$$
$$= (2.2 + 0.8 \cdot 0.83)/5.42 \cdot 75 = 40\ mg\ COD \cdot l^{-1} \tag{13.44}$$
$$S_{npu} = S_{pu} - S_{bpu} = 75 - 40 = 35\ mg\ COD \cdot l^{-1} \tag{13.45}$$

Table 13.14 summarizes the calculated UASB effluent COD composition for the conditions of Example 13.11. The COD fractions can be calculated by dividing with the COD concentration in the anaerobic effluent, which for the COD removal efficiency of 69.4% is $183\ mg\ COD \cdot l^{-1}$, when the methane concentration is ignored. Although residual methane will be present in the anaerobic effluent, it will not be detected in the COD analysis as it will be almost completely transferred to the atmosphere in the period between sampling and analysis. As also in reality the methane will be stripped from the UASB effluent during (pre-)aeration and hence does not constitute an oxygen demand, the methane fraction is not included in the load to the aerobic system.

Table 13.14 Characterization of the effluent organic material of Example 13.11

COD fraction in UASB effluent	Symbol	Value	Fraction of S_{tu}
Soluble (non settleable) COD:	S_{eu}/mS_{eu}	108	59%
– soluble non biodegradable	S_{nsu}/f_{nsu}	60	33%
– soluble/colloidal (easily) biodegradable COD	S_{bsu}/f_{bsu}	24	13%
– inorganic sulphide COD	S_{h2su}/f_{h2su}	24	13%
Particulate COD:	S_{pu}/f_{pu}	75	41%
– slowly biodegradable particulate COD	S_{bpu}/f_{bpu}	40	22%
– non-biodegradable part. COD	S_{npu}/f_{npu}	35	19%

The non-biodegradable fractions constitute more than 50% of the COD in the anaerobic effluent. Combined with the reduction in the COD concentration, it is obvious that the reduction in oxygen demand and excess sludge production in the aerobic system will be significant.

The definition of H_2S-COD as a separate COD fraction is very useful, as the sulphides will be partially oxidized in an (optional) pre-aeration step. If the activated sludge system is designed for secondary treatment with or without nitrification, then a further subdivision of the biodegradable COD in the anaerobic effluent is not needed. On the other hand, when biological nitrogen removal (i.e. by means of denitrification) is required, then a separation of biodegradable COD into easily and slowly biodegradable fractions becomes important.

Whereas the use of sulphides, which are rapidly biodegradable, as a source of COD for denitrification has been demonstrated (for example by Pantoja Filho et al., 2010), it is questionable whether this process will be useful in practice. The sulphides cannot be introduced directly into the anoxic zone of an activated sludge system, as this would result in a significant reduction of bacterial growth rates, especially for the nitrifiers. The sulphides can only be used for denitrification when a dedicated anoxic reactor (i.e. without sludge return from the main activated sludge system) is installed, which will be expensive. Thus, this will only be feasible if there is a large sulphide concentration present in the anaerobic effluent, but this would then render the anaerobic treatment process itself unattractive! Therefore the COD associated to sulphides is excluded from the easily biodegradable COD fraction (f_{sb}) available for denitrification. Thus, in the case of anaerobic effluent, f_{sb} is defined as:

$$f_{sb} = S_{bsu}/(S_{bpu} + S_{bsu} + S_{h2S}) = f_{bsu}/(f_{bpu} + f_{bsu} + f_{h2su}) \tag{13.47}$$

Once the biodegradable COD fraction $(1 - f_{nsu} - f_{npu})$ in the anaerobic effluent is known, then the oxygen demand and the aerobic excess sludge production can be calculated as usual with Eqs. (3.43 and 3.51). Figure 13.24 shows typical profiles of the oxidised COD fraction and the COD fraction discharged as aerobic excess sludge as a function of the *anaerobic* sludge age and the anaerobic hydraulic retention time, both for a low value (5 days – secondary treatment) and a high value (15 days – nitrification) of the aerobic sludge age. The normal operating range of the anaerobic sludge age for municipal UASB systems is also indicated.

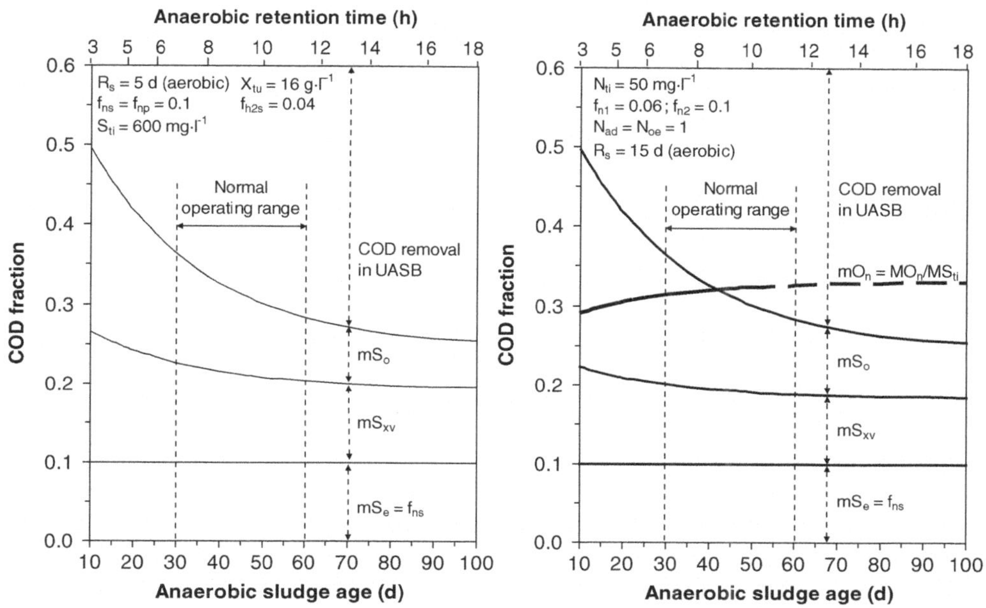

Figure 13.24 Fate of the sewage COD in the *aerobic* post-treatment system as function of the *anaerobic* sludge age: COD removal (left) and nitrification (right)

The anaerobic retention time (top scale) was calculated for conditions specified in Example 13.6 to Example 13.8, using Eq. (13.28) for an average biomass concentration in the UASB reactor X_{tu} of 16 g $TSS \cdot l^{-1}$. In Figure 13.24b the oxygen demand for nitrification is plotted for an influent TKN/COD ratio of 0.083 mg $N \cdot mg^{-1}$ COD. It can be observed that the oxygen demand for nitrification is several times higher than that for the oxidation of residual biodegradable COD, which is to be expected as most of the biodegradable COD has been removed in the UASB, whereas the removal efficiency of nitrogen is low.

As a result of the anaerobic pre-treatment, the composition of the sludge in the aerobic post-treatment unit will be very different from the composition of typical aerobic sludge generated from raw sewage. For instance, due to the increase of the TKN/COD ratio in the anaerobic effluent the nitrifier fraction the biomass will significantly increase. If it is required to include this fraction in the biomass, refer to Appendix A6.2.

More importantly, the activity of both heterotrophic and autotrophic bacteria tend to be significantly lower than the values measured in conventional activated sludge systems. Guimarães (2003) showed that in the case of nitrifiers the reduced metabolic capacity could be attributed to the presence of sulphide in the anaerobic effluent.

Table 13.15 shows some experimental results by Guimarães (2003) and Silva Filho (2005). Using a respirometer, the key kinetic constants were determined for heterotrophic and autotrophic bacteria in conventional activated sludge systems and in aerobic post-treatment systems operating under otherwise comparable conditions (temperature, sludge age and organic loading rate).

Table 13.15 Comparison of kinetic parameters (determined at 25°C) in a conventional activated sludge system and in an aerobic system following anaerobic pre-treatment

Parameter	UoM	Conventional	Post-treatment
Nitrifiers(1):			
– μ_m	d^{-1}	0.4–0.6	0.18–0.25
– K_n	$mg\ N \cdot l^{-1}$	< 1	< 1
Heterotrophs(2):			
– K_{ms}	$mg\ COD \cdot mg^{-1}\ VSS \cdot d^{-1}$	3–6	1.2–1.5
– K_{ss}	$mg\ COD \cdot l^{-1}$	< 1	< 1
Settleability:			
– k	$l \cdot g^{-1}\ TSS$	0.31–0.46	0.32–0.35
– v_0	$m \cdot h^{-1}$	6–12	9–22

Notes: (1) Batch fed with NH_4Cl and (2) fed with acetate.

The observed decrease in the values of μ_m and K_{ms} (by a factor 2 to 4) indicates that the metabolic activity of both autotrophic and heterotrophic bacteria in an aerobic post-treatment system is much smaller than in conventional aerobic systems. This in turn implies that the required sludge age for substantial complete removal of ammonium and organic material will increase compared to the sludge age required in a conventional aerobic system. However, even though the metabolic capacity of autotrophic and heterotrophic bacteria in aerobic post-treatment systems tends to be smaller than in comparable activated sludge systems treating raw sewage, it is perfectly possible to remove the relevant substrates (biodegradable organic material and ammonium) by means of an increase in the sludge age. The increase in sludge age does not result in a significant increase of the required aerobic reactor volume because:

- The biodegradable COD load to the aerobic post-treatment has been significantly reduced as a result of the anaerobic pre-treatment;
- The concentration of inert particulate material is low as most of the suspended solids present in the raw sewage are retained in the preceding anaerobic system;
- The settleability of the sludge in aerobic post treatment systems is very good, so that the post treatment system can be operated at an increased sludge concentration.

In any case, alternative aerobic systems such as the trickling filter also require operation at a long sludge age, if the same effluent quality is to be obtained. Furthermore, application of pre-aeration to reduce sulphide toxicity could also be considered, as it allows the sludge age in the aerobic system to be reduced to normal values.

On the other hand, the reduction of the nitrifier growth rate might be turned into an advantage if it is intended to reuse the effluent for irrigation: plants take up ammonium much better than nitrate. Furthermore, if only COD removal is desired, the reduction in nitrifier growth rate allows the aerobic post-treatment system at a higher sludge age before nitrification will develop. This will enhance the effluent quality because of the increased removal of surfactants, while the development of nitrification is prevented either partially or completely. This strategy will significantly reduce oxygen demand for nitrification while it also eliminates the risk of denitrification in the final settler, although this risk is small in any case (refer also to Appendix A8).

It is often assumed that anaerobic pre-treatment has a detrimental effect on the settling properties of the aerobic sludge in the post-treatment system. The (albeit scarce) experimental data seem to indicate the opposite: Guimarães (2003) operated a bench scale SBR treating UASB effluent for a period of 3 years and observed medium to good settleability with k-values ranging between 0.34 to 0.38 $l \cdot g^{-1}$ and v_0-values between 9 to 12 $m \cdot h^{-1}$ at sludge ages from 9 to 15 days. A pilot-scale Bardenpho system for post treatment of anaerobic effluent (Silva Filho et al., 2005) demonstrated even better sludge settleability: $k = 0.32$ to 0.38 $l \cdot g^{-1}$ and $v_0 = 18$ to 22 $m \cdot h^{-1}$ at an aerobic sludge age of 20 days. The values of the Vesilind constants indicate that the sludge settleability in post treatment systems tends to be better or at least equal to that in activated sludge systems treating raw sewage.

Furthermore, during the entire experimental research project of more than 4 years, the development of filamentous sludge was never observed. The presence of sulphides in the anaerobic effluent might be responsible for this effect, as the filaments extend outside of the sludge flocs and hence are more exposed. If this is true, then the sulphide in the anaerobic effluent has an effect similar to that of chlorination in the control of sludge bulking (refer to Section 9.3). Pre-aeration of anaerobic effluent might then actually have a detrimental effect on sludge settleability as it reduces sulphide toxicity to the filamentous bacteria. Thus, if nitrification is not required, it could be considered to operate the pre-aeration tank as a selector (i.e. with sludge return). However, should nitrification be required, one should then balance the costs of operating at a higher sludge age (i.e. without pre-aeration) against the benefits of a reduced final settler volume.

Another explanation for the improvement in settling characteristics of the aerobic sludge of post-treatment systems might be the nature of the feed, as the concentration of easily biodegradable COD in the anaerobic effluent will be low. In this case, the application of pre-aeration will not have a detrimental effect on sludge settleability.

EXAMPLE 13.12

Consider again the municipal sewage treatment plant of the previous examples (50,000 $m^3 \cdot d^{-1}$ and 30,000 kg $COD \cdot d^{-1}$). For the conditions specified in these examples, compare the total reactor volume, excess sludge production and energy consumption of an optimised conventional activated sludge system for secondary treatment with that of a combined anaerobic-aerobic treatment system. For the latter case, evaluate the performance with and without nitrification in the aerobic post-treatment system. The following additional data are given:

- Conventional activated sludge system:
  - $R_s = 5$ d; $f_v = 0.7$;
  - $k = 0.46$ $l \cdot g^{-1}$ and $v_0 = 144$ $m \cdot d^{-1}$;
  - The digester is operated at ambient temperature (25°C).
- Aerobic post-treatment system:
  - Secondary treatment: $R_s = 5$ d and $X_t = 3.0$ $g \cdot l^{-1}$;
  - Nitrification: $R_s = 15$ d; $X_t = 3.4$ $g \cdot l^{-1}$; $N_{ti} = 50$ and $N_{oe} + N_{ad} = 2$ $mg\ N \cdot l^{-1}$;
  - $f_v = 0.8$; $k = 0.4$ $l \cdot g^{-1}$ and $v_0 = 200$ $m \cdot d^{-1}$.

- General:
  - $H_d = 4$ m; $H_{th} = 3$ m; $s_{fd} = 2$; $s_{fth} = 1.5$;
  - $\eta_{aer} = 1.2$ kg $O_2 \cdot kWh^{-1}$; $\eta_{el} = 35\%$;
  - Energy content of methane is 14 kWh $\cdot$ kg^{-1} CH_4;
  - Assume all produced methane is recovered.

Solution

(a) Optimised design of conventional activated sludge system

The conventional activated sludge system consists of four main treatment units: biological reactor, final settler, sludge thickener and anaerobic sludge digester. The key parameter to be optimised is the aerobic sludge age: for secondary treatment, the lowest possible sludge age is selected that allows the removal of the biodegradable organic material. In principle, at the environmental temperature of 25°C, a sludge age as low as 2 days would suffice, but this would result in a poor effluent quality. Therefore a value of 5 days is selected in order to guarantee the removal of dispersed bacteria and surfactants. The raw sewage COD fractionisation will be:

$$mS_{te} = f_{ns} = 0.1 \quad (3.18)$$

$$mS_{xv} = f_{cv} \cdot (1 - f_{ns} - f_{np}) \cdot (1 + f \cdot b_h \cdot R_s) \cdot C_r/R_s + f_{np} = 0.38 \quad (3.38)$$

$$mS_o = (1 - f_{ns} - f_{np}) \cdot [(1 - f_{cv} \cdot Y) + f_{cv} \cdot (1 - f) \cdot b_h \cdot C_r] = 0.52 \quad (3.43)$$

Hence the organic and total excess sludge production will be:

$$mE_v = mS_{xv}/f_{cv} = 0.38/1.5 = 0.26 \text{ mg VSS} \cdot \text{mg}^{-1} \text{ COD} \quad \text{and}$$

$$mE_t = mE_v/f_v = 0.26/0.7 = 0.37 \text{ mg TSS} \cdot \text{mg}^{-1} \text{ COD}$$

The sludge mass that develops in the system per unit mass of applied COD is calculated as:

$$mX_t = mE_t \cdot R_s = 0.37 \cdot 5 = 1.83 \text{ mg TSS} \cdot \text{d} \cdot \text{mg}^{-1} \text{ COD} \quad (3.57)$$

Knowing that $mX_t = MX_t/MS_{ti} = (V_r \cdot X_t)/(Q_i \cdot S_{ti})$, the reactor volume required per unit influent flow can be expressed as a function of the sludge concentration:

$$v_r = V_r/Q_i = mX_t \cdot (S_{ti}/X_t)$$

Eq. (8.33) states that that when the critical sludge recycle rate s_c is applied (i.e. when clarification will be the limiting process in the final settler), the maximum allowable hydraulic loading rate of the final settler is given by:

$$T_{sm} = Q_i/A_{min} = v_0 \cdot \exp(-k \cdot X_t) \quad \text{so that} \quad V_d = s_{fd} \cdot H_d \cdot (Q_i/T_{sm}) \quad \text{and}$$

$$v_d = V_d/Q_i = (s_{fd} \cdot H_d/v_0) \cdot \exp(k \cdot X_t) \quad (8.33)$$

The reactor- and final settler volume are calculated for different X_t values in order to determine the optimal sludge concentration for which the total volume is minimum. If poor settleability is assumed ($k = 0.46$ l $\cdot$ g^{-1} and $v_0 = 6$ m $\cdot$ h^{-1}), then for the given conditions the minimal total volume is attained for $X_t = 3.2\ g \cdot l^{-1}$. The specific unit volumes are given as $v_r = 0.34\ m^3 \cdot m^{-3} \cdot d^{-1}$ and $v_d = 0.24\ m^3 \cdot m^{-3} \cdot d^{-1}$. From Figure 8.10, the critical sludge recirculation factor s_c is determined as 0.57 for $k \cdot Xt = 1.47$.

The optimal sludge thickener volume is calculated using the optimisation procedure demonstrated in Chapter 12 for the system consisting of a gravity thickener and an anaerobic digester, in which the solids flux design method is applied. The required thickener volume per kg COD applied is equal to:

$$v_{th} = (s_{fth} \cdot H_{th}/S_{ti}) \cdot (mE_t/F_{sol}) \quad (12.7a)$$

The maximum solids flux that can be transported depends on the value of the limiting flux, which is given by the limiting sludge concentration:

$$F_l = X_{th} \cdot v_0 \cdot (k \cdot X_l - 1) \cdot \exp(-k \cdot X_l) \quad (12.5a)$$

$$X_l = (X_{th}/2) \cdot \{1 + [1.4/(k \cdot X_{th})]^{0.5}\} \quad (12.5b)$$

When hydraulic sludge wasting is applied, the specific excess sludge production rate is equal to $mq = v_r/R_s$, which is reduced by thickening to a value $mq_{th} = mq \cdot (X_{th}/X_t)$. If anaerobic digestion is assumed for stabilisation of the thickened excess sludge, the required specific digester volume $v_{di} = mq_{th} \cdot R_{di}$, where $R_{di} = 20 \cdot 1 \cdot 1^{(20-25)}+5 = 17.4$ days. Under these conditions, the minimum total volume is found for a thickened sludge concentration of $X_{th} = 15\ g \cdot l^{-1}$, when $v_{th} = 0.06$ and $v_{di} = 0.27\ m^3 \cdot m^{-3} \cdot d^{-1}$. Hence the minimum volume of the entire wastewater treatment system is:

$$v_t = v_r + v_d + v_{th} + v_{di} = 0.34 + 0.24 + 0.06 + 0.27 = 0.92\, m^3 \cdot m^{-3} \cdot d^{-1}$$

For the given oxygen transfer efficiency of 1.2 kg $O_2 \cdot kWh^{-1}$, the power consumption per unit mass applied COD (mP_{aer}) is equal to $mS_o/1.2 = 0.43\ kWh \cdot kg^{-1}$ COD applied. The digester performance is calculated from the sludge composition and digester efficiency. The active- and non active volatile excess sludge production is calculated as:

$$mE_{vxa} = (1 - f_{ns} - f_{np}) \cdot C_r/R_s = 0.18\, mg\ X_a \cdot mg^{-1}\, COD$$

$$mE_{vxna} = mE_v - mE_{vxa} = 0.26 - 0.18 = 0.08\, mg\ X_v \cdot mg^{-1}\, COD$$

The solids conversion efficiency in the anaerobic digester at 25°C is (Eqs. 12.64 and 12.65):

$$\eta_{dp} = (0.67 \cdot 25 + 36)/100 = 53\% \quad \text{and} \quad \eta_{dn} = (0.19 \cdot 25 + 10)/100 = 15\%$$

Hence the digested sludge mass is $0.53 \cdot 0.18 + 0.15 \cdot 0.08 = 0.11$ mg VSS $\cdot\ mg^{-1}$ COD. If it is assumed that the inorganic sludge concentration is unaffected by the digestion process, the stabilised sludge production will be: $mE_{te} = 0.37 - 0.11 = 0.26$ kg TSS $\cdot\ kg^{-1}$ COD. The methane production can be estimated as $mM_d = mS_d/4 = 1.5 \cdot mE_d/4 = 0.04$ kg $CH_4 \cdot kg^{-1}$ COD. With the methane produced, the power production is $35\% \cdot 0.04 \cdot 14 = 0.197\ kW \cdot kg^{-1}$ COD applied, which is $0.197/0.43 = 45\%$ of the energy demand for aeration.

(b) The combined anaerobic-aerobic treatment system

To optimise the design of the anaerobic-aerobic treatment system, the optimal volume of the anaerobic pre-treatment system must be determined. As the anaerobic volume increases, so does the degree of solids digestion, resulting in increased COD removal and a reduction of required aerobic post treatment volume. However, the decrease in volume from the aerobic system at higher values of the anaerobic sludge age cannot compensate for the increase in anaerobic reactor volume. Therefore in principle the lowest possible anaerobic sludge age should be selected, which is often limited by a desired degree of UASB sludge stabilisation (e.g. $f_{pu} \leq 0.3$) or a certain desired minimum COD removal efficiency. On the other hand, it should also be considered that operating at a low anaerobic sludge age might result in operational instability, especially at lower temperatures. In practice the operational sludge age is often maintained between 30 and 60 days.

The volume of the main treatment units of the anaerobic-aerobic system can be evaluated as a function of the anaerobic sludge age or the anaerobic retention time:

- As a function of the anaerobic sludge age, calculate the anaerobic excess sludge production (Eq. 13.16) and the effluent COD concentration (Eq. 13.17);
- Determine the biodegradable COD concentration in the anaerobic effluent as a function of the anaerobic sludge age and proceed with the optimisation of the aerobic post treatment unit as in the example above, assuming that the aerobic excess sludge is mixed with the anaerobic excess sludge prior to discharge. In practice it might be possible to accommodate the aerobic excess sludge in the anaerobic pre-treatment reactor, as both the anaerobic- and the aerobic unit are frequently overdesigned. However, as demonstrated in Example 13.8, a better alternative might be to construct a dedicated heated digester;
- Determine the minimum total volume of the aerobic reactor and the final settler, assuming that the settleability of the aerobic sludge treating anaerobic effluent increases ($k = 0.4\ l \cdot g^{-1}$ TSS and $v_0 = 200\ m \cdot d^{-1}$);
- Plot the anaerobic reactor volume and the minimal volumes of the aerobic reactor and final settler as function of the anaerobic sludge age, as shown in Figure 13.25. The hydraulic residence time in the anaerobic reactor can be calculated with Eq. (13.28).

Figure 13.25 has been calculated for a secondary treatment system ($R_s = 5$ days, left) and for nitrification ($R_s = 15$ days, right).

The optimal aerobic sludge concentration has also been indicated. It turns out (not discussed here) that the applied anaerobic sludge age has only a small effect on the optimal value of aerobic sludge concentration. For example, for the aerobic system for secondary treatment the value of X_t increases from $2.8\ g \cdot l^{-1}$ at $R_{su} = 10$ days to $3.2\ g \cdot l^{-1}$ at $R_{su} = 100$ days. This difference is small and has only a minor impact on the overall results. Therefore, in order to simplify the example, average concentrations of $X_t = 3.0\ g \cdot l^{-1}$ at $R_s = 5$ days and $3.4\ g \cdot l^{-1}$ at 15 days have been used.

When the total volume of the anaerobic-aerobic system is compared with the minimal volume calculated for the optimised conventional activated sludge system, it is quite clear that there is a potential for a very substantial reduction in volume when anaerobic pre-treatment is applied. The minimum volume is obtained for an anaerobic sludge age of 10 days, but in practice a much longer anaerobic retention time is selected in order to ensure process stability and to stabilise the anaerobic excess sludge.

Figure 13.25 demonstrates that for the usual range of applied anaerobic sludge ages (30 to 60 days), the total volume of the anaerobic-aerobic system is 30 to 50% less than the volume of the optimised conventional aerobic system. Table 13.16 compares the performance of the combined anaerobic–aerobic system to that of the conventional aerobic system, for an anaerobic sludge age of 35 days, as applied in the previous examples.

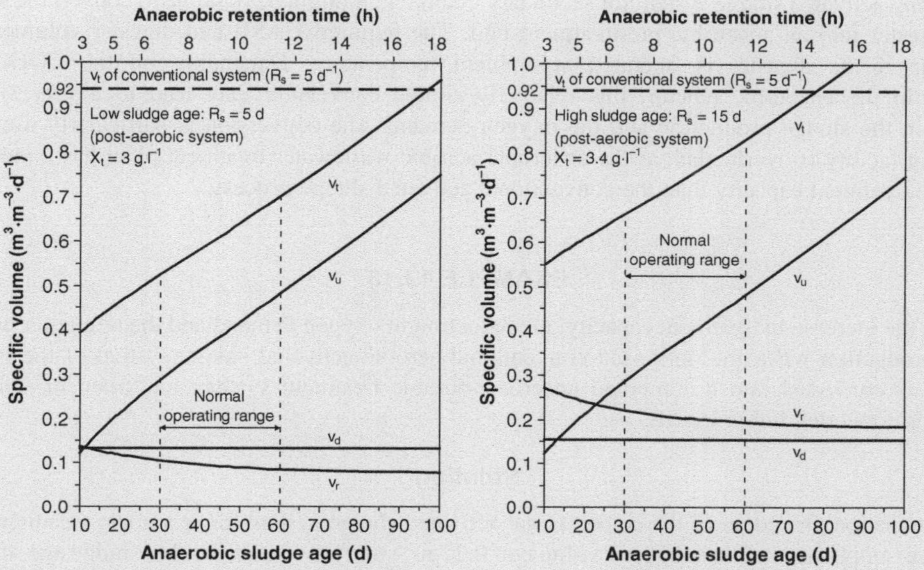

Figure 13.25 Specific volume of the combined aerobic-anaerobic system per unit influent flow as function of the anaerobic sludge age or -retention time. Left = secondary treatment and right = nitrification.

Table 13.16 Comparison of system volume and performance of the three configurations discussed in Example 13.12

Parameter	UoM	Conventional aerobic system		UASB + secondary treatment		UASB + nitrification	
Sludge age	days	$R_s = 5$		$R_{su} = 35/R_s = 5$		$R_{su} = 35/R_s = 15$	
System volume:							
– V_u	m^3	–		16,050		16,050	
– V_r	m^3	17,125		4999		10,723	
– V_d	m^3	12,105		6640		7792	
– V_{th}	m^3	2938		–		–	
– V_{di}	m^3	13,637		–		–	
– V_t	m^3	45,805		27,688		34,565	
Performance:							
– MS_{te}	kg COD · d^{-1}	3000	10%	3000	10%	3000	10%
– MS_{xve}	kg COD · d^{-1}	6683	22%	6728	23%	6046	20%
– MS_o	kg COD · d^{-1}	15492	52%	3696	12%	4378	15%
– MS_d	kg COD · d^{-1}	4826	16%	16,576	55%	16,576	55%
– MO_n	kg O_2 · d^{-1}	–		–		9507	
– P_{aer}	kW	538		128		482	
– P_{el}	kW	246		848		848	

In an existing activated sludge system for secondary treatment, it might be possible to convert the anaerobic sludge digester into an anaerobic pre-treatment unit. The required UASB and digester volumes will be comparable if the digester is operated at ambient temperature. Depending on the COD removal efficiency in the anaerobic reactor, this relatively simple conversion may lead to a very significant reduction in the sludge production and the oxygen demand. The conversion is particularly useful when the existing facility is overloaded, as the anaerobic-aerobic wastewater treatment plant has a much larger volumetric treatment capacity than the conventional activated sludge process.

EXAMPLE 13.13

Estimate the increase in treatment capacity, the reduction in oxygen demand and the reduction in excess sludge production when the optimised conventional aerobic activated sludge system of the previous example is converted into a combined anaerobic-aerobic treatment system: i.e. when the anaerobic digester is converted into a UASB.

Solution

The optimisation procedure of the conventional activated sludge system for secondary treatment in the previous example resulted in a digester volume of $0.27\ m^3 \cdot m^{-3} \cdot d^{-1}$ at an aerobic sludge age of 5 days. If this digester is to be used as a UASB reactor, the retention time will be $0.27 \cdot 24 = 7.2$ hours. For the specified conditions and using Eq. 13.28, this results in an anaerobic sludge age of 32 days, which is perfectly acceptable for a temperature of 25°C.

The anaerobic excess sludge production, the biodegradable COD fraction in the UASB effluent and the digested COD fraction can be calculated using Eqs. (13.16 to 13.18). Based on the composition of the anaerobic effluent, which can be defined with Eqs. (12.41 to 13.47), the fraction of the influent COD remaining in the final (aerobic) effluent and the COD fractions that are oxidised and transformed into aerobic excess sludge can be calculated.

UASB performance:

- Digested COD fraction: $mS_{du} = 0.53$
- COD fraction in the anaerobic sludge: $mS_{xvu} = 0.12$
- COD fraction in the anaerobic effluent: $mS_{seu} + mS_{peu} = 0.23 + 0.12 = 0.35$
  - Biodegradable COD fraction (incl. H_2S): $mS_{bu} = 0.13 + 0.07 = 0.20$
  - Non biodegradable COD fraction: $mS_{nu} = 0.10 + 0.05 = 0.15$

Overall system performance:

- COD fraction in the final effluent: $mS_{te} = 0.10$
- Digested COD fraction: $mS_{du} = 0.53$
- Oxidised COD fraction: $mS_o = 0.13$
- COD fraction in the combined excess sludge: $mS_{xvu} + mS_{xv2} = 0.12 + 0.12 = 0.24$

Under these conditions the (existing) aerobic reactor will be under loaded, so excess treatment capacity is available. Thus, if required, more influent can be treated, the extent of which is limited by the fact that an increase in influent flow rate results in a decrease in anaerobic sludge age and thus in a reduced anaerobic treatment efficiency. This in turn will increase the hydraulic- and organic loads to the aerobic post treatment system, resulting in a higher oxygen demand and more aerobic excess sludge production. Above a certain influent flow rate, the treatment system will become overloaded as one of the following three factors will be limiting:

- The increase of the influent COD load to the UASB reactor may result in an anaerobic sludge age that is too short to properly stabilise the anaerobic sludge;
- The oxygen demand in the aerobic reactor may exceed the available aeration capacity;
- The solids loading rate to the final settler may become excessive, resulting in problems with solid-liquid separation, as both the sludge concentration in the aerobic system and the flow rate to the final settler will increase.

It is difficult to predict in advance what will be the limiting constraint, but in any case there is a potential for a very substantial increase in the total treatment capacity by transforming the sludge digester into an anaerobic pre-treatment unit. An additional advantage is that the methane production in the UASB will be much higher than that in the sludge digester, so that the entire treatment system can be operated without an external energy source if the biogas is converted into electric power.

13.5.1.2 Stabilisation of aerobic excess sludge in the UASB reactor

In conventional aerobic sewage treatment plants, the produced excess sludge is often stabilised in a dedicated unit, often an anaerobic sludge digester (Section 12.4). When anaerobic pre-treatment is applied the excess sludge production in the aerobic post-treatment unit will be much smaller compared to that in a conventional aerobic system. Digestion of this excess sludge can take place in the pre-treatment unit itself, as indicated in Figure 13.23. While this simplifies the overall treatment scheme, there are some issues to consider.

First of all, the UASB reactor is not particularly suited to digest aerobic excess sludge, especially at lower temperatures. Only a part of the aerobic excess sludge will be retained in the UASB, while the conditions in the reactor (low mixing intensity and suboptimal temperatures) reduce the degree of solids degradation that can be attained. Second, the additional solids load to the UASB reduces the anaerobic sludge age and with it the COD removal efficiency.

As to the first subject, Silva Filho et al., (2005) investigated the influence of the addition of excess aerobic sludge on the performance of the UASB reactor. Aerobic excess sludge generated at a low sludge age was added in different proportions to raw sewage treated in four pilot scale UASB reactors (R_1 to R_4). Each 120 litre UASB reactor was fed with 480 $l \cdot d^{-1}$ of raw sewage with an average influent COD concentration of 706 $mg \cdot l^{-1}$. The hydraulic retention time was 6 hours and the applied organic load is 339 g $COD \cdot d^{-1}$ or 2.8 g $COD \cdot l^{-1} \cdot d^{-1}$, which may be considered as typical values for anaerobic sewage treatment at the operational temperature of 25°C. The first reactor (R_1) received only the raw sewage while the other three reactors also received aerobic sludge: an additional 20, 40 and 60% (as COD) of the organic sewage load in reactors R_2, R_3 and R_4 respectively. The aerobic excess sludge was generated in a separate aerobic system that was operated at a sludge age of 5 days. The UASB reactors were operated without intentional excess sludge discharge, so that the anaerobic sludge age was maximum. The anaerobic sludge production was calculated from the daily mass of sludge mass discharged with the UASB effluent. This effluent was received in 500 litre tanks, where the anaerobic sludge was allowed to settle. The COD concentration in the clarified effluent was used to evaluate the digestion efficiency of the UASB reactor, while the accumulated sludge was used to determine the sludge production and the properties of the produced sludge.

The collected effluent data in Table 13.17 show that an increase in the proportion of aerobic excess sludge in the influent to the UASB reactor did not have a negative effect on the performance. The pH in the effluent even increased, while the volatile fatty acids concentration in the effluent remained low (< 1 mmol HAc $\cdot$ l^{-1}). More importantly, because of the high operational temperature the removal efficiency of organic material remained high in all reactors, although a slight increase in the effluent total COD concentration was observed at increasing aerobic sludge loads.

Table 13.17 Sewage and settled effluent characteristics from 4 UASB reactors operated with different fractions of aerobic excess sludge in the feed (ranging from 0 to 60%) for $R_h = 6$ hrs and $T = 25°C$

Variables	Sewage	Effluent			
		R_1 (0%)	R_2 (20%)	R_3 (40%)	R_4 (60%)
pH	7.6	7.5	7.5	7.5	7.5
Alkalinity (mg $CaCO_3 \cdot l^{-1}$)	386	411	419	422	416
VFA (mg HAc $\cdot l^{-1}$)	113	47	48	50	51
Total COD (mg COD $\cdot l^{-1}$)	706	126	126	173	149
Settled COD (mg COD $\cdot l^{-1}$)	–	114	111	112	109
TSS (mg $\cdot l^{-1}$) (settled effluent)	305	29	38	79	62
VSS (mg $\cdot l^{-1}$) (settled effluent)	219	24	30	56	44

Table 13.18 shows the characteristics of the anaerobic excess sludge for each of the 4 pilot reactors. Regarding the influence of aerobic excess sludge on the properties of the stabilised anaerobic sludge, the following can be observed from the data:

(1) The sludge mass that developed was approximately equal in the four reactors (1700 g VSS or 14 g VSS $\cdot l^{-1}$), despite considerable differences in influent COD composition. This indicates that for this reactor configuration and upflow velocity it may be considered as the maximum sludge mass that can be maintained in the reactor;

(2) The sludge age is calculated as the ratio between the sludge mass present in the UASB and the daily discharge of sludge mass with the effluent. The sludge age decreases rapidly at increased loading of the UASB reactor with aerobic sludge, reaching a minimum value of only 11 days in reactor R_4. Due to the high operational temperature of 25°C, a collapse of UASB treatment efficiency did not occur;

(3) Despite the low value of the sludge age in reactor R_4, the stability of the final sludge was satisfactory. When the anaerobic excess sludge was subjected to further digestion during 30 days at 30°C, the mass of digested solids increased from reactor R_1 to R_4, but it did not exceed more than 16% of the incubated volatile sludge mass, which is considered by the EPA (1992) as an indication for sludge stability. Furthermore the sludge discharged by the UASB reactors did not have a bad odour;

(4) The daily mass of discharged (stabilised) sludge increased considerably with an increased proportion of aerobic excess sludge in the feed of the UASB reactors: in fact the daily mass of digested sludge only marginally increased at higher aerobic sludge loads. This is a very

important observation as it indicates that a significant part of the aerobic excess sludge is not retained in the UASB;

(5) The anaerobic sludge in the UASB reactors was more stable than the sludge discharged with the effluent, indicating that the aerobic excess sludge was more than proportionally washed-out of the UASB reactor;

(6) There was no clear effect on the specific methanogenic activity of the sludges: the measured values are all in the upper range usually found for raw sewage digestion;

(7) The settleability of the anaerobic sludge was very good; in particular, the compressibility (indicated by a low k value) was excellent, allowing a high UASB sludge concentration;

(8) The specific filter resistance values are rather high: this indicates that the sludge was not very well stabilised (Andreoli et al., 2002). The use of filtration aids (polyelectrolytes or metal salts) might be sufficient to prevent problems with mechanical sludge dewatering.

Table 13.18 Characterisation of the stabilised sludge in the 4 UASB reactors for an increasing fraction of aerobic excess sludge in the feed (0–60%)

Parameter	R_1 (0%)	R_2 (20%)	R_3 (40%)	R_4 (60%)
Solids load–aerobic sludge ($g\ VSS \cdot d^{-1}$)	0	45	90	135
Total solids load (sludge + sewage) ($g\ VSS \cdot d^{-1}$)	120	165	210	254
Stabilised sludge production ($g\ VSS \cdot d^{-1}$)	45	80	123	152
UASB sludge mass (g VSS)	1600	1710	1730	1600
Anaerobic sludge age (d)	36	21	14	11
Degradability of the excess sludge ($g\ VSS \cdot g^{-1}\ VSS$)	0.11	0.13	0.15	0.16
Specific methanogenic activity ($g\ COD \cdot g^{-1}\ VSS \cdot d^{-1}$)	0.17	0.24	0.19	0.18
Settleability constant k ($l \cdot g^{-1}\ TSS$)	0.03	0.03	0.06	0.05
Settleability constant v_0 ($m \cdot h^{-1}$)	4.9	6.0	8.4	10.0
Specific filter resistance ($m \cdot kg^{-1}\ TSS$)	$6.7 \cdot 10^{13}$	$5.9 \cdot 10^{13}$	$2.7 \cdot 10^{14}$	$2.5 \cdot 10^{14}$

In principle, the two main beneficial effects of UASB treatment of aerobic excess sludge are (I) stabilisation of the excess sludge through the reduction of the active volatile sludge fraction, and (II) an increase in the suspended solids concentration (thickening). However, part of the suspended solids present in the aerobic excess sludge is not retained by the sludge blanket and constitutes an additional TSS and COD load to the aerobic process. This is not apparent from Table 13.17, but it should be considered that the reactors were operated without intentional sludge discharge and that the data reported in this table refer to settled UASB effluent. Furthermore, in the experiment the UASB reactors were operated at 25°C whereas at temperatures lower than 20°C the digestion efficiency of the aerobic sludge will be significantly reduced. Non-degraded suspended solids will accumulate, which will reduce the anaerobic sludge age and with it the anaerobic COD removal efficiency.

Hence, combined with the "recirculation" of aerobic excess sludge, the COD load to the aerobic post-treatment will be increased even further. As discussed previously, an alternative approach is to digest the excess aerobic sludge with the anaerobic excess sludge in a small heated digester, as demonstrated in Example 13.8. Both the extent and the degree of solids digestion will increase. In Example 13.1 the influence of aerobic sludge recycle to the UASB reactor on the performance of an anaerobic – aerobic wastewater treatment plant is discussed.

EXAMPLE 13.14

An anaerobic – aerobic sewage treatment plant consisting of an UASB followed by an activated sludge system is operated under the following conditions:

- $R_{su} = 30$ days; $T = 20°C$;
- $Q_i = 50{,}000\ m^3 \cdot d^{-1}$; $S_{ti} = 600\ mg \cdot l^{-1}$; $f_{ns} = 0.05$ and $f_{np} = 0.15$;
- $V_u = 17{,}300\ m^3$; $X_{tu} = 17.7\ g \cdot l^{-1}$; $X_{tue} = 80\ mg \cdot l^{-1}$; $f_v = 0.65$.

Assess the impact of the recycle of aerobic excess sludge to the UASB on the performance of this system. To simplify the example, ignore the presence of sulphate in the raw sewage. The aerobic system is operated at a sludge age of five days ($V_r = 5410\ m^3$; $X_t = 3$ g $TSS \cdot l^{-1}$; $f_v = 0.85$). Assume that nitrification does not develop.

It is assumed that $\eta = 80\%$ of the suspended solids present in the aerobic sludge are retained in the UASB sludge blanket and that the relationships describing degradation of aerobic sludge in an anaerobic digester apply as well ($\eta_{dp} = 49.4\%$ (Eq. 12.64) and $\eta_{dn} = 13.8\%$ (Eq. 12.65) at $T = 20°C$). This might be to optimistic as the mixing intensity in the sludge blanket and hence the contact between biomass and solids will be less than in a sludge digester. The non-degraded organic suspended solids in the (retained) aerobic excess sludge will accumulate as an inert organic fraction in the UASB sludge.

Solution

First evaluate the performance of the existing anaerobic-aerobic system when aerobic sludge recycling is not applied. For $MS_{ti} = 50{,}000 \cdot 0.6 = 30{,}000$ kg $COD \cdot d^{-1}$, the "ideal" UASB effluent COD load (i.e. without suspended solids) is calculated as:

$$\begin{aligned} MS_{eu} &= [f_{ns} + 0.27 \cdot \exp[-0.04 \cdot (R_{su} - 4)]/1.067^{(T-25)}] \cdot MS_{ti} \\ &= [0.05 + 0.27 \cdot \exp[-0.04 \cdot (30 - 4)]/1.067^{(-5)}] \cdot 30{,}000 \\ &= 5460\ kg\ COD \cdot d^{-1} \end{aligned} \quad (13.16)$$

For $f_v = 0.65$ and $X_{tue} = 80$ mg $TSS \cdot l^{-1}$, the COD load associated to the suspended solids in the UASB effluent can be calculated as $50{,}000 \cdot 0.65 \cdot 0.08 \cdot 1.5 = 3900$ kg $COD \cdot d^{-1}$. The total effluent COD load in the anaerobic effluent is thus $5460 + 3900 = 9360$ kg $COD \cdot d^{-1}$ and the overall COD removal efficiency is $1 - 9360/30{,}000 = 68.8\%$, so $S_{tu} = 187.2$ mg $COD \cdot l^{-1}$.

Using Eqs. (13.20 to 13.27) to characterise the anaerobic biomass and assuming that the composition of the effluent suspended solids is equal to that in the UASB, the active fraction in the organic solids in anaerobic effluent can be calculated as $(mX_{bp} + (1-f) \cdot mX_{au})/mX_v = (3.0 + 0.8 \cdot 0.61)/6.64 = 52.5\%$. Now f_{npu} can be calculated as $(1-0.525) \cdot 3900/9360 = 19.8\%$. The inert soluble COD fraction f_{nsu} is equal to $f_{ns}/(1-\eta_{COD}) = 0.05/(1-0.688) = 16\%$. For $MS_{tu} = 9360$ kg $COD \cdot d^{-1}$ the performance of the aerobic reactor in terms of oxygen demand and sludge production can be calculated as:

$$C_r = Y \cdot R_s/(1 + b_h \cdot R_s) = 0.45 \cdot 5/(1 + 0.24 \cdot 5) = 1.02 \quad (3.30)$$

$$\begin{aligned} MS_o &= (1 - f_{npu} - f_{nsu}) \cdot (1 - f_{cv} \cdot Y + (1 - f) \cdot f_{cv} \cdot b_h \cdot C_r) \cdot MS_{tu} \\ &= (1 - 0.358) \cdot (1 - 1.5 \cdot 0.45 + (1 - 0.2) \cdot 1.5 \cdot 0.24 \cdot 1.02) \cdot 9360 \\ &= 3720\ kg\ O_2 \cdot d^{-1} \end{aligned} \quad (3.43)$$

$$\begin{aligned} ME_{t2} &= [(1 - f_{nsu} - f_{npu}) \cdot (1 + f \cdot b_h \cdot R_s) \cdot C_r/R_s + f_{np}/f_{cv}] \cdot MS_{tu}/f_v \\ &= [(1 - 0.358) \cdot (1 + 0.2 \cdot 0.24 \cdot 5) \cdot 1.02/5 + 0.198/1.5] \cdot 3720/0.85 \\ &= 3246 \text{ kg TSS} \cdot d^{-1} \end{aligned} \quad (3.50)$$

When hydraulic sludge wasting is assumed, this corresponds to an excess sludge flow of 1082 $m^3 \cdot d^{-1}$. When the aerobic excess sludge is sent to the UASB reactor, it is assumed that only 80% of the aerobic excess sludge is retained. Thus the TSS concentration in the UASB effluent will increase by $(1–0.8) \cdot 3246 \cdot 1000/(50{,}000 + 1082) = 13$ mg TSS $\cdot l^{-1}$. To simplify the example, assume that the composition of this fraction is equal to that of the other TSS in the UASB effluent. The organic excess sludge production in the activated sludge system $ME_{v2} = 3246 \cdot 0.85 = 2759$ kg VSS $\cdot d^{-1}$ of which a fraction $\eta = 80\%$ or 2207 kg $\cdot d^{-1}$ is retained in the UASB sludge blanket. To estimate the extent of solids degradation, first the active fraction of the aerobic excess sludge must be calculated:

$$\begin{aligned} f_{av} &= (1 - f_{ns} - f_{np}) \cdot C_r/[(1 \cdot f_{ns} \cdot f_{np}) \cdot (1 + f \cdot b_h \cdot R_s) \cdot C_r + f_{np} \cdot R_s/f_{cv}] \\ &= (1 - 0.358) \cdot 1.02/[(1 - 0.358) \cdot (1 + 0.2 \cdot 0.24 \cdot 5) \cdot 1.02 + 0.198 \cdot 5/1.5] \\ &= 45\% \end{aligned} \quad (3.52)$$

Now, for the given values of the digestion efficiencies η_{dp} and η_{dn}, the mass of digested aerobic excess sludge can be calculated as:

$$\begin{aligned} ME_d &= (f_{av} \cdot \eta_{dp} + (1 - f_{av}) \cdot \eta_{dn}) \cdot ME_{v2} \\ &= (0.45 \cdot 0.494 + (1 - 0.45) \cdot 0.138) \cdot 2207 = 655 \text{ kg VSS} \cdot d^{-1} \end{aligned} \quad (12.66)$$

$$ME_{ve} = ME_{v2} - ME_d = 2207 - 655 = 1553 \text{ kg VSS} \cdot d^{-1}$$

In order to evaluate the impact of aerobic excess sludge recycle on UASB performance, it will be necessary to recalculate the influent load and composition. The non-degrated organic solids remaining after digestion of the aerobic sludge in the UASB (i.e. ME_{ve} as calculated above) are added to the inert particulate COD load (MS_{np}) in the raw sewage, together with the non-active fraction in the part of the aerobic excess sludge that is not retained in the UASB.

As to the soluble inert COD load, the aerobic excess sludge will contain the same concentration of this material as the raw sewage (i.e. $0.05 \cdot 600 = 30$ mg COD $\cdot l^{-1}$). The combined feed (raw sewage + aerobic excess sludge) of the UASB is characterised as:

$$\begin{aligned} Q_i' &= Q_i + q_2 = 50{,}000 + 1082 = 51{,}082 \text{ m}^3 \cdot d^{-1} \\ MS_{ti}' &= MS_{ti} + f_{cv} \cdot ME_{v2} \\ &= 30{,}000 + 2759 \cdot 1.5 = 34{,}138 \text{ kg COD} \cdot d^{-1} \\ f_{np}' &= (f_{np} \cdot MS_{ti} + f_{cv} \cdot (ME_{ve} + (1 - \eta) \cdot (1 - f_{av}) \cdot ME_{v2})/MS_{ti}' \\ &= (0.15 \cdot 30{,}000 + 1.5 \cdot (1553 + (1 - 0.8) \cdot (1 - 0.45) \cdot 2759)/34{,}138 = 0.213 \\ f_{ns}' &= (f_{ns} \cdot MS_{ti} + q_2 \cdot S_{nsi})/MS_{ti}' \\ &= (0.05 \cdot 30{,}000 + 1082 \cdot 0.03)/34{,}138 = 0.045 \end{aligned}$$

For this influent load and composition the UASB performance can be recalculated (for $f_v = 0.65$). One important effect is the significant increase in the non-biodegradable particulate organic solids load to the UASB, which will result in an equally significant increase in anaerobic excess sludge production. In this specific case (for $R_{su} = 30$ days) the required UASB sludge mass will increase from 307,000 to 420,000 kg TSS. For the average UASB sludge concentration of 17.7 g $TSS \cdot l^{-1}$, this will require a reactor volume of 23,730 m^3, an increase of 37%. However, as the UASB is only 17,300 m^3 in size, the result would be an increase in the average sludge concentration to 24.3 g $TSS \cdot l^{-1}$.

As the UASB reactor is already operating at the maximum sludge inventory, this increase cannot be accommodated and there is no alternative but to decrease the sludge age to $R_{su} = 18.7$ days. However, as a result the COD removal efficiency in the UASB is reduced from 68.8% to 61.2%, which will in turn result in an increase in the COD load to the aerobic reactor. The performance of the aerobic system is calculated as $ME_{t2} = 4015$ kg $TSS \cdot d^{-1}$, $ME_{v2} = 3413$ kg $VSS \cdot d^{-1}$ and $MS_o = 6675$ kg $O_2 \cdot d^{-1}$. The total sludge mass in the aerobic reactor MX_t is equal to $5 \cdot 4015 = 20{,}075$ kg TSS, which for the existing aeration tank volume of 5410 m^3 will require an increase in the sludge concentration from 3.0 to 3.5 g $TSS \cdot l^{-1}$ (or almost 15%). Furthermore, as the aerobic excess sludge flow to the UASB increases, a further reduction in COD removal efficiency will result. After a series of iterations, equilibrium is obtained for an anaerobic sludge age R_{su} of 17.9 days. The performance of the combined anaerobic-aerobic system with and without aerobic sludge recycle is summarized in Table 13.19.

Table 13.19 Comparison of overall system performance with and without aerobic sludge recycle

Parameter	UoM	No recycle	With recycle
Anaerobic sludge age	d	30	17.9
UASB COD removal efficiency	%	68.8	61.2
UASB excess sludge production	kg $TSS \cdot d^{-1}$	6222	12,343
Putrescible fraction UASB sludge	g $X_{bp} \cdot g^{-1}$ VSS	0.45	0.49
Aerobic excess sludge production	kg $TSS \cdot d^{-1}$	3246	0(1)
Oxygen demand	kg $O_2 \cdot d^{-1}$	3721	7059
Aerobic sludge concentration	g $TSS \cdot l^{-1}$	3.0	3.5

Note (1): 4031 kg TSS · d⁻¹ of aerobic excess sludge is sent to the UASB

Perhaps counter intuitively, the recycle of aerobic excess sludge to the UASB has resulted in an *increase* of the total excess sludge production. This is due to the steep reduction in anaerobic sludge age, which decreases the conversion efficiency of slowly biodegradable organic solids. It can also be observed that both aerobic excess sludge production and oxygen demand increase significantly, which is caused by the increase in COD load in the UASB effluent resulting from the decrease in COD removal efficiency. Apart from the increase in operational costs (aeration), if the aerobic system has insufficient capacity for this increase, then operational problems might result.

On the other hand, when aerobic sludge is not recycled to the UASB, the problem of poorly stabilised aerobic sludge remains, requiring a small thickener (± 300 m^3, for a thickened sludge concentration of 30–35 $g \cdot l^{-1}$) plus a dedicated digester (± 1000–1500 m^3, depending on the applied digestion temperature). However, solids destruction and methane production will increase as a result of the higher digestion temperature.

From the above example it can be concluded that while at first glance treatment of aerobic excess sludge in the UASB seems very attractive, there are some issues to consider. Most importantly, in the design phase the COD and TSS load contained in the aerobic excess sludge must be explicitly taken into account. The UASB is a poor solids digester, especially in winter time as the operational temperature will be significantly lower than in a heated digester. Furthermore, the mixing intensity in an UASB will be less than in a digester, in order to maintain a sludge blanket in the reactor with a proper filtering action, which is essential for a satisfactory UASB performance.

An 30–40% increase in required UASB volume should be expected, as well as a slight increase in aerobic volume. The latter is due to an increased suspended solids content in the UASB effluent, though the increase in required volume will not be so accentuated as in the case of Example 13.1, where the UASB reactor became overloaded, reducing COD removal efficiency. The increase in UASB volume should be compared to the additional costs for installing an aerobic sludge thickener and a small (heated) digester. An additional benefit of the latter solution is a higher degree of solids digestion, which results in increased gas production and a reduction in stabilised excess sludge production, whereas the putrescible fraction of the stabilised sludge will be lower as well.

Finally, if the additional load to the UASB is not explicitly taken into account during the design phase, then the actual UASB performance is likely to be well below expected performance. This will result in a significant increase in the COD load to the aerobic post-treatment system and might seriously compromise overall system performance, as demonstrated in Example 13.14.

13.5.2 Nitrogen removal from anaerobic effluent

When biological nitrogen removal is required, the efficiency of the anaerobic reactor is limited by the constraint that sufficient organic material must be left in the effluent to effect the removal of nitrate in the downstream activated sludge process. In Chapter 5 expressions were derived for the maximum TKN/COD ratio allowing complete nitrate removal in a Bardenpho configuration (Eq. 5.88) and for the calculation of the optimum size of the pre-D zone (Eq. 5.92), resulting in the highest nitrate removal in either Bardenpho of pre-D configuration, if complete denitrification is not possible. The model developed for nitrogen removal remains valid for anaerobically pre-treated wastewater. However, the TKN/COD ratio of the digested sewage increases due to the removal of a large part of the organic material in the anaerobic unit, which reduces the maximum extent of denitrification considerably. There are several options that can be applied to limit the extent of this increase:

- Operate the anaerobic reactor at a lower hydraulic retention time (and sludge age), resulting in a reduced anaerobic treatment efficiency;
- Split the raw sewage flow and treat only a part of the wastewater anaerobically. The rest of the flow is sent to the anoxic reactor;
- Operate the anaerobic treatment system at a low methanogenic efficiency;
- Consider the application of alternative methods for nitrogen removal, for instance nitritation and Anammox, which will reduce the COD demand considerably, or the use of physical-chemical methods (e.g. struvite precipation).

Reduction of the anaerobic retention time is only possible to a limited extent, because at short retention times the anaerobic reactor tends to become unstable, which represents a major operational problem. In practice, the hydraulic retention time in anaerobic pre-treatment units is never less than 5–6 hours, which is equivalent to an anaerobic sludge age of 20 to 40 days, depending on the temperature and the efficiency of sludge retention in the solid-liquid separator of the anaerobic reactor. Bypassing a part of the raw sewage flow to the activated sludge system is a better alternative, but in this case the advantage of the anaerobic pre-treatment unit will be reduced as well. The feasibility of using a raw sewage bypass will be discussed in Section 13.5.2.1, while the third and fourth alternatives are discussed in Sections 13.5.2.2 and 13.5.2.3.

Another problem related to biological nitrogen removal in an aerobic post-treatment system is that the metabolic activity of both autotrophic and heterotrophic bacteria is reduced when anaerobic pre-treatment is applied (refer also to Table 13.15), so that both the nitrification- and denitrification rates are lower than in a conventional aerobic process. Most likely the presence of highly toxic sulphide in the anaerobic effluent is responsible for this effect. Thus a longer sludge age is necessary and this creates additional difficulties for the removal of nitrogen and phosphorus. However, as discussed before, the installation of a small pre-aeration step between the UASB and the aerobic post-treatment system is sufficient to eliminate the sulphide toxicity.

13.5.2.1 Bypass of raw sewage to the activated sludge system

Complete denitrification is no longer a realistic objective when anaerobic pre-treatment is applied, unless the COD/N ratio in the raw sewage is very favourable, which applies to certain industrial wastewaters. Due to the removal of a significant part of the biodegradable COD in the anaerobic reactor, a very large bypass would be required for full nitrogen removal. Instead, the objective should be to remove only so much nitrate as needed to comply to the effluent limits. As full denitrification is no longer an objective, a pre-D configuration is selected. This configuration maximises the extent of nitrogen removal at the low values of the COD/TKN ratio typically observed in anaerobic effluent. A second advantage of using a pre-D configuration is that the mixed liquor flow to the final settler will be aerobic, with a low background concentration of dissolved nitrogen gas. Hence, the risk of floating sludge in the final settler because of denitrification will be low: refer also to Appendix A8.

The most important design issue to be evaluated is the quantity of raw sewage that needs to be bypassed to the anoxic reactor. This is determined by:

- Raw sewage composition in terms of COD (f_{ns}, f_{np} and f_{sb});
- COD composition and -concentration in the UASB effluent, which depends on the removal efficiency in the UASB;
- Ratio between COD and nitrogen in the raw sewage.

Once these data are known, the composition and load of COD and nitrogen in the combined feed to the aerobic post-treatment system can be determined. The nitrogen removal performance can be calculated according to the general procedure outlined in Section 5.4. Using the anaerobic model that was developed in Section 13.3, the effluent of the UASB can be characterised. The fractions f_{nsu}, f_{npu} and f_{sb} are calculated with Eqs. (13.42 to 13.47).

It was already highlighted in Section 13.5.1.1 that the determination of the value of f_{sb} is problematic, as it is a priori difficult to estimate which part of the soluble and colloidal COD in the UASB effluent will be easily biodegradable. This is further complicated when pre-aeration is applied to remove sulphide toxicity. As only the sulphides are toxic, not the intermediate products in the oxidation reaction to sulphate, it is recommended to design the pre-aeration reactor for at least partial oxidation to either elemental sulphur or thiosulphates, using a molar O_2/S ratio of 0.5 to 1.0 mg $O_2 \cdot mg^{-1}$ S. There will still be residual COD (50–75%) associated to these partially oxidized sulphur components, but this fraction is not available for rapid denitrification.

As to the soluble/colloidal biodegradable organic COD in the effluent, a significant part of it will consist of residual VFA (typically less than 0.5–1.0 mmol $\cdot$ l^{-1} or <60 mg COD $\cdot$ l^{-1}). In any case, the contribution from the COD in the anaerobic effluent to the denitrification capacity will be small compared to that supplied by the raw sewage bypass.

As an example, Figure 13.26 shows the minimum aerobic sludge age required to meet an total nitrogen limit of 10 mg N $\cdot$ l^{-1} in the effluent as a function of the COD/N ratio in the raw sewage for different bypass fractions (25 to 50%).

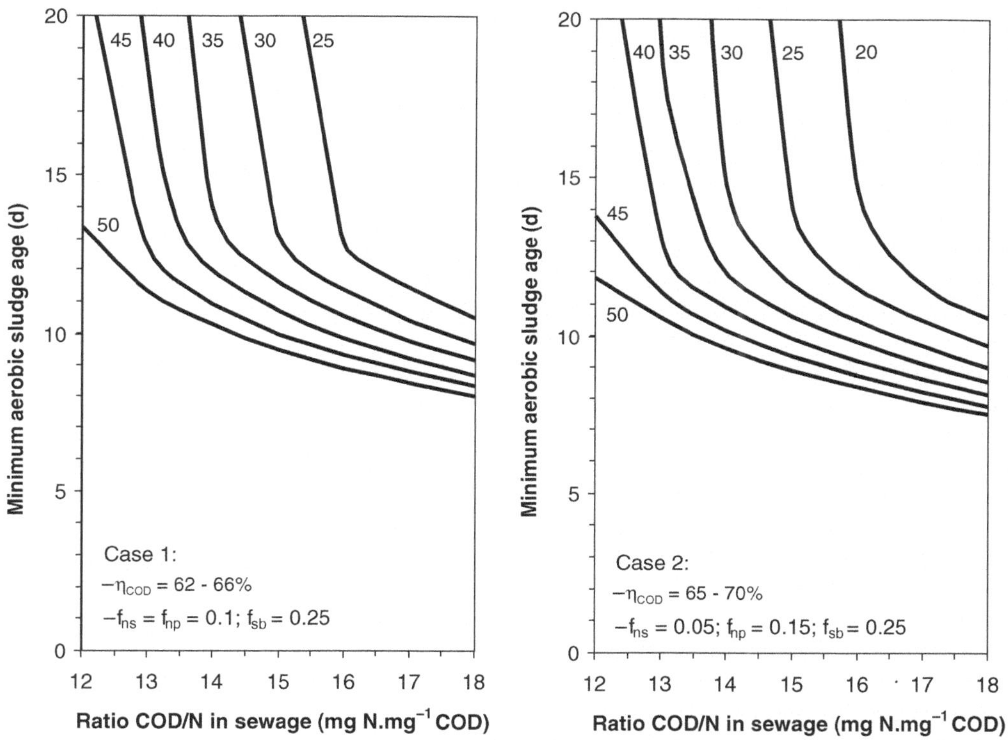

Figure 13.26 Minimum sludge age in the activated sludge system required to meet a total nitrogen effluent limit of less than 10 mg N $\cdot$ l^{-1}, as function of the influent COD/N ratio, for different fractions of the raw sewage flow bypassed (from 20–50%). The effect of differences in raw sewage composition is also indicated

Figure 13.26 has been calculated for the following conditions:

- The anaerobic sludge age R_{su} is 35 days;
- The sewage temperature is 20°C;
- The produced anaerobic- and aerobic excess sludge is subjected to secondary digestion in a dedicated heated digester: the released nitrogen is returned to the aerobic system;
- The nitrogen concentration in the raw sewage N_{ti} is 50 mg N $\cdot$ l^{-1};
- Effluent TKN concentration: $N_{ad} = N_{oe} = 1$ mg N $\cdot$ l^{-1};

- COD composition of the raw sewage:
  - Case 1: $f_{ns} = f_{np} = 10\%$ and $f_{sb} = 25\%$;
  - Case 2: $f_{ns} = 5\%$; $f_{np} = 15\%$ and $f_{sb} = 25\%$;
- The sulphate concentration is $70\ mg \cdot l^{-1}$, which is completely converted into H_2S. The anaerobic effluent is pre-aerated at an applied molar O_2/S ratio of $1\ mg\ O_2 \cdot mg^{-1}\ S$;
- All soluble organic biodegradable COD (i.e. excluding the COD associated to reduced sulphur compounds) in the anaerobic effluent is considered easily biodegradable;
- Nitrifier kinetics: $\mu_m = 0.4\ d^{-1}$; $b_n = 0.04\ d^{-1}$; $K_n = 1.0\ mg\ N \cdot l^{-1}$;
- Denitrification rate: $K_2 = 0.12\ mg\ N \cdot mg^{-1}\ X_a\text{-}VSS \cdot d^{-1}$; $f_{max} = 50\%$;
- The bypass fraction of raw sewage was varied between 20 to 50%.

As expected, the COD/N ratio in the raw sewage has a significant effect on the required minimum sludge age in the activated sludge system and the required bypass flow. Furthermore, when a large enough proportion of the raw sewage is bypassed to the aerobic post-treatment system (e.g. 50% for the conditions in the example), this will ensure that the required nitrogen effluent limits are met, even for a COD/TKN ratio in the raw sewage of 12. The resulting minimum required sludge age of 12–14 days is in itself acceptable. However, the selection of a large bypass flow approximately doubles the COD load to the aerobic post-treatment system and requires a proportional increase of the aerobic volume and the oxygen demand. Hence, although the UASB volume would be reduced as well, the overall attractiveness of combined anaerobic-aerobic treatment is reduced.

The reduction in the COD load to the aerobic post-treatment system for a bypass of 50% and for a COD removal efficiency of 65% in the UASB is equal to $(1–0.5) \cdot 0.65 = 32.5\%$. This is comparable to the degree of COD removal obtained in a primary settler, which is a much simpler (and cheaper) unit operation. Thus in general, whenever a bypass of more than 35% is required, the combination of UASB and aerobic treatment ceases to be attractive: the reduction in COD load is then equal to $(1–0.35) \cdot 0.65 = 42\%$. For the conditions specified in the examples shown in Figure 13.26 (i.e. at 20°C), this situation occurs for a COD/N ratio in the raw sewage lower than 14.

Table 13.20 summarises acceptable solutions for the two cases considered. To indicate the effect of temperature, a third case is added, which is equal to Case 1 (i.e. $f_{ns} = f_{np} = 10\%$), but now calculated at a temperature of 25°C. The beneficial effect of operating at higher temperature is due to the fact that nitrogen removal efficiency increases with temperature, as both nitrifier growth rate and denitrification rate are temperature dependent.

Furthermore there is a clear influence of the raw sewage composition. In Table 13.20 it can be observed that an increase in f_{ns} (for the same total non-biodegradable fraction of 20%) is detrimental to nitrogen removal performance. The excess sludge production in the UASB will be reduced and consequently less organic nitrogen is removed. From the results presented in Figure 13.26 and Table 13.20 it can be concluded that for $T = 20°C$ in all cases considered a suitable nitrogen removal performance can be obtained for $COD/TKN > 14$, for raw sewage bypass percentages ranging from 20 to 35%. Operation at 25°C significantly reduces the sludge age and hence the size of the aerobic post-treatment system.

13.5.2.2 Anaerobic digestion with reduced methanogenic efficiency

In the previous sections it was tacitly assumed that the methanogenic efficiency in the anaerobic digester was high, i.e. resulting in a low concentration of volatile fatty acids in the effluent. However, depending on the operational conditions (mainly temperature and sludge age) and the process configuration (separated acid and methanogenic fermentation), the methanogenic fermentation efficiency may be low or even nonexistent. Even under these conditions there may be a significant advantage in applying anaerobic pre-treatment, or rather acid fermentation. The COD produced in the anaerobic reactor will be easily biodegradable and may be used to enhance nutrient removal in activated sludge processes: i.e. to effect high rate denitrification and biological phosphorus removal.

Table 13.20 Required values of the raw sewage bypass and the sludge age in the aerobic post-treatment system (to comply with $N_{te} \leq 10$ mg N $\cdot$ l^{-1}) as function of the COD/N ratio in the raw sewage for the conditions specified in Figure 13.26. Case 3 indicates the effect of operation at higher temperature

COD/TKN	Case 1 (20°C) ($f_{ns} = f_{np} = 0.1$)		Case 2 (20°C) ($f_{ns} = 0.05$; $f_{np} = 0.15$)		Case 3 (25°C) ($f_{ns} = f_{np} = 0.1$)	
	% bypass	R_s	% bypass	R_s	% bypass	R_s
13	40%	15.6	40%	12.1	40%	9.8
14	35%	13.2	35%	11.8	35%	9.5
15	30%	12.5	30%	11.6	30%	9.6
16	25%	12.5	25%	11.5	30%	8.8
17	20%	13.2	25%	10.5	25%	9.0
18	20%	11.6	20%	10.6	20%	9.4

Figure 13.27 shows the flow scheme of an activated sludge process in which acid fermentation is applied to increase the nutrient removal capacity. In the first pre-treatment step the anaerobic sludge in the hydrolytic rector is used to convert soluble- and part of the particulate influent COD into volatile fatty acids. This acid fermentation may be enhanced by subjecting the excess sludge from the hydrolytic reactor to further hydrolysis in a sludge digester and by applying heating to enhance the conversion efficiency. Addition of metal salts may be required to prevent solubilisation of the phosphate released to the liquid phase.

The fermented wastewater is introduced into the activated sludge process, for example a UCT configuration, where the nutrients are removed by the same mechanisms as discussed in Chapters 5 and 7, i.e. nitrification-denitrification for nitrogen removal and luxury uptake by bio-P organisms in the case of phosphorus removal. If insufficient VFA are available for complete bio-P removal, it may be supplemented by the addition of metal-salts.

As a result of the anaerobic pre-fermentation, the composition of the influent COD is enhanced, so that the phosphorus and nitrogen removal capacity are higher than in a conventional nutrient removal activated sludge system treating raw or settled sewage. Hence, the post-treatment nutrient removal system will perform better at lower temperatures and/or at shorter sludge ages than conventional nutrient removal plants. Another advantage of this hydrolytic anaerobic pre-treatment is that the non biodegradable particulate material is largely retained in the pre-fermenter, so that the accumulation of inert particulate solids in the activated sludge system is much smaller and the reactor volume may be reduced.

A practical problem in regions with a cold to moderate climate is that heating of the influent stream is not economically feasible. Furthermore, at lower temperatures the rate of the hydrolysis process tends to be smaller than the rates of the processes of acidogenesis, acetogenesis and methanogenesis. After solubilisation of the particulate organic material, the conversion to methane is almost unavoidable, unless the activity of methanogenic bacteria is suppressed by the addition of acid to lower the pH or by the introduction of a limited amount of oxygen, which is toxic to methanogens.

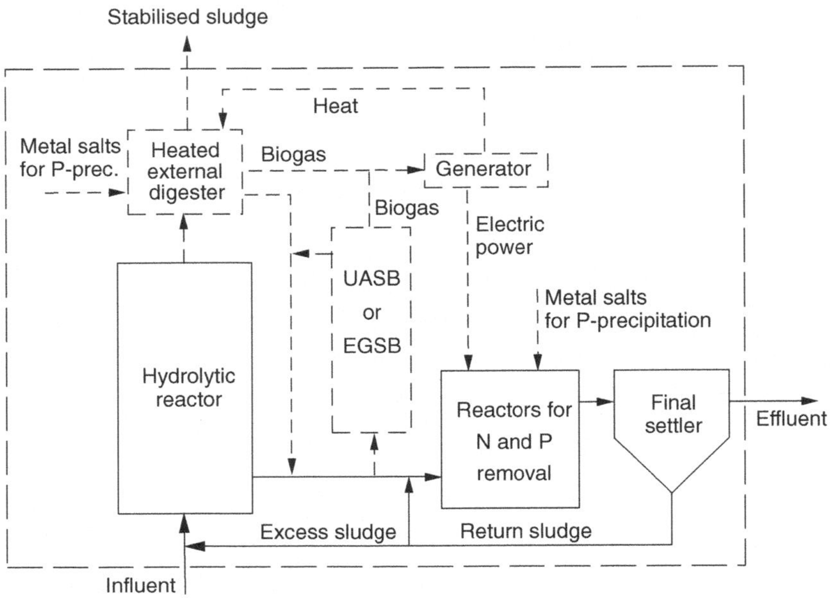

Figure 13.27 Hydrolytic pre-treatment as a means to enhance the nutrient removal capacity of the activated sludge process

Although substantial research effort has been allocated in the 1990s to the pre-fermentation of raw and pre-settled sewage at cold to moderate temperatures (between 8 and 15°C), the results were disappointing and the concept has never been applied in full-scale. Refer also to the section dealing with the improvement of substrate availability for nutrient removal in Chapter 7. Another disadvantage of the configuration in Figure 13.27 is that the influent COD is converted but not removed in the pre-treatment zone, so that the entire organic load must be treated aerobically. However, it is possible to create an intermediate situation in which part of the influent is pre-fermented while the rest receives full anaerobic treatment: this scenario is indicated in Figure 13.27 as well (Coelho et al., 2005).

13.5.2.3 Application of innovative nitrogen removal configurations

In Chapter 6 the anaerobic ammonium oxidation process (Anammox) was discussed. When the Anammox process is combined with nitritation, it becomes feasible to remove nitrogen without COD requirement and at a significantly reduced oxygen consumption. So far at full scale the combined process has only been demonstrated for high strength nitrogen wastewater streams (300–1500 mg $N \cdot l^{-1}$), for example digester reject water. However, there is no fundamental reason why it could not be applied to low strength (or low temperature) wastewater, for example for the post-treatment of anaerobic effluent where the bulk of the COD has already been removed.

The theoretical concept is schematically presented in the form of a mass balance in Figure 13.28. This figure was constructed based on the following data: $f_{ns} = f_{np} = 0.1$; $f_{sb} = 0.25$; $T = 20°C$; $SO_4^{2-} = 50 \text{ mg} \cdot l^{-1}$; $f_{max} = 0.6$. The COD and nitrogen concentrations in the different streams are all expressed in mg per litre influent. In a two reactor configuration, part of the anaerobic effluent is subjected to nitritation, i.e. the oxidation of ammonium to nitrite. Subsequently the nitrite-rich stream is contacted in the Anammox reactor with the remainder of the anaerobic effluent. Alternatively, all of the anaerobic effluent can be treated in a single reactor configuration. Both alternatives have their pros and cons. Whereas the single reactor will be less expensive, the two reactor configuration might allow for a higher degree of control over the system, for instance in the control of the nitrite to ammonium ratio.

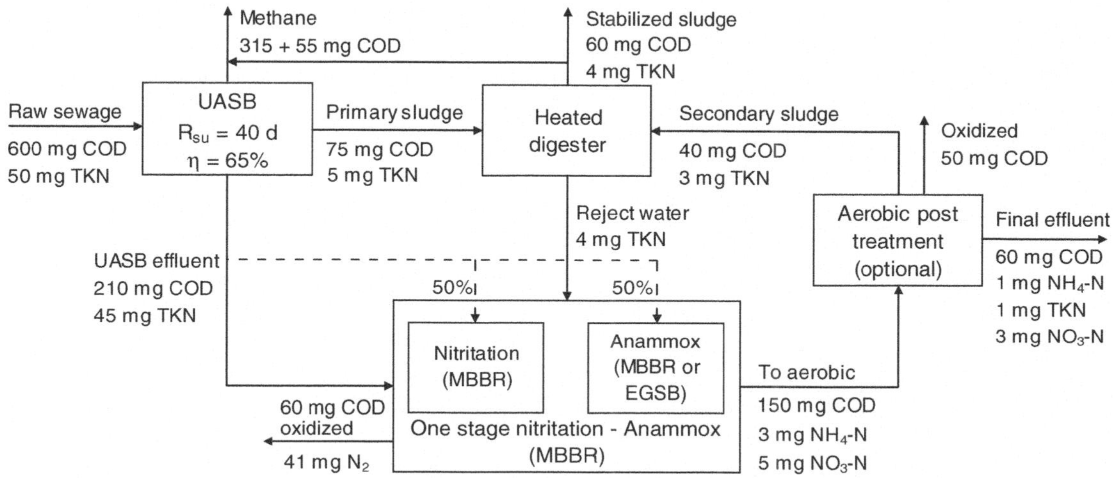

Figure 13.28 Sustainable wastewater treatment by combination of anaerobic pre-treatment and innovative nitrogen removal without COD requirement. Mass balance shows the concentrations of COD and N per litre influent

While this combination of UASB and sustainable nitrogen removal is very promising due to the possibility to produce a high quality effluent at low operational costs, the feasibility of this concept has yet to be proven in practice. But as more and more wastewater engineering firms develop and implement their own variant of the combined nitritation/Anammox process, the likelihood of success will increase.

At present, when nitrogen removal is required this effectively prevents anaerobic pre-treatment from being considered, due to the unfavourable COD/N ratio in the anaerobic effluent. When a bypass is considered to supply the required COD, the UASB treatment option quickly ceases to be attractive, unless the sewage contains a high COD/N ratio. Thus, should it it become feasible to achieve nitrogen removal without the need of organic material, this will be a tremendous push towards the further dissemination of anaerobic systems. The combination of anaerobic pre-treatment followed by nitritation – Anammox will allow for truly sustainable wastewater treatment, although initially this treatment option will be confined predominantly to regions with a warm climate. One could consider hydrolytic pre-treatment or a high loaded UASB combined with heated anaerobic digestion of the UASB excess sludge to allow application of the basic concept in regions with a colder climate, but again this will require a substantial R&D effort.

As an intermediate step the system configuration presented in Figure 13.29 might be considered. The COD/N ratio in the feed to the aerobic post-treatment system is increased by the removal of the nitrogen released during anaerobic digestion by means of the nitritation-Anammox process. In Figure 13.29 the quantitative effect on nitrogen removal is indicated. As can be observed, the treatment of the digester reject water reduces the nitrogen concentration in the feed to the aerobic system by almost 10%. In the case of the example, for a TKN/COD ratio of 12 and $f_{max} = 0.6$, this allows the nitrogen discharge limit of $<10\ mg\ N \cdot l^{-1}$ to be met for a bypass percentage of 35%. This certainly enhances the attractiveness of combined anaerobic-aerobic treatment. All elements in this configuration have already been implemented at full-scale.

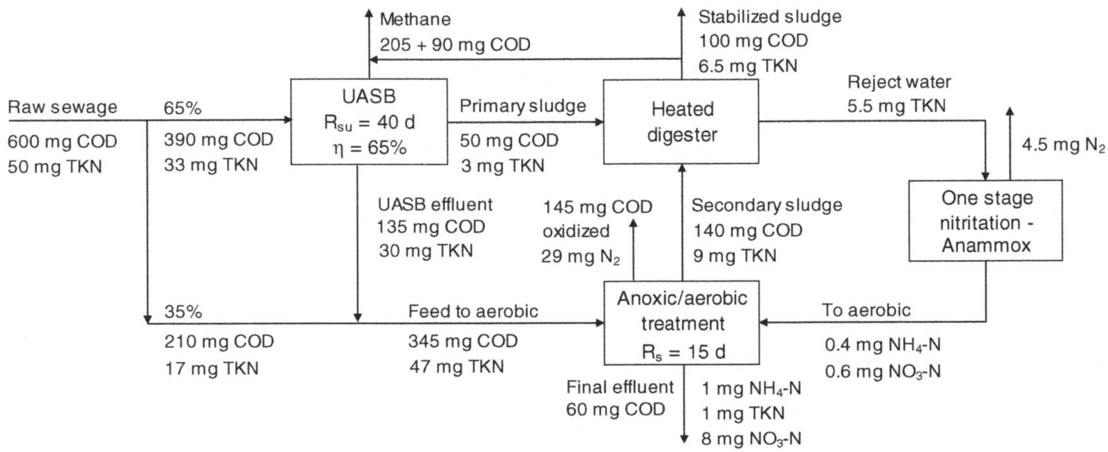

Figure 13.29 Intermediate solution: enhanced nitrogen removal capacity in a classic anaerobic-aerobic configuration using sustainable nitrogen removal. Concentrations of COD and N are expressed in mg per litre influent

13.5.3 Future developments

13.5.3.1 Two stage anaerobic digestion

When treating wastewater with a large particulate organic fraction like sewage, it may be advantageous to apply a two stage anaerobic process. In the first stage, the particulate organic matter is entrapped and partially converted into soluble compounds, which are then digested in a subsequent second reactor. This configuration is shown in Figure 13.30.

The first, hydrolytic reactor will typically contain a flocculent sludge and is operated at a relatively low upflow velocity. Most of the particulate influent matter is retained in the flocculent sludge layer and will be partially converted into soluble products by means of the process of hydrolysis. Methanogenesis will not or only partly develop in the hydrolytic reactor, because the environmental and operational conditions are not very suitable for this process. Moreover, the development of acid fermentation may tend to depress pH to a value below the optimal range. Due to the accumulation of solids in this reactor and the fact that only part of the entrapped organic material will be hydrolysed, a significant mass of excess sludge will have to be withdrawn from the reactor. As a result, the sludge age will remain relatively low and consequently the slow growing methanogenic bacteria cannot develop.

In the effluent of the hydrolytic reactor the organic matter will be present predominantly in a soluble form and thus can be treated conveniently in an EGSB reactor. A disadvantage of the two stage system may be the excessive accumulation of solids in the hydrolytic reactor, which will occur when the hydrolysis rate in this reactor is too low, as is the case at low temperatures. Under these conditions, the anaerobic sludge retention time may become too low to achieve the required degree of liquefaction to stabilise the excess sludge.

However, even in such a case it is still possible to produce a satisfactory excess sludge quality by applying sludge stabilisation in a heated digester. In such an auxiliary hydrolytic digester, the stabilised sludge can be separated in a liquid-solid separation step while the liquid phase, enriched with soluble organic material, is injected into the effluent of the unheated hydrolytic reactor and submitted to methanogenic treatment in the second reactor, which is an EGSB.

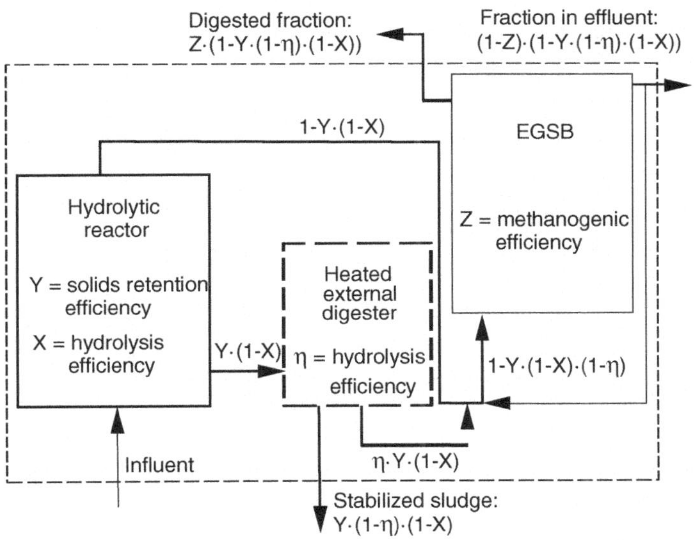

Figure 13.30 Schematic representation of a two stage anaerobic treatment system with auxiliary sludge digestion

Figures 13.30 and 13.31 show the configuration of such a two stage anaerobic digestion system, including the distribution of the influent organic matter over three fractions in such a two stage system: i.e. converted into (I) excess sludge, (II) methane and (III) remaining in the effluent. The effect of the temperature with respect to the variation of these fractions is indicated qualitatively in Figure 13.31.

13.5.3.2 Psychrophilic anaerobic wastewater treatment

So far, most engineered wastewater systems are operated in the mesophilic range (15–40°C), whereas currently there is a trend to apply thermophilic treatment (45–55°C) to high strength wastewaters or slurries (e.g. thin stillage from bio-ethanol production). On the contrary, psychrophilic treatment has never received a lot of attention due to the low growth rate of methanogenic bacteria at temperatures <10–15°C. The division between mesophilic and thermophilic bacteria is very clear, because mesophilic bacterial rapidly decay at temperatures higher than 40°C, as the genetic modifications required to protect their enzymes at high temperatures are lacking. On the other hand it has always been a bit unclear whether pyschrophilic bacteria really existed or whether observed psychrophilic activity actually involved mesophilic bacteria adapted to lower temperatures.

However, recently the research group of Curtis from the University of Newcastle presented some interesting findings. Methanogenic bacteria were identified from the sediments of arctic swamps that exhibited a similar methanogenization rate as their mesophilic counterparts. Whereas it proved perfectly possible to achieve a high methanogenisation rate at 20°C using the psychrophilic sludge, the opposite, i.e. a high methanogenisation rate at 4°C using the mesophilic sludge was not possible, even after a prolonged period of testing.

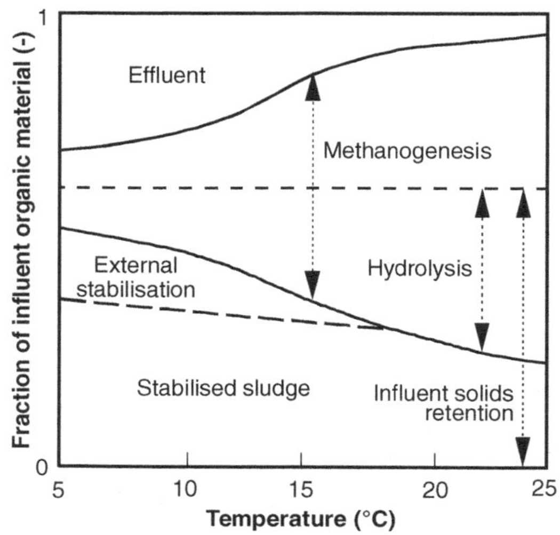

Figure 13.31 Influence of temperature on the division of the influent organic material into methane, sludge and the effluent

These phenomena were explained by the application of advanced population dynamic models (Curtis et al., 2006; Sloan et al., 2007). Basically, any operating biological reactor can be considered as a large established community receiving a steady flux of immigrants that are present in the influent. As the number of "immigrants" is very small compared to the number of "residents", the impact they have on the microbial composition will be small.

The hypothesis is thus that if a mesophilic system is fed with cold wastewater (even for a long time), the mesophilic bacteria will still tend to dominate the microbial community. Hence a reactor started up with mesophilic sludge is very unlikely to become a good psychrophilic reactor. On the other hand, if the reactor is started up with pschychrophilic sludge instead it is likely it will remain psychrophilic even when periodically subjected to conditions favouring mesophilic bacteria (for instance higher temperatures during summer time). If this theory is true, it might also have implications on the theories and practices used for starting up low-growth bacterial systems (Anammox, granular sludge etc).

Although the use of psychrophilic bacteria might allow the application of anaerobic treatment to colder areas (or dispense with the need for heating), much more research effort is required before all of this can be of practical use. One important issue is whether there is also psychrophilic hydrolysis, being the rate limiting step rather than methanogenesis in the treatment of complex substrates such as the particulate organic material present in raw sewage.

13.6 ANAEROBIC TREATMENT OF INDUSTRIAL WASTEWATER

As discussed in the previous sections, the application of anaerobic treatment to municipal sewage has always been rather difficult, due to the diluted nature of the wastewater, the high fraction of suspended solids it contains and the unfavourable temperature in a large part of the world. On the other hand, anaerobic treatment of *industrial* wastewater has proven an exceptional success, precisely because a lot of industrial wastewaters contain an elevated COD concentration and a high temperature. The latter is due to the fact that the factory wastewater originates from industrial processes that often involve heating as a process step. Even if the wastewater is outside of the recommended range for mesophilic digestion (25–37°C), it can often be heated with the energy released upon combustion of the produced biogas. From the energy content of methane (50,400 $MJ \cdot kg^{-1}$), it can be calculated that one kg of methane has the potential to heat one cubic meter of wastewater by 10.2°C, when a heating efficiency of 85% is assumed. Or alternatively, 98 g of methane is sufficient to heat a cubic meter of wastewater by one degree Celsius. Table 13.21 shows additional examples for different temperature intervals. Thus it is not surprising that it is often possible to operate the anaerobic system at or close to the optimum mesophilic temperature range (35–37°C).

Table 13.21 Heating potential of high strength wastewater streams

Required ΔT (°C)	Required methane production (g $CH_4 \cdot l^{-1}$)	Digested COD concentration (g $COD \cdot l^{-1}$)
1	0.1	0.4
5	0.5	2.0
10	1.0	3.9
15	1.5	5.9
20	1.95	7.8
25	2.45	9.8
30	2.95	11.8

Another factor that favours the application of anaerobic treatment to industrial wastewater is that these often have a high COD/N ratio or are even nutrient deficient. In the former case, full denitrification requires only a small bypass, whereas in the latter case nitrogen (and phosphorus) has to be added. This is very different for municipal sewage treatment, where the requirement for nitrogen removal renders the application of anaerobic pre-treatment very difficult. A final advantage is the fact that due to the low flow rate (and the high temperature) the fraction of methane lost with the effluent is insignificant. Consider a reactor operated at 37°C and receiving a wastewater with 10,000 mg $COD \cdot l^{-1}$, of which 85% can be removed. Assuming a yield of 0.05 mg $VSS \cdot mg^{-1}$ COD, the methane production can be calculated as $(1–1.5 \cdot 0.05) \cdot 0.85 \cdot 10{,}000/4 = 1966$ mg $CH_4 \cdot l^{-1}$. The effluent methane concentration at 37°C and 80% CH_4 content in the biogas is 14.2 mg $CH_4 \cdot l^{-1}$. Thus the non-recovered methane fraction is equal to $1–14.2/1966 = 0.7\%$, which is very low indeed. It is therefore not surprising that for industrial wastewater an anaerobic pre-treatment step is almost always implemented, if there are no specific reasons not to do so, for example due to toxicity.

The fact that conditions are so much more favourable for industrial wastewater allowed for the development of high rate anaerobic reactors, based on the discovery of granulated sludge. An interesting feature of the UASB process is that under specific conditions a granular type of sludge with a diameter of 1 to 5 mm can develop. These granules have a high density and mechanical strength and combine a high settling velocity with a high specific methanogenic activity. These features allow a high and active biomass concentration to be maintained in the reactor, up to 50–80 g $TSS \cdot l^{-1}$ in the granular sludge bed, which in turns makes it possible to apply a high volumetric loading rate of up to 10–15 g $COD \cdot m^{-3} \cdot d^{-1}$. Figure 13.32 shows microscopic and electro-microscopic of anaerobic sludge granules and the bacteria of which it consists. The granulation process has been studied by several researchers (De Zeeuw, 1987 and Hulshoff Pol, 1986 and 1987) and it seems that the formation of granules is related to the selection pressure resulting from the operational conditions prevailing in the UASB reactor (e.g. a high upflow velocity) and to the characteristics of the treated wastewater, where the most important one is a low content of suspended solids. So far, granulation has not been observed in any of the existing full-scale UASB reactors treating raw sewage: in all cases a flocculent type of sludge has developed. Probably this is due to the low upflow velocity and because a large fraction of the influent COD consists of suspended solids. Nevertheless, excellent COD and TSS removal efficiencies were achieved in these reactors, demonstrating that sludge granulation certainly is not a prerequisite for successful anaerobic sewage treatment.

Figure 13.32 Biogas producing sludge granules (left) and electron microscopic view of anaerobic bacteria inside biomass granule (mix of methanogenic and acidogenic bacteria). Courtesy of Biothane Systems International

The finding that granular sludge could be cultivated under specific conditions for soluble wastewaters, led to the development of the EGSB (Expanded Granular Sludge Bed) reactor (Van der Last, 1991). The EGSB is characterised by the fact that the granular sludge bed is operated in expanded mode as a result of the high upward velocity that is applied, i.e. 3 to 5 $m \cdot h^{-1}$ compared to less than 1 $m \cdot h^{-1}$ in a UASB reactor. The EGSB reactor has been shown to be quite efficient in the removal of soluble organic matter, even at lower temperatures, which can be attributed to the intensive contact between the incoming organic matter and the sludge granules. Compared to the granulated UASB systems, the volumetric loading rate of EGSB reactors is even higher, up to 25 $kg\ COD \cdot m^{-3} \cdot d^{-1}$. Due to the reduction in volume and footprint the EGSB reactors have replaced the UASB systems in popularity, despite the fact that due to poorer solids retention the COD removal efficiency is actually slightly lower. However, there certainly remains a market for industrial UASB systems, especially for wastewater with a higher content of suspended solids. Due to the increase in hydraulic- and gas loading rates in the granular UASB and EGSB reactors the design of the GLS separator had to be modified. Some examples of current designs are shown in Figure 13.33.

Table 13.22 summarizes some key design characteristics of both types of reactors. Both the high rate UASB and EGSB have been successfully implemented at several hundreds of industrial installations. The higher tolerance to toxic components in the influent results from the dilution of influent resulting from the recirculation of anaerobic effluent over the reactor, which is required in order to maintain the high upflow velocity that drives the selection on granular bacteria. As to reactor pressurisation, typically the large surface area of the UASB makes it difficult and expensive to pressurise the entire reactor, contrary to the much smaller area of the EGSB. The risk of odour emissions is thus prevented in an EGSB reactor.

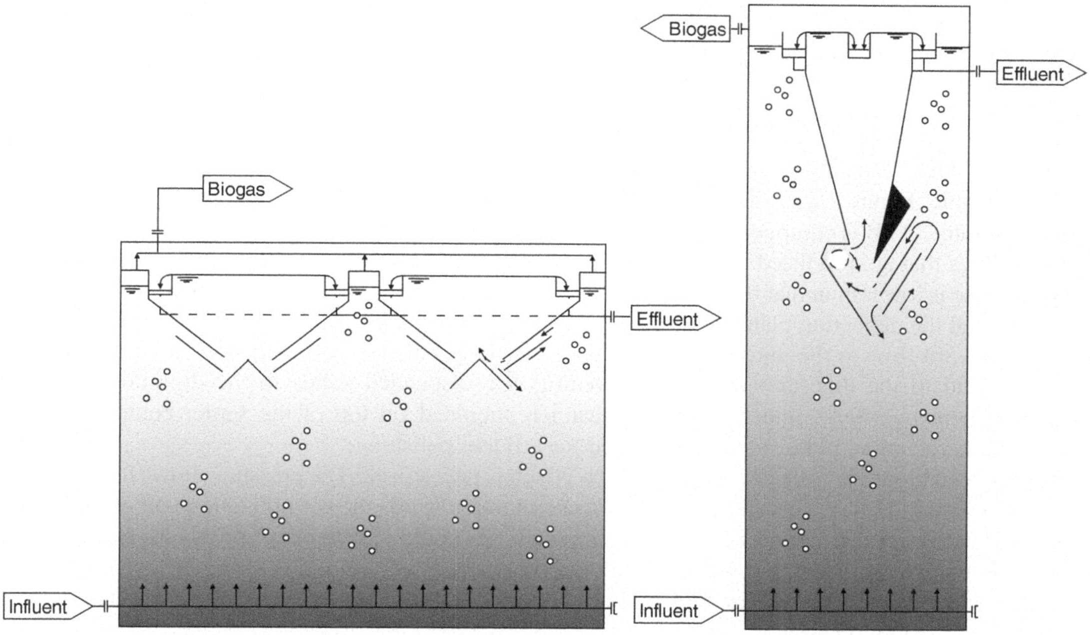

Figure 13.33 Schematic representation of an industrial (granulated) UASB reactor (Biothane® UASB – left) and an EGSB reactor (Biobed® EGSB – right). Additionally, Figure A3.4 shows a detailed artist impression of an Biobed EGSB reactor including internals.

Table 13.22 Comparison of typical values of design parameters of two high rate anaerobic systems: granular UASB- and EGSB reactors

Parameter	UoM	EGSB	UASB
Reactor upflow velocity	$m \cdot h^{-1}$	3–5	<1.5
Settler upflow velocity	$m \cdot h^{-1}$	5–10	1.0–1.2
Gas velocity reactor	$m \cdot h^{-1}$	<5	<1.0
Reactor height	m	12–18	6
Height of sludge bed	m	3–5	1–2
Volumetric loading rate	$g\ COD \cdot l^{-1} \cdot d^{-1}$	15–25	5–12
Tolerance for influent TSS	(–)	$<500\ mg \cdot l^{-1}/<10\%\ S_{pi}$	$<20\%\ S_{pi}$
Tolerance for toxicity[(1)]	(–)	high	medium
Reactor pressurised	(–)	yes (10–50 mbarg)	only gas hoods (10–50 mbarg)

Note (1) For organic biodegradable toxic components

The UASB market is characterized by many suppliers, some of them operating in local markets but with some global players as well. On the other hand, the market for expanded granular bed systems is much more exclusive as it is dominated by few companies (mainly Biothane (Biobed® EGSB) and Paques (IC system) with some niche players as well (e.g. Voight in the paper industry). The main reason is the increased technical complexity of the reactor internals compared to those for UASBs, for instance the GLS separator and the influent distribution system. Biothane and Paques have both patent protected designs and supply the key internal parts.

Figure 13.33 shows a schematic representation of the two main reactor types used by Biothane: the granulated UASB (Biothane® – left) and the granulated EGSB (Biobed® – right). Due to the high reactor upflow velocity applied in the Biothane® UASB reactor, especially in the constricted aperture area between two GLS separators, a conventional GLS design as used for the municipal UASB is no longer applicable. Figure 14.16 shows an example of a full-scale Biobed® reactor.

The high rate UASB is equipped with deflection baffles mounted parallel to the gas collection plates. As the gas bubbles that are produced can only travel upwards and the density of the rising gas-liquid mixture below the plate is lower than that of the gas-free liquid above the plate, a recirculation flow (gas lift) will be induced around the deflection plate, depositing the entrained suspended solids back on top of the deflector. As the downward flow to the digestion zone is generally larger than the effluent flow, there will be a net return of liquid to the digestion zone, which returns the suspended solids to the digestion zone. The polished effluent overflows into the effluent channels mounted on top of the settler compartment. To improve solids retention, plate packs can be used for effluent polishing.

The produced biogas is collected in the gas box, which is pressurised. The pressure allows for the flow of sour (i.e. containing H_2S) biogas through a gas treatment unit towards the gas utilization unit or to the flare. The UASB reactor itself is not pressurized, which potentially could lead to odour problems. Hence the gas volume is normally ventilated and sent to a vent gas treatment unit.

Another feature that distinguishes industrial UASB reactors from municipal UASB reactors is the design of the influent distribution system. The municipal UASB is gravity-fed and relies on an elaborate system of influent divisions culminating in the final distribution box mounted on top of the reactor, from which individual feed lines extend to the bottom of the reactor. On the other hand, industrial reactors always have a pumped feed through a distribution system mounted on the bottom of the reactor. This distribution system consists of a header system connected to influent feed pipes extending into the reactor with influent nozzles placed at regular intervals.

The change to a pumped feed allows for much more efficient distribution and efficient mixing and is made possible due to the nature of the wastewater (less fibres, debris etc.) and the much smaller flow rate, which allows for more extensive pre-treatment if required.

The Biobed® EGSB reactor is basically a high rate variant of the granular UASB. The GLS separator is optimised for operation under very high liquid and gas loading rates. Contrary to the UASB not all of the surface area is occupied by GLS separators, as there is no integrated gas collection box in the settler. Due to the high volumetric organic loading rate, the volume of an EGSB is quite small compared to an UASB reactor. This is even more accentuated in the case of the surface area, due to the increase in height that follows from the high upflow velocity. This allows the total reactor to be covered and pressurised, thereby eliminating the risk of odour emissions. In Figure 13.33 it can be observed that the number of deflection plates is increased. Furthermore on the left hand side of the settler a dedicated zone for energy dissipation is created, which allows entrained granules to escape the upward flow and settle back to the digestion compartment.

An EGSB reactor is not suitable for the removal of particulate organic matter due to the high upflow liquid velocity. The influent suspended solids are not (or only partly) retained by the granular bed nor by the GLS settler and will leave the reactor with the effluent. As methanogenic bacteria tend to grow on the suspended solids as well as on the granules, as a consequence part of the methanogens are washed out of the reactor together with the suspended solids. Under these circumstances, it will be very difficult to maintain a sufficient quantity of granular sludge in the reactor.

Figure 13.34 shows a typical configuration of an industrial anaerobic wastewater treatment plant, in this case an EGSB. However, it very similar to the UASB configuration. Additionally to the units shown, a number of pre-treatment steps might be required to remove the following constituents from the wastewater:

- Free oil, grease & fat: oil trap, oil separator or dissolved air flotation (DAF) unit;
- Sand and grit: grit removal unit;
- Fibrous material: static or vibrating screens, drum filters;
- Suspended solids: primary settler or DAF.

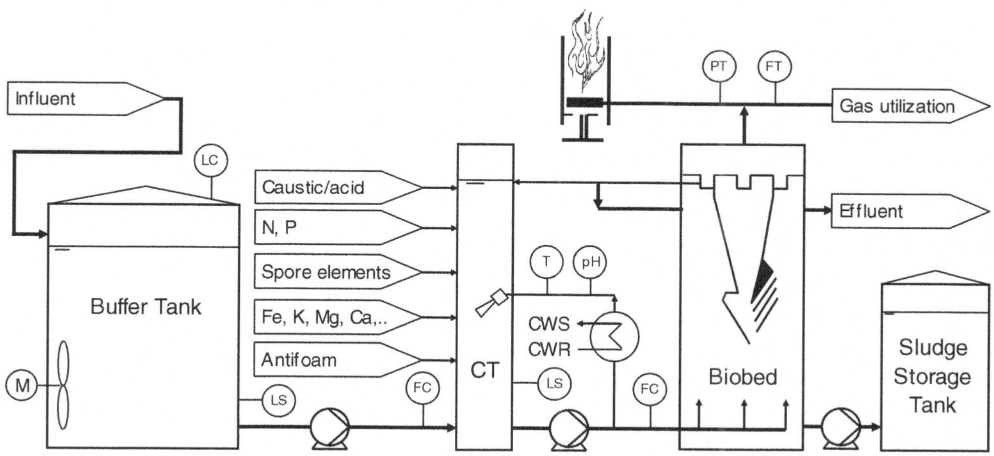

Figure 13.34 Typical configuration of an industrial high rate anaerobic wastewater treatment plant (Biobed® EGSB configuration). Figure A1.7 shows a picture of two Biobed EGSB reactors plus conditioning tank.

The raw wastewater enters into a buffer tank, which equalises both flow- and concentration fluctuations. The buffer tank should also provide suffient buffer volume to handle peak flows, which requires it to be operated only partially full. Proper selection of the size of the buffer tank is important: if it is too small the biological reactors are subjected to large variations in concentration and flow, but on the other hand if it too large then acidification might occur in the buffer tank. While sometimes this is actively encouraged in order to enhance biodegradability, it also results in the production of a significant mass of suspended solids. These solids will not be retained in the granulated sludge systems and because part of the substrate has been removed it will decrease granule production as well.

From the buffer tank the wastewater is sent to the conditioning tank (CT), a small but nonetheless crucial unit with a typical residence time of 10–20 minutes. In the conditioning tank the wastewater is contacted with anaerobic effluent recycled from the EGSB reactor. There are two reasons for this: (I) it reduces the demand for caustic (or acid) for pH correction due to recirculation of the bicarbonate buffer produced in the anaerobic reactor and (II) it reduces (dilutes) the concentration of toxic biodegradable components if present. Apart from the addition of caustic (NaOH) or acid (HCl) for pH control to a value between 6.8 to 7.2, the following chemicals might have to be added if a nutrient deficient wastewater is treated:

- The macro-nutrients nitrogen and phosphorus, possibly also Fe, K, Mg, Ca and S that are minor constituents of bacterial cells;
- Micro-nutrients (trace elements and vitamins) that are required for the synthesis or proper functioning of enzymes;
- If the wastewater has a tendency to induce foaming, then antifoam can be added as well.

Heating is applied (or sometimes cooling) to adjust the temperature to the optimal value for mesophilic operation of 35–37°C. Often a heat exchanger is placed in the recirculation (mixing) loop around the conditioning tank, or alternatively direct steam injection (in the conditioning tank!) can be applied. If the required degree of heating is very large, it is advantageous to install an influent-effluent heat exchanger to pre-heat the buffered wastewater in order to reduce the energy requirements for heating.

The reactor itself has been discussed earlier in this section. The produced excess sludge granules are taken from the reactor and stored in the sludge buffer tank. Under the right conditions this sludge can be kept for many months, which allows for a rapid re-start in the event of a serious process upset (e.g. toxic shock). Additionally (or alternatively) the produced granular biomass can be sold commercially as seed sludge for start up of new installations and in fact is thus a valuable commodity, compared to flocculent excess sludge, which constitutes a significant part of the operational cost of a municipal sewage treatment plant.

The collected biogas is sent to a gas treatment unit for H_2S removal (for instance in a caustic scrubber or biogas scrubber) and drying, prior to utilization in gas boilers, gas motors or cogeneration units. Alternatively or in case of process upsets, the biogas can be flared. To conclude this chapter, Table 13.23 provides typical COD loading rates for high-rate anaerobic reactors in different sectors of industry.

Table 13.23 Typical organic loading rates (in kg COD $\cdot$ m^{-3} $\cdot$ d^{-1}) for different industry sectors (based on total reactor volume)

Industry sector	Selected reactor type	Organic loading rate
Paper – pulp	EGSB	10–15
Paper – recycle paper	EGSB	15–20
Bakery yeast	UASB/EGSB	7.5
Beer brewery	UASB/EGSB	7.5–15
PTA	EGSB	15
Soda drinks/fruit juice	EGSB	15–20
Potatoes (chips and fries)	UASB	10
Bio-ethanol	EGSB	15–20

Chapter 14

Integrated cost-based design and operation

14.0 INTRODUCTION

In the preceding chapters a steady state model of the activated sludge system has been presented, which can be used for optimised design and operation of wastewater treatment plants. Obviously, the design of activated sludge systems has already received a large amount of research attention. Several design software packages have been developed, most of which were based on the Activated Sludge Models from the IWA (Henze et al. 1986, 1994 and 1998). The theoretical concepts used in the IWA models are for a large part based on the research done at the University of Cape Town (UCT) in South Africa (Water Research Commission, 1984). The same research also forms the basis of the general and ideal steady state models presented in this book. However, the format of the IWA models is not particularly suitable for application as a design tool. One should consider that the main objective of these dynamic models is to increase our knowledge of activated sludge system behaviour, for instance by allowing this behaviour to be simulated. For this purpose, a large number of variables and parameters are included. These are indispensable when studying system reactions to disturbances or to process control measures, but can be considered as unnecessary ballast from a design viewpoint. In fact, the IWA models are of such a complexity that an analytical optimised design solution is not possible.

An example is the dissolved oxygen (DO) concentration, which is included in the IWA models as one of many state variables, all having their own separate mass balance. Furthermore, the concentration of dissolved oxygen is included in nearly all reaction rate equations, in the form of a control function. This Monod type control function is either in the form $DO_l/(K_{DO}+DO_l)$ or $(K_{DO}+DO_l)/DO_l$ and thus "switches" a particular process on or -off, depending on the dissolved oxygen concentration. This is a crucial feature when simulating the behaviour of activated sludge systems. However, it is not required for system design, where sufficient availability of oxygen in the aerobic reactors and the absence of oxygen in anoxic- and anaerobic reactors are presupposed. Proper aeration control, including installation of sufficient aeration capacity and a suitable process control system, is there to ensure that oxygen will be present at the right time, location, quantity and concentration.

A second example is alkalinity. In the design of an activated sludge system, it is only important to know whether sufficient alkalinity will be available in the influent to maintain the pH at an adequate value, given the design requirements. If not, then provisions will be made for pH control.

Apart from model complexity, another issue is that most design models and -books only take into consideration the processes that develop in the biological reactor, such as metabolisation of organic material, nitrogen removal and biological excess phosphorus removal. The design of auxiliary units such as final settlers, thickeners and digesters is either excluded or it is treated as a separate, stand-alone unit operation.

The focus in this chapter is therefore on *integrated* cost-based design. First the basis of design for a project is discussed: this includes influent- and site characteristics as well as the relevant costing data, project limitations and the selection of an appropriate system configuration. In the second step an integrated design approach is presented that also includes the other main treatment units of the activated sludge system: i.e. final settler, sludge thickener, sludge digester and optionally pre-treatment units as the primary settler and the UASB reactor. The use of this integrated design approach will be demonstrated in several detailed examples, in a step-by-step determination of the optimal activated sludge system configuration.

In the optimised design of a new wastewater treatment plant the different parts of the systems can be optimised based on information or presuppositions about the composition and characteristics of the influent and the kinetic constants. However, the presented theory can also be applied to the operation of existing systems, to ensure the desired effluent criteria are met at minimum costs and under conditions of good operational stability. In many cases, existing activated sludge systems are considerably overdesigned when compared to the actual applied average flow and load. This is mainly caused by (I) the necessity to take into account peak organic and hydraulic loads and (II) a conservative design including safety margins, as there is uncertainty both with regard to expected loadings and to equipment performance. Some practical examples dealing with this subject will be discussed later in this chapter.

An important aspect in the design of wastewater treatment systems is the dynamic nature of the quality and quantity of the incoming wastewater flow. In many cases the wastewater flow increases in time, because the number of contributors and/or the fraction of the population served by a sewer system will increase. However, in a number of developed countries the total sewage flow is decreasing due to (I) stabilisation of the population, (II) the tendency to construct separate sewer systems for rainfall and municipal sewage in new residential areas, (III) the reduction of water consumption by industry and (IV) the construction of wastewater treatment plants by industry. However, whenever an increase is expected, a staged approach might be advantageous. This could be in the form of a modular approach with several parallel identical treatment "trains", constructed when additional treatment capacity is required and/or when funding becomes available. Another possibility is the construction in parts, first preliminary treatment followed at a later stage by secondary treatment, which will then deliver the desired effluent quality. The staged approach has the advantages that the initial investment is reduced and that possible shortcomings in system design or equipment performance can be corrected in the subsequent modules.

14.1 PREPARATIONS FOR SYSTEM DESIGN

Before starting the optimised design procedure for activated sludge systems four fundamental questions should have been answered:

- What is the design basis? For instance, what are the quantity and composition of the wastewater to be treated, environmental and climatic conditions etc.?
- What are the treatment objectives?

- What are the available treatment options?
- What are possible limitations and constraints?

14.1.1 The basis of design

In the design of wastewater treatment plants, often neither the quantity nor the characteristics of the wastewater to be treated are known and therefore these have to be estimated, for instance based on the size of the population and its expected growth, the fraction of the population served by a sewer system and the expected development of organic- and hydraulic contributions per capita. For any rational design approach, it will be necessary to attribute values to all of these parameters. In practice, sometimes there will be no alternative than to refer to "expert judgment" or "common sense" in order to come up with a value at all. In addition to quantitative data concerning the flow rate and the concentration of the wastewater, it will also be necessary to obtain data about the values of the kinetic parameters of the processes and of the settleability constants. Finally the relevant costing data have to be obtained, which will ultimately define the investment and operational costs.

14.1.1.1 Wastewater characteristics

A prerequisite for any rational process design is the availability of information about the quality and quantity of the wastewater to be treated. However, in many cases at the start of the design phase the wastewater to be treated does not even exist yet. In that case it will be necessary to refer to similar projects or to design literature to obtain the required data. In the case of the activated sludge system the most important parameters are:

(1) Influent flow rate;
(2) Concentration and composition of the organic material in the influent;
(3) Concentration of the inorganic suspended solids in the influent;
(4) Temperature profile;
(5) Value and stability of pH;
(6) Nutrient concentration;
(7) Especially in the case of industrial wastewaters: the presence of toxic components.

(1) Influent flow rate
In the case of domestic sewage, the flow rate depends on five main factors:

- Size of the contributing population;
- Per capita water use;
- Type of sewer system: combined or separate;
- Contribution of industries;
- Quality of the sewer network and the quantity of infiltration water;

Each of these factors tends to change in time: the population will increase or decrease and the per capita water consumption increases when economic conditions improve. The industrial contribution might increase together with the level of industrial production, but this might be partly offset by more efficient use of water in the production process (e.g. the use of water in the paper industry has been reduced by a factor of ten in the last fifty years). Furthermore it might be advantageous (or obligatory) for companies to treat their own wastewater. When the quality of the sewer system deteriorates, the infiltration rate of ground water into the sewer system will increase.

Apart from determining the average expected flow it is also necessary to consider daily and seasonal variations. This is particularly important for the determination of the OUR and its variation in time in the aeration tanks: the installed oxygen transfer capacity should be sufficient to meet the oxygen demand of the process under maximum loading conditions. When the daily variation in OUR is substantial, it will be necessary to install process control on the aerators in order to maintain a stable and adequate dissolved oxygen concentration in the reactor without having excessive aeration during periods of low load.

An alternative is the construction of a buffer tank to equalise variations in concentration and flow. Especially for industrial wastewaters, where large fluctuations in flow, temperature and/or pH occur, construction of a buffer tank may be required anyway to ensure operational stability. The average flow rate and its variation are also of major importance for the design of units in which size is primarily determined by hydraulic considerations, such as primary- and final settlers. On the other hand, the influence of the flow rate on the design of thickener and digester will be limited, as the design of these units is based on the organic loading rate.

(2) Concentration and composition of the organic material in the influent

Together with the influent flow rate, the concentration and composition of the organic material in the influent are the most important parameters in the design of an activated sludge system, as together they determine the organic load to be removed. The organic load determines the production of excess sludge and the oxygen consumption for the removal of carbonaceous material. The value of these variables also depends on the sludge age, the main operational parameter to be optimised.

Both the concentration and the flow rate of municipal sewage (and of most industrial wastewaters) tend to fluctuate in time. In many cases there is a distinct daily pattern as well. The amplitude of the variation depends on local topographic conditions and the residence time of sewage in the sewer system. In larger sewer systems there tends to be some equalisation. It is important to note that the organic load to the activated sludge system is defined as the average of the product of the variable flows and the associated concentrations and is (generally) *not* equal to the average flow multiplied by the average concentration. For example, in the case of municipal sewage the concentration in general tends to be highest at the hours of peak dry weather flow rate. Therefore the average organic load to the activated sludge system will be significantly larger than the product of average flow and concentration.

The opposite can also occur, especially for industrial wastewater, i.e. that maximum concentration and peak flow do not coincide. If the design is based on the product of maximum flow and concentration this might result in significant overdesign. It is therefore concluded that it is very important to acquire information about the expected daily profile of both influent flow and -concentration. As an example, in Table 14.1 and Figure 14.1 the influent flow and -concentration data of a typical municipal sewage treatment plant are listed. It can be observed that there is significant variation during the day in both flow and -load.

The average daily COD load can be calculated from Table 14.1 as $24 \cdot 2465 = 59{,}166\ \text{kg COD} \cdot \text{d}^{-1}$. However, when the product of the average values of Q_i and S_{ti} from Table 14.1 is used (bottom row), then the COD load is calculated as $24 \cdot 4691 \cdot 0.488 = 54{,}941\ \text{kg COD} \cdot \text{d}^{-1}$, which is significantly less than the actual value (–7.1%). In practice, almost invariably the flow-weighted average value of the product $Q_i \cdot S_{ti}$ will be larger than the product of the average values of Q_i and S_{ti}. This is due to the fact that at increasing flow rate the organic material concentration generally also increases. Especially during the morning peak flow, when the sewer system is flushed and accumulated solids are transported to the sewage treatment plant.

Table 14.1 Typical example of the variation of influent flow and -load over a day. The value of the correction factor is indicated as well

Time period (hrs)	Flow ($m^3 \cdot h^{-1}$)	[COD] ($mg\,COD \cdot l^{-1}$)	COD load ($kg\,COD \cdot h^{-1}$)	Correction factor (–)
00:00–02:00	3450	415	1432	1.27
02:00–04:00	2530	330	835	1.59
04:00–06:00	1920	305	586	1.72
06:00–08:00	2700	460	1242	1.14
08:00–10:00	5250	767	4027	0.69
10:00–12:00	7500	801	6008	0.66
12:00–14:00	6905	570	3936	0.92
14:00–16:00	4890	490	2396	1.07
16:00–18:00	5200	476	2475	1.10
18:00–20:00	7150	430	3075	1.22
20:00–22:00	4851	401	1945	1.31
22:00–24:00	3950	412	1627	1.28
Average hourly	4691	488	2465(1)	–

Note: (1) This is the value of the flow-weighted average COD load, i.e. not *equal to the product of average flow and average COD concentration, which is 2289 kg COD $\cdot$ d^{-1}, or 7.1% less than the flow-weighted average.*

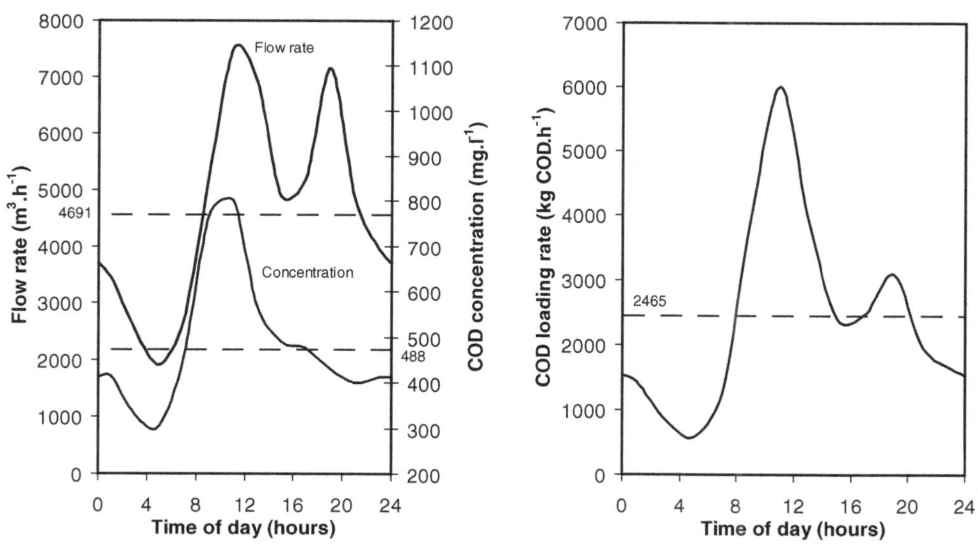

Figure 14.1 Daily variations in flow rate and concentration (left) and in the resulting organic load to the sewage treatment plant (right), based on the data of Table 14.1

It is not practical to obtain profiles of the flow and load by determining the flow rate and concentration over entire 24-hour periods as indicated in Table 14.1, at least not on a daily basis. Normally it will be necessary to estimate the daily organic load on the basis of a single grab sample taken at a specific time of day (often in the early morning, for instance 09:00) and the recorded cumulative daily flow. A correction factor must then be applied to the results of the analysis so that the product of the corrected value and the average flow reflects the best estimate of the organic load. To find this correction factor, the COD concentration profile from Table 14.1 can used to calculate the ratio between the COD concentration measured at a particular time and the average concentration. The average concentration can be calculated from Table 14.1 as $2465/4691 = 525$ mg COD $\cdot l^{-1}$ (i.e. not 488 mg $\cdot l^{-1}$!). So, for example at t = 09:00 hrs the value of the correction factor is $525/767 = 0.69$. Thus if on a particular day the measured COD value of a sample taken in the morning at 09:00 hrs has a value of 680 mg $\cdot l^{-1}$, then the equivalent concentration of this day is $0.69 \cdot 680 = 466$ mg $\cdot l^{-1}$. If the average influent flow of the same day is 4800 $m^3 \cdot d^{-1}$, then the best estimate of the organic load to the sewage treatment plant would be $0.466 \cdot 4800 = 2186$ kg COD $\cdot d^{-1}$. Table 14.1 lists the correction factors as a function of time over a 24 hr period for the data reported. It can be observed that the variation of the correction factor of the day is very large and that the value of the correction factor is either much lower or much higher than one for a large part of the time. Hence if the COD value of a sample taken at any time of the day is considered to be typical for the average over a 24 hour period, a very serious error can be made indeed.

A preferred, yet more expensive alternative to the use of a correction factor is an automatic influent sampling device, which allows composite samples to be taken. In this case, care should be taken that the sampler is operating according to a flow proportional sampling scheme (e.g. after every 200 m^3 of wastewater flow) and not on a time proportional basis. Time proportional sampling will attribute to much relative weight to periods with low flow.

The example discussed here only relates to the data presented in Table 14.1 and has no general validity. However, for many municipal sewage treatment plants the daily trend will be similar. Therefore, in every case where the organic load has to be estimated and the actual wastewater is available, it is recommended to establish the daily concentration profiles of COD, N and P. In general, larger sewer systems tend to have smaller amplitudes in both flow and load variations, as some degree of equalization and buffering takes place in the sewer system itself. Therefore, the value of the correction factor will be closer to unity. The above considerations, made for COD, are applicable to BOD and TSS as well. In practice the same correction factor is often used.

To conclude this section, the composition of the organic material, i.e. the value of the three important fractions f_{ns}, f_{np} and f_{sb} is very important as well. The influent composition will directly affect effluent quality, excess sludge production and oxygen consumption, as has been discussed extensively elsewhere in this book. If the values of these fractions cannot be determined experimentally then either default values from the literature or measured values from similar projects should be used.

(3) Concentration of inorganic suspended solids in the influent

The concentration of inorganic suspended solids is important, not only because it affects the concentration and production of excess sludge but also because it tends to influence the settling characteristics of the sludge.

(4) Temperature profile

Temperature is a very important parameter as it has a large influence on the values of the kinetic coefficients in the activated sludge system. Determining the minimum expected operational temperature is critical: if the system will function satisfactory at its lowest expected temperature then performance at higher temperature will not be a problem. The nitrification process is highly affected by a decrease in temperature. Apart from the direct influence on nitrifier growth rate, seasonal dynamics play a role as well: at the beginning of the cold season nitrification capacity is not yet negatively affected as the nitrifiers fraction in the sludge is still high. However, as the cold season proceeds, the nitrifier fraction will gradually decrease and so will the nitrification capacity. At the beginning of spring a similar (but reverse) effect will occur: although the temperature increases the nitrification capacity will tend to lag behind, as it will take longer for the nitrifier fraction in the sludge to be restored.

Apart from the obvious absolute lower limit of zero degrees Celsius, activated sludge systems are also limited with respect to the allowable upper temperature limit. Especially when nitrification is involved, the temperature in the reactor should not exceed a maximum value of 40°C. The highest average temperature to be expected for municipal sewage is about 30°C around the equator decreasing to about 20°C at the tropics of Cancer (23.5°N latitude) and Capricorn (23.5°S latitude). In regions with a moderate and cold climate the average temperature tends to be much more dependent on the change of seasons. Large seasonal temperature fluctuations can cause problems as processes might develop at higher temperatures that do not occur at lower temperature, for instance nitrification. If the system has not been designed for nitrification, serious problems may develop in the warm season.

In the case of hot industrial wastewaters it may be required to install heat exchangers in order to avoid the temperature rising above 40°C. The large wastewater treatment plant operated by CETREL at Camaçari successfully treats petrochemical wastewater at a temperature range varying between 35 and 39°C in the aeration tanks, producing an effluent with very low concentrations of biodegradable organic material and ammonium. Large short-term temperature variations should be avoided, as these will have an adverse effect on biomass activity. When these kinds of temperature variations are expected, the use of an influent buffer tank is recommended. Finally, temperature also affects the installed aeration capacity. The reactor temperature influences the equilibrium dissolved oxygen concentration but also the value of the oxygen transfer rate coefficient, while in the case of diffused aeration the ambient temperature influences the volume of the air flow to be compressed. At higher temperatures the difference between the equilibrium oxygen concentration and the oxygen setpoint, the driving force for oxygen transfer to the mixed liquor, will be smaller. At the same time, a high ambient temperature will increase the required compression power of the blower.

(5) Value and stability of pH

The value and stability of pH is mainly determined by the alkalinity present in the influent and the alkalinity produced or consumed in the activated sludge system. If the alkalinity in the reactor is not maintained at a minimum value of 35 $mg \cdot l^{-1}$ $CaCO_3$, external alkalinity should be added to prevent operational instability, as the pH might at least occasionally be reduced to values much below the optimum of 7.0 to 7.5.

(6) Nutrient concentration

A certain minimum concentration of nutrients (mainly nitrogen and phosphorus) is necessary to cover the requirements for the production of excess sludge. If nutrients are not available in sufficient quantity, the bacteria will still grow but the settleability of the sludge will decrease. In case of industrial wastewaters, the nutrient concentration in the influent might not be enough to meet the demand, making nutrient addition necessary. For domestic wastewater and some industrial wastewaters, the nutrient concentration will be much higher than the requirements for excess sludge production and in these cases the system should be designed for nutrient removal.

(7) Toxic components

Industrial wastewaters (or domestic sewage with industrial contributions) might contain toxic components, either having an impact on kinetic parameters such as growth and decay rates or on sludge properties such as sedimentability. Depending on the nature of the material the effect may be either acute or chronic. In the former case there is a temporary presence of toxic material. The toxic material may be removed in the treatment system by metabolism or desorption. Examples are volatile organic compounds (VOCs) such as light aromatics and chlorinated organics in oil refinery or petrochemical wastes. In the case of chronic toxicity, the toxic material tends to be retained in the treatment system and accumulates in the sludge, e.g. heavy metals. It is important to distinguish inhibitory compounds, which will only have an effect while they are present, from real toxic materials that will have an irreversible effect, even after having been eliminated.

14.1.1.2 Kinetic parameters and settleability of the sludge

In addition to the qualitative and quantitative characteristics of the wastewater to be treated, it will be necessary to determine the value of kinetic parameters, as these are important in determining both the performance of the system and the dimensions of the main treatment units. The parameters that determine the minimum sludge age for nitrification and the denitrification rate are especially important. If biomass is already available (for instance in projects for expansion of treatment capacity), the settleability should be established. This can be done either by means of an experimental procedure allowing direct determination of the Vesilind constants or through determination of a parameter that allows estimation of these constants. Often the value of these constants fluctuates in time: in this case the most unfavourable values should be adopted. When no activated sludge is available, literature values or measured values from similar wastewater treatment plants can be used. It is recommended to be somewhat conservative in the design, as underperformance of the final settler may otherwise cause sludge loss with the effluent.

14.1.2 Costing data

The costs involved in constructing and operating a wastewater treatment plant can be broadly divided into two categories: investment costs and operating expenses. The investment costs are all costs required for the construction of the wastewater treatment plant and include items such as:

- Preliminary studies;
- Acquisition of a construction site and preparation for building;
- Project costs such as design, engineering and legal affairs: e.g. permit applications and environmental impact assessments;
- Civil work, equipment & instrumentation costs, including process control systems;
- Construction and installation;
- Commissioning (i.e. testing and accepting the installation) and start-up costs.

The operational costs are all costs incurred to maintain and operate the wastewater treatment plant and include items such as:

- Personnel;
- Maintenance costs;
- Operational costs: chemicals, utilities, lab supplies, office supplies etc.;
- Aeration costs (electricity);
- Sludge disposal costs.

In this chapter the focus will be on the optimisation of the (annualised) investment and operational costs, as these are the costs that can be influenced by the optimised design of the wastewater treatment plant.

14.1.2.1 Investment costs

In general, construction costs are the main item in the total investment costs, although the contribution of desing & engineering may be significant as well, especially for smaller projects. In Table 14.2 a typical example is given of a breakdown of investment costs for a wastewater treatment plant.

Table 14.2 Typical division of cost items as proportion of total investment costs, excluding the costs of site acquisition and infrastructure

Cost item	Fraction	Description
1. Preparation		
1.1 Site acquisition	Location dependent	Acquisition of building plot, brokers, notaries, taxes
1.2 Infrastructure	Location dependent	Access roads, sewer lines and effluent discharge pipelines, power supply
1.3 Site preparation	0.5–2%	Demolishing, ground work, rerouting pipes & cables, roads
2. Construction	**70–85%**	
2.1 Civil	23–29%	Construction of tanks, buildings, foundations
2.2 Mechanical	21–27%	Equipment costs incl. installation, local piping etc.
2.3 E&I$^{(1)}$	10–16%	Local instrumentation and electro-technical equipment
2.4 Piping	2–5%	Interconnecting piping, utilities, sewers, including insulation and tracing
2.4 Central PC/E&I$^{(1)}$	2–5%	Central process control incl. software, motor cabinet, substation, frequency converters, cable work
2.5 Contingency	10–20%	Allowance for unforeseen expenses
3. Start-up	**1–3%**	
3.1 Equipment	Included in item 3	Maintenance and laboratory equipment
3.2 Start-up supplies/spares	Included in item 3	Chemicals, first fills (activated carbon, filter material). Fittings, cables, etc.
3.3 Personnel	Included in item 3	Hiring & training employees
4. Additional	**10–20%**	
4.1 Initial studies	Included in item 4	Feasibility study, system selection, geotechnical survey
4.2 Design and engineering	Included in item 4	Basic & detailed design and engineering, requisitions and tender process, procurement
4.3 Project management	Included in item 4	Planning and budget control
4.4 Construction management	Included in item 4	Site supervision, testing and commissioning
4.5 Miscellaneous	Included in item 4	Permits, taxes, insurance

Note: (1) Process Control / Electrical & Instrumentation

In the example the cost of site acquisition and infrastructure works (items 1.1 and 1.2) have not been included as these are very site-specific. For municipal wastewater treatment plants in particular, the costs of constructing a sewer system, pumping stations and pressure lines can be very high and might even be higher than the costs of constructing the wastewater treatment plant itself. To get an idea of the construction effort required,

Figure A7.7 shows the pressure pipeline connecting the existing municipal sewer system with the new Harnaschpolder STP near The Hague, The Netherlands. A second, similarly sized pipeline directs the treated sewage to a subsea outfall. These costs are less important for industrial wastewater treatment plants, which are often located on the factory site and near the source of the wastewater.

The total construction costs or TCC (i.e. cost item 2.1 to 2.3 in Table 14.2) of the main treatment units of the wastewater treatment plant discussed in this book typically comprise between 50–70% of the total investment costs, depending on the size of the wastewater treatment plant. In order to estimate the construction costs in the early stages of the wastewater treatment plant design, unit volume costs can be attributed to the different main treatment units. Similarly, unit costs per installed kW are assigned to aeration equipment and equipment for energy generation. These costs cover the expenditure of raw material and equipment, all civil and mechanical construction work and the installation of local electrical equipment and instrumentation. In the case of diffused aeration, apart from the blower package this includes the costs of the aeration elements, connecting pipelines, valves and instrumentation.

As can be observed in Table 14.3, costs per unit volume and per kW installed tend to be higher for smaller wastewater treatment capacities. This is one of the reasons (apart from personnel costs) why, for instance in the Netherlands, the trend is to construct very large wastewater treatment plants, (>0.5 to 1.0 million P.E.), replacing several smaller installations. However, the reduction in construction costs for the wastewater treatment plant has to be balanced with the increase in construction costs for the sewage supply and effluent discharge network.

Table 14.3 Typical value ranges of costing parameters for different WWTP sizes (expressed in Population Equivalents (P.E.)), based on the price level in 2006 – use cumulative inflation to adapt to present values

Parameter/Capacity	25,000 P.E.	50,000 P.E.	100,000 P.E.	200,000 P.E.
Cost per unit volume ($US\$ \cdot m^{-3}$):				
C_{d1} – primary settler	600–900	400–650	300–450	200–350
C_u – UASB	600–1000	500–700	350–500	250–400
C_r – aeration tank	220–300	180–250	150–200	120–170
C_d – final settler	350–550	300–400	250–330	200–260
C_{th} – sludge thickener	700–1000	500–800	300–500	250–400
C_{di} – anaerobic digester	600–1000	450–700	300–400	250–350
Cost of installed kW ($US\$ \cdot kW^{-1}$):				
C_{ae} – surface aeration	4500–7000	4000–5200	3200–4000	2800–3500
C_{ae} – diffused aeration	6000–9500	5500–8000	5000–7200	4000–6000
C_{gen} – power generation	2500–5000	1700–3500	1500–2500	1000–2000

The upper range values in Table 14.3 correspond to a well-equipped, quality constructed modern municipal WWTP in developed countries. For instance the construction of the basins will be mainly in concrete and pumping stations, control rooms and equipment are all located in concrete buildings. Furthermore the level of automation will be quite high. The lower end corresponds to industrial installations and municipal installations in developing countries, where equipment is often skid mounted and located outside under a shelter. To determine the total construction costs, use the costing data and the optimised unit volumes:

$$TCC = C_{d1} \cdot V_{d1} + C_u \cdot V_u + C_r \cdot V_r + C_d \cdot V_d + C_{th} \cdot V_{th} + C_{di} \cdot V_{di} + C_{ae} \cdot P_{aerm} + C_{gen} \cdot P_{el} \tag{14.1}$$

However, in addition to the main treatment units discussed in this book, a typical (municipal) wastewater treatment plant will contain additional units that will contribute significantly to the total construction costs as well:

(1) Influent lifting station, where all sewage streams are received and lifted to a hydraulic level that permits the flow of the wastewater through the water line (pre-treatment, activated sludge treatment and clarification) by gravity;
(2) Screening: for the removal of large debris, such as leaves, plastic, rags etc;
(3) Sand trap: removes sand, gravel and potentially also oil, fat and grease;
(4) Off-gas treatment: e.g. lava filters that treat contaminated air from covered units such as primary settling tanks, screening operations, sand trap and dewatering units;
(5) Control and maintenance buildings, laboratory and storage facilities, including equipment for heating, ventilation and air conditioning (HVAC);
(6) Chemical dosing units, e.g. for metal salts and polyelectrolyte, including storage facilities;
(7) Sludge dewatering: filter presses, decanter centrifuges or sludge drying beds, sludge buffer tank and dewatered sludge storage.

In the case of industrial treatment, the wastewater is often delivered trough a single pressure line and received in a buffer tank. An influent lifting station, screening units and a sand trap are normally not required. However, other units may be required such as plate pack separators and dissolved air flotation (DAF) units if oil is present in the wastewater or additional polishing steps such as advanced oxidation, sand filtration and activated carbon adsorption. For municipal wastewater treatment or industrial installations with a similar configuration, the construction costs of the additional units listed above (i.e. items 1 to 7) are dependent on the treatment capacity and range between 20% (large capacity) and 35% (small capacity) of the total construction costs. Thus once the construction costs of the main treatment units are estimated, the additional construction costs can be calculated by multiplication with a costing factor f_{ac}. Values of f_{ac} as a function of the size of the wastewater treatment plant are listed in Table 14.4.

Finally, to obtain the total investment costs, the remaining cost items listed in Table 14.2 will have to be accounted for: i.e. items 1.3, 2.4, 2.5, 3.1 to 3.3 and 4.1 to 4.5. A rough estimate of these costs is between 50% (large capacity) and 100% (small capacity) of the total construction costs. To include these cost items in the total investment costs, a second costing factor f_i is introduced. Typical value ranges of f_i are shown in Table 14.4. To obtain the total investment costs TIC, estimate the costs of additional units that have not been specified (f_{ac}) and calculate the additional project investment costs (f_i):

$$TIC = f_{ac} \cdot f_i \cdot TCC \tag{14.2}$$

A final note about the data presented in Table 14.3 and Table 14.4: as local requirements, conditions and prices differ for each project, the accuracy of the unit cost based estimate of the total investment costs will in general not be higher than +/–30%. To obtain a more accurate estimate, a proper design and engineering study will be required, including priced vendor quotations for the main equipment. On the other hand, the costing approach introduced in this section can be very useful in the optimised design procedure as it allows evaluation and comparison of different system configurations on a different criterion than total treatment volume alone. As can be observed in Table 14.3, the main treatment units differ considerably in construction cost per m^3 volume.

Table 14.4 Indicative values of the costing factors f_i and f_{ac} used to estimate the total investment costs of a STP from the construction costs of the main treatment units, as function of the treatment capacity in people equivalents (P.E.)

Capacity	25,000 P.E.	50,000 P.E.	100,000 P.E.	200,000 P.E.
$f_{ac}^{(1)}$	1.4–1.5	1.35–1.45	1.3–1.4	1.25–1.35
f_i	1.6–1.9	1.5–1.8	1.5–1.7	1.4–1.6

Note: [1] *The values given are applicable for municipal wastewater treatment plants with a similar configuration as discussed in this section, i.e. with the additional units listed in items (1) to (7) on the previous page*

EXAMPLE 14.1

The optimized design for a 100,000 P.E. plant yields the following results:

- $V_r = 8000\ m^3$ and $V_d = 2500\ m^3$
- $V_{th} = 300\ m^3$ and $V_{di} = 2000\ m^3$

Estimate the total investment costs, using the following additional data:

- Installed aeration capacity (surface aeration): $P_{el} = 250$ kW;
- Gas is used in existing boilers;
- Assume a "typical" municipal design using Dutch standards.

Solution

Retrieve from Table 14.3 and Table 14.4 the appropriate unit cost factors:

- $C_r = 200$ and $C_d = 330\ US\$ \cdot m^{-3}$;
- $C_{th} = 500$ and $C_{di} = 400\ US\$ \cdot m^{-3}$;
- $C_{ae} = 4000\ US\$ \cdot kW^{-1}$;
- $f_i = 1.7$ and $f_{ac} = 1.4$.

Total construction costs TCC (of the main units) is calculated as:

$$TCC = 8000 \cdot 200 + 2500 \cdot 330 + 300 \cdot 500 + 2000 \cdot 400 + 250 \cdot 4000$$
$$= 4.4 \text{ million US\$}$$
$$TIC = f_i \cdot f_{ac} \cdot TCC = 1.4 \cdot 1.7 \cdot 4.4 = 10.4 \text{ million US\$}$$

14.1.2.2 Operational costs

The total operational costs (TOC) are all costs incurred to maintain and operate the wastewater treatment plant and include items such as:

- Personnel, maintenance and insurance;
- Operational costs (chemicals, utilities, lab supplies, office supplies etc);

– Aeration costs (electricity) and heating costs (for the anaerobic digester);
– Costs for sludge disposal;
– Effluent discharge costs.

The value of the first items depends for a large part on the efficiency of the operating company, but for the purpose of cost estimation is taken as a percentage of the total investment costs. Often different maintenance percentages are used for civil works and for mechanical, electrical & instrumentation equipment, as the latter generally requires more maintenance. Insurance costs are also taken as a percentage of total investment costs.

For the aeration costs, the price per kWh is required. This price not only covers the production costs of electricity, but also that of transport, taxes and connection. For heating costs the price of the appropriate fuel can be used (natural gas, gasoil). Depending on the process configuration, sufficient energy may be generated from the digested sludge to cover the heating requirements.

The costs of sludge disposal may vary considerably, depending on the possibilities of reuse and the legal requirements for sludge disposal. Often the dewatered sludge has a dry solids content of 15–25%, although this can be much higher (up to 90%) when drying beds are applied. If direct application of stabilized sludge on farmland is allowed, then the costs of disposal will be very low (i.e. only transportation costs), however this is more and more prohibited due to healthcare concerns. In the case of disposal at a landfill, the cost per ton dry sludge are still relatively low, e.g. between 60–100 US$ ton dry sludge. However, if this is not allowed, then the dewatered sludge will require more costly disposal. Several methods are available: by far the most common ones are drying and incineration or composting. In this case the costs of disposal may easily be as high as 300 $US\$ \cdot ton\ solids^{-1}$. Transport costs will further increase this amount. Sometimes the dewatered sludge can be reused, for example as an additive in cement production.

Finally, if applicable, discharge levies might be payable for the residual organic- and nutrient load discharged with the effluent to municipal sewer systems. This does not apply to municipal wastewater treatment plants or to industrial treatment plants discharging directly to surface water. Table 14.5 shows typical ranges for the different items that make up the operational costs. Once the activated sludge system has been designed, the total operational costs can easily be calculated (assuming the price of produced and consumed electricity are equal):

$$\begin{aligned} TOC = {} & (p + o + m + n) \cdot TIC + 365 \cdot (24 \cdot ((P_{aer} - P_{el}) \cdot C_{el} + P_h \cdot C_h) \\ & + C_{sd} \cdot ME_{te}/1000 + PE_{res} \cdot C_{dl}) \end{aligned} \tag{14.3}$$

Table 14.5 Typical value ranges for the items included in the operational costs

Operational cost item	Symbol	Range	UoM
Personnel	P	2–5%	$US\$ \cdot year^{-1}$
Operation	o	0.5–1.5%	$US\$ \cdot year^{-1}$
Maintenance – civil	m	0.5–1.0%	$US\$ \cdot year^{-1}$
– mechanical/E&I		1–2.5%	$US\$ \cdot year^{-1}$
Insurance	n	0.2–0.4%	$US\$ \cdot year^{-1}$
Electrical energy	C_{el}	0.10–0.20	$US\$ \cdot kWh^{-1}$
Heating energy	C_h	0.2–0.5	$US\$ \cdot m^3$ gas/...
Sludge transport & disposal	C_{sd}	60–500	$US\$ \cdot ton^{-1}$ TSS
Discharge levies	C_{dl}	20–70	$US\$ \cdot P.E.^{-1}$

EXAMPLE 14.2

For the previous example, estimate the annual operational costs (TOC). Use the higher values from the ranges listed in Table 14.5. The following additional data are given:

- Average electrical power consumption: 150 kW;
- Daily stabilised sludge discharge: 1.5 ton dry solids;
- No costs for heating and discharge levies;
- The division between civil- and mechanical + E&I construction costs is 30 vs 70%.

Solution

Cost of personnel, maintenance, operation and insurance:

$$TOC = (p + o + n + 30\% \cdot m_{civ} + 70\% \cdot m_{me\&i}) \cdot TIC$$

$$= (5 + 1.5 + 0.4 + 0.3 \cdot 1 + 0.3 \cdot 2.5) \cdot 10.4 = 0.9 \text{ million US\$.yr}^{-1}$$

Costs of aeration and sludge disposal:

$$TOC = 365 \cdot (24 \cdot P_{aer} \cdot C_{el} + C_{sd} \cdot ME_{te})$$

$$= 365 \cdot (24 \cdot 150 \cdot 0.2 + 500 \cdot 1.5) = 0.5 \text{ million US\$.yr}^{-1}$$

Total operational costs $TOC = 0.9 + 0.5 = 1.4$ million US$ $\cdot$ yr^{-1}

14.1.2.3 Annualised investment costs

A fundamental aspect in the design optimisation procedure is the calculation of the total annual treatment costs, composed of the *annualized* investment costs (AIC) and the operational costs (TOC), which are annual by nature. However, it will be necessary to annualise the total investment costs (TIC) over the expected lifetime of the treatment plant, transforming them into net present value (financing or annualized investment costs). To transform the total investment costs into annualized costs, the following formula can be applied (De Faro, 1986):

$$AIC = TIC/a_{i,n} \qquad (14.4)$$

where:

AIC = annualized investment costs (depreciation) over the expected economic lifetime
TIC = total investment costs in present value
$a_{i,n} = [(1+i)^n - 1]/[i \cdot (1+i)^n]$ = annualization factor (14.5)
i = interest rate (annual) and n = economic lifetime of the treatment plant in years

Often a distinction is made between the economic lifetime of the civil part of the wastewater treatment plant and that of the mechanical and E&I part: for instance 30 years for the civil part and 15 to 20 years for the mechanical and E&I part. If this is the case, the total investment costs TIC are divided into different fractions. A typical division of the TIC is about 35–60% for the civil works, 25–40% for mechanical equipment and 15–25% for E&I equipment.

Equation (14.1) is a simplified expression in which it is assumed that the treatment plant will have no residual value after the expected economic lifetime, inflation is disregarded and the annual terms are equal until the end of the economic lifetime. However, the expression is adequate enough for an first analysis of the generated design. The calculated annual financing costs are added to the expected operational costs and the total costs can be used to compare design alternatives in various formats such as annual costs, costs per unit volume of treated wastewater and cost per people equivalent.

EXAMPLE 14.3

Calculate the annualized costs for sewage treatment plant of the previous examples. The following additional data are given: i = 8%, the expected economical lifetime of the civil works is 30 years and that of mechanical + E&I installations 17 years.

Solution

Calculate both annualization factors with Eq. (14.4). For the civil works:

$$a_{i,n} = [(1+i)^n - 1]/[i \cdot (1+i)^n]$$
$$= [(1.08^{30}) - 1]/[0.08 \cdot (1.08)^{30}] = 11.3$$

For the mechanical and E&I installations:

$$a_{i,n} = [(1.08^{17}) - 1]/[0.08 \cdot (1.08)^{17}] = 9.1$$

Calculate the annualized investment costs. For the civil works:

$$AIC = 0.3 \cdot 10.4/11.3 = 0.3 \text{ million US\$} \cdot yr^{-1}$$

For the mechanical and E&I installations:

$$AIC = 0.7 \cdot 10.4/8.6 = 0.8 \text{ million US\$} \cdot yr^{-1}$$

The total annualized investment costs AIC = 1.1 million US$ · yr^{-1}

Total annualized costs TAC = AIC + TOC = 1.1 + 1.4 = 2.5 million US$ · yr^{-1}

14.1.3 Performance objectives

A wastewater treatment plant is in general constructed with the objective to produce an effluent with a certain desired quality at minimum total or annualized investment and operational costs. In general the desired effluent quality is specified by environmental legislation, setting limits to operational parameters such as the concentration of suspended solids, organic material and nutrients. In the case of industrial wastewater, other parameters such as temperature, pH, oil, grease and fats and floating material could be important as well. In many developing countries effluent standards with respect to nutrient concentrations do not yet exist or are not yet applied or maintained. However, even if no legal requirements exist, it will be advantageous to apply tertiary treatment for the following reasons:

- Removal of nutrients (and particularly of nitrogen) is a necessary condition to obtain proper operational stability, especially in regions with a warm climate;
- The oxygen consumption will be reduced because of the recovery of equivalent oxygen by denitrification;
- The tendency of environmental legislation is to become stricter with time, making it very probable that nutrient removal will be required within the expected lifetime of the activated sludge system yet to be constructed.

During the last three decades in Europe and the US, huge investments have been made to upgrade existing activated sludge systems, first to achieve nitrogen removal and later to achieve phosphorus removal as well.

In developing countries, the lessons learned can be immediately applied to the design of nutrient removal systems. Apart from legislation, an important consideration to define the type of treatment is the final destination of the effluent. Many cities are located in coastal zones, in which case the effluent of wastewater treatment plants will be discharged directly into the sea (often by means of a pipeline outfall). The beneficial effects of nutrient removal (particularly phosphorus) may not be very significant, as the discharged nutrients will be instantly diluted and therefore do not significantly alter the existing background concentration. However, if the effluent is discharged into a surface water (such as a lake or a river) or into a bay, nutrient removal can be extremely important to maintain the quality of the receiving water. The following examples from Brazil will clarify this issue.

In the case of the São Paulo, the largest Brazilian city with a population of 18–20 million inhabitants in the metropolitan area and an enormous industrial complex, a large part of the treated effluent ends up in the rivers Tietê and Paraíba, originating in the metropolitan area. The rivers run through hundreds of kilometres of land before reaching the sea and in doing so serve as a major source of potable water in many cities (for instance Rio de Janeiro). It is easy to see that maintaining and improving the quality of these rivers is of great importance and for this reason tertiary treatment of all wastewaters in the São Paulo metropolitan area is necessary.

In the Rio de Janeiro metropolitan area, a large part of the sewage generated by its 12–13 million inhabitants and one of the largest industrial complexes in the country is discharged directly into the Bay of Guanabara. As a result a large part of the Bay, which in the past was an important area for fishing and tourism now has a septic aspect and a very low dissolved oxygen level. In this case, wastewater treatment without nutrient removal will not resolve the problem as the load of nitrogen and phosphorus (estimated at 50 ton $TKN \cdot d^{-1}$ and 10 ton $P \cdot d^{-1}$) results in eutrophication of the bay and still seriously limits the use for fishing and tourism. Therefore, in this case tertiary treatment will also be necessary.

In the city of Salvador da Bahia (2.5 to 3 million inhabitants) a wastewater collection system is in place which directs almost all sewage produced in the city to a primary treatment unit, from which it is sent into the Atlantic Ocean by means of a 5 km subsurface pipeline outfall. The primary treatment is applied to protect the outfall from accumulation of solids. Various studies have shown that the quality of the beach water is not significantly influenced by the discharge of the pre-treated sewage in the sea. It is concluded that in the case of Salvador, the subsurface disposal of sewage in the sea is a viable solution from an environmental viewpoint, although the initial investment cost were very high and a more economic alternative could have been implemented.

The city of Manaus in the heart of the Amazon Region is located at the bank of the Rio Negro, just upstream of the confluence with the Amazon River. The average flow of the Amazon River is 100,000 $m^3 \cdot sec^{-1}$. Assuming the population of Manaus (1.7 million inhabitants) produces a wastewater flow of

230,000 $m^3 \cdot d^{-1}$ or approximately 2.7 $m^3 \cdot sec^{-1}$, the discharge of raw sewage in the river will dilute the sewage with a factor 100,000/2.7 = 37,000. The pollutants will therefore automatically be diluted to a very low concentration, possibly even lower than the existing background concentration in the river. As there is a legal requirement to reduce the concentration of suspended solids and of biodegradable organic material, in the case of Manaus it may be considered to only install an efficient anaerobic treatment unit such as a UASB reactor (especially taking into account that the average water temperature in the region approaches 30°C year round), which will easily remove more than 70% of these pollutants.

The above examples clearly show that the geographical and demographical conditions have a decisive influence on the decision on the nature of the treatment to be applied and the required effluent quality.

14.1.4 Applicable system configurations

Once the influent flow and -composition and the required effluent composition are known, the selected configuration and main unit volumes will be determined by the ambient conditions and particularly by the temperature.

In the previous chapters it was shown that nitrification at very low temperature is practically impossible, allowing only secondary treatment with removal of suspended solids and organic material. At higher temperature nitrification will be possible and in regions with a warm climate it will practically be inevitable. In this case, secondary treatment systems with nitrification or tertiary treatment systems with nitrogen – and possibly also phosphorus removal might be selected. Depending on the situation, pre-treatment units such as primary settlers and anaerobic reactors can be included as well. In Section 14.2, five basic configurations of the activated sludge process will be discussed. Table 14.6 lists the selected configurations and their performance in the removal of several wastewater constituents. Figures 14.2 to Figure 14.4 show the basic process flow schemes.

Table 14.6 Treatment efficiency of the five basic configurations of the activated sludge system with respect to suspended solids (SS), organic material (COD), total organic nitrogen (TKN), total nitrogen (N_t) and phosphorus (P_t)

Configuration		Removal efficiency				
		SS	COD	TKN	N_t	P_t
A1	Secondary treatment (AS) with sludge thickening and anaerobic digestion	+	+	–	–	–
A2	Primary settling + secondary treatment (AS) + sludge thickening and anaerobic digestion	+	+	–	–	–
B1	Anaerobic pre-treatment with secondary post-treatment (AS) plus sludge thickening and anaerobic digestion	+	+	–	–	–
C1	Tertiary treatment (BDP) with nitrogen removal plus sludge thickening and anaerobic digestion	+	+	+	+	–
C2	Tertiary treatment with nitrogen and phosphorus removal (UCT) plus sludge thickening and anaerobic digestion	+	+	+	+	+

Starting from these basic configurations, hybrids can be constructed. For example, when the wastewater has a high COD/N ratio, it may be possible to include primary clarification (or even anaerobic pre-treatment) in configuration C1 or C2. The organic load to the activated sludge system will be reduced, while enough organic material will remain in the pre-treated influent for biological nutrient removal. Other types of treatment may be used as well. For instance in the Netherlands, configuration C2 is often preceded by primary clarification and supplemented by the addition of metal salts to remove any phosphate not taken up by the bio-P organisms. This configuration is only possible due to the relatively low N_{ti} values (40–60 $mg N \cdot l^{-1}$) and the fact that complete nitrogen removal is not required ($N_{te} < 10\ mg N \cdot l^{-1}$). If required, part of the raw sewage flow is bypassed around the primary settler.

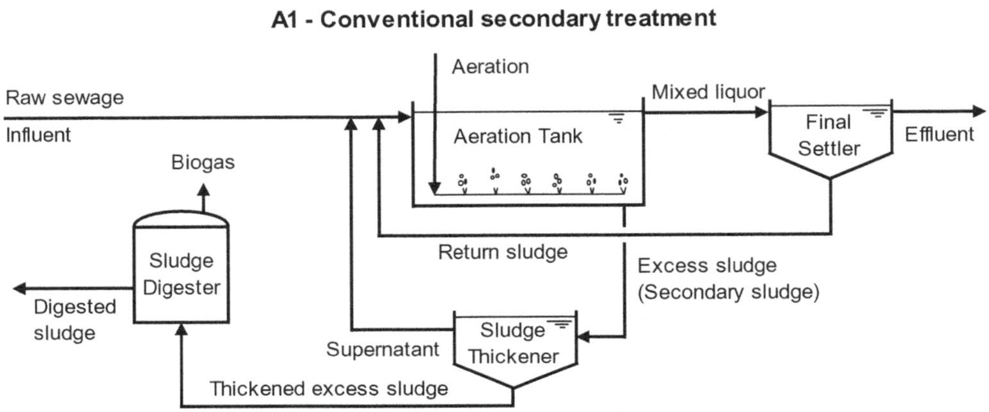

Figure 14.2 Basic process flow diagram of system configuration A1

In Section 14.2 the optimised design procedure will be restricted to the five basic configurations presented above. Additional design examples will be presented in Section 14.5.

14.1.5 Limitations and constraints

The selected solution in a specific design case depends on legal requirements, but also on other considerations such as the possible reuse of the effluent, the available area for construction, available funds (also taking into account the cost of the sewer system), the environmental and economical impact of the treatment plant and the availability of skilled employees.

Together with the legal requirements, the possible reuse of the effluent can determine the effluent standards. If for example the effluent is to be used for irrigation, it is advantageous to leave the nutrients in the effluent and thus to reduce the utilisation of chemical fertilisers. On the other hand, if the effluent is discharged to surface waters then nutrient removal may be of critical importance. In developing countries many people still do not have access to a public drinking water network. Therefore it should be taken into account that surface waters might be used for human consumption, in which case the hygienic quality of the water is very important. It might be necessary to remove pathogens and include post-treatment steps such as disinfection (chlorine, ozone, UV), filtration or biological units such as stabilisation ponds and helophyte filters.

Limited availability of funding can also influence the design. If the available resources are insufficient to comply with the legal requirements, it may be necessary to construct an imperfect system from a legal perspective, but within the financial possibilities and thus avoiding the possibility that nothing is done at

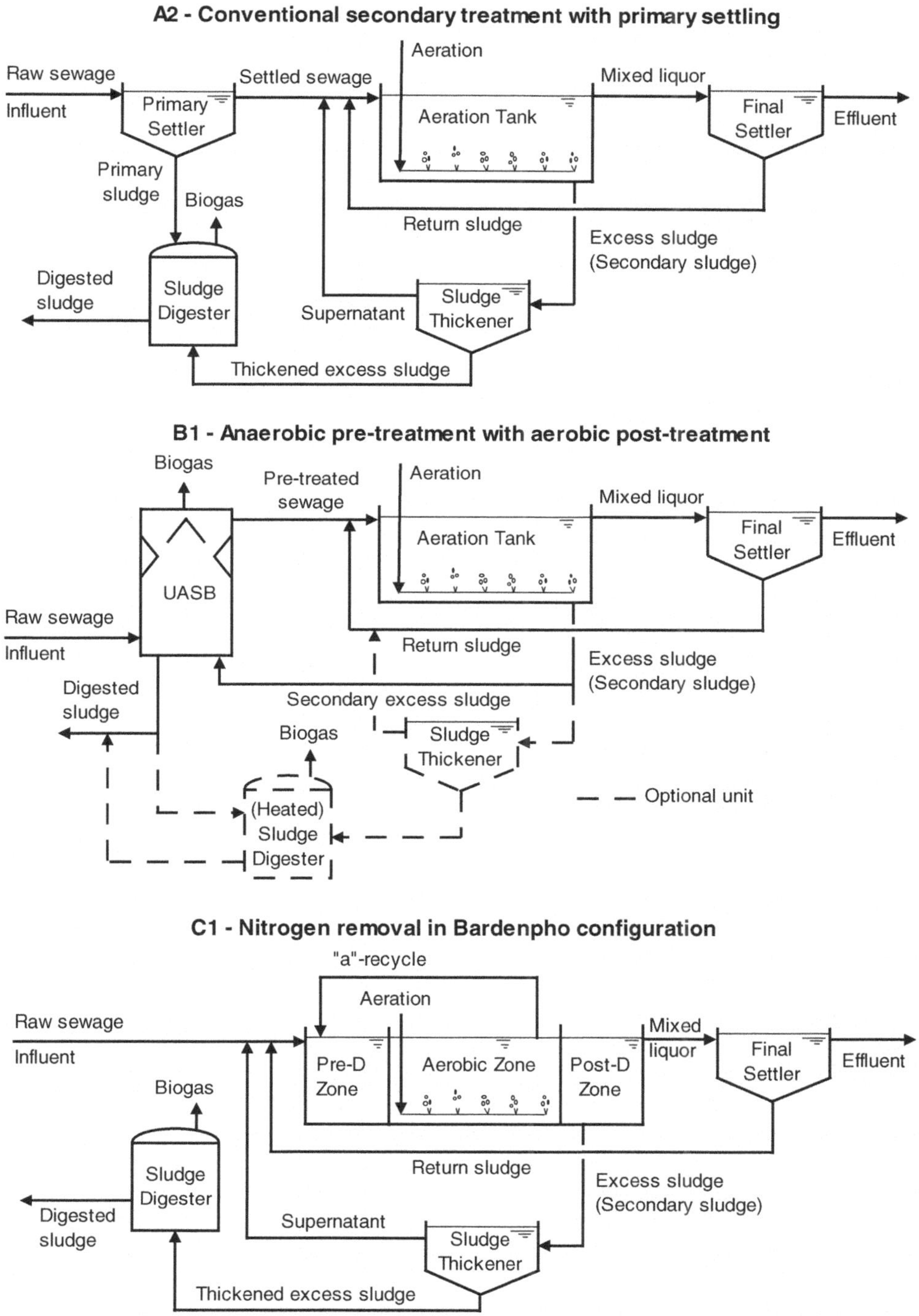

Figure 14.3 Basic process flow diagrams of system configurations A2, B1 and C1

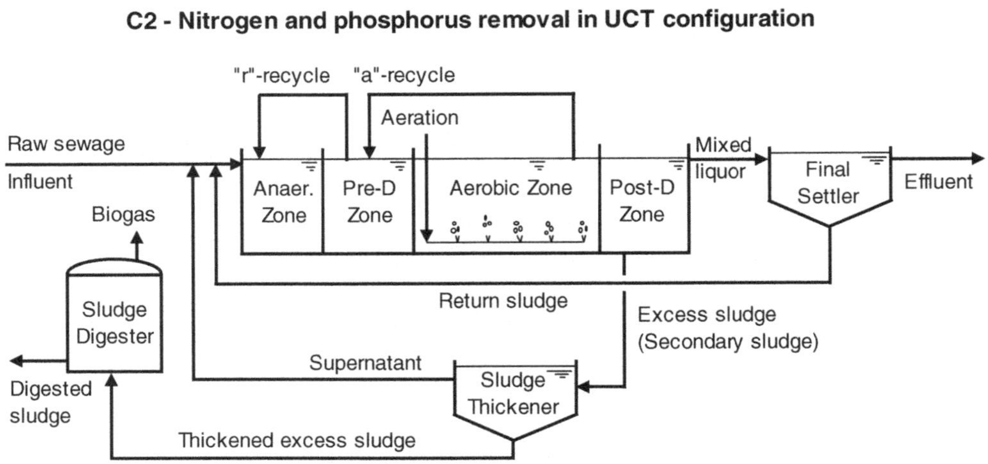

Figure 14.4 Basic process flow diagram of system configuration C2

all. In this chapter it will be demonstrated that pre-treatment of sewage with a UASB reactor will considerably lower the total investment cost of an activated sludge system (although only at sufficiently high ambient temperature). Thus, if financial resources are limited, construction could start with a UASB reactor removing most of the organic material in the influent, leaving for a later time the post treatment in an activated sludge system. Even in the event that at a later stage the anaerobic pre-treatment step is decommissioned (e.g. to allow biological nutrient removal), the UASB unit can still be useful as a sludge digester.

It is important to realise that the construction costs of a sewer system are considerable and might actually be higher than that of the actual wastewater treatment plant. Therefore in many cases it is possible to reduce investment costs significantly when several smaller decentralised treatment plants are built in a city instead of a single central treatment plant. The installation of high cost large diameter sewage pipes and pumping stations can be avoided.

To illustrate this point, in the Netherlands recently a large wastewater treatment plant was constructed near the city of The Hague, replacing several smaller treatment plants (Harnaschpolder – 1.3 million P.E.). In the same project, a second treatment plant (Houtrust) was extensively refurbished. The total investment costs were approximately 700 M€, divided into 270 M€ for the new STP; 90 M€ for renovation of the old STP and the remaining 350 M€ for the pressure lines and pumping stations transporting the raw sewage to the STP and discharging the treated effluent to sea. Thus in this case the costs of the sewer network represented 50% of the total investment and this does not even include the costs of the sewer collection system in the city itself, as this was already in place. On the other hand, the cost of purchasing building area inside the city limits would have been very high as well and locating decentralised wastewater treatment plants inside the city limits would not have been easy.

An alternative is to centralise only certain operations (e.g. sludge treatment). In this case all local wastewater treatment plants may pump or transport their excess sludge to a central sludge treatment unit. An interesting concept, applied by SANEPAR in the Brazilian state of Paraná, is to treat the wastewater locally by UASB reactors and to leave the post treatment (which will be centralised) for a later stage.

Frequently in developing countries it is decided to implement secondary treatment only, as tertiary treatment is considered to be an expensive sophistication only applicable for regions with ample resources. This point of view is disputed for the following reasons: (I) the cost difference between a system for secondary treatment and a system designed for nutrient removal is not very large, (II) secondary treatment may result in many operational problems in regions with a hot climate and (III) the nutrient removal system will have a superior effluent quality. For these reasons in general the construction of a tertiary treatment system for nitrogen removal is justified.

Another misconception is that the activated sludge system is too complex to be operated in developing countries. Although the presence of qualified operators and maintenance technicians is required, the degree of technical difficulty of an activated sludge system is in fact not very high. In a country like Brazil with a large industrial base there are literally thousands of factories that are much more complex. The frequent failures of activated sludge systems are not because of their technical complexity, but due to other factors such as inadequate design and lack of priority given to treatment systems by water and wastewater treatment companies, resulting in inadequate operation and maintenance. However, at present there are many large wastewater treatment systems in Brazil with adequate and stable performance. These systems have been in operation for more than three decades and have demonstrated excellent operational stability and a high quality effluent.

14.2 OPTIMISED DESIGN PROCEDURE

In the previous chapters various examples of the design and optimisation of the different units of the activated sludge systems have been discussed. In this section a conceptual method is presented, which can be used for the design optimisation of the activated sludge treatment configurations presented in Section 14.1.4. This optimisation method uses the same body of theory as already presented in the earlier chapters, but for benefit of the reader the whole procedure is presented in an integrated form, considering all steps of the design process.

14.2.1 System A1: Conventional secondary treatment

The most elementary configuration of the activated sludge system consists of a completely mixed aerobic reactor treating the influent, followed by a final settler and equipped with a gravity thickener and anaerobic digester for stabilisation of the produced excess sludge.

In practice this system will also be equipped with a pre-treatment capable of removing large debris (rags, paper, plastics), sand and if required oil, fat and grease. For the design of these units refer for instance to Metcalf & Eddy (2003). For optimised system design the following data are required:

(1) Sludge age at which the system should be operated;
(2) Values of the parameters of the steady state model of the activated sludge system;
(3) Influent characteristics;
(4) Costing and financial parameters.

(1) The design sludge age

The selected operating sludge age depends on the minimum sludge age required to:

- Obtain a substantially complete removal of organic material (incl. detergents and soaps);
- Allow development of protozoa predating on free bacteria.

When the above conditions are satisfied, a clean effluent with a low content of suspended solids and biodegradable organic material can be produced. In Chapter 3 and Appendix 3 it is demonstrated that a minimum sludge age of 2 to 3 days should be sufficient for temperatures >14°C. The sludge age may be marginally higher if complete removal of detergents is required to avoid foaming problems. On the other hand, when the sludge age is selected too high, this will result in growth of nitrifiers followed by the associated cycles of system instability.

(2) Parameters of the steady state model of the activated sludge system
With respect to the ideal model of the activated sludge system, the following parameters are distinguished:

- Characteristics defining sludge production (Y, f_{cv} and f);
- Characteristics defining sludge composition (f_v, f_n and f_p);
- The decay rate constant of the active sludge (b_h), which is temperature dependent;
- The Vesilind constants defining sludge settleability (k and v_0).

In Table 14.7 the default parameter values of the ideal activated sludge model are given. The values of the settleability constants correspond to sludge with poor settling characteristics. Also included in Table 14.7 is the oxygen transfer efficiency of the surface aerators, required to estimate the use of electricity for aeration.

In Table 14.8 the data required for the characterisation of the influent is given. The values of flow and concentration of organic material, as well as the composition in terms of non-biodegradable particulate and soluble fractions are required. To estimate whether nutrient addition (N, P) is required and if so, how much, the influent nitrogen and phosphorus concentrations should also be determined.

Table 14.7 Default parameter values of the activated sludge system for secondary treatment

Parameter	Symbol	Default value	Unit of measure
Sludge related parameters:			
– sludge yield	Y	0.45	$g\ VSS \cdot g^{-1}\ COD$
– ratio COD/VSS	f_{cv}	1.50	$g\ COD \cdot g^{-1}\ VSS$
– endogenous residue	f	0.20	$g\ VSS \cdot g^{-1}\ VSS$
– decay rate	b_h	$0.24 \cdot 1.04^{(T-20)}$	d^{-1}
Settleability parameters:			
– Vesilind parameter	k	0.46	$l \cdot g^{-1}\ TSS$
– Vesilind parameter	v_0	6.0	$m \cdot h^{-1}$
– diluted sludge volume index	I_{DSV}	120	$ml \cdot g^{-1}\ TSS$
– settler – thickener height	H_d/H_{th}	4/3	m
– settler – thickener safety factor	s_{fd}/s_{fth}	2/1.5	(–)
Sludge composition:			
– organic fraction	f_v	0.70	$g\ VSS \cdot g^{-1}\ TSS$
– nitrogen fraction	f_n	0.10	$g\ N \cdot g^{-1}\ VSS$
– phosphorus fraction	f_p	0.025	$g\ P \cdot g^{-1}\ VSS$
Efficiency of power generation	η_{el}	35	%
Oxygen transfer efficiency (actual)	OT_a	$1.2 \cdot 1.03^{(T-20)}$	$kg\ O_2 \cdot kWh^{-1}$

Table 14.8 Wastewater characteristics

Parameter	Symbol	Unit of measure
Influent flow rate	Q_i	$m^3 \cdot d^{-1}$
Organic material:		
– COD concentration	S_{ti}	$mg\ COD \cdot l^{-1}$
– non biodegradable soluble fraction	f_{ns}	$mg\ COD \cdot mg^{-1}\ COD$
– non biodegradable particulate fraction	f_{np}	$mg\ COD \cdot mg^{-1}\ COD$
Nutrients:		
– nitrogen	N_{ti}	$mg\ N \cdot l^{-1}$
– phosphorus	P_{ti}	$mg\ P \cdot l^{-1}$

The activated sludge system should be designed for the lowest expected wastewater temperature. Temperatures above this minimum will result in improved performance. As for industrial wastewaters, the temperature of the influent is often a result of upstream process operations. Here it is important that large fluctuations in temperature are avoided (consider the implementation of a feed buffer tank) and that the temperature in the bioreactor will not exceed 40°C (<35–38°C is recommended for nitrification), as this will result in decay of biomass. In regions with a hot climate, intensive solar radiation or when concentrated wastewater is treated (generation of reaction heat), the temperature in the bioreactor can increase several degrees above the influent temperature and influent coolers might have to be installed.

As for the digester, this unit can be operated either at or above ambient temperature (when the combustion energy in the produced methane is used for heating). At higher temperatures, the rate and extent of solids removal in the anaerobic digestion process will increase and the required digestion volume will be reduced. However, this will be at the expense of additional investment costs for a gas motor, heat exchangers and control & safety equipment.

When the required data in Table 14.7 and Table 14.8 have been gathered, as well as the financial and costing data in Table 14.3 to Table 14.5, the design optimisation of system A1 is performed according to the following procedure:

(1) For the selected sludge age, determine the sludge mass that will develop in the system and the daily excess sludge production:

$$C_r = Y \cdot R_s/(1 + b_h \cdot R_s) \quad (3.30)$$

$$MX_v = [(1 - f_{ns} - f_{np}) \cdot (1 + f\ \cdot b_h \cdot R_s) \cdot C_r + f_{np} \cdot R_s/f_{cv}] \cdot Q_i \cdot S_{ti} \quad (3.48)$$

$$MX_t = MX_v/f_v \quad (3.49)$$

$$ME_v = MX_v/R_s \quad (3.50)$$

$$ME_t = MX_t/R_s = ME_v/f_v \quad (3.51)$$

(2) Determine the optimal sludge concentration in the reactor for which the combined construction costs of reactor and settler MC_{rd} will be minimised and calculate the resulting unit volumes. That is minimise the following equation:

$$MC_{rd} = C_r \cdot V_r + C_d \cdot V_d \quad (8.39)$$

Where the volumes of the units are given by:

$$V_r = MX_t/X_t \quad (3.54)$$
$$V_d = s_{fd} \cdot (H_d/v_0) \cdot \exp(k \cdot X_t) \cdot Q_i \quad (8.32)$$

Determine the value of the critical sludge recycling factor s_c and select an appropriate value for s ($s \geq s_c$). Verify the hydraulic residence time in the settler and adapt the value of s if necessary (Section 8.3). Determine the excess sludge flow:

$$q = V_r/R_s \text{(hydraulic sludge wasting)}$$
$$q = V_r \cdot X_t/(R_s \cdot X_r) \text{ (sludge wasting from return sludge line)} \quad (3.16)$$

(3) Determine the optimal thickened sludge concentration for which the construction costs of thickener and digester C_{thdi} are minimal and calculate the resulting volumes. In the optimisation procedure the following parameters are calculated as a function of the thickened sludge concentration (solids flux design is assumed):

$$X_l = (X_{th}/2) \cdot [1 + (1 - 4/(k \cdot X_{th}))^{0.5}] \quad (12.5b)$$
$$F_l = X_{th} \cdot v_0 \cdot (k \cdot X_l - 1) \cdot \exp(-k \cdot X_l) \quad (12.5a)$$
$$A_{th} = s_{fth} \cdot ME_t/F_l \quad (12.7b)$$
$$V_{th} = H_{th} \cdot A_{th} \quad (12.8b)$$
$$q_{th} = q \cdot (X_t/X_{th}) \quad (12.9b)$$
$$V_{di} = q_{th} \cdot (20 \cdot 1.1^{(20-T)} + 5) \quad (12.63)$$
$$MC_{thdi} = C_{th} \cdot V_{th} + C_{di} \cdot V_{di} \quad (12.80)$$

The minimum value C_{thdi} indicates the optimal concentration that minimises the total construction costs of thickener and digester. For the optimal value, check if the hydraulic residence time in the thickener is adequate (<1 day for short sludge ages). If this is not the case, decrease the selected thickened sludge concentration. For the hydraulic residence time in the digester R_{di} (which depends on the temperature), calculate the digestion efficiency of active and inactive sludge, the digested sludge mass and the stabilised excess sludge production (both total and volatile):

$$\eta_{dp} = (36 + 0.67 \cdot T)/100 \quad (12.64)$$
$$\eta_{dn} = (10 + 0.19 \cdot T)/100 \quad (12.65)$$
$$MS_d = ME_v \cdot f_{cv} \cdot (f_{av} \cdot \eta_{dp} + (1 - f_{av}) \cdot \eta_{dn}) \quad (12.66)$$
$$ME_{ve} = ME_v - MS_d/f_{cv} \quad (12.74)$$
$$ME_{te} = ME_t - MS_d/f_{cv} \quad (12.75)$$

Determine if the nutrient concentration in the influent is sufficient to cover the demand for nitrogen and phosphorus for the production of excess sludge, using Eqs. (3.59 and 3.60). Take into account that in the digester part of the nutrients contained in the excess sludge are mineralised and released into the liquid phase. After solids–liquid separation this quantity will be returned to the aeration tank. Use $f_n = 0.1$ kg N and $f_p = 0.025$ kg P per kg of digested excess sludge. If the nutrient concentration in the influent is insufficient, calculate the quantity to be added. If not, then the nutrient effluent concentrations can be calculated. The actual nutrient demand will have a value between a maximum corresponding to the nutrient demand of the produced excess sludge and a minimum corresponding to that of the produced stabilised sludge. The maximum nutrient demand of the activated sludge system (in the secondary excess sludge) and the minimum nutrient demand of the activated sludge process (i.e. considering only the nutrients present in the stabilised sludge) can be calculated as respectively:

$$N_l = f_n \cdot ME_v/Q_i \text{ and } N_{le} = f_n \cdot ME_{ve}/Q_i \quad (3.59)/(12.78)$$

$$P_l = f_p \cdot ME_v/Q_i \text{ and } P_{le} = f_p \cdot ME_{ve}/Q_i \quad (3.60)/(12.79)$$

In practice the maximum demand for nutrients will be used to calculate the addition of nutrients for systems without nutrient removal (if required), while the minimum demand is used to size nutrient removal systems.

(4) Determine the different fluxes of organic material: effluent, oxidised, digested and transformed into stabilised sludge. Furthermore define the oxygen demand (MO_t), the aeration capacity to be installed and the potential energy that can be generated from the methane produced during anaerobic sludge digestion.

$$MS_{te} = f_{ns} \cdot MS_{ti} \quad (3.18)$$

$$MS_o = (1 - f_{ns} - f_{np}) \cdot (1 - f_{cv} \cdot Y + f_{cv} \cdot b_h \cdot (1 - f) \cdot C_r) \cdot MS_{ti} \quad (3.43)$$

$$MS_d = f_{cv} \cdot (f_{av} \cdot \eta_{dp} + (1 - f_{av}) \cdot \eta_{dn}) \cdot ME_v \quad (12.66)$$

$$MS_{xve} = f_{cv} \cdot [(1 - \eta_{dp}) \cdot f_{av} \cdot mE_v + (1 - \eta_{dn}) \cdot (1 - f_{av}) \cdot mE_v]$$

Check if the mass balance for organic material closes, i.e. whether the value of $B_o = (MS_{te} + MS_o + MS_d + MS_{xve})/MS_{ti}$ is close to unity. The average oxygenation capacity will be equal to the average consumption of oxygen. However, in order to be able to meet fluctuating oxygen demands, the installed capacity will have to be larger than the average capacity. The variation in oxygen demand depends on the characteristics of the wastewater (variation in influent COD and flow rate) and on the system configuration (e.g. the presence of a feed buffer tank and the flow pattern in the aeration tank). In general the maximum oxygen demand rarely exceeds more than 1.5 times the average demand:

$$MO_t = MS_o \text{ and } MO_{tm} = 1.5 \cdot MS_o$$

For a more accurate estimate of the required maximum oxygenation capacity, it will be necessary to perform dynamic simulations using the estimated variations in influent flow and organic load as input. For surface aeration, with the assumed oxygen transfer efficiency of the aerator in terms of mass of oxygen transferred per unit of energy consumed (OT_a), the average and maximum (installed) power can be calculated. For diffused aeration the calculation of the power requirements is demonstrated in Section 4.1.5 and Example 14.15.

$$P_{aer} = MO_t/(24 \cdot OT_a) \quad \text{and} \quad P_{aerm} = MO_{tm}/(24 \cdot OT_a) \tag{4.12}$$

The mass of produced methane will be equal to 25% of MS_d, the mass of digested COD. When the methane production is known, the electrical energy that may potentially be generated can be calculated. The value of this potential can be compared with the demand calculated above.

$$P_{el} = \eta_{el} \cdot 5.25/24 \cdot ME_d \tag{12.72}$$

(5) For the optimised unit volumes, use the costing data to determine the construction costs:

$$TCC = C_r \cdot V_r + C_d \cdot V_d + C_{th} \cdot V_{th} + C_{di} \cdot V_{di} + C_{ae} \cdot P_{aerm} \tag{14.1}$$

To obtain the total investment costs TIC, estimate the costs of additional units that have not been specified (using f_{ac}) and the additional project costs (using f_i):

$$TIC = f_{ac} \cdot f_i \cdot TCC \tag{14.2}$$

(6) Estimate the factors that contribute to financial, or annualized investment costs (i.e. interest rate and economic lifetime) and operational costs (personnel, operation, maintenance, insurance, energy consumption and sludge disposal). The annualized investment costs (AIC) are calculated from the total investment costs TIC, where AIC is given by:

$$AIC = TIC/a_{i,n} \tag{14.4}$$

The total operational costs are composed of several factors, which can be expressed as:

$$\begin{aligned} TOC = {} & (p + o + m + n) \cdot TIC + 365 \cdot ((24 \cdot (P_{aer} - P_{el}) \cdot C_{el} + P_h \cdot C_h) \\ & + C_{sd} \cdot ME_{te}/1000) \end{aligned} \tag{14.3}$$

where:

TOC = total operational costs ($US\$ \cdot year^{-1}$)
TIC = total investment costs (US$)
p = ratio between personnel and investment costs (% of TIC per year)
o = ratio between operational and investment costs (% of TIC per year)
m = ratio between maintenance and investment costs (% of TIC per year)
n = ratio between insurance and investment costs (% TIC per year)
C_{el} = cost of electrical energy ($US\$ \cdot kWh^{-1}$)
C_h = cost of heating medium ($US\$ \cdot m^3$ gas)
C_{sd} = cost of sludge disposal ($US\$ \cdot ton^{-1}$ TSS)

Finally, calculate the total annualized costs TAC as the sum of the annualised investment costs AIC and the operational costs TOC.

EXAMPLE 14.4

Optimise the design of an activated sludge system for secondary treatment (A1) of sewage from a city of 100,000 inhabitants. The average wastewater production is $120\ l \cdot d^{-1} \cdot hab^{-1}$ ($Q_i = 12{,}000\ m^3 \cdot d^{-1}$). The characteristics of the sewage and the costing data are presented in Table 14.9 and Table 14.10. The design sludge age is 3 days and hydraulic wasting is assumed. The methane that is produced will be flared.

Table 14.9 Wastewater characteristics of Example 14.4

Parameter	Value	Unit of measure
Influent flow rate:	12,000	$m^3 \cdot d^{-1}$
Organic material:		
– COD concentration	650	$mg\ COD \cdot l^{-1}$
– non biodegradable soluble fraction	0.10	–
– non biodegradable particulate fraction	0.08	–
Nutrients:		
– nitrogen	50	$mg\ N \cdot l^{-1}$
– phosphorus	15	$mg\ P \cdot l^{-1}$
Temperature:		
– minimum/digester	20	°C

Table 14.10 Costing data of Example 14.4

Parameter	Symbol	Value	Unit of measure
Cost per unit measure of volume/kW installed:			
– aeration tank	C_r	175	$US\$ \cdot m^{-3}$
– final settler	C_d	300	$US\$ \cdot m^{-3}$
– sludge thickener	C_{th}	400	$US\$ \cdot m^{-3}$
– digester	C_{di}	350	$US\$ \cdot m^{-3}$
– surface aeration	C_{ae}	3500	$US\$ \cdot kW^{-1}$
Cost annualization:			
– expected lifetime/interest rate	n/i	20/6%	year/per year
Multipliers:			
– construction costs of additional units	f_{ac}	1.35	(–)
– additional investment costs	f_i	1.6	(–)
Operational cost parameters:			
– electrical energy	C_{el}	0.15	$US\$ \cdot kWh^{-1}$
– sludge disposal	C_{sd}	200	$US\$ \cdot ton^{-1}$ dry sludge
– personnel/operation	p/o	3%/1%	per year of TIC
– maintenance/insurance	m/n	2%/0.3%	per year of TIC

Solution

(1) Determine the sludge mass in the system that will develop
For the given characteristics and conditions:

$$Cr = Y \cdot R_s/(1 + b_h \cdot R_s) = 0.45 \cdot 3/(1 + 0.24 \cdot 3) = 0.78 \quad (3.30)$$

$$MX_a = [(1 - f_{ns} - f_{np}) \cdot C_r] \cdot MS_{ti} \quad (3.46)$$

$$= (1 - 0.10 - 0.08) \cdot 0.78] \cdot 12{,}000 \cdot 0.65 = 5020 \text{ kg VSS}$$

$$MX_v = [(1 - f_{ns} - f_{np}) \cdot (1 + f \cdot b_h \cdot R_s) \cdot C_r + f_{np} \cdot R_s/f_{cv}] \cdot MS_{ti} \quad (3.48)$$

$$= [(1 - 0.10 - 0.08) \cdot (1 + 0.2 \cdot 0.24 \cdot 3) \cdot 0.78 + 0.08 \cdot 3/1.5] \cdot 12{,}000 \cdot 0.65$$

$$= 6991 \text{ kg VSS}$$

$$MX_t = MX_v/f_v = 6991/0.7 = 9987 \text{ kg TSS} \quad (3.49)$$

$$ME_v = MX_v/R_s = 6991/3 = 2330 \text{ kg VSS} \cdot d^{-1} \quad (3.50)$$

$$ME_t = ME_v/f_v = 2330/0.7 = 3329 \text{ kg TSS} \cdot d^{-1} \quad (3.51)$$

$$f_{av} = MX_a/MX_v = 5020/6991 = 0.72 \quad (3.52)$$

(2) Sizing the reactor – settler system
For the calculated sludge mass, optimise the volumes of reactor and settler using the minimal cost criterion, taking the sludge concentration in the reactor as the independent variable. Check the hydraulic residence time in the settler for the critical recirculation factor. Calculate the biological reactor and settler volumes and hence the construction costs as a function of X_t:

$$V_r = MX_t/X_t = 9987/X_t \quad (3.54)$$

$$V_d = (s_{fd} \cdot H_d/v_0) \cdot \exp(k \cdot X_t) \cdot Q_i = 2 \cdot 4/6 \cdot \exp(0.46 \cdot X_t) \cdot 12{,}000 \quad (8.32)$$

$$MC_{rd} = C_r \cdot V_r + C_d \cdot V_d = 175 \cdot V_r + 300 \cdot V_d \quad (8.39)$$

In Figure 14.5a the calculated construction costs of Example 14.4 are shown. It can be observed that the minimum value of the total costs is obtained for a sludge concentration $X_t = 2.4$ g TSS $\cdot$ l^{-1}, where $V_r =$ 4161 m^3 and $V_d = 2011$ m^3. The sludge volume is $2.4 \cdot 120 = 288$ ml: i.e. sufficient for zone settling. For the optimum concentration, calculate with the aid of Eq. (8.38) and/or 8.10 that for $k \cdot X_t = 1.1$ the critical recirculation factor $s_c = 0.35$. This results in a contact time (*not* hydraulic retention time) in the final settler of:

$$V_d/[Q_i \cdot (1 + s_c)] = 2011/(500 \cdot 1.35) = 3.0 \text{ h} \quad (8.36)$$

This value is at the higher limit but acceptable, especially as in practice the applied sludge recycle factor s will often be larger than the critical recirculation rate s_c, e.g. for $s = 1.0$ the contact time will be 2 hours.

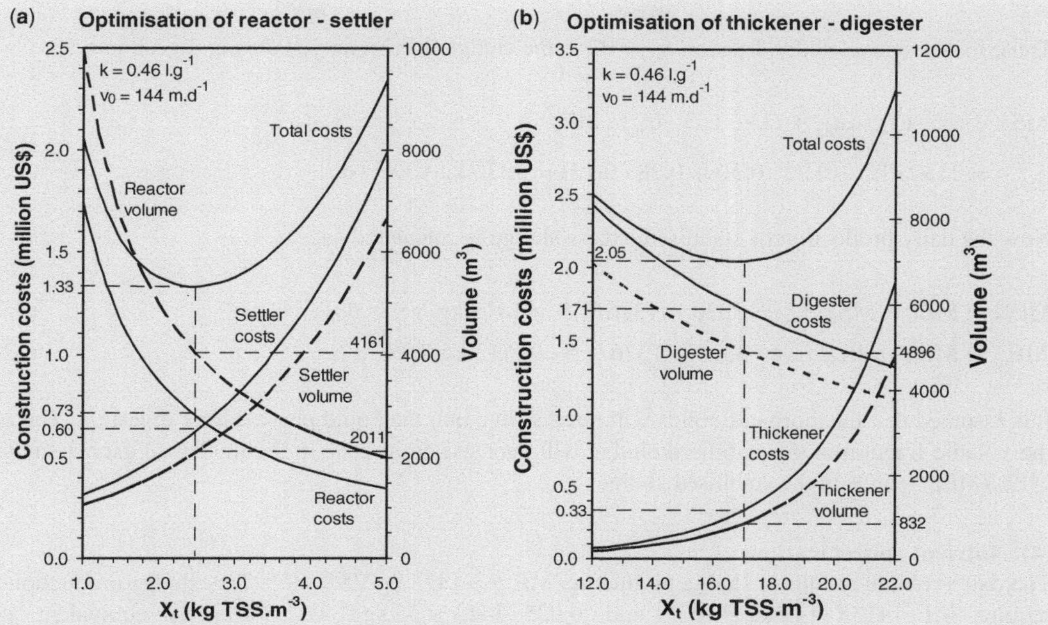

Figure 14.5 Graphical cost optimisation of system configuration A1

(3) Sizing the thickener – digester system

For the calculated excess sludge production optimise the volumes of thickener and anaerobic digester using the minimal cost criterion, taking the thickened sludge concentration as the independent variable.

$$X_l = (X_{th}/2) \cdot [1 + (1 - 4/(k \cdot X_{th}))^{0.5}] = (X_{th}/2) \cdot [1 + (1 - 4/(0.46 \cdot X_{th}))^{0.5}] \quad (12.5b)$$

$$F_l = X_{th} \cdot v_0 \cdot (k \cdot X_l - 1) \cdot \exp(-k \cdot X_l)$$

$$= X_{th} \cdot 6 \cdot (0.46 \cdot X_l - 1) \cdot \exp(-0.46 \cdot X_l) \quad (12.5a)$$

$$V_{th} = s_{fth} \cdot H_{th} \cdot A_{th} = s_{fth} \cdot H_{th} \cdot ME_t/F_l = 1.5 \cdot 4 \cdot 3329/F_l \quad (12.8b)$$

$$q_{th} = q \cdot (X_t/X_{th}) = 1387 \cdot (2.4/X_{th}) \quad (12.9b)$$

$$V_{di} = q_{th} \cdot (20 \cdot 1.1^{(20-T)} + 5) = q_{th} \cdot 25 \quad (12.63)$$

$$MC_{thdi} = C_{th} \cdot V_{th} + C_{di} \cdot V_{di} = 400 \cdot V_{th} + 350 \cdot V_{di} \quad (12.80)$$

In Figure 14.5b it can be observed that for the conditions specified in Example 14.4, the combined construction costs of thickener and digester have a minimum value for a thickened excess sludge concentration $X_{th} = 17$ g $TSS \cdot l^{-1}$. For this value of X_{th}, the volumes of thickener and digester are $V_{th} = 832$ m^3 and $V_{di} = 4896$ m^3 respectively. The excess sludge flow from the aerobic reactor is equal to $V_r/R_s = 4.161/3 = 1387$ m$^3 \cdot d^{-1}$. The hydraulic residence time in the thickener will be $24 \cdot 832/1387 = 14$ hours, an acceptable value. As for the efficiency of the anaerobic sludge digestion:

$$\eta_{dp} = (36 + 0.67 \cdot T)/100 = 0.49 \quad (12.64)$$

$$\eta_{dn} = (10 + 0.19 \cdot T)/100 = 0.14 \quad (12.65)$$

Thus, for the active sludge fraction $f_{av} = 0.72$, the sludge mass removed during digestion is:

$$MS_d = f_{cv} \cdot (f_{av} \cdot \eta_{dp} + (1 - f_{av}) \cdot \eta_{dn}) \cdot ME_v$$
$$= 2330 \cdot 1.5 \cdot (0.72 \cdot 0.50 + 0.28 \cdot 0.13) = 1376 \text{ kg COD} \cdot d^{-1} \quad (12.66)$$

Now the daily production of stabilised excess sludge is calculated as:

$$ME_{ve} = ME_v - MS_d/f_{cv} = 2330 - 1376/1.5 = 1413 \text{ kg VSS} \cdot d^{-1} \quad (12.74)$$

$$ME_{te} = ME_t - MS_d/f_{cv} = 3329 - 1376/1.5 = 2412 \text{ kg TSS} \cdot d^{-1} \quad (12.75)$$

It is assumed that the inorganic solids will not dissolve into the liquid phase during digestion. Therefore the volatile fraction in the stabilised sludge will decrease from 70% in the thickened excess sludge to $ME_{ve}/ME_{te} = 58\%$ in the stabilised sludge.

(4) Nutrient concentrations

The daily volatile stabilised sludge production $ME_{ve} = 1413 \text{ kg VSS} \cdot d^{-1}$. This sludge production will require $0.1 \cdot 1413 = 141 \text{ kg N} \cdot d^{-1}$ and $0.025 \cdot 1413 = 35 \text{ kg P} \cdot d^{-1}$. This is equivalent to a concentration of 11.8 mg N and 2.9 mg P per litre influent. Thus $38.2 \text{ mg N} \cdot l^{-1}$ will remain in the effluent. With respect to phosphorus, the effluent concentration $P_{te} = P_{ti} - P_{le} = 15–2.9 = 12.1 \text{ mg P} \cdot l^{-1}$, mainly in the form of phosphate. Therefore the extent of nutrient removal in this process configuration is very limited: 76% of the influent nitrogen load and 81% of the influent phosphorus load will leave with the effluent. However, in practice part of the phosphate released during digestion may precipitate on the sludge: $P_{ti} - P_l \leq P_{te} \leq P_{ti} - P_{le}$.

(5) Process fluxes

The oxygen consumption for the specified sludge age is calculated as:

$$MO_c = (1 - f_{ns} - f_{np}) \cdot (1 - f_{cv} \cdot Y + f_{cv} \cdot (1 - f) \cdot b_h \cdot C_r) \cdot Q_i \cdot S_{ti}$$
$$= (1 - 0.10 - 0.08) \cdot (1 - 1.5 \cdot 0.45 + 1.5 \cdot 0.8 \cdot 0.24 \cdot 0.78) \cdot 12{,}000 \cdot 0.65 \quad (3.43)$$
$$= 3524 \text{ kg} \cdot d^{-1}$$

For the assumed (actual) oxygen transfer capacity (AOTR) of $1.2 \text{ kg } O_2 \cdot kWh^{-1}$, this represents an electrical power consumption of $3524/1.2 = 2926 \text{ kWh} \cdot d^{-1}$ and an average net power demand of $2926/24 = 122$ kW. It can be verified that the dissipated energy ($122{,}000/4161 = 29 \text{ W} \cdot m^{-3}$) is more than sufficient to maintain the sludge in suspension. The average oxygen uptake rate is calculated as the ratio between the total consumption of oxygen and the aerated reactor volume: $3524/4161 = 0.85 \text{ kg } O_2 \cdot m^{-3} \cdot d^{-1}$ or $35.3 \text{ mg } O_2 \cdot l^{-1} \cdot h^{-1}$. In practice the installed aeration capacity will be larger to cover peak oxygen demands. An additional 50% capacity results in an installed capacity of 184 kW.

The mass of digested COD is equal to $MS_d = 1376$ kg COD $\cdot$ d^{-1} or $1376/4 = 344$ kg $CH_4 \cdot d^{-1}$, which according to Eq. (12.72) allows a potential for power generation of $P_{el} = 5.25/24 \cdot \eta_{el} \cdot MS_d/f_{cv} = 70$ kW at 35% efficiency. The potential for power generation represents 57% of the average energy demand for aeration (122 kW). Knowing that the mass of organic stabilised sludge (1413 kg VSS $\cdot$ d^{-1}) has a COD value of $1.5 \cdot 1413 = 2120$ kg COD $\cdot$ d^{-1}, the fluxes of organic material as fraction of the applied organic load MS_{ti} can be summarised as:

- Effluent COD = 780 kg COD $\cdot$ d^{-1} (10%)
- Oxidised COD = 3524 kg COD $\cdot$ d^{-1} (45%)
- Digested COD = 1376 kg COD $\cdot$ d^{-1} (18%)
- Excess sludge COD = 2120 kg COD $\cdot$ d^{-1} (27%)

(6) Determine the total investment costs

The construction costs include the construction of the four main units and the acquisition and installation of aeration equipment. The minimum value of the construction costs can be determined from the calculated volumes and the unit price of each unit type, as well as the price per kW of installed aeration capacity. Using the factor f_{ac}, the construction costs of the units not explicitly considered are estimated. Finally, using factor f_i, the additional project costs (listed in Table 14.2) are accounted for. The results are shown in Table 14.11.

Table 14.11 Total investment costs of system configuration A1

Equipment	Construction costs (US$)	Volume fraction	Fraction of TIC
Reactor	730,000	32%	8%
Settler	600,000	19%	7%
Thickener	330,000	8%	4%
Digester	1,710,000	41%	20%
Aeration	620,000	–	7%
Subtotal main equipment	4,010,000	–	46%
Additional unit costs	1,400,000	–	16%
Other project costs	3,250,000	–	38%
Total investment (TIC)	8,660,000	–	100%

(7) Determine the cost of financing, operational costs and total costs

The investment value can be annualised and transformed into net present value. Under the conditions of Example 14.4 (i.e. interest rate $i = 6\%$ and expected lifetime $n = 20$ years), the annual costs will be 8.7% of the total investment costs (TIC) for a period of 20 years. Thus an annualized investment cost of $0.087 \cdot 8{,}660{,}000 =$ US$ 760,000 per year or US$ 7.6 per capita per year. For the daily per capita contribution of 120 litres, the costs per unit volume of treated wastewater is $7.6/(0.12 \cdot 365) = 17.4$ US$ct per m^3. In the absence of a more detailed analysis, the annual costs of personnel, operation, maintenance and insurance can be estimated as a percentage of the TIC. The other operational costs include costs of energy for aeration, sludge disposal, heating of the digester and discharge levies (the latter two are not required in this example). In Table 14.12 the annualised investment and operational costs are presented.

Although in practice the actual numbers may deviate from those adopted in the example, invariably the annualised investment costs represent the largest cost component (in this case 46%). On the other hand, the costs of aeration are in reality a less important factor in the operational costs than generally perceived. Other operational costs, especially the costs of sludge disposal and personnel (for smaller installations) may be equally or more important.

Table 14.12 Annualised investment and operational costs of system configuration A1

Cost item	Annual costs US$ · year^{-1}	Costs per m^3 US$ct · m^{-3}	Cost per inh. US$ · inh^{-1} · yr^{-1}	Fraction %
Investment costs	760,000	17.4	7.6	46
Operational costs	890,000	20.3	8.9	54
– aeration	160,000	3.7	1.6	10
– sludge disposal	180,000	4.1	1.8	11
– personnel	260,000	5.9	2.6	16
– operation	90,000	2.1	0.9	5
– maintenance	170,000	3.9	1.7	11
– insurance	30,000	0.7	0.3	2
Total annualized costs	1,650,000	37.7	16.5	100

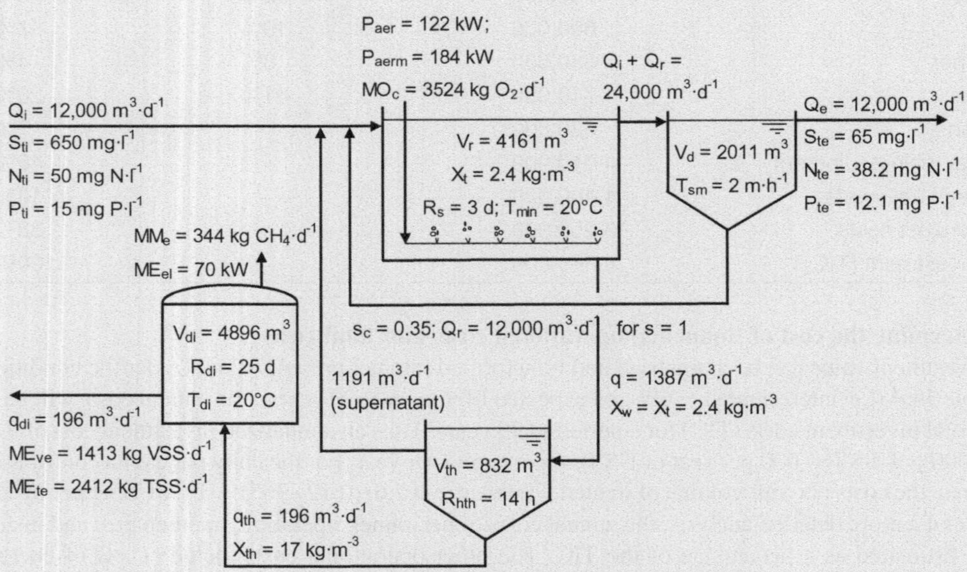

Figure 14.6 Schematic flow diagram of the optimised design of configuration A1

(8) Schematic diagram of the optimized system design

Figure 14.6 gives a schematic overview of the design of the optimised wastewater treatment plant of Example 14.4. In this figure all the main parameters and fluxes calculated in steps 1 to 7 above are indicated. Finally, in Table 14.16 located at the end of Section 14.2.3, the annualized treatment costs of the conventional configurations for secondary treatment (A1 and A2) are compared with those of configuration B1, combined anaerobic-aerobic treatment.

14.2.2 System A2: Secondary treatment with primary settling

In order to optimise system configuration A2, consisting of an activated sludge system with secondary treatment preceded by primary settling, essentially the same procedure is followed as for system A1. Table 14.13 list the additional data required to specify the conditions and efficiency of the pre-treatment. With these values, the characteristics of the settled sewage are calculated and the rest of the system will be optimised according to the procedure described in the previous section for system A1.

Table 14.13 Design data for primary settling in Example 14.5

Parameter	Symbol	Value	Unit of measure
Hydraulic residence time	R_{h1}	2	hours
COD removal efficiency	η_1	30–40	%
Primary sludge concentration	X_{d1}	30–50	$g\ TSS \cdot l^{-1}$
Organic fraction of primary sludge	f_v	0.6–0.7	$g\ VSS \cdot g^{-1}\ TSS$
Particulate, non biodegradable COD fraction in pre-settled sewage	f'_{np}	0.01–0.02	(–)

EXAMPLE 14.5

Optimise the design of an activated sludge system consisting of:

- Primary settler;
- Completely mixed aerated reactor for secondary treatment (no nitrification);
- Final settler;
- Thickener (for the excess sludge);
- Anaerobic digester for the stabilisation of primary and secondary sludge.

Adopt the influent characteristics and costing values as specified in Table 14.9 and Table 14.10 (i.e. the same values adopted in Example 14.4). For the primary clarification the following additional data are specified:

- Residence time in the primary settler $R_{h1} = 2$ hours;
- Removal efficiency of the organic material $\eta_1 = 40\%$;
- Solids concentration of the primary sludge $X_{d1} = 40\ g \cdot l^{-1}$;

- Residual fraction of particulate, non biodegradable organics in the pre-settled sewage $f'_{np} = 0.02$;
- The cost per unit volume of the primary settler $C_{d1} = 375$ US$ $\cdot m^{-3}$;
- The value of operational cost factor "p" (personnel) is increased to 4.5% to reflect the additional work load resulting from the addition of a new unit;
- Volatile solids fraction $f_v = 0.65$ in the primary sludge and 0.8 in the secondary excess sludge.

Solution

(1) Characterisation of the settled sewage and the primary sludge

Calculate the concentration and composition of the organic material in the settled sewage and the flow and composition of the primary sludge. For the assumed COD removal efficiency of 40% in the primary clarifier, the concentration in the settled influent will be:

$$S'_{ti} = (1 - \eta_1) \cdot S_{ti} = 0.60 \cdot 650 = 390 \text{ mg COD} \cdot l^{-1}$$

It is assumed that the concentration of non-biodegradable, soluble material in the influent will not change in the primary settler:

$$S'_{ns} = 0.10 \cdot S_{ti} = 0.10 \cdot 650 = 65 \text{ mg COD} \cdot l^{-1} \quad \text{and}$$

$$f'_{ns} = f_{ns}/(1 - \eta_1) = 0.1/0.6 = 0.167$$

The non-biodegradable fraction in the settled influent is given as $f'_{np} = 0.02$. The mass of primary sludge is calculated as:

$$ME_{t1} = MS_{ti} \cdot \eta_1/(f_{cv} \cdot f_v) = 7800 \cdot 0.4/(1.5 \cdot 0.65) = 3200 \text{ kg TSS} \cdot d^{-1} \tag{12.1}$$

This results in a primary excess sludge production of $3200/40 = 80 \text{ m}^3 \cdot d^{-1}$ for the assumed concentration of 40 g $TSS \cdot l^{-1}$.

(2) to (9) Optimize system configuration A2

Using the values of S'_{ti}, f'_{ns} and f'_{np}, the system parameters are calculated as shown in steps 1 to 8 of Example 14.4. The only difference is that in the current example, the excess sludge flow to the digester will be composed of two parts:

- Primary sludge, with a flow rate $q_1 = 80 \text{ m}^3 \cdot d^{-1}$ and a concentration of 40 g $TSS \cdot l^{-1}$;
- Thickened excess sludge from the activated sludge system, with a flow rate $q_{th} = 88 \text{ m}^3 \cdot d^{-1}$ and concentration $X_{th} = 17$ g $TSS \cdot l^{-1}$.

It is assumed that the efficiency of the anaerobic digestion of primary sludge is equal to that of the active fraction of the secondary excess sludge ($\eta_{dp} = 0.49$). The total volatile excess sludge mass $ME_{v1} + ME_{v2}$ is 2080 kg $VSS \cdot d^{-1}$ of primary sludge and 1202 kg $VSS \cdot d^{-1}$ of secondary sludge. After anaerobic digestion 1734 kg $VSS \cdot d^{-1}$ remains as stabilised sludge. The required volume of the digester $V_{di} = (80 + 88) \cdot (20 \cdot 1.1^{(20-20)} + 5) = 4209 \text{ m}^3$.

The concentration of nutrients in the effluent is assumed to be at the maximum value possible: i.e. all nutrients contained in the excess sludge are released to the liquid phase during anaerobic sludge digestion. The mass of nitrogen and phosphorus removed with the stabilised sludge from the digester is equal to $0.1 \cdot 1734 = 173$ kg N $\cdot$ d^{-1} and $0.025 \cdot 1734 = 43.3$ kg P $\cdot$ d^{-1}. This amounts to $N_{le} = 14.4$ mg N $\cdot$ l^{-1} and $P_{le} = 3.6$ mg P $\cdot$ l^{-1}, leaving in the effluent $50–14.4 = 35.6$ mg N $\cdot$ l^{-1} and $15–3.6 = 11.4$ mg P $\cdot$ l^{-1}.

In Table 14.14 the annualised investment costs of system configuration A2 are listed, while Figure 14.7 presents the configuration of the optimised design, including the fluxes of organic material and the system costs. In Table 14.16 at the end of Section 14.2.3, the annualized treatment costs of configurations A1, A2 and B1 are compared.

Table 14.14 Annualised investment and operational costs of system configuration A2

Cost item	Annual costs US\$ $\cdot$ year^{-1}	Costs per m^3 US\$ct $\cdot$ m^{-3}	Cost per inh. US\$ $\cdot$ inh^{-1} $\cdot$ yr^{-1}	Fraction %
Investment costs	620,000	14.2	6.5	41
Operational costs	880,000	20.1	8.8	59
– aeration	100,000	2.3	1.0	7
– sludge disposal	230,000	5.3	2.3	15
– personnel	320,000	7.3	3.2	21
– operation	70,000	1.6	0.7	4
– maintenance	140,000	3.2	1.4	9
– insurance	20,000	0.5	0.2	1
Total costs	1,500,000	34.2	15.0	100

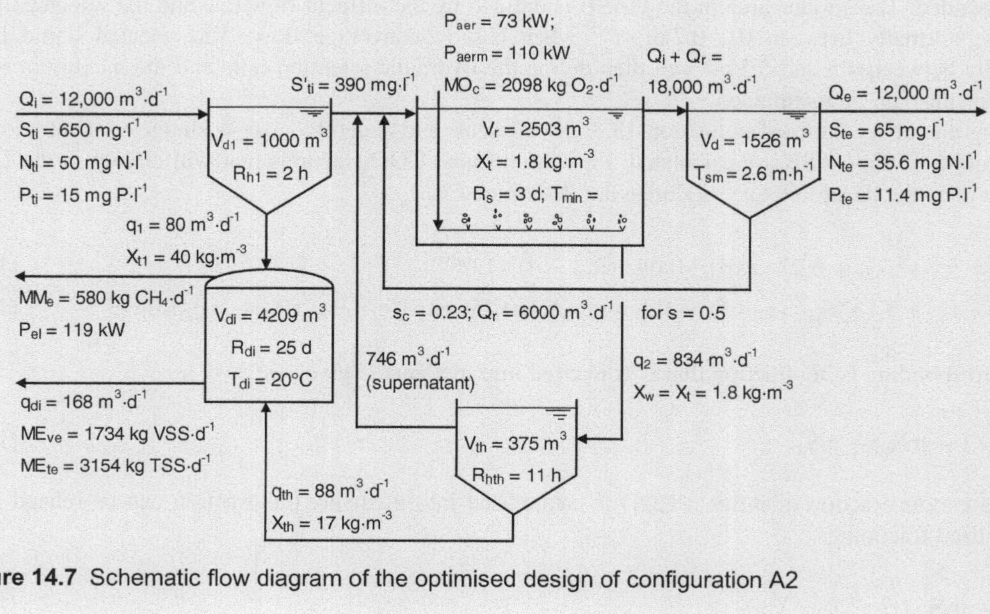

Figure 14.7 Schematic flow diagram of the optimised design of configuration A2

14.2.3 System B1: Combined anaerobic-aerobic treatment

In system B1 a large part of the organic material present in the influent will be removed in the UASB reactor. However, aerobic post treatment will be required to in order remove the residual organic material in the anaerobic effluent. As the growth rate of the nitrifiers is reduced due to the presence of sulphides in the UASB effluent, it is possible to increase the sludge age to 5 days while nitrification is still avoided. The increase in sludge age results in higher removal of surfactants and thereby improves the effluent quality.

For the optimisation of system B1 it will be necessary to determine the appropriate value for the hydraulic residence time or rather the anaerobic sludge age in the UASB reactor. These values are determined partly by economic constraints, but other important considerations are that the sludge retention time in the UASB is sufficient to remove a significant proportion of the applied COD load while simultaneously a well stabilised sludge is produced that can be dewatered directly. Both requirements require application of a high value of the sludge age. Alternatively, the anaerobic excess sludge can be digested in a small heated anaerobic digester together with the aerobic sludge, which will increase both rate and extent of solids degradation. This subject has been discussed extensively in Section 13.3.3 (Example 13.8) and Section 13.5.1.2 (Example 13.14). In this example only the configuration with a heated digester will be discussed.

An important question is what constitutes a well stabilised sludge. The anaerobic excess sludge from a UASB may contain a relatively high fraction of non metabolised particulate organic material if the sludge retention time is inadequate for the operational temperature. In the absence of further data, it is proposed that a reduction of f_{pu} in the anaerobic excess sludge to a value of less than 0.3 is sufficient for proper stabilisation. The fraction f_{pu} is defined here as the sum of the non-metabolised particulate COD fraction and the active fraction of the secondary excess sludge (if it is recycled to the UASB). This implies that the active fraction of the anaerobic biomass is excluded, as the decay rate of these micro-organisms is very low.

As to the hydraulic residence time, Van Haandel and Lettinga (1995) recommend a minimum hydraulic residence time of 6 hours in UASB reactors with a standard GLS design operating at a temperature around 25°C. To ensure proper operational stability, a hydraulic residence time of at least 8 hours at 20°C is recommended. The surface area of the UASB is defined by the influent flow rate and the selected upflow velocity, normally between 0.6–0.7 $m \cdot h^{-1}$ when based on average flow. The selected liquid height (typically between 4.5 and 5.5 m) will then define the hydraulic retention time and the maximum sludge inventory that can be maintained.

Using the theory outlined in Section 13.3, the effluent- and anaerobic excess sludge concentration and composition of the UASB are calculated. First the influent COD fractions that will end up in the UASB effluent or in the anaerobic excess sludge are determined:

$$mS_{eu} = f_{ns} + f_{h2s} + 0.27 \cdot \exp[-0.04 \cdot (R_{su} - 4)]/1.067^{(T-25)} \tag{13.16}$$

$$mS_{xvu} = f_{np} + f_{cv} \cdot Y_{an} \cdot (1 - f_{ns} - f_{np} - f_{h2s}) + 0.25 \cdot \exp[(-0.04 \cdot (R_{su} - 4)]/1.067^{(T-25)} \tag{13.17}$$

The corresponding COD fraction that is converted into methane is given as:

$$mS_{du} = 1 - mS_{eu} - mS_{xu} \tag{13.18}$$

When a certain fraction of influent COD is catabolised into methane, this fraction can be related to the metabolised fraction:

$$mS_{mb} = mS_{du}/(1 - f_{cv} \cdot Y_{an}) \tag{13.19}$$

Using the metabolised COD fraction, the active and endogenous COD fractions of the excess sludge can be calculated (mg VSS · d · mg^{-1} COD):

$$mX_{au} = Y_{an} \cdot R_{su}/(1 + b_{an} \cdot R_{su}) \cdot mS_{mb} \tag{13.20}$$

$$mX_{eu} = f \cdot b_{an} \cdot R_{su} \cdot mX_{au} \tag{13.21}$$

The non biodegradable particulate COD fraction in the influent f_{np} accumulates as an inert organic fraction in the anaerobic excess sludge (mg VSS · d · mg^{-1} COD):

$$mX_{iu} = f_{np} \cdot R_{su}/f_{cv} \tag{13.22}$$

Now the non-metabolised biodegradable particulate organic fraction mX_{bpu} in the volatile sludge is calculated from mS_{xvu} and the three previously calculated sludge fractions:

$$mX_{bpu} = mS_{xvu} \cdot R_{su}/f_{cv} - mX_{au} - mX_{eu} - mX_{iu} \tag{13.24}$$

$$mX_{vu} = mX_{au} + mX_{eu} + mX_{iu} + mX_{bpu} \tag{13.23}$$

Using the COD mass fractions defined above and including the mineral fraction mX_{mu}, the anaerobic sludge mass MX_{tu} that will develop can be calculated as a function of the anaerobic sludge age. Knowing that the average solids concentration X_{tu} in a properly operated UASB reactor equipped with a conventional GLS separator will be around 15–20 kg TSS · m^{-3} (this includes the settling compartment), the required UASB volume can be determined.

The selected value of the anaerobic sludge age depends on the desired COD removal efficiency, the maximum allowed f_{pu} value (defined as mX_{bpu}/mX_{vu}) and on the decision whether or not the excess anaerobic sludge will be further digested in an anaerobic digester, together with the aerobic excess sludge. In the subsequent step, the post-treatment system consisting of conventional aerobic treatment, final settler, thickener and heated digester is optimised according to the procedure outlined for system A1.

The COD removal efficiency in an UASB reactor can also be estimated as a function of the retention time using the empirical relationship given by Eq. (13.5), although the predicted COD removal is probably too optimistic. Furthermore, Eq. (13.5) is a simplified empirical relationship that does not take into account that a large part of the influent COD removal is actually due to accumulation of non-metabolised organic material in the anaerobic sludge and furthermore does not predict the UASB effluent quality in terms of COD composition.

EXAMPLE 14.6

Optimise the design of an activated sludge system composed of an UASB, a completely mixed aerobic reactor for secondary treatment of the effluent, a final settler, sludge thickener and heated anaerobic digester. Adopt the influent characteristics and costing data specified in Table 14.9 and Table 14.10 (i.e. the same values used in Example 14.9 and Example 14.5). For the UASB reactor the following additional data is specified:

- Design reactor upflow velocity is 0.7 m · h^{-1}
- Average UASB sludge concentration is 17.5 g TSS · l^{-1}, effluent TSS = 75 mg · l^{-1};

- $Y_{an} = 0.05$ mg VSS · mg^{-1} COD; $b_{an} = 0.01$ d^{-1};
- Anaerobic sludge concentration in the upper part of the UASB digestion zone (from where the excess sludge should be discharged preferably) is 30 g TSS · l^{-1};
- Inert suspended solids content (minerals) in the influent is 45.5 mg TSS · l^{-1}, so $f_{mi} = 0.07$ g ISS · g^{-1} COD;
- SO_4^{2-} concentration is 65 mg · l^{-1}, of which 80% is converted to H_2S in the UASB;
- Methane concentration in the UASB effluent is 20.4 mg CH_4 · l^{-1} (70% CH_4 in biogas);
- The value of the cost factor for operations "p" is increased to 5% to reflect the additional workload from operating the UASB.

For the aerobic post-treatment system the following additional data are given:

- Pre-aeration is not applied, instead the anaerobic effluent is contacted with the aerobic return sludge in an aerated selector, which is a part of the aeration tank volume. Hence the nitrification rate is significantly reduced: $\mu_m = 0.15$ d^{-1}, $b_n = 0.04$ d^{-1} and $K_n = 1.0$ mg N · l^{-1}
- Aerobic sludge age is 5 days; $X_t = 3$ g · l^{-1}; $f_v = 0.85$;
- Settleability of aerobic sludge following anaerobic pre-treatment is good: $k = 0.35$ l · g^{-1} TSS, $v_0 = 216$ m · d^{-1} and $I_{dsv} = 80$–100 ml · g^{-1} TSS.
- The attainable thickened sludge concentration is 35 g TSS · l^{-1} for a solids loading rate of 30 kg TSS · m^{-2} · d^{-1}. The digester temperature $T_{di} = 37$°C (waste heat from gas utilization is used for heating) and $\eta_{el} = 35\%$.

As the volume of the aerobic post-treatment units will be reduced compared to that in system configuration A1 and A2, the values of the relevant cost factors will increase. Furthermore, for the anaerobic-aerobic configuration it makes sense now to evaluate the costs and benefits of power generation. It is assumed that the unit benefits of the surplus power sold to the power company is approximately half the purchasing cost (purchasing price 0.15 US$ · kWh^{-1} and selling price 0.07 US $ · kWh^{-1}).

- $C_u = 400$ US$ · m^{-3}
- $C_r = 225$ US$ · m^{-3}
- $C_d = 300$ US$ · m^{-3}
- $C_{th} = 600$ US$ · m^{-3}
- $C_{di} = 500$ US$ · m^{-3}
- $C_{aer} = 4500$ US$ · kW^{-1}
- $C_{gen} = 2000$ US$ · kW^{-1}

Solution

(1) Determine the UASB performance as function of the anaerobic sludge age

As a first step in the calculation procedure, the anaerobic performance is characterised as a function of the anaerobic sludge age. For the applied influent flow rate of 12,000 m^3 · d^{-1} and design upflow velocity of 0.7 m · h^{-1}, the required UASB surface area is 714 m^2 (26.7 × 26.7 m). The UASB sludge mass and hence the anaerobic sludge age will then vary as a function of the selected UASB height. Using (Eqs. 13.16 and 13.17), the fractions of influent COD that will end up in the effluent and in the excess sludge are calculated, whereas Eqs. (13.20 to 13.27) can be used to characterise the UASB sludge mass. The calculations are demonstrated here for a reactor liquid height of 5.15 m (5.5 m total height), corresponding to an anaerobic sludge age of 37 days. First the equivalent H_2S fraction remaining in the UASB effluent is determined. For $[SO_4] = 65\ mg \cdot l^{-1}$ and 80% conversion to H_2S in the UASB, f_{h2s} is calculated as 0.058.

$$\begin{aligned} mS_{eu} &= f_{ns} + f_{h2s} + 0.27 \cdot \exp[-0.04 \cdot (R_{su} - 4)]/1.067^{(T-25)} \\ &= 0.1 + 0.06 + 0.27 \cdot \exp(-0.04 \cdot 33)/1.067^{(-5)} = 0.26 \end{aligned} \quad (13.16)$$

$$\begin{aligned} mS_{xvu} &= f_{np} + f_{cv} \cdot Y_{an} \cdot (1 - f_{ns} - f_{np} - f_{h2s}) \\ &\quad + 0.25 \cdot \exp[(-0.04 \cdot (R_{su} - 4)]/1.067^{(T-25)} \\ &= 0.1 + 1.5 \cdot 0.05 \cdot (1 - 0.1 - 0.08 - 0.06) \\ &\quad + 0.25 \cdot \exp(-0.04 \cdot 33)/1.067^{(-5)} = 0.23 \end{aligned} \quad (13.17)$$

The digested (methanised) fraction $mS_{du} = 1 - mS_{eu} - mS_{xvu} = 1 - 0.26 - 0.23 = 0.51$. The corresponding metabolised fraction can be calculated with Eq. (13.19):

$$mS_{mb} = mS_d/(1 - f_{cv} \cdot Y_{an}) = 0.51/(1 - 1.5 \cdot 0.05) = 0.55 \quad (13.19)$$

From the metabolised COD fraction, the active, endogenous and inert fractions of the volatile sludge are calculated with Eqs. (13.20 to 13.22) as:

$$\begin{aligned} mX_{au} &= Y_{an} \cdot R_{su}/(1 + b_{an} \cdot R_{su}) \cdot mS_{mb} \\ &= 0.05 \cdot 37/(1 + 0.01 \cdot 37) \cdot 0.55 = 0.75\ g\ VSS \cdot d \cdot g^{-1}\ COD \end{aligned} \quad (13.20)$$

$$mX_{eu} = f \cdot b_{an} \cdot R_{su} \cdot mX_{au} = 0.2 \cdot 0.01 \cdot 37 \cdot 0.75 = 0.06\ g\ VSS \cdot d \cdot g^{-1}\ COD \quad (13.21)$$

$$mX_{iu} = f_{np} \cdot R_{su}/f_{cv} = 0.08 \cdot 37/1.5 = 1.97\ g\ VSS \cdot g^{-1}\ COD \quad (13.22)$$

Now, from the difference between the value of mS_{xvu} and the sum of the sludge mass fractions determined above, mX_{bpu} can be calculated:

$$\begin{aligned} mX_{bpu} &= mS_{xvu} \cdot R_{su}/f_{cv} - mX_{au} - mX_{eu} - mX_{iu} \\ &= 0.23 \cdot 37/1.5 - 0.75 - 0.06 - 1.97 = 2.88\ g\ VSS \cdot g^{-1}\ COD \end{aligned} \quad (13.24)$$

$$\begin{aligned} mX_{vu} &= mX_{au} + mX_{eu} + mX_{iu} + mX_{bpu} \\ &= 0.75 + 0.06 + 1.97 + 2.88 = 5.66\ g\ VSS \cdot g^{-1}\ COD \end{aligned} \quad (13.23)$$

$$mX_{mu} = f_{mi} \cdot R_s \quad \text{or} \quad 0.07 \cdot 37 = 2.59\ g\ ISS \cdot g^{-1}\ COD$$

$$\begin{aligned} mX_{tu} &= mX_v + mXm_u = 5.66 + 2.59 = 8.25\ g\ TSS \cdot g^{-1}\ COD \quad \text{and} \quad f_{vu} \\ &= 5.66/8.25 = 0.69 \end{aligned}$$

For the applied organic load of $7800\ kg\ COD \cdot d^{-1}$ the total sludge mass is equal to $8.25 \cdot 7800 = 64{,}370$ kg TSS. The required (liquid) UASB reactor volume is calculated with Eq. (13.28) as $MX_{tu}/X_{tu} = 64{,}370/17.5 = 3680\ m^3$, i.e. $714\ m^2$ times 5.15 m. Figure 14.8 shows the UASB volume and the anaerobic sludge age as a function of the selected liquid height in the UASB for the conditions specified in this example.

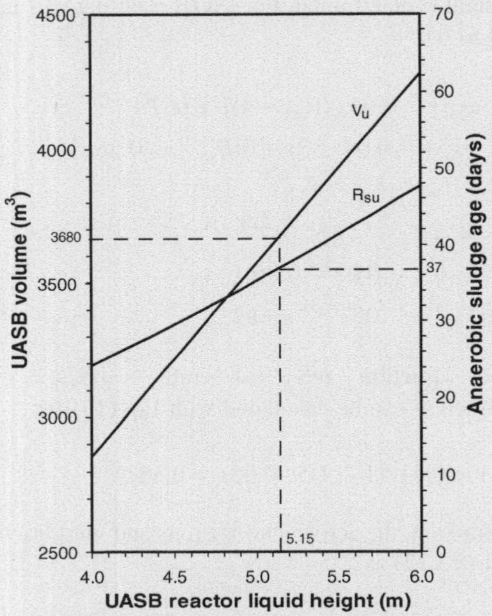

Figure 14.8 Anaerobic sludge age and reactor volume as function of the UASB liquid height

Figure 14.9 shows the COD removal efficiency and the specific excess sludge production (i.e. excess sludge production excluding the effluent solids). As to the composition of the UASB effluent, the soluble COD concentration in the UASB effluent is $0.26 \cdot 650 = 167\ mg \cdot l^{-1}$ (of which a fraction $0.06 \cdot 650 = 38\ mg \cdot l^{-1}$ consists of inorganic H_2S-COD. The particulate effluent COD concentration is equal to $f_{cv} \cdot f_{vu} \cdot X_{tue} = 1.5 \cdot 0.69 \cdot 75 = 77$ mg $COD \cdot l^{-1}$. The total effluent COD concentration is $244\ mg \cdot l^{-1}$, which defines the UASB COD removal efficiency η_{COD} as $1 - 244/650 = 62.4\%$.

The methane production is equal to $mS_d \cdot MS_{ti} = 0.51 \cdot 7800 = 4000$ kg $COD \cdot d^{-1}$ or 1000 kg $CH_4 \cdot d^{-1}$. Part of the produced methane will leave with the effluent as this methane fraction is not recovered: for $S_{ch4u} = 20.4\ mg \cdot l^{-1}$, this is calculated as 979 kg $COD \cdot d^{-1}$ or 245 kg $CH_4 \cdot d^{-1}$. The recovered methane production is 1000–245 kg = 755 $CH_4 \cdot d^{-1}$, so methane recovery is 75.5%. The recovered methane is used for energy generation: $MP_{el} = 50{,}400/(24 \cdot 3600) \cdot \eta_{el} \cdot MS_{du}/4 = 0.583 \cdot 0.35 \cdot 755 = 154$ kW.

From Figure 14.9 the average UASB excess sludge production is determined as $0.108 \cdot 7800 = 842$ kg $TSS \cdot d^{-1}$, equivalent to a volumetric production of $28\ m^3 \cdot d^{-1}$ when the sludge is wasted at a concentration of 30 g $TSS \cdot l^{-1}$. Note that the anaerobic sludge age is not defined by sludge wasting alone, but that the loss of suspended solids with the effluent should be included as well ($= 12{,}000 \cdot 0.075 = 900$ kg $TSS \cdot d^{-1}$). It can be confirmed that the sludge age is indeed $64{,}370/(842 + 900) = 37$ days. The putrescible fraction of the UASB sludge $f_{pu} = mX_{bpu}/mX_{vu} = 2.88/5.66 = 0.5$, while $f_{avu} = [(1 - f) \cdot mX_{au} + mX_{bpu}]/mX_{vu} = [(1 - 0.2) \cdot 0.75 + 2.88]/5.66 = 0.62$.

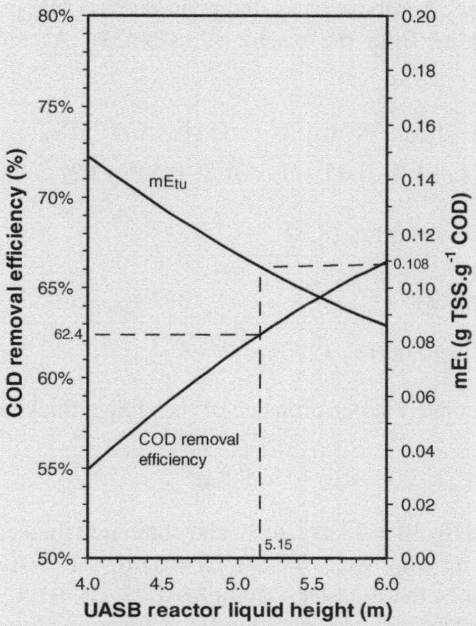

Figure 14.9 COD removal efficiency and specific (total) excess sludge production as function of the UASB liquid height

(2) Calculate the performance of the aerobic post-treatment system

Once the design of the UASB system has been defined, the performance of the post-treatment system can be evaluated, once again as a function of the anaerobic sludge age R_{su}. As a first step the UASB effluent composition is established. The COD load in the UASB effluent MS_{tu} is equal to $(1-\eta_{COD}) \cdot MS_{ti} = (1-0.624) \cdot 7800 = 2933$ kg COD $\cdot$ d^{-1}. Assuming the soluble inert COD in the influent is not retained in the UASB, $f_{nsu} = f_{ns}/(1-\eta_{COD}) = 0.1/(1\text{–}0.624) = 0.266$. To calculate the inert particulate COD fraction f_{npu}, it is assumed that the effluent TSS composition is equal to that in the reactor. The particulate COD load in the UASB effluent is $12{,}000 \cdot 0.077 = 926$ kg COD $\cdot$ d^{-1}. Assuming the active fraction is biodegradable, $MS_{npu} = (1-f_{avu}) \cdot MS_{pu} = (1-0.62) \cdot 926 = 352$ kg COD $\cdot$ d^{-1} and f_{npu} is $352/2933 = 0.12$.

The composition of the aerobic biomass can be calculated according to the procedure outlined in Example 14.4. Compared to configurations A1 and A2, a higher sludge age can be applied in the aerobic system, as the nitrifier growth rate is reduced due to the H_2S in the anaerobic effluent. It can be verified with Eq. (5.38) that the minimum sludge age for nitrification has increased to $1/(\mu_m - b_n) = 1/(0.15\text{–}0.04) = 9$ d.

The increase of R_s to 5 days will improve effluent quality significantly in terms of COD and BOD, due to the increased conversion of surfactants. For $R_{su} = 37$ days and $R_s = 5$ days, the aerobic sludge mass that will develop is equal to 3467 kg VSS ($f_{av2} = 53\%$) and $3467/0.85 = 4079$ kg TSS. For the selected aerobic sludge concentration of 3.0 g TSS $\cdot$ l^{-1}, the required aerobic reactor volume is calculated as $4079/3 = 1360\ m^3$. Aerobic excess sludge production $ME_{t2} = 4079/5 = 816\ kg\ TSS \cdot d^{-1}$ ($q_2 = 272\ m^3 \cdot d^{-1}$ when wasting from the reactor is assumed). As nitrification does not develop, $MO_t = MO_c$.

$$\begin{aligned} MO_c &= (1 - f_{nsu} - f_{npu}) \cdot (1 - f_{cv} \cdot Y + f_{cv} \cdot (1 - f) \cdot b_h \cdot C_r) \cdot MS_{tu} \\ &= (1 - 0.266 - 0.12) \cdot (1 - 1.5 \cdot 0.45 + 1.5 \cdot 0.8 \cdot 0.24 \cdot 1.02) \cdot 2933 = 1113\ kg\ O_2 \cdot d^{-1} \end{aligned}$$

The final settler volume follows from Eq. (8.32)

$$\begin{aligned} V_d &= (s_{fd} \cdot H_d/v_0) \cdot \exp(k \cdot X_t) \cdot Q_i \\ &= 2 \cdot 4/216 \cdot \exp(0.35 \cdot 3) \cdot 12{,}000 = 1270\ m^3 \end{aligned} \quad (8.32)$$

Now all that remains is to calculate the performance of the sludge thickener and heated digester.

$$V_{th} = H_{th} \cdot A_{th} = H_{th} \cdot ME_{t2}/F_{sol} = 3 \cdot 816/30 = 82\ m^3$$

The thickened excess flow is $816/30 = 23\ m^3 \cdot d^{-1}$. The combined thickened excess sludge flow to the anaerobic digester is equal to $q_w = q_1 + q_{th2} = 28 + 23 = 51\ m^3 \cdot d^{-1}$. At 30°C, the required retention time is 12.7 days, which defines the digester volume as $12.7 \cdot 51 = 652\ m^3$. Figure 14.10 shows the volume of the treatment units of system configuration B1 as function of the UASB reactor liquid height.

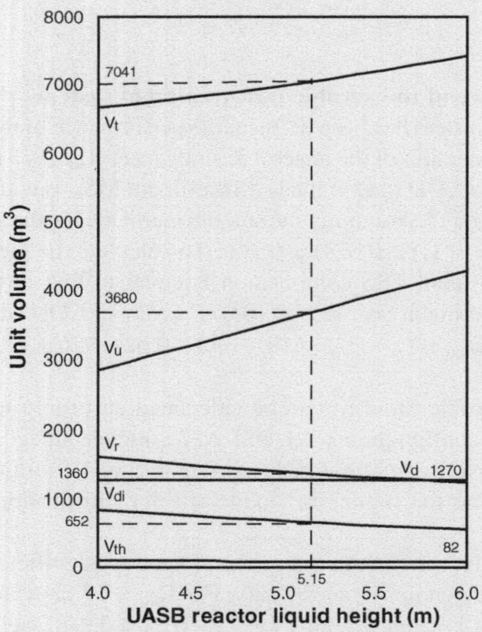

Figure 14.10 Volume of the treatment units in configuration B1 as function of the UASB liquid height

When the UASB liquid height increases (and with it the anaerobic sludge age), it can be observed that the decrease in aeration tank and digester volume does not compensate for the increase in UASB reactor volume. As shown in Figure 14.11, this is even further accentuated when the cost factors are included (as $C_u > C_r$). Thus one might be tempted to minimise the UASB liquid height, as long as stable performance is guaranteed. However, the investment costs are only one part of the equation, as will be discussed below. To finalise the example, the digester performance is calculated. The combined feed flow to the digester is characterized by:

$$ME_t = ME_{tu} + ME_{t2} = 842 + 816 = 1656\,kg\,TSS \cdot d^{-1}$$

$$ME_v = f_{vu} \cdot ME_{tu} + f_{v2} \cdot ME_{t2} = 0.69 \cdot 842 + 0.85 \cdot 816$$
$$= 576 + 693 = 1270\,kg\,VSS \cdot d^{-1}$$

$$ME_{va} = f_{avu} \cdot ME_{vu} + f_{av2} \cdot ME_{v2} = 0.62 \cdot 576 + 0.53 \cdot 693$$
$$= 723\,kg\,VSS \cdot d^{-1},\ so\ f_{av} = 57\%$$

$$ME_{vna} = ME_v - ME_{va} = 1270 - 723 = 547\,kg\,VSS \cdot d^{-1}$$

For $T = 37°C$ and using Eqs. (12.64 and 12.65), the anaerobic digestion efficiency is estimated as $\eta_{dp} = (36 + 0.67 \cdot 37)/100 = 0.61$ and $\eta_{dn} = (10 + 0.19 \cdot 37)/100 = 0.17$.

$$ME_d = \eta_{dp} \cdot ME_{va} + \eta_{dn} \cdot (ME_v - ME_{va}) = 0.61 \cdot 723 + 0.17 \cdot 547 = 532\,kg\,VSS \cdot d^{-1}$$

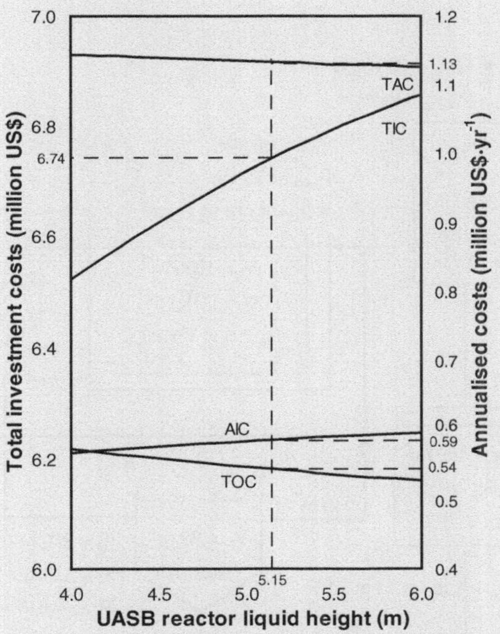

Figure 14.11 Total and annualised investment costs and yearly operational costs as function of the UASB liquid height

The stabilised excess sludge production is 1656–532 = 1123 kg TSS · d^{-1}, of which 737 kg is organic (f_v = 0.66). The active sludge mass decreases to (1–0.61) · 723 = 283 kg VSS · d^{-1}, so f_{ave} = 283/737 = 34%. The digested sludge mass ME_d is equivalent to MS_d = 798 kg COD · d^{-1} and 200 kg CH_4 · d^{-1}. The power that can be generated from the digested methane is 41 kW, which brings the total power production to 195 kW. When compared to the power demand for aeration $P_{aer} = MO_t/(24 \cdot \eta_{aer}) = 1113/(24 \cdot 1.2) = 39$ kW, it is concluded that there is ample potential for energy self-sufficiency, even when energy demand for pumping and dewatering is considered.

Figure 14.11 shows the annualised investment costs (AIC – including the additional investment costs for energy generation), the total yearly operational costs (TOC – reduced with the benefits due to the sales of surplus electricity) and the total annualised costs (TAC). It can be observed that the operational costs significantly impact the relative attractiveness of the different configurations. In general the trend is towards lower total annualised costs at increased reactor height (and anaerobic sludge age), but the difference is so small (only a 12,000 US$ · yr^{-1} decrease in costs over the liquid height interval between 4.0 and 6.0 m) that given the uncertainties and assumptions made for all practical purposes it may be considered as nonexistent.

As most customers focus more on investment costs, this might induce the selection of a lower reactor. However, when this results in an UASB sludge age of less than 25 days, this is not recommended, as it reduces the operational stability of the anaerobic process. Figure 14.12 shows the overall schematic flow diagram of system configuration B1, while Table 14.15 summarizes the annualised investment and operational costs.

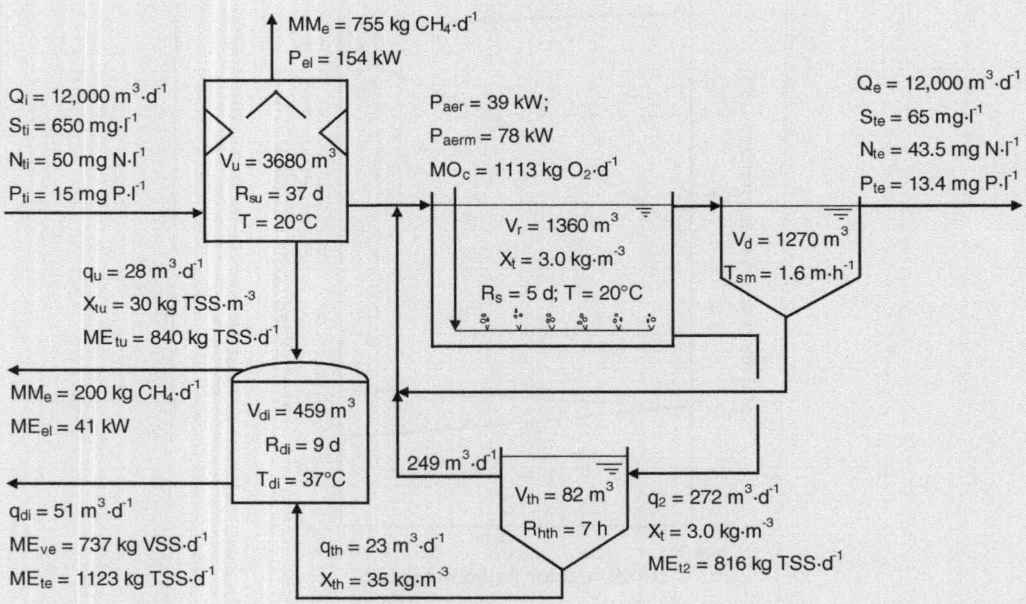

Figure 14.12 Schematic flow diagram of the optimised design of configuration B1

Table 14.15 Annualised investment and operational costs of system configuration B1

Cost item	Annual costs US$ · year^{-1}	Costs per m^3 US $ct · m^{-3}	Cost per inh. US$ · inh^{-1} · yr^{-1}	Fraction %
Investment costs	570,000	13.0	5.7	53
Operational costs	510,000	11.7	5.1	47
– aeration	50,000	1.1	0.5	5
– electricity generation	−150,000	−3.4	−1.5	−14
– sludge disposal	80,000	1.9	0.8	7
– personnel	330,000	7.3	3.2	30
– operation	60,000	1.4	0.6	6
– maintenance	130,000	3.0	1.3	12
– insurance	20,000	0.5	0.2	2
Total costs	1,080,000	24.7	10.8	100

In Table 14.16 and Table 14.17 the performance and costs of configurations A1, A2 and B1 are compared. Table 14.17 shows an interesting and perhaps surprising aspect of the design of configuration A2. For the specified conditions, the mass of digested organic material (30%) is actually larger than the mass of oxidised organic material (27%). Configuration A2 can thus be considered to be predominantly anaerobic, although it is a variant of the activated sludge process. This is also demonstrated when the potential energy generation (119 kW) is compared with the average energy demand for aeration (73 kW).

Table 14.16 Total annualized treatment costs of system A1, A2 and B1

System costs	UoM	A1	A2	B1
Annual costs:				
– annual. investment costs	US$ · yr^{-1}	760,000	620,000	570,000
– operational costs	US$ · yr^{-1}	890,000	880,000	510,000
– total	US$ · yr^{-1}	1,650,000	1,500,000	1,080,000
Costs per people equivalent per year:				
– annual. investment costs	US$ · inh^{-1} · yr^{-1}	7.6	6.2	5.7
– operational costs	US$ · inh^{-1} · yr^{-1}	8.9	8.8	5.1
– total	US$ · inh^{-1} · yr^{-1}	16.5	15.0	10.8
Costs per m^3 treated sewage:				
– annual. investment costs	US$ct · m^{-3}	17.4	14.2	13.0
– operational costs	US$ct · m^{-3}	20.3	20.1	11.7
– total	US$ct · m^{-3}	37.7	34.2	24.7

When configuration A1 and A2 are compared, it can be concluded that the two systems produce an effluent of similar quality. However, under the specified conditions, the unit treatment costs of configuration A2 are 8.3% lower than those of configuration A1 (37.3 versus 34.2 US$ct per m^3 wastewater treated).

Table 14.17 Division of influent COD into fractions for system A1, A2 and B1

COD fraction	UoM	A1	A2	B1
In effluent	%	10	10	10
In stabilised excess sludge	%	27	33	15
Oxidized	%	45	27	14
Digested	%	18	30	61(1)

Note: (1) Only 80% of the produced methane is recovered: 200 kg $CH_4 \cdot d^{-1}$ lost to the environment

It should be noted that in the examples of configuration A1 and A2 the additional investment costs and the reduced operational costs related to the generation of electricity from the produced biogas have not been included. The difference in costs between A1 and A2 might therefore be further increased if the potential capacity for electricity generation from biogas is used, provided that the price of internally generated electricity is smaller than that of externally procured electricity. However, the difference in costs between the two systems may not always be sufficient justification for the decision to implement system A2 instead of A1. Other factors may be important such as:

- The risk of odour problems is increased in systems with primary settling, as anaerobic digestion may develop at the bottom of the primary settler, resulting in the production of biogas (often with traces of H_2S). This can only be prevented by covering the primary settler and providing it with off-gas treatment (e.g. lava filters);
- The cost advantage of system A2 is largely based on efficient performance of the primary settler. If the settler is less efficient the cost advantage may be reduced or even inverted;
- When nutrient removal is required, the effect of primary settling on the availability of easily biodegradable organic material for denitrification and biological phosphorus removal should be considered;
- The potential for energy self-sufficiency of system A2 (under the specified conditions) with respect to the consumption of electricity may be considered a strategic advantage, as it will not be necessary to procure energy from a local supplier of electricity.

It is important not to generalise the results obtained in Example 14.4 and Example 14.5: primary clarification can be advantageous in some cases but might not be in others. Using the type of analysis as presented in the previous sections, the suitability or otherwise of including such a unit in the system design can be determined.

When system B1 is included in the comparison, it can be concluded that it is possible to produce an effluent with equal quality at lower costs, as the annualised treatment costs per m^3 treated sewage (0.247 US$) are significantly less than that of system A1 (0.377 US$) and A2 (0.342 US$).

Under the specified conditions the application of combined anaerobic- aerobic treatment has various advantages:

- A large part (> 60%) of the organic material contained in the influent is removed anaerobically, resulting in a significant reduction of energy demand for aeration;
- The mass of produced methane is more than sufficient to cover the energy demand: in the example presented here, the potential energy production is 5 times larger than the demand for aeration;
- The total volume of system B1 (7040 m^3) is much smaller than that of systems A1 (11,900 m^3) and A2 (8600 m^3). Although the UASB unit itself is large, it is more than compensated for by the decrease in volume of the aeration tank, final settler, thickener and digester. Note that the reduction in final settler volume will not materialize if enhanced sludge settleability does not occur, for instance if pre-aeration is required to remove the produced sulphides. Furthermore, if a heated digester is selected, the total volume of systems A1 and A2 will significantly decrease;
- The stabilised excess sludge production (51 $m^3 \cdot d^{-1}$) is significantly smaller than that of system A1 (196 $m^3 \cdot d^{-1}$) and A2 (168 $m^3 \cdot d^{-1}$), facilitating dewatering and final disposal. This is mainly due to the increase in thickened sludge concentration compared to that of configuration A1 and A2. Furthermore, due to the increased digestion temperature the extent of solids degradation is higher as well. However, under comparable conditions (i.e. when the excess sludge is subjected to heated digestion), the stabilised excess sludge production in terms of kg TSS $\cdot d^{-1}$ perhaps does not differ as much between the three alternatives as expected, as will be explained below.

An often cited advantage of anaerobic pre-treatment is the significant reduction in excess sludge production (up to 80–90%), compared to a conventional activated sludge system. However, whereas in industrial anaerobic systems the excess sludge production may indeed be very low, provided that the wastewater is mainly soluble in nature, in a municipal UASB a significant portion of the influent particulate material is not degraded and will leave as part of the excess sludge. Furthermore, if a heated anaerobic digester is included in the design of the conventional system, then a substantial part of the produced excess sludge will be degraded. Therfore, a more realistic value for the reduction in *stabilised* excess sludge production is between 25 and 50%.

UASB reactors should preferably be operated at temperatures above 15–18°C, to obtain sufficient digestion efficiency and to properly stabilise the anaerobic excess sludge. If the expected minimum temperature is below this value, it is recommended to digest the excess sludge in a separate anaerobic digester, as demonstrated in Example 14.6. The produced methane can be directly used for digester heating. However, if electricity is produced the excess heat from the generation of electricity can be utilised as well.

Finally, system configurations A1, A2 and B1 are not suited for biological nutrient removal (nitrogen and/or phosphorus). The aerobic reactors in systems A1 and A2 can be modified into a Bardenpho- or UCT type of configuration, although in the case of system A2 the availability of COD might be restricted. As for system B1, a partial bypass is required as demonstrated in Example 14.6. Alternatively it might be considered to use one of the new system configurations as discussed in Section 13.5.2.3, which are currently being researched by several R&D groups. However, no full-scale installations have been constructed to date.

14.2.4 System C1: Nitrogen removal

If nitrogen removal is to be achieved, the system configuration has to be modified to include non-aerated zones for denitrification. Another important change is that the sludge age is no longer set by the requirements for efficient organic material removal: now the sludge age will depend on the constraints given by the nitrification and denitrification processes. In Chapter 5, a method was presented to calculate the minimum sludge age for complete removal of nitrate in an activated sludge system (Eqs. 5.86 or 5.88). This minimum sludge age depends on various parameters, whose values are required for the design process:

- Nitrification constants: μ_m, K_n, b_n, f_{max} and N_{ad};
- Denitrification parameters: K_2 and K_3;
- Values of recirculation factors "a" and "s";
- Fraction of biodegradable COD that is easily degradable: f_{sb}.

The procedure to optimise the design of activated sludge systems with nitrification and denitrification is as follows:

Step 1 Determine the minimum sludge age and the aerobic/anoxic sludge mass fractions

Determine the minimum sludge age required for a Bardenpho configuration to achieve both efficient nitrification (i.e. the concentration of residual ammonia should be equal to N_{ad}) and complete removal of nitrate. Use either an iterative calculation (Eqs. 5.86 or 5.88) or a graphic analysis (Figure 5.27). If complete nitrogen removal is not possible for a sludge age lower than a specified maximum value (for example $R_s = 20$ days), then use Eq. (5.93 or 5.94) to calculate for this sludge age the largest extent of nitrogen removal possible and the resulting residual nitrate effluent concentration. Once the sludge age is established, the anoxic pre-D sludge mass fraction (f_{x1}) and post-D sludge mass fraction (f_{x3}) are calculated, using Eqs. (5.83 and 5.84) for complete removal of nitrate or Eq. (5.92) for incomplete removal.

Steps 2 – 9 Optimise the system design

For the sludge age determined in the previous step, the system design is finalised with essentially the same procedure as used in Example 14.4. A problem appears in that the nitrogen removal is calculated for the concentration present in the influent. However, the excess sludge will be digested and in the process of volatile sludge mineralisation a considerable quantity of nutrients will be released to the liquid phase, to be returned to the biological reactor. The effect of the nutrient solubilisation can be included in the calculation, when the mass of nitrogen released in the anaerobic digester is added to the influent nitrogen load. Therefore the optimisation of the system design will take place in two steps:

(1) The system is optimised for the nitrogen concentration present in the influent and the mass of nitrogen liberated during digestion is calculated;
(2) The calculations are repeated with the estimated nitrogen mass released during digestion added to the influent nitrogen load. If the resulting values of the optimal sludge ages from step 1 and 2 deviate significantly, then a third iteration may be required.

EXAMPLE 14.7

For the same conditions as in the previous examples (specified in Table 14.9 and Table 14.10), determine the optimal Bardenpho configuration for complete removal of nitrate. The following additional data are given:

(1) Nitrification parameters:

- $\mu_m = 0.3\ d^{-1}$
- $b_n = 0.04\ d^{-1}$
- $K_n = 1.0\ mg\ N \cdot l^{-1}$
- $f_{max} = 0.6$.
- $N_{ad} = 2\ mg\ N \cdot l^{-1}$
- $N_{oe} = 2\ mg\ N \cdot l^{-1}$

(2) Denitrification parameters:

- $K_2 = 0.1\ mg\,N \cdot mg^{-1}\,VSS \cdot d^{-1}$
- $K_3 = 0.08\ mg\,N \cdot mg^{-1}\,VSS \cdot d^{-1}$

(3) Values of recirculation factors

- $a = 4$
- $s = 1$

(4) Fraction of biodegradable COD that is easily biodegradable: $f_{sb} = 0.2$.

Solution

(1) Determine the sludge age and the division into aerobic and anoxic zones

Solving Eq. (5.86) for the specified conditions yields a minimum sludge age of 10.5 days for efficient nitrogen removal: $N_{ae} = N_{ad} = 2\ mg\,N \cdot l^{-1}$, the effluent organic nitrogen concentration $N_{oe} = 2\ mg\,N \cdot l^{-1}$ and nitrate removal is complete. For this sludge age value the activated sludge system parameters can be calculated as follows:

(a) Nitrification capacity:

$$C_r = Y \cdot R_s/(1 + b_h \cdot R_s) = 0.45 \cdot 10.5/(1 + 0.24 \cdot 10.5) = 1.34 \qquad (3.30)$$

$$\begin{aligned} N_c &= N_{ti} - N_l - N_{oe} - N_{ad} \qquad (5.51) \\ &= N_{ti} - f_n \cdot [(1 - f_{ns} - f_{np}) \cdot (1 + f \cdot b_h \cdot R_s) \cdot C_r/R_s + f_{np}/f_{cv}] \cdot S_{ti} - N_{oe} - N_{ad} \\ &= 50 - 0.1 \cdot [(1 - 0.18) \cdot (1 + 0.2 \cdot 0.24 \cdot 10.5) \cdot 1.34/10.5 + 0.08/1.5] \cdot 650 - 2 - 2 \\ &= 32.3\ mg\,N \cdot l^{-1} \end{aligned}$$

(b) Biodegradable organic material in the influent (Eq. 3.3):

$$S_{bi} = S_{ti} \cdot (1 - f_{ns} - f_{np}) = 650 \cdot (1 - 0.18) = 533\ mg\,COD \cdot l^{-1} \qquad (3.3)$$

(c) Anoxic sludge mass fractions f_{x1} and f_{x3}, for $f_{dn} = (1 - f_{cv} \cdot Y)/2.86 = 0.11$:

$$\begin{aligned} f_{x1} &= [(N_c/S_{bi}) \cdot a/(a + s + 1) - f_{dn} \cdot f_{sb}]/(K_2 \cdot C_r) \qquad (5.83) \\ &= [(32.3/533) \cdot 4/6 - 0.11 \cdot 0.2]/(0.1 \cdot 1.34) = 0.13 \end{aligned}$$

$$\begin{aligned} f_{x3} &= (N_c/S_{bi}) \cdot (s + 1)/(a + s + 1)/(K_3 \cdot C_r) \qquad (5.84) \\ &= (32.3/533) \cdot (2/6)/(0.08 \cdot 1.34) = 0.19 \end{aligned}$$

(d) Calculate the aerated sludge mass fraction:

$$f_{ae} = 1 - f_{x1} - f_{x3} = 1 - 0.13 - 0.19 = 0.68$$

(e) Verify if nitrate removal is complete by calculation of the denitrification capacity:

$$\begin{aligned} D_{c1} &= (f_{dn} \cdot f_{sb} + K_2 \cdot C_r \cdot f_{x1}) \cdot S_{bi} \qquad (5.68) \\ &= (0.11 \cdot 0.2 + 0.1 \cdot 1.34 \cdot 0.13) \cdot 533 = 21.5\ mg\,N \cdot l^{-1} \end{aligned}$$

$$\begin{aligned} D_{c3} &= K_3 \cdot C_r \cdot f_{x3} \cdot S_{bi} \qquad (5.69) \\ &= 0.08 \cdot 1.34 \cdot 0.19 \cdot 533 = 10.8\ mg\,N \cdot l^{-1} \end{aligned}$$

Thus $D_c = D_{c1} + D_{c3} = 21.5 + 10.8 = 32.3\ mg\ N \cdot l^{-1}$, which is indeed equal to the value of the nitrification capacity.

(f) Estimate the quantity of nitrogen liberated during the anaerobic sludge digestion:

For the conditions above, the excess volatile sludge mass before and after sludge digestion is calculated ($f_{av} = 0.50$)

$$ME_v = 1646\ kg\ VSS \cdot d^{-1}$$

$$ME_{ve} = 1128\ kg\ VSS \cdot d^{-1}$$

Therefore the maximum solubilisation of nitrogen in the digester is $f_n \cdot (ME_v - ME_{ve}) = 0.1 \cdot (1646–1128) = 52\ kg\ N \cdot d^{-1}$, which amounts to an equivalent influent nitrogen concentration of $4.3\ mg\ N \cdot l^{-1}$. Based on these preliminary results all calculations are repeated, based on an influent nitrogen concentration of $N'_{ti} = N_{ti} + 4.3 = 54.3\ mg\ N \cdot l^{-1}$. Complete nitrogen removal is still possible, but it will be necessary to increase the design sludge age to 12 days:

$$C_r = Y \cdot R_s/(1 + b_h \cdot R_s) \quad (3.30)$$
$$= 0.45 \cdot 12/(1 + 0.24 \cdot 12) = 1.39$$

$$N_c = N_{ti} - N_l - N_{oe} - N_{ad} \quad (5.51)$$
$$= 54.3 - 13.2 - 2 - 2 = 37.1 mg\ N \cdot l^{-1}$$

$$f_{x1} = [(N_c/S_{bi}) \cdot a/(a + s + 1) - f_{dn} \cdot f_{bs}]/(K_2 \cdot C_r) \quad (5.83)$$
$$= [(36.8/533) \cdot 4/6 - 0.11 \cdot 0.2]/(0.1 \cdot 1.39) = 0.17$$

$$f_{x3} = (N_c/S_{bi}) \cdot (s + 1)/(a + s + 1)/(K_3 \cdot C_r) \quad (5.84)$$
$$= 36.8/533 \cdot (2/6)/(0.08 \cdot 1.39) = 0.21$$

$$f_{ae} = 1 - f_{x1} - f_{x3} = 1 - 0.17 - 0.21 = 0.62$$

$$D_{c1} = (f_{dn} \cdot f_{sb} + K_2 \cdot C_r \cdot f_{x1}) \cdot S_{bi} \quad (5.68)$$
$$= (0.11 \cdot 0.2 + 0.1 \cdot 1.39 \cdot 0.17) \cdot 533 = 24.7\ mg\ N \cdot l^{-1}$$

$$D_{c3} = K_3 \cdot C_r \cdot f_{x3} \cdot S_{bi} \quad (5.69)$$
$$= 0.08 \cdot 1.39 \cdot 0.21 \cdot 533 = 12.4\ mg\ N \cdot l^{-1}$$

Thus $D_c = D_{c1} + D_{c3} = 24.5 + 12.3 = 37.1\ mg\ N \cdot l^{-1}$, which is equal to the nitrification capacity.

The excess sludge production ME_v is $1585\ kg\ VSS \cdot d^{-1}$, which is reduced after digestion to $ME_{ve} = 1102\ kg\ VSS \cdot d^{-1}$. So the maximum extent of solubilisation of nitrogen that can occur in the digester is equal to $f_n \cdot (ME_v - ME_{ve}) = 0.1 \cdot (1585–1102) = 48\ kg\ N \cdot d^{-1}$ which amounts to an equivalent influent nitrogen concentration of $4.0\ mg\ N \cdot l^{-1}$. So it appears that the increase of the nitrogen influent concentration by $4.3\ mg\ N \cdot l^{-1}$ in this second iteration has been a little too much and a third iteration could be executed, this time with an initial nitrogen concentration of $N'_{ti} = 50 + 4.0 = 54.0\ mg\ N \cdot l^{-1}$. However, the difference in results between the second and a third iteration will be minimal and therefore it can be concluded that when a sludge age of $R_s = 12$ days is adopted, the system will be able to remove all nitrate generated in the nitrification process, not only from the nitrogen in the influent, but also from the nitrogen released in the digester. The effluent nitrogen concentration N_{te} is equal to $4\ mg\ N \cdot l^{-1}$, consisting of a residual organic fraction N_{oe} of $2\ mg\ N \cdot l^{-1}$ and a residual ammonia concentration N_{ad} of $2\ mg\ N \cdot l^{-1}$. The nitrogen present in the influent is divided into different fractions as follows:

- Nitrogen in the effluent: $N_{te} = N_{oe} + N_{ad} = 2 + 2 = 4\ mg\ N \cdot l^{-1}$
- Nitrogen in the stabilised excess sludge: $N_{le} = f_n \cdot ME_{ve}/Q_i = 0.1 \cdot 1102/12 = 9.2\ mg\ N \cdot l^{-1}$
- Denitrified nitrogen: $N_d = N_{ti} - N_{le} - N_{te} = 50–9.2–4.0 = 36.8\ mg\ N \cdot l^{-1}$

(2) to (9) Optimise the system configuration C1

Once the minimum required sludge age and the division of the reactor into anoxic and aerobic zones has been established, the calculations for optimisation of the system can be started, according to the procedure of Example 14.4. In Table 14.18 the annualised costs of configuration C1 are listed, while in Figure 14.13 the optimised solution is specified.

Table 14.18 Annualised investment and operational costs of system configuration C1

Cost item	Annual costs $US\$ \cdot year^{-1}$	Costs per m^3 $US\$ct \cdot m^{-3}$	Cost per inh. $US\$ \cdot inh^{-1} \cdot yr^{-1}$	Fraction %
Investment costs	890,000	20.3	8.9	47
Operational costs	1,020,000	23.3	10.2	53
– aeration	250,000	5.7	2.5	13
– sludge disposal	130,000	3.0	1.3	7
– personnel	310,000	7.1	3.1	16
– operation	100,000	2.3	1.0	5
– maintenance	200,000	4.6	2.0	10
– insurance	30,000	0.7	0.3	2
Total costs	1,910,000	43.6	19.1	100

As to the design calculations, the following additional remarks can be made:

(1) Volumes of the anoxic and aerobic zones: For the sludge age $R_s = 12$ days, the optimal volume of the biological reactor is determined as $V_r = 8234\ m^3$. As the anoxic sludge mass fractions were determined as $f_{x1} = 0.17$ and $f_{x3} = 0.21$, the volumes of the pre-D, post D and aerated zones can be calculated as:

$$V_1 = f_{x1} \cdot V_r = 0.17 \cdot 8234 = 1400\ m^3$$

$$V_3 = f_{x3} \cdot V_r = 0.21 \cdot 8234 = 1715\ m^3$$

$$V_{aer} = (1 - f_{x1} - f_{x3}) \cdot V_r = 0.62 \cdot 8234 = 5119\ m^3$$

(2) Oxygen demand: The mass of oxygen required for the process is given as the sum of the oxygen required for the oxidation of organic material and the oxygen required for nitrification. The mass of organic material oxidised by nitrate is subtracted from this quantity (as equivalent oxygen):

$$MO_t = MO_c + MO_n - MO_{eq} \tag{5.13}$$

$$MO_c = (1 - f_{ns} - f_{np}) \cdot (1 - f_{cv} \cdot Y + f_{cv} \cdot (1 - f) \cdot b_h \cdot C_r) \cdot MS_{ti} = 4642\ kg\ O_2.d^{-1} \tag{3.43}$$

$$MO_n = 4.57 \cdot MN_c = 2034\ kg\ O_2.d^{-1} \tag{5.11}$$

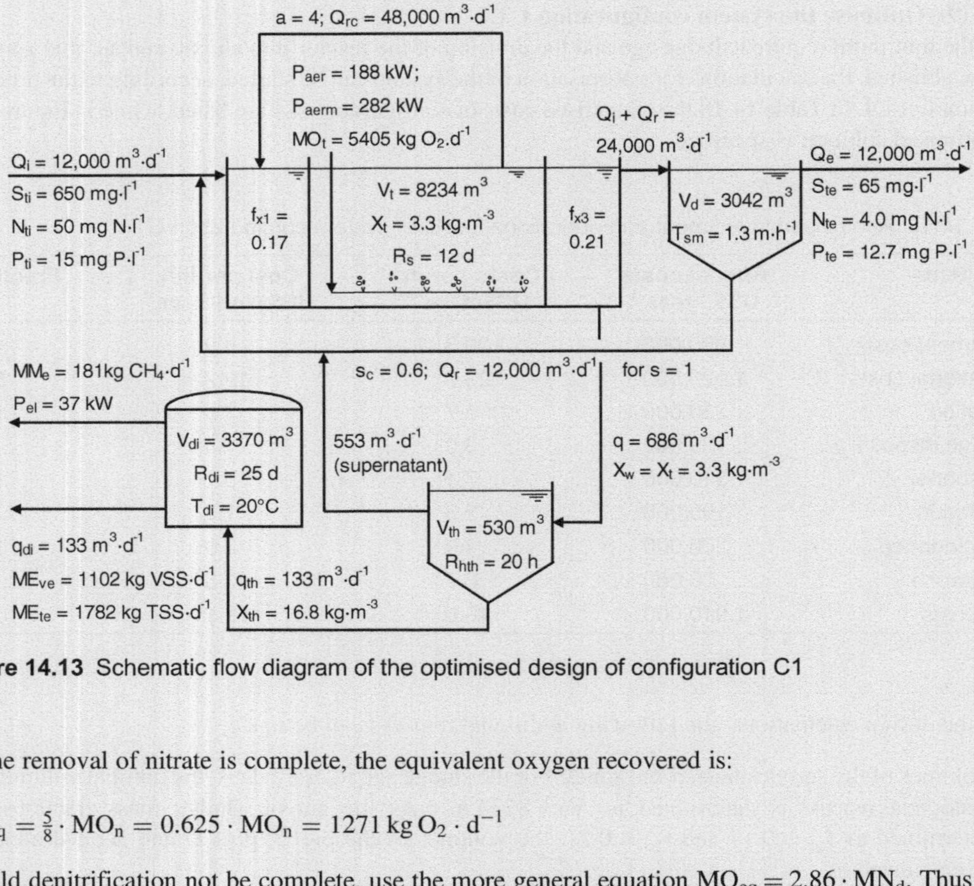

Figure 14.13 Schematic flow diagram of the optimised design of configuration C1

As the removal of nitrate is complete, the equivalent oxygen recovered is:

$$MO_{eq} = \tfrac{5}{8} \cdot MO_n = 0.625 \cdot MO_n = 1271\ kg\ O_2 \cdot d^{-1} \tag{5.12}$$

Should denitrification not be complete, use the more general equation $MO_{eq} = 2.86 \cdot MN_d$. Thus in this example: $MO_t = 4642 + 2034–1271 = 5405\ kg\ O_2 \cdot d^{-1}$. Oxygen is only consumed in the aerobic zone, so the OUR is given by:

$$O_t = MO_t/V_{aer} = 5405/5119 = 1.05\ kg\ O_2 \cdot m^{-3} \cdot d^{-1} \text{ or } 44\ mg\ O_2 \cdot l^{-1} \cdot h^{-1}$$

14.2.5 System C2: Nitrogen and phosphorus removal

In order to effect biological removal of phosphorus, an anaerobic zone will have to be incorporated in the system configuration. Based on the analysis made in Chapter 7 (Figure 7.7), it was concluded that a relatively small anaerobic zone will be sufficient and that preferably the sludge age should be low. On the other hand, the necessity to remove nitrogen requires a relatively long sludge age. There is no single optimized analytical solution for this problem: one has to find a solution iteratively while using expert judgment regarding the values of several operational and design variables. The most important ones are:

- The size of the anaerobic sludge fraction f_{an}, typically between 0.10 to 0.15;
- The recirculation rate from the anoxic zone to the anaerobic zone "r", where a default value of one is recommended;
- The hydraulic regime characterised by the number of completely mixed reactors in series in the anaerobic zone N. It has been demonstrated in Chapter 7 that a value of $N > 2$ increases phosphorus removal performance only marginally;
- The sludge age.

The optimised design procedure can be summarised as:

(1) Select appropriate values for N and r (e.g. $N = 2$, $r = 1$);
(2) Select a value for the anaerobic sludge mass fraction, e.g. $f_{an} = 0.10-0.15$;
(3) Determine the value of the sludge age that permits removal of both phosphorus and nitrogen. For phosphorus removal use the procedure described in Section 7.1.3. For nitrogen removal refer to Example 14.7;
(4) Adapt the values of f_{an} and when required also of N and r and determine the minimum sludge age that permits removal of the nutrients. For the selected sludge age, optimise the activated sludge system according to the procedure explained in Example 14.4;
(5) If it is not possible to comply with both the effluent phosphorus and nitrogen limits, optimise the system for compliance to the nitrogen limits and apply supplementary simultaneous precipitation for removal of the excess phosphate, as will be demonstrated in Example 14.14.

EXAMPLE 14.8

For the same conditions as specified in Table 14.9 and Table 14.10, determine an UCT configuration that complies with the following effluent criteria:

- $N_{ad} = 2\ mg\ N \cdot l^{-1}$;
- $P_{td} = 2\ mg\ P \cdot l^{-1}$
- $N_{td} = 10\ mg\ N \cdot l^{-1}$.

The following additional data are given:

- $f_{ep} = 0.25$
- $f_{vp} = 0.46$
- $f_{pd} = 0.8$
- $K_c = 60\ m^3 \cdot kg^{-1}\ VSS \cdot d^{-1}$
- $f_{pp} = 0.38\ mg\ P \cdot mg^{-1}\ VSS$
- $b_p = 0.04\ d^{-1}$

Solution

(1) Establish the sludge age for complete or maximum nutrient removal
The following values relevant for phosphorus removal are selected: $N = 2$, $r = 1$ and $f_{an} = 0.15$.

(a) Phosphorus removal
Follow the procedure presented in Section 7.1.3:

1. Calculate the concentration of organic material sequestered by the bio-P organisms;
2. Use this value to calculate the production of bio-P organisms;
3. Determine the "normal" active heterotrophic sludge concentration, based on the mass of non-sequestered organic material available in the effluent of the anaerobic zone;
4. Calculate the concentration of the other sludge fractions: i.e. the endogenous residue of both bio-P organisms and normal active sludge and the inert sludge;
5. Calculate the removal of phosphorus with the excess sludge.

(b) Nitrogen removal
Follow the procedure outlined in Example 14.7 and calculate as a function of the sludge age the following parameters:

1. Nitrification capacity;
2. Values of the different anoxic sludge mass fractions;
3. Denitrification capacity

Compare the denitrification capacity with the nitrification capacity and determine the residual concentration of nitrogen in the effluent (ammonium, organic nitrogen and nitrate). With trial and error (using a spreadsheet), it can be established that for a sludge age $R_s = 12$ days it is possible to comply with both the nitrogen and phosphorus effluent criteria. For this sludge age all main system parameters are calculated below.

The mass of nitrogen released to the liquid phase during anaerobic digestion is added to the nitrogen load to the activated sludge system. As for phosphorus, it is assumed that the phosphorus released during digestion (consisting of organic phosphorus and internally stored poly-phosphate) will be precipitated with $FeCl_3$ added to the digester. Due to the high phosphate concentration, the formation of (undesired) metal hydroxides will be significantly reduced, so that the required molar Me/P dosing ratio will be much lower than during pre-precipitation of the raw wastewater. The chemical precipitation of phosphate will not be discussed here: refer to Example 14.14. As an alternative, it could also be considered to directly dewater the bio-P sludge (without digestion).

The first step in the design process is to calculate the mass of COD sequestered by the bio-P organisms in the anaerobic zone. Initially it is assumed that nitrate is not being returned to the anaerobic zone (this will be confirmed later in this example). Using the iterative procedure presented in Section 7.1.3, the residual concentration in the anaerobic zone S_{bsN} is calculated as $9.3 \text{ mg COD} \cdot l^{-1}$ (for $N = 2$). The sequestered COD concentration is:

$$S_{seq} = S_{bsi} - (r + 1) \cdot S_{bsN} = 0.2 \cdot (1 - 0.10 - 0.08) \cdot 650 - 2 \cdot 9.3 = 88 \text{ mg COD} \cdot l^{-1} \quad (7.6)$$

This is the concentration of influent COD used for the growth of bio-P organisms, which is therefore no longer available for the production of "normal" biomass. The residual concentration of biodegradable

COD available for the normal biomass is $S_{bi}' = S_{bi} - S_{seq} = 533 - 88 = 445$ mg $COD \cdot l^{-1}$. Now the different sludge fractions can be calculated:

$$C_{rh} = Y \cdot R_s/(1 + b_h \cdot R_s) = 0.45 \cdot 12/(1 + 0.24 \cdot 12) = 1.39 \quad (3.30)$$

$$C_{rp} = Y \cdot R_s/(1 + b_p \cdot R_s) = 0.45 \cdot 12/(1 + 0.04 \cdot 12) = 3.65 \quad (3.30)$$

$$MX_{ap} = C_{rp} \cdot MS_{seq} = 3.65 \cdot 0.088 \cdot 12{,}000 = 3853 \text{ kg VSS} \quad (7.7)$$

$$MX_{ep} = f_{ep} \cdot b_p \cdot R_s \cdot MX_{ap} = 0.25 \cdot 0.04 \cdot 12 \cdot 3853 = 462 \text{ kg VSS} \quad (7.9)$$

$$MX_{ah} = C_{rh} \cdot MS'_{bi} = 1.39 \cdot 0.445 \cdot 12{,}000 = 7432 \text{ kg VSS} \quad (7.8)$$

$$MX_{eh} = f \cdot b_h \cdot R_s \cdot MX_{ah} = 0.2 \cdot 0.24 \cdot 12 \cdot 7432 = 4281 \text{ kg VSS} \quad (7.10)$$

$$MX_i = f_{ns} \cdot R_s/f_{cv} \cdot MS_{ti} = 0.08 \cdot 12/1.5 \cdot 0.650 \cdot 12{,}000 = 4992 \text{ kg VSS} \quad (3.45)$$

$$MX_v = MX_{ap} + MX_{ep} + MX_{ah} + MX_{eh} + MX_i = 21{,}020 \text{ kg VSS} \quad (7.11)$$

$$MX_t = MX_{ap}/f_{vp} + (MX_{ep} + MX_{ah} + MX_{eh} + MX_i)/f_v = 32{,}900 \text{ kg TSS} \quad (7.12)$$

The biological removal of phosphorus can be calculated as:

$$MP_l = 0.38 \cdot MX_{ap}/R_s + 0.025 \cdot (MX_v - MX_{ap})/R_s = 122 + 36 = 158 \text{ kg P} \cdot d^{-1} \quad (7.13)$$

$$P_l = MP_l/Q_i = 158/12{,}000 = 13.1 \text{ mg P} \cdot l^{-1} \quad (3.60)$$

It can be observed that the phosphorus demand for the production of excess sludge is somewhat smaller than the concentration available in the influent ($P_{ti} = 15$ mg $P \cdot l^{-1}$). Should the influent phosphate concentration be insufficient, then there would have been a concomitant decrease in the production of bio-P organisms and more non-sequestered material would have been available for the "normal" sludge. The residual concentration of phosphate in the effluent is equal to:

$$P_{te} = P_{ti} - P_l = 15.0 - 13.1 = 1.9 \text{ mg P} \cdot l^{-1}$$

It can be concluded that the residual phosphorus concentration meets the specified effluent limit. Should this not be the case, various measures can be considered to increase the biological removal of phosphorus:

- Increase the anaerobic sludge fraction f_{an} or increase the recirculation rate "r";
- Increase the number of anaerobic reactors N;
- Reduce the sludge age (at the expense of nitrogen removal) or a combination of these actions.

As the viability of biological phosphorus removal for $R_s = 12$ days has been demonstrated, the next step is to validate the nitrogen removal performance. For the specified sludge age, the additional nitrogen load returned to the activated sludge system from the anaerobic digester can be calculated. The volatile excess sludge production $ME_v = 21{,}020/12 = 1752$ kg $VSS \cdot d^{-1}$, so the nitrogen demand for excess sludge production is equal to:

$$MN_l = f_n \cdot ME_v = 0.1 \cdot 1752 = 175 \text{ kg N} \cdot d^{-1} \quad (3.59)$$

In the sludge digester 577 kg $VSS \cdot d^{-1}$ of volatile sludge is digested. The production of stabilised excess sludge is equal to $ME_{ve} = 1752 - 577 = 1175$ kg $VSS \cdot d^{-1}$. The mass of nitrogen disposed with the stabilised excess sludge $MN_{le} = 0.1 \cdot 1175 = 118$ kg $N \cdot d^{-1}$, which represents an influent concentration of $N_{le} = MN_{le}/Q_i = 118/12{,}000 = 9.8$ mg $N \cdot l^{-1}$. For the specified residual organic nitrogen- and ammonium concentrations of 2 mg $N \cdot l^{-1}$ each, the required nitrification capacity can thus be calculated as:

$$N_c = N_{ti} - N_{le} - N_{oe} - N_{ad} = 50 - 9.8 - 2 - 2 = 36.2\,mg\,N \cdot l^{-1} \tag{5.50}$$

For the sludge age of 12 days and $N_{ad} = 2$ mg $N \cdot l^{-1}$, the maximum allowed non-aerated sludge mass fraction is calculated as:

$$\begin{aligned} f_m &= 1 - (1 + K_n/N_{ad}) \cdot (b_n + 1/R_s)/\mu_m \\ &= 1 - (1 + 1/2) \cdot (0.04 + 1/12)/0.3 = 0.38 \end{aligned} \tag{5.47}$$

So the minimum aerated sludge fraction at $R_s = 12$ days is equal to $f_{ae} = 1 - f_m = 0.62$. As the value of the anaerobic mass fraction f_{an} has been specified as 0.15, the maximum anoxic mass fraction $f_x = f_m - f_{an} = 0.38 - 0.15 = 0.23$. For $f_{x1} = 0.18$ and $f_{x3} = 0.05$ it can be demonstrated that the nitrogen effluent limits are met. In this example it is assumed (conservatively) that only 80% of the bio-P organisms are capable of denitrification. Therefore, in order to evaluate the extent of denitrification, it will be necessary to differentiate between the influent COD fraction available for the bio-P organisms and for the normal heterotrophic biomass:

$$S_{bsp} = 88\,mg\,COD \cdot l^{-1} \longrightarrow f_{bsp} = 88/107 = 0.83 \tag{7.16}$$

$$S_{bsh} = 107 - 88 = 19\,mg\,COD \cdot l^{-1} \longrightarrow f_{bsh} = 19/107 = 0.17 \tag{7.17}$$

$$S_{bp} = 88\,mg\,COD \cdot l^{-1} \longrightarrow f_{bp} = 88/533 = 0.17 \tag{7.18}$$

$$S_{bh} = 533 - 88 = 445\,mg\,COD \cdot l^{-1} \longrightarrow f_{bh} = 445/533 = 0.83 \tag{7.19}$$

Now the denitrification capacity can be calculated with the equations specifically developed for denitrification in bio-P systems:

$$\begin{aligned} D_{c1} &= [f_{dn} \cdot (f_{bsp} \cdot f_{pd} + f_{bsh}) \cdot f_{sb} + K_2 \cdot f_{x1} \cdot (C_{rh} \cdot f_{bh} + C_{rp} \cdot f_{bp} \cdot f_{pd})] \cdot S_{bi} \\ &= [0.11 \cdot (0.83 \cdot 0.8 + 0.17) \cdot 0.2 + 0.10 \cdot 0.18 \cdot (1.39 \cdot 0.83 \\ &\quad + 3.65 \cdot 0.17 \cdot 0.8)] \cdot 533 \\ &= 27.5\,mg\,N \cdot l^{-1} \end{aligned} \tag{7.14}$$

$$\begin{aligned} D_{c3} &= K_3 \cdot f_{x3} \cdot (C_{rh} \cdot f_{bh} + C_{rp} \cdot f_{bp} \cdot f_{pd}) \cdot S_{bi} \\ &= 0.08 \cdot 0.05 \cdot (1.39 \cdot 0.83 + 3.65 \cdot 0.17 \cdot 0.8) \cdot 533 \\ &= 3.6\,mg\,N \cdot l^{-1} \end{aligned} \tag{7.15}$$

$$D_c = D_{c1} + D_{c3} = 27.5 + 3.6 = 31.1\,mg\,N \cdot l^{-1}$$

It can be checked that the concentration of nitrate available for denitrification (N_{av}) is larger than the denitrification capacity for both anoxic zones so that the effluent nitrate concentration can be calculated as:

$$N_{ne} = N_c - D_c = 36.2 - 31.1 = 5.1\,mg\,N \cdot l^{-1} \tag{5.78}$$

However, if $N_{av1}>D_{c1}$, not all nitrate is removed in the pre-D zone. Hence, nitrate will be recirculated from the pre-D to the anaerobic zone, which reduces the mass of VFA available to the bio-P organisms. In order to prevent this, the value of the recirculation factor "a" must be reduced. It can be verified that for $a = 3.25$ the value of N_{av1} is equal to D_{c1}:

$$N_{av1} = a/(a+s+1) \cdot N_c + s \cdot N_{ne} = D_{c1} \quad (5.89)$$

$$= 3.25/(3.25+1+1) \cdot 36.2 + 1 \cdot 5.1 = 27.5\ mg\,N \cdot l^{-1}$$

Finally the effluent nitrogen concentration is equal to:

$$N_{te} = N_{ne} + N_{oe} + N_{ad} = 5.1 + 2 + 2 = 9.1\ mg\,N \cdot l^{-1}$$

(2) To (9) Optimise the system C2 (UCT configuration)

Once the sludge age and the division of the sludge mass into aerobic, anoxic and anaerobic fractions has been determined, the system can be further optimised using the procedure outlined in Example 14.4.

Due to the fact that the sludge concentration in the anaerobic zone is a factor $r/(1+r)$ smaller than in the other reactors, the volumetric fraction will have to be a factor $(1+r)/r$ larger than the specified anaerobic sludge mass fraction f_{an}. Thus the total volume of a UCT system will be a factor $(1+f_{an}/r)$ larger than a system with an equivalent sludge mass but with a uniform concentration. In the example at hand, with $f_{an} = 0.15$ and $r = 1$, the volume of the UCT system will be a factor $(1+0.15/1) = 1.15$ or 15% larger.

Nutrients will be released during anaerobic digestion of the produced excess sludge. In the case of nitrogen, this effect can be compensated for by an increase of the sludge age. However, the mass of phosphorus released is too large to be handled in the same way. The daily excess sludge production for a sludge age of 12 days is $1752\ kg\,VSS \cdot d^{-1}$, of which $1175\ kg\,VSS \cdot d^{-1}$ remains as stabilised sludge mass after digestion. The volatile excess sludge contains $158\ kg\,P \cdot d^{-1}$, representing a concentration of $13.1\ mg\ P \cdot l^{-1}$ in the influent. Assuming that during anaerobic digestion the bio-P organisms are degraded in the same way as the other biomass and that the stabilised sludge will have a phosphorus content of $0.025\ mg\,P \cdot mg^{-1}\,VSS$ (i.e. all stored poly-P is released), the mass of phosphorus in the stabilised sludge will be only $29.4\ kg\,P \cdot d^{-1}$, resulting in a potential release of $158-29.7 = 128.3\ kg\,P \cdot d^{-1}$, representing $10.7\ mg\ P \cdot l^{-1}$ influent. This release of phosphate is partly compensated by the formation of metal-phosphate complexes that precipitate on the sludge. However, as it concerns such a large part of the phosphorus removed from the influent (10.7 of the $13.1\ mg\ P \cdot l^{-1}$ removed), often chemical precipitation with metal salts is applied either to the influent or the effluent of the digester, so that all phosphorus released is precipitated and is not returned to the activated sludge system. Taking into account that the flow of digested sludge will be small (in the example $q_{di} = 166\ m^3 \cdot d^{-1}$) and that the dosing ratio between metal and phosphorus will be close to the stoichiometrical ratio of one, this will be a relatively small-scale operation. Alternatively, the bio-P sludge can be directly dewatered, this is perhaps the best solution.

In Table 14.19 the annualised costs of system configuration C2 are shown, while in Figure 14.14 the optimised solution is further specified. In Table 14.20 the optimised design of system configurations C1 and C2 are compared in terms of the division of influent COD, while in Table 14.21 the annualized costs are compared to that of configuration C1.

Table 14.19 Annualised investment and operational costs of system configuration C2

Cost item	Annual costs $US\$ \cdot year^{-1}$	Costs per m^3 $US\$ct \cdot m^{-3}$	Cost per inh. $US\$ \cdot inh^{-1} \cdot yr^{-1}$	Fraction %
Investment costs	1,030,000	23.5	10.3	47
Operational costs	1,140,000	26.0	11.4	53
– aeration	230,000	5.3	2.3	11
– sludge disposal	160,000	3.7	1.6	7
– personnel	350,000	8.0	3.5	16
– operation	120,000	2.7	1.2	6
– maintenance	240,000	5.5	2.4	11
– insurance	40,000	0.9	0.4	2
Total costs	2,170,000	49.5	21.7	100

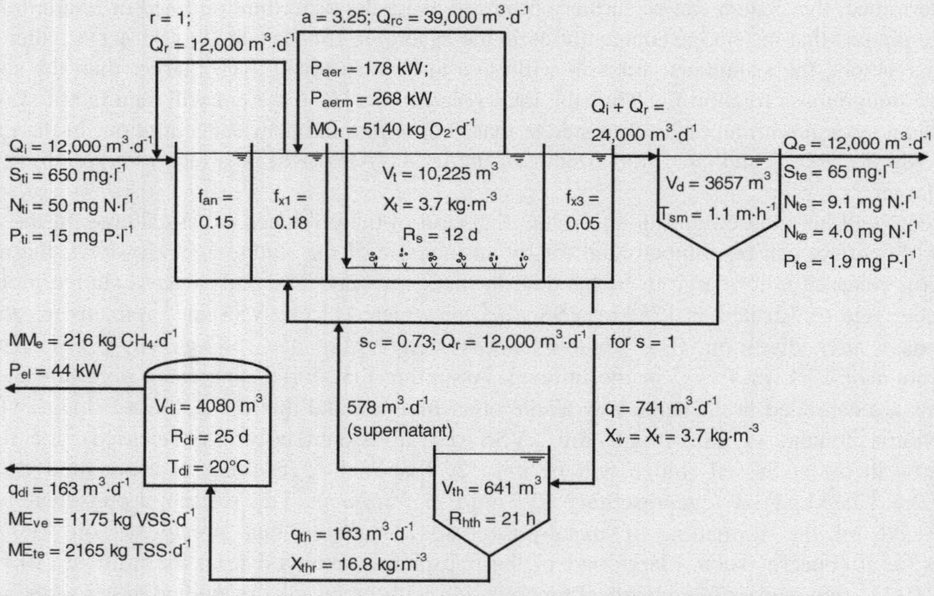

Figure 14.14 Schematic flow diagram of the optimised design of configuration C2

Table 14.20 Division of influent COD into fractions for system C1 and C2

COD fraction	UoM	C1	C2
In effluent	%	10	10
In stabilised excess sludge	%	21	23
Oxidized	%	60	56
Digested	%	9	11

Table 14.21 Total annualised costs of system configurations C1 and C2

System cost	UoM	C1	C2
Annual costs:			
– annualised investment costs	$US\$ \cdot year^{-1}$	890,000	1,030,000
– operational costs	$US\$ \cdot year^{-1}$	1,020,000	1,140,000
– total	$US\$ \cdot year^{-1}$	1,910,000	2,170,000
Costs per inhabitant:			
– annualised investment costs	$US\$ \cdot inh^{-1} \cdot yr^{-1}$	8.9	10.3
– operational costs	$US\$ \cdot inh^{-1} \cdot yr^{-1}$	10.2	11.4
– total	$US\$ \cdot inh^{-1} \cdot yr^{-1}$	19.1	21.7
Costs per m^3 treated wastewater			
– annualised investment costs	$US\$ct \cdot m^{-3}$	20.3	23.5
– operational costs	$US\$ct \cdot m^{-3}$	23.3	26.0
– total	$US\$ct \cdot m^{-3}$	43.6	49.5

14.2.6 System comparison

The main objective of the optimisation of the different activated sludge system configurations was to establish the minimum costs required to produce a specified effluent quality. In Table 14.22 the main effluent quality parameters are evaluated for the five configurations presented in the previous sections, while also key parameters in process performance are compared.

Some of these effluent quality parameters cannot be calculated using the theory derived in this book and have to be estimated based on operational experience of current operational systems. This applies to the effluent suspended solids concentration and the BOD concentration, which theoretically should be zero but in practice will have higher values due to imperfect phase separation in the final settler. The other parameters have been calculated in the optimised designs of Example 14.4 to Example 14.8. The following comments can be made relating to effluent quality and treatment costs:

- The secondary treatment systems (A1, A2 and B1) in general tend to have slightly higher TSS and BOD values in the effluent compared to the tertiary treatment systems, as these are operated at a higher sludge age and therefore will have less free bacteria in the effluent (the main source of suspended solids and BOD);
- The secondary treatment systems have a relatively small carbonaceous BOD value in the effluent, but if one considers the oxygen demand for the oxidation of ammonium, it is clear that this represents a considerable oxygen demand, up to half the BOD value in the influent;
- The higher ammonia- and phosphorus effluent concentrations for UASB pre-treatment (B1) are a result of more complete digestion in the heated digester operated at 37°C.
- Secondary treatment with primary clarification (A2) is less expensive than direct secondary treatment of the raw sewage (A1): 34.2 $US\$ct \cdot m^{-3}$ versus 37.7 $US\$ct \cdot m^{-3}$ or a 9% reduction in treatment costs;
- The costs of the secondary treatment also depend on the selected configuration. System B1 consisting of a UASB reactor + aerobic reactor + final settler + heated digester is much cheaper than A1 and A2.

It should however be noted that at lower temperatures this may be different: the operational costs of alternative B1 will rapidly increase due to the need to heat the UASB influent flow, which is clearly not economically feasible;

- At lower temperatures the anaerobic digesters will need to be heated as well, however for these units a much smaller flow will have to be heated: the total influent flow is 12,000 $m^3 \cdot d^{-1}$ compared to thickened excess flows of 196 $m^3 \cdot d^{-1}$ and 168 $m^3 \cdot d^{-1}$ for configurations A1 and A2 respectively. An alternative to heating would be to install a much larger digester volume. However, in principle it is advantageous to heat the digester, as not only the rate (= less volume) but also the extent of digestion (= less stabilised sludge produced and more methane generated) will increase;
- The nitrogen removal system C1 is approximately 15% more expensive than A1 and about 27% more expensive than A2. This should be balanced against the significant operational advantages (e.g. operational stability by eliminating the risk of denitrification in the final settler) and the superior effluent quality (nitrogen removal);
- The additional costs of adding biological phosphorus removal (C2) in this example are only 14% compared to a configuration with nitrogen removal only (C1). However, whereas nitrogen removal is complete in system C1, the effluent nitrogen limits are only just complied with in configuration C2. This is due to the conflicting requirements for optimal biological phosphorus removal (low sludge age) and optimal nitrogen removal (high sludge age). However, it has been demonstrated in Example 7.4 that the residual phosphorus concentration in a bio-P removal system only increases gradually with increasing sludge age. Therefore, in practice priority would be given to nitrogen removal, while bio-P removal would be supplemented by chemical precipitation: the quantity of chemical required to meet the effluent phosphorus limits would probably be modest. Refer also to Example 14.14 for an additional design example involving different methods for combined phosphorus and nitrogen removal.

Table 14.22 Comparative analysis of effluent quality, values of key operational and design parameters and cost implications of the five basic configurations discussed in Example 14.4 to Example 14.8

System		A1	A2	B1	C1	C2
TSS	mg TSS · l^{-1}	<20	<20	<20	<10	<10
BOD	mg $O_2 \cdot l^{-1}$	<20	<20	<20	<10	<10
BOD for nitrification	mg $O_2 \cdot l^{-1}$	160	150	180	15	15
COD	mg COD · l^{-1}	65	65	65	65	65
TKN	mg N · l^{-1}	38.2	35.6	43.5	4.0	4.0
Total nitrogen	mg N · l^{-1}	38.2	35.6	43.5	4.0	9.1
Total phosphorus	mg P · l^{-1}	12.1	11.4	13.4	12.7	1.9
Total volume	m^3	11,900	9614	7040	15,176	18,439
ME_{te}	ton TSS · d^{-1}	2.4	3.2	1.2	1.8	2.2
Energy consumption	kW	122	73	39	188	178
Potential self-reliance	%	57	161	500	20	25
Financial costs	US\$ct · m^{-3}	17.4	14.2	13.0	20.3	23.5
Operational costs	US\$ct · m^{-3}	20.3	20.1	11.7	23.3	26.0
Total annualized costs	US\$ct · m^{-3}	37.7	34.2	24.7	43.6	49.5

14.3 FACTORS INFLUENCING DESIGN

14.3.1 Influence of the wastewater temperature

One of the most important characteristics influencing the design of wastewater treatment systems is the wastewater temperature. The temperature influences the rate of all biological processes, which will influence the size of the treatment units. Furthermore, the oxygen transfer efficiency will be affected as well, as both the equilibrium dissolved oxygen concentration and the oxygen transfer rate are temperature dependent.

EXAMPLE 14.9

Compare the volume of the main units, the excess sludge production, the energy demand and the estimated costs of a Bardenpho system for the conditions specified in Table 14.9 and Table 14.10, for minimum temperatures of 14, 20 and 26°C respectively.

Solution

For the different temperatures follow the procedure outlined in Example 14.7. First the minimum sludge age for complete nitrogen removal will be determined. Once the value of the sludge age is known, the other operational parameters can be calculated. In Table 14.23 the results are summarised.

Table 14.23 Effect of the wastewater temperature on the main parameters and dimensions of an activated sludge system (C1 – Bardenpho configuration)

Parameters	UoM	Temperature		
		14°C	20°C	26°C
Minimum sludge age	d	22.5	12	7
Volume				
– reactor	m^3	11,775	8234	5960
– settler	m^3	4009	3042	2531
– thickener	m^3	510	530	455
– digester	m^3	3001⁽¹⁾	3370	2515
– total	m^3	19,295	15,176	11,461
Stabilised sludge production	ton TSS $\cdot d^{-1}$	1.65	1.78	1.86
Energy demand	kW	196	188	179
Potential for energy production	kW	N.A.⁽¹⁾	20	26
Annualised treatment costs				
– financing	US\$ct $\cdot m^{-3}$	23.8	20.3	16.4
– operational	US\$ct $\cdot m^{-3}$	24.8	23.3	21.7
– total	US\$ct $\cdot m^{-3}$	48.6	43.6	38.1

Note: (1) Anaerobic digester heated to 30°C. Methane gas is used for heating of the excess sludge flow and not available for power generation.

It can be observed that the excess sludge production and the energy demand of the system are comparable for all temperatures. However, the minimum required sludge age and thus also the biological reactor volume depends heavily on the temperature. Both nitrification and anaerobic digestion proceed slowly at low temperatures. It can clearly be seen that the financial costs are temperature dependent, due to the effect on the biological treatment volume to be installed. The operational costs are also temperature dependent, although less than the annualised financing costs. This is due to the fact that maintenance, insurance and operational costs are calculated as a percentage of total investment costs (personnel costs are assumed to be equal for all temperatures). It can be concluded that the selection of the correct design temperature is very important, as a value that is too low will result in overdesign and unnecessary costs while a value that is too high might result in non-compliance with the effluent limits.

A typical phenomenon observed in countries with a strong seasonal temperature variation is that the nitrification capacity tends to lag behind the changes in temperature. In autumn the nitrification capacity is higher and in the spring it is lower than expected for the wastewater temperature. The reason is that the nitrifier fraction in the mixed liquor responds to the temperature but it will take significant time (several sludge ages) before a new state of equilibrium is reached. Furthermore, it is important to note that the minimum sludge age in a nitrogen removal system depends not only on the temperature but also on the specific growth rate of nitrifiers. A realistic estimate of the value of this parameter is therefore very important and in Appendix 4 an experimental procedure will be presented to determine its value.

14.3.2 Influence of the sludge age

In the optimised design procedure presented in the previous examples, it was assumed that the selected sludge age should be equal to the minimum required sludge age. The design procedure will now be used to demonstrate that this is in fact a correct assumption. For several values of the sludge age the system parameters will be calculated in order to determine the quantitative effect the sludge age will have on system design and costs.

EXAMPLE 14.10

Use the data of Table 14.9 and Table 14.10 to validate that the optimal sludge age for system A1, conventional secondary treatment (without nitrification) is indeed equal to the minimum required sludge age.

Solution

The calculations made for the optimal sludge age are repeated for other values of the sludge age, in order to assess what the influence of this variable is on the performance and the costs of the system. In Figure 14.15 the following parameters are plotted as function of the sludge age:

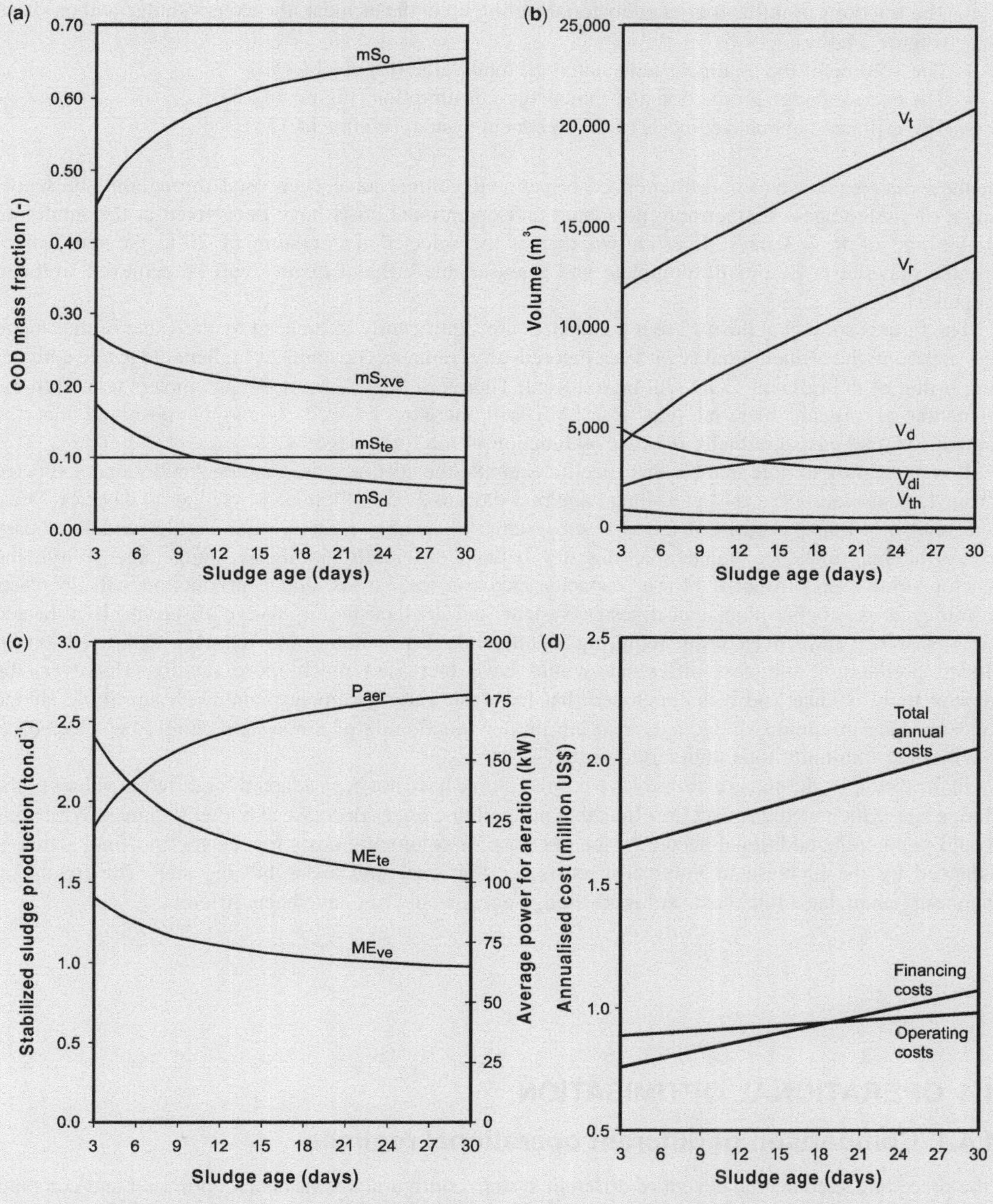

Figure 14.15 Influence of the sludge age on activated sludge system performance and annualised treatment costs

- The fractions of influent organic material ending up in the effluent, the excess sludge and oxidised (Figure 14.15a);
- The volume of the treatment units and their total value (Figure 14.15b);
- The excess sludge production and the energy consumption (Figure 14.15c);
- The estimated annualized costs of the treatment system (Figure 14.15d).

In the example, constant construction costs per unit volume have been used throughout the whole range of sludge ages. Furthermore personnel and operational costs have been fixed at the minimum sludge age of $R_s = 3$ days. It is known that at the selected temperature of 20°C the removal of organic material is essentially complete and a reasonable effluent quality will be achieved (refer to Appendix 3).

The figures show that the different parameters are significantly influenced by the value of the sludge age, although the effluent quality (at least theoretically) remains constant. At higher sludge age a higher proportion of the influent COD will be oxidised. Therefore the values of the parameters related to the oxidation of organic material (e.g. P_{aer}, V_r) will increase as well. It can be observed that the annualised total costs gradually increase as function of the sludge age.

It is interesting to note that for this specific example the cost increase may be smaller than expected (from 1.65 million US$ $\cdot$ yr^{-1} at a sludge age of 3 days to 2.05 million US$ $\cdot$ yr^{-1} at 30 days, or 25%). The reason is that the costs of the two subsystems of aerobic reactor + final settler and secondary excess sludge thickener + anaerobic digester behave differently at higher sludge age. While the reactor volume and installed blower capacity increase, the excess sludge production will decrease, resulting in a smaller thickener-digester system and less costs for sludge disposal. If a heated digester would have been used, requiring significantly less volume and reducing stabilised excess sludge production, the cost difference would have increased much more rapidly. However, the general trend is clear and it is concluded that for wastewater treatment plants with anaerobic sludge digestion, the minimum sludge age ensuring proper functioning of the system should be selected, as it will yield minimum total annualised costs.

In the example, the construction costs per unit volume have not been adapted for different values of the sludge age, which would reflect the effect that unit volume prices decrease at higher volumes. While this would yield some additional accuracy, the decrease in volumetric costs for reactor and final settler is balanced by the increase in volumetric costs for thickener and anaerobic digester. The trend, i.e. increasing annualised total costs at higher sludge ages, would not have been different.

14.4 OPERATIONAL OPTIMISATION

14.4.1 Comparison of different operational regimes

In the preceding sections the design of different system configurations has been optimised and compared. Another application of the optimisation procedure is to use it to compare the operational costs of a single system configuration under different operational regimes. This application will be demonstrated here with some examples.

EXAMPLE 14.11

Determine whether it is worthwhile in economic terms to increase the temperature in the anaerobic digester in Example 14.4 (system configuration A1, secondary treatment of raw sewage in a conventional activated sludge system). Assume that the costs of a heating system are US$ 150,000 (including installation and instrumentation) for an expected methane production of 344 $kg\,CH_4 \cdot d^{-1}$. The costs of operation and personnel increase by US$ 50,000 per year.

Solution

First it has to be determined whether the potential heat generation is sufficient to increase the temperature of the digester from its ambient value of 20°C to 30°C, the optimal value for anaerobic digestion. In Example 14.4 it was determined that the methane production was 344 $kg\,CH_4 \cdot d^{-1}$, with a thickened excess sludge flow of 196 $m^3 \cdot d^{-1}$ to be heated.

Knowing that the combustion value of methane is 12,000 $kCal \cdot kg^{-1}$ CH_4 and that it takes 1000 kCal to raise the temperature of one m^3 of sludge (mainly water) by 1°C, the mass of methane required to raise the temperature of 1 m^3 of sludge by 10°C is $10{,}000/12{,}000 = 0.83$ $kg\,CH_4$. The mass of methane required to increase the temperature of the thickened excess sludge flow is equal to $0.83 \cdot 198 = 163$ $kg\,CH_4 \cdot d^{-1}$. Hence, the energy contained in the produced biogas is more than sufficient to meet the energy demand for heating, even taking into consideration that heat transfer will not be 100% efficient.

Having established that heating of the digester with the produced methane is feasible, it remains to be determined whether it is an economically viable option. For this, the costs of the system with and without heating are compared. The benefits of heating the digester are a reduction of the hydraulic residence time (resulting in a reduction in volume) and an increase in digestion efficiency. Using Eq. (12.63) it is estimated that the hydraulic residence time will be reduced from 25 days at 20°C to 13 days at 30°C, resulting in a volume reduction of 44%.

Disadvantages are the costs associated with the acquisition, operation and maintenance of the heat exchange equipment. Maintaining the same conditions as before (in Table 14.9 and Table 14.10), a total cost of US$ 0.329 per m^3 with heating and US$ 0.377 per m^3 without heating can be calculated. Thus a significant cost reduction (12%) can be obtained when the digester is heated to 30°C.

In the above evaluation it has been tacitly assumed that the temperature is constant. If this is not so, heating will become less attractive as can be seen from the following. If the wastewater temperature changes over the year from, say, 17 to 23°C, the heating capacity will have to be designed for the lowest temperature (highest difference: $30-17 = 13$°C, so that in this case the installed heat exchange equipment will have to be larger and therefore more costly. The fact that during the warm season less energy is consumed does not offset the higher investment costs.

In practice the decision whether or not to implement digester heating not only depends on the results of the cost analysis. Other factors to be considered are (I) the possibility to generate electricity from the produced methane and to use the residual heat (energy losses) to heat the digester and (II) the added complexity of the system, where potential instability may be introduced in the case of failure of the heating system or operating errors.

EXAMPLE 14.12

A beer brewery is considering treatment of its effluent. The initial design resulted in the selection of a conventional activated sludge system, to be operated at a sludge age of 4 days and at an oxygen concentration of 1.0 mg $O_2 \cdot l^{-1}$.

From experiences at other beer breweries it is known that conventional wastewater treatment of this effluent frequently leads to serious problems with sludge settleability. For example, at another brewery with a similar activated sludge system (also operated at $R_s = 4$ days and at an oxygen concentration of 1.0 mg $O_2 \cdot l^{-1}$) the values of the settleability constants were $k = 0.5\ l \cdot g^{-1}$ and $v_0 = 5\ m \cdot h^{-1}$.

Several methods are considered to improve the settleability of the sludge:

1. Addition of $FeSO_4$ at a concentration of 150 mg $\cdot l^{-1}$ mixed liquor modifies the values of the Vesilind constants to $k = 0.30\ l \cdot g^{-1}$ and $v_0 = 10\ m \cdot h^{-1}$ for the same sludge age and dissolved oxygen concentration. An additional 100 mg $\cdot l^{-1}$ of mineral sludge is generated. The $FeSO_4$ can be obtained as a by-product of another factory at a price of US$ 60 $\cdot$ ton^{-1};
2. Increase the DO concentration from 1.0 to 2.5 mg $O_2 \cdot l^{-1}$ in order to suppress the presence of bulking sludge and to increase the settleability to $k = 0.40\ l \cdot g^{-1}$ and $v_0 = 8\ m \cdot h^{-1}$;
3. Installation of a selector (US$ 50,000), which is expected to improve the values of the settleability constants to $k = 0.35\ l \cdot g^{-1}$ and $v_0 = 10\ m \cdot h^{-1}$.

Calculate for the brewery what the costs per m^3 of effluent will be for the conventional activated sludge treatment (base alternative) and whether any of the above alternatives will result in a substantial reduction of the operating- and investment costs. The characteristics of the wastewater are listed in Table 14.24. Use the costs coefficients of the previous examples.

Table 14.24 Wastewater characterisation of the beer brewery of Example 14.12

Parameter	Value	UoM	Parameter	Value	UoM
Q_i	250	$m^3 \cdot h^{-1}$	X_{ti}	300	mg TSS $\cdot l^{-1}$
S_{ti}	2000	mg COD $\cdot l^{-1}$	N_{ki}	10	mg N $\cdot l^{-1}$
f_{np}	0.02	(−)	P_{ti}	2	mg P $\cdot l^{-1}$
f_{ns}	0.02	(−)	T	35	°C
f_{sb}	0.6	(−)	Alk_i	400	mg $\cdot l^{-1}$ $CaCO_3$

Solution

Using the procedure demonstrated in Example 14.4 calculate for all cases the optimised configuration and the values of the main process parameters. In Table 14.25 the results are presented for the base case and the three alternatives. All four alternatives will result in the same effluent quality and an equal demand for nitrogen and phosphorus.

With respect to the operational parameters, the addition of $FeSO_4$ will result in an additional production of 100 mg $\cdot l^{-1}$ or 0.6 $t \cdot d^{-1}$ of mineral sludge. In the alternative with increased oxygen concentration, the oxygen transfer efficiency (OT_a) is reduced. As the saturation constant at 35°C is

7.1 mg $O_2 \cdot l^{-1}$, the ratio between the OT_a of an aerator operating at 2.5 mg $O_2 \cdot l^{-1}$ and at 1.0 mg $O_2 \cdot l^{-1}$ is $(7.1-2.5)/(7.1-1.0) = 0.75$. Thus, the energy requirements and installed aeration capacity for the system operated at 2.5 mg $O_2 \cdot l^{-1}$ will be $1/0.75 = 33\%$ larger.

Table 14.25a Comparison of the alternatives for treatment of beer brewery effluent

Parameter	Alternative			
	1 – Basis	**2 – $FeSO_4$**	**3 – DO = 2.5**	**4 – Selector**
Effluent:				
– COD	40	40	40	40
– TKN (demand)	35	35	35	35
– P (demand)	9	9	9	9
Volume (m^3):				
– Reactor	4851	3504	3696	3234
– Settler	1981	1139	1341	1073
– Thickener	600	159	408	280
– Digester	2814	2426	2110	1853
– Total	10,245	7228	7555	6440

Table 14.25b Comparison of alternatives for treatment of beer brewery WW

Parameter	Alternative			
	1 – Basis	**2 – $FeSO_4$**	**3 – DO = 2.5**	**4 – Selector**
Excess sludge production				
– TSS	2.61	3.24	2.61	2.61
– VSS	1.45	1.45	1.45	1.45
Energy demand	267	267	356	267
Potential energy production	97	97	97	97
Costs (US\$ct $\cdot$ m^{-3}):				
– operational	50.2	47.9	53.4	43.8
– financial	34.7	27.9	32.9	26.5
– total	84.9	75.8	86.3	69.9
Reduction	–	11%	–2%	18%

The total volumes of all three alternatives will be considerably reduced compared to the base alternative, due to a smaller required settling area and the possibility of maintaining a higher sludge concentration in the reactor. When the total annualised treatment costs are evaluated, it is clear from a cost perspective that the installation of the selector is the most financially interesting alternative, resulting in a reduction of treatment costs from US$ 0.85 m^{-3} to US$ 0.70 m^{-3} (or 18%). Dosing of $FeSO_4$ might be considered as well, while operation at a higher oxygen concentration is not attractive.

14.4.2 Optimised operation of existing treatment plants

The previous examples in this chapter all refer to the design of an activated sludge system based on an expected or experimentally determined wastewater flow or composition. Once the treatment system has been constructed, the actual quantity and quality of the wastewater will probably differ from those expected, as well as the values of the operational parameters. In this case the theory presented in this book can be used for another type of optimisation: for a given configuration determine the optimal operational conditions, characterised by production of the specified effluent quality at minimal costs.

EXAMPLE 14.13

After construction of the Bardenpho activated sludge system designed in Example 14.7, it is established that the flow and composition of the wastewater differ from the design values, i.e. the actual values are $Q_i = 10{,}000\ m^3 \cdot d^{-1}$, $S_{ti} = 975\ mg \cdot l^{-1}$, $N_{ti} = 40\ mg \cdot l^{-1}$ and an average temperature in wintertime of 24°C. Define the optimal operational conditions that will produce an effluent without nitrate and with maximum 2 mg ammonium in the effluent. Estimate the treatment costs. Assume that the size of the anoxic reactors cannot be increased or reduced.

Solution

The changes in actual loading and temperature compared to the design values will have the following effects on the activated sludge system:

- The minimum sludge age required for full denitrification will be smaller than the one calculated in Example 14.7 (12 days) because the ratio N_{ti}/S_{ti} has decreased and the temperature is higher, which will increase the values of the kinetic parameters;
- However, if the sludge age is reduced, a new limiting factor in this example will be the effluent ammonium concentration, as it depends on the applied aerobic sludge age and it is not possible to increase the aerobic fraction in the system as the anoxic volumes are fixed;
- The increase in organic load and the reduction in sludge age will increase the solids load to the digester, although this is partly compensated by the increase in digestion efficiency at higher temperature;
- The hydraulic loading rate of the final settler will be lower as a result of the decrease in influent flow. However, depending on the actual sludge concentration, the solids loading rate might increase.

Taking the above constraints into consideration a solution will have to be determined. It is not possible to adapt the size of the anoxic zones in the existing activated sludge system (i.e. $f_x = 0.38$). It can be calculated with Eq. (5.44) that a minimum sludge of 8.25 days is required to comply with the limit of $N_{ae} < 2\ mg\ N \cdot l^{-1}$. At this sludge age denitrification will be complete. However, it can be demonstrated that the digester will be slightly overloaded. For the temperature of 24°C the required sludge retention time in the digester is equal to:

$$R_{di} = 20 \cdot 1.1^{(T-20)} + 5 = 18.7\,\text{days} \tag{12.63}$$

As the volume of the aerobic reactor does not change, the value of the excess sludge flow rate q will be equal to $8234/8.25 = 998\ m^3 \cdot d^{-1}$. MX_t is equal to $24{,}838\ kg\ TSS \cdot m^{-3}$. For an excess sludge production of $ME_t = MX_t/R_s = 24{,}838/8.24 = 3011\ kg\ TSS \cdot d^{-1}$ and an available thickener surface $A_{th} = V_{th}/(H_{th} \cdot s_{fth}) = 530/(3 \cdot 1.5) = 118\ m^2$, the applied solids flux is:

$$F_{sol} = ME_t/A_{th} = 3011/118 = 25.6\,kg\,TSS \cdot m^{-2} \cdot d^{-1}$$

Furthermore it is know that the solids flux is minimum for the limiting flux. Therefore:

$$F_l = F_{sol} = 25.6 = X_{th} \cdot v_0 \cdot (k \cdot X_{th} - 1) \cdot \exp(-k \cdot X_{th}) \tag{12.5a}$$

This can be solved iteratively for $X_{th} = 15.7\ kg\ TSS \cdot m^{-3}$. The thickened excess flow rate $q_{th} = 3011/15.7 = 192\ m^3 \cdot d^{-1}$, resulting in a sludge retention time in the digester of $R_{di} = V_{di}/q_{th} = 3370/192 = 17.6$ d, which is indeed slightly smaller than the required digestion time of 18.7 days. A solution complying with all restrictions can be found for $R_s = 9.75$ days.

$$N_{ae} = 1.6\ mg\ N \cdot l^{-1}$$
$$D_c = D_{c1} + D_{c3} = 57.5\ mg\ N \cdot l^{-1}$$
$$R_{di} = 18.8\ \text{days}$$

The combined denitrification capacity is much larger than the nitrification capacity N_c of $26.7\ mg\ N \cdot l^{-1}$, so denitrification is complete (it can be verified that $D_c > N_{av}$ for both anoxic zones).

To evaluate the performance of the final settler, the applied solids loading rate must be compared with the solids flux that can be transported in the settler for all values of X between $X_t = 3.4$ and $X_r = (s+1)/s \cdot 3.4 = 6.8\ g\ TSS \cdot l^{-1}$. The solids loading rate is calculated as

$$F_{sol} = (s+1) \cdot X_t \cdot Q_i/(A_d/s_{fd}) = 143\,kg\,TSS \cdot m^{-2} \cdot d^{-1} \tag{8.15}$$

The solids flux due to abstraction of return sludge is $s \cdot Q_i \cdot X/(A_d/s_{fd})$ and the flux due to settling is $X \cdot v_0 \cdot \exp(-k \cdot X)$. The combined solids flux has a lowest value of $149\ kg\ TSS \cdot m^{-2} \cdot d^{-1}$ for $X = 6$ $g\ TSS \cdot l^{-1}$, which is larger than the applied solids loading rate and thus problems with solids-liquid separation are not to be expected in the final settler.

In Table 14.26 the main operational values and the costs of the original design are compared with the actual situation. The data show that using the theory presented in this book, it is possible to accommodate the 25% increase in daily organic load (from 7800 kg COD to 9750 kg COD per day) without additional investments and only a limited increase in operational costs.

Table 14.26 Main operational parameters and annualized costs of the optimised system under design and actual conditions

Parameter	UoM	Original design	Actual situation
Influent parameters:			
– Q_i	$m^3 \cdot d^{-1}$	12,000	10,000
– $S_{ti}/N_{ti}/P_{ti}$	$mg \cdot l^{-1}$	650/50/15	975/40/10
– T (min)	°C	20	24
Operational R_s	days	12	9.75
Sludge production:			
– total	ton $TSS \cdot d^{-1}$	1.8	2.2
– volatile	ton $VSS \cdot d^{-1}$	1.1	1.4
Energy:			
– demand/pot. generation	kW	188/37	213/50
Costs:			
– financial	US$ $\cdot$ year^{-1}	890,000	890,000
– operation	US$ $\cdot$ year^{-1}	1,020,000	1,080,000
– total	US$ $\cdot$ year^{-1}	1,910,000	1,970,000

14.5 INTEGRATED DESIGN EXAMPLES

14.5.1 Nutrient removal in different configurations

In Chapter 7 various methods for phosphorus removal were discussed: pre-precipitation, simultaneous precipitation and bio-P removal. In Example 14.14 these configurations will be compared in more detail.

EXAMPLE 14.14

A Bardenpho system treats 15,000 m^3 of municipal wastewater per day. Table 14.27 lists the influent characteristics, the main process conditions and the estimated values of the kinetic- and stoichiometric parameters. For convenience, the values of all kinetic parameters have been adjusted to the design temperature of 15°C. The excess sludge is thickened and sent to a central sludge treatment unit for digestion and final disposal. The effluent limits for nitrogen are:

- Total nitrogen: $N_{td} < 10$ mg $\cdot$ l^{-1} total nitrogen;
- Total Kjeldahl Nitrogen $N_{kd} < 5$ mg TKN $\cdot$ l^{-1};
- Ammonium nitrogen $N_{ad} < 1$ mg NH_4-N $\cdot$ l^{-1}.

Future legislation is expected in which an effluent total phosphorus concentration of < 2 mg P $\cdot$ l^{-1} ($= P_{td}$) is required.

Table 14.27 Wastewater- and process characteristics of Example 14.14

Par.	Value	Unit	Par.	Value	Unit
Q_i	15,000	$m^3 \cdot d^{-1}$	b_h	0.197	d^{-1}
V_r	17,500	m^3	K_2	0.082	$mg\,N \cdot mg^{-1}\,X_a \cdot d^{-1}$
S_{ti}	600	$mg\,COD \cdot l^{-1}$	K_3	0.069	$mg\,N \cdot mg^{-1}\,X_a \cdot d^{-1}$
N_{ti}	65	$mg\,N \cdot l^{-1}$	μ_m	0.224	d^{-1}
P_{ti}	15	$mg\,P \cdot l^{-1}$	b_n/b_p	0.033	d^{-1}
f_{sb}	0.30	(−)	K_n	0.560	$mg\,N \cdot l^{-1}$
f_{np}	0.15	(−)	f_n	0.1/0.06[(2)]	$mg\,N \cdot mg^{-1}\,VSS$
f_{ns}	0.08	(−)	f_p	0.025/0.015[(2)]	$mg\,P \cdot mg^{-1}\,VSS$
T	15	°C	f_{pp}	0.38	$mg\,P \cdot mg^{-1}\,VSS$
f_v	0.7/0.8[(1)]	mg VSS/mg TSS	f	0.2	$mg\,VSS \cdot mg^{-1}\,VSS$
f_{vp}	0.46	mg VSS/mg TSS	f_{ep}	0.25	$mg\,VSS \cdot mg^{-1}\,VSS$
f_{x1}	0.225	(−)	f_{pd}	0.7	(−)
f_{x3}	0.200	(−)	f_{an}	0.125	(−)
R_s	24	d	K_c	0.06	$l \cdot mg^{-1}\,VSS \cdot d^{-1}$
a	4	(−)	η_{x1}	0.65[(3)]	(−)
s	1	(−)	X_{te}	20	$mg\,TSS \cdot l^{-1}$
r	1	(−)	N_{ose}	0.5	$mg\,N \cdot l^{-1}$
N	2	(−)	P_{ose}	0.125	$mg\,P \cdot l^{-1}$

Notes: (1) When pre-precipitation is applied.
(2) Fraction of N and P in the organic solids of the primary sludge removed during pre-precipitation.
(3) Removal efficiency of particulate + colloidal organic material when pre-precipiation is applied.

The following issues are to be evaluated:

(1) Determine whether the existing system can produce an effluent quality that meets the current nitrogen standards;
(2) To comply with the new effluent phosphorus limit, consider the following measures:
  a. Simultaneous precipitation (using $FeCl_3$);
  b. Installation of a primary settler and pre-precipitation (using $FeCl_3$);
  c. Biological phosphorus removal (UCT configuration), if required this can be supplemented with addition of $FeCl_3$.

Indicate for all options:

- The system performance with respect to phosphorus removal;
- The expected daily $FeCl_3$ consumption;
- The effect on the effluent nitrogen quality.

Correct N_l and P_l for the loss of particulate organic solids with the effluent (refer to the procedure described in Appendix A6).

Solution

(1) Determine the performance of the existing Bardenpho system
Calculate the sludge mass that develops in the Bardenpho system:

$$C_r = 0.45 \cdot 24/(1 + 0.197 \cdot 24) = 1.88 \quad (3.30)$$

$$MX_v = [(1 - f_{np} - f_{ns}) \cdot (1 + f \cdot b_h \cdot R_s) \cdot C_r + f_{np} \cdot R_s/f_{cv}] \cdot MS_{ti} \quad (3.48)$$

$$= [(1 - 0.15 - 0.08) \cdot (1 + 0.2 \cdot 0.197 \cdot 24) \cdot 1.88 + 0.15 \cdot 24/1.5] \cdot 15{,}000 \cdot 0.6$$

$$= 47{,}010\,\text{kg VSS}$$

$$MX_t = MX_v/f_v = 47{,}010/0.7 = 67{,}158\,\text{kg TSS} \quad (3.49)$$

$$X_t = MX_t/V_r = 67{,}158/17{,}500 = 3.84\,\text{kg TSS} \cdot m^{-3}$$

$$ME_v = MX_v/R_s = 47{,}010/24 = 1959\,\text{kg VSS} \cdot d^{-1} \quad (3.50)$$

Calculate the effluent ammonium concentration:

$$N_{ae} = K_n - (b_n + 1/R_s)/[(1 - f_x) \cdot \mu_m - b_n - 1/R_s] \quad (5.44)$$

$$= 0.56 \cdot (0.033 + 1/24)/[(1 - 0.425) \cdot 0.224 - 0.033 - 1/24)] = 0.8\,\text{mg N} \cdot l^{-1}$$

To calculate the nitrification capacity, remember that in this example N_l is corrected for the loss of N_{ope} with the effluent. The value of $N_{ope} = f_n \cdot f_v \cdot X_{te} = 0.1 \cdot 0.7 \cdot 20 = 1.4$ mg $N \cdot l^{-1}$.

$$N_{lx} = f_n \cdot ME_v/Q_i - N_{ope} = 0.1 \cdot 1958 \cdot 1000/15{,}000 - 1.4 = 11.7\,\text{mg N} \cdot l^{-1} \quad (A6.2)$$

$$N_c = N_{ti} - N_{lx} - N_{oe} - N_{ae} = 65 - 11.7 - (1.4 + 0.5) - 0.8 = 50.7\,\text{mg N} \cdot l^{-1} \quad (A6.3)$$

The denitrification capacity is calculated as:

$$S_{bi} = (1 - 0.15 - 0.08) \cdot 600 = 462\,\text{mg COD} \cdot l^{-1} \quad (3.3)$$

$$D_{c1} = (f_{sb} \cdot f_{dn} + K_2 \cdot C_r \cdot f_{x1}) \cdot S_{bi} \quad (5.68)$$

$$= (0.3 \cdot 0.114 + 0.082 \cdot 1.88 \cdot 0.225) \cdot 462 = 31.7\,\text{mg N} \cdot l^{-1}$$

$$D_{c3} = K_3 \cdot C_r \cdot f_{x3} \cdot S_{bi} \quad (5.69)$$

$$= 0.069 \cdot 1.88 \cdot 0.2 \cdot 462 = 12.0\,\text{mg N} \cdot l^{-1}$$

In each anoxic zone, either the available denitrification capacity (D_c) or the nitrate available for denitrification (N_{av}) can be the limiting factor. Assuming D_c is limiting, N_{ne} is equal to $50.7{-}31.7{-}12.0 = 6.9$ mg $N \cdot l^{-1}$. Check the assumption by calculating N_{av1} and N_{av3} as well:

$$N_{av1} = a/(a + s + 1) \cdot N_c + s \cdot N_{ne} \quad (5.74)$$

$$= 4/(4 + 1 + 1) \cdot 50.7 + 1 \cdot 6.9 = 40.7\,\text{mg N} \cdot l^{-1}$$

Indeed N_{av1} is larger than D_{c1}. Now N_{av3} can be simply calculated as:

$$\begin{aligned} N_{av3} &= N_c - D_{c1} \\ &= 50.7 - 31.7 = 18.9\,\text{mg N} \cdot l^{-1} \end{aligned} \tag{5.76}$$

Again the denitrification capacity is limiting ($N_{av3} > D_{c3}$). It may be concluded that in both zones sufficient nitrate is available to ensure maximum denitrification and thus N_{ne} is indeed equal to $N_c - D_c = 6.9$ mg $N \cdot l^{-1}$. The total effluent nitrogen concentration N_{te} is 9.6 mg $N \cdot l^{-1}$ and is composed of 6.9 mg $N \cdot l^{-1}$ of nitrate and 2.7 mg $N \cdot l^{-1}$ of Kjeldahl nitrogen ($N_{ae} = 0.8$ mg $N \cdot l^{-1}$ and $N_{oe} = 1.9$ mg $N \cdot l^{-1}$). It can be concluded that the nitrogen effluent criteria are met.

To calculate the phosphorus removal with the excess sludge, P_l is corrected for the loss of particulate organic phosphorus with the effluent. $P_{ope} = f_p \cdot f_v \cdot X_{te} = 0.35$ mg $P \cdot l^{-1}$.

$$P_{lx} = P_l - P_{ope} = 0.025 \cdot 1959 \cdot 1000/15{,}000 - 0.35 = 2.9\,\text{mg P} \cdot l^{-1} \tag{7.22}$$

Thus approximately 20% of the influent phosphorus concentration is removed together with the excess sludge. The total effluent phosphorus concentration $P_{te} = 15-2.9 = 12.1$ mg $P \cdot l^{-1}$, with a soluble phosphate concentration P_{pe} equal to $P_{ti} - P_l - P_{oe} = 15.0-2.9-0.5 = 11.6$ mg $P \cdot l^{-1}$

(2a) Simultaneous precipitation

It is assumed that the sludge age of the BDP system will not change as a result of simultaneous precipitation. In order to meet the total phosphorus effluent discharge limit, the phosphate effluent concentration must be smaller than:

$$P_{pe} < P_{td} - P_{oe} < 2.0 - 0.475 < 1.525\,\text{mg P} \cdot l^{-1}$$

MP_{te} is $15{,}000 \cdot 2/1000 = 30$ kg $P \cdot d^{-1}$. The mass of phosphorus removed together with the biological excess sludge is:

$$\begin{aligned} MP_{lx} &= f_p \cdot ME_v - Q_i \cdot f_v \cdot f_p \cdot X_{te} \quad \text{or} \\ Q_i \cdot P_{lx}/1000 &= 15{,}000 \cdot 2.9/1000 = 44\,\text{kg P} \cdot d^{-1} \end{aligned} \tag{7.34}$$

Using Eq. (7.42), the required chemical removal of phosphorus can be calculated:

$$\begin{aligned} MP_{chem} &= MP_{ti} - MP_{lx} - MP_{te} \\ &= 225 - 44 - 30 = 151\,\text{kg P} \cdot d^{-1} \quad \text{or} \quad 4.9\,\text{kmol P} \cdot d^{-1} \\ P_{chem} &= 151 \cdot 1000/15{,}000 = 10.1\,\text{mg P} \cdot l^{-1} \end{aligned} \tag{7.42}$$

From Figure 7.14 one can determine that to lower P_{pe} to a value of 1.525 mg $P \cdot l^{-1}$ a molar Me/P_{ti} dosing ratio between 0.8 and 1.3 is required. In our case the conservative estimate of 1.3 (i.e. the "max" curve) is selected. Multiplying this ratio with the phosphorus influent load MP_{ti} (on a molar basis) yields the daily consumption of $FeCl_3$:

$$M_{ME} = 225/31 \cdot 1.3 = 9.4\,\text{kmol FeCl}_3 \cdot d^{-1}$$

Using the data on $FeCl_3$ in Table 7.3, this amounts to 1531 kg $FeCl_3 \cdot d^{-1}$ or 2.73 $m^3 \cdot d^{-1}$ 40% wt $FeCl_3$. In order to calculate the production of chemical sludge, Eqs. (7.45 and 7.46) are used:

$$ME_{mp} = MP_{chem}/31 \cdot mw_{mp} = 4.9 \cdot (55.8 + 31.0 + 4 \cdot 16.0)$$
$$= 736\ \text{kg FePO}_4 \cdot d^{-1} \tag{7.45}$$

$$ME_{mo} = (M_{ME} - MP_{chem}) \cdot mw_{mo} = (9.4 - 4.9) \cdot (55.8 + 3 \cdot 17)$$
$$= 487\ \text{kg Fe(OH)}_3 \cdot d^{-1} \tag{7.46}$$

The total daily chemical excess sludge production ME_{chem} equals 736 + 487 = 1222 kg TSS $\cdot d^{-1}$. If the sludge age R_s is maintained at 24 days, the daily excess sludge production will increase as a result of chemical dosing from a value of $MX_t/R_s = 67{,}168/24 = 2798$ kg TSS $\cdot d^{-1}$ to a value of 4021 kg TSS $\cdot d^{-1}$. Clearly simultaneous precipitation will have implications for the sludge concentration in the system. For $R_s = 24$ days, MX_t increases to $24 \cdot 4021 = 96{,}496$ kg TSS, an increase in sludge concentration X_t from 3.84 to 5.51 kg TSS $\cdot m^{-3}$.

In practice, the produced Fe(III)salts will act as a flocculant and sludge settleability will increase. Depending on the design of the final settler and the applied solids load, operating the activated sludge system at the increased concentration of 5.51 kg TSS $\cdot m^{-3}$ might well be possible. Then, as the sludge age does not change, the effluent nitrogen concentration and composition will not change either.

(2b) Pre-precipitation

In the case of pre-precipitation a primary settler needs to be installed. The hydraulic retention time is typically around 2 hours, so the volume of the primary settler will be $2/24 \cdot 15{,}000 = 1250\ m^3$. The calculation of the pre-precipitation example proceeds according to the following steps:

- Calculate the pre-settled influent composition and recalculate the activated sludge system performance based on the adapted influent composition;
- Take appropriate action if nitrogen removal is no longer guaranteed;
- Calculate the required chemical phosphorus removal in the primary clarifier including chemical consumption.

To assess the effect of pre-precipitation on nitrogen removal, several issues are to be considered:

- Both the organic- and solids load to the biological reactor are reduced. As the reactor volume will not change, a significant increase in sludge age will be the result;
- The organic fraction of the sludge in the biological reactor (f_v) will increase due to removal of a large part of the inorganic (mineral) suspended solids in the primary settler;
- The N_{ti}/S_{ti} ratio increases as the removal of COD is higher than that of TKN (most of the influent nitrogen is present in the form of soluble ammonium).

The quantitative effect on nitrogen removal performance can be calculated as follows. As a first step, the influent composition after pre-precipitation is defined:

$$f_{bs} = f_{sb} \cdot (1 - f_{np} - f_{ns}) \tag{7.38}$$
$$= 0.3 \cdot (1 - 0.15 - 0.08) = 0.23$$

$$f_{bp} = (1 - f_{sb}) \cdot (1 - f_{np} - f_{ns}) \tag{7.37}$$
$$= (1 - 0.3) \cdot (1 - 0.15 - 0.08) = 0.54$$

$$f'_{ns} = f_{ns}/[1 - \eta_{x1} \cdot (f_{np} + f_{bp})] \quad (7.35)$$
$$= 0.08/[1 - 0.65 \cdot (0.15 + 0.54)] = 0.14$$
$$f'_{np} = (1 - \eta_{x1}) \cdot f_{np}/[1 - \eta_{x1} \cdot (f_{np} + f_{bp})] \quad (7.36)$$
$$= (1 - 0.65) \cdot 0.15/[1 - 0.65 \cdot (0.15 + 0.54)] = 0.10$$
$$f'_{bs} = f_{bs}/[1 - \eta_{x1} \cdot (f_{np} + f_{bp})] \quad (7.40)$$
$$= 0.23 \cdot [1 - 0.65 \cdot (0.15 + 0.54)] = 0.42$$
$$f'_{bp} = (1 - \eta_{x1}) \cdot f_{bp}/[1 - \eta_{x1} \cdot (f_{np} + f_{bp})] \quad (7.41)$$
$$= (1 - 0.65) \cdot 0.54/[1 - 0.65 \cdot (0.15 + 0.54)] = 0.34$$
$$f'_{sb} = f'_{bs}/(f'_{bs} + f'_{bp}) \quad (7.39)$$
$$= 0.42/(0.42 + 0.34) = 0.55$$

The overall COD removal efficiency is $\eta_{x1} \cdot (f_{np} + f_{bp}) = 0.65 \cdot (0.15 + 0.54) = 45\%$. The pre-settled total influent COD concentration $S'_{ti} = (1-0.45) \cdot S_{ti} = 0.55 \cdot 600 = 331 \text{ mg COD} \cdot l^{-1}$. Assuming the sludge concentration in the reactor is maintained at $3.84 \text{ kg TSS} \cdot m^{-3}$, the sludge age R_s that will develop after pre-precipitation can be determined from the value of mX_t, which is given by two independent equations:

$$mX_t = MX_t/MS'_{ti} = V_r \cdot X_t/(Q_i \cdot S'_{ti}) \quad (3.49)$$
$$= 67{,}158/(15{,}000 \cdot 0.331) = 13.51 \text{ mg VSS} \cdot d \cdot mg^{-1} \text{ COD}$$
$$mX_t = mX_v/f_v = [(1 - f'_{ns} - f'_{np}) \cdot (1 + f \cdot b_h \cdot R_s) \cdot C_r + f'_{np} \cdot R_s/f_{cv}]/f_v \quad (3.49)$$
$$= [(1 - 0.14 - 0.10) \cdot (1 + 0.2 \cdot 0.197 \cdot R_s) \cdot 0.45 \cdot R_s/$$
$$(1 + 0.197 \cdot R_s) + 0.10 \cdot R_s/1.5]/0.80$$

The equations are solved for $R_s = 72$ days. Clearly this is an excessive sludge age and well outside the range of sludge ages for which the model has been validated (2 to 50 days). The sludge will have a very low active fraction, which may result in the presence of many small, poorly settling sludge flocs in the effluent of the final settler. A second effect of this high sludge age is a reduction in phosphorus removal with the secondary excess sludge, which will have to be compensated for by an increase of the chemical dosing to the primary settler.

To determine whether denitrification will be complete, the limiting ratio for complete denitrification in a Bardenpho configuration $(N_{ti}/S_{ti})_o$, must be compared to the ratio N'_{ti}/S'_{ti} in the pre-settled influent. S'_{ti} has been calculated already as $269 \text{ mg COD} \cdot l^{-1}$. N'_{ti} can be calculated by subtracting the nitrogen removed in the primary settler from N_{ti} (remember that the nitrogen fraction in the primary sludge $f_n = 0.06$, as most of the N_{ki} is present in the form of ammonium).

$$N'_{ti}/S'_{ti} = [65 - 0.06 \cdot (600 - 331)/1.5]/331 = 54.3/331 = 0.164 \text{ mg N} \cdot mg^{-1} \text{ COD}$$

To calculate $(N_{ti}/S_{ti})_o$ the following procedure is followed:

$$C_r = 0.45 \cdot 72/(1 + 0.197 \cdot 72) = 2.13 \quad (3.30)$$

$$
\begin{aligned}
(N_c/S_{bi})_o &= [(a+s+1)\cdot(f'_{sb}\cdot f_{dn}+K_2\cdot C_r\cdot f_m)]/[a+(K_2/K_3)\cdot(s+1)] && (5.86)\\
&= [(4+1+1)\cdot(0.55\cdot 0.114+0.082\cdot 2.13\cdot 0.425)]/\\
&\quad [4+(0.082/0.069)\cdot(1+1)]\\
&= 0.129\ \text{mg N}\cdot\text{mg}^{-1}\ \text{COD}\\
N_{lx} &= MN_{lx}/Q_i && (A6.2)\\
&= (f_n\cdot ME_v-MN_{ope})/Q_i=(0.1\cdot 0.8\cdot 67{,}168/72-21)\cdot 1000/15{,}000\\
&= 3.6\ \text{mg N}\cdot\text{l}^{-1}\\
N_{ae} &= K_n\cdot(b_n+1/R_s)/[(1-f_m)\cdot\mu_m-b_n-1/R_s] && (5.44)\\
&= 0.56\cdot(0.033+1/72)/[(1-0.425)\cdot 0.224-0.033-1/72]\\
&= 0.3\ \text{mg N}\cdot\text{l}^{-1}\\
(N_{ti}/S_{ti})_o &= (1-f'_{ns}-f'_{np})\cdot(N_c/S_{bi})_o+(N_{lx}+N_a+N_{oe})/S'_{ti} && (5.88)\\
&= (1-0.14-0.10)\cdot 0.129+(3.6+0.3+1.9)/331\\
&= 0.115\ \text{mg N}\cdot\text{mg}^{-1}\ \text{COD}
\end{aligned}
$$

As expected, $N_{ti}/S_{ti} \gg (N_{ti}/S_{ti})_o$ ($0.164 \gg 0.115$) and complete denitrification is no longer possible. The effluent nitrogen concentration can be calculated as $N_{te} = N_{ne} + N_{ae} + N_{oe}$.

$$
\begin{aligned}
S'_{bi} &= (1-f'_{ns}-f'_{np})\cdot S'_{ti}=(1-0.14-0.10)\cdot 331=252\ \text{mg COD}\cdot\text{l}^{-1} && (3.3)\\
D_{c1} &= (f'_{sb}\cdot f_{dn}+K_2\cdot C_r\cdot f_{x1})\cdot S'_{bi} && (5.68)\\
&= (0.55\cdot 0.114+0.082\cdot 2.13\cdot 0.225)\cdot 252=25.7\ \text{mg N}\cdot\text{l}^{-1}\\
D_{c3} &= K_3\cdot C_r\cdot f_{x3}\cdot S'_{bi} && (5.69)\\
&= 0.069\cdot 2.13\cdot 0.2\cdot 252=7.4\ \text{mg N}\cdot\text{l}^{-1}\\
N_{ne} &= N'_{ti}-N_{oe}-N_{lx}-N_{ae}-D_{c1}-D_{c3} && (5.78)\\
&= 54.3-1.9-3.6-0.3-25.7-7.4=15.4\ \text{mg N}\cdot\text{l}^{-1}\\
N_{te} &= 15.4+0.3+1.9=17.6\ \text{mg N}\cdot\text{l}^{-1}
\end{aligned}
$$

The effluent total nitrogen limit is not met, requiring an additional 7.6 mg of NO_3-N $\cdot\, l^{-1}$ influent to be removed (114.1 kg N $\cdot\, d^{-1}$). Clearly a further increase of the sludge age is not a viable option here. Instead, as the availability of biodegradable substrate for denitrification is limiting, an easily biodegradable carbon source such as methanol could be added. Alternatively, part of the raw sewage flow could be bypassed around the primary settler.

For a quick (and conservative) indicative estimate of the methanol consumption, the ratio between the utilisation rates of nitrate and -substrate can be used: $r_{dn} = (1-f_{cv}\cdot Y)/2.86 \cdot r_{us} = f_{dn}\cdot r_{us}$. This figure only takes into account the nitrate demand for exogenous respiration. The estimated methanol consumption is the equal to 114.1/0.114 = 951 kg COD $\cdot\, d^{-1}$ or 63 mg COD $\cdot\, l^{-1}$ influent. For a more reliable estimate, the required methanol consumption can be calculated considering that there are three parameters that can be optimised (X_t, R_s and the methanol dosing) while there are a number of constraints to be complied to:

- Select a target sludge concentration X_t. In this example, the main objective will be to keep $X_t > 3.0$ kg TSS · m^{-3} in order to avoid problems with clarification in the final settler: the sludge volume ($X_t \cdot I_{DSV}$) should preferably be larger than 200 ml;
- Select an operating sludge age, taking into account that a significant reduction of the sludge age will increase methanol consumption;
- Select the quantity of methanol to be added;
- Constraint 1: $N_{ne} < N_{td} - N_{oe} - N_{ae} < 10 - 1.9 - N_{ae}$;
- Constraint 2: $N_{ae} < 1$ (which will generally be true at the sludge age under consideration).

Multiple solutions exist that satisfy the constraints. One of them is $R_s = 50$ days; $X_t = 3.06$ kg TSS · m^{-3}, requiring a methanol consumption of 615 kg COD · d^{-1} (410 kg methanol · d^{-1}) or 41 mg COD · l^{-1}. This is significantly lower than the initial estimate of the methanol consumption, as is to be expected because:

- The additional endogenous denitrification resulting from the increase in active biomass, due to the increase in COD load and the decrease in sludge age, was not considered in the initial indicative estimate;
- Due to the reduce in sludge age, more nitrogen will be removed with the secondary excess sludge.

The effluent nitrogen concentration and composition are calculated below:

$$S_{bi} = 252 + 41 = 293\ \text{mg COD} \cdot l^{-1}$$
$$S_{ti} = 331 + 41 = 372\ \text{mg COD} \cdot l^{-1}$$
$$f_{ns} = 48/372 = 0.13$$
$$f_{np} = 32/372 = 0.08$$
$$f_{sb} = (139 + 41)/293 = 0.61$$
$$C_r = 0.45 \cdot 50/(1 + 0.197 \cdot 50) = 2.07 \quad (3.30)$$
$$ME_{v2} = [(1 - 0.13 - 0.08) \cdot (1 + 0.2 \cdot 0.197 \cdot 50) \cdot 2.07/50 + 0.08/1.5] \cdot 0.372$$
$$= 856\ \text{kg VSS} \cdot d^{-1} \quad (3.50)$$
$$N_{lx} = 0.1 \cdot 856 \cdot 1000/15{,}000 - 1.4 = 4.3\ \text{g N} \cdot l^{-1} \quad (A6.2)$$
$$N_{ae} = 0.56 \cdot (0.033 + 1/50)/[(1 - 0.425) \cdot 0.224 - 0.033 - 1/50] = 0.4\ \text{mg N} \cdot l^{-1} \quad (5.44)$$
$$N_c = 54.3 - 4.3 - 1.9 - 0.4 = 47.7\ \text{mg N} \cdot l^{-1} \quad (5.50)$$
$$D_{c1} = (0.114 \cdot 0.61 + 0.082 \cdot 2.07 \cdot 0.225) \cdot 293 = 31.6\ \text{mg N} \cdot l^{-1} \quad (5.68)$$
$$D_{c3} = 0.069 \cdot 2.07 \cdot 0.2 \cdot 293 = 8.4\ \text{mg N} \cdot l^{-1} \quad (5.69)$$
$$N_{ne} = N_c - (D_{c1} + D_{c3}) = 47.7 - (31.6 + 8.4) = 7.7\ \text{mg N} \cdot l^{-1} \quad (5.78)$$

As the actual denitrification capacity is equal to the required denitrification capacity the effluent nitrogen criteria are met. The effluent nitrogen composition will be: $N_{oe} = 1.9$ mg N · l^{-1}; $N_{ae} = 0.4$ mg N · l^{-1} and $N_{ne} = 7.7$ mg N · l^{-1}. To finalise the design, it is only required to calculate how much phosphorus needs to be pre-precipitated in order to meet the effluent phosphorus limit:

$$MP_{lx} = f_p \cdot ME_v - MP_{ope} = 0.025 \cdot 856 - 5.3 = 16.1\,kg\,P \cdot d^{-1} \quad (7.34)$$

$$MP_{chem} = MP_{ti} - MP_{lx} - MP_{te} = 225 - 16.1 - 30$$
$$= 178.9\,kg\,P \cdot d^{-1} \quad or \quad P_{chem} = 11.9\,mg\,P \cdot l^{-1} \quad (7.42)$$

This amounts to $178.9/31 = 5.8$ kmol $P \cdot d^{-1}$. The phosphorus fraction to be removed in the primary settler is $11.9/15 = 79\%$, which requires a molar Me/P_{ti} dosing ratio of 2 as can be determined from Figure 7.12. The daily $FeCl_3$ consumption is 11.5 kmol $\cdot d^{-1}$, which amounts to 1873 kg $FeCl_3 \cdot d^{-1}$ or 3.34 m^3 40% wt $FeCl_3 \cdot d^{-1}$. The production of chemical sludge can be calculated after correcting for the mass of phosphorus removed with the primary sludge:

$$ME_{v1} = Q_i \cdot (S_{ti} - S'_{ti})/f_{cv} = 15{,}000 \cdot (0.6 - 0.33)/1.5 = 2687\,kg\,VSS \cdot d^{-1}$$

$$MP_{l1} = f_p \cdot ME_{v1} = 0.015 \cdot 2687 = 40.3\,kg\,P \cdot d^{-1} \quad (7.43)$$

$$ME_{mp} = (178.9 - 40.3)/31 \cdot mw_{mp} = 4.5 \cdot 151 = 674\,kg\,FePO_4 \cdot d^{-1} \quad (7.47)$$

$$ME_{mo} = (11.5 - 4.5) \cdot mw_{mo} = 7.1 \cdot 107 = 755\,kg\,Fe(OH)_3 \cdot d^{-1} \quad (7.48)$$

$$ME_t = 2687/0.7 + 674 + 755 + 856/0.8 = 6337\,kg\,TSS \cdot d^{-1}$$

If pre-precipitation is selected, a digester could be employed for sludge volume reduction and energy generation. However, as ammonium is released during the sludge digestion process, the N_{ki} load to the activated sludge system will increase, resulting in a higher methanol consumption for denitrification.

(2c) Biological excess phosphorus removal + supplementary chemical dosing

In this configuration, the existing Bardenpho system will be expanded with an anaerobic zone, modifying it from a Bardenpho into an UCT configuration. The main performance objective of the UCT system will be to ensure that nitrogen effluent limits are complied with, while simultaneously maximising the extent of phosphorus removal by biological means. If insufficient phosphorus is removed, then supplementary simultaneous precipitation will be employed. Note that nitrification may become a critical process, as the aerobic sludge mass fraction will decrease when an anaerobic zone is included: this should be checked.

The calculation proceeds in an iterative manner. Based on a first estimate of the operational sludge age and of the concentration of nitrate in the recycle stream "r" to the anaerobic reactor, the proportion of COD available to bio-P organisms and to normal heterotrophic biomass is calculated. Using these values, the performance of the activated sludge system is evaluated with respect to nitrogen removal as a function of the sludge age. If the selected sludge age and actual nitrate concentration in the "r" recirculation flow differ significantly from the first estimate, they are adapted and the proportion of COD available to bio-P organisms and heterotrophs is recalculated. As a starting value, use $R_s = 25$ days and assume that the nitrate concentration in the "r" recycle $(N_{n1}) = 0\,N \cdot l^{-1}$. With Eq. (7.2) the initial estimate of the mass of easily biodegradable COD available to the bio-P organisms is calculated:

$$S_{seq} = S'_{bsi} = S_{bsi} - r \cdot N_{ne} \cdot 2.86/(1 - f_{cv} \cdot Y) \cdot K_1/(K_1 + K_2) \quad (7.2)$$
$$= 138.6 - 0 = 138.6\,mg\,N \cdot l^{-1}$$

The residual, non-sequestered biodegradable COD in the influent will sustain an active "normal" heterotrophic sludge mass equal to:

$$MX_{ah} = C_{rh} \cdot (MS_{bi} - MS_{seq}) = 1.9 \cdot 15{,}000 \cdot (462 - 138.6) = 9201 \text{ kg VSS} \quad (7.8)$$

Now the estimate for S_{seq} can be revised, using Eq. (7.5) to calculate the concentration of S_{bsN} in the effluent of the second anaerobic reactor (as not all S_{bsi} is fermented into VFA):

$$\begin{aligned} S_{bsN} &= S'_{bsi}/(1+r)/[1 + f_{an} \cdot K_c \cdot MX_{ah}/(Q_i \cdot N \cdot (1+r))]^N \quad (7.5) \\ &= 138.6/(1+1)/\{1 + 0.125 \cdot 60 \cdot 9201/[15{,}000 \cdot 2 \cdot (1+1)]\}^2 \\ &= 15.0 \text{ mg COD} \cdot l^{-1} \end{aligned}$$

This reduces the concentration of easily biodegradable COD that will be sequestered by the bio-P organisms to:

$$S_{seq} = S'_{bsp} = 138.6 - (1+1) \cdot 15.0 = 108.6 \text{ mg COD} \cdot l^{-1} \quad (7.6)$$

This value can be used for a new calculation of MX_a, leading to a revised estimate for S_{bsN} and so on. After 5 iterations equilibrium is established for $MX_a = 9981$ kg VSS, $S_{bsN} = 13.7$ mg COD $\cdot l^{-1}$ and $S_{bsp} = 111.2$ mg COD $\cdot l^{-1}$. The influent COD composition can be summarised as:

$$S_{bsp} = S_{seq} = 111.2 \text{ mg COD} \cdot l^{-1} \longrightarrow f_{bsp} = 0.80 \quad (7.16)$$

$$S_{bsh} = 138.6 - 111.2 = 27.4 \text{ mg COD} \cdot l^{-1} \longrightarrow f_{bsh} = 0.20 \quad (7.17)$$

$$S_{bp} = S_{seq} = 111.2 \text{ mg COD} \cdot l^{-1} \longrightarrow f_{bp} = 0.24 \quad (7.18)$$

$$S_{bh} = 462 - 111.2 = 350.8 \text{ mg COD} \cdot l^{-1} \longrightarrow f_{bh} = 0.76 \quad (7.19)$$

Having established the amount of biodegradable COD available to bio-P organisms and to normal heterotrophs, it is now possible to evaluate the performance of the UCT system in terms of nitrogen- and phosphorus removal. The first step is to calculate the effect of the inclusion of an anaerobic reactor on the total reactor volume and on the values of the anoxic- and aerobic sludge mass fractions

- Calculate the volume of the anoxic- and aerobic zones in the original BDP system from the data provided (i.e. $V_r = 17{,}500$ m^3; $f_{x1} = 0.225$ and $f_{x3} = 0.2$);
- Determine the mass fractions f'_{x1}, f'_{x3} and f'_{aer} after conversion to an UCT configuration. As there is now a fraction $f_{an} = 0.125$ of the sludge mass in the anaerobic reactor, the other fractions decrease correspondingly. For example in the post-D reactor: $f'_{x3} = f_{x3} \cdot (1 - f_{an}) = 0.175$;
- Calculate the equivalent new volume: this is the volume the UCT reactor would have if the concentration was uniform in all reactors: $V_{eq} = V_3/f'_{x3} = 3500/0.175 = 20{,}000$ m^3 and hence the equivalent volume of the anaerobic reactor would be 12.5% of this: 2500 m^3;
- However, as the sludge concentration in the anaerobic zone is a factor $r/(r+1)$ lower, the anaerobic volume must be a factor $(r+1)/r$ times larger. Hence the real anaerobic volume is: $2 \cdot 2500 = 5000$ m^3;
- Calculate the new total volume and the resulting volume fractions. The results are listed in Table 14.28):

Table 14.28 Division of volume- and mass fractions over the different zones of the UCT system of Example 14.14

Reactor	Volume (m^3)	Mass fraction	Vol. fraction
Pre-D	$= 0.225 \cdot 17{,}500 = 3937$	$f'_{x1} = 0.20$	0.18
Post-D	$= 0.200 \cdot 17{,}500 = 3500$	$f'_{x3} = 0.175$	0.16
Aerobic	$= 0.575 \cdot 17{,}500 = 10{,}063$	$f'_{aer} = 0.50$	0.45
Anaerobic	$= (3500/0.175 - 17{,}500) \cdot (1+1) = 5000$	$f_{an} = 0.125$	0.22
Total	$= 22{,}500$	$= 1.00$	1.00

The composition of the mixed culture biomass is calculated as follows:

$$\begin{aligned} MX_{vh} &= MX_{ah} + MX_{eh} \\ &= [f_{bh} \cdot (1 - f_{ns} - f_{np}) \cdot (1 + f \cdot b_h \cdot R_s) \cdot C_{rh}] \cdot MS_{ti} \text{ or} \\ &= (1 + f \cdot b_h \cdot R_s) \cdot C_{rh} \cdot (MS_{bi} - MS_{seq}) \qquad (7.8 + 7.10) \\ &= [0.76 \cdot (1 - 0.08 - 0.15) \cdot (1 + 0.2 \cdot 0.197 \cdot 25) \cdot 1.90] \cdot 9000 \\ &= 19{,}826 \text{ kg VSS} \end{aligned}$$

$$C_{rp} = 0.45 \cdot 25/(1 + 0.033 \cdot 25) = 6.17 \qquad (3.30)$$

$$MX_{ap} = f_{bp} \cdot (1 - f_{ns} - f_{np}) \cdot C_{rp} \cdot MS_{ti} \quad \text{or} \quad MX_{ap} = C_{rp} \cdot MS_{seq} \qquad (7.7)$$

$$= 0.24 \cdot (1 - 0.08 - 0.15) \cdot 6.17 \cdot 9000 = 10{,}296 \text{ kg VSS}$$

$$MX_{ep} = f_{ep} \cdot b_p \cdot R_s \cdot MX_{ap} = 0.25 \cdot 0.033 \cdot 25 \cdot 10{,}296 = 2116 \text{ kg VSS} \qquad (7.9)$$

$$MX_i = f_{np} \cdot R_s/f_{cv} \cdot MS_{ti} = 0.15 \cdot 25/1.5 \cdot 9000 = 22{,}500 \text{ kg VSS} \qquad (3.43)$$

$$\begin{aligned} MX_v &= MX_{vh} + MX_{ap} + MX_{ep} + MX_i \\ &= 19{,}826 + 10{,}296 + 2116 + 22{,}500 = 54{,}738 \text{ kg VSS} \qquad (7.11) \end{aligned}$$

$$\begin{aligned} MX_t &= MX_{ap}/f_{vp} + (MX_{vh} + MX_{ep})/f_v \qquad (7.12) \\ &= 10{,}296/0.46 + (54{,}838 - 10{,}296)/0.70 = 85{,}871 \text{ kg TSS} \end{aligned}$$

The average biomass concentration in the aerobic and anoxic reactors is equal to:

$$\begin{aligned} X_t &= MX_t/[V_{x1} + V_{aer} + V_{x3} + V_{an}/(r+1)] \\ &= 85{,}871/[17{,}500 + 2500/(1+1)] = 4.3 \text{ kg TSS} \cdot m^{-3} \end{aligned}$$

The bio-P phosphorus removal can be calculated as:

$$\begin{aligned} MP_{lx} &= [f_{pp} \cdot MX_{ap} + f_p \cdot (mX_v - mX_{ap})]/R_s - MP_{ope} \qquad (7.34) \\ &= (0.38 \cdot 10{,}296 + 0.025 \cdot (54{,}738 - 10{,}296))/25 - 5.3 = 195.7 \text{ kg P} \cdot d^{-1} \end{aligned}$$

$$P_{lx} = MP_{lx}/Q_i = 13.0 \text{ mg P} \cdot l^{-1}$$

Compliance with the effluent total phosphorus limit P_{td} is ensured when $P_{ti} - P_{lx} - P_{chem} < 2$ mg $P \cdot l^{-1}$. As it happens, the bio-P removal is such that supplementary chemical phosphorus removal is not required, with $P_{ti} - P_{lx} = 15.0 - 13.0 = 2.0$ mg $P \cdot l^{-1}$. In practice a $FeCl_3$ dosing installation might be installed anyway, to be used in the event of process upsets, peak influent phosphorus loads and rainwater flows. The effluent phosphorus composition can be calculated as $P_{oe} = 0.5$ mg $P \cdot l^{-1}$ and $P_{pe} = 1.5$ mg $P \cdot l^{-1}$. Now all that remains is to validate the nitrogen removal performance of the system.

$$MN_{lx} = f_n \cdot MX_v/R_s - MN_{ope} = 0.1 \cdot 54{,}738/25 - 21 = 198\,\text{kg N} \cdot d^{-1} \quad \text{(A6.2)}$$

$$N_{lx} = MN_{lx}/Q_{ti} = 13.2\,\text{mg N} \cdot l^{-1}$$

$$N_{ae} = K_n \cdot (b_n + 1/R_s)/[(1 - f_x) \cdot \mu_m - b_n - 1/R_s] \quad (5.44)$$

$$= 0.56 \cdot (0.033 + 1/25)/(0.50 \cdot 0.224 - 0.033 - 1/25) = 1.0\,\text{mg N} \cdot l^{-1}$$

$$N_c = N_{ti} - N_{lx} - N_{oe} - N_{ae} = 65 - 13.2 - 1.9 - 1.0 = 48.9\,\text{mg N} \cdot l^{-1} \quad (5.50)$$

$$D_{c1} = [f_{dn} \cdot (f_{bsp} \cdot f_{pd} + f_{bsh}) \cdot f_{sb} + K_2 \cdot f_{x1} \cdot (C_{rh} \cdot f_{bh} + C_{rp} \cdot f_{bp} \cdot f_{pd})] \cdot S_{bi} \quad (7.14)$$

$$= [0.11 \cdot (0.80 \cdot 0.7 + 0.20) \cdot 0.3 + 0.082 \cdot 0.20 \cdot (1.90 \cdot 0.76 + 6.17 \cdot 0.24 \cdot 0.7)] \cdot 462$$

$$= 30.4\,\text{mg N} \cdot l^{-1}$$

$$D_{c3} = K_3 \cdot f_{x3} \cdot (C_{rh} \cdot f_{bh} + C_{rp} \cdot f_{bp} \cdot f_{pd}) \cdot S_{bi} \quad (7.15)$$

$$= 0.068 \cdot 0.175 \cdot (1.90 \cdot 0.76 + 6.17 \cdot 0.24 \cdot 0.7) \cdot 462 = 13.8\,\text{mg N} \cdot l^{-1}$$

$$D_c = D_{c1} + D_{c3} = 30.4 + 13.8 = 44.2\,\text{mg N} \cdot l^{-1}$$

This is less than the nitrification capacity, so that the effluent nitrate concentration is *at least* equal to (assuming the denitrification capacities are fully utilised):

$$N_{ne} \geq N_c - D_c = 48.9 - 44.2 = 4.7\,\text{mg N} \cdot l^{-1} \quad (5.78)$$

The available nitrate in the pre-D reactor may be estimated as:

$$N_{av1} = a/(a + s + 1) \cdot N_c + s \cdot N_{ne} = 4/6 \cdot 48.9 + 1 \cdot 4.7 = 37.2\,\text{mg N} \cdot l^{-1} \quad (5.89)$$

Hence D_{c1} is smaller than N_{av1}, resulting in recycling of nitrate to the anaerobic zone. As this will have a negative effect on bio-P removal, it should be avoided. It is possible to reduce the value of the "a" recirculation factor until N_{av1} is equal to D_{C1}.

$$a = (s + 1)(D_{c1} - s \cdot N_{ne})/(N_c - D_{c1} + s \cdot N_{ne}) = 2 \cdot (30.4 - 4.7)/(48.9 - 30.4 + 4.7) = 2.22$$

It can be verified that this indeed reduces N_{av1} to the value of D_{c1}:

$$N_{av1} = (2.22/4.22) \cdot 48.9 + 1 \cdot 4.7 = 30.4\,\text{mg N} \cdot l^{-1} \quad (5.89)$$

Now, having prevented the recycle of nitrate from the anoxic- to the anaerobic reactor (as $D_{c1} = N_{av1}$), the available nitrate in the post-D reactor N_{av3} is calculated as:

$$N_{av3} = N_c - D_{c1} \quad (5.76)$$

$$= 48.9 - 30.4 = 18.5\,\text{mg N} \cdot l^{-1}$$

As N_{av1} is equal to D_{c1} and N_{av3} is larger than D_{c3}, the assumption of an effluent nitrate concentration of 4.7 mg $N \cdot l^{-1}$ was indeed correct.

$$N_{te} = N_{oe} + N_{ne} + N_{ae} = 1.9 + 4.7 + 1.0 = 7.6\,mg\,N \cdot l^{-1}$$

It is concluded that for the given situation it will be possible to produce an effluent with a more than acceptable quality in terms of nutrient concentrations. Table 14.29 compares the effluent characteristics and sludge production for the different configurations of Example 14.14.

Table 14.29 Comparison of effluent quality and sludge production of the different configurations in Example 14.14

Parameter	UoM	BDP	Sim-P	Pre-P	UCT
Sludge concentration	$kg\,TSS \cdot m^{-3}$	3.86	5.51	3.06	4.29
Sludge age	days	24	24	50	25
Effluent composition:					
– total nitrogen	$mg\,N \cdot l^{-1}$	9.6	9.6	10.0	7.6
– nitrate	$mg\,N \cdot l^{-1}$	6.9	6.9	7.7	4.7
– TKN	$mg\,N \cdot l^{-1}$	2.7	2.7	2.3	2.9
– total phosphorus	$mg\,P \cdot l^{-1}$	12.0	2.0	2.0	2.0
– ortho-phosphate	$mg\,P \cdot l^{-1}$	11.5	1.5	1.5	1.5
Sludge production:					
– primary	$kg\,TSS \cdot d^{-1}$	–	–	2687	–
– biological	$kg\,TSS \cdot d^{-1}$	2798	2798	856	3435
– chemical	$kg\,TSS \cdot d^{-1}$	–	1212	1429	–
– total	$kg\,TSS \cdot d^{-1}$	2798	4021	6337	3435
Volatile sludge fraction	(–)	0.7	0.49	0.56	0.64
Chemical dosing:					
– $FeCl_3$	$kg\,FeCl_3 \cdot d^{-1}$	–	1531	1873	–
– methanol	$kg\,CH_3OH \cdot d^{-1}$	–	–	410	–
Additional units:					
– primary clarifier	m^3	–	–	1250	–
– anaerobic reactor	m^3	–	–	–	5000

As can be observed, the performance of the UCT system is superior to that of all other phosphorus removal configurations with respect to the effluent quality, the mass of excess sludge produced and the requirement for chemicals. However, this will be at the expense of additional anaerobic reactor volume. On the other hand, the costs of constructing an anaerobic reactor are smaller than the construction costs of a primary clarifier, as the latter unit is fitted with an inclined bottom and an expensive scraper mechanism. Furthermore, air tight sealing or forced ventilation of the surface area of the primary settler might be requird to prevent odour problems. Furthermore, the downstream biological nitrogen removal process is seriously affected. Methanol dosing might be required to meet the nitrogen effluent standards, as in the example, further adding to the costs and operational complexity. As an alternative raw sewage flow might be bypassed directly to the pre-D anoxic zone.

Pre-precipitation is therefore only an interesting option if the existing activated sludge system is overloaded and a primary settler is already installed. In regions with a hot climate, the combination with primary sludge hydrolysis might be interesting, in order to increase the VFA content in the wastewater, but this requires an assessment of the potential VFA production for the wastewater at hand.

The main disadvantages of simultaneous precipitation are the high consumption of chemicals and the reduction of biological treatment capacity, should it be required to reduce the sludge age. Meeting nitrogen effluent limits may then be hard.

14.5.2 Membrane bioreactor design – case study

In Example 14.15 the application of submerged and cross-flow membranes to municipal sewage will be discussed in an extensive design case. As will be shown, application of cross-flow membranes is not a suitable alternative in this specific situation, but it is included here nonetheless for tutorial purposes.

EXAMPLE 14.15

An old municipal WWTP is facing strict effluent nitrogen limits (total nitrogen < 5 mg N $\cdot$ l^{-1} and TKN < 2.2 mg N $\cdot$ l^{-1}) and will have to be replaced. As available space is limited, an MBR is considered. Influent wastewater characteristics and kinetic/stoichiometric parameters are listed in Table 14.30, while MBR related parameters and aeration characteristics are listed in Table 14.31.

Table 14.30 Wastewater- and process characteristics for Example 14.15

Influent characteristics			Kinetic- and settling parameters		
Par.	Value	Unit	Par.	Value	Unit
Q_i	3500	$m^3 \cdot d^{-1}$	b_h	0.162	d^{-1}
Q_{pf}	19200	$m^3 \cdot d^{-1}$	K_2	0.074	$mg\,N \cdot mg^{-1}\,X_a \cdot d^{-1}$
S_{ti}	860	$mg\,COD \cdot l^{-1}$	K_3	0.060	$mg\,N \cdot mg^{-1}\,X_a \cdot d^{-1}$
N_{ti}/P_{ti}	80/15	$mg\,N/P \cdot l^{-1}$	μ_m	0.125	d^{-1}
f_{sb}	0.32	(–)	b_n	0.027	d^{-1}
f_{np}	0.15	(–)	K_n	0.313	$mg\,N \cdot l^{-1}$
f_{ns}	0.06	(–)	f_n/f_p	0.1/0.025	$mg\,N/P \cdot mg^{-1}\,VSS$
T	10	°C	f	0.2	$mg\,VSS \cdot mg^{-1}\,VSS$
f_v	0.7	(–)	I_{dsv}	100	$ml \cdot g^{-1}\,TSS$
$X_t^{(1)}$	4/10/12	$kg\,TSS \cdot m^{-3}$	v_0	178	$m \cdot d^{-1}$
$N_{oe}^{(2)}$	1.2	$mg\,N \cdot l^{-1}$	k	0.42	$l \cdot g^{-1}\,TSS$
P_{oe}	0.25	$mg\,P \cdot l^{-1}$	H_d	2	m
X_{te}	15	$mg\,TSS \cdot l^{-1}$	s_{fd}	2	(-)
			α	0.08	m/m

Notes: (1) Mixed liquor concentration in conventional system/submerged MBR/cross-flow MBR
(2) N_{oe} is composed of $N_{ose} + N_{ope}$. $N_{ope} = 0.1 \cdot 0.7 \cdot 15 = 1.05$ and $N_{ose} = 1.2–1.05 = 0.15$ mg N $\cdot$ l^{-1}

Table 14.31 Aeration and -membrane characteristics for Example 14.15

Aeration characteristics			Membrane characteristics		
Par.	**Value**	**Unit**	**Par.**	**Value**	**Unit**
$p,\ p_w$	1.013/0.017	bar	$F_{m,avg}$ (10°C)	15	$litre \cdot m^{-2} \cdot h^{-1}$
$\alpha \cdot F^{(1)}$	0.6/0.5/0.4	(–)	$F_{m,avg}$ (16°C)	20	$litre \cdot m^{-2} \cdot h^{-1}$
β	0.95	(–)	$F_{m,pf}$ (10°C)	45	$litre \cdot m^{-2} \cdot h^{-1}$
θ	1.024	(–)	$F_{m,cf}$	100	$litre \cdot m^{-2} \cdot h^{-1}$
$OT_{4.5}$	28%	(–)	η_m	0.9	(–)
DO_{ss}/DO_{sp}	9.1/2.0	$mg\ O_2 \cdot l^{-1}$	$A_{mod}^{(4)}$	1250/500/33	m^2
T_{air} (avg./max)	12/35	°C	$V_{mod}^{(5)}$	10/10/0.1	m^3
$H_{liq}^{(2)}$	5/3.5	m	Spec. aer.$^{(6)}$	0.3/0.4	$Nm^3 \cdot m^{-2}$
$H_{dif}^{(2)}$	0.3/0.5	m	F/P ratio (CF)	$4/2^{(7)}$	(–)
$\Delta p^{(3)}$	0.05/0.15	bar	CF module flow	212	$m^3 \cdot h^{-1}$
Peak factor	1.5	(–)	Min. TMP (CF)	1.4	bar
η_{aer}	70%	(–)	F/P ratio (subm)	5	(–)
			DO_{mt}	6	$mg\ O_2 \cdot l^{-1}$

Notes: (1) $\alpha \cdot$ F-factor for different suspended solids concentrations, respectively 4, 12 and 15 kg TSS $\cdot\ m^{-3}$
(2) Height of liquid/diffusers in reactor resp. membrane tank
(3) Pressure loss in lines and aerators for coarse bubble resp. fine bubble aeration
(4) Membrane surface area per fibre/plate/cross-flow module
(5) Volume per fibre/plate/cross-flow module. The tank volume will be larger than the combined module volume, taking into account requirements for distance between modules, to walls etc.
(6) Specific aeration requirements for fibre/plate membranes, taking into account the fraction of time the membranes are aerated
(7) Operation in feed & bleed configuration under average flow resp. peak flow conditions

Design an MBR in both submerged and cross-flow configuration and compare the design results with that of a conventional activated sludge system. For all configurations, take into account the reduction of the denitrification capacity due to the recycle of oxygen to the anoxic zones. The produced excess sludge is transported to a central sludge treatment facility located off-site. Finally consider the contribution of the (aerated) return sludge flow from the submerged membrane tank to total oxygen supply.

Solution

(1) Design of the conventional Bardenpho system

(A1) Design of the biological treatment

The system complies with the nitrogen effluent standards for $R_s = 31$ days, $a = 4$, $s = 1$, $f_{x1} = 0.15$ and $f_{x3} = 0.225$ ($f_x = 0.375$). The total sludge mass $MX_t = 29{,}100$ kg TSS, requiring a reactor volume $V_r = MX_t/X_t = 29{,}100/4 = 7275\ m^3$.

$$S_{bi} = (1 - 0.15 - 0.06) \cdot 860 = 679\ mg\,COD \cdot l^{-1} \quad (3.3)$$

$$C_r = 0.45 \cdot 31/(1 + 0.162 \cdot 31) = 2.32 \quad (3.30)$$

$$N_l = 1000 \cdot f_n \cdot f_v \cdot MX_t/(R_s \cdot Q_i) \quad (3.59)$$
$$= 1000 \cdot 0.1 \cdot 0.7 \cdot 29{,}100/(31 \cdot 3500) = 18.8\,mg\,N \cdot l^{-1}$$
$$N_{ae} = K_n \cdot (b_n + 1/R_s)/[(1 - f_m) \cdot \mu_m - b_n - 1/R_s] \quad (5.44)$$
$$= 0.313 \cdot (0.027 + 1/31)/[(1 - 0.375) \cdot 0.125 - 0.027 - 1/31] = 1.0\,mg\,N \cdot l^{-1}$$
$$N_c = N_{ti} - N_{ose} - N_{ae} - N_l = 80 - 0.15 - 1.0 - 18.8 = 60.1\,mg\,N \cdot l^{-1} \quad (5.50)$$

To be able to compare the effluent quality of an MBR and a conventional activated sludge system, a differentiation has been made between the soluble (N_{ose}) and particulate (N_{ope}) organic nitrogen concentration in the effluent, as N_{ope} will be retained by the MBR together with all other suspended solids.

$$D_{c1} = (f_{sb} \cdot f_{dn} + K_2 \cdot C_r \cdot f_{x1}) \cdot S_{bi} \quad (5.68)$$
$$= (0.32 \cdot 0.114 + 0.074 \cdot 2.32 \cdot 0.15) \cdot 679 = 42.2\,mg\,N \cdot l^{-1}$$
$$D_{c3} = K_3 \cdot C_r \cdot f_{x3} \cdot S_{bi} \quad (5.69)$$
$$= 0.060 \cdot 2.32 \cdot 0.225 \cdot 679 = 21.2\,mg\,N \cdot l^{-1}$$

The reduction of the denitrification capacity due to oxygen recycle can be calculated as:

$$\Delta D_{c1} = a \cdot DO_{sp}/2.86 \quad (5.95)$$
$$= 4 \cdot 2/2.86 = 2.8\,mg\,N \cdot l^{-1}$$
$$\Delta D_{c3} = (s + 1) \cdot DO_{sp}/2.86 \quad (5.96)$$
$$= 2 \cdot 2/2.86 = 1.4\,mg\,N \cdot l^{-1}$$

The corrected values of the denitrification capacities are $D_{c1} = 39.4$ mg $N \cdot l^{-1}$ and $D_{c3} = 19.8$ mg $N \cdot l^{-1}$. As long as D_{c1} is limiting ($D_{c1} < N_{av1}$), the following equation can be used to calculate N_{ne}

$$N_{ne} = N_c - D_{c1} - D_{c3} \quad (5.78)$$
$$= 60.1 - 39.4 - 19.8 = 0.9\,mg\,N \cdot l^{-1}$$
$$N_{av1} = a/(a + s + 1) \cdot N_c + s \cdot N_{ne} \quad (5.74)$$
$$= 4/(4 + 1 + 1) \cdot 60.1 + 1 \cdot 0.9 = 40.9\,mg\,N \cdot l^{-1}$$

N_{av1} is larger than D_{c1} so Eq. (5.78) could indeed be used for calculation of N_{ne}. The estimate of the effluent nitrogen composition can now be finalized as:

$$N_{ke} = N_{ae} + N_{oe} = 1.0 + 1.2 = 2.2\,mg\,N \cdot l^{-1}$$

$$N_{te} = N_{ke} + N_{ne} = 2.2 + 0.9 = 3.1\,mg\,N \cdot l^{-1}$$

(A2) Calculation of the oxygen demand

The average daily oxygen demand MO_t is equal to $MO_c + MO_n - MO_{eq}$

$$MX_a = C_r \cdot (1 - f_{ns} - f_{np}) \cdot MS_{ti} \tag{3.46}$$
$$= 2.32 \cdot (1 - 0.06 - 0.15) \cdot 3500 \cdot 0.86 = 5508\,kg\,VSS$$

$$MO_c = MO_{ex} + MO_{en} = (1 - f_{cv} \cdot Y) \cdot (1 - f_{ns} - f_{np}) \cdot MS_{ti} + f_{cv} \cdot (1 - f) \cdot b_h \cdot MX_a \tag{3.43}$$
$$= (1 - 1.5 \cdot 0.45) \cdot (1 - 0.06 - 0.15) \cdot 3500 \cdot 0.86 + 1.5 \cdot (1 - 0.2) \cdot 0.162 \cdot 5508$$
$$= 1844\,kg\,O_2.d^{-1}$$

$$MO_n = 4.57 \cdot N_c \cdot Q_i/1000 \tag{5.11}$$
$$= 4.57 \cdot 60.1 \cdot 3500/1000 = 961\,kg\,O_2.d^{-1}$$

$$MO_{eq} = 2.86 \cdot (D_{c1} + D_{c3}) \cdot Q_i \cdot 1000 \tag{5.12}$$
$$= 2.86 \cdot (39.4 + 19.8) \cdot 3500/1000 = 607\,kg\,O_2.d^{-1}$$

$$MO_t = 1844 + 961 - 607 = 2198\,kg\,O_2.d^{-1} = 91.6\,kg\,O_2.h^{-1} \tag{5.13}$$

It is not necessary to compensate for the oxygen consumption in the anoxic reactors due to oxygen recycle, as this effect has already been included in the reduction of MO_{eq} (less denitrification). The calculation of the average power requirement for aeration proceeds as follows. To adapt the oxygen transfer efficiency ($OT_{4.5}$) at 4.5 m submergence to the actual liquid height in the aeration tank, Eq. (4.13) can be used:

$$OT_s = OT_{4.5} \cdot \left[(H_{liq} - H_{dif})/4.5\right]^{0.8}$$
$$= 28\% \cdot [(5 - 0.3)/4.5]^{0.8} = 29\% \tag{4.13}$$

In the second step the relationship between the oxygen transfer efficiency under actual and standard conditions is determined using Eq. (4.16). In this example it is not necessary to correct for elevation above sea level.

$$OT_a = OT_s \cdot \alpha \cdot F \cdot \theta^{(T-20)} \cdot \left[\frac{(p - p_w + (H_{liq} - H_{dif})/20) \cdot 51.6 \cdot \beta \cdot DO_{ss}}{(p_s - p_w) \cdot (31.6 + T)} - DO_l\right]/DO_{ss} \tag{4.16}$$

$$OT_a = 29\% \cdot 0.6 \cdot 1.024^{(16-20)} \cdot \left[\frac{(1.013 - 0.017 + (5 - 0.3)/20) \cdot 51.6 \cdot 0.95 \cdot 9.1}{(1.013 - 0.017) \cdot (31.6 + 16)} - 2.0\right]/9.1 = 16.5\%$$

Now the required air mass flow rate can be calculated for the average ambient temperature in order to obtain the average power consumption and for the maximum temperature (and peak oxygen demand) to size the electrical motor:

$$Q_{air,av} = MO_t \cdot (29/32) \cdot (1/0.209) \cdot (1/OT_a)$$
$$= 91.4 \cdot (29/32) \cdot (1/0.209) \cdot (1/0.174) = 2400\text{ kg air} \cdot h^{-1} \tag{4.17}$$

$$Q_{air,max} = 1.5 \cdot 2400 = 3600\text{ kg air} \cdot h^{-1}$$

The blower discharge pressure is calculated with Eq. (4.18) as:

$$p_{dis} = 1.013 + (H_{liq} - H_{dif})/10 + \Delta p = 1.633 \tag{4.18}$$

The required blower power can be calculated with Eq. (4.19) as:

$$P_{aer} = Q_{air} \cdot R \cdot T_{in} \cdot [(p_{dis}/p_s)^{0.283} - 1]/(3600 \cdot 29.7 \cdot 0.283 \cdot \eta_{aer}) \tag{4.19}$$

So for average and peak conditions respectively:

$$P_{aer} = [(2400/3600) \cdot 8.314 \cdot 285.3 \cdot ((1.633/1.013)^{0.283} - 1)]/(29.7 \cdot 0.283 \cdot 0.7)$$
$$= 39 \text{ kW}$$

$$P_{aerm} = [(3600/3600) \cdot 8.314 \cdot 308.3 \cdot ((1.633/1.013)^{0.283} - 1)]/(29.7 \cdot 0.283 \cdot 0.7)$$
$$= 63 \text{ kW}$$

(A3) Design of the final settler

Given the large ratio between peak- and average flow and the prolonged peak flow duration of 24 hours, the design of the final settler will be based on the peak (rainfall) flow. The alternative would be to design for average flow and to increase the settler height to 4–5 m, thus increasing the buffer volume. However, during the peak flow, part of the biomass will be transferred to the final settler, so that the mixed liquor concentration X_t during peak flow will be lower than the normal mixed liquor reactor concentration of 4.0 kg TSS $\cdot$ m^{-3}: in practice up to 10 to 30% lower, depending on the ratio between peak- and average flow rate and the duration of the peak flow. Apart from a reduction in treatment capacity, this will have two other effects: the solids loading rate to the settler is decreased and at the same time the solids flux that can be transported in the final settler increases.

In this example a maximum 30% decrease in mixed liquor concentration is assumed, from 4.0 to 2.7 kg TSS $\cdot$ m^{-3}. When the critical sludge recycle ratio is applied, the maximum hydraulic overflow rate in the final settler can be calculated using Eq. (8.28) as:

$$T_{sm} = v_0 \cdot \exp(-k \cdot X_t) \tag{8.28}$$
$$= 178 \cdot \exp(-0.42 \cdot 2.7) = 54.9 \text{ m} \cdot \text{d}^{-1}$$

$$A_d = s_{fd} \cdot Q_{i,max}/T_{sm} \tag{8.31}$$
$$= 2 \cdot 19{,}200/54.9 = 699 \text{ m}^2$$

$$D_d = (4 \cdot 699/\pi)^{0.5} = 29.8 \text{ m}$$

When calculating the volume of the settler, the cone volume is often omitted, although for a typical bottom gradient α of 0.08 m $\cdot$ m^{-1} the contribution to total settler volume is not insignificant. The volume of the cone can be calculated as:

$$V_c = 0.33 \cdot (\alpha \cdot D_d/2) \cdot A_d$$
$$= 0.33 \cdot (0.08 \cdot 29.8/2) \cdot 699 = 275 \text{ m}^3$$

$$V_d = 275 + 2 \cdot 699 = 1674 \text{ m}^3$$

Assuming $s = 1$, then the retention time (or rather: contact time) at average flow ($Q_i = 146\ m^3 \cdot h^{-1}$) is $1674/(2 \cdot 146)$ or about 5.7 hrs and 1.0 hrs during peak flow. The retention time during average flow is far above the recommended limit of 1–3 hrs, while the retention time during peak flow is at the lower end. In this particular case a high retention time might be acceptable given the moderate temperature and the high sludge age applied. However, two other options could be considered as well: (I) construct two settlers and use the second one only during peak flows and (II) operate during average flow with a value of $s \gg 1$ (the pumps are sized for peak flow anyway). For $s = 3$ ($Q_{rs} = 438\ m^3 \cdot h^{-1}$) the retention time in the final settler is reduced to 2.9 hrs.

The power requirement of a pump with water service can be approximated with the following formula:

$$P_{el} = (Q/3600) \cdot \Delta p \cdot 100/\eta_{el} \qquad (10.2)$$

Assuming $\eta_{el} = 0.7$ and Δp of 0.15 bar for the nitrate recirculation pump and $\Delta p = 1.0$ bar for the sludge recycle pumps, the estimated power consumption can be calculated. The results are listed in Table 14.32. It is assumed that $a = 5$ during average flow and $a = 2$ during rainwater flow conditions (as the influent concentrations will be significantly diluted). Furthermore $s = 3$ during average flow and $s = 1.0$ during rainwater flow.

Table 14.32 Power requirements of recirculation and sludge recycle pumps

Flow rate	Recirculation pump		Sludge recycle pump		Overall
	$m^3 \cdot h^{-1}$	kW	$m^3 \cdot h^{-1}$	kW	kW
Average	730	4	438	17	21
Peak	1600	9	800	32	41

(B) Design of the MBR Bardenpho system

(B1) Design of the biological treatment

When designing a Bardenpho MBR system, the following issues need to be considered:

- If nitrate removal is complete, $N_{av1} = a/(a + s + 1) \cdot N_c$;
- If nitrate removal is incomplete, $N_{av1} = a \cdot (N_c + s \cdot N_{ne})/(a + s + 1) \longrightarrow$ this will reduce the value of N_{av1} in a MBR configuration compared to a conventional configuration and makes a larger post-D zone a necessity;
- The return sludge is not sent to the pre-D anoxic zone as it contains oxygen, hence to compensate for the dilution that will result, the volume of this zone will have to be increased by a factor $(a + 1)/a$;
- The effluent TKN concentration will decrease as organic particulate nitrogen (N_{ope}) is retained by the membranes;
- For the submerged MBR, the minimum value of the sludge recycle factor s will be equal to 4 in order to limit the biomass concentration in the membrane tank (X_r) to a maximum value of 12.5 kg $TSS \cdot m^{-3}$. For the cross-flow system, the membrane feed to permeate ratio is equal to 3, so in effect $s = 2$;
- The effect of oxygen recirculation will be much more noticeable than in a conventional BDP system, as the recirculation flow rates will be higher. The aeration intensity in the last part of the nitrification zone could be decreased to reduce the DO load returned to the anoxic zones.

However, nitrification capacity is reduced at lower DO values, so in practice operation of the nitrification zone at a value lower than 1.5 mg $O_2 \cdot l^{-1}$ is not recommended (unless excess nitrification capacity is available);
- The aerated return sludge flow from the submerged membrane tanks will contribute to the oxygen supply.

As a result of above factors, both MBR systems will require a higher sludge age to meet the desired effluent criteria. The following solutions satisfy the effluent requirements:

- Submerged MBR: $R_s = 36$ days, $a = 6$, $s = 4$; $f_{x1} = 0.10$ and $f_{x3} = 0.32$
- Cross-flow MBR: $R_s = 33$ days, $a = 5$, $s = 2$; $f_{x1} = 0.13$ and $f_{x3} = 0.265$

The calculation of the nitrogen removal performance in the submerged MBR system is demonstrated below ($C_r = 2.37$ and $N_l = 18.3$ mg $N \cdot l^{-1}$).

$$N_{ae} = 0.313 \cdot (0.027 + 1/36)/[(1 - 0.42) \cdot 0.125 - 0.027 - 1/36] = 1.0\,\text{mgN} \cdot l^{-1} \quad (5.44)$$

$$N_c = 80 - 0.15 - 1.0 - 18.3 = 60.6\,\text{mgN} \cdot l^{-1} \quad (5.50)$$

$$D_{c1} = (0.32 \cdot 0.114 + 0.074 \cdot 2.37 \cdot 0.10) \cdot 679 = 36.6\,\text{mgN} \cdot l^{-1} \quad (5.68)$$

$$\Delta D_{c1} = a \cdot DO_{sp}/2.86 \quad (5.95)$$
$$= 6 \cdot 2/2.86 = 4.2\,\text{mgN} \cdot l^{-1}$$

$$D_{c3} = 0.060 \cdot 2.37 \cdot 0.32 \cdot 679 = 30.9\,\text{mgN} \cdot l^{-1} \quad (5.69)$$

$$\Delta D_{c3} = (s + 1) \cdot DO_{sp}/2.86 \quad (5.96)$$
$$= (4 + 1) \cdot 2/2.86 = 3.5\,\text{mgN} \cdot l^{-1}$$

$$N_{ne} = N_c - D_{c1} - D_{c3} \quad (5.78)$$
$$= 60.6 - (36.6 - 4.2) - (30.9 - 3.5) = 0.7\,\text{mgN} \cdot l^{-1}$$

$$N_{avl} = a \cdot (N_c + s \cdot N_{ne})/(a + s + 1) \quad (10.10)$$
$$= 6 \cdot (60.6 + 1 \cdot 0.7)/(6 + 4 + 1) = 34.6\,\text{mgN} \cdot l^{-1}$$

It can be observed that the reduction of D_c due to oxygen recycle is quite significant: (4.2 + 3.5)/(36.6 + 30.9) = 11.4%. For the cross-flow system, the reduction in the value of D_{c1} (−3.5) and D_{C3} (−2.1) is somewhat smaller, as the value of "s" is only increased to 2 instead of 4. This limits the reduction in N_{avl} and reduces oxygen recirculation over the post-D zone, compared to the submerged membranes.

Table 14.33 Nitrogen effluent composition of the configurations of Example 14.15

Effluent composition	Conventional	Submerged	Cross-flow
– Total nitrogen	3.1	1.9	1.8
– Nitrate	0.9	0.7	0.6
– Ammonia	1.0	1.0	1.0
– Organic nitrogen	1.2	0.2	0.2
– TKN	2.2	1.2	1.2

Table 14.33 summarizes the nitrogen effluent composition of the three configurations. It can be observed that the required effluent quality can be obtained with all configurations. However, proper final settler performance will be required for the conventional activated sludge system to comply, as an increase in TSS in the effluent will quickly result in a violation of the TKN limit of 2.2 mg $N \cdot l^{-1}$.

(B2) Calculation of oxygen demand

Calculation of the average electrical power consumption for the aeration blower proceeds as shown in Section (a2). The following issues should be considered:

- The oxygen transfer efficiency is reduced due to effect of operation at higher biomass concentrations. For the submerged system ($\alpha \cdot F = 0.5$), OT_a is lowered to 13.9% and for the cross-flow system ($\alpha \cdot F = 0.4$), OT_a is as low as 11.1%;
- Oxygen demand in both systems will be slightly higher, due the operation at increased sludge age;
- For the submerged MBR, this increase is partly compensated by the contribution of the membrane tank aeration to the total oxygen supply in the aeration tank. However, this contribution is small, i.e. $s \cdot Q_i \cdot DO_{mt}/1000 = 4 \cdot 3500 \cdot 6/1000 = 84$ kg $O_2 \cdot d^{-1}$, or about 4%.

In Table 14.34 the main parameters relating to oxygen demand and aeration capacity are summarized.

Table 14.34 Aeration capacity of the configurations of Example 14.15

Parameter	Conventional	Submerged	Cross-flow
Oxygen demand			
– MO_c	1844	1869	1855
– MO_n	961	969	965
– MO_{eq}	–607	–599	–597
– Contribution MT blowers	–	–84	–
– MO_t	2198	2155	2222
Aeration capacity			
– average consumption	39	44	58
– installed capacity	63	72	93

(B3) Calculation of membrane unit configuration

The submerged membranes must be able to process the average influent flow rate at average net flux and the peak influent flow at peak net flux, at the *lowest* reactor temperature expected (i.e. 10°C). The largest calculated surface area is adopted:

$$\text{Average flow: } A_m = Q_i \cdot 1000/(F_m \cdot \eta_m) = (1000 \cdot 3500/24)/(15 \cdot 0.9) = 7800 \text{ m}^2 \quad (10.7)$$

$$\text{Peak flow: } A_m = Q_{pf} \cdot 1000/(F_m \cdot \eta_m) = (1000 \cdot 800)/(45 \cdot 0.9) = 19{,}750 \text{ m}^2 \quad (10.7)$$

As expected, the peak flow scenario is limiting. Taking into account the standard sizes of the vendor modules, the required number of modules is 19,750/1250 = 15.8 = 16 fibre membrane modules or 19,750/500 = 39.5 = 40 plate membrane modules. Because of the large variation in influent flow rate, the submerged membranes are installed in several parallel tanks. The number of tanks is a trade-off between the cost of additional equipment- and instrumentation versus the reduction in aeration costs and increased flexibility of operation. In this example, four membrane tanks are selected, each with 5000 m^2 of membrane surface area.

At the *average* reactor temperature of 16°C, the selected gross flux will be higher, i.e. 20 $litre \cdot m^{-2} \cdot h^{-1}$ The net average operating flux is $0.9 \cdot 20 = 18$ $litre \cdot m^{-2} \cdot h^{-1}$, so one membrane tank will produce $5000 \cdot 18/1000 = 90\ m^3 \cdot h^{-1}$. This means that for the average influent flow of 146 $m^3 \cdot h^{-1}$, on average only one membrane unit will be in constant operation while a second unit will be switched on or off as required, depending on the liquid level in the biological reactor. The other membrane units will be in stand-by mode.

Using the membrane module dimensions specified in Table 14.31 and assuming the membrane tank volume is twice that of the modules, the required membrane tank volume can be estimated at $16 \cdot 2 \cdot 10 = 320\ m^3$ for the submerged membranes and $40 \cdot 2 \cdot 10 = 800\ m^3$ for the plate membranes. Additional building volume will be required to install the permeate pumps and buffer, the membrane feed pumps, the CIP tank, the chemical dosing units and the membrane aeration blowers.

The net aeration requirement is 0.3 $Nm^3 \cdot m^{-2}$ for the fibre membranes and 0.4 $Nm^3 \cdot m^{-2}$ for the plate membranes. The average membrane surface area in operation equals $(146{,}000)/18 = 8100\ m^2$. In Table 14.31 the average- and peak membrane aeration power requirements are listed for plate- and fibre membranes. The power requirements are calculated using the following parameters: $\Delta p = 0.05$ bar (coarse bubble aeration), $H_{liq} = 3.5$ m and $H_{dif} = 0.5$ m, T = 12°C (average) and T = 35°C (maximum). Other energy consumers are the membrane feed- and permeate pumps. Assuming $\eta_{el} = 0.7$ and $\Delta p = 0.15$ bar for the membrane feed pump; $\Delta p = 1.0$ bar for the permeate pumps at average flow and $\Delta p = 1.5$ bar at peak flow, the estimated power consumption of the membrane section can be calculated. The results are listed in Table 14.35.

Table 14.35 Power requirements of (submerged) membrane section

Membrane Type	Average flow		Peak flow	
	$(N)m^3 \cdot h^{-1}$	kW	$(N)m^3 \cdot h^{-1}$	kW
Blower fibre membranes	2430	31	6000	82
Blower plate membranes	3240	41	8000	110
Feed pump	730	4	4000	24
Permeate pump	146	6	800	48
Total fibres/plates		41/51		154/181

As for the cross-flow membranes, the membrane flux will decline during operation (for equal TMP). In this example the average flux is 100 $litre \cdot m^{-2} \cdot h^{-1}$, so assuming that the decrease in flux is linear, then for example the initial flux is 125 $litre \cdot m^{-2} \cdot h^{-1}$ and the membranes will be cleaned when the flux decreases to 75 $litre \cdot m^{-2} \cdot h^{-1}$. It is further assumed that cleaning of the cross-flow modules will be sequentially. As no back flushing or relaxation is applied, the value of $\eta = 1.0$.

$$A_m = Q_{pf} \cdot 1000/F_m = (146{,}000)/100 = 8000\ m^2 \qquad (10.7)$$

For the standard 8″ mm module of 33 m^2, 242 modules will be required. Assuming the modules are stacked in series of 7 modules, then similar to the submerged membranes, the modules will be arranged in several parallel units to allow for flexible operation, for example $4 \times 9 \times 7 = 252$ modules. The combined skid volume can be estimated at 300 m^3. Additional volume will be required for installation of the pumps and the CIP unit.

The energy consumption is composed of two main factors: the energy required by the membrane feed pump to pressurise the feed and the energy required by the recirculation pump. In Table 14.36 the power requirement for both average- and peak influent flow is given for $\eta_{el} = 70\%$. The following assumptions have been made:

- The required differential pressure of the recirculation pump is equal to the differential pressure over the series of 7 modules: i.e. $7 \cdot 0.8 = 5.6$ bar, while the required discharge pressure of the feed pump is $5.6 + 1.4 = 7.0$ bar;
- The membrane feed/permeate ratio under average flow conditions is 4: a 7 module skid produces 23.1 $m^3 \cdot h^{-1}$ of permeate, so the required feed flow per skid is 92.4 $m^3 \cdot h^{-1}$
- As the required cross-flow through a module is 212 $m^3 \cdot h^{-1}$, an additional 119.6 $m^3 \cdot h^{-1}$ of recirculation flow is required (the recirculation flow per skid equals the flow per module as the modules are placed in series);
- Under average flow conditions 6.3 racks ($6.3 \cdot 7 \cdot 33 \cdot 0.1 = 146$ $m^3 \cdot h^{-1}$ of permeate) are in operation. The feed flow is $4 \cdot 146 = 583$ $m^3 \cdot h^{-1}$ and the required recirculation flow rate is $6.3 \cdot 212 = 1338 - 583 = 755$ $m^3 \cdot h^{-1}$
- At peak flow rate (800 $m^3 \cdot h^{-1}$) all 36 skids are operational. As this is a temporary situation, a lower feed/permeate ratio is accepted (f/p = 2). The feed flow is $2 \cdot 800 = 1600$ $m^3 \cdot h^{-1}$ and the recirculation flow is $36 \cdot 212 - 1600 = 6032$ $m^3 \cdot h^{-1}$

Table 14.36 Power requirements of feed & bleed CF membranes

Flow rate	Membrane feed pumps		Recirculation pumps		Overall
	$m^3 \cdot h^{-1}$	kW	$m^3 \cdot h^{-1}$	kW	kW
Average	583	162	755	168	330
Peak	1600	445	6030	1340	1785

It can be observed that, in this example, the average energy requirements of feed & bleed cross-flow membrane configuration will be about 6–8 times higher than those of the submerged membrane configuration.

It is also interesting to consider the effect of the cross-flow filtration energy input on the reactor temperature: at average flow the energy input equals $500 \cdot 3600/1000 = 1800$ MJ. Reactor volume is 2620 m^3 and hydraulic residence time is 2620/146 is 18 hrs. Total energy input in 18 hours equals 21,330 MJ. Assuming all energy is released as heat, the increase in reactor temperature would be $21{,}330/(4.2 \cdot 2620) = 1.9°C$.

In this specific example cross-flow MBR cannot be considered as a viable alternative due to the high energy demand and the high ratio between average and peak flow. As discussed in Section 10.2.3, the cross-flow membranes have to be sized for maximum flow for a fixed membrane flux value, while the submerged membranes are adaptable and can increase the membrane flux during peak flow conditions. Thus the required submerged membrane area is only 19,753 m^2 at 45 litre · m^{-2} · h^{-1} instead of $(1000 \cdot 800)/(15 \cdot 0.9) = 59{,}260$ m^2 at 15 litre · m^{-2} · h^{-1} (at 10°C).

In Table 14.37 the design summary of the treatment configurations is given. Both MBR and conventional treatment meet the imposed nitrogen effluent limits. From the point of construction, the reduction in volume should be balanced against the costs for membranes, additional equipment etc.

Table 14.37 Design comparison of the different configurations of Example 14.15

Parameter	UoM	Conventional		Submerged		Cross-flow	
Volume/fraction:							
– pre-D	m^3/(–)	1091	0.15	384	0.10	398	0.13
– nitrification	m^3/(–)	4547	0.675	1910	0.58	1545	0.605
– post-D	m^3/(–)	1637	0.225	1054	0.32	677	0.265
– total	m^3/(–)	7276	1.0	3348	1.0	2620	1.0
Clarifier:							
– diameter	M	29.8		–		–	
– surface area	m^2	700		–		–	
– volume	m^3	1674		–		–	
Membranes:				Fibres	Plates		
– tank volume	m^3	–		320	800	300	
– surface area	m^2	–		20,000		8000	
Nitrification blower:							
– avg/peak air flow	kg · h^{-1}	2400	3600	2800	4206	3614	5421
– avg/peak power	kW	39	63	44	72	58	93
Membrane aeration				Fibres	Plates		
– avg air flow	$Nm^3 \cdot h^{-1}$	–		2430	3240	–	
– peak air flow	$Nm^3 \cdot h^{-1}$	–		6000	8000	–	
– avg power	kW	–		31	41	–	
– peak power	kW	–		82	110	–	
Pump power							
– avg./peak flow	kW	21	41	10/72	10/72	330	1785
Total power				Fibres	Plates		
– avg/peak flow	kW	60	104	86/226	97/253	387	1878
Pre-treatment:							
– coarse screens	(–)	Yes		Yes		Yes	
– sand/fat trap	(–)	only sand trap		sand/fat trap		sand/fat trap	
– fine screens	(–)	No		Yes		Yes	

The energy use of the submerged MBR system will be approximately 1.5 times higher than that of a conventional system. However, as has been demonstrated in Example 14.4 to Exampl 14.8, the cost impact of the energy consumption for aeration is often overestimated, as it represents only a minor fraction of the annualised *total* costs of a wastewater treatment plant (typically around 10–15%). Thus the annualized total costs would increase by a much lower percentage. However, the investment costs for MBR are higher as well.

It is clear that the cross-flow system will be very uneconomical in operation at more than 6 times the energy demand of the conventional system and 4.5 times the energy demand of the submerged MBR system. The reduction in investment costs compared to submerged MBR will not compensate for the very large increase in energy consumption. This is why cross-flow systems are only used for much smaller applications.

In both MBR configurations, the introduction of the aerobic sludge recycle results in a decrease in nitrate removal. As was demonstrated in step (b1) of the example, this can be resolved by increasing the anoxic sludge mass fraction at the expense of an increased ammonium effluent concentration. If this is unacceptable, the sludge age will have to be increased, which allows a larger anoxic mass fraction while maintaining the same effluent ammonium concentration.

As for the conventional system, it clearly requires more space than the other configurations and is more vulnerable with regard to the concentration of (particulate) organic nitrogen in the effluent. Due to the organic nitrogen contained in the effluent suspended solids, a disturbance in final settler operation might result in an immediate violation of the effluent TKN limit. Other alternatives to MBR might be considered as well, such as post treatment with a sand-filter, but also these units will result in additional costs.

14.6 FINAL REMARKS

In this text it has been demonstrated that the activated sludge system is an efficient process to remove suspended solids, organic material and the macronutrients nitrogen and phosphorus from wastewater. The removal of suspended solids and organic material is virtually complete if there are no imperfections in the separation between the solid and liquid phases in the final settler. These imperfections may have their origin in the settleability of the sludge (pin-point flocs, free bacteria, sludge bulking) or in the design and operation of the settler (flotation due to denitrification, inadequate sludge recycle rate, insufficient settler area). However, when the activated sludge system is designed and operated correctly, it is possible to consistently maintain low concentrations of suspended solids and organic material in the effluent.

Removal of nutrients is particularly desirable when the effluent of the treatment system is discharged into surface water. The nutrients present in the effluent will contribute to eutrophication, which might affect the water quality. Complete removal of nitrogen is feasible for most wastewaters (depending on the TKN/COD ratio), using sequentially the processes of nitrification and denitrification. In Chapter 5 it was shown that in regions with a hot climate the application of biological nitrogen removal is a factor that will contribute significantly to operational stability, preventing both the flotation of sludge in the final settler and the formation of filamentous sludge. In this chapter it was demonstrated that the removal of nitrogen will increase the treatment costs by 15–40% depending on the configuration of the conventional activated sludge system. However, it should be taken into account that the removal of nitrogen is of great advantage in the prevention of eutrophication, contributes to the operational stability of the system and is a legal requirement in many cases anyway. Therefore often it is recommended to include nitrogen removal in the system configuration. New developments in nitrogen removal (nitrogen removal over nitrite and anaerobic ammonium oxidation) have been discussed in Chapter 6, where it was shown that their application may be very advantageous. However, the suitability of these new technologies should be evaluated for each specific application.

As for phosphorus removal, biological excess phosphorus removal is possible for many wastewaters and offers a cost advantage compared to chemical phosphorus removal. However, inclusion of biological phosphorus removal will increase treatment costs slightly and will add to the complexity of the treatment system (for instance, it complicates sludge treatment). It may conflict with the objective of nitrogen removal, as for optimal phosphorus removal a low sludge age is required, while for nitrification a high sludge age is necessary.

However, depending on the wastewater composition, for most types of municipal sewage it is very well possible to deliver an effluent within the specified nitrogen and phosphorus limits, as was demonstrated in Example 14.8 and Example 14.14. Furthermore, phosphorus removal might be of critical importance to prevent eutrophication of a receiving surface water. When phosphorus removal is required, it is concluded from Example 14.14 that, in principle, biological phosphorus removal should be preferred over chemical phosphorus removal. If insufficient easily biodegradable COD is available, chemical phosphorus removal might be selected, but whenever possible only as a backup process, should the biological phosphorus removal capacity be insufficient.

New activated sludge configurations have been developed in the last decades: the most important one is the membrane bioreactor, in which the traditional gravitational settler has been replaced by micro- or ultra-filtration membranes. The advantages are the production of an effluent that is free of suspended solids and the possibility of operating at higher sludge concentrations, thereby reducing the required treatment area. Although many small-scale industrial installations have been built, for municipal applications the MBR is currently not yet cost competitive. However, this might change as membrane prices go down and if effluent legislation becomes stricter.

Independent of the objective or configuration of the wastewater treatment plant, control of the sludge age is of crucial importance in the operation of activated sludge systems. This control is aimed at aligning the total sludge mass in the system with the applied organic- and nitrogen load, by means of continuous or daily discharge of excess sludge from the system. This ensures treatment objectives will be met, while at the same time the solid-liquid separation step (final settler and thickener) and the sludge treatment units will not be overloaded. Traditionally the excess sludge is taken from the return sludge flow, but in this text it has been demonstrated that direct discharge from the aeration tank is preferable. The reason is that the concentration in the biological reactor is relatively constant, even when the influent flow varies. In contrast, the sludge concentration in the return sludge flow varies markedly with hydraulic fluctuations, making it very difficult to relate the volume of the excess sludge flow to the mass of suspended solids actually discharged from the system.

The recent developments in anaerobic treatment will certainly have important implications for the future development of the activated sludge process, particularly in regions with a warm climate where anaerobic treatment is very efficient. In Chapters 13 and 14 it was demonstrated that under identical conditions, the combined anaerobic-aerobic system may require only half of the costs of a conventional activated sludge system. However, an important disadvantage of anaerobic digestion of wastewaters is that it increases the ratios between nitrogen and organic material (N_{ti}/S_{ti}) and phosphorus and organic material (P_{ti}/S_{ti}). Often this prevents nutrient removal in the subsequent activated sludge treatment process, if this is required. This said, anaerobic pre-treatment (particularly the UASB/EGSB reactor types) is very suitable for wastewaters with a low nutrient content as produced in factories processing vegetable products (beer breweries, paper factories etc).

Independent from the selected configuration, an important part of the cost of activated sludge system refers to investment costs. When the wastewater flow is expected to increase, the use of a modular construction should always be considered, preventing investment in systems that will not be used immediately. An important operational cost item is personnel expenses, resulting in an ever increasing use of process automation in wastewater treatment plants. While energy prices remain at their present level (on average about US$ 0.15 per kWh), aeration costs will not play a very important role in the operational costs. Sludge disposal costs however may be much more significant, although in principle activated sludge could be applied in fertilisation of poor soils, if this is permitted.

Figure 14.16 Arrangement of Biobed®EGSB reactor, conditioning tank and reactor feed and mixing pumps. Courtesy of Bulmers Ltd, Ireland – member of C&C Group PLC

Reference List

Abwasser Technische Vereinigung (ATV) (1976). Erlaüterungen unde Ergänzungen zum Arbeitsbericht des ATV Fachausschusses: Die bemessungen des Nachklärbecken von Belebungsanlagen (Explanatory notes and additions to the report of the ATV Technical Committee: design of the final clarifier of activated sludge systems), Korrespondenz Abwasser, vol. 23, pp. 231–235.

Abwasser Technische Vereinigung (ATV) (1992). Bemessung von Belebungsanlagen ab 5000 insgesamt Einwohner und Einwohnergleichwerte (Dimensioning of single stage activated sludge plants upwards from 5000 total inhabitants and population equivalents). *ATV Rules and Standards*, **11**, A131.

Abwasser Technische Vereinigung (ATV) (1993). Anwendungshinweise zum ATV arbeidsblatt A131 – Teil Nächklärbecken (Application information for ATV guideline A131 – part Final Settlers). *Arbeitsbericht des ATV Fachausschusses 2.5 – Absetzverfahrenn, Korrespondenz Abwasser*, **40**(2), 224–225.

Aisse M. M. and Bollmann H. A. (1989). Avaliaçâo da aplicaçâo de reatores anaeróbios de fluxo ascendente no tratamento de esgotos domêsticos de pequenas comunidades (Evaluation of the performance of upflow anaerobic reactors for the treatment of domestic sewage for small communities). Proceedings of 15th Congress of the Brazilian Sanitary Engineering Organisation, Belém, 140.

Anderson J. A. (1981). Thickening performance of activated sludge settlement tanks. *Water Pollution Control*, **4**, 521–528.

Andrews J. F. and Busby J. R. (1973). Dynamic modelling and control strategies for the activated sludge process, Research report, Department of Environmental Systems Engineering, Clemson University, South Carolina, U.S.A.

Anthonissen A. C., Loehr R. C., Prakasam T. B. S. and Srinath E. G. (1976). Inhibition of nitrification. *Journal Water Control Pollution Federation*, **48**(5), 835–852.

Antoniou P., Hamilton J., Koopman B., Jain R., Holloway B., Lyberatos G. and Svoronos S. A. (1990). Effect of temperature and pH on the effective maximum specific growth rate of nitrifying bacteria. *Water Research*, **24** (1), 97–101.

Araújo L., Catunda P. F. C. and Van Haandel A. C. (1994). Influence of the composition of waste activated sludge on its anaerobic stabilization. *Water SA*, **24**(3), 231–236.

Araújo L. and Van Haandel A. C. (1993). Influência da composição de lodo ativado sobre a sua estabilização anaeróbia (Influence of the activated sludge composition on its anaerobic stabilisation). Proceedings of 15th Congress of the Brazilian Sanitary Engineering Organisation, Natal, pp. 513–523.

Arkley J. J. and Marais G. v. R. (1981). Effect of anoxic sludge mass fraction on the activated sludge process, Research report W38, Department of Civil Engineering, University of Cape Town.

Arvin E. (1985). Biological removal of phosphorus from wastewaters. *CRC Critical Rev. Environmental Control*, **15**, 24–94.

Baker D. R., Loehr R. C. and Anthonissen A. C. (1975). Oxygen transfer at high solids concentrations. Journal Environmental Engineering Division, ASCE EE5, 759–774.

Balakrishnan and Eckenfelder (1970). Nitrogen removal by modified activated sludge process. *ASCE Journal of the Sanitary Engineering Division*, **96**, 501–512.

Barbosa R. A. and Santa Anna G. L. Jr. (1989). Treatment of raw domestic sewage in a UASB reactor. *Water Research*, **23**(12), 1483.

Barnard J. L. (1973). Biological denitrification. *Water Pollution Control*, **72**(6), 705–720.

Barnard J. L. (1975). Nutrient removal in biological systems. *Water Pollution Control*, **74**, 143–154.

Barnard J. L. (1976). A review of biological phosphorus removal in the activated sludge process. *Water SA*, **2**(3), 136–144.

Barnard J. L. (1991). Entwurf einer Belebungsanlage für nitrifikation/denitrification (Design of an activated sludge system for nitrification/denitrification). Institut für Siedlungswasserwirtschaft, University of Braunschweig, Germany, pp. 9–18.

Bart Brenner and Lewis (1969). Chemical-biological control of nitrogen and phosphorus in wastewater effluent. *Journal Water Pollution Control Federation*, **40**, 2040–2054.

Berends D., Salem S., Van Loosdrecht M. C. M. and Van de Kuij R. (2002). De BABE-technologie – BABE combineert rejectiewaterbehandeling met de enting van nitrificerende organismen (The BABE-technology – BABE combines reject water treatment with seeding of nitrifying organisms), STOWA report no. 2002-42.

Berkhof D., de Bruin B., Kerstholt M, Kraan R., Miska V., Peeters T., van der Roest H., Verschoor J., de Kreuk M. and van Loosdrecht M. (2010). Nereda pilot onderzoeken 2003–2010 (Nereda pilot trials 2003–2010), STOWA report no. 2010-29.

Beun J. J., Hendriks A., Morgenroth E., Wilderer P. A., Van Loosdrecht M. C. M and Heijnen J. J. (1999). Aerobic granulation in a sequencing batch reactor. *Water Research*, **33**(10), 2283–2290.

Billmeier E. (1993). Verbesserung des Feststofretention in Nachklärbecken (Improvement of suspended solids retention in final settlers). *AWT-Abwassertechnik*, **2**, 41–45.

Blackbeard J. R. and Ekama G. A. (1984). Survey of activated sludge bulking and foaming in Southern Africa. *IMIESA*, **9**(3), 20–25.

Blackbeard J. R, Ekama G. A. and Marais G. v. R. (1986). A survey of filamentous bulking and foaming in activated sludge plants in South Africa. *Water SA*, **14**(1), 29–34.

Blackwell L. G. (1971). A theoretical and Experimental Investigation of the Transient Response of the Activated Sludge Process. PhD thesis, Clemson University, South Carolina, USA.

Bock E., Stüven R. Schmidt I. and Zart D. (1995). Nitrogen loss caused by denitrifying Nitrosomonas cells using ammonium or hydroxylamine as electron donors and nitrite as electron acceptor. *Archives of Microbiology*, **163**, 16–20.

Broda E. (1977). Two kinds of lithotrophs missing in nature. *Z. F. Allgemeine Mikrobiologie*, **17**, 491–493.

Brouwer M., Van Loosdrecht M. C. M. and Heijnen J. J. (1996). Behandeling van stikstofrijke retourstromen op huishoudelijke afvalwaterzuiveringen in enkelvoudige reactor systemen (Treatment of nitrogen-rich reject water on municipal wastewater treatments plants in single reactor systems), STOWA report no. 96-21.

Buchan L. (1981). The location and nature of accumulated phosphorus in seven sludges from activated sludge plants which exhibited enhanced phosphorus removal. *Water SA*, **7**, 1–7.

Burke R. A., Dold P. L and Marais G. v. R. (1986). Biological phosphorus removal in short sludge age activated sludge systems, Research report W58, Department of Civil Engineering, University of Cape Town, South Africa.

Cakir F. Y. and Stenstrom M. K. (2005). Greenhouse gas production: a comparison between aerobic and anaerobic wastewater treatment. *Water Research*, **39**(17), 4197–4203.

Casey T. G., Wentzel M. C., Loewenthal R. E., Ekama G. A. and Marais G. v. R. (1994). A hypothesis for the causes and control of anoxic–aerobic (AA) filament bulking in nutrient removal activated sludge systems. *Water Science and Technology*, **29**(7), 203–212.

Casey T. G., Wentzel M. C. and Ekama G. A. (1999). Filamentous organism bulking in nutrient removal activated sludge systems. Paper 11: a biochemical/microbiological model for proliferation of anoxic–aerobic (AA) filamentous organisms. *Water SA*, **25**(4), 425–442.

Casey T. G., Ekama G. A., Wentzel M. C. and Marais G. v. R. (2003). A hypothesis for the causes and control of anoxic-aerobic (AA) filament bulking in nutrient removal activated sludge systems. *Water Science and Technology*, **29**(7), 203–212.

Catunda P. F. C., Van Haandel A. C. and Lettinga G. (1994). Post treatment of anaerobically digested sewage in stabilisation ponds. Proceedings of 7th Anaerobic Digestion IAWQ Conference, Cape Town, South Africa.

Catunda P. F. C., Van Haandel A. C. and Sousa J. T. (1990). Dimensionamento e otimização de decantadores de lodo ativado (Design and optimisation of activated sludge settlers). Revista Bio. Ano II, no. 1, pp. 35–42.

Cavalcanti P. F. F. (2003). Integrated Application of the UASB Reactor and Ponds for Domestic Sewage Treatment in Tropical Regions. Doctoral thesis, University of Wageningen, The Netherlands.

Chernicharo C. A. L. (2006). Post-treatment options for the anaerobic treatment of domestic wastewater. Rev. Environ. *Science Biotechnology*, **5**(1), 73–92.

Chernicharo C. A. L. (2007). Princípios do tratamento biologic de agues residuárias – reatores anaeróbios (Fundamentals of biological wastewater treatment – anaerobic reactors), vol. 5, 2nd edn, Editora UFMG, Belo Horizonte, MG, Brasil.

Chernicharo C. A. L. (2007). Biological Wastewater Treatment Series: vol. 4 – Anaerobic Reactors. IWA Publishing, London, UK.

Chernicharo C. A. L., Almeida P. G. S., Lobato L. C. S., Couto T. C., Borges J. M. and Lacerda Y. S. (2009). Experience with the design and start up of two full scale UASB plants: enhancements and feedback. *Water Science and Technology*, **60**(2), 507–515.

Christensen M. H. and Harremoes P. (1977). Biological denitrification of sewage: a literature review. *Progress in Water Technology*, **8**(4/5), 509–555.

Chudoba J., Grau P. and Ottova V. (1973). Control of activated sludge bulking: II – Selection of micro-organisms by means of a selector. *Water Research*, **7**(10), 1399–1406.

Chudoba J., Ottova V. and Madera V. (1973). Control of activated sludge filamentous bulking: I – Effect of hydraulic regime and degree of mixing in the aeration tank. *Water Research*, **7**(8), 1163–1182.

Clayton J. A., Ekama G. A., Wentzel M. C. and Marais G. v. R. (1991). Denitrification kinetics in biological nitrogen and phosphorus removal activated sludge systems treating municipal wastewater. *Water Science and Technology*, **23** (4–6), 1025–1035.

Collazos C. J. (1990). Operación de la planta piloto de La Rosita, informacion de la 2° etapa por la agencia de protección de la cuenca Bucaramanga (Operation of the La Rosita pilot plant, 2nd stage report for the protection agency of the Bucaramanga catchment area), Bucaramanga, Colombia.

Comeau Y., Hall K. J. and Oldham W. K. (1985). A biochemical model for biological enhanced phosphorus removal. *Water Research*, **20**(12), 1511–1521.

Cookney J., McAdam E. J., Cartmell E. and Jefferson B. (2010). Recovery of methane from anaerobic process effluent using poly-di-methyl-siloxane membrane contactors. Proceedings of the 12th IWA Anaerobic Digestion Conference (AD12), Guadalajara, Mexico.

Costa Silva J. C. (1990). Remoção de nutrientes do efluente de um DAFA, tratando esgoto doméstico através de um sistema Bardenpho (Nutrient Removal from the Effluent of an UASB Reactor Using a Bardenpho System). MSc thesis, Federal University of Paraíba, Campina Grande, Brazil.

Coulter J. B., Soneda S. and Ettinger M. B. (1957). Anaerobic contact process for sewage disposal. *Sewage and Industrial Wastes*, **29**(4), 468–477.

Coura Dias M, Catunda P. F. C and Van Haandel A. C. (1983). O sistema de lodo ativado parte I: O estado estacionário (The activated sludge system part I – the steady state). Proceedings of the 12th Brazilian Sanitary Engineering Congress, Camburiu.

Curtis T. P., Head I. M., Lunn M., Woodcock S., Schloss P. D. and Sloan W. T. (2006). What is the extent of prokaryotic diversity? *Philosophical Transactions of the Royal Society of Biological Sciences*, **361**, 2023–2037.

Dangcong P., Bernet M., Delgenes J. P. and Moletta R. (1999). Aerobic granulated sludge – a case report. *Water Research*, **33**(3), 890–893.

De Bruin L. M. M., Kraan M. W., Verkuijlen J., Van der Roest H. F., de Kreuk M. K. and Van Loosdrecht M. C. M. (2005). Aëroob korrelslibtechnologie – pilot onderzoek naar de toepassing voor de behandeling van huishoudelijk afvalwater (Granulated aerobic sludge – pilot plant research for the application in municipal wastewater treatment), STOWA report no. 2005-35.

De Jong P., Van de Pluijm J. P. L. M, Slange J. and Visser F. A. (1996). Varianten op voorbezinking – een haalbaarheidsstudie (Alternatives to primary settling – a feasibility study), STOWA report no. 96-20.

De Jong P., Voors E. H. and van der Pluijm J. L. P. M. (1993). Handleiding chemische P-verwijdering (Manual on chemical phosphorus removal), STOWA report no. 93-06.

De Kreuk M. K. (2006). Aerobic Granular Sludge – Scaling-Up of a New Technology. PhD thesis, Delft University of Technology, The Netherlands.

De Kreuk M. K. and Van Loosdrecht M. C. M. (2004). Selection of slow growing organisms as a means for improving aerobic granulated sludge stability. *Water Science and Technology*, **49**(11–12), 9–17.

De Kreuk M. K., Van Loosdrecht M. C. M. and De Bruin B. (2003). Development of the aerobic granule reactor technology, STOWA report no. 2003-07.

De Kreuk M. K., Van Loosdrecht M. C. M. and Heijnen J. J. (2005). Simultaneous COD, nitrogen and phosphate removal by aerobic granulated sludge. *Biotechnology and Bioengineering*, **90**(6), 761–769.

De Man A. W. A. (1990). Anaerobe zuivering van ruw afvalwater met behulp van korrelslib in UASB reactoren – eindrapport (Anaerobic purification of raw sewage with the aid of granular sludge in UASB reactors – final report), Department of Environmental Studies, Agricultural University Wageningen, The Netherlands.

De Zeeuw W. (1987). Granular sludge in UASB reactors. Proceedings of the GASMAT Workshop, Lunteren, The Netherlands.

Dick R. I. (1970). Role of activated sludge final settling tanks. *Journal of the Sanitary Engineering Division, ASCE*, **96**(2), 423–436.

Dold P. L., Ekama G. A. and Marais G. v. R. (1980). A general model for the activated sludge process. *Progress in Water Technology*, **12**(6), 47–77.

Dos Santos E. V. M. (2009). Estequiometria e cinética da desnitrificação em sistemas de lodo ativado (Stoichiometrics and Kinetics of Denitrification in Activated Sludge Systems). Masters thesis, UFCG-DEC, Brazil.

Downing A. L., Painter H. A. and Knowles G. (1964). Nitrification in the activated sludge process. *Journal and Proceedings of the Institution of Sewage Purification*, **64**(2), 130–158.

Downing A. L., Tomlinson T. G. and Truesdale G. A. (1964). Effects of inhibitors on nitrification in the activated sludge process. *Journal and Proceedings of the Institution of Sewage Purification*, **6**, 537–550.

Eastman J. A. and Ferguson J. F. (1971). Solubilisation of particulate organic carbon during the acid phase of anaerobic digestion. *Journal Water Pollution Control Federation*, **53**(3), 352–366.

Eckenfelder W. W. Jr. (1991). Berechnung einer Belebungsanlage zur Stickstoff Elimination (Design of an activated sludge systems for nitrogen removal), Institut für Siedlungwasserwirtschaft, Univ. Braunschweig 33–45.

Eikelboom D. H. (1993). Invloed van P- en N-verwijdering op de bezinkeigenschappen van actief slib (Influence of N- and P removal on the settling characteristics of activated sludge), STOWA report no. 93-10.

Eikelboom D. H. (1977). Identification of filamentous organisms in bulking activated sludge. *Progress in Water Technology*, **8**, 153.

Eikelboom D. H. (2000). Process Control of Activated Sludge Plants by Microscopic Investigation. IWA Publishing, London, UK.

Eikelboom D. H. and van Buijsen H. J. J. (1981). Microscopic sludge investigation manual. TNO – Research Institute for Environmental Hygiene, The Netherlands.

Ekama G. A., Barnard J. L., Günthert F. W., Krebs P., McCorquedale J. A., Parker D. S. and van Wahlberg E. J. (1997). Secondary settling tanks – theory, modelling, design and operation, IAWQ Scientific and Technical Report no. 6.

Ekama G. A. and Marais G. v. R (1986). Sludge Settleability of Secondary Sludge and Settling Tank Design Procedures. *Water Pollution Control*, **1**, 101–113.

Ekama G. A. and Marais G. v. R. (1984). Two improved sludge settleability parameters. *IMIESA*, **9**(6), 20–27.

Ekama G. A. and Marais G. v. R. (1985). Exploratory study on activated sludge bulking and foaming problems in South Africa (1983–1984), Final report to Water Research Commission, UCT Report no. 54, University of Cape Town, South Africa.

Ekama G. A. and Marais G. v. R. (1986). Sludge settleability and secondary settling design procedures. *Water Pollution Control*, **5**(1), 101–113.

Ekama G. A. and Wentzel M. C. (2008). Biological wastewater treatment – principles, modelling and design – chapter 5: nitrogen removal. In: M. Henze, M. C. M. van Loosdrecht, G. A. Ekama and D. Brdjanovic (eds), IWA Publishing, London, UK.

EPA (1974). Process Design Manual for Sludge Treatment and Disposal.

Filipe C. D. M., Daigger G. T. and Leslie Grady C. P. Jr. (2001). A metabolic model for acetate uptake under anaerobic conditions by glycogen accumulating organisms: stoichiometrics, kinetics and the effect of pH. *Biotechnology and Bioengineering*, **76**(1), 17–31.

Filipe C. D. M., Daigger G. T. and Leslie Grady C. P. Jr. (2001). pH as a key factor in the competition between glycogen accumulating organisms and phosphorus accumulating organisms. *Water Environment Research*, **73** (1), 223–232.

Fuhs G. W. and Chen M. (1975). Microbiological basis for phosphate removal in the activated sludge process for the treatment of wastewater. *Microbial Ecology*, **2**, 119–138.

Fukase T., Shibata M. and Mijayi (1982). Studies on the mechanism of biological phosphorus removal. *Japanese Journal Water Pollution Resource*, **5**, 309–317.

Fux C., Boehler M., Huber F., Brunner I. and Siegrist H. (2002). Biological treatment of ammonium-rich wastewater by partial nitritation and subsequent anaerobic ammonium oxidation (Anammox) in a pilot plant. *Journal of Biotechnology*, **99**(3), 295–306.

Garret M. T. Jr. and Sawyer C. N. (1952). Kinetics of removal of soluble BOD by activated sludge. Proceedings of the 7th Industrial Waste Conference, Purdue University Extension Series, vol. 79, pp. 51–77.

Genung R. K., Donaldson T. L. and Reed G. D. (1985). Pilot scale development of anaerobic filter technology for municipal wastewater treatment. Proceedings of the Anaerobic Sewage Treatment Seminar, M. S. Schwitzenbaum (ed.), Univ. of Massachusetts, Amherst, US.

Giesen A. Advancements in aerobic granular biomass processes. Presentation at the Neptune and Innowatch end user conference: innovative and sustainable technologies for urban and industrial wastewater treatment; Ghent – Belgium; 27th of January 2010.

Gloyna E. F. (1971). Waste stabilisation ponds. WHO monograph series, Geneva, Switserland.

Gloyna E. F. and Aguirra J. (1972). New experimental ponds data. R. E. McKinney (ed.), 2nd International Symposium on Waste Treatment Lagoons, Lawrence, University of Kansas, US.

Gomes C. S. (1985). Research at SANEPAR and State of Parana, Brazil, with anaerobic treatment of domestic sewage in full scale and pilot plants. Proceedings of the Anaerobic Sewage Treatment Seminar, M. S. Schwitzenbaum (ed.), Univ. of Massachusetts, Amherst, US.

Gujer W. (1977). Design of nitrifying activated sludge process with the aid of dynamic simulation. *Progress in Water Technology*, **9**, 323–336.

Gujer W. and Jenkins D. (1975). A nitrification model for the contact-stabilisation activated sludge process. *Water Research*, **9**(5/6), 561–566.

Gujer W. and Zehnder A. J. B. (1983). Conversion processes in anaerobic digestion. *Water Science and Technology*, **15** (8–9), 127–167.

Harremoes P., Bundgaard E. and Henze M. (1991). Developments in Wastewater Treatment for Nutrient Removal. *European Water Pollution Control*, **1**(1), 19–23.

Hartley K. J. (2008). Controlling sludge settleability in the oxidation ditch process. *Water Research*, **42**(6–7), 1459–1466.

Haskoning (1989). Anaerobic treatment of domestic wastewater under tropical conditions – final report pilot plant research.

Haskoning and Euroconsult (1990). Monitoring report of the 5 MLD UASB treatment plant at Kanpur – Technical report.

Haug R. T. and McCarty P. L. (1971). Nitrification with the submerged filter. Report prepared by the Department of Civil Engineering, Stanford University for the US Environmental Protection Agency, Research Grant No. 17010 EPM.

Heffernan B., Otten M. and Van der Lubbe J. G. M. (2010). Design and operational aspects of UASB sewage treatment plants. Singapore International Water Week; Singapore; 28th June–2nd July 2010.

Heffernan B., Van der Lubbe J. G. M. and Van Lier J. B. (2010). Performance review of large scale up-flow anaerobic sludge blanket sewage treatment plants. Proceedings of Bi-Annual IWA World Water Congress – Montreal, Canada, 19th–24th September 2010.

Heide B. A. (1975). Aerobe en verdergaande zuivering van afvalwater in zeer laagbelaste actief slib systemen, deel II (Advanced aerobic treatment of wastewater in very low loaded activated sludge systems – part II), Report no. A77, Inst. voor milieuhygiene en gezondheidstechniek, TNO, Delft.

Heidman J. A. (1979). Sequential nitrification - denitrification in a plug flow activated sludge system. U. S. Environmental Protection Agency, Cincinnati, Ohio, report no. EPA 600/2-79-159.

Heijnen J. J. (1991). Bioenergetics and biokinetics of microbial growth and product formation – Lecture notes. TU Delft, The Netherlands.

Hellinga C., Van Loosdrecht M. C. M. and Heijnen J. J. (1999). Model based design of a novel process for nitrogen removal from concentrated flows. *Mathematical and Computer Modelling of Dynamic Systems*, **5**(4), 1–13.

Helmer C., Tromm C., Hippen A., Rosenwinkel K. H., Seyfried C. F. and Kunst S. (2001). Single stage biological nitrogen removal by nitritation and anaerobic ammonium oxidation in biofilm systems. *Water Science and Technology*, **43**(1), 311–320.

Henze M., Dupont R., Grau P. and De la Sota A. (1993). Rising sludge in secondary settlers due to denitrification. *Water Research*, **27**(2), 231–236.

Henze M., Grady C. P. L., Gujer W., Marais G. v. R. and Matsuo T. (1986). Activated sludge model no. 1. Scientific and Technical reports no 1, IAWPRC, London, U.K.

Henze M., Grady C. P. L., Gujer W., Marais G. v. R. and Matsuo T. (1994). Activated sludge model No. 2. IAWQ, London, U.K.

Heukelekian H. (1958). Basic principles of sludge digestion. In: Biological Treatment of Sewage and Industrial Wastes, J. NcCabe and W. W. Eckenfelder Jr. (eds), Reinhold, New York, US.

Hiao X., Heijnen J. J. and van Loosdrecht M. C. M. (2002). Sensitivity analysis of a biofilm model describing a one stage completely autotrophic nitrogen removal process (CANON). *Biotechnology and Bioengineering*, **77**(3), 266–277.

Higgins A. J., Kaplovsky A. J. and Hunter J. V. (1982). Organic composition of aerobic, anaerobic and compost-stabilised Sludges. *Journal of Water Pollution Control Federation*, **54**(5), 466–473.

Hippen A., Rosenwinkel K. H., Baumgarten G. and Seyfried C. F. (1997). Aerobic deammonification: a new experience in the treatment of wastewater. *Water Science and Technology*, **35**(10), 111–120.

Hoover S. R. and Porges N. (1952). Assimilation of dairy waste by activated sludge – II: The equations of synthesis and rate of oxygen utilisation. *Sewage and Industrial Wastes Journal*, **24**(3), 306–312.

Hulshof Pol L. W., Heynekamp K. and Lettinga G. (1987). The selection pressure as driving force behind the granulation of sludge. Proceedings of the GASMAT Workshop, Lunteren, The Netherlands.

Hulshof Pol L. W. and Lettinga G. (1986). New technologies for anaerobic wastewater treatment. *Water Science and Technology*, **18**(12), 41–53.

IPCC (2007). III. I. Avoidance of methane production in wastewater treatment through replacement of anaerobic systems by aerobic systems (version 6).

Janssen P. M. J., Meinema K. and Van de Roest H. F. (2002). Biological phosphorus removal – manual for design and operation, STOWA Report, Published by IWA.

Jenkins D. (1980). The control of activated sludge bulking. Presented at the 52nd annual California water pollution control association conference, Monterey, California, US.

Jenkins D., Richard M. G. and Daigger G. T. (2004). Manual on the Causes and Control of Activated Sludge Bulking, Foaming and Other Separation Problems, 3rd edn, IWA Publishing, London, UK.

Jeris J. S. (1982). Industrial wastewater treatment using anaerobic fluidised bed reactors. Proceedings of IAWPRC Seminar on Anaerobic Treatment, Copenhagen, Denmark.

Jeris J. S., Ehlers R. and Witkowski P. (1985). Anaerobic treatment of municipal wastewater using the fluidised bed process. Proceedings of the Anaerobic Sewage Treatment Seminar, M. S. Schwitzenbaum (ed.), Univ. of Massachusetts Amherst, US.

Jetten M. S. M., Van Gerven E., Ping Zheng and Strous M. (1996). Verwijdering van ammonium uit slibgistingswater met het Anammox process – een haalbaarheids studie (Removal of ammonium from digester reject water with the Anammox process – a feasibility study), STOWA report no. 96-21.

Jetten M., Van Loosdrecht M. C. M. and Van Dongen U. (2000). Het gecombineerd Sharon – Anammox process – een duurzame methode voor stikstof verwijdering uit slibgistingswater (The combined SHARON – Anammox process – a sustainable method for N-removal from digestion reject water), STOWA report no. 2000-25.

Jewell W. J. (1982). Development of the attached microbial film expanded bed process for anaerobic waste treatment. Biological fluidised bed treatment of water and wastewater, P. E. Cooper and B. Atkinson (eds.), Ellis Horwood, Chichester, U.K.

Jewell W. J., Schwitzenbaum M. S. and Morris J. W. (1981). Municipal wastewater treatment with the anaerobic attached microbial film expanded bed process. *Journal Water Pollution Control Federation*, **53**(4), 482–490.

Kalbskopf K. H. (1971). Theoretisch Grundlagen, bemessung und Verfahrensweise der Schlammeindickung (Theoretical foundations, design and methodology of sludge thickening). Gewässerschutz – Wasser – Abwasser.

Katz W. J. and Rohlich G. A. (1956). A study of the equilibria and kinetics of adsorption by activated sludge. In: Biological Treatment of Sewage and Industrial Wastes, B. J. Mc. Cabe and W. W. Eckenfelder Jr. (eds), Reinhold Publishing Co., New York, USA.

Kayser R. (1991). Berechnungsbeispiel für Stickstoffentfernung (Sample calculation for nitrogen removal). Institut fur Siedlungs wasserwirtschatft, University of Braunschweig, Germany.

Keinath T. M. (1985). Operational control dynamics of secondary clarifiers. *Journal of Water Pollution Control Federation*, **57**(7), 770–776.

Koopman B. and Cadee K. (1983). Prediction of thickening capacity using diluted sludge volume index. *Water Research*, **17**(10), 1427–1431.

Koot A. C. J. (1980). Behandeling van afvalwater (Treatment of wastewater). Lecture notes, Technical University of Delft.

Kruit J., Hulsbeek J. J. W. and Visser A. (2001). Beheersing van lichtslib bij de behandeling van stedelijk afvalwater met biologische nutrientenverwijdering (Control of bulking sludge in nutrient removal systems for the treatment of municipal sewage), STOWA report no. 2001-02.

Kuai L. and Verstraete W. (1998). Ammonium removal by the oxygen limited autotrophic nitrification – denitrification system. *Applied and Environmental Microbiology*, **64**(11), 4500–4506.

Lakshminarayana J. S. S. (1972). Prevention of sewage pollution by stabilisation ponds. *Environmental Letters*, **8**(2), 121–134.

Landberg G. G., Graulich B. P. and Kipple W. H. (1969). Experimental problems associated with the testing of surface aeration equipment. *Water Research*, **3**(6), 445–455.

Lawrence A. W. and Brown C. G. (1973). Biokinetic approach to optimal design and control of nitrifying activated sludge systems. Annual meeting of the New York water pollution control association, New York City.

Lawrence A. W. and McCarty P. L. (1970). Unified basis for biological treatment design and operation. *Journal of Sanitary Engineering Division, ASCE*, **96**(SA3), 757–778.

Lechevalier H. A. and Lechevalier M. P. (1975). Actinomycetes of sewage treatment plants, EPA Report no. 600/2-75/031.

Lee S. E., Koopman B. L., Bode H. and Jenkins D. (1983). Evaluation of alternative sludge settleability indices. *Water Research*, **17**(10), 1421–1426.

Levenspiel O. (1977). Engenharia das reacoes químicas (Chemical reaction engineering). Edgard Blucher Ed., São Paulo, Brazil.

Lijklema L. (1973). Model for nitrification in the activated sludge process. *Environmental Science and Technology*, **7**(5), 428–433.

Lin C-Y., Sato K., Noike T. and Matsumoto J. Methanogenic digestion using mixed substrate of acetic, propionic and butyric acids. *Water Research*, **20**(3), 385–394.

Liu Y. and Tay J. H. (2004). State of the art of biogranulation technology for wastewater treatment. *Biotechnology Advances*, **22**(7), 553–563.

Loewenthal R. E. and Marais G. v. R. (1976). Carbonate Chemistry of Aquatic Systems: Theory and Application. Ann Arbor Science, Michigan, U.S.A.

Lopez-Vazquez C. M., Hooijmans C. M., Brdjanovic D., Gijzen H. J. and van Loosdrecht M. C. M. (2008). Factors affecting the microbial populations at full scale enhanced biological phosphorus removal (EPBR) wastewater treatment plants in the Netherlands. *Water Research*, **42**, 2349–2360.

Lopez-Vazquez C. M., Hooijmans C. M., Brdjanovic D., Gijzen H. J. and van Loosdrecht M. C. M. (2009). Temperature effects on glycogen accumulating organisms. *Water Research*, **43**(11), 2852–2864.

Lopez-Vazquez C. M., Oehmen A., Zhiguo Y., Hooijmans C. M., Brdjanovic D., Gijzen H. J., Yuan Z. and van Loosdrecht M. C. M. (2009). Modelling the PAO-GAO competition: effects of carbon source, temperature and pH. *Water Research*, **43**(2), 450–462.

Ludzack F. J. and Ettinger M. G. (1962). Controlling operation to minimise activated sludge effluent nitrogen. *Journal of Water Pollution Control Federation*, **34**(9), 920–931.

Macedo Alves R. F. C. (1990). Tratamento biológico da lixívia (Biological Treatment of Leachate). MSc thesis – University of Paraíba – Campina Grande, Brazil.

Malan W. N. and Gouws E. P. (1966). Geaktiveerde slyk vir rioolwatersuivering op Bellville (Activated sludge systems for sewage treatment at Belville), Research report, Council for Scientific and Industrial research.

Mara D. D. (1976). Sewage Treatment in Hot Climates. John Wiley & Sons, Chichester, U.K.

Marais G. v. R. and Ekama G. A. (1976). The activated sludge process: steady state behaviour. *Water SA*, **2**(4), 163–200.

Marais G. v. R. and Shaw (1961). Rational theory for design of waste stabilisation ponds in South Africa. *Transactions of the South African Institute of Civil Engineers*, **3**(11), 205.

Marsden M. G. and Marais G. v. R. (1976). The role of the primary anoxic reactor in denitrification and biological phosphorus removal, Research report. no. 19, Department of Civil Engineering, University of Cape Town, South Africa.

Martins A. M. P., van Loosdrecht M. C. M. and Heijnen J. J. (2004). Bulking sludge in biological nutrient removal systems. *Biotechnology and Bioengineering*, **86**(2), 125–135.

Matsche (1972). The elimination of nitrogen in the treatment plant of Vienna-Blumenthal. *Water Research*, **6**, 485–486.

McCarty P. L. (1964). Anaerobic waste treatment fundamentals – part II: environmental requirements and control. *Public Works*, **95**, 123–126.

McCarty P. L. and Brodersen C. F. (1962). Theory of extended aeration activated sludge. *Journal of Water Pollution Control Federation*, **34**(11), 1095–1103.

McGarry M. G. and Pescod M. B. (1970). Stabilisation pond design criteria for tropical Asia. In: Proceedings 2nd International Symposium on Waste Treatment Lagoons, R. E. McKinney (ed.), University of Kansas, USA.

McKinney R. E. and Symons J. M. (1968). Growth and endogenous metabolism a discussion. Proceedings 1st International Conference on Water Pollution Control, Pergamon Press, London, U.K.

Medeiros U. T. de P., Cavalcanti P. F. F. and Van Haandel A. C. (2007). The use of respirometry to determine nitrification kinetics in activated sludge systems. Proceedings of the Conference on Design, Operation & Economics of Large Wastewater Treatment Plants, Vienna, September 2007.

Meiring P. G., Drews R. J., Van Eck H. and Stander G. J. (1968). A guide to the use of pond systems in South Africa for the purification of raw and partially treated sewage, CSIR Special Reports WAT, National Institute for Water Research, Pretoria, Republic of South Africa.

Melamed A., Saliternik C. and Wachs A. M. (1970). BOD removal and nitrification of anaerobic effluent by activated sludge. Presented at the 5th IAWPR conference, San Francisco, California, USA.

Mohlman F. W. (1934). The sludge index. *Sewage Works*, **6**, 119–126.

Morgenroth E., Sherden T., Van Loosdrecht M. C. M., Heijnen J. J. and Wilderer P. A. (1997). Aerobic granular sludge in a sequencing batch reactor. *Water Research*, **31**(12), 3191–3194.

Mulder A., van de Graaf A. A., Robertson L. A. and Kuenen J. G. (1995). Anaerobic ammonium oxidation discovered in a denitrifying fluidised bed reactor. *FEMS Microbiology Letters*, **16**, 177–184.

Mulder J. W., Van Loosdrecht M. C. M., Hellinga C. and Van Kempen R. (2001). Full-scale application of the SHARON process for treatment of rejection water of digested sludge dewatering. *Water Science and Technology*, **43**(11), 127–134.

Neto D. B. and Van Haandel A. C. (1993). Viabilidade da digestão anaeróbia do lodo de excesso gerado no complexo petroquímioco em Camaçari (Viability of anaerobic sludge digestion of the excess sludge produced at the petrochemical complex in Camaçari). Proceedings of 17th Seminar of the Brazilian Sanitary Engineering Organisation, Natal, pp. 362–372.

Nicholls (1975). Modification of extended aeration plants in Johannesburg (South Africa) to achieve denitrification. Presented at the conference on nitrogen as a water pollutant, Copenhagen, Denmark.

Nicolella C., van Loosdrecht M. C. M. and Heijnen J. J. (2000). Wastewater treatment with particulate biofilm reactors. *Journal of Biotechnology*, **80**, 1–33.

Nicolella C., van Loosdrecht M. C. M. and Heijnen J. J. (2000). Particle-based biofilm reactors. *Trends in Biotechnology*, **18**(7), 312–320.

Nobre C. A. and Guimaraes M. O. (1987). Experiências do tratamento anaeróbio de esgotos urbanos (Experiences with anaerobic sewage treatment). *Revista DAE*, **47**(75), 75–85.

Oliveira S. C. and Von Sperling M. (2009). Performance evaluation of UASB systems with and without post-treatment. *Water Science and Technology*, **59**(7), 1299–1306.

O'Rourke J. T. (1968). Kinetics of Anaerobic Treatment at Reduced Temperatures. PhD thesis, Stanford University, Stanford, California.

Pantoja Filho J. L. R. and Foresti E. (2010). Denitrification of domestic sewage anaerobic effluent with biogas: preliminary results. Proceedings of the 12th IWA Anaerobic Digestion Conference (AD12), Guadalajara, Mexico.

Paques (2011a) Reference sheet: PHOSPAQ – one step ANAMMOX®, Waterstromen Olburg.

Paques (2011b). PHOSPAQ – one step ANAMMOX – Case study Aviko Steenderen (Potato Processing Industry).

Pasveer (1965). Beitrag uber Stickstoffbeseitigung aus Abwasser (Contribution about nitrogen removal from wastewater). Munchner Beitrage zur Abwasser – Fischerei – und Flussbiologie. Liebmann (ed.), Bd 12, pp. 197–200.

Pavlostathis S. G. (1985). A Kinetic Model for Anaerobic Digestion of Waste Activated Sludge. PhD thesis, Cornell University, Ithaca, NY., USA.

Phelps E. B. (1944). Stream Sanitation. Wiley, New York, USA.

Pitman A. R. (1980). Settling properties of extended aeration sludge. *Journal of Water Pollution Control Federation*, **52** (3), 524–536.

Poduska R. A. and Andrews J. F. (1974). Dynamics of Nitrification in the Activated Sludge Process. Department of Environmental Systems Engineering, Clemson University, Clemson, South Carolina, USA.

Pretorius W. A. (1971). Anaerobic digestion of raw sewage. *Water Research*, **5**(9), 681–687.

Pynaert K., Smets B. F., Wyffels S., Beheydt D., Siciliano S. D. and Verstraete W. (2003). Characterization of an autotrophic nitrogen removing biofilm from a highly loaded lab-scale rotating biological contactor. *Applied Environmental Microbiology*, **69**(6), 3626–3635.

Rabinowitz B. and Marais G. v. R. (1981). Chemical and biological phosphorus removal in the activated sludge process, Research report no. W32, Department of Civil Engineering, University of Cape Town, Rondebosch, South Africa.

Rachwal A. J., Johnstone D. W. M., Hanbury M. J. and Critchard D. J. (1982). The application of settleability tests for control of activated sludge plants. In: Bulking of Activated Sludge – Preventative and Remedial Methods (editors Chambers B. and Tomlinson E. J.), Ellis Hopwood Chichester, U.K.

Rajapakse J. P. and Scutt J. E. (1999). Denitrification with natural gas and various new growth media. *Water Research*, **33**(18), 3723–3734.

Richard M. G., Shimizu G. P. and Jenkins D. (1984). The growth physiology of the filamentous organism Type 021N and its significance to activated sludge bulking. Presented at the 57th annual conference, Water Pollution Control Federation, New Orleans, USA.

Rinzema A. (1989). Anaerobic Treatment of Wastewaters with High Concentrations of Lipids or Sulphate. PhD thesis, Wageningen University, The Netherlands.

Roeleveld P. J. and van Loosdrecht M. C. M. (2002). Experience with guidelines for wastewater characterisation in The Netherlands. *Water Science and Technology*, **45**(6), 77–87.

Rössle W. H. and Pretorius W. A. (2001). A review of characterisation requirements for in-line prefermenters – paper 1: Wastewater characterisation. *Water SA*, **27**(3), 405–412.

Rössle W. H. and Pretorius W. A. (2001). A review of characterisation requirements for in-line prefermenters – paper 2: Process characterisation. *Water SA*, **27**(3), 413–423.

Sastry C. A. and Mohanras G. J. (1976). Waste stabilisation pond design and experiences in India. In Ponds as a wastewater treatment alternative, Water Resources Symposium no. 9, University of Texas, Austin, Texas, USA.

Schellinkhout A. and Collazos C. J. (1991). Full scale application of the UASB technology for sewage treatment. Presented at 6th international IAWPRC symposium, São Paulo, Brazil.

Schmid M., Twachtmann U., Klein M., Strous M., Juretschko S., Jetten M. S. M., Metzger J. W., Schleifer K. H. and Wagner M. (2000). Molecular evidence for genus level diversity of bacteria capable of catalyzing anaerobic ammonium oxidation. *Systematic and Applied Microbiology*, **23**(1), 93–106.

Sezgin M., Jenkins D. and Parker D. S. (1978). A unified explanation of filamentous activated sludge bulking. *Journal of Water Pollution Control Federation*, **50**(2), 362–381.

Siebritz I. P, Ekama G. A. and Marais G. v. R. (1982). A parametric model for biological excess phosphorus removal. *Water Science and Technology*, **15**(3/4), 127–152.

Siegrist H., Renggli D. and Gujer W. (1993). Mathematical modelling of anaerobic mesophilic sewage sludge treatment. *Water Science and Technology*, **27**(2), 25–36.

Silva S. A. (1989). Stabilisation pond series with UASB reactors for the treatment of domestic waste in NE Brazil. Proceedings of 15th Seminar of Brazilian Sanitary Engineers, vol. 1, pp. 86, Belem, Brazil.

Sloan W. T., Woodcock S., Lunn M., Head I. M. and Curtis T. P. (2007). Modelling taxa-abundance distributions in microbial communities using environmental sequence data. *Microbial Ecology*, **53**, 443–455.

Srinath E. G., Sastry C. B. and Pillai S. G. (1959). Rapid removal of phosphorus from sewage by activated sludge. *Cellular and Molecular Life Sciences*, **15**(9), 339–340.

Smollen M. and Ekama G. A. (1984). Comparison of empirical settling velocity equations in flux theory for secondary settling tanks. *Water SA*, **10**, 175–184.

Souza C. L., Chernicharo C. A. L. and Aquino S. F. (2010a). Quantification of dissolved methane in UASB reactors treating domestic wastewater under different operating conditions. Proceedings of the 12th IWA Anaerobic Digestion Conference (AD12), Guadalajara, Mexico.

Souza C. L., Chernicharo C. A. L. and Melo G. C. B. (2010b). Methane and hydrogen sulphide measurements in UASB reactors treating domestic wastewater. Proceedings of the 12th IWA Anaerobic Digestion Conference (AD12), Guadalajara, Mexico.

Spanjers H., Vanrolleghem P. A., Olsson G. and Dold P. (1996). Respirometry in control of activated sludge processes. *Water Science and Technology*, **34**(3–4), 117–126.

Spanjers H., Vanrolleghem P. A., Olsson G. and Dold P. (1998). Respirometry in control of activated sludge processes – principles, IAWQ Scientific and Technical Report no. 7, IAWQ.

Speece R. E (1983). Anaerobic biotechnology for industrial wastewater treatment. *Environmental Science and Technology*, **17**(9), 416–427.

Stenstrom M. K. and Poduska R. A. (1980). The effect of dissolved oxygen concentration on nitrification. *Water Research*, **14**(6), 645–650.

Stern L. B. and Marais G. v. R. (1974). Sewage as electron donor in biological denitrification, Research report. no. 7, Department of Civil Engineering, University of Cape Town, South Africa.

Stichting Toegepast Onderzoek Reiniging Afvalwater (STORA) (1981). Hydraulische en technologische aspecten van het nabezinkproces – literatuuronderzoek (Hydraulic and technological aspects of the settling process – literature review), STORA report no. 1981-10.

Stichting Toegepast Onderzoek Reiniging Afvalwater (STORA) (1981). Hydraulische en technologische aspecten van het nabezinkproces – Ronde nabezinktanks: ontwerpgegevens en bedrijfservaring (Hydraulic and technological aspects of the settling process – circular settling tanks: design criteria and operational experiences), STORA report no. 1981-11.

Stichting Toegepast Onderzoek Reiniging Afvalwater (STORA) (1981). Hydraulische en technologische aspecten van het nabezinkproces – Ronde nabezinktanks: praktijkonderzoek (Hydraulic and technological aspects of the settling process – circular settling tanks: research findings), STORA report no. 1981-12.

Stobbe C. T. (1964). Über das Verhalten von belebtes Schlammes in aufsteigender Wasserbewegung (About the behaviour of the upflow activated sludge process) Publication of the sanitary engineering institute of the University of Hannover, Germany, vol. 18.

Stofkoper J. A. and Trentelman C. C. M. (1982). Richtlijnen voor het dimensioneren van ronde nabezinktanks voor actiefslibinstallaties (Guidelines for design of circular final settling tanks for activated sludge systems). *H_2O*, **15** (14), 344–354.

Strous M., Fuerst J. A., Kramer E. H. M., Logeman S., Muyzer G., van de Pas-Schoonen K. T., Webb R., Kuenen J. G. and Jetten M. S. M. (1999). Missing lithotroph identified as new planctomycete. *Nature*, **400**, 446–449.

Strous M., Heijnen J. J., Kuenen J. G. and Jetten M. S. M. (1998). The sequencing batch reactor as a powerful tool for the study of slowly growing anaerobic ammonium oxidising micro-organisms. *Applied Microbiology and Biotechnology*, **50**(5), 589–596.

Sutton P. M., Jank B. E., Monaghan B. A. and Murphy K. L. (1979). Single sludge nitrogen removal systems, Research report no. 88, Environmental Protection Service, Canada.

Switzenbaum M. S. and Jewell W. J. (1980). Anaerobic attached-film and expanded-bed reactor treatment. *Journal of Water Pollution Control Federation*, **52**(7), 1953–1965.

Tandoi V., Jenkins D. and Wanner J. (2006). Activated sludge separation problems – scientific and technical report no. 16. IWA publishing.

Ten Have C. and Van Kempen R. (2004). Rejectiewaterbehandeling geëvalueerd: SHARON, effluentkwaliteit, alternatieven en marktpotentie (Evaluation of reject water treatment – SHARON, effluent quality, alternatives and market potential), STOWA report no. 2004-20.

Tsai M. W., Wentzel M. C. and Ekama G. A. (2003). The effect of residual ammonia concentration under aerobic conditions on the growth of Microthrix parvicella in biological nutrient removal plants. *Water Research*, **37** (12), 3009–3015.

Tuntoolavest M. and Grady C. P. L. J. (1980). Effect of activated sludge operational conditions on sludge thickening characteristics. *Journal of Water Pollution Control Federation*, **54**(7), 1112–1117.

U. S. EPA (1975). Design manual for suspended solids removal. EPA 625/1-75-003; Office of Technology Transfer, Washington DC, USA.

U. S. EPA (1975). Process design manual for nitrogen control. Office of Technology Transfer, Washington DC, USA.

U. S. EPA (1985). Health effects of land application of municipal sludge, Report no. 1-85/015; Office of Technology Transfer, Washington, DC, USA.

Van Bentem A. G. N., Verkuijlen J. and Van de Roest H. F. (2002). MBR: Inleiding tot de bedrijfsvoering (MBR: Introduction to process operation), STOWA report no. 2002-12.

Van Betuw W., Wiegant W. and Kruit J. (2008). SHARON-Anammox systemen: evaluatie van rejectiewaterbehandeling op slibverwerkingsbedrijf Sluisjesdijk (SHARON – Anammox systems: evaluation of reject water treatment on sludge treatment plant Sluisjesdijk), STOWA report no. 2008-18.

Van de Graaf A. A., de Bruijn P., Robertson L. A., Jetten M. S. M. and Kuenen J. G. (1996). Autotrophic growth of anaerobic ammonium oxidising bacteria in a fluidised bed reactor. *Microbiology*, **142**(8), 2187–2196.

Van der Lubbe J. G. M., Heffernan B. and Van Lier J. B. (2010). Engineering and operational issues in UASB reactors treating municipal sewage. Anaerobic digestion conference (AD12), Guadalajara – Mexico.

Van der Lubbe J. G. M., Otten M. and Heffernan B. (2010). UASB – How to get it right. *Asian Water*, **26**(6), 13–16.

Van de Roest H. F., Van Bentem A. G. N., Lawrence D. P., Van der Herberg P. and Kraan R. (2002). MBR for municipal wastewater treatment – pilot plant research Beverwijk, STOWA report no. 11A.

Van de Roest H. F., Van Bentem A. G. N., Lawrence D. P., Van der Herberg P. and Kraan R. (2002). MBR for municipal wastewater treatment – pilot side studies, STOWA report no. 11B.

Van Dijk L., Postma H. J. W. and Roncken G. C. G. (1995). Behandeling van stikstofrijke retourstromen op rwzi's – praktijkonderzoek met de membraan-bioreactor naar nitrificatie/denitrificatie over nitriet bij het slibverwerkingsbedrijf Sluisjesdijk (Treatment of nitrogen-rich reject water on municipal wastewater treatment plants – use of an MBR for nitrification/denitrification over nitrite at sludge treatment plant Sluisjesdijk), STOWA report no. 95-15.

Van Dongen U., Jetten M. and Van Loosdrecht M. C. M. (2001). The SHARON – Anammox process for treatment of ammonium rich wastewater. *Water Science and Technology*, **44**(1), 153–160.

Van Haandel A. C. and Catunda P. F. C. (1986). Dimensionamento e Otimização de Digestores Aeróbios de Lodo Ativado (Design and optimisation of aerobic sludge digesters). *Engenharia Sanitária*, **25**(2), 181–190.

Van Haandel A. C., Catunda P. F. C., Araújo L. S. and Vilar A. (1989). Determinação da Sedimentabilidade de Lodo Ativado (Determination of activated sludge settleability). Presented on the 15th sanitary engineering congress, Belém, Brazil.

Van der Last A. R. M. (1991). Anaerobe behandeling van voorbezonken afvalwater met EGSB en FB processen (Anaerobic treatment of settled sewage with the EGSB and the FB processes), Final report, Environmental Technology Department, Wageningen University.

Van Haandel A. C. (1992). Activated sludge settling – part II: Settling theory and application to design. *Water SA*, **18**(3), 173–180.

Van Haandel A. C. (1994). Alkalinity requirements in anaerobic digesters. 17th biennial international IAWQ conference, Budapest, Hungary.

Van Haandel A. C. and Catunda P. F. C. (1982). Determinação da taxa de consumo de oxigênio (Determination of the oxygen uptake rate). *Revista Engenharia Sanitária*, **21**(4), 481–488.

Van Haandel A. C. and Catunda P. F. C. (1983). O balanço de massa em sistemas de tratamento com lodo em suspensão (Mass balance in treatment systems with suspended sludge). *Revista Engenharia Sanitária*, **22**(4), 409–413.

Van Haandel A. C. and Catunda P. F. C. (1986). Dimensionamento e otimização de digestores anaeróbios de lodo ativado (Design and optimisation of anaerobic digesters for activated sludge). *Engenharia Sanitária*, **25**(2), 181–190.

Van Haandel A. C. and Catunda P. F. C. (1987). Alkalinity and pH changes in the activated sludge process. *Water Science and Technology*, **19**, 1263–1265.

Van Haandel A. C. and Catunda P. F. C. (1992). Activated sludge settling – part I: Experimental determination of settling characteristics. *Water SA*, **18**(3), 165–172.

Van Haandel A. C., Catunda P. F. C., Araújo L. S. and Vilar A. (1989). Determinação da Sedimentabilidade de Lodo Ativado (Determination of the settleability of activated sludge). Presented at the 15th Congress of the Brazilian sanitary engineering organisation; Belém.

Van Haandel A. C., Catunda P. F. C. and Neiva M. R. (1997). Um método experimental para a determinação da eficiênciade aeradores superficiais em sistemas de lodo ativado (An experimental method to determine the aeration efficiency of surface aerators in activated sludge systems). Proceedings of 19th Congress of the Brazilian Sanitary Engineering Organisation, Foz de Iguaçu, Brazil.

Van Haandel A. C. and Lettinga G. (1994). Anaerobic Sewage Treatment. J. Wiley, Chichester, U.K.

Van Haandel A. C. and Marais G. v. R. (1981). Nitrification and denitrification kinetics in the activated sludge process, Research report no. W41, Department of Civil Engineering, University of Cape Town, South Africa.

Van Haandel A. C., Tenório M. A. A., Oliveira R. and Catunda P. F. C. (1986). A cinética de digestão anaeróbia de lodo ativado (The kinetics of anaerobic digestion of activated sludge). *Engenharia Sanitária*, **25**(2), 191–200.

Van Kempen R., De Mooij H. W. and Draaijer H. (2004). SHARON – N-removal over nitrite of nitrogen rich wastewaters. Grontmij Water & Reststoffen.

Van Kempen R., Mulder J. W., Uiterlinde C. A. and Van Loosdrecht M. C. M. (2001). Overview: full scale experience of the SHARON process for treatment of rejection water of digested sludge dewatering. *Water Science and Technology*, **44**(1), 145–152.

Van Lier J. B., Vashi A., Van der Lubbe J. G. M. and Heffernan B. (2010). Anaerobic technology for sustainable environment – chapter 4: anaerobic sewage treatment using UASB reactors: engineering and operational aspects. In: H. Fang (ed.). Imperial College Press, London, UK.

Van Loosdrecht M. C. M. (2008). Biological wastewater treatment – principles, modelling and design – chapter 6: innovative nitrogen removal. In: M. Henze, M. C. M. van Loosdrecht, G. A. Ekama and D. Brdjanovic (eds), IWA Publishing, London, UK.

Van Starkenburg W. and Kruit J. (1994). Dimensionering van selectoren – the rol van influent karakterisering (Guidelines for selector dimensioning – the role of influent characterization), STOWA report no. 94-16.

Van Veldhuizen H. and Van Loosdrecht M. C. M. (1997). Stikstofverwijdering bij lage BZV/stikstof verhoudingen – stikstof verwijdering over nitrite in de hoofdstroom van actief-slib processen (Nitrogen removal at low BOD/N ratio – nitrogen removal over nitrite in the main process stream of activated sludge processes), STOWA report no. 97-34.

Vesilind P. A. (1968). Theoretical Considerations: Design of Prototype Thickeners from Batch Settling Tests. *Water and Sewage Works*, **115**, 302.

Vieira S. M. M. and Garcia A. D. Jr. (1991). Sewage treatment by UASB reactor – operational results and recommendations for design and utilisation. Presentation at the 6th IAWPRC congress, Sao Paulo, Brazil.

Vieira S. M. M. and Souza M. E. (1986). Development of technology for use of the UASB reactor in domestic sewage treatment. Proceedings of IAWPRC Seminar on Anaerobic Treatment in Tropical Countries, Sao Paulo, Brazil.

Visscher R. (2010). Gas uit water – Royal Haskoning wint methane terug (Gas from water – Royal Haskoning recovers methane). *De Ingenieur*, **122**(8).

Vlaeminck S. E., Cloetens L. F. F., De Clippeleir H., Carballa M. and Verstraete W. (2009). Granular biomass capable of partial nitritation and Anammox. *Water Science and Technology*, **58**(5), 1113–1120.

Voors E. H. and De Jong P. (1993). Bij-effecten van ijzer sulfaat dosering in actief-slib installaties (Side effects of iron sulphate dosing in activated sludge systems). *H_2O*, **26**, 769–789.

Wagner F. (1982). Study of the causes and prevention of sludge bulking in Germany. Chapter 2 in Bulking of activated sludge – preventative and remedial methods. In: B. Chambers and E. J. Tomlinson (eds), Ellis, Horwood Ltd., Chichester, U.K.

Wang K. (1994). Integrated Anaerobic and Aerobic Treatment of Sewage. PhD thesis, University of Wageningen.

Washington D. R. and Hetling L. J. (1965). Volatile sludge accumulation in activated sludge plants. *Journal of Water Pollution Control Federation*, **38**(4), 499–507.

Water Pollution Control Federation (1977). Wastewater treatment plant design – manual of practice no. 8. WPCF and ASCE, Lancaster Press Inc., Lancaster, USA.

Water Pollution Control Federation (1985). Manual of practice no. 8: clarifier design. WPCF, Washington D.C., USA.

Wentzel M. C., Dold P. L., Ekama G. A. and Marais G. v. R. (1985). Kinetics of biological phosphorus release. *Water Science and Technology*, **17**(11/12), 57–71.

Wentzel M. C., Ekama G. A., Dold P. L. and Marais G. v. R. (1990). Biological excess phosphorus removal – steady state process design. *Water SA*, **16**(1), 29–47.

Wentzel C. M., Ekama G. A. and Marais G. v. R. (1992). Processes and modelling nitrification – denitrification, biological excess phosphorus removal systems – a review. *Water Science and Technology*, **25**(6), 59–82.

Wentzel M. C., Loewenthal R. E., Ekama G. A. and Marais G. v. R. (1988). Enhanced polyphosphate organism cultures in activated sludge systems – part II: Experimental behaviour. *Water SA*, **15**(2), 71–88.

Wentzel M. C., Lötter L. H., Loewenthal R. E. and Marais G. v. R. (1986). Metabolic behaviour of Acinetobacter spp. in enhanced biological phosphorus removal: a biochemical model. *Water SA*, **12**(4), 209–224.

Werner M. and Kayser R. (1991). Denitrification with biogas as external carbon source. *Water Science and Technology*, **23**(4–6), 701–708.

Wett B. (2006). Solved upscaling problems for implementing deammonification of rejection water. *Water Science and Technology*, **53**(12), 121–128.

Wett B. (2007). Development and implementation of a robust deammonification process. Proceedings of IWA Leading Edge Technology Conference, Singapore.

Wett B., Murthy S., Takács I., Hell M., Bowden G., Deur and O'Shaughnessy M. (2007). Key parameters for control of DEMON deammonification process. Proceedings of WEF-IWA Conference on Nutrient Removal, Baltimore.

White M. J. D. (1975). Settling of activated sludge, Technical Report TR11, WRC, Stevanhage, UK.

Wilson D. E. and Marais G. v. R. (1976). Adsorption phase in biological denitrification, Research Report. no. 11, Department of Civil Engineering, UCT, South Africa.

WRC (1984). Theory, design and operation of biological nutrient removal activated sludge processes. Water Research Commission; South Africa.

Wyffels S., Boeckx P., Pynaert K., Verstraete W. and Van Cleemput O. (2004). Sustained nitrite accumulation in a membrane-assisted bioreactor (MBR) for the treatment of ammonium-rich wastewater. *Journal of Chemical Technology and Biotechnology*, **78**(4), 412–419.

Wuhrmann K. (1964). Nitrogen removal in sewage treatment processes. *Verhandlungen der Internationale Vereinigung fuer Theoretische und Angewandte Limnologie*, **15**, 580–596.

Yoda M., Hattori M. and Mijaji Y. (1985). Treatment of municipal wastewater by the anaerobic fluidised bed – behaviour of organic suspended solids in an anaerobic reactor. Proceedings of the Seminar of Anaerobic Treatment of Sewage, M. S. Schwitzenbaum (ed.), University of Massachusetts, Amherst. USA.

Yoshioka N., Holta Y., Tanaka S., Naito S. and Tsugami S. (1957). Continuous thickening of homogeneous flocculated slurries. *Chemical Engineering*, **21**, 66.

Young J. C. (1990). Summary of design and operating factors for upflow anaerobic filters. Proceedings of International Workshop on Anaerobic Treatment Technology for Municipal and Industrial Wastewater, Valladolid, Spain.

Young J. C. and McCarty P. L. (1969). The anaerobic filter for waste treatment. *Journal of Water Pollution Control Federation*, **41**(5), research supplement, R160–R173.

Zeng R. J., Van Loosdrecht M. C. M., Yuan Z. and Keller J. (2003). Metabolic model for glycogen accumulating organisms in anaerobic/aerobic activated sludge systems. *Biotechnology and Bioengineering*, **81**(1), 92–105.

Appendix 1

Determination of the oxygen uptake rate

The oxygen uptake rate (OUR, or in this book O_t) is a parameter used to evaluate the rate at which metabolic processes take place in activated sludge treatment processes, characterised by the presence of sludge in suspension. The main uses of the OUR test are to:

- Estimate the values of kinetic and stoichiometric parameters;
- Obtain data required to set up a mass balance of organic- and/or nitrogenous material;
- Evaluate the sludge activity in terms of the maximum substrate utilisation rate;
- Determine the degree of sludge stabilisation after aerobic digestion.

The OUR test procedure is simple: the aeration of a mixed liquor batch is interrupted and the resulting decrease in the oxygen concentration is measured as a function of time. Preferably the oxygen sensor is connected to a recorder or a computer, so that a continuous profile can be obtained. The results of the OUR test are given in terms of mg $O_2 \cdot l^{-1} \cdot h^{-1}$. If the volatile sludge concentration is determined at the same time, the OUR per unit mass of sludge or the specific OUR (mg $O_2 \cdot g^{-1}$ VSS $\cdot h^{-1}$) can be calculated. For a reliable determination of the OUR, the oxygen consumption rate during the test should be at least 2 to 3 mg $O_2 \cdot l^{-1}$. In most activated sludge processes, the OUR will be in the order of 20 to 100 mg $O_2 \cdot l^{-1} \cdot h^{-1}$ for low and high rate systems respectively, so the time required to carry out the test will be only a few minutes.

The OUR test may be carried out in the aeration tank itself or in bench scale reactors. The first option is only feasible if it is possible to keep the sludge in suspension while the aeration is interrupted (e.g. when mixers are installed), or if the time required for the test is so short that the effect of sludge settling may be ignored. However, in most aeration tanks the sludge is maintained in suspension by the agitation introduced by the aeration equipment.

Therefore, usually bench scale reactors or even beakers are used. Sludge is taken from the aeration tank and influent is fed to this sample at proportionally the same rate as to the aeration tank. The OUR determination is sometimes carried out without simultaneous feeding of influent to the sludge sample (Standard methods, 1993), but this procedure leads to a significant underestimate of the actual OUR in the activated sludge system. This is due to the fact that part of the influent material (easily biodegradable material and ammonia) is utilised at a high rate, so that its concentration is low at any time. Therefore, if influent is not continuously supplied, it will rapidly be depleted in the batch reactor and the corresponding OUR will not be measured.

On the other hand, a batch OUR determination without continuous (proportional) feeding can be used if the objective is to measure the maximum specific activity of sludge, as in general a large amount of substrate will be added to the sludge batch. Similarly, when the endogenous respiration rate is to be determined, no external substrate is allowed at all and again the batch OUR measurement can be used.

The three methods for OUR measurement as described above are schematically represented in Figure A1.1. Concerning the third method, presented in Figure A1.1c, it is important that the OUR reactor is operated at the same hydraulic residence time as the aeration tank and that the proportion between influent and return sludge is correct: i.e. $Q_i/q_i = Q_r/q_r$

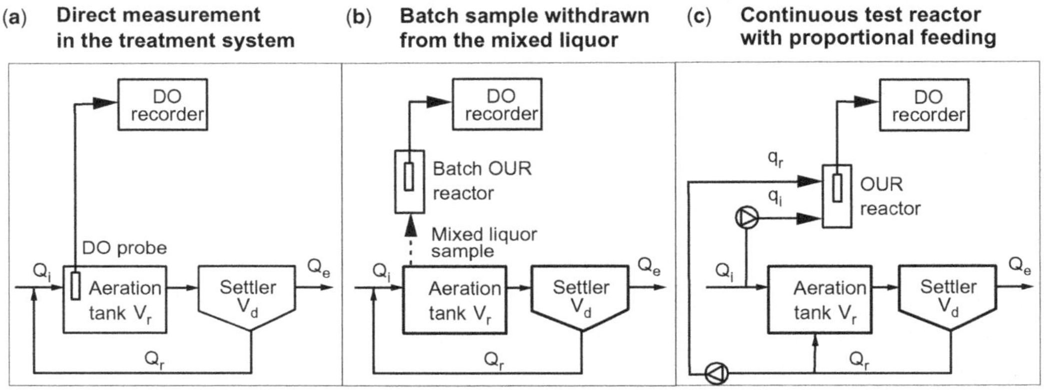

Figure A1.1 Schematic representation of different experimental set-ups for the determination of the OUR

A1.1 DETERMINATION OF THE APPARENT OUR

In the simplest version of the OUR test, the only materials required are a dissolved oxygen analyzer with sensor and a stopwatch. The time period required for a decrease in the dissolved oxygen concentration from an initial value DO_1 to a lower value DO_2 is determined and the corresponding oxygen uptake rate is calculated as:

$$OUR_a = (DO_1 - DO_2)/(t_2 - t_1) \qquad (A1.1)$$

where:

OUR_a = apparent OUR, the observed rate of change of the dissolved oxygen concentration after the aeration of the mixed liquor is interrupted
$DO_{1,2}$ = initial and final dissolved oxygen concentration respectively
$t_2 - t_1$ = time interval

As the measured dissolved oxygen concentration tends to oscillate slightly, it is preferable to record the dissolved oxygen concentration in time and subsequently draw the "best-fit" straight line. Alternatively, the data can be stored in a computer and standard software can be used to calculate the best-fit straight line through the data points.

A1.2 CORRECTION FACTORS OF THE APPARENT OUR

In reality, the interpretation of the OUR test is less straightforward than suggested in the preceding section. Along with metabolic activity, there are several other factors that may influence the mixed liquor dissolved oxygen concentration and these must be taken into consideration. These factors will now be discussed.

A1.2.1 Representativeness of mixed liquor operational conditions

In order to determine the OUR, the sludge must have a uniform concentration in the aeration tank. Therefore the applicability of the test is restricted to aerobic processes with sludge in suspension, such as the conventional activated sludge process, the oxidation ditch and the aerated lagoon.

The sludge mass must be stirred continuously to avoid sludge settling during the test. As in most aeration tanks the mixed liquor cannot be stirred independently of aeration, the determination must be carried out in a small reactor (several litres) in which a sludge sample is placed and fed with influent in conformity with Figure A1.1c. In most cases, the duration of the OUR test is so short that the effect of the introduction of influent and the resulting dilution of the sludge in the reactor vessel can be ignored.

A1.2.2 Critical dissolved oxygen concentration

Oxygen consumption mainly occurs within the sludge flocs, so there is a tendency for the dissolved oxygen concentration to decrease from the periphery of the floc (where it is assumed to be equal to the bulk concentration) towards the floc centre. Figure A1.2 schematically shows the resulting profile of the dissolved oxygen concentration (a spherical floc is assumed). Depending on the applied bulk dissolved oxygen concentration, a zone without oxygen (anoxic or anaerobic) may develop in the central region of the flocs. As no oxygen is consumed in this region the overall specific OUR will decrease.

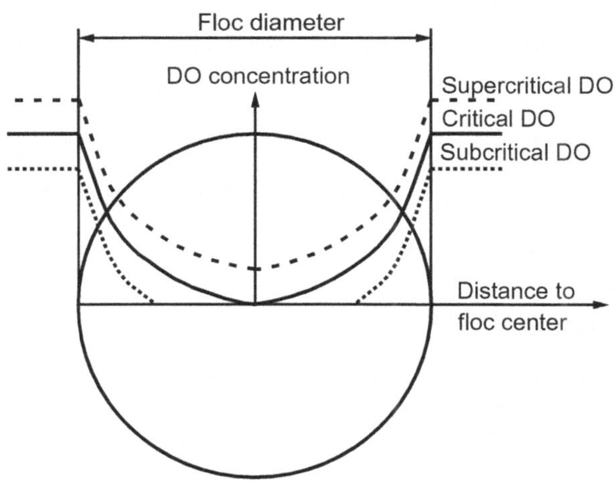

Figure A1.2 Oxygen profiles within sludge flocs as function of the distance to the floc centre

The minimum concentration of dissolved oxygen in the bulk mixed liquor required to avoid oxygen becoming a limiting factor inside the sludge flocs is called the critical dissolved oxygen concentration. Its numerical value depends on several factors, of which the OUR value and the stirring intensity (which influences the floc size) are the most important.

The critical dissolved oxygen concentration may be determined experimentally by observing the decrease in the dissolved oxygen concentration with time until all oxygen is consumed: provided that the availability of external substrate does not change during the test, there will be a constant oxygen consumption rate as long as the dissolved oxygen concentration is above the critical value. At the critical concentration the rate of oxygen consumption starts to decrease gradually, resulting in a subsequent decrease in the slope of the line of dissolved oxygen concentration versus time.

Figure A1.3 shows how the critical dissolved oxygen concentration can be determined in practice. Naturally, the OUR should be determined in the dissolved oxygen concentration range above the critical value.

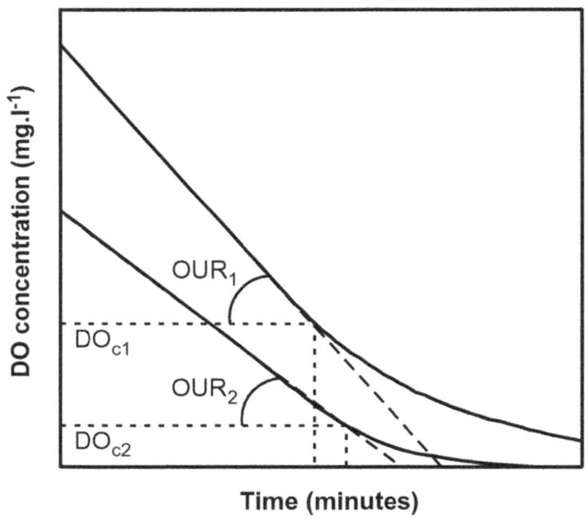

Figure A1.3 Schematic representation of the determination of the critical dissolved oxygen concentration

In practice this critical value is seldom more than 0.5 to 1 mg $O_2 \cdot l^{-1}$ for organic material utilisation and 1 to 2 mg $O_2 \cdot l^{-1}$ for nitrification. The value of the critical dissolved oxygen concentration is of considerable practical importance, as it is in principle the optimal dissolved oxygen concentration in the aeration tank. A smaller concentration will reduce the available treatment capacity of the plant, while a higher dissolved oxygen concentration will result in unnecessary aeration costs.

A1.2.3 Hydraulic effects

Apart from metabolic processes, the oxygen concentration may change because of other factors. In the case of a reactor with influent entering and mixed liquor leaving continuously, the rate of change of the dissolved oxygen concentration due to this hydraulic effect must be taken into consideration:

$$OUR_h = (dDO/dt)_h = (DO_e - DO_i)/R_h \qquad (A1.2)$$

EXAMPLE A1.1

Determine the hydraulic effect in a 10 litre reactor receiving $20\,l \cdot d^{-1}$ of influent ($DO = 1$ mg $O_2 \cdot l^{-1}$) and $10\,l \cdot d^{-1}$ of return sludge ($DO = 0.4$ mg $O_2 \cdot l^{-1}$), if a dissolved oxygen concentration of 4 mg $O_2 \cdot l^{-1}$ is maintained in the reactor.

Solution

First the weighted average of dissolved oxygen in the inlet (influent+return sludge) is calculated as:

$$DO_i = (20 \cdot 1 + 10 \cdot 0.4)/(20 + 10) = 24/30 = 0.8\,mg\,O_2 \cdot l^{-1}$$

Now Eq. (A1.2) is applied:

$$OUR_h = (dDO/dt)_h = (DO_e - DO_i)/R_h = (4 - 0.8)/(10/30) = 9.6\,mg\,O_2 \cdot l^{-1} \cdot d^{-1}$$
$$= 0.4\,mg\,O_2 \cdot l^{-1} \cdot h^{-1}.$$

It is concluded that the dissolved oxygen concentration in the reactor decreases at a rate of 0.4 mg $O_2 \cdot l^{-1} \cdot h^{-1}$ due to the hydraulic effect, independent of any biological oxygen consumption. In most cases the hydraulic effect is very small compared to the OUR required for metabolism. However, when the biological OUR is low, for example in aerated lagoons, correction for the hydraulic effect may be important.

A1.2.4 Absorption of atmospheric oxygen

In the case of a low OUR, absorption of atmospheric oxygen may interfere with the determination process. The rate of oxygen absorption from the air depends on several factors:

- Size of the liquid-air interface: this area may be reduced by placing floating material on the surface or using a closed vessel;
- Dissolved oxygen concentration of the mixed liquor;
- Mixing intensity: if mixing is intense and surface renewal frequent, more oxygen will be absorbed.

The absorption effect on the dissolved oxygen concentration can be determined when water without dissolved oxygen is placed in the OUR reactor and the increase in the dissolved oxygen concentration in time is observed. The rate of change of the dissolved oxygen concentration due to absorption can be expressed as:

$$OUR_{abs} = (dDO/dt)_a = k_{abs} \cdot (DO_s - DO) \quad (A1.3)$$

The solution of the differential equation is:

$$\ln\left[(DO_s - DO)/(DO_s - DO_0)\right] = k_{abs} \cdot t \quad (A1.4)$$

where:

OUR_{abs} = rate of change of the DO concentration due to absorption of atmospheric oxygen
DO_s, DO_0 = saturated and initial dissolved oxygen concentration in the water
DO = dissolved oxygen concentration at time "t"
t = absorption time (h)
k_{abs} = absorption constant (1/h)

With the aid of Eq. (A1.4) the value of the absorption constant k_{abs} can be determined with the following steps:

- Remove the dissolved oxygen in the water in the OUR reactor with Na_2SO_3 and a trace ($10\ mg \cdot l^{-1}$) of $CoCl_2$;
- While mixing at the same intensity as during a regular OUR determination, determine the increase of the dissolved oxygen concentration in time;
- Plot data of $\ln[(DO_s - DO)/(DO_s - DO_0)]$ as a function of the absorption time: the slope of the resulting straight line is equal to k_{abs}.

As the rate of oxygen absorption is higher at increasing mixing intensity, it is important that during the OUR determination the applied mixing intensity is low while still sufficient to maintain the sludge in suspension. The absorption effect is more pronounced for smaller reactors, because of the relatively large ratio of surface area to rector volume. In Figure A1.4 an example is given of the graphical determination of the constant k_{abs}. The experimental dissolved oxygen concentrations are plotted as a function of time in the left-hand graph of Figure A1.4. The saturation concentration was determined as $7.7\ mg\ O_2 \cdot l^{-1}$. In the right-hand graph of Figure A1.4 the corresponding values of $\ln[(DO_s - DO)/(DO_s - DO_0)]$ are plotted as a function of time (the value of $DO_0 = 0$).

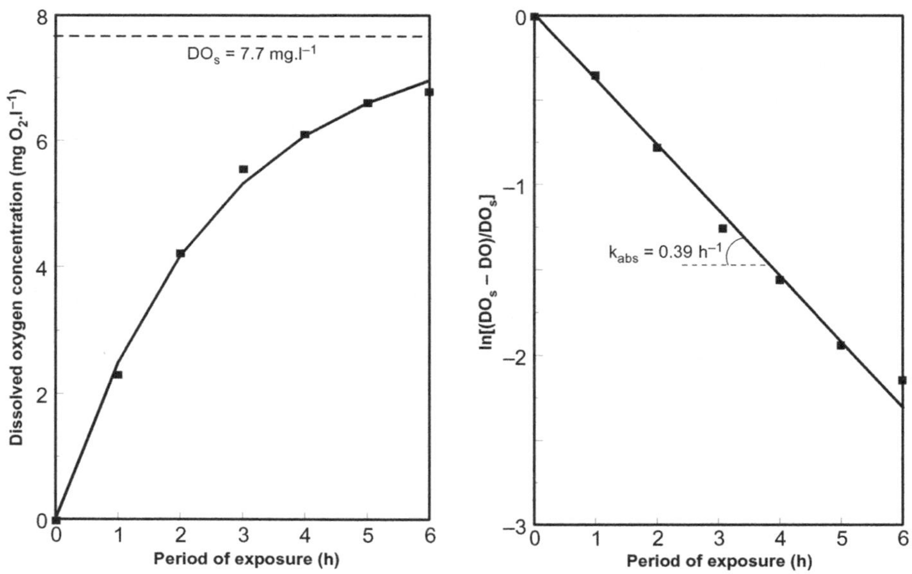

Figure A1.4 Experimental determination of the absorption constant

The value of k_{abs} is calculated from the slope of the best-fit straight line through the experimental values: $k_{abs} = 0.39\ h^{-1}$. So, when this test reactor is operating at a dissolved oxygen of 3 mg $O_2 \cdot l^{-1}$, under the specified conditions the rate of oxygen absorption is calculated as $OUR_{abs} = 0.39 \cdot (7.7 - 3) = 1.6$ mg $O_2 \cdot l^{-1} \cdot h^{-1}$.

The apparent OUR (the *observed* rate of change of the dissolved oxygen concentration in the reactor) is a result of the combination of the three effects described above:

$$OUR_a = OUR + OUR_h - OUR_{abs} \quad \text{or}$$

$$OUR = OUR_a - (DO_i - DO)/R_h + k_{abs} \cdot (DO_s - DO) \qquad (A1.5)$$

Equation (A1.5) shows how the observed or apparent OUR value should be corrected for the hydraulic effects and the absorption of atmospheric oxygen. The importance of these corrections in practice will depend on the relative values of OUR, OUR_h and OUR_{abs}.

EXAMPLE A1.2

In a 10 litre reactor with an absorption constant of $k_{abs} = 0.39\ h^{-1}$ and receiving 20 $l \cdot d^{-1}$ influent (dissolved oxygen = 1.0 mg $O_2 \cdot l^{-1}$ and 10 $l \cdot d^{-1}$ return sludge (dissolved oxygen = 0.4 mg $O_2 \cdot l^{-1}$), the aeration is interrupted and it is observed that the dissolved oxygen concentration decreases from 5 to 3 mg $O_2 \cdot l^{-1}$ in 320 seconds. Determine the apparent OUR and the value corrected for the hydraulic effect and the absorption of oxygen.

Solution

The value of the apparent OUR is calculated as:

$$OUR_a = (DO_1 - DO_2)/(t_2 - t_1) = (5 - 3)/320 = 22.5\ mg\ O_2.l^{-1} \cdot h^{-1}$$

The hydraulic effect was determined in Example A1.1 for the average dissolved oxygen concentration of 4 mg $O_2 \cdot l^{-1}$ in the reactor:

$$OUR_h = 0.8\ mg\ O_2\ cot\, l^{-1} \cdot h^{-1}$$

The OUR effect of the absorption of oxygen can be calculated for $k_{abs} = 0.39\ h^{-1}$ and the average dissolved oxygen concentration of 4 mg $\cdot\ l^{-1}$ in the reactor:

$$OUR_{abs} = 0.39 \cdot (7.7 - 4) = 1.44\ mg\ O_2 \cdot l^{-1} \cdot h^{-1}$$

Hence the OUR due to metabolism in the reactor can now be calculated as:

$$OUR = OUR_a - (DO_i - DO)/R_h + k_{abs} \cdot (DO_s - DO) = 22.5 - 0.8 + 1.4 = 23.1\ mgO_2 \cdot l^{-1} \cdot h^{-1}$$

It is concluded that under the specified conditions the apparent OUR value (22.5 mg $O_2.l^{-1} \cdot h^{-1}$) is virtually identical to the OUR due to metabolism (23.1 mg $O_2.l^{-1} \cdot h^{-1}$).

A1.2.5 The relaxation effect

When the sensor of a dissolved oxygen analyser is suddenly transferred from a liquid with a particular dissolved oxygen concentration (for example saturated water) to a different environment (for example water without dissolved oxygen), it can be observed that a relatively long period is required before the reading has adapted to the new situation and indicates the true dissolved oxygen concentration in the new environment. This time period is called the relaxation period of the sensor. The relatively low response rate of the dissolved oxygen sensor may be a problem for OUR measurements. It can be observed experimentally that the rate of change of the analyzer reading is proportional to the difference that exists between the indicated value and the true dissolved oxygen concentration in the water, so that:

$$(DO_m/dt)_r = k_r \cdot (DO_m - DO) \quad (A1.6)$$

where:

DO_m, DO = dissolved oxygen reading on the analyzer respectively the true oxygen concentration
k_r = relaxation constant of the sensor

If the true dissolved oxygen concentration is constant, Eq. (A1.6) can be solved as:

$$(DO_m - DO)/(DO_{m0} - DO) = \exp(-k_r \cdot t) \quad (A1.7)$$

where DO_{m0} = initial dissolved oxygen reading on the analyzer

The value of the relaxation constant can be determined as follows:

(1) Fill a beaker with water depleted of dissolved oxygen ($DO = 0\ mg \cdot l^{-1}$) and another one with saturated water;
(2) Place the electrode in one of the beakers and wait until a constant reading is obtained;
(3) Transfer the electrode suddenly to the other beaker and record the dissolved oxygen reading on the meter as a function of time;
(4) Determine the relaxation time t_r, which is reached when the difference between the reading and the true value is a factor 0.37 of initial difference (the value of 0.37 corresponds to $e^{-1} = 1/2.72$). At this moment the value of $k_r \cdot t_r = 1$ and $k_r = 1/t_r$.

In Figure A1.5 two examples of the experimental determination of the k_r value are shown, for intense and for slow mixing of the beaker. Using the above procedure, values of $k_r = 520\ h^{-1}$ and $340\ h^{-1}$ were obtained for fast and slow mixing respectively. It should be noted that apart from the mixing intensity, other factors also influence the value of the relaxation constant, such as the type of electrode, the condition of the membrane and the temperature. When the true dissolved oxygen concentration changes (as will be the case during OUR determinations), the differential equation can be written as:

$$(dDO_m/dt)_r = k_r \cdot (DO_m - DO) = k_r \cdot (DO_m - DO_0 - OUR_a \cdot t) \quad (A1.8)$$

where DO_0 = initial value of the actual dissolved oxygen concentration.

The solution to Eq. (A1.8) is:

$$DO_m = DO + (OUR_t/k_r) \cdot (1 - \exp(-k_r \cdot t)) + (DO_{m0} - DO_0) \cdot \exp(-k_r \cdot t) \quad \text{or} \tag{A1.9}$$

$$DO_m - DO = (OUR_t/k_r) \cdot (1 - \exp(-k_r \cdot t)) + (DO_{m0} - DO_0) \cdot \exp(-k_r \cdot t) \tag{A1.10}$$

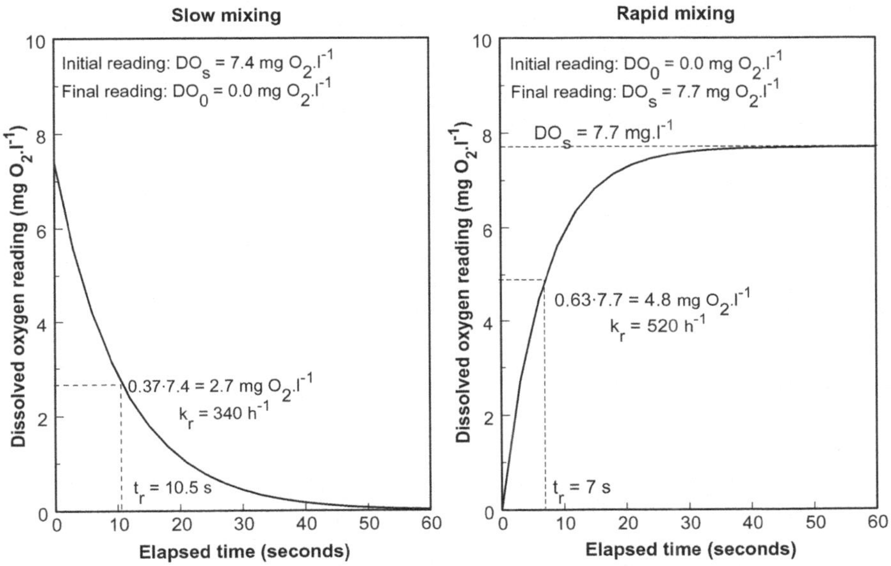

Figure A1.5 Experimental determination of the relaxation constant of the oxygen sensor

Equation (A1.10) shows that during the OUR test the measured dissolved oxygen concentration will never be equal to the true value of the dissolved oxygen concentration in the mixed liquor: there will always remain a difference between these two parameters. However, this does not invalidate the test. As the relaxation constant k_r generally has a large value (normally in the range of 5 to 10 min^{-1} or 0.08 to 0.17 s^{-1}), the exponential factors in Eq. (A1.10) quickly become insignificant. For example: in order to decrease the relaxation effect by 99% [i.e. $\exp(-k_r \cdot t) < 0.01$], it is required that $t > 4.6/k_r = 20$ to 40 seconds. Consequently after a period of 20 to 40 seconds 99% of the relaxation effect will have been eliminated. After this initial period a difference will remain between the reading on the meter and the true value:

$$\Delta DO = OUR_t/k_r \tag{A1.11}$$

In Figure A1.6 true and measured profiles of the dissolved oxygen concentration are shown as a function of time for two values of the relaxation constant. In both cases an OUR of 1 mg $O_2 \cdot l^{-1} \cdot min^{-1}$ and a true initial dissolved oxygen concentration of 5 mg $O_2 \cdot l^{-1}$ have been assumed. The profiles have been drawn for three initial readings: (1) $DO_{m0} = 2$ mg $O_2 \cdot l^{-1}$; (2) $DO_{m0} = 5$ mg $O_2 \cdot l^{-1}$ and (3) $DO_{m0} = 8$ mg $O_2 \cdot l^{-1}$. It can be observed that the OUR determination may easily lead to a considerable error when the relaxation constant has a low value. Even when the initial dissolved oxygen analyser reading is equal to the true dissolved oxygen concentration, there is a danger of misinterpretation of the apparent straight line that is obtained. In the case of Figure A1.6 (right), the curve beginning at DO = 5 mg $O_2 \cdot l^{-1}$ could be interpreted as a straight line: the reduction of the dissolved oxygen concentration from 5 to 3 mg $O_2 \cdot l^{-1}$ in a period of 3.2 minutes would then lead to an assumed OUR value of 2/3.2 = 0.6 mg $O_2 \cdot l^{-1} \cdot min^{-1}$ instead of the true value of 1 mg $O_2 \cdot l^{-1} \cdot min^{-1}$ (i · e. a decrease of 5 mg $O_2 \cdot l^{-1}$ in 5 minutes).

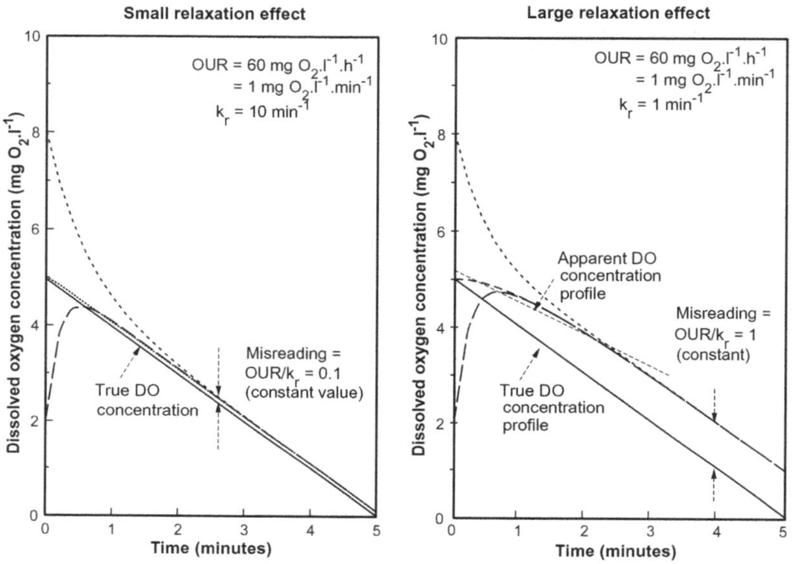

Figure A1.6 Relaxation effect: true and measured dissolved oxygen profiles for different values of the relaxation constant k_r

Figure A1.7 Two Biobed®EGSB reactors installed at a cider factory, with on the background one of the conditioning tanks. On the foreground the flare. Courtesy of Bulmers Ltd, Ireland - member of C&C Group PLC

Appendix 2

Calibration of the general model

The general model can be used to develop a series of differential equations describing the behaviour of the activated sludge system for any kind of feeding pattern and process configuration. The rate of change of the concentration of the different variables per unit time is expressed as the sum of the reaction rate and the rate of change due to hydraulic effects:

$$r_{C,n} = r_{rC,n} + r_{hC,n} \tag{A2.1}$$

where:

C = variable linked to the metabolism of organic material (S_{sbs}, S_{sbp}, S_{pa}, X_a or X_e)
$r_{C,n}$ = rate of change of variable C in reactor "n"
$r_{rC,n}$ = rate of change of variable C in reactor "n", due to biological reactions
$r_{hC,n}$ = rate of change of variable C in reactor "n", due to hydraulic effects

The value of the rate of change due to hydraulic effects ($r_{hC,n}$) can be calculated from a simple mass balance over reactor "n". Using Figure A2.1 one has:

$$V_{r,n} \cdot r_{hC,n} = Q_{n-1} \cdot C_{n-1} + f_{i,n} \cdot Q_i \cdot C_i + f_{r,n} \cdot Q_r \cdot C_r - Q_n \cdot C_n \tag{A2.2}$$

where:

$V_{r,n}$ = volume of reactor "n"
$f_{i,n}$ = fraction of the influent flow to reactor "n"
$f_{r,n}$ = fraction of the return mixed liquor flow to reactor "n"
$C_{i,\ n,\ n-1,\ r}$ = respectively concentration of component C in the influent, reactor "n", entering reactor "n" and in the return sludge
$Q_{i,\ n,\ n-1,\ r}$ = respectively influent flow, flow from- and to reactor "n" and return sludge flow

In Eq. (A2.2) other internal recycles can be included as required, by adding the appropriate term. Having established the differential equations for the parameters involved in the general model, calibration will consist of generating theoretical profiles as a function of place (reactor) and time.

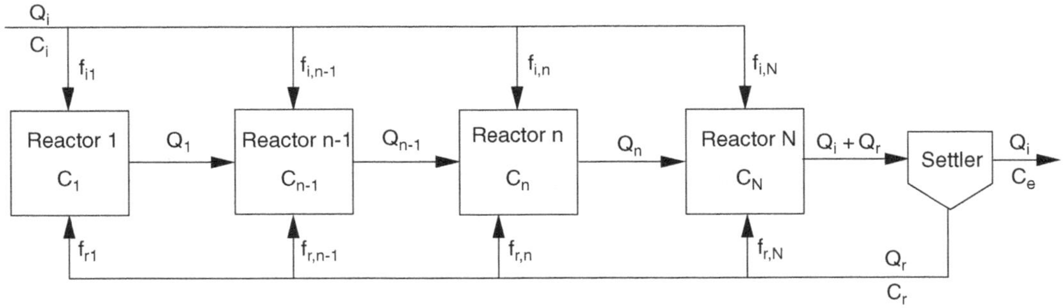

Figure A2.1 Schematic representation of the incoming and outgoing fluxes in an activated sludge system composed of several reactors in series

Different sets of stoichiometric- and kinetic coefficients can be used. The generated theoretical profiles will be compared to experimental values. If no full-scale activated sludge process is available, the values may be obtained by operating a lab- or pilot scale unit. In general, the validity of the calibration will increase when the parameter values are varied over a larger range. In the following sections two simple techniques for calibration of the general model will be discussed: (I) calibration with cyclic loading and (II) calibration with batch loading.

A2.1 CALIBRATION WITH CYCLIC LOADING

In this calibration method a bench-scale activated sludge system is operated until steady state conditions are achieved. The reactor is then subjected to a series of repeated influent loadings. In the demonstration example below a system consisting of a 12 litre batch reactor, a final settler and a sludge return pump was fed at a constant rate with 3 litres of sewage for a period of 1¼ hours. Thereafter aeration was continued for another 4¾ hours after which a new influent feed period was started. This was repeated 6 times. During the whole period the OUR was regularly measured. Allyl-thio-urea (ATU) was added to suppress nitrification. The reactor was operated at 26°C and at a sludge age R_s of 20 days. In Table A2.1 the main experimental data of the test are shown.

The data in Table A2.1 clearly indicate that the COD of the liquid phase of the mixed liquor (S_m), the effluent COD (S_{te}) and the volatile solids concentration (X_v) are essentially constant during the experiment. The average values of these parameters may therefore be used for the provisional determination of the non-biodegradable influent COD fractions f_{ns} and f_{np}. If it is assumed that the effluent COD is composed of non-biodegradable dissolved material, then for the example of Table A2.1 one has:

$$f_{ns} = S_{te}/S_{ti} = 66/677 = 0.10$$

Furthermore, when it is assumed that the concentration of non degraded particulate biodegradable material in the reactor is low, the volatile sludge concentration will be only composed of inert, active and endogenous sludge.

Table A2.1 Experimental results of 6 experiments with cyclic flow and load variations

Test	S_{ti} mg COD·l^{-1}	S_m (mg COD·l^{-1})		S_{te} mg COD·l^{-1}	X_v (mg VSS·l^{-1})	
		Start	End		Start	End
1	712	61	87		2610	2260
2	618	107	67	80	2250	2250
3	658	72	73		2410	2180
4	672	61	61	54	2340	2380
5	718	53	61		2260	2350
6	685	97	82	65	2510	2550
Avg	677	76	72	66	2396	2323

Notes: Index "m" refers to the liquid phase of the mixed liquor.
Start = just before start of feed period; end = directly after feed period.

The non-biodegradable particulate influent COD fraction (f_{np}) can be calculated with the aid of Eq. (3.35) of the simplified model:

$$X_v = [(1 - f_{ns} - f_{np}) \cdot (1 + f \cdot b_h \cdot R_s) \cdot C_r + f_{np} \cdot R_s/f_{cv}] \cdot Q_i \cdot S_{ti}/V_r$$

In this example Q_i is 12 litres when the 6 hour test period is extrapolated to 24 hours (as $4 \cdot 3 = 12$ litres). Using the data of Table A2.1 [$X_v = (2396 + 2323)/2 = 2360$ mg VSS·l^{-1} and $C_r = Y_h \cdot R_s/(1 + b_h \cdot R_s) = 0.45 \cdot 20/(1 + 0.30 \cdot 20) = 1.27$], f_{np} can be solved as:

$$2360 = [(1 - 0.10 - f_{np}) \cdot (1 + 0.2 \cdot 0.3 \cdot 20) \cdot 1.27) + f_{np} \cdot 20/1.5] \cdot 12 \cdot 677/12 \quad \text{or} \quad f_{np} = 0.09$$

With the provisional estimates of f_{ns} and f_{np} at hand, the OUR profile can now be used to estimate the kinetic constants for the utilisation of organic material under non-ideal conditions. Figure A2.2 shows the behaviour of the OUR profile under cyclic flow and load conditions:

- In the period before feeding, the OUR decreases slowly and tends to a more or less constant value, presumably corresponding to the endogenous respiration rate (level I in Figure A2.2);
- As soon as feeding begins, there is a rapid increase of the OUR, tending later to a constant (maximum) value at a higher level (level III in Figure A2.2);
- Almost immediately after the end of the feeding period, there is a precipitous drop in the OUR (to level II in Figure A2.2) and then a gradual decrease until the lowest level (level I) is reached again just before the next feeding period.

The OUR profile allows easy determination of the value of two parameters: the fraction of biodegradable COD that is easily degraded (f_{sb}) and the specific utilisation rate of slowly biodegradable (adsorbed) organic material (K_{mp}).

If no feed is supplied, the concentration of biodegradable organic material will eventually be depleted and oxygen consumption will then only be due to endogenous respiration. To estimate the endogenous respiration rate, again assume provisionally that the stored material concentration is low and use

Eq. (3.41). In the case of this example:

$$
\begin{aligned}
O_{en} &= f_{cv} \cdot r_o = f_{cv} \cdot (1 - f) \cdot b_h \cdot X_a \\
&= (1 - f_{ns} - f_{np}) \cdot (1 - f) \cdot f_{cv} \cdot b_h \cdot C_r \cdot Q_i \cdot S_{ti}/V_r \\
&= (1 - 0.10 - 0.09) \cdot (1 - 0.2) \cdot 1.5 \cdot 0.30 \cdot 1.27 \cdot 12 \cdot 677/12 \\
&= 255\ \mathrm{mg\ O_2 \cdot l^{-1} \cdot d^{-1}} \quad \text{or} \quad 10.6\ \mathrm{mg\ O_2 \cdot l^{-1} \cdot h^{-1}}
\end{aligned}
$$

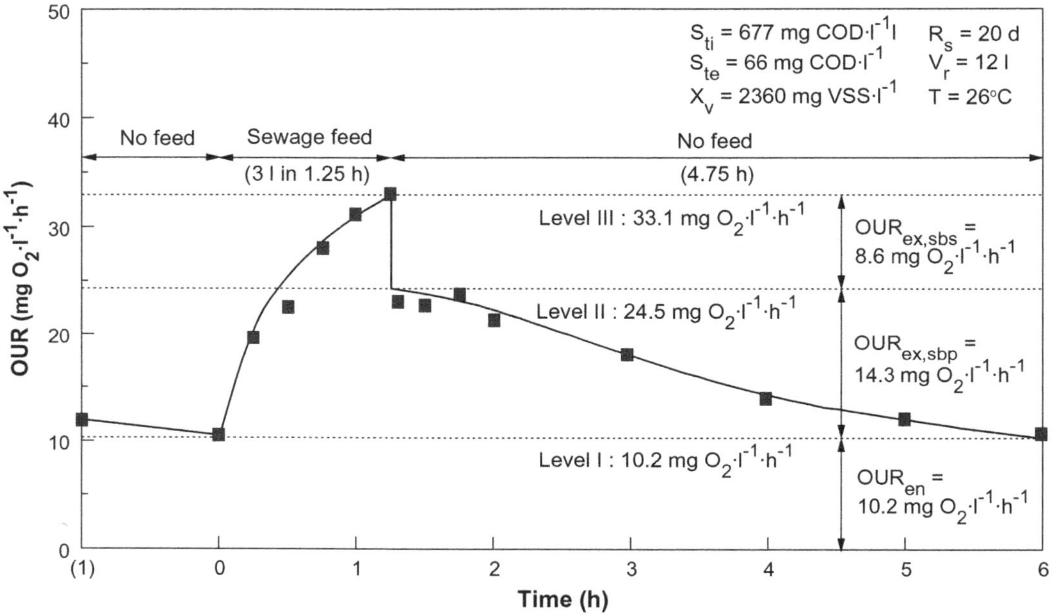

Figure A2.2 Typical OUR profile during the experiments with cyclic flow and load conditions: measured values and theoretical profile

In Figure A2.2 it can be observed that the OUR just before the start of the feeding period tends toward a value of $O_c = 10.2$ mg $O_2 \cdot l^{-1} \cdot h^{-1}$, which is quite close to the value calculated above for the endogenous respiration rate of 10.6 mg $O_2 \cdot l^{-1} \cdot h^{-1}$. Therefore it can be concluded that the utilisation rate of organic material at the end of the feed interruption was indeed very low. This means that our earlier assumption was correct that the biodegradable organic material S_{bp} stored during the feed period has been almost completely hydrolysed and utilised in the period after feed interruption.

The sudden drop of the OUR value at the end of the feed period can be attributed to the rapid depletion of easily biodegradable material. As this material is utilised at a high rate, at any moment during the feed period its concentration will be low, as is corroborated by the fact that the COD of the liquid phase of the mixed liquor does not increase during the feed period. Hence, during the feed period the utilisation rate of the easily biodegradable material is practically equal to the feeding rate of this material and the corresponding oxygen consumption rate can be expressed as:

$$O_{ex,sbs} = (1 - f_{cv} \cdot Y) \cdot r_{us} \quad \text{and} \tag{A2.3}$$

$$r_{us} = Q_i \cdot S_{bsi}/V_r \tag{A2.4}$$

Note that the value of the influent flow Q_i in Eq. (A2.4) should be the *actual* flow during the feed period and not the *average* flow! The actual flow during the feed period of 75 minutes is $60/75 \cdot 3$ l/h or 2.4 l/h. Combining Eqs. (A2.3) and (A2.4) one has:

$$O_{ex,sbs} = (1 - f_{cv} \cdot Y_h) \cdot Q_i \cdot S_{bsi}/V_r \quad \text{or} \tag{A2.5}$$

$$S_{bsi} = V_r \cdot O_{ex,sbs}/((1 - f_{cv} \cdot Y_h) \cdot Q_i) \tag{A2.6}$$

As very little easily biodegradable material S_{bs} will be present in the reactor at any given moment, it is quickly depleted after interruption of the feed and the OUR will decrease by an amount corresponding to $O_{ex,sbs}$. In Figure A2.2:

$$O_{ex,sbs} = \text{OUR (level III)} - \text{OUR (level II)} \tag{A2.7}$$

In the case of the example: $O_{ex,sbs} = 33.1 - 24.5 = 8.6$ mg $O_2 \cdot l^{-1} \cdot h^{-1}$. The easily biodegradable influent COD concentration can now be calculated as:

$$S_{bsi} = 12 \cdot 8.6/((1 - 1.5 \cdot 0.45) \cdot 2.4) = 132 \text{ mg COD} \cdot l^{-1}$$

Hence, on the basis of the OUR profile in Figure A2.2, the easily biodegradable COD fraction (of the total biodegradable COD) is estimated as:

$$f_{sb} = S_{bsi}/S_{bi} = 138/((1 - 0.10 - 0.09) \cdot 677) = 0.25$$

Using this figure, a preliminary estimate of the constant K_{mp} can be obtained from the value of the maximum OUR (level III) at the end of the feed period. The maximum OUR of 33 mg $O_2 \cdot l^{-1} \cdot h^{-1}$ as measured in the example is composed of:

– Endogenous respiration: $O_{en} = 10.2 \text{ mg } O_2 \cdot l^{-1} \cdot h^{-1}$;
– Utilisation of easily biodegr. material: $O_{ex,sbs} = 8.6 \text{ mg } O_2 \cdot l^{-1} \cdot h^{-1}$;
– Utilisation of stored/adsorbed material: $O_{ex,sbp} = O_c - O_{en} - O_{ex,sbs}$
$= 33.1 - 10.2 - 8.6 = 14.3 \text{ mg } O_2 \cdot l^{-1} \cdot h^{-1}$

Combining Eqs. (3.70) and (A2.3) one has:

$$O_{ex,sbp} = (1 - f_{cv} \cdot Y_h) \cdot r_{hi}$$

$$= (1 - f_{cv} \cdot Y_h) \cdot K_{mp} \cdot X_a \cdot S_{pa}/(S_{pa} + K_{sp} \cdot X_a) \tag{A2.8}$$

where:

S_{pa} = adsorbed slowly biodegradable organic material
K_{sp} = half reaction rate (Monod) constant for growth on S_{pa}

If it is now assumed that at the end of the feed period $S_{pa} \gg K_{sp} \cdot X_a$ (i.e. the active sludge has almost become saturated with adsorbed slowly biodegradable organic material during the feed period), one has $S_{pa}/(S_{pa} + K_{sp} \cdot X_a) \rightarrow 1$ and thus (for $X_a = C_r \cdot Q_i \cdot S_{bi}/V_r = 1.27 \cdot 12 \cdot 0.81 \cdot 677/12 = 696$ mg VSS $\cdot$ l^{-1}):

$$O_{ex,sbp} = (1 - f_{cv} \cdot Y_h) \cdot K_{mp} \cdot X_a \text{ or } 24 \cdot 14.3 = (1 - 1.5 \cdot 0.45) \cdot K_{mp} \cdot 696$$

Hence the K_{mp} value is estimated as $24 \cdot 14.3/(0.33 \cdot 696) = 1.5$ mg COD $\cdot$ mg^{-1} $X_a \cdot d^{-1}$ (at T = 26°C). When the temperature dependency according to Dold (1980) is assumed (Table A2.2), the value of K_{mp} is corrected to 0.85 at T = 20°C. Once the values of f_{ns}, f_{np}, f_{sb} and K_{mp} have been estimated, a theoretical OUR profile can be generated, by attributing values to the other kinetic parameters. This is repeated until there is a good correlation between the theoretical and the measured OUR profiles during the whole period of feeding and feed interruption. Table A2.2 shows the parameter values for which in Figure A2.2 the closest correlation between the measured OUR values and the theoretical OUR profile was obtained.

Table A2.2 Values of the kinetic constants calculated from the OUR profile of Figure A2.2. Determined at T = 26°C; temperature dependency applied according to Dold et al. (1980)

Symbol	Description	Value at 20°C	Temp. dep.
K_{ms}	Specific utilisation rate of easily biodegradable organic material	20 mg COD $\cdot$ mg^{-1} $X_a \cdot d^{-1}$	$1.2^{(T-20)}$
K_{mp}	Specific utilisation rate of slowly bio-degradable (adsorbed) organic material	0.85 mg COD $\cdot$ mg^{-1} $X_a \cdot d^{-1}$	$1.1^{(T-20)}$
K_{ss}	Half rate (Monod) constant (S_{bs})	5.0 mg COD $\cdot$ l^{-1}	–
K_{sp}	Half rate (Monod) constant (S_{pa})	0.04 mg COD $\cdot$ mg^{-1} X_a	$1.1^{(T-20)}$
K_a	Adsorption rate constant	0.25 litre $\cdot$ mg^{-1} $X_a \cdot d^{-1}$	$1.1^{(T-20)}$
K_{ap}	Adsorption saturation constant	1.5 mg COD $\cdot$ mg^{-1} X_a	–

A2.2 CALIBRATION WITH BATCH LOADING

The second method of calibration is the batch loading technique, which is very simple: sewage is added to the pilot SBR reactor almost instantaneously in a single batch. In between batch feeding there is no feeding at all. Nitrification is suppressed with ATU. The experimental procedure can be summarised as:

- Aeration and agitation of the mixed liquor in the activated sludge reactor is interrupted for a short period (15 minutes), prior to addition of a batch of sewage, allowing the sludge to settle;
- After settling, a volume of supernatant, equal to the batch volume to be added, is siphoned off and the concentrated sludge is aerated. This takes another 15 minutes;
- The batch of sewage, previously aerated, is added to the concentrated sludge;
- This procedure is repeated with a certain frequency (for example once per 12 hours).

Table A2.3 shows experimental values of the influent COD (S_{ti}), the COD in the liquid phase (S_m) and the volatile sludge concentration (X_v), directly before and directly after addition of a batch of sewage. The 6 litre batches of sewage were added every 12 hours (aeration time 11.5 hrs) to a mixed liquor reactor operating at a sludge age of 20 days. Temperature was maintained at 26°C. As in the cyclic loading experiments, the COD and VSS concentrations during the batch loading experiments remain almost constant.

Table A2.3 Experimental results of a series of tests with batch loading of an activated sludge reactor (refer also Figure A2.3). T = 26°C and R_s = 20 days

Test	S_{ti} mg COD · l^{-1}	S_m (mg COD · l^{-1})		X_v (mg VSS · l^{-1})	
		Before	After	Before	After
1		90	67		
2	736	97	120	2490	2460
3		61	83		
4	752	73	67	2530	2490
Average	744	80	84	2510	2475

Notes: –before = 15 minutes before addition of the sewage batch
–after = 1 minute after addition

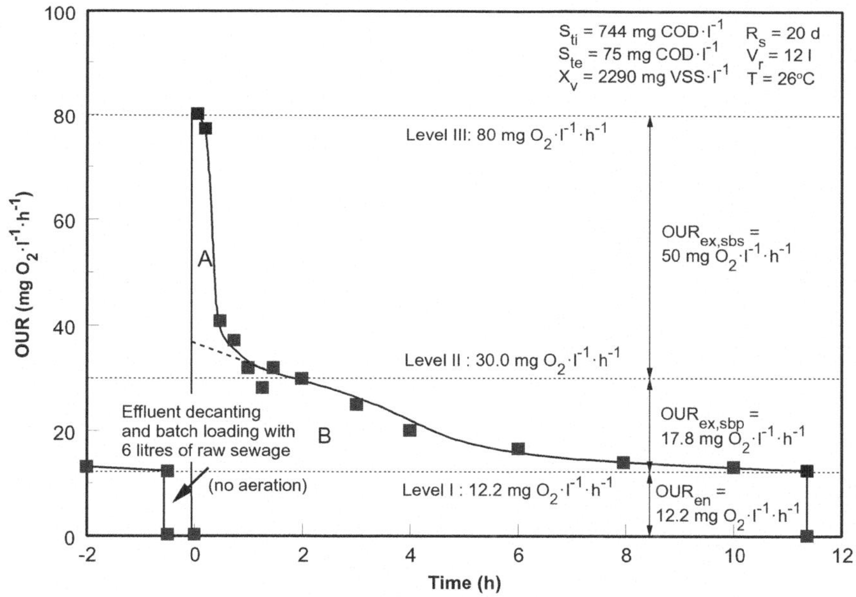

Figure A2.3 OUR profile during the experiments with batch loadings (Table A2.3)

Figure A2.3 shows the OUR profile as a function of time during the batch loading experiments. Using the experimental data, the non-biodegradable soluble fraction can be estimated as described in Section A2.1: f_{ns} = 0.10 and f_{np}=0.08. In this case the determination of f_{sb} is different from that described in Section A2.1, as the substrate feeding rate can now no longer be calculated (the feed was added instantaneously). The alternative approach is to calculate the oxygen demand that can be attributed to oxidation of easily biodegradable substrate (area A in Figure A2.3) and to oxidation of slowly biodegradable COD (area B in Figure A2.3). MO_t can be estimated as 6145 mg $O_2 \cdot d^{-1}$. The demand for exogenous oxidation is estimated by subtracting from MO_t the oxygen consumption for endogenous respiration: $MO_{en} = 2 \cdot 11.5 \cdot 12 \cdot 12.2 = 3367$ mg $O_2 \cdot d^{-1}$. $MO_{ex} = 6145 - 3367 = 2777$ mg $O_2 \cdot d^{-1}$ and this oxygen demand is by definition related to the oxidation of biodegradable COD. The surface area of A is estimated as 27.4 mg $O_2 \cdot l^{-1}$, so the oxygen demand for the respiration of S_{bs} is equal to $MO_{ex,sbs} = 2 \cdot 12 \cdot 27.4 = 658$ mg $O_2 \cdot d^{-1}$. Finally f_{sb} is given as $MO_{ex,sbs}/MO_{ex} = 658/2777 = 0.24$.

As in the case of the cyclic flow and -load experiments, it is possible to use OUR profiles to calibrate the general model by trial and error, generating theoretical profiles for different values of the constants until the closest correlation between the theoretical and experimental values is obtained. The theoretical profile of Figure A2.3 was obtained using the values of the kinetic constants listed in Table A2.2: i.e. the values that were derived from the cyclic loading experiments. It is important to consider that the same set of kinetic constants is capable of accurately predicting the behaviour of the activated sludge process with cyclic loading (Table A2.1 and Figure A2.2) and with batch loading (Table A2.3 and Figure A2.3), even though the actual process conditions in these experiments were very different.

Having established that the general model is able to predict the actual behaviour of the activated sludge process under extreme conditions of flow and load, it is to be expected that it can also be successfully applied to full-scale installations, where changes in flow- and loading patterns are much more gradual.

Appendix 3

The non-ideal activated sludge system

The main advantage of the non-ideal model of the activated sludge system as presented in Section 3.4 is that it can be used to describe the activated sludge system under a wide variety of process conditions, reactor configurations and influent characteristics.

At the same time it has an important disadvantage: the model has such a complexity that optimisation (often only possible by means of simulation) requires the use of specialist software. However, it is possible to estimate the deviation from ideality for a simplified situation, assuming a constant flow and load and a completely mixed reactor.

When the kinetic constants for the utilisation of biodegradable material are known, it is possible to estimate the residual concentration of the biodegradable material (both dissolved and particulate) and of the stored material. If the utilisation of the influent organic material is incomplete, a certain concentration S_b is not metabolised and this material will then leave the system either in dissolved form in the effluent and/or the excess sludge flow (S_{bs}), or in particulate form in the excess sludge, either enmeshed (S_{bp}) or adsorbed (S_{pa}). The daily mass of biodegradable non-metabolised material leaving the reactor can be expressed as:

$$MS_b = Q_i \cdot S_b = Q_i \cdot S_{bs} + q \cdot (S_{bp} + S_{pa}) \tag{A3.1}$$

where:

MS_b = mass of discharged biodegradable material
S_b = non-metabolised biodegradable COD concentration
S_{bs} = concentration of easily biodegradable material in the effluent
S_{bp} = concentration of slowly biodegradable material in the excess sludge
S_{pa} = concentration of stored material in the excess sludge.

Dividing by Q_i and applying Eqs. (3.15 and 3.17) one has:

$$S_b = S_{bs} + (R_h/R_s) \cdot (S_{bp} + S_{pa}) \tag{A3.2}$$

The utilisation rate of the easily biodegradable material is very high and its metabolism is therefore virtually complete. Hence, deviations of the ideal behaviour of the activated sludge process are mainly due to incomplete utilisation of slowly biodegradable material. When perfect solid-liquid separation is assumed in the final settler, this material can only leave the activated sludge system with the discharged excess sludge. The flow rate of non-metabolised organic material in the waste sludge is a factor (R_s/R_h) greater than in the influent flow, where it originates.

In general in an activated sludge system (but not in an aerated lagoon) the value of $R_s/R_h \gg 1$ and therefore only a small non-metabolised S_{bp} concentration is required to accumulate into a significant concentration in the sludge. The concentration of non-metabolised material can be estimated as follows:

(1) Calculate the active sludge concentration for the effectively metabolised COD concentration, i.e. $S_{bi} - S_b$:

$$X_a = Y \cdot R_s/(1 + b_h \cdot R_s) \cdot (S_{bi} - S_b)/R_h \tag{A3.3}$$

(2) Use the kinetic expressions of Table 3.10 to develop expressions for the concentration of easily and slowly biodegradable material and for the stored material concentration:

$$r_{sbs} = (S_{bs} - S_{bsi})/R_h = -r_{us} + r_{hi}$$

$$= -K_{ms} \cdot X_a \cdot S_{bs}/(S_{bs} + K_{ss}) + K_{mp} \cdot X_a \cdot S_{pa}/(S_{pa} + K_{sp} \cdot X_a) \tag{A3.4}$$

$$r_{sbp} = S_{bp}/R_s - S_{bpa}/R_h = -r_a$$

$$= -K_a \cdot X_a \cdot S_{bp} \cdot (K_{ap} - S_{ba}/X_a) \tag{A3.5}$$

$$r_{spa} = S_{pa}/R_s = (r_a - r_{hi})$$

$$= K_a \cdot X_a \cdot S_{bp} \cdot (K_{ap} - S_{ba}/X_a) - K_{mp} \cdot X_a \cdot S_{pa}/(S_{pa} + K_{sp} \cdot X_a) \tag{A3.6}$$

Rearranging the three equations above:

$$S_{bs} = A - K_{ss} - R_h \cdot K_{ms} - ((A - K_{ss} - R_h \cdot K_{ms})^2 + 4 \cdot K_{ss} \cdot A) \tag{A3.7}$$

$$S_{bp} = (R_s/R_h) \cdot S_{bpa}/[1 + R_s \cdot K_a \cdot S_{pa} \cdot S_a \cdot (K_{ap} - S_{pa}/X_a)] \tag{A3.8}$$

$$S_{pa} = [-b + (b^2 - 4 \cdot a \cdot c)^{0.5}]/(2 \cdot a) \tag{A3.9}$$

where:

$$A = S_{bsa} + R_h \cdot K_{ms} \cdot S_{pa} \cdot X_a/(K_{sp} + S_{pa)} \tag{A3.10a}$$

$$a = 1 + R_s \cdot K_a \cdot S_{bp} \tag{A3.10b}$$

$$b = K_{sp} \cdot X_a + R_s \cdot K_{mp} \cdot X_a - R_s \cdot K_a \cdot S_{bp} \cdot X_a \cdot K_{ap} - K_{sp} \tag{A3.10c}$$

$$c = R_s \cdot K_a \cdot S_{bp} \cdot X_a^2 \cdot K_{ap} \cdot K_{sp} \tag{A3.10d}$$

Equations (A3.7 to A3.9) do not allow immediate determination of the concentration of non-metabolised biodegradable material, because this would require the value of X_a to be known, while the value of X_a itself is dependent on the efficiency of the utilisation of the biodegradable material. However, the solution can be found by using an iterative procedure that can be summarised as follows:

(a) Initially assume ideal behaviour (i.e. $S_b = S_{bs} = S_{bp} = S_{pa} = 0$ and calculate X_a from Eq. (A3.3);
(b) Using the value of X_a from step (a), calculate the resulting values of S_{bs}, S_{bp} and S_{pa} in Eqs. (A3.7 to A3.9) and use the sum of these (S_b) to calculate a new value for X_a;
(c) With the new value of X_a, repeat the calculations of step (b) until the differences in the values for S_{bs}, S_{bp} and S_{pa} are smaller than a specified minimum (for example 0.1%). Usually only a few iterations are necessary.

In order to carry out the iterative calculation procedure outlined above, the values of the kinetic constants must be known. Table A3.1 shows the temperature dependent expressions and the numerical values for several temperatures.

Table A3.1 Values of the kinetic constants at different temperatures (Dold et al., 1980)

Symbol	Temp. dep.	Unit of measure	8°C	14°C	20°C	26°C
K_{ms}	$20 \cdot 1.2^{(T-20)}$	$mg\ COD \cdot mg^{-1}\ X_a \cdot d^{-1}$	2.24	6.70	20	59.7
K_{ss}	5.0	$mg\ COD \cdot l^{-1}$	5.0	5.0	5.0	5.0
K_a	$0.25 \cdot 1.1^{(T-20)}$	$litre \cdot mg^{-1}\ X_a \cdot d^{-1}$	0.080	0.141	0.25	0.443
K_{mp}	$3.0 \cdot 1.1^{(T-20)}$	$mg\ COD \cdot mg^{-1}\ X_a \cdot d^{-1}$	0.956	1.693	3.0	5.315
K_{sp}	$0.04 \cdot 1.1^{(T-20)}$	$mg\ COD \cdot mg^{-1}\ X_a$	0.013	0.023	0.04	0.071
b_h	$0.24 \cdot 1.04^{(T-20)}$	d^{-1}	0.150	0.190	0.24	0.304

Once the value of the non-metabolised biodegradable COD concentration has been determined, its influence on other process variables can be calculated. The formation of endogenous residue (and consequently that of volatile sludge) and the oxygen uptake rate are directly affected:

$$X_e = f \cdot b_h \cdot R_s \cdot Y/[(1 + b_h \cdot R_s) \cdot R_h] \cdot (S_{bi} - S_b) \quad (A3.11)$$

Hence:

$$X_v = (1 + f \cdot b_h \cdot R_s) \cdot Y \cdot R_s/(1 + b_h \cdot R_s) \cdot (S_{ba} - S_b)/R_h + R_s \cdot R_h \cdot [f_{np} + (S_b - S_{bs})]/f_{cv} \quad (A3.12)$$

and

$$OUR_c = [1 - f_{cv} \cdot Y + f_{cv} \cdot (1 - f) \cdot b_h \cdot Y \cdot R_s/(1 + b_h \cdot R_s)] \cdot (S_{bi} - S_b)/R_h \quad (A3.13)$$

To illustrate the calculation method, in Figure A3.1 typical removal efficiencies of the biodegradable organic material and the concentrations of easily and slowly biodegradable material in the effluent and excess sludge are given. In Figure A3.2 the fractions of the influent organic material in the effluent (mS_{te}), the excess sludge (mS_{xv}) and oxidised (mS_o) are shown plotted as functions of the sludge age for temperatures of 14, 20 and 26°C.

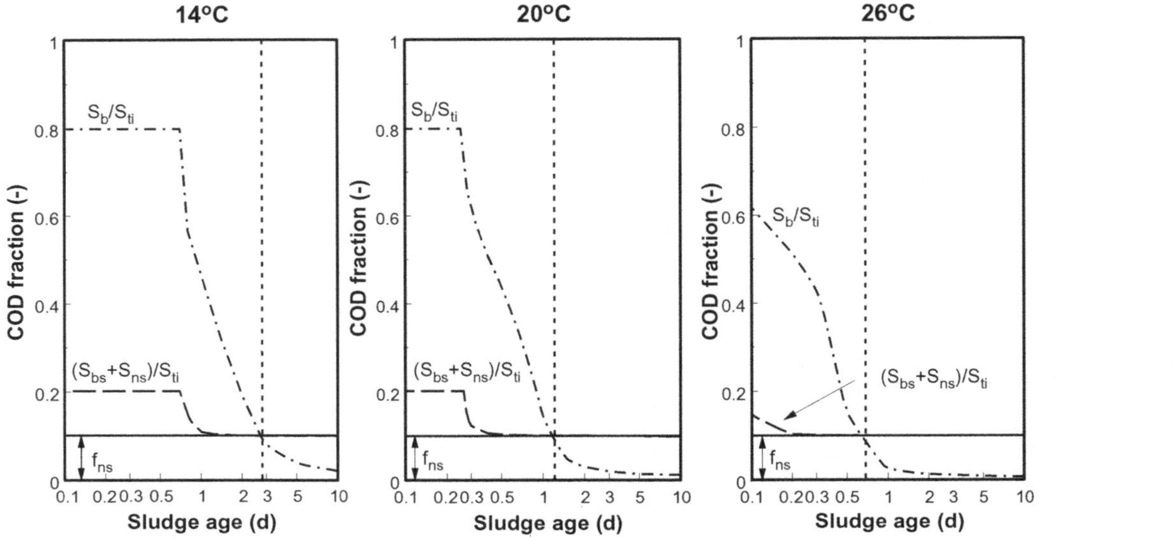

Figure A3.1 Predicted concentration of biodegradable material for the ideal and non-ideal models as a function of the sludge age for different temperatures

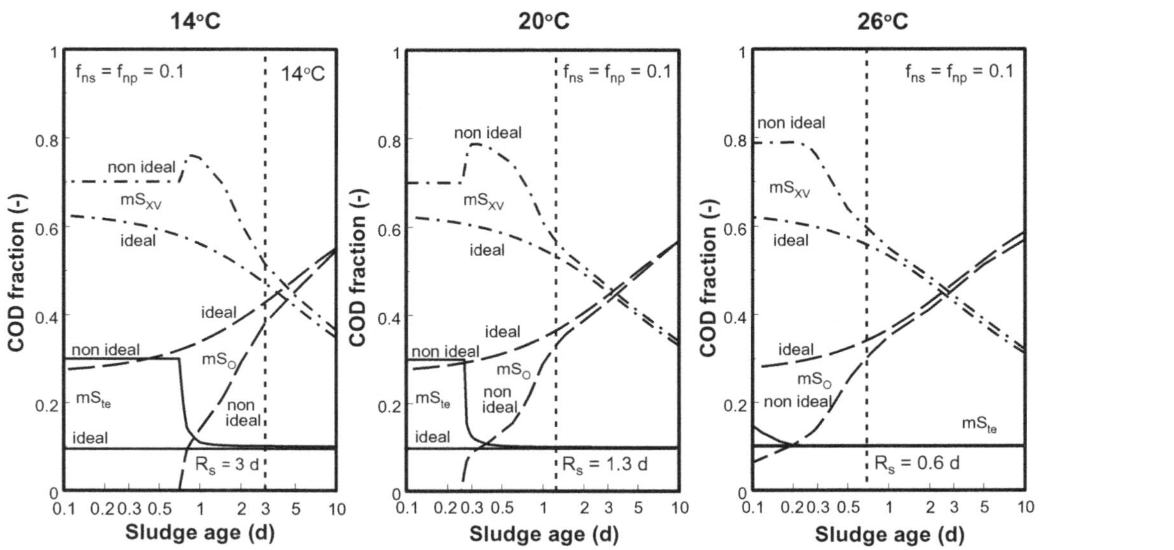

Figure A3.2 Division of influent COD over the fractions in the effluent, oxidised and in the excess sludge predicted according to the ideal and non-ideal model as a function of the sludge age for different temperatures

From the simulations that generated Figure A3.1 and Figure A3.2 the following can be concluded:

(1) At decreasing sludge age, long before the efficiency of the easily biodegradable material removal starts to decline, the removal efficiency of slowly biodegradable material has already collapsed;
(2) The minimum sludge age for a particular required efficiency is heavily influenced by the temperature. In Figure A3.1 the sludge age at which the removal efficiency of organic material removal starts to deviate from its minimum value has been indicated. It can be noted that this sludge age is *very short* at any temperature:

 - 3 days at 14°C;
 - 1 days at 20°C;
 - 0.6 days at 26°C.

Figure A3.3 One of the autors checking the biomass quality in one of the UASB reactors of the Onça STP, Belo Horizonte-Brazil. Courtesy of B. Heffernan

Figure A3.4 Artist impression of a Biobed® EGSB reactor showing the influent distribution system, the granular sludge bed and the GLS separators - courtesy of Biothane Systems International

Appendix 4

Determination of nitrification kinetics

In Chapter 5 the equation of Downing et al. (1964) was used to model the growth of nitrifiers with a Monod relationship:

$$(dX_n/dt) = Y_n \cdot r_n = [\mu_m \cdot (N_a/(N_a + K_n) - b_n] \cdot X_n \qquad (4.28)$$

where:

(dX_n/dt) = net growth rate of the nitrifying bacteria ($mg\ VSS \cdot l^{-1} \cdot d^{-1}$)
r_n = nitrification rate ($mg\ N \cdot l^{-1} \cdot d^{-1}$)
Y_n = yield coefficient of nitrifiers ($mg\ VSS \cdot mg^{-1}\ N$)
μ_m = maximum specific growth constant of nitrifiers (d^{-1})
N_a = ammonium concentration ($mg\ N \cdot l^{-1}$)
K_n = half saturation constant for ammonium ($mg\ N \cdot l^{-1}$)
b_n = autotrophic decay constant (d^{-1})
X_n = nitrifier concentration ($mg\ VSS \cdot l^{-1}$)

Stenstrøm and Poduska (1980) have shown that also the dissolved oxygen concentration has a significant effect on the nitrifier growth rate, which can be described with a Monod expression. As explained in Appendix 1, the effect is caused by oxygen limitation inside the sludge floc that will occur when the bulk dissolved oxygen concentration decreases below the critical concentration. In the general model this has not been included, as oxygen concentration is considered more as an operational than a design parameter: typically a (bulk) dissolved oxygen concentration around $2\ mg \cdot l^{-1}\ O_2$ is assumed, which is sufficient for nitrification. However, during an OUR test the effect of a low oxygen concentration cannot be ignored. Thus for nitrifiers (i.e. the ammonium oxidizers being responsible for the rate limiting step):

$$dX_n/dt = r_n \cdot Y_n = [\mu_m \cdot N_a/(N_a + K_n) \cdot DO/(DO + K_o) - b_n] \cdot X_n \qquad (A4.1)$$

where:

μ_m = maximum specific growth rate for nitrifiers
K_o = half saturation constant of dissolved oxygen for nitrification

Three different approaches have been developed to determine the value of the kinetic parameters of nitrification:

(1) A method in which the sludge age of an activated sludge system is gradually lowered while observing the resulting change in the ammonium concentration in the effluent. This method is not particularly accurate and is very laborious; it may take several weeks of even months to obtain reliable values for the constants;
(2) A method proposed by Van Haandel and Marais (1981), in which an activated sludge process operating under constant flow and load conditions was submitted to alternating anoxic- and aerobic periods. During the aerobic periods the variation of the nitrate concentration was determined and used to calculate the nitrification constants. This method is much more accurate, but it still requires significant effort to carry out;
(3) The third method uses respirometrics. Oxygen uptake rate (OUR) tests are extremely simple to carry out as demonstrated in Appendix 1. The experimentally determined OUR is taken as a measure for the nitrification rate and from it the nitrification constants can readily be calculated. When the OUR determination is repeated at different dissolved oxygen concentrations, the half saturation value for dissolved oxygen may also be determined.

Only the respirometric method will be discussed here. Stoichiometrically there is a consumption of two moles of oxygen per mol of nitrified ammonium or (equivalent) per mol of produced nitrate. Hence there is a proportionality of 4.57 mg O_2 per mg nitrate N produced, so that:

$$O_n = 4.57 \cdot r_n = [\mu_m \cdot N_a/(N_a + K_n) \cdot DO/(DO + K_o) - b_n] \cdot X_n/Y_n \quad (A4.2)$$

where O_n = oxygen uptake rate for nitrification

As a prerequisite for the determination of the nitrification kinetics, first a nitrifying sludge must be generated under steady state conditions, which allows the nitrifier concentration of the sludge to be determined. The concentration of nitrifiers is given by Marais and Ekama, (1976):

$$X_n = Y_n \cdot R_s \cdot N_c/[(1 + b_n \cdot R_s) \cdot R_h] \quad (A4.3)$$

where N_c = nitrified ammonium concentration (nitrification capacity) of the activated sludge system under consideration

The procedure starts when a nitrifying sludge batch is aerated (without feed) until the OUR has declined to a more or less constant value, equivalent to the endogenous respiration rate. In general this takes about 0.5 to 1 hour. At this point a known quantity of ammonium is added, e.g. in the form of an ammonium chloride solution. As a result of the ammonium addition, the OUR will increase steeply. After some time the OUR will decrease again to approximately the same value as before the ammonium addition. Figure A4.1 in Example A4.1 represents a typical curve of the OUR after addition of a batch-load of ammonium. This curve can be used to determine the kinetic nitrification constants as will be explained in the subsequent sections.

(a) Mass balance check

The area between the OUR curve and the base endogenous respiration line represents the oxygen consumption per litre of reactor resulting from the addition of ammonium. When the surface area is determined, it can be compared with the observed nitrate concentration increase. If the oxygen consumption is approximately 4.57 times higher than the produced nitrate concentration (in mg NO_3-N $\cdot$ l^{-1}), then it can be concluded that the experimental data are reliable and may be used for calculation of the kinetic constants.

(b) Estimate of the maximum specific growth rate constant μ_m

From the observation that the OUR is almost constant in the period directly after the addition of ammonium, it can be concluded that during this phase no ammonium limitation exists and that nitrification is proceeding at the maximum possible rate (i.e. $N_a \gg K_n$). With the aid of Eq. (A4.3) the concentration of nitrifiers can be calculated. In the next step, the maximum specific growth rate μ_m can be determined from:

$$OUR_m = 4.57 \cdot r_n = 4.57 \cdot (\mu_m - b_n) \cdot X_n/Y_n \qquad (A4.4)$$

where OUR_m = maximum OUR due to nitrification during the respiration test

In general, the numerical value of the maximum specific growth rate μ_m is much larger than the value of the decay constant b_n, so that a good approximation is $(\mu_m - b_n) \approx \mu_m$.

(c) Half saturation constant of ammonium

To determine the value of the half saturation constant of ammonium K_n, once again the OUR profile is evaluated. When the OUR starts to decrease, ammonium is becoming a rate limiting factor. In accordance with Monod kinetics, when the OUR is equal to half the maximum OUR value, the ammonium concentration at that particular moment is equal to the value of K_n. Hence K_n is determined as follows:

(1) When it is observed that the OUR for nitrification is reduced to half of the maximum OUR value, a sample is withdrawn and the ammonium concentration is determined. ATU (Allyl-Thio-Urea) is added to suppress further nitrification in the withdrawn sample;
(2) Alternatively the oxygen consumption for nitrification is determined in the time period required to decrease the OUR from $0.5 \cdot OUR_m$ to OUR_{en}: the amount of oxygen consumed is equivalent to the grey area indicated in Figure A4.1. The ammonium concentration present in the batch at the moment that the OUR was equal to $0.5 \cdot OUR_m$, can be estimated by dividing the oxygen consumption by 4.57.

(d) Half saturation constant of dissolved oxygen

To evaluate the effect of dissolved oxygen limitation on the nitrification rate, several batch tests are carried out, each at a different dissolved oxygen concentration. For each of these tests, the μ_m value is calculated by the procedure outlined above. The values of μ_m are plotted as a function of the average dissolved oxygen concentration in the different batch tests (see Figure A4.2). From the observation that the μ_m value increases at higher dissolved oxygen concentrations, it is concluded that the oxygen concentration is a limiting factor in the nitrification rate. A true maximum value for the specific growth rate of nitrifiers (μ_{max}) can now be calculated using Eq. (A4.5):

$$OUR_m = 4.57 \cdot r_{Nm} = 4.57 \cdot \mu_m \cdot [DO_{av}/(DO_{av} + K_o)] \cdot X_n/Y_n \qquad (A4.5)$$

Alternatively written, $\mu_{max} = \mu_m \cdot [DO_{av}/(DO_{av} + K_o)]$. As both the μ_m value and the DO_{av} concentration are different for each of the batch tests, the best estimate for μ_{max} is the mean of all values calculated from Eq. (A4.5) for a particular K_o value. For different values of K_o a plot is made of μ_m as a function of the dissolved oxygen concentration. The K_o value that results in the closest correlation between the experimental values and the theoretical curves is selected. This will be explained in detail in Example A4.1. Note that in the model presented in this book, K_o has not been included as a parameter, as in the optimised design procedure the applied oxygen concentration is assumed to be higher than the critical concentration, i.e. oxygen limitation will not occur.

(e) Determination of the decay constant

The decay rate of nitrifiers is small and for that reason the test to determine the value of b_n is time consuming as well. A nitrifying sludge batch is aerated without feed for a period sufficient to remove any remaining substrate from the sludge batch and to ensure the virtual elimination of heterotrophic biomass (as heterotrophs have a much higher decay rate than nitrifiers). The decay of the heterotrophic biomass may take between 2 to 6 weeks, depending on the temperature of the sludge batch. When the decay of the heterotrophs is virtually complete, the decay rate of nitrifiers can be evaluated by adding ammonium chloride at regular intervals (several days) and observe the decrease in OUR_m in time.

An alternative is to feed the sludge batch with ammonium only, and to periodically discharge excess sludge. After a period equal to two or three sludge ages, the elimination of heterotrophic biomass can be considered almost complete. Theoretically there should be an exponential decay of OUR_m with time, as the concentration of nitrifiers decreases exponentially. Hence there will be a linear relationship between $\ln(OUR_m)$ and time and the slope of the straight line is equal to the decay constant b_n. As the decay rate of nitrifiers is very slow, the test will take at least a month. In practice it is not very important to know the exact value of the decay constant, as invariably $\mu_m \gg b_n$ and for that reason the influence of b_n on nitrification kinetics is very limited. For practical purposes one can adopt a value already reported by other authors, for example: $b_n = 0.04 \cdot 1.03^{(T-20)}$ (Marais and Ekama, 1976).

EXAMPLE A4.1

To illustrate the methodology presented in the previous section, an experiment will be analysed where three pilot-scale activated sludge systems were operated with influent received on a large industrial wastewater treatment plant (CETREL, Brazil). The dissolved oxygen concentration was controlled by means of a respirometer. The three reactors were operated at different dissolved oxygen concentrations and were identified as R_1 (DO between 3.5 and 4.5 mg $O_2 \cdot l^{-1}$, R_2 (DO between 1.5 and 2.5 mg $O_2 \cdot l^{-1}$) and R_3 (DO between 0.5 and 1.5 mg $O_2 \cdot l^{-1}$). In Table A4.1 the influent and effluent characteristics of the reactors are listed, along with the operational conditions during the experiment.

Table A4.1 Experimental results and operational conditions from the three pilot-scale activated sludge systems operated at different dissolved oxygen concentrations (V_r = 25 litres; Q_i = 17 $l \cdot d^{-1}$; R_s = 20 d and T = 26°C)

Parameter	UoM	Influent	Effluent R_1	Effluent R_2	Effluent R_3
Avg. DO	mg $O_2 \cdot l^{-1}$		4	2	1
COD	mg $\cdot l^{-1}$	1482	296	288	282
TKN/NH_4^+	mg N $\cdot l^{-1}$	104/58	11.6/0.9	14.1/ 0.8	13.3/1.2
NO_3^-	mg N $\cdot l^{-1}$	<1	51.7	45.0	40.3
X_t/X_v	mg SS $\cdot l^{-1}$		4841/3098	4530/2900	4663/2864

It can be observed that the effluent ammonium concentration in all three systems was low. However, the effluent nitrate concentration decreased at lower average dissolved oxygen concentration. This can be attributed to simultaneous denitrification: at a dissolved oxygen bulk concentration below the critical concentration, anoxic conditions are established inside (part of the) sludge floc and denitrification will occur.

Using the data in Table A4.1 and Eq. (A4.3), the concentration of nitrifying bacteria X_n can be calculated. The excess sludge produced from the reactors was used to determine the kinetics of the nitrification process. Figure A4.1 presents a respirogram that may be considered as typical for the experiment and that was generated under the following conditions: (I) a high dissolved oxygen concentration (i.e. sludge from system R_1) and (II) addition of sufficient NH_4Cl to obtain an ammonium concentration of 5 mg $N \cdot l^{-1}$ in the reactor. The interpretation of the respirogram will now be discussed.

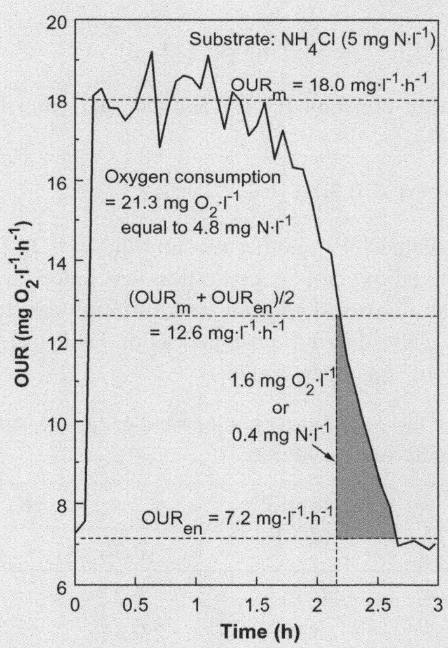

Figure A4.1 Typical respirogram after feeding of a nitrifying sludge batch with ammonium chloride

Solution

(1) Verification of the mass balance

In order to confirm that the data is reliable, it is first verified whether the mass balance closes. The area between the maximum OUR curve and the endogenous respiration level is related to the concentration of ammonium added. For the test displayed in Figure A4.1, this area is equal to 21.3 mg $O_2 \cdot l^{-1}$, which corresponds to the oxidation of $21.3/4.56 = 4.8$ mg $N \cdot l^{-1}$. As this value is very close to the added ammonium concentration of 5 mg $N \cdot l^{-1}$, the experimental data is considered to be reliable.

(2) Determination of the maximum nitrifier specific growth rate μ_m and K_o
The nitrifier concentration in the batch is calculated with the data from Table A4.1. The value of N_c is equal to the nitrate concentration produced and a nitrifier yield of 0.1 mg VSS · mg^{-1} N was assumed. The nitrifier concentration should be reduced by 50% of the calculated value because of the dilution of the sludge sample with influent (1:1) prior to the test:

$$X_n = Y_n \cdot R_s \cdot N_c/((1 + b_n \cdot R_s) \cdot R_h)$$

$$= 0.1 \cdot 20 \cdot 51.7/((1 + 0.046 \cdot 20) \cdot 25/17)/2 = 19.0\,mg\ X_n \cdot l^{-1}$$

Furthermore the maximum OUR for nitrification can be determined from Figure A4.1 as:

$$OUR_n = (OUR_m - OUR_{en}) = (18 - 7.2) = 10.8\,mg\ O_2 \cdot l^{-1} \cdot h^{-1}$$

Hence the maximum nitrification rate is:

$$r_n = 10.8/4.57 = 2.36\,mg\ N \cdot l^{-1}.h^{-1} = 56.7\,mg\ N \cdot l^{-1} \cdot d^{-1}$$

Now the value of the nitrification constant μ_m (or actually $\mu_m - b_n$) for the sludge in reactor R_3 is determined as:

$$\mu_m = Y_n \cdot r_n/X_n = 0.1 \cdot 56.7/19.0 = 0.30\ d^{-1}$$

When the values of μ_m are calculated for the other reactors ($\mu_m = 0.27$ for R_2 and $\mu_m = 0.19$ for R_1), it becomes apparent that the dissolved oxygen concentration was indeed a limiting factor during the tests. To eliminate the influence of the dissolved oxygen concentration and to determine the true maximum specific growth rate constant μ_m, the data are arranged as in Table A4.2. In Table A4.2 an additional batch test with $DO_{av} = 0.5$ mg · l^{-1} has been added.

Table A4.2 Calculated values of the maximum specific nitrifier growth rate (μ_m) without ammonium or dissolved oxygen limitation, for different K_o values

Reactor	DO_{av} (mg $O_2 \cdot l^{-1}$)	μ_m (exp.)[1] (d^{-1})	K_o values (mg $O_2 \cdot l^{-1}$)			
			0.25	0.5	1.0	2.0
–	0.5	0.15	0.22	0.30	0.45	0.75
R_3	1	0.19	0.24	0.28	0.38	0.57
R_2	2	0.27	0.30	0.34	0.40	0.54
R_1	4	0.30	0.32	0.34	0.37	0.45
Average μ_m value for each K_o value:			0.28	0.32	0.40	0.59
Standard deviation of μ_m:			0.05	0.03	0.03	0.12

Note: (1) From step (2)

The values of μ_m as calculated above for each batch are listed in column 3. The average dissolved oxygen concentration applied during the batch test is indicated as well (column 2). As the oxygen concentration affects the calculated value of μ_m, it has to be corrected for the factor $DO_{av}/(K_o + DO_{av})$. Therefore in the next step, for each of the four K_o values, the μ_m value is calculated for each value of DO_{av} using

Eq. (A4.5). The results are listed in columns 4 to 7. As an example, the calculation is presented for a K_o value of 1.0 mg $O_2 \cdot l^{-1}$ and an average dissolved oxygen concentration of 4 mg $O_2 \cdot l^{-1}$:

$$OUR_m = 4.57 \cdot [DO_{av}/(DO_{av} + K_o)] \cdot \mu_m \cdot X_n/Y_n \quad \text{(A4.5)}$$
$$10.8 \cdot 24 = 4.57 \cdot (4/(4+1)) \cdot \mu_m \cdot 19 \cdot 1/0.1 \quad \text{or} \quad \mu_m = 0.3/0.8 = 0.37$$

For each K_o value an average value for μ_m is calculated. The results already indicate that the true value of K_o will be located somewhere between 0.5 and 1.0 mg $O_2 \cdot l^{-1}$, as the standard deviation of the average μ_m value is small in both cases and the individual μ_{max} values do not deviate significantly from the average.

Using the average values of μ_m as calculated in Table A4.2, for each K_o value a theoretical curve is generated of μ_m as function of the dissolved oxygen concentration, as shown in Figure A4.2. Finally, the theoretical curves are compared with the experimental data (in column 3 in Table A4.2) and the K_o value resulting in the closest correlation between experiment and theory is selected as the true half saturation value for dissolved oxygen. In Figure A4.2 it can be observed that the best fit is obtained for $K_o = 1.0$ mg $\cdot l^{-1}$ and $\mu_m = 0.4$ d^{-1}. This implies that the growth rate of the nitrifiers is reduced by 50% of its maximum value, if the aeration tank is operated at a dissolved oxygen concentration of 1 mg $\cdot l^{-1}$.

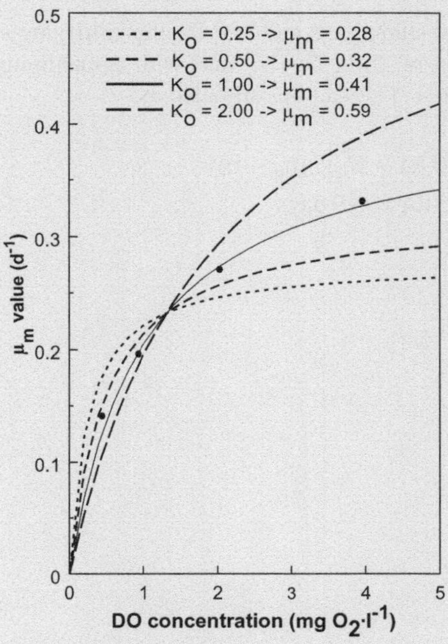

Figure A4.2 Theoretical curves and experimental values of μ_m of the nitrifiers as a function of the dissolved oxygen concentration

(3) Determination of the half saturation constant for ammonium K_n

To estimate the value of the half saturation constant for ammonium K_n, the surface of the grey area in Figure A4.1 is measured at the moment that $\mu = ½ \cdot \mu_m$ or $OUR_n = OUR_m/2 = (18+7)/2 = 12.5$ mg

$O_2 \cdot l^{-1} \cdot h^{-1}$. The surface of the grey area is equal to 1.6 mg $O_2 \cdot l^{-1}$, which is equivalent to 1.6/4.57 = 0.4 mg $N \cdot l^{-1}$. It is concluded that the growth rate μ is reduced to 50% of its maximum value when the ammonium concentration has decreases to 0.4 mg $N \cdot l^{-1}$, so $K_n = 0.4$ mg $N \cdot l^{-1}$.

The experiment described above was also carried out with sludge batches taken from systems operating at lower dissolved oxygen concentrations. This did not influence the values of μ_m and K_n. This indicates that the decrease of the maximum growth rate μ_m as observed at low dissolved oxygen concentrations is only due to the reduced availability of dissolved oxygen and not to generation of a sludge with a lower substrate affinity or metabolic capacity. The value of the nitrifier decay constant has not been determined, instead the value suggested by Marais and Ekama (1976) was accepted: $b_n = 0.04 \cdot 1.03^{(T-20)} = 0.046\ d^{-1}$ at 26°C.

For the industrial effluent treated by CETREL and for a temperature of 26°C, the nitrification kinetics can thus be summarised as:

- $b_n = 0.046\ d^{-1}$
- $\mu_m = 0.40\ d^{-1}$
- $K_n = 0.4$ mg $N \cdot l^{-1}$
- $K_o = 1.0$ mg $O_2 \cdot l^{-1}$

Hence the minimum operational sludge age required for this nitrifying system, operating at an average dissolved oxygen concentration of 2 mg $O_2 \cdot l^{-1}$ and with a maximum allowed effluent ammonium concentration of less than 1 mg $N \cdot l^{-1}$, can be estimated as:

$$R_s = 1/[(N_a/(N_a + K_n) \cdot DO/(DO + K_o) \cdot \mu_m - b_n]$$
$$= 1/[1/1.4 \cdot 2/3 \cdot 0.4 - 0.046] = 6.9 \text{ days}$$

Appendix 5

Determination of denitrification kinetics

In Chapter 5 a model for denitrification was presented. It was demonstrated that the rate of denitrification depends on the type of substrate available and the type of configuration used (pre-D or post-D). The following parameters were introduced:

K_1 – denitrification rate constant for easily biodegradable material in a pre-D reactor;
K_2 – denitrification rate constant for slowly biodegradable material in a pre-D reactor;
K_3 – denitrification rate constant for slowly biodegradable material in a post-D reactor.

The effect of endogenous respiration is included in the values of K_2 and K_3. As long as the pre-D anoxic mass fraction is large enough to allow for complete metabolisation of the easily biodegradable COD fraction f_{sb} (i.e. $f_{x1} > f_{min}$), the value of K_2 and K_3 together with the available active sludge mass and the value of f_{sb} define the extent of denitrification possible. Although in municipal wastewater the values of K_2 and K_3 are not expected to differ significantly from the default values, this might very well be the case for wastewater of industrial origin.

In this Appendix an experimental method is presented that can be used to determine the denitrification capacity in an activated sludge system and to establish the values of the denitrification constants.

EXAMPLE A5.1

A Bardenpho system with a total reactor volume of 235 l (Figure A5.1) was operated on raw sewage (300 $l \cdot d^{-1}$) from the city of Campina Grande, Brazil. The pre-D zone is divided into two parts to induce plug flow behaviour. The main system parameters are given in Table A5.1, while the experimental results are given in Table A5.2. The BDP system is operated with a sludge recycle flow that is equal to the influent flow rate ($s = 1$), while the mixed liquor recycle to the second anoxic reactor is equal to $a = 3$. Calculate the denitrification constants K_2 and K_3 from the data presented in this example.

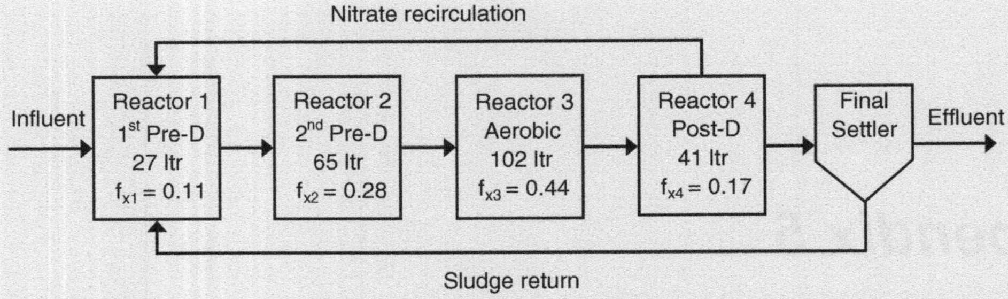

Figure A5.1 Schematic layout of the pilot system used for the determination of the denitrification kinetics (Example A5.1)

Table A5.1 System parameters of the BDP pilot plant used for determination of the denitrification kinetics (Example A5.1)

Parameter	Value	UoM	Parameter	Value	UoM
Q_i	300	$l \cdot d^{-1}$	a	3	(-)
V_t	235	litre	s	1	(-)
R_s	20	d	X_t	2900	mg TSS.l^{-1}
R_h	0.78	d^{-1}	OUR	50	mg $O_2 \cdot l^{-1} \cdot h^{-1}$
T	25	°C	f_{sb}	0.06	(-)
f_v	0.7	(−)	b_{h25}	0.29	d^{-1}

Table A5.2 Experimental results of the BDP pilot plant used for determination of the denitrification kinetics (Example A5.1)

Parameter	UoM	Infl.	R1	R2	R3	R4	Effl.
COD	$mg \cdot l^{-1}$	456	22	28	28	30	23
TKN	mg $N \cdot l^{-1}$	57	14	13	1.9	3.3	3.5
NH_4-N	mg $N \cdot l^{-1}$	42	10	11	0.8	1.6	1.6
NO_3-N	mg $N \cdot l^{-1}$	0	4.4	0.5	9.6	5.3	4.5
NO_2-N	mg $N \cdot l^{-1}$	0	0.5	0.2	0.4	0.2	0.4

Solution

(1) Concentration of nitrate and nitrite

The experimental data demonstrate that both nitrification and denitrification in the modified BDP system are efficient, as indicated by the total nitrogen effluent concentration $N_{ti} = N_{ke} + N_{Nn}$ of $3.5 + 4.5 + 0.4 = 8.4$ mg $N \cdot l^{-1}$, which is only 15% of the influent nitrogen concentration of 57 mg $N \cdot l^{-1}$.

Denitrification was not complete in any of the anoxic reactors, as nitrate was present in all reactors and in the effluent. Therefore, it can be concluded that nitrate was never limiting and that the measured nitrate removal reflects the available denitrification capacity. The presence of nitrate in all reactors also indicates that it is impossible to have full denitrification under the applied operational conditions. The nitrite concentration is so low (always less than 0.5 mg $N \cdot l^{-1}$) that it for all practical purposes can be ignored. In the calculation below the nitrite concentration has been added to the nitrate concentration.

(2) Check the nitrogen mass balance

To create the nitrogen mass balance, the daily masses of nitrogen leaving with the effluent, discharged with the excess sludge and denitrified to nitrogen gas are compared with the daily applied nitrogen load.

(a) Influent nitrogen load:

$$MN_{ti} = N_{ti} \cdot Q_i$$
$$= 57 \cdot 300 = 16{,}100\,\text{mg N} \cdot d^{-1}$$

(b) Effluent nitrogen load:

$$MN_{te} = N_{te} \cdot (Q_i - q)$$
$$= (3.5 + 4.9) \cdot (300 - 11.8) = 2420\,\text{mg N} \cdot d^{-1}$$

where $q = V_r/R_s = 235/20 = 11.8$ litres $\cdot\, d^{-1}$

(c) Nitrogen discharged with the excess sludge:

$$MN_l = f_n \cdot MX_v/R_s + q \cdot N_{te} = f_n \cdot f_v \cdot X_t \cdot V_r/R_s$$
$$= 0.1 \cdot 0.7 \cdot 2900 \cdot 235/20 + 11.8 \cdot 8.4 = 2485\,\text{mg N} \cdot d^{-1}$$

(d) Denitrified nitrogen:

To determine the mass of nitrogen removed by denitrification one calculates the sum of the nitrate removal observed in the anoxic reactors and in the final settler, where denitrification takes place as well, as indicated by the reduction of nitrate in this unit. The mass of nitrogen removed in the first reactor is given by the difference between the concentration in the in- and outgoing flows, multiplied with the flow rate passing through the reactor. The flow averaged nitrate concentration of the incoming streams is equal to:

$$N_{n0} = (s \cdot N_{ne} + a \cdot N_{n3} + N_{ni})/(a + s + 1)$$
$$= (1 \cdot 4.9 + 3 \cdot 10 + 0)/5 = 7.0\,\text{mg N} \cdot l^{-1}$$

The nitrate concentration leaving reactor 1 is equal to $N_{n1} = 4.9$ mg $N \cdot l^{-1}$. The mass of nitrate removed in reactor 1 is:

$$MN_{d1} = (a + s + 1) \cdot Q_i \cdot (N_{n0} - N_{n1})$$
$$= 300 \cdot 5 \cdot (7.0 - 4.9) = 3120\,\text{mg N} \cdot d^{-1}$$

In the same way the daily mass of denitrified nitrogen is calculated in the other anoxic reactors:

$$MN_{d2} = (a + s + 1) \cdot Q_i \cdot (N_{n1} - N_{n2})$$
$$= 5 \cdot 300 \cdot (4.9 - 0.7) = 6300\,\text{mg N} \cdot d^{-1}$$

$$MN_{d4} = (s + 1) \cdot Q_i \cdot (N_{n3} - N_{n4})$$
$$= 2 \cdot 300 \cdot (10 - 5.5) = 2700\,\text{mg N} \cdot d^{-1}$$
$$MN_{dd} = (s + 1) \cdot Q_i \cdot (N_{n4} - N_{ne})$$
$$= 2 \cdot 300 \cdot (5.5 - 4.9) = 360\,\text{mg N} \cdot d^{-1}$$
$$MN_d = MN_{d1} + MN_{d2} + MN_{d4} + MN_{dd} = 12{,}480\,\text{mg N} \cdot d^{-1}$$

(e) Nitrogen mass balance check:

$$B_n = (MN_{te} + MN_l + MN_d)/MN_{ti}$$
$$= (2420 + 2485 + 12{,}480)/17,\ 100 = 17{,}385/17{,}100 = 1.01$$

As the measured value of B_n is very close to the theoretical value of unity, it can be concluded that the nitrogen mass balance closes and that the experimental data is probable reliable. If the value of B_n would have been less than 0.9 or higher than 1.1, then the data should be rejected as the mass balance does not close. A possible reason could be that the system is not operated under rigorous steady state conditions.

(3) Check the mass balance of organic material

Once again, check whether the data are reliable, now by closing the COD mass balance (i.e. by calculating the recovery factor B_o):

$$MS_{ti} = Q_i \cdot S_{ti}$$
$$= 300 \cdot 456 = 136{,}800\,\text{mg COD} \cdot d^{-1}$$
$$MS_{te} = (Q_i - q) \cdot S_{te}$$
$$= (300 - 11.8) \cdot 23 = 6630\,\text{mg COD} \cdot d^{-1}$$
$$MS_{xv} = q \cdot (f_{cv} \cdot f_v \cdot X_t + S_{te})$$
$$= 11.8 \cdot (1.5 \cdot 0.7 \cdot 2900 + 23) = 36{,}050\,\text{mg COD} \cdot d^{-1}$$

When the oxidized COD mass is determined, it should be considered that part of the measured oxygen demand is due to nitrification, while on the other hand part of the organic material is oxidized with nitrate instead of oxygen.

$$MS_o = MO_c + MO_{eq}$$
$$MO_{eq} = 2.86 \cdot MN_d = 2.86 \cdot 12{,}480 = 35{,}700\,\text{mg } O_2 \cdot d^{-1}$$
$$MO_c = MO_t - MO_n$$
$$MO_t = 24 \cdot V_r \cdot O_t$$
$$= 24 \cdot 102 \cdot 50 = 122{,}400\,\text{mg } O_2 \cdot d^{-1}$$
$$MO_n = 4.57 \cdot MN_c = 4.57 \cdot (MN_{ki} - MN_{ke} - MN_l)$$
$$= 4.57 \cdot 300 \cdot (57 - 3.5) - 2485 = 62{,}000\,\text{mg } O_2 \cdot d^{-1}$$

$$MO_c = 122{,}400 - 62{,}000 = 60{,}400\,\text{mg } O_2 \cdot d^{-1}$$
$$MS_o = 60{,}400 + 35{,}700 = 96{,}100\,\text{mg } O_2 \cdot d^{-1}$$
$$B_o = (MS_{te} + MS_{xv} + MS_o)/MS_{ti}$$
$$= (6630 + 36{,}050 + 96{,}100)/136{,}800 = 1.01$$

Again the value of the recovery factor B_o is very close to the theoretical value of unity.

(4) Composition of the organic material in the influent

Having established that the analytical data is reliable, the soluble non biodegradable fraction can now be calculated from the COD values in the influent and effluent:

$$f_{ns} = S_{te}/S_{ti} = 23/456 = 0.05$$

The particulate non biodegradable fraction can be calculated from the volatile sludge concentration, which is the sum of the active sludge (X_a), the endogenous residue (X_e) and the inert sludge (X_i):

$$X_v = [(1 + f \cdot b_h \cdot R_S) \cdot (1 - f_{ns} - f_{np}) \cdot C_r + f_{np} \cdot R_s/f_{cv}] \cdot S_{ti}/R_h$$
$$= f_v \cdot X_t = 2900 \cdot 0.7 = 2030\,\text{mg VSS} \cdot l^{-1}$$

where $C_r = Y \cdot R_s/(1 + b_h \cdot R_s) = 0.45 \cdot 20/(1 + 0.29 \cdot 20) = 1.32$

$$2030 = [(1 + 0.2 \cdot 0.29 \cdot 20) \cdot (1 - 0.05 - f_{np}) \cdot 1.32 + f_{np} \cdot 20/1.5] \cdot 456/0.78$$

This equation can be solved for $f_{np} = 0.0725$. The easily biodegradable fraction was determined as $f_{sb} = 0.06$ using respirometric methods (not discussed here, for an example refer to Appendices A1 and A2), which is very low compared to the values generally found in domestic sewage ($0.2 < f_{sb} < 0.25$). The low concentration of easily biodegradable COD limits the extent of denitrification that can be obtained, as this COD fraction is used in the pre-D reactor(s) for the reduction of nitrogen nitrate.

(5) Calculation of the denitrification constants in pre-D and post-D reactors:

From the previous calculations it is possible to determine the denitrification constants of the activated sludge system. As nitrate was present in the effluent of all anoxic reactors, the availability of nitrate (N_{av}) was never limiting. Therefore the available denitrification capacity is equal to the mass of nitrogen removed. Considering that in the post-D reactor (fourth reactor) there is no easily biodegradable material left, one has:

$$S_{bi} = (1 - f_{ns} - f_{np}) \cdot S_{ti} = (1 - 0.05 - 0.0725) \cdot 456 = 400\,\text{mg COD} \cdot l^{-1}$$
$$D_{c4} = MN_{d4}/Q_i = k_3 \cdot C_r \cdot f_{x4} \cdot S_{bi} \quad \text{or}$$

$$k_3 = D_{c4}/(C_r \cdot f_{x4} \cdot S_{bi}) = (2700/300)/(1.32 \cdot 0.17 \cdot 400) = 0.098\,\text{mg N} \cdot \text{mg}^{-1} X_a \cdot d^{-1}$$

Similarly in the second reactor (where there is also no easily biodegradable COD left):

$$D_{c2} = MN_{d2}/Q_i = k_2 \cdot C_r \cdot f_{x4} \cdot S_{bi} \quad \text{or}$$

$$k_2 = D_{c2}/(C_r \cdot f_{x2} \cdot S_{bi}) = (6300/300)/(1.32 \cdot 0.28 \cdot 400) = 0.144\,\text{mg N} \cdot \text{mg}^{-1} X_a \cdot d^{-1}$$

The denitrification capacity of the first anoxic reactor is composed of two parts: the oxidation of easily biodegradable organic material D_{c1s} and the oxidation of particulate organic material plus endogenous respiration D_{c1p}. The value of D_{c1s} is directly proportional to the concentration of the easily biodegradable COD in the influent:

$$D_{c1s} = (1 - f_{cv} \cdot Y)/2.86 \cdot f_{sb} \cdot S_{bi}$$

$$= 0.33/2.86 \cdot 0.06 \cdot 400 = 2.7\,\text{mg N} \cdot l^{-1}$$

$$D_{c1p} = k_2 \cdot C_r \cdot f_{x1} \cdot X_a$$

So with the value of k_2 already calculated for reactor 2:

$$D_{c1p} = k_2 \cdot C_r \cdot f_{x1} \cdot S_{bi}$$

$$= 0.144 \cdot 1.32 \cdot 0.11 \cdot 400 = 8.7\,\text{mg N} \cdot l^{-1}$$

$$D_{c1} = D_{c1s} + D_{c1p}$$

$$= 2.7 + 8.7 = 11.4\,\text{mg N} \cdot l^{-1}$$

It can be concluded that the calculated denitrification capacity in reactor 1 is comparable to the observed capacity:

$$D_{c1} = MN_{d1}/Q_i = 3120/300 = 10.4\,\text{mg N} \cdot l^{-1}$$

The values of the denitrification constants are almost identical to the ones found earlier in Campina Grande (Brazil) and South Africa by Van Haandel et al. (1999): $k_2 = 0.10 \cdot 1.08^{(T-20)}$ and $k_3 = 0.08 \cdot 1.04^{(T-20)}$. Considering the effect of the operational temperature of 25°C, the temperature corrected values are $k_2 = 0.146$ and $k_3 = 0.097$ mg $N \cdot mg^{-1}\ X_a \cdot d^{-1}$, almost identical to the values determined above.

The easily biodegradable fraction ($f_{sb} = 0.06$) is much smaller than the usual value of 0.20 to 0.25. This is one of the reasons that complete denitrification was not possible in this experiment. Another reason was the ratio between TKN and COD in the influent, which had a value of $57/456 = 0.125$. This is high for raw sewage and indicates that the availability of organic material for denitrification is limited. Such a high ratio between TKN and COD in a reasonably poor city such as Campina Grande, where the consumption of proteins (meat) is limited, might be explained by the extensive use of septic tanks. The raw sewage produced at households passes through septic tanks, where part of the biodegradable COD is removed, prior to discharge to the sewer.

Appendix 6

Extensions to the ideal model

The most accurate model may not always the most suitable model, as it should be fit to purpose. Although added complexity may make a model much more true to reality, it will also be increasingly cumbersome to use. Calibration of a complex model is very difficult, as is the interpretation of the results, while it is also difficult to avoid errors as the number of equations grows.

When the ideal steady state activated sludge model was developed, it was with the specific objective of being used as a tool for the design and optimisation of activated sludge systems, not as a scientific tool. Therefore a number of assumptions and simplifications have been implemented to reduce the calculation requirements at the expense of a small decrease in accuracy. Notwithstanding, the model is complex enough as it stands. Two of the model simplifications that have been applied are the following:

- N_{oe} and P_{oe} are modelled as a single fraction containing both soluble and particulate nitrogen or phosphorus. This slightly overestimates the amount of nitrogen/phosphorus discharged with the excess sludge;
- Nitrifiers are not included as a separate fraction in the biomass, which is perfectly justified for municipal wastewater with a relatively low TKN/COD ratio. However, this is not the case for some industrial wastewaters and for some of the the new nitrogen removal configurations that have been discussed in Chapter 6.

In this appendix the following (optional) extensions to the ideal model will be discussed:

- The effect of imperfect solid-liquid separation in the final settler on nitrogen- and phosphorus mass balances and the method to correct for this effect;
- The incorporation of the nitrifiers as a mass fraction in the volatile sludge, which is required at a high N_{ki}/S_{ti} ratio.

A6.1 IMPERFECT SOLID-LIQUID SEPARATION IN FINAL SETTLER

In the ideal model presented in this book it is assumed that solid-liquid separation in the final settler is perfect. However, in practice some suspended solids will always be present in the effluent, the concentration of which depend on the type of wastewater treated, the applied sludge age and the design and operational conditions of the final settler. In general X_{te} is lower than 20 mg TSS · l^{-1} and often a value between 10–15 mg TSS · l^{-1} is feasible. However, as both organic nitrogen and phosphorus are present in the suspended solids, this will have an effect on the mass balance of these compounds and their effluent concentrations.

A6.1.1 Particulate organic nitrogen and phosphorus in the effluent

In the ideal model, the organic nitrogen present in the effluent is modelled as a single fraction N_{oe} that contains both particulate- and soluble organic nitrogen. The particulate nitrogen N_{ope} is present in the suspended solids leaving with the effluent. As such, this nitrogen fraction is actually part of N_l, but instead of ending up in the excess sludge, it is discharged with the effluent. Assuming the composition of the effluent suspended solids is equal to that of the mixed liquor biomass, the concentration of N_{ope} is equal to:

$$N_{ope} = f_n \cdot f_v \cdot X_{te} \qquad (A6.1)$$

As f_v ranges between 0.6 and 0.8 for most activated sludge processes, N_{ope} will have a value in the range of 0.6 to 1.8 mg N · l^{-1}. With regards to the soluble organic nitrogen concentration N_{ose}, this value depends also on the wastewater, but for municipal wastewater it is often less than 0.5 mg N · l^{-1}. In the ideal model the value of N_l is not adjusted for the loss of N_{ope} with the effluent, as $N_{ope} \ll N_l$. In effect, this results in "double-counting" of N_{ope}: once in the effluent and once in the excess sludge. The same also applies to the particulate organic phosphorus in the effluent. The repercussions, although they are often small, are best explained with a numerical example. In Table A6.1 the true (measured) values and model (predicted) values of the different nitrogen fractions are compared for three cases:

Table A6.1 Effect of model simplifications on the predicted effluent nitrate concentration when N_l is not corrected for N_{ope} (= 1 mg N · l^{-1}). Predicted (model) and actual values of the different nitrogen fractions for three cases in a BDP configuration

	Influx	Outgoing nitrogen flux					Nitrogen removal		
	N_{ti}	N_l	N_{ose}	N_{ope}	N_{ae}	N_{ne}	N_d	N_c	D_c
Complete denitrification ($N_c - D_c < 0$)									
True	50	14	0.5	1.0	1.0	0	33.5	33.5	35.0
Model	50	15	1.5		1.0	0	32.5	32.5	35.0
Incomplete denitrification ($N_c - D_c > 0$)									
True	50	14	0.5	1.0	1.0	6	27.5	33.5	27.5
Model	50	15	1.5		1.0	5	27.5	32.5	27.5
Apparent complete denitrification ($N_c - D_c > 0$ and $N_c - D_c < N_{ope}$)									
True	50	14	0.5	1.0	1.0	1	32.5	33.5	32.5
Model	50	15	1.5		1.0	0	32.5	32.5	32.5

- Complete denitrification ($D_c > N_c$);
- Incomplete denitrification ($D_c < N_c$);
- Apparent complete denitrification ($N_c - D_c < N_{ope}$).

The following effluent nitrogen concentrations are given for the actual (true) case: $N_{ae} = 1$ mg N · l^{-1} and $N_{oe} = N_{ope} + N_{ose} = 1.0 + 0.5 = 1.5$ mg N · l^{-1}. When the calculated model results for each case are compared with the actual values, the following observations can be made:

(a) Predicted model results for complete denitrification

- Assume that $D_c = 35$ mg N · l^{-1} for both the true and model cases. The denitrification capacity is not influenced by the assumptions for model ideality, as it is determined by the influent COD load and -composition and by the applied sludge age;
- The fraction N_{oe} is not differentiated into N_{ope} and N_{ose} and the value used for N_{oe} is equal to 1.5 mg N · l^{-1}. Furthermore, the value of N_l is not reduced to compensate for the loss of 1 mg N · l^{-1} of organic particulate nitrogen in the effluent and N_l is thus overestimated by 1 mg N · l^{-1} or 7%;
- This is partly compensated if anaerobic digestion is applied: between 20 - 40% of the "additional" nitrogen present in the excess sludge will be released during digestion and returned to the activated sludge system;
- $N_c = N_{ti} - N_l - N_{oe} - N_a = 50 - 15 - 1.5 - 1 = 32.5$ mg N · l^{-1}. Therefore, the nitrification capacity is slightly underestimated, with 1 mg N · l^{-1} or 3%, resulting in a similar reduction in the calculated value O_n, the aeration demand for nitrification;
- The concentration of denitrified nitrogen $N_d = N_c - D_c$ is reduced by 1 mg N · l^{-1} as well because of the decrease in N_c, resulting in a similar reduction in O_{eq}, the equivalent oxygen recovery;
- The model effluent nitrogen composition is equal to the actual composition. As D_c is larger than N_c for both true and model values, all nitrate will be denitrified and $N_{ne} = 0$ mg N · l^{-1} for both cases.

(b) Predicted model results for incomplete denitrification

- Assume that $D_c = 27.5$ mg N · l^{-1} for both the true and model cases, so $D_c < N_c$;
- The value of N_l is overestimated by 1 mg N · l^{-1} or 7%;
- The nitrification capacity is slightly underestimated by 1 mg N · l^{-1} or 3%;
- The model effluent nitrogen composition is *not* equal to the true composition. In this case N_c is larger than D_c. As the predicted value of N_c is 1 mg N · l^{-1} lower than the true value of N_c, while the D_c value is equal in both cases, then according to the equation $N_{ne} = N_c - D_c$, the predicted value of N_{ne} will also be 1 mg N · l^{-1} smaller than the true value;

(c) Predicted model results for incomplete denitrification (but apparently complete!)

- $D_c = 32.5$ mg N · l^{-1} for both the true and model cases
- The value of N_l is overestimated by 1 mg N · l^{-1} or 7%;
- The nitrification capacity is slightly underestimated by 1 mg N · l^{-1} or 3%;
- Denitrification is only *apparently* complete ($N_{ne} = 0$ mg N · l^{-1}). For $(N_c - D_c) < N_{ope}$ (in the example $32.5 < D_c < 33.5$ mg N · l^{-1}), the model erroneously calculates complete denitrification, while in reality the N_{ne} value will be between 0 and 1 mg N · l^{-1}.

It can be observed that the errors resulting from the model simplifications are small and moreover that the effects tend to cancel each other partly out. Furthermore, in practice the design errors resulting from inaccurate attribution of values to the kinetic parameters (e.g. μ_m, K_2 and K_3) and to the influent COD fractions (f_{ns}, f_{np} and f_{sb}) are many times larger. However, the inaccuracy resulting from considering N_{ope} and N_{ose} as a single fraction N_{oe}, while not correcting the value of N_l, can easily be corrected if desired. As a first step, the value of N_l is adapted to include only the nitrogen discharged with the excess sludge:

$$N_{lx} = N_l - N_{ope} \quad \text{(A6.2)}$$

where

N_{lx} = nitrogen influent concentration discharged with the excess sludge, corrected for N_{ope}
N_{ope} = $f_n \cdot f_v \cdot X_{te}$ (or a typical default value can be used, for instance 1.5 mg $N \cdot l^{-1}$)

$$\begin{aligned} N_c &= N_{ti} - N_{ae} - N_{oe} - N_{lx}, \text{which is equivalent to} \\ &= N_{ti} - N_{ae} - N_{ose} - N_l \end{aligned} \quad \text{(A6.3)}$$

In general it is recommended to design an activated sludge system for full nitrogen removal. The potential underestimate of N_{ne} resulting from the model simplification as demonstrated in this section will then not influence the compliance of the treatment plant to the imposed effluent nitrogen limits. Therefore the differentiation between N_{ose} and N_{ope} and the correction of N_l to N_{lx} has in general not been applied throughout this book, with the exception of Example 14.14.

A similar correction could be applied to the particulate phosphorus in the effluent. When phosphorus removal is required, typical effluent limits of $<1 - 2$ mg $P \cdot l^{-1}$ apply. The P_{ope} concentration may then be a significant part of the total effluent phosphorus concentration. Not correcting P_l for the value of P_{ope} might result in violation of the effluent limits. Therefore in the case of phosphorus removal (often combined with nitrogen removal), the values of N_l and P_l can be compensated for the loss of N_{ope} and P_{ope} with the effluent. The phosphorus concentration discharged with the excess sludge is equal to:

$$P_{ope} = f_p \cdot f_v \cdot X_{te} (\text{or a typical default value can be used, e.g. } 0.35\,\text{mg}\,P \cdot l^{-1}) \quad \text{(7.20)}$$

$$P_{lx} = P_l - P_{ope} \quad \text{(7.22)}$$

For a more detailed discussion about the effects of the particulate organic phosphorus concentration in the effluent refer to Section 7.1.3.4. In general, the distinction between P_{ose} and P_{ope} and the correction of P_l has not been applied throughout this book, with the exception of Example 14.14 and Example 14.15.

A6.1.2 Excess sludge production and composition

For consistency, the excess sludge production can also be corrected for the loss of suspended material with the effluent. Assuming a X_{te} value of 20 mg TSS $\cdot l^{-1}$, then from a simple mass balance calculation it can be determined that at a shorter sludge age (e.g. 8–12 days) the mass of suspended solids leaving with the effluent is approximately 10 percent of the mass discharged with the excess sludge. At higher sludge ages, this value will increase: for example to 12–13 percent for a sludge age between 20 and 30 days. However, as far as the design of the sludge treatment units is concerned, this deviation can be ignored as it is important to have sufficient sludge processing capacity available. Proper performance of sludge thickener and -digester is crucial for the overall treatment performance of the STP. Given the small flow rate of the (thickened) excess sludge compared to the size of the influent flow, a certain degree of overdesign will thus not be very expensive and provides some operational flexibility. Therefore in this book the excess sludge production has not been corrected for the loss of suspended solids with the effluent. However, if desired the true daily excess sludge production ME_{tx} can be calculated as:

$$ME_{tx} = ME_t - (Q_i - q) \cdot X_{te} \quad (A6.4)$$

When N_l and P_l are not corrected for the loss of N_{ope} and P_{ope} with the effluent, the calculated effect will be a small overestimate in the nitrogen and phosphorus loads returned to the main activated sludge system. Should larger accuracy be required, then for the calculation of all unit operations from the thickener to the sludge digester, the adapted values ME_{tx}, N_{lx} and P_{lx} can be used, as discussed in the previous section.

A6.2 NITRIFIER FRACTION IN THE VOLATILE SLUDGE MASS

In municipal sewage treatment plants the nitrifier fraction in the volatile biomass will be small: typically between 2 and 3%. The exact value of the nitrifier fraction depends on the N_{ti}/S_{ti} ratio and the applied sludge age.

For municipal sewage, excluding the nitrifiers from the model results in a very small overestimate of the nitrification capacity (i.e. with a value equal to N_{ln}, the nitrogen content of the nitrifiers discharged with the excess sludge) and thus also of the resulting oxygen demand for nitrification. On the other hand, the excess sludge production is slightly underestimated. Again, the impact of these inaccuracies on the resulting design will be many times smaller than for instance an erroneous estimate of f_{np} or K_2.

An important reason not to include the nitrifier biomass in the simplified model is that to do so would significantly complicate calculations. The quantity of nitrifier biomass formed is dependent on N_c, the nitrified ammonium concentration. As N_c itself is dependent on the mass of nitrogen discharged with the excess sludge, the values of N_c and N_{ln} can only be calculated iteratively.

On the other hand, for industrial wastewaters and/or under different operational conditions the inclusion of nitrifiers might be recommended or even necessary. Examples are the treatment of sludge reject water (discussed in Chapter 6) and the aerobic post-treatment (including nitrification) of UASB effluent (discussed in Example 14.6). Sludge digestion reject water is characterized by a high N_{ki} concentration (500–1500 mg $N \cdot l^{-1}$) and a very low COD/N ratio. For such a wastewater it makes sense to include the nitrifiers explicitly in the model equations, as they may become a significant fraction of the volatile sludge. In those cases where only nitritation and no denitritation is applied (such as in the nitritation reactors detailed in Chapter 6), the volatile sludge mass will predominantly consist of nitrifiers. In order to include the nitrifying biomass in the general model, as a first step the specific active nitrifier production of per unit mass of daily applied nitrifiable nitrogen is calculated:

$$C_{rn} = Y_n \cdot R_s/(1 + b_n \cdot R_s) \quad (A6.5)$$

where Y_n = overall nitrifier yield = 0.12 mg VSS $\cdot$ mg^{-1} N converted

For heterotrophic bacteria mX_a is equal to the product of C_r and the biodegradable COD fraction in the influent. However, in the case of nitrifiers it is not possible to divide the influent nitrogen a priori into a "nitrifiable"- and a "non-nitrifiable" fraction. Both the nitrification potential (N_p) and the nitrification capacity (N_c) are dependent on the applied sludge age: N_p through the value of N_l, while N_c is dependent on the value of N_{ae}. As N_l and N_c are thus interdependent, the calculation proceeds iteratively. When the nitrifiers are included as a sludge fraction in the model, the volatile sludge X_v will consist of the following fractions:

- X_a: active heterotrophs (X_{ah}) and active nitrifiers (X_{an});
- X_e: endogenous residue from heterotrophs (X_{eh}) and nitrifiers (X_{en});
- X_i: inert particulate organic material.

The nitrifier biomass can be calculated as:

$$MX_{an} = C_{rn} \cdot MN_c \quad \text{(A6.6)}$$
$$MX_{en} = f \cdot b_n \cdot R_s \cdot C_{rn} \cdot MN_c \quad \text{(A6.7)}$$
$$MX_{vn} = C_{rn} \cdot (1 + f \cdot b_n \cdot R_s) \cdot MN_c \quad \text{(A6.8)}$$

where $MN_c = Q_i \cdot (N_{ti} - N_l - N_{oe} - N_{ae})$

In Eqs. (A6.6 to A6.8) the only unknown variable is the value of N_l:

$$N_l = N_{lh} + N_{ln} \quad \text{(A6.9)}$$

N_{lh} can be calculated as usual with Eq. (3.59):

$$N_{lh} = f_n \cdot ME_{vh}/Q_i = f_n \cdot [(1 - f_{np} - f_{ns}) \cdot (1 + f \cdot b_h \cdot R_s) \cdot C_{rh} + f_{np} \cdot R_s/f_{cv}] \cdot S_{ti}/R_s$$

while

$$N_{ln} = f_n \cdot MX_{vn}/(R_s \cdot Q_i) = f_n \cdot C_{rn} \cdot (1 + f \cdot b_n \cdot R_s) \cdot N_c/R_s \quad \text{(A6.10)}$$

As discussed in Section A6.1.1, the concentration of nitrogen discharged with the excess sludge (N_l) is slightly overestimated as N_{ope} is included in both the value of N_l and that of N_{oe}. If this is not desired, the value of N_{lx} (Eq. A6.2) can be used instead of that of N_l (Eq. 3.59). As concerns the demand for oxygen, MO_{en} is adapted to include the endogenous demand for respiration by the nitrifiers.

$$\begin{aligned} MO_c &= MO_{ex} + MO_{en} \\ &= (1 - f_{cv} \cdot Y_h) \cdot (1 - f_{ns} - f_{np}) \cdot MS_{ti} + f_{cv} \cdot (1 - f) \cdot (b_h \cdot MX_{ah} + b_n \cdot MX_{an}) \end{aligned} \quad \text{(A6.11)}$$

In this book the contribution of nitrifiers to the total volatile sludge mass is generally ignored, unless otherwise specified.

EXAMPLE A6.1

Consider a wastewater with the following composition:

- $Q_i = 1200\ m^3 \cdot d^{-1}$
- $S_{ti} = 1000$ mg COD $\cdot l^{-1}$
- $T = 30°C$
- $f_{ns} = 0.3$
- $f_{np} = 0.1$
- $N_{ki} = 600$ mg N $\cdot l^{-1}$
- $N_{oe} = 30$ mg N $\cdot l^{-1}$
- $b_h = 0.36\ d^{-1}$
- $f_v = 0.85$
- $\mu_m = 0.6\ d^{-1}$
- $b_n = 0.06\ d^{-1}$
- $K_n = 3$ mg N $\cdot l^{-1}$
- $Y_n = 0.12$ mg VSS $\cdot mg^{-1}$ N

It is required to reduce the effluent ammonia concentration to ≤ 2 mg N $\cdot l^{-1}$. Determine the sludge mass that will develop, when the nitrifiers are included. What will the nitrifier fraction be?

Solution

The minimum sludge age required to comply to the specified value of $N_{ad} = 2$ mg N $\cdot l^{-1}$ can be calculated with Eq. (5.39) as:

$$\begin{aligned} R_{sm} &= 1/[\mu_m/(1 + K_n/N_{ad}) - b_n] \qquad (5.39) \\ &= 1/[0.6/(1 + 3/2) - 0.06] \\ &= 5.6\ \text{days} \end{aligned}$$

The heterotrophic biomass is calculated as usual with Eqs. (3.30 and 3.48)

$$C_{rh} = Y_h \cdot R_s/(1 + b_h \cdot R_s) = 0.45 \cdot 5.6/(1 + 0.36 \cdot 5.6) = 0.84 \qquad (3.30)$$

$$\begin{aligned} MX_{vh} &= [(1 - f_{ns} - f_{np}) \cdot (1 + f \cdot b_h \cdot R_s) \cdot C_r + f_{np} \cdot R_s/f_{cv}] \cdot MS_{ti} \qquad (3.48) \\ &= [(1 - 0.4) \cdot (1 + 0.2 \cdot 0.36 \cdot 5.6) \cdot 0.84 + 0.1 \cdot 5.6/1.5] \cdot 1200 \\ &= 1289\ \text{kg VSS} \end{aligned}$$

$$MN_{lh} = f_n \cdot ME_{vh} = 0.1 \cdot 1289/5.6 = 23.2\ \text{kg N} \cdot d^{-1} \qquad (3.59)$$

The mass of nitrified nitrogen depends on the discharge of organic nitrogen with the excess sludge mass, a value not know a priori as the nitrifier mass is not yet known. Therefore MN_c and MN_l have to be calculated iteratively. Initially assume that $MN_{ln} = 0$ kg N $\cdot d^{-1}$

$$\begin{aligned} MN_c &= MN_{ti} - MN_{lh} - MNl_n - MN_{oe} - MN_{ad} \qquad (A6.10) \\ &= 720 - 23.2 - 0 - 36 - 2.4 = 658.4\ \text{kg N} \cdot d^{-1} \end{aligned}$$

$$\begin{aligned} C_{rn} &= Y_n \cdot R_s/(1 + b_n \cdot R_s) \qquad (A6.5) \\ &= 0.12 \cdot 5.6/(1 + 0.06 \cdot 5.6) = 0.5 \end{aligned}$$

The nitrifier sludge mass can now be calculated as:

$$MX_{vn} = C_{rn} \cdot (1 + f \cdot b_n \cdot R_s) \cdot Q_i \cdot N_c \quad \text{(A6.8)}$$

$$= 0.5 \cdot (1 + 0.2 \cdot 0.06 \cdot 5.6) \cdot 658.4 = 351\,\text{kg VSS} \cdot d^{-1}$$

$$MNl_n = f_n \cdot ME_{vn} = 0.1 \cdot 351/5.6 = 6.3\,\text{kg N} \cdot d^{-1}$$

$$MN_c = 658.4 - 6.3 = 652.1\,\text{kg N} \cdot d^{-1}$$

After recalculation, the new value of the nitrifier sludge mass $MX_{vn} = 348$ kg VSS and $MN_{ln} = 6.26$ kg $N \cdot d^{-1}$, which is so close to the previous value that a further iteration is not required. The total biomass concentration is 1289 + 348 = 1637 kg VSS, of which 21% consists of nitrifiers (active + endogenous residue). It can be concluded that in wastewaters with a low COD/N ratio, the contribution of the nitrifiers is indeed considerable and should not be ignored.

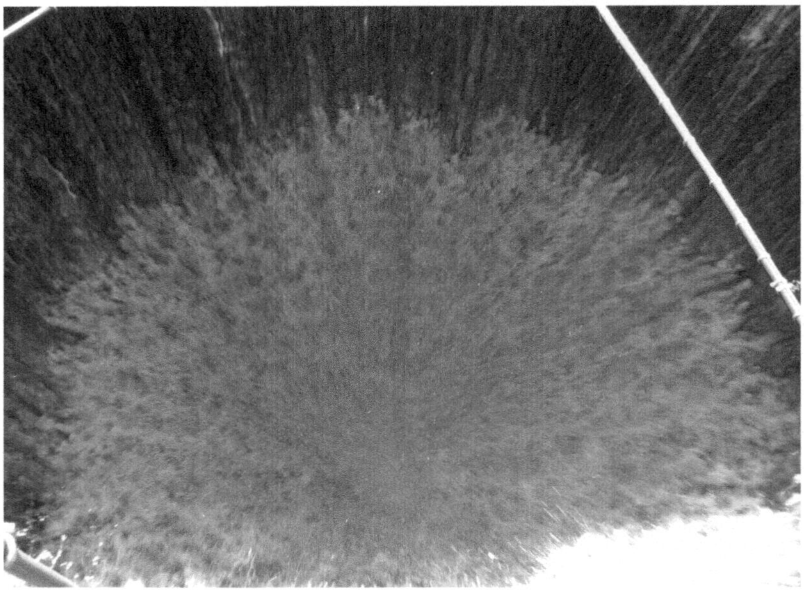

Figure A6.1 Impressive UASB effluent overflow, discharging the effluent of 20 UASB reactors at the Onça STP, Belo Horizonte - Brazil. Courtesy of B. Heffernan

Appendix 7

Empiric methods for final settler sizing

In this appendix two of the empiric design guidelines that have been briefly discussed in Section 8.3.5 will be reviewed in more detail: the guidelines of the Dutch foundation for water research (STORA) in 1981, which have recently been re-evaluated (STOWA, 2002) and those by the German foundation for water research (ATV, 1976 and 1991). These design procedures are based on the application of a maximum sludge volume loading rate, which is a function of the sludge concentration, the sludge volume and the surface area of the final settler.

A7.1 STORA DESIGN GUIDELINES (1981)

In the original research study (STORA, 1981) an extensive series of practical measurements were performed on 21 full-scale final settlers. The applied hydraulic loading rate was increased until settler failure was observed. The final settlers selected in the research study shared the following characteristics:

- Circular settler with a diameter between 30–50 m;
- Sidewall depth between 1.5 and 2.5 m;
- Bottom slope $0.08\ m \cdot m^{-1}$;
- Central inlet flocculation chamber and effluent weir/channel on the perimeter;
- Sludge removal by means of a bottom scraper.

Sidewall settler depth in the Netherlands is often limited to a relative shallow 1.5 to 2.5 m as the groundwater level is rather high, while even a shallow final settler with a sidewall depth of 2.5 m and a diameter of 30 m will have a lowest point of at least 3.5 m deep.

A7.1.1 Theoretical aspects

A typical characteristic of the Dutch situation is that almost all sewer systems are combined. Therefore the ratio between peak (rainfall) and average (dry weather) flow rate is quite high, typically between 2.5 and 4. Therefore, one of the most important differences between a design based on the solids flux theory and that according to the STORA guidelines is that the former uses the average influent flow rate (sufficient buffer volume is assumed to be present as the sidewall depth of the settler is high, at values between 4–6 m), while the latter is based on the peak (rainwater) influent flow rate. The main design parameter in the STORA method is the sludge volume loading rate. The sludge volume loading rate ($l \cdot m^{-2} \cdot h^{-1}$) is defined as:

$$T_{vx} = T_s \cdot v_x = Q_{pf} \cdot X_t \cdot I_{dsv}/A_d \qquad \text{(A7.1)}$$

where

T_s = hydraulic overflow rate ($m^3 \cdot m^{-2} \cdot h^{-1}$ or $m \cdot h^{-1}$)
v_x = sludge volume ($l \cdot m^{-3}$)
Q_{pf} = peak- or rainwater flow ($m^3 \cdot h^{-1}$)
X_t = sludge concentration in the mixed liquor flow to the final settler ($kg\ TSS \cdot m^{-3}$)
I_{dsv} = diluted sludge volume index (DSVI in $ml \cdot g^{-1}$ TSS)
A_d = net final settler surface area (m^2), i.e corrected for the area occupied by inlet chamber and effluent channel, typically between 5–7% of the total area for $d_d = 25–50$ m

The T_{vx} value at the maximum wastewater flow rate is not allowed to exceed the permissible value of T_{vxm}, which is a function of the sludge volume v_x. The applied T_{vx} value is defined as the product of the peak overflow rate and the sludge concentration X_{tpf} in the reactor, which will be established during the peak flow event. This concentration will be lower than the design reactor concentration X_t, as part of the sludge inventory will have been transferred to the final settler. In Fig. A7.1 the maximum hydraulic loading rate T_{sm} (= T_{vxm}/v_x) is given as function of the sludge volume. The value of T_{vxm} that is allowed depends on the sludge volume v_x and varies between 300 and 400 $l \cdot m^{-2}$, i.e. four different regions are defined:

- $v_x < 200\ l \cdot m^{-3}$: the sludge volume is to low, fines will escape with the effluent;
- $200 < v_x < 300\ l \cdot m^{-3}$: maximum value of $T_{vx} = 300\ l \cdot m^{-2} \cdot h^{-1}$;
- $300 < v_x < 600\ l \cdot m^{-3}$: transition range, maximum value of $T_{vx} = \frac{1}{3} \cdot v_x + 200$;
- $v_x > 600\ l \cdot m^{-3}$: maximum value of $T_{vx} = 400\ l \cdot m^{-2} \cdot h^{-1}$.

During the peak flow, part of the sludge mass in the reactor will be transferred to the final settler where it will accumulate, resulting in an increase of the level of the sludge blanket. In the experiments of 1981 it was established that if the sludge blanked level did not stabilize at 0.3 m above the sidewall depth, it would continue to rise until finally sludge was discharged with the effluent. Therefore this value was accepted as the maximum height of the sludge buffer zone H_{db}. The rate of increase of the sludge blanket level was found to be less than 0.5 $m \cdot h^{-1}$ in all cases.

The equilibrium sludge concentration in the reactor during peak flow can be estimated from the mass of sludge transferred to the final settler during a period of peak flow. The following assumptions are made:

(1) Under average flow conditions the sludge mass in the final settler will be very small;
(2) The average concentration X_{av} at which the sludge will accumulate in the final settler is dependent on the value of I_{dsv} and is calculated with the following equation:

$$X_{av} = 480/I_{dsv} \qquad \text{(A7.2)}$$

If the calculated value of X_{av} is larger than that of X_t, the value of X_t will be selected.

(3) The available sludge buffer volume V_{db} in the final settler includes the volume of the cone and an additional height H_{db} (= 0.3 m) above the side wall depth:

$$V_{db} = A_d \cdot (H_{db} + \tfrac{1}{3} \cdot \alpha \cdot d_d/2) \quad (A7.3)$$

where α = slope of final settler bottom ($m \cdot m^{-1}$)

(4) The available sludge buffer capacity MX_{tba} is equal to the product of maximum buffer volume and the average sludge concentration in the final settler: i.e. when the sludge blanket level rises above H_{db} the washout of sludge is imminent (within a couple of hours):

$$MX_{tba} = V_{db} \cdot X_{av} \quad (A7.4)$$

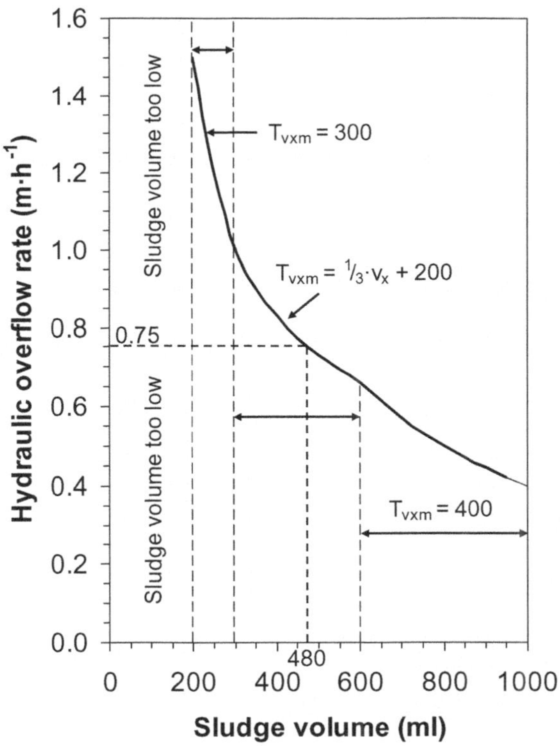

Figure A7.1 STORA guidelines (1981) for final settler design: T_{sm} value as function of v_x

Several restrictions apply to the maximum sludge quantity that can be buffered in the settler:

- The value of X_{tpf} is not allowed to decrease below 2.0 g $TSS \cdot l^{-1}$ and the value of v_x not below 200 $ml \cdot l^{-1}$, whichever is limiting, to prevent excessive washout of sludge fines with the effluent;

- No more than 30% of the total sludge mass MX_t may be transferred to the final settler, in order to maintain adequate treatment capacity in the reactor;
- The maximum quantity of buffered sludge may not exceed the available sludge buffer capacity MX_{tb}.

The transfer of sludge mass to the final settler will only result in a equivalent rapid decrease in the sludge concentration in the mixed liquor flow to the final settler, if the reactor is operated in a completely mixed flow configuration (for example a single non-compartmentalised aeration tank or circuit systems such as carrousels). In the case of plug flow reactors there will be a delay, so that it may take several hours before the decrease is observed. During this time, the final settler will be significantly overloaded with solids that can result in a rapid increase in sludge blanket level.

As concerns the return sludge concentration and -flow, the maximum return sludge concentration was found to depend on the diluted sludge volume index and is defined as (valid for $90 < I_{dsv} < 150$ $ml \cdot g^{-1}$ TSS):

$$X_{rm} = 1200/I_{dsv} \tag{A7.5}$$

The sludge recycle ratio required to maintain equilibrium over the final settler is given by Eq. (8.11) as $s = X_t/(X_t - X_r)$.

A7.1.2 Application of the STORA 1981 design guidelines

The design procedure for final settlers according to the STORA guidelines from 1981 can be summarized as follows:

(1) Select the target operational sludge concentration in the reactor X_t;
(2) Calculate the sludge volume v_x from the values of X_t and I_{dsv},
(3) For this sludge volume v_x, determine the maximum value of the sludge volume loading rate T_{vxm} from Figure A7.1 and determine the maximum hydraulic loading rate T_{sm};
(4) For the peak influent flow rate Q_{pf}, calculate the required surface area of the final settler (compensate for the loss of area due to inlet chamber and effluent channel);
(5) Calculate the diameter of the final settler;
(6) Calculate the available sludge buffer volume V_{db} and the maximum sludge mass that can be buffered in the final settler MX_{tba};
(7) Evaluate whether one of the restrictions regarding the extent of sludge transfer of reactor to final settler listed in Section A7.1.1 applies, i.e. whether:
    (a) $X_{tpf} < 2.0$ kg TSS $\cdot m^{-3}$ or $v_x < 200\ l \cdot m^{-3}$;
    (b) The decrease in MX_t in the reactor is more than 30%;
    (c) No more sludge buffer volume is available in the final settler. i.e. $MX_{tbr} > MX_{tba}$.
(8) If none of the above restrictions apply, reduce the value of the peak flow equilibrium mixed liquor concentration X_{tpf} and repeat the procedure;
(9) Once one of the restrictions applies, the procedure is stopped and the calculated diameter of the final settler is accepted as the design value.

EXAMPLE A7.1

Design a final settler for a 15.000 m^3 carrousel operated at 4 $g \cdot l^{-1}$. Assume a prolonged peak flow rate of 1.500 $m^3 \cdot h^{-1}$. The value of the diluted sludge volume index $I_{dsv} = 120$ $ml \cdot g^{-1}$ TSS. Only one settler is to be installed with the following characteristics: $H_d = 2$ m; $H_{db} = 0.3$ m and $\alpha = 0.08$ $m \cdot m^{-1}$. The loss of surface area from inlet chamber and effluent weir/channel is 6%.

Solution

Start the calculation for the initial reactor solids concentration of $X_t = X_{tpf} = 4$ $g \cdot l^{-1}$. The sludge volume v_x of the mixed liquor flow to the final settler is equal to $4 \cdot 120 = 480$ ml. From Figure A7.1 it can be established that for this sludge volume the maximum sludge volume loading rate is equal to:

$$T_{vxm} = 1/3 \cdot v_x + 200 = 1/3 \cdot 480 + 200 = 360 l \cdot m^{-2} \cdot h^{-1}$$

The maximum allowed hydraulic overflow rate is equal to:

$$T_{sm} = T_{vx}/v_x = 360/480 = 0.75 \, m \cdot h^{-1}$$

The required surface area of the final settler can now be determined as:

$$A_d = Q_{pf}/(T_{sm} \cdot (1 - 0.06)) = 1500/(0.75 \cdot (1 - 0.06)) = 2128 \, m^2$$

The final settler diameter d_d is 52.1 m. Alternatively two smaller settlers could be constructed. The available buffer volume can now be calculated as:

$$V_{db} = A_d \cdot (H_{db} + 1/3 \cdot \alpha \cdot d_d/2)$$

$$= 2128 \cdot (0.3 + 1/3 \cdot 0.08 \cdot 52.1/2) = 2115 \, m^3$$

$$X_{av} = 480/I_{dsv} = 480/120 = 4 \, kg \, TSS \cdot m^{-3}$$

The average sludge concentration in the buffer zone of the final settler is (coincidentally) equal to the sludge concentration in the reactor X_t. The maximum sludge mass that can be buffered in the final settler is calculated as:

$$MX_{tba} = V_{db} \cdot X_{av} = 2115 \cdot 4 = 8416 \, kg \; TSS$$

It can be established that, as expected for the first iteration, none of the four criteria limiting the transfer of sludge to the final settler yet applies:

- $X_{tpf} > 2$ kg TSS $\cdot m^{-3}$ and $v_x > 200$ $ml \cdot l^{-1}$
- $MX_{tbr} = 15000 \cdot (X_t - X_{tpf}) = 15.000 \cdot (4 - 4) = 0$ kg TSS, so $MX_{tbr} < MX_{tba}$
- $\Delta MX_t = (X_t - X_{tpf})/X_t < 30\%$

Consequently, in the next step the concentration of X_{tpf} is lowered and the calculations are repeated. The results are shown in Table A7.1. As can be observed, the available sludge buffer capacity in the settler is limiting once the sludge concentration in the reactor is reduced to less than 3.5 kg TSS $\cdot$ m^{-3}. Therefore the final settler dimensions for $X_{tpf} = 3.5$ kg TSS $\cdot$ m^{-3} are accepted as design values. The sludge concentration in the return sludge flow X_r can be calculated as $1200/I_{dsv} = 1200/120 = 10$ kg TSS $\cdot$ m^{-3}. Using Eq. (A7.5) the maximum return sludge concentration is calculated as $1200/I_{dsv} = 10\ g \cdot l^{-1}$. Therefore the required recycle ratio s during peak flow $= 3.5\ /(10 - 3.5) = 0.54$.

Table A7.1 Results of the final settler design procedure of Example A7.1, using the STORA guidelines from 1981 (settler volume includes cone volume)

Par.	UoM	X_{tpf}–equilibrium peak flow concentration in aeration tank					
		4.0	3.8	3.7	3.6	3.5	3.4
ΔMX_t	%	0%	-5%	-8%	-10%	-13%	-14%
v_x	$l \cdot m^3$	480	456	444	432	420	414
T_{vxm}	$l \cdot m^{-2} \cdot h^{-1}$	360	352	348	344	340	338
T_{sm}	$m \cdot h^{-1}$	0.75	0.77	0.78	0.80	0.81	0.82
A_d	m^2	2128	2067	2036	2004	1971	1955
d_d	m	52.1	51.3	50.9	50.5	50.1	49.9
V_{db}	m^3	2115	2035	1993	1951	1908	1887
X_{av}	kg TSS $\cdot m^{-3}$	4.0	4.0	4.0	4.0	4.0	4.0
MX_{tba}	kg TSS	8461	8138	7973	7805	***7634***	7547
MX_{tbr}	kg TSS	0	3000	4500	6000	***7500***	8250
V_d	m^3	4654	4501	4423	4342	4261	4219
V_t	m^3	19,654	20,291	20,639	21,009	21,404	21,611

A7.1.3 Modifications to the STORA 1981 design guidelines

When the original STORA guidelines were developed in 1981, the settling characteristics of the activated sludge were certainly worse than those prevalent in today's operation. The average I_{dsv} value of the activated sludge systems included in the study without primary settling was 140 ml $\cdot$ g^{-1} TSS, while the average I_{dsv} value of the systems with primary settling was even worse at 190 ml $\cdot$ g^{-1} TSS. The latter values might even be considered to be indicative of a certain degree of sludge bulking. This is not unlikely, as at the time none of the considered activated sludge systems was provided with measures against sludge bulking (e.g. a selector). As the settling characteristics have since improved to average I_{dsv} values between 70 – 150 ml $\cdot$ g^{-1} TSS, it might therefore be assumed that is now possible to apply a higher sludge volume loading rate to a final settler.

In a recent investigation the original STORA guidelines were compared with the measured zone settling velocities at different concentrations of sludge taken from 11 full-scale sewage treatment plants (STOWA, 2002). The results are displayed in Figure A7.2, where the results of comparable sludges have been clustered and exponential trend lines added. Also indicated are the original STORA guidelines from 1981.

At first glance the results seem to indicate that indeed much higher sludge volume loading rates can be applied over the whole range of sludge volumes investigated. The only exception are the results for the sludge that exhibited a very low I_{dsv} value of 55 ml · g^{-1} TSS, perhaps due to the high suspended solids concentration of this sludge.

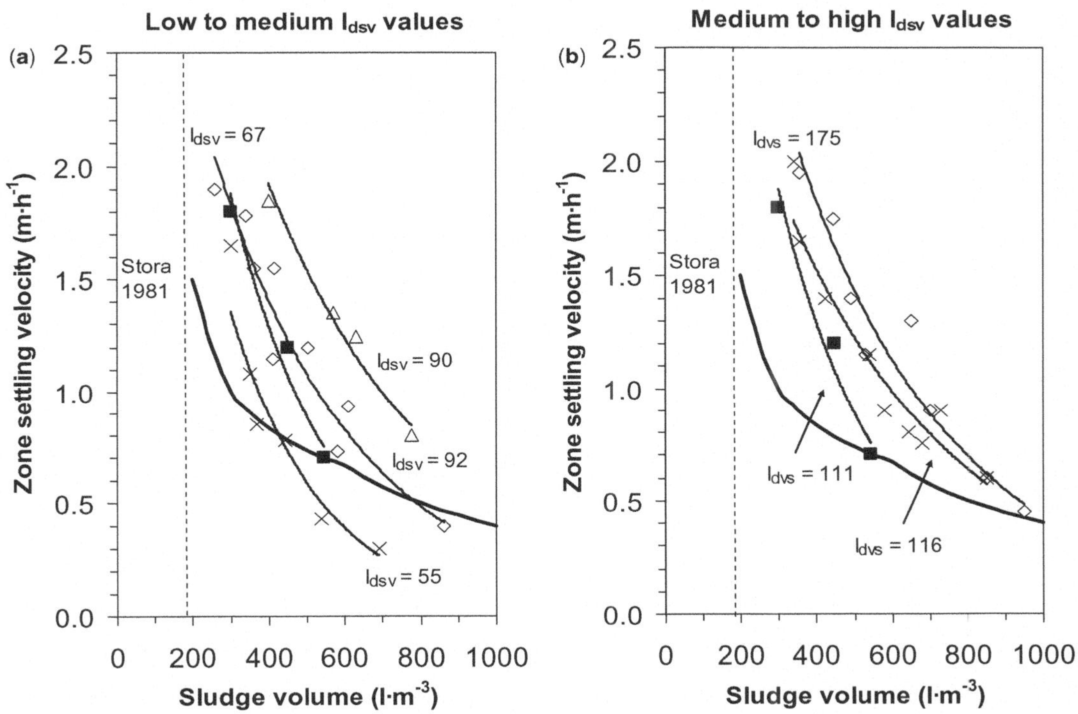

Figure A7.2 Zone settling velocity of mixed liquor from several full-scale activated sludge systems and for different I_{dsv} values (adapted from STOWA, 2002)

However, it should be considered that the zone settling velocities of the different sludges were determined from lab-scale experiments, where the non-idealities associated to full-scale settlers are absent (i.e. no dead volume). Otherwise stated, the data points in Figure A7.2 should be corrected with a safety factor, before the observed batch zone settling velocities can be translated into the maximum allowed hydraulic overflow rate of a full-scale settler.

Unfortunately in this study the safety factors of the final settlers have not been determined. To indicate the effect that dead volume, density currents et cetera will have, the same data set is presented again in Figure A7.3 for an assumed safety factor s_{fd} of 1.5 and 2.0. From these two graphs it can be concluded that when realistic values are selected for s_{fd}, the T_{sm} curves correspond to sludge volume loading rates between 300 and 500 l · m^{-2} · h^{-1} for $s_{fd} = 1.5$ and between 225–400 l · m^{-2} · h^{-1} for $s_{fd} = 2.0$. When these results are compared with the original Stora guidelines from 1981, also indicated in Figure A7.3, then it can be observed that this line is located approximately in the middle of the two extremes. For $s_{fd} = 1.5$ and for lower sludge volumes ($v_x < 600$ ml · l^{-1}) the Stora guideline underestimates the value of T_{vxm}, while for $s_{fd} = 2.0$ and higher sludge volumes ($v_x > 600$ ml · l^{-1}) the value of T_{vxm} is overestimated. As s_{fd} values of 1.5 and 2.0 may be considered as the lower respectively higher extremes one might find in practice for well-designed final settlers, it can be concluded that the original STORA guidelines of 1981 are still valid.

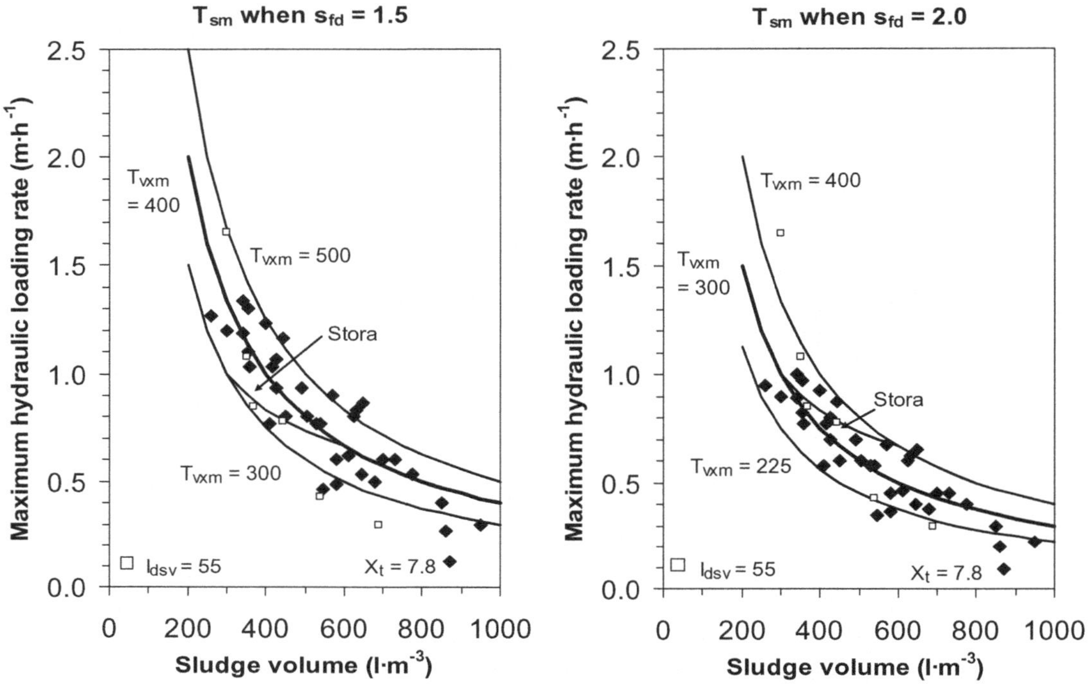

Figure A7.3 ZSV measurements (STOWA, 2002) compared to T_{sm} curves for different T_{vxm} values when the individual data points are corrected with $s_{fd} = 1.5$ and $= 2.0$, transforming ZSV into T_{sm}

Several other findings of the STOWA 2002 research project are briefly summarized below:

- Concerning the required sludge recycle ratio s, the maximum return sludge concentration given by Eq. (A7.5) is $X_{rmax} = 1200/I_{dsv}$. Using this value the minimum value of the sludge recycle ratio is obtained (at maximum influent flow rate). However, according to the findings from the a more realistic value of s is calculated when Eq. (A7.5) is adapted to $X_{rmax} = 1200/(1\frac{1}{3} \cdot I_{dsv})$;
- For a T_{vxm} value of $300\,l \cdot m^{-2} \cdot h^{-1}$ (and for shallow final settlers), the recommended sidewall depth is equal to $0.05 \cdot d_d$;
- The final settler buffer volume available during peak flow for buffering was considered to be conservative. Consequently the value of H_{db} can be increased from its original value of 0.3 m, where deeper tanks allow for a higher amount of sludge buffering. The minimum distance between the level of the sludge blanket and the surface level is 1.0 m, while 1.5 m is recommended to prevent the risk of high effluent suspended solids concentrations.

A7.2 FINAL SETTLER DESIGN COMPARISON METHODOLOGY

When comparing the solids flux design method with the empiric methods, it should be considered that identical T_{sm} values for a certain sludge concentration do not imply an identical final settler design. As explained above, the design according to the ATV and STORA guidelines are based on the equilibrium *peak flow* situation, while the solids flux design is based on equilibrium at *average flow*. In both ATV and STOWA design procedures a maximum transfer of 30% of the sludge mass from the reactor to the settler is accepted during peak flow.

For example, assume that in case of medium settleability, T_{sm} according to the flux theory is $1.11\ m \cdot h^{-1}$ for $X_t = 4\ g \cdot l^{-1}$. Assuming a 30% reduction in sludge concentration at peak flow, this would be $T_{sm} = 1.85\ m \cdot h^{-1}$ at $X_t = 2.7\ g \cdot l^{-1}$ according to the maximum value allowed by the STOWA guidelines (for $T_{vxm} = 500\ l \cdot m^{-2} \cdot h^{-1}$). The required surface area of the final settler depends on the values of Q_{iav} (DWF) and Q_{pf} (RWF). For a Q_{ipf}/Q_{iav} ratio of $2\frac{1}{2}$, not uncommon in the Netherlands, the calculated surface area of the settler according to the STOWA guidelines will be a factor $(Q_{pf}/Q_{av})/(T_{smstowa}/T_{smsf}) = 2.5/(1.85/1.11) = 2.5/1.67 = 1.5$ larger. On the other hand, the sidewall depth of the final settler designed according to the solids flux theory will be higher: for instance 4–6 m compared to 1.5–2.5 m often applied in the Netherlands. In this section an example will be presented in which the design of a final settler according to empirical methods and to the solids flux theory are compared for a specific situation.

EXAMPLE A7.2

For the same data as in Example A7.1, compare the optimal design of the reactor - final settler system according the revised STORA guidelines with that according to the solids flux theory. Use the data summarized in Table A7.2.

Table A7.2 Design data for Example A7.2

General data:			**Data specific for solids flux design:**		
MX_t	60,000	kg TSS	s_{fd}	2	(-)
Q_i/Q_{pf}	800/1500	$m^3 \cdot h^{-1}$	k	0.4	$l \cdot g^{-1}$ TSS
I_{dsv}	120	$ml \cdot g\ TSS^{-1}$	v_0	8	$m \cdot h^{-1}$
α	0.08	$m \cdot m^{-1}$	H_d	4	m
C_r	200	$US\$ \cdot m^{-3}$	C_d	275	$US\$ \cdot m^{-3}$
Data specific for revised STORA design (i.e. a fair to good settleability is assumed):					
T_{vx} $(200 < v_x < 300)$	500	$l \cdot m^{-2} \cdot h^{-1}$	H_d	2.75	m
T_{vx} $(300 < v_x < 600)$	$600 - \frac{1}{3} \cdot v_x$	$l \cdot m^{-2} \cdot h^{-1}$	H_{db}	1.25	m
T_{vx} $(v_x > 600)$	400	$l \cdot m^{-2} \cdot h^{-1}$	C_d	400	$US\$ \cdot m^{-3}$

Note that the settleability of the sludge is assumed to be fair to good, which is reflected in the values of the Vesilind constants k and v_0. Therefore, for fair comparison, the maximum allowable sludge volume loading rate in the Stowa design has also been increased: starting at a maximum of $500\ l \cdot m^{-2} \cdot h^{-1}$ for sludge volumes $< 300\ ml \cdot l^{-1}$ and gradually decreasing to $400\ l \cdot m^{-2} \cdot h^{-1}$ for sludge volumes $> 600\ ml \cdot l^{-1}$

The costs per unit volume of final settler are higher for the STORA design method than for an equivalent unit volume in the solids flux design method. The reason is that in a shallow settler the costs of civil works, settler bottom and scraper mechanism will be divided over less cubic meters. The loss of surface area due to inlet chamber and effluent weir is given as 6%.

Solution

For the STOWA guidelines (2002), the calculation proceeds identical to that presented in Example A7.1 for the STORA 1981 guidelines with the following modifications:

- An increase in maximum sludge volume loading rate T_{vxm} (in accordance with Table A7.2);
- An increase in available buffer volume ($H_{db} = 1.25$ m)

Selected results of the optimisation of the final settler according to the STOWA recommendations are presented in Table A7.3, while the cost optimisation curves are displayed in Figure A7.4. It can be observed in Table A7.3 that for values of $X_t > 6.5$ g $TSS \cdot l^{-1}$ the maximum allowed transfer of sludge to the final settler has become limiting (30% of 60,000 = 18,000 kg TSS). The available sludge buffer volume is therefore only partly used and this explains why the total cost rapidly increases for values of $X_t > 6.5$ g $TSS \cdot l^{-1}$.

Table A7.3 Optimized design according to the revised STORA procedure (2002)

Par.	X_t - Selected aeration tank sludge concentration (kg TSS $\cdot$ m^{-3})						
	3.0	**4.0**	**5.0**	**6.0**	**6.5**	**7.0**	**8.0**
X_{tpf}	2.65	3.34	3.94	4.37	4.55	4.90	5.60
V_x	318	401	473	524	546	588	672
ΔMX_t	-12%	-16%	-21%	-27%	***-30%***	-30%	-30%
T_{vxm}	495	468	444	427	420	406	400
T_{sm}	1.56	1.17	0.94	0.82	0.77	0.69	0.60
A_d	1025	1368	1700	1958	2075	2311	2681
d_d	36.1	41.7	465	49.9	51.4	54.3	58.4
MX_{tba}	7101	9888	12,722	16,380	***18,277***	22,351	30,464
MX_{tbr}	7000	9875	12,700	16,333	***18,000***	18,000	18,000
V_r	3308	15,000	12,000	10,000	9231	8571	7500
V_d	20,000	4517	5.720	6675	7116	8012	9440
V_t	23,308	19,517	17,720	16,675	16,347	16,583	16,940

The solids flux design method has been extensively discussed elsewhere so only the graphical results are displayed in Figure A7.4. It can be observed that in this particular case the solids flux theory results in a cheaper design (3.5 million US$ versus 4.67 million US$ for the STOWA 2002 method). It should however be checked with the static point procedure whether the selected depth of 4 m is sufficient to sustain a prolonged peak flow.

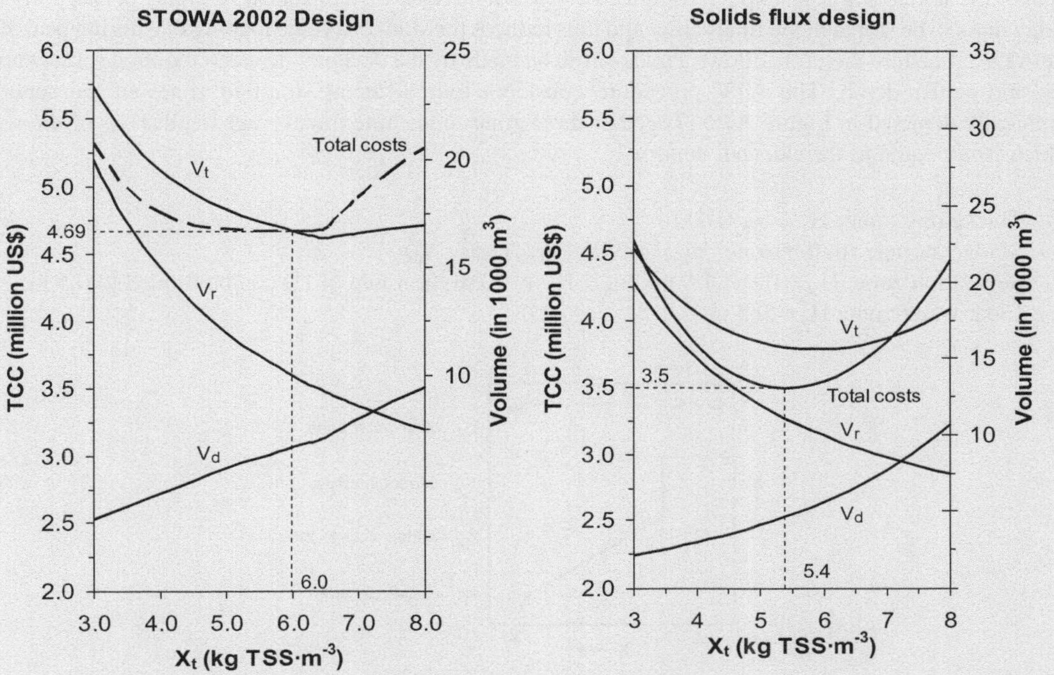

Figure A7.4 Unit volumes and total construction costs (TCC) of the reactor-settler system according to the STOWA 2002 (revised STORA 1981) guidelines and according to the solids flux theory for the conditions of Example A7.2

A7.3 ATV DESIGN GUIDELINES (1976)

The German ATV design procedure from 1976 precedes the STORA guidelines from 1981 but is similar to the STORA method in many aspects. Most importantly, the surface area of the final settler is defined by the maximum sludge volume loading rate.

A7.3.1 Theoretical aspects

In the case of the ATV guidelines the maximum value of T_{vxm} is approximately $400\ l \cdot m^{-2} \cdot h^{-1}$ for $v_x = 200\ ml \cdot l^{-1}$ and decreases slowly to a value of $200\ l \cdot m^{-2} \cdot h^{-1}$ for $v_x = 1000\ ml \cdot l^{-1}$. For these values it was observed that full-scale settlers complied to the effluent limit of 30 mg $TSS \cdot l^{-1}$, applicable in Germany in the seventies. The allowed maximum hydraulic loading rate T_{sm} is calculated as:

$$T_{sm} = 2400/(v_x)^{1.34} \tag{8.42}$$

Similar to the STORA guideline, Eq. (8.42) is valid for $X > 2.0\ g \cdot l^{-1}$ and $v_x > 200\ ml \cdot l^{-1}$ and sizing is based on the sustained peak flow. The main distinguishing feature from the STORA guidelines is that the depth of the settler is now an explicit design criterion. An increase in depth allows a higher proportion of the sludge mass to be stored in the final settler and thus reduces the sludge volume loading rate during peak flow. The ATV procedure therefore allows a trade-off to be made by the designer between required settler surface area and settler depth. The ATV procedure considers four different stratified zones in the setter, as graphically depicted in Figure A7.5. Together these zones constitute the average depth H_{dav} of the settler (which is *not* equal to the sidewall depth):

- Thickening zone: $H_1 = v_x/1000$ (A7.6)
- Sludge storage (buffer)zone: $H_2 = \Delta MX_t \cdot I_{dsv}/(500 \cdot A_d)$ (A7.7)
- Separation zone: $H_3 = 0.8 - 1.0$ m, but if $H_2 > 1.0$ m the value of H_3 can be limited to 0.5 m;
- Clear water zone: $H_4 = 0.5$ m.

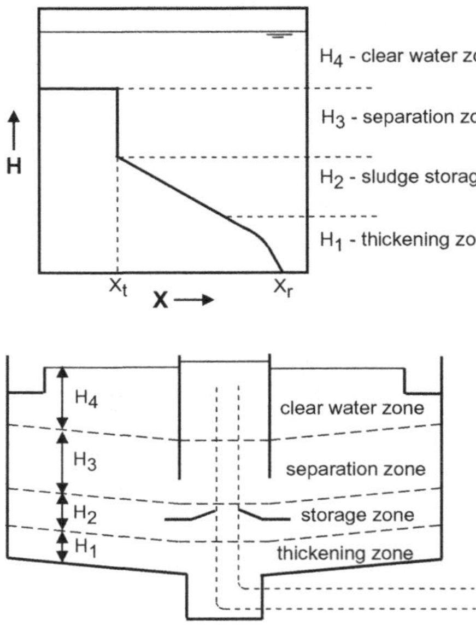

Figure A7.5 Schematic sludge profile in a final settler (not to scale)

The maximum average settler depth H_{dav} is determined when one of the four limiting conditions applies: i.e. $\Delta MX_t > 30\%$, $\Delta X_t > 1.3\ kg \cdot m^{-3}$, $X_{tpf} < 2.0\ g \cdot l^{-1}$ and $v_x < 200\ ml \cdot l^{-1}$. The minimum average depth H_{dav} is 2.0 m. The resulting sidewall depth can be calculated as $H_d = H_{dav} - 0.25 \cdot \alpha \cdot d_d$.

EXAMPLE A7.3

Using the same data as in Example A7.1, i.e. $V_t = 15.000\ m^3$, $X_t = 4$ kg TSS $\cdot m^{-3}$, $I_{dsv} = 120\ ml \cdot g^{-1}$ TSS and $Q_{pf} = 1500\ m^3 \cdot h^{-1}$, calculate the required final settler surface area and -volume as function of the selected average depth according to the AVG design guidelines.

Solution

It can be calculated that the limiting constraint for the maximum average settler depth is the maximum amount of sludge mass that is allowed to be transferred to the final settler: i.e. $0.3 \cdot 60{,}000 = 18{,}000$ kg TSS. The minimum peak flow sludge concentration X_{tpf} is $0.7 \cdot 4.0 = 2.8$ kg TSS $\cdot$ m^{-3}, which corresponds to a maximum allowed hydraulic loading rate of (Eq. 8.42):

$$T_{sm} = 2400/(v_x)^{1.34} = 2400/(2.8 \cdot 120)^{1.34} = 0.99\,\text{m} \cdot \text{h}^{-1}$$

Thus the required surface area of the settler is $Q_{pf}/T_{sm} = 1512$ m^2. The height of the different zones in the final settler can be calculated as:

$$H_1 = v_x/1000 = 2.8 \cdot 120/1000 = 0.34\,\text{m}$$

$$H_2 = \Delta MX_t \cdot I_{dsv}/(500 \cdot A_d) = 18{,}000 \cdot 120/(500 \cdot 1512) = 2.85\,\text{m}$$

$$H_3 = 0.5\,\text{m as } h_2 > 1.0\,\text{m and } H_4 = 0.5\,\text{m}$$

$H_{dav} = 2.85 + 0.34 + 0.5 + 0.5 = 4.18$ m. For the bottom inclination α of 0.08 m $\cdot$ m^{-1}, the sidewall depth H_d can be calculated as $H_d = H_{dav} - 0.25 \cdot 0.08 \cdot d_d = 4.18 - 0.25 \cdot 0.08 \cdot 44 = 3.3$ m.

In Table A7.4 the results of the calculations are presented in a tabulated form for different values of X_{tpf} in between the boundary conditions of $X_t = 4.0$ kg TSS $\cdot$ m^{-3} and $X_{tpf} = 2.8$ kg TSS $\cdot$ m^{-3}. Note that the ATV guideline does not specify an optimal solution but rather a range of possible solutions. In practice the total construction costs of the different design alternatives will not differ much, as the increase in volume at larger depth will be compensated by the decrease in required civil works and the diameter of the sludge raking mechanism.

Table A7.4 Final settler design as function of the selected *average* settler depth according to the ATV 1976 method

Par.	UoM	X_{tpf} - equilibrium peak flow sludge concentration in the aeration tank						
		4.0	3.8	3.6	3.4	3.2	3.0	2.8
ΔMX_t	%	0	3000	6000	9000	12,000	15,000	18,000
v_x	l $\cdot$ m^3	480	456	432	408	384	360	336
T_{sm}	m $\cdot$ h^{-1}	0.61	0.66	0.71	0.76	0.83	0.90	0.99
A_d	m^2	2448	2285	2125	1969	1815	1665	1518
H_1	m	0.48	0.46	0.43	0.41	0.38	0.36	0.34
H_2	m	0	0.32	0.68	1.10	1.59	2.16	2.85
H_3	m	1.0	1.0	1.0	0.5	0.5	0.5	0.5
H_4	m	0.5	0.5	0.5	0.5	0.5	0.5	0.5
H_{dav}	m	1.98	2.27	2.61	2.51	2.97	3.52	4.18
H_d	m	0.86	1.19	1.57	1.50	2.01	2.60	3.30
d_d	m	56	54	52	50	48	46	44
V_d	m^3	3935	4368	4809	4275	4810	5353	5903
V_t	m^3	18,935	19,368	19,809	19,275	19,810	20,353	20,903

The maximum return sludge concentration during dry weather flow is calculated with:

$$X_{rm} = 1200/I_{dsv} \tag{A7.8}$$

This value is increased with $2\ kg \cdot m^{-3}$ during RWF. The sludge recycle ratio required to maintain equilibrium over the final settler is determined by Eq. (8.11) as $s = X_t/(X_t - X_r)$ and is calculated both for DWF and RWF conditions. Finally the ATV recommends a maximum weir loading rate of 5 – 10 $m^3 \cdot m^{-1} \cdot h^{-1}$, while for diffuse sludges (i.e. containing many pinpoint flocs) a maximum between 3–5 $m^3 \cdot m^{-1} \cdot h^{-1}$ is recommended.

A7.3.2 Modifications to the ATV 1976 design guidelines

In 1991 the ATV guideline was revised. The main changes were that a higher sludge volume loading rate was allowed ($T_{vx} < 450\ l \cdot m^{-2} \cdot h^{-1}$) and that the depth of the final settler was increased. The latter was a consequence from the application of stricter effluent limits ($X_{te} < 20$ mg TSS $\cdot\ l^{-1}$). The value of $T_{vxm} < 450\ l \cdot m^{-2} \cdot h^{-1}$ is significantly less conservative than the values used by the earlier ATV and STORA guidelines. However, the maximum hydraulic overflow rate T_{sm}, while still calculated based on the peak flow rate, now no longer uses the value of X_{tpf} instead of that of X_t. Depending on the ratio between RWF and DWF, the area requirements compared to the ATV 1976 guidelines may thus actually be larger. The applicability of the revised ATV guidelines is limited to:

- Final settlers with a length < 60 m (rectangular tanks) or a diameter between 30–50 m;
- $I_{dsv} \leq 180\ ml \cdot g^{-1}$ TSS;
- $v_x \leq 600\ l \cdot m^{-3}$;
- $s < 1.5$ for DWF and $s < 0.75$ for RWF.

According to Billmeier (1993) and Ekama et al. (1996), the final settler depth influences the effluent suspended solids concentration. The value of T_{vxm} that complies to $X_{te} = 20$ mg TSS $\cdot\ l^{-1}$ can be calculated from the following equation:

$$T_{vxm} = 190 \cdot H_{dav}/(1 + s_{pf}) \tag{A7.9}$$

where s_{pf} is the return sludge ratio during peak flow conditions

This equation is valid for $3 < H_{dav} < 4.5$ m. According to Billmeier (1993), for $H_{dav} < 3.0$ m the value of T_{vxm} is reduced to $350\ l \cdot m^{-2} \cdot h^{-1}$ while T_{sm} is limited to $1.1\ m \cdot h^{-1}$. A further change compared to the 1976 ATV procedure is that the maximum return sludge concentration (for both DWF and RWF) is calculated as:

$$X_{rm} = \eta_d \cdot X_f = \eta \cdot (1000/I_{dsv}) \cdot R_{hth}^{1/3} \tag{A7.10}$$

where:

X_f = average suspended solids concentration on final settler bottom
η_d = efficiency factor to account for short-circuiting between sludge inlet and outlet
– 0.7 for bottom scrapers
– between 0.5 – 0.7 for suction nozzles

Figure A7.6 Final settler of WWTP Beverwijk - courtesy of DHV BV

R_{hth} = thickening time t_{th}

1.5 – 2.0 h without denitrification
2.0 – 2.5 h with denitrification
1.0 – 1.5 h with nitrification
1.5 – 2.0 h for bio-P removal

As for the calculation of the height of the different zones in the settlers, the calculation proceeds as follows:

(1) Thickening zone – h_1

The recommended retention times of the sludge in the thickening zone have already been indicated above. Therefore:

$$h_1 = v_x \cdot (1 + s_{pf}) \cdot R_{hth}/(300 \cdot R_{hth} + 500) \quad \text{(A7.11)}$$

(2) Storage zone – h_2

The height of the storage zone is defined by the condition that the minimum storage time is 1.5 hours. The storage sludge concentration is $500/I_{dsv}$. Furthermore a maximum decrease of 30% of X_t is allowed during peak flow. Hence:

$$h_2 = 1.5 \cdot \Delta X_t \cdot I_{dsv} \cdot (1 + s_{pf}) \cdot T_{spf}/500 \quad \text{(A7.12)}$$

where T_{spf} = hydraulic loading rate during peak flow

(3) Separation zone – h_3

The retention time of the "free water fraction" of the sludge at peak flow is to be at least 30 minutes. The free water fraction is defined as $1 - v_x$ for $X = X_{tpf} = X_t$ (the influence of a decrease of the sludge concentration is not taken into account, except for the calculation of the height of the sludge storage zone h_3). Thus:

$$h_3 = 0.5 \cdot T_{spf} \cdot (1 + s_{pf})/(1 - (v_x/1000)) \quad (A7.13)$$

The height of the clear water zone h_4 remains unchanged at 0.5 m.

Regarding the depth of the settler, the following additional comments are made in the new guidelines:

- $H_{dav} > 3$ m over at least 2/3rd of the tank length (rectangular tanks);
- $H_d > 2.5$ m for circular settlers with 30 m $< d_d <$ 50 m

Figure A7.7 Large diameter sewer pipes are laid during construction of the Harnaschpolder STP - courtesy of Delfland Waterboard

Appendix 8

Denitrification in the final settler

In Section 13.5.2 the design of an anaerobic–aerobic system with nitrification was discussed. It might be argued that this creates a problem, as in principle the produced nitrate is available for denitrification in the final settler. Subsequently, the produced nitrogen gas can result in a floating sludge blanket and hence in potential loss of biomass and poor effluent quality. The instability resulting from uncontrolled denitrification has previously been discussed in the introduction of Chapter 5. However, it will be demonstrated in this section that in the case of treatment of anaerobic effluent, the biodegradable COD concentration is so low that, if the nitrogen effluent limits permit this, there is a strong incentive not to denitrify. If denitrification is required the conventional solution is to bypass part of the raw sewage to the pre-D zone of the aerobic system. However, not only will this reduce the potential for energy generation, but also the required activated sludge system volume will increase.

Like methane, nitrogen gas (N_2) is poorly soluble in water as it is neither polar nor reactive, contrary to for instance H_2S, which dissociates into HS^- and H^+. When the nitrogen gas concentration in the mixed liquor exceeds the equilibrium concentration, micro-bubbles of nitrogen gas will be formed, predominantly inside of the sludge floc, causing it to float. As shown in Table A8.1, the equilibrium concentration (solubility) of nitrogen gas at a given depth in the final settler is determined by the local pressure at this depth and the gas composition.

As the return sludge travels downward, the bottom of the settler is the location where the cumulative nitrogen gas production will be maximum. Furthermore, at the bottom of the final settler all oxygen will have been consumed. Therefore it seems reasonable to accept the nitrogen gas concentration in equilibrium with pure nitrogen at the sidewall depth as the maximum allowable value. Another factor to consider is that the nitrogen gas will be produced inside the sludge floc and will have to diffuse out of the floc. Thus a concentration gradient is formed from the centre of the floc to the bulk liquid. Therefore it seems prudent to introduce a safety margin to compensate for locally increased nitrogen gas concentrations inside the sludge flocs. Without firm data it is recommended to use a safety factor of $s_d = 1.5$.

The nitrogen gas concentration in the mixed liquor flow into the final settler is determined by the operational conditions of the reactor preceding it. In the case of an aerated reactor and assuming that the flow to the final settler is taken from a surface overflow, the equilibrium concentration of nitrogen gas at atmospheric conditions is applicable. Although part of the oxygen present in the aeration air will have been consumed during its upward passage through the mixed liquor (typically from 21 to 18%), this will largely have been replaced by the carbon dioxide formed in the respiration process.

Table A8.1 Solubility of nitrogen gas in water as function of the liquid depth and gas composition: pure nitrogen and atmospheric air (78% nitrogen). Adapted from Henze et al. (1993)

Temperature	N_2 concentration in mg $N \cdot l^{-1}$ equilibrium with atmosphere		N_2 concentration in mg $N \cdot l^{-1}$ equilibrium with pure nitrogen		
	0.0 m	3.0 m	0.0 m	3.0 m	4.0 m
10°C	17.9	23.3	22.6	29.4	31.7
15°C	15.9	20.7	20.4	26.5	28.6
20°C	14.5	18.9	18.6	24.2	26.0
25°C	13.2	17.2	16.9	22.0	23.6
30°C	12.1	15.7	15.6	20.2	21.8

On the other hand, if the mixed liquor is taken from the post-D anoxic reactor, then oxygen will be absent from the gas phase and equilibrium with a gas phase of pure nitrogen at atmospheric pressure may be assumed (although actually some carbon dioxide will be present). In this case, re-aeration of the mixed liquor is recommended, which will have the following beneficial effects:

- Nitrogen gas will be stripped, thereby reducing the concentration to the equilibrium value with atmospheric air: for example at 25°C from 16.9 to 13.2 mg $N_2 \cdot l^{-1}$;
- Oxygen will be introduced into the mixed liquor, which will reduce the denitrification capacity with $DO_l/2.86$, as oxygen will be preferentially used (refer to Section 5.4.2.3).

Due to the low respiration rate in a post-D zone, the aeration capacity required for re-aeration will be small and it may be easy to reach a dissolved oxygen concentration of at least 30 to 50% of the equilibrium value. If the hydraulic profile allows this then, instead of aerators, a series of cascades can be used.

If biological nitrogen removal (denitrification) is required, the attractiveness of combined anaerobic-aerobic treatment will decrease, so often the aerobic post-treatment system is only designed for COD removal and nitrification. In this case the mixed liquor flow to the final settler is taken from an aerated reactor. For plug flow systems, the oxygen concentration in this flow will often be higher than the setpoint value, as the oxygen demand is significantly reduced towards the end of the aerobic reactor.

The following equation defines the maximum allowable production of nitrogen gas in the return sludge flow during its passage through the final settler to the abstraction point (N_{ddmax} expressed in mg $N \cdot l^{-1}$ *return sludge* flow):

$$N_{ddmax} = (N_{N2eq} + DO_l/2.86 - N_{N2in})/s_d \qquad (A8.1)$$

where:

N_{N2eq} = equilibrium dissolved nitrogen gas concentration at the maximum liquid depth of the final settler, assuming an atmosphere of 100% nitrogen

DO_l = dissolved oxygen concentration in the mixed liquor flow to the final settler, which is equal to DO_{sp} in the case of nitrification and equal to zero in the case of mixed liquor from a post-D zone

N_{N2in} = dissolved nitrogen gas concentration in the incoming mixed liquor flow. This is equal to the equilibrium dissolved nitrogen gas concentration at 78% nitrogen in the case of aerated or re-aerated mixed liquor and 100% nitrogen in the case of mixed liquor taken from a post-D zone, in both cases at atmospheric pressure
s_d = safety factor to compensate for the locally increased nitrogen gas concentration inside the sludge floc

Having established the maximum allowable nitrogen gas production per liter return sludge flow during its passage through the final settler, we can now focus on the other important factor: the expected denitrification and hence nitrogen gas production. In Chapter 5 the following preconditions for denitrification were identified:

- The availability of nitrate (or nitrite), which will obviously not be a limiting factor when the activated sludge system has only been designed for nitrification;
- The availability of biodegradable substrate: either exogenous (influent COD) or endogenous (respiration of active cell mass).

The biodegradable COD concentration in the mixed liquor is very low after passing the biological reactors and taking into account that the temperature is at least 15°C when anaerobic pre-treatment is applied, so that the metabolisation rate will be high. The endogenous respiration rate is dependent on the temperature and the active sludge concentration, which is a function of the sludge age, the influent COD concentration and the COD composition. To describe the denitrification process in the final settler, it is assumed that this unit may be considered as a (poorly mixed) post-D reactor. The equation defining the post-D denitrification capacity is given by Eq. (5.69) and can be adapted as follows:

$$D_{cd} = K_3 \cdot C_r \cdot f_{xd} \cdot S_{bi} \tag{A8.2}$$

where:

D_{cd} = denitrification capacity in the final settler (mg $N \cdot l^{-1}$ influent)
f_{xd} = sludge mass fraction located in final settler

After anaerobic pre-treatment, the concentration of biodegradable COD ($= S_{bu}$) will be low and is defined by the overall COD removal percentage η_{cod} and the fractions f_{nsu} and f_{npu} in the anaerobic effluent. The value of f_{xd} is defined as the ratio between the average sludge mass present in the final settler and the total sludge mass in the aerobic system. Under conditions of average flow and load, the mass of sludge in the final settler will be low. An indicative estimate of the mass of sludge in the final settler can be made with the following equation:

$$MX_{td} = f_{xvd} \cdot V_d \cdot (X_t + X_r)/2 \tag{A8.3}$$

where f_{xvd} = fraction of final settler volume filled with sludge (typically 5–10%, max. 20%)

$$f_{xd} = MX_{td}/(MX_{td} + MX_t) \tag{A8.4}$$

When the final settler receives a peak flow, the sludge blanket level will rise, but on the other hand the return sludge concentration and the contact time in the settler will decrease, so the net effect on the extent of denitrification that occurs will be small.

On the other hand, when a very low sludge age is applied and/or when the influent COD concentration is low, then the reactor volume might be small compared to the final settler volume and in these cases the value of f_{xd} may then be considerable. In any case, the true sludge age, based on the total sludge mass rather than the sludge mass present in the aeration tanks, will be higher than the design sludge age:

$$R_{sa} = R_s/(1 - f_{xd}) \tag{A8.5}$$

Note that this is the sludge age to be used in Eq. (A8.2). This non-ideal behaviour of the activated sludge system is discussed in Section 3.4 and Appendix. In the ideal steady state model the presence of biomass in the final settler is ignored as due to the absence of substrate and the low mixing intensity the reaction rates will be low. Furthermore, the effect on the treatment performance of the activated sludge system will actually be beneficial as some additional denitrification will occur, typically between 1–3 mg $N \cdot l^{-1}$. This is similar in magnitude to the reduction of the denitrification capacity resulting from oxygen recirculation to the anoxic zones: at least when reasonable values are selected for the "a"- and "s" recycles. Therefore, the sum of these two effects tends to be close to zero.

To assess whether the occurrence of denitrification in the final settler might result in the formation of nitrogen gas bubbles, the following factors should be considered:

- The value of D_{cd} should be increased by a factor $1/v_d$, where v_d is the hydraulic retention time of the final settler in days ($= V_d/Q_i$) to reflect that the hydraulic retention time in the final settler will be (much) less than one day;
- The value of D_{cd} should be expressed per litre return sludge rather than per litre influent, so it has to be divided by the value of the sludge recycle factor "s";
- The concentration of nitrate denitrified in the return sludge *per passage through the final settler* will be significantly lower than D_{cd}, as the contents of the final settler will be refreshed $24 \cdot v_d/(s+1)$ times per day.

When the above factors are taken into consideration, the value of N_{dd}, the concentration of nitrate that will be denitrified in the return sludge stream per passage through the final settler, can now be calculated as:

$$N_{dd} = [D_{cd}/(v_d \cdot s)]/[24 \cdot v_d/(s + 1)] = (s + 1) \cdot D_{cd}/(24 \cdot s \cdot v_d^2) \tag{A8.6}$$

The factors influencing the risk of denitrification in the final settler can be summarized as:

- The dissolved nitrogen gas concentration in the incoming mixed liquor flow N_{2in}, normally equal to the equilibrium with atmospheric air. Anoxic effluent can be re-aerated;
- The dissolved oxygen concentration in the mixed liquor flow to the final settler DO_l, as the oxygen will be consumed preferentially by the heterotrophic biomass and reduces the denitrification capacity;
- The maximum dissolved nitrogen gas solubility at the bottom of the final settler N_{2eq}, which depends on the settler sidewall depth H_d and the temperature;
- Locally increased dissolved nitrogen gas concentrations inside the sludge floc, which can be accounted for by means of a safety factor s_d;
- The COD concentration- and composition in the feed of the activated sludge system;

- The applied sludge age R_s;
- The biomass volume fraction in the final settler f_{xvd};
- The biomass concentration in the return sludge X_r, which is a function of X_t and s;
- The hydraulic retention time in the final settler v_d, which is a function of the sludge settleability (k and v_0), the settler height H_d, the safety factor s_{fd} and the value of the sludge recirculation factor "s".

EXAMPLE A8.1

For an activated sludge system treating UASB effluent (COD removal + nitrification only), evaluate whether denitrification in the final settler might constitute a problem. Compare the maximum nitrogen gas production with the maximum allowable dissolved nitrogen gas concentration at a depth of 4.0 m for the following conditions:

- $Q_i = 50{,}000\ m^3 \cdot d^{-1}$; $S_{tu} = 164$ mg COD $\cdot\ l^{-1}$; $f_{nsu} = f_{npu} = 20\%$;
- $R_s = 10$ d (design value); $b_h = 0.292\ d^{-1}$; $f_v = 0.8$ mg VSS $\cdot$ mg^{-1} TSS
- $k = 0.46\ l \cdot g^{-1}$; $v_0 = 144\ m \cdot d^{-1}$ (poor settleability); $s_{fd} = 2$; $H_d = 4$; $s = 0.85$;
- $f_{xvd} = 15\%$; $s_d = 1.5$ and $H_i = 2.5$ m;
- $T = 25°C$; $DO_l = 2.07\ mg \cdot l^{-1}$ (25% of DO_{sa}) and $X_t = 4\ g \cdot l^{-1}$.

Solution

From Table A8.1 it can be determined that at T = 25°C the mixed liquor dissolved nitrogen gas concentration in the inlet to the final settler is equal to 13.2 mg $N \cdot l^{-1}$ and that at a liquid depth of 4.0 m the dissolved nitrogen gas concentration in equilibrium with an atmosphere of pure nitrogen will be 23.6 mg $N \cdot l^{-1}$. As a first step, determine the maximum allowable nitrogen gas production per liter return sludge flow during the passage through the final settler:

$$N_{ddmax} = (N_{N2eq} + DO_l/2.86 - N_{N2in})/s_d$$
$$= (23.6 + 2.07/2.86 - 13.2)/1.5 = 7.4\ mg\ N_2 \cdot l^{-1} \quad \text{(A8.1)}$$

In order to be able to calculate D_{cd}, first f_{xd} must be determined. For this we need to define the division of the total sludge mass that develops into its constituents MX_t and MX_{td}.

$$C_r = 0.45 \cdot 10/(1 + 0.292 \cdot 10) = 1.15\ mg\ VSS \cdot d \cdot mg^{-1}\ COD \quad (3.30)$$
$$MX_t = [(1 - f_{ns} - f_{np}) \cdot (1 + f \cdot b_h \cdot R_s) \cdot C_r + f_{np} \cdot R_s/f_{cv}] \cdot MS_{ti}/f_v \quad (3.49)$$
$$= [0.6 \cdot (1 + 0.2 \cdot 0.292 \cdot 10) \cdot 1.15 + 0.2 \cdot 10/1.5] \cdot 8200/0.8 = 24{,}910\ kg\ TSS$$

To calculate MX_{td}, we need the final settler volume. Assuming clarification is limiting ($s = s_c$) the volume of the final settler can be calculated as:

$$V_d = s_{fd} \cdot Q_i \cdot H_d/[v_0 \cdot \exp(-k \cdot X_t)] \quad (8.33)$$
$$= 2 \cdot 50{,}000 \cdot 4/[144 \cdot \exp(-0.46 \cdot 4)]$$
$$= 17{,}490\ m^3$$

For $s = 0.85$, $X_r = (1 + 0.85)/0.85 \cdot 4 = 8.7$ g TSS $\cdot$ l^{-1}

$$MX_{td} = f_{xvd} \cdot V_d \cdot (X_t - X_r)/2 = 0.15 \cdot 17{,}490 \cdot (4 + 8.7)/2 = 16{,}667 \text{ kg TSS} \quad \text{(A8.3)}$$

Now f_{xd} is given as $MX_{td}/(MX_t + MX_{td}) = 16{,}667/(24{,}910 + 16{,}667) = 0.40$ (A8.4)

$$R_{sa} = R_s/(1 - f_{xd}) = 10/(1 - 0.40) = 16.7 \text{ d} \quad \text{(A8.5)}$$

For $R_{sa} = 16.7$ days, C_r increases to 1.28. The denitrification capacity in the final setter per litre influent flow can now be calculated as:

$$D_{cd} = K_3 \cdot C_r \cdot f_{xd} \cdot S_{bu} = 0.08 \cdot 1.04^{(25-20)} \cdot 1.28 \cdot 0.4 \cdot 0.6 \cdot 164 = 4.9 \text{ mg N} \cdot \text{l}^{-1} \text{influent} \quad \text{(A8.2)}$$

The hydraulic retention time in the final settler is:

$$v_d = V_d/Q_i = 17{,}490/50{,}000 = 0.35 \text{ d}$$

Now finally the expected nitrogen gas production per litre return sludge passing from the inlet to the final settler to the return sludge abstraction point can be calculated as:

$$N_{dd} = (s + 1) \cdot D_{cd}/(24 \cdot s \cdot v_d^2) = 1.85 \cdot 4.9/(24 \cdot 0.85 \cdot 0.35^2) = 3.6 \text{ mg N} \cdot \text{l}^{-1} \quad \text{(A8.6)}$$

When the value of N_{dd} is compared to the maximum allowable nitrogen production N_{ddmax} of 7.4 mg N $\cdot$ l^{-1} return sludge, it can be concluded that it is indeed very unlikely that a floating sludge blanket will develop due to excessive denitrification, even with the adverse parameter selection in this example: i.e poor sludge settleability and a sludge fraction in the final settler f_{xvd} of 15%.

For the conditions of Example A8.1, Figure A8.1a compares as function of the *design* sludge age the potential nitrogen gas production to the maximum acceptable nitrogen gas concentration in the bottom part of the final settler ($N_{ddmax} - N_{dd}$). For anaerobically pre-treated sewage typically only secondary treatment and/or nitrification is applied and the mixed liquor flow to the final settler will thus always be aerobic. It can be observed in Figure A8.1a that, over the whole range of design sludge ages considered, the production of nitrogen gas in the return sludge flow will be significantly lower than the critical limit $N_{ddmax} = N_{N2eq} + DO_l/2.86 - N_{N2in}$. Therefore, for aerobic systems treating anaerobic effluent it can be concluded that under normal conditions the risk of floating sludge due to denitrification in the final settler is indeed very low. It should be stressed that this is not just a theoretical exercise: several full-scale aerobic systems are operated at high temperatures and with high mixed liquor nitrate concentrations, without any sludge separation problems resulting from denitrification in the final settler.

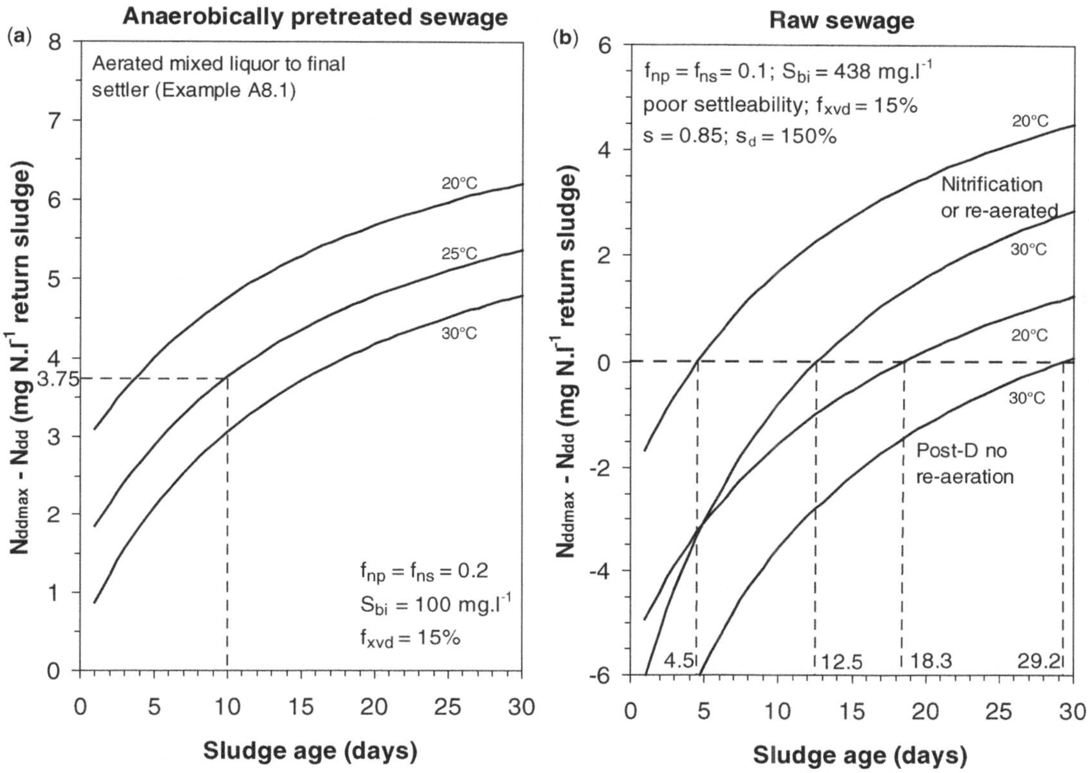

Figure A8.1 Difference between maximum allowable and predicted nitrogen gas production ($N_{ddmax} - N_{dd}$) in the return sludge flow as a function of the *design* sludge age in the activated sludge system for anaerobically pre-treated sewage (a) and raw sewage (b)

Figure A8.1b shows the curves of $N_{dd} - N_{ddmax}$ for typical raw sewage. Two cases are considered: (I) when the mixed liquor flow to the final settler is aerobic (i.e. nitrification or post-D re-aerated) and (II) when it is anoxic (post-D without re-aeration). The data used to calculate the curves in Figure A8.1b are indicated as well. When Figure A8.1a and b are compared, it can be observed that the nature of the wastewater has a significant impact.

With regards to the first case, nitrification, this requires an aerobic sludge age of at least 6.5 days at 20°C and 5 days at 25°C in order to reduce the ammonia concentration to a value of 1 mg $N \cdot l^{-1}$ or less. It can be observed that for the conditions specified in Figure A8.1b the application of a higher sludge age is recommended, especially at higher temperatures.

When denitrification is required as well, the sludge age will consequently be increased. However, if the mixed liquor is taken directly from the post-D zone (without re-aeration), the risk of development of a floating sludge blanket in the final settler will still be significant, but only if denitrification is incomplete. In that case, re-aeration is recommended, as it is such a cheap and effective measure.

The evaluation in the raw sewage example (Figure A8.1b) was performed for rather conservative assumptions: $f_{xvd} = 15\%$, poor sludge settleability and $s_d = 1.5$. Therefore in practice operation at a lower sludge age might be feasible: however this should be evaluated on a case by case basis. On the other hand, as sufficient biodegradable COD is available in the raw sewage, there is no good reason *not* to apply denitrification.

A final remark: as the final settler was designed for poorly settling sludge, it was necessarily quite large. If now during operation actual sludge settleability turns out to be fair instead, then it is possible to reduce the sludge recycle rate. However, this will also increase the retention time of the return sludge in the final settler. Furthermore, the return sludge concentration will increase, but this effect will be (partly) cancelled by a reduction of the sludge volume fraction in the final settler (f_{xvd}). However, should problems arise, this can easily be remedied by an increase in the sludge recycle flow rate.

Appendix 9

Aerobic granulated sludge

The authors would like to acknowledge the contributions of Tom Peeters (Tom.Peeters@DHV.com) and Merle de Kreuk (m.dekreuk@wshd.nl) to this appendix.

The feasibility of cultivating granular biomass under anaerobic conditions has been extensively investigated since the early 1980s. Since then, many full-scale anaerobic granular wastewater treatment systems have been constructed worldwide, e.g. the Expanded Granular Sludge Bed Reactor (EGSB), refer to also Section 13.6. The main driver for the development of granular sludge systems has always been the reduction in reactor volume compared to conventional suspended growth systems, resulting in lower investment costs and a reduced footprint. It should be emphasised that these anaerobic granular sludge systems differ from expanded or fluidized bed systems, in which a carrier material (e.g. sand grains, plastic supports) is used on which a biofilm will grow. These systems are notoriously hard to operate and control, as the growth of biofilm will change the apparent density of the granule (Nicolella et al., 2000).

Already in the 1990s, spontaneous formation of aerobic sludge granules was observed at several full-scale activated sludge systems for biological phosphorus removal. However, due to the shear stress induced by recirculation- and return sludge pumps, the size of these granules was never very large. The formation of these granules was mainly linked to those bio-P systems equipped with a true plug-flow anaerobic zone (Martins et al., 2004).

In the late nineties of the last century aerobic granular sludge (AGS) was formed for the first time under controlled laboratory conditions. This research was initiated by the Delft University of Technology and the Technical University of Munich and fundamental research has continued ever since throughout the world by many universities and companies. From 2003 to date an intensive applied research program was conducted in the Netherlands to develop a new process technology based on this aerobic granular sludge. The research into the so-called Nereda® technology was executed by the international consultancy and engineering firm DHV in close cooperation with the Delft University of Technology, the Dutch Foundation for Applied Water Research (STOWA) and six water boards, supported by the Ministry of Economic Affairs. It consisted of five pilot-scale investigations, the realization of multiple full-scale municipal and industrial installations and the support of fundamental research.

One of the problems that arises when research results on this subject are compared is the use of different methods for characterizing aerobic granular sludge. Therefore, during the first IWA aerobic granule workshop in Munich in 2004 a general definition was agreed upon:

- The granules that make up aerobic granular activated sludge are to be understood as aggregates of microbial origin, which do not coagulate under reduced hydrodynamic shear, and which subsequently settle significantly faster than activated sludge flocs;
- An additional characteristic is that the (diluted) sludge volume index after 5 minutes ($DSVI_5$) is almost equal to that after 30 minutes ($DSVI_{30}$).

Furthermore granular sludge is characterized by a large fraction of particles larger than 0.212 mm. In a full-scale AGS installation the value of this granular fraction would typically range between 70–95%. Due to the granular nature of the AGS sludge, settling velocities typically range between 10 and 50 $m \cdot h^{-1}$, which is very high when compared with the activated sludge flocs from conventional systems, which typically settle at velocities of 1 to 3 $m \cdot h^{-1}$.

In Table A9.1 results on sludge settleability are presented from two pilot AGS systems (De Bruin et al., 2005 and Berkhof et al., 2010), which were operated in parallel to the full-scale plants, which allowed for comparison of the obtained results. After start-up of an AGS reactor, in time the diluted sludge volume index of the aerobic granulated sludge will decrease, while furthermore the $DSVI_5$ value will tend towards the $DSVI_{30}$ value, which generally is in the range of 20 to 60 $ml \cdot g^{-1}$ TSS. The difference in sludge settleability is graphically displayed in Figure A9.1.

Table A9.1 Settleability characteristics of the aerobic granular sludge from two Nereda® pilot plants, located at respectively the Ede (2003 – 2005) and Epe (2006 – 2010) municipal sewage treatment plants, both situated in the Netherlands (adapted from De Bruin et al., 2005 and Berkhof et al., 2010)

Parameter	UoM	Ede		Epe	
		CAS	AGS	CAS	AGS
Type of sewage	(–)	Pre-settled		Raw	
Pre-treatment	(–)	Coarse screening Primary settling		Coarse screening Aerated sand trap Fine screen (2 mm)	
$DSVI_5$	$mg \cdot l^{-1}$	160–180	70–80	200–240	40–50
$DSVI_{30}$	$mg \cdot l^{-1}$	100–120	55–60	100–120	30–40

Due to the high settling velocities it is possible to achieve and maintain sludge concentrations in the reactor as high as 20 $kg\ TSS \cdot m^{-3}$. However, for design purposes a value of 8–12 $kg\ TSS \cdot m^{-3}$ is generally applied. When this is compared to the design sludge concentration typically used for conventional activated sludge systems, between 2.5–5.0 $kg\ TSS \cdot m^{-3}$, it becomes obvious that there is a potential for a significant reduction in system volume, especially taking into account that a final settler will no longer be required.

In the 1990s, it was determined experimentally by Beun et al. (1999), Dangcong et al. (1999) and Morgenroth et al. (1997) that feast-famine conditions are instrumental in aerobic granule formation and that these can be obtained in a sequencing batch reactor (SBR) when applying a short pulse feed, inducing sufficient shear stress and selecting on good settleability of the sludge (i.e. by imposing a short effluent withdrawal period). Later research by de Kreuk et al. (2004) and Liu et al. (2004) showed that the main aspects in granule formation are:

- Hydraulic selection pressure, resulting in process conditions which outcompete poor settling biomass in favour of biomass with excellent settling properties;
- Initial high substrate concentrations in order to apply high gradients;
- The conversion of easily biodegradable substrate into slowly biodegradable intermediate products, which stimulates the growth of slow growing organisms;
- High shear forces, stimulating the growth of smooth and dense granules.

Figure A9.1 Difference in sludge settleability (i.e. the sludge volume resulting after 5 minutes at 4.0 g TSS $\cdot$ l^{-1}) between aerobic granulated sludge and conventional activated sludge. Courtesy of DHV BV

The initial research and pilot studies focused on the crucial role that phosphate accumulating bacteria played in the formation of aerobic granules (De Kreuk, 2006), but later research (and practical findings from full-scale installations) have shown that this is certainly not a perquisite.

A9.1 BENEFITS OF AEROBIC GRANULAR SLUDGE SYSTEMS

The aerobic granulated sludge system has the following benefits compared to conventional aerobic treatment:

- Easy and efficient nutrient removal;
- Lower energy requirements;
- Reduced footprint, due to the high settling velocity of the sludge granules;
- Reduction in investment and operational costs;
- Increased sustainability.

(a) Efficient nutrient removal

An important feature of granular biomass is the enhanced capacity for biological nutrient removal. Even when the AGS system is not explicitly design for nutrient removal, performance will be better than in a conventional activated sludge system. The reason for this is that a dissolved oxygen (DO) gradient will develop during aeration, from the bulk liquid towards the centre of the sludge granule. Oxygen will penetrate only partly into the granule, as it is consumed by autotrophic- and heterotrophic organisms in the outside layer. The oxygen penetration depth depends on the DO concentration in the bulk liquid, the granule diameter and the oxygen uptake rate. In the outer (aerobic) layers of the granule, nitrate and nitrite are produced, which diffuse towards the anoxic centre of the sludge particle where denitrification takes place. This process is called simultaneous denitrification and can also be observed in conventional activated sludge systems containing zones with low DO concentrations (e.g. in Carrousel® systems). In Figure A9.2 this process is schematically demonstrated.

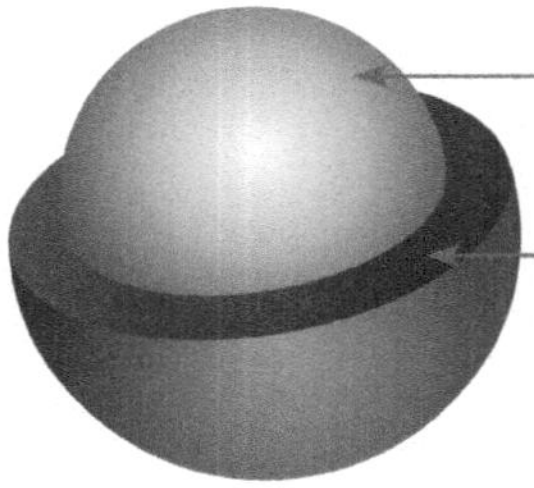

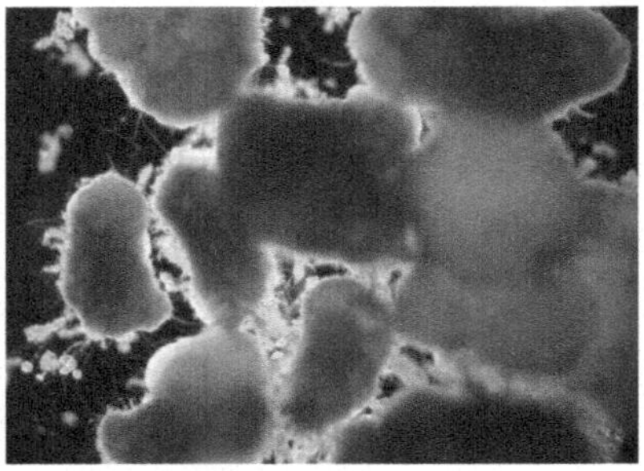

Figure A9.2 Schematical representation of the different zones in an aerobic sludge granule and the processes that occur in each zone (left) and close-up of an aerobic granule (right). Courtesy of DHV BV

As to aerobic granular sludge systems with biological phosphorus removal, during the anaerobic phase, typically pre-settled or raw influent is distributed over the bottom of the reactor, moving upward through the packed sludge bed in a plug-flow manner. The easily (soluble) biodegradable substrate (S_{bs}) diffuses into the sludge granules and is fermented into VFA. As described in Chapter 7, the bio-P organisms store the VFA as internal cell products, mainly PHB. The reduction equivalents required for the conversion of VFA into PHB are supplied by the conversion of glycogen, while internally stored polyphosphate is first split into ortho-phosphate and then released to the liquid phase of the reactor. In the anoxic- and aerobic phases, PHB will be used as a substrate for biomass growth and for the regeneration of glycogen and polyphosphate. If not all of the VFA has been absorbed during the anaerobic phase, it will be used for direct growth of other heterotrophs, for growth of bio-P biomass, and partly for PHB production.

During the aerobic period, an oxygen gradient is formed from the bulk liquid towards the centre of the sludge granule and the processes of nitrification and simultaneous denitrification as discussed above take place. Similar to the conventional bio-P system, during the anoxic- and aerobic phases "normal" heterotrophic biomass will grow on the slowly biodegradable organic substrate S_{bp}. In Figure A9.3 typical concentration profiles in the sludge granules are shown that develop during the anaerobic feed period and the subsequent aerobic period.

The anaerobic sludge mass fraction f_{an} in an AGS reactor can be defined as the length of the anaerobic feed time divided by the total cycle time and has a larger value (typically around 0.25) than the anaerobic sludge mass fraction of a conventional bio-P removal system ($f_{an} = 0.1-0.15$). The hydrolysis under anaerobic conditions of slowly biodegradable COD (S_{bp}) into easily biodegradable COD (S_{bs}) and VFA will be more complete and the fraction of influent COD available to bio-P bacteria in an AGS system will thus be higher than in a conventional bio-P system.

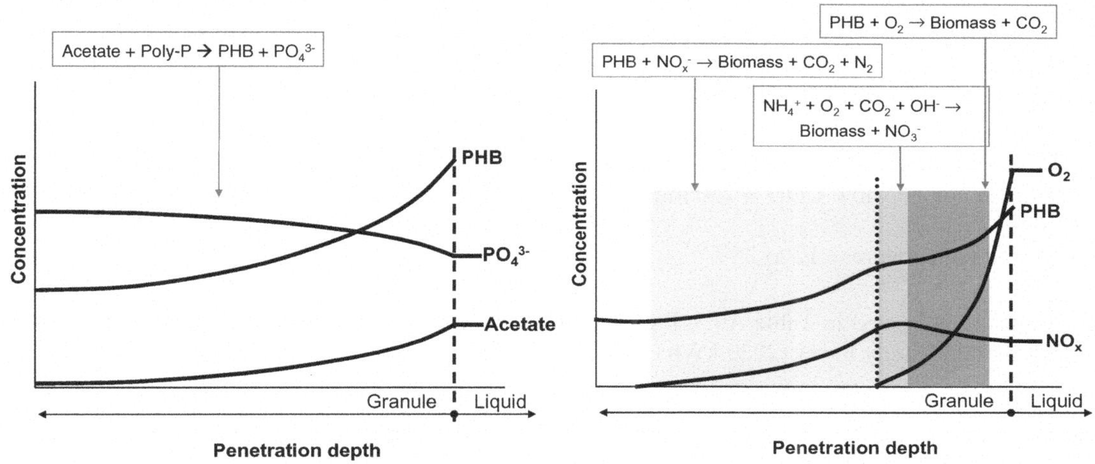

Figure A9.3 Concentration profiles of selected key components during the anaerobic feed phase (left) and the aerobic phase (right) of an AGS process cycle, De Kreuk (2006)

The high concentration of nitrifying, denitrifying and bio-P organisms in the aerobic granule result in an improved biological nutrient removal capacity compared to a conventional activated sludge system with a comparable sludge mass. Table A9.2 shows the treatment performance of a Nereda® pilot plant treating a mixture of industrial (slaughterhouse) and municipal wastewater under typical Dutch climate conditions, where the sewage temperature varied between 7–20°C.

Table A9.2 Average treatment performance of the Nereda® pilot plant at STP Epe, the Netherlands - 2009 (based on data reported by Berkhof et al., 2010)

Parameter	UoM	Influent	Effluent	Limit	Removal
COD	$mg \cdot l^{-1}$	585	55	125	>90%
Total nitrogen	$mg\ N \cdot l^{-1}$	>75	5	5–8	>90%
TKN	$mg\ N \cdot l^{-1}$	>75	1	–	>95%
NH_4-N	$mg\ N \cdot l^{-1}$	52	0.5	–	>99%
NO_x-N	$mg\ N \cdot l^{-1}$	0	4	–	–
Total phosphorus	$mg\ P \cdot l^{-1}$	>10	<1	0.5	>90%
PO_4-P	$mg\ P \cdot l^{-1}$	6	0.4	–	>90%
Suspended solids	$mg \cdot l^{-1}$	193	<15	<20	>90%

(b) Lower energy requirements

In comparison with the conventional activated sludge (CAS) process, the aerobic granular sludge process can have a better aeration efficiency, i.e. when efficient fine bubble aeration is used, due to the possibility of operation at increased height. Furthermore there are neither return sludge or nitrate recycle streams nor mixing and propulsion requirements. Therefore significantly less energy is required for plant operation. Depending on the site specific conditions, the energy requirements for AGS systems may be 20% to 50% lower compared to a CAS system. In an internal evaluation performed by DHV the energy consumption of a Nereda® for biological nutrient removal (100,000 P.E. at 136 g TOD per P.E.) was compared to that of a conventional Carrousel® for the following design values:

- Daily flow rate = 110,000 $m^3 \cdot d^{-1}$, peak flow rate = 11,500 $m^3 \cdot h^{-1}$;
- Influent composition: COD = 750 $mg \cdot l^{-1}$, TSS = 300 $mg \cdot l^{-1}$ TKN = 55 $mg\ N \cdot l^{-1}$ and total-P = 10 $mg\ P \cdot l^{-1}$;
- Design temperature = 12 to 25°C

The results are presented in Table A9.3. It can be observed that the predicted energy consumption of the Nereda® is significantly lower (2220 $kWh \cdot d^{-1}$ or 38%) than that of the Carrousel® configuration.

Table A9.3 Comparison of energy requirements of a conventional Carrousel® system and Nereda® in a nutrient removal configuration treating municipal sewage

Parameter	Carrousel®		Nereda®	
	$kWh \cdot d^{-1}$	%	$kWh \cdot d^{-1}$	%
Influent lifting station	150	3%	262	7%
Screen and sand/grit removal	73	1%	73	2%
Biological reactor	4972	85%	2397	67%
– mixers anaerobic zone	192	3%	–	–
– mixers pre-denitrification	318	5%	–	–
– nitrate recirculation	648	11%	–	–
– propulsors aerobic zone	848	15%	–	–
– aeration	2534	44%	2397	67%
– final settlers	60	1%	–	–
– return sludge pumping station	372	6%	–	–
Sludge dewatering	93	2%	383	11%
Miscellaneous small equipment	228	4%	228	6%
Cable/frequency converter losses	291	5%	244	7%
Total energy consumption	**5807**	**100%**	**3587**	**100%**

(c) Reduced footprint

The increase in sludge concentration that is possible because of the high settling velocity of the aerobic sludge granules (as discussed earlier in this appendix) and the absence of a final settler result in a significant reduction in the required footprint of the treatment plant. In Table A9.4 the difference in footprint between a Carrousel® and Nereda® for the case study discussed above are shown: the Nereda®

is respectively 100,000 m^3 (54%) and 30,000 m^2 (71%) smaller than the Carrousel® system. Note that the total required sludge mass is slightly larger for the Nereda®, to compensate for the time required for settling and decanting.

Table A9.4 Comparison of footprint and system volume of a conventional Carrousel® system and Nereda® in a nutrient removal configuration treating municipal sewage

Parameter	UoM	Carrousel®	Nereda®
Design reactor sludge concentration	$g\,TSS \cdot l^{-1}$	4	8
Aeration tank volume	m^3	138,000	84,000
Additional volume of anoxic/anaerobic zones	m^3	13,000	–
Settler area	m^2	11,000	–
Total volume	m^3	184,000	84,000
Total area	m^2	42,000	12,000

(d) Reduced investment and operational costs

The concentrated biomass substantially reduces tank volume and easily reduces the plant footprint to 50%, which is further accentuated by the elimination of the final settler from the design. Furthermore the number of mechanical equipment is reduced, as for example mixers, recirculation pumps and return sludge pumps are not required. This lowers the direct investment costs for green field, retrofit or capacity extension applications and the existing treatment site might be utilized rather than having to purchase new land.

Operation and maintenance costs are lower as well, due to the reduction in mechanical equipment, reduction of chemical usage and the high energy efficiency of the process. On the other hand, at present there are few suppliers of aerobic granular sludge systems, as the technology is still largely patent protected. This implies that cost-competition between AGS vendors will not be very large, which partly negates the potential savings in investment costs.

(e) Sustainable

The AGS technology may be considered as more sustainable than the conventional activated sludge system, mainly because energy demands are significantly lower and less equipment and construction material are required. The Swedish Research Institute IVL executed an extensive lifecycle assessment (LCA) study in which the environmental impact of Nereda® was compared to that of a conventional activated sludge system for the aerobic polishing of anaerobic pre-treated wastewater from breweries (Giesen, 2010). It was concluded that the aerobic granular system was more sustainable for all investigated environmental parameters.

A9.2 SYSTEM DESIGN AND OPERATION

A9.2.1 Process configurations

Different system configurations for aerobic granulated systems may be selected, depending on the situation at hand:

(a) Greenfield applications (Figure A9.5);
(b) Retrofit/upgrades of existing installations (Figure A9.6);
(c) Hybrid capacity extension of existing installations (Figure A9.7).

Figure A9.4 Examples of some full-scale Nereda® applications: edible oil industry (left) and Nereda® under construction at the STP Epe - The Netherlands (right). Courtesy of DHV BV

(a) Greenfield applications

Depending on the wastewater flow and characteristics, a typical greenfield AGS plant comprises multiple modular reactors (often three), operating out of sequence so that there is always one reactor available to receive the incoming wastewater. Alternatively, if the quantity of wastewater to be treated is low, as will the case for many industrial applications, a single reactor preceded with a buffer tank can be selected as well. Full-scale examples of Nereda® systems are shown in Figure A9.4.

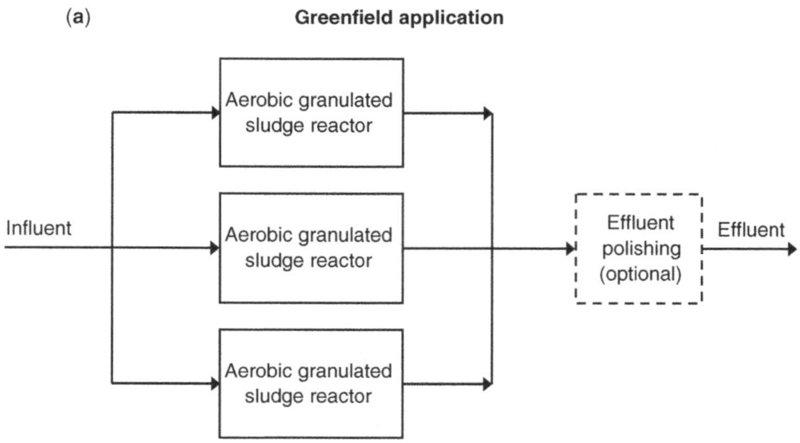

Figure A9.5 Schematic representation of a greenfield AGS configuration

(b) Retrofit/upgrades of existing installations

As the physical design of an aerobic granular sludge reactor is quite flexible, it is often possible to convert the reactors of existing conventional continuous activated sludge systems or Sequencing Batch Reactors

(SBRs) into an AGS reactor. As the application of aerobic granulated sludge allows an increase of the biomass concentration with (typically) a factor two, after the retrofit the treatment capacity of the original plants will be significantly increased and/or the effluent quality will be considerably improved.

(c) Hybrid capacity extension of existing installations

In this application, the AGS reactor receives only part of the raw wastewater flow while the remaining part is treated by the existing conventional treatment system. Depending on the specific requirements, several variants of this configuration can be distinguished:

- If both the hydraulic- and biological capacity of the existing conventional treatment plant are to be increased, one or more AGS reactors can be operated in parallel with the existing biological reactors. Depending on the local circumstances and effluent requirements, a post treatment step might be required for extensive phosphorus- and suspended solids removal;
- If only the biological capacity needs to be increased, this can be easily implemented by constructing only one AGS reactor in parallel to the existing biological reactors. If necessary, a storm water buffer for hydraulic optimisation can be considered;
- Another variant of the hybrid capacity extension can be applied when a part of the organic load originates from a concentrated wastewater flow. In this case it can be considered to treat the concentrated flow in a compact AGS system in parallel with the existing activated sludge reactors, although the application of AGS should be compared to other alternatives, such as anaerobic treatment.

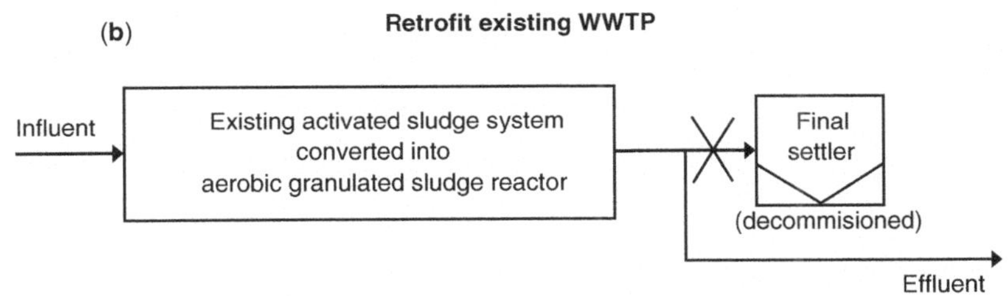

Figure A9.6 Schematic representation of retrofit/upgrade of existing activated sludge plants into an AGS configuration

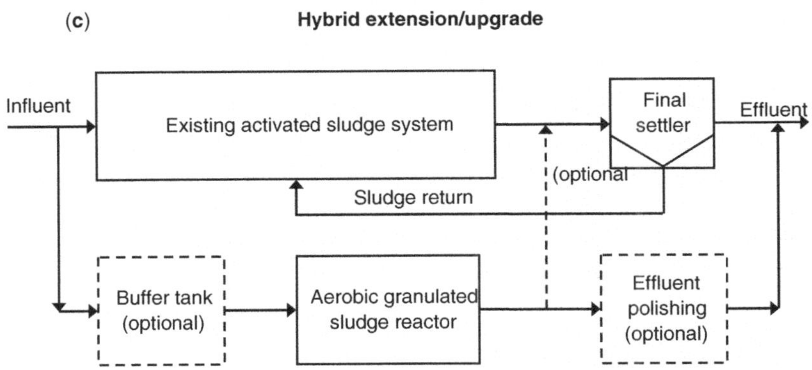

Figure A9.7 Schematic representation of an hybrid capacity extension of existing activated sludge plants using an AGS reactor

An important additional advantage of a hybrid capacity extension with an aerobic granular sludge reactor is that the conventional activated sludge system can be seeded with granular surplus sludge, either with the AGS effluent or -waste sludge. Because of this inoculation process, the sludge characteristics and settling performance of the existing treatment plant will gradually improve, resulting in increased capacity and improved treatment performance.

A9.2.2 Reactor configuration

A schematic layout of a typical greenfield aerobic granulated sludge reactor is shown in Figure A9.8. When the fill- and draw stages are combined, then the critical engineering aspects are the proper design of the inlet distribution system and that of the effluent removal section. In this design, the influent requires distribution over the entire reactor bottom to prevent short-circuiting between reactor inlet and -outlet. The valve in the main effluent discharge line is only open during the aerobic fill phase, when effluent is simultaneously discharged. The valve in the secondary line is opened directly after the end of the feed phase, in order to create some empty volume in the reactor to allow for expansion of the liquid column when aeration is resumed. An alternative would be the use of moving or floating decanters as commonly used in SBR systems. In general, the physical design of an AGS reactor is quite flexible, which allows the reuse of existing reactors in plant retrofits. In Table A9.5 typical values of several AGS design and performance characteristics are summarised.

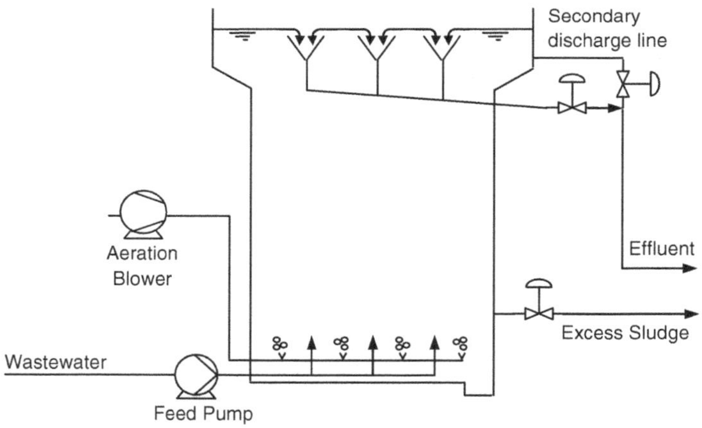

Figure A9.8 Schematic layout of a typical (greenfield) aerobic granulated sludge reactor

The specific organic loading rates (in kg $COD \cdot kg^{-1} TSS \cdot d^{-1}$) used for the design of aerobic granular sludge systems and conventional activated sludge systems are comparable. However, the high biomass settling rate and the increased biomass concentration will result in an increase in volumetric loading rate (kg $COD \cdot m^{-3} \cdot d^{-1}$) compared to conventional treatment systems.

A9.2.3 Operation of AGS systems

The aerobic granular sludge process is operated in a sequencing batch cycle mode comprising the following process stages:

(1) Pulse feed or anaerobic fill, optionally combined with effluent discharge;
(2) Draining (optional);
(3) Aeration;
(4) Anoxic (optional);
(5) Settling;
(6) Effluent discharge (optional);
(7) Sludge discharge.

An example of a typical process cycle for nutrient removal is shown in Table A9.6.

Table A9.5 Typical value ranges for several design and process parameters in aerobic granulated sludge systems

Parameter	Value	Parameter	Value
Avg. upflow velocity[1]	2–3 $m \cdot h^{-1}$	$DSVI_5$	40–80 $ml \cdot g^{-1}$
Max. upflow velocity[1]	5 $m \cdot h^{-1}$	$DSVI_{30}$	30–60 $ml \cdot g^{-1}$
Settling velocity	10–50 $m \cdot h^{-1}$	DO setpoint (bulk)	2–2.5 mg $O_2 \cdot l^{-1}$
Reactor sludge concentration	8–12 g $TSS \cdot l^{-1}$	Reactor height	4–12 m
Batch size	5–70% of reactor volume	Type of aeration	Fine bubble
Organic loading rate	0.1–0.3 g COD. g^{-1} $TSS \cdot d^{-1}$	Effluent TSS concentration	10–20 mg $TSS \cdot l^{-1}$

Note: (1) only relevant when operated under simultaneous filling and effluent withdrawal

Table A9.6 Example of the process cycle of an aerobic granulated sludge system designed for nutrient removal (4-hour process cycle: in practice the length of the process cycle may range from 2–9 hrs)

No.	Description	Duration	
		Minutes	Fraction of cycle time
1	Anaerobic fill	60	25%
2	Drain	5	2%
3	Aerobic process	165	69%
4	Anoxic process	0	0%
5	Settling	10	4%
6	Effluent discharge[1]	0	0%
7	Sludge discharge[1]	0	0%
Total	Cycle time	240	100%

Note: (1) in this example fill and draw takes place simultaneously while excess sludge is withdrawn during the aerobic process stage

(1) Pulse feed or anaerobic fill

In this phase the wastewater is pumped into the reactor. The high organic loading rate applied in this phase favours the slow-growing organisms, which are often capable of storing easily biodegradable substrate as cell-internal polymer, whereas the fast growing organisms are not. When the AGS system is operated according to an optimized SBR cycle, which is typically recommended for greenfield applications, the effluent will be withdrawn simultaneously, while in retrofit situations this may not be the case. The influent needs to be carefully and evenly distributed over the reactor bottom to prevent short-circuiting between reactor inlet and -outlet. In practice the filling rate will be limited to a value of $2-3\ m \cdot h^{-1}$. Effluent discharge take places using floating decanters or fixed effluent collection channels or pipes, as shown in Figure A9.8.

(2) Draining (optional - depending on reactor configuration)

When the reactor is not equipped with a floating decanter but with a fixed effluent pipe and simultaneous fill and draw is applied, then an additional drain period is required. Aeration without a prior drain discharge period would result in the discharge of mixed liquor due to the expansion ('gas hold-up') of the liquid volume. Therefore prior to resuming aeration, a small volume of additional effluent is withdrawn from a secondary discharge pipe located beneath the effluent weir in order to create volume for liquid expansion during aeration without spills.

(3) Aeration

During the aeration phase the biological conversion processes take place. The outer layer of the granules will be aerobic and here nitrification will occur. The produced nitrate is denitrified in the anoxic core of the granules, where substrate is available as a result of the feed/fill phase. Biological phosphorus uptake takes place in large part of the granule, both under aerobic and anoxic conditions.

Depending on the influent characteristics and applied process conditions, part of the biomass may be present in traditional flocculent form, i.e. outside the granules. In the flocculent sludge, the same processes occur that take place in the aerobic zone of a conventional activated sludge system.

(4) Anoxic reaction phase (optional)

Depending on the TKN/COD ratio in the influent, the concentration of easily biodegradable COD in the influent (S_{bsi}) and the applied bulk oxygen concentration, it might be necessary to include an anoxic stage to meet the the nitrate or total nitrogen effluent limits. In this (optional) anoxic phase, all readily biodegradable material present in the bulk liquid (S_{bs}) and a large part of the internal cell polymers will have been depleted. Denitrification will thus proceed at much lower rate. This phase may be considered as equivalent to a post-D zone in a Bardenpho or UCT configuration.

(5) Settling phase

The settling phase in an AGS is relatively short and is applied mainly to allow for liquid/solid separation in advance of effluent discharge, i.e. in order to ensure that the liquid phase in the upper part of the reactor is free of suspended solids before the fill and discharge phase starts.

(6) Effluent discharge (optional)

Instead of simultaneous fill and draw, effluent discharge can also be applied as a separate phase after the settling period. In SBR systems, the use of floating decanters is effective in reducing the effluent discharge period compared to fixed decanters, as effluent discharge may already start during the sludge settling phase, as soon as the distance between the sludge bed and the decanter device has become large enough to prevent sludge entrainment with the effluent. Due to the high settling velocity of the granulated sludge, it will often be possible to reduce the duration of the dedicated effluent discharge period (i.e. after the settling phase) to a very low value.

(7) Sludge discharge
Sludge will generally be discharged at the end of the settling phase or during the aeration phase. The advantage of the first option is the selective sludge discharge, which enables the retention of the aerobic granules with the highest settling velocities. However, an advantage of the latter option is the more constant concentration of the discharged sludge, which makes it much easier to determine the actual quantity of solids discharged and to control the sludge age: i.e. by means of hydraulic sludge control as explained in Section 3.3.3.6.

A general disadvantage of batch operated systems is the difficulty in handling of peak flows. However, due to the high settling velocity of the aerobic granular sludge it is possible to increase the hydraulic loading rate during peak flow without compromising effluent quality, also because, contrary to conventional activated sludge systems, all biomass will remain in the reactor (no sludge transfer to the final settler). However, in some cases it may be beneficial to reduce the length of the process cycle during peak flow periods.

A9.2.4 Start-up of aerobic granular sludge reactors

Depending on the biomass growth rate and the size of the AGS system, the start up period might take several months, required to transform the original suspended activated biomass into granules. This is a general disadvantage of all granular biomass systems. However, due to the significantly higher sludge growth rate of aerobic biomass, the start up period of aerobic granular sludge is much lower compared to that of anaerobic granular sludge (i.e. when no or insufficient anaerobic granular biomass is available to seed the reactor).

The transformation from activated sludge flocs into aerobic sludge granules requires an initial high sludge selection pressure, e.g. through the application of high upflow velocities during the filling phase, in order to selectively retain those organisms in the system that settle well. As illustrated in Figure A9.9, for the Nereda® pilot in Ede it took only a few months to increase the aerobic sludge fraction to more than 70% using only a small quantity of granular seed biomass from lab reactors. Generally, the biomass in a fully adapted AGS system will contain between 70–95% of granules.

An alternative and preferred option to accelerate startup is to supply seed sludge, as is customary with the slow growing anaerobic granular sludge systems. Granular biomass is stable and can be stored easily for long periods. While the granular structure will survive for years, the effect of storage on the biological activity is dependent on the method and duration of storage. When aerobic granular biomass is kept under cooled conditions (about 4°C), it can be stored for several weeks without losing its biological activity . When sufficient high quality aerobic granular seed sludge is available, then the start-up period can be reduced to two weeks.

A9.3 GRANULAR BIOMASS: EVALUATION AND POTENTIAL

Since the first release of this book in 2007, the development of the aerobic granular biomass technology has advanced considerably. Was the technology at that stage a promising emerging technology, nowadays it can be considered as "proven", although the number of installed systems is at present still limited compared to other systems. However, since 2005 the AGS technology has been implemented successfully more than ten times for both industrial and municipal wastewater. Furthermore, at the time of publication (2011) more systems were in the design or construction phase. An important finding was that the AGS system has been proven under harsh African conditions in which it has been demonstrated that due to extensive degree of automation that can be applied (i.e. fully automated process cycles), also relatively low-skilled staff can operate the treatment works. The benefits of using aerobic granular biomass are evident: lower costs, excellent treatment performance, reduced energy consumption and easy operation.

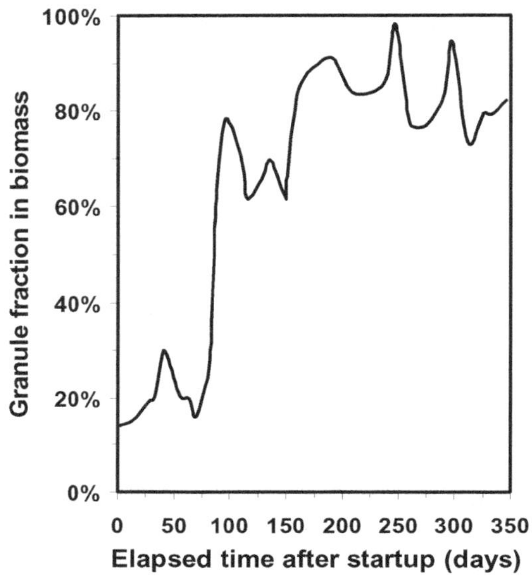

Figure A9.9 Observed degree of granulation in the biomass after start-up of the first Nereda® pilot installation located at the Ede STP, The Netherlands (De Bruin et al., 2005)

The research field is still relatively new and not all interactions between wastewater composition, operational conditions and system performance are yet fully understood. However, the available research clearly shows that the aerobic granular sludge technology is capable of handling both dissolved and particulate organic pollutants. Furthermore, it seems the granules are less vulnerable to toxic compounds than the suspended flocs in conventional activated sludge systems. This is due (I) to the relative small penetration depth of toxic compounds into the granule, so that the bacteria in inner parts are partly protected and (II) to the relatively large and heterogeneous micro-organism population, in which latent available capacity inside the granule can (partly) replace the affected micro-organism population at the outside of the granule.

In South Africa, a Nereda® installation has been in operation from 2009 on wastewater containing very high levels of particulate matter, due to the contribution of septic tank waste. The treatment plant is designed for a capacity of 4,000 $m^3 \cdot d^{-1}$ and easily meets local discharge limits with an average effluent quality of NH_4-N < 1.5 mg $N \cdot l^{-1}$, NO_3-N between 5–10 mg $N \cdot l^{-1}$, PO_4-P < 5 mg $P \cdot l^{-1}$ and SS < 10 mg $\cdot l^{-1}$, which allows it to be directly reused for irrigation purposes. Detailed performance data of this plant can be found in Table A9.7. The wastewater plant itself is shown in Figure A9.10.

Figure A9.10 Full-scale municipal sewage Nereda® application (4000 $m^3 \cdot d^{-1}$) at the Gansbaai STP in South Africa. Courtesy of DHV BV

Table A9.7 Average performance of the 4000 $m^3 \cdot d^{-1}$ municipal sewage Nereda® located at Gansbaai, South Africa (based on data provided by DHV BV)

Parameter	UoM	Influent	Effluent	Limit	Removal %
COD	$mg \cdot l^{-1}$	1240	42	75	96.6%
NH_4-N(1)	$mg\ N \cdot l^{-1}$	81	1.4	6	98.3%
NO_x-N	$mg\ N \cdot l^{-1}$	–	8.7	15	–
Total phosphorus	$mg\ P \cdot l^{-1}$	19.2	–	–	–
PO_4-P	$mg\ P \cdot l^{-1}$	–	4.7	10	–
Suspended solids	$mg\ TSS \cdot l^{-1}$	690	5.5	25	99.2%

Note (1): The NH_4-N/TKN ratio was 0.75

When comparing the aerobic granular sludge technology to another (more or less) recent development, the membrane bioreactor (MBR), the following remarks can be made:

- Though the footprint of both systems is comparable, the aerobic granular sludge system is technologically much simpler than the MBR, requiring less instrumentation and process control. Furthermore the skill level required for operation is lower, better meeting the capacity in the developing countries in for example Asia, Africa and Eastern Europe;

- The investment- and operating costs, energy consumption and environmental profile for the AGS system are all much more favourable. Furthermore the AGS system will often be able to achieve the required effluent quality at much lower investments and operational costs than the MBR;
- Mainly because of the retention of suspended solids, including bacteria and viruses, the MBR yields a better effluent quality, whereas AGS requires additional effluent disinfection or polishing. However, especially when the AGS is equipped with relatively simple mechanical polishing filters or a reed bed, the removal of suspended solids and the associated COD, nitrogen and phosphorus content for both technologies is quite similar and in any case effluent quality is better than limits currently applied in many countries (COD $< 50\ mg \cdot l^{-1}$, total nitrogen $< 10\ mg\ N \cdot l^{-1}$, total phosphorus < 1 to $2\ mg\ P \cdot l^{-1}$);
- For difficult wastewaters (e.g. of a predominantly industrial origin: high strength, saline or with toxic contributions) the MBR might be more stable as the membrane will prevent biomass loss even in the effect of toxic shocks, whereas for the AGS under certain conditions there might be a risk of degranulation. On the other hand, as mentioned earlier, the bacteria inside the granule are partially protected against toxicity;
- Both technologies can be considered important instruments in upgrading or extending existing wastewater treatment facilities without extending the actual plant footprint;

IWA Publishing's authorised EU representative for General Product Safety Regulations is Diane D'Arras, 15 rue Duret, 75116 Paris, France, e-mail: safety@iwap.co.uk.

Printed and bound by CPI Group (UK) Ltd, Croydon, CR0 4YY
12/05/2026
02108863-0016